Intermediate Algebra

Seventh Edition

John Tobey
North Shore Community College
Danvers, Massachusetts

Jeffrey Slater
North Shore Community College
Danvers, Massachusetts

Jamie Blair
Orange Coast College
Costa Mesa, California

Jennifer Crawford
Normandale Community College
Bloomington, Minnesota

PEARSON

Boston Columbus Indianapolis New York San Francisco Upper Saddle River
Amsterdam Cape Town Dubai London Madrid Milan Munich Paris Montréal Toronto
Delhi Mexico City São Paulo Sydney Hong Kong Seoul Singapore Taipei Tokyo

Editorial Director, Mathematics: *Christine Hoag*
Editor in Chief: *Paul Murphy*
Acquisitions Editor: *Dawn Giovanniello*
Executive Content Editor: *Kari Heen*
Senior Content Editor: *Lauren Morse*
Editorial Assistant: *Chelsea Pingree*
Executive Director of Development: *Carol Trueheart*
Senior Development Editor: *Elaine Page*
Senior Managing Editor: *Karen Wernholm*
Senior Production Supervisor: *Ron Hampton*
Design Manager: *Andrea Nix*
Interior Design: *Tamara Newnam*
Senior Design Specialist: *Barbara Atkinson*
Digital Assets Manager: *Marianne Groth*
Supplements Production Project Manager: *Katherine Roz*
Content Development Manager: *Rebecca E. Williams*
Senior Content Developer: *Mary Durnwald*
Executive Manager, Course Production: *Peter Silvia*
Media Producers: *Audra Walsh and Vicki Dreyfus*
Executive Marketing Manager: *Michelle Renda*
Marketing Manager: *Rachel Ross*
Marketing Assistant: *Ashley Bryan*
Senior Author Support/Technology Specialist: *Joe Vetere*
Procurement Manager/Boston: *Evelyn M. Beaton*
Procurement Specialist: *Debbie Rossi*
Media Procurement Specialist: *Ginny Michaud*
Permissions Project Supervisor: *Michael Joyce*
Production Management, Composition, and Answer Art: *Integra*
Text Art: *Scientific Illustrators*
Cover Images: *Illustration by Amy DeVoog*

Library of Congress Cataloging-in-Publication Data

Intermediate algebra / John Tobey ... [et al.]. —7th ed.
p. cm.
Includes index.
ISBN 978-0-321-76950-3 (alk. paper)
1. Algebra—Textbooks. I. Tobey, John.
QA154.3.T64 2013
512.9—dc23 2011024002

1 2 3 4 5 6 7 8 9 10—DOW—16 15 14 13 12

ISBN-10: 0-321-76950-3 (paperback)
ISBN-13: 978-0-321-76950-3 (paperback)

PEARSON

pearsonhighered.com

This book is dedicated to three amazing couples
of the next generation:
John and Stephanie Tobey
Greg and Marcia Salzman
Phil and Melissa LaBelle
They excel at marriage, parenthood, work, and
graduate work. They are a blessing to us all.

Contents

CHAPTER 9

The Conic Sections 485

CHAPTER 10

Additional Properties of Functions 539

CHAPTER 11

Logarithmic and Exponential Functions 583

Preface

TO THE INSTRUCTOR

One of the hallmark characteristics of *Intermediate Algebra* that makes the text easy to learn and teach from is the building-block organization. Each section is written to stand on its own, and every homework set is completely self-testing. Exercises are paired and graded and are of varying levels and types to ensure that all skills and concepts are covered. As a result, the text offers students an effective and proven learning program suitable for a variety of course formats—including lecture-based classes; discussion-oriented classes; modular, self-paced courses; distance learning; mathematics laboratories; and computer-supported centers.

We have visited and listened to teachers across the country and have incorporated a number of suggestions into this edition to help you with the particular learning delivery system at your school.

WHAT'S NEW IN THE SEVENTH EDITION?

- Chapter Organizers have been updated to include a You Try It column that provides additional opportunity for students to practice relevant chapter topics and procedures.
- A solid correlation has been made between the material on the **How Am I Doing? Chapter Test** and the examples, exercises, Chapter Review, and Cumulative Review. Each Chapter Test problem has at least one example, two Chapter Review exercises, and two Cumulative Review exercises that represent the same problem type. **New assessment check boxes** allow students to tally their answers and gauge their preparedness for the actual test.
- Following each Chapter Test, the new **Math Coach** provides students with a personal office-hour experience by walking them through some helpful hints to keep them from making common errors on test problems. For additional help, students can also watch the authors work through these problems on the accompanying Math Coach videos, available on YouTube and in MyMathLab.
- Select **Examples and Student Practice** problems, representing some of the most difficult concepts for students to master in a chapter, have been placed side by side to encourage students to work through each step of these problems to gain further understanding. These concepts are also covered on the Chapter Test and in the Math Coach.
- **Enhanced emphasis on Steps to Success boxes** (formerly Developing Your Study Skills) have been integrated throughout the text to provide students with more guided techniques for improving their study skills and succeeding in math.
- The **Use Math to Save Money** features are now assignable so that students can apply this new knowledge to their everyday lives. All of the topics have been chosen based on a student survey of over 1000 developmental math college students. These give practical, realistic examples of how students can use math to cut costs and spend less.
- Ten percent of the exercises throughout the text have been refreshed.
- All real-world application problems have been updated.
- ***New* The Lecture Series on DVD** has been completely revised to provide students with extra help for each section of the textbook. The Lecture Series DVD includes
    - **Interactive Lectures** that highlight key examples and exercises from every section of the textbook. A new interface allows for easy navigation to sections, objectives, and examples.
    - **Math Coach Videos**, featuring the text authors (John Tobey, Jeffrey Slater, Jamie Blair, and Jennifer Crawford), coach students in avoiding the most commonly made mistakes in a particular problem when students need the most help: the night before an exam.
    - **Chapter Test Prep Videos** provide step-by-step video solutions to every problem in each How Am I Doing? Chapter Test in the textbook.

Student and Instructor Resources

Student Resources

Worksheets with the Math Coach
Provides extra vocabulary and practice exercises for every section of the text. Each chapter also includes the Math Coach problems with ample space for students to show their work. The worksheets can be packaged with the textbook or with the MyMathLab access kit.

Student Solutions Manual
Provides worked-out solutions to all odd-numbered section exercises, even and odd exercises in the Quick Quiz, mid-chapter reviews, chapter reviews, chapter tests (including problems covered in the Math Coach), and cumulative reviews.

Lecture Series on DVD Featuring Math Coach and Chapter Test Prep Videos
Provides students with extra help for each section of the textbook. The videos include

- A complete lecture for each section of the textbook. The new interface allows easy navigation to objectives and examples.
- Math Coach videos that coach students in avoiding the most commonly made mistakes in a particular problem.
- Step-by-step video solutions to every problem in each How Am I Doing? Chapter Test.

Math Coach and Chapter Test Videos are also available in MyMathLab and on YouTube.

All Student Resources are available for purchase at www.mypearsonstore.com

Instructor Resources

Annotated Instructor's Edition
Contains all of the content found in the student edition, plus the following:

- Answers to all practice problems, section exercises, mid-chapter reviews, chapter reviews, chapter tests, cumulative tests, and practice final exam
- Teaching Tips placed in the margin at key points where students historically need extra help
- Teaching Examples placed in the margins to accompany each example

Instructor's Solutions Manual

- Detailed step-by-step solutions to the even-numbered section exercises
- Solutions to every exercise (odd and even) in the Classroom Quiz, mid-chapter reviews, chapter reviews, chapter tests, cumulative tests, and practice final

(Available for download from the Instructor's Resource Center)

Instructor's Resource Manual with Tests and Mini-Lectures

- Mini-lecture for each text section
- Two short group activities per chapter
- Three forms of additional practice exercises
- Two pretests per chapter—free response and multiple-choice
- Six tests per chapter—free response and multiple-choice
- Two cumulative tests per even-numbered chapter—free response and multiple choice
- Two final exams—free response and multiple choice
- Answers to all items

(Available for download from the Instructor's Resource Center)

Online Resources

MyMathLab® Online Course (access code required)

MathXL® Online Course (access code required)

TestGen® (Available for download from the Instructor's Resource Center)

Diagnostic Pretest: Intermediate Algebra

1. ______
2. ______
3. ______
4. ______
5. ______
6. ______
7. ______
8. ______
9. ______
10. ______
11. ______
12. ______
13. ______
14. ______
15. ______

Chapter 1

1. Evaluate. $3 - (-4)^2 + 16 \div (-2)$

2. Simplify. $(3xy^{-2})^3(2x^2y)$

3. Simplify. $3x - 4x[x - 2(3 - x)]$

4. Evaluate $F = \frac{9}{5}C + 32$ when $C = -35°$.

Chapter 2

Solve the following:

5. $-8 + 2(3x + 1) = -3(x - 4)$

6. When ice floats in water, approximately $\frac{8}{9}$ of the height of the ice lies under water. If the tip of an iceberg is 23 feet above the water, how deep is the iceberg below the waterline?

7. $4 + 5(x + 3) \geq x - 1$

8. $\left|3\left(\frac{2}{3}x - 4\right)\right| \leq 12$

Chapter 3

9. Find the slope and the y-intercept of $3x - 5y = -7$.

10. Find an equation of the line that passes through $(-5, 6)$ and $(2, -3)$.

11. Is this relation also a function? $\{(5, 6), (-6, 5), (6, 5), (-5, 6)\}$

12. If $f(x) = -2x^2 - 6x + 1$, find $f(-3)$.

Chapter 4

Solve the following:

13.
$$\begin{aligned} 3x + 5y &= 30 \\ 5x + 3y &= 34 \end{aligned}$$

14.
$$\begin{aligned} 2x - y + 2z &= 8 \\ x + y + z &= 7 \\ 4x + y - 3z &= -6 \end{aligned}$$

15. A speedboat can travel 90 miles with the current in 2 hours. It can travel upstream 105 miles against the current in 3 hours. How fast is the boat in still water? How fast is the current?

16. Graph the system.

$$x - y \le -4$$
$$2x + y \le 0$$

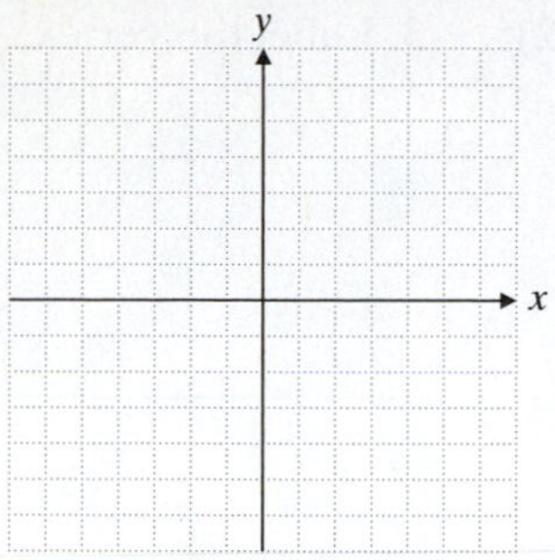

Chapter 5

17. Multiply. $(3x - 4)(2x^2 - x + 3)$

18. Divide. $(2x^3 + 7x^2 - 4x - 21) \div (x + 3)$

19. Factor. $125x^3 - 8y^3$

20. Solve. $2x^2 - 7x - 4 = 0$

Chapter 6

21. Simplify. $\dfrac{10x - 5y}{12x + 36y} \cdot \dfrac{8x + 24y}{20x - 10y}$

22. Combine. $2x - 1 + \dfrac{2}{x + 2}$

23. Simplify. $\dfrac{\dfrac{1}{x + h} - \dfrac{1}{x}}{h}$

24. Solve for x. $\dfrac{x}{x - 2} + \dfrac{3x}{x + 4} = \dfrac{6}{x^2 + 2x - 8}$

Chapter 7

Assume that all expressions under radicals represent nonnegative numbers.

25. Multiply and simplify. $(\sqrt{3} + \sqrt{2x})(\sqrt{7} - \sqrt{2x^3})$

26. Rationalize the denominator. $\dfrac{3\sqrt{x} + \sqrt{y}}{\sqrt{x} - \sqrt{y}}$

27. Solve and check your solutions. $2\sqrt{x - 1} = x - 4$

Chapter 8

Solve.

28. $x^2 - 2x - 4 = 0$

29. $x^4 - 12x^2 + 20 = 0$

16. ______

17. ______

18. ______

19. ______

20. ______

21. ______

22. ______

23. ______

24. ______

25. ______

26. ______

27. ______

28. ______

29. ______

30.

31.

32.

33.

34.

35.

36.

37.

38.

39.

40.

41.

30. Graph $f(x) = (x - 2)^2 + 3$. Label the vertex.

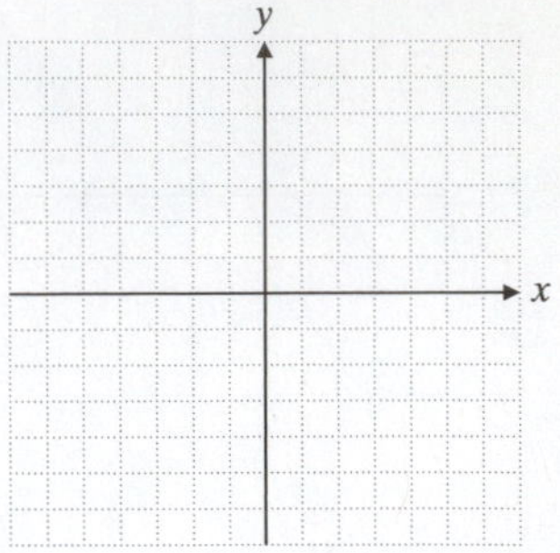

Chapter 9

31. Write in standard form the equation of the circle with center at $(5, -2)$ and a radius of 6.

32. Write in standard form the equation of the ellipse whose center is at $(0, 0)$ and whose intercepts are at $(3, 0)$, $(-3, 0)$, $(0, 4)$, and $(0, -4)$.

33. Solve the following nonlinear system of equations.

$$\begin{aligned} x^2 + 4y^2 &= 9 \\ x + 2y &= 3 \end{aligned}$$

Chapter 10

34. If $f(x) = 2x^2 - 3x + 4$, find $f(a + 2)$.

35. Graph on one axis $f(x) = |x + 3|$ and $g(x) = |x + 3| - 3$.

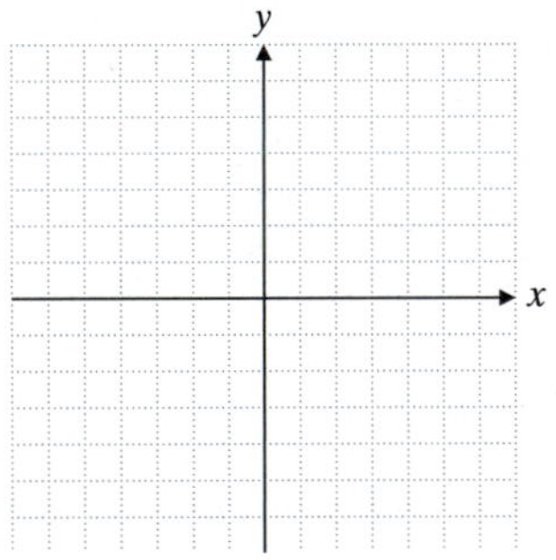

36. If $f(x) = \dfrac{3}{x + 2}$ and $g(x) = 3x^2 - 1$, find $g[f(x)]$.

37. If $f(x) = -\dfrac{1}{2}x - 5$, find $f^{-1}(x)$.

Chapter 11

38. Find y if $\log_5 125 = y$.

39. Find b if $\log_b 4 = \dfrac{2}{3}$.

40. What is log 10,000?

41. Solve for x. $\log_6(5 + x) + \log_6 x = 2$

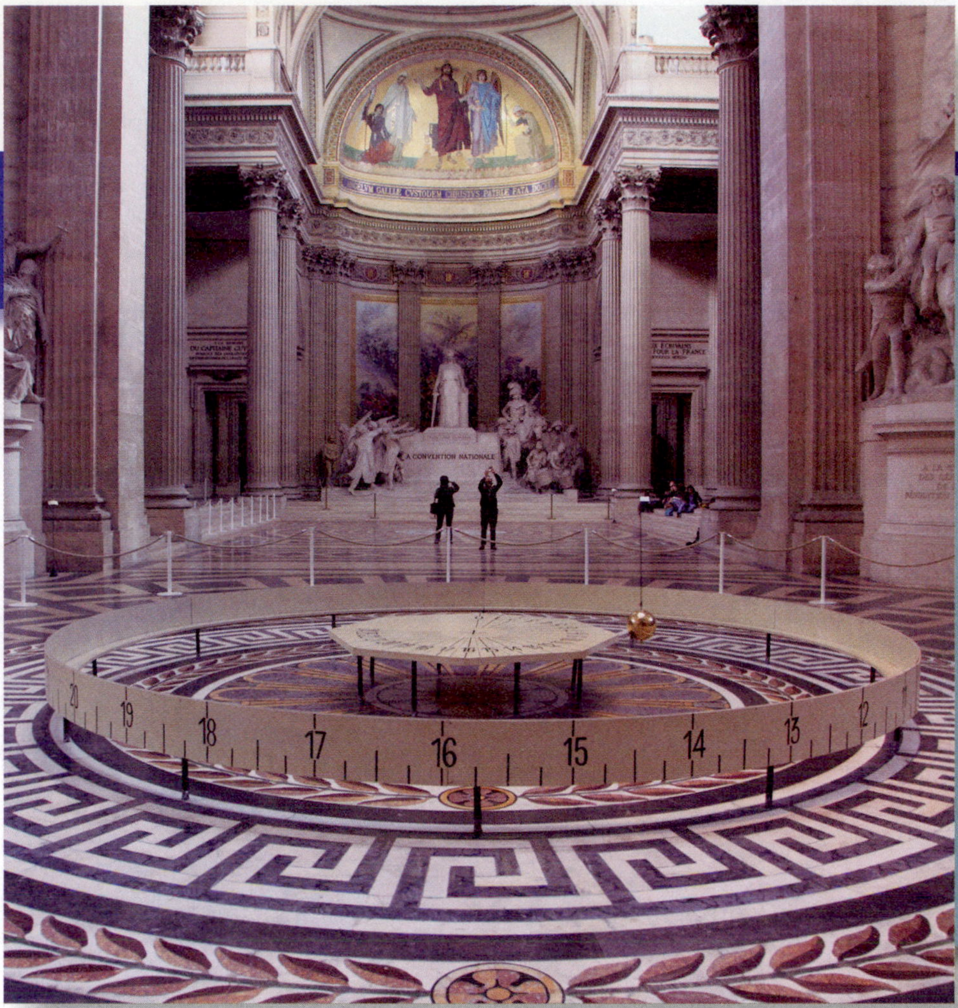

CHAPTER 1

Mathematicians and scientists are able to use formulas to predict a number of things. This is a photograph of the famous Pantheon Pendulum located in the Pantheon building in Paris, France. Here in 1851, the physicist Jean Foucault demonstrated the rotation of Earth by constructing a 67-meter pendulum suspended from the dome. We will study the use of pendulums in this chapter.

Basic Concepts

1.1 The Real Number System

Student Learning Objectives

After studying this section, you will be able to:

1. **Identify the subsets of the real numbers to which a given number belongs.**
2. **Identify which property of the real numbers is used in a given equation.**

① Identifying Subsets of the Real Numbers

A **set** is a collection of objects called **elements.** A set of numbers is simply a listing, within braces { }, of the numbers (elements) in the set. For example,

$$S = \{1, 3, 5, \ldots\}$$

is the set of positive odd integers. The three dots (called an *ellipsis*) here mean that the set is **infinite;** in other words, we haven't written all the possible elements of the set. A set that contains no elements is called the **empty set** and is symbolized by $\varnothing$ or { }.

Some important sets of numbers that we will study are the following:

- Natural numbers
- Whole numbers
- Integers
- Rational numbers
- Irrational numbers
- Real numbers

DEFINITION

The **natural numbers** N (also called the *positive integers*) are the counting numbers:

$$N = \{1, 2, 3, \ldots\}.$$

The **whole numbers** W are the natural numbers plus 0:

$$W = \{0, 1, 2, 3, \ldots\}.$$

The **integers** I are the whole numbers plus the *negatives* of all natural numbers:

$$I = \{\ldots, -3, -2, -1, 0, 1, 2, 3, \ldots\}.$$

The **rational numbers** Q include the integers and all *quotients* of integers (but division by zero is not allowed):

$$Q = \left\{\frac{a}{b} \,\middle|\, a \text{ and } b \text{ are integers but } b \neq 0\right\}.$$

The last expression means "the set of all fractions a divided by b, such that a and b are integers but b is not equal to zero." (The | is read as "such that.") The letters a and b are **variables;** that is, they can represent different numbers.

A rational number can be written as a **terminating decimal,** $\frac{1}{8} = 0.125$, or as a **repeating decimal,** $\frac{2}{3} = 0.6666\ldots$. For repeating decimals we often use a bar over the repeating digits. Thus, we write $0.232323\ldots$ as $0.\overline{23}$ to show that the digits 23 repeat indefinitely. A terminating decimal has a finite number of digits. A repeating decimal goes on forever, but the digits repeat in a definite pattern.

Some numbers in decimal notation are nonterminating and nonrepeating. In other words, we can't write them as quotients of integers. Such numbers are called **irrational numbers.**

The **irrational numbers** are numbers whose decimal forms are nonterminating and nonrepeating.

For example, $\sqrt{3} = 1.7320508\ldots$ can be carried out to an infinite number of decimal places with no repeating pattern of digits. In Chapter 7 we will study irrational numbers extensively. We can now describe the set of real numbers.

The set R of **real numbers** is the set of all numbers that are rational or irrational.

The following figure will help you see the relationships among the sets of numbers that we have described.

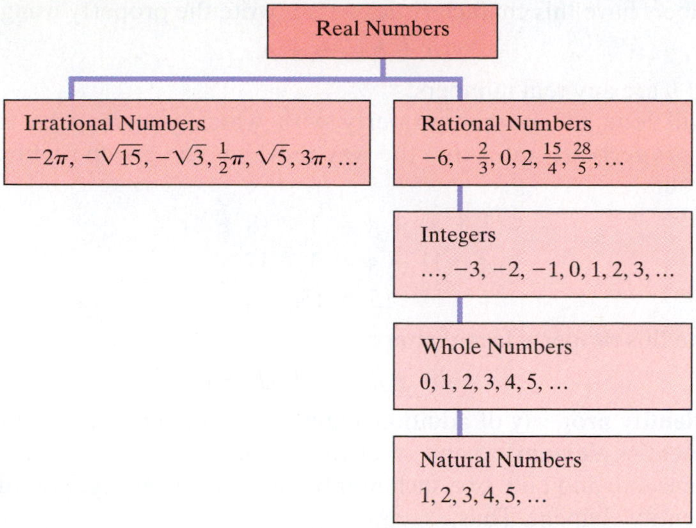

The figure shows that the natural numbers are contained within the set of whole numbers (or we could say that the set of whole numbers contains the set of natural numbers). If all the members of one set are also members of another set, then the first set is a **subset** of the second set. The natural numbers are a subset of the whole numbers. The natural numbers are also a subset of the integers, rational numbers, and real numbers.

EXAMPLE 1 Name the sets to which each of the following numbers belongs.

(a) 5 **(b)** 0.2666 . . . **(c)** 1.4371826138526 . . .

(d) 0 **(e)** $\frac{1}{9}$

Solution

(a) 5 is a natural number. Thus, we can say that it is also a whole number, an integer, a rational number, and a real number.

(b) 0.2666 . . . is a repeating decimal. Thus, it is a rational number and a real number.

(c) 1.4371826138526 . . . doesn't have any repeating pattern. Therefore, it is an irrational number and a real number.

(d) 0 is a whole number, an integer, a rational number, and a real number.

(e) $\frac{1}{9}$ is a quotient of integers and can be written as 0.111 . . . (or $0.\overline{1}$), so it is a rational number and a real number.

Student Practice 1 Name the sets to which the following numbers belong.

(a) 1.26 **(b)** 3 **(c)** $\frac{3}{7}$ **(d)** -2 **(e)** 5.182671 . . .

***NOTE TO STUDENT:** Fully worked-out solutions to all of the Student Practice problems can be found at the back of the text starting at page SP-1.*

② Identifying the Properties of Real Numbers

A property of something is a characteristic that we know to be true. For example, a property of road surfaces is that if they are not treated with chemicals, water on them will form a skin of ice when the temperature drops to 32°F.

Real numbers also have properties. There are some characteristics of real numbers that we have found to be true. For example, we know that we can add any two real numbers and the order in which we add the numbers does not affect the sum: $6 + 8 = 8 + 6$. Mathematicians call this property the **commutative property.** Since all real numbers have this characteristic, we can write the property using letters:

$$a + b = b + a,$$

where a and b are any real numbers.

We will look at another property with which you may be familiar, the **associative property.** It states that the way in which you group numbers does not affect their sum:

$$(3 + 8) + 4 = 3 + (8 + 4)$$
$$11 + 4 = 3 + 12$$
$$15 = 15.$$

We can write this property using letters:

$$(a + b) + c = a + (b + c).$$

The **identity property** of addition states that if we add a unique real number (called an **identity element**) to any other real number, that number is not changed. Unique means one and only one such number has this property. For addition, zero (0) is the identity element. Thus,

$$99 + 0 = 99,$$

and, in general, $a + 0 = a$.

The last property we will describe is the additive **inverse property.** For any real number a there is a unique real number $-a$ such that if we add them we obtain the identity element. For example,

$$8 + (-8) = 0 \quad \text{and} \quad -1.5 + 1.5 = 0.$$

In general, $a + (-a) = 0$.

These are the properties of real numbers for addition. These properties apply to multiplication as well. Multiplication of real numbers is commutative: $3 \cdot 8 = 8 \cdot 3$. It is also associative: $3 \cdot (5 \cdot 6) = (3 \cdot 5) \cdot 6$. Note that the dot $\cdot$ indicates multiplication. Thus, $3 \cdot 5$ means 3 times 5; $a \cdot b$ means a times b. These properties as well as a few others are summarized in the following table.

Properties of Real Numbers

Addition	Property	Multiplication
	For all real numbers a, b, c:	
$a + b = b + a$	Commutative properties	$a \cdot b = b \cdot a$
$a + (b + c) = (a + b) + c$	Associative properties	$a \cdot (b \cdot c) = (a \cdot b) \cdot c$
$a + 0 = a$	Identity properties	$a \cdot 1 = a$
$a + (-a) = 0$	Inverse properties	$a \cdot \frac{1}{a} = 1$ when $a \neq 0$
	Distributive property of multiplication over addition	
	$a(b + c) = a \cdot b + a \cdot c$	

Note that the additive inverse of a number is the **opposite** of the number. The additive inverse of 6 is -6. The additive inverse of $-\frac{2}{3}$ is $\frac{2}{3}$. This idea is different from the idea of a multiplicative inverse. The multiplicative inverse of a number is the **reciprocal** of a number. The multiplicative inverse of 7 is $\frac{1}{7}$. The multiplicative inverse of $-\frac{8}{9}$ is $-\frac{9}{8}$.

TO THINK ABOUT: Multiplication Properties What is the identity element of multiplication? Why? Give a numerical example of the inverse property of multiplication. What is $\frac{1}{a}$ if a is $\frac{2}{5}$?

We will use each of these properties to simplify expressions and to solve equations. It will be helpful for you to become familiar with these properties and to recognize what these properties enable you to do.

EXAMPLE 2 State the name of the property that justifies each statement.

(a) $5 + (7 + 1) = (5 + 7) + 1$ **(b)** $x + 0.6 = 0.6 + x$

Solution

(a) Associative property of addition **(b)** Commutative property of addition

Student Practice 2 State the name of the property that justifies each statement.

(a) $9 + 8 = 8 + 9$ **(b)** $17 + 0 = 17$

The distributive property links multiplication with addition. It states that

$$a(b + c) = (a \cdot b) + (a \cdot c).$$

We can illustrate this property with a numerical example.

$$\begin{aligned} 5(2 + 7) &= (5 \cdot 2) + (5 \cdot 7) \\ 5(9) &= 10 + 35 \\ 45 &= 45 \end{aligned}$$

EXAMPLE 3 State the name of the property that justifies each statement.

(a) $5 \cdot y = y \cdot 5$ **(b)** $6 \cdot \frac{1}{6} = 1$ **(c)** $5(9 + 3) = 5 \cdot 9 + 5 \cdot 3$

Solution

(a) Commutative property of multiplication

(b) Inverse property of multiplication

(c) Distributive property of multiplication over addition

Student Practice 3 State the name of the property that justifies each statement.

(a) $6 \cdot (2 \cdot w) = (6 \cdot 2) \cdot w$ **(b)** $4 \cdot \frac{1}{4} = 1$ **(c)** $6(8 + 7) = 6 \cdot 8 + 6 \cdot 7$

A word about percents: Throughout this book the topic of percent will be used. If you remember percents but need a brief refresher, here are three basic facts.

1. A **percent** means parts per hundred. So 30% means 30 parts per hundred. We write this as 30% = 0.30.
2. To find a percent of a number, change the percent to a decimal and multiply the number by the decimal. To find 30% of 67, we calculate $0.30 \times 67 = 20.1$.
3. To answer such questions as "what percent is 36 out of 48?" we write the fractional form of "36 out of 48," which is $\frac{36}{48}$. We then change the fraction to a decimal by calculating $36 \div 48 = 0.75$. Finally, we change the decimal to percent form, 0.75 = 75%. Thus, 36 out of 48 is 75%.

If this brief review is not sufficient for you to recall the basic ideas of percent, you should refer immediately to a book on beginning algebra or basic college mathematics and study these topics carefully. You will frequently see exercises involving percent in the section exercises.

1.1 Exercises

MyMathLab®

Watch the videos in MyMathLab

Download the MyDashBoard App

Verbal and Writing Skills, Exercises 1–4

1. How do integers differ from whole numbers?

2. How do rational numbers differ from integers?

3. What is a terminating decimal?

4. What is a repeating decimal?

	Indicate the set(s) to which the number belongs.	Natural Numbers	Whole Numbers	Integers	Rational Numbers	Irrational Numbers	Real Numbers
5.	13,001						
6.	0						
7.	-42						
8.	$-\frac{144}{4}$						
9.	-6.1313						
10.	$10.\overline{59}$						
11.	$-\frac{8}{7}$						
12.	$-5\frac{1}{2}$						
13.	$\frac{\pi}{5}$						
14.	$\sqrt{7}$						
15.	0.79						
16.	1.314278619 . . . (no discernible pattern)						
17.	7.040040004 . . . (pattern of increasing numbers of zeros between the 4s)						
18.	54.989898 . . .						

Exercises 19–26 refer to the set $\left\{-25, -\frac{28}{7}, -\frac{18}{5}, -\pi, -0.763, -0.333\ldots, 0, \frac{1}{10}, \frac{2}{7}, \frac{\pi}{4}, \sqrt{3}, 9, 52.8, \frac{283}{5}\right\}$.

19. List the negative integers.

20. List the rational numbers.

21. List the negative real numbers.

22. List the irrational numbers.

23. List the positive rational numbers.

24. List the negative rational numbers.

25. List the negative rational numbers that are not integers.

26. List the positive rational numbers that are integers.

List all the elements of each set.

27. $\{a|a \text{ is a positive odd integer}\}$

28. $\{b|b \text{ is a natural number less than } 10\}$

Name the property that justifies each statement. All variables represent real numbers.

29. $8 + (-5) = (-5) + 8$

30. $(6 + 1) + 3 = 6 + (1 + 3)$

31. $\frac{1}{14} \cdot 14 = 1$

32. $4\left(\frac{1}{3} + 6\right) = 4 \cdot \frac{1}{3} + 4 \cdot 6$

33. $5.6 + 0 = 5.6$

34. $5.6 + (-5.6) = 0$

35. $-\frac{1}{6} \cdot \frac{5}{9} = \frac{5}{9} \cdot \left(-\frac{1}{6}\right)$

36. $(-2.6) + 7 = 7 + (-2.6)$

37. $\frac{3}{7} \cdot 1 = \frac{3}{7}$

38. $\left(-\frac{1}{5}\right) + \left(\frac{1}{5}\right) = 0$

39. $-6(4.5 + 3) = -6(4.5) + (-6)(3)$

40. $0 + \pi = \pi$

41. $2.4 + (3 + \sqrt{5}) = (2.4 + 3) + \sqrt{5}$

42. $3\left(-\frac{x}{2}\right) = -\frac{x}{2}(3)$

43. $\left(-\frac{1}{2}\right) + 9 = 9 + \left(-\frac{1}{2}\right)$

44. $2 \cdot (5 \cdot x) = (2 \cdot 5) \cdot x$

45. $-\sqrt{3}\left(5 \cdot \frac{1}{2}\right) = \left(-\sqrt{3} \cdot 5\right) \cdot \frac{1}{2}$

46. $y \cdot \frac{1}{y} = 1$

47. What is the additive inverse of -12?

48. What is the additive inverse of -1.3?

49. What is the multiplicative inverse of $-\frac{5}{3}$?

50. What is the multiplicative inverse of -10?

To Think About *Use real numbers to visualize an applied situation.*

51. ***Siberian Tiger*** The Siberian tiger is the largest living cat in the world, an endangered subspecies of the tiger family. It is estimated that there are no more than 210 of these animals left in the wild, and approximately the same number in captivity. Of the eight subspecies of tiger, three are extinct, and the remaining five are endangered.

(a) What is the percentage of subspecies of tiger that have become extinct?

(b) What is the percentage of subspecies of tiger that are endangered?

(c) How many subspecies of tiger are not extinct and are not endangered?

52. ***Mountain Height*** According to the world's geologists, the tallest mountain in the world, measured from its base, is the island of Hawaii. The entire island is one big mountain. The tallest peak is Mauna Kea, 13,784 feet above sea level and rising from a sea bottom that is 18,000 feet below sea level. Mount Everest's peak is 29,022 feet above sea level, but it rises from a plateau 12,000 feet high. How high is each mountain measured from its base to its peak? (*Source:* U.S. Department of the Interior.) How many of these mountains measure more than 40,000 feet from base to peak?

Game Show Winnings *In a game show the incremental winnings are \$100, \$200, \$300, \$500, \$1000, \$2000, \$4000, \$8000, \$16,000, \$32,000, \$64,000, \$125,000, \$250,000, \$500,000, and \$1,000,000.*

53. Where does the pattern not follow a logical progression?

54. What number would you obtain if you followed the original progression of the first three numbers—\$100, \$200, \$300—and followed that pattern twenty times after the \$1 million mark?

55. Suppose a new television game show is established where the top prize is \$1,000,000. However, the incremental winnings (listed in reverse order) are \$500,000, \$250,000, \$125,000, \$62,500, \$31,250, etc. Would the set of all numbers that are possible winnings be a set of whole numbers?

56. ***Pass Completions*** During seven games of the 2010 football season, Drew Brees of the New Orleans Saints was the leader in the NFL for most pass completions. He completed 200 out of 287 pass attempts. What percent of his pass attempts were completed? Round to the nearest whole percent.

57. ***Fire Deaths*** The regional fire marshall said that in the past 20 years, smoke detectors have reduced deaths in home fires by about 60%. Twenty years ago 154 people were killed in home fires. If the fire marshall is correct, how many people per year are killed in house fires?

58. ***Gas Mileage*** In 2007, President Bush signed into law a requirement that by the year 2020, cars manufactured in the United States must achieve an average gas mileage of 35 miles per gallon. This will be about a 27% increase over the average gas mileage in 2007. What was the average gas mileage in 2007? Round to the nearest tenth.

Quick Quiz 1.1

Consider the set of numbers $\{-2\pi, -0.5333\ldots, -\frac{3}{11}, 0, 2, \sqrt{7}, \frac{55}{7}, 23.5, 77.222\}$.

1. List the irrational numbers.

2. List the negative real numbers.

3. Name the property that justifies this statement. $3(-3.56 + 9) = 3(-3.56) + 3(9)$

4. **Concept Check** Explain what properties would be needed to justify this statement.
$(4)(3.5 + 9.3) = (9.3 + 3.5)(4)$

1.2 Operations with Real Numbers

Student Learning Objectives

After studying this section, you will be able to:

1. **Find the absolute value of any real number.**
2. **Add, subtract, multiply, and divide real numbers.**
3. **Perform mixed operations of addition, subtraction, multiplication, and division in the proper order.**

① Finding the Absolute Value of a Real Number

We can think of real numbers as points on a line, called the **real number line,** where positive numbers lie to the right of zero and negative numbers lie to the left of zero.

A number line helps us to understand the important concept of absolute value and lets us see how to add and subtract real numbers.

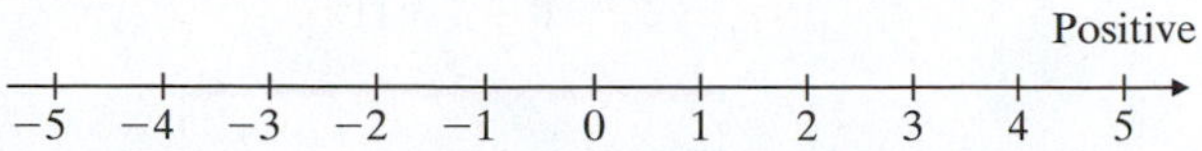

All real numbers can be placed in order on a number line. For any two numbers graphed on a number line, the number to the left is less than the number to the right. This means that the number to the right is greater than the number on the left.

The symbol $<$ means **is less than.** Since -5 is to the left of -3 on the number line, we can write this as $-5 < -3$. The symbol $>$ means **is greater than.** Since -3 is to the right of -5 on the number line, we can write this as $-3 > -5$.

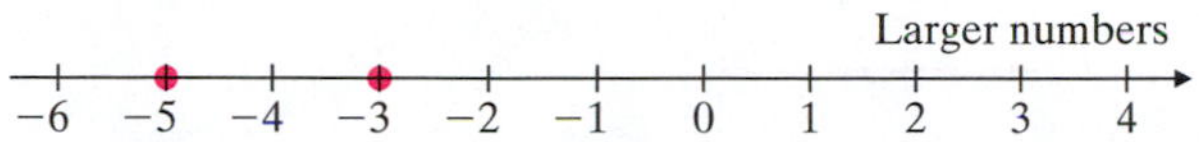

If we are considering some number x that is greater than 4, we would say that x is greater than 4 and write $x > 4$. But if we wanted to write some number that could be 4 or greater than 4, the approach is a little different. We would say that x is greater than or equal to 4 and write $x \geq 4$. In a similar way, we can say "x is less than or equal to 4."

We will discuss inequalities in great detail in Section 2.6. We provide this brief introduction so that we can refer to inequalities while discussing the concept of absolute value.

The **absolute value** of a number x, written $|x|$, can be visualized as its distance from 0 on a number line.

For example, the absolute value of 5 is 5, because 5 is located 5 units from 0 on a number line. The absolute value of -5 is also 5 because -5 is located 5 units from 0 on a number line. We illustrate this concept on the number line below.

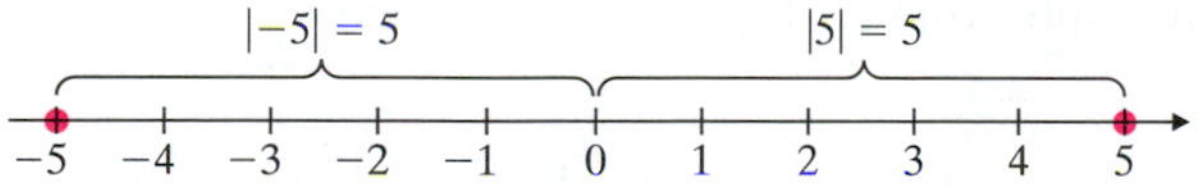

Even though -5 and 5 are located on opposite sides of 0, we are concerned only with distance, which is always positive. (We never say, for instance, that we traveled -10 miles.) Thus, we see that the absolute value of any number is always positive or zero. We formally define absolute value as follows:

DEFINITION OF ABSOLUTE VALUE

Absolute value of x: $$|x| = \begin{cases} x, & \text{if } x \geq 0 \\ -x, & \text{if } x < 0 \end{cases}$$

EXAMPLE 1 Evaluate.

(a) $|6|$ **(b)** $|-8|$ **(c)** $|0|$ **(d)** $|5 - 3|$ **(e)** $|-1.9|$

Solution

(a) $|6| = 6$ **(b)** $|-8| = 8$ **(c)** $|0| = 0$

(d) $|5 - 3| = |2| = 2$ **(e)** $|-1.9| = 1.9$

Student Practice 1 Evaluate.

(a) $|-4|$ **(b)** $|3.16|$ **(c)** $|8 - 8|$ **(d)** $|12 - 7|$ **(e)** $\left|-2\frac{1}{3}\right|$

***NOTE TO STUDENT:** Fully worked-out solutions to all of the Student Practice problems can be found at the back of the text starting at page SP-1.*

② Adding, Subtracting, Multiplying, and Dividing Real Numbers

Adding Real Numbers

Addition of real numbers can be pictured on a number line. For example, to add +6 and +5, we start at 6 on a number line and move 5 units in the positive direction (to the right).

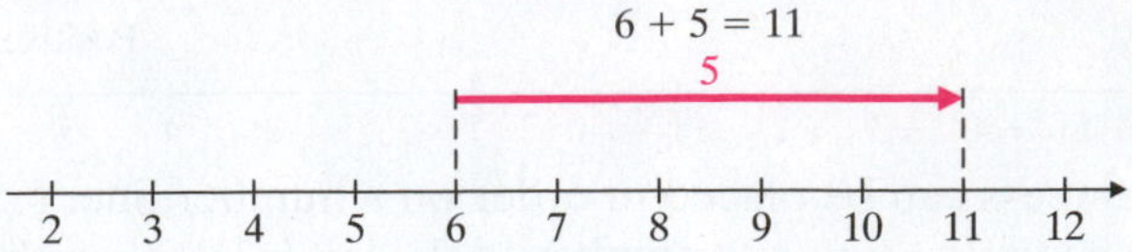

To add two negative numbers, say −4 and −3, we start at −4 and move 3 units in the negative direction (to the left). In other words, we add their absolute values and assign their common sign to the sum. We state this procedure as a rule.

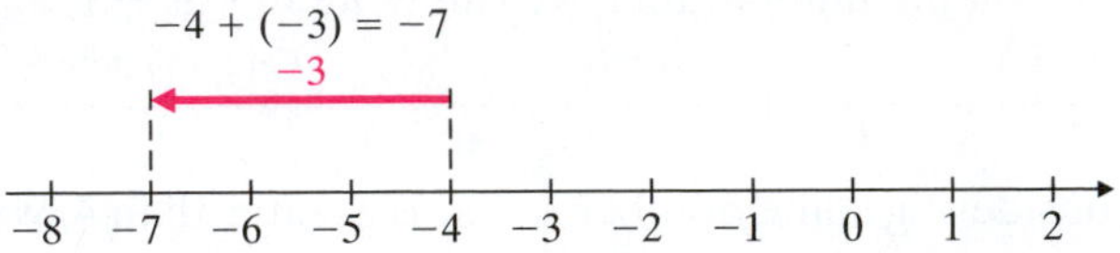

RULE 1.1

To add two real numbers with the *same* sign, add their absolute values. The sum takes the common sign.

EXAMPLE 2 Add.

(a) $-5 + (-0.6)$ **(b)** $-\frac{1}{2} + \left(-\frac{1}{3}\right)$ **(c)** $\frac{2}{5} + \frac{3}{7}$

Solution We apply Rule 1.1 to all three problems.

(a) $-5 + (-0.6) = -5.6$

(b) $-\frac{1}{2} + \left(-\frac{1}{3}\right) = -\frac{3}{6} + \left(-\frac{2}{6}\right) = -\frac{5}{6}$

Obtain a common denominator before adding.

(c) $\frac{2}{5} + \frac{3}{7} = \frac{14}{35} + \frac{15}{35} = \frac{29}{35}$

Student Practice 2 Add.

(a) $3.4 + 2.6$ **(b)** $-\frac{3}{4} + \left(-\frac{1}{6}\right)$ **(c)** $-5 + (-37)$

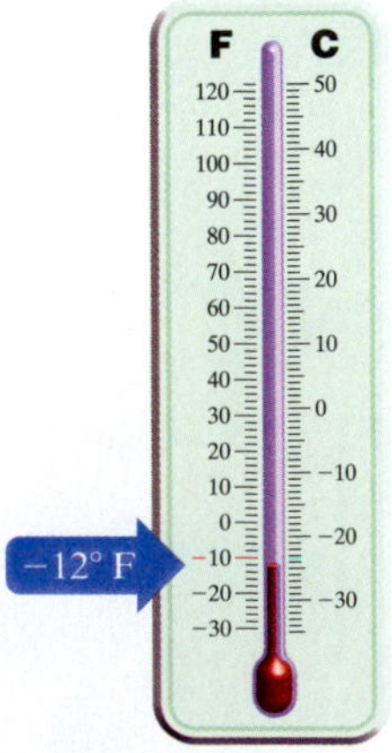

One place where we routinely see positive and negative numbers is a thermometer. If the temperature is −12°F, and then there is a drop of −1.5°F, the new temperature is $-12 + (-1.5) = -13.5$°F.

What do we do if the numbers have different signs? Let's add −4 and 3. Again using a number line, we start at −4 and move 3 units in the positive direction.

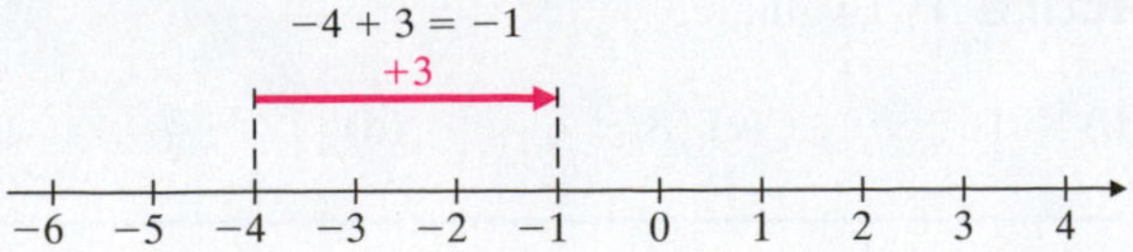

RULE 1.2

To add two real numbers with different signs, find the difference between their absolute values. The answer takes the sign of the number with the larger absolute value.

EXAMPLE 3 Add.

(a) $12 + (-5)$ **(b)** $-\frac{1}{3} + \frac{1}{4}$

Solution

(a) $12 + (-5) = 7$ **(b)** $-\frac{1}{3} + \frac{1}{4} = -\frac{4}{12} + \frac{3}{12} = -\frac{1}{12}$

Student Practice 3 Add.

(a) $24 + (-30)$ **(b)** $-\frac{1}{5} + \frac{2}{3}$

Notice that if you add two numbers that are opposites, the answer will *always* be zero. The **opposite** of a number is the number with the same absolute value but a different sign. The opposite of -6 is 6. The opposite of $\frac{2}{3}$ is $-\frac{2}{3}$.

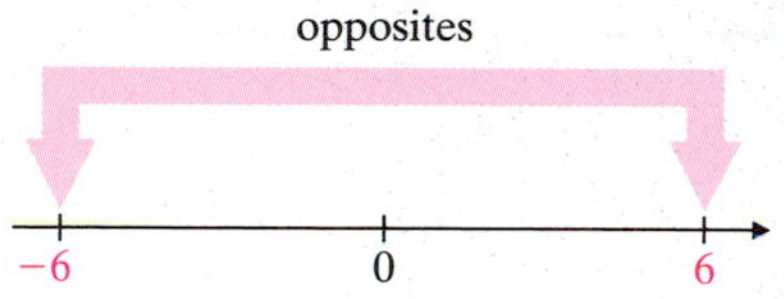

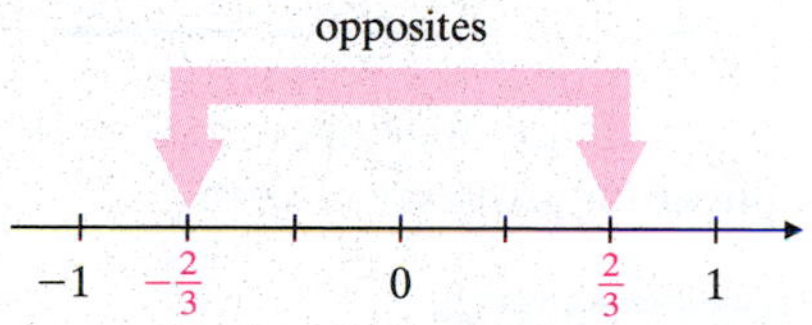

The opposite of a number is also called the **additive inverse.**

TO THINK ABOUT: Additive Inverse Why is the opposite of a number called the additive inverse of the number?

Subtracting Real Numbers

We define subtraction in terms of addition and use our rules for adding real numbers. So actually, you already know how to subtract if you understand how to add real numbers.

RULE 1.3

To subtract b from a, add the opposite (additive inverse) of b to a. Thus, $a - b = a + (-b)$.

EXAMPLE 4 Use Rule 1.3 to subtract.

(a) $5 - 7$ **(b)** $-12 - (-3)$ **(c)** $-0.06 - 0.55$

Solution

(a) $5 - 7 = 5 + (-7) = -2$

(b) $-12 - (-3) = -12 + 3 = -9$

(c) $-0.06 - 0.55 = -0.06 + (-0.55) = -0.61$

Student Practice 4 Use Rule 1.3 to subtract.

(a) $-8 - (-3)$ **(b)** $\frac{1}{2} - \left(-\frac{1}{4}\right)$ **(c)** $-0.35 - 0.67$

Calculator

Negative Numbers

You can use a scientific calculator to find $-32 + (-46)$.

The [+/−] key changes the sign of a number from positive to negative or negative to positive.

Press these keys:

32 [+/−] [+] 46 [+/−] [=]

The display should read:

[−78]

On a graphing calculator, use the [(−)] key to indicate a negative number. Press this before you enter the number. Note that this is not the same as pressing the subtraction key [−]. In general, operations on a graphing calculator are entered in the same order as written.

Try the following:

(a) $-256 + 184$

(b) $94 + (-51)$

(c) $-18 - (-24)$

(d) $-6 + (-10) - (-15)$

Multiplying and Dividing Real Numbers

Recall that the product or quotient of two positive numbers is positive.

$$(+2)(+5) = (2)(5) = 2 + 2 + 2 + 2 + 2 = 10$$

Thus,

$$(+10) \div (+5) = +2 \quad \text{because} \quad (+2)(+5) = +10.$$

Suppose that the signs of the numbers we want to multiply or divide are different. What will be the sign of the answer? We again use repeated addition for multiplication.

$$-2(5) = (-2) + (-2) + (-2) + (-2) + (-2) = -10$$

Similarly,

$$-5(2) = (-5) + (-5) = -10.$$

Thus,

$$-10 \div 5 = -2 \quad \text{and} \quad -10 \div 2 = -5.$$

Notice that multiplying or dividing a negative number by a positive number gives a negative answer.

RULE 1.4
When you multiply or divide two real numbers with different signs, the answer is a negative number.

If b is any real number, then $b \cdot 0 = 0$. Also, $0 \cdot b = 0$. This is sometimes called the **multiplication property of zero.**

EXAMPLE 5 Evaluate.

(a) $5(-8)$ **(b)** $(-5)\left(\frac{2}{17}\right)$ **(c)** $\frac{1.6}{-0.08}$ **(d)** $\frac{-50}{10}$

Solution

(a) $5(-8) = -40$ **(b)** $(-5)\left(\frac{2}{17}\right) = \left(-\frac{5}{1}\right)\left(\frac{2}{17}\right) = -\frac{10}{17}$

(c) $\frac{1.6}{-0.08} = -20$ **(d)** $\frac{-50}{10} = -5$

Student Practice 5 Evaluate.

(a) $\left(-\frac{2}{5}\right)\left(\frac{3}{4}\right)$ **(b)** $\frac{150}{-30}$ **(c)** $\frac{0.27}{-0.003}$ **(d)** $\frac{-12}{24}$

Now let us look at multiplication and division of two negative numbers. We begin by letting a and b be two positive numbers. Then we have the following:

$-a(0) = 0$	The multiplication property of zero states that any number times 0 is zero.
$-a(-b + b) = 0$	Rewrite 0 as $-b + b$ using the additive inverse property.
$(-a)(-b) + (-a)(b) = 0$	Use the distributive property.
$ab \quad + \quad -ab \quad = 0$	By the additive inverse property.

We know that $(-a)(b) = -ab$, a negative number. For the last statement above to be true, $(-a)(-b)$ must be the positive number ab. Thus, we state that $(-a)(-b) = ab$.

RULE 1.5

When you multiply or divide two real numbers with like signs, the answer is a positive number.

EXAMPLE 6 Evaluate.

(a) $\frac{-3}{5} \cdot \frac{-2}{11}$ **(b)** $\frac{-15}{-3}$ **(c)** $\frac{2}{3} \div \frac{5}{7}$

Solution

(a) $\frac{-3}{5} \cdot \frac{-2}{11} = \frac{(-3)(-2)}{(5)(11)} = \frac{6}{55}$ **(b)** $\frac{-15}{-3} = 5$

(c) $\frac{2}{3} \div \frac{5}{7} = \frac{2}{3} \cdot \frac{7}{5} = \frac{(2)(7)}{(3)(5)} = \frac{14}{15}$

Student Practice 6 Evaluate.

(a) $-\frac{2}{7} \cdot \left(-\frac{3}{5}\right)$ **(b)** $-60 \div (-5)$ **(c)** $\frac{3}{7} \div \frac{1}{5}$

Before we leave this topic, we take one more look at division by zero.

$$-\frac{5}{3} \div 0 = ? \quad \text{means} \quad -\frac{5}{3} = ? \times 0.$$

In words, "What times 0 is equal to $-\frac{5}{3}$?" By the multiplication property of zero, any number times zero is zero. Thus, our division problem is impossible to solve. We say that division by 0 is undefined.

Division by 0 is undefined.

③ Performing Mixed Operations

It is very important to perform all mathematical operations in the proper order. Otherwise, you will get wrong answers.

If addition, subtraction, multiplication, and division are written horizontally, do the operations in the following order.

1. Do all multiplications and divisions from left to right.
2. Do all additions and subtractions from left to right.

EXAMPLE 7 Evaluate. $20 \div (-4) \times 3 + 2 + 6 \times 5$

Solution

1. Beginning at the left, we do multiplication and division as we encounter it. Here we encounter division first and then multiplication.

$$\begin{aligned} 20 \div (-4) \times 3 + 2 + 6 \times 5 &= -5 \times 3 + 2 + 6 \times 5 \\ &= -15 + 2 + 6 \times 5 \\ &= -15 + 2 + 30 \end{aligned}$$

Continued on next page

Calculator

Order of Operations

Suppose you need to evaluate the following on a scientific calculator.

$$-3.56 \div 8.9(-6.2) + 5.5(-8.34)$$

Press these keys:

3.56 [+/−] [÷] 8.9

[×] 6.2 [+/−] [+] 5.5

[×] 8.34 [+/−] [=]

The display should read:

[−43.39]

2. Next we add and subtract from left to right.

$$\begin{aligned} -15 + 2 + 30 &= -13 + 30 \\ &= 17 \end{aligned}$$

Student Practice 7 Evaluate. $5 + 7(-2) - (-3) + 50 \div (-2)$

You may encounter a fraction with several operations in the numerator or denominator or both. In such cases complete the operations in the numerator and then complete the operations in the denominator before simplifying the fraction.

EXAMPLE 8 Evaluate.

(a) $-20 \div (-5)(3) + 6 - 5(-2)$ **(b)** $\dfrac{13 - (-3)}{5(-2) - 6(-3)}$

Solution

(a) $-20 \div (-5)(3) + 6 - 5(-2)$

$= 4(3) + 6 - 5(-2)$

$= 12 + 6 + 10$

$= 28$

(b) $\dfrac{13 - (-3)}{5(-2) - 6(-3)} = \dfrac{13 + 3}{-10 + 18} = \dfrac{16}{8} = 2$

Student Practice 8 Evaluate.

(a) $6(-2) + (-20) \div (2)(3)$ **(b)** $\dfrac{7 + 2 - 12 - (-1)}{(-5)(-6) + 4(-8)}$

STEPS TO SUCCESS Doing Homework for Each Class is Critical

Usually every student in the course has to ask the question, "Is homework really that important? Do I actually have to do it?"

You learn by doing. It really makes a difference. Mathematics involves mastering a set of skills that you learn by practicing, not by watching someone else do it. Your instructor may make solving a mathematics problem look very easy, but for you to learn the necessary skills, you must practice them over and over.

The key to success in practice. Learning mathematics is like learning to play a musical instrument, to type, or to play a sport. No matter how much you watch someone else do mathematical calculations, no matter how many books you read on "how to" do it, no matter how easy it appears to be, the key to success in mathematics is practice on each homework set.

Do each kind of problem. Some exercises in a homework set are more difficult than others. Some stress different concepts. Usually you need to work at least all the odd-numbered problems in the exercise set. This allows you to cover the full range of skills in the problem set. Remember, the more exercises you do, the better you will become in your mathematical skills.

Making it personal: Write down your personal reason for why you think doing the homework in each section is very important for success. Which of the three points made do you find is the most convincing? ▼

1.2 Exercises

MyMathLab®

Watch the videos in MyMathLab

Download the MyDashBoard App

Verbal and Writing Skills, Exercises 1 and 2

1. Explain in your own words how to add two real numbers with the same sign or with different signs.

2. Explain in your own words how to multiply or divide two real numbers with the same sign or with different signs.

Evaluate. Assume a and b are greater than 0.

3. $\left|-\frac{2}{3}\right|$

4. $|-27|$

5. $|8.3|$

6. $\left|3\frac{1}{2}\right|$

7. $|9 - 14|$

8. $|2 - 6|$

9. $|-b|$

10. $|a|$

Perform the operations indicated. Write your answer in simplest form.

11. $-6 + (-12)$

12. $-17 + (-3)$

13. $-3 - (-5)$

14. $-9 - 6$

15. $5\left(-\frac{1}{3}\right)$

16. $(-16)(-2)$

17. $\frac{-18}{-2}$

18. $(-42) \div 7$

19. $(-0.3)(0.1)$

20. $(1.2)(-5)$

21. $-4.9 + 10.5$

22. $1.4 - (-3.6)$

23. $-\frac{5}{12} + \frac{7}{18}$

24. $-\frac{7}{9} + \frac{1}{2}$

25. $(-2.4) \div 6$

26. $3.6 \div (-3)$

27. $\left(-\frac{2}{3}\right)\left(-\frac{7}{4}\right)$

28. $\frac{4}{5}\left(-\frac{15}{11}\right)$

29. $(-4)\left(\frac{1}{2}\right) + (-7)(-3)$

30. $(9)\left(-\frac{1}{3}\right) + (-4)(3)$

Perform each of the following operations, if possible.

31. $12 + (-12)$

32. $-5.9 + 5.9$

33. $\frac{-5}{0}$

34. $\frac{-12}{0}$

35. $\frac{0}{-3}$

36. $\frac{0}{4}$

37. $\frac{3 - 3}{-8}$

38. $\frac{-5 + 5}{6}$

39. $\frac{-7 + (-7)}{-14}$

40. $\frac{-4 + (-4)}{-20}$

Mixed Practice

Perform the following operations in the proper order. Write your answer in simplest form.

41. $\frac{17}{18} + \left(-\frac{5}{6}\right)$

42. $\frac{9}{20} + \left(-\frac{1}{5}\right)$

43. $\frac{5}{6} \div \left(-\frac{3}{4}\right)$

44. $-\frac{4}{5} \div \frac{7}{10}$

45. $5 + 6 - (-3) - 8 + 4 - 3$

46. $12 - 3 - (-4) + 6 - 5 - 8$

47. $\dfrac{9(-3) + 7}{3 - 7}$

48. $\dfrac{12 - 2(6)}{1 - 5}$

49. $6(-2) + 3 - 5(-3) - 4$

50. $-7(-3) - 10 + 3(-1) + 8$

51. $15 + 20 \div 2 - 4(3)$

52. $10(0.2) + 6 \div (-0.1)$

53. $\dfrac{6 - 2(7)}{5 - 6}$

54. $\dfrac{3(5) + 1}{4(-1) - 2}$

55. $\dfrac{1 + 49 \div (-7) - (-3)}{-1 - 2}$

56. $\dfrac{72 \div (-4) + 3(-4)}{5 - (-5)}$

57. $4\left(-\dfrac{1}{2}\right) + \dfrac{2}{3}(9)$

58. $-2.4(5) - 1.6(2)$

59. $\dfrac{1.63482 - 2.48561}{(16.05436)(0.07814)}$

60. $(1.783)(2.5725) - (1.0526)(-5.9812)$

To Think About

61. Three numbers are multiplied: $a \cdot b \cdot c = d$. The value of d is a negative number. What are the possible signs of a, b, and c?

62. Three numbers are multiplied: $a \cdot b \cdot c = d$. The value of d is a positive number. What are the possible signs of a, b, and c?

Cumulative Review *Name the property illustrated by each equation.*

63. [1.1.2] $5 + 17 = 17 + 5$

64. [1.1.2] $4 \cdot (3 \cdot 6) = (4 \cdot 3) \cdot 6$

Refer to the set of numbers $\left\{-16, -\frac{1}{2}\pi, 0, \sqrt{3}, 9.36, \frac{19}{2}, 10.\overline{5}\right\}$.

65. [1.1.1] List the irrational numbers.

66. [1.1.1] List the rational numbers.

Quick Quiz 1.2 *Perform the following operations in the proper order. Write your answer in simplest form.*

1. $-8 + (3)(-4)$

2. $5.6 - (-3.8)$

3. $-12 + 15 \div (-5) + 3(6)$

4. Concept Check Explain what steps need to be taken in what order to perform the following operations.

$$\frac{3(-2) + 8}{5 - 9}$$

1.3 Powers, Square Roots, and the Order of Operations

① Raising a Number to a Power

Exponents or powers are used to indicate repeated multiplication. For example, we can write $6 \cdot 6 \cdot 6 \cdot 6$ as 6^4. In the expression 6^4, 4 is the **exponent** or **power** that tells us how many times the **base,** 6, appears as a factor. This is called **exponential notation.**

$$\underset{\uparrow \text{ base}}{6}^{\overset{\text{exponent}}{4}} = \overbrace{6 \cdot 6 \cdot 6 \cdot 6}^{4 \text{ factors}}$$

> **EXPONENTIAL NOTATION**
>
> If x is a real number and n is a positive integer, then
>
> $$x^n = \underbrace{x \cdot x \cdot x \cdot x \cdots}_{n \text{ factors}}$$

EXAMPLE 1 Write in exponential notation.

(a) $(-2)(-2)(-2)(-2)(-2)$ **(b)** $x \cdot x \cdot x \cdot x \cdot x \cdot x \cdot x \cdot x$

Solution

(a) $(-2)(-2)(-2)(-2)(-2) = (-2)^5$ **(b)** $x \cdot x \cdot x \cdot x \cdot x \cdot x \cdot x \cdot x = x^8$

Student Practice 1 Write in exponential notation.

(a) $(-4)(-4)(-4)(-4)$ **(b)** $z \cdot z \cdot z \cdot z \cdot z \cdot z \cdot z$

EXAMPLE 2 Evaluate.

(a) $(-2)^4$ **(b)** -2^4 **(c)** 3^5 **(d)** $(-5)^3$ **(e)** $\left(\frac{1}{3}\right)^3$

Solution

(a) $(-2)^4 = (-2)(-2)(-2)(-2) = (4)(-2)(-2) = (-8)(-2) = 16$

Notice that we are raising -2 to the fourth power. That is, the base is -2. We use parentheses to clearly indicate that the base is negative.

(b) $-2^4 = -(2 \cdot 2 \cdot 2 \cdot 2) = -16$

Here the base is 2. The base is not -2. We wish to find the negative of 2 raised to the fourth power.

(c) $3^5 = 3 \cdot 3 \cdot 3 \cdot 3 \cdot 3 = 243$

(d) $(-5)^3 = (-5)(-5)(-5) = (25)(-5) = -125$

(e) $\left(\frac{1}{3}\right)^3 = \left(\frac{1}{3}\right)\left(\frac{1}{3}\right)\left(\frac{1}{3}\right) = \frac{1}{27}$

Student Practice 2 Evaluate.

(a) $(-3)^5$ **(b)** $(-3)^6$ **(c)** $(-4)^4$

(d) -4^4 **(e)** $\left(\frac{1}{5}\right)^2$

Student Learning Objectives

After studying this section, you will be able to:

① **Raise a number to a positive integer power.**

② **Find square roots of numbers that are perfect squares.**

③ **Evaluate expressions by using the proper order of operations.**

***NOTE TO STUDENT:** Fully worked-out solutions to all of the Student Practice problems can be found at the back of the text starting at page SP-1.*

Calculator

Exponents

You can use a scientific calculator to evaluate $(-5)^3$.

Press these keys:

5 [+/−] [y^x] 3 [=]

The display should read:

[−125]

On a graphing calculator, use the [∧] key to raise to a power. Also, it is necessary to use the parentheses around a negative number in order to raise the entire number to the power.

TO THINK ABOUT: Raising Negative Numbers to a Power Look at Student Practice 2. What do you notice about raising a negative number to an even power? To an odd power? Will this always be true? Why?

② Finding Square Roots

We say that a square root of 16 is 4 because $4 \cdot 4 = 16$. You will note that, since $(-4)(-4) = 16$, another square root of 16 is -4. For practical purposes, we are usually interested in the nonnegative square root. We call this the **principal square root.** $\sqrt{}$ is the principal square root symbol and is called a **radical.**

radical $\rightarrow \sqrt{9} = 3$

radicand (the 9) ... principal square root (the 3)

The number or expression under the radical sign is called the **radicand.** Both 16 and 9 are called *perfect squares* because their square roots are integers.

> If x is an integer and a is a positive real number such that $a = x^2$, then x is a **square root** of a, and a is a **perfect square.**

EXAMPLE 3 Find the square roots of 25. What is the principal square root?

Solution Since $(-5)^2 = 25$ and $5^2 = 25$, the square roots of 25 are 5 and -5. The principal square root is 5.

Student Practice 3 What are the square roots of 49? What is the principal square root of 49?

EXAMPLE 4 Evaluate.

(a) $\sqrt{81}$ **(b)** $\sqrt{0}$ **(c)** $-\sqrt{49}$

Solution

(a) $\sqrt{81} = 9$ because $9^2 = 81$. **(b)** $\sqrt{0} = 0$ because $0^2 = 0$.

(c) $-\sqrt{49} = -\left(\sqrt{49}\right) = -(7) = -7$

Remember: A principal square root is *positive* or *zero.*

Student Practice 4 Evaluate.

(a) $\sqrt{100}$ **(b)** $\sqrt{1}$ **(c)** $-\sqrt{36}$

We can find the square root of a positive number. However, there is no real number for $\sqrt{-4}$ or $\sqrt{-9}$. The square root of a negative number is not a real number.

EXAMPLE 5 Evaluate.

(a) $\sqrt{0.04}$ **(b)** $\sqrt{\dfrac{25}{36}}$ **(c)** $\sqrt{-16}$

Solution

(a) $(0.2)^2 = (0.2)(0.2) = 0.04$. Therefore, $\sqrt{0.04} = 0.2$.

Calculator

Square Roots

You can use a scientific calculator to evaluate $\sqrt{625}$.

Press these keys:

625 [$\sqrt{}$]

The display should read:

[25]

On a graphing calculator, the square root operation is usually accessed by pressing a [SHIFT] or [2nd] key followed by pressing the [x^2] key. This should precede the entry of the number you are taking the square root of.

(b) We can write $\sqrt{\frac{25}{36}}$ as $\frac{\sqrt{25}}{\sqrt{36}}$, and $\frac{\sqrt{25}}{\sqrt{36}} = \frac{5}{6}$. Thus, $\sqrt{\frac{25}{36}} = \frac{5}{6}$.

(c) This is not a real number.

Student Practice 5 Evaluate.

(a) $\sqrt{0.09}$ **(b)** $\sqrt{\frac{4}{81}}$ **(c)** $\sqrt{-25}$

The square roots of some numbers are irrational numbers. For example, $\sqrt{3}$ and $\sqrt{7}$ are irrational numbers. We often use rational numbers to *approximate* square roots that are irrational. They can be found using a calculator with a square root key. To approximate $\sqrt{3}$ on most calculators, we enter the 3 and then press the [√] key. Using a calculator, we might get 1.7320508 as our approximation. If you do not have a calculator, you can use the square root table at the back of the book. From the table we have $\sqrt{3} \approx 1.732$.

③ The Order of Operations of Real Numbers

Parentheses are used in numerical and algebraic expressions to group numbers and variables. When evaluating an expression containing parentheses, evaluate the numerical expressions inside the parentheses first. When we need more than one set of parentheses, we may also use brackets. To evaluate such an expression, work from the inside out.

EXAMPLE 6 Evaluate.

(a) $2(8 + 7) - 12$ **(b)** $12 - 3[7 + 5(6 - 9)]$

Solution

(a)

$2(8 + 7) - 12 = 2(15) - 12$	Work inside the parentheses first.
$= 30 - 12$	Multiply.
$= 18$	Subtract.

(b)

$12 - 3[7 + 5(6 - 9)] = 12 - 3[7 + 5(-3)]$	Begin with the innermost grouping symbols.
$= 12 - 3[7 + (-15)]$	Multiply inside the grouping symbols.
$= 12 - 3(-8)$	Add inside the grouping symbols.
$= 12 + 24$	Multiply.
$= 36$	Add.

Student Practice 6 Evaluate.

(a) $6(12 - 8) + 4$ **(b)** $5[6 - 3(7 - 9)] - 8$

A fraction bar acts like a grouping symbol. We must evaluate the expressions above and below a fraction bar before we divide.

EXAMPLE 7 Evaluate. $\frac{(5)(-2)(-3)}{6 - 8 + 4}$

Solution $\frac{(5)(-2)(-3)}{6 - 8 + 4} = \frac{30}{2} = 15$

Student Practice 7 Evaluate. $\frac{(-6)(3)(2)}{5 - 12 + 3}$

Calculator

Parentheses and Exponents

A scientific calculator will perform calculations in the right order for expressions involving exponents and parentheses. Let us evaluate

$(4.89 - 8.34)^3$
$- 7(2.87 + 9.05) + 12.78.$

On most scientific calculators we use the following keystrokes:

[(] 4.89 [−] 8.34 [)]

[y^x] 3 [−] 7 [×]

[(] 2.87 [+] 9.05 [)]

[+] 12.78 [=]

The display will show the following value:

[−111.723625]

A radical or absolute value bars group the quantities within them. Thus, we simplify the numerical expressions within these grouping symbols before we find the square root or the absolute value.

EXAMPLE 8 Evaluate.

(a) $\sqrt{(-3)^2 + (4)^2}$ **(b)** $|5 - 8 + 7 - 13|$

Solution

(a) $\sqrt{(-3)^2 + (4)^2} = \sqrt{9 + 16} = \sqrt{25} = 5$

(b) $|5 - 8 + 7 - 13| = |-9| = 9$

Student Practice 8 Evaluate.

(a) $\sqrt{(-5)^2 + 12^2}$ **(b)** $|-3 - 7 + 2 - (-4)|$

When many arithmetic operations or grouping symbols are used, we use the following order of operations for calculations with real numbers.

ORDER OF OPERATIONS FOR CALCULATIONS

1. Combine numbers inside grouping symbols.
2. Raise numbers to their indicated powers and take any indicated roots.
3. Multiply and divide numbers from left to right.
4. Add and subtract numbers from left to right.

MC **EXAMPLE 9** Evaluate. $(4 - 6)^3 + 5(-4) + 3$

Solution

$(4 - 6)^3 + 5(-4) + 3 = (-2)^3 + 5(-4) + 3$ Combine $4 - 6$ in parentheses.

$= -8 + 5(-4) + 3$ Cube -2. $(-2)^3 = -8$.

$= -8 - 20 + 3$ Multiply $5(-4)$.

$= -25$ Combine $-8 - 20 + 3$.

Student Practice 9 Evaluate.

$-7 - 2(-3) + (4 - 5)^3$

MC **EXAMPLE 10** Evaluate. $2 + 66 \div 11 \cdot 3 + 2\sqrt{36}$

Solution

$2 + 66 \div 11 \cdot 3 + 2\sqrt{36} = 2 + 66 \div 11 \cdot 3 + 2 \cdot 6$ Evaluate $\sqrt{36}$.

$= 2 + 6 \cdot 3 + 2 \cdot 6$ Divide $66 \div 11$.

$= 2 + 18 + 12$ Multiply $6 \cdot 3$ and $2 \cdot 6$.

$= 32$ Add 2, 18, and 12.

Student Practice 10 Evaluate.

$5 + 6 \cdot 2 - 12 \div (-2) + 3\sqrt{4}$

Students often find that they make errors in exercises like Example 10 when they try to omit steps. Therefore, we recommend that when doing exercises in this section, you should make a separate step for each of the four priorities listed in the preceding box.

EXAMPLE 11 Evaluate. $\dfrac{2 \cdot 6^2 - 12 \div 3}{4 - 8}$

Solution We evaluate the numerator first.

$$\begin{aligned} 2 \cdot 6^2 - 12 \div 3 &= 2 \cdot 36 - 12 \div 3 && \text{Raise to a power.} \\ &= 72 - 4 && \text{Multiply and divide from left to right.} \\ &= 68 && \text{Subtract.} \end{aligned}$$

Next we evaluate the denominator.

$$4 - 8 = -4$$

Thus,

$$\frac{2 \cdot 6^2 - 12 \div 3}{4 - 8} = \frac{68}{-4} = -17$$

Student Practice 11 Evaluate.

$$\frac{2(3) + 5(-2)}{1 + 2 \cdot 3^2 + 5(-3)}$$

As we become a more technologically-oriented society, it is becoming more necessary to understand in what order a computer or a scientific calculator performs arithmetic operations. Be sure you take the time to master this procedure. If you have a scientific calculator, be sure to become familiar with how it works for these types of problems.

STEPS TO SUCCESS What Good Is It to Study Mathematics?

Students often question the value of mathematics. They see little real use for it in their everyday lives. Let us think about three things.

Get a good job. Mathematics is often the key that opens the door to a better paying job or just to get a job if you are unemployed. In our present-day technological world, many people use mathematics daily. Many vocational and professional areas—such as the fields of business, statistics, economics, psychology, finance, computer science, chemistry, physics, engineering, electronics, nuclear energy, banking, quality control, nursing, medical technology, and teaching—require a certain level of expertise in mathematics. Those who want to work in these fields must be able to function at a given mathematical level. Those who cannot will not be able to enter these job areas.

Save money. These are challenging financial times. We are all looking for ways to save money. The more mathematics you learn, the more you will be able to find ways to save money. Several suggestions for saving money are given in this book. Be sure to read over each one and think how it might apply to your life.

Make decisions. Should I buy a car or lease one? Should I buy a house or rent an apartment? Should I drive to work or take public transportation? What career field should I pick if I want to increase my chances of getting a good job? Mathematics will help you to think more clearly and make better decisions because it will help you collect all the facts.

Making it personal: Which of these three paragraphs is the most relevant to you? Which one helps you to see why mathematics can really help you? Write down what you think is the most important reason for you to study mathematics. ▼

1.3 Exercises

MyMathLab®   Watch the videos in MyMathLab   Download the MyDashBoard App

Verbal and Writing Skills, Exercises 1–6

1. In the expression a^3, identify the base and the exponent.

2. When a negative number is raised to an odd power, is the result positive or negative?

3. When a negative number is raised to an even power, is the result positive or negative?

4. Will $-a^n$ always be negative? Why or why not?

5. What are the square roots of 121? Why are there two answers?

6. What is the principal square root?

Write in exponential form.

7. $9 \cdot 9 \cdot 9 \cdot 9$

8. $12 \cdot 12 \cdot 12 \cdot 12 \cdot 12 \cdot 12 \cdot 12$

9. $(-6)(-6)(-6)(-6)(-6)$

10. $(-8)(-8)(-8)(-8)$

11. $x \cdot x \cdot x \cdot x \cdot x \cdot x \cdot y \cdot y \cdot y$

12. $a \cdot a \cdot a \cdot a \cdot b \cdot b \cdot b$

Evaluate.

13. 2^5

14. 7^3

15. $(-5)^2$

16. $(-4)^3$

17. -6^2

18. -3^4

19. -1^4

20. $(-3)^2$

21. $\left(\frac{2}{3}\right)^2$

22. $\left(-\frac{1}{5}\right)^3$

23. $\left(-\frac{1}{4}\right)^4$

24. $\left(\frac{2}{3}\right)^4$

25. $(0.7)^2$

26. $(-0.5)^2$

27. $(0.04)^3$

28. $(0.03)^3$

Evaluate each expression containing a principal square root.

29. $\sqrt{81}$

30. $\sqrt{121}$

31. $-\sqrt{16}$

32. $-\sqrt{64}$

33. $\sqrt{\frac{4}{9}}$

34. $\sqrt{\frac{1}{36}}$

35. $\sqrt{0.09}$

36. $\sqrt{0.25}$

37. $\sqrt{9 + 7}$

38. $\sqrt{12 + 24}$

39. $\sqrt{3599 + 1}$

40. $\sqrt{420 - 20}$

41. $\sqrt{\frac{5}{36} + \frac{31}{36}}$

42. $\sqrt{\frac{1}{9} + \frac{3}{9}}$

43. $\sqrt{-36}$

44. $\sqrt{-49}$

45. $-\sqrt{-0.36}$

46. $-\sqrt{-0.49}$

Follow the proper order of operations to evaluate each of the following.

47. $5(3 - 9) + 7$

48. $4(3 - 5) + 9$

49. $-15 \div 3 + 7(-4)$

50. $16 \div (-8) - 6(-2)$

51. $(-2)(-10) + (-4)^2$

52. $(-8)(-5) + (7)^2$

53. $(5 + 2 - 8)^3 - (-7)$

54. $(8 - 6 - 7)^2 \div 5 - 6$

55. $(-2)^3 + (-4)^2 - 3$

56. $-5(-10) + (-4)^3 - (-20)$

57. $-5^2 + 3(1 - 8)$

58. $-8^2 - 4(1 - 12)$

Mixed Practice *Follow the proper order of operations to evaluate each of the following.*

59. $5[(1.2 - 0.4) - 0.8]$

60. $-3[(4.2 + 0.5) - 0.7]$

61. $4(-6) - 3^2 + \sqrt{25}$

62. $10(-1) - 2^5 + \sqrt{144}$

63. $\dfrac{7 + 2(-4) + 5}{8 - 6}$

64. $\dfrac{15 + 5^2 - 10}{3 + 2}$

65. $\dfrac{-3(2^3 - 1)}{3 - 10}$

66. $\dfrac{4 + 2(3^2 - 12)}{4 - 6}$

67. $\dfrac{|2^2 - 5| - 3^2}{-5 + 3}$

68. $\dfrac{-3 + |3^3 - 30|}{2 - 6}$

69. $\dfrac{\sqrt{(-5)^2 - 3 + 14}}{|19 - 6 + 3 - 25|}$

70. $\dfrac{\sqrt{(-2)^2 - 3 + 3}}{6 - |3 \cdot 2 - 8|}$

71. $\dfrac{\sqrt{6^2 - 3^2 - 2}}{(-3)^2 - 4}$

72. $\dfrac{\sqrt{4 \cdot 7 + 2^3}}{3^2 - 5}$

73. $(5.986)^5$

74. $(0.325)^4$

To Think About

Coin Toss *When a coin is tossed n times, there are 2^n possible results. For example, if a coin is tossed three times, there are $2^3 = 8$ possible results. If we record the outcome of a head as H and the outcome of a tail as T, we can list the eight results as: HHH, HHT, HTH, THH, HTT, TTH, THT, and TTT.*

75. How many more different results are there when a coin is tossed eight times than when it is tossed six times?

76. How many more different results are there when a coin is tossed twelve times than when it is tossed ten times?

Cumulative Review *State the property illustrated by each equation.*

77. **[1.1.2]** $a \cdot \frac{1}{a} = 1$

78. **[1.1.2]** $b + (-b) = 0$

79. **[1.2.2]** ***Polar Bears*** Due to conservation projects and regulated hunting, the polar bear population has risen from five thousand to forty thousand. What is the percent of increase in the polar bear population? (*Hint:* The percent of increase is obtained by dividing the amount of increase by the original amount.)

80. **[1.2.2]** ***Astronomy*** The pressure at the center of the planet Jupiter is 81,000 tons per square inch, and the pressure at the center of Earth is 27,000 tons per square inch. How many times greater is the pressure at the center of Jupiter than the pressure at the center of Earth?

81. **[1.2.2]** ***Professional Basketball*** During the 2005–06 basketball season, LeBron James of the Cleveland Cavaliers scored 2478 points. During the 2006–07 season, he scored 2132 points. This was his greatest drop in points scored in a season between 2003 and 2010. What was the percent of decrease in points scored? (*Hint:* The percent of decrease is obtained by dividing the amount of decrease by the original amount.) Round to the nearest whole percent.

82. **[1.2.2]** ***Professional Basketball*** Kareem Abdul-Jabar is the all-time leading scorer in the NBA. During his 20-year career, he attempted 28,307 field goals (two-point shots). He made approximately 56% of these shots. How many field goals did Abdul-Jabar make during his career? Round to the nearest whole number.

Quick Quiz 1.3

1. Evaluate. $\left(\frac{2}{3}\right)^5$

Follow the proper order of operations to evaluate each of the following.

2. $\dfrac{8 + 3(-2) + 4}{9 - 12}$

3. $(8 - 10)^3 - 30 \div (-3) + \sqrt{17 + 8}$

4. **Concept Check** Explain what operations need to be done in what order to evaluate the following.

$$\frac{\sqrt{(-3)^3 - 6(-2) + 15}}{|3 - 5|}$$

How Am I Doing? Sections 1.1–1.3

How are you doing with your homework assignments in Sections 1.1 to 1.3? Do you feel you have mastered the material so far? Do you understand the concepts you have covered? Before you go further in the textbook, take some time to do each of the following problems.

1.1

Exercises 1 and 2 refer to the set $\{\pi, \sqrt{9}, \sqrt{7}, -5, 3, \frac{6}{2}, 0, \frac{1}{2}, 0.666\ldots\}$.

1. List the irrational real numbers.

2. List the integers.

3. What number sets does $\sqrt{3}$ belong to?

4. What is the property of real numbers that justifies the following equation?

$$(x + y) + z = x + (y + z)$$

5. What is the property of real numbers that justifies the following equation?

$$12\left(\frac{1}{12}\right) = 1$$

1.2

Simplify.

6. $30 \div (-6) + 3 - 2(-5)$

7. $6\left(-\frac{2}{3}\right) + (-5)(-2)$

8. $\dfrac{20 + (5)(-2)}{3 - 7}$

9. $\dfrac{-5 + (-5)}{-15}$

10. $-9 + 6(-2) - (-3)$

1.3

Evaluate.

11. $\sqrt{\dfrac{16}{49}}$

12. $\sqrt{0.81}$

13. 4^4

14. $12 - \sqrt{3^3 + 6(-3)}$

15. $(-4)^3 + 2(3^2 - 2^2)$

16. $\dfrac{4 - 5^2}{14 - \sqrt{16 + 9}}$

17. $|2^2 - 5 - 6|$

18. $\dfrac{\sqrt{(-2)^2 + 5}}{|12 - 15|}$

Now turn to page SA-1 for the answers to each of these problems. Each answer also includes a reference to the objective in which the problem is first taught. If you missed any of these problems, you should stop and review the Examples and Student Practice problems in the referenced objective. A little review now will help you master the material in the upcoming sections of the text.

1. ______
2. ______
3. ______
4. ______
5. ______
6. ______
7. ______
8. ______
9. ______
10. ______
11. ______
12. ______
13. ______
14. ______
15. ______
16. ______
17. ______
18. ______

1.4 Integer Exponents and Scientific Notation

Student Learning Objectives

After studying this section, you will be able to:

1. **Rewrite expressions with negative exponents as expressions with positive exponents.**
2. **Use the product rule of exponents.**
3. **Use the quotient rule of exponents.**
4. **Use the power rules of exponents.**
5. **Express numbers in scientific notation.**

***NOTE TO STUDENT:** Fully worked-out solutions to all of the Student Practice problems can be found at the back of the text starting at page SP-1.*

① Rewriting Expressions with Negative Exponents as Expressions with Positive Exponents

Before we formally define the meaning of a negative exponent, let us look for a pattern.

On this side we decrease each exponent by 1 to obtain the expression on the next line.	$3^4 = 81$ $3^3 = 27$ $3^2 = 9$ $3^1 = 3$ $3^0 = 1$ $3^{-1} = ?$ $3^{-2} = ?$	On this side we divide each number by 3 to obtain the number on the next line.

What results would you expect on the last two lines? $3^{-1} = \frac{1}{3}$? Then $3^{-2} = \frac{1}{3^2} = \frac{1}{9}$. Do you see the pattern? Then we would have

$$3^{-3} = \frac{1}{3^3} = \frac{1}{27} \quad \text{and} \quad 3^{-4} = \frac{1}{3^4} = \frac{1}{81}.$$

Now we are ready to make a formal definition of a negative exponent.

DEFINITION OF NEGATIVE EXPONENTS

If x is any nonzero real number and n is an integer,

$$x^{-n} = \frac{1}{x^n}.$$

EXAMPLE 1 Simplify. Do not leave negative exponents in your answers.

(a) 2^{-5} **(b)** w^{-6}

Solution

(a) $2^{-5} = \frac{1}{2^5} = \frac{1}{32}$ **(b)** $w^{-6} = \frac{1}{w^6}$

Student Practice 1 Simplify. Do not leave negative exponents in your answers.

(a) 3^{-2} **(b)** z^{-8}

EXAMPLE 2 Simplify. $\left(\frac{2}{3}\right)^{-4}$

Solution $\left(\frac{2}{3}\right)^{-4} = \frac{1}{\left(\frac{2}{3}\right)^4} = \frac{1}{\frac{16}{81}} = (1)\left(\frac{81}{16}\right) = \frac{81}{16}$

Student Practice 2 Simplify.

$$\left(\frac{3}{4}\right)^{-2}$$

② The Product Rule of Exponents

Numbers and variables with exponents can be multiplied quite simply if *the base is the same*. For example, we know that

$$(x^3)(x^2) = (x \cdot x \cdot x)(x \cdot x).$$

Since the factor x appears five times, it must be true that

$$x^3 \cdot x^2 = x^5.$$

Hence we can state a general rule.

RULE 1.6 PRODUCT RULE OF EXPONENTS

If x is a real number and n and m are integers, then

$$x^m \cdot x^n = x^{m+n}.$$

Remember that we don't usually write an exponent of 1. Thus, $3 = 3^1$ and $x = x^1$.

EXAMPLE 3 Multiply. Leave your answers in exponential form.

(a) $4^3 \cdot 4^{10}$ **(b)** $y \cdot y^6 \cdot y^3$ **(c)** $(a + b)^2(a + b)^3$

Solution

(a) $4^3 \cdot 4^{10} = 4^{3+10} = 4^{13}$

(b) $y \cdot y^6 \cdot y^3 = y^{1+6+3} = y^{10}$

(c) $(a + b)^2(a + b)^3 = (a + b)^{2+3} = (a + b)^5$ (The base is $a + b$.)

Student Practice 3 Multiply. Leave your answers in exponential form.

(a) $2^8 \cdot 2^{15}$ **(b)** $x^2 \cdot x^8 \cdot x^6$ **(c)** $(x + 2y)^4(x + 2y)^{10}$

EXAMPLE 4 Multiply.

(a) $(3x^2)(5x^6)$ **(b)** $(5x^2y)(-2xy^3)$

Solution

(a) $(3x^2)(5x^6) = (3 \cdot 5)(x^2 \cdot x^6) = 15x^8$

(b) $(5x^2y)(-2xy^3) = (5)(-2)(x^2 \cdot x^1)(y^1 \cdot y^3) = -10x^3y^4$

Student Practice 4 Multiply.

(a) $(7w^3)(2w)$ **(b)** $(-5xy)(-2x^2y^3)$

Using the Product Rule with Negative Exponents

Rule 1.6 says that the exponents are integers. Thus, they can be negative.

EXAMPLE 5 Multiply, then simplify. Do not leave negative exponents in your answer. $(8a^{-3}b^{-8})(2a^5b^5)$

Solution

$$\begin{aligned}(8a^{-3}b^{-8})(2a^5b^5) &= 16a^{-3+5}b^{-8+5} \\ &= 16a^2b^{-3} \\ &= 16a^2\left(\frac{1}{b^3}\right) = \frac{16a^2}{b^3}\end{aligned}$$

Continued on next page

Student Practice 5 Multiply, then simplify. Do not leave negative exponents in your answer.

$$(7xy^{-2})(2x^{-5}y^{-6})$$

③ The Quotient Rule of Exponents

We now develop the rule for dividing numbers with exponents. We know that

$$\frac{x^5}{x^3} = \frac{x \cdot x \cdot \cancel{x} \cdot \cancel{x} \cdot \cancel{x}}{\cancel{x} \cdot \cancel{x} \cdot \cancel{x}} = x \cdot x = x^2.$$

Note that $x^{5-3} = x^2$. This leads us to the following general rule.

RULE 1.7 QUOTIENT RULE OF EXPONENTS

If x is a nonzero real number and m and n are integers,

$$\frac{x^m}{x^n} = x^{m-n}.$$

EXAMPLE 6 Divide. Leave your answers in exponential form with no negative exponents.

(a) $\frac{x^{12}}{x^3}$ **(b)** $\frac{5^{16}}{5^7}$ **(c)** $\frac{y^3}{y^{20}}$ **(d)** $\frac{2^{20}}{2^{30}}$

Solution

(a) $\frac{x^{12}}{x^3} = x^{12-3} = x^9$

(b) $\frac{5^{16}}{5^7} = 5^{16-7} = 5^9$

(c) $\frac{y^3}{y^{20}} = y^{3-20} = y^{-17} = \frac{1}{y^{17}}$

(d) $\frac{2^{20}}{2^{30}} = 2^{20-30} = 2^{-10} = \frac{1}{2^{10}}$

Student Practice 6 Divide. Leave your answers in exponential form with no negative exponents.

(a) $\frac{w^8}{w^6}$ **(b)** $\frac{3^7}{3^3}$ **(c)** $\frac{x^5}{x^{16}}$ **(d)** $\frac{4^5}{4^8}$

Our quotient rule leads us to an interesting situation if $m = n$.

$$\frac{x^m}{x^m} = x^{m-m} = x^0$$

But what exactly is x^0? Whenever we divide any nonzero value by itself we always get 1, so we would therefore expect that $x^0 = 1$. But can we prove that? Yes.

Since $x^{-n} = \frac{1}{x^n}$,

Then $x^{-n} \cdot x^n = 1$

$x^{-n+n} = 1$ Using the product rule.

$x^0 = 1$ Since $-n + n = 0$.

RULE 1.8 RAISING A NUMBER TO THE ZERO POWER

For any nonzero real number x, $x^0 = 1$.

EXAMPLE 7 Simplify. Do not leave negative exponents in your answers.

(a) $3x^0$ **(b)** $(3x)^0$ **(c)** $(-2x^{-5})(y^3)^0$

Solution

(a) $3x^0 = 3(1)$ Since $x^0 = 1$.

$= 3$

(b) $(3x)^0 = 1$ Note that the entire expression is raised to the zero power.

(c) $(-2x^{-5})(y^3)^0 = (-2x^{-5})(1)$

$$= (-2)\left(\frac{1}{x^5}\right) = \frac{-2}{x^5} \quad \text{or} \quad -\frac{2}{x^5}$$

Student Practice 7 Simplify. Do not leave negative exponents in your answers.

(a) $6y^0$ **(b)** $(3xy)^0$ **(c)** $(5^{-3})(2a)^0$

EXAMPLE 8 Divide. Then simplify your answers. Do not leave negative exponents in your answers.

(a) $\dfrac{26x^3y^4}{-13xy^8}$

(b) $\dfrac{-150a^3b^4c^2}{-300abc^2}$

Solution

(a) $\dfrac{26x^3y^4}{-13xy^8} = \dfrac{26}{-13} \cdot \dfrac{x^3}{x} \cdot \dfrac{y^4}{y^8} = -2x^2y^{-4} = -\dfrac{2x^2}{y^4}$

(b) $\dfrac{-150a^3b^4c^2}{-300abc^2} = \dfrac{-150}{-300} \cdot \dfrac{a^3}{a} \cdot \dfrac{b^4}{b} \cdot \dfrac{c^2}{c^2} = \dfrac{1}{2} \cdot a^2 \cdot b^3 \cdot c^0 = \dfrac{a^2b^3}{2}$

Student Practice 8 Divide. Then simplify your answers. Do not leave negative exponents in your answers.

(a) $\dfrac{30x^6y^5}{20x^3y^2}$ **(b)** $\dfrac{-15a^3b^4c^4}{3a^5b^4c^2}$

For the remainder of this chapter, we will assume that for all exercises involving exponents, a simplified answer should not contain negative exponents.

EXAMPLE 9 Divide, then simplify your answer. $\dfrac{3x^{-5}y^{-6}}{27x^2y^{-8}}$

Solution

$$\frac{3x^{-5}y^{-6}}{27x^2y^{-8}} = \frac{1}{9}x^{-5-2}y^{-6-(-8)} = \frac{1}{9}x^{-5-2}y^{-6+8} = \frac{1}{9}x^{-7}y^2 = \frac{y^2}{9x^7}$$

Student Practice 9 Divide, then simplify your answer.

$$\frac{2x^{-3}y}{4x^{-2}y^5}$$

④ The Power Rules of Exponents

Note that $(x^4)^3 = x^4 \cdot x^4 \cdot x^4 = x^{4+4+4} = x^{4\cdot 3} = x^{12}$. In the same way we can show

$$(xy)^3 = x^3y^3$$

$$\text{and } \left(\frac{x}{y}\right)^3 = \frac{x^3}{y^3} \quad (y \neq 0).$$

Therefore, we have the following rules.

RULE 1.9 POWER RULES OF EXPONENTS

If x and y are any real numbers and n and m are integers,

$$(x^m)^n = x^{mn}, \quad (xy)^n = x^n y^n, \text{ and}$$

$$\left(\frac{x}{y}\right)^n = \frac{x^n}{y^n}, \qquad \text{if } y \neq 0.$$

EXAMPLE 10 Use the power rules of exponents to simplify.

(a) $(x^6)^5$ **(b)** $(2^8)^4$ **(c)** $[(a + b)^2]^4$

Solution

(a) $(x^6)^5 = x^{6\cdot 5} = x^{30}$

(b) $(2^8)^4 = 2^{32}$ Careful. Don't change the base of 2.

(c) $[(a + b)^2]^4 = (a + b)^8$ The base is $a + b$.

Student Practice 10 Use the power rules of exponents to simplify.

(a) $(w^3)^8$ **(b)** $(5^2)^5$ **(c)** $[(x - 2y)^3]^3$

EXAMPLE 11 Simplify.

(a) $(3xy^2)^4$ **(b)** $\left(\frac{2a^2b^3}{3ab^4}\right)^3$ **(c)** $(2a^2b^{-3}c^0)^{-4}$

Solution

(a) $(3xy^2)^4 = 3^4x^4y^8 = 81x^4y^8$

(b) $\left(\frac{2a^2b^3}{3ab^4}\right)^3 = \frac{2^3a^6b^9}{3^3a^3b^{12}} = \frac{8a^3}{27b^3}$

(c) $(2a^2b^{-3}c^0)^{-4} = 2^{-4}a^{-8}b^{12} = \frac{b^{12}}{2^4a^8} = \frac{b^{12}}{16a^8}$

Student Practice 11 Simplify.

(a) $(4x^3y^4)^2$ **(b)** $\left(\frac{4xy}{3x^5y^6}\right)^3$ **(c)** $(3xy^2)^{-2}$

We need to derive one more rule. You should be able to follow the steps.

$$\frac{x^{-m}}{y^{-n}} = \frac{\frac{1}{x^m}}{\frac{1}{y^n}} = \frac{1}{x^m} \cdot \frac{y^n}{1} = \frac{y^n}{x^m}$$

RULE 1.10 RULE OF NEGATIVE EXPONENTS

If n and m are positive integers and x and y are nonzero real numbers, then

$$\frac{x^{-m}}{y^{-n}} = \frac{y^n}{x^m}.$$

For example, $\frac{x^{-5}}{y^{-6}} = \frac{y^6}{x^5}$ and $\frac{2^{-3}}{x^{-4}} = \frac{x^4}{2^3} = \frac{x^4}{8}$.

EXAMPLE 12 Simplify.

(a) $\frac{3x^{-2}y^3z^{-1}}{4x^3y^{-5}z^{-2}}$ **(b)** $\left(\frac{5xy^{-3}}{2x^{-4}yz^{-3}}\right)^{-2}$

Solution

(a) First remove all negative exponents.

$$\frac{3x^{-2}y^3z^{-1}}{4x^3y^{-5}z^{-2}} = \frac{3y^3y^5z^2}{4x^3x^2z^1}$$ Only variables with negative exponents will change their position.

$$= \frac{3y^8z^2}{4x^5z^1}$$

$$= \frac{3y^8z}{4x^5}$$

(b) First remove the parentheses by using the power rules of exponents.

$$\left(\frac{5xy^{-3}}{2x^{-4}yz^{-3}}\right)^{-2} = \frac{5^{-2}x^{-2}y^6}{2^{-2}x^8y^{-2}z^6}$$

$$= \frac{2^2y^6y^2}{5^2x^8x^2z^6}$$

$$= \frac{4y^8}{25x^{10}z^6}$$

Student Practice 12 Simplify.

(a) $\frac{7x^2y^{-4}z^{-3}}{8x^{-5}y^{-6}z^2}$

(b) $\left(\frac{4x^2y^{-2}}{x^{-4}y^{-3}}\right)^{-3}$

EXAMPLE 13 Simplify $(-3x^2)^{-2}(2x^3y^{-2})^3$. Express your answer with positive exponents only.

Solution

$$(-3x^2)^{-2}(2x^3y^{-2})^3 = (-3)^{-2}x^{-4} \cdot 2^3x^9y^{-6}$$

$$= \frac{2^3x^9}{(-3)^2x^4y^6} = \frac{8x^9}{9x^4y^6} = \frac{8x^5}{9y^6}$$

Student Practice 13 Simplify $(2x^{-3})^2(-3xy^{-2})^{-3}$. Express your answer with positive exponents only.

⑤ Scientific Notation

Scientific notation is a convenient way to write very large or very small numbers. For example, we can write 50,000,000 as 5×10^7 since $10^7 = 10{,}000{,}000$, and we can write

$$0.0000000005 = 5 \times 10^{-10} \text{ since } 10^{-10} = \frac{1}{10^{10}} = \frac{1}{10{,}000{,}000{,}000} = 0.0000000001.$$

In scientific notation, the first factor is a number that is greater than or equal to 1, but less than 10. The second factor is a power of 10.

SCIENTIFIC NOTATION

A positive number written in **scientific notation** has the form $a \times 10^n$, where $1 \leq a < 10$ and n is an integer.

Decimal form and scientific notation are just equivalent forms of the same number. To change a number from decimal notation to scientific notation, follow the steps below. Remember that the first factor must be a number between 1 and 10. This determines where to place the decimal point.

RULE 1.11 CONVERTING FROM DECIMAL NOTATION TO SCIENTIFIC NOTATION

1. Move the decimal point from its original position to the right of the first nonzero digit.
2. Count the number of places that you moved the decimal point. This number is the absolute value of the power of 10 (that is, the exponent).
3. If you moved the decimal point to the right, the exponent is negative; if you moved it to the left, the exponent is positive.

EXAMPLE 14 Write in scientific notation.

(a) 7816 **(b)** 15,200,000 **(c)** 0.0123 **(d)** 0.00046

Solution

(a) $7816 = 7.816 \times 10^3$ We moved the decimal point three places to the left, so the power of 10 is 3.

(b) $15{,}200{,}000 = 1.52 \times 10^7$

(c) $0.0123 = 1.23 \times 10^{-2}$ We moved the decimal point two places to the right, so the power of 10 is -2.

(d) $0.00046 = 4.6 \times 10^{-4}$

Student Practice 14 Write in scientific notation.

(a) 128,320 **(b)** 476 **(c)** 0.0786 **(d)** 0.007

We can also change a number from scientific notation to decimal notation. We simply move the decimal point to the right or to the left the number of places indicated by the power of 10.

EXAMPLE 15 Write in decimal form.

(a) 8.8632×10^4 **(b)** 6.032×10^{-2} **(c)** 4.4861×10^{-5}

Solution

(a) $8.8632 \times 10^4 = 88{,}632$

Move the decimal point two places to the left.

(b) $6.032 \times 10^{-2} = 0.06032$

(c) $4.4861 \times 10^{-5} = 0.000044861$

Student Practice 15 Write in decimal form.

(a) 4.62×10^6 **(b)** 1.973×10^{-3} **(c)** 4.931×10^{-1}

Using scientific notation and the laws of exponents can greatly simplify calculations.

EXAMPLE 16 Evaluate using scientific notation. $\dfrac{(0.000000036)(0.002)}{0.000012}$

Solution Rewrite the expression in scientific notation.

$$\frac{(3.6 \times 10^{-8})(2 \times 10^{-3})}{1.2 \times 10^{-5}}$$

Now rewrite using the commutative property.

$$\frac{\overset{3}{\cancel{(3.6)}}(2)(10^{-8})(10^{-3})}{\underset{1}{\cancel{(1.2)}}(10^{-5})} = \frac{6.0}{1} \times \frac{10^{-11}}{10^{-5}}$$ Simplify and use the laws of exponents.

$$= 6.0 \times 10^{-11-(-5)}$$

$$= 6.0 \times 10^{-6}$$

Student Practice 16 Evaluate using scientific notation.

$$\frac{(55{,}000)(3{,}000{,}000)}{5{,}500{,}000}$$

EXAMPLE 17 In a scientific experiment, a scientist stated that a proton is theoretically traveling at 3.36×10^5 meters per second. If that is the correct speed, how far would the proton travel in 2×10^4 seconds?

Solution Here we use the idea that the rate times the time equals the distance. So we multiply 3.36×10^5 by 2×10^4. Using the commutative and associative properties, we can write this as

$$3.36 \times 2 \times 10^5 \times 10^4 = 6.72 \times 10^9.$$

The proton would travel 6.72×10^9 meters.

Student Practice 17 The mass of Earth is considered to be 6.0×10^{24} kilograms. A scientist is studying a star whose mass is 3.4×10^5 times larger than the mass of Earth. If the scientist is correct, what is the mass of this star?

Calculator

Scientific Notation

Most scientific calculators can display only eight digits at one time. Numbers with more than eight digits are shown in scientific notation. [1.12 E 08] means 1.12×10^8. Note that the display on your calculator may be slightly different.

You can use the calculator to compute large numbers by entering the numbers using scientific notation. For example, to compute

$$(7.48 \times 10^{24}) \times (3.5 \times 10^8)$$

on a scientific calculator, press these keys:

7.48 [EE] 24 [×]

3.5 [EE] 8 [=]

The display should read:

[2.618 E 33]

1.4 Exercises

MyMathLab® Watch the videos in MyMathLab Download the MyDashBoard App

Simplify. Rewrite all expressions with positive exponents only.

1. 3^{-2} **2.** 4^{-3} **3.** x^{-5} **4.** y^{-4}

5. $(-7)^{-2}$ **6.** $(-2)^{-5}$ **7.** $\left(-\frac{1}{9}\right)^{-1}$ **8.** $\left(-\frac{1}{2}\right)^{-4}$

Multiply. Leave your answer in exponential form.

9. $x^4 \cdot x^8$ **10.** $y^{10} \cdot y$ **11.** $17^4 \cdot 17$

12. $12^5 \cdot 12^9$ **13.** $(3x)(-2x^5)$ **14.** $(4y^2)(2y)$

15. $(-11x^2y^2)(-x^4y^7)$ **16.** $(-15x^4y)(-6xy^5)$ **17.** $4x^0y$

18. $-6a^2b^0$ **19.** $(3xy)^0(7xy)$ **20.** $-9a^3b^5(-ab)^0$

21. $(-6x^2yz^0)(-4x^0y^2z)$ **22.** $(5^0a^3b^4)(-2a^3b^0)$

23. $\left(-\frac{3}{5}m^{-2}n^4\right)(5m^2n^{-5})$ **24.** $\left(\frac{3}{4}mn^{-2}\right)(8m^{-4}n^3)$

Divide. Simplify your answers.

25. $\frac{x^{16}}{x^5}$ **26.** $\frac{x^{17}}{x^3}$ **27.** $\frac{a^{20}}{a^{25}}$ **28.** $\frac{x^4}{x^7}$

29. $\frac{2^8}{2^5}$ **30.** $\frac{3^{16}}{3^{18}}$ **31.** $\frac{2x^3}{x^8}$ **32.** $\frac{4y^3}{8y}$

33. $\frac{-15x^4yz}{3xy}$ **34.** $\frac{40a^3b}{-5a^3}$ **35.** $\frac{-20a^{-3}b^{-8}}{14a^{-5}b^{-12}}$ **36.** $\frac{-24x^5y^8}{-9x^{-1}y^{-2}}$

Use the power rules to simplify each expression.

37. $(x^2)^8$ **38.** $(a^5)^7$ **39.** $(3a^5b)^4$

40. $(2xy^6)^5$ **41.** $\left(\frac{x^2y^3}{z}\right)^6$ **42.** $\left(\frac{x^3}{y^5z^8}\right)^4$

43. $\left(\frac{3ab^{-2}}{4a^0b^4}\right)^2$ **44.** $\left(\frac{5a^3b}{-3a^{-2}b^0}\right)^3$

45. $\left(\frac{2xy^2}{x^{-3}y^{-4}}\right)^{-3}$ **46.** $\left(\frac{3x^{-4}y}{x^{-3}y^2}\right)^{-2}$

47. $(x^{-1}y^3)^{-2}(2x)^2$ **48.** $(x^2y^{-1})^{-2}(3x^{-3})^2$

49. $\frac{(-3m^5n^{-1})^3}{(mn)^2}$ **50.** $\frac{(m^4n^3)^{-1}}{(-5m^{-3}n^4)^2}$

Mixed Practice *Simplify. Express your answers with positive exponents only.*

51. $\dfrac{2^{-3}a^2}{2^{-4}a^{-2}}$

52. $\dfrac{3^4a^{-3}}{3^3a^4}$

53. $\left(\dfrac{1}{3}y\right)^{-3}$

54. $\left(\dfrac{2}{5}x^3\right)^{-2}$

55. $\left(\dfrac{y^{-3}}{x}\right)^{-2}$

56. $\left(\dfrac{y}{z^{-4}}\right)^{-3}$

57. $\dfrac{a^0b^{-4}}{a^{-3}b}$

58. $\dfrac{c^{-3}d^{-2}}{c^{-4}d^{-5}}$

59. $\left(\dfrac{14x^{-3}y^{-3}}{7x^{-4}y^{-3}}\right)^{-2}$

60. $\left(\dfrac{25x^{-1}y^{-6}}{5x^{-4}y^{-6}}\right)^{-2}$

61. $\dfrac{7^{-8}\cdot 5^{-6}}{7^{-9}\cdot 5^{-5}}$

62. $\dfrac{9^{-2}\cdot 8^{-10}}{9^{-1}\cdot 8^{-9}}$

63. $(9x^{-2}y)\left(-\dfrac{2}{3}x^3y^{-2}\right)$

64. $(-12x^5y^{-2})\left(\dfrac{3}{4}x^{-6}y^3\right)$

65. $(-3.6982x^3y^4)^7$

66. $\dfrac{1.98364\times 10^{-14}}{4.32571\times 10^{-16}}$

Write in scientific notation.

67. 38

68. 759

69. 1,730,000

70. 405,300,000

71. 0.83

72. 0.0654

73. 0.0008125

74. 0.0000048

Write in decimal notation.

75. 7.13×10^5

76. 4.006×10^6

77. 3.07×10^{-1}

78. 7.07×10^{-3}

79. 9.01×10^{-7}

80. 6.668×10^{-9}

Perform the calculations indicated. Express your answers in scientific notation.

81. $(3.1\times 10^{-4})(1.5\times 10^{-2})$

82. $(1.8\times 10^{-3})(4.0\times 10^8)$

83. $\dfrac{3.6\times 10^{-5}}{1.2\times 10^{-6}}$

84. $\dfrac{10.5\times 10^{-10}}{2.1\times 10^2}$

To Think About

85. ***Amazon River*** The Amazon River is famous for sending forth one-fifth of all the moving freshwater on Earth, amounting to 7,200,000 cubic *feet* per second. How many cubic *meters* per second pour out of the mouth of the Amazon River? (Use 1 foot $\approx$ 0.305 meter.)

86. ***Sensory Perception*** Certain moths and butterflies can detect sweetness in a solution when the ratio of sugar to water is 1:300,000. Humans, on the other hand, are considered very sensitive if they are able to detect sweetness in a solution of one part sugar to two hundred parts water. How much more sensitive are moths and butterflies than the most sensitive humans?

Applications

87. ***Sound from Bats*** A bat emits a sound at a very high frequency that humans cannot hear. The frequency is approximately 5.1×10^4 cycles per second. How many cycles would occur in 1.5×10^2 seconds?

88. ***Speed of Light*** In one year, a beam of light travels 5.87×10^{12} miles. How far would that light travel in 5×10^3 years?

89. ***Oxygen Molecules*** The weight of one oxygen molecule is 5.3×10^{-23} gram. How much would 2×10^4 molecule of oxygen weigh?

90. ***Solar Probe*** The average distance from Earth to the sun is 4.90×10^{11} feet. If a solar probe is launched from Earth and travels at 2×10^4 feet per second, how long would it take to reach the sun?

Cumulative Review *Evaluate.*

91. **[1.4.5]** ***Planet Mass*** The mass of Mercury is about 3.64×10^{20} tons. The mass of Jupiter, the largest planet, is about 2.09×10^{24} tons. How many times greater than the mass of Mercury is the mass of Jupiter? Round to the nearest whole number.

92. **[1.4.5]** ***Calories*** One calorie is equal to 2.78×10^{-7} kilowatt-hours. How many calories are in 5.56×10^3 kilowatt-hours?

93. **[1.3.3]** $-9 + 14 \div (-2) + 5^2$

94. **[1.3.3]** $-6^2 + 16 \div 2$

Quick Quiz 1.4 *Simplify. Do not leave negative exponents in your answers.*

1. $\dfrac{20x^4y^3}{25x^{-2}y^6}$

2. $\left(\dfrac{3a^{-4}b^2}{a^3}\right)^2$

3. Write in scientific notation. 0.000578

4. **Concept Check** Explain how you would simplify the following.

$$(3x^2y^{-3})(2x^4y^2)$$

1.5 Operations with Variables and Grouping Symbols

① Combining Like Terms in an Algebraic Expression

Student Learning Objectives

After studying this section, you will be able to:

1. **Combine like terms in an algebraic expression.**
2. **Multiply algebraic expressions using the distributive property.**
3. **Remove grouping symbols in their proper order to simplify algebraic expressions.**

A collection of numerical values, variables, and operation signs is called an **algebraic expression.** An algebraic expression sometimes contains the sum or difference of several *terms*. A **term** is a real number, a variable, or a product or quotient of numbers and variables.

> Terms can *always* be separated by + signs.

EXAMPLE 1 List the terms in each algebraic expression.

(a) $5x + 3y^2$ **(b)** $5x^2 - 3xy - 7$

Solution

(a) $5x$ is a product of a real number (5) and a variable (x), so $5x$ is a term. $3y^2$ is also a product of a real number and a variable, so $3y^2$ is a term.

(b) Note that we can write $5x^2 - 3xy - 7$ in an equivalent form with plus signs: $5x^2 + (-3xy) + (-7)$. This second form helps us to identify more readily the three terms. The terms are $5x^2$, $-3xy$, and -7.

Student Practice 1 List the terms in each algebraic expression.

(a) $7x - 2w^3$ **(b)** $5 + 6x + 2y$

NOTE TO STUDENT: Fully worked-out solutions to all of the Student Practice problems can be found at the back of the text starting at page SP-1.

Any factor in a term is the **coefficient** of the product of the remaining factors. For example, $4x^2$ is the coefficient of y in the term $4x^2y$, and $4y$ is the coefficient of x^2 in $4x^2y$. The numerical coefficient of a term is the numerical value multiplied by the variables. The numerical coefficient of $4x^2y$ is 4. Mathematicians often use "coefficient" to mean the "numerical coefficient." We will use it this way for the remainder of the book. Thus, in the expression $-5x^2y$, the coefficient will be considered to be -5. If no numerical coefficient appears before a variable in a term, the coefficient is understood to be 1. For example, the coefficient of xy is 1. The coefficient of $-x$ is -1.

EXAMPLE 2 Identify the coefficient of each term.

(a) $5x^2 - 2x + 3xy$ **(b)** $8x^3 - 12xy^2 + y$

(c) $\frac{1}{2}x^2 + \frac{1}{4}x$ **(d)** $3.4ab - 0.5b$

Solution

(a) The coefficient of the x^2 term is 5. The coefficient of the x term is -2. The coefficient of the xy term is 3.

(b) The coefficient of the x^3 term is 8. The coefficient of the xy^2 term is -12. The coefficient of the y term is 1.

(c) The coefficient of the x^2 term is $\frac{1}{2}$. The coefficient of the x term is $\frac{1}{4}$.

(d) The coefficient of the ab term is 3.4. The coefficient of the b term is -0.5.

Student Practice 2 Identify the coefficient of each term.

(a) $5x^2y - 3.5w$ **(b)** $\frac{3}{4}x^3 - \frac{5}{7}x^2y$ **(c)** $-5.6abc - 0.34ab + 8.56bc$

We can add or subtract terms if they are **like terms;** that is, if the terms have the same variables and the same exponents. When we combine like terms, we are using the following form of the distributive property:

$$ba + ca = (b + c)a.$$

EXAMPLE 3 Combine like terms by using the distributive property.

(a) $8x + 2x$ **(b)** $5x^2y + 12x^2y$

Solution

(a) $8x + 2x = (8 + 2)x = 10x$

(b) $5x^2y + 12x^2y = (5 + 12)x^2y = 17x^2y$

Student Practice 3 Combine like terms by using the distributive property.

(a) $9x - 12x$ **(b)** $4ab^2c + 15ab^2c$

EXAMPLE 4 Combine like terms.

(a) $7x^2 - 2x - 8 + x^2 + 5x - 12$ **(b)** $\frac{1}{3}x^2 + \frac{1}{4}x - \frac{1}{6}x^2$

(c) $2.3x^3 - 5.6x + 5.8x^3 - 7.9x$

Solution

(a) $7x^2 - 2x - 8 + x^2 + 5x - 12 = 8x^2 + 3x - 20$

Remember that the coefficient of x^2 is 1, so you are adding $7x^2 + 1x^2$.

(b) $\frac{1}{3}x^2 + \frac{1}{4}x - \frac{1}{6}x^2 = \frac{2}{6}x^2 - \frac{1}{6}x^2 + \frac{1}{4}x = \frac{1}{6}x^2 + \frac{1}{4}x$

(c) $2.3x^3 - 5.6x + 5.8x^3 - 7.9x = 8.1x^3 - 13.5x$

Student Practice 4 Combine like terms.

(a) $12x^3 - 5x^2 + 7x - 3x^3 - 8x^2 + x$

(b) $\frac{1}{3}a^2 - \frac{1}{5}a - \frac{4}{15}a^2 + \frac{1}{2}a + 5$

(c) $4.5x^3 - 0.6x - 9.3x^3 + 0.8x$

② Multiplying Algebraic Expressions Using the Distributive Property

We can use the distributive property $a(b + c) = ab + ac$ to multiply algebraic expressions. The kinds of expressions usually encountered are called *polynomials*. **Polynomials** are variable expressions that contain terms with *nonnegative* integer exponents. Some examples of polynomials are $6x^2 + 2x - 8$, $5a + b$, $16x^3$, and $5x + 8$.

EXAMPLE 5 Use the distributive property to multiply. $-2x(x^2 + 5x)$

Solution The distributive property tells us to multiply each term in the parentheses by the term outside the parentheses.

$$-2x(x^2 + 5x) = (-2x)(x^2) + (-2x)(5x) = -2x^3 - 10x^2$$

Student Practice 5 Use the distributive property to multiply.

$$-3x^2(2x - 5)$$

There is no limit to the number of terms we can multiply. For example, $a(b + c + d + \cdots) = ab + ac + ad + \cdots$.

EXAMPLE 6 Multiply.

(a) $7x(x^2 - 3x - 5)$ **(b)** $5ab(a^2 - ab + 8b^2 + 2)$

Solution

(a) $7x(x^2 - 3x - 5) = 7x^3 - 21x^2 - 35x$

(b) $5ab(a^2 - ab + 8b^2 + 2) = 5a^3b - 5a^2b^2 + 40ab^3 + 10ab$

Student Practice 6 Multiply.

(a) $-5x(2x^2 - 3x - 1)$ **(b)** $3ab(4a^3 + 2b^2 - 6)$

A parenthesis preceded by no sign or a positive sign $(+)$ can be considered to have a numerical coefficient of 1. A parenthesis preceded by a negative sign $(-)$ can be considered to have a numerical coefficient of -1.

EXAMPLE 7 Simplify.

(a) $(3x^2 + 2)$ **(b)** $-(2x + 3)$

(c) $-4x(x - 2y)$ **(d)** $\frac{2}{3}(6x^2 - 2x + 3)$

Solution

(a) $(3x^2 + 2) = 1(3x^2 + 2) = 3x^2 + 2$

(b) $-(2x + 3) = -1(2x + 3) = -2x - 3$

(c) $-4x(x - 2y) = -4x^2 + 8xy$

(d)
$$\frac{2}{3}(6x^2 - 2x + 3) = \frac{2}{3}(6x^2) - \frac{2}{3}(2x) + \frac{2}{3}(3)$$
$$= 4x^2 - \frac{4}{3}x + 2$$

Student Practice 7 Simplify.

(a) $(7x^2 - 8)$ **(b)** $-(3x + 2y - 6)$

(c) $-5x^2(x + 2xy)$ **(d)** $\frac{3}{4}(8x^2 + 12x - 3)$

3 Removing Grouping Symbols to Simplify Algebraic Expressions

To simplify an expression that contains parentheses that are not placed within other parentheses, multiply and then combine like terms.

EXAMPLE 8 Simplify. $5(x - 2y) - (y + 3x) + (5x - 8y)$

Solution

$$5(x - 2y) - (y + 3x) + (5x - 8y)$$
$$= 5x - 10y - y - 3x + 5x - 8y$$
$$= 7x - 19y$$

Remember that you are really multiplying by -1, so don't forget to change the signs of the terms inside the parentheses.

Student Practice 8 Simplify. $-7(a + b) - 8a(2 - 3b) + 5a$

To simplify an expression that contains grouping symbols within grouping symbols, work from the inside out. The grouping symbols [ ] and { } are used like parentheses.

EXAMPLE 9 Simplify. $-2\{3 + 2[x - 4(x + y)]\}$

Solution

$-2\{3 + 2[x - 4(x + y)]\}$

$= -2\{3 + 2[x - 4x - 4y]\}$ Remove the parentheses by multiplying each term of $x + y$ by -4.

$= -2\{3 + 2[-3x - 4y]\}$ Combine like terms inside the brackets.

$= -2\{3 - 6x - 8y\}$ Remove the brackets by multiplying each term of $-3x - 4y$ by 2.

$= -6 + 12x + 16y$ Remove the braces by multiplying each term by -2.

Student Practice 9 Simplify.
$-2\{4x - 3[x - 2x(1 + x)]\}$

STEPS TO SUCCESS What Is the Best Way to Review Before a Test?

Here is what students have found.

1. Read over your textbook again. Make a list of any terms, rules, or formulas you need to know for the exam. Make sure you understand them all.
2. Go over your notes. Look back at your homework and quizzes. Redo the problems you missed. Make sure you can get the right answer.
3. Practice some of each type of problem covered in the chapter(s) you are to be tested on.
4. At the end of the chapter are special sections to help you review. Be sure to do the Chapter Review problems. Study each part of the Chapter Organizer and do the You Try It problems.
5. When you think you are ready, take the How Am I Doing? Chapter Test. Make sure you study the Math Coach notes right after the test.
6. Get help for those concepts that are giving you difficulty. Don't be afraid to ask for help. Teachers, tutors, friends in class, and other friends are ready to help you. Don't wait. Get help now.

Making it personal: Which of the six steps given do you most need to follow? Take some time today to start doing these things. These methods of review have helped thousands of students! They can help you—NOW! ▼

1.5 Exercises

MyMathLab® Watch the videos in MyMathLab Download the MyDashBoard App

Verbal and Writing Skills, Exercises 1–4

1. Explain what the coefficient of y is in $-5x^2y$.

2. Explain what the coefficient of x is in $3xy^3z$.

3. What are the terms in the expression $5x^3 - 6x^2 + 4x + 8$?

4. What are the terms in the expression $5x^3 + 3x^2 - 2y - 8$?

In exercises 5–10, list the numerical coefficient of each term.

5. $x^5 - 3x - 8y$

6. $-4xy + 9x^2 + y$

7. $5x^3 - 3x^2 + x$

8. $6x^2 - x - 6y$

9. $6.5x^3y^3 - 0.02x^2y + 3.05y$

10. $-\frac{1}{2}a^2b^2 - \frac{10}{3}a^2b - \frac{4}{5}ab$

Combine like terms.

11. $3ab + 8ab$

12. $7ab - 5ab$

13. $4y - 7x + 2x - 6y$

14. $10a + 6b - 7a + 2b$

15. $-4x^2 + 3 + x^2 - 7$

16. $-6y^2 - y + 3y + 2y^2$

17. $4ab - 3b^3 - 5ab + 3b^3$

18. $2a - ab - 2a - ab$

19. $0.1x^2 + 3x - 0.5x^2$

20. $1.2x^2 - x + 0.3x^2$

21. $\frac{2}{3}m + \frac{5}{6}n - \frac{1}{3}m + \frac{1}{3}n$

22. $\frac{5}{8}m + \frac{4}{5}n + \frac{1}{2}m - \frac{1}{5}n$

23. $\frac{2}{3}a^2 + 2b + \frac{1}{3}a^2 - 8b$

24. $\frac{3}{2}x^2 + 7x + \frac{1}{2}x^2 - 10y$

25. $1.2x^2 - 5.6x - 8.9x^2 + 2x$

26. $6y^2 - 1.6y - 3.2y - 3.8y^2$

Multiply. Simplify your answers wherever possible.

27. $6x(3x + y)$

28. $7y(4x - 3)$

29. $-y(y^2 - 3y + 5)$

30. $-3x(x^3 + 2x^2 - x)$

31. $-2a^2(a - 3a^2 + 2ab)$

32. $-4m^3(2m + 6 - 5mn)$

33. $2xy(x^2 - 3xy + 4y^2)$

34. $3ab(a^2 - ab - 4b^2)$

35. $\frac{3}{4}(8x^2 - 4x + 2)$

36. $\frac{1}{2}(5x - 8y + 4)$

37. $\frac{x}{5}(5x^2 - 2x + 1)$

38. $\frac{x}{4}(x^2 + 5x - 12)$

39. $3ab(a^4b - 3a^2 + a - b)$

40. $5xy^2(y^3 - y^2 + 3x + 1)$

41. $0.5x^2(2x - 4y + 3y^2)$

42. $1.2x^2(4x - 2xy + 5)$

Remove grouping symbols and simplify.

43. $2(x - 1) - x(x + 1) + 3(x^2 + 2)$

44. $5(x - 2) + x(3x - 8) - (x - 2)$

45. $2\{3x - 2[x - 4(x + 1)]\}$

46. $-3\{3y + 2[y + 2(y - 4)]\}$

Mixed Practice *Remove grouping symbols if necessary and simplify.*

47. $a(a - 4b) - 5a(a + b)$

48. $4(xy - 1) - (x - 3)$

49. $6x^3 - 2x^2 - 9x^3 - 12x^2$

50. $4x^4 - 7x^3 - 12x^4 + 6x^3$

51. $2[-3(2x + 4) + 8(2x - 4)]$

52. $3[3(3x + 6) - 5(5x - 9)]$

53. $3y[y - (x - 5)]$

54. $2x[4 - (3x + 2y)]$

Cumulative Review *Evaluate.*

55. **[1.3.3]** $3(-2)^3 - 5(-6)$

56. **[1.3.3]** $\sqrt{81} - 5(3 - 5 + 2)$

57. **[1.2.3]** $\dfrac{5(-2) - 8}{3 + 4 - (-3)}$

58. **[1.3.3]** $(-3)^5 + 2(-3)$

59. **[1.4.5]** ***Bacteria Size*** The smallest organism known to contain all the chemicals needed to sustain independent life is a bacterium called the *pleuropneumonia* organism. It would take 1,893,500 of them, touching side by side, to span an inch. How many *pleuropneumonia* organisms would be found if you put them in a line 1 kilometer long? Express your answer in scientific notation. (Use 1 inch = 0.0254 meter.)

60. **[1.2.2]** ***Efficiency of Internal Combustion Engines*** Efficiency of internal combustion engines is lost at the rate of 2% for every 1000 feet of altitude above sea level. What percentage of efficiency would be lost by powerboats and cars at Lake Titicaca, 4167 *meters* above sea level? (Use 1 foot = 0.305 meter.)

Quick Quiz 1.5 *Simplify.*

1. $2xy(-3x^2 + 4xy - 5x)$

2. $3x^2(x + 4y) - 2(5x^3 - 2x^2y)$

3. $3[-4(x + 2) + 3(5x - 1)]$

4. **Concept Check** Explain how to simplify the following. $2x^2 - 3x + 4y - 2x^2y - 8x - 5y$

1.6 Evaluating Variable Expressions and Formulas

① Evaluating a Variable Expression

We need to know how to **evaluate**—compute a numerical value for—variable expressions when we know the values of the variables.

Student Learning Objectives

After studying this section, you will be able to:

① **Evaluate a variable expression when the value of the variable is given.**

② **Evaluate a formula to determine a given quantity.**

EVALUATING A VARIABLE EXPRESSION

1. Replace each variable (letter) by its numerical value. Put parentheses around the value (watch out for negative values).
2. Carry out each step, using the correct order of operations.

EXAMPLE 1 Evaluate $x^2 - 5x - 6$ when $x = -4$.

Solution

$(-4)^2 - 5(-4) - 6$	Replace x by -4 and put parentheses around it.
$= 16 - 5(-4) - 6$	Square -4.
$= 16 - (-20) - 6$	Multiply $5(-4)$.
$= 16 + 20 - 6$	Multiply $-1(-20)$.
$= 30$	Combine $16 + 20 - 6$.

Student Practice 1 Evaluate $2x^2 + 3x - 8$ when $x = -3$.

NOTE TO STUDENT: Fully worked-out solutions to all of the Student Practice problems can be found at the back of the text starting at page SP-1.

MC **EXAMPLE 2** Evaluate $(5 - x)^2 + 3xy$ when $x = -2$ and $y = 3$.

Solution $[5 - (-2)]^2 + 3(-2)(3)$

$= [5 + 2]^2 + 3(-2)(3) = [7]^2 + 3(-2)(3)$

$= 49 + 3(-2)(3) = 49 - 18 = 31$

Student Practice 2 Evaluate $(x - 3)^2 - 2xy$ when $x = -3$ and $y = 4$.

EXAMPLE 3 Evaluate when $x = -3$.

(a) $(-2x)^2$ **(b)** $-2x^2$

Solution

(a) $[-2(-3)]^2 = [6]^2 = 36$ Multiply $(-2)(-3)$ and then square the result.

(b) $-2(-3)^2 = -2(9)$ Square -3.

$= -18$ Multiply $-2(9)$.

Student Practice 3 Evaluate when $x = -4$.

(a) $(-3x)^2$ **(b)** $-3x^2$

TO THINK ABOUT: Importance of Grouping Symbols Why are the answers to parts **(a)** and **(b)** different in Example 3? What does $(-2x)^2$ mean? What does $-2x^2$ mean? Why are the parentheses so important in this situation?

② Evaluating Formulas

A **formula** is a rule for finding the value of a variable when the values of other variables in the expression are known. The word *formula* is usually applied to some physical situation, much like a recipe. For example, we can determine the Fahrenheit temperature F for any Celsius temperature C from the formula

$$F = \frac{9}{5}C + 32.$$

EXAMPLE 4 Find the Fahrenheit temperature when the Celsius temperature is −30°C.

Solution $F = \frac{9}{5}(-30) + 32$ Substitute the known value −30 for the variable C. Then evaluate.

$= \frac{9}{\cancel{5}_1}(-\cancel{30}^{6}) + 32 = 9(-6) + 32$

$= -54 + 32 = -22$

Thus, when the temperature is −30°C, the equivalent Fahrenheit temperature is −22°F.

Student Practice 4 Find the Fahrenheit temperature when the Celsius temperature is 70°C.

EXAMPLE 5 The period T of a pendulum (the time in seconds for the pendulum to swing back and forth one time) is $T = 2\pi\sqrt{\frac{L}{g}}$, where L is the length of the pendulum in feet and g is the acceleration due to gravity in feet/second^2. Find the period when $L = 288$ ft and $g = 32$ ft/sec^2. Approximate the value of π as 3.14. Round your answer to the nearest tenth of a second.

Solution $T = 2(3.14)\sqrt{\frac{288}{32}} = 6.28\sqrt{9} = 6.28(3)$

$= 18.84$

The time for a 288-foot-long pendulum to swing back and forth once is approximately 18.84 seconds.

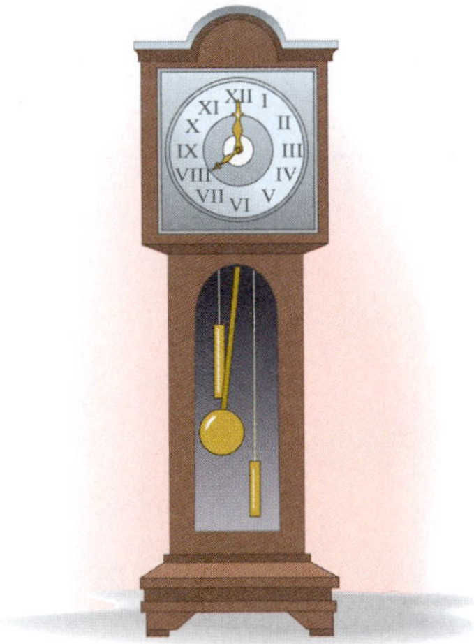

Student Practice 5 Tarzan is swinging back and forth on a 128-foot-long rope. Use the above formula to find out how long it takes him to swing back and forth one time. Round your answer to the nearest tenth of a second.

EXAMPLE 6 An amount of money invested or borrowed (not including interest) is called the *principal*. Find the amount A to be repaid on a principal p of \$1000 borrowed at a simple interest rate r of 8% for a time t of 2 years. The formula is $A = p(1 + rt)$.

Solution $A = 1000[1 + (0.08)(2)]$ Change 8% to 0.08.

$= 1000[1 + 0.16]$

$= 1000(1.16) = 1160$

The amount to be repaid (principal plus interest) is \$1160.

Student Practice 6 Find the amount to be repaid on a loan of \$600 at a simple interest rate of 9% for a time of 3 years.

One of the areas of mathematics where formulas are very helpful is geometry. We use formulas to find the perimeters, areas, and volumes of common geometric figures.

▲ **EXAMPLE 7** Find the perimeter of a rectangular school playground with length 28 meters and width 16.5 meters. Use the formula $P = 2l + 2w$.

Solution We draw a picture to get a better idea of the situation.

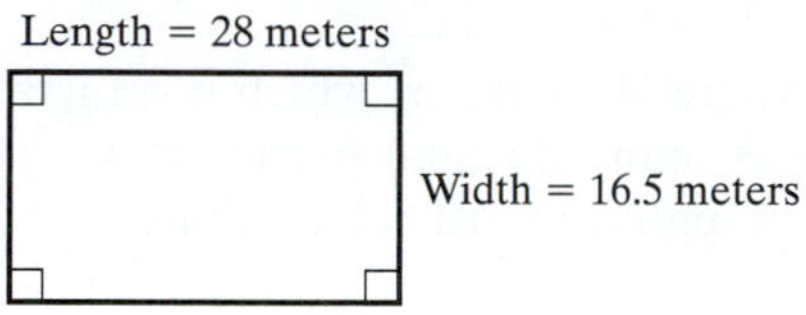

$$P = 2l + 2w = 2(28) + 2(16.5) = 56 + 33 = 89$$

The perimeter of the playground is 89 meters.

▲ **Student Practice 7** Find the perimeter of a rectangular computer chip. The length of the chip is 0.76 centimeter and the width is 0.38 centimeter.

The most commonly used formulas for geometric figures are listed in the following box.

AREA AND PERIMETER FORMULAS

In the following formulas, A = area, P = perimeter, and C = circumference. The "squares" in the figures like ⊿ mean that the angle formed by the two lines is 90°.

Rectangle

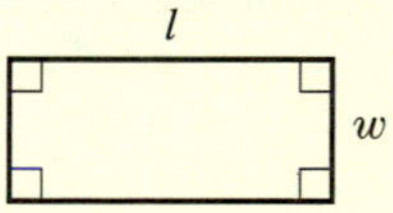

$A = lw$

$P = 2l + 2w$

Triangle

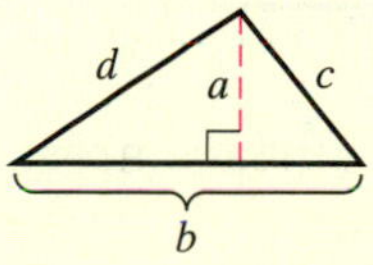

$A = \frac{1}{2}ab$

$P = b + c + d$

Parallelogram

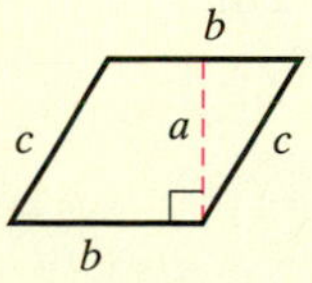

$A = ab$

$P = 2b + 2c$

Rhombus

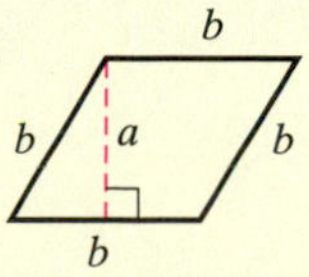

$A = ab$

$P = 4b$

Trapezoid

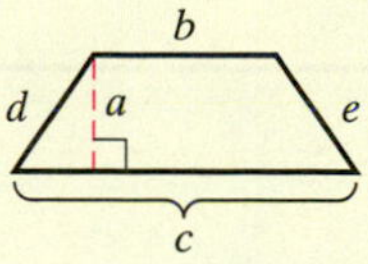

$A = \frac{1}{2}a(b + c)$

$P = b + c + d + e$

AREA AND PERIMETER FORMULAS (*continued*)

Circle

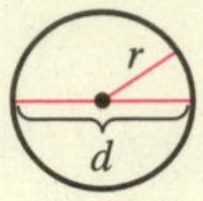

$A = \pi r^2$

$C = 2\pi r$, where r is the radius

$C = \pi d$, where d is the diameter

$\pi \approx 3.14$

In circle formulas we need to use π, which is an irrational number. Its value can be approximated to as many decimal places as we like, but we will use 3.14 because that is accurate enough for most calculations.

▲ **EXAMPLE 8** Find the area of a trapezoid that has a height of 6 meters and bases of 7 and 11 meters.

Solution The formula is $A = \frac{1}{2}a(b + c)$. We are told that $a = 6$, $b = 7$, and $c = 11$, so we put those values into the formula.

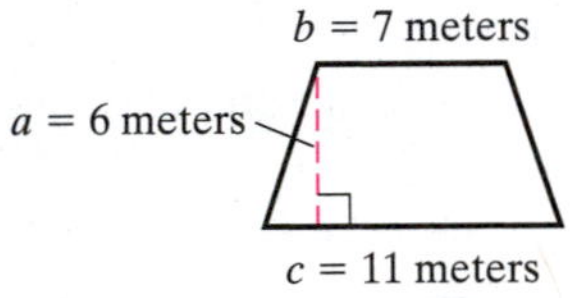

$$A = \frac{1}{2}(6)(7 + 11) = \frac{1}{2}(6)(18) = 54$$ The area is 54 square meters or 54 m^2.

▲ **Student Practice 8** Find the area of a triangle with an altitude of 12 meters and a base of 14 meters.

VOLUME AND SURFACE AREA FORMULAS

In the following formulas, V = volume and S = surface area.

Rectangular solid

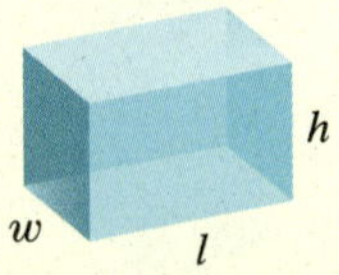

$V = lwh$

$S = 2lw + 2wh + 2lh$

Sphere

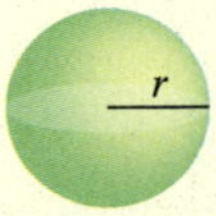

$V = \frac{4}{3}\pi r^3$

$S = 4\pi r^2$

Right circular cylinder

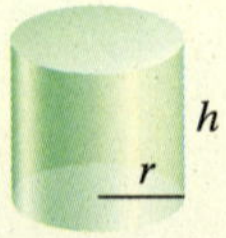

$V = \pi r^2 h$

$S = 2\pi rh + 2\pi r^2$

EXAMPLE 9 Find the volume of a sphere with a radius of 3 centimeters.

Solution The formula is $V = \frac{4}{3}\pi r^3$.

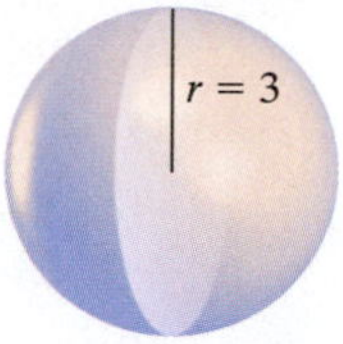

Therefore, we have $V \approx \frac{4}{3}(3.14)(3)^3 = \frac{4}{3}(3.14)(27)$

$$= \frac{4}{\overset{}{\cancel{3}}_{1}}(3.14)(\overset{9}{\cancel{27}}) = 113.04$$

The volume is approximately 113.04 cubic centimeters or 113.04 cm^3.

Student Practice 9 Find the volume of a right circular cylinder of height 10 meters and radius 6 meters.

STEPS TO SUCCESS

If you attend a traditional mathematics class that meets one or more times each week:

Faithful Class Attendance Is Well Worth It.

Get started in the right direction. Make a personal commitment to attend class every day, beginning with the first day of class. Teachers and students all over the country have discovered that faithful class attendance and good grades go together.

The vital content of class. What goes on in class is designed to help you learn more quickly. Each day, significant information is given that will truly help you to understand concepts. There is no substitute for this firsthand learning experience.

Meet a friend. You will soon discover that other students are also coming to class every single class period. It is easy to strike up a friendship with students who share this common commitment. They will usually be available to answer a question after class and give you an additional source of help when you encounter difficulty.

Making it personal: Write down what you think is the most compelling reason to attend every class meeting. Make that commitment and see how much it helps you. ▼

If you are enrolled in an online mathematics class, a self-paced mathematics class taught in a math lab, or some other type of nontraditional class:

Keep Yourself on Schedule.

The key to success is to keep on schedule. In a class where you determine your own pace, you will need to commit yourself to following the suggested pace provided in your course materials. Check off each assignment as you do it so you can see your progress.

Make sure all your class materials are organized and available. Keep all course schedules and assignments right where you can quickly find them. Review them often to be sure you are doing everything that you should.

Discipline yourself to follow the detailed schedule exactly for the first six weeks. Professor Tobey and Professor Slater have both taught online classes for several years. They have found that students usually succeed in the course as long as they do every suggested activity for the first six weeks!

Making it personal: Are you good at following schedules and keeping track of details? Which of the suggestions above do you find the most helpful? ▼

1.6 Exercises

MyMathLab®

Watch the videos in MyMathLab

Download the MyDashBoard App

In this exercise set, round all answers to the nearest hundredth unless otherwise stated.
Evaluate each expression for the values given.

1. $14 + 6x; x = -2$

2. $11x - 7; x = 3$

3. $x^2 + 2x - 9; x = -4$

4. $x^2 + 3x - 12; x = -5$

5. $3 + 7x - x^2; x = 1$

6. $-4x - x^2 + 7; x = 3$

7. $-2x^2 + 5x - 3; x = -4$

8. $6x^2 - 3x + 5; x = 5$

9. $(-3y)^4; y = -1$

10. $(-5a)^2; a = -2$

11. $-3y^4; y = -1$

12. $-5a^2; a = -2$

13. $-ay + 4bx - b; a = -1, b = 3, x = 2, y = -4$

14. $-2ay + 6ab - y; a = -3, b = 2, y = -1$

15. $\sqrt{b^2 - 4ac}; b = 5, a = 1, c = -14$

16. $\sqrt{b^2 - 4ac}; b = 3, a = -1, c = -2$

17. Evaluate $2x^2 - 5x + 6$ when $x = -3.52176$. (Round to five decimal places.)

18. Evaluate $3x^2 - 7x - 2$ when $x = -0.56736$. (Round to five decimal places.)

Applications

Temperature Conversion *For exercises 19 and 20, use the formula* $F = \frac{9}{5}C + 32$.

19. Find the Fahrenheit temperature when the Celsius temperature is −60°C.

20. Find the Fahrenheit temperature when the Celsius temperature is 30°C.

For exercises 21 and 22, use the formula $C = \dfrac{5F - 160}{9}$.

21. Find the Celsius temperature if the Fahrenheit temperature is 122°F.

22. Find the Celsius temperature if the Fahrenheit temperature is −40°F.

Swinging of a Pendulum *For exercises 23 and 24, use* $T = 2\pi\sqrt{\dfrac{L}{g}}$. *Let* $\pi \approx 3.14$ *and* $g = 32$ *feet per second*^{2}.

23. A child is swinging on a rope 32 feet long over a river swimming hole. How long does it take (in seconds) to complete one swing back and forth?

24. A cable swinging from a skyscraper under construction acts as a pendulum. The cable is 512 feet long. How many seconds will it take for the cable to swing back and forth one time?

Simple Interest *In exercises 25–28, use the simple interest formula* $A = p(1 + rt)$.

25. Find A if $p = \$4800$, $r = 12\%$, and $t = 1.5$ years.

26. Find A if $p = \$5000$, $r = 4\%$, and $t = 4$ years.

27. Find the amount to be repaid on a loan of \$2200 at a simple interest rate of 4% for 4 years.

28. Find the amount to be repaid on a loan of \$3500 at a simple interest rate of 7% for 6 years.

Falling Time in Gravity *In exercises 29–32, use the fact that the distance S in feet an object falls in t seconds is given by* $S = \frac{1}{2}gt^2$*, where* $g = 32$ *feet per second*^{2}.

29. Find S if $t = 3$ seconds.

30. Find S if $t = 6$ seconds.

31. A piece of a window ledge fell from the sixty-eighth floor of the Texas Commerce Tower in Houston. It took 7 seconds to hit the ground. How far did the piece of window ledge fall?

32. A bolt fell out of a window frame on the twenty-third floor of the John Hancock Tower in Boston. The bolt took 4 seconds to hit the ground. How far did the bolt fall?

33. Find $z = \dfrac{Rr}{R + r}$ if $R = 36$ and $r = 4$.

34. Find $z = \dfrac{Rr}{R + r}$ if $R = 35$ and $r = 15$.

Dosage of Medicine *The approximate number of milligrams m of a medicine that should be given a child of age c when the usual adult dosage is x milligrams can be obtained from the following equation. Use the equation to solve exercises 35 and 36. Round your answers to the nearest whole number.*

$$m = \frac{cx}{c + 12}$$

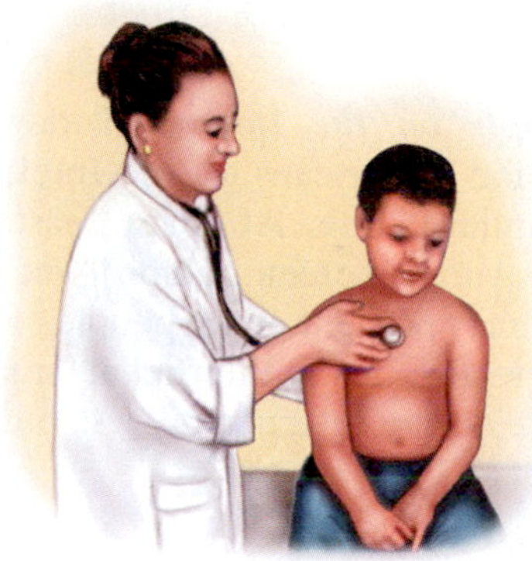

35. Find the amount of ibuprofen to give a child of age 8 if the usual adult dosage is 400 milligrams.

36. Find the amount of aspirin to give a child of age 10 if the usual adult dosage is 325 milligrams.

Geometry *In exercises 37–49, use the geometry formulas on pages 45 and 46 to find the quantities specified. Use* $\pi \approx 3.14$.

▲ **37.** Find the area of a circle with a radius of 0.5 inches.

▲ **38.** Find the circumference of a circle with a diameter of 0.2 meters.

▲ **39.** Find the area of a triangle with a base of 16 centimeters and a height of 7 centimeters.

▲ **40.** Find the area of a triangle with a base of 12 meters and an altitude of 14 meters.

▲ **41.** Find the area of a parallelogram with an altitude of 4 yards and a base of $\frac{7}{8}$ yards.

▲ **42.** Find the area of a rhombus with an altitude of 8 centimeters and a base of $10\frac{3}{4}$ centimeters.

▲ **43.** ***Graphing Calculator Display Window*** A graphing calculator has a display window 6.1 centimeters wide and 4.05 centimeters high. What is the area of the display window?

▲ **44.** Find the surface area of an ice cube that is 1.8 inches long, 1 inch wide, and 1 inch high.

▲ **45.** A trapezoid has a height of 6 centimeters and parallel bases measuring 5 centimeters and 7 centimeters. Find its area.

▲ **46.** A trapezoid has sides 5.2, 6.1, 3.5, and 2.2 meters long. Find its perimeter.

▲ **47.** ***Soup Can*** A can of soup has a height of 11 centimeters and a radius of 4 centimeters.

(a) Find its volume. (Use $\pi \approx 3.14$.)

(b) Find its total surface area.

▲ **48.** ***Baseball*** A baseball has a radius of 2.9 centimeters.

(a) Find its volume. (Use $\pi \approx 3.14$.)

(b) Find its surface area.

▲ **49.** ***Telephone Cable*** A cross section of telephone cable is a circle that has a radius of 8 centimeters. What is the area of the cross section? What is the circumference of this circular cross section? (Use $\pi \approx 3.14$.)

▲ **50.** ***Stainless Steel Circle*** An engineer must make a stainless steel circle with a radius of 6 centimeters. What is the area of the circle? What is the circumference of the circle? (Use $\pi \approx 3.14$.)

Cumulative Review *Simplify.*

51. **[1.4.4]** $(6x^{-4}y^3z^0)^2$

52. **[1.4.4]** $\left(\dfrac{2x^3}{3y}\right)^3$

53. **[1.5.3]** $2\{5 - 2[x - 3(2x + 1)]\}$

54. **[1.3.3]** $2^3 - 4^2 + \sqrt{9 \cdot 2 - 2}$

55. **[1.2.2]** ***People on the Internet*** Today there were twelve million Americans on the Internet at 9 P.M., eastern standard time. At 2 P.M., eastern standard time, there were nineteen million Americans on the Internet. Thirty percent of the people on the Internet at 2 P.M. were also on the Internet at 9 P.M. How many people were on the Internet at 9 P.M. but were not on the Internet at 2 P.M.?

56. **[1.2.2]** ***High School Reunion*** An invitation was sent out to Beverly High School alumni from the class of 1961. The graduating class that year had 180 students. A total of 69% of the living students went to the reunion. A total of 5% of the graduates of the class of 1961 are no longer living. Seventy-seven percent of all the students attending the reunion brought a spouse to the gathering. In addition, twenty-two faculty members returned for this big event. What was the total number of people at the reunion? Round your answer to the nearest whole number.

Quick Quiz 1.6

1. Evaluate $3x^2 - 5x - 2$ when $x = -3$.

2. Evaluate $-2x^2 + 4xy - y^2$ when $x = -2$ and $y = 4$.

3. Find the area of a circle with a radius of 5 centimeters. Use $\pi \approx 3.14$.

4. **Concept Check** Explain how you would find the amount A to be repaid on a principal of \$5000 borrowed at a simple interest rate r of 8% for a time t of 2 years. The formula is $A = p(1 + rt)$.

Did You Know...

That When Buying a Vehicle You Can Save Hundreds of Dollars Off the Asking Price by Negotiating?

NEGOTIATING FOR THE BEST PRICE

Understanding the Problem:
Rachael is looking to purchase a new vehicle. She wants to negotiate the best deal she can. She knows she should research the car she wants and find out the manufacturer's suggested retail price (MSRP), invoice price, rebates, and the reasonable profit margin on the model she wants.

Making a Plan:
First Rachael needs to determine what model and options she wants. She needs to be very specific because pricing varies widely from model to model, and additional options can add thousands of dollars to the price. Once she has picked out a specific vehicle, she can use auto pricing and consumer information websites to research the vehicle and find out prices and rebates.

Step 1: After she has determined which vehicle she wants, Rachael does some research and finds that the invoice price is $12,360 and the MSRP is $13,710.

Task 1: *What is the difference between the invoice price and the MSRP?*

Step 2: The difference between invoice price and MSRP gives the amount of profit that the dealer will make on a full price sale. The profit margin is the amount of profit divided by the expense. For a car dealer, the expense is the invoice price, which he pays to the car manufacturer.

Task 2: *What is the profit margin on a full price sale, rounded to the nearest percent?*

Step 3: During her research, Rachael found out that a 5% profit margin is a reasonable amount for this vehicle in the current market.

Task 3: *What is 5% of $12,360?*

Task 4: *Using the 5% profit margin, how much should Rachael plan on paying for the vehicle?*

Step 4: As part of her research, she looked up rebates. She found that there is a $500 manufacturer's rebate on the vehicle.

Task 5: *If Rachael buys the vehicle for the price found previously, how much will it cost her after the rebate?*

Finding a Solution:
Rachael goes to the dealership with printouts of her research information so she can refer to it and use it to back up her claim that the price she is offering is reasonable. If they don't accept her offer she is willing to leave and try another dealership.

Applying the Situation to Your Life:
A vehicle is a very expensive purchase. Make sure you do your research and don't go to the dealership until you've determined what you are willing to pay. Some car sales representatives can be very intimidating. It may be helpful to bring a friend or family member with you for moral support. Let them know how much you plan to spend, and they can encourage you to stick to that amount. If you can't get the price you want, leave and try a different dealership.

Chapter 1 Organizer

Topic and Procedure	Examples	You Try It
Commutative property of addition, p. 4 $a + b = b + a$	$12 + 13 = 13 + 12$	**1.** Name the property that justifies each statement. **(a)** $0.5 + 0 = 0.5$
Commutative property of multiplication, p. 4 $a \cdot b = b \cdot a$	$11 \cdot 19 = 19 \cdot 11$	**(b)** $(1 + 9) + 3 = 1 + (9 + 3)$
Associative property of addition, p. 4 $a + (b + c) = (a + b) + c$	$4 + (3 + 6) = (4 + 3) + 6$	**(c)** $10 \cdot 1 = 10$
Associative property of multiplication, p. 4 $a \cdot (b \cdot c) = (a \cdot b) \cdot c$	$7 \cdot (3 \cdot 2) = (7 \cdot 3) \cdot 2$	**(d)** $10 + 4 = 4 + 10$
Identity property of addition, p. 4 $a + 0 = a$	$9 + 0 = 9$	**(e)** $2 \cdot 9 = 9 \cdot 2$
Identity property of multiplication, p. 4 $a \cdot 1 = a$	$7 \cdot 1 = 7$	**(f)** $5 \cdot (4 \cdot 6) = (5 \cdot 4) \cdot 6$
Inverse property of addition, p. 4 $a + (-a) = 0$	$8 + (-8) = 0$	**(g)** $8\left(\frac{1}{8}\right) = 1$
Inverse property of multiplication, p. 4 If $a \neq 0, a\left(\frac{1}{a}\right) = 1$	$15\left(\frac{1}{15}\right) = 1$	**(h)** $5(2 + 6) = 5 \cdot 2 + 5 \cdot 6$
Distributive property of multiplication over addition, p. 4 $a(b + c) = a \cdot b + a \cdot c$	$7(9 + 4) = 7 \cdot 9 + 7 \cdot 4$	**(i)** $-14 + 14 = 0$
Addition of real numbers, p. 11 To add two real numbers with the *same sign,* add their absolute values and use the common sign. To add two real numbers with *different signs,* find the difference between their absolute values. The answer takes the sign of the number with the larger absolute value.	**(a)** $9 + 5 = 14$ **(b)** $-7 + (-3) = -10$ **(c)** $-\frac{1}{5} + \frac{3}{5} = \frac{2}{5}$ **(d)** $-42 + 19 = -23$	**2.** Add. **(a)** $8 + 3$ **(b)** $-5 + (-9)$ **(c)** $-\frac{3}{4} + \frac{1}{4}$ **(d)** $-18 + 25$
Subtraction of real numbers, p. 11 To subtract b from a, *add the opposite* of b to a. $a - b = a + (-b)$	**(a)** $12 - (-3) = 12 + (+3) = 15$ **(b)** $-7.2 - (+1.6) = -7.2 + (-1.6) = -8.8$	**3.** Subtract. **(a)** $19 - (-6)$ **(b)** $-8.4 - 2.1$
Multiplication and division of real numbers, p. 12 When you multiply or divide two real numbers with like signs, the answer is a *positive* number. When you multiply or divide two real numbers whose signs are *different,* the answer is a *negative* number.	**(a)** $(-6)(-3) = 18$ **(b)** $-20 \div (-4) = 5$ **(c)** $(-8)(5) = -40$ **(d)** $16 \div (-2) = -8$	**4.** Multiply. **(a)** $(-8)(-6)$ **(b)** $-36 \div (-9)$ **(c)** $3(-6)$ **(d)** $-40 \div 5$
Absolute value of a number, p. 9 $\lvert x \rvert = \begin{cases} x, & \text{if } x \geq 0 \\ -x, & \text{if } x < 0 \end{cases}$	**(a)** $\lvert 6 \rvert = 6$ **(b)** $\lvert -2 \rvert = 2$ **(c)** $\left\lvert -\frac{4}{7} \right\rvert = \frac{4}{7}$ **(d)** $\lvert 0 \rvert = 0$ **(e)** $\lvert -3.6 \rvert = 3.6$	**5.** Find each absolute value. **(a)** $\lvert 7 \rvert$ **(b)** $\lvert -10 \rvert$ **(c)** $\left\lvert -\frac{3}{5} \right\rvert$ **(d)** $\lvert 1.25 \rvert$

Topic and Procedure	Examples	You Try It
Order of operations of real numbers, p. 20 To simplify numerical expressions, use this order of operations. **1.** Combine numbers inside grouping symbols. **2.** Raise numbers to their indicated powers and take any indicated roots. **3.** Multiply and divide numbers from left to right. **4.** Add and subtract numbers from left to right.	Simplify. $5 + 2(5 - 8)^3 - 12 \div (-4)$ $= 5 + 2(-3)^3 - 12 \div (-4)$ $= 5 + 2(-27) - 12 \div (-4)$ $= 5 + (-54) - (-3)$ $= -49 + 3$ $= -46$	**6.** Simplify. $9 - 3(1 - 3)^2 + 21 \div (-3)$
Negative exponents, p. 26 $x^{-n} = \frac{1}{x^n} \qquad \frac{x^{-n}}{y^{-m}} = \frac{y^m}{x^n}$	**(a)** $x^{-6} = \frac{1}{x^6}$ **(b)** $2^{-8} = \frac{1}{2^8}$ **(c)** $\frac{x^{-4}}{y^{-5}} = \frac{y^5}{x^4}$	**7.** Simplify. **(a)** a^{-5} **(b)** 3^{-3} **(c)** $\frac{a^{-2}}{b^{-4}}$
Rules of Exponents for multiplication and division, pp. 27 and 28 If x and y are any nonzero real numbers and m and n are integers, $x^m x^n = x^{m+n}$ and $\frac{x^m}{x^n} = x^{m-n}$.	**(a)** $(2x^5)(3x^6) = 6x^{11}$ **(b)** $\frac{15x^8}{5x^3} = 3x^5$	**8.** Simplify. **(a)** $(3x^4)(4x^3)$ **(b)** $\frac{24x^6}{8x}$
Zero exponent, p. 28 $x^0 = 1$ when $x \neq 0$.	**(a)** $x^0 = 1$ **(b)** $5^0 = 1$ **(c)** $(3ab)^0 = 1$	**9.** Simplify. **(a)** m^0 **(b)** 8^0 **(c)** $(3x)^0$
Power rules, p. 30 $(x^m)^n = x^{mn}$ $(xy)^n = x^n y^n$ $\left(\frac{x}{y}\right)^n = \frac{x^n}{y^n}$, if $y \neq 0$	**(a)** $(7^3)^4 = 7^{12}$ **(b)** $(3x^{-2})^4 = 3^4x^{-8} = \frac{3^4}{x^8} = \frac{81}{x^8}$ **(c)** $\left(\frac{2a^2}{b^3}\right)^{-4} = \frac{2^{-4}a^{-8}}{b^{-12}} = \frac{b^{12}}{2^4a^8} = \frac{b^{12}}{16a^8}$	**10.** Simplify. **(a)** $(3^4)^2$ **(b)** $(2a^{-5})^3$ **(c)** $\left(\frac{x^2}{4y^3}\right)^{-3}$
Scientific notation, p. 31 A positive number is written in scientific notation if it is in the form $a \times 10^n$, where $1 \leq a < 10$ and n is an integer.	**(a)** $128 = 1.28 \times 10^2$ **(b)** $2{,}568{,}000 = 2.568 \times 10^6$ **(c)** $13{,}200{,}000{,}000 = 1.32 \times 10^{10}$ **(d)** $0.16 = 1.6 \times 10^{-1}$ **(e)** $0.00079 = 7.9 \times 10^{-4}$ **(f)** $0.0000034 = 3.4 \times 10^{-6}$	**11.** Write in scientific notation. **(a)** 3124 **(b)** 18,250,000 **(c)** 27,800,000,000 **(d)** 0.039 **(e)** 0.00021 **(f)** 0.0000007
Combining like terms, p. 38 Combine terms that have identical variables and exponents.	$7x^2 - 3x + 4y + 2x^2 - 8x - 9y = 9x^2 - 11x - 5y$	**12.** Simplify. $-6a + 12b - 15a^2 - a + 3b + 9a^2$
Using the distributive property, p. 38 Use the distributive property to remove parentheses. $a(b + c) = ab + ac$	$3(2x^2 - 6x - 8) = 6x^2 - 18x - 24$	**13.** Multiply. $5(3x^2 - x + 5)$

Topic and Procedure	Examples	You Try It
Removing grouping symbols, p. 39 **1.** Remove grouping symbols from the inside out. **2.** Combine like terms inside grouping symbols whenever possible. **3.** Continue until all grouping symbols are removed. **4.** Combine like terms.	$5\{3x - 2[4 + 3(x - 1)]\} = 5\{3x - 2[4 + 3x - 3]\}$ $= 5\{3x - 2[1 + 3x]\}$ $= 5\{3x - 2 - 6x\}$ $= 5\{-3x - 2\}$ $= -15x - 10$	**14.** Simplify. $-2\{3x - 3[6 - 2(x - 1)]\}$
Evaluating variable expressions, p. 43 **1.** Replace each variable by the numerical value given. Put parentheses around the value. **2.** Follow the order of operations in evaluating the expression.	Evaluate $2x^3 + 3xy + 4y^2$ for $x = -3$ and $y = 2$. $2(-3)^3 + 3(-3)(2) + 4(2)^2$ $= 2(-27) + 3(-3)(2) + 4(4)$ $= -54 - 18 + 16$ $= -56$	**15.** Evaluate $3x^2 - 5xy + y^2$ for $x = -2$ and $y = 3$.
Using formulas, p. 44 **1.** Replace the variables in the formula by the given values. **2.** Evaluate the expression. **3.** Label units carefully.	Find the area of a circle with a radius of 4 feet. Use $A = \pi r^2$ and $\pi \approx 3.14$. $A \approx (3.14)(4\text{ ft})^2$ $= (3.14)(16\text{ ft}^2)$ $= 50.24\text{ ft}^2$ The area of the circle is approximately 50.24 square feet.	**16.** Find the area of a triangle with a height of 16 feet and a base of 5 feet.

Chapter 1 Review Problems

Indicate the set(s) to which the number belongs.

		Natural Numbers	Whole Numbers	Integers	Rational Numbers	Irrational Numbers	Real Numbers
1.	-5						
2.	$\frac{7}{8}$						
3.	3						
4.	$0.\overline{3}$						
5.	2.1652384 . . . (no discernible pattern)						

In exercises 6 and 7, name the property of real numbers that justifies each statement.

6. $4 + a = a + 4$

7. $5(2 \cdot x) = (5 \cdot 2) \cdot x$

8. Are all rational numbers also real numbers?

Compute, if possible.

9. $-15 - (-20)$

10. $-7.3 + (-16.2)$

11. $-8(-6)$

12. $-\frac{4}{5} \div \left(-\frac{12}{5}\right)$

13. $-\frac{5}{6}\left(\frac{7}{10}\right)$

14. $5 + 6 - 2 - 5$

15. $-3.6(-1.5)$

16. $0 \div (-14)$

17. $7 \div 0$

18. $-17 + (+17)$

19. $17 - 3(6)$

20. $\dfrac{5 - 8}{2 - 7 - (-2)}$

21. $2\sqrt{49} - 3^2 + 5$

22. $4(6) - |-8| + (-1)^3$

23. $\sqrt{(-1)^2 + 6(4)} + 8 \div (-2)$

24. $\sqrt{\frac{25}{36}} - 2\left(\frac{1}{12}\right)$

25. $6|-3 - 1| + 5(-3)(0) - 4^2$

26. $(-0.4)^3$

Simplify. Do not leave negative exponents in your answers.

27. $(3xy^2)(-2x^0y)(4x^3y^3)$

28. $(5a^4bc^2)(-6ab^2)$

29. $\frac{16abc^0}{48ab^4c^2}$

30. $\left(\frac{-3x^3y}{2x^4z^2}\right)^4$

31. $(-2xy^6z^0)^3$

32. $(2x^2y^{-4})(-5x^{-1}y)$

33. $\frac{3x^5y^{-6}}{12x^{-2}y}$

34. $\frac{(5^{-1}x^{-2})^{-1}}{(2^{-2}y)^{-3}}$

35. $\left(\frac{x^3y^4}{5x^6y^8}\right)^3$

36. Write in scientific notation. 0.00721

37. Change to scientific notation and multiply. Express your answer in scientific notation.
(5,300,000) (2,000,000,000)

38. Combine like terms. $-x + 8 + 6x^2 + 7x - 4$

39. Multiply. $-5ab^2(a^3 + 2a^2b - 3b - 4)$

In exercises 40 and 41, remove grouping symbols and simplify.

40. $3x(x - 7) - (x^2 + 1)$

41. $2x^2 - \{2 + x[3 - 2(x - 1)]\}$

42. Evaluate $5x^2 - 3xy - 2y^3$ when $x = 2$ and $y = -1$.

▲ 43. A coffee can has a height of 8 inches and a radius of 3 inches. Find its volume. (Use $V = \pi r^2 h$ and $\pi \approx 3.14$.)

▲ 44. ***Park Measurements*** A triangular park in a Minneapolis neighborhood has an altitude of 88 yards and a base of 52 yards. Find the area of this triangular region.

▲ 45. ***Aircraft Windshield*** The front view of the windshield of a stealth bomber appears to the observer to be shaped like a trapezoid. The altitude of the trapezoid is 14 inches. The bases of the trapezoid are 26 inches and 34 inches. What is the area of this apparent trapezoid?

Mixed Practice

46. Evaluate. $9(-2) + (-28 \div 7)^3 - 5$

Simplify each expression.

47. $(-7a^2b)(-2a^0b^3c^2)$

48. $\frac{(3x^{-1}y^2)^3}{(4x^2y^{-2})^2}$

49. $\left(\frac{3a^{-5}b^0}{2a^{-2}b^3}\right)^2$

50. Write 0.000058 in scientific notation.

51. Write 8.95×10^7 in decimal form.

Simplify.

52. $4x^2 - x^3 + 7x - 5x^2 + 6x^3 - 2x$

53. $2a^3b(5a - ab - 3)$

54. $-2\{x + 3[y - 5(x + y)]\}$

55. Evaluate $5a^2 - 3ab + 4b$ when $a = -3$ and $b = -2$.

▲ 56. ***Geometry*** Use the formula $A = \pi r^2$ to find the area of a circle with a radius of 4 meters. (Use $\pi \approx 3.14$.)

57. ***Swinging Pendulum*** Use the formula $T = 2\pi\sqrt{L/g}$ to find the period of a pendulum T in seconds where the length of the pendulum $L = 512$ feet and gravity $g = 32$ ft/sec^2. (Use $\pi \approx 3.14$.)

58. Simplify. $\frac{2}{3}(6x - 9y) - (x - 2y)$

59. Find the amount to be repaid on a loan of \$3200 at a simple interest rate of 9% for 2 years.

How Am I Doing? Chapter 1 Test

1. ______
2. ______
3. ______
4. ______
5. ______
6. ______
7. ______
8. ______
9. ______
10. ______
11. ______
12. ______
13. ______
14. ______
15. ______
16. ______
17. ______
18. ______
19. ______
20. ______
21. ______

Total correct: ______

CHAPTER Test Prep VIDEOS MATH COACH MyMathLab® YouTube™

After you take this test read through the Math Coach on pages 57–58. Math Coach videos are available via MyMathLab and YouTube. Step-by-step test solutions in the Chapter Test Prep Videos are also available via MyMathLab and YouTube. (Search "TobeyInterAlg" and click on "Channels.")

Exercises 1 and 2 refer to the set $\left\{-2, 12, \frac{9}{3}, \frac{25}{25}, 0, 2.585858, \ldots, \pi, \sqrt{4}, 2\sqrt{5}\right\}$.

1. List the real numbers that are not rational numbers.

2. List the integers.

3. Name the property that justifies the statement $(8 \cdot x)3 = 3(8 \cdot x)$.

In exercises 4–11, do not leave negative exponents in your answers.

MC **4.** Evaluate. $(7 - 5)^3 - 18 \div (-3) + \sqrt{10 + 6}$

5. Evaluate. $(4 - 5)^2 - 3(-2) \div 3$

6. Simplify. $\dfrac{16x^3y}{20x^{-1}y^5}$

7. Simplify. $(5x^{-3}y^{-5})(-2x^3y^0)$

MC **8.** Simplify. $\left(\dfrac{5a^{-2}b}{a}\right)^2$

9. Simplify. $7x - 9x^2 - 12x - 8x^2 + 5x$

10. Simplify. $5a + 4b - 6a^2 + b - 7a - 2a^2$

11. Simplify. $3xy^2(4x - 3y + 2x^2)$

12. Write in scientific notation. 0.000002186

13. Write in decimal form. 2.158×10^9

14. Perform the calculation indicated. Write your answer in scientific notation. $(3.8 \times 10^{-5})(4 \times 10^{-2})$

15. Simplify. $2x^2(x - 3y) - x(4 - 8x^2)$

MC **16.** Simplify. $2[-3(2x + 4) + 8(3x - 2)]$.

17. Evaluate $2x^2 - 3x - 6$ when $x = -4$.

MC **18.** Evaluate $5x^2 + 3xy - y^2$ when $x = 3$ and $y = -3$.

▲ **19.** Find the area of a trapezoid with an altitude of 12 meters and bases of 6 and 7 meters.

▲ **20.** Find the area of a circle with a radius of 6 meters. Use $\pi \approx 3.14$.

21. Find the amount A to be repaid on a principal of \$8000 borrowed at a simple interest rate r of 5% for a time t of 3 years. The formula is $A = p(1 + rt)$.

MATH COACH

Mastering the skills you need to do well on the test.

Students often make the same types of errors when they do the Chapter 1 Test. Here are some helpful hints to keep you from making these common errors on test problems.

Using the Order of Operations with Grouping Symbols and Many Operations—Problem 4

Evaluate. $(7 - 5)^3 - 18 \div (-3) + \sqrt{10 + 6}$

Helpful Hint First combine any numbers inside grouping symbols (parentheses, radicals). Next, evaluate exponents and evaluate roots.

Did you first simplify the expression to $(2)^3 - 18 \div (-3) + \sqrt{16}$?

Yes ____ No ____

If you answered No, please go back and carefully combine the numbers inside the grouping symbols.

Did you next simplify the expression to $8 - 18 \div (-3) + 4$?

Yes ____ No ____

If you answered No, please go back and raise the number 2 to the third power. Then find the square root of 16.

If you answered Problem 4 incorrectly, go back and rework the problem using these suggestions.

Using the Power Rule with Negative Exponents—Problem 8

Simplify. $\left(\dfrac{5a^{-2}b}{a}\right)^2$

Helpful Hint First remove the parentheses by using the power rule for exponents. Then write the expression using only positive exponents.

When you used the power rule of exponents, did you get $\dfrac{5^2a^{-4}b^2}{a^2}$?

Yes ____ No ____

If you answered No, remember that the power rule allows you to multiply the exponents.

When you used the rule of negative exponents, did you get a^6 as your denominator?

Yes ____ No ____

If you answered No, remember that $a^4(a^2)$ requires you to use the product rule of exponents.

Now go back and rework the problem using these suggestions.

Need help? Watch the MATH COACH videos in MyMathLab® or on YouTube™.

Simplifying Algebraic Expressions with Many Grouping Symbols—Problem 16

Simplify. $2[-3(2x + 4) + 8(3x - 2)]$

Helpful Hint Work from the inside out by removing the innermost parentheses first. Then work within the resulting expression inside the brackets.

Did you first apply the distributive property to remove the innermost parentheses and obtain the expression $2[-6x - 12 + 24x - 16]$?

Yes ____ No ____

If you answered No, stop and perform that step again. Be careful to avoid sign errors.

Were you able to combine like terms to get $2[18x - 28]$?

Yes ____ No ____

If you answered No, go back and take the time to combine the x terms and then combine the constant terms. Then finish the problem on your own.

If you answered Problem 16 incorrectly, go back and rework the problem using these suggestions.

Evaluating a Variable Expression Given the Values of the Variables—Problem 18

Evaluate $5x^2 + 3xy - y^2$ when $x = 3$ and $y = -3$.

Helpful Hint First, replace each variable with its given value. Place parentheses around each numerical value when you make the substitution. Then use the proper order of operations.

In your first step, did you write $5(3)^2 + 3(3)(-3) - (-3)^2$?

Yes ____ No ____

If you answered No, stop and carefully make this substitution. Be sure to surround the number you are substituting with parentheses.

After evaluating the exponents, did you get the expression $5(9) + 3(3)(-3) - (9)$?

Yes ____ No ____

If you answered No, go back and square 3 and then square -3. Remember that $(-3)^2 = 9$.

Now go back and rework the problem using these suggestions.

CHAPTER 2

How do you count the number of walruses? Are these ice-dependent animals being endangered by global warming? Recently scientists from Russia and Alaska were involved in a joint venture to count the number of these animals located on ice floes between these two regions. The count used infrared technology and the kind of mathematical equations you will study in this chapter.

Linear Equations and Inequalities

2.1 First-Degree Equations with One Unknown

Student Learning Objective

After studying this section, you will be able to:

① **Solve first-degree equations with one unknown.**

① Solving First-Degree Equations with One Unknown

An **equation** is a mathematical statement that two quantities are equal. A **first-degree equation with one unknown** is an equation in which only one kind of variable appears, and that variable has an exponent of 1. This is also called a **linear equation with one unknown.** The variable itself may appear more than once. The equation $-6x + 7x + 8 = 2x - 4$ is a first-degree equation with one unknown because there is one variable, x, and that variable has an exponent of 1. Other examples are $5(6y + 1) = 2y$ and $5z + 3 = 10z$.

To **solve** a first-degree equation with one unknown, we need to find the value of the variable that makes the equation a true mathematical statement. This value is the **solution** or **root** of the equation.

EXAMPLE 1 Is $\frac{1}{3}$ a solution of the equation $2a + 5 = a + 6$?

Solution We replace a by $\frac{1}{3}$ in the equation $2a + 5 = a + 6$.

$$2\left(\frac{1}{3}\right) + 5 \stackrel{?}{=} \frac{1}{3} + 6$$

$$\frac{2}{3} + 5 \stackrel{?}{=} \frac{1}{3} + 6$$

$$\frac{17}{3} \neq \frac{19}{3}$$

This last statement is not true. Thus, $\frac{1}{3}$ is not a solution of $2a + 5 = a + 6$.

Student Practice 1 Is $-\frac{3}{2}$ a solution of the equation $6a - 5 = -4a + 10$?

NOTE TO STUDENT: Fully worked-out solutions to all of the Student Practice problems can be found at the back of the text starting at page SP-1.

This procedure of verifying that a value is indeed a root of an equation is very valuable. We will use it throughout Chapter 2 to check our answers when solving equations.

Equations that have the same solution are said to be **equivalent.** The equations

$$7x - 2 = 12, \quad 7x = 14, \quad \text{and} \quad x = 2$$

are equivalent because the solution of each equation is $x = 2$.

To solve an equation, we perform algebraic operations on it to obtain a simpler, equivalent equation of the form variable = constant or constant = variable.

PROPERTIES OF EQUIVALENT EQUATIONS

1. If $a = b$, then $a + c = b + c$ and $a - c = b - c$.

   If the same number is added to or subtracted from both sides of an equation, the result is an equivalent equation.

2. If $a = b$ and $c \neq 0$, then $ac = bc$ and $\frac{a}{c} = \frac{b}{c}$.

   If both sides of an equation are multiplied or divided by the same nonzero number, the result is an equivalent equation.

To solve a first-degree equation, we isolate the variable on one side of the equation and the constants on another, using the properties of equivalent equations.

EXAMPLE 2 Solve. $x - 8.2 = 5.0$

Solution $x - 8.2 + 8.2 = 5.0 + 8.2$ Add 8.2 to each side.

$x = 13.2$

Check.

$$x - 8.2 = 5.0$$
$$13.2 - 8.2 \stackrel{?}{=} 5.0$$
$$5.0 = 5.0 \checkmark$$

The statement is valid, so our answer is correct. 13.2 is the root of the equation.

Student Practice 2 Solve. $x + 5.2 = -2.8$

EXAMPLE 3 Solve. $\frac{1}{3}y = -6$

Solution $3\left(\frac{1}{3}\right)y = 3(-6)$ Multiply each side by 3 to eliminate the fraction.

$y = -18$

The solution is -18. Be sure to check this value.

Student Practice 3 Solve.

$$\frac{1}{5}w = -6$$

EXAMPLE 4 Solve. $6x - 2 - 4x = 8x + 3$

Solution

$2x - 2 = 8x + 3$ Combine like terms.

$2x - 8x - 2 = 8x - 8x + 3$ Subtract $8x$ from (or add $-8x$ to) each side.

$-6x - 2 = 3$

$-6x - 2 + 2 = 3 + 2$ Add 2 to each side.

$-6x = 5$

$\frac{-6x}{-6} = \frac{5}{-6}$ Divide each side by -6.

$x = -\frac{5}{6}$

When checking fractional values like $x = -\frac{5}{6}$, take extra care in order to perform the operations correctly.

Check. $6\left(-\frac{5}{6}\right) - 2 - 4\left(-\frac{5}{6}\right) \stackrel{?}{=} 8\left(-\frac{5}{6}\right) + 3$ Replace x by $-\frac{5}{6}$ in the *original* equation.

$$-5 - 2 + \frac{10}{3} \stackrel{?}{=} \frac{-20}{3} + 3$$
$$\frac{-21}{3} + \frac{10}{3} \stackrel{?}{=} \frac{-20}{3} + \frac{9}{3}$$
$$-\frac{11}{3} = -\frac{11}{3} \checkmark$$

Thus, $-\frac{5}{6}$ is the solution.

Student Practice 4 Solve.

$$8w - 3 = 2w - 7w + 4$$

When an equation contains grouping symbols and fractions, use the following procedure to solve it.

PROCEDURE FOR SOLVING FIRST-DEGREE EQUATIONS WITH ONE UNKNOWN

1. Remove grouping symbols in the proper order.
2. If fractions exist, multiply both sides of the equation by the least common denominator (LCD) of all the fractions.
3. Combine like terms if possible.
4. Add or subtract a variable term on both sides of the equation to collect all variable terms on one side of the equation.
5. Add or subtract a numerical value on both sides of the equation to collect all numerical values on the other side of the equation.
6. Divide both sides of the equation by the coefficient of the variable.
7. Simplify the solution (if possible).
8. Check your solution.

Steps 4 and 5 are interchangeable. You may do Step 5 first and then do Step 4, if you wish.

It is not necessary to memorize this eight-step procedure. However, you should refer to it often as you work the exercises in Section 2.1 until the sequence of steps is second nature to you.

EXAMPLE 5 Solve and check. $3(3x + 2) - 4x = -2(x - 3)$

Solution

$$9x + 6 - 4x = -2x + 6 \quad \text{Remove the parentheses.}$$
$$5x + 6 = -2x + 6 \quad \text{Combine like terms.}$$
$$5x + 2x + 6 = -2x + 2x + 6 \quad \text{Add } 2x \text{ to each side.}$$
$$7x + 6 = 6$$
$$7x + 6 - 6 = 6 - 6 \quad \text{Subtract 6 from each side.}$$
$$7x = 0$$
$$\frac{7x}{7} = \frac{0}{7} \quad \text{Divide each side by 7.}$$
$$x = 0$$

Check.
$$3[3(0) + 2] - 4(0) \stackrel{?}{=} -2(0 - 3)$$
$$3(0 + 2) - 0 \stackrel{?}{=} -2(-3)$$
$$6 - 0 \stackrel{?}{=} 6$$
$$6 = 6 \checkmark$$

Thus, 0 is the solution.

Student Practice 5 Solve and check. $a - 4(2a - 7) = 3(a + 6)$

EXAMPLE 6 Solve. $\frac{x}{5} + \frac{1}{2} = \frac{4}{5} + \frac{x}{2}$

Solution

$$10\left(\frac{x}{5} + \frac{1}{2}\right) = 10\left(\frac{4}{5} + \frac{x}{2}\right)$$ Multiply each term by the LCD 10.

$$10\left(\frac{x}{5}\right) + 10\left(\frac{1}{2}\right) = 10\left(\frac{4}{5}\right) + 10\left(\frac{x}{2}\right)$$ Use the distributive property.

$$2x + 5 = 8 + 5x$$ Simplify.

$$2x - 2x + 5 = 8 + 5x - 2x$$ Subtract $2x$ from each side.

$$5 = 8 + 3x$$

$$5 - 8 = 8 - 8 + 3x$$ Subtract 8 from each side.

$$-3 = 3x$$

$$\frac{-3}{3} = \frac{3x}{3}$$ Divide each side by 3.

$$-1 = x$$

Check. See if you can verify this solution.

Student Practice 6 Solve and check.

$\frac{y}{3} + \frac{1}{2} = 5 + \frac{y-9}{4}$ $\left(\textit{Hint: } \text{You can write } \frac{y-9}{4} \text{ as } \frac{y}{4} - \frac{9}{4}.\right)$

An equation that contains many decimals can be multiplied by an appropriate power of 10 to clear it of decimals.

EXAMPLE 7 Solve and check. $0.9 + 0.2(x + 4) = -3(0.1x - 0.4)$

Solution

$$0.9 + 0.2x + 0.8 = -0.3x + 1.2$$ Remove the parentheses.

$$10(0.9) + 10(0.2x) + 10(0.8) = 10(-0.3x) + 10(1.2)$$ Multiply each side of the equation by 10 and use the distributive property.

$$9 + 2x + 8 = -3x + 12$$ Simplify.

$$2x + 17 = -3x + 12$$ Combine like terms.

$$2x + 3x + 17 = -3x + 3x + 12$$ Add $3x$ to each side.

$$5x + 17 = 12$$

$$5x + 17 - 17 = 12 - 17$$ Add -17 to each side.

$$5x = -5$$

$$x = -1$$ Divide each side by 5.

Check. $0.9 + 0.2(-1 + 4) \stackrel{?}{=} -3[0.1(-1) - 0.4]$

$$0.9 + 0.2(3) \stackrel{?}{=} -3(-0.1 - 0.4)$$

$$0.9 + 0.6 \stackrel{?}{=} -3(-0.5)$$

$$1.5 = 1.5 \checkmark$$

Thus, -1 is the solution.

Student Practice 7 Solve and check.

$$4(0.01x + 0.09) - 0.07(x - 8) = 0.83$$

Not every equation has a solution. Some equations have no solution at all.

EXAMPLE 8 Solve. $7x + 3 - 9x = 14 - 2x + 5$

Solution

$$-2x + 3 = -2x + 19 \quad \text{Combine like terms.}$$
$$-2x + 3 + 2x = -2x + 19 + 2x \quad \text{Add } 2x \text{ to each side.}$$
$$3 = 19 \quad \text{Simplify. We obtain a false equation.}$$

No matter what value we use for x in this equation, we get a false sentence. This equation has **no solution.**

Student Practice 8 Solve. $7 + 14x - 3 = 2(x - 4) + 12x$

Now we examine a totally different situation. There are some equations for which any real number is a solution.

EXAMPLE 9 Solve. $5(x - 2) + 3x = 10x - 2(x + 5)$

Solution

$$5x - 10 + 3x = 10x - 2x - 10 \quad \text{Remove parentheses.}$$
$$8x - 10 = 8x - 10 \quad \text{Combine like terms.}$$
$$8x - 10 - 8x = 8x - 10 - 8x \quad \text{Subtract } 8x \text{ from each side of the equation.}$$
$$-10 = -10 \quad \text{Simplify. This is an equation that is always true.}$$

Replacing x in this equation by any real number will always result in a true sentence. This is an equation for which **any real number is a solution.**

Student Practice 9 Solve. $13x - 7(x + 5) = 4x - 35 + 2x$

As you work through the exercises in this section you will notice that solutions to equations can be integers, fractions, or decimals. Recall from page 2 that a **terminating decimal** is one that has a definite number of digits. Throughout this book, the decimal form of a solution will be given only if is a terminating decimal.

STEPS TO SUCCESS Be Involved.

If you are in a traditional class:
Don't just sit on the sidelines of the class and watch. Take part in the classroom discussion. People learn mathematics best through active participation. Whenever you are not clear about something, ask a question. Usually your questions will be helpful to other students in the room. When the teacher asks for suggestions, be sure to contribute your own ideas. Sit near the front where you can see and hear well. This will help you to focus on the material being covered.

Making it personal: Which of these suggestions is the one you most need to follow? Write down what you need to do to improve in this area. ▼

If you are in an online class or a nontraditional class:
Be sure to e-mail the teacher. Talk to the tutor on duty. Ask questions. Think about concepts. Make your mind interact with the textbook. Be mentally involved. This active mental interaction is the key to your success.

Making it personal: Which of these suggestions is the one you most need to follow? Write down what you need to do to improve in this area. ▼

2.1 Exercises

MyMathLab®

Watch the videos in MyMathLab

Download the MyDashBoard App

Verbal and Writing Skills, Exercises 1–8

1. Is -20 a root of the equation $3x - 15 = 45$? Why or why not?

2. Is 21 a root of the equation $2x + 12 = -30$? Why or why not?

3. Is $\frac{2}{7}$ a solution to the equation $7x - 8 = -6$? Why or why not?

4. Is $\frac{3}{5}$ a solution to the equation $5y + 9 = 12$? Why or why not?

5. What is the first step in solving the equation $\frac{x}{3} + \frac{3}{4} = 2 - \frac{x}{2}$? Why?

6. What is the first step in solving the equation $0.7 + 0.03x = 4$? Why?

7. When solving the equation $x + 3.6 = 8$, would you clear it of decimals by multiplying each term by 10? Why or why not?

8. When solving the equation $x - \frac{1}{4} = 3$, would you clear it of fractions by multiplying each term by 4? Why or why not?

Solve exercises 9–44. Check your solutions.

9. $-11 + x = -3$

10. $26 + x = -35$

11. $-8x = 56$

12. $-16x = -64$

13. $-14x = -70$

14. $-12x = 72$

15. $8x - 1 = 11$

16. $10x + 3 = 15$

17. $8x + 5 = 2x - 13$

18. $16x + 5 = 10x - 1$

19. $16 - 2x = 5x - 5$

20. $-12x - 8 = 10 - 3x$

21. $3a - 5 - 2a = 2a - 3$

22. $5a - 2 + 4a = 2a + 12$

23. $4(y - 1) = -2(3 + y)$

24. $3(5 - y) = 3(y + 4)$

25. $3 + y = 10 - 3(y + 1)$

Solve the following exercises. You may leave answers in either decimal or fraction form.

26. $4y + 5 = 6(y + 3) - y$

27. $\frac{2}{3}x = 8$

28. $-\frac{5}{6}x = 5$

29. $\frac{y}{2} + 4 = \frac{1}{6}$

30. $\frac{y}{3} + 2 = \frac{4}{5}$

31. $\frac{2}{3} - \frac{x}{6} = 1$

32. $\frac{4x}{5} + \frac{3}{2} = 2x$

33. $\frac{1}{2}(x + 3) - 2 = 1$

34. $5 - \frac{2}{3}(x + 2) = 3$

35. $5 - \frac{2x}{7} = 1 - (x - 4)$

36. $6 + 2(x - 1) = \frac{3x}{5} + 4$

37. $0.3x + 0.4 = 0.5x - 0.8$

38. $0.8x - 0.1 = 0.4x + 0.7$

39. $0.3 - 0.05x = 0.2x - 0.1$

40. $0.1x - 0.12 = 0.04x + 0.03$

41. $0.2(x - 4) = 3$

42. $0.5(3x + 5) = 1$

43. $0.05x - 2 = 0.3(x - 5)$

44. $0.3(x + 2) - 2 = 0.05x$

Mixed Practice *Solve each of the following.*

45. $2x + 12 = 3 + 4x - 7$

46. $4y - 11 - 7y = 22 - 3$

47. $\frac{1}{2} - \frac{x}{3} = \frac{2x - 3}{3}$

48. $\frac{1}{6} - \frac{x}{2} = \frac{x - 5}{3}$

49. $\frac{1}{3} - \frac{x + 1}{5} = \frac{x}{3}$

50. $\frac{y + 5}{12} = \frac{3}{4} - \frac{y + 1}{8}$

51. $2 + 0.1(5 - x) = 1.3x - (0.4x - 2.5)$

52. $3(0.4 - x) + 2 = x + 0.4(x + 8)$

To Think About

Some of the equations in this section have no solution. For some of the equations, any real number is a solution. Other equations have one real solution. Work carefully and solve each of the following equations.

53. $6x - 8 = 9 - 2x + 5 - 3x$

54. $7x - 5 = -2x - 15 + 10x + 6$

55. $2x - 4(x + 1) = -2x + 14$

56. $3x - 17 = 8x - 5(x - 2)$

57. $-(x - 2) + 1 = 2x + 3(1 - x)$

58. $8(x + 2) - 7 = 3(x + 3) + 5x$

59. $6 + 8(x - 2) = 10x - 2(x + 4)$

60. $2x + 4(x - 5) = -x + 7(x - 1) + 3$

61. $x - 2 + \frac{2x}{5} = -2 + \frac{7x}{5}$

62. $x + \frac{2x + 8}{3} = \frac{5x + 5}{3} + 1$

Cumulative Review

Simplify. Do not leave negative exponents in your answers.

63. **[1.3.3]** $5 - (4 - 2)^2 + 3(-2)$

64. **[1.3.3]** $(-2)^4 - 12 - 6(-2)$

65. **[1.4.4]** $\left(\frac{3xy^2}{2x^2y}\right)^3$

66. **[1.4.2]** $(2x^{-2}y^{-3})^2(4xy^{-2})^{-2}$

Quick Quiz 2.1

Solve for x.

1. $4(7 - 3x) = 16 - 3(3x + 1)$

2. $\frac{2}{3}(2x - 1) + 3 = 4(2x - 4)$

3. $0.7x + 1.3 = 5x - 2.14$

4. **Concept Check** Explain how you would solve the following equation for x.

$$\frac{3x + 1}{2} + \frac{2}{3} = \frac{4x}{5}$$

2.2 Literal Equations and Formulas

Student Learning Objective

After studying this section, you will be able to:

1. Solve literal equations for the desired unknown.

***NOTE TO STUDENT:** Fully worked-out solutions to all of the Student Practice problems can be found at the back of the text starting at page SP-1.*

1 Solving Literal Equations for the Desired Unknown

A first-degree **literal equation** is an equation that contains at least one variable other than the variable that we are solving for. When you solve for an unknown in a literal equation, the final expression will contain these other variables. We use this procedure to deal with formulas in applied problems.

EXAMPLE 1 Solve for x. $5x + 3y = 2$

Solution

$$5x = 2 - 3y \quad \text{Subtract } 3y \text{ from each side.}$$

$$\frac{5x}{5} = \frac{2 - 3y}{5} \quad \text{Divide each side by 5.}$$

$$x = \frac{2 - 3y}{5} \quad \text{The solution is a fractional expression.}$$

Student Practice 1 Solve for W. $P = 2L + 2W$

Where possible, combine like terms as you solve the equation.

When solving more complicated first-degree literal equations, use the following procedure.

PROCEDURE FOR SOLVING FIRST-DEGREE (OR LINEAR) LITERAL EQUATIONS

1. Remove grouping symbols in the proper order.
2. If fractions exist, multiply all terms on both sides by the LCD.
3. Combine like terms if possible.
4. Add or subtract a term with the desired unknown on both sides of the equation to collect all terms with the desired unknown on one side of the equation.
5. Add or subtract appropriate terms on both sides of the equation to collect all other terms on the other side of the equation.
6. Divide each side of the equation by the coefficient of the desired unknown.
7. Simplify the solution (if possible).

Some equations appear difficult to solve because they contain fractions and parentheses. Immediately remove the parentheses. Then multiply each term by the LCD. The equation will then appear less threatening.

MC **EXAMPLE 2** Solve for b. $A = \frac{2}{3}(a + b + 3)$

Solution

$$A = \frac{2}{3}a + \frac{2}{3}b + 2 \quad \text{Remove parentheses.}$$

$$3A = 3\left(\frac{2}{3}a\right) + 3\left(\frac{2}{3}b\right) + 3(2) \quad \text{Multiply all terms by the LCD 3.}$$

$$3A = 2a + 2b + 6 \quad \text{Simplify.}$$

$$3A - 2a - 6 = 2b \quad \text{Subtract } 2a \text{ from each side. Subtract 6 from each side.}$$

$$\frac{3A - 2a - 6}{2} = \frac{2b}{2} \quad \text{Divide each side of the equation by the coefficient of } b.$$

$$\frac{3A - 2a - 6}{2} = b \quad \text{Simplify.}$$

Student Practice 2 Solve for a.

$$H = \frac{3}{4}(a + 2b - 4)$$

Be sure to combine like terms after removing the parentheses. This will simplify the equation and make it much easier to solve.

EXAMPLE 3 Solve for x. $5(2ax + 3y) - 4ax = 2(ax - 5)$

Solution

$10ax + 15y - 4ax = 2ax - 10$ Remove parentheses.

$6ax + 15y = 2ax - 10$ Combine like terms.

$6ax - 2ax + 15y = -10$ Subtract $2ax$ from each side to obtain terms containing x on one side.

$4ax = -10 - 15y$ Simplify and subtract $15y$ from each side.

$\frac{4ax}{4a} = \frac{-10 - 15y}{4a}$ Divide each side by the coefficient of x.

$$x = \frac{-10 - 15y}{4a}$$

Student Practice 3 Solve for b. $-2(ab - 3x) + 2(8 - ab) = 5x + 4ab$

Marathons (26.2-mile road races) are run all over the world. The five marathons with the most participants are held in New York City, Boston, Chicago, London, and Berlin. As world-class distance runners continue to get faster, the world record for the marathon continues to drop. The most recent world-record breaker in the men's marathon is Haile Gebrselassie of Ethiopia. His time of 2 hours 3 minutes 59 seconds won the Berlin Marathon in 2008. If we round to the nearest minute, this time is approximately equal to 124 minutes. The bar graph below shows the approximate world record times for the men's marathon for selected years from 1960 to 2010.

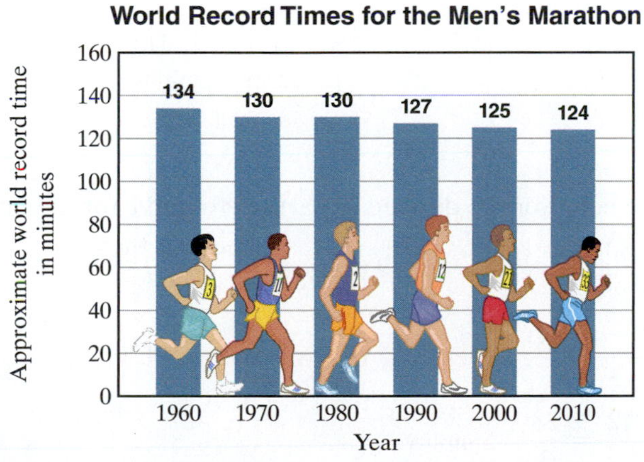

EXAMPLE 4 The world record times in minutes for the men's marathon are approximated by the equation $t = -0.2x + 135$, where x is the number of years since 1960. Solve this equation for x. Determine approximately what year it will be when the world record time for the men's marathon is 122 minutes.

Solution

$t = -0.2x + 135$

$0.2x = 135 - t$ Add $0.2x - t$ to each side.

$2x = 1350 - 10t$ Multiply each side by 10 to clear decimals.

$x = \frac{1350 - 10t}{2}$ Divide each side by 2.

Continued on next page

Now we will use this equation, which has been solved for x (the number of years since 1960), to find when the winning time is 122 minutes.

$$x = \frac{1350 - 10(122)}{2} \quad \text{We substitute 122 for } t.$$

$$x = \frac{1350 - 1220}{2}$$

$$x = \frac{130}{2} = 65$$

It will be 65 years from 1960. Thus, we estimate that this winning time will occur in the year 2025.

Student Practice 4 The winning time in minutes for a 15-mile race to benefit breast cancer research is given by the equation $t = -0.4x + 81$, where x is the number of years since 1990. Solve this equation for x. Determine approximately what year it will be when the winning time for this 15-mile race is 71 minutes.

▲ **EXAMPLE 5**

(a) Solve the formula for the area of a trapezoid, $A = \frac{1}{2}a(b + c)$, for c.

(b) Find c when $A = 20$ square inches, $a = 3$ inches, and $b = 4$ inches.

Solution

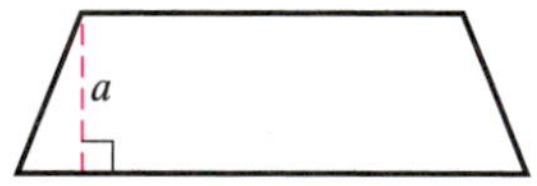

(a)

$$A = \frac{1}{2}ab + \frac{1}{2}ac \quad \text{Remove the parentheses.}$$

$$2A = 2\left(\frac{1}{2}ab\right) + 2\left(\frac{1}{2}ac\right) \quad \text{Multiply each term by 2.}$$

$$2A = ab + ac \quad \text{Simplify.}$$

$$2A - ab = ac \quad \text{Subtract } ab \text{ from each side to isolate the term containing } c.$$

$$\frac{2A - ab}{a} = c \quad \text{Divide each side by } a.$$

(b) We use the equation we derived in part **(a)** to find c for the given values.

$$c = \frac{2A - ab}{a}$$

$$= \frac{2(20) - (3)(4)}{3} \quad \text{Substitute the given values of } A, a, \text{ and } b \text{ to find } c.$$

$$= \frac{40 - 12}{3} = \frac{28}{3} \quad \text{Simplify.}$$

Thus, side $c = \frac{28}{3}$ inches or $9\frac{1}{3}$ inches.

▲ **Student Practice 5**

(a) Solve for h. $A = 2\pi rh + 2\pi r^2$

(b) Find h when $A = 100$ and $r = 2.0$. Use $\pi \approx 3.14$. Round your answer to the nearest hundredth.

2.2 Exercises

MyMathLab® Watch the videos in MyMathLab Download the MyDashBoard App

Solve for x.

1. $6x + 5y = 3$

2. $7x + 2y = 5$

3. $4x + y = 18 - 3x$

4. $7x - 9 = 6y - x$

5. $y = \frac{2}{3}x - 4$

6. $y = -\frac{1}{5}x + 3$

Solve for the variable specified.

7. $x = -\frac{3}{4}y + \frac{2}{3}$; for y

8. $x = \frac{5}{2}y - \frac{1}{5}$; for y

▲**9.** $A = lw$; for l

▲**10.** $V = lwh$; for w

▲**11.** $A = \frac{h}{2}(B + b)$; for B

12. $C = \frac{5}{9}(F - 32)$; for F

▲**13.** $A = 2\pi rh$; for r

▲**14.** $V = \pi r^2 h$; for h

15. $H = \frac{2}{3}(a + 2b)$; for b

16. $H = \frac{3}{4}(5a + b)$; for a

17. $2(2ax + y) = 3ax - 4y$; for x

18. $4(-ax + 2y) = 5ax + y$; for x

Follow the directions given.

▲**19.** **(a)** Solve for b. $A = \frac{1}{2}ab$

(b) Evaluate when $A = 18$ and $a = \frac{3}{2}$.

20. **(a)** Solve for C. $F = \frac{9}{5}C + 32$

(b) Evaluate when $F = 23°$.

21. **(a)** Solve for n. $A = a + d(n - 1)$

(b) Evaluate when $A = 28$, $a = 3$, and $d = 15$.

▲**22.** **(a)** Solve for h. $V = \frac{1}{3}\pi r^2 h$

(b) Evaluate when $V = 6.28$, $r = 3$, and $\pi \approx 3.14$.

Applications

23. ***Oil Imported from Canada*** The amount of oil imported to the United States from Canada in millions of barrels per day can be approximated by the equation $y = 0.076x + 1.35$, where x is the number of years since 2000. Solve this equation for x. Use this new equation to determine in which year the approximate number of oil barrels imported from Canada per day will be 2.49 million. (*Source:* www.eia.doe.gov)

24. ***Life Expectancy*** The number of years a person born in the United States is expected to live can be approximated by the equation $y = 0.189x + 70.8$, where x is the number of years since 1970. Solve this equation for x. Use this new equation to determine in which year the approximate life expectancy will be 80.2 years. (*Source:* www.cdc.gov)

25. ***Mariner's Formula*** The mariner's formula can be written in the form $\frac{m}{1.15} = k$, where m is the speed of a ship in miles per hour and k is the speed of the ship in knots (nautical miles per hour).

(a) Solve the formula for m.

(b) Use this result to find the number of miles per hour a ship is traveling if its speed is 29 knots.

26. ***Patient Appointments*** Some doctors use the formula $ND = 1.08T$ to relate the variables N (the number of patient appointments the doctor schedules in one day), D (the duration of each patient appointment), and T (the total number of minutes the doctor can use to see patients in one day).

(a) Solve the formula for N.

(b) Use this result to find the number of patient appointments N a doctor should make if she has 6 hours available for patients and each appointment is 15 minutes long. (Round your answer to the nearest whole number.)

In exercises 27 and 28, the variable C represents the consumption of products in the United States in billions of dollars, and D represents disposable income in the United States in billions of dollars.

27. ***An Economic Model*** Suppose economists use as a model of the country's economy the equation $C = 0.6547D + 5.8263$.

(a) Solve the equation for D.

(b) Use this result to determine the disposable income D if the consumption C is \$9.56 billion. Round your answer to the nearest tenth of a billion.

28. ***An Economic Model*** Suppose economists use as a model of the country's economy the equation $C = 0.7649D + 6.1275$.

(a) Solve the equation for D.

(b) Use this result to determine the disposable income D if the consumption C is \$12.48 billion. Round your answer to the nearest tenth of a billion.

Cumulative Review *Write with positive exponents in simplest form.*

29. **[1.4.4]** $(2x^{-3}y)^{-2}$

30. **[1.4.4]** $\left(\frac{5x^2y^{-3}}{x^{-4}y^2}\right)^{-3}$

31. **[1.3.3]** Simplify. $1 + 16 \div (2 - 4)^3 - 3$

32. **[1.5.3]** Simplify. $2[a - (3 - 2b)] + 5a$

33. **[1.2.2]** ***Education Fund*** Sharon and James want to begin an education fund for their two daughters. They invest $5000 in a certificate of deposit for one year, with an annual return of 5%. $4000 is invested in a more risky venture that they hope will have an annual return of 9%. How much money will they have at the end of 1 year if their risky investment performs as they hope?

(*Hint:* Use $I = prt$.)

34. **[1.2.2]** ***Automobile Costs*** Drew wants to go to college in Pennsylvania. He and his parents take a long weekend to drive there from Kansas City, Missouri. His odometer read 45,711.3 when he left the college in Pennsylvania and 46,622.1 when he arrived back home. He started and ended his trip on a full tank of gas. He made gas purchases of 9.9 gallons, 11.7 gallons, 10.6 gallons, 5.8 gallons, and 8 gallons during the trip. How many miles per gallon did the car get on the trip?

Quick Quiz 2.2

1. Solve for b. $A = d + g(b + 3)$

2. Solve for x. $V = \frac{1}{3}xy$

3. Solve for w. $G = 4x + \frac{1}{2}w + \frac{3}{4}$

4. **Concept Check** Explain how you would solve for x in the following equation. $3(3ax + y) = 2ax - 5y$

STEPS TO SUCCESS Taking Good Notes Each Class Session

Don't Copy Down Everything the Teacher Says. You will get overloaded with facts. Instead write down the important ideas and examples as the instructor lectures. Be sure to include any helpful hints or suggestions that your instructor gives you. You will be amazed at how easily you forget these if you do not write them down.

Be an Active Listener. Keep your mind on what the instructor is saying. Be ready with questions whenever you do not understand something. Stay alert in class. Keep your mind on mathematics.

Preview the Lesson Material. Before class glance over the topics that will be covered. You can take much better notes if you know what the topics of the lecture will be.

Look Back at Your Notes. Try to review them the same day sometime after class. You will find the content of your notes much easier to understand if you read them again within a few hours of class.

Making it personal: Which of these suggestions are the most helpful to you? How can you improve your skills in note taking?

2.3 Absolute Value Equations

Student Learning Objectives

After studying this section, you will be able to:

1. Solve absolute value equations of the form $|ax + b| = c$.
2. Solve absolute value equations of the form $|ax + b| + c = d$.
3. Solve absolute value equations of the form $|ax + b| = |cx + d|$.

1 Solving Absolute Value Equations of the Form $|ax + b| = c$

From Section 1.2, you know that the absolute value of a number x can be pictured as the distance between 0 and x on the number line. Let's look at a simple absolute value equation, $|x| = 4$, and draw a picture.

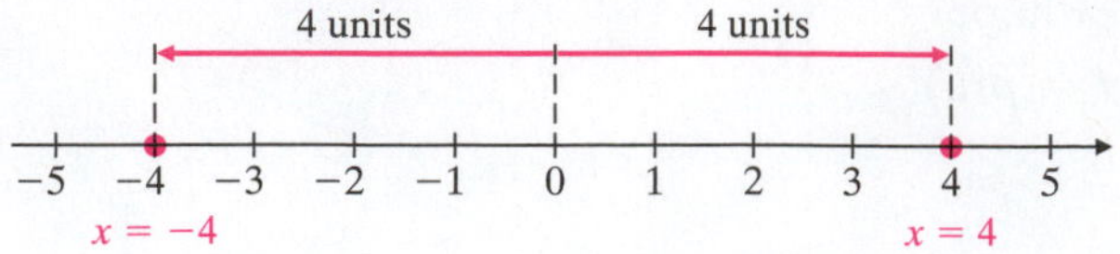

Thus, the equation $|x| = 4$ has two solutions, $x = 4$ or $x = -4$. Let's look at another example.

$$\text{If} \quad |x| = \frac{2}{3},$$
$$\text{then} \quad x = \frac{2}{3} \quad \text{or} \quad x = -\frac{2}{3},$$
$$\text{because} \quad \left|\frac{2}{3}\right| = \frac{2}{3} \quad \text{and} \quad \left|-\frac{2}{3}\right| = \frac{2}{3}.$$

We can solve these relatively simple absolute value equations by recalling the definition of absolute value.

$$|x| = \begin{cases} x, & \text{if } x \geq 0 \\ -x, & \text{if } x < 0 \end{cases}$$

Now let's take a look at a more complicated absolute value equation: $|ax + b| = c$.

> The solutions of an equation of the form $|ax + b| = c$, where $a \neq 0$ and c is a positive number, are those values that satisfy
> $$ax + b = c \quad \textit{or} \quad ax + b = -c.$$

EXAMPLE 1 Solve and check your solutions. $|2x + 5| = 11$

Solution Using the rule established in the box, we have the following:

$$\begin{aligned} 2x + 5 &= 11 \quad &\text{or} \quad 2x + 5 &= -11 \\ 2x &= 6 & 2x &= -16 \\ x &= 3 & x &= -8 \end{aligned}$$

The two solutions are 3 and −8.

Check. **if $x = 3$** **if $x = -8$**

$$\begin{aligned} |2x + 5| &= 11 \\ |2(3) + 5| &\stackrel{?}{=} 11 \\ |6 + 5| &\stackrel{?}{=} 11 \\ |11| &\stackrel{?}{=} 11 \\ 11 &= 11 \quad ✓ \end{aligned} \qquad \begin{aligned} |2x + 5| &= 11 \\ |2(-8) + 5| &\stackrel{?}{=} 11 \\ |-16 + 5| &\stackrel{?}{=} 11 \\ |-11| &\stackrel{?}{=} 11 \\ 11 &= 11 \quad ✓ \end{aligned}$$

Student Practice 1 Solve and check your solutions. $|3x - 4| = 23$

***NOTE TO STUDENT:** Fully worked-out solutions to all of the Student Practice problems can be found at the back of the text starting at page SP-1.*

EXAMPLE 2 Solve and check your solutions. $\left|\frac{1}{2}x - 1\right| = 5$

Solution The solutions of the given absolute value equation must satisfy

$$\frac{1}{2}x - 1 = 5 \quad \text{or} \quad \frac{1}{2}x - 1 = -5.$$

If we multiply each term of both equations by 2, we obtain the following:

$$x - 2 = 10 \quad \text{or} \quad x - 2 = -10$$

$$x = 12 \qquad\qquad x = -8$$

Check. **if $x = 12$**

$$\left|\frac{1}{2}(12) - 1\right| \stackrel{?}{=} 5$$
$$|6 - 1| \stackrel{?}{=} 5$$
$$|5| \stackrel{?}{=} 5$$
$$5 = 5 \checkmark$$

if $x = -8$

$$\left|\frac{1}{2}(-8) - 1\right| \stackrel{?}{=} 5$$
$$|-4 - 1| \stackrel{?}{=} 5$$
$$|-5| \stackrel{?}{=} 5$$
$$5 = 5 \checkmark$$

Student Practice 2 Solve and check your solutions.

$$\left|\frac{2}{3}x + 4\right| = 2$$

② Solving Absolute Value Equations of the Form $|ax + b| + c = d$

Notice that in each of the previous examples the absolute value expression is on one side of the equation and a positive real number is on the other side of the equation. What happens when we encounter an equation of the form $|ax + b| + c = d$?

MC **EXAMPLE 3** Solve $|3x - 1| + 2 = 5$ and check your solutions.

Solution First we will rewrite the equation so that the absolute value expression is alone on one side of the equation.

$$|3x - 1| + 2 - 2 = 5 - 2$$

$$|3x - 1| = 3$$

Now we solve $|3x - 1| = 3$.

$$3x - 1 = 3 \quad \text{or} \quad 3x - 1 = -3$$
$$3x = 4 \qquad\qquad 3x = -2$$
$$x = \frac{4}{3} \qquad\qquad x = -\frac{2}{3}$$

Check. **if $x = \frac{4}{3}$**

$$\left|3\left(\frac{4}{3}\right) - 1\right| + 2 \stackrel{?}{=} 5$$
$$|4 - 1| + 2 \stackrel{?}{=} 5$$
$$|3| + 2 \stackrel{?}{=} 5$$
$$3 + 2 \stackrel{?}{=} 5$$
$$5 = 5 \checkmark$$

if $x = -\frac{2}{3}$

$$\left|3\left(-\frac{2}{3}\right) - 1\right| + 2 \stackrel{?}{=} 5$$
$$|-2 - 1| + 2 \stackrel{?}{=} 5$$
$$|-3| + 2 \stackrel{?}{=} 5$$
$$3 + 2 \stackrel{?}{=} 5$$
$$5 = 5 \checkmark$$

Student Practice 3 Solve and check your solutions. $|2x + 1| + 3 = 8$

③ Solving Absolute Value Equations of the Form $|ax + b| = |cx + d|$

Let us now consider the possibilities for a and b if $|a| = |b|$.

$$\text{Suppose } a = 5\text{; then } b = 5 \text{ or } -5.$$

$$\text{If } a = -5\text{, then } b = 5 \text{ or } -5.$$

To generalize, if $|a| = |b|$, then $a = b$ or $a = -b$.

We now apply this property to solve more complex equations.

EXAMPLE 4 Solve and check. $|3x - 4| = |x + 6|$

Solution The solutions of the given equation must satisfy

$$3x - 4 = x + 6 \quad \text{or} \quad 3x - 4 = -(x + 6).$$

Now we solve each equation in the normal fashion.

$$\begin{aligned} 3x - 4 &= x + 6 & \text{or} \quad 3x - 4 &= -x - 6 \\ 3x - x &= 4 + 6 & 3x + x &= 4 - 6 \\ 2x &= 10 & 4x &= -2 \\ x &= 5 & x &= -\frac{1}{2} \end{aligned}$$

We will check each solution by substituting it into the *original equation.*

Check. **if $x = 5$**

$$|3(5) - 4| \stackrel{?}{=} |5 + 6|$$
$$|15 - 4| \stackrel{?}{=} |11|$$
$$|11| \stackrel{?}{=} |11|$$
$$11 = 11 \ ✓$$

if $x = -\frac{1}{2}$

$$\left|3\left(-\frac{1}{2}\right) - 4\right| \stackrel{?}{=} \left|-\frac{1}{2} + 6\right|$$
$$\left|-\frac{3}{2} - 4\right| \stackrel{?}{=} \left|-\frac{1}{2} + 6\right|$$
$$\left|-\frac{3}{2} - \frac{8}{2}\right| \stackrel{?}{=} \left|-\frac{1}{2} + \frac{12}{2}\right|$$
$$\left|-\frac{11}{2}\right| \stackrel{?}{=} \left|\frac{11}{2}\right|$$
$$\frac{11}{2} = \frac{11}{2} \ ✓$$

Student Practice 4 Solve and check. $|x - 6| = |5x + 8|$

TO THINK ABOUT: Two Other Absolute Value Equations Explain how you would solve an absolute value equation of the form $|ax + b| = 0$. Give an example. Does $|3x + 2| = -4$ have a solution? Why or why not?

2.3 Exercises

MyMathLab®

Watch the videos in MyMathLab

Download the MyDashBoard App

Verbal and Writing Skills, Exercises 1–4

1. The equation $|x| = b$, where b is a positive number, will always have how many solutions? Why?

2. The equation $|x| = b$ might have only one solution. How could that happen?

3. To solve an equation like $|x + 7| - 2 = 8$, what is the first step that must be done? What will be the result?

4. To solve an equation like $|3x - 1| + 5 = 14$, what is the first step that must be done? What will be the result?

Solve each absolute value equation. Check your solutions for exercises 5–24.

5. $|x| = 30$

6. $|x| = 14$

7. $|x + 4| = 10$

8. $|x + 6| = 13$

9. $|2x - 5| = 13$

10. $|7x - 3| = 11$

11. $|5 - 4x| = 11$

12. $|2 - 3x| = 13$

13. $\left|\frac{1}{2}x - 3\right| = 2$

14. $\left|\frac{1}{4}x + 5\right| = 3$

15. $|1.8 - 0.4x| = 1$

16. $|0.9 - 0.7x| = 4$

17. $|x + 2| - 1 = 7$

18. $|x + 3| - 4 = 8$

19. $\left|\frac{1}{2} - \frac{3}{4}x\right| + 1 = 3$

20. $\left|\frac{2}{3} - \frac{1}{2}x\right| - 2 = -1$

21. $\left|2 - \frac{2}{3}x\right| - 3 = 5$

22. $\left|5 - \frac{7}{2}x\right| + 1 = 11$

23. $\left|\frac{1-3x}{2}\right| = \frac{4}{5}$

24. $\left|\frac{2x-1}{4}\right| = \frac{1}{3}$

Solve each absolute value equation.

25. $|x+4| = |2x-1|$

26. $|x-7| = |3x-1|$

27. $\left|\frac{x-1}{2}\right| = |2x+3|$

28. $\left|\frac{2x+3}{3}\right| = |x+4|$

29. $|1.5x-2| = |x-0.5|$

30. $|2.2x+2| = |1-2.8x|$

31. $|3-x| = \left|\frac{x}{2}+3\right|$

32. $\left|\frac{2x}{5}+1\right| = |1-x|$

Solve for x. Round to the nearest hundredth.

33. $|1.62x+3.14| = 2.19$

34. $|-0.74x-8.26| = 5.36$

Mixed Practice *Solve each equation, if possible. Check your solutions.*

35. $|3(x+4)| + 2 = 14$

36. $|4(x-2)| + 1 = 19$

37. $\left|\frac{5x}{3}-1\right| = 0$

38. $\left|\frac{3}{4}x+9\right| = 0$

39. $\left|\frac{4}{3}x-\frac{1}{8}\right| = -5$

40. $\left|\frac{3}{4}x-\frac{2}{3}\right| = -8$

41. $\left|\frac{3x-1}{3}\right| = \frac{2}{5}$

42. $\left|\frac{4x+5}{2}\right| = \frac{1}{3}$

Cumulative Review

43. [1.4.2] Simplify. $(3x^{-3}yz^0)\left(\frac{5}{3}x^4y^2\right)$

44. [1.3.3] Evaluate. $\frac{\sqrt{3 - 2 \cdot 1^2 + 5}}{4^2 - 2 \cdot 3}$

Quick Quiz 2.3 *Solve for x.*

1. $|3x - 4| = 49$

2. $\left|\frac{2}{3}x + 1\right| - 3 = 5$

3. $|2x + 5| = |x - 4|$

4. Concept Check Explain how you would solve for x in the following equation.

$$|2x + 4| = \frac{1}{2}$$

STEPS TO SUCCESS Getting the Greatest Value from Your Homework

Read the textbook first before doing the homework. Take some time to read the text and study the sample examples. Try working out the Student Practice problems. You will be amazed at the amount of understanding you will obtain by studying the book before jumping into the homework exercises.

Take your time. Read the directions carefully. Be sure you understand what is being asked. Check your answers with those given in the back of the textbook. If your answer is incorrect, study similar sample examples in the text. Then redo the problem, watching for errors.

Make a schedule. You will need to allow two hours outside of class for each hour of actual class time. Make a weekly schedule of the times you have class. Now write down the times each day you will devote to doing math homework. Then write down the times you will spend doing homework for your other classes. If you have a job be sure to write down all your work hours.

Making it personal: Write down your own schedule of class, work, and study time. ▼

Sunday	Monday	Tuesday	Wednesday	Thursday	Friday	Saturday

2.4 Using Equations to Solve Word Problems

Student Learning Objective

After studying this section, you will be able to:

① **Solve applied problems by using equations.**

① Solving Applied Problems by Using Equations

The skills you have developed in solving equations will allow you to solve a variety of applied problems. The following steps may help you to organize your thoughts and provide you with a procedure to solve such problems.

1. **Understand the problem.**
   (a) Read the word problem carefully to get an overview.
   (b) Determine what information you will need to solve the problem.
   (c) Draw a sketch. Label it with the known information. Determine what needs to be found.
   (d) Choose a variable to represent one unknown quantity.
   (e) If necessary, represent other unknown quantities in terms of the same variable.
2. **Write an equation.**
   (a) Look for key words to help you translate the words into algebraic symbols.
   (b) Use a relationship given in the problem or an appropriate formula in order to write an equation.
3. **Solve the equation and state the answer.**
4. **Check.**
   (a) Check the solution in the original situation.
   (b) Be sure the solution to the equation answers the question in the word problem. You may need to do some additional calculations if it does not.

Symbolic Equivalents of English Phrases

English Phrase	Mathematical Symbol
and, added to, increased by, greater than, plus, more than, sum of	$+$
decreased by, subtracted from, less than, diminished by, minus, difference between	$-$
product of, multiplied by, of, times	$\cdot$ or ( )( ) or $\times$
divided by, quotient of, ratio of	$\div$ or fraction bar
equals, are, is, will be, yields, gives, makes, is the same as, has a value of	$=$

Although we often use x to represent the unknown quantity when we write equations, any letter can be used. It is a good idea to use a letter that helps us remember what the variable represents. (For example, we might use s for speed or h for hours.) We now look at some translations of English sentences into algebraic equations.

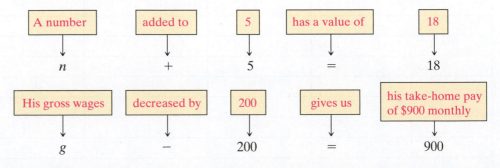

A number	added to	5	has a value of	18
↓	↓	↓	↓	↓
n	$+$	5	$=$	18

His gross wages	decreased by	200	gives us	his take-home pay of $900 monthly
↓	↓	↓	↓	↓
g	$-$	200	$=$	900

Five	times	the speed	yields	300 feet per second
5	×	s	=	300

The annual salary	divided by	52	will be	$380
a	÷	52	=	380

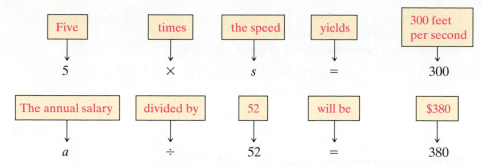

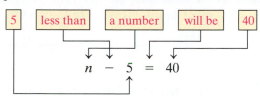

Be careful when translating the expressions "more than" and "less than." The order in which you write the symbols does not follow the order found in the English sentence. For example, "5 less than a number will be 40" is written as follows:

5	less than	a number	will be	40

$$n - 5 = 40$$

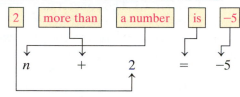

"Two more than a number is −5" is written as shown next.

2	more than	a number	is	−5

$$n + 2 = -5$$

Since addition is commutative, we can also write $2 + n = -5$ for "2 more than a number is −5." Be careful, however, with subtraction. "5 less than a number will be 40" must be written as $n - 5 = 40$.

We will now illustrate the use of the four-step procedure in solving a number of word problems.

EXAMPLE 1 Nancy works for an educational services company that provides computers and software to local public schools. She went to a local truck rental company to rent a truck to deliver some computers to the Newbury Elementary School. The truck rental company has a fixed rate of $40 per day plus 20¢ per mile. Nancy rented a truck for 3 days and was billed $177. How many miles did she drive?

Solution

1. *Understand the problem.* Let $n =$ the number of miles driven.

   Since each mile driven costs 20¢, we multiply the 20¢ (or $0.20) per-mile cost by the number of miles n.

   Thus, $0.20n =$ the cost of driving n miles at 20¢ per mile.

2. *Write an equation.*

Fixed costs for 3 days	plus	mileage charge	equals	$177
$(40)(3)$	+	$(0.20)(n)$	=	177

3. *Solve the equation and state the answer.*

$$120 + 0.20n = 177 \quad \text{Multiply (40)(3).}$$

$$0.20n = 57 \quad \text{Subtract 120 from each side.}$$

$$\frac{0.20n}{0.20} = \frac{57}{0.20} \quad \text{Divide each side by 0.20.}$$

$$n = 285 \quad \text{Simplify.}$$

Nancy drove the truck for 285 miles.

Continued on next page

4. *Check.* Does a truck rental for 3 days at \$40 per day plus 20¢ per mile for 285 miles come to a total of \$177?

We will check our values in the original equation.

$$\begin{aligned}(40)(3) + (0.20)(n) &= 177\\ 120 + (0.20)(285) &\stackrel{?}{=} 177\\ 120 + 57 &\stackrel{?}{=} 177\\ 177 &= 177 \quad ✓\end{aligned}$$

It checks. Our answer is correct.

***NOTE TO STUDENT:** Fully worked-out solutions to all of the Student Practice problems can be found at the back of the text starting at page SP-1.*

Student Practice 1 Western Laboratories rents a computer terminal for \$400 per month plus \$8 per hour of computer use time. The bill for 1 year's computer use was \$7680. How many hours did Western Laboratories actually use the computer?

EXAMPLE 2 The Acetones are a barbershop quartet. They travel across the country in a special bus and usually give six concerts a week. This popular group always sings to a sell-out crowd. The concert halls have an average seating capacity of three thousand people each. Concert tickets average \$12 per person. The onetime expenses for each concert are \$15,000. The cost of meals, motels, security and sound people, bus drivers, and other expenses totals \$100,000 per week. How many weeks per year will the Acetones need to be on tour if each member wants to earn \$71,500 per year?

Solution

1. *Understand the problem.* The item we are trying to find is the number of weeks the quartet needs to be on tour. So we let

$$x = \text{the number of weeks on tour.}$$

Now we need to find an expression that describes the quartet's income. Each week there are six concerts with three thousand people paying \$12 per person. Thus,

$$\begin{aligned}\text{weekly income} &= (6)(3000)(12) = \$216{,}000, \quad \text{and}\\ \text{income for } x \text{ weeks} &= \$216{,}000x.\end{aligned}$$

Now we need to find an expression for the total expenses. The expenses for each concert are \$15,000, and there are six concerts per week. Thus, concert expenses will be $(6)(15{,}000) = \$90{,}000$ per week. Now the cost of meals, motels, security and sound people, bus drivers, and other expenses total \$100,000 per week. Thus,

$$\begin{aligned}\text{total weekly expenses} &= \$90{,}000 + \$100{,}000 = \$190{,}000, \quad \text{and}\\ \text{total expenses for } x \text{ weeks} &= \$190{,}000x.\end{aligned}$$

The four Acetones each want to earn \$71,500 per year, so the group will need $4 \times \$71{,}500 = \$286{,}000$ for the year.

2. *Now we can write an equation.*

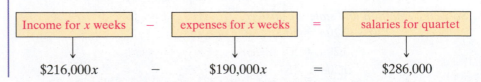

Income for *x* weeks	−	expenses for *x* weeks	=	salaries for quartet
↓		↓		↓
$\$216{,}000x$	−	$\$190{,}000x$	=	$\$286{,}000$

3. *Solve the equation and state the answer.*

$$26{,}000x = 286{,}000 \quad \text{Combine like terms.}$$

$$\frac{26{,}000x}{26{,}000} = \frac{286{,}000}{26{,}000} \quad \text{Divide each side of the equation by 26,000.}$$

$$x = 11$$

Thus, the Acetones will need to be on tour for 11 weeks to meet their salary goal.

4. *Check.* The check is left to the student.

Student Practice 2 A group of five women in a rock band plans to go on tour. They are scheduled to give five concerts a week with an average audience of six thousand people at each concert. The average ticket price for a concert is \$14. The onetime expenses for each concert are \$48,000. The additional costs per week are \$150,000. How many weeks per year will this group need to be on tour if each member wants to earn \$60,000 per year?

Sometimes we need a simple formula from geometry or some other science to write the original equation.

▲ **EXAMPLE 3** An astronaut's space suit contains a small rectangular steel plate that supports the breathing control valve. The length of the rectangle is 3 millimeters more than double its width. Its perimeter is 108 millimeters. Find the width and length.

Solution

1. *Understand the problem.* We draw a sketch to assist us. The formula for the perimeter of a rectangle is $P = 2w + 2l$, where $w =$ the width and $l =$ the length. Since the length is compared to the width, we will begin with the width. Let $w =$ the width. Then $2w + 3 =$ the length.

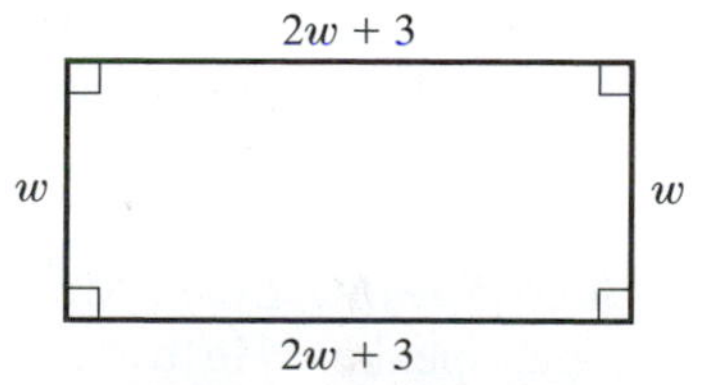

2. *Write an equation.*

$$P = 2w + 2l$$

$$108 = 2w + 2(2w + 3)$$

3. *Solve the equation and state the answer.*

$$108 = 2w + 4w + 6$$

$$108 = 6w + 6$$

$$102 = 6w$$

$$17 = w$$

Because $w = 17$, we have $2w + 3 = 2(17) + 3 = 37$. Thus, the rectangle is 17 millimeters wide and 37 millimeters long.

4. *Check.*

$$P = 2w + 2l$$

$$108 \stackrel{?}{=} 2(17) + 2(37)$$

$$108 \stackrel{?}{=} 34 + 74$$

$$108 = 108 \quad ✓$$

▲ **Student Practice 3** The perimeter of a triangular lawn is 162 meters. The length of the first side is twice the length of the second side. The length of the third side is 6 meters shorter than three times the length of the second side. Find the dimensions of the triangle.

2.4 Exercises

MyMathLab® Watch the videos in MyMathLab Download the MyDashBoard App

Write an algebraic equation and use it to solve each problem.

1. Three-fifths of a number is −54. What is the number?

2. Five-sixths of a number is −60. What is the number?

Applications

3. ***Gym Membership Costs*** Allyson can pay for her gym membership on a monthly basis, but if she pays for an entire year's membership in advance, she'll receive a \$50 discount. Her discounted bill for the year would then be \$526. What is the monthly membership fee at her gym?

4. ***Parking Garage Fees*** The monthly parking fee at the Safety First Parking Garage is one and one-half times last year's monthly fee, but compact cars will now receive a \$10 per month discount. If the owner of a compact car pays \$98 per month this year, what was his monthly fee last year?

5. ***Computer Shipments*** Barbara Stormer's computer company ships its packages through Reliable Shipping Express. Reliable charges \$3 plus \$0.80 per pound to ship each package. If the charge for shipping a package to Idaho was \$17.40, how much did the package weigh?

6. ***Parking Garage Fees*** At New York City's La Guardia airport, one parking garage charges a onetime "transportation fee" of \$2.75 plus \$22 for indoor parking per day. How many days has a car been parked at this location if the parking charge is \$134.75?

7. ***Laundry Expenses*** Roberto and Maria Santanos spend approximately \$11.75 per week to wash and dry their family's clothes at a local coin laundry. A new washer and dryer would cost them a total of \$846. How many weeks will it take for the laundromat cost to equal the cost of a new washer and dryer?

8. ***Personal Banking*** Paytrust.com is an online bill-paying service that charges \$5 per month plus \$0.50 per paid bill. Thomas used this service for 6 months and paid a total of \$48.50. How many bills were paid through Paytrust.com during that period?

9. ***Guest Speaker Costs*** Mr. Ziglar is a famous motivational speaker. He travels across the country by jet and gives an average of eight presentations a week. He gives his speeches in auditoriums that are always full and hold an average of 2000 people each. He charges \$15 per person at each of these speeches. His average expenses for each presentation, including renting the auditorium, are about \$14,000. The cost of meals, motels, jet travel, and support personnel totals \$85,000 per week. How many weeks will Mr. Ziglar need to travel if he wants to earn \$129,000 per year?

10. ***Music Tour Costs*** The Three Tenors are planning a nationwide tour. They will travel across the country by jet and give four concerts a week. They will sing in concert halls that hold an average of 5000 people each. Concert tickets average \$18 per person, and most concerts are totally sold out. The advance expenses for each concert are \$55,000. The cost of meals, motels, jet travel, and support personnel totals \$110,000 per week. How many weeks will the Three Tenors need to be on tour if each of them wants to earn \$120,000 per year?

▲ 11. ***Geometry*** A new Youth Opportunity Center is being built in Roxbury. The perimeter of the rectangular playing field is 340 yards. The length of the field is 6 yards less than triple the width. What are the dimensions of the playing field?

▲ 12. ***Geometry*** Dave and Jane Wells have a new rectangular driveway. The perimeter of the driveway is 168 feet. The length is 12 feet longer than three times the width. What are the dimensions of the driveway?

▲ **13.** ***Geometry*** A vacant city lot is being turned into a neighborhood garden. The neighbors want to fence in a triangular section of the lot and plant flowers there. The longest side of the triangular section is 7 feet shorter than twice the shortest side. The third side is 6 feet longer than the shortest side. The perimeter is 59 feet. How long is each side?

▲ **14.** ***Geometry*** A leather coin purse has the shape of a triangle. Two sides are equal in length and the third side is 3 centimeters shorter than one and one-half times the length of the equal sides. The perimeter is 28.5 centimeters. Find the lengths of the sides.

Cumulative Review *Name the property that justifies each statement.*

15. **[1.1.2]** $57 + 0 = 57$

16. **[1.1.2]** $(2 \cdot 3) \cdot 9 = 2 \cdot (3 \cdot 9)$

17. **[1.2.3]** Evaluate. $7(-2) \div 7(-3) - 3$

18. **[1.3.3]** Evaluate. $(7 - 12)^3 - (-4) + 3^3$

Quick Quiz 2.4

1. Three-fourths of a number is -69. What is the number?

2. A triangle has a perimeter of 100 yards. The first side is twice the length of the second side. The third side is 8 yards longer than the second side. How long is each side?

3. The Manchester Airport Central Garage charges \$6 to park for the first hour and then \$3.50 an hour for any additional hours. Ben paid \$65.50 to park there one day when he flew to Philadelphia. How many hours was he parked at the airport?

4. **Concept Check** Explain how you would set up an equation to solve the following problem. A driveway has a perimeter of 212 feet. The length of the driveway is 7 feet longer than four times the width. Find the width and the length of the driveway.

How Am I Doing? Sections 2.1–2.4

How are you doing with your homework assignments in Sections 2.1 to 2.4? Do you feel you have mastered the material so far? Do you understand the concepts you have covered? Before you go further in the textbook, take some time to do each of the following problems.

1. ______
2. ______
3. ______
4. ______
5. ______
6. ______
7. ______
8. ______
9. ______
10. ______
11. ______
12. ______
13. ______
14. ______
15. ______
16. ______

2.1

Solve for x.

1. $2x - 1 = 12x + 36$

2. $\frac{x - 2}{4} = \frac{1}{2}x + 4$

3. $4(x - 3) = x + 2(5x - 1)$

4. $0.6x + 3 = 0.5x - 7$

2.2

5. Solve for y. $-5x + 9y = 18$

6. Solve for a. $5ab - 2b = 16ab - 3(8 + b)$

7. Solve for r. $A = P + Prt$

8. Use your results from problem 7 to find r when $P = 100$, $t = 3$, and $A = 118$.

2.3

Solve for x.

9. $|5x + 8| = 3$

10. $|9 - x| + 2 = 5$

11. $\left|\frac{2x + 3}{4}\right| = 2$

12. $|5x - 8| = |3x + 2|$

2.4

Use an algebraic equation to find a solution for each exercise.

▲ **13.** The length of a queen-size mattress is 20 inches longer than the width, and the perimeter is 280 inches. Find the dimensions of a queen-size mattress.

14. Eastern Bank charges its customers a flat fee of \$6 per month for a checking account plus 12¢ for each check. The bank charged Jose \$9.12 for his checking account last month. How many checks did he use?

▲ **15.** Alan and Cindi pick up food donations for their local food bank from grocery stores. Alan picked up 80 more than half as many pounds of food as Cindi did. Together they picked up 455 pounds of food donations. How many pounds did each pick up?

16. In Freeport, Maine, the north end of L. L. Bean has a triangular parking lot for bicycles. The longest side is 5 feet shorter than twice the shortest side. The third side is 9 feet longer than the shortest side. The perimeter is 62 feet. How long is each side?

Now turn to page SA-3 for the answers to each of these problems. Each answer also includes a reference to the objective in which the problem is first taught. If you missed any of these problems, you should stop and review the Examples and Student Practice problems in the referenced objective. A little review now will help you master the material in the upcoming sections of the text.

2.5 Solving More-Involved Word Problems

① Solving More-Involved Word Problems by Using an Equation

To solve some word problems, we might need to understand percents, simple interest, mixtures, or some other concept before we can use an algebraic equation as a model. We review several of these concepts in this section.

From arithmetic you know that to find a percent of a number we write the percent as a decimal and multiply the decimal by the number.

Thus, to find 36% of 85, we calculate $(0.36)(85) = 30.6$. If the number is not known, we can represent it by a variable.

Student Learning Objective

After studying this section, you will be able to:

① **Solve more-involved word problems by using an equation.**

EXAMPLE 1 The Wildlife Refuge Rangers tagged 144 deer. They estimate that they have tagged 36% of the deer in the refuge. If they are correct, approximately how many deer are in the refuge?

Solution

1. *Understand the problem.* Let $n =$ the number of the deer in the refuge.

Then $0.36n = 36\%$ of the deer in the refuge.

2. *Write an equation.*

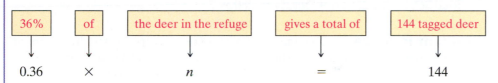

36%	of	the deer in the refuge	gives a total of	144 tagged deer
↓	↓	↓	↓	↓
0.36	×	n	=	144

3. *Solve the equation and state the answer.*

$$0.36n = 144$$

$$\frac{0.36n}{0.36} = \frac{144}{0.36} \quad \text{Divide each side by 0.36.}$$

$$n = 400 \quad \text{Simplify.}$$

There are approximately 400 deer in the refuge.

4. *Check.* Is it true that 36% of 400 is 144?

$$(0.36)(400) \stackrel{?}{=} 144$$

$$144 = 144 \checkmark$$

It checks. Our answer is correct.

Student Practice 1 Technology Resources, Inc. sold 6900 computer workstations, a 15% increase in sales over the previous year. How many computer workstations were sold last year? (*Hint*: Let $x =$ the amount of sales last year, then $0.15x =$ the increase in sales over last year.)

NOTE TO STUDENT: Fully worked-out solutions to all of the Student Practice problems can be found at the back of the text starting at page SP-1.

Adding two numbers yields a total. We can call one of the numbers x and the other number (total $- x$). We will use this concept in Example 2.

EXAMPLE 2 Bob's and Marcia's weekly salaries total \$265. If they both went from part-time to full-time employment, their combined weekly income would be \$655. Bob's salary would double, while Marcia's would triple. How much do they each make now?

Solution

1. *Understand the problem.* Let $b =$ Bob's part-time salary.

Since the total of the two part-time weekly salaries is \$265, we can write $265 - b =$ Marcia's part-time salary.

Continued on next page

2. *Write an equation.*

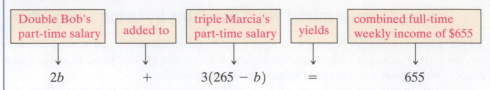

Double Bob's part-time salary	added to	triple Marcia's part-time salary	yields	combined full-time weekly income of $655
$2b$	$+$	$3(265 - b)$	$=$	655

3. *Solve the equation and state the answer.*

$$2b + 3(265 - b) = 655$$
$$2b + 795 - 3b = 655 \quad \text{Remove parentheses.}$$
$$795 - b = 655 \quad \text{Simplify.}$$
$$-b = -140 \quad \text{Subtract 795 from each side.}$$
$$b = 140 \quad \text{Multiply each side by } -1.$$

If $b = 140$, then $265 - b = 265 - 140 = 125$. Thus, Bob's present part-time weekly salary is $140, and Marcia's present part-time weekly salary is $125.

4. *Check.* Do their present weekly salaries total $265? Yes: $140 + 125 = 265$. If Bob's income is doubled and Marcia's is tripled, will their new weekly salaries total $655? Yes: $2(140) + 3(125) = 280 + 375 = 655$.

Student Practice 2 Together Alicia and Heather sold forty-three cars at Prestige Motors last month. If Alicia doubles her sales and Heather triples her sales next month, they will sell 108 cars. How many cars did each person sell this month?

Simple interest is an income from investing money or a charge for borrowing money. It is computed by multiplying the amount of money borrowed or invested (called the *principal*) by the rate of interest and by the period of time it is borrowed or invested (usually measured in years unless otherwise stated). Hence

$$\text{interest} = \text{principal} \times \text{rate} \times \text{time}.$$
$$I = prt$$

All interest problems in this chapter involve simple interest.

EXAMPLE 3 Maria has a job as a financial advisor in a bank. She advised a customer to invest part of his money in a money market fund earning 12% simple interest and the rest in an investment fund earning 14% simple interest. The customer had $6000 to invest. If he earned $772 in interest in 1 year, how much did he invest in each fund?

Solution

1. *Understand the problem.* Let $x =$ the amount of money invested at 12% interest. The other amount of money is (total $- x$).
   Thus $6000 - x =$ the amount of money invested at 14% interest.
2. *Write an equation.*

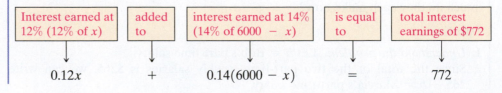

Interest earned at 12% (12% of x)	added to	interest earned at 14% (14% of 6000 − x)	is equal to	total interest earnings of $772
$0.12x$	$+$	$0.14(6000 - x)$	$=$	772

3. *Solve the equation and state the answer.*

$$0.12x + 0.14(6000 - x) = 772$$
$$0.12x + 840 - 0.14x = 772 \quad \text{Remove parentheses.}$$
$$840 - 0.02x = 772 \quad \text{Add } 0.12x \text{ to } -0.14x.$$
$$-0.02x = -68 \quad \text{Subtract 840 from each side.}$$
$$\frac{-0.02x}{-0.02} = \frac{-68}{-0.02} \quad \text{Divide each side by } -0.02.$$
$$x = 3400 \quad \text{Simplify.}$$

If $x = 3400$, then $6000 - x = 6000 - 3400 = 2600$. Thus, \$3400 was invested in the money market fund earning 12% interest, and \$2600 was invested in the investment fund earning 14% interest.

4. *Check.* Do the two amounts of money total \$6000?

Yes: \$3400 + \$2600 = \$6000.

Does the total interest amount to \$772?

$$(0.12)(3400) + (0.14)(2600) \stackrel{?}{=} 772$$
$$408 + 364 \stackrel{?}{=} 772$$
$$772 = 772 \checkmark$$

Our answers are correct.

Student Practice 3 Tricia received an inheritance of \$5500. She invested part of it at 8% simple interest and the remainder at 12% simple interest. At the end of the year she had earned \$540. How much did Tricia invest at each interest amount?

Sometimes we encounter a situation in which two or more items are combined to form a mixture or solution. These types of problems are called **mixture problems.**

EXAMPLE 4 A small truck has a radiator that holds 20 liters. A mechanic needs to fill the radiator with a solution that is 60% antifreeze. He has 70% and 30% antifreeze solutions. How many liters of each should he use to achieve the desired mix?

Solution

1. *Understand the problem.* Let $x =$ the number of liters of 70% antifreeze to be used.

Since the total amount of solution must be 20 liters, we can use $20 - x$ for the other part. So $20 - x =$ the number of liters of 30% antifreeze to be used.

In this problem a chart or table is very helpful. We will multiply the entry in column (A) by the entry in column (B) to obtain the entry in column (C).

	(A) Number of Liters of the Solution	(B) Percent Pure Antifreeze	(C) Number of Liters of Pure Antifreeze
70% antifreeze solution	x	70%	$0.70x$
30% antifreeze solution	$20 - x$	30%	$0.30(20 - x)$
Final 60% solution	20	60%	$0.60(20)$

Continued on next page

2. *Write an equation.* Now we form an equation from the entries in column (C).

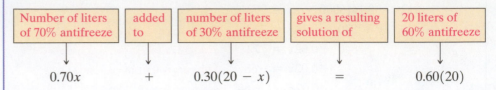

Number of liters of 70% antifreeze	added to	number of liters of 30% antifreeze	gives a resulting solution of	20 liters of 60% antifreeze
↓	↓	↓	↓	↓
$0.70x$	$+$	$0.30(20 - x)$	$=$	$0.60(20)$

3. *Solve the equation and state the answer.*

$$\begin{aligned} 0.70x + 0.30(20 - x) &= 0.60(20) \\ 0.70x + 6 - 0.30x &= 12 \\ 6 + 0.40x &= 12 \\ 0.40x &= 6 \\ \frac{0.40x}{0.40} &= \frac{6}{0.40} \\ x &= 15 \end{aligned}$$

If $x = 15$, then $20 - x = 20 - 15 = 5$. Thus, the mechanic needs 15 liters of 70% antifreeze solution and 5 liters of 30% antifreeze solution.

4. *Check.* Can you verify that this answer is correct?

Student Practice 4 A jeweler wishes to prepare 200 grams of 80% pure gold from sources that are 68% pure gold and 83% pure gold. How many grams of each should he use?

Some problems involve the relationship distance = rate × time or $d = rt$.

EXAMPLE 5 Frank drove at a steady speed for 3 hours on the turnpike. He then slowed his traveling speed by 15 miles per hour on the secondary roads. The entire trip took 5 hours and covered 245 miles. What was his speed on the turnpike? What was his speed on the secondary road?

Solution

1. *Understand the problem.* Let x = speed on the turnpike in miles per hour.

So $x - 15$ = speed on the secondary roads in miles per hour.

Again, as in Example 4, a chart or table is very helpful. We will use a chart that records values of $(r)(t) = d$.

	Rate (miles per hour) r	Time (hours) t	Distance (miles) $(r)(t) = d$
Turnpike	x	3	$3x$
Secondary roads	$x - 15$	2	$2(x - 15)$
Entire trip	Not appropriate	5	245

2. *Write an equation.* Using the distance entries (third column of the table), we have

Distance on turnpike	+	Distance on secondary roads	=	Total distance
$3x$	+	$2(x - 15)$	=	245

3. *Solve the equation and state the answer.*

$$3x + 2(x - 15) = 245$$
$$3x + 2x - 30 = 245$$
$$5x - 30 = 245$$
$$5x = 275$$
$$x = 55$$

Thus, Frank traveled at an average speed of 55 miles per hour on the turnpike. Therefore his speed on the secondary road was 40 miles per hour.

4. *Check.* Can you verify that this answer is correct?

Student Practice 5 Wally drove for 4 hours at a steady speed. He slowed his speed by 10 miles per hour for the last part of the trip. The entire trip took 6 hours and covered 352 miles. How fast did he drive on each portion of the trip?

2.5 Exercises

MyMathLab® Watch the videos in MyMathLab Download the MyDashBoard App

Write an algebraic equation for each problem and solve it. When necessary round your answer to the nearest tenth.

1. ***U.S. Population*** The population of the United States in 2010 was estimated to be 310.7 million people. This was an increase of 24% from the population in 1990. What was the population of the United States in 1990? (*Source:* www.census.gov)

2. ***National Debt*** The U.S. national debt on November 9, 2010, was approximately \$13.7 trillion. This was an increase of 71% from the national debt on November 9, 2005. What was the national debt on November 9, 2005? (*Source:* www.treasurydirect.gov)

3. ***Consumer Purchases*** Eastwing dormitory just purchased a new Sony color television on sale for \$340. The sale price was 80% of the original price. What was the original price of the television set?

4. ***Health Club Membership*** In 2010, the number of Americans who were members of a health club was 45.3 million. This was an increase of 119% from the number of health club members in 1990. How many Americans were health club members in 1990? (*Source:* www.numberof.net)

5. ***Unemployment Rates*** Allentown's employment statistics show that 969 of its residents were unemployed last month. This is a decrease of 15% from the previous month. How many residents were unemployed in the previous month?

6. ***Lyme Disease*** A wildlife expert at the Crane Wildlife Reserve has found fifteen deer with ticks that are carrying Lyme disease. She estimates that this number is about 60% of the total number of deer carrying infected ticks. Approximately how many such deer are there on the Reserve?

7. ***Reforestation Program*** In a reforestation program, 1400 seedling trees were planted. Twice as many spruce as hemlocks were planted. The number of balsams planted was twenty more than triple the number of hemlocks. How many of each type of seedling was planted?

8. ***Summer Rental*** Lynn and Judy are pooling their savings to rent a cottage in Maine for a week this summer. The rental cost is \$950. Lynn's family is joining them, so she is paying a larger part of the cost. Her share of the cost is \$250 less than twice Judy's. How much of the rental fee is each of them paying?

9. ***Salary Increases*** When Angela and Walker first started working for the supermarket, their weekly salaries totaled $500. Now during the last 25 years Walker has seen his weekly salary triple. Angela has seen her weekly salary become four times larger. Together their weekly salaries now total $1740. How much did they each make 25 years ago?

10. ***Salary Increases*** When Grace and Tony started as junior engineers at a manufacturing company, their weekly salaries totaled $1300. Now ten years later they have both become senior engineers. Grace has had her salary double. Tony has had his salary triple. Together their weekly salaries now total $3200. How much did they each make 10 years ago?

11. ***Medicine Dosage*** A hospital received a shipment of 8-milligram doses of a medicine. Each 8-milligram package was repacked into two smaller doses of unequal size and labeled packet A and packet B. The hospital then used 17 doses of packet A and 14 doses of packet B in one week. The hospital used a total of 127 milligrams of the medicine during that week. How many milligrams of the medicine are contained in each A packet? In each B packet?

12. ***Cookie Sales*** This year, two Girl Scout troops together sold 460 boxes of cookies. Half of the Rockland troop's sales were Thin Mints and $\frac{2}{5}$ of the Harrisville troop's sales were Thin Mints. Together they sold 205 boxes of Thin Mints. How many boxes of cookies did each troop sell?

13. ***Interest on a Loan*** Rebecca and Peter borrowed $1500 at a simple interest rate of 9% for a period of 2 years. What was the interest?

14. ***Interest on a Loan*** Juanita and Carlos borrowed $4800 at a simple interest rate of 11% for a period of 2 years. What was the interest?

15. ***College Fund Investment*** The Vegas want to invest $5000 of their daughter's college fund for 18 months in a certificate of deposit that pays a simple interest rate of 3.1%. How much interest will they earn?

16. ***Earned Interest*** David had $4000 invested for one-fourth of a year at a simple interest rate of 6.1%. How much interest did he earn?

17. ***Investment Income*** Cynthia is a hardworking single mother on a very limited budget. She invests her savings only in safe investments because she is not in a position to take risks and lose any of her hard-earned money. She invested $6400 in two types of accounts for one year. The first type earned 5% simple interest, and the second type earned 8% simple interest. At the end of the year, Cynthia had earned $395 in interest. How much did she invest at each rate?

18. ***Investment Income*** The Johnsons' family business did well *this* year due to their investments *last* year. The business earned $6570 on an investment of $45,000 in mutual funds. There were two types of funds that Jenna Johnson invested in. The first was a pharmaceutical fund, which paid out simple interest of 13%. The second was a biotech fund, which paid out simple interest of 16%. How much did Jenna Johnson invest in each fund last year for her family's business?

19. ***Retirement Fund Investment*** Jim Jacobs decided to invest $18,000 of his retirement fund conservatively. He invested part of this money in a certificate of deposit that pays 3.5% simple interest and part in a fixed interest account that pays 2.2% simple interest. Last year he earned $552 in interest. How much did he invest in each type of account?

20. ***Investment Income*** Tori invested $8000 in money market funds. Part was invested at 5% simple interest, and the rest was invested at 7% simple interest. At the end of 1 year, Tori had earned $496 in interest. How much did she invest in each fund?

21. ***Fat Content of Food*** A chef has one cheese that contains 45% fat and another cheese that contains 20% fat. How many grams of each cheese should she use in order to obtain 30 grams of a cheese mixture that is 30% fat?

22. ***Chemical Mixtures*** Becky was awarded a college internship at a local pharmaceutical research corporation. She has been given her own corner of the laboratory to try her hand at pharmacology. One of her assignments is to mix two solutions, one that is 16% strength and the other that is 9% strength. How many milliliters of each should she use in order to obtain 350 milliliters of a 12% strength solution?

23. ***Fat Content of Food*** The meat department manager at a large food store wishes to mix some hamburger with 30% fat content and some hamburger that has 10% fat content in order to obtain 100 pounds of hamburger with 25% fat content. How much of each type of hamburger should she use?

24. ***Cost of Tea*** The manager of a gourmet coffee and tea store is mixing two green teas, one worth $7 per pound and the other worth $9 per pound. He wants to obtain 32 pounds of tea worth $8.50 per pound. How much of each tea should he use?

25. ***Fertilizer Mix*** A landscaping company needs 150 gallons of 18% fertilizer to fertilize the shrubs in an office park. They have in stock 25% fertilizer and 15% fertilizer. How much of each type should they mix together?

26. ***Insecticide Manufacturing*** Most mosquito repellents contain the ingredient DEET. A manufacturer needs to produce 10-ounce spray cans of 40% DEET. How much 100% DEET and how much 25% DEET should be mixed to produce each of these spray cans?

27. ***Auto Travel*** Susan drove for 4 hours on secondary roads at a steady speed. She completed her trip by driving 2 hours on an interstate highway. Her total trip was 250 miles. Her speed on the interstate highway portion of the trip was 20 miles per hour faster than her speed on the secondary roads. How fast did she travel on the secondary roads?

28. ***Airplane Flight*** Alice and Wendy flew a small plane for 930 miles. For the first 3 hours they flew at maximum speed. After refueling they finished the trip at a cruising speed that was 60 miles per hour slower than maximum speed. The entire trip took 5 hours of flying time. What is the maximum flying speed of the plane?

29. ***Walking on a Treadmill*** Yissania and Charlotte walked on treadmills at the gym for the same amount of time. Yissania walks at 5 miles per hour, and Charlotte walks at 4.2 miles per hour. If Yissania walked 0.6 miles farther than Charlotte, how long did they use the treadmills?

30. ***Boating on a Lake*** The Clarke family went sailing on a lake. Their boat averaged 6 kilometers per hour. The Rourke family took their outboard runabout for a trip on the lake for the same amount of time. Their boat averaged 14 kilometers per hour. The Rourke family traveled 20 kilometers farther than the Clarke family. How many hours did each family spend on their boat trip?

Cumulative Review *Evaluate.*

31. **[1.6.1]** $5a - 2b + c$ when $a = 1, b = -3, c = -4$

32. **[1.6.1]** $2x^2 - 3x + 1$ when $x = -2$

33. **[1.3.3]** $\dfrac{5 + 8(-2) + 2^4}{|2 - 7|}$

34. **[1.3.3]** $\dfrac{\sqrt{7^2 - 24}}{2^3(-1) + 7(4)}$

Quick Quiz 2.5

1. Eastview dormitory purchased a new 52-inch HD television for $2208 this year. This was a price that was 8% less than the one they purchased last year. How much was the television that they purchased last year?

2. A landscaping company needs 200 gallons of 45% fertilizer to fertilize the shrubs at North Shore Community College. They have in stock 50% fertilizer and 30% fertilizer. How much of each type should they mix together?

3. Lexi invested $4000 in two mutual funds. In one year part of her investment earned 5% simple interest and the other part earned 7% simple interest. At the end of that year Lexi had earned $250. How much did she invest at each rate?

4. **Concept Check** How would you set up an equation to solve the following problem? A new sports car sold for a certain amount of money. This year the price went up 12%. The new price is $39,200. What was the price the previous year?

2.6 Linear Inequalities

Student Learning Objectives

After studying this section, you will be able to:

1. **Determine whether one number is less than or greater than another number.**
2. **Graph linear inequalities in one variable.**
3. **Solve linear inequalities in one variable.**

1 Determining Whether One Number Is Less Than or Greater Than Another Number

We briefly introduced inequalities in Chapter 1. Now we will discuss this concept more completely.

A **linear inequality** is a statement that describes how two numbers or linear expressions are related to one another. We can use a number line to visualize the concept of inequality.

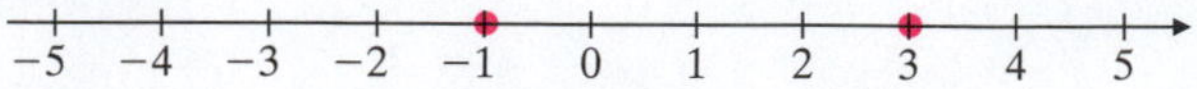

It is a mathematical property that -1 is less than 3. We write this in the following way:

$$-1 < 3.$$

Notice the position of these numbers on the number line. -1 is to the *left* of 3 on the number line. We say that a first number *is less than* a second number if the first number is *to the left* of the second number on a number line.

We could also say that 3 is greater than -1 (because it is *to the right* of -1 on the number line). We then write this in the following way:

$$3 > -1.$$

An inequality symbol can face either right or left, but its opening must face the larger number.

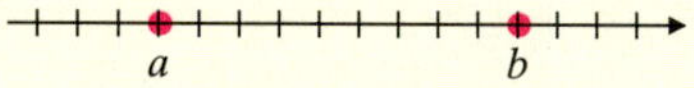

In general, if $a < b$, then it is also true that $b > a$.

EXAMPLE 1 Insert the proper symbol between the numbers.

(a) 8 _____ 6 **(b)** -2 _____ 3 **(c)** -4 _____ -2

(d) $\frac{1}{2}$ _____ $\frac{1}{3}$ **(e)** -0.033 _____ -0.0329

Solution

(a) $8 > 6$ because 8 is to the right of 6 on a number line.

(b) $-2 < 3$ because -2 is to the left of 3 on a number line.

(c) $-4 < -2$ because -4 is to the left of -2 on a number line.

(d) When comparing two fractions, rewrite them with a common denominator.

$\frac{1}{2}$ ___ $\frac{1}{3}$ Rewrite fractions with the same denominator.

$\frac{3}{6}$ ___ $\frac{2}{6}$ Compare the numerators: $3 > 2$.

$\frac{3}{6} > \frac{2}{6}$

Thus, $\frac{1}{2} > \frac{1}{3}$ because $\frac{3}{6} > \frac{2}{6}$.

(e) $-0.033 < -0.0329$ because $-0.0330 < -0.0329$.

Notice that since both numbers are negative, -0.0330 is to the left of -0.0329 on a number line.

Student Practice 1 Insert the proper symbol between the two numbers.

(a) -1 ______ -2 **(b)** $\frac{2}{3}$ ______ $\frac{3}{4}$ **(c)** -0.561 ______ -0.5555

NOTE TO STUDENT: *Fully worked-out solutions to all of the Student Practice problems can be found at the back of the text starting at page SP-1.*

Numerical or algebraic expressions as well as numbers can be compared.

EXAMPLE 2 Insert the proper symbol between the expressions.

(a) $(5 - 8)$ ______ $(2 - 3)$ **(b)** $|1 - 7|$ ______ $|-4 - 12|$

Solution

(a) $(5 - 8)$ ______ $(2 - 3)$ Evaluate each expression and compare.

$-3 < -1$

Thus, $(5 - 8) < (2 - 3)$ because $-3 < -1$.

(b) $|1 - 7|$ ______ $|-4 - 12|$ Evaluate each expression and compare.

$6 < 16$

Thus, $|1 - 7| < |-4 - 12|$ because $6 < 16$.

Student Practice 2 Insert $<$ or $>$ between the expressions.

(a) $(-8 - 2)$ ______ $(-3 - 12)$ **(b)** $|-15 + 8|$ ______ $|7 - 13|$

In addition to the $<$ and $>$ symbols, we will also encounter two other notations when we deal with inequalities. We briefly mentioned these in Chapter 1. If we want to say that a number x is greater than or equal to 5, we would write this using the following notation:

$$x \geq 5.$$

Likewise, if we want to say that a number x is less than or equal to 8, we would write it using the following notation:

$$x \leq 8.$$

The symbol $\leq$ means **less than or equal to,** and the symbol $\geq$ means **greater than or equal to.** We say that the inequality symbols $\leq$ and $\geq$ "contain the equals sign."

② Graphing Linear Inequalities in One Variable

Inequality symbols are often used with variables. For example,

$$x > 4$$

means that x can be any number that is greater than 4. The variable x cannot equal 4. We can use a number line to graph this inequality.

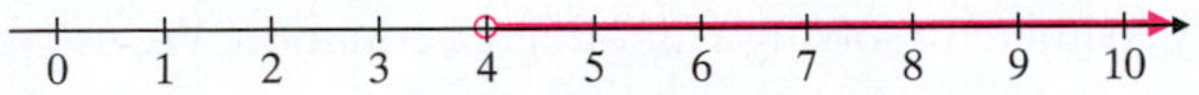

On the number line, we shade the portion that is to the right of 4. Any point in the shaded portion will satisfy the inequality since all points to the right of 4 are greater than 4. The open circle at 4 means that x cannot be 4.

Let's look at the inequality $x < -1$. This means that x can be any number that is less than -1. To graph the inequality, we will shade all points to the left of -1 on a number line.

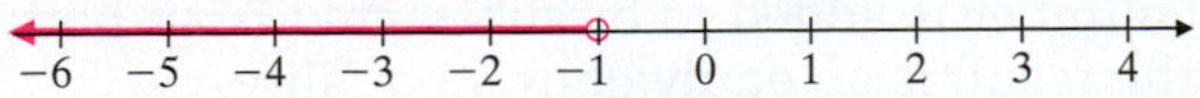

$x < -1$

Inequality symbols used with variables can contain the equals sign. The inequality $x \geq -2$ means all numbers greater than or equal to -2. We graph this inequality as follows.

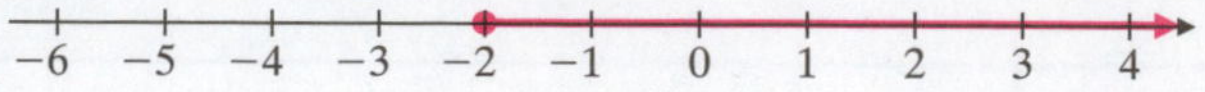

$x \geq -2$

The shaded circle at -2 means that the graph includes the point -2. -2 is a solution to the inequality.

Similarly, $x \leq 1$ is graphed as follows.

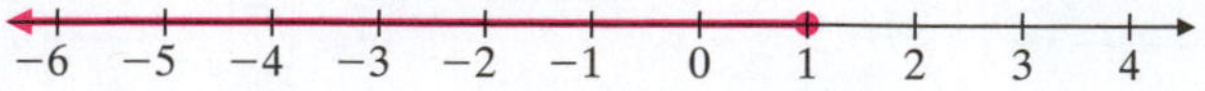

$x \leq 1$

EXAMPLE 3 Graph each inequality.

(a) $x < 0$ **(b)** $x \leq 0$ **(c)** $x > -5$ **(d)** $x \geq -5$ **(e)** $2 < x$

Solution

(a) $x < 0$

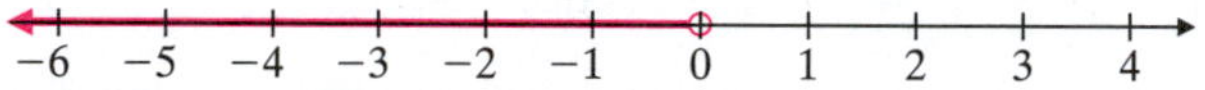

(b) $x \leq 0$

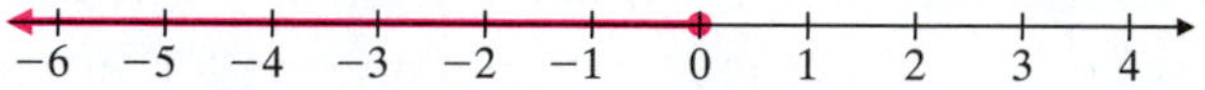

(c) $x > -5$

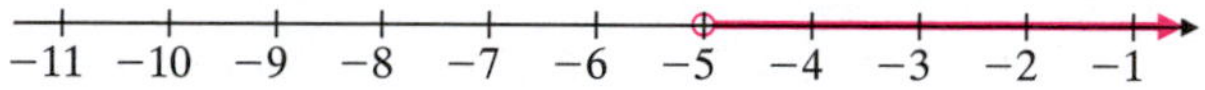

(d) $x \geq -5$

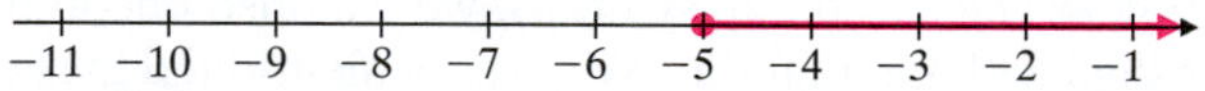

(e) We read an inequality starting with the variable. Thus, we read $2 < x$ as "x is greater than 2" and graph the expression accordingly.

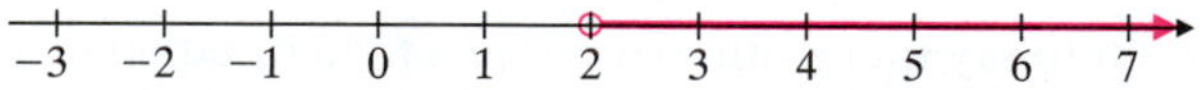

Student Practice 3 Graph each inequality.

(a) $x > 3.5$

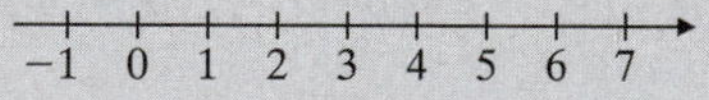

(b) $x \leq -10$

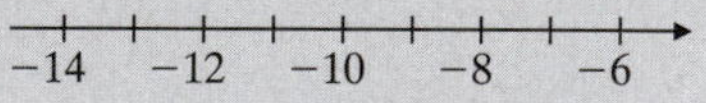

(c) $x \geq -3$

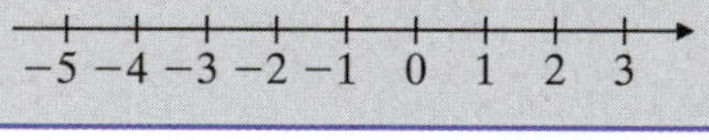

(d) $-4 > x$

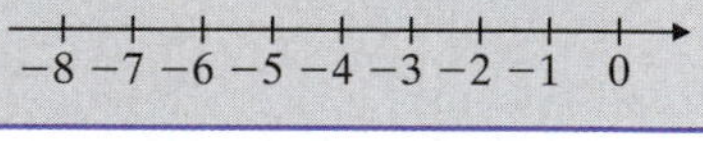

③ Solving Linear Inequalities in One Variable

Inequalities that have the same solution are said to be **equivalent.** Solving a first-degree inequality is similar to solving a first-degree equation. We use various properties of real numbers.

ADDITION AND SUBTRACTION PROPERTY FOR INEQUALITIES

For all real numbers a, b, and c, if $a < b$, then

$$a + c < b + c \quad \text{and} \quad a - c < b - c.$$

If the same number is added to or subtracted from both sides of an inequality, the result is an equivalent inequality.

The same number can be added to or subtracted from both sides of an inequality without affecting the direction of the inequality. (Any inequality symbol can be used. We used $<$ for convenience.)

EXAMPLE 4 Solve the inequality. Graph and check the solution. $x - 8 < 15$

Solution

$$x - 8 < 15 \quad \text{Add 8 to each side.}$$
$$x - 8 + 8 < 15 + 8 \quad \text{Simplify.}$$
$$x < 23$$

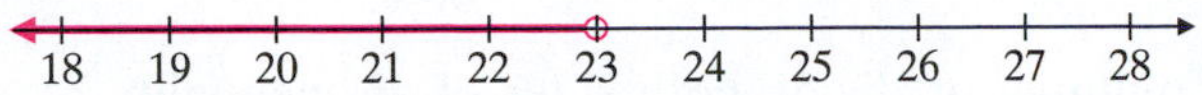

To check, choose any numerical value that lies in the region indicated by the red arrow on the number line. See whether a true statement results when you substitute it into the inequality. We will choose 22.5.

$$x - 8 < 15 \quad \text{Substitute 22.5 for } x \text{ in the original inequality.}$$
$$22.5 - 8 \stackrel{?}{<} 15$$
$$14.5 < 15 \ ✓$$

Student Practice 4 Solve the inequality. Graph and check the solution. $x + 2 > -12$

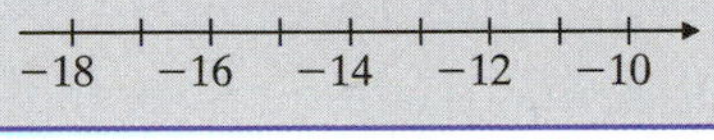

MULTIPLICATION OR DIVISION BY A POSITIVE NUMBER

For all real numbers a, b, and c when $c > 0$, if $a < b$, then

$$ac < bc \quad \text{and} \quad \frac{a}{c} < \frac{b}{c}.$$

If both sides of an inequality are multiplied or divided by the same *positive* number, the result is an equivalent inequality.

When we multiply or divide both sides of an inequality by a positive number, the direction of the inequality is not changed. That is, if $5x < -15$ and we divide both sides by 5, we obtain $x < -3$.

To check this solution, we choose the value -4. Is $5(-4) < -15$ true? Yes.

EXAMPLE 5 Solve and graph your solution.

$6x + 3 \leq 2x - 5$.

Solution

$$6x + 3 - 3 \leq 2x - 5 - 3 \quad \text{Subtract 3 from each side.}$$
$$6x \leq 2x - 8 \quad \text{Simplify.}$$
$$6x - 2x \leq 2x - 2x - 8 \quad \text{Subtract } 2x \text{ from each side.}$$
$$4x \leq -8 \quad \text{Simplify.}$$
$$\frac{4x}{4} \leq \frac{-8}{4} \quad \text{Divide each side by 4. The inequality is not changed.}$$
$$x \leq -2$$

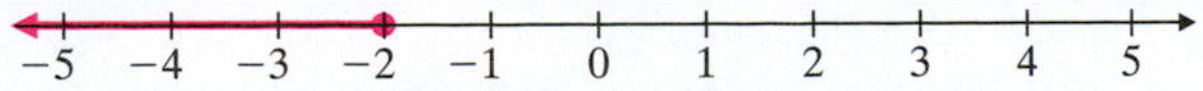

Student Practice 5 Solve and graph your solution. $8x - 8 \geq 5x + 1$.

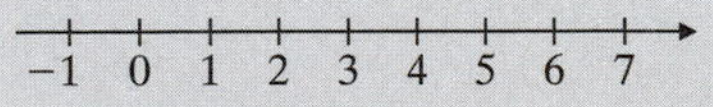

Continued on next page

To check, we choose −2 and −3.

$$6x + 3 \le 2x - 5$$
$$6(-2) + 3 \overset{?}{\le} 2(-2) - 5$$
$$-9 \le -9 \checkmark$$
$$6x + 3 \le 2x - 5$$
$$6(-3) + 3 \overset{?}{\le} 2(-3) - 5$$
$$-15 \le -11 \checkmark$$

When we multiply or divide both sides of an inequality by a *negative number,* the direction of the inequality is *reversed.* That is, if we divide both sides of $-3x < 21$ by -3, we obtain $x > -7$. If we divide both sides of $-4x \ge -16$ by -4, we obtain $x \le 4$.

To check the solution of the first inequality, we choose the value 1. Is $-3(1) < 21$? Yes. To reverse the inequality symbol might seem to be an unusual move. Let's see what would happen if we did not reverse the symbol. If we did not, the solution would be $x < -7$. To check this solution, we choose -8. Is $-3(-8) < 21$ true? No, 24 is not less than 21.

MULTIPLICATION OR DIVISION BY A NEGATIVE NUMBER

For all real numbers a, b, and c when $c < 0$, if $a < b$, then

$$ac > bc \quad \text{and} \quad \frac{a}{c} > \frac{b}{c}.$$

If both sides of an inequality are multiplied or divided by the same *negative* number *and the inequality symbol is reversed,* the result is an equivalent inequality.

EXAMPLE 6 Solve. $-8x - 12 < -4(x - 4) + 8$

Solution

$-8x - 12 < -4x + 16 + 8$ Use the distributive property to remove the parentheses.

$-8x - 12 < -4x + 24$ Simplify.

$-8x + 4x - 12 < -4x + 4x + 24$ Add $4x$ to each side.

$-4x - 12 < 24$ Simplify.

$-4x - 12 + 12 < 24 + 12$ Add 12 to each side.

$-4x < 36$ Simplify.

$\frac{-4x}{-4} > \frac{36}{-4}$ Divide each side by −4 and reverse the inequality.

$x > -9$ Simplify.

Student Practice 6 Solve. $2 - 12x > 7(1 - x)$

An inequality that contains all decimal terms is best handled by first multiplying both sides of the inequality by the appropriate power of 10.

EXAMPLE 7 Solve. $-0.3x + 1.0 \leq 1.2x - 3.5$

Solution

$10(-0.3x + 1.0) \leq 10(1.2x - 3.5)$ Multiply each side by 10.

$-3x + 10 \leq 12x - 35$

$-3x - 12x + 10 \leq 12x - 12x - 35$ Subtract $12x$ from each side.

$-15x + 10 \leq -35$ Simplify.

$-15x + 10 - 10 \leq -35 - 10$ Subtract 10 from each side.

$-15x \leq -45$ Simplify.

$\frac{-15x}{-15} \geq \frac{-45}{-15}$ Divide each side by -15 and reverse the direction of the inequality.

$x \geq 3$

Student Practice 7 Solve. $-0.8x + 0.9 \geq 0.5x - 0.4$

To solve an inequality that contains fractions, multiply both sides of the inequality by the LCD to clear the fractions.

MC **EXAMPLE 8** Solve. $\frac{1}{7}(x + 5) > \frac{1}{5}(x + 1)$

Solution

$\frac{x}{7} + \frac{5}{7} > \frac{x}{5} + \frac{1}{5}$ Using the distributive property, remove the parentheses.

$35\left(\frac{x}{7}\right) + 35\left(\frac{5}{7}\right) > 35\left(\frac{x}{5}\right) + 35\left(\frac{1}{5}\right)$ Multiply each term by the LCD.

$5x + 25 > 7x + 7$ Simplify.

$5x + 25 - 25 > 7x + 7 - 25$ Subtract 25 from each side.

$5x > 7x - 18$ Simplify.

$5x - 7x > 7x - 7x - 18$ Subtract $7x$ from each side.

$-2x > -18$ Simplify.

$\frac{-2x}{-2} < \frac{-18}{-2}$ Divide each side by -2 and reverse the direction of the inequality.

$x < 9$

Student Practice 8 Solve.

$\frac{1}{5}(x - 6) < \frac{1}{3}(x - 2)$

TO THINK ABOUT: Multiplying by the LCD Another approach to solving the inequality in Example 8 would be to multiply each side of the inequality by the LCD before removing the parentheses. Try it. Think about the pros and cons of this approach. Choose the method you like best.

EXAMPLE 9 Lexi and her mother are using a public phone to make a long distance phone call from Honolulu, HI, to West Chicago, IL. The charge is $4.50 for the first minute and 85¢ for each additional minute. Any fractional part of a minute will be rounded up to the nearest whole minute.

What is the maximum time that Lexi and her mother can talk if they have $15.55 in change to make the call?

Solution Let $x =$ the number of minutes they talk after the first minute. The cost must be less than or equal to $15.55. So we write the inequality

$$4.50 + 0.85x \leq 15.55$$
$$0.85x \leq 11.05 \quad \text{We subtract 4.50 from each side}$$
$$x \leq 13 \quad \text{We divide each side by 0.85}$$

Now we add the 13 minutes to the one minute that cost $4.50. This gives us 14 minutes. Thus the maximum amount of time they can talk is 14 minutes.

Student Practice 9 Olivia and her mother are making a long distance phone call from Anchorage, AK, to Southborough, MA. The charge is $3.50 for the first minute and 65¢ for each additional minute. Any fractional part of a minute will be rounded up to the nearest whole minute.

What is the maximum time that Olivia and her mother can talk if they have $13.90 in change to make the call?

Solving inequalities is a very important skill. Take some extra time to review Examples 1–9. Carefully work out the solutions to the Student Practice problems 1–9. Turn to page SP-5 and and check to be sure you have done the Student Practice problems correctly. Some extra time spent carefully studying the Examples and Student Practice problems will make the homework exercises much easier to complete.

STEPS TO SUCCESS Review a Little Every Day

Successful students find that review is not something you do the night before the test. Take time to review a little each day. When you are learning new material, take a little time to look over the concepts previously learned in the chapter.

By this continual review you will find the pressure to prepare for a test is reduced. You need time to think about what you have learned and make sure you really understand it. This will help to tie together the different topics in the chapter.

A little review of each idea and each kind of problem will enable you to feel confident. You will think more clearly and have less tension when it comes to test time.

Making it personal: Which of these suggestions is the one you most need to follow? Write down what you need to do to improve in this area. ▼

2.6 Exercises

MyMathLab®

Watch the videos in MyMathLab

Download the MyDashBoard App

Verbal and Writing Skills, Exercises 1–6

True or false?

1. The statement $6 < 8$ conveys the same information as $8 > 6$.
2. Adding $-5x$ to each side of an inequality reverses the direction of the inequality.
3. Dividing each side of an inequality by -4 reverses the direction of the inequality.
4. The graph of $x > -2$ is the set of all points to the right of -2 on a number line.
5. The graph of $x \leq 6$ does not include the point at 6 on a number line.
6. To solve the inequality $\frac{2}{3}x + \frac{3}{4} \geq \frac{1}{2}x - 4$, multiply only the fractions by the LCD.

Insert the symbol < or > between each pair of numbers.

7. 6 ______ -3
8. -15 ______ 4
9. -7 ______ -2
10. -5 ______ -9
11. $\frac{3}{4}$ ______ $\frac{2}{3}$
12. $\frac{5}{6}$ ______ $\frac{5}{7}$
13. $-\frac{2}{9}$ ______ $-\frac{3}{14}$
14. $-\frac{7}{16}$ ______ $-\frac{6}{13}$
15. -3.5 ______ -3.41
16. -2.69 ______ -2.7
17. $|3 - 7|$ ______ $|9 - 2|$
18. $|-8 + 2|$ ______ $|6 - 13|$

Graph each inequality.

19. $x \geq -2$
20. $x \geq -4$
21. $x < 15$
22. $x < 80$

Solve for x and graph your solution.

23. $2x - 7 \leq -5$
24. $3 + 5x \geq 18$
25. $3x - 7 > 9x + 5$
26. $2x + 5 > 4x - 5$
27. $0.5x + 0.1 < 1.1x + 0.7$
28. $1.7 - 0.6x \leq x + 0.1$

Solve for x.

29. $4x - 1 > 15$
30. $5x - 1 > 29$
31. $5x + 3 \leq 2x - 9$
32. $8x - 7 \leq 4x - 19$
33. $2x + \frac{5}{3} > \frac{2}{5}x - 1$
34. $2x + \frac{5}{2} > \frac{3}{2}x - 2$
35. $3x - 11 + 4(x + 8) < 0$
36. $4x + 7 + 5(x - 5) < 0$
37. $\frac{3}{5}x - (x + 2) \geq -2$
38. $-3(x + 1) - \frac{x}{2} + \frac{3}{2} < 0$

Mixed Practice *Solve for x.*

39. $0.4x + 1 \leq 2.6$
40. $0.3x + 1.2 \geq 3.8 - x$
41. $0.1(x - 2) \geq 0.5x - 0.2$
42. $1.2 - 0.8x \leq 0.3(4 - x)$
43. $2 - \frac{1}{5}(x - 1) \geq \frac{2}{3}(2x + 1)$
44. $\frac{3}{4} + \frac{1}{2}(x - 7) \leq 1 - \frac{x}{4}$
45. $\frac{2x - 3}{5} + 1 \geq \frac{1}{2}x + 3$
46. $1 - \frac{2x + 1}{2} > \frac{x}{4} + \frac{4}{3}$

Applications *For exercises 47–52, describe the situation with a linear inequality and then solve the inequality.*

47. ***Tip Income*** Malcolm is a waiter and earns \$5.50 per hour plus an average tip of \$6 for every table served. How many tables must he serve to earn more than \$100 for a 4-hour shift?

48. ***Telemarketing*** A phone solicitor selling long-distance services earns \$7.50 an hour plus \$25 for every new customer she signs up. How many customers must she sign up to earn more than \$400 during the next 30 working hours?

49. ***Telephone Rates*** Rusty Slater is making a long-distance phone call to Orlando, Florida, from a pay phone. The operator informs him that the charge will be \$3.95 for the first minute and 55¢ for each additional minute. Any fractional part of a minute used will be rounded up to the nearest whole minute. What is the maximum time Rusty can talk if he has \$13.30 in change in his pocket?

50. ***Aircraft Cargo Capacity*** A small plane carrying packages takes off from Beverly Airport. Each package weighs 68.5 pounds. The plane has a carrying capacity for people and packages of 2395 pounds. The plane is carrying a pilot who weighs 180 pounds and a passenger who weighs 160 pounds. How many packages can be safely carried?

51. ***Elevator Capacity*** Molly and Denton from the computer services department are delivering several new computers to faculty offices using the college elevator. The elevator has a maximum capacity of 1100 pounds. Molly weighs 130 pounds, and Denton weighs 155 pounds. Each computer weighs 59 pounds. How many computers can Molly and Denton place in the elevator and then safely ride with the computers up to the next floor?

52. ***Mailing Costs*** DeWolf Associates sent out several boxes of literature by Priority Mail. They planned for a mailing budget of \$8.46 per box. The post office charged \$0.41 for the first ounce and \$0.23 for each additional ounce. What was the most a box could weigh and still be mailed at a cost that did not exceed the budget?

Cumulative Review *Simplify.*

53. **[1.5.3]** $3xy(x + 2) - 4x^2(y - 1)$

54. **[1.5.2]** $\frac{2}{3}ab(6a - 2b + 9)$

55. **[1.4.4]** $\left(\frac{4x^2}{3yw^{-1}}\right)^3$

56. **[1.4.4]** $(-3a^0b^{-3}c^5)^{-2}$

Quick Quiz 2.6 *Solve each inequality.*

1. $5x + 7 > 3x - 9$

2. $-3(x + 4) < 6x - 8$

3. $\frac{1}{5}(x - 3) \geq \frac{1}{4}(x - 5) + 1$

4. **Concept Check** If you were to solve the inequalities $-3x < 9$ and $3x < 9$, in one case you would have to reverse the direction of the inequality and in the other case you would not. Explain how you can tell which case is which.

2.7 Compound Inequalities

① Graphing Compound Inequalities That Use *and*

Some inequalities consist of two inequalities connected by the word *and* or the word *or*. They are called **compound inequalities.** The solution of a compound inequality using the connective *and* includes all the numbers that make both parts true at the same time.

EXAMPLE 1 Graph the values of x where $7 < x$ *and* $x < 12$.

Solution We read the inequality starting with the variable. Thus, we graph all values of x, where x is greater than 7 and where x is less than 12. All such values must be between 7 and 12. Numbers that are greater than 7 and less than 12 can be written as $7 < x < 12$.

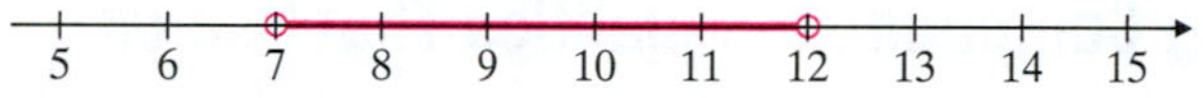

Student Practice 1 Graph the values of x where $-8 < x$ *and* $x < -2$.

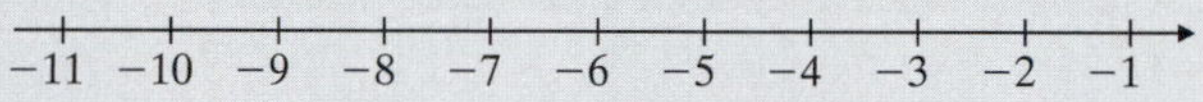

EXAMPLE 2 Graph the values of x where $-6 \leq x \leq 2$.

Solution Here we have that x is greater than or equal to -6 and that x is less than or equal to 2. We remember to include the points -6 and 2 since the inequality symbols contain the equals sign.

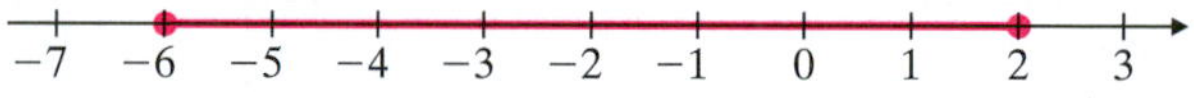

Student Practice 2 Graph the values of x where $-1 \leq x \leq 5$.

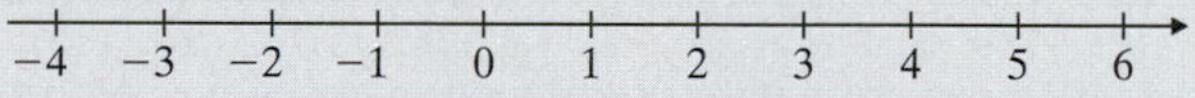

EXAMPLE 3 Graph the values of x where $-8.5 \leq x < -1$.

Solution Note the shaded circle at -8.5 and the open circle at -1.

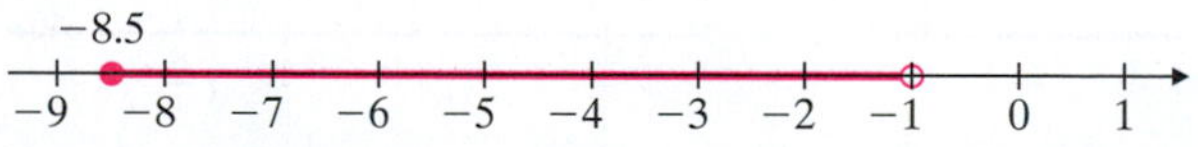

Student Practice 3 Graph the values of x where $-10 < x \leq -5.5$.

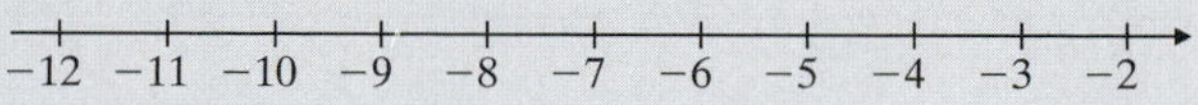

EXAMPLE 4 Graph the salary range (s) of the full-time employees of Tentron Corporation. Each person earns at least \$190 weekly, but not more than \$800 weekly.

Continued on next page

Student Learning Objectives

After studying this section, you will be able to:

① **Graph compound inequalities that use *and*.**

② **Graph compound inequalities that use *or*.**

③ **Solve compound inequalities and graph their solutions.**

***NOTE TO STUDENT:** Fully worked-out solutions to all of the Student Practice problems can be found at the back of the text starting at page SP-1.*

Solution "At least \$190" means that the weekly salary of each person is greater than or equal to \$190 weekly. We write $s \geq \$190$. "Not more than" means that the weekly salary of each person is less than or equal to \$800. We write $s \leq \$800$. Thus, s may be between 190 and 800 and may include those endpoints.

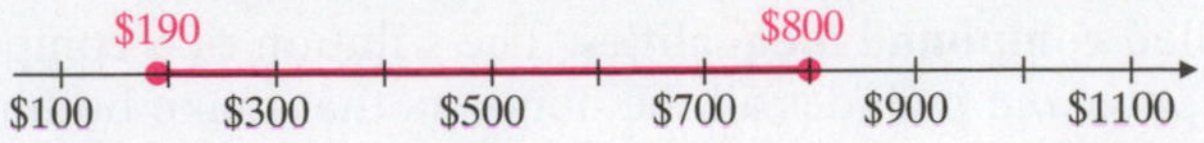

Student Practice 4 Graph the weekly salary range of a person who earns at least \$200 per week, but never more than \$950 per week.

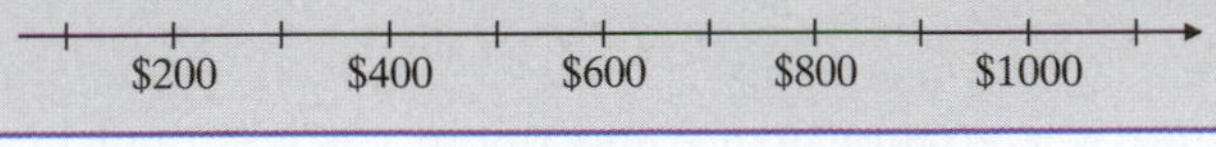

② Graphing Compound Inequalities That Use *or*

The solution of a compound inequality using the connective *or* includes all the numbers that are solutions of either of the two inequalities.

EXAMPLE 5 Graph the region where $x < 3$ *or* $x > 6$.

Solution Notice that a solution to this inequality need not be in both regions at the same time.

Read the inequality as "x is less than 3 or x is greater than 6." Thus, x can be less than 3 or x can be greater than 6. This includes all values to the left of 3 as well as all values to the right of 6 on a number line. We shade these regions.

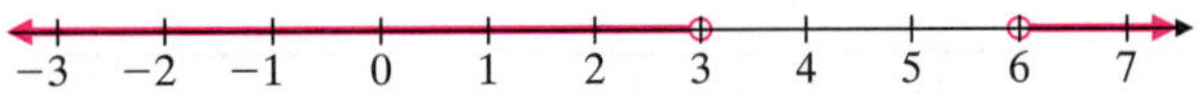

Student Practice 5 Graph the region where $x < 8$ *or* $x > 12$.

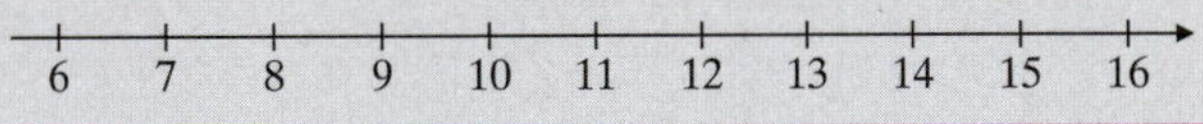

EXAMPLE 6 Graph the region where $x > -2$ *or* $x \leq -5$.

Solution Note the shaded circle at -5 and the open circle at -2.

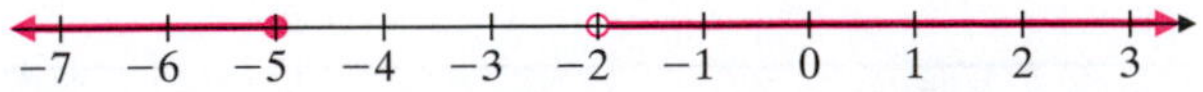

Student Practice 6 Graph the region where $x \leq -6$ *or* $x > 3$.

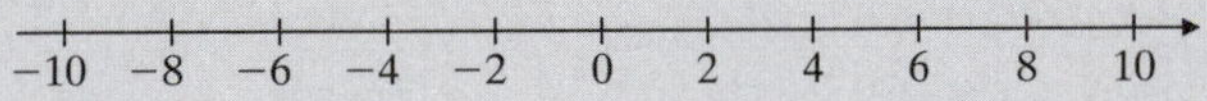

EXAMPLE 7 Male applicants for the state police force in Fred's home state are ineligible for the force if they are shorter than 60 inches or taller than 76 inches. Graph the range of rejected applicants' heights.

Solution Each rejected applicant's height h will be less than 60 inches ($h < 60$) or will be greater than 76 inches ($h > 76$).

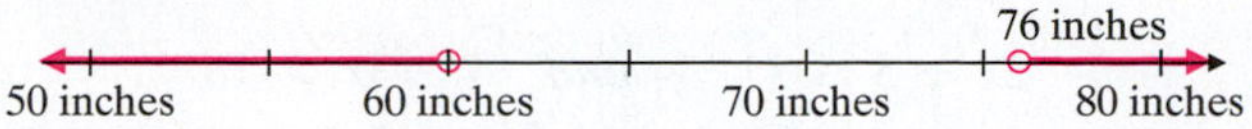

Student Practice 7 Female applicants are ineligible if they are shorter than 56 inches or taller than 70 inches. Graph the range of rejected applicant's heights.

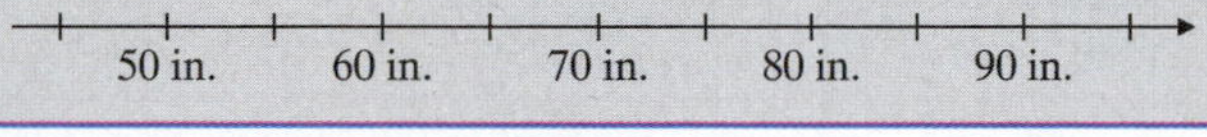

③ Solving Compound Inequalities

When asked to solve a more complicated compound inequality for x, we normally solve each individual inequality separately.

EXAMPLE 8 Solve for x and graph the compound solution. $3x + 2 > 14$ *or* $2x - 1 < -7$

Solution We solve each inequality separately.

$$\begin{aligned} 3x + 2 &> 14 \quad \textit{or} \quad & 2x - 1 &< -7 \\ 3x &> 12 & 2x &< -6 \\ x &> 4 & x &< -3 \end{aligned}$$

The solution is $x < -3$ *or* $x > 4$.

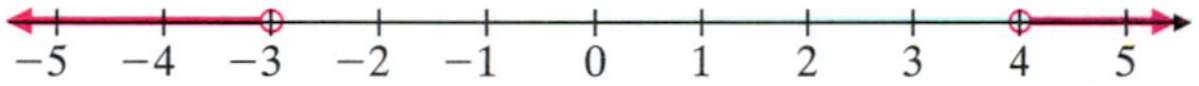

Student Practice 8 Solve for x and graph the compound solution. $3x - 4 < -1$ *or* $2x + 3 > 13$

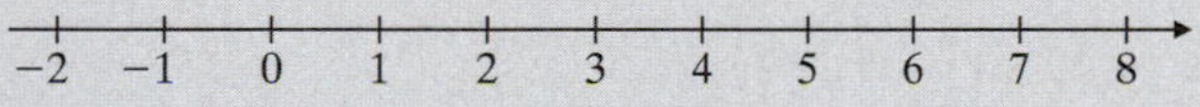

EXAMPLE 9 Solve for x and graph the compound solution. $5x - 1 > -2$ *and* $3x - 4 < 8$

Solution We solve each inequality separately.

$$\begin{aligned} 5x - 1 &> -2 \quad \textit{and} \quad & 3x - 4 &< 8 \\ 5x &> -1 & 3x &< 12 \\ x &> -\frac{1}{5} & x &< 4 \end{aligned}$$

The solution is the set of numbers between $-\frac{1}{5}$ and 4, not including the endpoints.

$$-\frac{1}{5} < x < 4$$

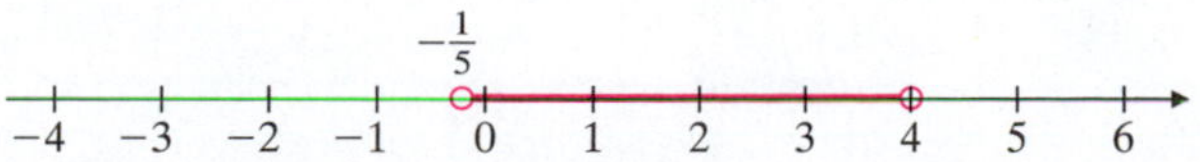

Student Practice 9 Solve for x and graph the compound solution. $3x + 6 > -6$ *and* $4x + 5 < 1$

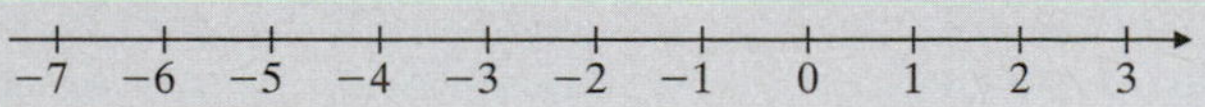

EXAMPLE 10 Solve and graph. $2x + 5 \leq 11$ *and* $-3x > 18$

Solution We solve each inequality separately.

$$2x + 5 \leq 11 \quad \textit{and} \quad -3x > 18$$

$$2x \leq 6 \qquad \frac{-3x}{-3} < \frac{18}{-3}$$

$$x \leq 3 \qquad x < -6$$

The separate solutions are $x < -6$ *and* $x \leq 3$.

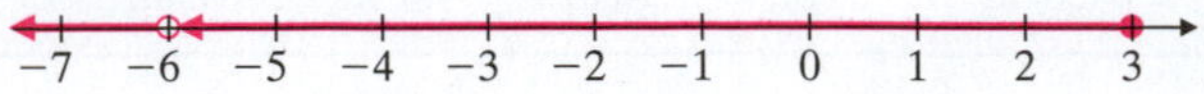

The only numbers that satisfy the statements $x \leq 3$ *and* $x < -6$ at the same time are $x < -6$. Thus, $x < -6$ is the solution to the compound inequality.

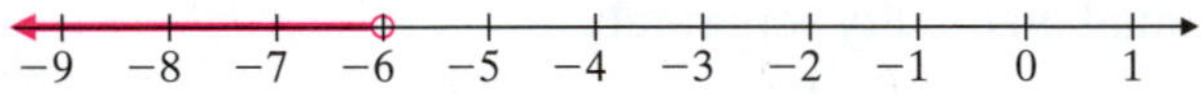

Student Practice 10 Solve and graph. $-2x + 3 < -7$ *and* $7x - 1 > -15$

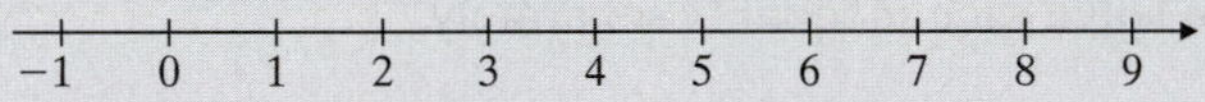

EXAMPLE 11 Solve. $-3x - 2 < -5$ *and* $4x + 6 < -12$

Solution We solve each inequality separately.

$$-3x - 2 < -5 \quad \textit{and} \quad 4x + 6 < -12$$

$$-3x < -3 \qquad 4x < -18$$

$$\frac{-3x}{-3} > \frac{-3}{-3} \qquad \frac{4x}{4} < \frac{-18}{4}$$

$$x > 1 \qquad x < -4\frac{1}{2}$$

Now, clearly it is impossible for one number to be greater than 1 *and* at the same time be less than $-4\frac{1}{2}$.

Thus, there is *no solution*. We can express this by the notation $\emptyset$, which is the **empty set.** Or we can just state, "There is no solution."

Student Practice 11 Solve. $-3x - 11 < -26$ *and* $5x + 4 < 14$

2.7 Exercises

MyMathLab®

Watch the videos in MyMathLab

Download the MyDashBoard App

Graph the values of x that satisfy the conditions given.

1. $3 < x$ *and* $x < 8$

2. $5 < x$ *and* $x < 10$

3. $-4 < x$ *and* $x < 2$

4. $-7 < x$ *and* $x < 1$

5. $7 < x < 9$

6. $3 < x < 5$

7. $-2 < x \leq \frac{1}{2}$

8. $-\frac{3}{2} \leq x \leq 4$

9. $x > 8$ *or* $x < 2$

10. $x \geq 2$ *or* $x \leq 1$

11. $x \leq -\frac{5}{2}$ *or* $x > 4$

12. $x < 3$ *or* $x > \frac{11}{2}$

13. $x \leq -10$ *or* $x \geq 40$

14. $x \leq -6$ *or* $x \geq 2$

Solve for x and graph your results.

15. $2x + 3 \leq 5$ *and* $x + 1 \geq -2$

16. $4x - 1 < 7$ *and* $x \geq -1$

17. $2x - 3 > 0$ *or* $x - 2 < -7$

18. $x + 1 \geq 5$ *or* $x + 5 < 2.5$

19. $x < 8$ *and* $x > 10$

20. $x < 6$ *and* $x > 9$

Applications *Express as a compound inequality.*

21. ***Toothpaste*** A tube of toothpaste is not properly filled if the amount of toothpaste t in the tube is more than 11.2 ounces or less than 10.9 ounces.

22. ***Clothing Standards*** The width of a seam on a pair of blue jeans is unacceptable if it is narrower than 10 millimeters or wider than 12 millimeters.

23. ***Interstate Highway Travel*** The number of cars c driving over Interstate 91 during the evening hours in January is always at least 5000, but never more than 12,000.

24. ***Campsite Capacity*** The number of campers c at a campsite during the Independence Day weekend is always at least 490, but never more than 2000.

Temperature Conversion *Solve the following application problems by using the formula* $C = \frac{5}{9}(F - 32)$. *Round to the nearest tenth.*

25. When visiting Montreal this spring, Marcos had been advised that the temperature could range from −20°C to 11°C. Find an inequality that represents the range in Fahrenheit temperatures.

26. The temperature in Sao Paulo, Brazil, during January can range from 16°C to 24°C. Find an inequality that represents the range in Fahrenheit temperatures.

Exchange Rates *At one point in 2010, the exchange rate for converting American dollars into Japanese yen was $Y = 81(d - 5)$. In this equation, d is the number of American dollars, Y is the number of yen, and \$5 represents a onetime fee that banks sometimes charge for currency conversion. Use the equation to solve the following problems. (Round answers to the nearest cent.)*

27. Frank is traveling to Tokyo, Japan, for two weeks, and he has been advised to have between 20,000 yen and 34,000 yen for spending money for each week he is there. Including the conversion charge, write an inequality that represents the number of American dollars he will need to exchange at the bank for this two-week period.

28. Carrie is traveling to Osaka, Japan, for three weeks. Her friend told her she should plan to have between 23,000 yen and 28,000 yen for spending money for each week she is there. Including the conversion charge, write an inequality that represents the number of American dollars she will need to exchange at the bank for the three-week period.

Mixed Practice *Solve each compound inequality.*

29. $x - 3 > -5$ *and* $2x + 4 < 8$

30. $x - 2 < 9$ *and* $x + 3 < 6$

31. $-6x + 5 \geq -1$ *and* $2 - x \leq 5$

32. $5x + 6 \geq -9$ *and* $10 - x \geq 3$

33. $4x - 3 < -11$ *or* $7x + 2 \geq 23$

34. $5x + 1 < 1$ *or* $3x - 9 > 9$

35. $-0.3x + 1 \geq 0.2x$ *or* $-0.2x + 0.5 > 0.7$

36. $-0.3x - 0.4 \geq 0.1x$ *or* $0.2x + 0.3 \leq -0.4x$

37. $\frac{5x}{2} + 1 \geq 3$ *and* $x - \frac{2}{3} \geq \frac{4}{3}$

38. $\frac{5x}{3} - 2 < \frac{14}{3}$ *and* $3x + \frac{5}{2} < -\frac{1}{2}$

39. $2x + 5 < 3$ *and* $3x - 1 > -1$

40. $6x - 10 < 8$ *and* $2x + 1 > 9$

41. $2x - 3 \geq 7$ *and* $5x - 8 \leq 2x + 7$

42. $7x + 2 \geq 11x + 14$ *and* $x + 9 \geq 6$

To Think About *Solve the compound inequality.*

43. $\frac{1}{4}(x + 2) + \frac{1}{8}(x - 3) \leq 1$ *and* $\frac{3}{4}(x - 1) > -\frac{1}{4}$

44. $\frac{x - 4}{6} - \frac{x - 2}{9} \leq \frac{5}{18}$ *or* $-\frac{2}{5}(x + 3) < -\frac{6}{5}$

Cumulative Review

45. **[1.5.3]** Simplify. $-3(x + 5) + 2(2x - 1)$

46. **[1.6.2]** Find the area of a circle with a diameter of 6 inches. Use $\pi \approx 3.14$.

47. **[2.2.1]** Solve for x. $3y - 5x = 8$

48. **[2.2.1]** solve for y. $7x + 6y = -12$

Quick Quiz 2.7 *Find the values of x that satisfy the given conditions.*

1. $3x + 2 < 8$ *and* $3x > -16$

2. $x > 5$ *and* $2x - 1 < 23$

3. $x - 7 \leq -15$ *or* $2x + 3 \geq 5$

4. **Concept Check** Explain why there are no values of x that satisfy these given conditions.

$$x + 8 < 3 \quad \textit{and} \quad 2x - 1 > 5$$

2.8 Absolute Value Inequalities

Student Learning Objectives

After studying this section, you will be able to:

1. Solve absolute value inequalities of the form $|ax + b| < c$.
2. Solve absolute value inequalities of the form $|ax + b| > c$.

① Solving Absolute Value Inequalities of the Form $|ax + b| < c$

We begin by looking at $|x| < 3$. What does this mean? The inequality $|x| < 3$ means that x is less than 3 units from 0 on a number line. We draw a picture.

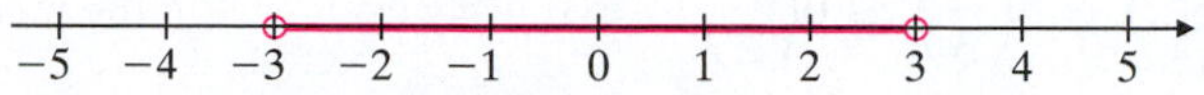

This picture shows all possible values of x such that $|x| < 3$. We see that this occurs when $-3 < x < 3$. We conclude that $|x| < 3$ and $-3 < x < 3$ are equivalent statements.

DEFINITION OF $|x| < a$

If a is a positive real number and $|x| < a$, then $-a < x < a$.

EXAMPLE 1 Solve. $|x| \leq 4.5$

Solution The inequality $|x| \leq 4.5$ means that x is less than or equal to 4.5 units from 0 on a number line. We draw a picture.

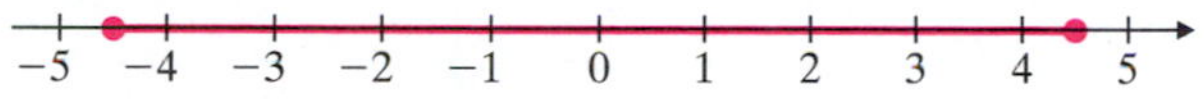

Thus, the solution is $-4.5 \leq x \leq 4.5$.

Student Practice 1 Solve and graph. $|x| < 2$

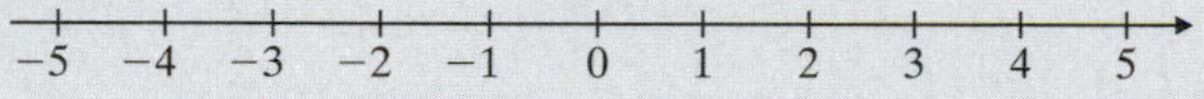

NOTE TO STUDENT: Fully worked-out solutions to all of the Student Practice problems can be found at the back of the text starting at page SP-1.

This same technique can be used to solve more complicated inequalities.

EXAMPLE 2 Solve and graph the solution. $|x + 5| \leq 10$

Solution We want to find the values of x that make $-10 \leq x + 5 \leq 10$ a true statement. We need to solve the compound inequality.

To solve this inequality, we subtract 5 from each part.

$$-10 - 5 \leq x + 5 - 5 \leq 10 - 5$$
$$-15 \leq x \leq 5$$

Thus, the solution is $-15 \leq x \leq 5$. We graph this solution.

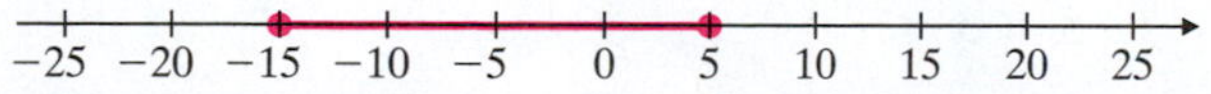

Student Practice 2 Solve and graph the solution. (*Hint:* Choose a convenient scale.) $|x - 6| < 15$

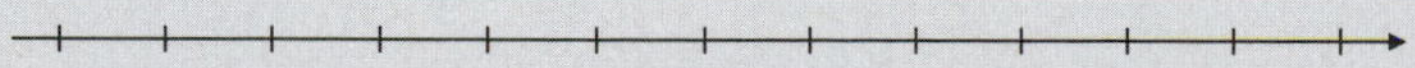

EXAMPLE 3 Solve and graph the solution. $\left|x - \frac{2}{3}\right| \leq \frac{5}{2}$

Solution

$$-\frac{5}{2} \leq x - \frac{2}{3} \leq \frac{5}{2} \qquad \text{If } |x| < a, \text{ then } -a < x < a.$$

$$6\left(-\frac{5}{2}\right) \leq 6(x) - 6\left(\frac{2}{3}\right) \leq 6\left(\frac{5}{2}\right) \qquad \text{Multiply each part of the inequality by 6.}$$

$$-15 \leq 6x - 4 \leq 15 \qquad \text{Simplify.}$$

$$-15 + 4 \leq 6x - 4 + 4 \leq 15 + 4 \qquad \text{Add 4 to each part.}$$

$$-11 \leq 6x \leq 19 \qquad \text{Simplify.}$$

$$-\frac{11}{6} \leq \frac{6x}{6} \leq \frac{19}{6} \qquad \text{Divide each part by 6.}$$

$$-1\frac{5}{6} \leq x \leq 3\frac{1}{6} \qquad \text{Change to mixed numbers to facilitate graphing.}$$

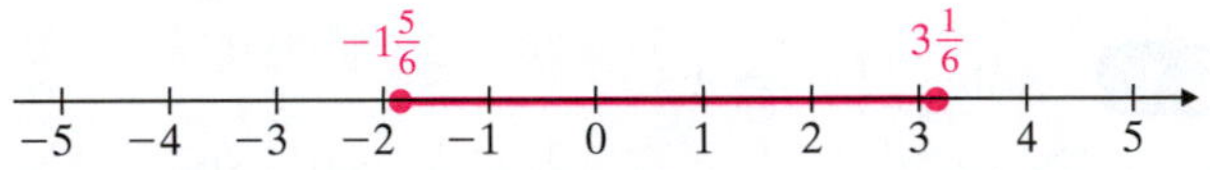

Student Practice 3 Solve and graph the solution.

$$\left|x + \frac{3}{4}\right| \leq \frac{7}{6}$$

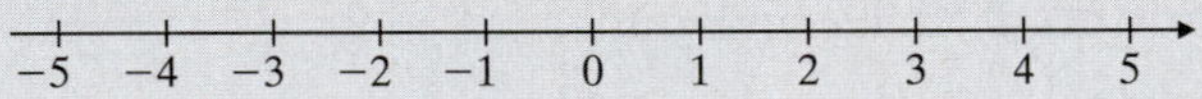

EXAMPLE 4 Solve and graph the solution. $|2(x - 1) + 4| < 8$.

Solution First we simplify the expression within the absolute value.

$$|2x - 2 + 4| < 8$$

$$|2x + 2| < 8$$

$$-8 < 2x + 2 < 8 \qquad \text{If } |x| < a, \text{ then } -a < x < a.$$

$$-8 - 2 < 2x + 2 - 2 < 8 - 2 \qquad \text{Subtract 2 from each part.}$$

$$-10 < 2x < 6 \qquad \text{Simplify.}$$

$$\frac{-10}{2} < \frac{2x}{2} < \frac{6}{2} \qquad \text{Divide each part by 2.}$$

$$-5 < x < 3$$

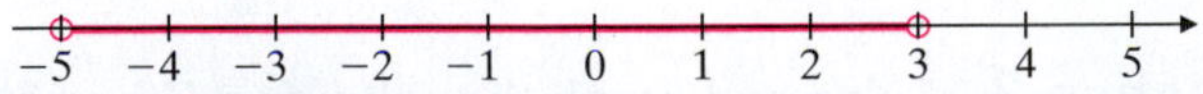

Student Practice 4 Solve and graph the solution. $|2 + 3(x - 1)| < 20$

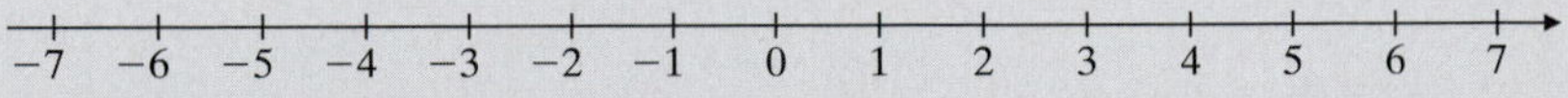

② Solving Absolute Value Inequalities of the Form $|ax + b| > c$

Now consider $|x| > 3$. What does this mean? This inequality $|x| > 3$ means that x is greater than 3 units from 0 on a number line. We draw a picture.

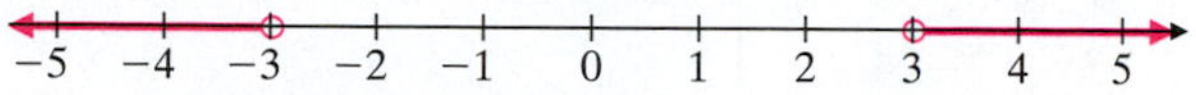

This picture shows all possible values of x such that $|x| > 3$. This occurs when $x < -3$ or when $x > 3$. (Note that a solution can be either in the region to the left of -3 on the number line or in the region to the right of 3 on the number line.) We conclude that the expression $|x| > 3$ and the expression $x < -3$ or $x > 3$ are equivalent statements.

DEFINITION OF $|x| > a$

If a is a positive real number and $|x| > a$, *then* $x < -a$ or $x > a$.

EXAMPLE 5 Solve and graph the solution. $|x| \geq 5\frac{1}{4}$

Solution The inequality $|x| \geq 5\frac{1}{4}$ means that x is more than $5\frac{1}{4}$ units from 0 on a number line. We draw a picture.

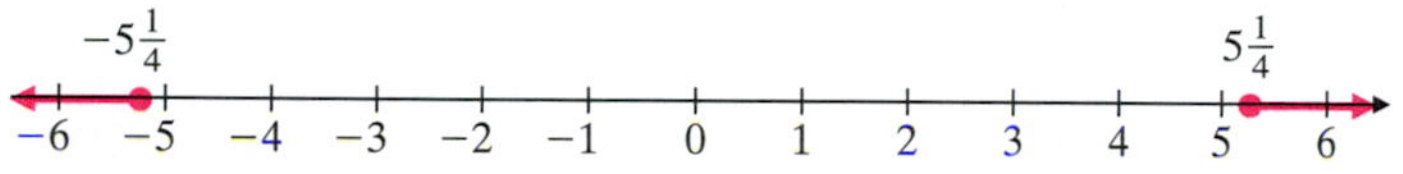

Thus, the solution is $x \leq -5\frac{1}{4}$ or $x \geq 5\frac{1}{4}$.

Student Practice 5 Solve and graph the solution. $|x| > 2.5$.

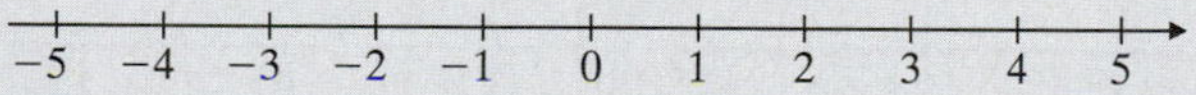

This same technique can be used to solve more complicated inequalities.

EXAMPLE 6 Solve and graph the solution. $|x - 4| > 5$

Solution We want to find the values of x that make $x - 4 < -5$ or $x - 4 > 5$ a true statement. We need to solve the compound inequality.

We will solve each inequality separately.

$$x - 4 < -5 \quad \textit{or} \quad x - 4 > 5$$
$$x - 4 + 4 < -5 + 4 \qquad x - 4 + 4 > 5 + 4$$
$$x < -1 \qquad x > 9$$

Thus, the solution is $x < -1$ *or* $x > 9$. We graph the solution on a number line.

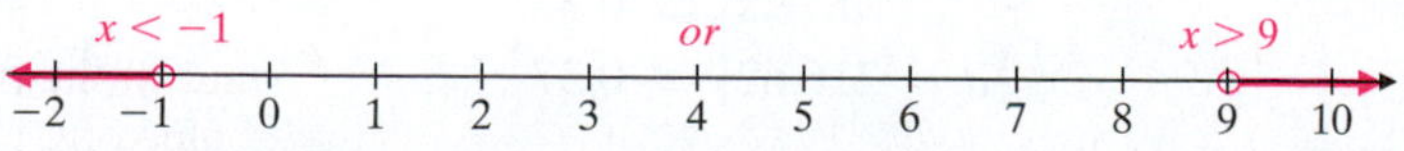

Student Practice 6 Solve and graph the solution. $|x + 6| > 2$

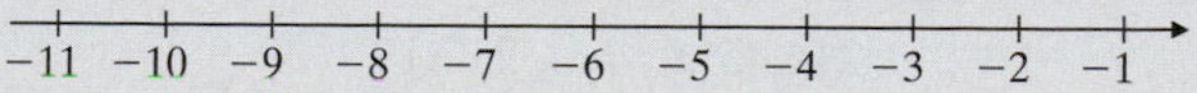

EXAMPLE 7 Solve and graph the solution. $|-3x + 6| > 18$

Solution By definition, we have the following compound inequality.

$$-3x + 6 > 18 \quad \textit{or} \quad -3x + 6 < -18$$

$$-3x > 12 \qquad\qquad -3x < -24$$

$$\frac{-3x}{-3} < \frac{12}{-3} \longleftarrow \qquad \longrightarrow \frac{-3x}{-3} > \frac{-24}{-3}$$

Division by a negative number reverses the inequality sign.

$$x < -4 \qquad\qquad x > 8$$

The solution is $x < -4$ *or* $x > 8$.

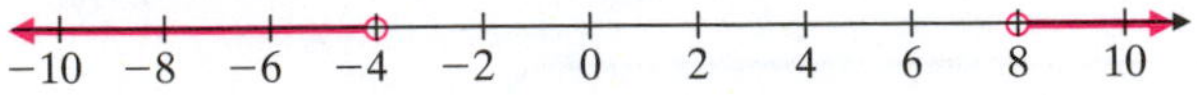

Student Practice 7 Solve and graph the solution. $|-5x - 2| > 13$.

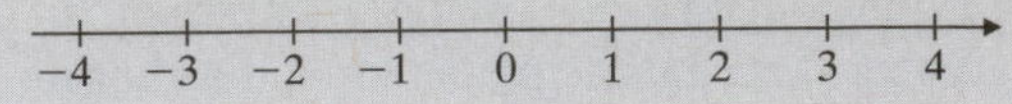

EXAMPLE 8 Solve and graph the solution. $\left|3 - \frac{2}{3}x\right| \geq 5$

Solution By definition, we have the following compound inequality.

$$3 - \frac{2}{3}x \geq 5 \quad \textit{or} \quad 3 - \frac{2}{3}x \leq -5$$

$$3(3) - 3\left(\frac{2}{3}x\right) \geq 3(5) \qquad 3(3) - 3\left(\frac{2}{3}x\right) \leq 3(-5)$$

$$9 - 2x \geq 15 \qquad 9 - 2x \leq -15$$

$$-2x \geq 6 \qquad -2x \leq -24$$

$$\frac{-2x}{-2} \leq \frac{6}{-2} \qquad \frac{-2x}{-2} \geq \frac{-24}{-2}$$

$$x \leq -3 \qquad x \geq 12$$

The solution is $x \leq -3$ *or* $x \geq 12$.

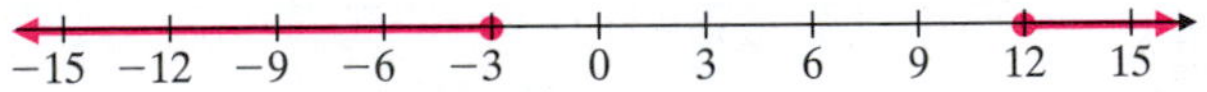

Student Practice 8 Solve and graph the solution.

$$\left|4 - \frac{3}{4}x\right| \geq 5$$

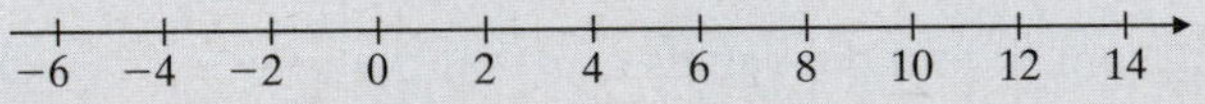

EXAMPLE 9 When a new car transmission is built, the diameter d of the transmission must not differ from the specified standard s by more than 0.37 millimeter. The engineers express this requirement as $|d - s| \leq 0.37$. If the standard s is 216.82 millimeters for a particular car, find the limits of d.

Solution

$$|d - s| \leq 0.37$$

$$|d - 216.82| \leq 0.37 \qquad \text{Substitute the known value of } s.$$

$$-0.37 \leq d - 216.82 \leq 0.37 \qquad \text{If } |x| \leq a, \text{ then } -a \leq x \leq a.$$

$$-0.37 + 216.82 \leq d - 216.82 + 216.82 \leq 0.37 + 216.82$$

$$216.45 \leq d \leq 217.19$$

Thus, the diameter of the transmission must be at least 216.45 millimeters, but not greater than 217.19 millimeters.

Student Practice 9 The diameter d of a transmission must not differ from the specified standard s by more than 0.37 millimeter. This is written as $|d - s| \leq 0.37$. Solve to find the allowed limits of d for a truck transmission for which the standard s is 276.53 millimeters.

SUMMARY OF ABSOLUTE VALUE EQUATIONS AND INEQUALITIES

It may be helpful to review the key concepts of absolute value equations and inequalities that we have covered in Sections 2.3 and 2.8. For real numbers a, b, and c, where $a \neq 0$ and $c > 0$, we have the following:

Absolute value form of the equation or inequality	Equivalent form without the absolute value	Type of solution obtained	Graphed form of the solution on a number line
$\lvert ax + b \rvert = c$	$ax + b = c$ or $ax + b = -c$	Two distinct numbers: m and n	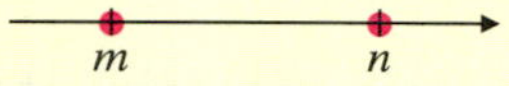
$\lvert ax + b \rvert < c$	$-c < ax + b < c$	The set of numbers between the two numbers m and n: $m < x < n$	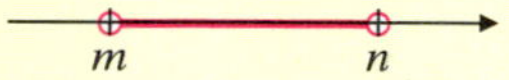
$\lvert ax + b \rvert > c$	$ax + b < -c$ or $ax + b > c$	The set of numbers less than m or the set of numbers greater than n: $x < m$ or $x > n$	

STEPS TO SUCCESS Helping Your Accuracy

It is easy to make a mistake. But here are five ways to cut down on errors. Look over each one and think about how each suggestion can help you.

1. Work carefully, and take your time. Do not rush through a problem just to get it done.
2. Concentrate on the problem. Sometimes your mind starts to wander. Then you get careless and will likely make a mistake.
3. Check your problem. Be sure you copied it correctly from the book.
4. Check your computations from step to step. Did you do each step correctly?
5. Check your final answer. Does it work? Is it reasonable?

Making it personal: Look over these five suggestions. Which one do you think will help you the most? Write down how you can use this suggestion to help you personally as you try to improve your accuracy. ▼

2.8 Exercises

MyMathLab®

Watch the videos in MyMathLab

Download the MyDashBoard App

Solve and graph the solutions.

1. $|x| \leq 8$

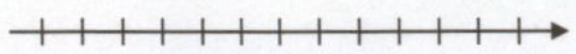

2. $|x| < 6$

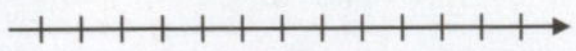

3. $|x + 4.5| < 5$

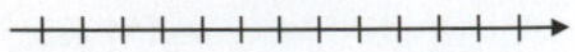

4. $|x + 6| < 3.5$

Solve for x.

5. $|x - 3| \leq 5$

6. $|x - 8| \leq 12$

7. $|2x - 1| \leq 5$

8. $|4x - 3| \leq 9$

9. $|5x - 2| \leq 4$

10. $|2x - 3| \leq 1$

11. $|0.5 - 0.1x| < 1$

12. $|0.9 - 0.2x| < 2$

13. $\left|\frac{1}{4}x + 2\right| < 6$

14. $\left|\frac{1}{3}x + 4\right| < 7$

15. $\left|\frac{2}{3}(x - 2)\right| < 4$

16. $\left|\frac{3}{5}(x - 1)\right| < 3$

17. $\left|\frac{3x - 2}{4}\right| < 3$

18. $\left|\frac{5x - 3}{2}\right| < 4$

Solve for x.

19. $|x| > 5$

20. $|x| \geq 7$

21. $|x + 2| > 5$

22. $|x + 4| > 7$

23. $|x - 1| \geq 2$

24. $|x - 6| \geq 4$

25. $|4x - 7| \geq 9$

26. $|6x - 5| \geq 7$

27. $|6 - 0.1x| > 5$

28. $|0.5 - 0.1x| > 6$

29. $\left|\frac{1}{5}x - \frac{1}{10}\right| > 2$

30. $\left|\frac{1}{4}x - \frac{3}{8}\right| > 1$

Mixed Practice

31. $\left|\frac{1}{3}(x - 2)\right| < 5$

32. $\left|\frac{2}{5}(x - 2)\right| \leq 4$

33. $|3x + 5| < 17$

34. $|2x + 3| < 5$

35. $|3 - 8x| > 19$

36. $|2 - 5x| > 2$

Applications

Manufacturing Standards *In a certain company, the measured thickness m of a helicopter blade must not differ from the standard s by more than 0.12 millimeter. The manufacturing engineer expresses this as* $|m - s| \leq 0.12$.

37. Find the limits of m if the standard s is 18.65 millimeters.

38. Find the limits of m if the standard s is 17.48 millimeters.

Computer Chip Standards *A small computer microchip has dimension requirements. The manufacturing engineer has written the specification that the new length n of the chip can differ from the previous length p by only 0.05 centimeter or less. The equation is* $|n - p| \leq 0.05$.

39. Find the limits of the new length if the previous length was 9.68 centimeters.

40. Find the limits of the new length if the previous length was 7.84 centimeters.

Cumulative Review

Perform the correct order of operations to simplify.

41. **[1.4.5]** Write in scientific notation. 0.000045

42. **[2.3.1]** Solve for x. $|2x - 1| = 8$

In exercises 43 and 44 use $\pi \approx 3.14$. *Round answers to the nearest hundredth.*

▲ **43.** **[1.6.2]** ***Geometry*** The Outward Bound program in the United States is famous for teaching self-esteem and personal achievement to young people. One of its physical challenges is for a student to hang on to a rope 19 meters long and swing from one shore to another and then back. The rope swings through a circular arc, measuring $\frac{1}{8}$ of the circumference of a circle. How many meters does the end of the rope travel in one *round-trip* swing?

▲ **44.** **[1.6.2]** ***Geometry*** The rigging on a sailboat comes loose from the mast. The end of the wire rigging that is hanging down is 30 feet from the top of the mast. This end swings through a circular arc, measuring $\frac{1}{6}$ of the circumference of a circle. How many feet does the end of the wire travel in one *round-trip* swing?

Quick Quiz 2.8 *Solve for x.*

1. $\left|\frac{1}{2}x + \frac{1}{4}\right| < 6$

2. $|8x - 4| \leq 20$

3. $|5x + 2| > 7$

4. **Concept Check** Explain what happens when you try to solve for x.

$$|7x + 3| < -4$$

Did You Know...

That Some Apartments Include Utilities, So You Will Need to Calculate Total Monthly Costs When Comparing Them?

SEARCHING FOR THE BEST DEAL ON AN APARTMENT

Understanding the Problem:
Mariam has decided it is time to move out of her parent's house and rent her first apartment. She is planning on rooming with Jen, her college roommate. They have been looking at apartments and have narrowed the choices down to three they both like. The apartments are each offering a deal if they sign a one-year lease. Apartment 1 is offering one month free so they would only pay rent for 11 months. Apartment 2 is offering free heat, and Apartment 3 is offering free electricity, cable TV, and Internet. They discussed the costs associated with the different apartments and came up with the table below.

Monthly bills	Monthly expenses Apartment 1	Monthly expenses Apartment 2	Monthly expenses Apartment 3
Rent	$800	$850	$900
Heat	$110	Included	$110
Electricity	$90	$90	Included
Cable and Internet	$90	$90	Included
Insurance	$25	$25	$25

Making a Plan:
The apartments are very similar so the deciding factor will be which apartment is the cheapest. They need to calculate what the total monthly costs are for each apartment.

Step 1: They total the monthly expenses for each apartment.

Task 1: *What is the total monthly cost for Apartment 1 (don't consider the one month free rent here)? Apartment 2? Apartment 3?*

Step 2: They calculate the monthly expenses for Apartment 1 taking into consideration the one month free rent.

Task 2: *What is the monthly cost for Apartment 1 taking into consideration the free month?*

Finding a Solution:
Now that they know the monthly cost for each apartment they can decide which apartment to rent.

Task 3: *Which apartment should they rent?*

Step 3: Since they are splitting the expenses of the apartment they need to divide the cost to determine each person's share.

Task 4: *What is the amount of each person's share of the total monthly costs to rent that apartment?*

Applying the Situation to Your Life:
When comparing apartment costs, there is more to consider than just the monthly rents. You need to take into consideration utilities like heat and electricity. The apartment with the cheapest rent isn't always the cheapest after these expenses. For apartments with included utilities, consider your lifestyle and determine if it is going to be saving you money. For example, if you are never home so you wouldn't use the cable or Internet, don't include it in the cost of comparable apartments since you wouldn't pay for it if it wasn't included. Also, when you sign a one-year lease, if you decide to move out early you will have to pay a fee, usually equal to one month's rent.

Chapter 2 Organizer

Topic and Procedure	Examples	You Try It
Solving first-degree equations, p. 62 **1.** Remove any grouping symbols in the proper order. **2.** If fractions exist, multiply all terms on both sides by the LCD of all the fractions. **3.** If decimals exist, multiply all terms on both sides by a power of 10. **4.** Combine like terms if possible. **5.** Add or subtract terms on both sides of the equation to get all terms with the variable on one side of the equation. **6.** Add or subtract a value on both sides of the equation to get all terms not containing the variable on the other side of the equation. **7.** Divide both sides of the equation by the coefficient of the variable. **8.** Simplify the solution (if possible). **9.** Check the solution.	Solve for x. $\frac{1}{3}(2x-3)+\frac{1}{2}(x+1)=3$ $\frac{2}{3}x-1+\frac{1}{2}x+\frac{1}{2}=3$ $6\left(\frac{2}{3}x\right)-6(1)+6\left(\frac{1}{2}x\right)+6\left(\frac{1}{2}\right)=6(3)$ $4x-6+3x+3=18$ $7x-3=18$ $7x=21$ $x=3$ *Check.* $\frac{1}{3}[2(3)-3]+\frac{1}{2}(3+1)\stackrel{?}{=}3$ $\frac{1}{3}(3)+\frac{1}{2}(4)\stackrel{?}{=}3$ $1+2=3$ ✓	**1.** Solve for x. $\frac{1}{4}(x+5)=6-\frac{1}{3}(2x-5).$
Equations and formulas with more than one variable, p. 68 If an equation or formula has more than one variable, you can solve for a particular variable by using the procedure for solving linear equations. Remember that your goal is to get all terms containing the desired variable on one side of the equation and all other terms on the opposite side of the equation.	Solve for r. $A=P(1+rt)$ $A=P+Prt$ $A-P=Prt$ $\frac{A-P}{Pt}=\frac{Prt}{Pt}$ $\frac{A-P}{Pt}=r$	**2.** Solve for b. $A=\frac{h}{2}(B+b)$
Absolute value equations, p. 74 To solve an equation that involves an absolute value, we rewrite the absolute value equation as two separate equations without the absolute value. We solve each equation. If $\lvert ax+b\rvert=c$ where $c>0$, then $ax+b=c$ or $ax+b=-c$.	Solve for x. $\lvert 4x-1\rvert=17$ $4x-1=17$ or $4x-1=-17$ $4x=17+1$ $4x=-17+1$ $4x=18$ $4x=-16$ $x=\frac{18}{4}$ $x=\frac{-16}{4}$ $x=\frac{9}{2}$ $x=-4$	**3.** Solve for x. $\lvert 3x+5\rvert=11$
Solving word problems, pp. 80 and 87 **1.** *Understand the problem.* **(a)** Read the word problem carefully to get an overview. **(b)** Determine what information you need to solve the problem. **(c)** Draw a sketch. Label it with the known information. Determine what needs to be found. **(d)** Choose a variable to represent one unknown quantity. **(e)** If necessary, represent other unknown quantities in terms of that same variable. **2.** *Write an equation.* **(a)** Look for key words to help you translate the words into algebraic symbols. **(b)** Use a relationship given in the problem or an appropriate formula in order to write an equation. **3.** *Solve the equation and state the answer.* **4.** *Check.* **(a)** Check the solution in the original equation. **(b)** Be sure the solution to the equation answers the question in the word problem. You may need some additional calculations if it does not.	Fred and Linda invested \$7000 for 1 year. Part was invested at 3% simple interest, and part was invested at 5% simple interest. If they earned a total of \$320 in interest, how much did they invest at each rate? **1.** *Understand the problem.* Let $x=$ the amount invested at 3% simple interest. Then $7000-x=$ the amount invested at 5% simple interest. **2.** *Write an equation.* $0.03x+0.05(7000-x)=320$ **3.** *Solve the equation and state the answer.* $0.03x+350-0.05x=320$ $350-0.02x=320$ $-0.02x=-30$ $x=1500$ $7000-x=5500$ Therefore, \$1500 was invested at 3% simple interest and \$5500 at 5% simple interest. **4.** *Check.* $0.03(1500)+0.05(7000-1500)\stackrel{?}{=}320$ $45+275\stackrel{?}{=}320$ $320=320$ ✓	**4.** When Jocelyn inherited \$12,000, she decided to invest it. Part was invested at 6% simple interest, and part was invested at 9% simple interest. She earned a total of \$960 in interest in one year. How much did she invest at each rate?

Topic and Procedure	Examples	You Try It

Solving linear inequalities, pp. 98–100

Topic and Procedure

1. If $a < b$, then for all real numbers a, b, and c,
$$a + c < b + c \text{ and } a - c < b - c.$$
2. If $a < b$, then for all real numbers a, b, and c when $c > 0$,
$$ac < bc \quad \text{and} \quad \frac{a}{c} < \frac{b}{c}.$$
Multiplying or dividing both sides of an inequality by a positive number does **not** reverse the inequality.
3. If $a < b$, then for all real numbers a, b, and c when $c < 0$,
$$ac > bc \quad \text{and} \quad \frac{a}{c} > \frac{b}{c}.$$
Multiplying or dividing both sides of an inequality by a negative number **reverses the direction** of the inequality symbol.

Examples

Solve and graph.

$$3(2x - 4) + 1 \geq 7$$
$$6x - 12 + 1 \geq 7$$
$$6x - 11 \geq 7$$
$$6x \geq 18$$
$$x \geq 3$$

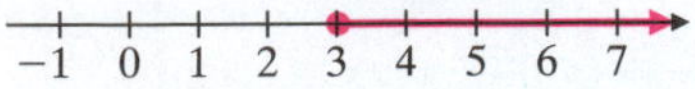

Solve and graph.

$$\frac{1}{4}(x + 3) \leq \frac{1}{3}(x - 2)$$
$$\frac{1}{4}x + \frac{3}{4} \leq \frac{1}{3}x - \frac{2}{3}$$
$$3x + 9 \leq 4x - 8$$
$$-1x + 9 \leq -8$$
$$-1x \leq -17$$
$$\frac{-1x}{-1} \geq \frac{-17}{-1}$$
$$x \geq 17$$

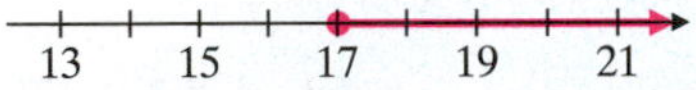

You Try It

5. Solve and graph.

(a) $8 - 2(3x + 1) \leq 18$

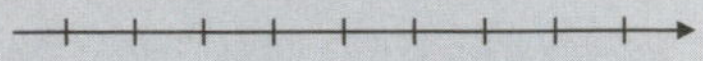

(b) $\frac{1}{2}(x - 6) < \frac{2}{5}(x - 2)$

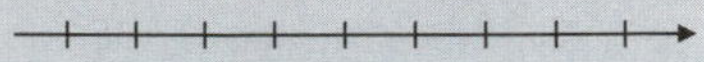

Solving compound inequalities containing and, p. 105

Topic and Procedure

The solution is the desired region containing all values of x that meet both conditions.

Examples

Graph the values of x satisfying $x + 6 > -3$ *and* $2x - 1 < -4$.

$$x + 6 - 6 > -3 - 6 \quad \textit{and} \quad 2x - 1 + 1 < -4 + 1$$
$$x > -9 \quad \textit{and} \quad x < -1.5$$

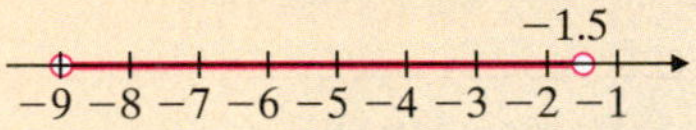

You Try It

6. Graph the values of x satisfying $x + 7 > -1$ *and* $3x + 4 < 10$.

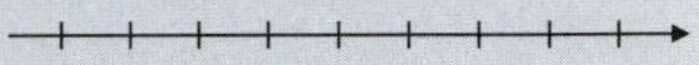

Solving compound inequalities containing or, p. 106

Topic and Procedure

The solution is the desired region containing all values of x that meet either of the two conditions.

Examples

Graph the values of x satisfying $-3x + 1 \leq 7$ *or* $3x + 1 \leq -11$.

$$-3x + 1 - 1 \leq 7 - 1 \quad \textit{or} \quad 3x + 1 - 1 \leq -11 - 1$$
$$-3x \leq 6 \quad \textit{or} \quad 3x \leq -12$$
$$\frac{-3x}{-3} \geq \frac{6}{-3} \quad \textit{or} \quad \frac{3x}{3} \leq \frac{-12}{3}$$
$$x \geq -2 \quad \textit{or} \quad x \leq -4$$

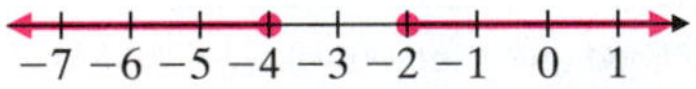

You Try It

7. Graph the values of x satisfying $5x + 2 \leq -8$ *or* $4x - 3 \geq 9$.

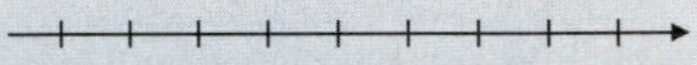

Solving absolute value inequalities involving < or ≤, p. 111

Topic and Procedure

Let a be a positive real number.

If $|x| < a$, then $-a < x < a$.

If $|x| \leq a$, then $-a \leq x \leq a$.

Examples

Solve and graph.

$$|3x - 2| < 19$$
$$-19 < 3x - 2 < 19$$
$$-19 + 2 < 3x - 2 + 2 < 19 + 2$$
$$-17 < 3x < 21$$
$$-\frac{17}{3} < \frac{3x}{3} < \frac{21}{3}$$
$$-5\frac{2}{3} < x < 7$$

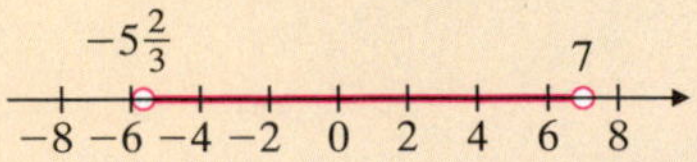

You Try It

8. Solve and graph. $|2x + 7| < 17$

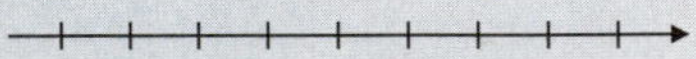

Topic and Procedure	Examples	You Try It
Solving absolute value inequalities involving $>$ or $\geq$, p. 113 Let a be a positive real number. If $\|x\| > a$, then $x < -a$ or $x > a$. If $\|x\| \geq a$, then $x \leq -a$ or $x \geq a$.	Solve and graph. $\left\|\frac{1}{3}(x-2)\right\| \geq 2$ $\frac{1}{3}(x-2) \leq -2$ *or* $\frac{1}{3}(x-2) \geq 2$ $\frac{1}{3}x - \frac{2}{3} \leq -2$ $\quad$ $\frac{1}{3}x - \frac{2}{3} \geq 2$ $x - 2 \leq -6$ $\quad$ $x - 2 \geq 6$ $x \leq -6 + 2$ $\quad$ $x \geq 6 + 2$ $x \leq -4$ *or* $x \geq 8$ 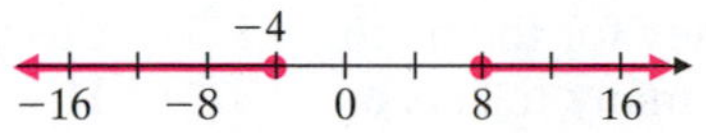	9. Solve and graph. $\left\|\frac{1}{4}(x+8)\right\| > 1$ 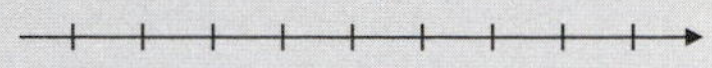

Chapter 2 Review Problems

Solve for x.

1. $7x - 3 = -5x - 18$

2. $8 - 2(x + 3) = 24 - (x - 6)$

3. $5(x - 2) + 4 = x + 9 - 2x$

4. $x - \frac{4}{3} = \frac{11}{12} + \frac{3}{4}x$

5. $\frac{1}{9}x - 1 = \frac{1}{2}\left(x + \frac{1}{3}\right)$

6. $5x = 3(1.6x - 4.2)$

7. Solve for a. $P = \frac{1}{2}ab$

8. Solve for a. $2(3ax - 2y) - 6ax = -3(ax + 2y)$

9. **(a)** Solve for F: $C = \frac{5F - 160}{9}$

(b) Now find F when $C = 10°$.

▲ **10.** **(a)** Solve for W: $P = 2W + 2L$

(b) Now find W when $P = 100$ meters and $L = 20.5$ meters.

Solve for x.

11. $|2x - 7| = 9$

12. $|5x + 2| = 7$

13. $|3 - x| = |5 - 2x|$

14. $|x + 8| = |2x - 4|$

15. $\left|\frac{1}{4}x - 3\right| = 8$

16. $|2x - 8| + 7 = 12$

Solve each problem.

▲ **17.** ***Geometry*** Jessica wants to fence her rectangular vegetable garden. The length of the garden is 3 feet longer than twice its width. The perimeter is 42 feet. What are the length and width of the garden?

18. ***Education*** The number of men attending Western Tech is 200 less than twice the number of women. The number of students at the school is 280. How many men attend? How many women attend?

19. ***Car Rental Costs*** Rent-It-Right rents compact cars for \$38 per day plus 15¢ per mile. Lucia rented a car for 3 days and was charged \$150. How many miles did she drive?

20. ***Withholding from Monthly Paycheck*** Alice's employer withholds from her monthly paycheck \$102 for federal and state taxes and for retirement. She noticed that the amount withheld for her state income tax is \$13 more than that withheld for retirement. The amount withheld for federal income tax is three times the amount withheld for the state tax. How much is withheld monthly for federal tax? State tax? Retirement?

21. ***Raffle Tickets*** Emma, Jackson, and Nicholas have been selling raffle tickets to raise money for the math club. Emma sold 5 less than twice as many tickets as Nicholas. Jackson sold 10 more than twice as many tickets as Nicholas. Together the three sold 180 tickets. How many did each student sell?

22. ***Education*** Valleyview College has 15% more students this year than five years ago. There are 2415 enrolled. How many students were enrolled five years ago?

23. ***Investments*** Huang invested \$9000 in mutual funds and bonds. The mutual fund earned 11% simple interest. The bonds earned 6% simple interest. At the end of one year, he had earned \$815 in interest. How much had he invested at each rate?

24. ***Chemical Mixtures*** To make 24 liters of a weak solution of 4% acid, a lab technician will use some premixed solutions: one is 2% acid and the other is 5% acid. How many liters of each type should he use to obtain the desired solution?

25. ***Coffee Costs*** A local specialty coffee shop wants to obtain 30 pounds of a mixture of coffee beans costing \$4.40 per pound. They have a mixture costing \$4.25 per pound and a mixture costing \$4.50 per pound. How much of each should be used?

26. ***Education*** When Eastern Slope Community College opened, the number of students (full-time and part-time) was 380. Since then the number of full-time students has doubled, and the number of part-time students has tripled. There are now 890 students at the school. How many of the present students are full-time? Part-time?

Solve for x.

27. $7x + 8 < 5x$

28. $9x + 3 < 12x$

29. $3(3x - 2) < 4x - 16$

30. $\frac{5}{3} - x \geq -\frac{1}{6}x + \frac{5}{6}$

31. $\frac{1}{3}(x - 2) < \frac{1}{4}(x + 5) - \frac{5}{3}$

32. $\frac{1}{3}(x + 2) > 3x - 5(x - 2)$

Graph the values of x that satisfy the conditions given.

33. $-3 \leq x < 2$

34. $-8 \leq x \leq -4$

35. $x < -2$ *or* $x \geq 5$

36. $x > -5$ *and* $x < -1$

37. $x > -8$ *and* $x < -3$

38. $x + 3 > 8$ *or* $x + 2 < 6$

Solve for x.

39. $x - 2 > 7$ *or* $x + 3 < 2$

40. $x + 3 > 8$ *and* $x - 4 < -2$

41. $-1 < x + 5 < 8$

42. $0 \leq 5 - 3x \leq 17$

43. $2x - 7 < 3$ *and* $5x - 1 \geq 8$

44. $4x - 2 < 8$ *or* $3x + 1 > 4$

Solve for x.

45. $|x + 7| < 15$

46. $|x + 9| < 18$

47. $\left|\frac{1}{2}x + 2\right| < \frac{7}{4}$

48. $|2x - 1| \geq 9$

49. $|3x - 1| \geq 2$

50. $|2(x - 5)| \geq 2$

51. ***Telephone Charges*** Greg Salzman is making a long-distance phone call to Chicago, Illinois, from a pay phone. The operator informs him that the charge will be \$3.95 for the first minute and 65¢ for each additional minute. Any fractional part of a minute used will be rounded up to the nearest whole minute. What is the maximum time Greg can talk if he has \$13.05 in change in his pocket?

52. ***Airplane Capacity*** A small plane takes off carrying packages from Manchester Airport. Each package weighs 77.5 pounds. The plane has a carrying capacity for people and packages of 1765 pounds. The plane is carrying a pilot who weighs 170 pounds and a passenger who weighs 200 pounds. How many packages can be safely carried?

53. ***Landscaping Costs*** Emmanuel Vargas wants to order bark mulch from a landscaping company. He does not want to spend more than \$250 for the order. The mulch costs \$28 per cubic yard and there is a \$40 delivery charge. Only whole cubic yards can be ordered. What is the maximum number of cubic yards of mulch he can order?

54. ***Census*** The Census Bureau has projected that the population of Nevada in the year 2025 may be as large as 2,854,000 or as small as 2,312,000. A group of real estate offices in Nevada has projected that the larger number should be 6% higher. They feel that the smaller number should be 4% higher. Assuming the Nevada real estate offices are correct, write an inequality that expresses the revised population projections for the year 2025. Use the variable x to represent the population of Nevada in the year 2025. (*Source:* U.S. Census Bureau)

Mixed Practice

55. Solve for x. $4 - 7x = 3(x + 3)$

56. Solve for B. $H = \frac{3}{4}B - 16$

Use an algebraic equation to find a solution.

57. ***Chemical Mixtures*** A technician needs 100 grams of an alloy that is 80% pure copper (Cu). She has one alloy that is 77% pure Cu and another that is 92% pure Cu. How many grams of each alloy should she use to obtain 80% pure Cu?

Solve for x. Graph your solution.

58. $7x + 12 < 9x$

59. $\frac{2}{3}x - \frac{5}{6}x - 3 \leq \frac{1}{2}x - 5$

Solve for x. Graph your solution.

60. $-2 \leq x + 1 \leq 4$

61. $2x + 3 < -5$ or $x - 2 > 1$

Solve for x.

62. $|2x - 7| + 4 = 5$

63. $\left|\frac{2}{3}x - \frac{1}{2}\right| \leq 3$

64. $|2 - 5x - 4| > 13$

How Am I Doing? Chapter 2 Test

Chapter Test Prep Videos | MATH COACH | MyMathLab® | YouTube

After you take this test read through the Math Coach on pages 125–126. Math Coach videos are available via MyMathLab and YouTube. Step-by-step test solutions in the Chapter Test Prep Videos are also available via MyMathLab and YouTube. (Search "TobeyInterAlg" and click on "Channels.")

Solve for x.

1. $5x - 8 = -6x - 10$

2. $3(7 - 2x) = 14 - 8(x - 1)$

3. $\frac{1}{3}(-x + 1) + 4 = 4(3x - 2)$

4. $0.5x + 1.2 = 4x - 3.05$

MC **5.** Solve for n. $L = a + d(n - 1)$

6. Solve for b. $A = \frac{1}{2}bh$

7. Use your answer for problem 6 to evaluate b when $A = 15 \text{ cm}^2$ and $h = 10$ cm.

8. Solve for r. $H = \frac{1}{2}r + 3b - \frac{1}{4}$

Solve for x.

9. $|5x - 2| = 37$

MC **10.** $\left|\frac{1}{2}x + 3\right| - 2 = 4$

Use an algebraic equation to find a solution.

▲ **11.** A triangle has a perimeter of 69 meters. The length of the second side is twice the length of the first side. The third side is 5 meters longer than the first side. How long is each side?

12. Mercy Hospital's electric bill for September was \$2489. This is a decrease of 5% from August's electric bill. What was the hospital's electric bill in August?

13. Linda needs 10 gallons of solution that is 60% antifreeze. She has a solution that is 90% antifreeze and another that is 50% antifreeze. How much of each should she use?

14. Lon Triah invested \$5000 at a local bank. Part was invested at 6% simple interest and the remainder at 10% simple interest. At the end of one year, Lon had earned \$428 interest. How much was invested at each rate?

Solve and graph.

15. $5 - 6x < 2x + 21$

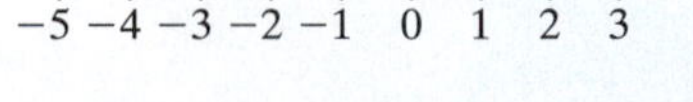

MC **16.** $-\frac{1}{2} + \frac{1}{3}(2 - 3x) \geq \frac{1}{2}x + \frac{5}{3}$

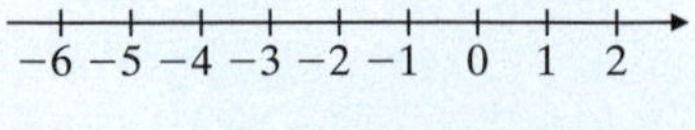

Find the values of x that satisfy the given conditions.

17. $-11 < 2x - 1 \leq -3$

18. $x - 4 \leq -6$ *or* $2x + 1 \geq 3$

Solve each absolute value inequality.

19. $|7x - 3| \leq 18$

MC **20.** $|3x + 1| > 7$

1. ______ ☐
2. ______ ☐
3. ______ ☐
4. ______ ☐
5. ______ ☐
6. ______ ☐
7. ______ ☐
8. ______ ☐
9. ______ ☐
10. ______ ☐
11. ______ ☐
12. ______ ☐
13. ______ ☐
14. ______ ☐
15. ______ ☐
16. ______ ☐
17. ______ ☐
18. ______ ☐
19. ______ ☐
20. ______ ☐

Total Correct: ☐

MATH COACH

Mastering the skills you need to do well on the test.

Students often make the same types of errors when they do the Chapter 2 Test. Here are some helpful hints to keep you from making those common errors on test problems.

Solving Literal Equations for a Specified Variable—Problem 5

Solve for n. $L = a + d(n - 1)$

Helpful Hint Use the distributive property to remove all parentheses before continuing with the other steps of the problem.

Did you obtain the equation $L = a + dn - d$ after removing the parentheses?

Yes ____ No ____

If you answered No, stop and carefully multiply each term inside the parentheses by the variable d.

You want to get the dn term on one side of the equation and all other terms on the other side.

If you added $-a + d$ to both sides, did you get $L - a + d = dn$?

Yes ____ No ____

If you answered No, go back and carefully perform the step of adding $-a + d$ to each side of the equation after the parentheses have been removed. Remember that in the final step, you must divide both sides of the equation by the variable d.

If you answered Problem 5 incorrectly, go back and rework the problem using these suggestions.

Solving Absolute Value Equations—Problem 10

Solve for x. $\left|\frac{1}{2}x + 3\right| - 2 = 4$

Helpful Hint Remember that you must add a number to each side of the equation to isolate the absolute value expression.

Did you add 2 to each side of the equation to obtain $\left|\frac{1}{2}x + 3\right| = 6$?

Yes ____ No ____

If you answered No, stop and complete that step before you do any other operations.

In your next step, did you write the two equations $\frac{1}{2}x + 3 = 6$ and $\frac{1}{2}x + 3 = -6$ and solve for x in each equation?

Yes ____ No ____

If you answered No, carefully perform those steps. Your answer should provide two potential solutions.

Now go back and rework the problem using these suggestions.

Need help? Watch the MATH COACH videos in MyMathLab® or on YouTube™.

Solving Linear Inequalities in One Varaible—Problem 16

Solve and graph. $-\frac{1}{2} + \frac{1}{3}(2 - 3x) \geq \frac{1}{2}x + \frac{5}{3}$

Helpful Hint Remove parentheses in linear inequalities as the first step. If there are fractions, multiply both sides of the inequality by the LCD to clear the fractions.

Did you first remove the parentheses to obtain the inequality $-\frac{1}{2} + \frac{2}{3} - x \geq \frac{1}{2}x + \frac{5}{3}$?

Yes ______ No ______

If you answered No, go back and carefully multiply the terms inside the parentheses by $\frac{1}{3}$.

Did you then multiply each term of the inequality by the LCD 6 to obtain the inequality $-3 + 4 - 6x \geq 3x + 10$?

Yes ______ No ______

If you answered No, consider why the LCD is 6 and carefully multiply all the terms of the equation by 6.

The next goal is to combine like terms on each side of the equation and get all the x terms on one side and all the numerical terms on the other side. If you divide by a negative number, you must reverse the inequality.

If you answered Problem 16 incorrectly, go back and rework the problem using these suggestions.

Solving Absolute Value Inequalities—Problem 20

Solve. $|3x + 1| > 7$

Helpful Hint If c is a positive real number and $|ax + b| > c$, then $ax + b > c$ or $ax + b < -c$. Use this first to create the two separate inequalities.

Did you apply the rule in the Helpful Hint to obtain the inequalities $3x + 1 > 7$ or $3x + 1 < -7$?

Yes ______ No ______

If you answered No, stop and perform those steps.

Did you add the correct number to each inequality to obtain the inequalities $3x > 6$ or $3x > -8$?

Yes ______ No ______

If you answered No, consider how to solve each inequality and perform each step carefully again.

In your final step, you will need to divide each inequality by 3 to solve for x.

Now go back and rework the problem using these suggestions.

Need more help? Look for section examples marked with MC to review.

CHAPTER 3

Successfully growing fruit involves more than just planting trees and waiting for the fruit to ripen for harvest. In addition to concerns over weather, pests, and disease, yield per acre is a variable that the farmer needs to consider. In this chapter you will develop your knowledge of functions to investigate some of these issues that might be used in managing an apple orchard.

Equations and Inequalities in Two Variables and Functions

3.1 Graphing Linear Equations with Two Unknowns

Student Learning Objectives

After studying this section, you will be able to:

1. **Graph a linear equation in two variables.**
2. **Use *x*- and *y*-intercepts to graph a linear equation.**
3. **Graph horizontal and vertical lines.**
4. **Graph a linear equation using different scales.**

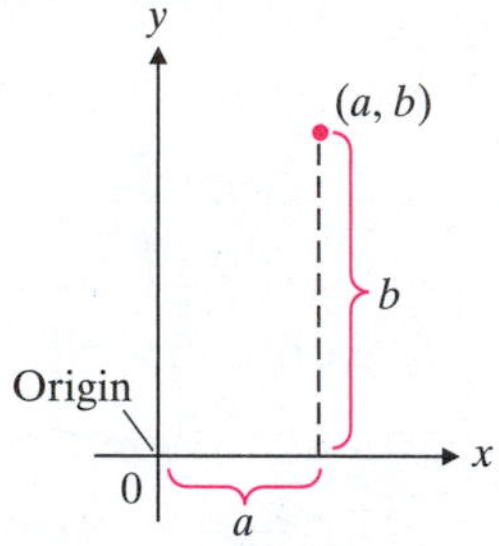

Graphs are often used to show the relationships among sets of data. You may be familiar with graphs that are used in applied mathematics, science, and business.

The following is a graph that could be found in a local newspaper. It shows the daily low temperature for the first 30 days of January 2011 and compares this to the normal (average) low temperature, as well as to the record low temperature for that month in that region of the country.

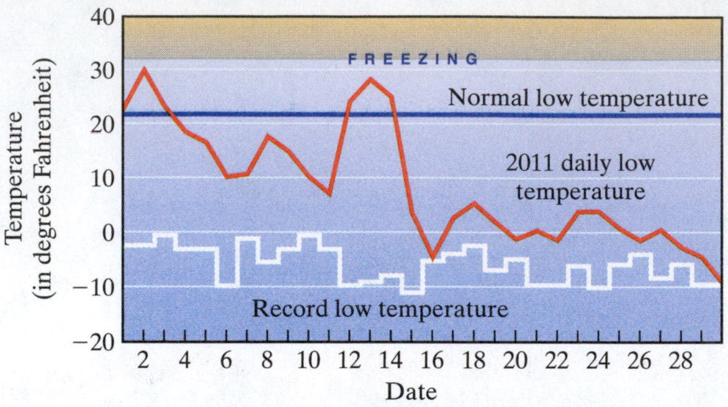

In mathematics we can also use graphs to show the relationships among the variables in an equation. To do so, we will use two number lines. For convenience we construct two real number lines—one horizontal and one vertical—that intersect to form a **rectangular coordinate system.** The horizontal line is the ***x*-axis.** The vertical line is the ***y*-axis.** They intersect at the **origin.**

An **ordered pair** of real numbers (a, b) represents a point on the rectangular coordinate system. See the graph in the left margin. To graph an ordered pair, we begin at the origin. The first coordinate is the x-coordinate, and the second coordinate is the y-coordinate. To locate the point described by the ordered pair (a, b), where $a, b > 0$, we move a units to the right of the origin along the x-axis. Then we move b units up parallel to the y-axis. Since (a, b) is an ordered pair, order is important. That is, if $a \neq b$, then $(a, b) \neq (b, a)$.

The next figure shows the graphs of (3, 4) and (4, 3).

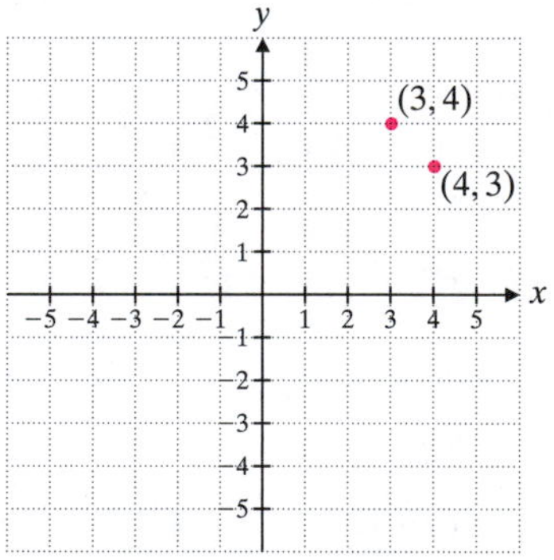

Graphing Calculator

Graphing Ordered Pairs

You can use a graphing calculator to graph ordered pairs. Most graphing calculators can plot statistical data. Use your calculator's statistical plot feature to plot the points $(3, 2)$; $(0, -4)$; $(-2, -1)$; $(-5, -4)$; and $(-3, 4)$. The display should be similar to the one below. Be sure to use an appropriate window.

Display:

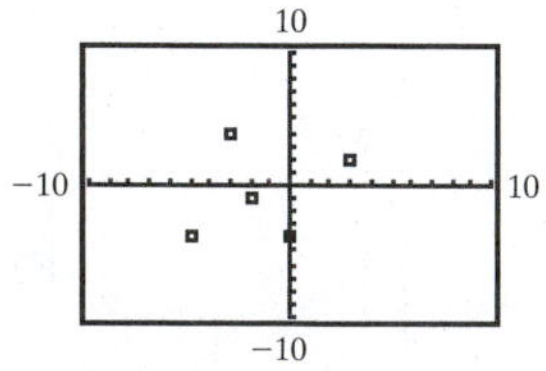

1 Graphing a Linear Equation in Two Variables

We now define a *linear equation in two variables.*

A **linear equation in two variables** is an equation that can be written in the form $Ax + By = C$, where A, B, and C are real numbers and A and B are not both zero. This form is called the **standard form** of a linear equation in two variables.

A **solution** of an equation in two variables is an ordered pair of real numbers that *satisfies* the equation. In other words, when we substitute the values of the coordinates into the equation, we get a true statement. For example, (6, 4) is a solution to $4x - 3y = 12$. The ordered pair (6, 4) means that $x = 6$ and $y = 4$.

$$4x - 3y = 12$$
$$4(6) - 3(4) \stackrel{?}{=} 12$$
$$24 - 12 \stackrel{?}{=} 12$$
$$12 = 12 \quad \checkmark$$

To graph the equation, we could graph all its solutions. However, this would be impossible, since there is an infinite number of solutions. It is a mathematical property that the graph of an equation of the form $Ax + By = C$, where A, B, and C are constants (A, B not both zero), is a straight line. Hence, to graph the equation we graph three ordered pair solutions and connect them with a straight line. (The third ordered pair solution is used to check the line.)

EXAMPLE 1 Graph the equation. $y = -3x + 2$

Solution We choose three values of x and then substitute them into the equation to find the corresponding values of y. Let's choose $x = -1$, $x = 1$, and $x = 2$.

For $x = -1$, $y = -3(-1) + 2 = 5$, so the first point, or solution, is $(-1, 5)$.

For $x = 1$, $y = -3(1) + 2 = -1$, and for $x = 2$, $y = -3(2) + 2 = -4$.

We can condense this information by using a table.

x	y
−1	5
1	−1
2	−4

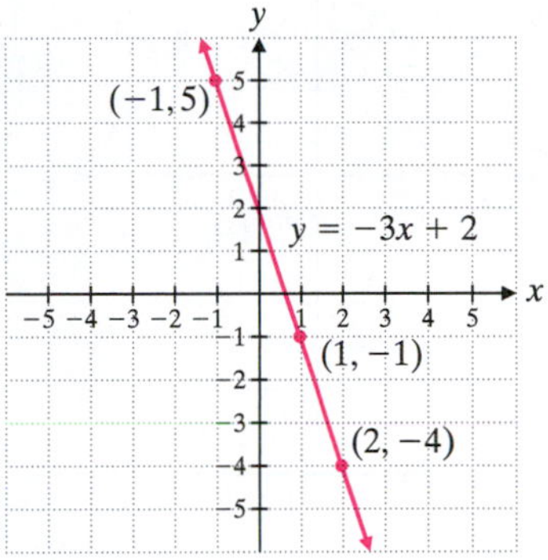

Student Practice 1 Graph the equation. $y = -4x + 2$

Student Practice 1

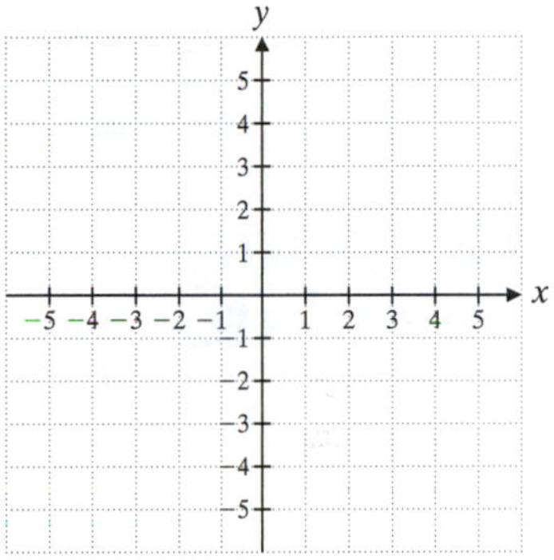

② Using *x*- and *y*-intercepts to Graph a Linear Equation

We can usually graph a straight line by using the x- and y-intercepts. A straight line that is not vertical or horizontal has these two intercepts.

The **x-intercept** of a line is the point where the line crosses the x-axis (that is, where $y = 0$). It is described by an ordered pair of the form $(a, 0)$.

The **y-intercept** of a line is the point where the line crosses the y-axis (that is, where $x = 0$). It is described by an ordered pair of the form $(0, b)$.

NOTE TO STUDENT: Fully worked-out solutions to all of the Student Practice problems can be found at the back of the text starting at page SP-1.

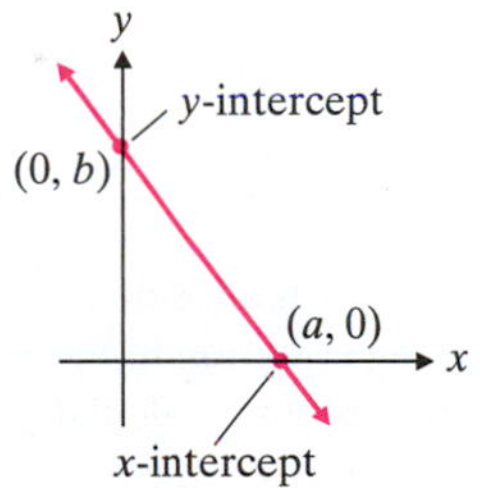

EXAMPLE 2 Find the x-intercept, the y-intercept, and one additional ordered pair that satisfies the equation. Then graph the equation $4x - 3y = -12$.

Solution

Find the x-intercept by using $y = 0$.

$$4x - 3(0) = -12$$
$$4x = -12$$
$$x = -3$$

The x-intercept is $(-3, 0)$.

Find the y-intercept by using $x = 0$.

$$4(0) - 3y = -12$$
$$-3y = -12$$
$$y = 4$$

The y-intercept is $(0, 4)$.

We can now pick any value of x or y to find our third point. Let's pick $y = 2$.

$$4x - 3(2) = -12$$
$$4x - 6 = -12$$
$$4x = -12 + 6$$

Student Practice 2 Find the x-intercept, the y-intercept, and one additional ordered pair that satisfies the equation. Then graph the equation $3x - 2y = -6$.

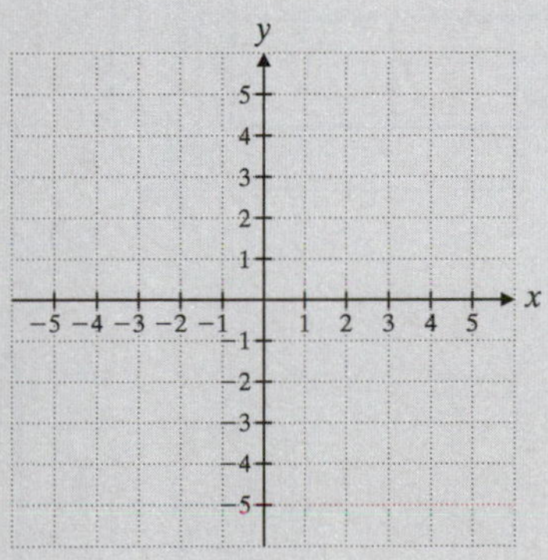

Continued on next page

$$4x = -6$$

$$x = -\frac{6}{4} = -\frac{3}{2}$$

Hence, the third point on the line is $\left(-\frac{3}{2}, 2\right)$. A table and the graph of the equation is shown below.

x	y
-3	0
0	4
$-\frac{3}{2}$	2

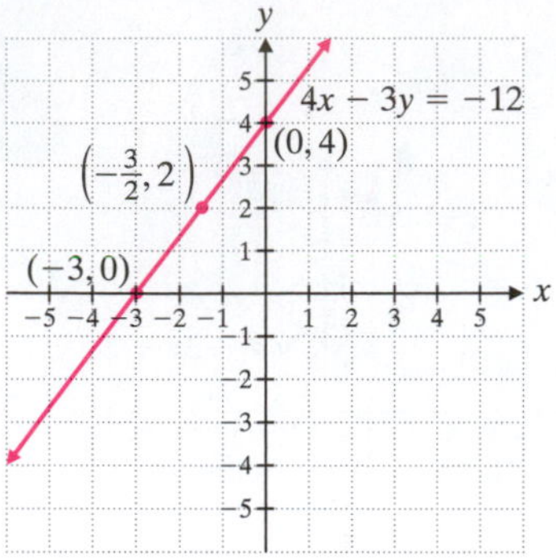

Graphing Calculator

Graphing a Line

You can graph a line given in the form $y = mx + b$ using a graphing calculator. For example, to graph the equation in Example 2, first rewrite the equation by solving for y.

$$4x - 3y = -12$$
$$-3y = -4x - 12$$
$$y = \frac{4}{3}x + 4$$

Enter the right-hand side of the resulting equation in the Y = editor of your calculator and graph. Choose an appropriate window to show all the intercepts. The following window is $[-10, 10]$ by $[-10, 10]$.

Display:

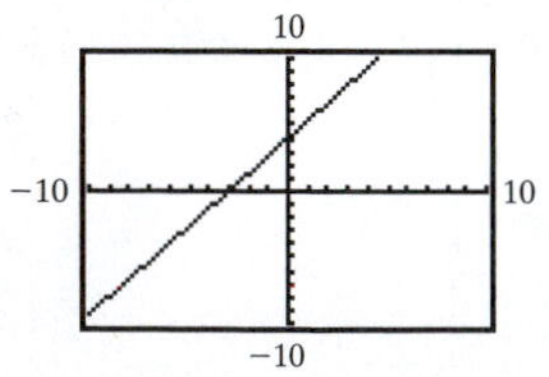

③ Graphing Horizontal and Vertical Lines

Let's look at the standard form of a linear equation, $Ax + By = C$, when $B = 0$.

$$Ax + (0)y = C$$
$$Ax = C$$
$$x = \frac{C}{A}$$

Notice that when we solve for x we get $x = \frac{C}{A}$, which is a constant. For convenience we will rename it a. The equation then becomes

$$x = a.$$

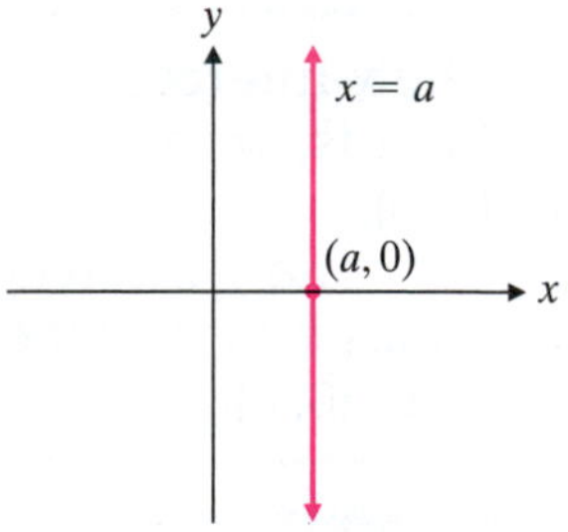

What does this mean? The equation $x = a$ means that for any value of y, x is a. The graph is a vertical line. What happens to $Ax + By = C$ when $A = 0$?

$$(0)x + By = C$$
$$By = C$$
$$y = \frac{C}{B}$$

We will rename the constant $\frac{C}{B}$ as b. The equation then becomes

$$y = b.$$

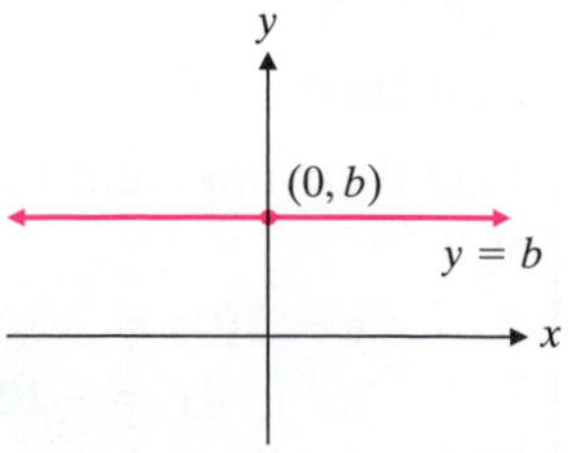

What does this mean? The equation $y = b$ means that for any value of x, y is b. The graph is a horizontal line.

The graph of the equation $x = a$, where a is any real number, is a **vertical line** through the point $(a, 0)$.

The graph of the equation $y = b$, where b is any real number, is a **horizontal line** through the point $(0, b)$.

EXAMPLE 3 Simplify and graph each equation.

(a) $x = -3$ **(b)** $2y - 4 = 0$

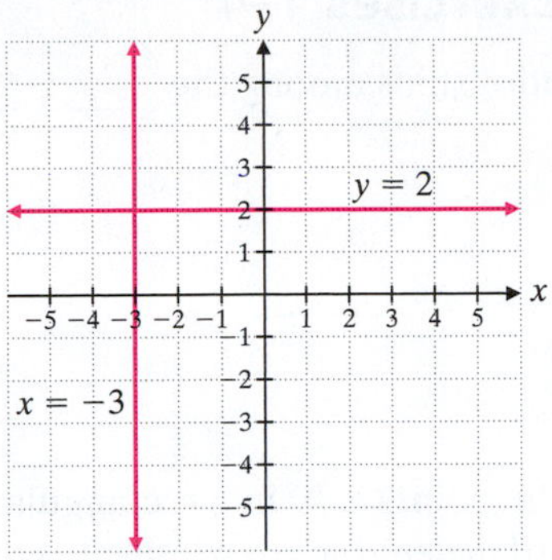

Solution

(a) The equation $x = -3$ means that for any value of y, x is -3. The graph of $x = -3$ is a vertical line 3 units to the left of the origin.

(b) The equation $2y - 4 = 0$ can be simplified.

$$2y - 4 = 0$$
$$2y = 4$$
$$y = 2$$

The equation $y = 2$ means that, for any value of x, y is 2. The graph of $y = 2$ is a horizontal line 2 units above the x-axis.

Student Practice 3 Simplify and graph each equation.

(a) $x = 4$ **(b)** $3y + 12 = 0$

Student Practice 3

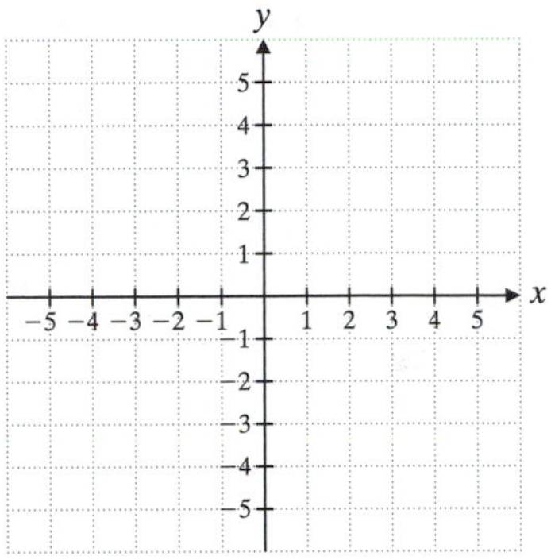

④ Graphing a Linear Equation Using Different Scales for the Axes

By common convention, each tick mark on a graph's axis indicates 1 unit, so we don't need to use a marked scale on each axis. But sometimes a different scale is more appropriate. This new scale must then be clearly labeled on each axis.

EXAMPLE 4 A company's finance officer has determined that the monthly cost in dollars for leasing a photocopier is $C = 100 + 0.002n$, where n is the number of copies produced in a month in excess of a specified number. Graph the equation using $n = 0$, $n = 30{,}000$, and $n = 60{,}000$. Let the n-axis be the horizontal axis.

Solution For each value of n we obtain C.

When $n = 0$,
then $C = 100 + 0.002(0) = 100 + 0 = 100$.
When $n = 30{,}000$,
then $C = 100 + 0.002(30{,}000)$
$= 100 + 60 = 160$.
When $n = 60{,}000$,
then $C = 100 + 0.002(60{,}000)$
$= 100 + 120 = 220$.

The table of values is shown next.

n	C
0	100
30,000	160
60,000	220

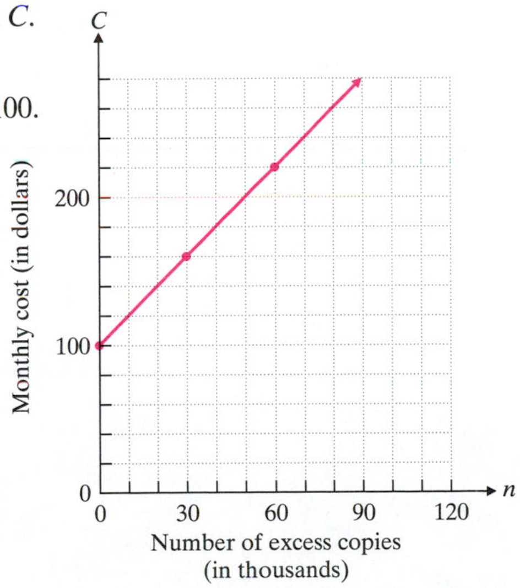

Since n varies from 0 to 60,000 and C varies from 100 to 220, we need different scales on the axes. We let each tick on the horizontal scale represent 10,000 excess copies and each tick on the vertical scale represent \$20.

Student Practice 4 The cost of a product in dollars is given by $C = 300 + 0.15n$, where n is the number of products produced. Graph the equation using an appropriate scale. Use $n = 0$, $n = 1000$, and $n = 2000$.

Student Practice 4

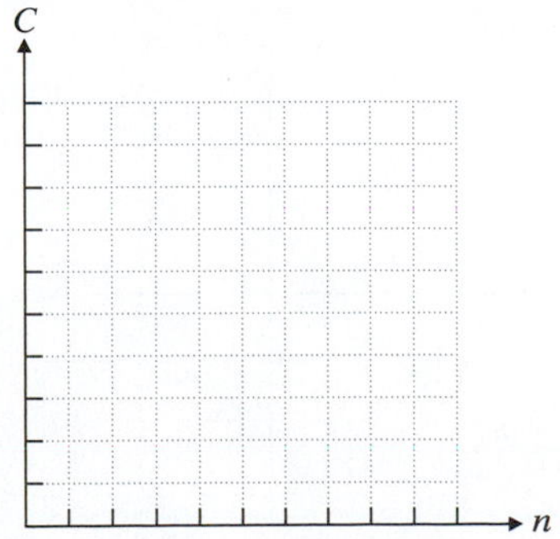

3.1 Exercises

MyMathLab®   Watch the videos in MyMathLab   Download the MyDashBoard App

Verbal and Writing Skills, Exercises 1–4

1. Graphs are used to show the relationships among the __________ in an equation.

2. The x-axis and the y-axis intersect at the __________.

3. Explain in your own words why the point (a, b) in a rectangular coordinate system is an *ordered* pair. In other words, what is the importance of the word *ordered* when we say it is an ordered pair? Give an example.

4. $(5, 1)$ is a solution to the equation $2x - 3y = 7$. What does this mean?

Find the missing coordinate.

5. $(-2,$ ______ $)$ is a solution of $y = 3x - 7$.

6. $(-3,$ ______ $)$ is a solution of $y = 4 - 3x$.

7. $\left(\_\_\_\_\_\_, \frac{1}{4}\right)$ is a solution of $5x + 12y = -17$.

8. $\left(\_\_\_\_\_\_, \frac{1}{2}\right)$ is a solution of $-x + 2y = -7$.

Graph each equation.

9. $y = 2x - 3$

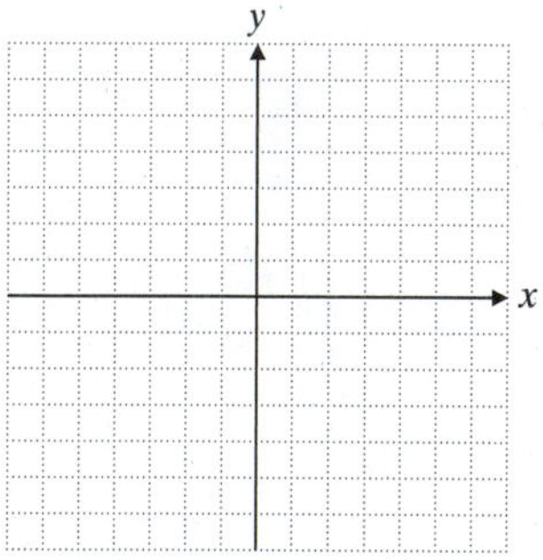

10. $y = 3x + 2$

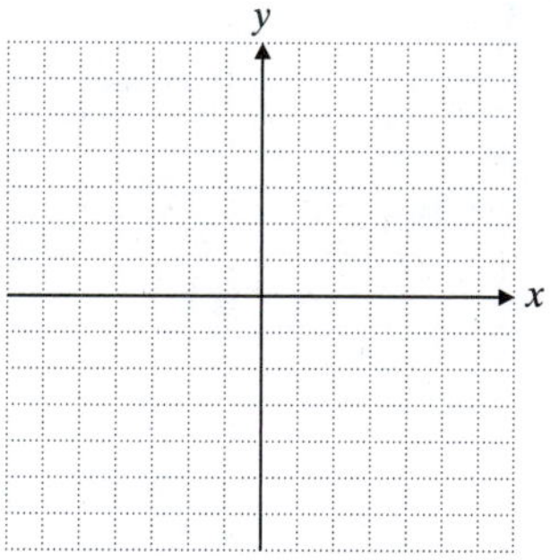

11. $y = 4 - 2x$

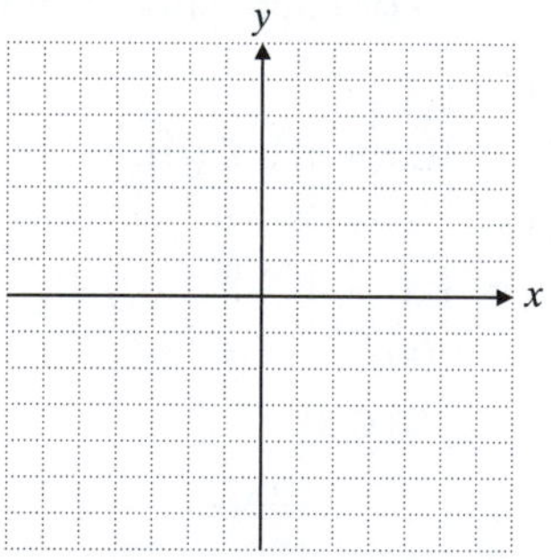

12. $y = -5x - 2$

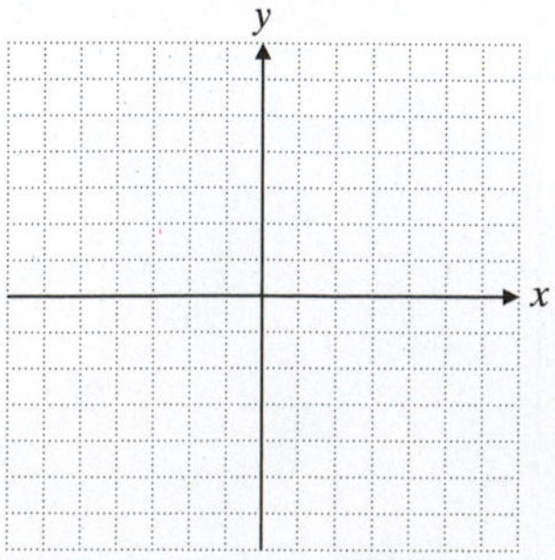

13. $y = \frac{2}{3}x - 4$

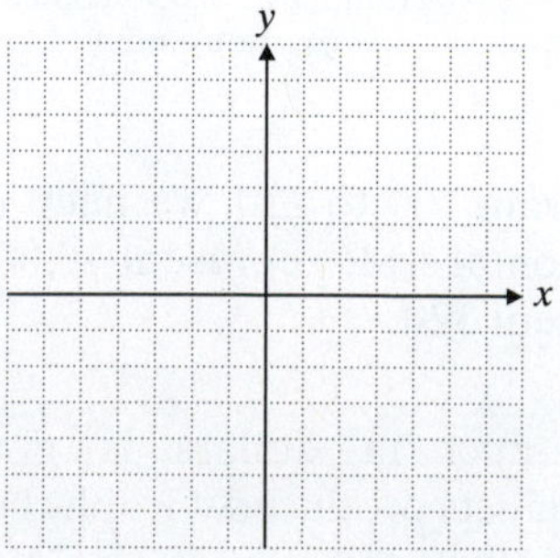

14. $y = \frac{5}{2}x + 1$

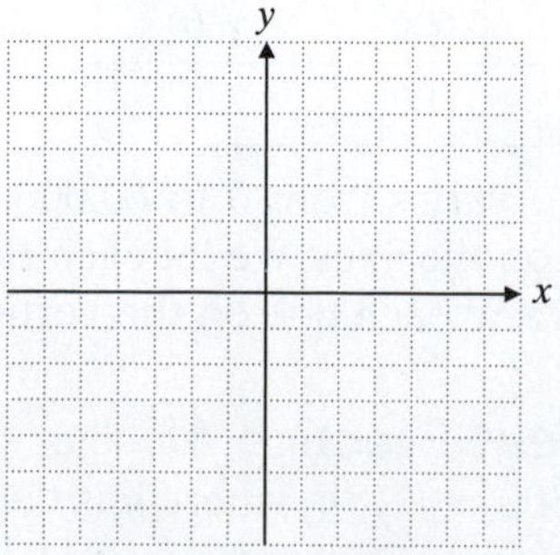

Simplify the equation if possible. Find the x-intercept, the y-intercept, and one or two additional ordered pairs that are solutions to the equation. Then graph the equation.

15. $2y - 3x = 6$

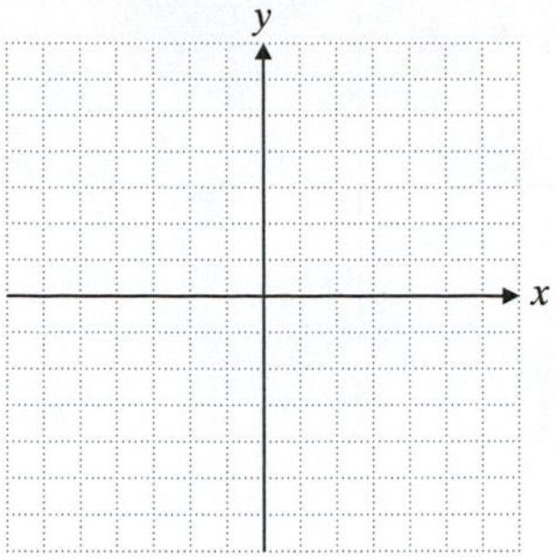

16. $2y + 5x = 10$

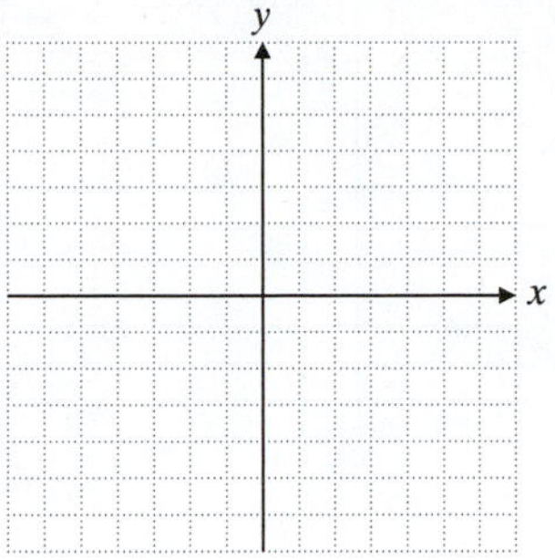

17. $2x - y = 6$

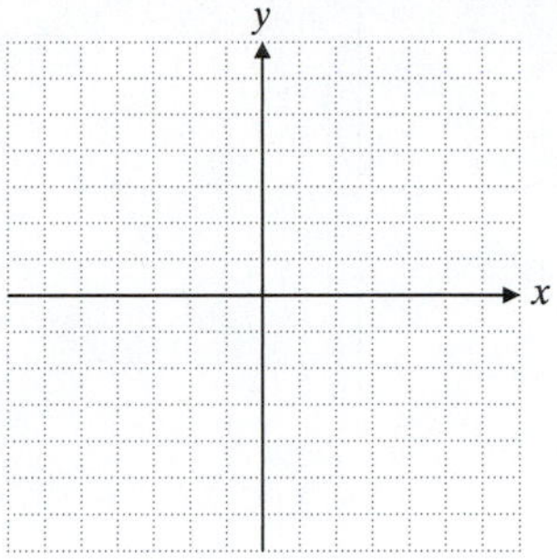

18. $4x - y = -4$

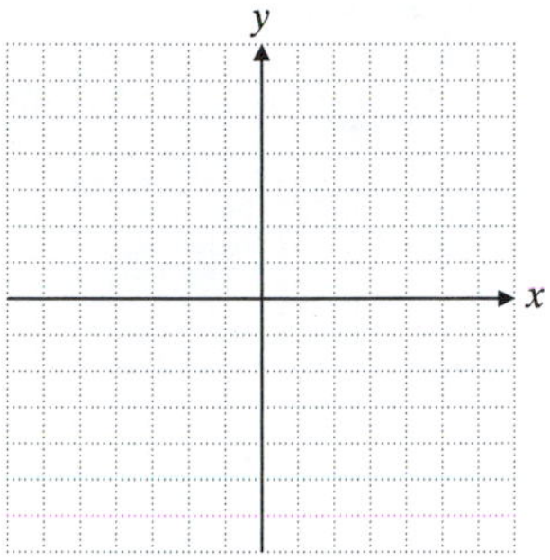

19. $-4x - 3y = 6$

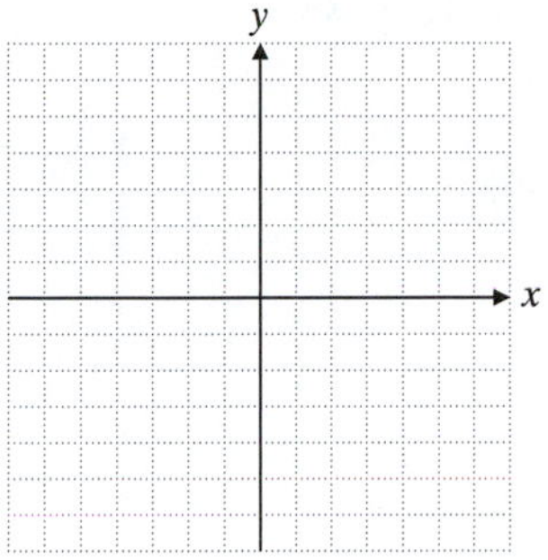

20. $5x - 2y = -12$

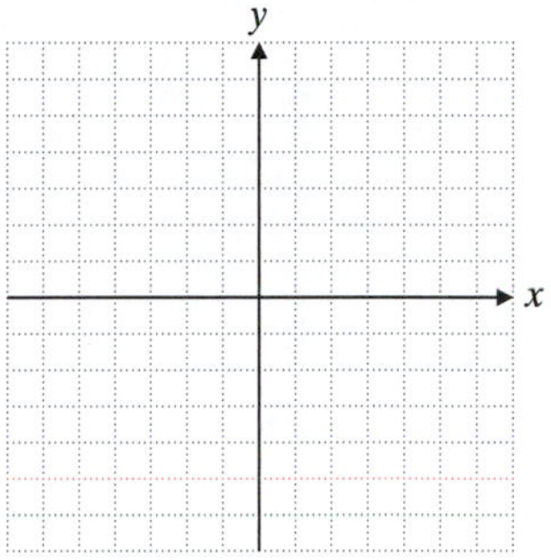

21. $5y - 4 = 3x - 4$

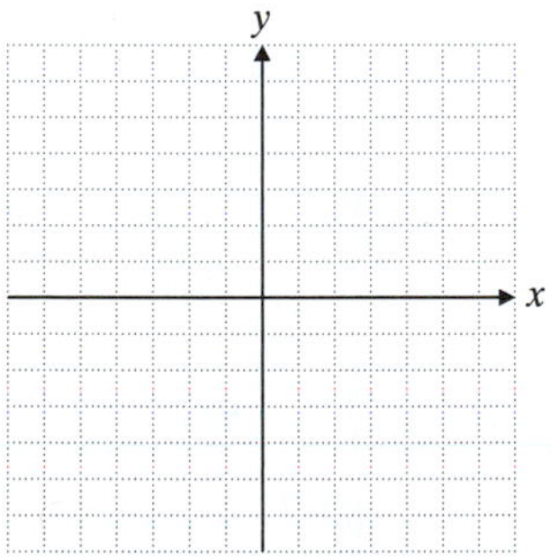

22. $4x + 6y + 2 = 2$

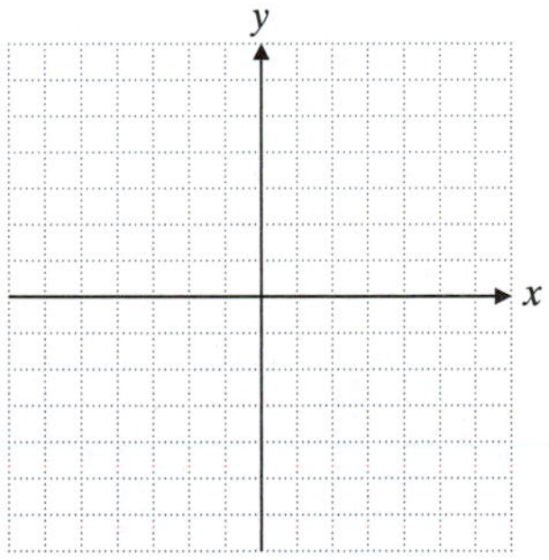

Simplify the equation if possible. State whether the equation represents a horizontal or a vertical line. Then graph the equation.

23. $x = -5$

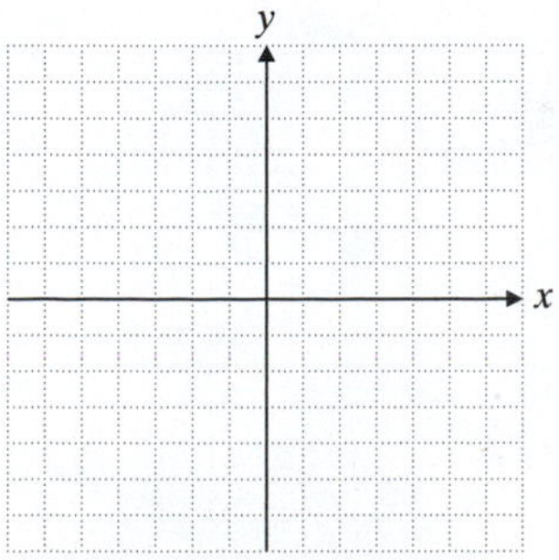

24. $x = 2$

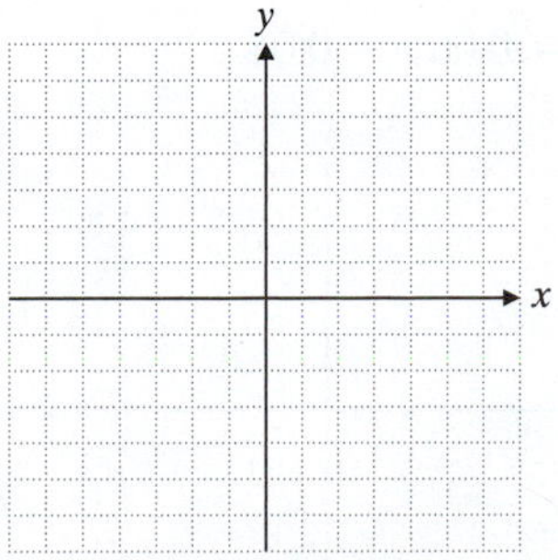

25. $4x - 16 = 0$

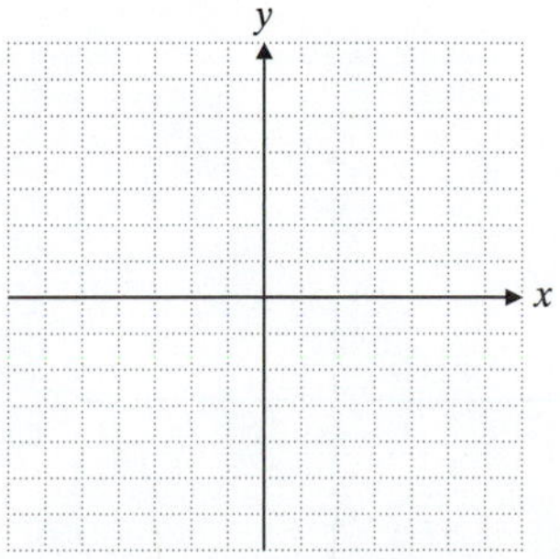

26. $2x - 3 = 3x$

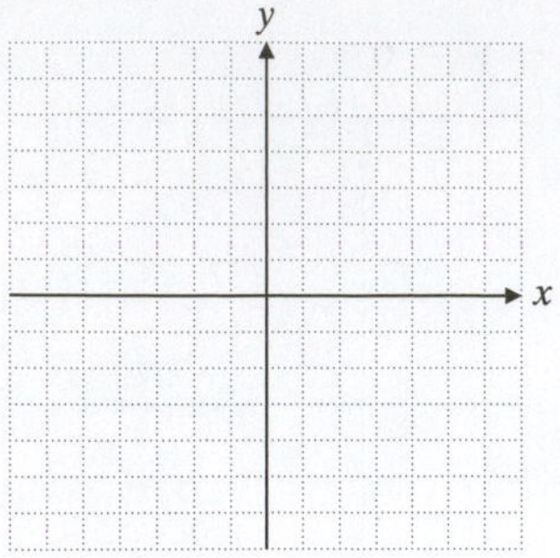

27. $2y + 8 = 0$

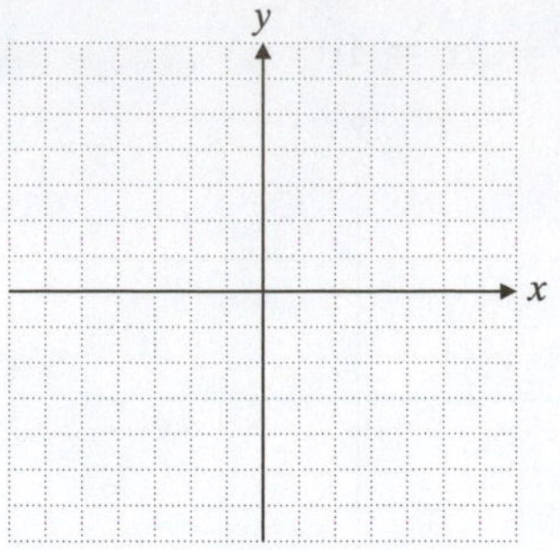

28. $5y + 6 = 2y$

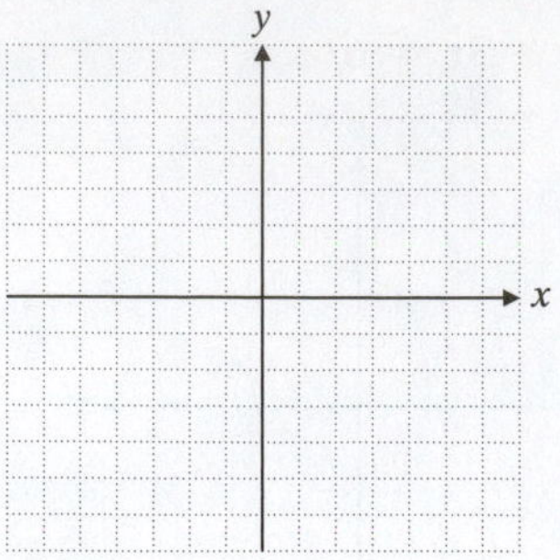

Mixed Practice *Simplify each equation if possible. Then graph the equation by any appropriate method.*

29. $y = -1.5x + 2$

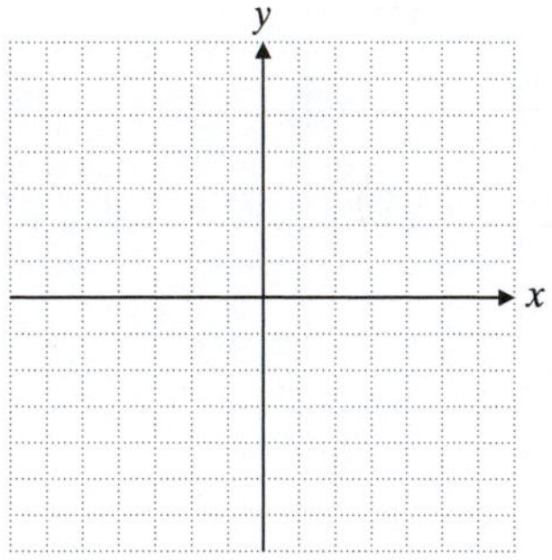

30. $y = 0.5x + 4$

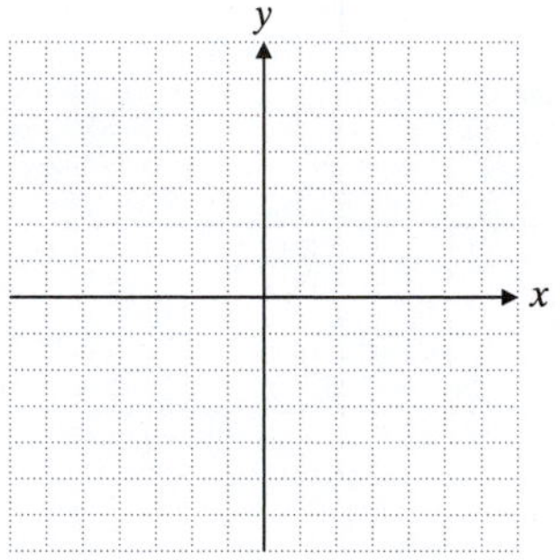

31. $2x + 5y = -5$

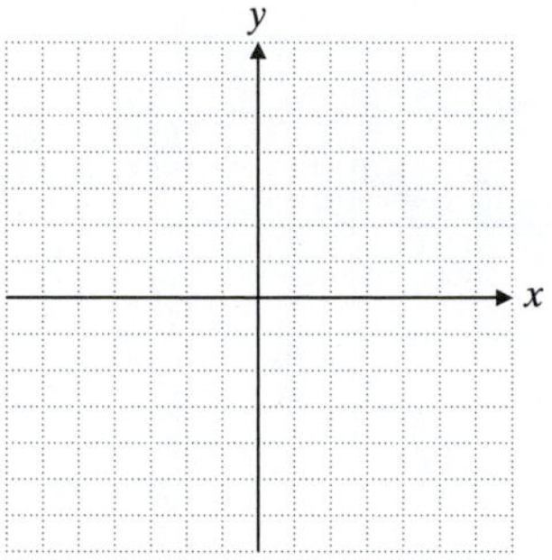

32. $4x - 3y = 6$

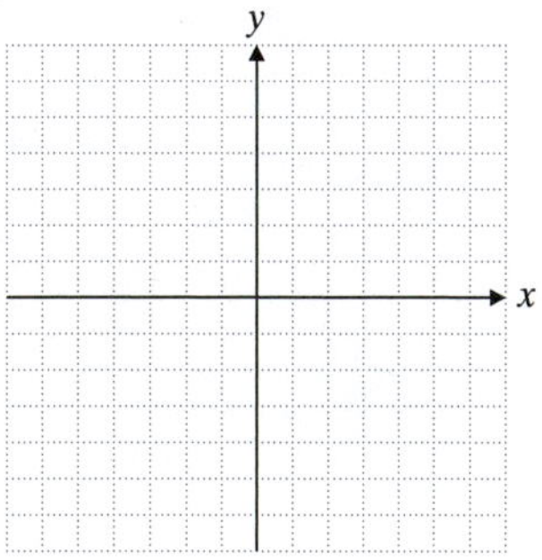

33. $5x + y + 4 = 8x$

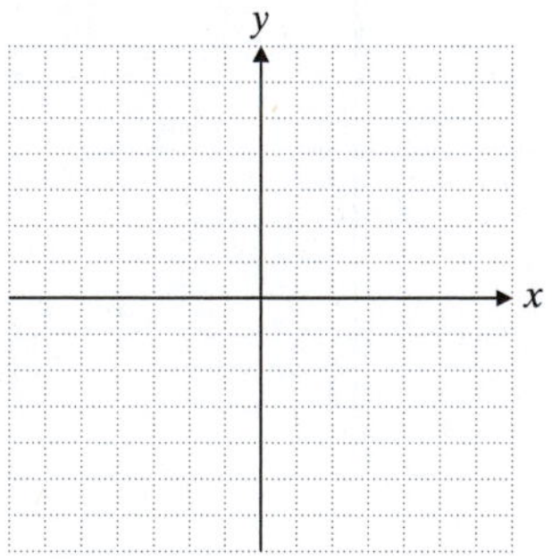

34. $5x - 4y - 4 = 4x$

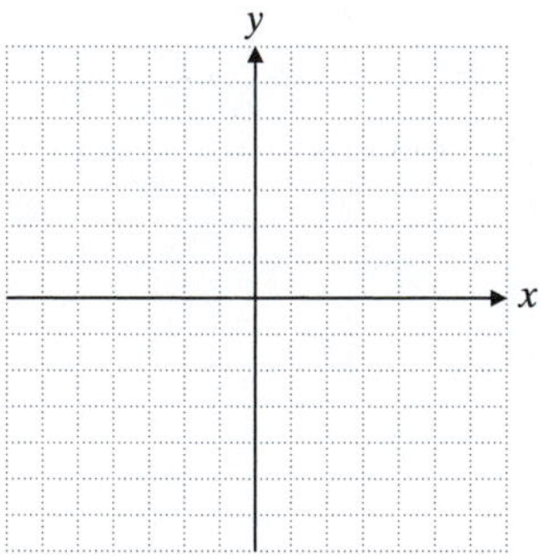

Graph each equation. Use appropriate scales on each axis.

35. $y = 82x + 150$

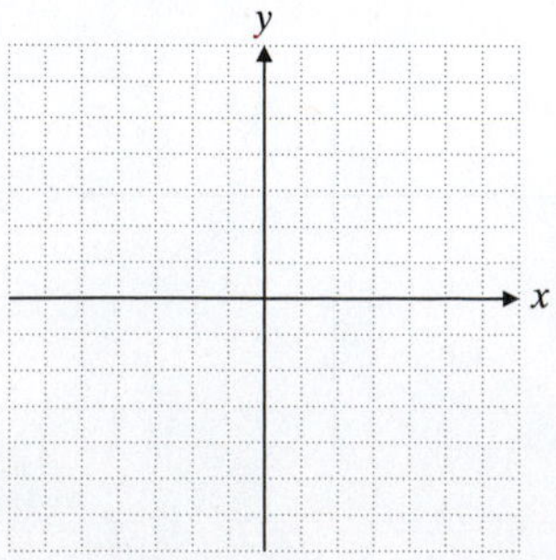

36. $y = 0.06x - 0.04$

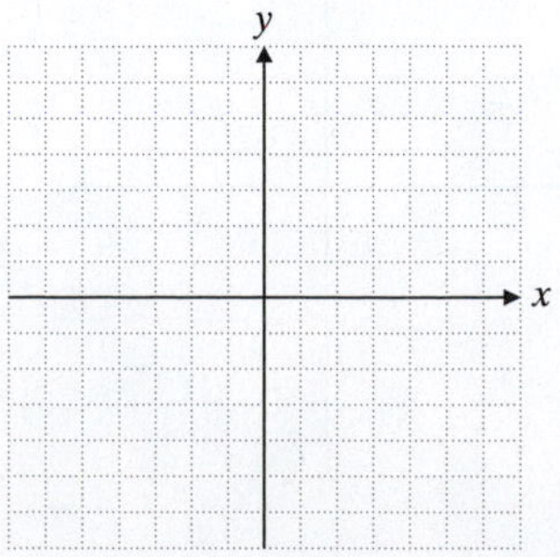

To Think About

Income of Men Versus Women *The following graph shows the median weekly earnings of men and women in the United States during the period 1999 to 2009. Use the graph to answer exercises 37–42.*

Median Weekly Earnings of Full-Time Workers in the U.S.

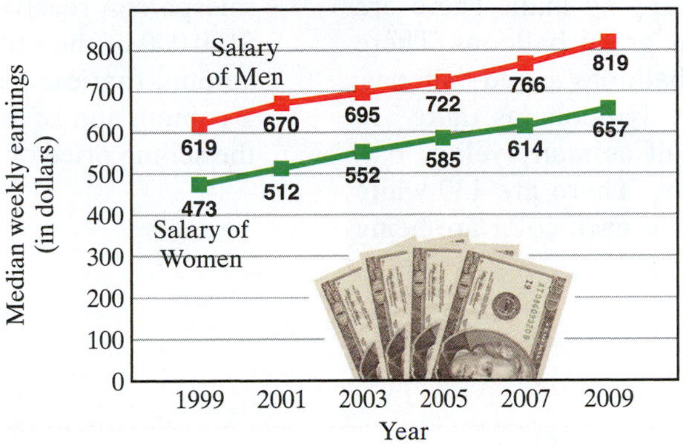

Source: www.bls.gov

37. During what 2-year period did the greatest increase in the median weekly earnings of men occur?

38. During what 2-year period did the greatest increase in the median weekly earnings of women occur?

39. In what year did the median weekly earnings of men and women have the largest difference?

40. In what year did the median weekly earnings of men and women have the smallest difference?

41. What was the percent of increase in earnings for women during the period 1999 to 2009? Round your answer to the nearest tenth of a percent.

42. What was the percent of increase in earnings for men during the period 1999 to 2009? Round your answer to the nearest tenth of a percent.

Applications

43. ***Baseball*** If a baseball is thrown vertically upward by Paul Frydrych when he is standing on the ground, the velocity of the baseball V (in feet per second) after T seconds is $V = 120 - 32T$.

(a) Find V for $T = 0, 1, 2, 3,$ and 4.

(b) Graph the equation, using T as the horizontal axis.

(c) What is the significance of the negative value of V when $T = 4$?

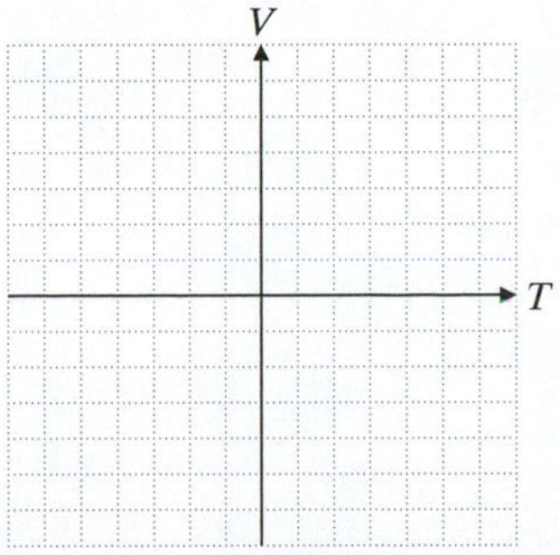

44. ***Gasoline Storage Tank*** A full storage tank on the Robinson family farm contains 900 gallons of gasoline. Gasoline is then pumped from the tank at a rate of 15 gallons per minute. The equation $G = 900 - 15m$ describes the number of gallons of gasoline G in the tank m minutes after the pumping began.

(a) Find G for $m = 0, 10, 20, 30,$ and 60.

(b) Graph the equation, using m as the horizontal axis.

(c) What happens when $m = 61$?

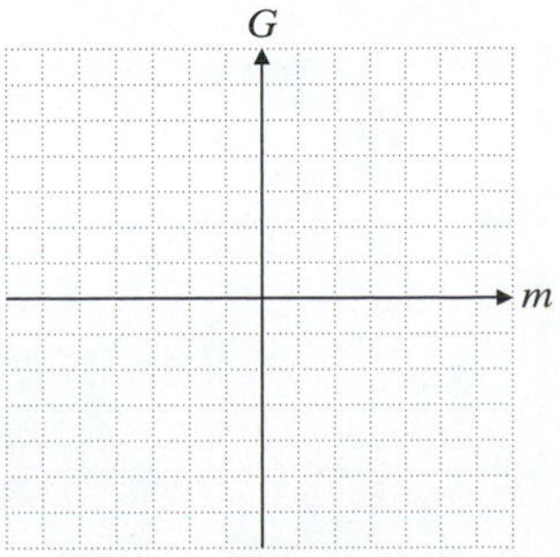

Cumulative Review

45. **[1.3.3]** Evaluate. $36 \div (8 - 6)^2 + 3(-4)$

46. **[2.6.3]** Solve for x. $3(x - 6) + 2 \leq 4(x + 2) - 21$

47. **[2.4.1.]** ***Balloon Giveaway*** A novelty company is giving away balloons in a shopping mall. There are twice as many red balloons as green balloons. There are three times as many blue balloons as red balloons. There are half as many white balloons as there are yellow balloons. There are half as many yellow balloons as there are red balloons. There are 130 white balloons. How many balloons of each color are being given away?

48. **[2.5.1]** ***Commission Sales*** At Greenland Realty, a salesperson receives a commission of 7% on the first \$100,000 of the selling price of a house and 3% on the amount that exceeds \$100,000. Ray Peterson received a commission of \$9100 for selling a house. What was the selling price of the house?

Quick Quiz 3.1

Graph each of the following. Plot at least three points.

1. $y = -\frac{2}{3}x + 4$

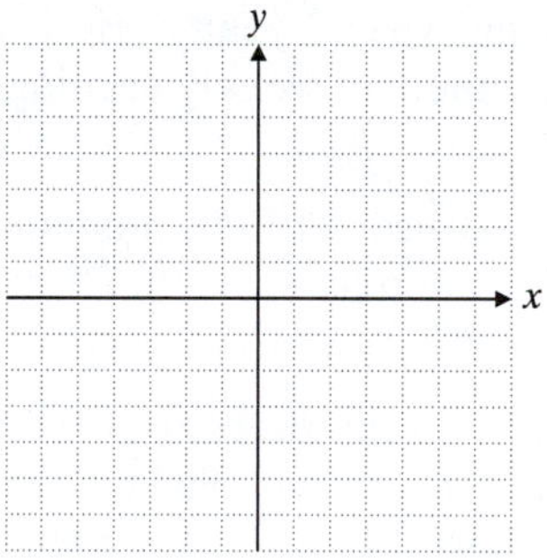

2. $7y - 5 = 16$

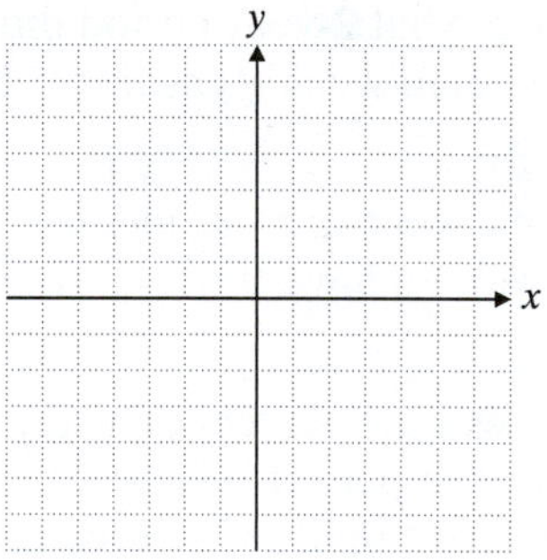

3. $-4x + 2y = -12$

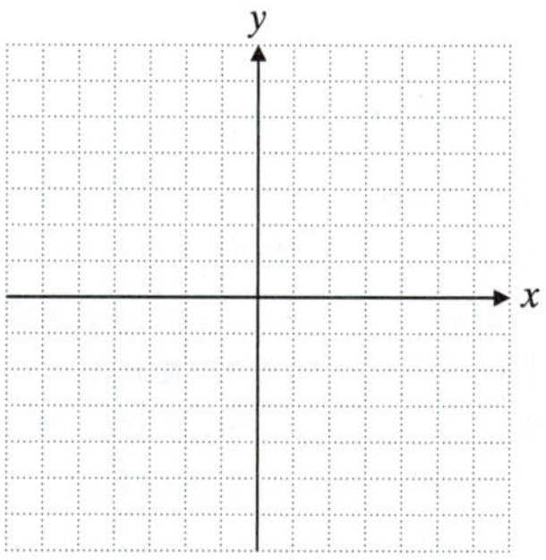

4. **Concept Check** Explain how you would find the x-intercept and the y-intercept for the following equation.

$$7x + 3y = -14$$

3.2 Slope of a Line

① Finding the Slope If Two Points Are Known

The concept of slope is one of the most useful in mathematics and has many practical applications. For example, a carpenter needs to determine the slope (or pitch) of a roof. (You may have heard someone say that a roof has a 5 : 12 pitch.) Road engineers must determine the proper slope (or grade) of a roadbed. If the slope is steep, you feel as if you're driving almost straight up. Simply put, slope is a measure of steepness. That is, slope measures the ratio of the vertical change *(rise)* to the horizontal change *(run)*.

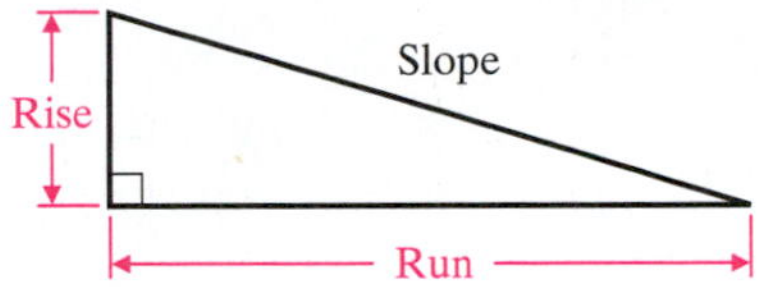

Mathematically, we define slope of a line as follows:

> The **slope of a straight line** containing points (x_1, y_1) and (x_2, y_2) is
> $$\text{slope} = m = \frac{y_2 - y_1}{x_2 - x_1} \qquad x_2 \neq x_1.$$

In the sketch, we see that the rise is $y_2 - y_1$ and the run is $x_2 - x_1$.

Student Learning Objectives

After studying this section, you will be able to:

① **Find the slope of any nonvertical straight line if two points are known.**

② **Determine whether two lines are parallel or perpendicular by comparing their slopes.**

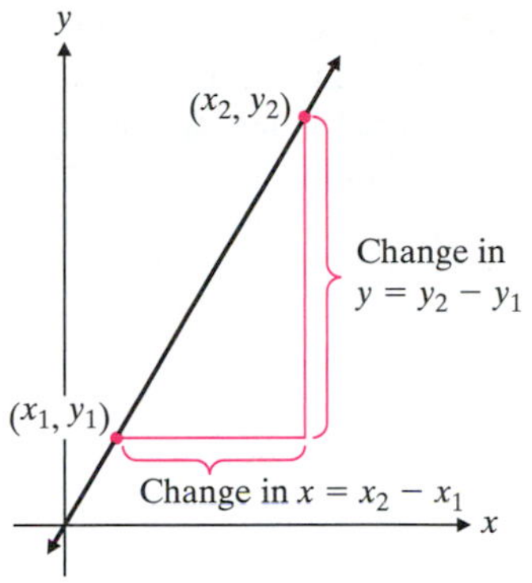

EXAMPLE 1 Find the slope of the line passing through $(-2, -3)$ and $(1, -4)$.

Solution Identify the y-coordinates and the x-coordinates for the points $(-2, -3)$ and $(1, -4)$.

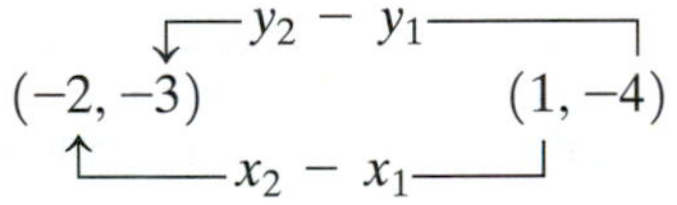

Use the formula.

$$\text{slope} = m = \frac{y_2 - y_1}{x_2 - x_1}$$
$$= \frac{-4 - (-3)}{1 - (-2)}$$
$$= -\frac{1}{3}$$

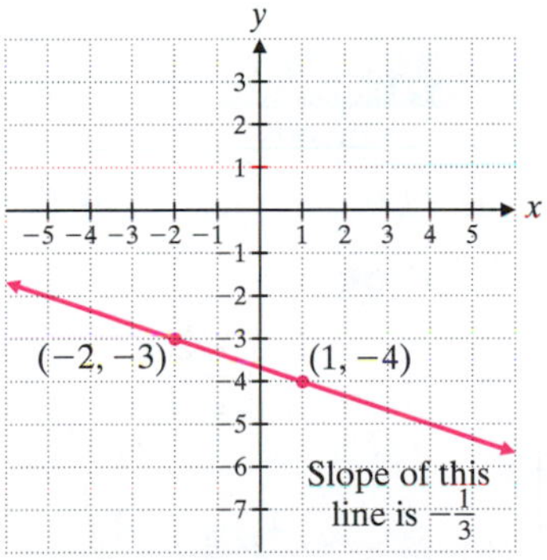

Student Practice 1 Find the slope of the line passing through $(-6, 1)$ and $(-5, -2)$.

***NOTE TO STUDENT:** Fully worked-out solutions to all of the Student Practice problems can be found at the back of the text starting at page SP-1.*

Notice that it does not matter which ordered pair we label (x_1, y_1) and which we label (x_2, y_2) as long as we subtract the x-coordinates in the same order that we subtract the y-coordinates. Let's redo Example 1.

$$\begin{array}{cc} (x_1, & y_1) \\ \downarrow & \downarrow \\ (1, & -4) \end{array} \qquad \begin{array}{cc} (x_2, & y_2) \\ \downarrow & \downarrow \\ (-2, & -3) \end{array} \qquad m = \frac{-3 - (-4)}{-2 - 1} = \frac{-3 + 4}{-3} = -\frac{1}{3}$$

Graphing Calculator

Slopes

Using a graphing calculator, graph

$y_1 = 3x + 1,$

$y_2 = \frac{1}{3}x + 1,$

$y_3 = -3x + 1,$ and

$y_4 = -\frac{1}{3}x + 1$

on the same set of axes. How is the coefficient of x in each equation related to the slope of the line? Will the graph of the line $y = -2x + 3$ slope upward or downward? How do you know? Verify using your calculator.

The Mount Washington Cog Railway in New Hampshire has a train track that in some places rises 37 feet for every 100 feet horizontally. This is a slope of $\frac{37}{100}$.

1. Lines sloping upward to the right have positive slopes.
2. Lines sloping downward to the right have negative slopes.

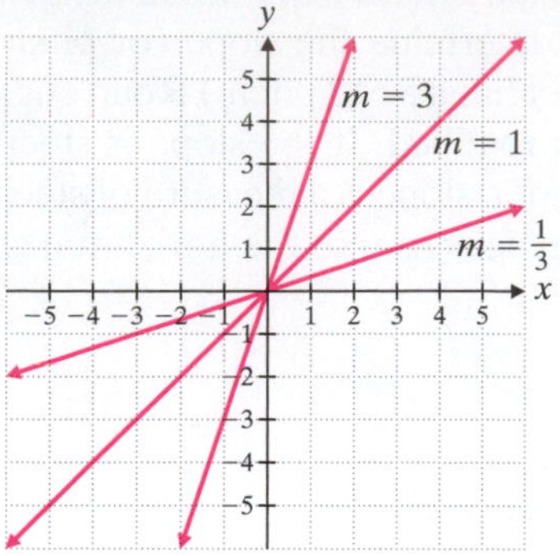

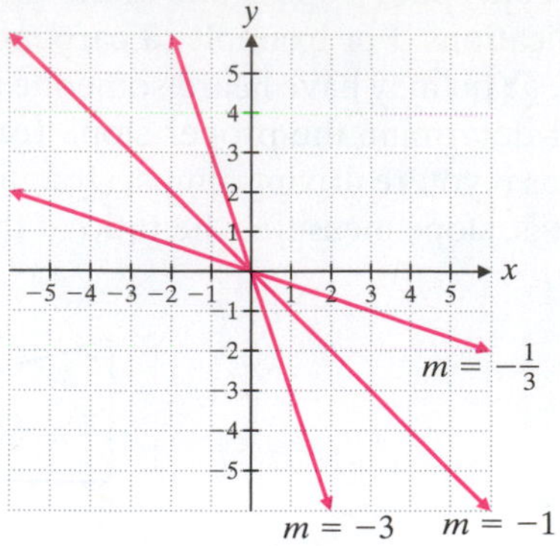

What is the slope of a horizontal line? Let's look at the equation $y = 4$. The graph of $y = 4$ is the horizontal line 4 units above the x-axis. It means that for any x, y is 4. The slope is

$$m = \frac{y_2 - y_1}{x_2 - x_1} = \frac{4 - 4}{x_2 - x_1} = \frac{0}{x_2 - x_1} = 0.$$

In general, for all horizontal lines, $y_2 - y_1 = 0$. Hence, the slope is 0.

What is the slope of a vertical line? The equation of the vertical line 4 units to the right of the y-axis is $x = 4$. It means that for any y, x is 4. The slope is

$$m = \frac{y_2 - y_1}{x_2 - x_1} = \frac{y_2 - y_1}{4 - 4} = \frac{y_2 - y_1}{0}$$

Because division by zero is not defined, we say that a vertical line has no slope or the slope is undefined.

The slope of a horizontal line is 0. The slope of a vertical line is undefined.

EXAMPLE 2 Find the slope, if possible, of the line passing through each pair of points.

(a) (1.6, 2.3) and (−6.4, 1.8) **(b)** $\left(\frac{5}{3}, -\frac{1}{2}\right)$ and $\left(\frac{2}{3}, -\frac{1}{4}\right)$

Solution

(a) $m = \dfrac{1.8 - 2.3}{-6.4 - 1.6} = \dfrac{-0.5}{-8.0} = 0.0625$

(b) $m = \dfrac{-\frac{1}{4} - \left(-\frac{1}{2}\right)}{\frac{2}{3} - \frac{5}{3}} = \dfrac{-\frac{1}{4} + \frac{2}{4}}{-\frac{3}{3}} = \dfrac{\frac{1}{4}}{-1} = -\dfrac{1}{4}$

Student Practice 2 Find the slope, if possible, of the line through each pair of points.

(a) (1.8, −6.2) and (−2.2, −3.4) **(b)** $\left(\frac{1}{5}, -\frac{1}{2}\right)$ and $\left(\frac{4}{15}, -\frac{3}{4}\right)$

When dealing with practical situations, such as the grade of a road or the pitch of a roof, we can find the slope by using the formula

$$\text{slope} = \frac{\text{rise}}{\text{run}}.$$

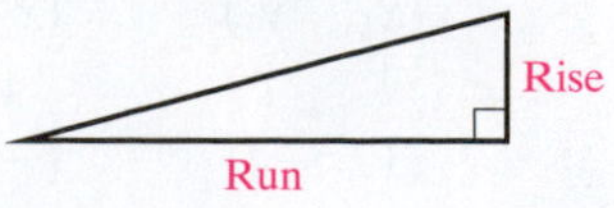

EXAMPLE 3 Find the pitch of a roof as shown in the sketch. Use only positive numbers in your calculation.

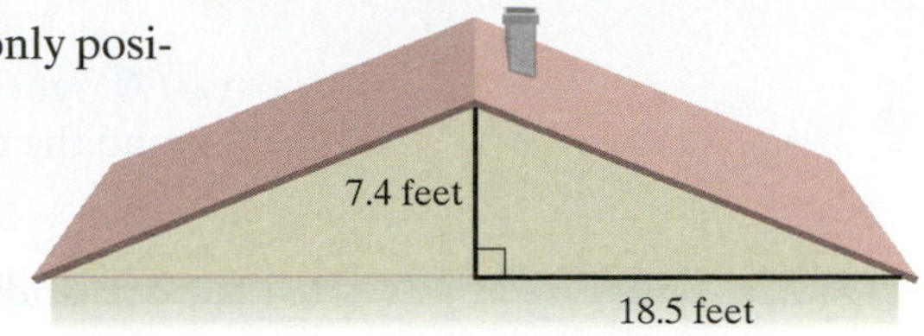

$$\text{slope} = \frac{\text{rise}}{\text{run}} = \frac{7.4}{18.5} = 0.4$$

Solution This could also be expressed as the fraction $\frac{2}{5}$. A builder might refer to this as a *pitch* (slope) of 2:5.

Student Practice 3 Find the slope of a river that drops 25.92 feet vertically over a horizontal distance of 1296 feet. (*Hint:* Use only positive numbers. In everyday use, the slope of a river or road is always considered to be a positive value.)

② Determining Whether Two Lines Are Parallel or Perpendicular

We can tell a lot about a line by looking at its slope. A positive slope tells us that the line rises from left to right. A negative slope tells us that the line falls from left to right. What might be true of the slopes of parallel lines? Determine the slope of each line in the following graphs, and compare the slopes of the parallel lines.

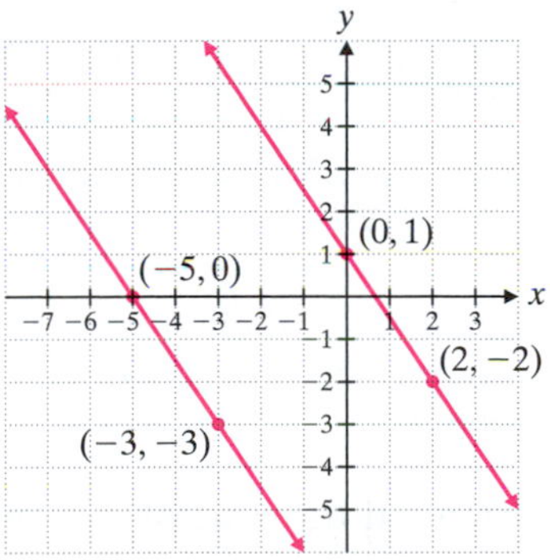

(a)

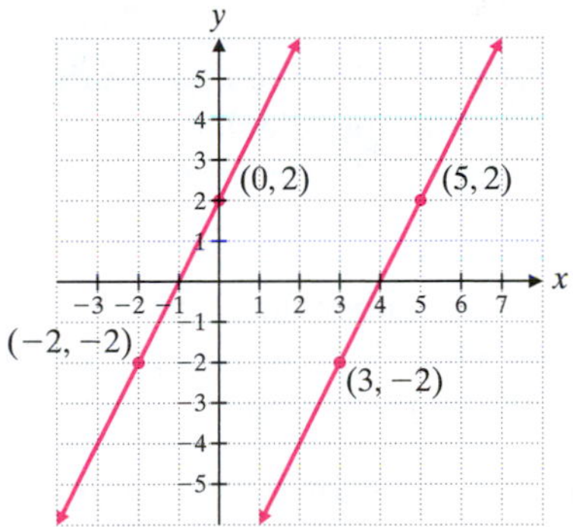

(b)

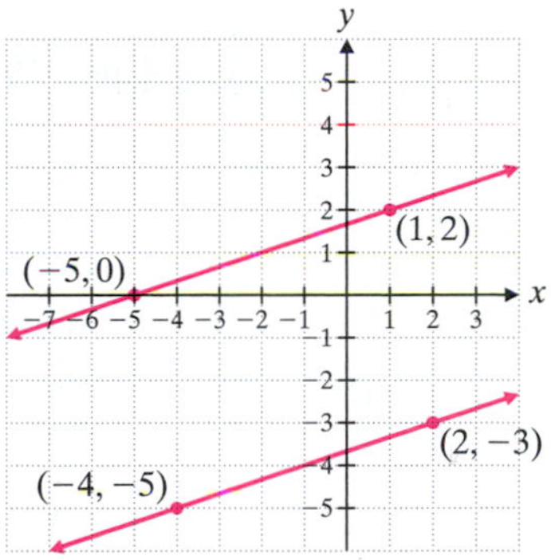

(c)

Since parallel lines are lines that never intersect, their slopes must be equal.

PARALLEL LINES

Two different lines with slopes m_1 and m_2 are *parallel* if $m_1 = m_2$.

Now take a look at perpendicular lines. Determine the slope of each line in the following graphs, and compare the slopes of the perpendicular lines. What do you notice?

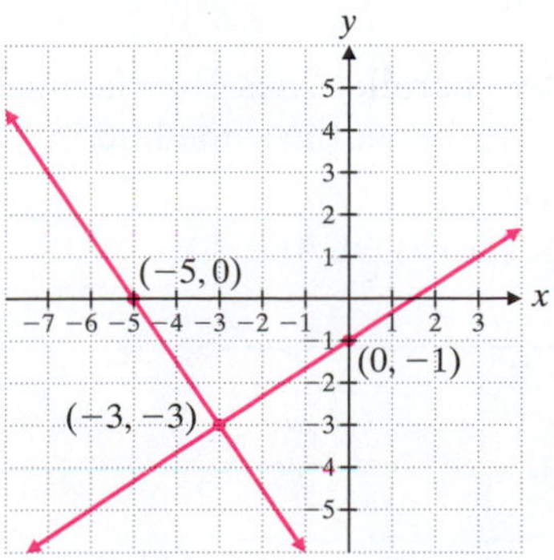

(a)

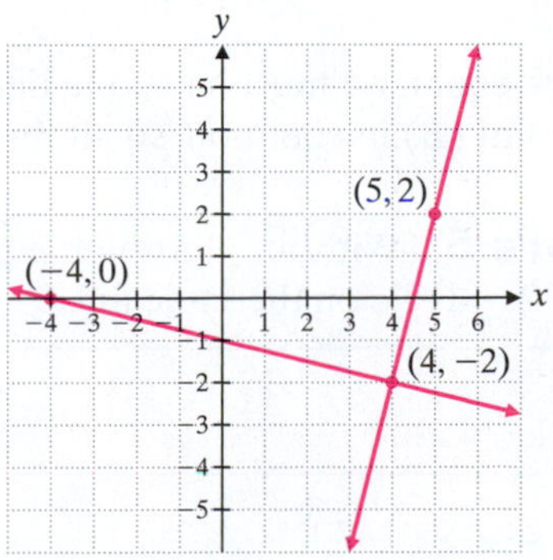

(b)

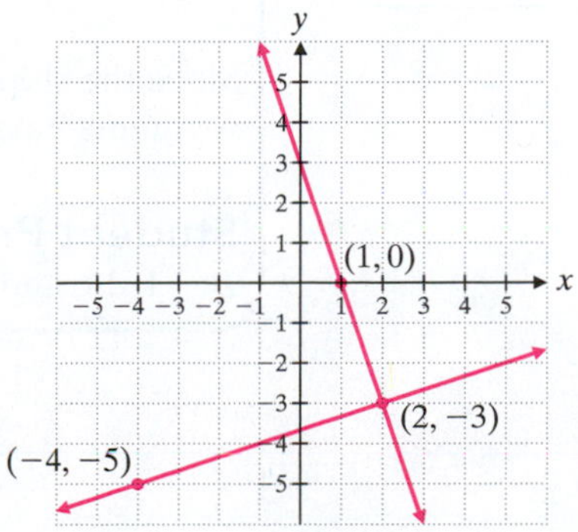

(c)

By definition, two lines are perpendicular if they intersect at right angles. (A right angle is an angle of 90°.) We would expect that one slope would be positive and the other slope negative.

In fact, it is true that if the slope of a line is 5, then the slope of a line that is perpendicular to it is $-\frac{1}{5}$.

PERPENDICULAR LINES

Two lines with slopes m_1 and m_2 are *perpendicular* if $m_1 = -\frac{1}{m_2}$ $(m_1, m_2 \neq 0)$.

EXAMPLE 4 Find the slope of a line that is perpendicular to the line l that passes through $(4, -6)$ and $(-3, -5)$.

Solution The slope of line l is

$$m_l = \frac{-5 - (-6)}{-3 - 4} = \frac{-5 + 6}{-7} = \frac{1}{-7} = -\frac{1}{7}.$$

The slope of a line perpendicular to line l must have a slope of 7.

Student Practice 4 If a line l passes through $(5, 0)$ and $(6, -2)$, what is the slope of a line h that is perpendicular to l?

It is helpful to remember that if two lines are perpendicular, their slopes are negative reciprocals. One slope will be positive and the other negative. If one line had a slope of $-\frac{3}{7}$, the slope of a line perpendicular to it would have a slope of $\frac{7}{3}$.

Now we will use our knowledge of slopes to determine if three points lie on the same line.

EXAMPLE 5 Without plotting any points, show that the points $A(-5, -1)$, $B(-1, 2)$, and $C(3, 5)$ lie on the same line.

Solution First we find the slope of the line segment from A to B and the slope of the line segment from B to C.

$$m_{AB} = \frac{2 - (-1)}{-1 - (-5)} = \frac{2 + 1}{-1 + 5} = \frac{3}{4}$$

$$m_{BC} = \frac{5 - 2}{3 - (-1)} = \frac{3}{3 + 1} = \frac{3}{4}$$

Since the slopes are equal, we must have one line or two parallel lines. But the line segments have a point (B) in common, so all three points lie on the same line.

Student Practice 5 Without plotting the points, show that $A(1, 5)$, $B(-1, 1)$, and $C(-2, -1)$ lie on the same line.

Graphing Calculator

Perpendicular Lines

In order to have perpendicular lines appear perpendicular to each other on a graphing calculator, you must use a "square" window setting. In other words, the actual unit length along both axes must be the same. You can produce square windows on most calculators with a few keystrokes.

Using a graphing calculator, graph the following equations on one coordinate system.

$$y_1 = 1.5625x + 1.8314$$
$$y_2 = -0.64x - 1.3816$$

Do the graphs appear to represent perpendicular lines? Are the lines really perpendicular? Why or why not?

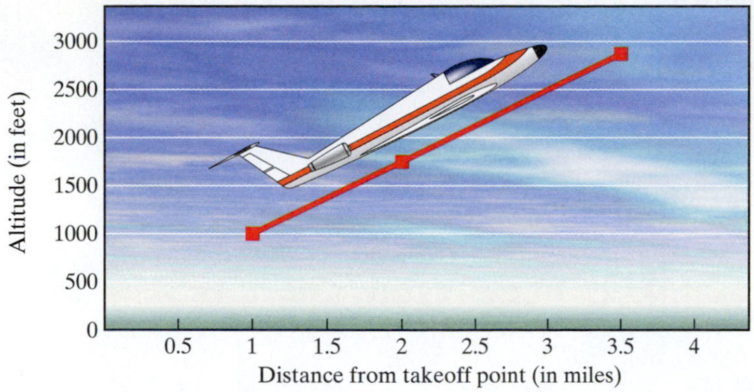

At takeoff an aircraft climbs into the sky at a certain rate of speed. This measurement of a slope is often called the *rate of climb*.

EXAMPLE 6 A gulfstream jet takes off from Orange County Airport in California. Look at the graph at the top of the page. At 1 mile from the takeoff point, the jet is 1000 feet above the ground and begins a specified rate of climb. At 2 miles from the takeoff point, it is 1750 feet above the ground. At 3.5 miles from the takeoff point, it is 2865 feet above the ground. Is the jet traveling in a straight line from the 1-mile point to the 3.5-mile point?

Solution From the 1-mile point to the 2-mile point, the jet traveled 750 feet upward over 1 horizontal mile. From the 2-mile point to the 3.5-mile point, the jet traveled 1115 feet upward over 1.5 horizontal miles. Thus, the first slope is 750 feet per mile, and the second slope is $743\frac{1}{3}$ feet per mile. These slopes are not the same, so the jet has not traveled in a straight line.

Student Practice 6 On the return flight, the jet is at an altitude of 4850 feet when it is 4.5 miles from the airport. Its altitude is 3650 feet when it is 3.5 miles from the airport and 1010 feet when it is 1.3 miles from the airport. Is the jet descending in a straight line?

STEPS TO SUCCESS Why Does Reviewing Make Such a Big Difference?

Students are often amazed that reviewing makes such a huge difference in helping them to learn. It is one of the most powerful tools that a math student can have.

Mathematics involves learning concepts one step at a time. Then the concepts are put together in a chapter. At the end of the chapter you need to know each of these concepts. Therefore, to succeed in each chapter you need to be able to put together all the pieces of each chapter. Reviewing in each section and reviewing at the end of each chapter are the amazing tools that help you to master the mathematical concepts.

As you review, if you find you cannot work out a problem be sure to study the Examples and the Student Practice problems very carefully. Then things will become more clear.

Making it personal: Start with this section. Do each Cumulative Review problem at the end of the exercise set. Check your answer for each one in the back of the book. If you miss any problem, go back to the appropriate section of the book for help.

For example, if you miss Problem 43 in Exercises 3.2 you see it is coded [1.4.3]. This means to go back to Section 1.4 of the book and look at objective 3. There you will find similar examples that explain this kind of problem.

In the shaded area below, write one thing that you learned by doing these review problems. ▼

3.2 Exercises

MyMathLab®

Watch the videos in MyMathLab

Download the MyDashBoard App

Verbal and Writing Skills, Exercises 1–6

1. Slope measures __________ change (rise) versus __________ change (run).

2. A positive slope indicates that the line slopes __________ to the right.

3. The slope of a horizontal line is __________.

4. Two different lines are parallel if their slopes are __________.

5. Does the line passing through $(-3, -7)$ and $(-3, 5)$ have a slope? Give a reason for your answer.

6. Let $(x_1, y_1) = (-6, -3)$ and $(x_2, y_2) = (-4, 5)$. Find $\dfrac{y_2 - y_1}{x_2 - x_1}$ and $\dfrac{y_1 - y_2}{x_1 - x_2}$. Are the results the same? Why or why not?

Find the slope, if possible, of the line passing through each pair of points.

7. $(2, 2)$ and $(6, -6)$

8. $(2, -1)$ and $(6, 3)$

9. $\left(\frac{3}{2}, 4\right)$ and $(-2, 0)$

10. $(5, 3)$ and $\left(1, \frac{3}{2}\right)$

11. $(6.8, -1.5)$ and $(5.6, -2.3)$

12. $(-2, 5.2)$ and $(4.8, -1.6)$

13. $\left(\frac{3}{2}, -2\right)$ and $\left(\frac{3}{2}, \frac{1}{4}\right)$

14. $\left(\frac{7}{3}, -6\right)$ and $\left(\frac{7}{3}, \frac{1}{6}\right)$

15. $(-7, -3)$ and $(10, -3)$

16. $(4, 12)$ and $(-5, 12)$

17. $\left(-\frac{1}{3}, -2\right)$ and $(1, -4)$

18. $\left(2, \frac{1}{3}\right)$ and $(4, 0)$

Applications

19. ***Snowboarding*** Find the slope (grade) of a snowboard recreation hill that rises 48 feet vertically over a horizontal distance of 80 feet.

20. ***Grade of a Road*** Find the grade of a mountain road that rises 22.5 feet vertically over a horizontal distance of 225 feet. (The grade of a road is usually expressed as a percent.)

21. ***Rock Formation*** Find the slope (pitch) of a perfectly smooth rock formation that rises 35.7 feet vertically over a horizontal distance of 142.8 feet.

22. ***Pitch of a Roof*** Find the slope (pitch) of a roof that rises 3.15 feet vertically over a horizontal distance of 10.50 feet.

23. ***Slope of Street*** Baldwin Street in Dunedin, New Zealand, is the world's steepest residential street. The slope of this street is 0.35. How many feet vertically does it rise over a horizontal distance of 120 feet?

24. ***Slope of a Roller Coaster*** T-Express at Everland in South Korea has the steepest drop of any wooden roller coaster in the world. The slope of the drop is $\frac{13}{3}$. The horizontal distance of the drop is 42 feet. How many feet does it fall vertically?

Find the slope of a line parallel to the line that passes through the following points.

25. $(6, 7)$ and $(24, 3)$

26. $(32, -5)$ and $(17, -10)$

27. $(5.5, 2)$ and $(5, 4)$

28. $(4, 0)$ and $(3.8, 2)$

29. $\left(-9, \frac{1}{2}\right)$ and $(-6, 5)$

30. $\left(1, \frac{5}{2}\right)$ and $\left(\frac{1}{3}, 2\right)$

Find the slope of a line perpendicular to the line that passes through the following points.

31. $(8, 12)$ and $(3, 9)$

32. $(4, 11)$ and $(-2, 1)$

33. $\left(3, \frac{1}{2}\right)$ and $\left(2, -\frac{3}{2}\right)$

34. $\left(\frac{1}{4}, -1\right)$ and $\left(\frac{5}{4}, \frac{1}{2}\right)$

35. $(-8.4, 0)$ and $(0, 4.2)$

36. $(0, -5)$ and $(-2, 0)$

To Think About

37. A line k passes through the points $(-3, -9)$ and $(1, 11)$. A second line h passes through the points $(-2, -13)$ and $(2, 7)$. Is line k parallel to line h? Why or why not?

38. A line k passes through the points $(4, 2)$ and $(-4, 4)$. A second line h passes through the points $(-8, 1)$ and $(8, -3)$. Is line k parallel to line h? Why or why not?

39. Show that $ABCD$ is a parallelogram if the four vertices are $A(2, 1)$, $B(-1, -2)$, $C(-7, -1)$, and $D(-4, 2)$. (*Hint:* A parallelogram is a four-sided figure with opposite sides parallel.)

40. Do the points $A(-1, -2)$, $B(2, -1)$, and $C(8, 1)$ lie on a straight line? Explain.

41. ***Ramp for the Disabled*** Most new buildings are required to have a ramp for the disabled that has a maximum vertical rise of 5 feet for every 60 feet of horizontal distance.

(a) What is the value of the slope of a ramp for the disabled?

(b) If the builder constructs a new building in which the ramp has a horizontal distance of 24 feet, what is the maximum height of the doorway above the level of the parking lot where the ramp begins?

(c) What is the shortest possible length of the ramp if the architect redesigns the building so that the doorway is 1.7 feet above the parking lot?

42. ***Rate of Climb of Aircraft*** A small Cessna plane takes off from Hyannis Airport. When the plane is 1 mile from the airport, it is flying at an altitude of 3000 feet. When the plane is 2 miles from the airport, it is flying at an altitude of 4300 feet. Round your answers to the following questions to the nearest tenth.

(a) If the plane continues flying at the same slope (the same rate of climb), what will its altitude be when it is 4.8 miles from the airport?

(b) If the plane continues flying at the same rate of climb, how many miles from the airport will it be when it reaches an altitude of 6000 feet?

(c) A Lear jet leaves the airport at the same time and has the same altitude (3000 feet) as the Cessna when each plane is 1 mile from the airport. When the jet is 1.8 miles from the airport, it is flying at an altitude of 4040 feet. Is the jet being flown at the same rate of climb as the Cessna?

Cumulative Review *Simplify.*

43. **[1.4.3]** $\dfrac{-15x^6y^3}{-3x^{-4}y^6}$

44. **[1.5.3]** $8x(x - 1) - 2(x + y)$

45. **[2.4.1]** Rama ordered a new custom-made rectangular rug for her living room. The length is 2 feet longer than the width. Its perimeter is 36 feet. Find the length and width of the rug.

46. **[2.7.1]** Solve for x. $3x + 1 > 10$ *or* $-2x - 3 > 1$

Quick Quiz 3.2

1. Find the slope of the line passing through $(-7, -2)$ and $\left(3, \dfrac{1}{2}\right)$.

2. Find the slope of the line passing through $(-8, -3)$ and $(5, -3)$.

3. Find the slope of a line that is *perpendicular* to the line passing through $(19, 4)$ and $(-3, -7)$.

4. **Concept Check** How would you find the grade (slope) of a driveway that rises 8.5 feet vertically over a horizontal distance of 120 feet?

3.3 Graphs and the Equations of a Line

① Using the Slope–Intercept Form of the Equation of a Line

Recall that the standard form of the equation of a line is $Ax + By = C$. Although the standard form tells us that the graph is a straight line, it reveals little about the line. A more useful form of the equation is the **slope–intercept form.** The slope–intercept form immediately reveals the slope of the line and where it intersects the y-axis. This is important information that will help us graph the line.

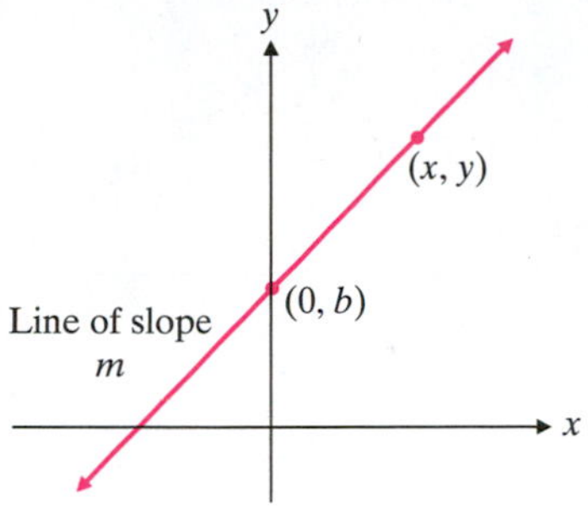

We will use the definition of slope to derive the equation. Let a nonvertical line with slope m cross the y-axis at some point $(0, b)$. Choose any other point on the line and label it (x, y). By the definition of slope, $\dfrac{y_2 - y_1}{x_2 - x_1} = m$. But here, $x_1 = 0, y_1 = b, x_2 = x$, and $y_2 = y$. So

$$\frac{y - b}{x} = m.$$

Now we solve this equation for y.

$$y - b = mx$$
$$y = mx + b$$

> **SLOPE–INTERCEPT FORM**
>
> The **slope–intercept form** of the equation of a line is $y = mx + b$, where m is the slope and $(0, b)$ is the y-intercept.

Write the Equation of a Line Given Its Slope and y-intercept

EXAMPLE 1 Write an equation of the line with the slope $-\frac{2}{3}$ and the y-intercept $(0, 5)$.

Solution

$$y = mx + b$$
$$y = \left(-\frac{2}{3}\right)x + (5) \quad \text{Substitute } -\tfrac{2}{3} \text{ for } m \text{ and 5 for } b.$$
$$y = -\frac{2}{3}x + 5$$

Student Practice 1 Write an equation of the line with slope 4 and y-intercept $\left(0, -\frac{3}{2}\right)$.

Write the Equation of a Line Given the Graph We can write an equation of a line if we are given its graph since we can determine the y-intercept and the slope from the graph.

Student Learning Objectives

After studying this section, you will be able to:

① **Use the slope–intercept form of the equation of a line.**

② **Use the point–slope form of the equation of a line.**

③ **Write the equation of the line passing through a given point that is parallel or perpendicular to a given line.**

Graphing Calculator

Exploring y-intercepts

Using a graphing calculator, graph

$$y_1 = 2x,$$
$$y_2 = 2x + 1,$$
$$y_3 = 2x - 1, \text{ and}$$
$$y_4 = 2x + 2$$

on the same set of axes. Where does each graph cross the y-axis? What effect does b have on the graph of $y = mx + b$? What would the graph of the line $y = 2x - 5$ look like? Use your graphing calculator to verify your conclusion.

NOTE TO STUDENT: Fully worked-out solutions to all of the Student Practice problems can be found at the back of the text starting at page SP-1.

Graphing Calculator

Exploring Slopes

Using a graphing calculator, graph

$$y_1 = x + 1,$$
$$y_2 = 3x + 1, \text{ and}$$
$$y_3 = \frac{1}{3}x + 1$$

on the same set of axes. What effect does m have on the graph of $y = mx + b$? What would the graph of the line $y = \frac{1}{2}x + 1$ look like?

EXAMPLE 2 Find the slope and y-intercept of the line whose graph is shown. Then use these to write an equation of the line.

Solution Looking at the graph, we can see that the y-intercept is at $(0, 5)$. That is, $b = 5$. If we can identify the coordinates of another point on the line, we will have two points and we can determine the slope. Another point on the line is $(3, 3)$.

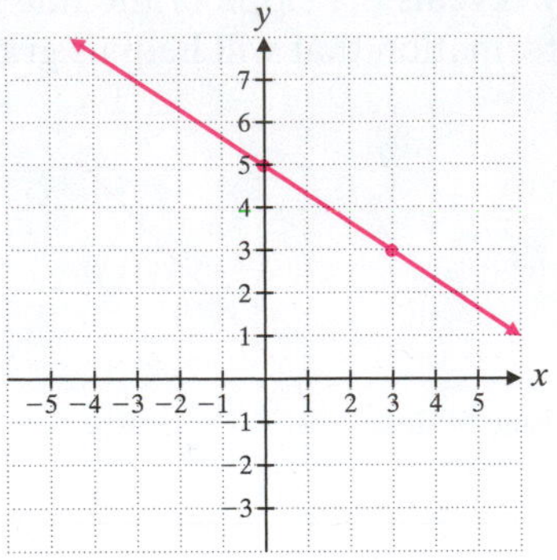

Thus,

$$(x_1, y_1) = (0, 5),$$
$$(x_2, y_2) = (3, 3), \text{ and}$$
$$m = \frac{y_2 - y_1}{x_2 - x_1} = \frac{3 - 5}{3 - 0} = -\frac{2}{3}.$$

We can now write the equation of the line in slope–intercept form.

$$y = mx + b$$
$$y = -\frac{2}{3}x + 5 \quad \text{Substitute } -\tfrac{2}{3} \text{ for } m \text{ and 5 for } b.$$

Student Practice 2 Find the slope and y-intercept of the line whose graph is shown in the margin on the left. Then use these to write an equation of this line.

Student Practice 2

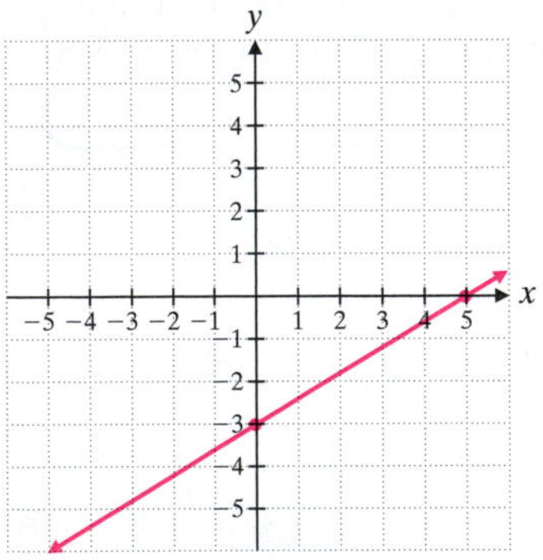

Use the Slope–Intercept Form to Graph an Equation We have just seen that given the graph of a line, we can determine its equation by identifying the y-intercept and finding the slope. We can also draw the graph of an equation without plotting points if we can write the equation in slope–intercept form and then locate the y-intercept on the graph.

EXAMPLE 3 Find the slope and the y-intercept. Then sketch the graph of the equation $28x - 7y = 21$.

Solution First we will change the standard form of the equation into slope–intercept form. This is a very important procedure. Be sure that you understand each step.

$$28x - 7y = 21$$
$$-7y = -28x + 21$$
$$\frac{-7y}{-7} = \frac{-28x}{-7} + \frac{21}{-7}$$
$$y = \overbrace{4x + (-3)}^{\text{slope}} \qquad \underbrace{y = mx + b}_{\text{gives } y\text{-intercept}}$$

Thus, the slope is 4, and the y-intercept is $(0, -3)$.

To sketch the graph, begin by plotting the point where the graph crosses the y-axis, $(0, -3)$. Plot the point. Now look at the slope. The slope, m, is 4 or $\frac{4}{1}$. This means there is a rise of 4 for every run of 1. From the point $(0, -3)$ go up 4 units and to the right 1 unit to locate a second point on the line. Draw the straight line that contains these two points, and you have the graph of the equation $28x - 7y = 21$.

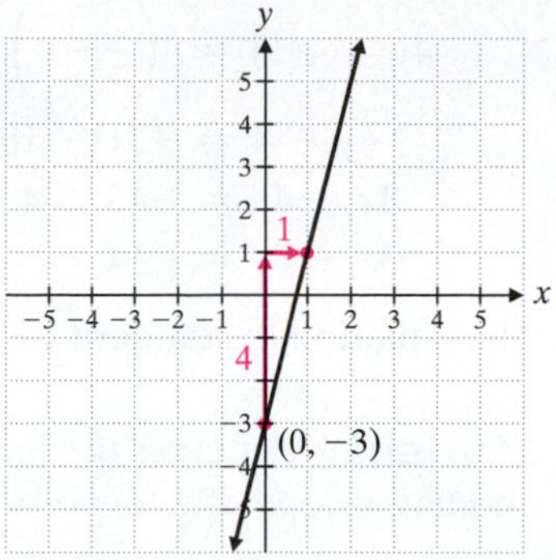

Student Practice 3 Find the slope and the y-intercept. Then sketch the graph of the equation $3x - 4y = -8$.

Student Practice 3

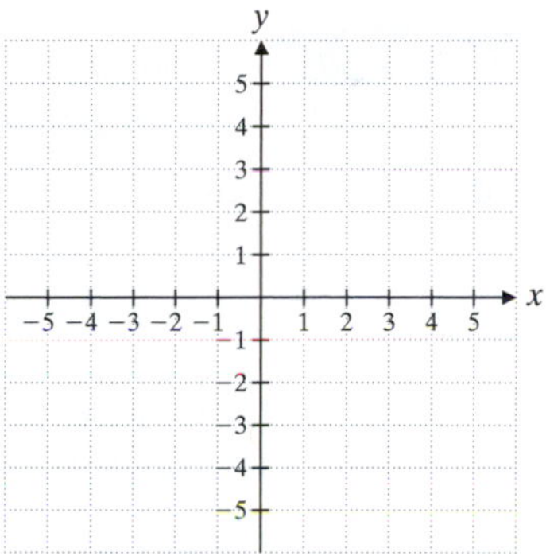

② Using the Point–Slope Form of the Equation of a Line

What happens if we know the slope of a line and a point on the line that is not the y-intercept? Can we write the equation of the line? By the definition of slope, we have the following:

$$m = \frac{y - y_1}{x - x_1}$$
$$m(x - x_1) = y - y_1$$
$$\text{That is, } y - y_1 = m(x - x_1).$$

This is the point–slope form of the equation of a line.

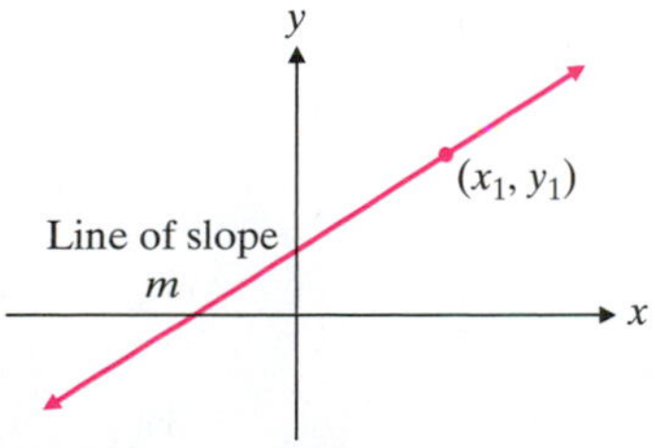

POINT–SLOPE FORM

The **point–slope form** of the equation of a line is $y - y_1 = m(x - x_1)$, where m is the slope and (x_1, y_1) are the coordinates of a known point on the line.

Write the Equation of a Line Given Its Slope and One Point on the Line

EXAMPLE 4 Find an equation of the line that has slope $-\frac{3}{4}$ and passes through the point $(-6, 1)$. Express your answer in standard form.

Solution Since we don't know the y-intercept, we can't use the slope–intercept form easily. Therefore, we use the point–slope form.

$$y - y_1 = m(x - x_1)$$

$$y - 1 = -\frac{3}{4}[x - (-6)] \quad \text{Substitute the given values.}$$

$$y - 1 = -\frac{3}{4}x - \frac{9}{2} \quad \text{Simplify. (Do you see how we did this?)}$$

Continued on next page

$$4y - 4(1) = 4\left(-\frac{3}{4}x\right) - 4\left(\frac{9}{2}\right)$$ Multiply each term by the LCD 4.

$$4y - 4 = -3x - 18$$ Simplify.

$$3x + 4y = -18 + 4$$ Add $3x + 4$ to each side.

$$3x + 4y = -14$$ Add like terms.

The equation in standard form is $3x + 4y = -14$.

Student Practice 4 Find an equation of the line that passes through $(5, -2)$ and has a slope of $\frac{3}{4}$. Express your answer in standard form.

Write the Equation of a Line Given Two Points on the Line We can use the point–slope form to find the equation of a line if we are given two points. Carefully study the following example. Be sure you understand each step. You will encounter this type of problem frequently.

EXAMPLE 5 Find an equation of the line that passes through $(3, -2)$ and $(5, 1)$. Express your answer in slope–intercept form.

Solution First we find the slope.

$$m = \frac{y_2 - y_1}{x_2 - x_1} = \frac{1 - (-2)}{5 - 3} = \frac{1 + 2}{2} = \frac{3}{2}$$

Now we substitute the value of the slope and the coordinates of either point into the point–slope equation. Let's use $(5, 1)$.

$$y - y_1 = m(x - x_1)$$

$$y - 1 = \frac{3}{2}(x - 5)$$ Substitute $m = \frac{3}{2}$ and $(x_2, y_2) = (5, 1)$.

$$y - 1 = \frac{3}{2}x - \frac{15}{2}$$ Remove parentheses.

$$y = \frac{3}{2}x - \frac{15}{2} + 1$$ Add 1 to each side of the equation.

$$y = \frac{3}{2}x - \frac{15}{2} + \frac{2}{2}$$ Add the two fractions.

$$y = \frac{3}{2}x - \frac{13}{2}$$ Simplify.

Student Practice 5 Find an equation of the line that passes through $(-4, 1)$ and $(-2, -3)$. Express your answer in slope–intercept form.

Before we go further, we want to point out that these various forms of the equation of a straight line are just that—*forms* for convenience. We are *not* using different equations each time, nor should you simply try to memorize the different variations without understanding when to use them. They can easily be derived from the definition of slope, as we have seen. And remember, you can *always* use the definition of slope to find the equation of a line. You may find it helpful to review Examples 4 and 5 for a few minutes before going ahead to Example 6. It is important to see how each example is different.

3 Writing the Equation of a Parallel or Perpendicular Line

Let us now look at parallel and perpendicular lines. If we are given the equation of a line and a point not on the line, we can find the equation of a second line that passes through the given point and is parallel or perpendicular to the first line. We can do this because we know that the slopes of parallel lines are equal and that the slopes of perpendicular lines are negative reciprocals of each other.

We begin by finding the slope of the given line. Then we use the point–slope form to find the equation of the second line. Study each step of the following example carefully.

EXAMPLE 6 Find an equation of the line passing through the point $(-2, -4)$ and parallel to the line $2x + 5y = 8$. Express the answer in standard form.

Solution First we need to find the slope of the line $2x + 5y = 8$. We do this by writing the equation in slope–intercept form.

$$5y = -2x + 8$$
$$y = -\frac{2}{5}x + \frac{8}{5}$$

The slope of the given line is $-\frac{2}{5}$. Since parallel lines have the same slope, the slope of the unknown line is also $-\frac{2}{5}$. Now we substitute $m = -\frac{2}{5}$ and the coordinates of the point $(-2, -4)$ into the point–slope form of the equation of a line.

$$y - y_1 = m(x - x_1)$$
$$y - (-4) = -\frac{2}{5}[x - (-2)] \quad \text{Substitute.}$$
$$y + 4 = -\frac{2}{5}(x + 2) \quad \text{Simplify.}$$
$$y + 4 = -\frac{2}{5}x - \frac{4}{5} \quad \text{Remove parentheses.}$$
$$5y + 5(4) = 5\left(-\frac{2}{5}x\right) - 5\left(\frac{4}{5}\right) \quad \text{Multiply each term by the LCD 5.}$$
$$5y + 20 = -2x - 4 \quad \text{Simplify.}$$
$$2x + 5y = -4 - 20 \quad \text{Add } 2x - 20 \text{ to each side.}$$
$$2x + 5y = -24 \quad \text{Simplify.}$$

$2x + 5y = -24$ is an equation of the line passing through the point $(-2, -4)$ and parallel to the line $2x + 5y = 8$.

Student Practice 6 Find an equation of the line passing through $(4, -5)$ and parallel to the line $5x - 3y = 10$. Express the answer in standard form.

One extra step is needed if the desired line is to be perpendicular to the given line. Carefully note the approach in Example 7.

EXAMPLE 7 Find an equation of the line that passes through the point $(2, -3)$ and is perpendicular to the line $3x - y = -12$. Express the answer in standard form.

Solution To find the slope of the line $3x - y = -12$, we rewrite it in slope–intercept form.

$$-y = -3x - 12$$
$$y = 3x + 12$$

This line has a slope of 3. Therefore, the slope of a line perpendicular to this line is the negative reciprocal $-\frac{1}{3}$.

Continued on next page

Graphing Calculator

Using Linear Regression to Find an Equation

Many graphing calculators, such as the TI-84 Plus, will find the equation of a line in slope–intercept form if you enter the points as a collection of data and use the Regression feature. We would enter the data from Example 5 as follows:

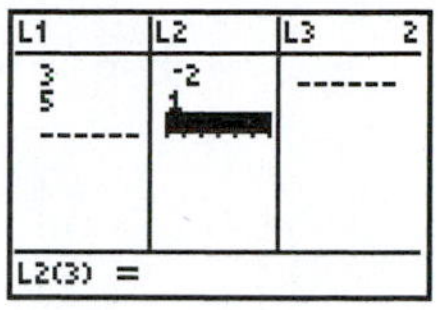

The output of the calculator uses the notation $y = ax + b$ instead of $y = mx + b$.

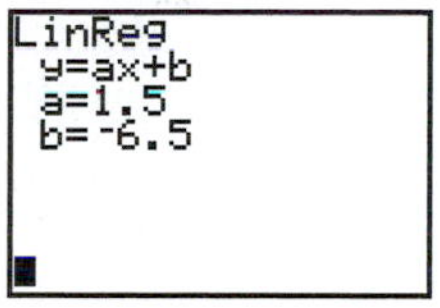

Thus, our answer to Example 5 using a graphing calculator would be $y = 1.5x - 6.5$.

Now substitute the slope $m = -\frac{1}{3}$ and the coordinates of the point $(2, -3)$ into the point–slope form of the equation.

$$y - y_1 = m(x - x_1)$$

$$y - (-3) = -\frac{1}{3}(x - 2) \quad \text{Substitute.}$$

$$y + 3 = -\frac{1}{3}(x - 2) \quad \text{Simplify.}$$

$$y + 3 = -\frac{1}{3}x + \frac{2}{3} \quad \text{Remove parentheses.}$$

$$3y + 3(3) = 3\left(-\frac{1}{3}x\right) + 3\left(\frac{2}{3}\right) \quad \text{Multiply each term by the LCD 3.}$$

$$3y + 9 = -x + 2 \quad \text{Simplify.}$$

$$x + 3y = 2 - 9 \quad \text{Add } x - 9 \text{ to each side.}$$

$$x + 3y = -7 \quad \text{Simplify.}$$

$x + 3y = -7$ is an equation of the line that passes through the point $(2, -3)$ and is perpendicular to the line $3x - y = -12$.

Student Practice 7 Find an equation of the line that passes through $(-4, 3)$ and is perpendicular to the line $6x + 3y = 7$. Express the answer in standard form.

STEPS TO SUCCESS Making a Friend in Class

Try to make a friend in your class. You may find that you enjoy sitting together and drawing support and encouragement from one another. Exchange phone numbers so you can call each other whenever you get stuck while doing your homework. Set up convenient times to study together on a regular basis, to do homework, and to review for exams.

You must not depend on a friend or fellow student to tutor you, do your work for you, or in any way be responsible for your learning. However, you will learn from one another as you seek to master the course. Studying with a friend and comparing notes, methods, and solutions can be very helpful. And it can make learning mathematics a lot more fun!

Making It personal: What can you do this week to help make a new friend in class? ▼

3.3 Exercises

MyMathLab®

Watch the videos in MyMathLab

Download the MyDashBoard App

Verbal and Writing Skills, Exercises 1 and 2

1. You are given two points that lie on a line. Explain how you would find an equation of the line.

2. Suppose $y = -\frac{2}{7}x + 5$. What can you tell about the graph by looking at the equation?

Write an equation of the line with the given slope and the given y-intercept. Leave the answer in slope–intercept form.

3. Slope $\frac{3}{4}$, y-intercept $(0, -9)$

4. Slope $-\frac{2}{3}$, y-intercept $(0, 5)$

Write an equation of the line with the given slope and the given y-intercept. Express the answer in standard form.

5. Slope $\frac{3}{4}$, y-intercept $\left(0, \frac{1}{2}\right)$

6. Slope $\frac{5}{6}$, y-intercept $\left(0, \frac{1}{3}\right)$

Find the slope and y-intercept of each of the following lines. Then use these to write an equation of the line.

7.

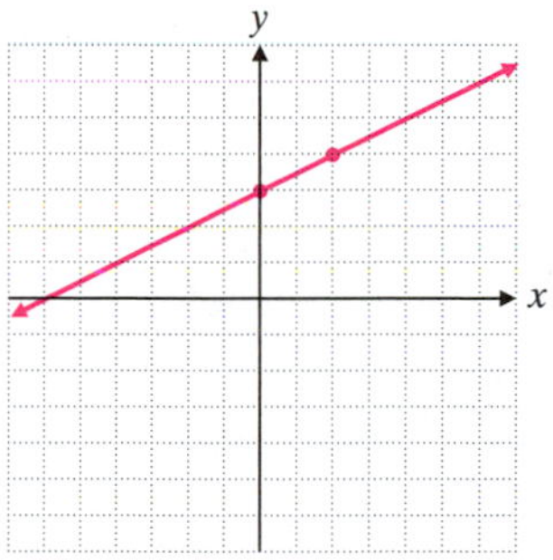

8.

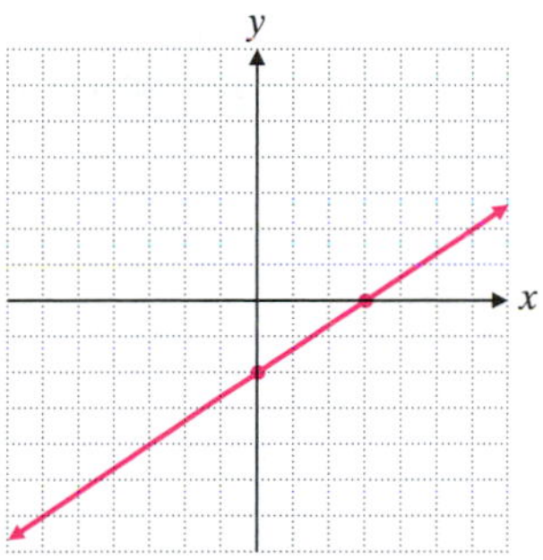

9.

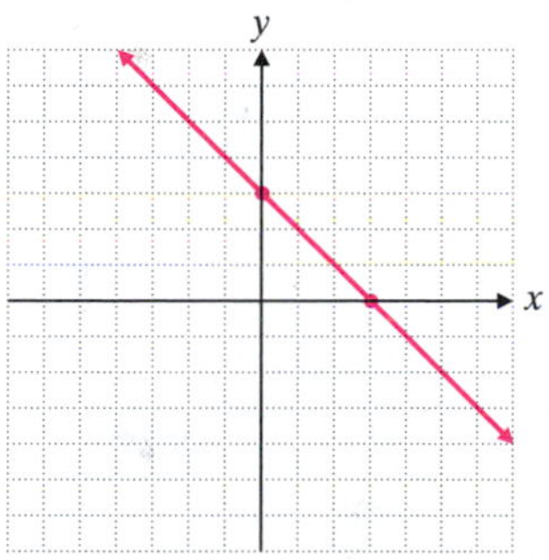

10.

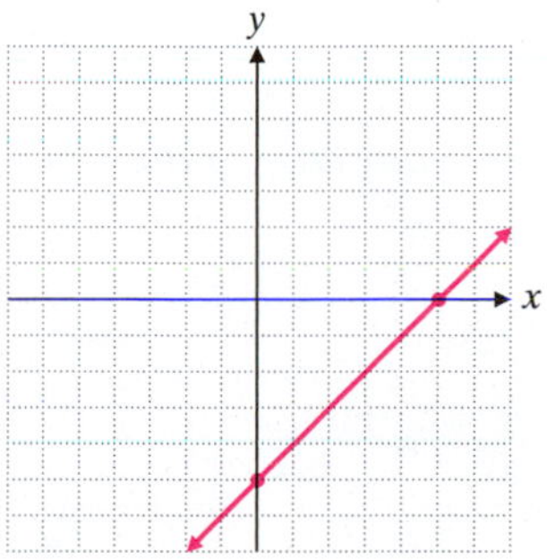

11.

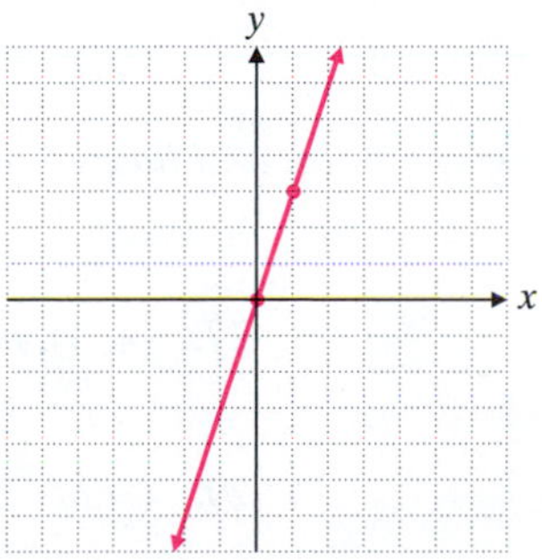

12.

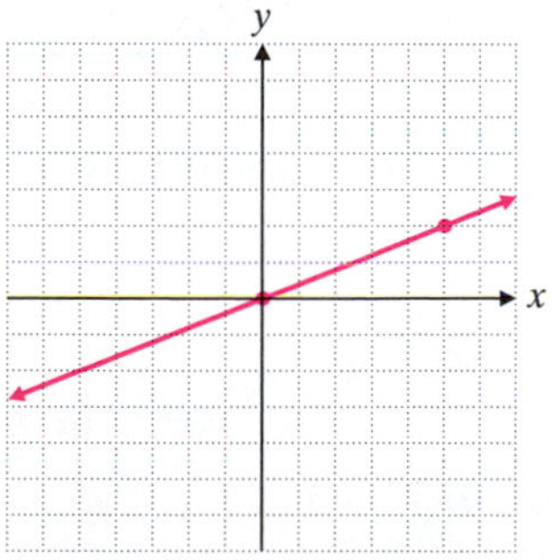

13.

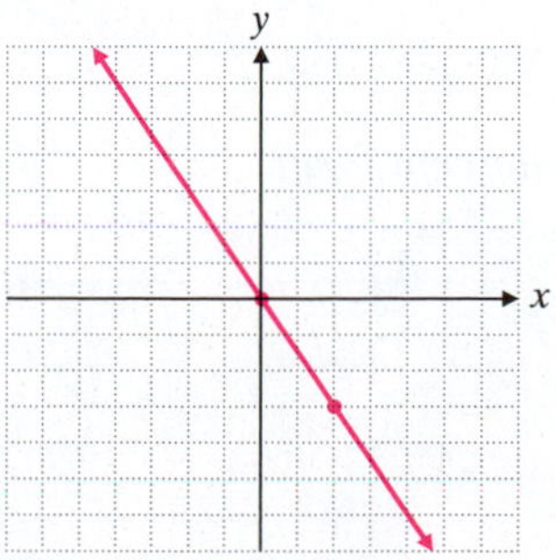

14.

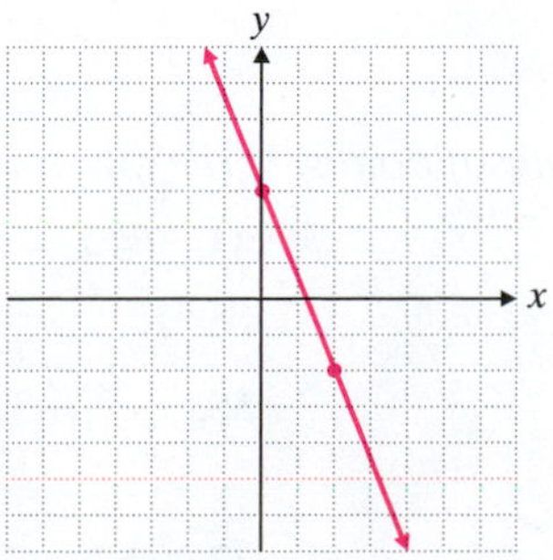

15.

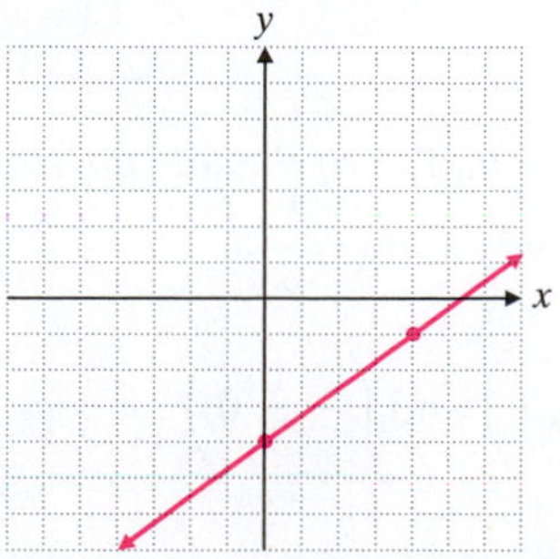

Write each equation in slope–intercept form. Then identify the slope and the y-intercept for each line.

16. $2x - y = 12$

17. $x - y = 5$

18. $2x - 3y = -8$

19. $5x - 4y = -20$

20. $\frac{1}{3}x + 2y = 7$

21. $2x + \frac{3}{4}y = -3$

For each equation find the slope and the y-intercept. Use these to graph the equation.

22. $y = 3x + 4$

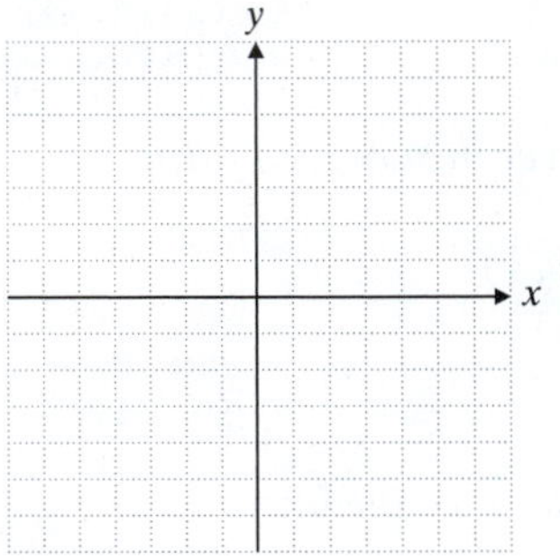

23. $y = \frac{1}{2}x - 3$

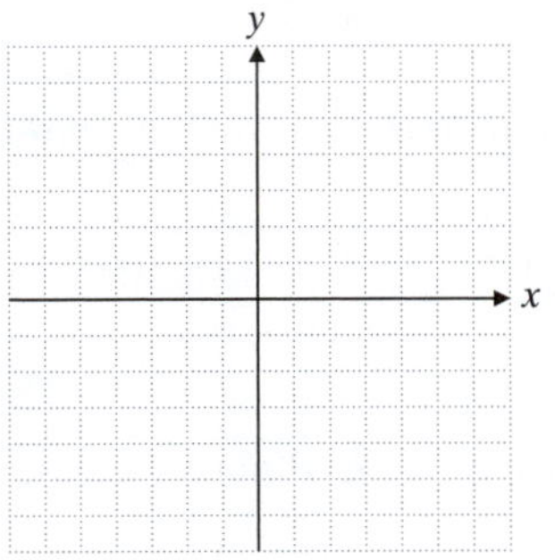

24. $5x - 4y = -20$

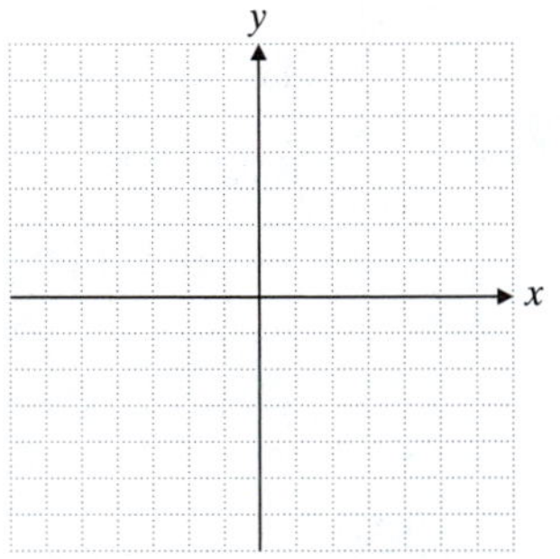

25. $5x + 3y = 18$

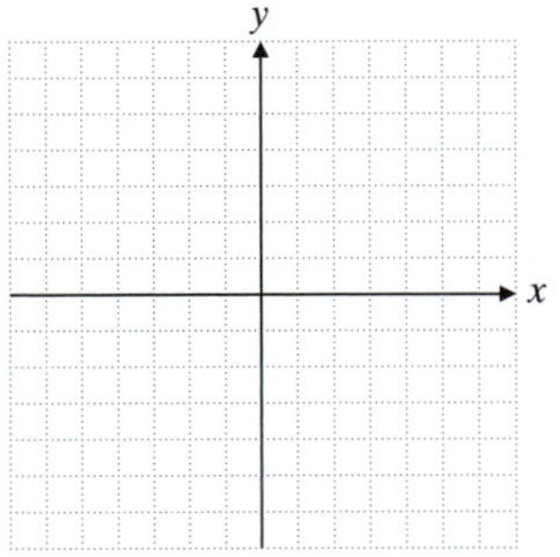

Find an equation of the line that passes through the given point and has the given slope. Express your answer in slope–intercept form.

26. $(6, 5), m = \frac{1}{3}$

27. $(4, 4), m = -\frac{1}{2}$

28. $(8, 0), m = -3$

29. $(-7, -2), m = 5$

30. $(0, -1), m = -\frac{5}{3}$

31. $(6, 0), m = -\frac{1}{5}$

Find an equation of the line passing through the pair of points. Write the equation in slope–intercept form.

32. $(5, -3)$ and $(1, -4)$

33. $(-4, -1)$ and $(3, 4)$

34. $\left(\frac{7}{6}, 1\right)$ and $\left(-\frac{1}{3}, 0\right)$

35. $\left(\frac{1}{2}, -3\right)$ and $\left(\frac{7}{2}, -5\right)$

36. $(4, 8)$ and $(-3, 8)$

37. $(12, -3)$ and $(7, -3)$

Find an equation of the line satisfying the conditions given. Express your answer in standard form.

38. Parallel to $x - y = 4$ and passing through $(-3, 2)$

39. Parallel to $5x - y = 4$ and passing through $(-2, 0)$

40. Parallel to $2y + x = 7$ and passing through $(-5, -4)$

41. Parallel to $x = 3y - 8$ and passing through $(5, -1)$

42. Perpendicular to $y = 3x$ and passing through $(-3, 2)$

43. Perpendicular to $2y = -3x$ and passing through $(6, -1)$

44. Perpendicular to $x - 4y = 2$ and passing through $(3, -1)$

45. Perpendicular to $x + 7y = -12$ and passing through $(-4, -1)$

To Think About

Without graphing, determine whether the following pairs of lines are (a) parallel, (b) perpendicular, or (c) neither parallel nor perpendicular.

46. $5x - 6y = 19$
$6x + 5y = -30$

47. $-3x + 5y = 40$
$5y + 3x = 17$

48. $y = \frac{2}{3}x + 6$
$-2x - 3y = -12$

49. $y = -\frac{3}{4}x - 2$
$6x + 8y = -5$

50. $y = \frac{3}{7}x - \frac{1}{14}$
$14y + 6x = 3$

51. $y = \frac{5}{6}x - \frac{1}{3}$
$6x + 5y = -12$

Optional Graphing Calculator Problems

If you have a graphing calculator, use it to graph each pair of equations. Do the graphs appear to be parallel?

52. $y = -2.39x + 2.04$ and $y = -2.39x - 0.87$

53. $y = 1.43x - 2.17$ and $y = 1.43x + 0.39$

Applications

Cost of Homes *The median sale price of single-family homes in the United States has been increasing steadily. The increase can be approximated by a linear equation of the form $y = mx + b$. The U.S. Census Bureau reported that in 2000 the median sale price of a single-family home in the United States was \$174,900. In 2010, the median sale price of a single-family home was \$223,900. We can record the data as follows:*

Number of Years Since 2000	Price of Home in Thousands of Dollars
0	174.9
10	223.9

Source: www.census.gov

Use the table of values on p. 153 for exercises 54–57.

54. Using these two ordered pairs, find the equation $y = mx + b$ where x is the number of years since 2000 and y is the median sale price of a single-family home in thousands of dollars.

55. Use the equation obtained in exercise 54 to find the expected median sale price of a single-family home in the year 2020 (20 years after 2000).

56. Graph the equation using the data for 2000, 2010, and 2020.

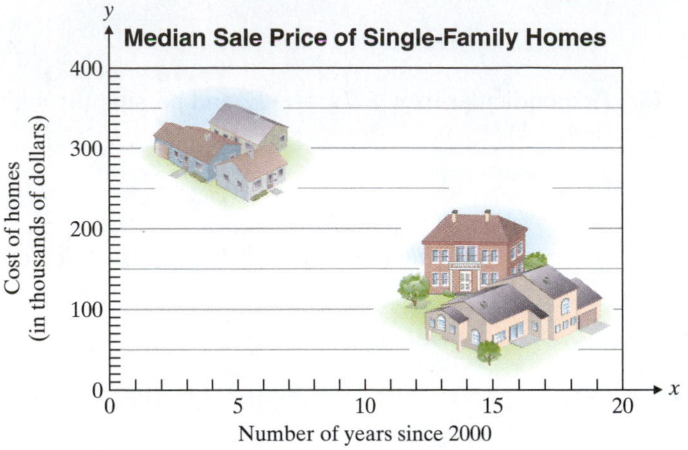

Source: www.census.gov

57. Use your graph to estimate the median cost of a home in 2015 (15 years after 2000).

Cumulative Review

58. [1.6.2] The face of a watch has a radius of 5 millimeters. Find the area of the face. Use $\pi \approx 3.14$.

59. [2.1.1] Solve for x. $0.3x + 0.1 = 0.27x - 0.02$

60. [2.2.1] Solve for x. $\frac{5}{4} - \frac{3}{4}(2x + 1) = x - 2$

61. [2.5.1] Geoff invested a total of \$9000. Part was invested at 4% simple interest and the rest was invested at 6% simple interest. After one year, he had made \$480 in interest. How much was invested at each rate?

Quick Quiz 3.3

1. Write the following equation in slope–intercept form. Then identify the slope and y-intercept.
$-3x + 7y = -9$

2. Find the standard form of the equation of the line that passes through $(-3, 2)$ and $(-4, -7)$.

3. Find an equation of the line with slope $-\frac{3}{4}$ and y-intercept $(0, -5)$. Write the answer in slope–intercept form.

4. Concept Check How would you find an equation of the horizontal line that passes through the point $(-4, -8)$?

How Am I Doing? Sections 3.1–3.3

How are you doing with your homework assignments in Sections 3.1 to 3.3? Do you feel you have mastered the material so far? Do you understand the concepts you have covered? Before you go further in the textbook, take some time to do each of the following problems.

3.1

1. Find the value of a if $(a, 6)$ is a solution to $5x + 2y = -12$.

Graph each equation.

2. $y = -\frac{1}{2}x + 5$

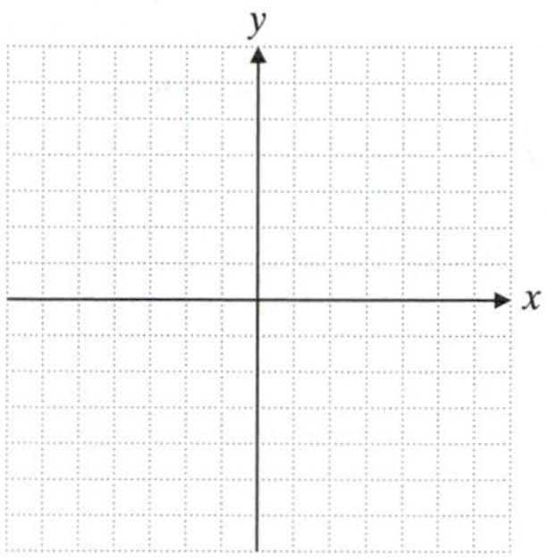

3. $5x + 3y = -15$

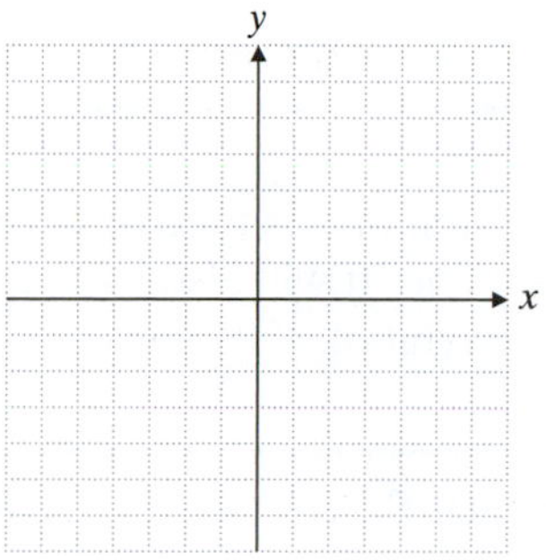

4. $4y + 6x = -8 + 9x$

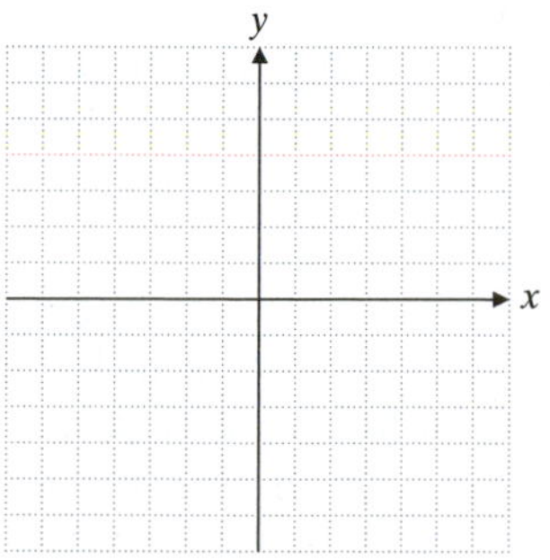

3.2

5. Find the slope of the line passing through the points $(-2, 3)$ and $(-1, -6)$.

6. Find the slope of a line parallel to the line that passes through $\left(\frac{2}{3}, 4\right)$ and $\left(\frac{5}{6}, -2\right)$.

7. Find the slope of a line perpendicular to the line passing through the points $(5, 0)$ and $(-13, -4)$.

8. A Colorado road has a slope of 0.13. How many feet does it rise vertically over a horizontal distance of 500 feet?

3.3

9. Find the slope and the y-intercept of $2x - 4y = 10$.

10. Write an equation of the line of slope -2 that passes through $(7, -3)$.

11. Write an equation of the line passing through $(-1, -2)$ and perpendicular to $3x - 5y = 10$.

12. Find an equation of the line that passes through $(-3, 1)$ and $(-1, 5)$.

Now turn to page SA-6 for the answers to each of these problems. *Each answer also includes a reference to the objective in which the problem is first taught. If you missed any of these problems, you should stop and review the Examples and Student Practice problems in the referenced objective. A little review now will help you master the material in the upcoming sections of the text.*

1. __________
2. __________
3. __________
4. __________
5. __________
6. __________
7. __________
8. __________
9. __________
10. __________
11. __________
12. __________

3.4 Linear Inequalities in Two Variables

Student Learning Objectives

After studying this section, you will be able to:

① **Graph a linear inequality in two variables.**

② **Graph a linear inequality for which the coefficient of one variable is zero.**

① Graphing a Linear Inequality in Two Variables

A linear inequality in two variables is similar to a linear equation in two variables. However, in place of the $=$ sign, there appears instead one of the following four inequality symbols: $<, >, \leq, \geq$.

LINEAR INEQUALITY IN TWO VARIABLES

A **linear inequality in two variables** is an inequality that can be written

$$\text{as} \quad Ax + By > C \quad \text{or} \quad Ax + By < C$$
$$\text{or} \quad Ax + By \geq C \quad \text{or} \quad Ax + By \leq C$$

where A, B, and C are real numbers and A and B are not both zero.

The graph of this type of linear inequality is a half-plane that lies on one side of a straight line. It will also include the boundary line if the inequality contains the $\leq$ or the $\geq$ symbol.

PROCEDURE FOR GRAPHING LINEAR INEQUALITIES

1. Replace the inequality symbol by an equals sign. This equation will be the boundary for the desired region.
2. Graph the line obtained in step 1. Use a dashed line if the original inequality contains a $<$ or $>$ symbol. Use a solid line if the original inequality contains a $\leq$ or $\geq$ symbol.
3. Choose a test point that does not lie on the boundary line. Substitute the coordinates into the original inequality. If you obtain an inequality that is true, shade the region on the side of the line containing the test point. If you obtain a false inequality, shade the region on the side of the line opposite the test point.

If the boundary line does not pass through (0, 0), that is usually a good test point to use.

EXAMPLE 1 Graph $y < 2x + 3$.

Solution

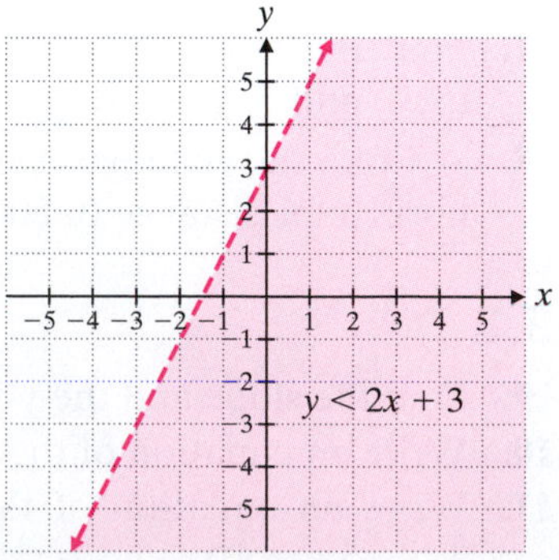

1. The boundary line is $y = 2x + 3$.
2. We graph $y = 2x + 3$ using a dashed line because the inequality contains $<$.
3. Since the line does not pass through (0, 0), we can use it as a test point. Substituting (0, 0) into $y < 2x + 3$, we have the following:

$$0 < 2(0) + 3$$
$$0 < 3$$

This inequality is true. We therefore shade the region on the same side of the line as (0, 0). See the sketch. The solution is the shaded region *not including* the dashed line.

Student Practice 1 Graph $y > 3x + 1$.

Student Practice 1

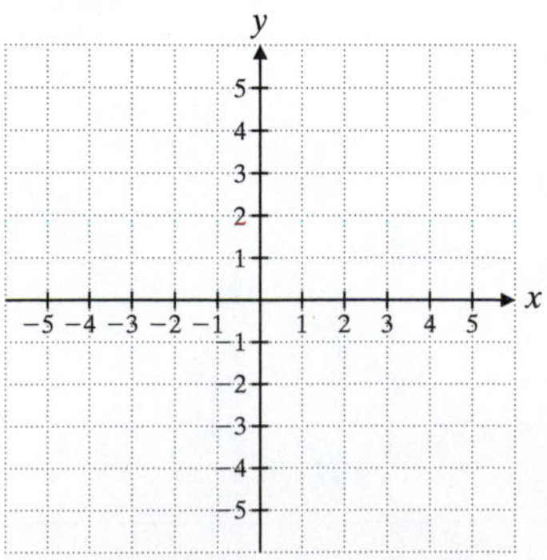

NOTE TO STUDENT: Fully worked-out solutions to all of the Student Practice problems can be found at the back of the text starting at page SP-1.

EXAMPLE 2 Graph $3x + 2y \geq 4$.

Solution

1. The boundary line is $3x + 2y = 4$.
2. We graph the boundary line with a solid line because the inequality contains $\geq$.
3. Since the line does not pass through (0, 0), we can use it as a test point. Substituting (0, 0) into $3x + 2y \geq 4$ gives the following:

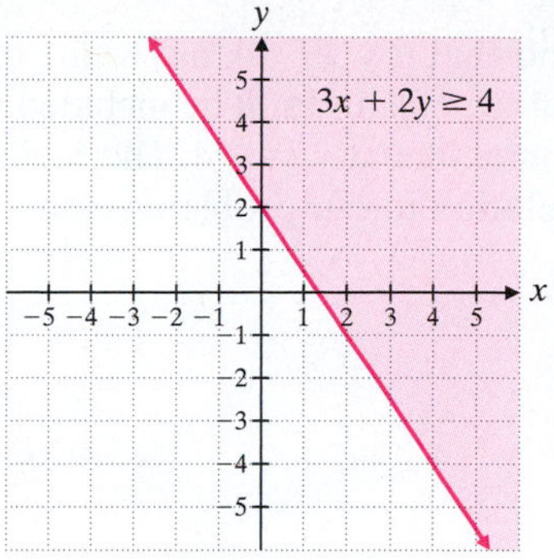

$$3(0) + 2(0) \geq 4$$
$$0 + 0 \geq 4$$
$$0 \geq 4$$

This inequality is false. We therefore shade the region on the side of the line opposite (0, 0). See the sketch. The solution is the shaded region *including* the boundary line.

Student Practice 2 Graph $-4x + 5y \leq -10$.

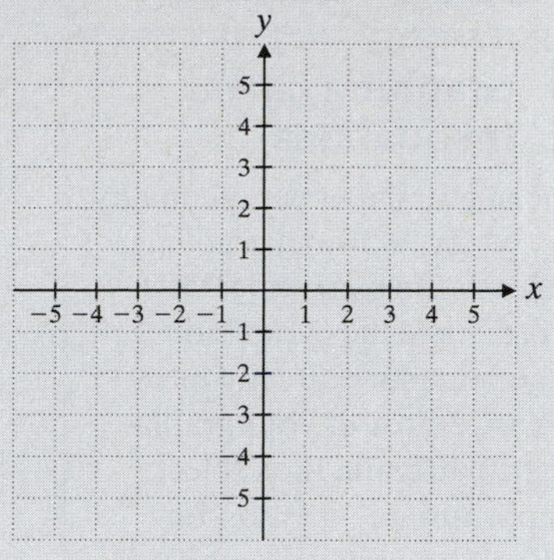

EXAMPLE 3 Graph $4x - y < 0$.

Solution

1. The boundary line is $4x - y = 0$.
2. We graph the boundary line with a dashed line.
3. Since (0, 0) lies on the line, we cannot use it as a test point. We must choose another point not on the line. We try to pick some point that is *not* close to the dashed boundary line. Let's pick $(-1, 5)$. Substituting $(-1, 5)$ into $4x - y < 0$ gives the following:

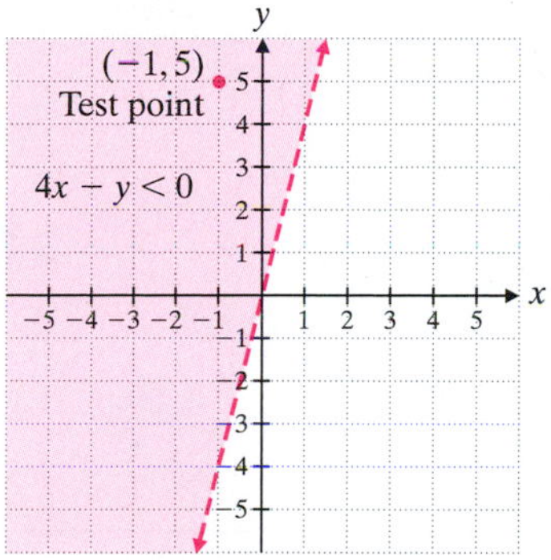

$$4(-1) - 5 < 0$$
$$-4 - 5 < 0$$
$$-9 < 0$$

This is true. Therefore, we shade the side of the boundary line that contains $(-1, 5)$. See the sketch. The solution is the shaded region above the dashed line but *not including* the dashed line.

Student Practice 3 Graph $3y + x < 0$.

Student Practice 3

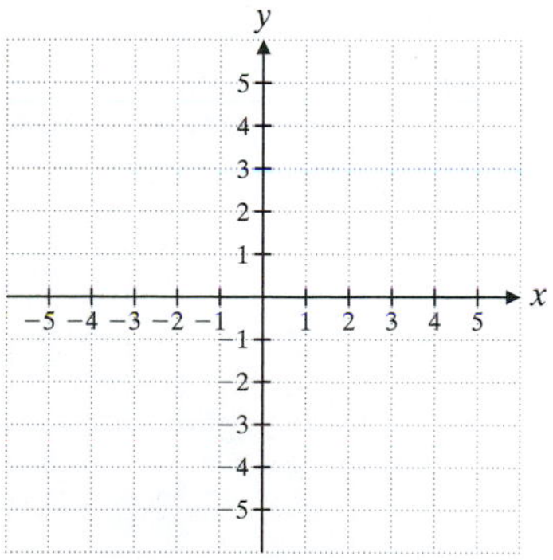

② Graphing a Linear Inequality for Which the Coefficient of One Variable Is Zero

EXAMPLE 4 Simplify and graph $3y \geq -12$.

Solution First we will need to simplify the original inequality by dividing each side by 3.

$$3y \geq -12$$
$$\frac{3y}{3} \geq \frac{-12}{3}$$
$$y \geq -4$$

Continued on next page

Graphing Calculator

Graphing Linear Inequalities

You can graph a linear inequality like $y \leq 3x + 1$ on a graphing calculator. On some calculators you can enter the expression for the boundary directly into the Y = editor of your graphing calculator and then select the appropriate direction for shading. Other calculators may require using a Shade command to shade the region. In general, most calculator displays do not distinguish between dashed and solid boundaries.

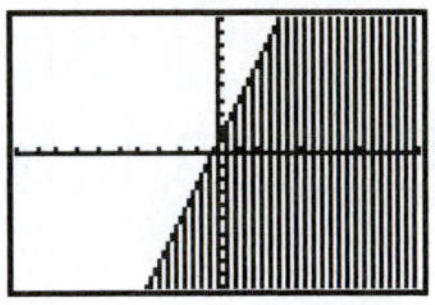

Graph the following:

$$y < 3.45x - 1.232$$

This inequality is equivalent to

$$0x + y \geq -4.$$

We find that if $y \geq -4$, any value of x will make the inequality true. Thus, x can be any value at all and still be included in our shaded region. Therefore, we draw a solid horizontal line at $y = -4$. The region we want to shade is the region above the line. The solution to our problem is the line and the shaded region above the line.

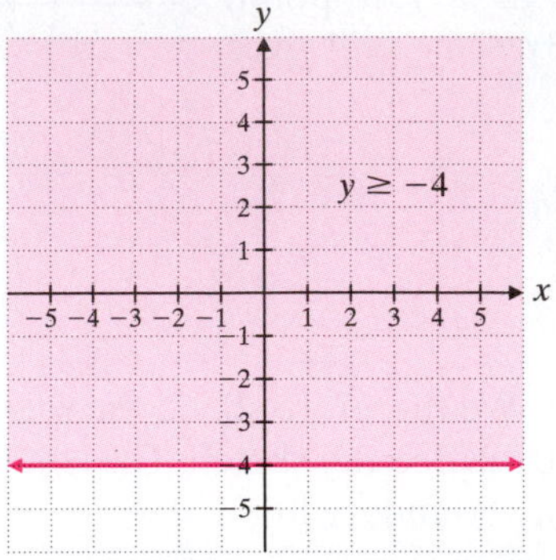

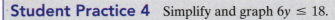

Student Practice 4 Simplify and graph $6y \leq 18$.

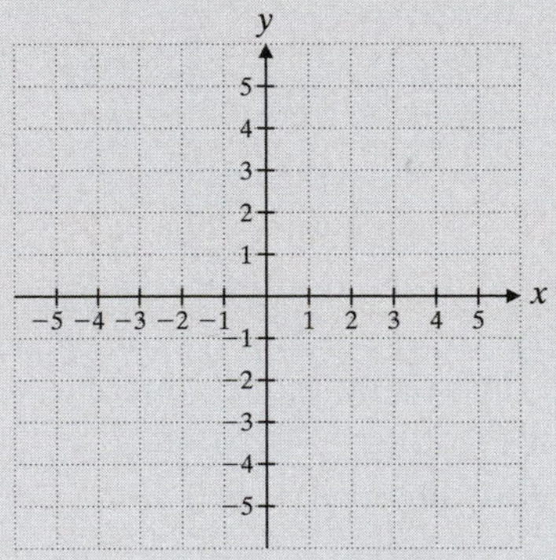

3.4 Exercises

MyMathLab®

Watch the videos in MyMathLab

Download the MyDashBoard App

Verbal and Writing Skills, Exercises 1–6

1. Explain when to use a dashed line as a boundary when graphing a linear inequality.

2. Explain when to use a solid line as a boundary when graphing a linear inequality.

3. When graphing $x > 5$, should the region to the left of the line $x = 5$ be shaded or should the region to the right of the line $x = 5$ be shaded?

4. When graphing $y < -6$, should the region below the line $y = -6$ be shaded or should the region above the line $y = -6$ be shaded?

5. When we graph the inequality $3x - 2y \geq 0$, why can't we use $(0, 0)$ as a test point? If we test the point $(-4, 2)$, do we obtain a false statement or a true one?

6. When we graph the inequality $4x - 3y \geq 0$, why can't we use $(0, 0)$ as a test point? If we test the point $(6, -5)$, do we obtain a false statement or a true one?

Graph each region.

7. $y > -2x + 4$

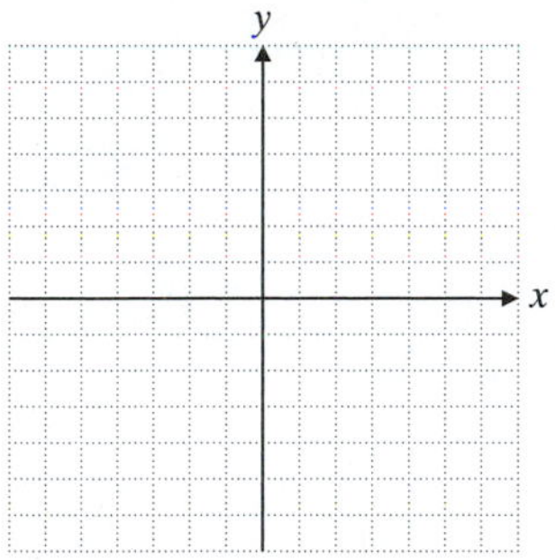

8. $y > -3x + 2$

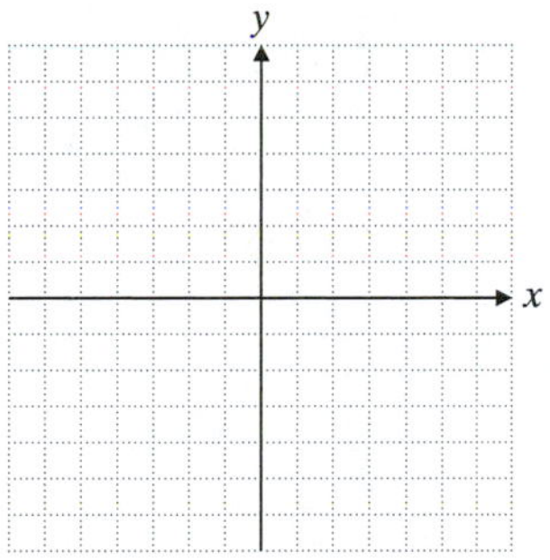

9. $y < \frac{2}{3}x - 2$

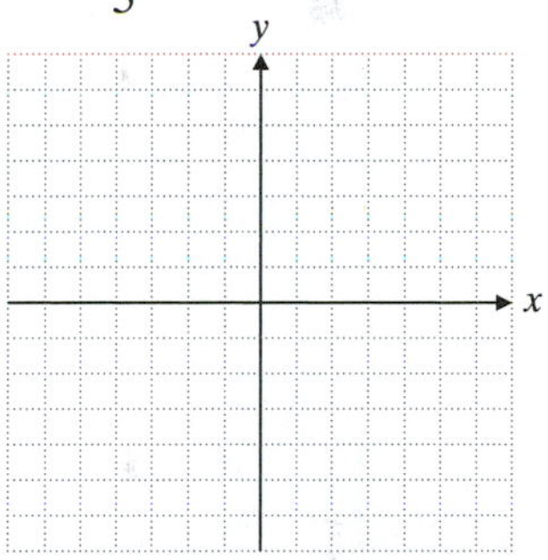

10. $y < \frac{3}{2}x - 1$

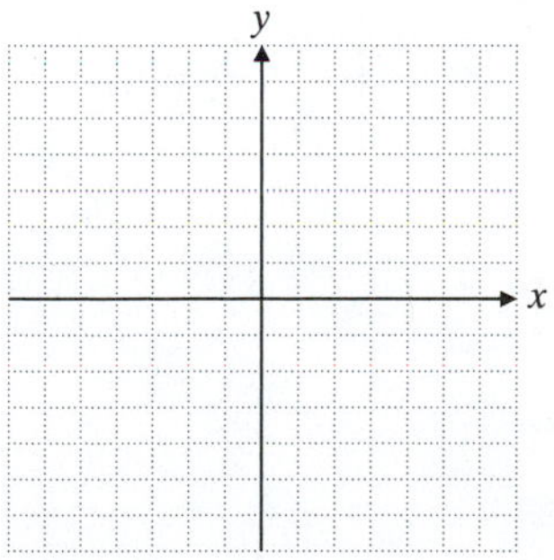

11. $y \geq \frac{3}{4}x + 4$

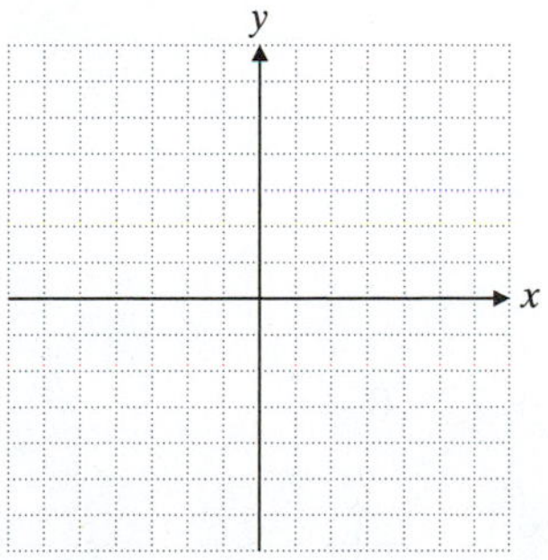

12. $y \geq -\frac{1}{5}x + 2$

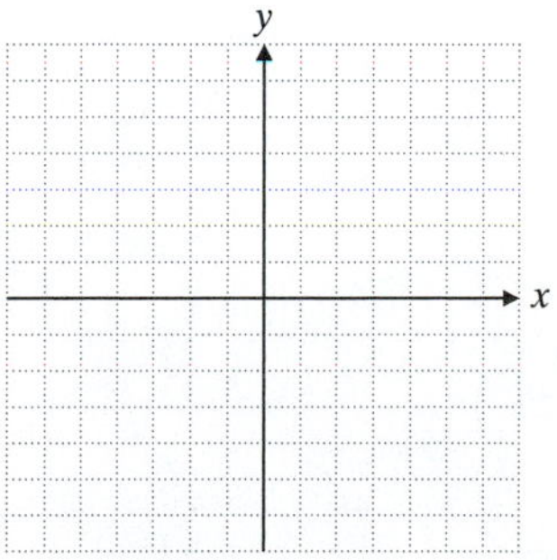

13. $-2x - y > 3$

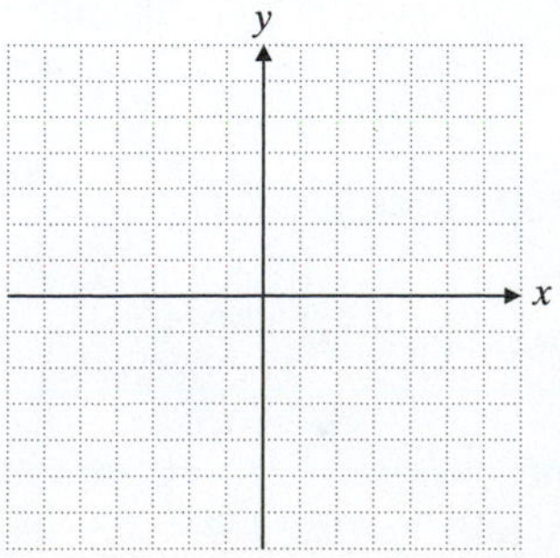

14. $6y - x \leq 12$

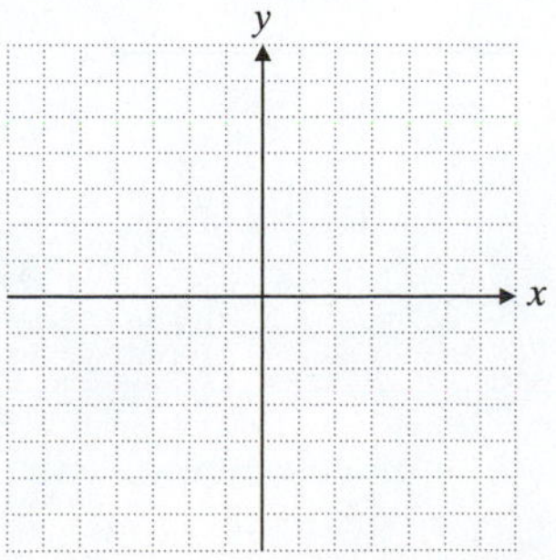

15. $-3x + y > 0$

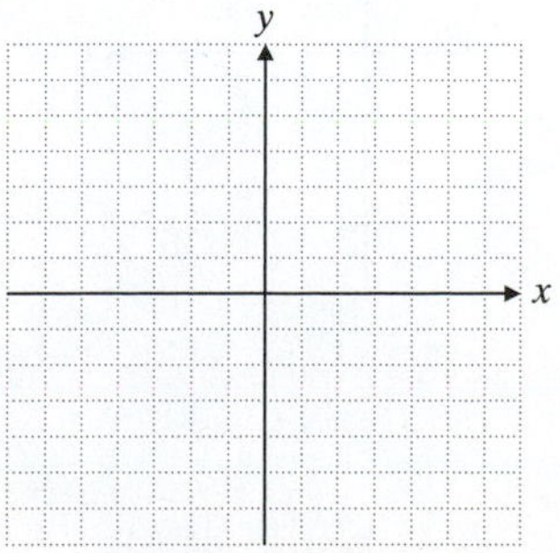

16. $x + y < 0$

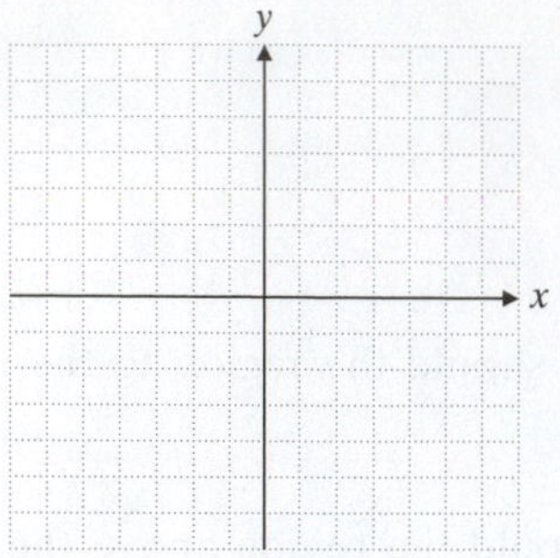

17. $5x - 2y \geq 0$

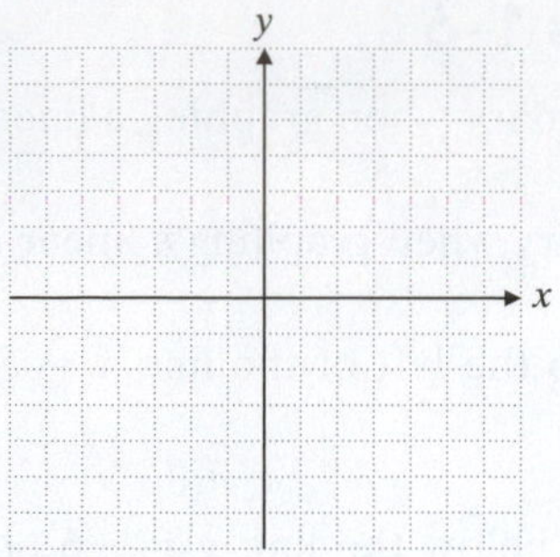

18. $x - 3y \geq 0$

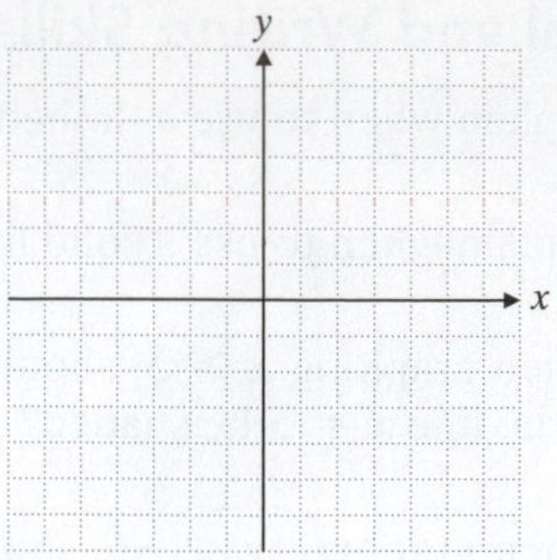

Simplify and graph each inequality in a rectangular coordinate system.

19. $x > -4$

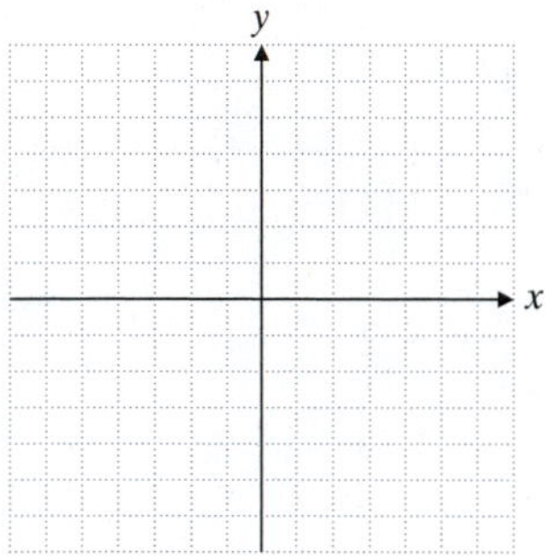

20. $x < 3$

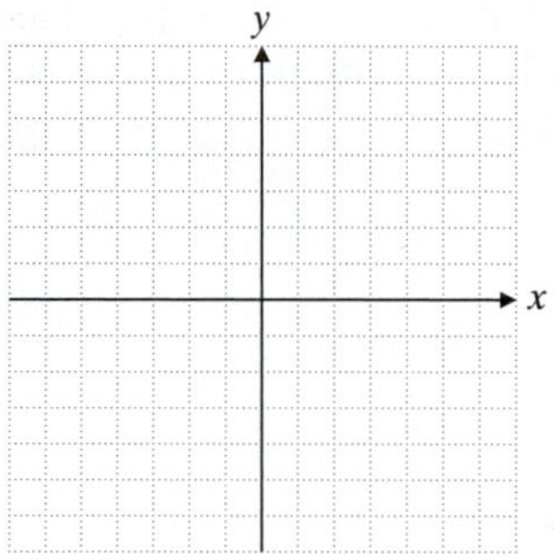

21. $y \leq -1$

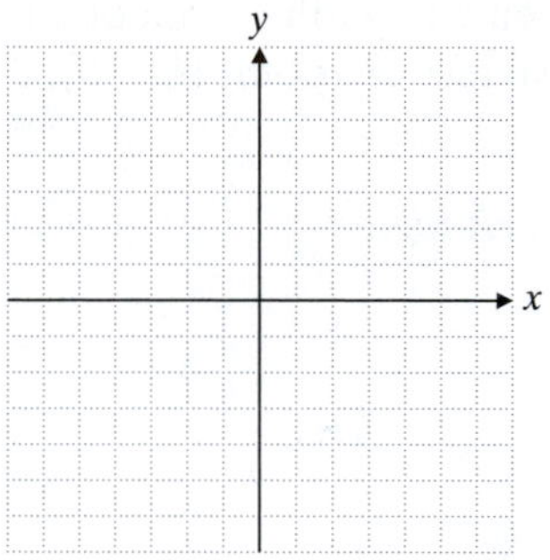

22. $y \geq -1$

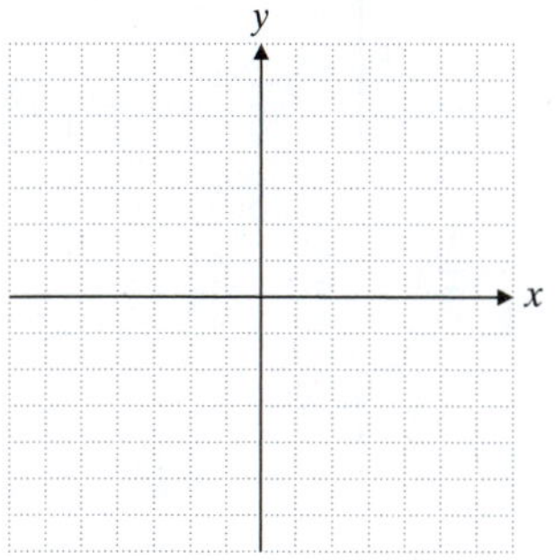

23. $-8x \leq -12$

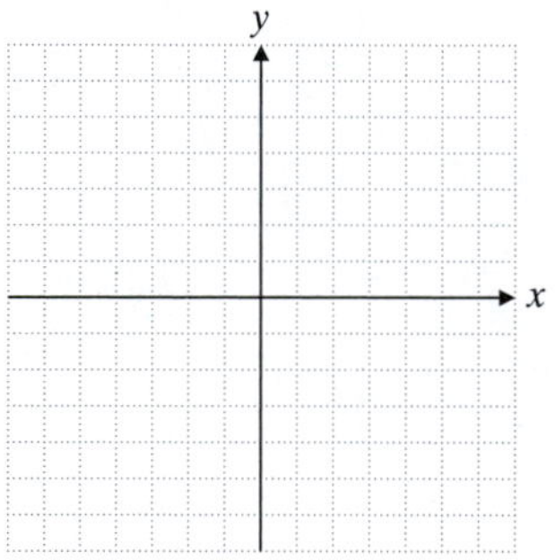

24. $-5x \leq -10$

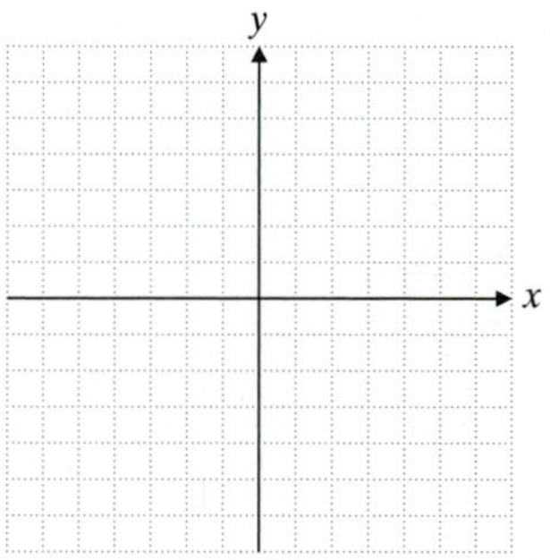

25. $4y \geq 2$

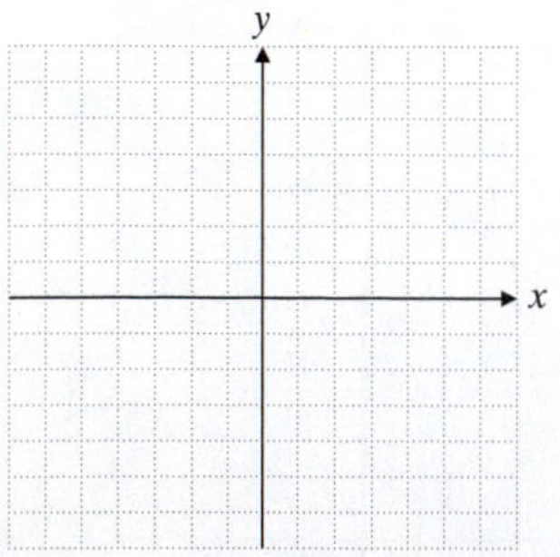

26. $3y + 2 > 0$

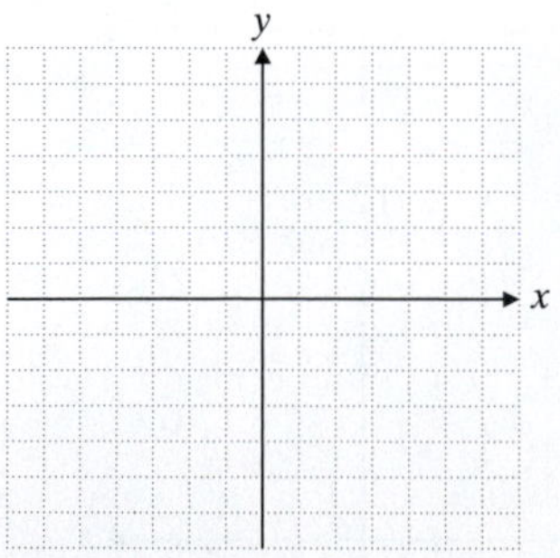

Cumulative Review

27. **[2.4.1]** Tina and Jerry ordered a carpet runner for their long upstairs hallway. The length of the runner is 5 feet longer than triple the width. If the perimeter of the runner is 34 feet, find its length and width.

28. **[2.7.3]** Solve for x. $x - 5 < 2$ *and* $3x + 1 > -2$

29. **[2.7.3]** Solve for x. $2x + 5 \leq 5$ *or* $-x + 4 \leq 2$

Quick Quiz 3.4

1. The point $(3, -4)$ lies below the line $y = 3x + 4$. Now suppose we want to graph the region where $y > 3x + 4$. Test the point $(3, -4)$ in the inequality. Should the graph of the region be shaded above the line or below the line?

2. Graph $y < -\frac{3}{5}x + 3$.

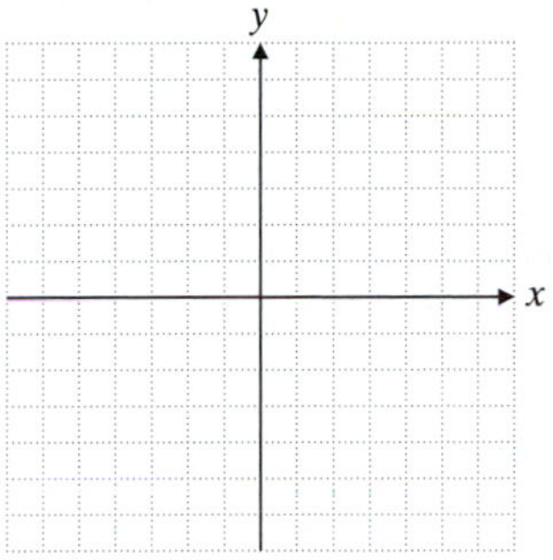

3. Graph $3x - 5y \leq -10$.

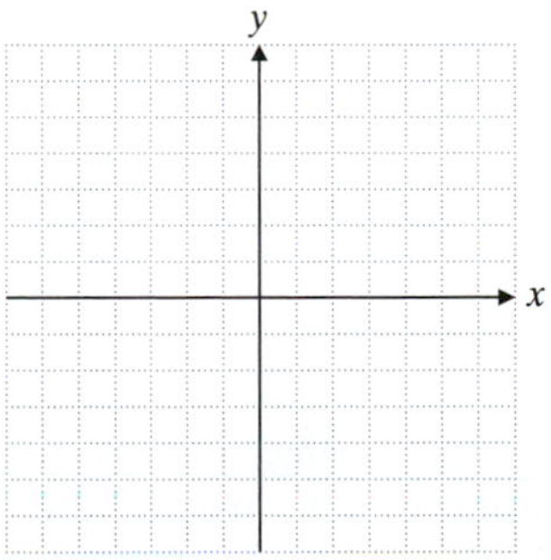

4. **Concept Check** Explain how you would graph the following region. $4y - 8 < 0$

3.5 Concept of a Function

Student Learning Objectives

After studying this section, you will be able to:

1. Describe a relation and determine its domain and range.
2. Determine whether a relation is a function.
3. Evaluate a function using function notation.

PIZZA PLUS

Size of Pizza	*Diameter of Pizza*	*Price of Pizza*
Small	5"	$4.75
Medium	10"	$7.00
Large	15"	$9.25
Party Size	20"	$11.50

① Describing a Relation and Determining Its Domain and Range

Whenever we collect and study data, we look for relationships. If we can determine a relationship between two sets of data, we can make predictions about the data. Let's begin with a simple finite example.

A local pizza parlor offers four different sizes of pizza and prices them as shown in the table on the left. By looking at the table, we can see that the price depends on the diameter of the pizza. The larger the size, the more expensive it will be. This appears to be an increasing relation. We can use a set of ordered pairs to show the correspondence between the diameter of the pizza and the price.

$$\{(5 \text{ in., } \$4.75), \quad (10 \text{ in., } \$7.00), \quad (15 \text{ in., } \$9.25), \quad (20 \text{ in., } \$11.50)\}$$

We can graph the ordered pairs to get a better picture of the relationship.

Following convention, we will assign the **independent variable** to the horizontal axis and the **dependent variable** to the vertical axis. Note that the dependent variable is price because price *depends* on size.

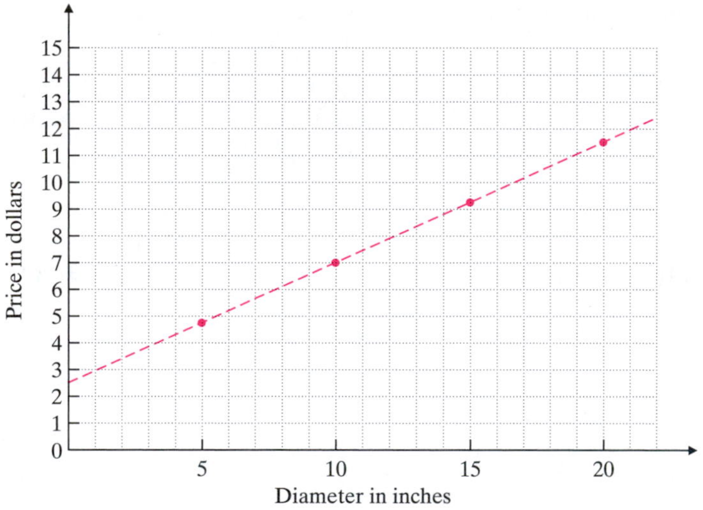

Just as we suspected, the relation is increasing. That is, the graph goes up as we move from left to right. We draw a line that approximately fits the data. This allows us to analyze the data more easily. Notice that if Pizza Plus decides to come out with a size between the small pizza and the medium pizza, it will probably be priced around \$6. Thus, we can see that there is a relation between the diameter of a pizza and its price and that we can express this relation as a table of values, as a set of ordered pairs, or in a graph.

Mathematicians have found it most useful to define a relation in terms of ordered pairs.

RELATION

A **relation** is any set of ordered pairs.

All the first items of each ordered pair in a relation can be grouped as a set called the **domain** of the relation. The domain of the pizza example is the set of sizes (diameters measured in inches) {5, 10, 15, 20}. These are all the possible values of the independent variable *diameter*. The domain is the input. It is the starting value.

All the second items of each ordered pair in a relation can be grouped as a set called the **range** of the relation. The range of the pizza example is the set of prices {\$4.75, \$7.00, \$9.25, \$11.50}. These are the corresponding values of the dependent variable *price*. The range is the output. The output is the result that comes out once you pick a diameter.

EXAMPLE 1 The information in the following table can be found in most almanacs.

Look at the data for the men's 100-meter race. Is there a relation between any two sets of data in this table? If so, describe the relation as a table of values, a set of ordered pairs, and a graph.

Olympic Games: 100-Meter Race for Men

Year	Winning runner, Country	Time
1900	Francis W. Jarvis, USA	11.0 s
1912	Ralph Craig, USA	10.8 s
1924	Harold Abrahams, Great Britain	10.6 s
1936	Jesse Owens, USA	10.3 s
1948	Harrison Dillard, USA	10.3 s
1960	Armin Harg, Germany	10.2 s
1972	Valery Borzov, USSR	10.14 s
1984	Carl Lewis, USA	9.99 s
1996	Donovan Bailey, Canada	9.84 s
2008	Usain Bolt, Jamaica	9.69 s

Source: The World Almanac

Solution A useful relation might be the correspondence between the year the event occurred and the winning time in the race. Let's see how this looks in a table, as a set of ordered pairs, and on a graph. Because we will choose the year as the independent variable, we will list it first in the table, as is customary.

Table.

Year	1900	1912	1924	1936	1948	1960	1972	1984	1996	2008
Time in Seconds	11.0	10.8	10.6	10.3	10.3	10.2	10.14	9.99	9.84	9.69

Ordered pairs.

$$\{(1900, 11.0), (1912, 10.8), (1924, 10.6), (1936, 10.3), (1948, 10.3), (1960, 10.2), (1972, 10.14), (1984, 9.99), (1996, 9.84), (2008, 9.69)\}$$

Graph. To save space, we draw the graph so that the time values on the vertical axis range from 9 seconds to 11 seconds rather than from 0 seconds to 11 seconds. We indicate a break like this in the scale on a vertical axis with the symbol ≀ on the axis. On a horizontal axis we use the symbol –∿–. By looking at the graph, we can see that the winning time usually decreases each year. This is a decreasing relation most of the time. What might we expect the winning time to be in 2012? Can we expect the time to decrease indefinitely? Why or why not?

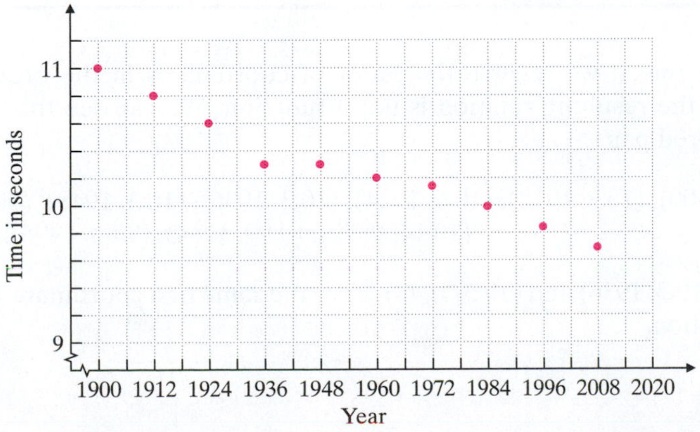

Continued on next page

***NOTE TO STUDENT:** Fully worked-out solutions to all of the Student Practice problems can be found at the back of the text starting at page SP-1.*

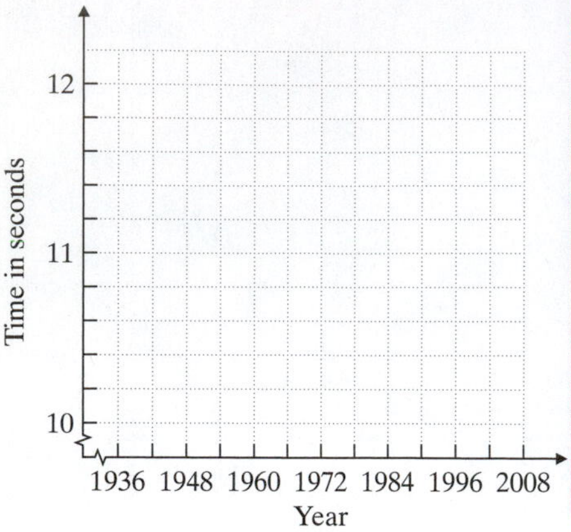

Student Practice 1

The data for the women's 100-meter race in the Olympic Games for selected years are given in the table on the right. Describe the relation between the year and the winning time in a table, as a set of ordered pairs, and as a graph.

Olympic Games: 100-Meter Race for Women

Year	Winning runner, Country	Time
1936	Helen Stephens, USA	11.5 s
1948	Francina Blankers-Koen, Netherlands	11.9 s
1960	Wilma Rudolph, USA	11.0 s
1972	Renate Stecher, East Germany	11.07 s
1984	Evelyn Ashford, USA	10.97 s
1996	Gail Devers, USA	10.94 s
2008	Shelly-Ann Fraser, Jamaica	10.78 s

Source: The World Almanac

Year							
Time in Seconds							

② Determining Whether a Relation Is a Function

The two relations we have just discussed have a special characteristic. Each value in the domain is matched with exactly one value in the range. This is true of our pizza example. One size of pizza does not have several different prices. This is also true of the relation between the year and winning time in the 100-meter race for men. No year has several different winning times.

Looking at this issue in terms of ordered pairs, we say that no two different ordered pairs have the same first coordinate. We will list each set of ordered pairs so that you can verify this characteristic.

{(5 in., \$4.75), (10 in., \$7.00), (15 in., \$9.25), (20 in., \$11.50)}

{(1900, 11.0), (1912, 10.8), (1924, 10.6), (1936, 10.3), (1948, 10.3), (1960, 10.2), (1972, 10.14), (1984, 9.99), (1996, 9.84), (2008, 9.69)}

All such relations with this special property are called *functions*. Each input has only one output.

FUNCTION

A **function** is a relation in which no two different ordered pairs have the same first coordinate.

Notice that if we reverse the order of coordinates in the ordered pairs in Example 1, the resulting relation is not a function. We can see this readily if we list the ordered pairs.

{(11.0, 1900), (10.8, 1912), (10.6, 1924), (10.3, 1936), (10.3, 1948), (10.2, 1960), (10.14, 1972), (9.99, 1984), (9.84, 1996), (9.69, 2008)}

Two pairs, (10.3, 1936) and (10.3, 1948), have the same first coordinate. This relation is not a function.

EXAMPLE 2 Give the domain and range of each relation. Indicate whether the relation is a function.

(a) $g = \{(2, 8), (2, 3), (3, 7), (5, 12)\}$

(b) Individuals' Incomes and Taxes

Income in Dollars	14,000	18,000	24,500	33,000	50,000	50,000
Income Tax in Dollars	2350	2800	2900	3750	1350	7980

(c) Women's Tibia Bone Lengths and Heights

Length of Tibia Bone in Centimeters	33	34	35	36
Height of the Woman in Centimeters	151	154	156	159

Source: National Center for Health Statistics.

Solution Recall that the domain of a function consists of all the possible values of the independent variable or input. In a set of ordered pairs, this is the first item in each ordered pair.

The range of a function consists of the corresponding values of the dependent variable. In a set of ordered pairs, this is the second item or output in each ordered pair.

(a) Domain $= \{2, 3, 5\}$

Range $= \{8, 3, 7, 12\}$

(2, 8) and (2, 3) have the same first coordinate. Thus, g is not a function.

(b) Domain $= \{14{,}000, 18{,}000, 24{,}500, 33{,}000, 50{,}000\}$

Range $= \{2350, 2800, 2900, 3750, 1350, 7980\}$

(50,000, 1350) and (50,000, 7980) have the same first coordinate. This relation is not a function.

(c) Domain $= \{33, 34, 35, 36\}$

Range $= \{151, 154, 156, 159\}$

No two different ordered pairs have the same first coordinate. This relation is a function.

Student Practice 2 Give the domain and range of the given relation. Indicate whether the relation is a function.

Car Performance

Horsepower	158	161	163	160	161
Top Speed (Mph)	98.6	89.2	101.4	102.3	94.9

If we are looking at the graph of a function, it will never have two different ordered pairs with the same x-value. We will examine some graphs in exercises 17–25.

Often we can tell a lot about the graph of a relation. Any graph that is not a function will have at least one region in which a vertical line will cross the graph more than once.

VERTICAL LINE TEST

If a vertical line can intersect the graph of a relation more than once, the relation is not a function. If no such line can be drawn, then the relation is a function.

EXAMPLE 3 Determine whether each of the following is a graph of a function.

(a)

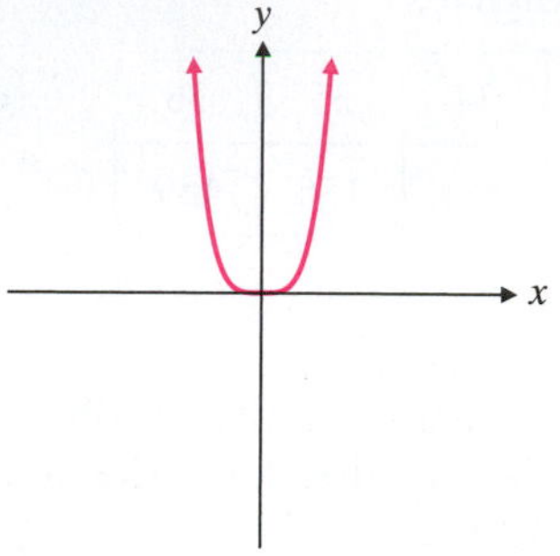

(b)

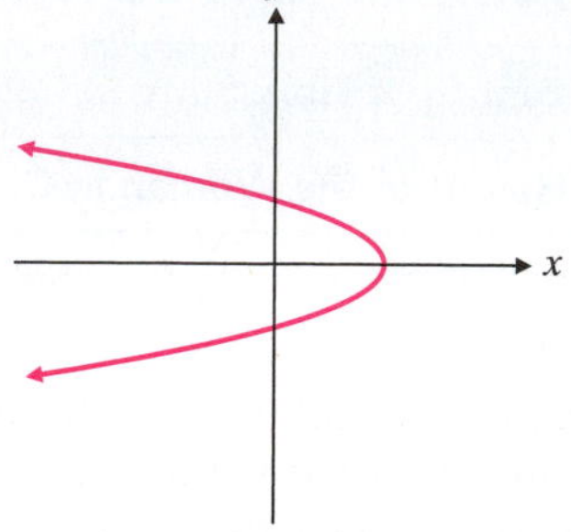

Solution

(a) This is a function. No vertical line will pass through more than one ordered pair on the curve.

(b) This is not a function. A vertical line could pass through two points on the curve.

Student Practice 3 Determine whether each of following is a graph of a function.

(a)

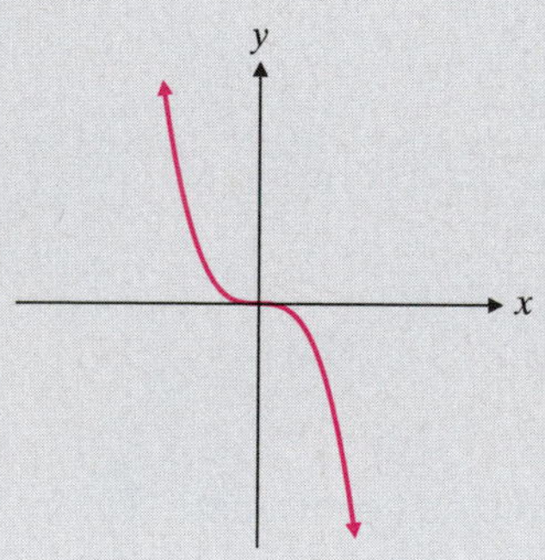

(b)

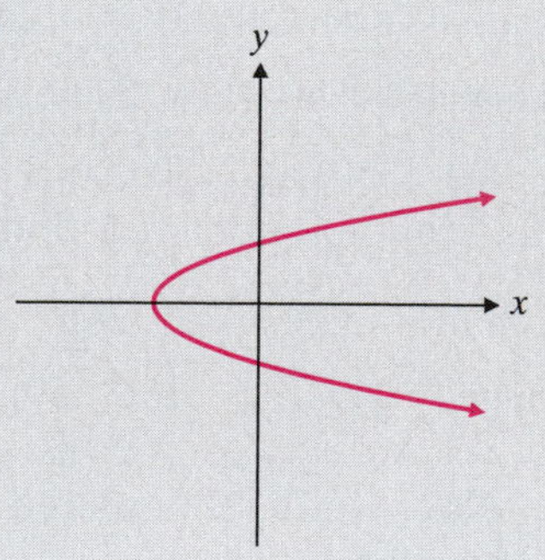

③ Evaluating a Function Using Function Notation

We can use **function notation** to indicate the relationship between the ordered pairs in a function. Looking back at our pizza example, we see that the diameter of a pizza determines its price. Since price is a function of diameter and a 5-inch-diameter pizza costs \$4.75 we can write

$$f(5) = 4.75.$$

Some equations describe functions. You are already familiar with linear equations. Let's look at one such equation.

$$y = 3x - 2$$

In this equation, y is a function of x. That is, for each value of x in the domain, there is a unique value of y in the range. We can use function notation when we determine the function value y for specific values of x. Let $x = 0$. Then we can write the following.

$$f(x) = 3x - 2$$
$$f(0) = 3(0) - 2 \quad \text{Substitute 0 for } x.$$
$$f(0) = -2 \quad \text{Evaluate.}$$

We see that when x is 0, $f(x)$ is -2. That is, the value of the function is -2 when x is 0. We can write this ordered pair solution as $(x, f(x)) = (0, f(0)) = (0, -2)$.

Note: The notation $f(x)$ does not mean that f is multiplied by x. It means that for any specific value of x, there is only one value for y. $f(x)$ is read as "f of x." Although we commonly use f as the function name, we can also use other variables, like g and h.

It may help you to understand the idea of a function by imagining a "function machine." An item from the domain enters as input into the machine—the function—and a member of the range results as output.

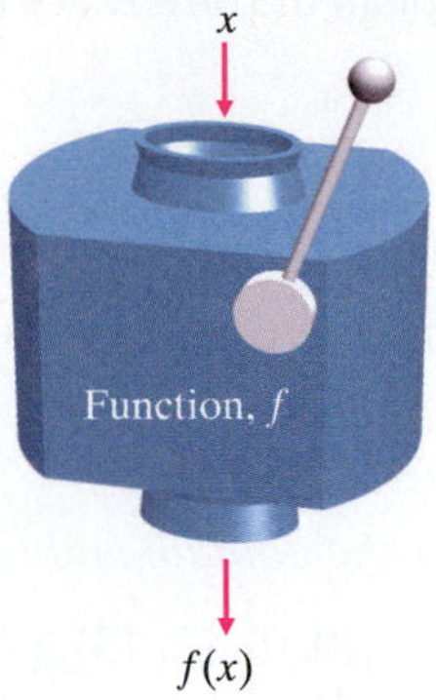

EXAMPLE 4 If $f(x) = 4 - 3x$, find each of the following.

(a) $f(2)$ **(b)** $f(-1)$ **(c)** $f\left(\frac{1}{3}\right)$

Solution In each case we replace x by the specific value from the domain.

(a)
$$f(2) = 4 - 3(2)$$
$$= 4 - 6$$
$$f(2) = -2$$

(b)
$$f(-1) = 4 - 3(-1)$$
$$= 4 + 3$$
$$f(-1) = 7$$

(c)
$$f\left(\frac{1}{3}\right) = 4 - 3\left(\frac{1}{3}\right)$$
$$= 4 - 1$$
$$f\left(\frac{1}{3}\right) = 3$$

Student Practice 4 If $f(x) = 2x^2 - 8$, find each of the following.

(a) $f(-3)$ **(b)** $f(4)$ **(c)** $f\left(\frac{1}{2}\right)$

3.5 Exercises

MyMathLab®

Watch the videos in MyMathLab

Download the MyDashBoard App

Verbal and Writing Skills, Exercises 1–4

1. Explain the difference between a relation and a function.

2. Explain the difference between the domain and the range of a relation.

3. What are the three ways you can describe a function?

4. Write the ordered pair for $f(-5) = 8$ and identify the x- and y-values.

What are the domain and range of each relation? Is the relation a function?

5. $D = \{(0, 0), (5, 13), (7, 11), (5, 0)\}$

6. $C = \{(\frac{1}{2}, 5), (-3, 7), (\frac{3}{2}, 5), (\frac{1}{2}, -1)\}$

7. $F = \{(85, -12), (16, 4), (-102, 4), (62, 48)\}$

8. $E = \{(40, 10), (-18, 27), (38, 10), (57, -15)\}$

9. ***Women's Dress Sizes***

USA	6	8	10	12	14
France	38	40	42	44	46

10. ***Women's Shoe Sizes***

USA	7	$7\frac{1}{2}$	8	$8\frac{1}{2}$	9
Europe	39	40	41	42	43

11. ***Average Monthly Fahrenheit Temperature: Pago Pago, Samoa***

Month	Jan.	Feb.	Mar.	Apr.	May	June	July	Aug.	Sept.	Oct.	Nov.	Dec.
Temperature	81	81	81	81	80	80	79	79	79	80	80	81

Source: United Nations Statistics Division

12. ***Tallest Mountain Peaks***

Country	Nepal/Tibet	Pakistan/China	India/Nepal	Nepal/Tibet	Nepal/Tibet
Height in Meters	8850	8611	8586	8516	8463

13. ***Tallest Buildings, USA***

City	Chicago	New York	New York	Chicago	Chicago	New York
Height in Feet	1454	1350	1250	1136	1127	1046

Source: World Almanac

14. ***Metric Conversion***

Miles	1	2	3	4	5
Kilometers	1.61	3.22	4.83	6.44	8.05

15. ***Mariner's Speed Conversion***

Knots	10	20	30	40	50
Miles per Hour	11.51	23.02	34.53	46.04	57.55

16. ***Metric Conversion***

Gallons	1	2	3	4	5	6
Liters	3.79	7.57	11.36	15.14	18.93	22.71

Determine whether the graph represents a function.

17.

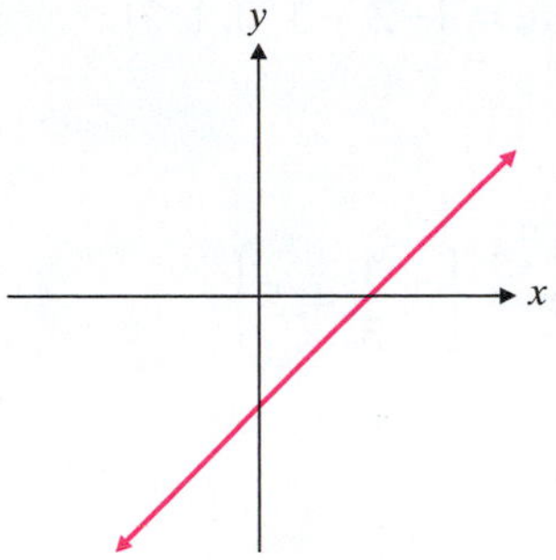

18.

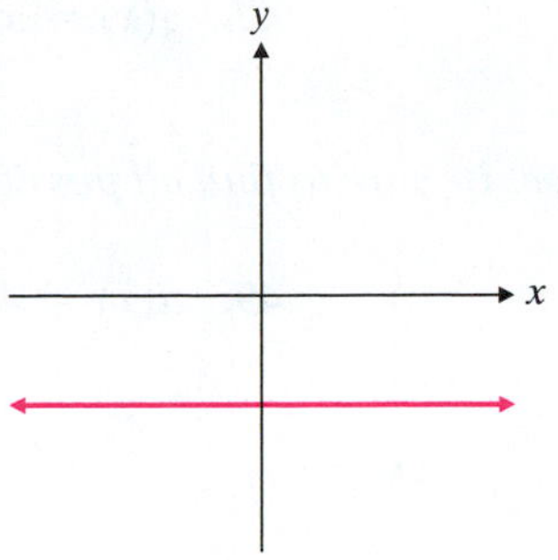

19.

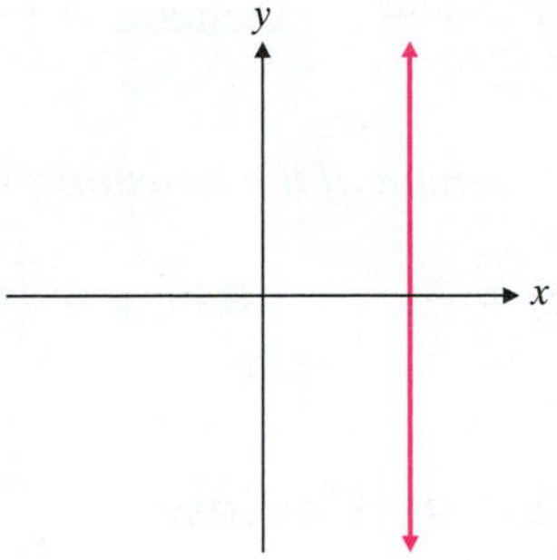

20.

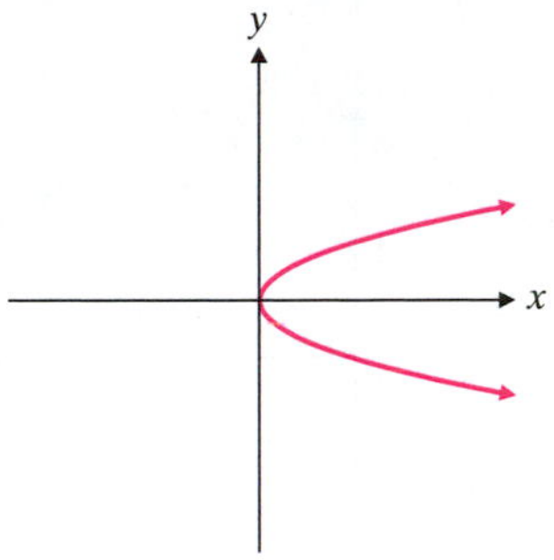

21.

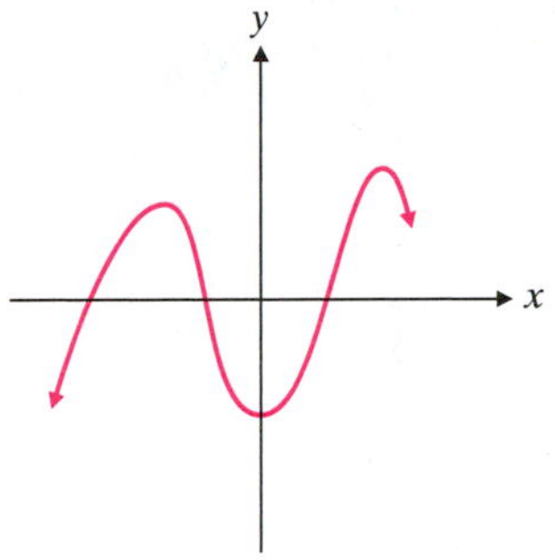

22.

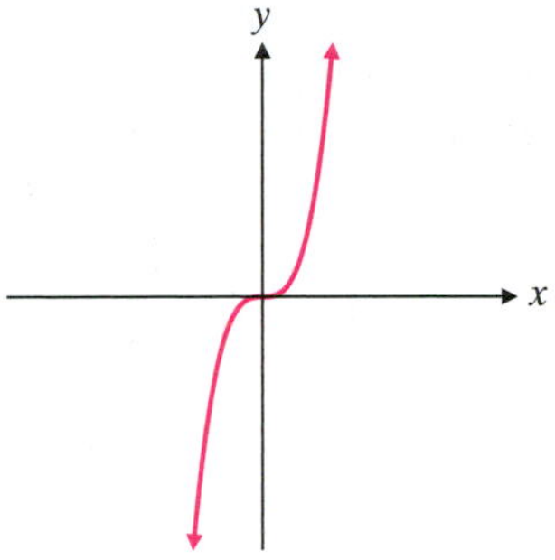

23.

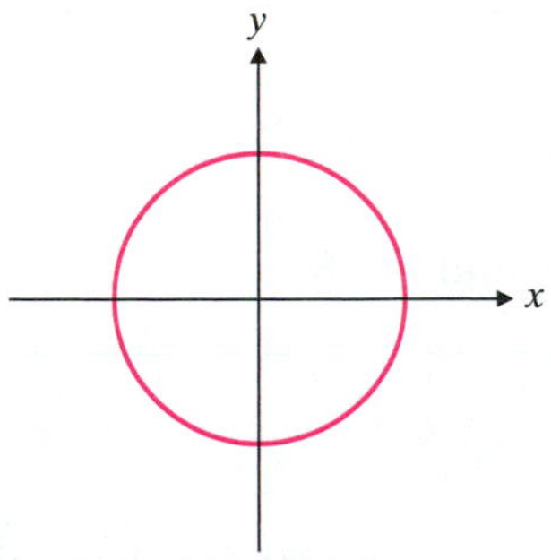

24.

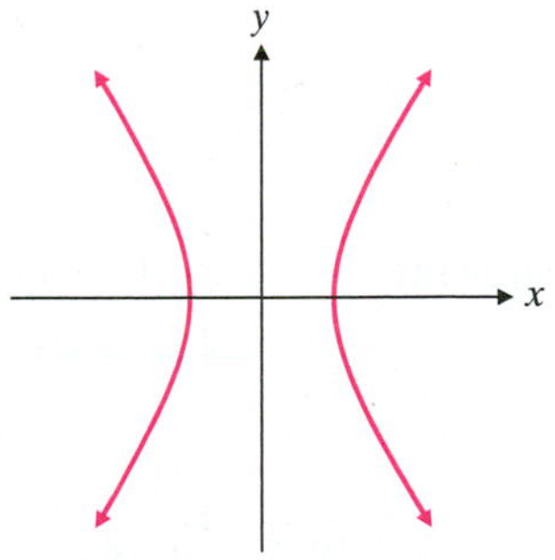

25.

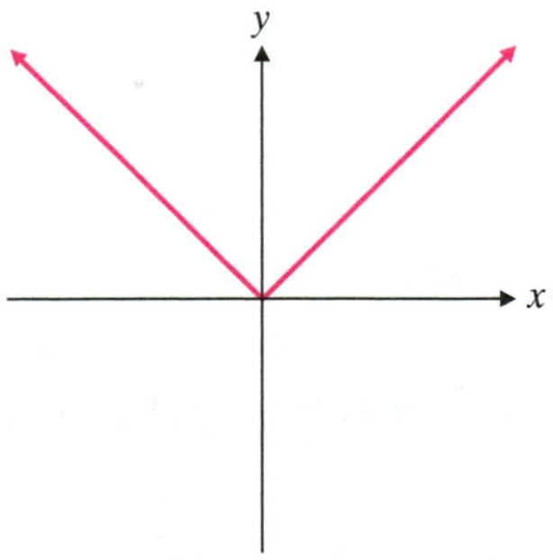

Given the function defined by $g(x) = 2x - 5$, *find the following.*

26. $g(-3)$ **27.** $g(-2.4)$ **28.** $g\left(\frac{1}{3}\right)$ **29.** $g\left(\frac{1}{2}\right)$

Given the function defined by $h(x) = \frac{3}{4}x + 1$, *find the following.*

30. $h(0)$ **31.** $h(-8)$ **32.** $h\left(\frac{1}{3}\right)$ **33.** $h(3)$

Given the function defined by $r(x) = 2x^2 - 4x + 1$, *find the following.*

34. $r(1)$ **35.** $r(-1)$ **36.** $r(0.1)$ **37.** $r(-0.1)$

Given the function defined by $t(x) = x^3 - 3x^2 + 2x - 3$, *find the following.*

38. $t(10)$ **39.** $t(5)$ **40.** $t(-1)$ **41.** $t(-2)$

42. If $f(x) = \sqrt{x + 10}$, find $f(-6)$.

43. If $f(x) = \sqrt{2 - x}$, find $f(-2)$.

44. If $g(x) = |x^2 + 5|$, find $g(-4)$.

45. If $g(x) = |6 - x^2|$, find $g(2)$.

To Think About

Find the range of the function for the given domain.

46. $f(x) = x + 3$ Domain $= \{-2, -1, 0, 1, 2\}$

47. $g(x) = x^2 + 3$ Domain $= \{-2, -1, 0, 1, 2\}$

Find the domain of the function for the given range. Be sure to find all possible values.

48. $h(x) = \frac{1}{2}x + 3$ Range $= \left\{0, 2, \frac{7}{2}, 9\right\}$

49. $d(x) = 4 - \frac{1}{3}x$ Range $= \left\{0, \frac{4}{3}, 4, 5\right\}$

Cumulative Review

50. [2.3.1] Solve. $|3x - 2| = 1$

51. [2.8.1] Solve. $|x - 5| \leq 3$

52. [3.1.1] Graph the line $2x - 5y = 10$.

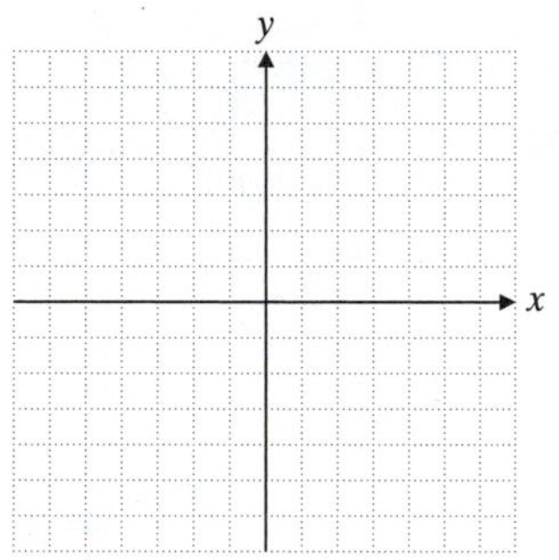

Quick Quiz 3.5

1. What are the domain and range of the following relation?

$\{(9, 3), (-2, 3), (4, 5), (-4, 5)\}$

2. If $p(x) = \left|-\frac{3}{4}x + 2\right|$, find $p(-8)$.

3. If $f(x) = -2x^3 + 4x^2 - x + 4$, find $f(-2)$.

4. Concept Check Explain how you can determine from a graph whether or not the graph represents a function.

3.6 Graphing Functions from Equations and Tables of Data

① Graphing a Function from an Equation

Frequently, we are given a function in the form of an equation and are asked to graph it. Each value of the function $f(x)$, often labeled y, corresponds to a value in the domain, often labeled x. This correspondence is the ordered pair (x, y) or $(x, f(x))$. The graph of the function is the graph of the ordered pairs.

If a function can be written in the form $f(x) = mx + b$, it is called a **linear function.** The graph of a linear function is a straight line.

If we can describe a real-life relationship with a function in the form of an equation, a table of values, and/or a graph, we can determine characteristics of the relationship and make predictions.

Student Learning Objectives

After studying this section, you will be able to:

① **Graph a function from an equation.**

② **Graph a function from a table of data.**

EXAMPLE 1 A salesperson earns \$15,000 a year plus a 20% commission on her total sales. Express her annual income in dollars as a function of her total sales. Determine values of the function for total sales of \$0, \$25,000, and \$50,000. Graph the function and determine whether the function is increasing or decreasing.

Solution Since we want to express income as a function of sales, let's see how income is determined.

$$\text{income} = \$15{,}000 + 20\% \text{ of total sales}$$

Income depends on total sales. Thus, total sales is the independent variable, and income is the dependent variable.

Let x = the amount of total sales in dollars and
$i(x)$ = income in dollars.
$i(x) = 15{,}000 + 0.20x$ Change 20% to 0.20.
$i(0) = 15{,}000 + 0.20(0) = 15{,}000$
$i(25{,}000) = 15{,}000 + 0.20(25{,}000) = 20{,}000$
$i(50{,}000) = 15{,}000 + 0.20(50{,}000) = 25{,}000$

We can put this information in a table of values.

x (dollars)	i(x) (dollars)
0	15,000
25,000	20,000
50,000	25,000

To facilitate the graphing, we will use a scale in thousands. Thus, we modify the table to make our task of graphing easier.

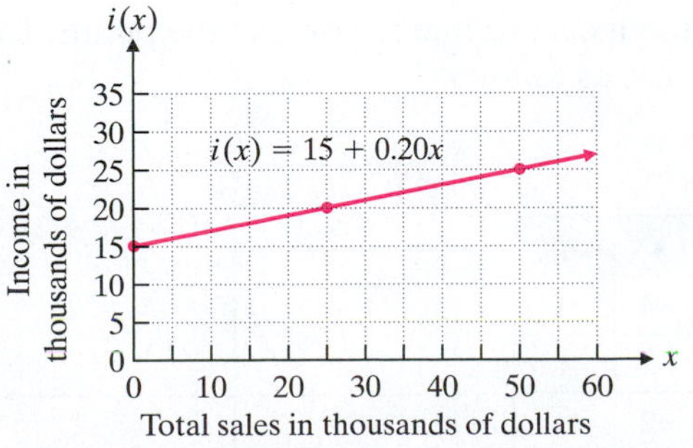

x (thousands of dollars)	i(x) (thousands of dollars)
0	15
25	20
50	25

The function is increasing.

Student Practice 1 A salesperson earns \$10,000 a year plus a 15% commission on her total sales. Express her annual income in dollars as a function d of her total sales x. Use sales amounts of \$0, \$40,000, and \$80,000 to graph her salary.

Student Practice 1

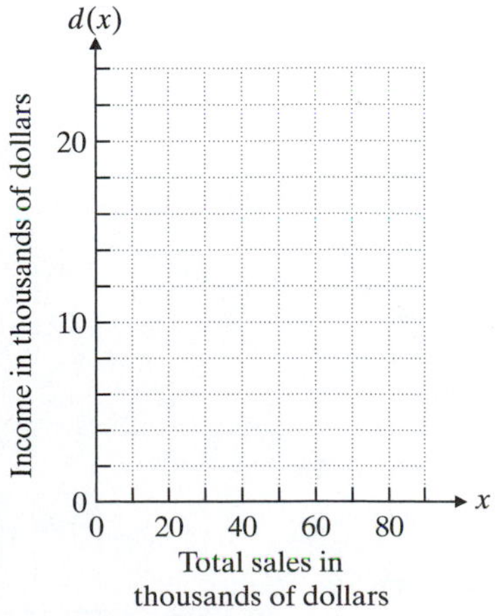

***NOTE TO STUDENT:** Fully worked-out solutions to all of the Student Practice problems can be found at the back of the text starting at page SP-1.*

Thus far, most of our work has been with linear functions. Let's look at some other functions and their graphs. We will begin with the **absolute value function** $f(x) = |x|$.

EXAMPLE 2 Graph $f(x) = |x|$.

Solution Let's find the function values for five values of the independent variable x. Because of the nature of the function, we will choose both negative and positive values for x. The table of values and the resulting graph are as follows:

x	f(x)
−2	2
−1	1
0	0
1	1
2	2

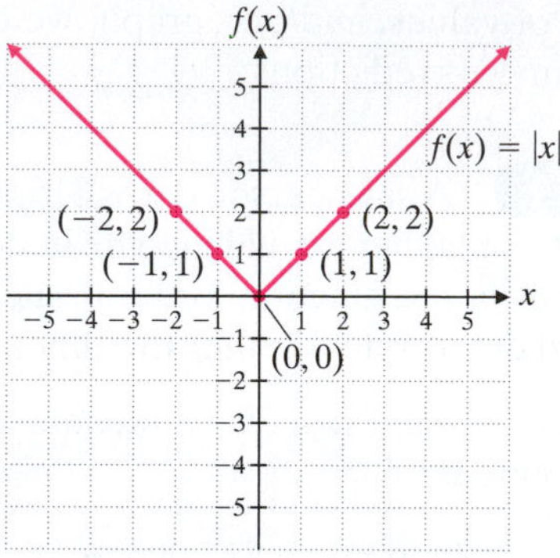

Notice that the graph is **symmetric about the *y*-axis.** That is, the graph on one side of the y-axis is a mirror image of the graph on the other side of the y-axis. This means that if the point (a, b) is on the graph, then the point $(-a, b)$ is also on the graph.

Student Practice 2 Graph the function $h(x) = |x + 2|$. Use integer values for x from -4 to 0.

Student Practice 2

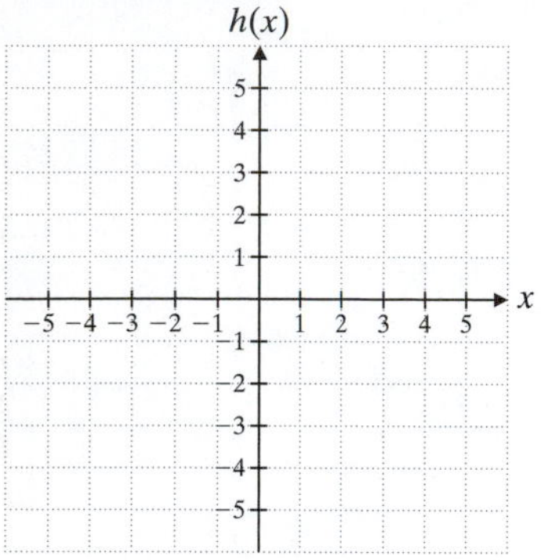

TO THINK ABOUT: Example 2 Follow-up How are the graphs of $f(x) = |x|$ and $h(x) = |x + 2|$ the same? How are they different? What would the graph of $k(x) = |x - 2|$ look like?

Now let's take a look at another function, $p(x) = x^2$. What happens when $x = 2$? When $x = -2$? Notice that when x is 2 or -2, the value of the function is 4. The ordered pairs are $(-2, 4)$ and $(2, 4)$. This graph is also symmetric about the y-axis as we shall see in the next example.

EXAMPLE 3 Graph. **(a)** $p(x) = x^2$ **(b)** $q(x) = (x + 2)^2$

Solution For each function we will choose both negative and positive values of x. Since these are *not* linear functions, we use a curved line to connect the points. The tables of values and the resulting graphs are as follows:

(a)

x	p(x)
−2	4
−1	1
0	0
1	1
2	4

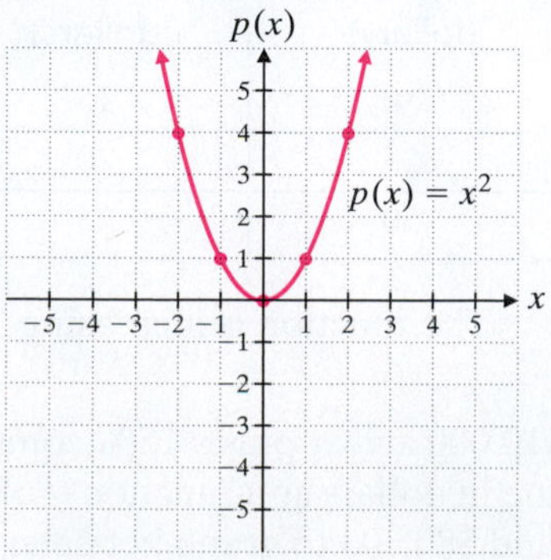

(b)

x	q(x)
−4	4
−3	1
−2	0
−1	1
0	4
1	9

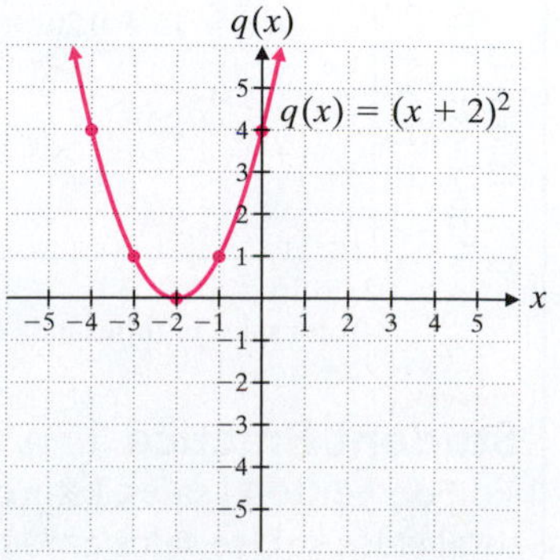

Student Practice 3 Graph **(a)** $r(x) = (x - 2)^2$ **(b)** $s(x) = x^2 + 2$

(a)

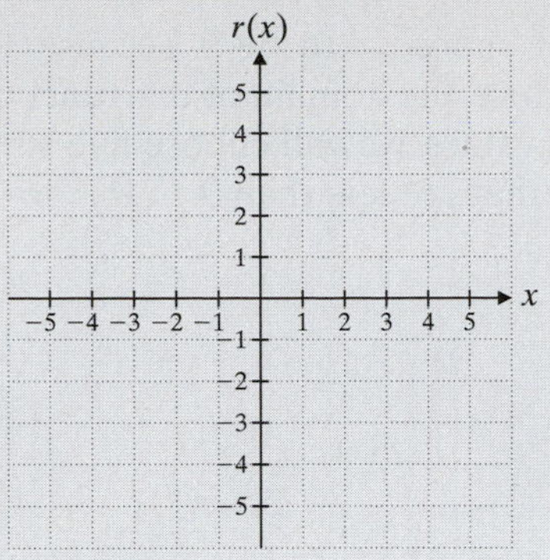

(b)

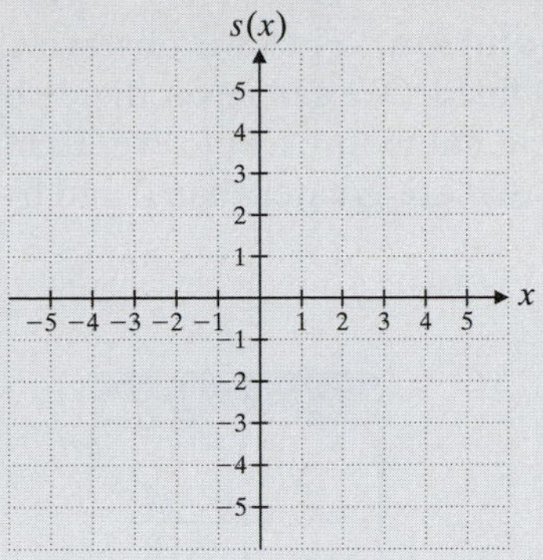

TO THINK ABOUT: Example 3 Follow-up How are the graphs of each of the previous functions the same? How are they different? Describe how changing $p(x) = x^2$ to $q(x) = (x + 5)^2$ and to $s(x) = x^2 + 5$ would affect the graph of the function.

Another interesting graph is the graph of the function $g(x) = x^3$. A quick look at $g(x)$ when x is 2 and -2 reveals that $g(x)$ is *not* symmetric about the y-axis. That is, there are different function values for 2 and -2. Let's see.

$$g(x) = x^3$$

$$g(2) = 2^3 = 8 \qquad g(-2) = (-2)^3 = -8$$

EXAMPLE 4 Graph. **(a)** $g(x) = x^3$ **(b)** $h(x) = x^3 + 1$

Solution We will pick five values for x, find the corresponding function values, and plot the five points to assist us in sketching the graph.

(a)

x	g(x)
−2	−8
−1	−1
0	0
1	1
2	8

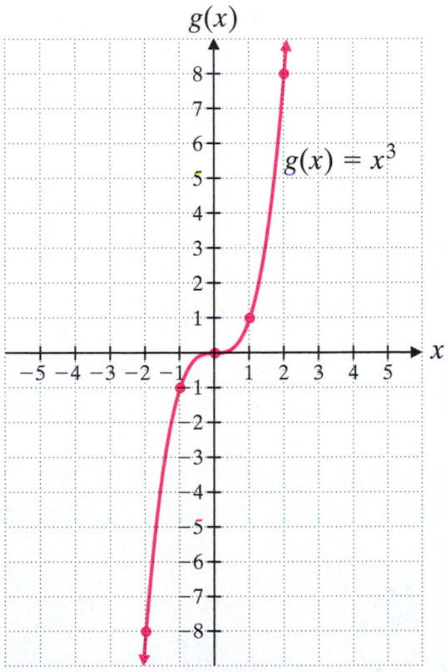

(b)

x	h(x)
−2	−7
−1	0
0	1
1	2
2	9

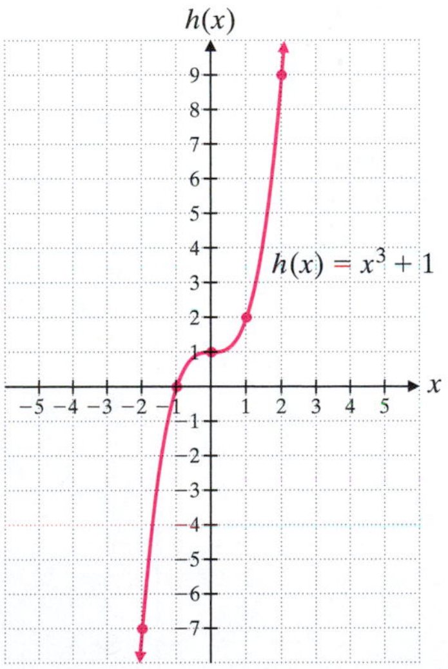

Student Practice 4 Consider the function $f(x) = x^3 - 2$. What do you think the graph of $f(x) = x^3 - 2$ will look like? Make a table of values for the function and draw the graph.

Student Practice 4

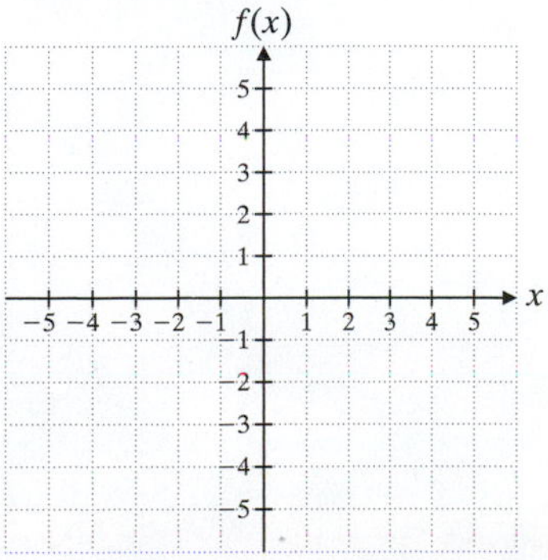

TO THINK ABOUT: Example 4 Follow-up Describe how the graph of $f(x) = x^3 - 2$ is related to the graph of $g(x) = x^3$.

In Examples 1–4 each function has had a domain of all real numbers. We were free to choose any real number for x. However, if a function is written in the form of a fraction and it contains a variable in the denominator, the domain of that function will not include any value for which the denominator becomes zero.

EXAMPLE 5 Graph $p(x) = \frac{4}{x}$.

Solution First we observe that we cannot choose x to be 0 because $\frac{4}{0}$ is not defined. (We can never divide by zero.) Therefore, the domain of this function is all real numbers except zero. To make a table of values, we will choose five values for x that are greater than 0 and five values for x that are less than 0.

x	$p(x)$
8	$\frac{1}{2}$
4	1
2	2
1	4
$\frac{1}{2}$	8
$-\frac{1}{2}$	-8
-1	-4
-2	-2
-4	-1
-8	$-\frac{1}{2}$

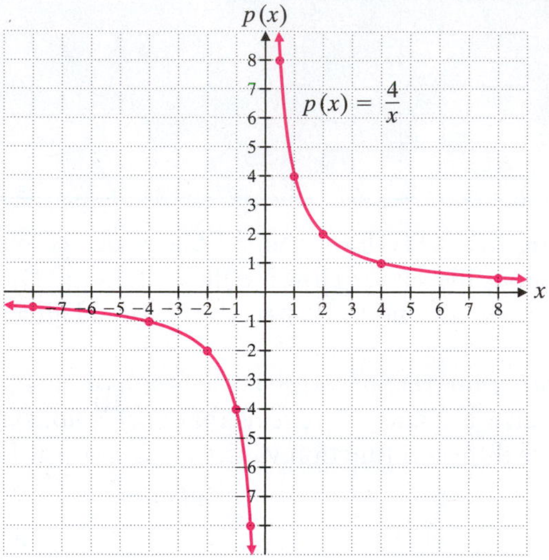

The study of this type of function, called a **rational function,** will be continued in a more advanced course, such as college algebra or precalculus.

Graphing Calculator

Horizontal Shifts

Using a graphing calculator, graph

$y_1 = 4/x$,
$y_2 = 4/(x + 2)$, and
$y_3 = 4/(x - 3)$

on the same set of axes. Use a window of $[-8, 8]$ by $[-8, 8]$.

What pattern do you observe?
What effect does c have on the graph of $y = 4/(x + c)$?
What would the graph of $y = 4/(x + 4)$ look like?
Verify using your calculator.

Student Practice 5 Graph.

$$q(x) = -\frac{4}{x}$$

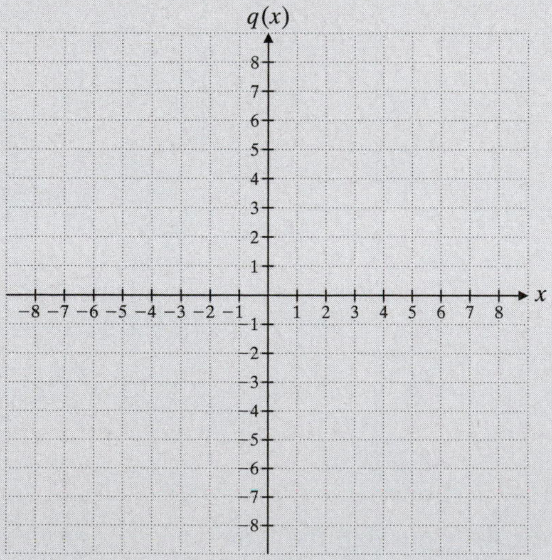

TO THINK ABOUT: Example 5 Follow-up Describe how the graph of $q(x) = -\frac{4}{x}$ is related to the graph of $p(x) = \frac{4}{x}$.

② Graphing a Function from a Table of Data

Sometimes when we study an event in daily life, we record data to better understand the event. In many cases we will not have an equation to work with, but rather a table of values that gives a general indication of some functional relationship. A graph of the values may help us to understand this underlying function.

In the following example, the number of items sold is the domain and the profit obtained from the sale is the range.

EXAMPLE 6 The marketing manager of a shoe company compiled the data in the table.

(a) Plot the data values and connect the points to see the graph of the underlying function.

(b) From the graph, determine the profit from selling 4000 pairs of shoes in one month.

(c) What kind of profit would you expect to make from selling 0 pairs of shoes in a month? What value do you obtain on the graph for $x = 0$? What does this mean?

x Pairs of Shoes Sold in a Month (In Thousands of Pairs)	$p(x)$ Monthly Profit from the Sales of Shoes (In Thousands of Dollars)
3	5
5	9
7	13

Solution

(a) We plot the three ordered pairs and connect the points by a straight line. This means the function is a linear function.

(b) We find that the value $x = 4$ on the graph corresponds to the value $y = 7$. Thus, we would expect that if we sold 4000 pairs of shoes, we would have a profit of $7000 for the month.

(c) In any business we would not expect to make a profit at all if we sold no items. Looking at our graph, we find that when $x = 0, y = -1$. Thus, we would predict a loss of $1000 in a month of zero sales of shoes. The loss would most likely be due to fixed expenses such as rent and supplies.

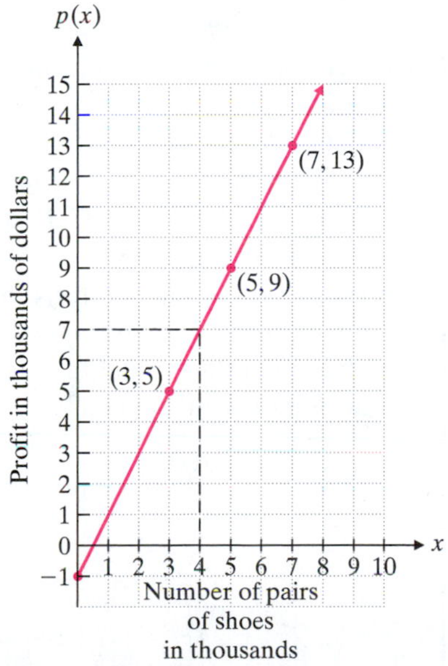

TO THINK ABOUT: Example 6 Follow-up Can x in Example 6 be any real number? Well, not really. Obviously the number of pairs of shoes sold monthly must be a whole number. You cannot sell 345.859 pairs of shoes. However, when we connect the points on the graph we use a *continuous* line (i.e., a line with no breaks in it). When we study the function for all real numbers greater than or equal to zero, it helps us better understand the relationship between sales and profit. Using a continuous line helps us make better predictions for the future. Now, of course, we know as well that the graphed line does not really extend forever. There are physical limitations on the number of shoes that can be made and sold by any one shoe manufacturer.

Continued on next page

x Average Number of Patients Per Hour	$g(x)$ Weekly Profit
6	$2000
9	$4000
12	$6000

Student Practice 6

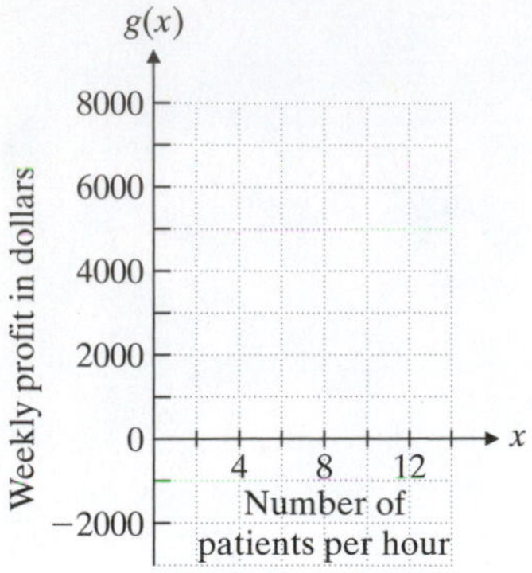

Student Practice 6 An accountant reviewed the profitability of a doctor's practice by comparing the number of patients the doctor saw with the profit. The data collected are in the table on the previous page.

(a) Plot the data values and connect the points to see the graph of the underlying function.
(b) Describe the function in as much detail as possible.
(c) What would the profit be if the doctor saw three patients per hour?
(d) What would the weekly loss be if the doctor saw zero patients per hour?
(e) Can we expect the function to increase indefinitely? Why or why not?

Given a set of data, we do not always know what the graph of the underlying function will look like. Not all functions are linear. When points do not appear to lie on a line, we connect them with a smooth curve.

Let's look at windchill, the effect of wind on the skin's reaction to the outside temperature. For example, if the temperature outside were 10° Fahrenheit (°F) and the wind were blowing at 20 miles per hour (mph), the effect would be the same as if the temperature were 24° below zero! A windchill table that describes this relationship is used in the next example.

EXAMPLE 7 The windchill values for various wind speeds when the temperature is 10° Fahrenheit are given in the table.

(a) Plot the points and connect them to see the graph of the underlying function.
(b) Estimate the windchill when the wind is blowing at 23 mph.
(c) Estimate the windchill when the wind is blowing at 40 mph.
(d) If the windchill is −15°F, at what speed is the wind blowing?
(e) Based on your analysis of the graph, does an increase in wind speed result in a greater change in the windchill at lower wind speeds or at higher wind speeds?

Wind Speed (mph)	Windchill (°F)
0	10
5	7
10	−9
15	−18
20	−24
25	−29
30	−33
35	−35

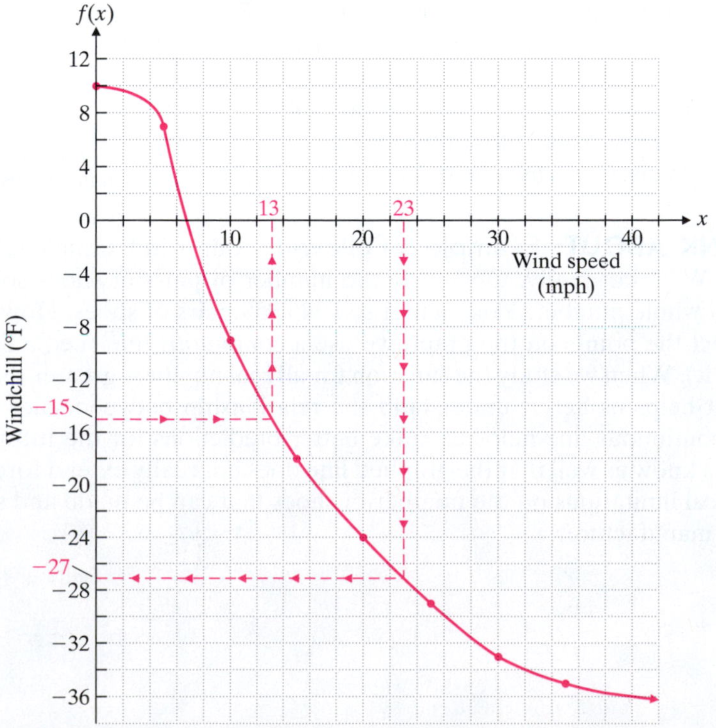

Solution

(a) First we need to determine which value is the independent variable and which value is the dependent variable. Since windchill depends on wind speed, wind speed is the independent variable and windchill is the dependent variable. Thus, the windchill values are the function values. Label the vertical axis "Windchill" and the horizontal axis "Wind Speed." Plot the points. Use a smooth curve to connect the points.

(b) Find 23 along the horizontal axis. Move down to the curve. Then move left until you intersect the vertical axis. The function value on the vertical axis for 23 mph is about −27°F.

(c) Since the last windchill number in the table is for a wind speed of 35 mph, we need to extend the curve. This is called *extrapolation*, and the function values obtained by this technique may be less accurate. At 40 mph, the windchill is about −36°F.

(d) Find −15 along the vertical scale. Move to the right until you intersect the curve. Then move up until you intersect the horizontal axis. This is the wind speed when the windchill is −15°F. This is about 13 mph.

(e) The curve goes downward more quickly at lower wind speeds. Thus, lower wind speeds have a greater effect on windchill. For example, when the wind speed goes from 0 to 15 mph, the windchill goes from 10°F to −18°F. This is a change of 28°F. A change in wind speed from 20 to 35 mph produces a change in windchill of only 11°F.

Student Practice 7

A table of windchill values at 0°F is given below.

Wind Speed (mph)	0	5	10	15	20	25	30	35
Windchill (°F)	0	−5	−22	−31	−39	−44	−48	−52

(a) Plot the points and connect them to graph the underlying function.

(b) Estimate the windchill when the wind is blowing at 32 mph.

(c) Estimate the windchill when the wind is blowing at 40 mph.

(d) If the windchill is −24°F, at what speed is the wind blowing?

(e) What is the domain of the function? What is the range?

Student Practice 7

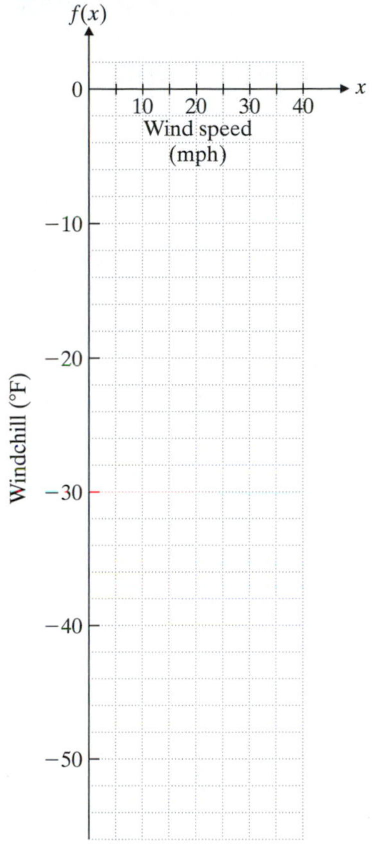

Scientists continue to study the present windchill table to see if it is truly accurate. Many scientists feel the figures need to be revised. Further research is needed to determine how the revision should be carried out. In addition to windchill tables, calculations can also be made using the Internet. There are several sites that feature a "windchill calculator."

An interesting example of a place where windchill is critical is on top of Mount Washington in New Hampshire. In 1934 the observatory recorded a wind gust of 231 miles per hour. You can learn more at www.mountwashington.org.

EXAMPLE 8 Graph the function $f(x)$ suggested by the data given in the following table. The domain is $x \geq 2$, and the range is $f(x) \geq 0$.

x	2	3	6	11
f(x)	0	1	2	3

(a) Determine an approximate value for $f(x)$ when $x = 8$.

(b) Determine an approximate value for x when $f(x) = 1.5$.

(c) What do you notice about the graph as the values of x get larger?

Continued on next page

Solution Assign the independent variable x to the horizontal axis. Assign the function values, the dependent variable, to the vertical axis. Plot the points and connect them with a smooth curve.

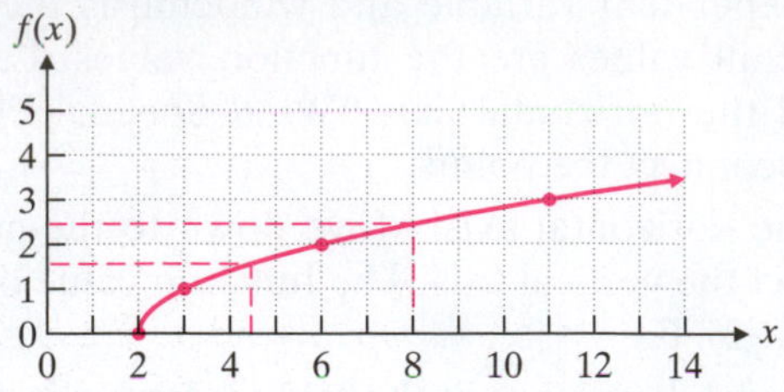

(a) Find $x = 8$ along the horizontal axis. Move up until you intersect the graph. Then move to the left until you intersect the vertical axis. This value is $f(x)$ when x is 8. Read the scale. Thus, $f(x)$ is about 2.5.

(b) Find $f(x) = 1.5$ along the vertical axis. Move to the right until you intersect the graph. Then move down until you intersect the horizontal axis. This value is x when $f(x)$ is 1.5. Read the scale. Thus, x is about 4.5.

(c) As the values of x get larger, the function values are increasing more slowly. In other words the *rate of change* decreases for larger values of x. Since the rate of change is the slope, we say that the slope is decreasing as x increases.

Student Practice 8 Graph the function $g(x)$ suggested by the data given in the following table. The domain is $x \leq 4$, and the range is $g(x) \geq 0$.

x	4	3	0	−5
g(x)	0	1	2	3

(a) Determine an approximate value for $g(x)$ when $x = -3$.

(b) Determine an approximate value for x when $g(x) = 1.5$.

(c) As x goes from −5 to 0 to 3, what do you observe about the curve?

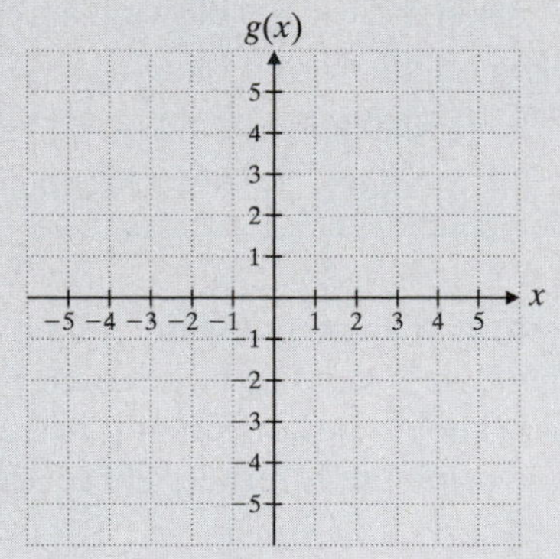

3.6 Exercises

MyMathLab®

Watch the videos in MyMathLab

Download the MyDashBoard App

Graph each function.

1. $f(x) = \frac{3}{4}x + 2$

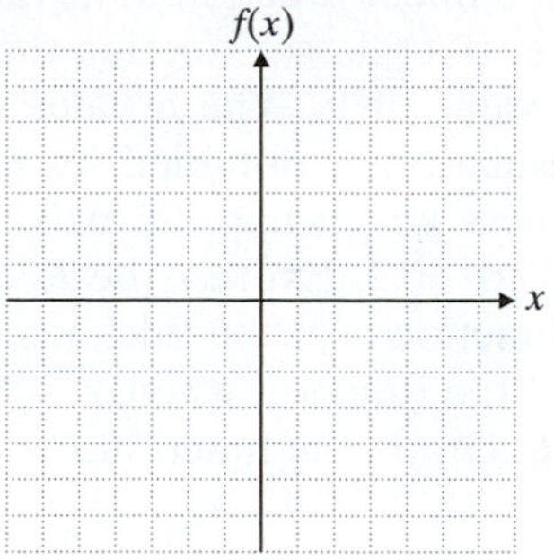

2. $f(x) = \frac{3}{2}x - 5$

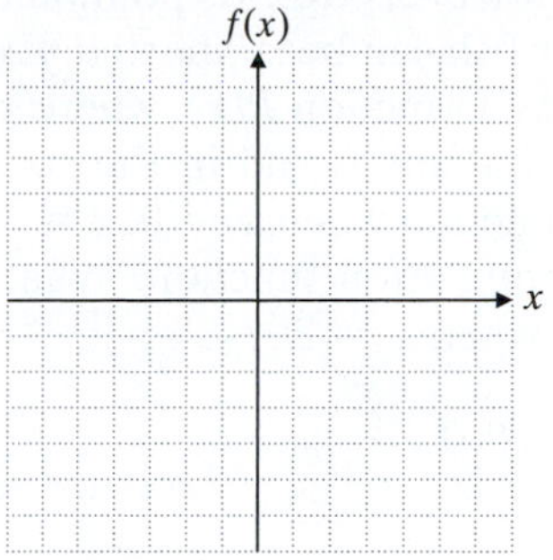

3. $f(x) = -3x - 1$

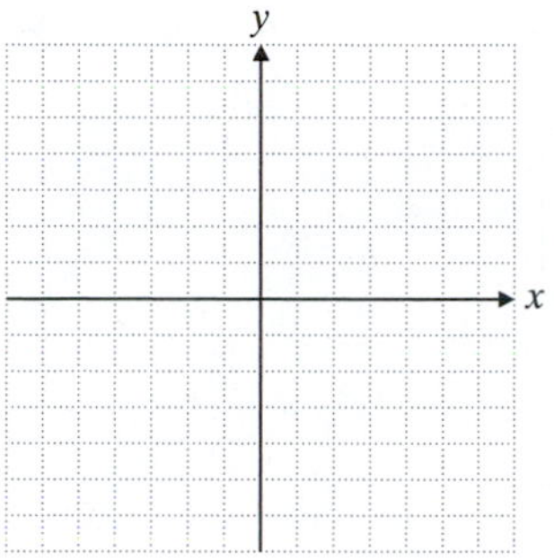

4. $f(x) = -2x + 3$

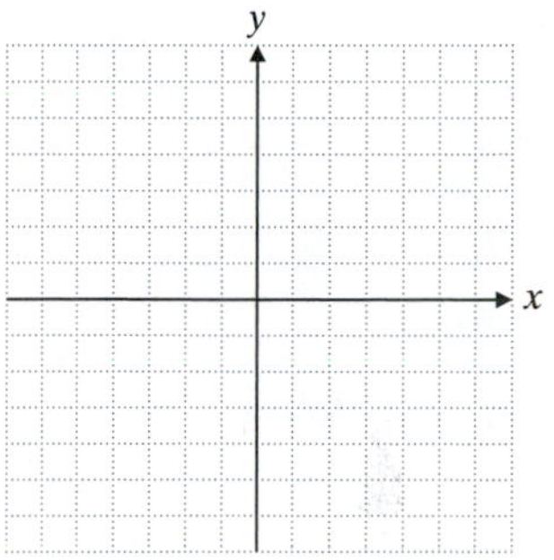

Applications

5. ***Cost of a Newsletter*** The cost of producing a newsletter is an initial charge of \$25 for formatting and editing plus \$0.15 for printing each copy. Express the cost $C(x)$ of printing the newsletter as a function of the number of copies printed x. Obtain values for the function when $x = 0, x = 100, x = 200,$ and $x = 300$. Graph the cost function.

6. ***Population Growth*** In 1996, the population of Paynesville was 18,000. The population has increased by 300 people each year since then and is expected to continue to do so. Express the population as a function $P(x)$, where x is the number of years since 1996. Obtain values for the function when $x = 0, x = 6, x = 10,$ and $x = 18$. Graph the population function.

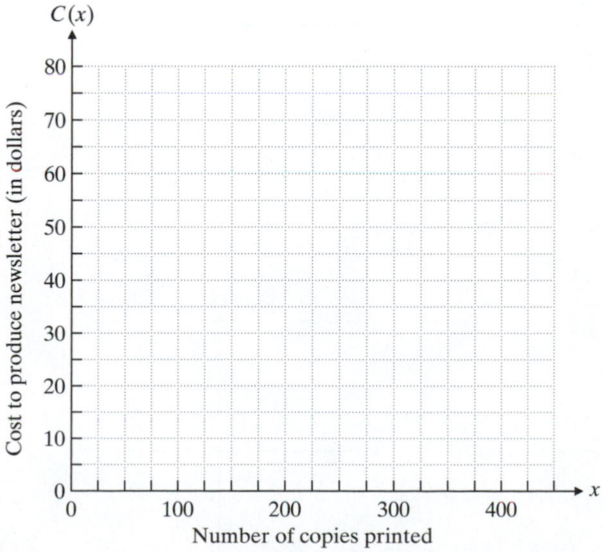

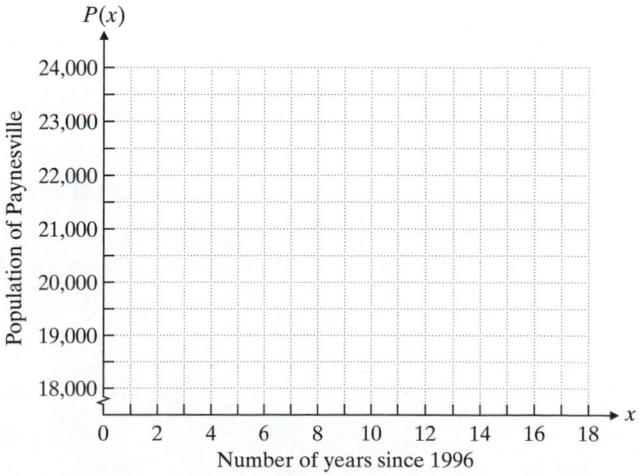

7. ***Water Pollution*** A wildlife biologist has determined that a certain stream in New Hampshire can support 45,000 fish if it is free of pollution. She has estimated that for every ton of pollutants in the stream, 1500 fewer fish can be supported. Express the fish population as a function $P(x)$, where x is the number of tons of pollutants found in the stream. Obtain values of the function when $x = 0$, $x = 10$, and $x = 30$. Graph the population function. What is the significance of $P(x)$ when $x = 30$?

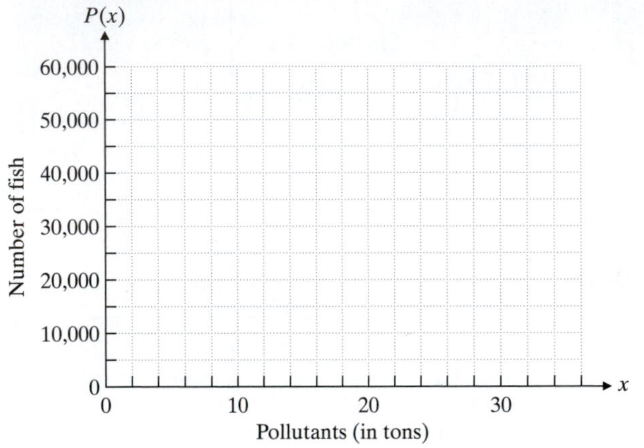

8. ***Insulation in a House*** The R value of insulation in a house is a measure of its ability to resist the loss of heat from the house. The R value of fiberglass insulation is a linear function of its thickness in inches. One type of fiberglass insulation that is 6 inches thick has an R value of 19. The R value in general of this type of insulation is obtained by multiplying 3.2 by the thickness x measured in inches and then adding the result to -0.2. Express the R value of this insulation as a function of x, the thickness in inches. Obtain values of the function when $x = 0$, $x = 1$, $x = 3.5$, and $x = 6$. Graph the function.

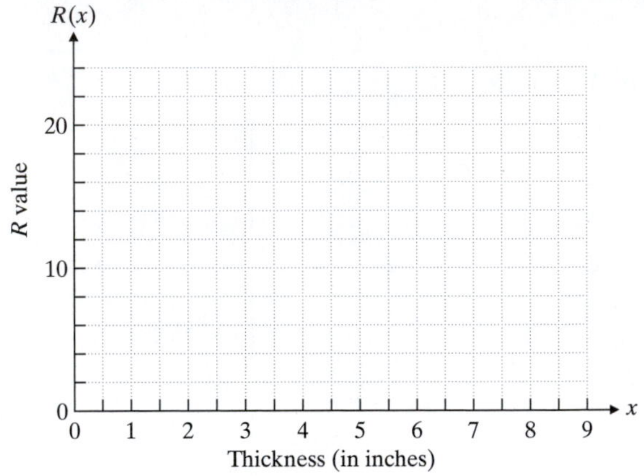

Graph each function.

9. $f(x) = |x - 1|$

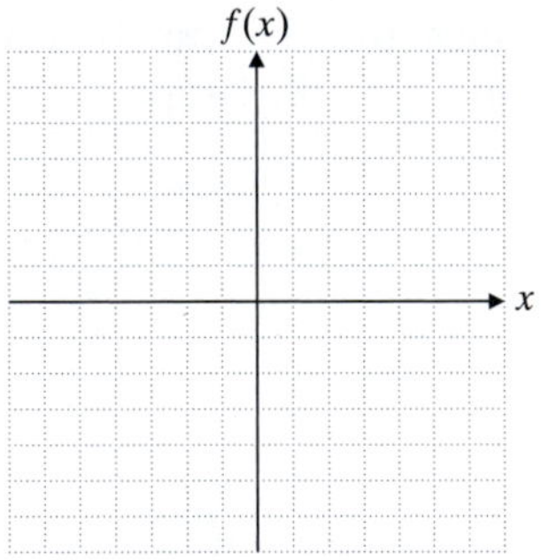

10. $g(x) = |x - 3|$

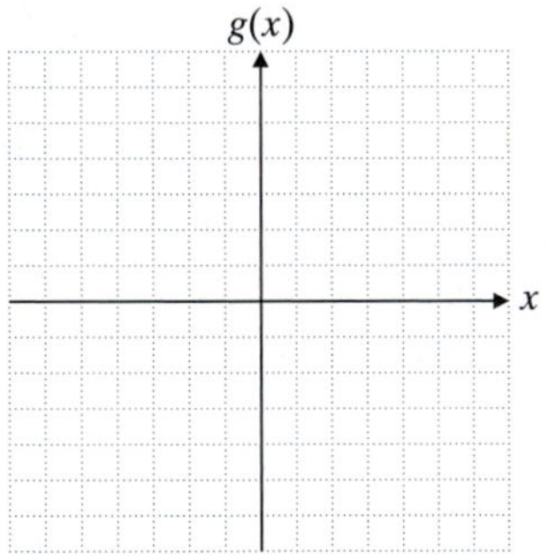

11. $g(x) = |x| - 5$

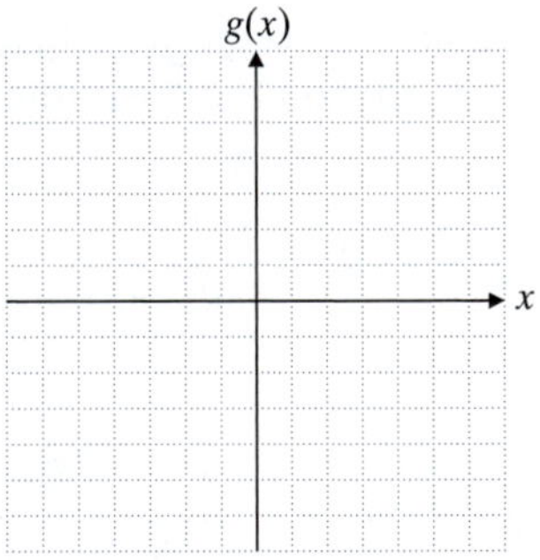

12. $f(x) = |x| + 2$

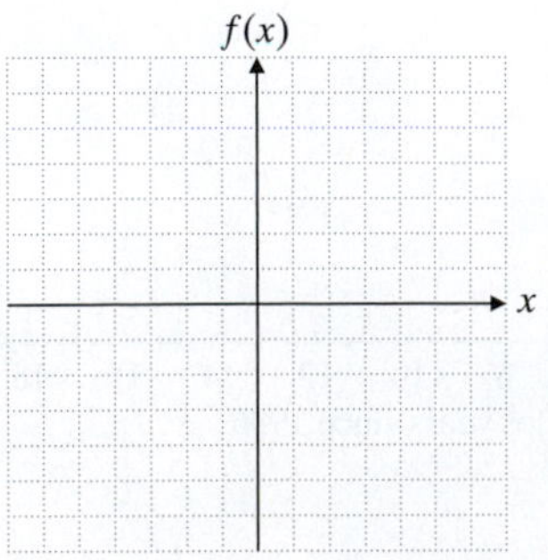

13. $g(x) = x^2 - 4$

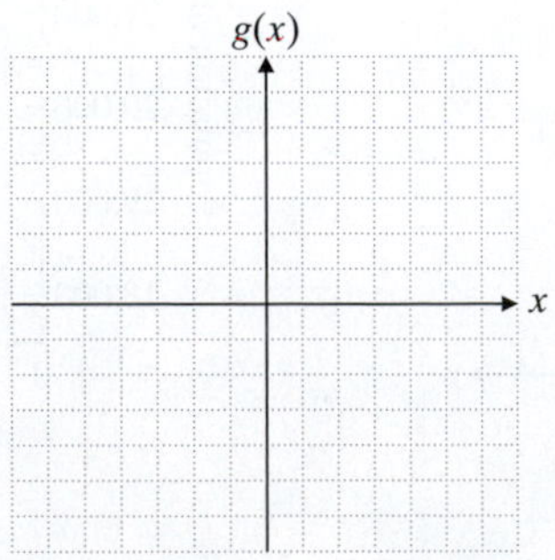

14. $f(x) = x^2 + 1$

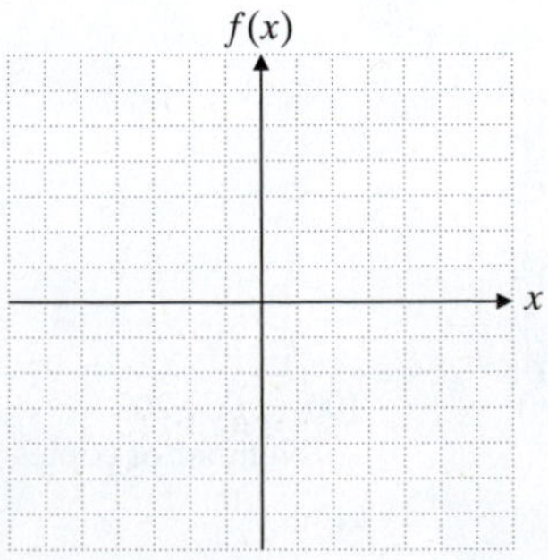

15. $g(x) = (x + 1)^2$

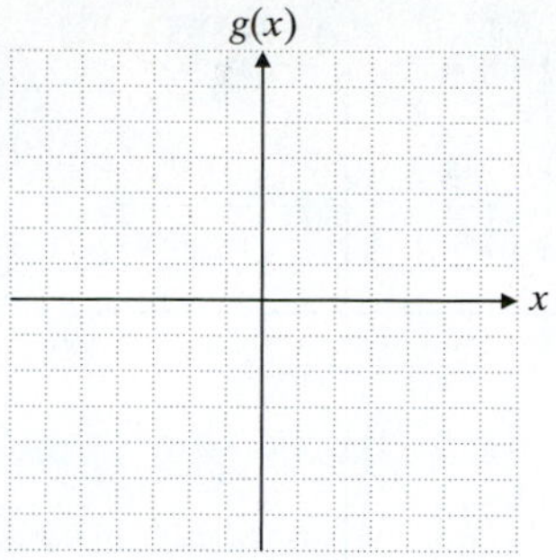

16. $f(x) = (x - 3)^2$

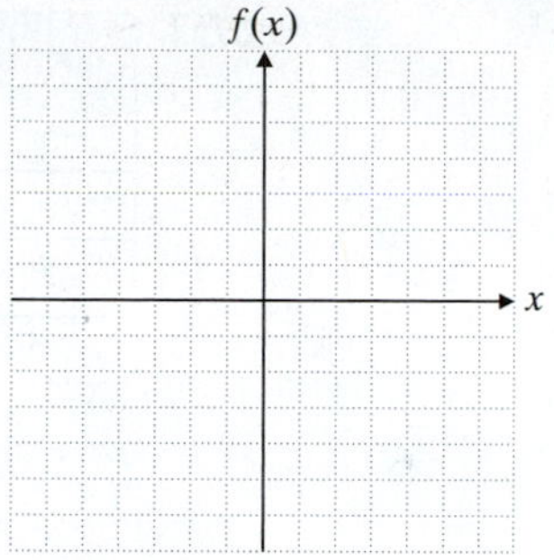

17. $g(x) = x^3 - 3$

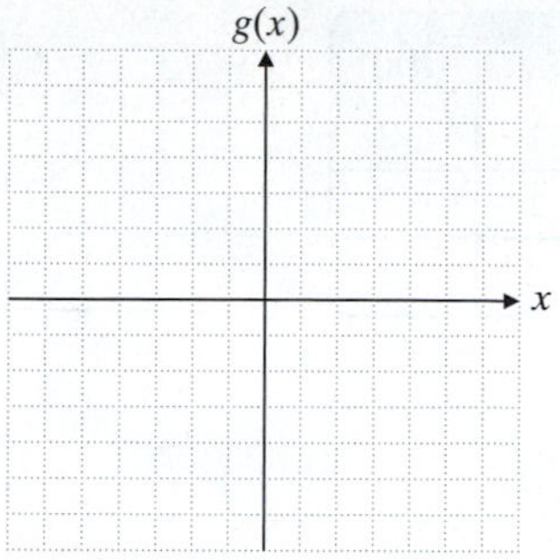

18. $f(x) = x^3 + 2$

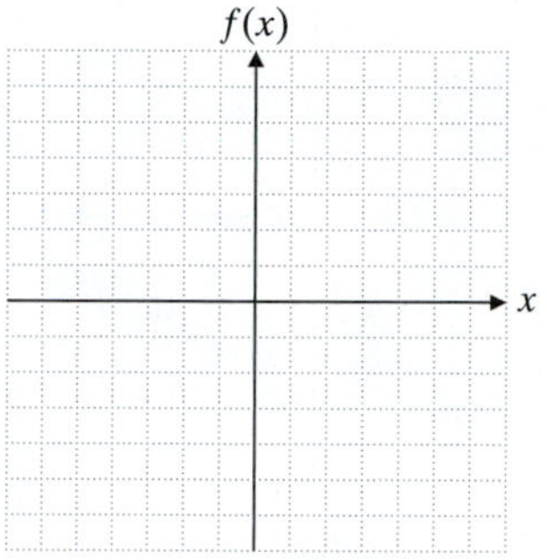

19. $p(x) = -x^3$

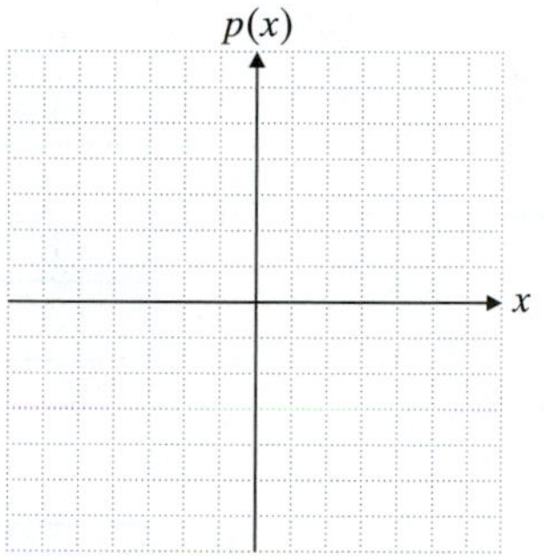

20. $s(x) = -x^3 + 2$

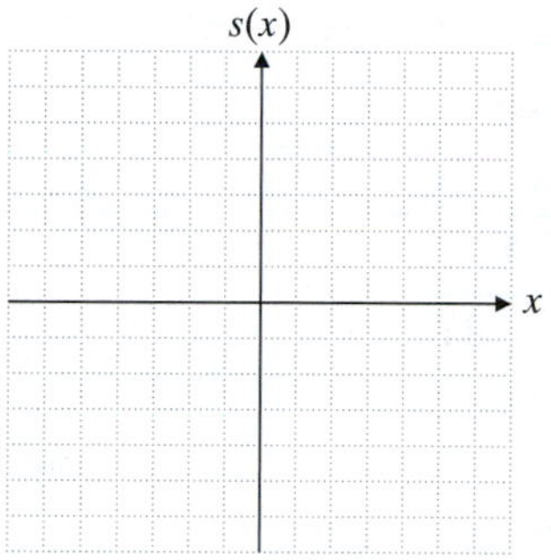

21. $f(x) = \dfrac{2}{x}$

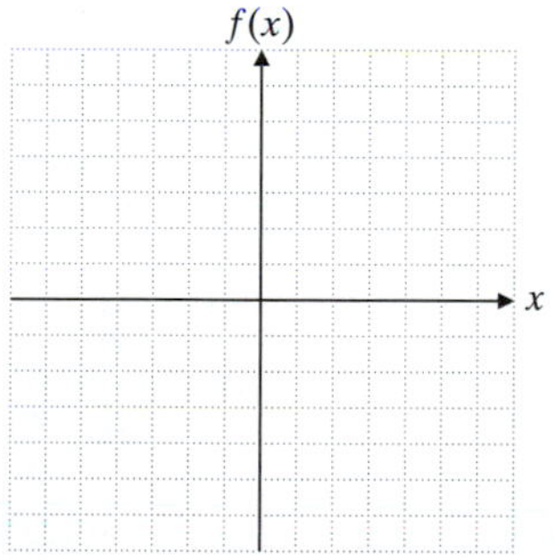

22. $g(x) = -\dfrac{3}{x}$

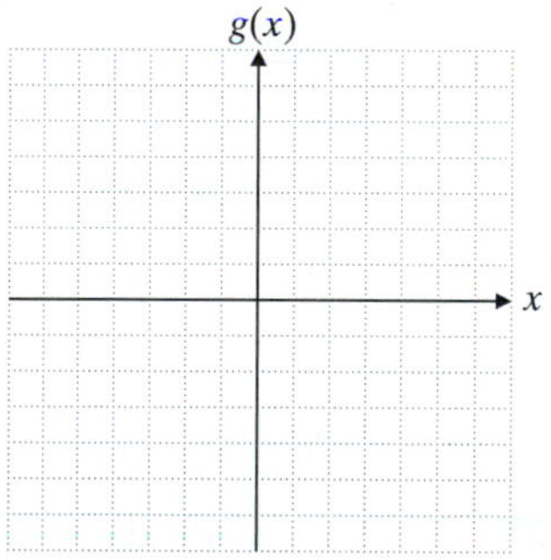

23. $h(x) = -\dfrac{6}{x}$

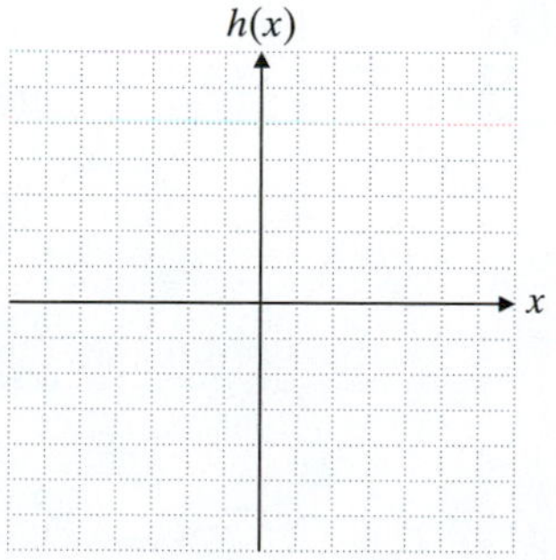

24. $t(x) = \dfrac{10}{x}$

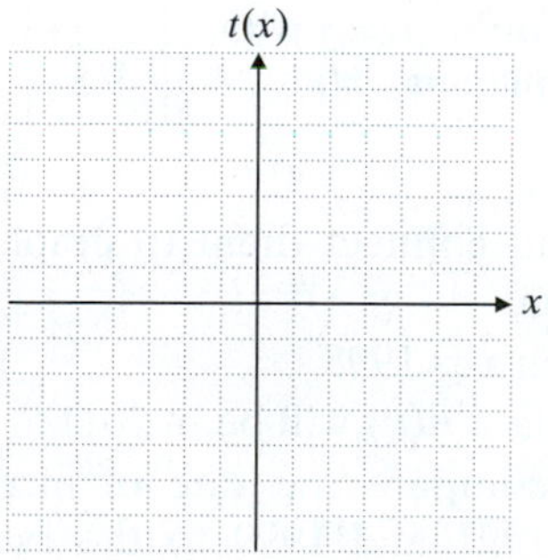

Optional Graphing Calculator Problem

25. On your graphing calculator, graph on the same set of axes: $y_1 = x^2$, $y_2 = 0.4x^2$, and $y_3 = 2.6x^2$. What effect does the coefficient have on the graph?

In each of the following problems, a table of values is given. Graph the function based on the table of values. Assume that both domain and range are all real numbers. After you have finished your graph in each case, estimate the value of f(x) when x = 2.

26.

x	f(x)
−1	2
1	−2
3	−6

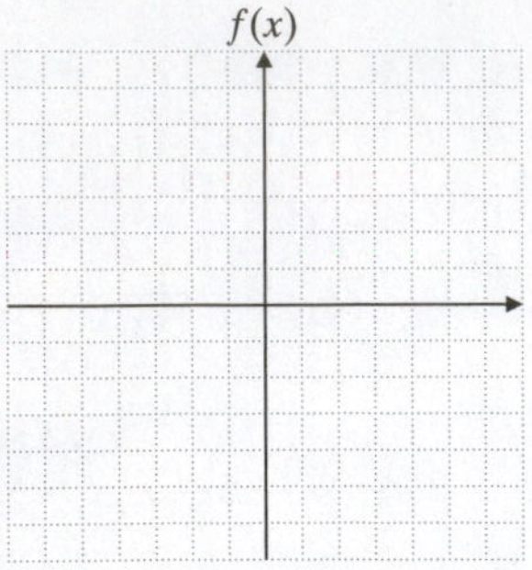

27.

x	f(x)
−5	−2
3	2
5	3

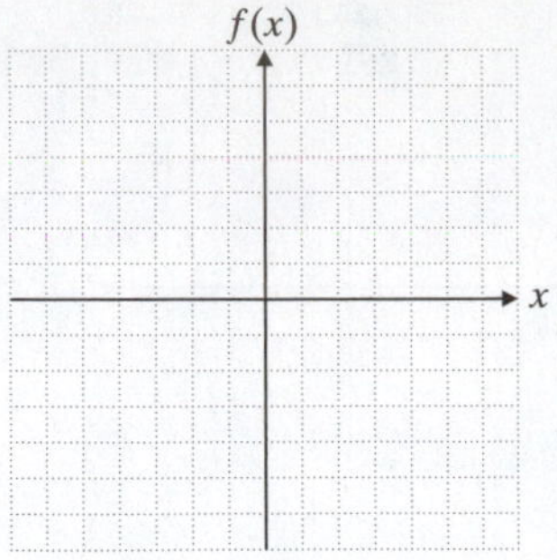

28.

x	f(x)
−2	0.25
−1	0.5
0	1
2.5	5.7

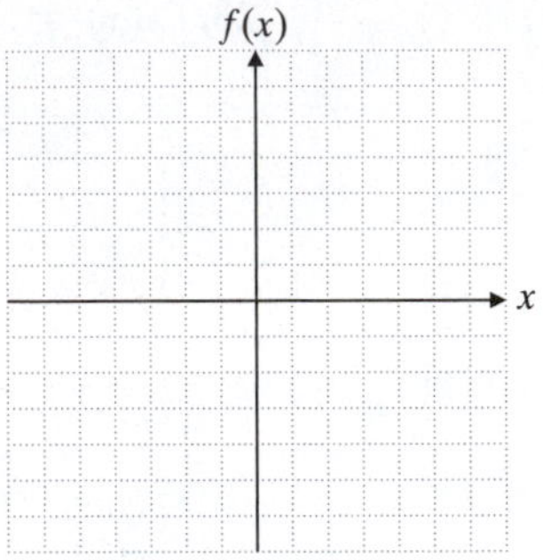

29.

x	f(x)
0	3
1	2.5
3	2.2
−1	4
−2	6

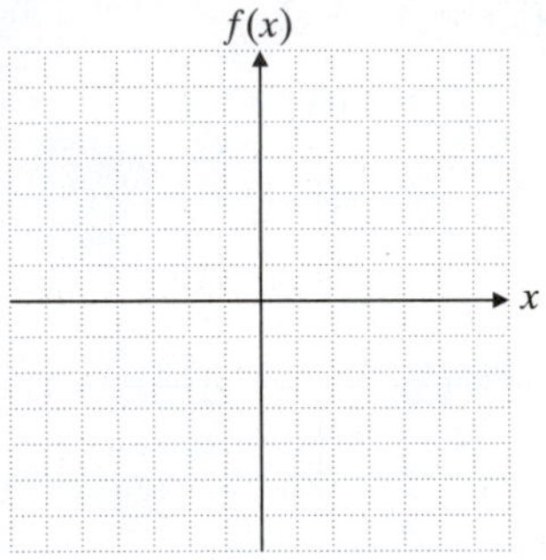

30.

x	f(x)
−3	−1
−2	2
−1	3
0	2
1	−1

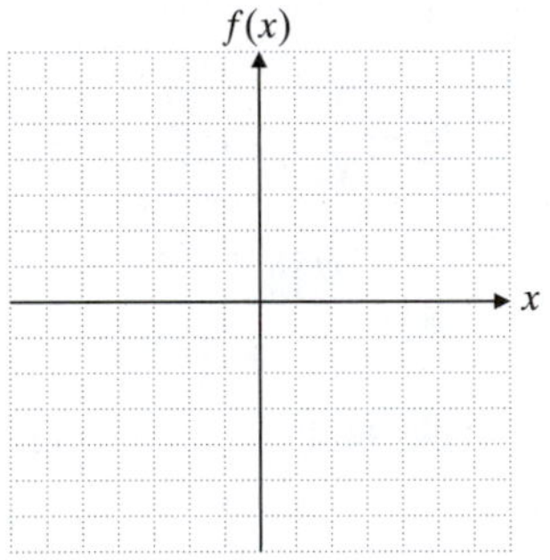

31.

x	f(x)
−1	−3
0	0
1	1
3	−3

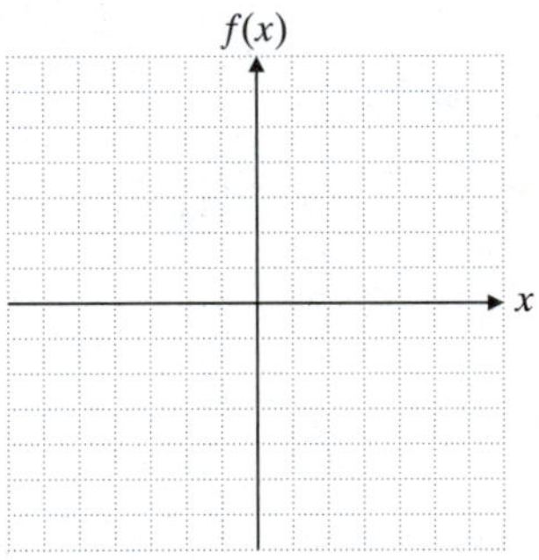

Applications

32. ***Living with HIV/AIDS*** The following table records the number of people worldwide living with HIV/AIDS for selected years.

Year, x	1991	1993	1995	1997	1999	2001	2003	2005	2007	2009
Number of People Living with HIV/AIDS (in millions), $n(x)$	9.5	13.5	18	22	25.5	28.5	31	32	33	33.5

Source: www.avert.org

(a) Plot the points and connect them to graph the underlying function.

(b) Estimate $n(x)$ when x is 1998.

(c) What do you estimate $n(x)$ will be in 2011?

(d) How would you compare the rate of increase in $n(x)$ between 1991 and 1999 to the rate of increase in $n(x)$ between 1999 and 2009?

(e) During what year did the number of people living with HIV/AIDS reach 20 million?

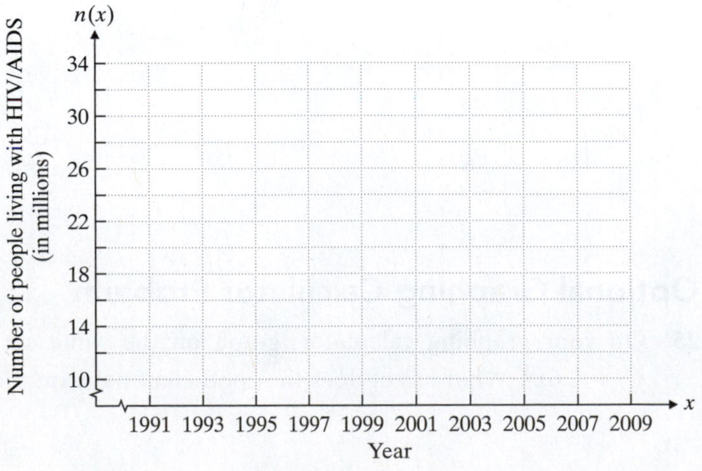

33. ***Computer Manufacturing*** The following table shows the relationship between the number of computers manufactured by a company each day and the company's profit for that day.

Number of Computers Made	20	30	40	50	60	70	80
Profit for the Company ($)	0	5000	8000	9000	8000	5000	0

(a) Plot the points and connect them to graph the underlying function.

(b) How many computers should be made each day to achieve the maximum profit?

(c) If the company wants to earn a profit of $8000 or more each day, how many computers should it manufacture each day?

(d) What will the profit picture be if the company manufactures eighty-two computers per day?

(e) Estimate the profit if the company manufactures forty-five computers per day.

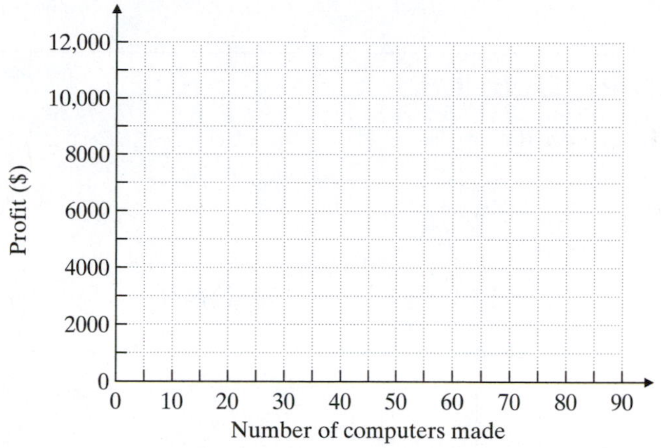

To Think About

34. ***Heat Index*** In hot weather, meteorologists use a heat index to indicate the relative comfort and safety of people subjected to high temperatures and humidity. The following table gives the heat index at 80°F for a relative humidity of 0% to 100%.

Relative Humidity	0%	20%	40%	60%	80%	100%
Heat Index at 80°F	73	77	79	82	86	91

Source: National Oceanic and Atmospheric Administration

(a) Graph the heat index values at 80°F and connect the points. Is the function linear? Why or why not?

(b) Estimate the value of the function when the relative humidity is 50%.

(c) Between what levels of humidity does the heat index increase the fastest?

(d) For what level of humidity does 80°F have a heat index of 80°F?

(e) What is the significance of the heat index being only 77°F when the relative humidity is 20%?

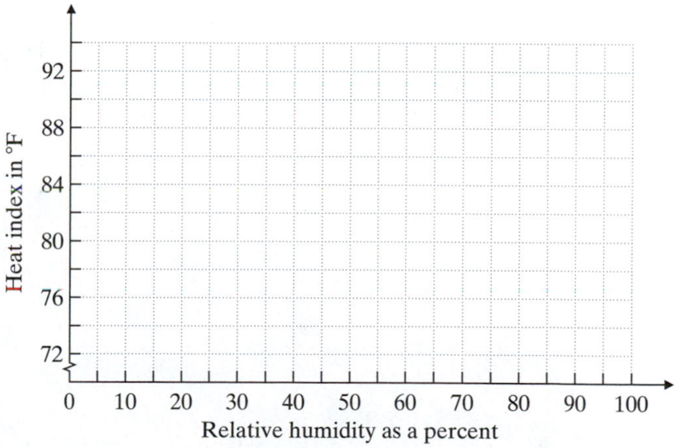

Cumulative Review *Solve for x.*

35. **[2.2.1]** $2(3ax - 4y) = 5(ax + 3)$

36. **[2.1.1]** $0.12(x - 4) = 1.16x - 8.02$

37. **[3.2.1]** Find the slope of the line passing through $(-1, 5)$ and $(2, -1)$.

38. **[3.3.3]** Write the standard form of the equation of the line that passes through $(2, -4)$ and is perpendicular to $y = 3x + 1$.

Quick Quiz 3.6

1. Graph $f(x) = |x - 4|$.

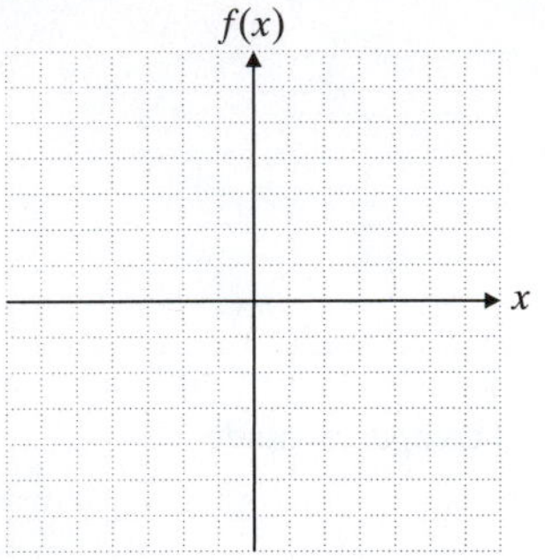

2. Graph $g(x) = 6 - x^2$.

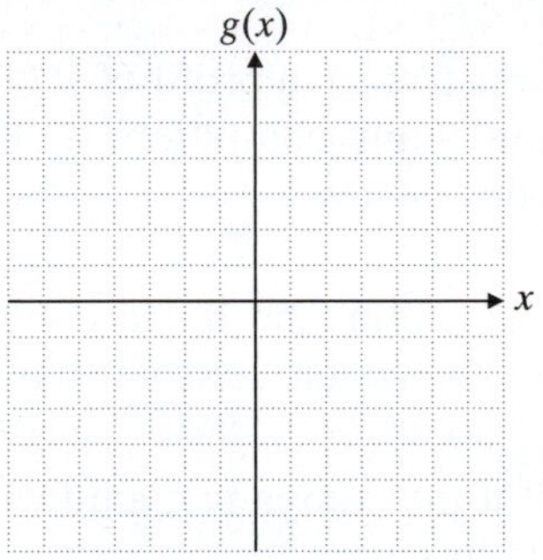

3. Graph $h(x) = x^3 - 2$.

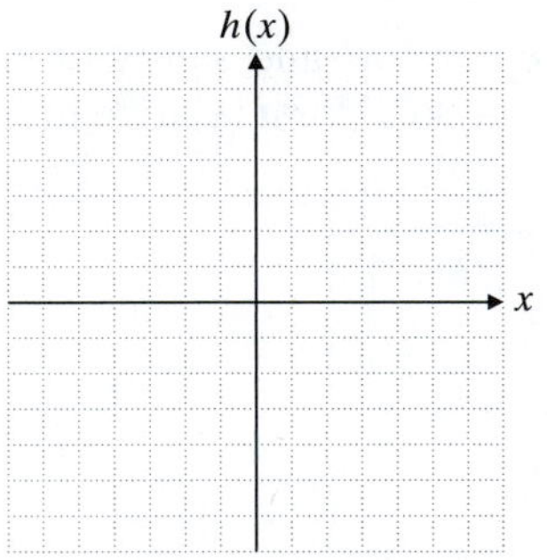

4. **Concept Check** Explain how you would find $f\left(\frac{1}{4}\right)$ for the following function.

$$f(x) = \frac{5}{8x - 5}$$

Did You Know...

That It May Be Better to Buy a Car Than Lease It?

DECIDING TO BUY OR LEASE A CAR

Understanding the Problem:

Louvy has his eye on a brand new car. He thinks he should lease the car because his best friend Tranh has a car lease and says he can get the same deal for Louvy. On the other hand, Louvy's girlfriend Allie says it is always better to buy the car and finance it by taking out a loan.

Louvy does some research and finds that it is not at all simple. While a lease offers lower monthly payments, at the end of the lease period you are left with nothing.

	Lease	Purchase
Automobile price	\$23,000	\$23,000
Interest rate	6%	6%
Length of loan	36 months	36 months
Down payment	\$1000	\$1000
Residual value (the value of the car you are turning in, or the price you would pay if you want to buy it)	\$11,000	Not applicable
Monthly payment	\$388.06	\$669.28

Making a Plan:

Step 1: Louvy needs to compare the total amounts he would pay for the entire loan.

Task 1: *Determine how much Louvy would pay to lease the car for three years.*

Task 2: *Determine how much Louvy would pay to buy the car with a three-year loan.*

Step 2: If Louvy wanted to buy the car at the end of the lease, he would have to pay an additional \$11,000.

Task 3: *Determine the total cost if Louvy wants to buy the car at the end of the lease.*

Task 4: *Determine the total overall savings if Louvy buys the car instead of leasing it.*

Making a Decision:

Step 3: Louvy is unsure what to do. He likes the fact that buying the car would give him an overall savings, but a lower monthly payment is also important to him.

Task 5: *How much will he save in payments each month if he leases the car?*

Task 6: *Which option should Louvy choose if he wants the best overall price? If he is concerned about his monthly payments?*

Applying the Situation to Your Life:

Task 7: *How much can you afford to pay for a car payment each month?*

Task 8: *How much would you pay in insurance, gas, and taxes?*

Task 9: *Would you rather buy or lease a car?*

Facts You Should Know:

You may wish to lease a car if

- you don't drive more than the specified number of miles in the lease on a yearly basis.
- you want a new vehicle every two to three years.

You may wish to buy a car if

- you intend to keep it a long time.
- you want to be debt-free after a time.

Chapter 3 Organizer

Topic and Procedure	Examples	You Try It
Graphing straight lines, p. 128 An equation of the form $$Ax + By = C$$ where A, B, and C are real numbers and A and B are not both zero, is a linear equation and has a graph that is a straight line. To graph such an equation, plot any three points—two give the line and the third checks it. (Where possible, use the x- and y-intercepts.)	$2x + 3y = 12$ Form is $Ax + By = C$, so it is a straight line. Let $x = 0$; then $y = 4$. Let $y = 0$; then $x = 6$. Let $x = 3$; then $y = 2$. x | y: 0 | 4; 6 | 0; 3 | 2 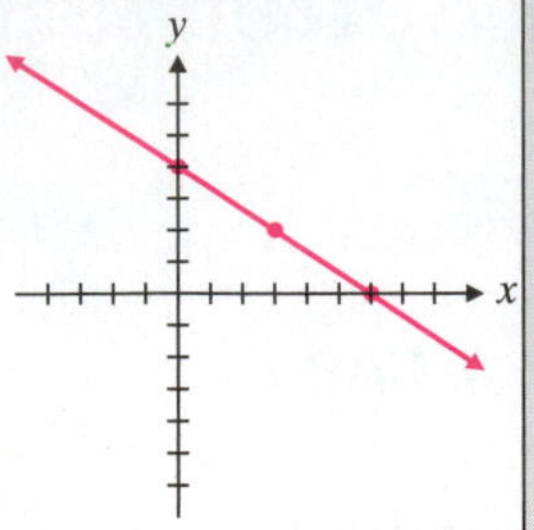	**1.** Graph $x - 2y = 8$. 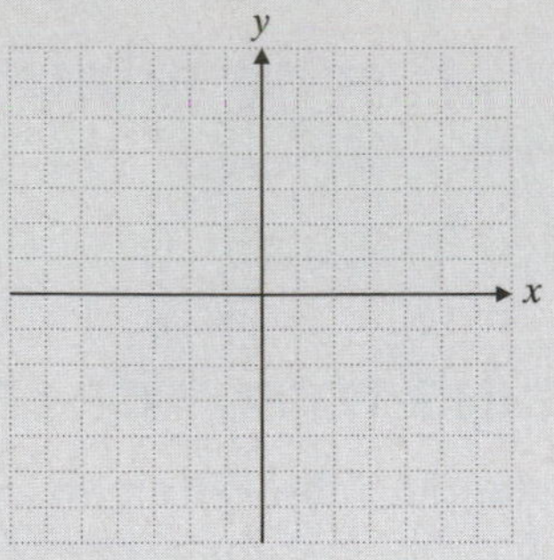
Intercepts, p. 129 A line crosses the x-axis at its x-intercept $(a, 0)$. A line crosses the y-axis at its y-intercept $(0, b)$.	Find the intercepts of $7x - 2y = -14$. If $x = 0$, $-2y = -14$ and $y = 7$. The y-intercept is $(0, 7)$. If $y = 0$, $7x = -14$ and $x = -2$. The x-intercept is $(-2, 0)$.	**2.** Find the intercepts of $-3x + 5y = -15$.
Slope, p. 137 The *slope* m of the straight line that contains the points (x_1, y_1) and (x_2, y_2) is defined by $$m = \frac{y_2 - y_1}{x_2 - x_1}, \quad \text{where } x_2 \neq x_1.$$	Find the slope of the line passing through $(6, -1)$ and $(-4, -5)$. $$m = \frac{-5 - (-1)}{-4 - 6} = \frac{-5 + 1}{-4 - 6} = \frac{-4}{-10} = \frac{2}{5}$$	**3.** Find the slope of the line passing through $(-5, 2)$ and $(-3, 1)$.
Zero slope, p. 138 All horizontal lines have *zero slope*. They can be described by equations of the form $y = b$, where b is a real number.	What is the slope of $2y = 8$? This equation can be simplified to $y = 4$. It is a horizontal line; the slope is zero.	**4.** What is the slope of $-3y = 3$?
Undefined slope, p. 138 The slopes of all vertical lines are undefined. The lines can be described by equations of the form $x = a$, where a is a real number.	What is the slope of $5x = -15$? This equation can be simplified to $x = -3$. It is a vertical line. This line has an undefined slope.	**5.** What is the slope of $4x = 2$?
Parallel and perpendicular lines, pp. 139–140 Two distinct lines with nonzero slopes m_1 and m_2 are **1.** parallel if $m_1 = m_2$. **2.** perpendicular if $m_1 = -\frac{1}{m_2}$.	**(a)** Find the slope of a line parallel to $y = -\frac{3}{2}x + 6$. $$m = -\frac{3}{2}$$ **(b)** Find the slope of a line perpendicular to $y = -4x + 7$. $$m = \frac{1}{4}$$	**6. (a)** Find the slope of a line parallel to $y = \frac{1}{2}x - 5$. **(b)** Find the slope of a line perpendicular to $y = 2x + 3$.
Standard form, p. 145 The equation of a line is in standard form when it is written as $Ax + By = C$, where A, B, and C are real numbers and A and B are not both zero.	Place this equation in standard form: $y = -5(x + 6)$. $$y = -5x - 30$$ $$5x + y = -30$$	**7.** Place this equation in standard form: $y = 2(x - 4)$.
Slope–intercept form, p. 145 The slope–intercept form of the equation of a line is $y = mx + b$, where the slope is m and the y-intercept is $(0, b)$.	Find the slope and y-intercept of each equation. **(a)** $y = -\frac{7}{3}x + \frac{1}{4}$ The slope is $-\frac{7}{3}$; the y-intercept is $\left(0, \frac{1}{4}\right)$. **(b)** $4x - 2y = 6$ First solve for y. $y = 2x - 3$ Now the equation is in slope–intercept form. The slope is 2; the y-intercept is $(0, -3)$.	**8.** Find the slope and y-intercept of each equation. **(a)** $y = \frac{1}{4}x - \frac{3}{4}$ **(b)** $-2x + y = 5$

Topic and Procedure	Examples	You Try It

Point–slope form, p. 147

The point–slope form of the equation of a line is $y - y_1 = m(x - x_1)$, where m is the slope and (x_1, y_1) are the coordinates of a point on the line.

Find an equation of the line passing through the points (6, 0) and (3, 4).

$$m = \frac{4 - 0}{3 - 6} = -\frac{4}{3}$$

Then use the point–slope form.

$$y - 0 = -\frac{4}{3}(x - 6)$$

$$y = -\frac{4}{3}x + 8$$

9. Find an equation of the line passing through the points $(-2, 4)$ and $(8, 0)$.

Graphing the solution of a linear inequality in two variables, p. 156

1. Replace the inequality symbol by an equals sign. This equation will be the boundary for the desired region.
2. Graph the line obtained in step 1, using a solid line if the original inequality contains a $\le$ or $\ge$ symbol and a dashed line if the original inequality contains a $<$ or $>$ symbol.
3. Pick any point that does not lie on the boundary line. Substitute the coordinates into the original inequality. If you obtain a true inequality, shade the region on the side of the line containing the point. If you obtain a false inequality, shade the region on the side of the line opposite the test point.

Graph the solution of $3y - 4x \ge 6$.

1. The boundary line is $3y - 4x = 6$.
2. The line passes through (0, 2) and (−1.5, 0). The inequality includes the boundary; draw a solid line.
3. Pick (0, 0) as a test point.

$$3y - 4x \ge 6$$
$$3(0) - 4(0) \stackrel{?}{\ge} 6$$
$$0 - 0 \stackrel{?}{\ge} 6$$
$$0 \ge 6$$

Our test point fails. We shade the side that does not contain (0, 0).

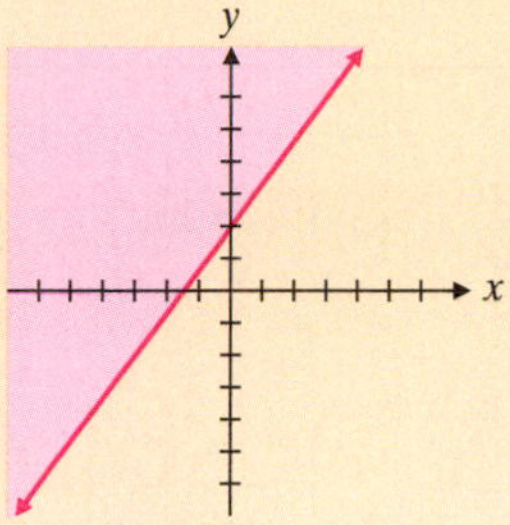

10. Graph the solution of $y - 2x \le 4$.

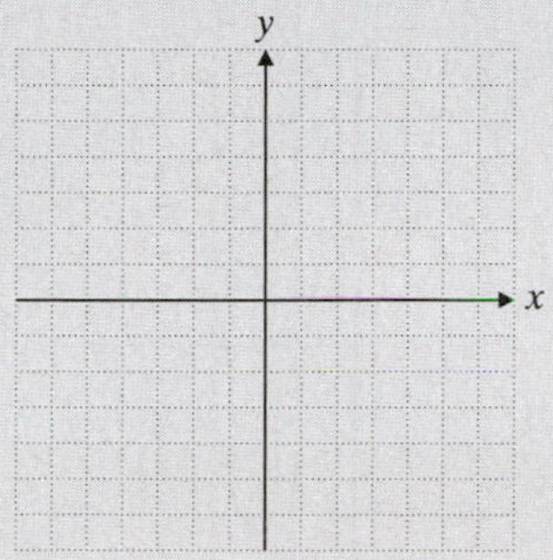

Finding the domain and the range of a relation, p. 162

The set of all the first items of the ordered pairs in a relation is called the domain. The set of all the second items of the ordered pairs in a relation is called the range.

Find the domain and range of the relation

$$A = \{(5, 6), (1, 6), (3, 4), (2, 3)\},$$
$$\text{Domain} = \{5, 1, 3, 2\}$$
$$\text{Range} = \{6, 4, 3\}$$

11. Find the domain and range of the relation $B = \{(0, 2), (-1, 3), (4, 1), (0, -1)\}$.

Determining whether a relation is a function, p. 164

A function is a relation in which no two different ordered pairs have the same first coordinate.

Determine whether each of the following is a function.

(a) $B = \{(6, 7), (3, 0), (6, 4)\}$

(b) $C = \{(-9, 3), (16, 4), (9, 3)\}$

(a) B is not a function. Two different pairs have the same first coordinate. They are (6, 7) and (6, 4).

(b) C is a function. There are no different pairs with the same first coordinate.

12. Determine whether each of the following is a function.

(a) $E = \{(6, 1), (2, -3), (4, 1), (2, 3)\}$

(b) $F = \{(2, 4), (5, 0), (0, -4)\}$

Determining whether a graph represents the graph of a function, p. 166

The graph of a function will have no two different ordered pairs with the same first coordinate.

If a vertical line can intersect a graph more than once, the graph does not represent a function. If no such line can be drawn, the graph represents a function.

(a)

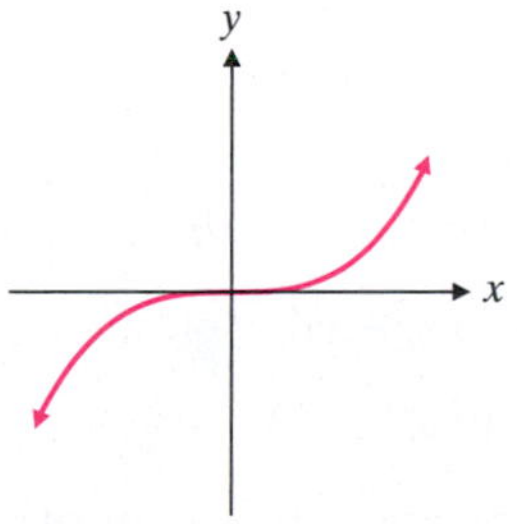

This is a function.

(b)

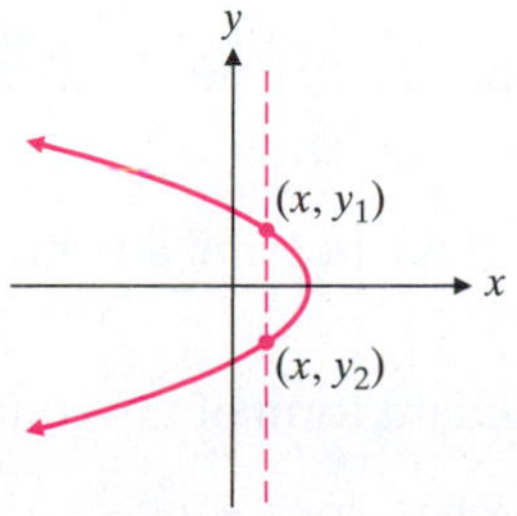

This is not a function. There are at least two different ordered pairs with the same first coordinate.

13. Determine whether each is a function.

(a)

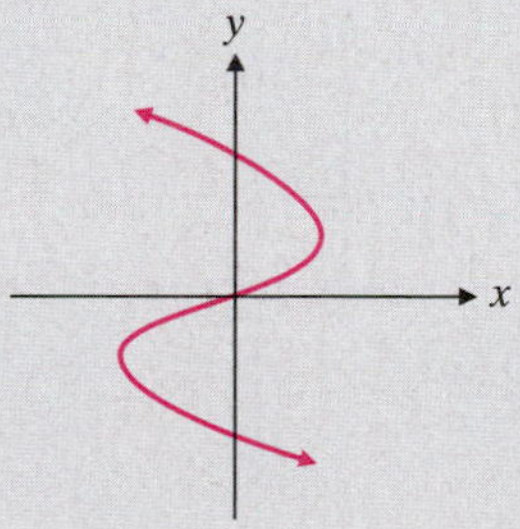

(b)

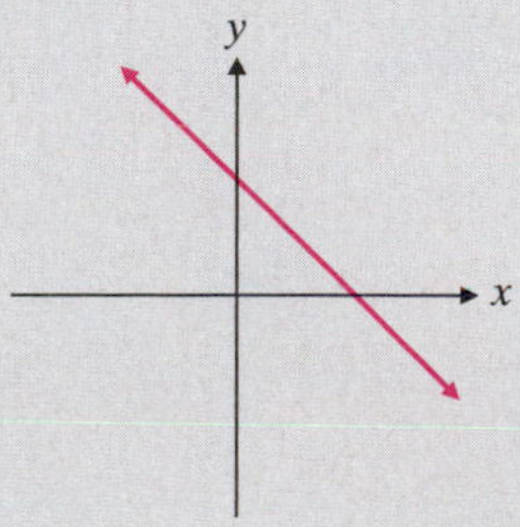

Topic and Procedure	Examples	You Try It
Function notation, p. 166 Use function notation to evaluate the function at a given value. Replace the x by the quantity within the parentheses and then simplify.	If $f(x) = 2x^2 - 3x + 4$, find $f(-2)$. $f(-2) = 2(-2)^2 - 3(-2) + 4$ $= 2(4) - 3(-2) + 4$ $= 8 + 6 + 4 = 18$	**14.** If $g(x) = x^2 - x + 6$, find $g(-1)$.
Graphing functions, p. 171 Prepare a table of ordered pairs (if one is not provided) that satisfy the function equation. Graph these ordered pairs. Connect the ordered pairs with a line or curve.	Graph $f(x) = \lvert x \rvert - 2$. First make a table. Then graph the ordered pairs and connect them.	**15.** Graph $f(x) = x^2 - 1$.

x	f(x)
−2	0
−1	−1
0	−2
1	−1
2	0

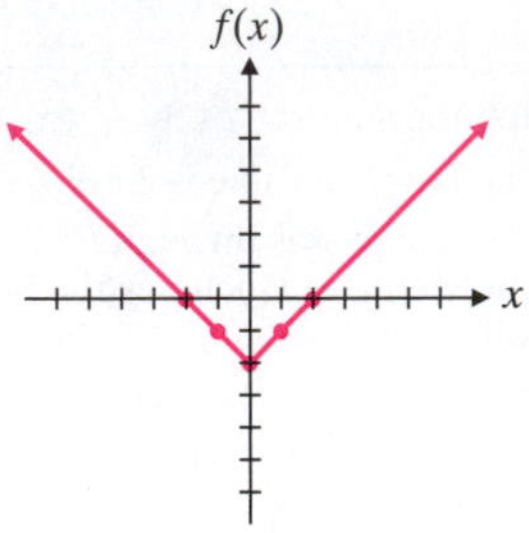

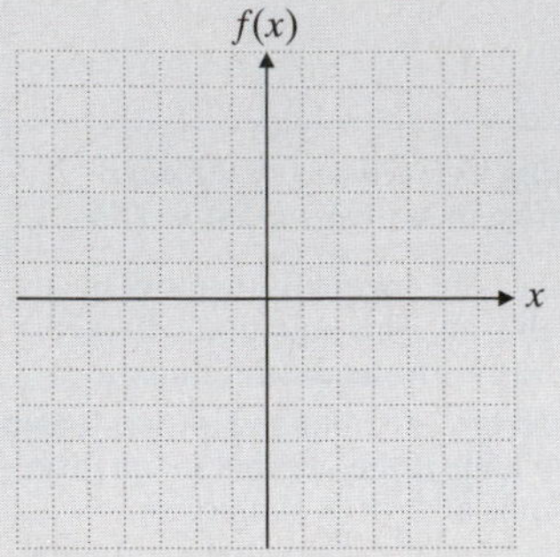

Chapter 3 Review Problems

Graph the straight line determined by each of the following equations.

1. $y = -\frac{3}{2}x + 5$

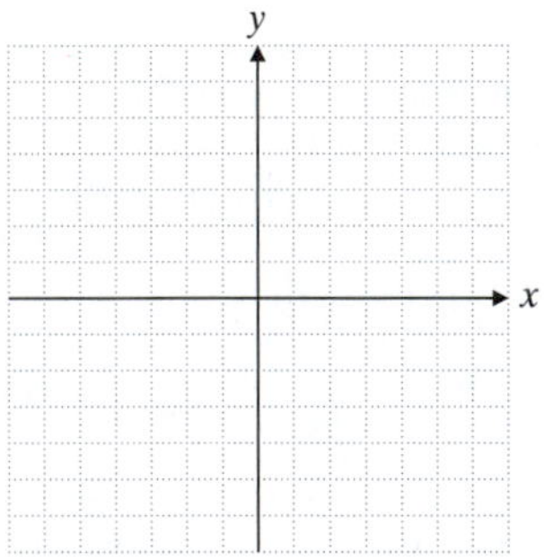

2. $y - 2x + 4 = 0$

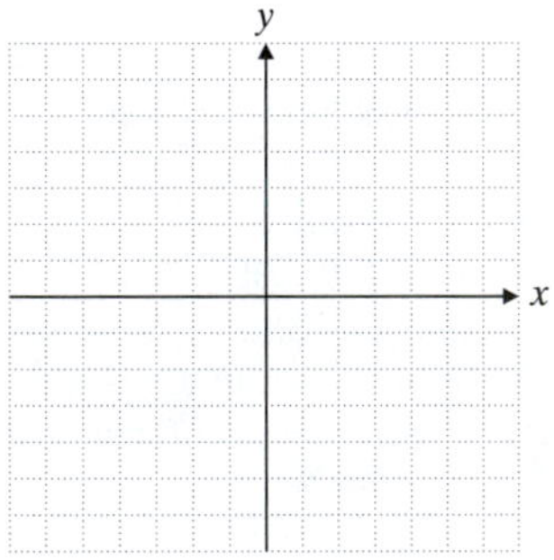

3. $5x - 3 = 9x + 13$

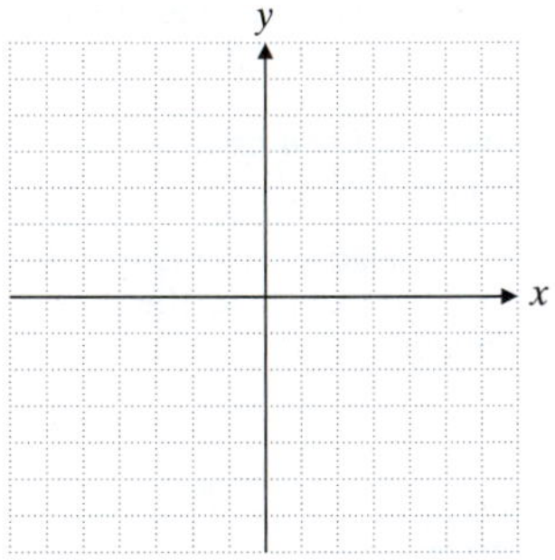

4. $8y + 5 = 10y + 1$

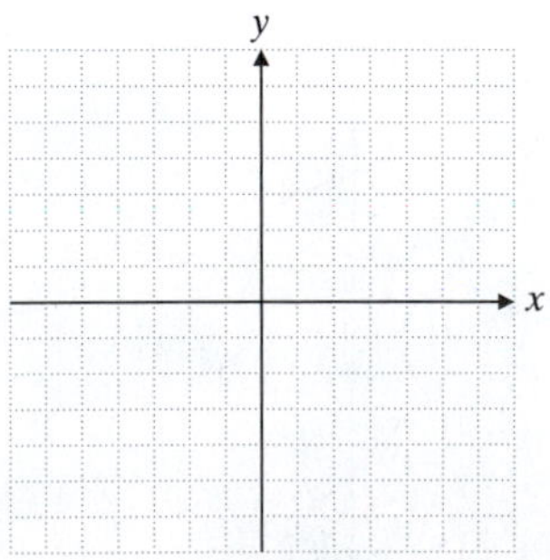

Find the slope, if possible, of the line connecting the given points.

5. $(-8, -4)$ and $(2, -3)$

6. $(7.5, -1)$ and $(0.3, -1)$

7. Find the slope and y-intercept of $-x + 3y = 12$.

8. Find the slope of a line perpendicular to the line passing through $\left(\frac{2}{3}, \frac{1}{3}\right)$ and $(4, 2)$.

9. A line has a slope of $\frac{1}{3}$ and a y-intercept of $(0, 5)$. Write its equation in standard form.

10. Find the standard form of the equation of the line that passes through $\left(\frac{1}{2}, -2\right)$ and has slope -4.

11. Find the standard form of the equation of the line that passes through $(-3, 1)$ and has slope 0.

12. ***Computer Company Profit*** A microcomputer company's profit in dollars is given by the equation $P = 140x - 2000$, where x is the number of microcomputers sold each day. **(a)** What is the slope of the equation $P = 140x - 2000$? **(b)** How many microcomputers must be sold each day for the company to make a profit?

In exercises 13–16, find the equation of the line satisfying the conditions given. Write your answer in standard form.

13. A line passing through $(5, 6)$ and $\left(-1, -\frac{1}{2}\right)$

14. A line that has an undefined slope and passes through $(-6, 5)$

15. A line perpendicular to $7x + 8y - 12 = 0$ and passing through $(-2, 5)$

16. A line parallel to $3x - 2y = 8$ and passing through $(5, 1)$

In exercises 17 and 18, find the slope and y-intercept of each of the following lines. Use these to write the equation of the line.

17.

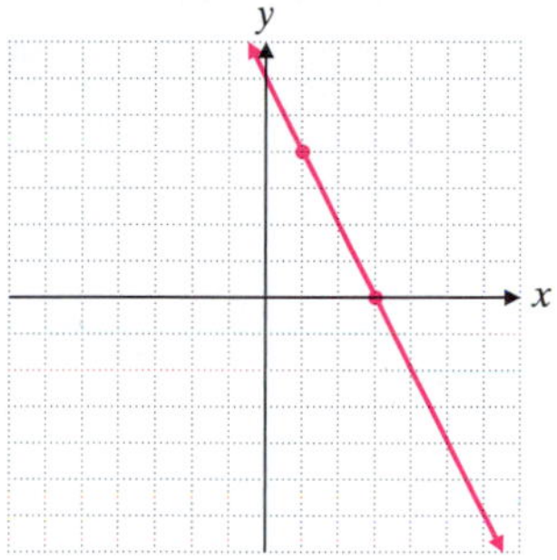

18.

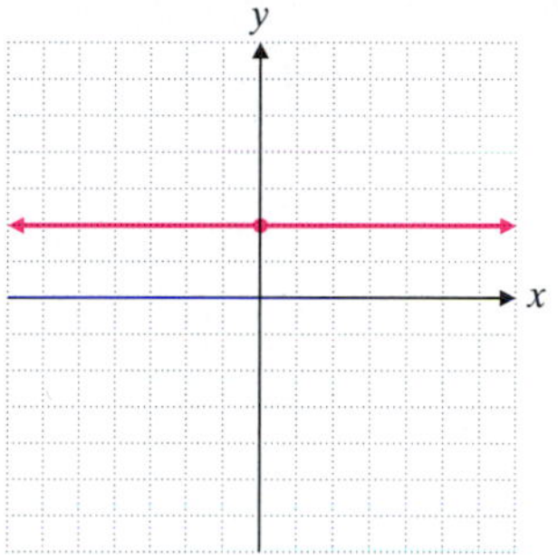

Graph the region described by the inequality.

19. $y < 2x + 4$

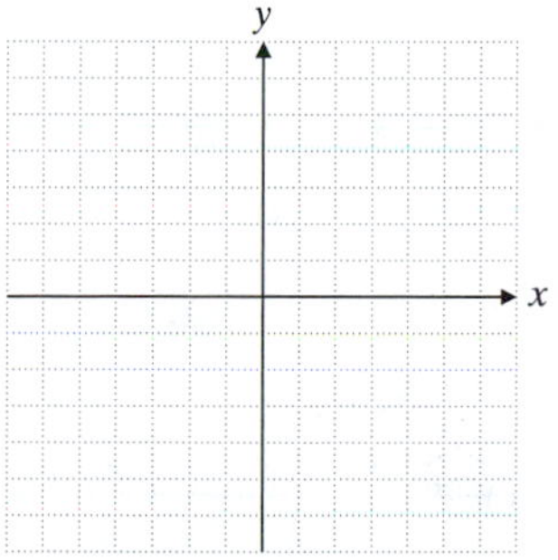

20. $y > -\frac{1}{2}x + 3$

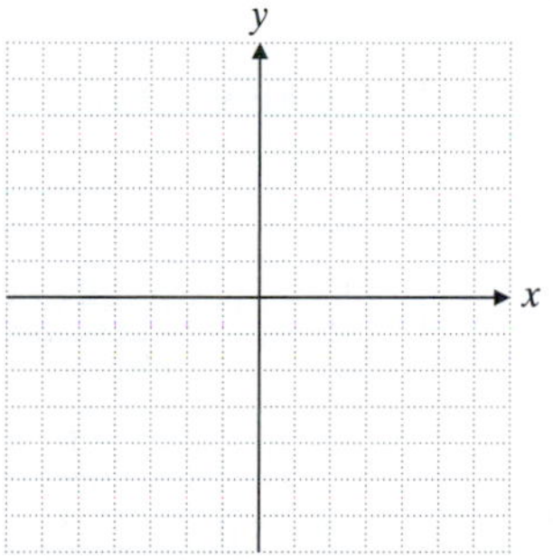

21. $3x + 4y \leq -12$

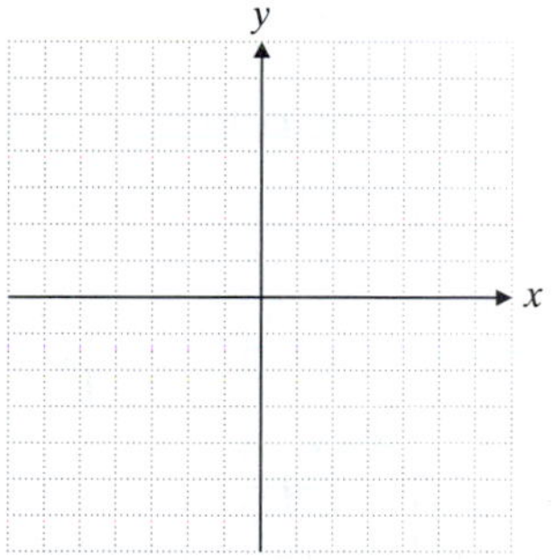

22. $x \leq 3y$

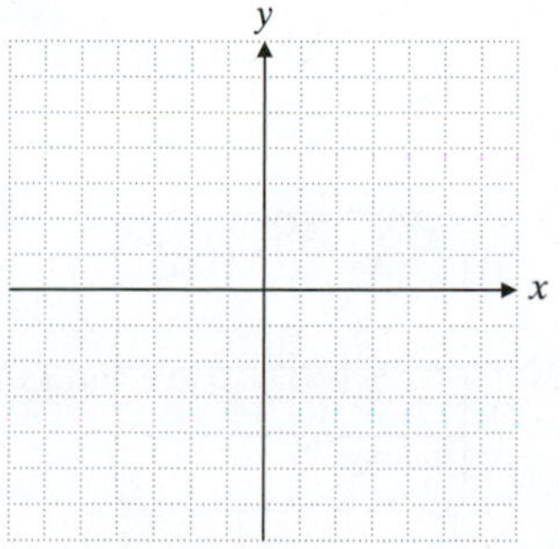

23. $3x - 5 < 7$

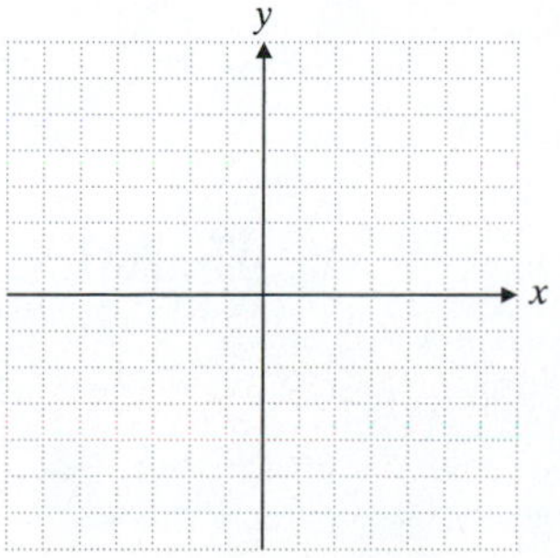

24. $5y - 2 > 3y - 10$

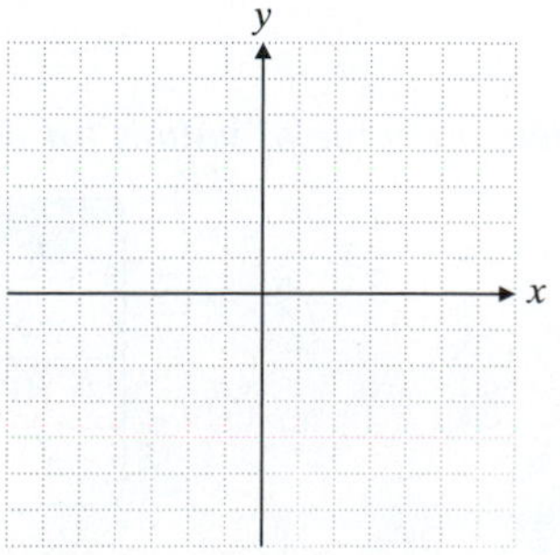

What is the domain and range of each relation? Is the relation a function?

25. $B = \{(-20, 18), (-18, 16), (-16, 14), (-12, 18)\}$

26. $A = \{(0, 0), (1, 1), (2, 4), (3, 9), (1, 16)\}$

Determine whether the graph represents a function.

27.

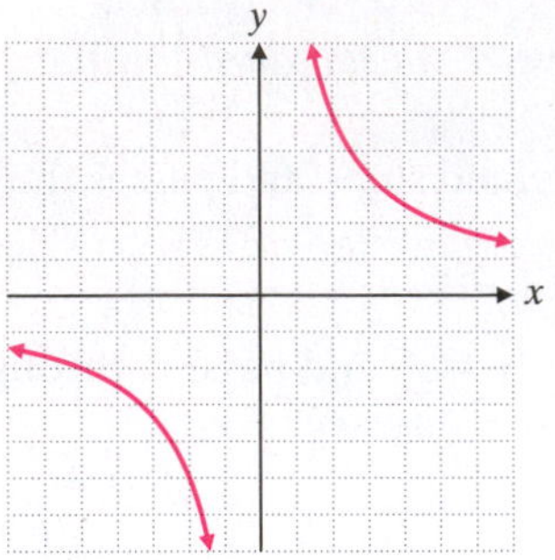

28.

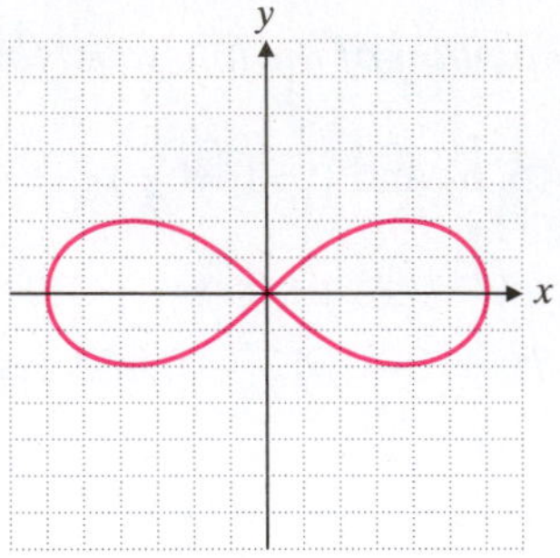

29. If $f(x) = -2x + 10$, find $f(-1)$ and $f(-5)$.

30. If $h(x) = x^3 + 2x^2 - 5x + 8$, find $h(-1)$.

Graph each function.

31. $f(x) = 2|x - 1|$

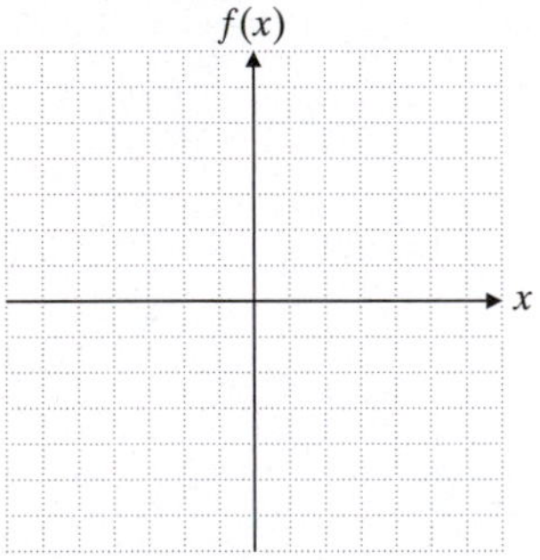

32. $g(x) = x^2 - 5$

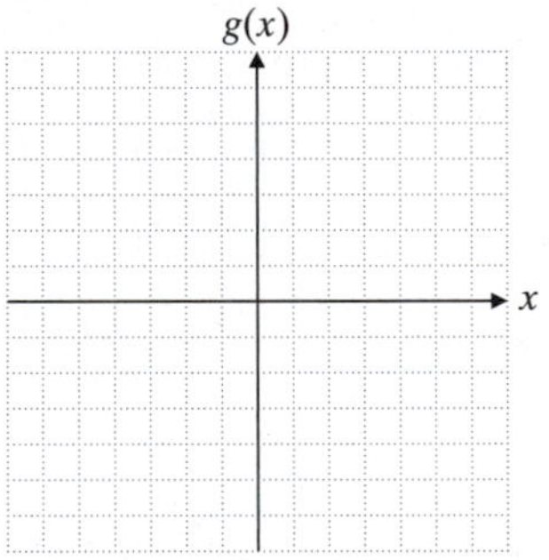

33. $h(x) = x^3 + 3$

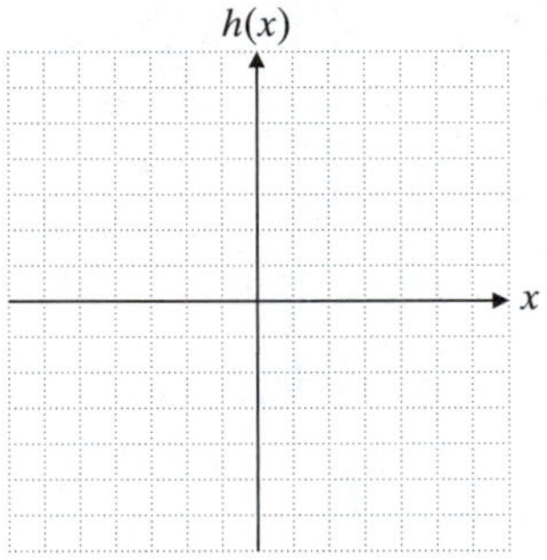

Plot the points and connect them to graph the underlying function. Estimate the value of $f(x)$ when $x = -2$.

34.

x	f(x)
−1	5
−3	−3
−4	−4
−5	−3
−6	0
−7	5

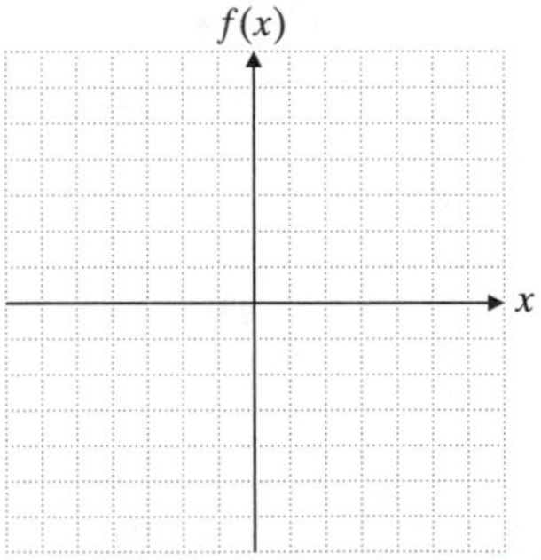

35.

x	f(x)
−3	4
−1	2
0	1
1	0
2	−1
3	0
4	1

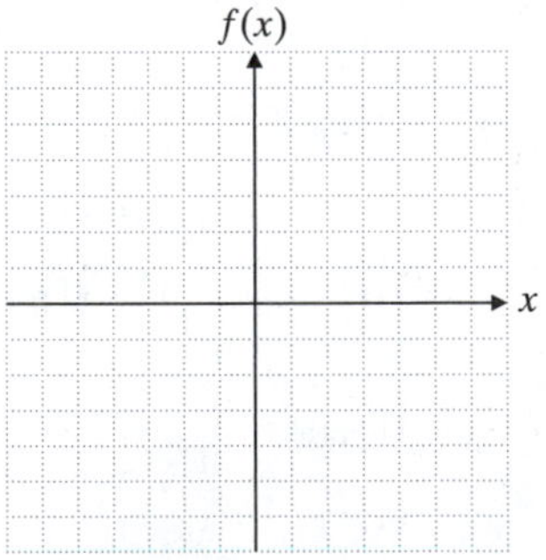

Complete the table of values for each function.

36. $f(x) = -\frac{4}{5}x + 3$

x	f(x)
−5	
0	
10	

37. $f(x) = 2x^2 - 3x + 4$

x	f(x)
−3	
0	
4	

Mixed Practice

38. Find the slope and y-intercept of $3x - 2y = 9$.

39. A line n is perpendicular to the line passing through $(-3, 6)$ and $(1, 9)$. What is the slope of line n?

40. A line p is parallel to the line passing through $(4.5, 8)$ and $(2.5, -6)$. What is the slope of line p?

41. Find the standard form of the equation of the line that has slope $\frac{5}{6}$ and y-intercept $(0, -5)$.

42. Find the equation of the line parallel to $y = 5x - 2$ and passing through $(4, 10)$. Write the answer in slope–intercept form.

43. Find the slope–intercept form of the equation of the line that passes through $(5, 6)$ and $(-7, 3)$.

44. Find the equation of the line perpendicular to $3x - 6y = 9$ and passing through $(-2, -1)$. Write the answer in standard form.

45. Find the equation of a vertical line passing through $(5, 6)$.

Applications

46. ***Car Rental Costs*** The cost of renting a full-size sedan at Palm Beach Car Rental is \$40 per day plus \$0.20 per mile that the car is driven. Express the cost of renting a full-size sedan for one day at Palm Beach Car Rental as a function $f(x)$, where x is the number of miles the car is driven.

47. ***Population Growth of a Town*** The population of Garrison was 24,000 in 1995. Since then the town has grown by 200 people each year. Express the population as a function $f(x)$, where x is the number of years since 1995.

48. ***Fish Population in a Stream*** A biologist in Montana has determined that a certain mountain stream initially had 18,000 fish on June 1. The number of fish in the stream has decreased by 65 fish each day since June 1. Express the number of fish in this stream as a function $f(x)$, where x is the number of days after June 1.

1.

2.

3.

4.

5.

6.

7.

8.

9.

10.

11.

How Am I Doing? Chapter 3 Test

CHAPTER Test Prep VIDEOS MATH COACH MyMathLab® YouTube™

After you take this test read through the Math Coach on pages 194–195. Math Coach videos are available via MyMathLab and YouTube. Step-by-step test solutions in the Chapter Test Prep Videos are also available via MyMathLab and YouTube. (Search "TobeyInterAlg" and click on "Channels.")

Graph each line. Plot at least three points.

1. $y = \frac{1}{3}x - 2$

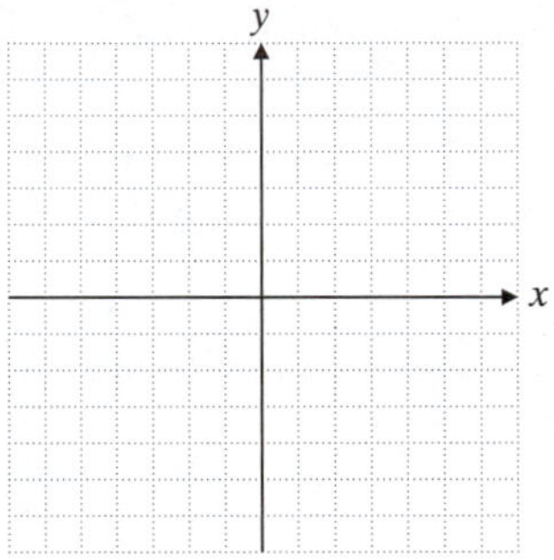

2. $2x - 3 = 1$

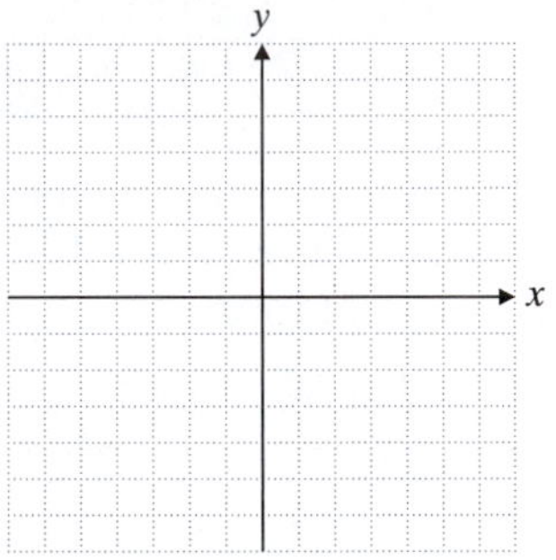

3. $5x + 3y = 9$

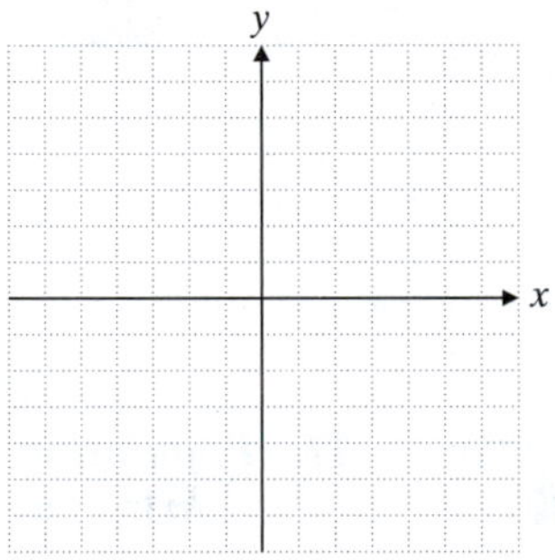

MC **4.** $2x + 3y = -10$

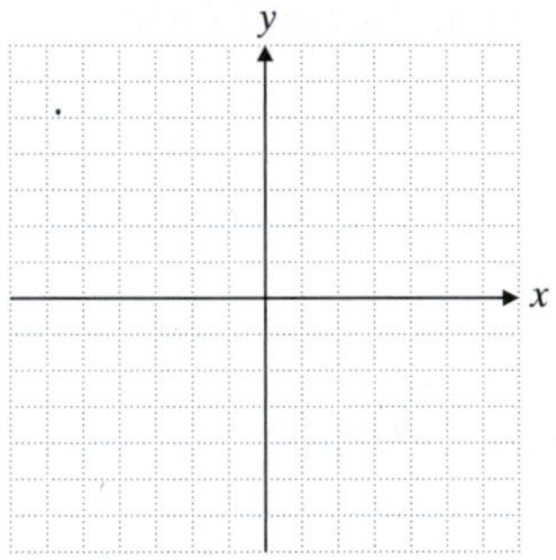

5. Find the slope of the line passing through $(2, -3)$ and $\left(\frac{1}{2}, -6\right)$.

6. Find the slope of a line passing through $(-7, 5)$ and $(6, 5)$.

7. Find the slope and y-intercept of the line $-2x + 5y = 15$.

8. Write the standard form equation of the line that is perpendicular to $6x - 7y - 1 = 0$ and passes through $(0, -2)$.

MC **9.** Write the standard form of the equation of the line that passes through $(5, -2)$ and $(-3, -1)$.

10. Write the equation of the horizontal line passing through $\left(-\frac{1}{3}, 2\right)$.

11. Write the equation of the line with slope -5 and y-intercept $(0, -8)$. Write the answer in slope–intercept form.

Graph the regions.

12. $y \geq -4x$

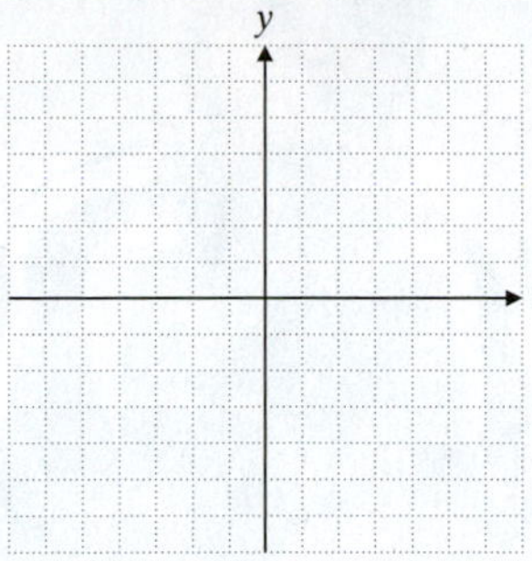

MC **13.** $4x - 2y < -6$

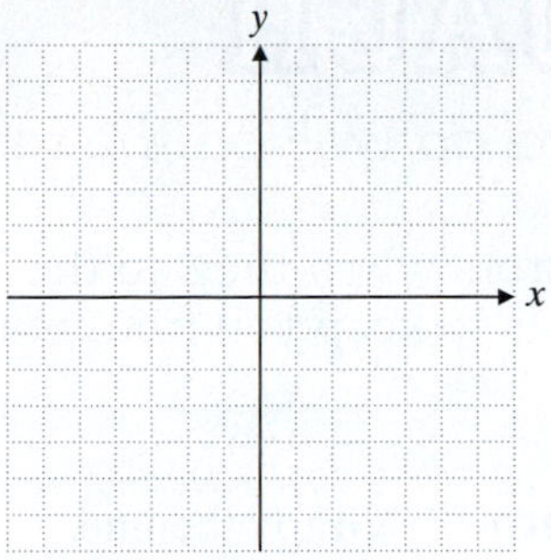

14. What are the domain and range of the following relation?
$A = \{(0, 0), (1, 1), (1, -1), (2, 4), (2, -4)\}$

15. If $f(x) = 2x - 3$, find $f\left(\frac{3}{4}\right)$.

16. If $g(x) = \frac{1}{2}x^2 + 3$, find $g(-4)$.

17. If $h(x) = \left|-\frac{2}{3}x + 4\right|$, find $h(-9)$.

MC **18.** If $p(x) = -2x^3 + 3x^2 + x - 4$, find $p(-2)$.

Graph each function.

19. $g(x) = 5 - x^2$

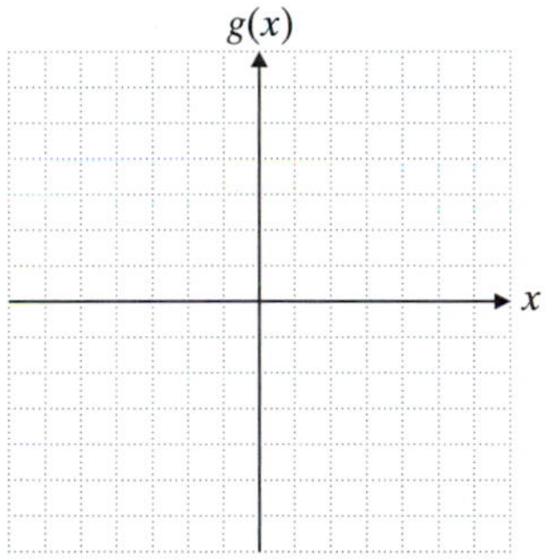

20. $h(x) = x^3 - 4$

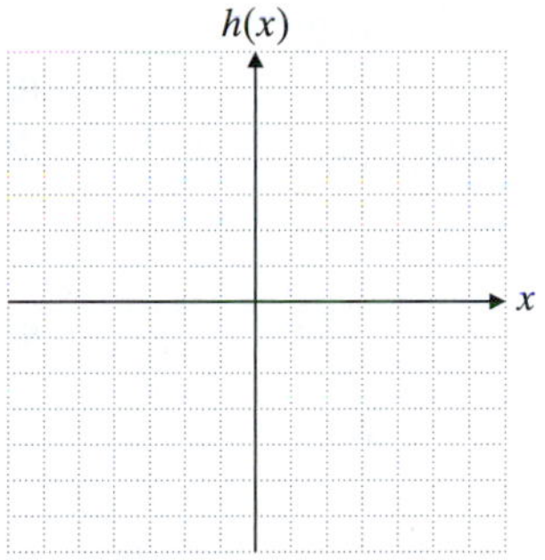

21. The following table describes the approximate distance $f(x)$ in miles that you can see across the ocean on a clear day if you are x feet above the water. Plot these points and connect them to graph the function.

Height in Feet, x	0	3	9	15
Distance in Miles, $f(x)$	0	6	54	150

Based on your graph, how many miles can you see if you are 4 feet above the water?

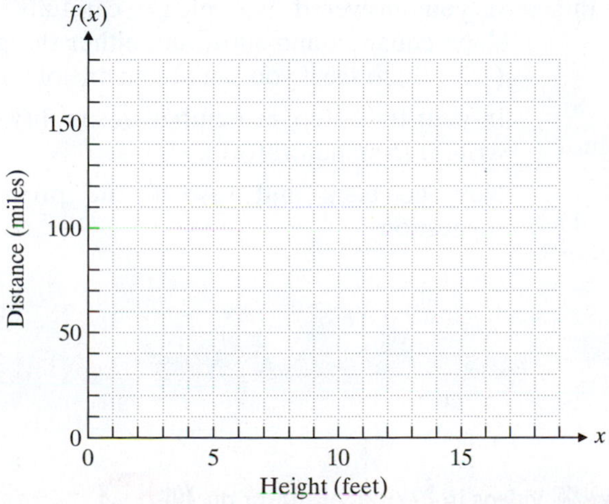

12. ______

13. ______

14. ______

15. ______

16. ______

17. ______

18. ______

19. ______

20. ______

21. ______

Total Correct: ______

MATH COACH

Mastering the skills you need to do well on the test.

Students often make the same types of errors when they do the Chapter 3 Test. Here are some helpful hints to keep you from making those common errors on test problems.

Graphing a Linear Equation—Problem 4

Graph the line. Plot at least three points. $2x + 3y = -10$

Helpful Hint If you carefully plot three points that form ordered-pair solutions to the given equation and then draw a straight line through these points, you can construct the graph of the line. If you cannot connect the three points with a straight line, then you know you have made a mistake.

Did you let $y = 0$ and substitute for y in the original equation to obtain -5 as the value for x? Did you plot $(-5, 0)$ on your graph paper?

Yes ____ No ____

If you answered No to either of these questions, substitute $y = 0$ into the original equation and solve for x.

Plot your results for that ordered pair on your graph paper.

Did you let $x = -2$ and substitute for x in the original equation to obtain -2 as the value for y?

Yes ____ No ____

Did you let $x = 1$ in the original equation and solve for y to get $y = -4$?

Yes ____ No ____

If you answered No to either of these questions, substitute these values for x into the original equation and solve for y in each case. Then plot each of these ordered pairs.

Remember to carefully draw a straight line that passes through each of the three points.

If you answered Problem 4 incorrectly, go back and rework the problem using these suggestions.

Write the Equation of a Line Given Two Points on the Line—Problem 9

Write the standard form of the equation of the line that passes through $(5, -2)$ and $(-3, -1)$.

Helpful Hint First, we can find the slope of a line when given two points on that line. Then we substitute the slope and the coordinates of one of the given points into the point-slope equation. Remember to simplify the equation if necessary.

Did you obtain a slope of $-\frac{1}{8}$ when you substituted the values of x and y for each point in the slope formula?

Yes ____ No ____

If you answered No, carefully substitute the x and y values into the slope formula.

Did you correctly substitute the slope and the point $(-3, -1)$ or $(5, -2)$ into the point-slope equation to obtain either $y + 1 = -\frac{1}{8}(x + 3)$ or $y + 2 = -\frac{1}{8}(x - 5)$?

Yes ____ No ____

If you answered No, please carefully write the point-slope equation and substitute either the point $(-3, -1)$ or $(-5, -2)$ to see if you can obtain one of these two equations.

In your final step, remember to simplify the equation and write it in standard form.

Now go back and rework the problem using these suggestions.

Need help? Watch the MATH COACH videos in MyMathLab® or on YouTube™.

Graphing a Linear Inequality in Two Variables—Problem 13

Graph the region. $4x - 2y < -6$

Helpful Hint Remember to use a dashed line for the graph of the boundary line if the inequality is written with a > or a < symbol.

Did you draw a dashed line for the graph of the boundary line $4x - 2y = -6$? Did it pass through the points (0, 3) and (−3, −3)?

Yes ____ No ____

If you answered No, remember to carefully plot points along the boundary line and connect these points with a dashed line.

If you use the test point (0, 0), does it satisfy the inequality $4x - 2y < -6$?

Yes ____ No ____

If you answered Yes, this is not correct. When you substitute (0, 0) into the inequality, you obtain $0 < -6$, which is not true. Does this mean that we shade above or below the dashed line? Remember that we shade the side of the boundary line where any test points chosen make the inequality true.

If you answered Problem 13 incorrectly, go back and rework the problem using these suggestions.

Evaluating a Function Using Function Notation—Problem 18

If $p(x) = -2x^3 + 3x^2 + x - 4$, find $p(-2)$.

Helpful Hint When you replace x with a numerical value in the function, place a parentheses around the number in the substitution step to avoid making mathematical errors.

Did you obtain $p(-2) = -2(-2)^3 + 3(-2)^2 + (-2) - 4$ in your first step?

Yes ____ No ____

If you answered No, stop and perform that step carefully again. Writing out this step will help you avoid errors.

Did you obtain $p(-2) = -2(-8) + 3(4) + (-2) - 4$ in your next step?

Yes ____ No ____

If you answered No, apply the order of operations by evaluating exponents next. Note that –2 to the third power is $(-2)(-2)(-2) = -8$ and –2 squared is $(-2)(-2) = 4$.

Now go back and rework the problem using these suggestions.

Cumulative Test for Chapters 1–3

This test provides a comprehensive review of the key objectives for Chapters 1–3.

1. Name the property that justifies $(-6) + 6 = 0$.

2. Evaluate the expression. $3(4 - 6)^2 + \sqrt{16} + 12 \div (-3)$

3. Simplify. $(2a^{-3}b^4)^{-2}$

4. Simplify. $\dfrac{20a^4b}{25ab^3}$

5. Simplify. $4(2x^2 - 1) - 2x(3x - 5y)$

6. Evaluate $2x^2 - 3xy - y^2$ for $x = -1$ and $y = 4$.

7. Write using scientific notation. 0.000437

▲ 8. Find the area of the semicircle to the right. Round your answer to the nearest hundredth. (Use $\pi \approx 3.14$.)

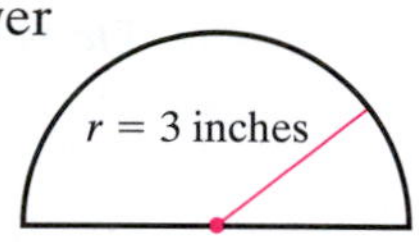

9. Solve for x. $\frac{1}{2}(x - 5) + \frac{2}{3}(x + 1) = \frac{1}{2}$

10. Solve for x. $|4x + 3| = 7$

11. Solve for x. $|x - 2| \leq 5$

12. Solve and graph your solution. $3(x - 2) > 6$ *or* $5 - 3(x + 1) > 8$

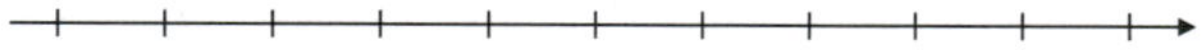

13. Solve for x. $2a - x = \frac{1}{3}(6x - y)$

▲ 14. A plastic insulator is made in a rectangular shape with a perimeter of 92 centimeters. The length is 1 centimeter longer than double the width. Find the dimensions of the insulator.

15. Marissa invested \$7000 for 1 year. Part was invested at 4% simple interest and part at 7% simple interest. At the end of 1 year, she had earned \$391 in interest. How much had she invested at each rate?

16. Graph the line $4x - 6y = 10$. Plot at least three points.

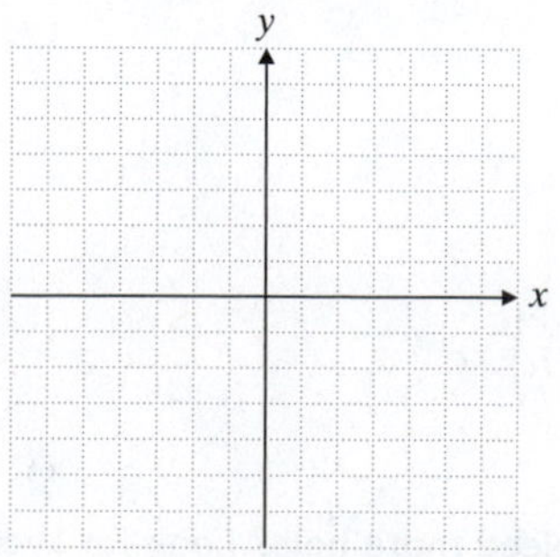

1. ______
2. ______
3. ______
4. ______
5. ______
6. ______
7. ______
8. ______
9. ______
10. ______
11. ______
12. ______
13. ______
14. ______
15. ______
16. ______

17. Find the slope of the line passing through (6, 5) and (−2, 1).

18. Write the standard form of the equation of the line that passes through (5, 1) and (4, 3).

19. Write the standard form of the equation of the line that passes through (−1, 4) and is perpendicular to $y = -\frac{1}{3}x + 6$.

20. What are the domain and range of the relation $\left\{(3, 7), (5, 8), \left(\frac{1}{2}, -1\right), (2, 2)\right\}$? Is the relation a function?

Graph the following functions.

21. $p(x) = -\frac{1}{3}x + 2$

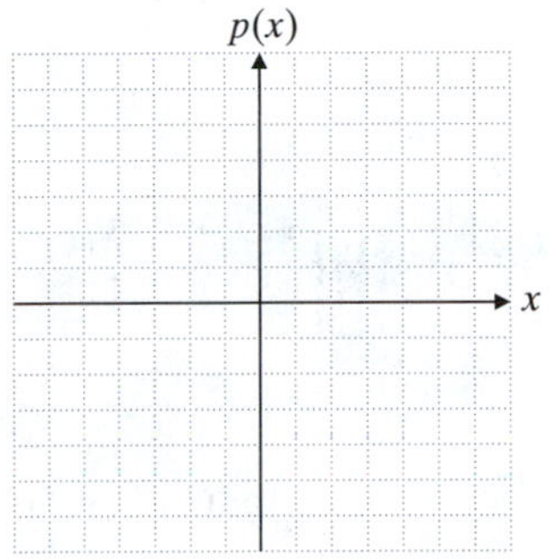

22. $h(x) = |x - 2|$

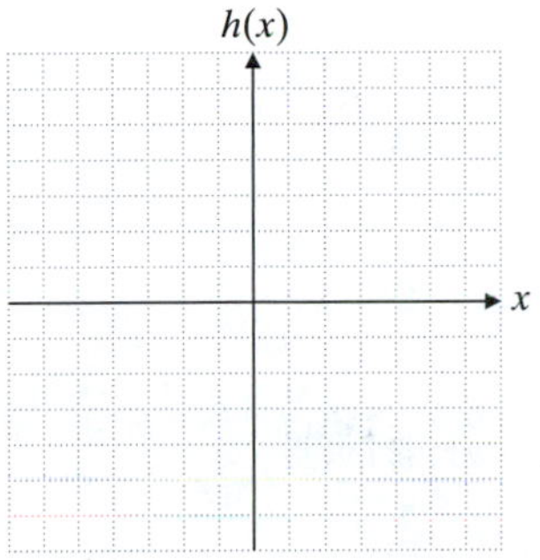

23. Graph the region. $y \leq -\frac{3}{2}x + 3$

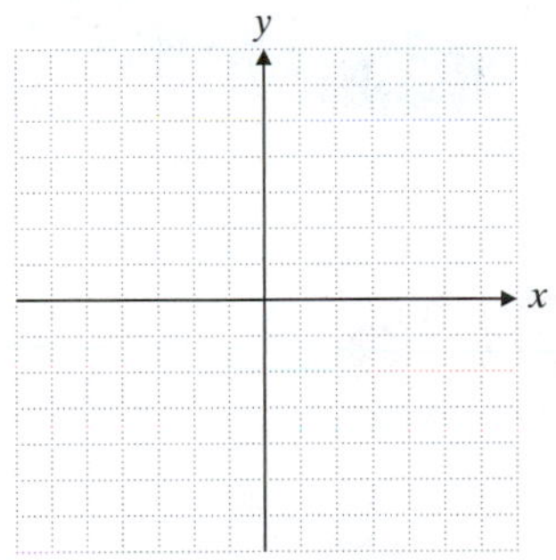

24. Complete the following table of values for $f(x)$. $f(x) = -2x^3 + 4$

x	f(x)
−2	
0	
3	

25. According to the U.S. Bureau of the Census, there were 31.1 million people living below the poverty level in 2000. The number has increased by 1.5 million each year since 2000.

(a) Express the number of people living below the poverty level, in millions, as a function $f(x)$ with x being the number of years since 2000.

(b) Using your answer from part **(a)**, predict how many people will be living below the poverty level in 2012.

17. ______

18. ______

19. ______

20. ______

21. ______

22. ______

23. ______

24. ______

25. (a) ______

(b) ______

CHAPTER 4

People who visit the Bay of Fundy in Nova Scotia, Canada, are often amazed at the huge difference between high tide and low tide. As the tide changes, huge volumes of moving water make a very strong tidal current. Trying to row a dory against the tide can be a challenging experience. Enjoying the brisk run of rowing a dory with the tide can be an exhilarating experience. The kind of mathematics you will learn in this chapter will allow you to solve a variety of problems regarding boating with and against the current.

Systems of Linear Equations and Inequalities

4.1 Systems of Linear Equations in Two Variables

Student Learning Objectives

After studying this section, you will be able to:

1. Determine whether an ordered pair is a solution to a system of two linear equations.
2. Solve a system of two linear equations by the graphing method.
3. Solve a system of two linear equations by the substitution method.
4. Solve a system of two linear equations by the addition (elimination) method.
5. Identify systems of linear equations that do not have a unique solution.
6. Choose an appropriate method to solve a system of linear equations algebraically.

① Determining Whether an Ordered Pair Is a Solution to a System of Two Linear Equations

In Chapter 3 we found that a linear equation containing two variables, such as $4x + 3y = 12$, has an unlimited number of ordered pairs (x, y) that satisfy it. For example, $(3, 0)$, $(0, 4)$, and $(-3, 8)$ all satisfy the equation $4x + 3y = 12$. We call *two* linear equations in two unknowns a **system of two linear equations in two variables.** Many such systems have exactly one solution. A **solution to a system** of two linear equations in two variables is an *ordered pair* that is a solution to *each* equation.

EXAMPLE 1 Determine whether $(3, -2)$ is a solution to the following system.

$$\begin{aligned} x + 3y &= -3 \\ 4x + 3y &= 6 \end{aligned}$$

Solution We will begin by substituting $(3, -2)$ into the first equation to see whether the ordered pair is a solution to the first equation.

$$\begin{aligned} 3 + 3(-2) &\stackrel{?}{=} -3 \\ 3 - 6 &\stackrel{?}{=} -3 \\ -3 &= -3 \checkmark \end{aligned}$$

Likewise, we will determine whether $(3, -2)$ is a solution to the second equation.

$$\begin{aligned} 4(3) + 3(-2) &\stackrel{?}{=} 6 \\ 12 - 6 &\stackrel{?}{=} 6 \\ 6 &= 6 \checkmark \end{aligned}$$

Since $(3, -2)$ is a solution to each equation in the system, it is a solution to the system itself.

It is important to remember that we cannot confirm that a particular ordered pair is in fact the solution to a system of two equations unless we have checked to see whether the solution satisfies both equations. Merely checking one equation is not sufficient. Determining whether an ordered pair is a solution to a system of equations requires that we verify that the solution satisfies *both* equations.

Student Practice 1 Determine whether $(-3, 4)$ is a solution to the following system.

$$\begin{aligned} 2x + 3y &= 6 \\ 3x - 4y &= 7 \end{aligned}$$

NOTE TO STUDENT: Fully worked-out solutions to all of the Student Practice problems can be found at the back of the text starting at page SP-1.

② Solving a System of Two Linear Equations by the Graphing Method

We can verify the solution to a system of linear equations by graphing each equation. If the lines intersect, the system has a unique solution. The point of intersection lies on both lines. Thus, it is a solution to each equation and the solution to the system. We will illustrate this by graphing the equations in Example 1. Notice that the coordinates of the point of intersection are $(3, -2)$. The solution to the system is $(3, -2)$.

This example shows that we can find the solution to a system of linear equations by graphing each line and determining the point of intersection.

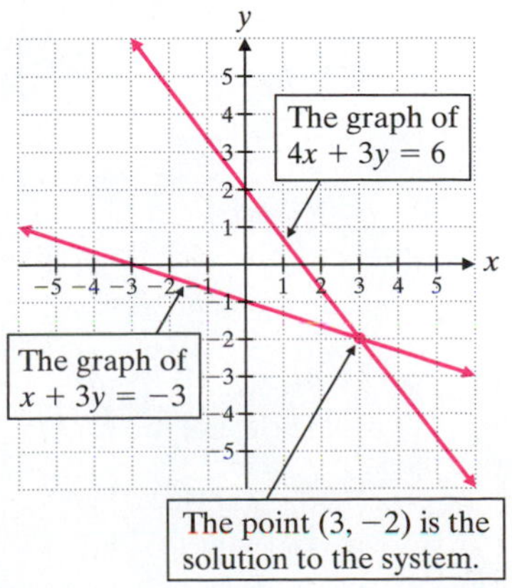

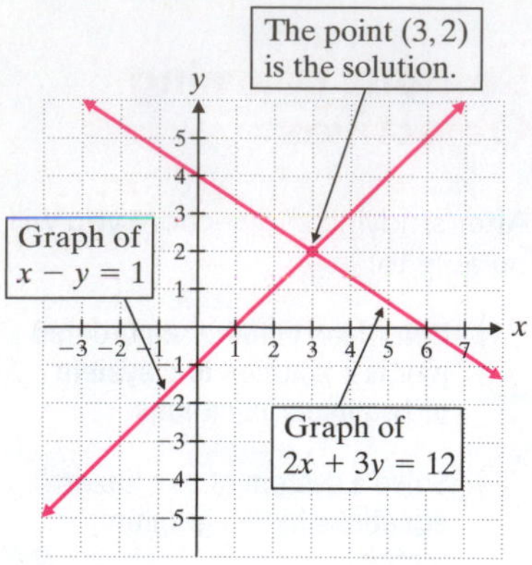

Student Practice 2

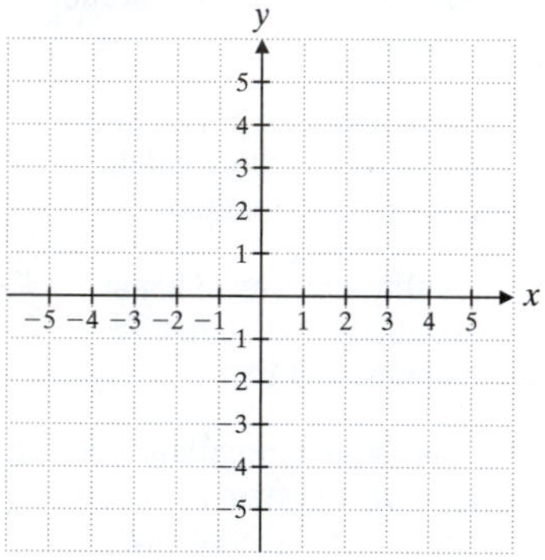

EXAMPLE 2 Solve this system of equations by graphing.

$$2x + 3y = 12$$
$$x - y = 1$$

Solution Using the methods that we developed in Chapter 3, we graph each line and determine the point at which the two lines intersect. The graph is to the left.

Finding the solution by the graphing method does not always lead to an accurate result, however, because it involves visual estimation of the point of intersection. Also, our plotting of one or more of the lines could be off slightly. Thus, we verify that our answer is correct by substituting $x = 3$ and $y = 2$ into the system of equations.

$$2x + 3y = 12 \qquad\qquad x - y = 1$$
$$2(3) + 3(2) \stackrel{?}{=} 12 \qquad\qquad 3 - 2 \stackrel{?}{=} 1$$
$$12 = 12 \checkmark \qquad\qquad 1 = 1 \checkmark$$

Thus, we have verified that the solution to the system is (3, 2).

Student Practice 2 Solve this system of equations by graphing. Check your solution.

$$3x + 2y = 10$$
$$x - y = 5$$

Many times when we graph a system, we find that the two lines intersect at one point. However, it is possible for a given system to have as its graph two parallel lines. In such a case there is no solution because there is no point that lies on both lines (i.e., no ordered pair that satisfies both equations). Such a system of equations is said to be **inconsistent.** Another possibility is that when we graph each equation in the system, we obtain one line. In such a case there is an infinite number of solutions. Any point (i.e., any ordered pair) that lies on the first line will also lie on the second line. A system of equations in two variables is said to have **dependent equations** if it has infinitely many solutions. We will discuss these situations in more detail after we have developed algebraic methods for solving systems of equations.

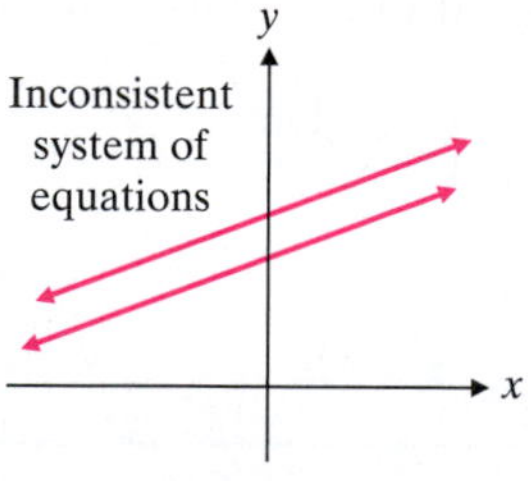

No solution

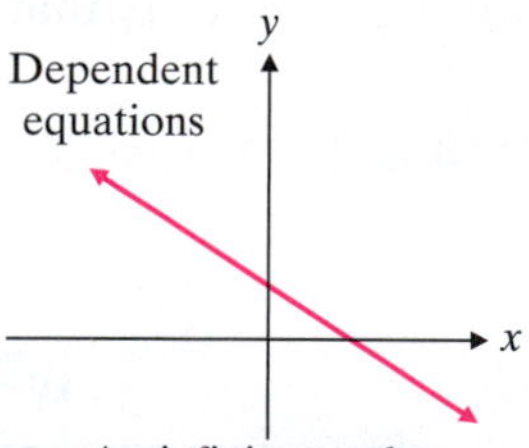

An infinite number of solutions

③ Solving a System of Two Linear Equations by the Substitution Method

One algebraic method of solving a system of linear equations in two variables is the **substitution method.** To use this method, we choose one equation and solve for one variable. It is usually best to solve for a variable that has a coefficient of +1 or −1, if possible. This will help us avoid introducing fractions. When we solve for one variable, we obtain an expression that contains the other variable. We *substitute* this expression into the second equation. Then we have one equation with one unknown, which we can easily solve. Once we know the value of this variable, we can substitute it into one of the original equations to find the value of the other variable.

EXAMPLE 3 Find the solution to the following system of equations. Use the substitution method.

$$x + 3y = -7 \quad \textbf{(1)}$$
$$4x + 3y = -1 \quad \textbf{(2)}$$

Solution We can work with equation **(1)** or equation **(2).** Let's choose equation **(1)** because x has a coefficient of 1. Now let us solve for x. This gives us equation **(3).**

$$x = -7 - 3y \quad \textbf{(3)}$$

Now we substitute this expression for x into equation **(2)** and solve the equation for y.

$$\begin{aligned} 4x + 3y &= -1 \quad \textbf{(2)} \\ 4(-7 - 3y) + 3y &= -1 \\ -28 - 12y + 3y &= -1 \\ -28 - 9y &= -1 \\ -9y &= -1 + 28 \\ -9y &= 27 \\ y &= -3 \end{aligned}$$

Now we substitute $y = -3$ into equation **(1)** or **(2)** to find x. Let's use **(1):**

$$\begin{aligned} x + 3(-3) &= -7 \\ x - 9 &= -7 \\ x &= -7 + 9 \\ x &= 2 \end{aligned}$$

Therefore, our solution is the ordered pair $(2, -3)$.
Check. We must verify the solution in both of the *original* equations.

$$\begin{aligned} x + 3y &= -7 \\ 2 + 3(-3) &\stackrel{?}{=} -7 \\ 2 - 9 &\stackrel{?}{=} -7 \\ -7 &= -7 \ \checkmark \end{aligned} \qquad \begin{aligned} 4x + 3y &= -1 \\ 4(2) + 3(-3) &\stackrel{?}{=} -1 \\ 8 - 9 &\stackrel{?}{=} -1 \\ -1 &= -1 \ \checkmark \end{aligned}$$

Student Practice 3 Use the substitution method to solve this system.

$$2x - y = 7$$
$$3x + 4y = -6$$

We summarize the substitution method here.

HOW TO SOLVE A SYSTEM OF TWO LINEAR EQUATIONS BY THE SUBSTITUTION METHOD

1. Choose one of the two equations and solve for one variable in terms of the other variable.
2. Substitute the expression from step 1 into the *other* equation.
3. You now have one equation with one variable. Solve this equation for that variable.
4. Substitute this value for the variable into one of the original or equivalent equations to obtain a value for the second variable.
5. Check the solution in both original equations.

Graphing Calculator

Solving Systems of Linear Equations

We can solve systems of equations graphically by using a graphing calculator. For example, to solve the system of equations in Example 2, first rewrite each equation in slope–intercept form.

$$y = -\frac{2}{3}x + 4$$
$$y = x - 1$$

Then graph $y_1 = -\frac{2}{3}x + 4$ and $y_2 = x - 1$ on the same screen.

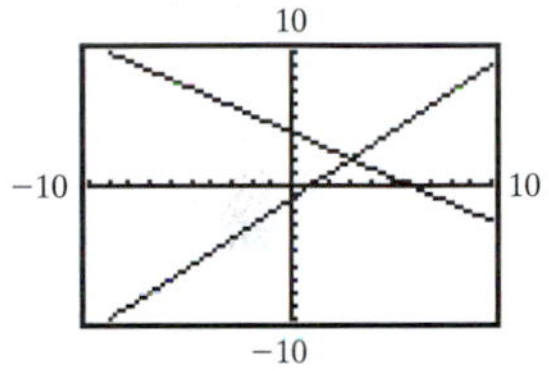

Next, use the Zoom and Trace features to find the intersection of the two lines.

Some graphing calculators have a command to find and calculate the intersection.

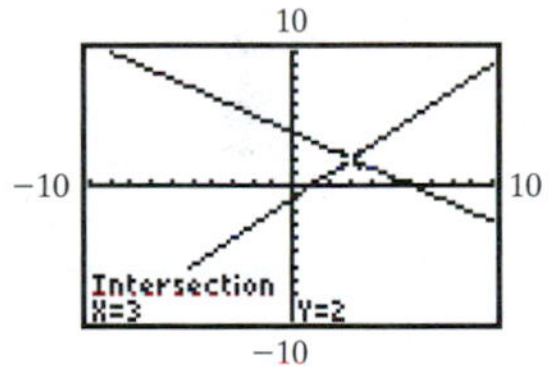

Try to find the solution to:

$$y_1 = -3x + 9$$
$$y_2 = 4x - 5$$

Optional Graphing Calculator Note. Before solving the system in Example 3 with a graphing calculator, you will first need to solve each equation for y. Equation **(1)** can be written as $y_1 = -\frac{1}{3}x - \frac{7}{3}$ or as $y_1 = \frac{-x-7}{3}$. Likewise, equation **(2)** can be written as $y_2 = -\frac{4}{3}x - \frac{1}{3}$ or as $y_2 = \frac{-4x-1}{3}$.

EXAMPLE 4 Solve the following system of equations.

$$\frac{1}{2}x - \frac{1}{4}y = -\frac{3}{4} \quad \textbf{(1)}$$
$$3x - 2y = -6 \quad \textbf{(2)}$$

Solution First clear equation **(1)** of fractions by multiplying each term by 4.

$$4\left(\frac{1}{2}x\right) - 4\left(\frac{1}{4}y\right) = 4\left(-\frac{3}{4}\right)$$
$$2x - y = -3 \quad \textbf{(3)}$$

We now have an equivalent system that does not contain fractions:

$$2x - y = -3 \quad \textbf{(3)}$$
$$3x - 2y = -6 \quad \textbf{(2)}$$

Step 1 Let's solve equation **(3)** for y. We select this because the y-variable has a coefficient of -1.

$$-y = -3 - 2x$$
$$y = 3 + 2x$$

Step 2 Substitute this expression for y into equation **(2).**

$$3x - 2(3 + 2x) = -6$$

Step 3 Solve this equation for x.

$$3x - 6 - 4x = -6$$
$$-6 - x = -6$$
$$-x = -6 + 6$$
$$-x = 0$$
$$x = 0$$

Step 4 To find the value of y, the easiest equation to use is our solution for y from step 1. Substitute $x = 0$ into this equation.

$$y = 3 + 2x$$
$$y = 3 + 2(0)$$
$$y = 3$$

So our solution is $(0, 3)$.

Step 5 We must verify the solution in both original equations.

$$\frac{1}{2}x - \frac{1}{4}y = -\frac{3}{4} \qquad 3x - 2y = -6$$
$$\frac{0}{2} - \frac{3}{4} \stackrel{?}{=} -\frac{3}{4} \qquad 3(0) - 2(3) \stackrel{?}{=} -6$$
$$-\frac{3}{4} = -\frac{3}{4} \checkmark \qquad -6 = -6 \checkmark$$

Student Practice 4 Use the substitution method to solve this system.

$$\frac{1}{2}x + \frac{2}{3}y = 1$$
$$\frac{1}{3}x + y = -1$$

④ Solving a System of Two Linear Equations by the Addition (Elimination) Method

Another way to solve a system of two linear equations in two variables is to add the two equations so that a variable is eliminated. This technique is called the **addition method** or the **elimination method.** We usually have to multiply one or both of the equations by suitable factors so that we obtain opposite coefficients on one variable (either x or y) in the equations.

EXAMPLE 5 Solve the following system by the addition method.

$$5x + 8y = -1 \quad \textbf{(1)}$$
$$3x + y = 7 \quad \textbf{(2)}$$

Solution We can eliminate either the x- or the y-variable. Let's choose y. We multiply equation **(2)** by -8.

$$-8(3x) + (-8)(y) = -8(7)$$
$$-24x - 8y = -56 \quad \textbf{(3)}$$

We now add equations **(1)** and **(3).**

$$5x + 8y = -1 \quad \textbf{(1)}$$
$$-24x - 8y = -56 \quad \textbf{(3)}$$
$$-19x = -57$$

We solve for x.

$$x = \frac{-57}{-19} = 3$$

Now we substitute $x = 3$ into either of the original equations **(1)** or **(2)**, or equivalent equation **(3).** We will use equation **(2).**

$$3(3) + y = 7$$
$$9 + y = 7$$
$$y = -2$$

Our solution is $(3, -2)$.

Check.
$$5(3) + 8(-2) \stackrel{?}{=} -1$$
$$15 + (-16) \stackrel{?}{=} -1$$
$$-1 = -1 \checkmark$$
$$3(3) + (-2) \stackrel{?}{=} 7$$
$$9 + (-2) \stackrel{?}{=} 7$$
$$7 = 7 \checkmark$$

Student Practice 5 Use the addition method to solve this system.

$$-3x + y = 5$$
$$2x + 3y = 4$$

For convenience, we summarize the addition method on the next page.

HOW TO SOLVE A SYSTEM OF TWO LINEAR EQUATIONS BY THE ADDITION (ELIMINATION) METHOD

1. Arrange each equation in the form $ax + by = c$. (Remember that a, b, and c can be any real numbers.)
2. Multiply one or both equations by appropriate numbers so that the coefficients of one of the variables are opposites.
3. Add the two equations from step 2 so that one variable is eliminated.
4. Solve the resulting equation for the remaining variable.
5. Substitute this value into one of the original or equivalent equations and solve to find the value of the other variable.
6. Check the solution in both of the original equations.

EXAMPLE 6 Solve the following system by the addition method.

$$\frac{x}{4} + \frac{y}{6} = -\frac{2}{3} \quad \textbf{(1)}$$
$$\frac{x}{5} + \frac{y}{2} = \frac{1}{5} \quad \textbf{(2)}$$

Solution Clear equation **(1)** of fractions by multiplying each term by 12.

$$12\left(\frac{x}{4}\right) + 12\left(\frac{y}{6}\right) = 12\left(-\frac{2}{3}\right)$$
$$3x + 2y = -8 \quad \textbf{(3)}$$

Clear equation **(2)** of fractions by multiplying each term by 10.

$$10\left(\frac{x}{5}\right) + 10\left(\frac{y}{2}\right) = 10\left(\frac{1}{5}\right)$$
$$2x + 5y = 2 \quad \textbf{(4)}$$

We now have an equivalent system that does not contain fractions.

$$3x + 2y = -8 \quad \textbf{(3)}$$
$$2x + 5y = 2 \quad \textbf{(4)}$$

To eliminate the variable x, we multiply equation **(3)** by 2 and equation **(4)** by -3. We now have the following equivalent system.

$$6x + 4y = -16$$
$$\underline{-6x - 15y = -6}$$
$$-11y = -22 \quad \text{Add the equations.}$$
$$y = 2 \quad \text{Solve for } y.$$

Since the original equations contain fractions, it will be easier to find the value of x if we use one of the equivalent equations. Substitute $y = 2$ into equation **(3)**.

$$3x + 2(2) = -8$$
$$3x + 4 = -8$$
$$3x = -12$$
$$x = -4$$

The solution to the system is $(-4, 2)$.

Check. Verify that this solution is correct.

Note. We could have easily eliminated the variable y in Example 6 by multiplying equation **(3)** by 5 and equation **(4)** by -2. Try it. Is the solution the same? Why?

Student Practice 6 Use the addition (elimination) method to solve this system.

$$\frac{x}{4} + \frac{y}{5} = \frac{23}{20}$$
$$\frac{7}{15}x - \frac{y}{5} = 1$$

⑤ Identifying Systems of Linear Equations That Do Not Have a Unique Solution

So far we have examined only those systems that have one solution. But other systems must also be considered. These systems can best be illustrated with graphs. In general, the system of equations

$$ax + by = c$$
$$dx + ey = f$$

may have one solution, no solution, or an infinite number of solutions.

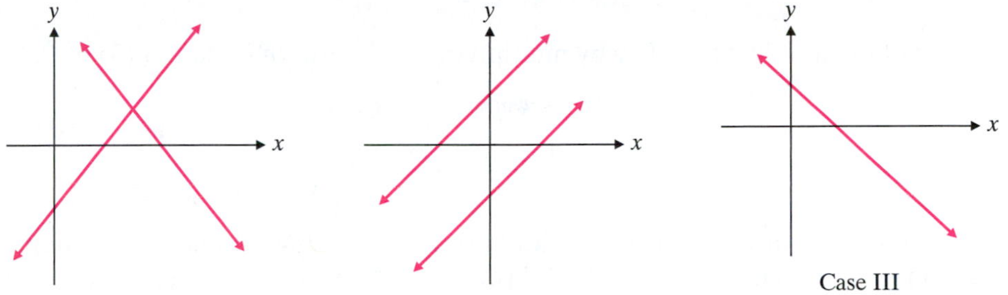

Case I (1 solution) | Case II (No solution) | Case III (Infinite number of solutions)

Case I: *One solution.* The two graphs intersect at one point, which is the solution. We say that the equations are **independent.** It is a **consistent system** of equations. There is a point (an ordered pair) *consistent* with both equations.

Case II: *No solution.* The two graphs are parallel and so do not intersect. We say that the system of equations is **inconsistent** because there is no point consistent with both equations.

Case III: *An infinite number of solutions.* The graphs of each equation yield the same line. Every ordered pair on this line is a solution to both of the equations. We say that the equations are **dependent.**

Graphing Calculator

Identifying Systems of Linear Equations

If there is a concern as to whether or not a system of equations has no solution, how can this be determined quickly using a graphing calculator? What is the most important thing to look for on the graph?

If the equations are dependent, how can we be sure of this by looking at the display of a graphing calculator? Why? Determine whether there is one solution, no solution, or an infinite number of solutions for each of the following systems.

1. $y_1 = -2x + 1$
$y_2 = -6x + 3$

2. $y_1 = 3x + 6$
$y_2 = 3x + 2$

3. $y_1 = x - 3$
$y_2 = -2x + 12$

EXAMPLE 7 If possible, solve the system.

$$2x + 8y = 16 \quad \textbf{(1)}$$
$$4x + 16y = -8 \quad \textbf{(2)}$$

Solution To eliminate the variable y, we'll multiply equation **(1)** by -2.

$$-2(2x) + (-2)(8y) = (-2)(16)$$
$$-4x - 16y = -32 \quad \textbf{(3)}$$

We now have the following equivalent system.

$$-4x - 16y = -32 \quad \textbf{(3)}$$
$$4x + 16y = -8 \quad \textbf{(2)}$$

When we add equations **(3)** and **(2),** we get

$$0 = -40,$$

which, of course, is false. Thus, we conclude that this system of equations is inconsistent, and **there is no solution.** Therefore, equations **(1)** and **(2)** do not intersect, as we can see on the graph to the right.

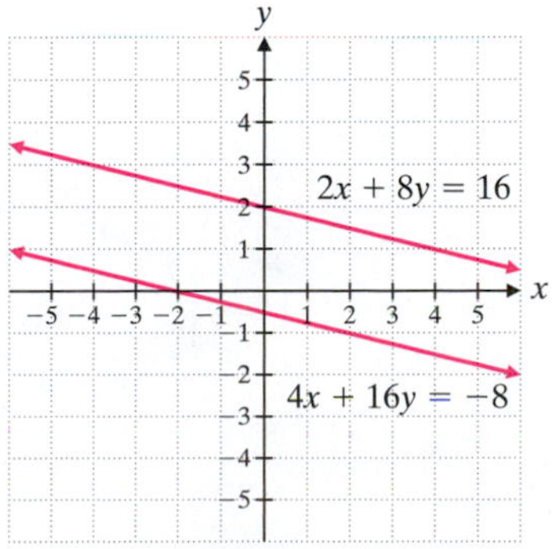

If we had used the substitution method to solve this system, we still would have obtained a false statement. When you try to solve an inconsistent system of linear equations by any method, you will always obtain a mathematical equation that is not true.

Student Practice 7 If possible, solve the system.

$$4x - 2y = 6$$
$$-6x + 3y = 9$$

EXAMPLE 8 If possible, solve the system.

$$0.5x - 0.2y = 1.3 \quad \textbf{(1)}$$
$$-1.0x + 0.4y = -2.6 \quad \textbf{(2)}$$

Solution Although we could work directly with the decimals, it is easier to multiply each equation by the appropriate power of 10 (10, 100, and so on) so that the coefficients of the new system are integers. Therefore, we will multiply equations **(1)** and **(2)** by 10 to obtain the following equivalent system.

$$5x - 2y = 13 \quad \textbf{(3)}$$
$$-10x + 4y = -26 \quad \textbf{(4)}$$

We can eliminate the variable y by multiplying each term of equation **(3)** by 2.

$$10x - 4y = 26 \quad \textbf{(5)}$$
$$\underline{-10x + 4y = -26} \quad \textbf{(4)}$$
$$0 = 0 \qquad \text{Add the equations.}$$

This statement is always true; it is an **identity.** Hence, the two equations are dependent, and there is an infinite number of solutions. Any solution satisfying equation **(1)** will also satisfy equation **(2).** For example, (3, 1) is a solution to equation **(1).** (Prove this.) Hence, it must also be a solution to equation **(2).** (Prove it). Thus, the equations actually describe the same line, as you can see on the graph.

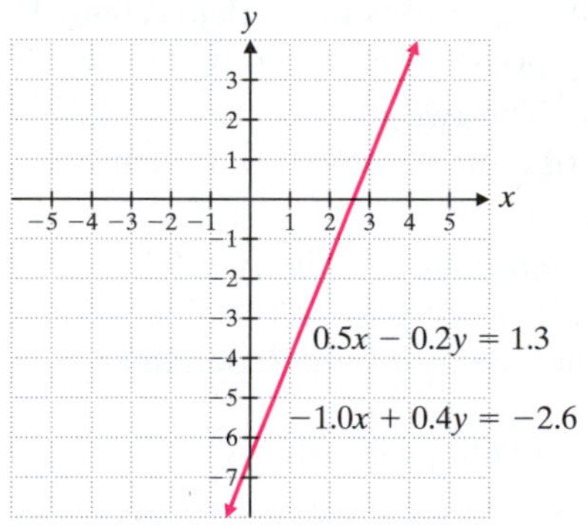

Student Practice 8 If possible, solve the system.

$$0.3x - 0.9y = 1.8$$
$$-0.4x + 1.2y = -2.4$$

⑥ Choosing an Appropriate Method to Solve a System of Linear Equations Algebraically

At this point we will review the algebraic methods for solving systems of linear equations and discuss the advantages and disadvantages of each method.

Method	Advantage	Disadvantage
Substitution	Works well if one or more variables has a coefficient of 1 or −1.	Often becomes difficult to use if no variable has a coefficient of 1 or −1.
Addition	Works well if equations have fractional or decimal coefficients. Works well if no variable has a coefficient of 1 or −1.	None

EXAMPLE 9 Select a method and solve each system of equations.

(a) $x + y = 3080$
$2x + 3y = 8740$

(b) $5x - 2y = 19$
$-3x + 7y = 35$

Solution

(a) Since there are x- and y-values that have coefficients of 1, we will select the substitution method.

$y = 3080 - x$ Solve the first equation for y.

$2x + 3(3080 - x) = 8740$ Substitute the expression into the second equation.

$2x + 9240 - 3x = 8740$ Remove parentheses.

$-1x = -500$ Simplify.

$x = 500$ Divide each side by -1.

Substitute $x = 500$ into the first equation.

$$x + y = 3080$$
$$500 + y = 3080$$
$$y = 3080 - 500$$
$$y = 2580 \quad \text{Simplify.}$$

The solution is (500, 2580).

Check. Verify that this solution is correct.

(b) Because none of the x- and y-variables has a coefficient of 1 or -1, we select the addition method. We choose to eliminate the y-variable. Thus, we would like the coefficients of y to be -14 and 14.

$7(5x) - 7(2y) = 7(19)$ Multiply each term of the first equation by 7.

$2(-3x) + 2(7y) = 2(35)$ Multiply each term of the second equation by 2.

$35x - 14y = 133$ We now have an equivalent system of equations.

$-6x + 14y = 70$

$29x = 203$ Add the two equations.

$x = 7$ Divide each side by 29.

Substitute $x = 7$ into one of the original equations. We will use the first equation.

$$5(7) - 2y = 19$$
$$35 - 2y = 19 \quad \text{Solve for } y.$$
$$-2y = -16$$
$$y = 8$$

The solution is (7, 8).

Check. Verify that this solution is correct.

Student Practice 9 Select a method and solve each system of equations.

(a) $3x + 5y = 1485$
$x + 2y = 564$

(b) $7x + 6y = 45$
$6x - 5y = -2$

TO THINK ABOUT: Two Linear Equations in Two Variables

Now is a good time to look back over what we have learned. When you graph a system of two linear equations, what possible kinds of graphs will you obtain?

What will happen when you try to solve a system of two linear equations using algebraic methods? How many solutions are possible in each case? The following chart may help you to organize your answers to these questions.

Graph	Number of Solutions	Algebraic Interpretation
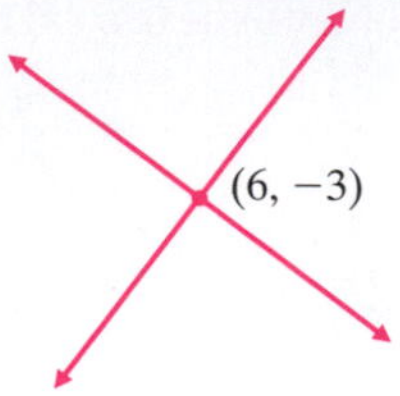 Two lines intersect at one point	**One unique solution**	You obtain one value for x and one value for y. For example, $x = 6, \quad y = -3.$
 Parallel lines	**No solution**	You obtain an equation that is inconsistent with known facts. For example, $0 = 6.$ The system of equations is inconsistent.
 Lines coincide	**Infinite number of solutions**	You obtain an equation that is always true. For example, $8 = 8.$ The equations are dependent.

4.1 Exercises

MyMathLab® Watch the videos in MyMathLab Download the MyDashBoard App

Verbal and Writing Skills, Exercises 1–4

1. Explain what happens when a system of two linear equations is inconsistent. What effect does it have in obtaining a solution? What would the graph of such a system look like?

2. Explain what happens when a system of two linear equations has dependent equations. What effect does it have in obtaining a solution? What would the graph of such a system look like?

3. How many possible solutions can a system of two linear equations in two unknowns have?

4. When you have graphed a system of two linear equations in two unknowns, how do you determine the solution of the system?

Determine whether the given ordered pair is a solution to the system of equations.

5. $\left(\frac{3}{2}, -1\right)$ $\quad 4x + 1 = 6 - y$
$2x - 5y = 8$

6. $\left(-4, \frac{2}{3}\right)$ $\quad 2x - 3(y - 5) = 5$
$6y = x + 8$

Solve the system of equations by graphing. Check your solution.

7. $3x + y = 2$
$2x - y = 3$

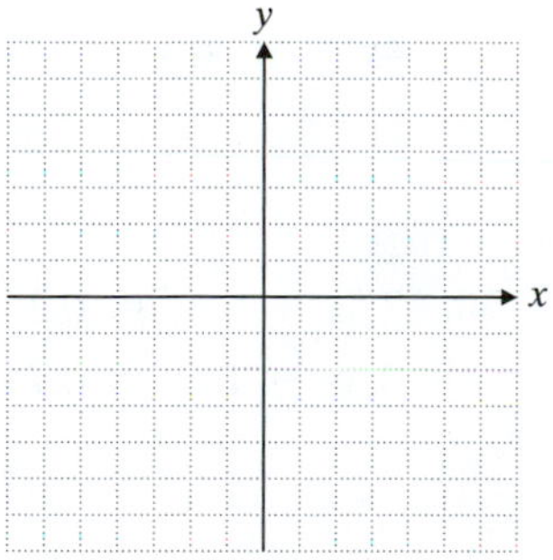

8. $3x + y = 5$
$2x - y = 5$

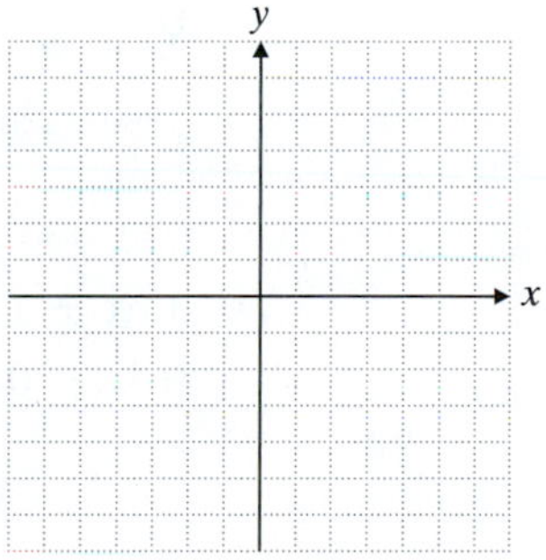

9. $3x - 2y = 6$
$4x + y = -3$

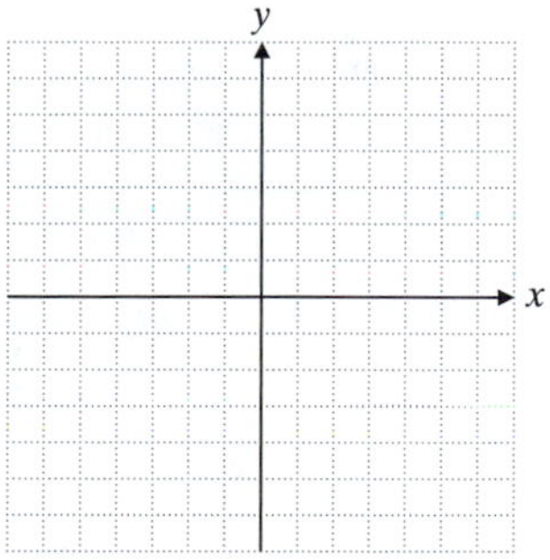

10. $3x - y = 5$
$2x - 3y = -6$

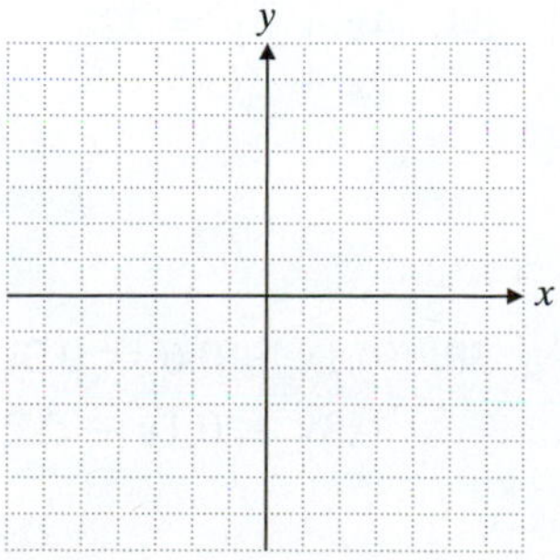

11. $y = -x + 3$
$x + y = -\frac{2}{3}$

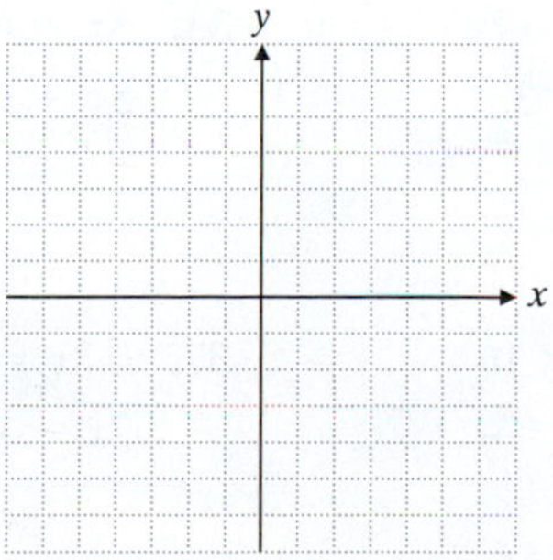

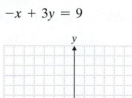

12. $y = \frac{1}{3}x - 2$
$-x + 3y = 9$

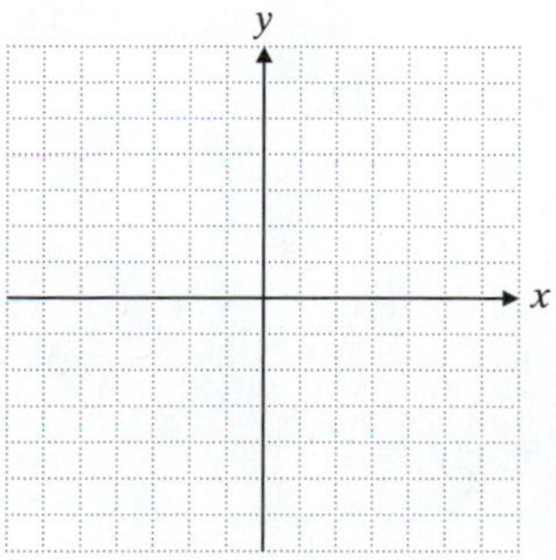

13. $y = -2x + 5$
$3y + 6x = 15$

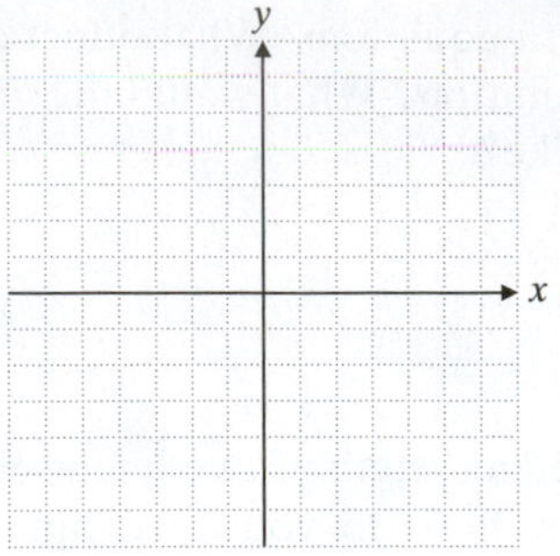

14. $x - 3 = 2y + 1$
$y - \frac{x}{2} = -2$

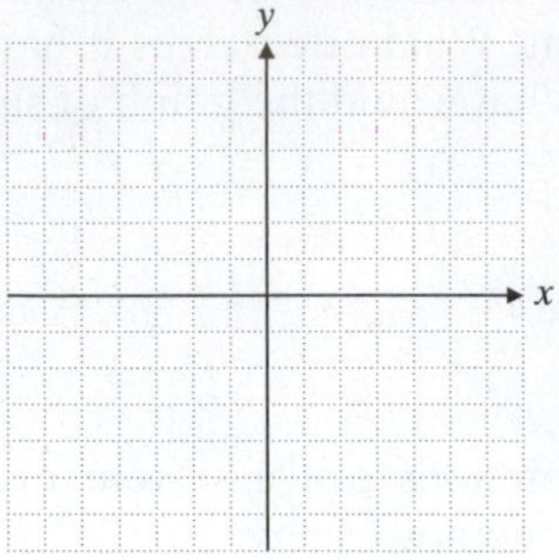

Find the solution to each system by the substitution method. Check your answers for exercises 15–18.

15. $3x + 4y = 14$
$x + 2y = -2$

16. $3x - 5y = 7$
$x - 4y = -7$

17. $-x + 3y = -8$
$2x - y = 6$

18. $10x + 3y = 8$
$2x + y = 2$

19. $2x - \frac{1}{2}y = -3$
$\frac{x}{5} + 2y = \frac{19}{5}$

20. $x - \frac{3}{2}y = 1$
$2x - 7y = 10$

21. $\frac{1}{2}x - \frac{1}{8}y = 3$
$\frac{2}{3}x + \frac{3}{4}y = 4$

22. $x + \frac{1}{4}y = 1$
$\frac{3}{8}x - y = -4$

Find the solution to each system by the addition (elimination) method. Check your answers for exercises 23–26.

23. $9x + 2y = 2$
$3x + 5y = 5$

24. $12x - 5y = -7$
$4x + 2y = 5$

25. $3s + 3t = 10$
$4s - 9t = -4$

26. $5s + 9t = 6$
$4s - 3t = 15$

27. $\frac{7}{2}x + \frac{5}{2}y = -4$
$3x + \frac{2}{3}y = 1$

28. $\frac{4}{3}x - y = 4$
$\frac{3}{4}x - y = \frac{1}{2}$

29. $1.6x + 1.5y = 1.8$
$0.4x + 0.3y = 0.6$

30. $2.5x + 0.6y = 0.2$
$0.5x - 1.2y = 0.7$

Mixed Practice

If possible, solve each system of equations. Use any method. If there is not a unique solution to a system, state a reason.

31. $7x - y = 6$
$3x + 2y = 22$

32. $8x - y = 17$
$4x + 3y = 33$

33. $3x + 4y = 8$
$5x + 6y = 10$

34. $4x + 5y = 22$
$9x + 2y = -6$

35. $2x + y = 4$
$\frac{2}{3}x + \frac{1}{4}y = 2$

36. $2x + 3y = 16$
$5x - \frac{3}{4}y = 7$

37. $0.2x = 0.1y - 1.2$
$2x - y = 6$

38. $0.1x - 0.6 = 0.3y$
$0.3x + 0.1y + 2.2 = 0$

39. $\begin{aligned} 5x - 7y &= 12 \\ -10x + 14y &= -24 \end{aligned}$

40. $\begin{aligned} 2x - 8y &= 5 \\ -6x + 24y &= 15 \end{aligned}$

41. $\begin{aligned} 0.8x + 0.9y &= 1.3 \\ 0.6x - 0.5y &= 4.5 \end{aligned}$

42. $\begin{aligned} 0.6y &= 0.9x + 1 \\ 3x &= 2y - 4 \end{aligned}$

43. $\begin{aligned} \frac{5}{3}b &= \frac{1}{3} + a \\ 9a - 15b &= 2 \end{aligned}$

44. $\begin{aligned} a - 3b &= \frac{3}{4} \\ \frac{a}{3} &= \frac{1}{4} + b \end{aligned}$

45. $\begin{aligned} \frac{2}{3}x - y &= 4 \\ 2x - \frac{3}{4}y &= 21 \end{aligned}$

46. $\begin{aligned} \frac{2}{9}x + y &= 20 \\ x - \frac{5}{4}y &= -2 \end{aligned}$

47. $\begin{aligned} 3.2x - 1.5y &= -3 \\ 0.7x + y &= 2 \end{aligned}$

48. $\begin{aligned} 3x - 0.2y &= 1 \\ 1.1x + 0.4y &= -2 \end{aligned}$

49. $\begin{aligned} 3 - (2x + 1) &= y + 6 \\ x + y + 5 &= 1 - x \end{aligned}$

50. $\begin{aligned} 2(y - 3) &= x + 3y \\ x + 2 &= 3 - y \end{aligned}$

To Think About

51. ***Bathroom Tile*** Wayne Burton is having some tile replaced in his bathroom. He has obtained an estimate from two tile companies. Old World Tile gave an estimate of \$200 to remove the old tile and \$50 per hour to place new tile on the wall. Modern Bathroom Headquarters gave an estimate of \$300 to remove the old tile and \$30 per hour to place new tile on the wall.

(a) Create a cost equation for each company where y is the total cost of the tile work and x is the number of hours of labor used to install new tile. Write a system of equations.

(b) Graph the two equations using the values $x = 0$, 4, and 8.

(c) Determine from your graph how many hours of installing new tile will be required for the two companies to cost the same.

(d) Determine from your graph which company costs less to remove old tile and to install new tile if the time needed to install new tile is 6 hours.

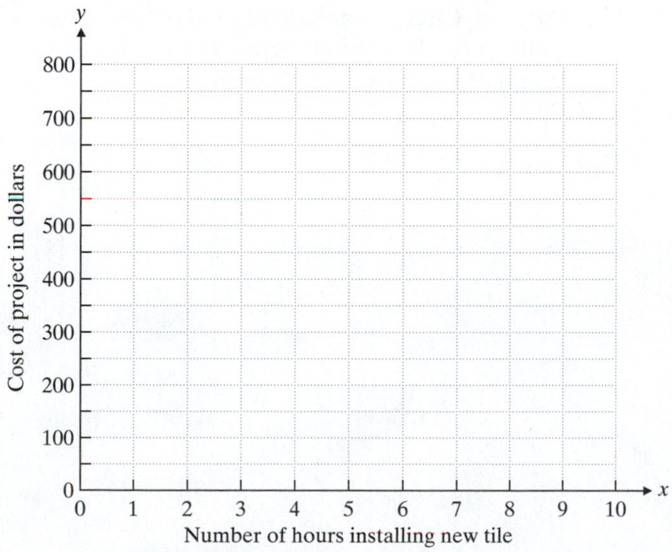

Optional Graphing Calculator Problems

On a graphing calculator, graph each system of equations on the same set of axes. Find the point of intersection to the nearest hundredth.

52. $y_1 = -0.81x + 2.3$
$y_2 = 1.6x + 0.8$

53. $y_1 = -1.7x + 3.8$
$y_2 = 0.7x - 2.1$

54. $5.86x + 6.22y = -8.89$
$-2.33x + 4.72y = -10.61$

55. $0.5x + 1.1y = 5.5$
$-3.1x + 0.9y = 13.1$

Cumulative Review

56. [1.2.2] ***Winter Road Salt*** Nine million tons of salt are applied to American highways for road deicing each year. The cost of buying and applying the salt totals $200 million. How much is this per pound? Round your answer to the nearest cent.

57. [1.2.2] ***City Parking Space*** Four-fifths of the automobiles that enter the city of Boston during rush hour will have to park in private or municipal parking lots. If there are 273,511 private or municipal lot spaces filled by cars entering the city during rush hour every morning, how many cars enter the city during rush hour? Round your answer to the nearest car.

Quick Quiz 4.1

1. Solve using the substitution method.
$5x - 3y = 14$
$2x - y = 6$

2. Solve using the addition method.
$6x + 7y = 26$
$5x - 2y = 6$

3. Solve by any method.
$\frac{2}{3}x + \frac{3}{5}y = -17$
$\frac{1}{2}x - \frac{1}{3}y = -1$

4. Concept Check Explain what happens when you go through the steps to solve the following system of equations. Why does this happen?
$6x - 4y = 8$
$-9x + 6y = -12$

4.2 Systems of Linear Equations in Three Variables

① Determining Whether an Ordered Triple Is the Solution to a System of Three Linear Equations in Three Variables

Student Learning Objectives

After studying this section, you will be able to:

① **Determine whether an ordered triple is the solution to a system of three linear equations in three variables.**

② **Find the solution to a system of three linear equations in three variables if none of the coefficients is zero.**

③ **Find the solution to a system of three linear equations in three variables if some of the coefficients are zero.**

We are now going to study **systems of three linear equations in three variables** (unknowns). A **solution** to a system of three linear equations in three unknowns is an **ordered triple** of real numbers (x, y, z) that satisfies each equation in the system.

EXAMPLE 1 Determine whether $(2, -5, 1)$ is a solution to the following system.

$$\begin{aligned} 3x + y + 2z &= 3 \\ 4x + 2y - z &= -3 \\ x + y + 5z &= 2 \end{aligned}$$

Solution How can we determine whether $(2, -5, 1)$ is a solution to this system? We will substitute $x = 2, y = -5$, and $z = 1$ into each equation. If a true statement occurs each time, $(2, -5, 1)$ is a solution to each equation and hence, a solution to the system. For the first equation:

$$\begin{aligned} 3(2) + (-5) + 2(1) &\stackrel{?}{=} 3 \\ 6 - 5 + 2 &\stackrel{?}{=} 3 \\ 3 &= 3 \ \checkmark \end{aligned}$$

For the second equation:

$$\begin{aligned} 4(2) + 2(-5) - 1 &\stackrel{?}{=} -3 \\ 8 - 10 - 1 &\stackrel{?}{=} -3 \\ -3 &= -3 \ \checkmark \end{aligned}$$

For the third equation:

$$\begin{aligned} 2 + (-5) + 5(1) &\stackrel{?}{=} 2 \\ 2 - 5 + 5 &\stackrel{?}{=} 2 \\ 2 &= 2 \ \checkmark \end{aligned}$$

Since we obtained three true statements, the ordered triple $(2, -5, 1)$ is a solution to the system.

Student Practice 1 Determine whether $(3, -2, 2)$ is a solution to this system.

$$\begin{aligned} 2x + 4y + z &= 0 \\ x - 2y + 5z &= 17 \\ 3x - 4y + z &= 19 \end{aligned}$$

***NOTE TO STUDENT:** Fully worked-out solutions to all of the Student Practice problems can be found at the back of the text starting at page SP-1.*

TO THINK ABOUT: Graphs in Three Variables
Can we graph an equation in three variables? How? What would the graph look like? What would the graph of the system in Example 1 look like? Describe the graph of the solution. At the end of Section 4.2 we will show you how the graphs might look.

② Finding the Solution to a System of Three Linear Equations in Three Variables If None of the Coefficients Is Zero

One way to solve a system of three equations with three variables is to obtain from it a system of two equations in two variables; in other words, we eliminate one variable from both equations. We can then use the methods of Section 4.1 to solve the resulting system. You can find the third variable (the one that was eliminated) by substituting the two variables that you have found into one of the original equations.

EXAMPLE 2 Find the solution to (that is, solve) the following system of equations.

$$\begin{aligned} -2x + 5y + z &= 8 \quad \textbf{(1)} \\ -x + 2y + 3z &= 13 \quad \textbf{(2)} \\ x + 3y - z &= 5 \quad \textbf{(3)} \end{aligned}$$

Solution Let's eliminate z because it can be done easily by adding equations **(1)** and **(3).**

$$\begin{aligned} -2x + 5y + z &= 8 \quad \textbf{(1)} \\ x + 3y - z &= 5 \quad \textbf{(3)} \\ \hline -x + 8y \quad &= 13 \quad \textbf{(4)} \end{aligned}$$

Now we need to choose a *different pair* from the original system of equations and once again eliminate the same variable. In other words, we have to use equations **(1)** and **(2)** or equations **(2)** and **(3)** and eliminate z. Let's multiply each term of equation **(3)** by 3 (and call it equation **(6)**) and add the result to equation **(2).**

$$\begin{aligned} -x + 2y + 3z &= 13 \quad \textbf{(2)} \\ 3x + 9y - 3z &= 15 \quad \textbf{(6)} \\ \hline 2x + 11y \quad &= 28 \quad \textbf{(5)} \end{aligned}$$

We now can solve the resulting system of two linear equations.

$$\begin{aligned} -x + 8y &= 13 \quad \textbf{(4)} \\ 2x + 11y &= 28 \quad \textbf{(5)} \end{aligned}$$

Multiply each term of equation **(4)** by 2.

$$\begin{aligned} -2x + 16y &= 26 \\ 2x + 11y &= 28 \\ \hline 27y &= 54 \quad \text{Add the equations.} \\ y &= 2 \quad \text{Solve for } y. \end{aligned}$$

Substituting $y = 2$ into equation **(4),** we have the following:

$$\begin{aligned} -x + 8(2) &= 13 \\ -x &= -3 \\ x &= 3. \end{aligned}$$

Now substitute $x = 3$ and $y = 2$ into one of the original equations (any one will do) to solve for z. Let's use equation **(1).**

$$\begin{aligned} -2x + 5y + z &= 8 \\ -2(3) + 5(2) + z &= 8 \\ -6 + 10 + z &= 8 \\ z &= 4 \end{aligned}$$

The solution to the system is (3, 2, 4).

Check. Verify that (3, 2, 4) satisfies *each* of the three *original* equations.

Student Practice 2 Solve this system.

$$\begin{aligned} x + 2y + 3z &= 4 \\ 2x + y - 2z &= 3 \\ 3x + 3y + 4z &= 10 \end{aligned}$$

Here's a summary of the procedure that we just used.

HOW TO SOLVE A SYSTEM OF THREE LINEAR EQUATIONS IN THREE VARIABLES

1. Use the addition method to eliminate any variable from any pair of equations. (The choice of variable is arbitrary.)
2. Use appropriate steps to eliminate the *same variable* from a *different pair* of equations. (If you don't eliminate the same variable, you will still have three unknowns.)
3. Solve the resulting system of two equations in two variables.
4. Substitute the values obtained in step 3 into one of the three original equations. Solve for the remaining variable.
5. Check the solution in all of the original equations.

It is helpful to write all equations in the form $Ax + By + Cz = D$ before using this five-step method.

③ Finding the Solution to a System of Three Linear Equations in Three Variables If Some of the Coefficients Are Zero

If a system of three linear equations in three variables contains one or more equations of the form $Ax + By + Cz = D$, where one of the values of A, B, or C is zero, then we will slightly modify our approach to solving the system. We will select one equation that contains only two variables. Then we will take the remaining system of two equations and eliminate the variable that is missing in the equation that we selected.

EXAMPLE 3 Solve the system.

$$\begin{aligned} 4x + 3y + 3z &= 4 \quad \textbf{(1)} \\ 3x \phantom{{}+3y} + 2z &= 2 \quad \textbf{(2)} \\ 2x - 5y \phantom{{}+3z} &= -4 \quad \textbf{(3)} \end{aligned}$$

Solution Note that equation **(2)** has no y-term and equation **(3)** has no z-term. Obviously, that makes our work easier. Let's work with equations **(2)** and **(1)** to obtain an equation that contains only x and y.

Step 1 Multiply equation **(1)** by 2 and equation **(2)** by -3 to obtain the following system.

$$\begin{aligned} 8x + 6y + 6z &= 8 \quad \textbf{(4)} \\ -9x \phantom{{}+6y} - 6z &= -6 \quad \textbf{(5)} \\ \hline -x + 6y \phantom{{}+6z} &= 2 \quad \textbf{(6)} \end{aligned}$$

Step 2 This step is already done, since equation **(3)** has no z-term.

Step 3 Now we can solve the system formed by equations **(3)** and **(6)**.

$$\begin{aligned} 2x - 5y &= -4 \quad \textbf{(3)} \\ -x + 6y &= 2 \quad \textbf{(6)} \end{aligned}$$

Continued on next page

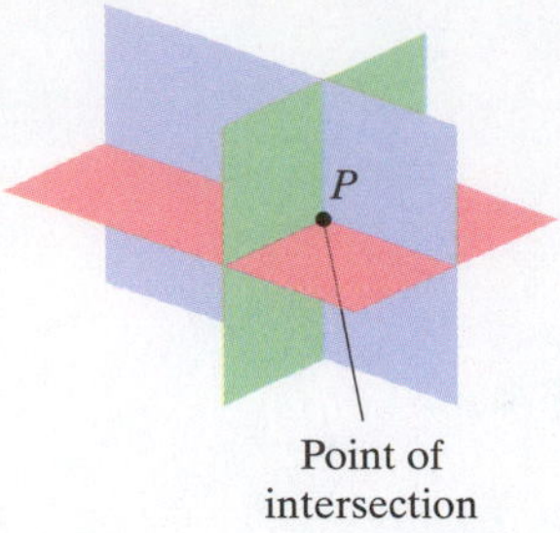

(a)

If we multiply each term of equation **(6)** by 2, we obtain the system

$$\begin{aligned} 2x - 5y &= -4 \\ -2x + 12y &= 4 \\ \hline 7y &= 0 \quad \text{Add.} \\ y &= 0. \quad \text{Solve for } y. \end{aligned}$$

Substituting $y = 0$ in equation **(6),** we find the following:

$$\begin{aligned} -x + 6(0) &= 2 \\ -x &= 2 \\ x &= -2. \end{aligned}$$

Step 4 To find z, we substitute $x = -2$ and $y = 0$ into one of the original equations containing z. Since equation **(2)** has only two variables, let's use it.

$$\begin{aligned} 3x + 2z &= 2 \\ 3(-2) + 2z &= 2 \\ 2z &= 8 \\ z &= 4 \end{aligned}$$

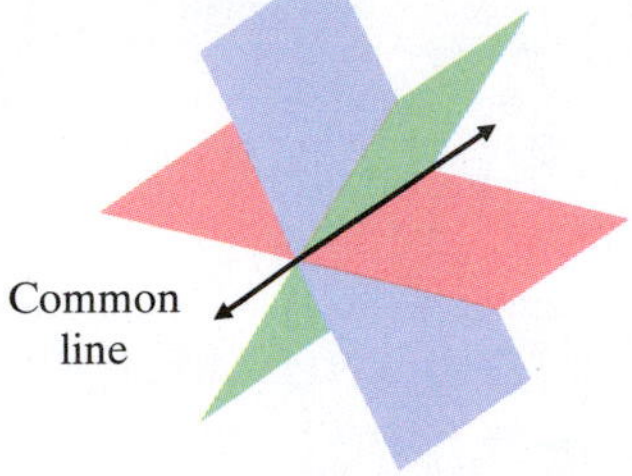

(b)

The solution to the system is $(-2, 0, 4)$.

Check. Verify this solution by substituting these values into equations **(1), (2),** and **(3).**

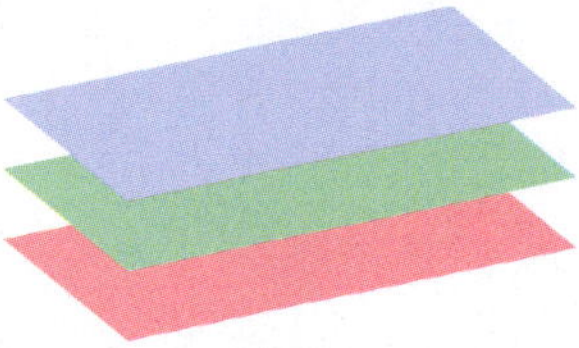

(c)

Student Practice 3 Solve the system.

$$\begin{aligned} 2x + y + z &= 11 \\ 4y + 3z &= -8 \\ x - 5y &= 2 \end{aligned}$$

A linear equation in three variables is a plane in three-dimensional space. A system of three linear equations in three variables is three planes. The solution to the system is the set of points at which all three planes intersect. There are three possible results. The three planes may intersect at one point. (See figure **(a)** in the margin.) This point is described by an ordered triple of the form (x, y, z) and lies in each plane.

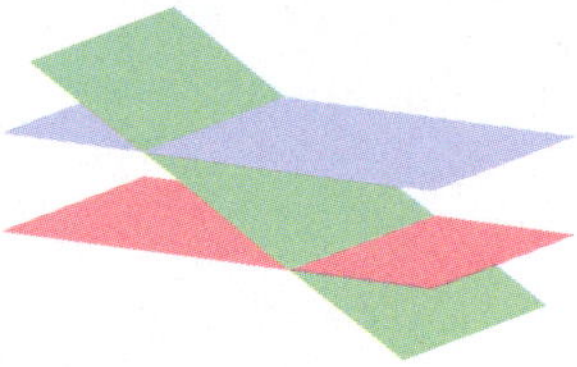

(d)

The three planes may intersect at a line. (See figure **(b)** in the margin.) In this case the system has an infinite number of solutions; that is, all the points on the line are solutions to the system.

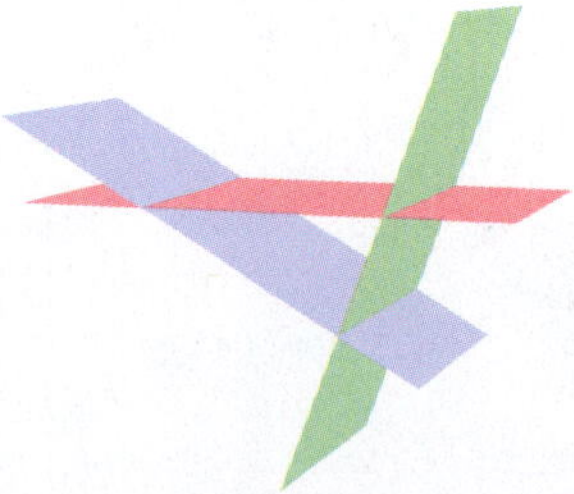

(e)

Finally, all three planes may not intersect at any points. It may mean that all three planes never share any point of intersection. (See figure **(c)** in the margin.) It may mean that two planes intersect. (See figures **(d)** and **(e)** in the margin.) In all such cases there is no solution to the system of equations.

4.2 Exercises

MyMathLab®

Watch the videos in MyMathLab

Download the MyDashBoard App

1. Determine whether $(2, 1, -4)$ is a solution to the system.

$$\begin{aligned} 2x - 3y + 2z &= -7 \\ x + 4y - z &= 10 \\ 3x + 2y + z &= 4 \end{aligned}$$

2. Determine whether $(-3, 0, 1)$ is a solution to the system.

$$\begin{aligned} 2x + 5y - z &= -7 \\ x - 11y + 4z &= 1 \\ -5x + 8y - 12z &= 3 \end{aligned}$$

3. Determine whether $(-1, 5, 1)$ is a solution to the system.

$$\begin{aligned} 3x + 2y - z &= 6 \\ x - y - 2z &= -8 \\ 4x + y + 2z &= 5 \end{aligned}$$

4. Determine whether $(3, 2, 1)$ is a solution to the system.

$$\begin{aligned} x + y + 2z &= 7 \\ 2x + y + z &= 9 \\ 3x + 4y - 2z &= 13 \end{aligned}$$

Solve each system.

5.
$$\begin{aligned} x + y + 2z &= 0 \\ 2x - y - z &= 1 \\ x + 2y + 3z &= 1 \end{aligned}$$

6.
$$\begin{aligned} 2x + 2y + z &= -6 \\ x - y + 3z &= 1 \\ x + 4y + z &= 8 \end{aligned}$$

7.
$$\begin{aligned} -x + 2y - z &= -1 \\ 2x + y + z &= 2 \\ x - y - 2z &= -13 \end{aligned}$$

8.
$$\begin{aligned} -4x + 2y + z &= 1 \\ x - y + 3z &= -5 \\ 3x + y - 4z &= 10 \end{aligned}$$

9.
$$\begin{aligned} 8x - 5y + z &= 15 \\ 3x + y - z &= -7 \\ x + 4y + z &= -3 \end{aligned}$$

10.
$$\begin{aligned} -3x + y - z &= 3 \\ 4x - 3y + 2z &= -2 \\ 3x + 2y + 4z &= 6 \end{aligned}$$

11.
$$\begin{aligned} x + 4y - z &= -5 \\ -2x - 3y + 2z &= 5 \\ x - \frac{2}{3}y + z &= \frac{11}{3} \end{aligned}$$

12.
$$\begin{aligned} x - 4y + 4z &= -1 \\ -x + \frac{y}{2} - \frac{5}{2}z &= -3 \\ -x + 3y - z &= 5 \end{aligned}$$

13.
$$\begin{aligned} 2x \quad + 2z &= -7 + 3y \\ \frac{3}{2}x + y + \frac{1}{2}z &= 2 \\ x + 4y \quad &= 10 + z \end{aligned}$$

14.
$$\begin{aligned} 8x - y &= 2z \\ y &= 3x + 4z + 2 \\ 5x + 2z &= y - 1 \end{aligned}$$

15.
$$\begin{aligned} a &= 8 + 3b - 2c \\ 4a + 2b - 3c &= 10 \\ c &= 10 + b - 2a \end{aligned}$$

16.
$$\begin{aligned} a &= c - b \\ 3a - 2b + 6c &= 1 \\ c &= 4 - 3b - 7a \end{aligned}$$

17. $$\begin{aligned} 0.2a + 0.1b + 0.2c &= 0.1 \\ 0.3a + 0.2b + 0.4c &= -0.1 \\ 0.6a + 1.1b + 0.2c &= 0.3 \end{aligned}$$

18. $$\begin{aligned} -0.1a + 0.2b + 0.3c &= 0.1 \\ 0.2a - 0.6b + 0.3c &= 0.5 \\ 0.3a - 1.2b - 0.4c &= -0.4 \end{aligned}$$

Find the solution for each system of equations. Round your answers to five decimal places.

19. $$\begin{aligned} x - 4y + 4z &= -3.72186 \\ -x + 3y - z &= 5.98115 \\ 2x - y + 5z &= 7.93645 \end{aligned}$$

20. $$\begin{aligned} 4x + 2y + 3z &= 9 \\ 9x + 3y + 2z &= 3 \\ 2.987x + 5.027y + 3.867z &= 18.642 \end{aligned}$$

Solve each system.

21. $$\begin{aligned} x - y &= 5 \\ 2y - z &= 1 \\ 3x + 3y + z &= 6 \end{aligned}$$

22. $$\begin{aligned} -x + 4z &= -3 \\ -y + 2z &= 5 \\ 3x + y + 2z &= 4 \end{aligned}$$

23. $$\begin{aligned} -y + 2z &= 1 \\ x + y + z &= 2 \\ -x + 3z &= 2 \end{aligned}$$

24. $$\begin{aligned} -2x + y - 3z &= 0 \\ -2y - z &= -1 \\ x + 2y - z &= 5 \end{aligned}$$

25. $$\begin{aligned} x - 2y + z &= 0 \\ -3x - y &= -6 \\ y - 2z &= -7 \end{aligned}$$

26. $$\begin{aligned} 2x - 4z &= -1 \\ 5x + y - 2z &= 1 \\ 2y + 8z &= 4 \end{aligned}$$

27. $$\begin{aligned} \frac{3a}{2} - b + 2c &= 2 \\ \frac{a}{2} + 2b - 2c &= 4 \\ -a + b &= -6 \end{aligned}$$

28. $$\begin{aligned} \frac{2a}{3} - \frac{b}{3} &= -1 \\ 2a + b + \frac{c}{3} &= -3 \\ -3b - c &= 3 \end{aligned}$$

Try to solve the system of equations. Explain your result in each case.

29.
$$\begin{aligned} 2x + y &= -3 \\ 2y + 16z &= -18 \\ -7x - 3y + 4z &= 6 \end{aligned}$$

30.
$$\begin{aligned} 6x - 2y + 2z &= 2 \\ 4x + 8y - 2z &= 5 \\ -2x - 4y + z &= -2 \end{aligned}$$

31.
$$\begin{aligned} 3x + 3y - 3z &= -1 \\ 4x + y - 2z &= 1 \\ -2x + 4y - 2z &= -8 \end{aligned}$$

32.
$$\begin{aligned} -3x + 4y - z &= -4 \\ x + 2y + z &= 4 \\ -12x + 16y - 4z &= -16 \end{aligned}$$

Cumulative Review

33. [2.3.1] Solve for x. $|2x - 1| = 7$

34. [1.4.5] Write using scientific notation. 76,300,000

35. [3.3.2] Explain the steps to find the standard form of the equation of the line that passes through $(1, 4)$ and $(-2, 3)$.

36. [3.3.3] Explain the steps to find the standard form of the equation of the line that is perpendicular to $y = -\frac{2}{3}x + 4$ and passes through $(-4, 2)$.

Quick Quiz 4.2 *Solve each system.*

1.
$$\begin{aligned} 4x - y + 2z &= 0 \\ 2x + y + z &= 3 \\ 3x - y + z &= -2 \end{aligned}$$

2.
$$\begin{aligned} x + 2y + 2z &= -1 \\ 2x - y + z &= 1 \\ x + 3y - 6z &= 7 \end{aligned}$$

3.
$$\begin{aligned} 4x - 2y + 6z &= 0 \\ 6y + 3z &= 3 \\ x + 2y - z &= 5 \end{aligned}$$

4. Concept Check Explain how you would eliminate the variable z and obtain two equations with only the variables x and y in the following system.
$$\begin{aligned} 2x + 4y - 2z &= -22 \\ 4x + 3y + 5z &= -10 \\ 5x - 2y + 3z &= 13 \end{aligned}$$

How Am I Doing? Sections 4.1–4.2

1. ______

2. ______

3. ______

4. ______

5. ______

6. ______

7. ______

8. ______

9. ______

10. ______

11. ______

12. ______

How are you doing with your homework assignments in Sections 4.1 and 4.2? Do you feel you have mastered the material so far? Do you understand the concepts you have covered? Before you go further in the textbook, take some time to do each of the following problems.

4.1

1. Solve by the substitution method.

$$\begin{aligned} 4x - y &= -1 \\ 3x + 2y &= 13 \end{aligned}$$

2. Solve by the addition method.

$$\begin{aligned} 5x + 2y &= 0 \\ -3x - 4y &= 14 \end{aligned}$$

Find the solution to each system of equations by any method. If there is no single solution to a system, state the reason.

3. $$\begin{aligned} 5x - 2y &= 27 \\ 3x - 5y &= -18 \end{aligned}$$

4. $$\begin{aligned} 7x + 3y &= 15 \\ \frac{1}{3}x - \frac{1}{2}y &= 2 \end{aligned}$$

5. $$\begin{aligned} 2x &= 3 + y \\ 3y &= 6x - 9 \end{aligned}$$

6. $$\begin{aligned} 0.2x + 0.7y &= -1 \\ 0.5x + 0.6y &= -0.2 \end{aligned}$$

7. $$\begin{aligned} 6x - 9y &= 15 \\ -4x + 6y &= 8 \end{aligned}$$

4.2

8. Determine whether $(3, -4, 0)$ is a solution to the system.

$$\begin{aligned} 3x + y - 2z &= 5 \\ 4x + 3y - z &= 12 \\ -2x - y + 3z &= -2 \end{aligned}$$

Find the solution to each system of equations.

9. $$\begin{aligned} 5x - 2y + z &= -1 \\ 3x + y - 2z &= 6 \\ -2x + 3y - 5z &= 7 \end{aligned}$$

10. $$\begin{aligned} 2x - y + 3z &= -1 \\ 5x + y + 6z &= 0 \\ 2x - 2y + 3z &= -2 \end{aligned}$$

11. $$\begin{aligned} x + y + 2z &= 9 \\ 3x + 2y + 4z &= 16 \\ 2y + z &= 10 \end{aligned}$$

12. $$\begin{aligned} x - 2z &= -5 \\ y - 3z &= -3 \\ 2x - z &= -4 \end{aligned}$$

Now turn to page SA-11 for the answers to each of these problems. Each answer also includes a reference to the objective in which the problem is first taught. If you missed any of these problems, you should stop and review the Examples and Student Practice problems in the referenced objective. A little review now will help you master the material in the upcoming sections of the text.

4.3 Applications of Systems of Linear Equations

Student Learning Objectives

After studying this section, you will be able to:

1. **Solve applications requiring the use of a system of two linear equations in two unknowns.**
2. **Solve applications requiring the use of a system of three linear equations in three unknowns.**

① Solving Applications Requiring the Use of a System of Two Linear Equations in Two Unknowns

We will now examine how a system of linear equations can assist us in solving applied problems.

EXAMPLE 1 For the paleontology lecture on campus, advance tickets cost \$5 and tickets at the door cost \$6. The ticket sales this year came to \$4540. The department chairman wants to raise prices next year to \$7 for advance tickets and \$9 for tickets at the door. He said that if exactly the same number of people attend next year, the ticket sales at these new prices will total \$6560. If he is correct, how many tickets were sold in advance this year? How many tickets were sold at the door?

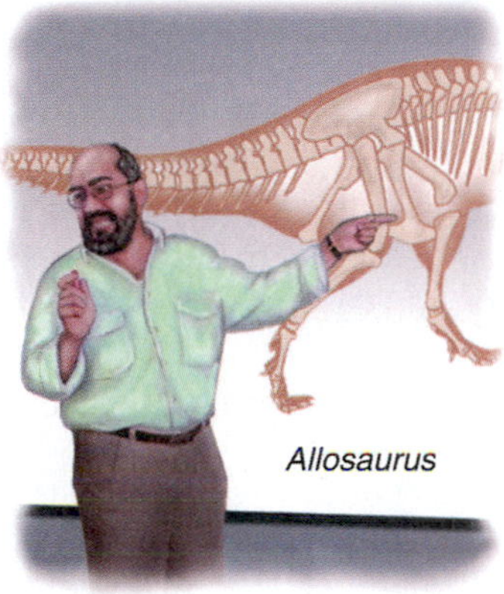

Solution

1. ***Understand the problem.*** Since we are looking for the number of tickets sold, we let

x = number of tickets bought in advance and

y = number of tickets bought at the door.

2. ***Write a system of two equations in two unknowns.*** If advance tickets cost \$5, then the total sales will be $5x$; similarly, total sales of door tickets will be $6y$. Since the total sales of both types of tickets was \$4540, we have

$$5x + 6y = 4540.$$

By the same reasoning, we have

$$7x + 9y = 6560.$$

Thus, our system is as follows:

$$5x + 6y = 4540 \quad \textbf{(1)}$$
$$7x + 9y = 6560 \quad \textbf{(2)}$$

3. ***Solve the system of equations and state the answer.*** We multiply each term of equation **(1)** by -3 and each term of equation **(2)** by 2 to obtain the following equivalent system.

$$\begin{array}{rcr} -15x - 18y &=& -13{,}620 \\ 14x + 18y &=& 13{,}120 \\ \hline -x \qquad &=& -500 \end{array}$$

Therefore, $x = 500$. Substituting $x = 500$ into equation **(1),** we have the following:

$$5(500) + 6y = 4540$$
$$6y = 2040$$
$$y = 340$$

Thus, 500 advance tickets and 340 door tickets were sold.

4. ***Check.*** We need to check our answers. Do they seem reasonable?

Would 500 advance tickets at \$5 and 340 door tickets at \$6 yield \$4540?

$$5(500) + 6(340) \stackrel{?}{=} 4540$$
$$2500 + 2040 \stackrel{?}{=} 4540$$
$$4540 = 4540 \checkmark$$

Would 500 advance tickets at \$7 and 340 door tickets at \$9 yield \$6560?

$$7(500) + 9(340) \stackrel{?}{=} 6560$$
$$3500 + 3060 \stackrel{?}{=} 6560$$
$$6560 = 6560 \checkmark$$

Continued on next page

***NOTE TO STUDENT:** Fully worked-out solutions to all of the Student Practice problems can be found at the back of the text starting at page SP-1.*

Student Practice 1 Coach Perez purchased baseballs at \$6 each and bats at \$21 each last week for the college baseball team. The total cost of the purchase was \$318. This week he noticed that the same items are on sale. Baseballs are now \$5 each and bats are \$17. He found that if he made the same purchase this week, it would cost only \$259. How many baseballs and how many bats did he buy last week?

EXAMPLE 2 An electronics firm makes two types of switching devices. Type A takes 4 minutes to make and requires \$3 worth of materials. Type B takes 5 minutes to make and requires \$5 worth of materials. When the production manager reviewed the latest batch, he found that it took 35 hours to make these switches with a materials cost of \$1900. How many switches of each type were produced for this latest batch?

Solution

1. ***Understand the problem.*** We are given a lot of information, but the major concern is to find out how many of the type A devices and the type B devices were produced. This becomes our starting point to define the variables we will use. Let

$$A = \text{the number of type } A \text{ devices produced and}$$
$$B = \text{the number of type } B \text{ devices produced.}$$

2. ***Write a system of two equations.*** How should we construct the equations? What relationships exist between our variables (or unknowns)? According to the problem, the devices are related by time and by cost. So we set up one equation in terms of time (minutes in this case) and one in terms of cost (dollars). Each type A took 4 minutes to make, each type B took 5 minutes to make, and the total time was $60(35) = 2100$ minutes. Each type A used \$3 worth of materials, each type B used \$5 worth of materials, and the total materials cost was \$1900. We can gather this information in a table. Making a table will help us form the equations.

	Type A Devices	Type B Devices	Total
Number of Minutes	$4A$	$5B$	2100
Cost of Materials	$3A$	$5B$	1900

$$4A + 5B = 2100$$
$$3A + 5B = 1900$$

Therefore, we have the following system.

$$4A + 5B = 2100 \quad \textbf{(1)}$$
$$3A + 5B = 1900 \quad \textbf{(2)}$$

3. ***Solve the system of equations and state the answer.*** Multiplying equation **(2)** by -1 and adding the equations, we find the following:

$$4A + 5B = 2100$$
$$\underline{-3A - 5B = -1900}$$
$$A = 200$$

Substituting $A = 200$ into equation **(1)**, we have the following:

$$800 + 5B = 2100$$
$$5B = 1300$$
$$B = 260$$

Thus, 200 type A devices and 260 type B devices were produced.

4. ***Check.*** If each type A requires 4 minutes and each type B requires 5 minutes, does this amount to a total time of 2100 minutes?

$$4A + 5B = 2100$$
$$4(200) + 5(260) \stackrel{?}{=} 2100$$
$$800 + 1300 \stackrel{?}{=} 2100$$
$$2100 = 2100 \checkmark$$

If each type A costs \$3 and each type B costs \$5, does this amount to a total cost of \$1900?

$$3A + 5B = 1900$$
$$3(200) + 5(260) \stackrel{?}{=} 1900$$
$$600 + 1300 \stackrel{?}{=} 1900$$
$$1900 = 1900 \checkmark$$

Student Practice 2 A furniture company makes both small and large chairs. It takes 30 minutes of machine time and 1 hour and 15 minutes of labor to build the small chair. The large chair requires 40 minutes of machine time and 1 hour and 20 minutes of labor. The company has 57 hours of labor time and 26 hours of machine time available each day. If all available time is used, how many chairs of each type can the company make?

When we encounter motion problems involving rate, time, or distance, it is useful to recall the formula $D = RT$ or distance $=$ (rate)(time).

EXAMPLE 3 An airplane travels between two cities that are 1500 miles apart. The trip against the wind takes 3 hours. The return trip with the wind takes $2\frac{1}{2}$ hours. What is the speed of the plane in still air (in other words, how fast would the plane travel if there were no wind)? What is the speed of the wind?

Solution

1. ***Understand the problem.*** Our unknowns are the speed of the plane in still air and the speed of the wind.

Let

$x =$ the speed of the plane in still air and

$y =$ the speed of the wind.

Let's make a sketch to help us see how these speeds are related to one another. When we travel against the wind, the wind is slowing us down. Since the wind speed opposes the plane's speed in still air, we must subtract: $x - y$.

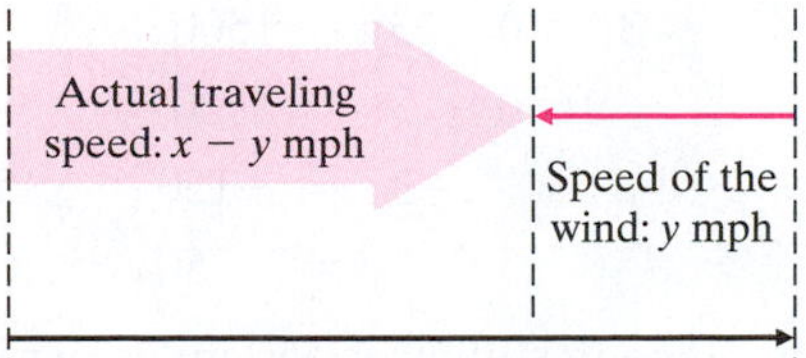

Continued on next page

Graphing Calculator

Exploration

A visual interpretation of two equations in two unknowns is sometimes helpful. Study Example 3. Graph the two equations.

$$3x - 3y = 1500$$
$$2.5x + 2.5y = 1500$$

What is the significance of the point of intersection? If you were an air traffic controller, how would you interpret the linear equation $3x - 3y = 1500$? Why would this be useful? How would you interpret $2.5x + 2.5y = 1500$? Why would this be useful?

When we travel with the wind, the wind is helping us travel forward. Thus, the wind speed is added to the plane's speed in still air, and we add: $x + y$.

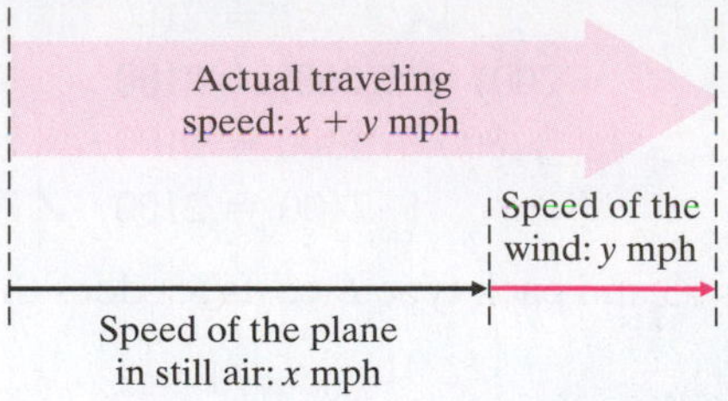

2. ***Write a system of two equations.*** To help us write our equations, we organize the information in a chart. The chart will be based on the formula $RT = D$, which is (rate)(time) = distance.

	R	· T	= D
Flying against the wind	$x - y$	3	1500
Flying with the wind	$x + y$	2.5	1500

Using the rows of the chart, we obtain a system of equations.

$$(x - y)(3) = 1500$$
$$(x + y)(2.5) = 1500$$

If we remove the parentheses, we will obtain the following system.

$$3x - 3y = 1500 \quad \textbf{(1)}$$
$$2.5x + 2.5y = 1500 \quad \textbf{(2)}$$

3. ***Solve the system of equations and state the answer.*** It will be helpful to clear equation **(2)** of decimal coefficients. Although we could multiply each term by 10, doing so will result in large coefficients on x and y. For this equation, multiplying by 2 is a better choice.

$$3x - 3y = 1500 \quad \textbf{(1)}$$
$$5x + 5y = 3000 \quad \textbf{(3)}$$

If we multiply equation **(1)** by 5 and equation **(3)** by 3, we will obtain the following system.

$$\begin{aligned} 15x - 15y &= 7500 \\ \underline{15x + 15y} &= \underline{9000} \\ 30x &= 16{,}500 \\ x &= 550 \end{aligned}$$

Substituting this result in equation **(1),** we obtain the following:

$$\begin{aligned} 3(550) - 3y &= 1500 \\ 1650 - 3y &= 1500 \\ -3y &= -150 \\ y &= 50 \end{aligned}$$

Thus, the speed of the plane in still air is 550 miles per hour, and the speed of the wind is 50 miles per hour.

Student Practice 3 An airplane travels west from city A to city B against the wind. It takes 3 hours to travel 1950 kilometers. On the return trip the plane travels east from city B to city C, a distance of 1600 kilometers in a time of 2 hours. On the return trip the plane travels with the wind. What is the speed of the plane in still air? What is the speed of the wind?

② Solving Applications Requiring the Use of a System of Three Linear Equations in Three Unknowns

EXAMPLE 4 A trucking firm has three sizes of trucks. The biggest truck holds 10 tons of gravel, the next size holds 6 tons, and the smallest holds 4 tons. The firm has a contract to provide fifteen trucks to haul 104 tons of gravel. To reduce fuel costs, the firm's manager wants to use two more of the fuel-efficient 10-ton trucks than the 6-ton trucks. How many trucks of each type should she use?

Solution

1. ***Understand the problem.*** Since we need to find three things (the numbers of 10-ton trucks, 6-ton trucks, and 4-ton trucks), it would be helpful to have three variables. Let

$x =$ the number of 10-ton trucks used,

$y =$ the number of 6-ton trucks used, and

$z =$ the number of 4-ton trucks used.

2. ***Write a system of three equations.*** We know that fifteen trucks will be used; hence, we have the following:

$$x + y + z = 15 \quad \textbf{(1)}$$

How can we get our second equation? Well, we also know the *capacity* of each truck type, and we know the total tonnage to be hauled. The first type of truck hauls 10 tons, the second type 6 tons, and the third type 4 tons, and the total tonnage is 104 tons. Hence, we can write the following:

$$10x + 6y + 4z = 104 \quad \textbf{(2)}$$

We still need one more equation. What other given information can we use? The problem states that the manager wants to use two more 10-ton trucks than 6-ton trucks. Thus, we have the following:

$$x = 2 + y \quad \textbf{(3)}$$

(We could also have written $x - y = 2$.) Hence, our system of equations is as follows:

$$\begin{aligned} x + y + z &= 15 \quad \textbf{(1)} \\ 10x + 6y + 4z &= 104 \quad \textbf{(2)} \\ x - y &= 2 \quad \textbf{(3)} \end{aligned}$$

3. ***Solve the system of equations and state the answer.*** Equation **(3)** doesn't contain the variable z. Let's work with equations **(1)** and **(2)** to eliminate z. First, we multiply equation **(1)** by -4 and add it to equation **(2)**.

$$\begin{aligned} -4x - 4y - 4z &= -60 \quad \textbf{(4)} \\ 10x + 6y + 4z &= 104 \quad \textbf{(2)} \\ \hline 6x + 2y &= 44 \quad \textbf{(5)} \end{aligned}$$

Continued on next page

Student Practice 4 A factory uses three machines to wrap boxes for shipment. Machines A, B, and C working together can wrap 260 boxes in 1 hour. If machine A runs 3 hours and machine B runs 2 hours, they can wrap 390 boxes. If machine B runs 3 hours and machine C runs 4 hours, 655 boxes can be wrapped. How many boxes per hour can each machine wrap?

Make sure you understand how we got equation **(5).** Dividing each term of equation **(5)** by 2 and adding to equation **(3)** gives the following:

$$\begin{array}{rl} 3x + y = 22 & \textbf{(6)} \\ \underline{x - y = 2} & \textbf{(3)} \\ 4x \qquad = 24 & \\ x = 6 & \end{array}$$

For $x = 6$, equation **(3)** yields the following:

$$6 - y = 2$$
$$4 = y$$

Now we substitute the known x- and y-values into equation **(1).**

$$6 + 4 + z = 15$$
$$z = 5$$

Thus, the manager needs six 10-ton trucks, four 6-ton trucks, and five 4-ton trucks.

4. *Check.* The check is left to the student.

STEPS TO SUCCESS Why Do We Have to Learn to Solve Word Problems?

Many students ask that question. It is a fair question and it deserves an honest answer.

Applications or word problems are the very life of mathematics! They are the reason for doing mathematics. They teach you how to put into use the mathematical skills you have developed.

Almost all branches of mathematics are studied because they solve problems in real life. You will find in this textbook that studying the Use Math to Save Money problems will help you manage your finances. Many decisions in daily life can be made more easily if mathematics is used to evaluate the possible options.

How can I learn to solve word problems more easily? Many students ask that question also. The key to success is practice. Make yourself do as many problems as you can. You may not be able to do them all correctly at first, but keep trying. If you cannot solve a problem, try another one. Ask for help from your teacher or someone in the tutoring lab. Ask other classmates how they solved the problem. Soon you will see great progress in your own problem-solving ability.

Making it personal: Which of these statements best answers the question "Why do we have to learn to solve word problems?" What suggestion do you think is the most helpful to help you to learn to solve word problems more easily? Write the suggestion in your own words. Start to apply it today to help you with the homework in Exercises 4.3. Remember, much of life is like a word problem. ▼

4.3 Exercises

MyMathLab®

Watch the videos in MyMathLab

Download the MyDashBoard App

Applications

Use a system of two linear equations to solve each exercise.

1. The sum of 2 numbers is 87. If twice the smaller number is subtracted from the larger number, the result is 12. Find the two numbers.

2. The difference between two numbers is 10. If three times the larger is added to twice the smaller, the result is 70. Find the two numbers.

3. ***Temporary Employment Agency*** An employment agency specializing in temporary construction help pays heavy equipment operators \$140 per day and general laborers \$90 per day. If thirty-five people were hired and the payroll was \$3950, how many heavy equipment operators were employed? How many laborers?

4. ***Broadway Ticket Prices*** A Broadway performance of *Les Misérables* had a paid attendance of 308 people. Balcony tickets cost \$38 and orchestra tickets cost \$60. Ticket sales receipts totaled \$15,576. How many tickets of each type were sold?

5. ***Amtrak Train Tickets*** Ninety-eight passengers rode in an Amtrak train from Boston to Denver. Tickets for regular coach seats cost \$120. Tickets for sleeper car seats cost \$290. The receipts for the trip totaled \$19,750. How many passengers purchased each type of ticket?

6. ***Farm Operations*** The Tupper Farm has 450 acres of land allotted for raising corn and wheat. The cost to cultivate corn is \$42 per acre. The cost to cultivate wheat is \$35 per acre. The Tuppers have \$16,520 available to cultivate these crops. How many acres of each crop should the Tuppers plant?

7. ***Computer Training*** A large company wants to train its human resources department to use new spreadsheet software and a new payroll program. Experienced employees can learn the spreadsheet in 3 hours and the payroll in 4 hours. New employees need 4 hours to learn the spreadsheet and 7 hours to learn the payroll. The company can afford to pay for 115 hours of spreadsheet instruction and 170 hours of payroll instruction. How many of each type of employee can the company train?

8. ***Radar Detector Manufacturing*** Ventex makes auto radar detectors. Ventex has found that its basic model requires 3 hours of manufacturing for the inside components and 2 hours for the housing and controls. Its advanced model requires 5 hours to manufacture the inside components and 3 hours for the housing and controls. This week, the production division has available 1050 hours for producing inside components and 660 hours for housing and controls. How many detectors of each type can be made?

9. ***Farm Management*** A farmer has several packages of fertilizer for his new grain crop. The old packages contain 50 pounds of long-term-growth supplement and 60 pounds of weed killer. The new packages contain 65 pounds of long-term-growth supplement and 45 pounds of weed killer. Using past experience, the farmer estimates that he needs 3125 pounds of long-term-growth supplement and 2925 pounds of weed killer for the fields. How many old packages of fertilizer and how many new packages of fertilizer should he use?

10. ***Hospital Dietician*** A staff hospital dietician has two prepackaged mixtures of vitamin additives available for patients. Mixture 1 contains 5 grams of vitamin C and 3 grams of niacin; mixture 2 contains 6 grams of vitamin C and 5 grams of niacin. On an average day she needs 87 grams of niacin and 117 grams of vitamin C. How many packets of each mixture will she need?

11. ***Coffee and Snack Expenses*** On Monday, Luis picked up five scones and six large coffees for the office staff. He paid \$18.39. On Tuesday, Rachel picked up four scones and seven large coffees for the office staff. She paid \$17.99. What is the cost of one scone? What is the cost of one large coffee?

12. ***Advertising Costs*** A local department store is preparing four-color sales brochures to insert into the *Salem Evening News*. The printer has a fixed charge to set up the printing of the brochure and a specific per-copy amount for each brochure printed. He quoted a price of \$1350 for printing five thousand brochures and a price of \$1750 for printing seven thousand brochures. What is the fixed charge to set up the printing? What is the per-copy cost for printing a brochure?

13. ***Airspeed*** Against the wind a small plane flew 210 miles in 1 hour and 10 minutes. The return trip took only 50 minutes. What was the speed of the wind? What was the speed of the plane in still air?

14. ***Aircraft Operation*** Against the wind, a commercial airline in South America flew 630 miles in 3 hours and 30 minutes. With a tailwind, the return trip took 3 hours. What was the speed of the wind? What was the speed of the plane in still air?

15. ***Fishing and Boating*** Don Williams uses his small motorboat to go 8 miles upstream to his favorite fishing spot. Against the current, the trip takes $\frac{2}{3}$ hour. With the current, the trip takes $\frac{1}{2}$ hour. How fast can the boat travel in still water? What is the speed of the current?

16. ***Canoe Trip*** It look Lance and Ivan 6 hours to travel 33 miles downstream by canoe on the Mississippi River. The next day they traveled for 8 hours upstream for 20 miles. What was the rate of the current? What was their average speed in still water?

17. ***Basketball*** In 1962, Wilt Chamberlain set an NBA single-game scoring record. He scored 64 times for a total of 100 points. He made no 3-point shots (these were not part of the professional basketball game until 1979), but made several free throws worth 1 point each and several regular shots worth 2 points each. How many free throws did he make? How many 2-point shots did he make?

18. ***Basketball*** Dirk Nowitzki of the Dallas Mavericks scored 44 points in a recent basketball game. He scored no 3-point shots, but made a number of 2-point field goals and 1-point free throws. He scored 30 times. How many 2-point shots did he make? How many free throws did he make?

19. ***Telephone Charges*** Nick's telephone company charges \$0.05 per minute for weekend calls and \$0.08 for calls made on weekdays. This month Nick was billed for 625 minutes. The charge for these minutes was \$43.40. How many minutes did he talk on weekdays and how many minutes did he talk on the weekends?

20. ***Office Supply Costs*** A new catalog from an office supply company shows that some of its prices will increase next month. This month, copier paper costs \$2.70 per ream and printer cartridges cost \$15.50. If Chris submits his order this month, the cost will be \$462. Next month, when paper will cost \$3.00 per ream and the cartridges will cost \$16, his order would cost \$495. How many reams of paper and how many printer cartridges are in his order?

21. ***Highway Department Purchasing*** This year the state highway department in Montana purchased 256 identical cars and 183 identical trucks for official use. The purchase price was \$5,791,948. Due to a budget shortfall, next year the department plans to purchase only 64 cars and 107 trucks. It will be charged the same price for each car and for each truck. Next year it plans to spend \$2,507,612. How much does the department pay for each car and for each truck?

22. ***Concert Ticket Prices*** A recent concert at Gordon College had a paid audience of 987 people. Advance tickets were \$9.95 and tickets at the door were \$12.95. A total of \$10,738.65 was collected in ticket sales. How many of each type of ticket were sold?

Use a system of three linear equations to solve exercises 23–31.

23. ***Bookstore Supplies*** Johanna bought 15 items at the college bookstore. The items cost a total of \$23.00. The pens cost \$0.50 each, the notebooks were \$3.00 each, and the highlighters cost \$1.50 each. She bought 2 more notebooks than highlighters. How many of each item did she buy?

24. ***Basketball*** In January 2006, Kobe Bryant of the Los Angeles Lakers had the highest-scoring game of his career. He made 46 shots for a total of 81 points. He made several free throws worth 1 point each, several regular shots worth 2 points each, and several 3-point shots. He made three times more regular shots than 3-point shots. How many of each type of shot did he make?

25. ***Holiday Concert*** A total of three hundred people attended the holiday concert at the city theater. The admission prices were \$12 for adults, \$9 for high school students, and \$5 for children. The ticket sales totaled \$2460. The concert coordinator suggested that next year they raise the prices to \$15 for adults, \$10 for high school students, and \$6 for children. She said that if the same number of people attend next year, the ticket sales at the higher prices will total \$2920. How many adults, high school students, and children attended this year?

26. ***CPR Training*** The college conducted a CPR training class for students, faculty, and staff. Faculty were charged \$10, staff were charged \$8, and students were charged \$2 to attend the class. A total of four hundred people came. The receipts for all who attended totaled \$2130. The college president remarked that if he had charged faculty \$15 and staff \$10 and let students come free, the receipts this year would have been \$2425. How many students, faculty, and staff came to the CPR training class?

27. ***City Subway Token Costs*** A total of twelve thousand passengers normally ride the green line of the MBTA during the morning rush hour. The token prices for a ride are \$0.25 for children under 12, \$1 for adults, and \$0.50 for senior citizens, and the revenue from these riders is \$10,700. If the token prices were raised to \$0.35 for children under 12 and \$1.50 for adults, and the senior citizen price were unchanged, the expected revenue from these riders would be \$15,820. How many riders in each category normally ride the green line during the morning rush hour?

28. ***Commission Sales*** The owner of Danvers Ford found that he sold a total of 520 cars, Escapes, and Explorers last year. He paid the sales staff a commission of \$100 for every car, \$200 for every Escape, and \$300 for every Explorer sold. The total of these commissions last year was \$87,000. In the coming year he is contemplating an increase so that the commission will be \$150 for every car and \$250 for every Escape, with no change in the commission for Explorer sales. If the sales are the same this year as they were last year, the commissions will total \$106,500. How many vehicles in each category were sold last year?

29. ***Pizza Costs*** The Hamilton House of Pizza delivered twenty pepperoni pizzas to Gordon College on the first night of final exams. The cost of these pizzas totaled \$181. A small pizza costs \$5 and contains 3 ounces of pepperoni. A medium pizza costs \$9 and contains 4 ounces of pepperoni. A large pizza costs \$12 and contains 5 ounces of pepperoni. The owner of the pizza shop used 5 pounds 2 ounces of pepperoni in making these twenty pizzas. How many pizzas of each size were delivered to Gordon College?

30. ***Roast Beef Sandwich Costs*** One of the favorite meeting places for local college students is Nick's Roast Beef in Beverly, Massachusetts. Last night from 8 P.M. to 9 P.M. Nick served twenty-four roast beef sandwiches. He sliced 15 pounds 8 ounces of roast beef to make these sandwiches and collected \$131 for them. The medium roast beef sandwich has 6 ounces of beef and costs \$4. The large roast beef sandwich has 10 ounces of beef and costs \$5. The extra large roast beef sandwich has 14 ounces of beef and costs \$7. How many of each size of roast beef sandwich did Nick sell from 8 P.M. to 9 P.M.?

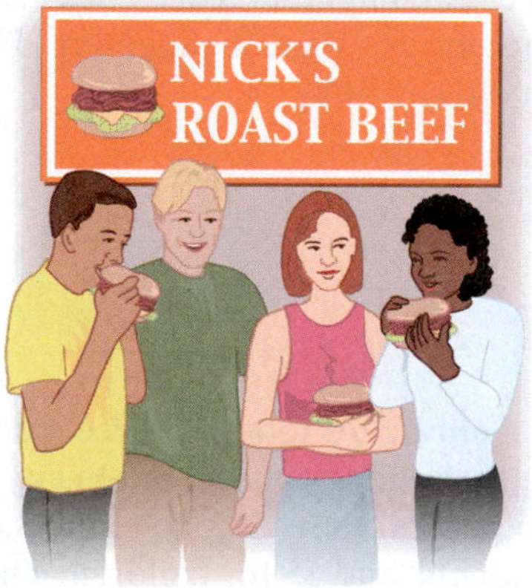

31. ***Packing Fruit*** Florida Fruits packs three types of gift boxes containing oranges, grapefruit, and tangerines. Box A contains 12 oranges, 5 grapefruit, and 3 tangerines. Box B contains 10 oranges, 6 grapefruit, and 4 tangerines. Box C contains 5 oranges, 8 grapefruit, and 5 tangerines. The shipping manager has available 91 oranges, 63 grapefruit, and 40 tangerines. How many gift boxes of each type can she prepare?

Cumulative Review *Solve for the variable indicated.*

32. **[2.1.1]** Solve for x. $\frac{1}{3}(4 - 2x) = \frac{1}{2}x - 3$

33. **[2.1.1]** Solve for x. $0.06x + 0.15(0.5 - x) = 0.04$

34. **[2.1.1]** Solve for y. $2(y - 3) - (2y + 4) = -6y$

35. **[2.2.1]** Solve for x. $6a(2x - 3y) = 7ax - 3$

Quick Quiz 4.3

1. A plane flew 1200 miles with a tailwind in 2.5 hours. The return trip against the wind took 3 hours. Find the speed of the wind. Find the speed of the plane in still air.

2. A man needed a car in Alaska to travel to fishing sites. He found a company that rented him a car but charged him a daily rental fee and a mileage charge. He rented the car for 8 days, drove 300 miles, and was charged \$355. Next week his friend rented the same car for 9 days, drove 260 miles, and was charged \$380. What was the daily rental charge? How much did the company charge per mile?

3. Three friends went on a trip to Uganda. They went to the same store and purchased the exact same items. Nancy purchased 3 drawings, 2 carved elephants, and 1 set of drums for \$55. John purchased 2 drawings, 3 carved elephants, and 1 set of drums for \$65. Steve purchased 4 drawings, 3 carved elephants, and 2 sets of drums for \$85. What was the price of each drawing, carved elephants, and set of drums?

4. **Concept Check** Explain how you would set up two equations using the information in problem 1 above if the plane flew 1500 miles instead of 1200 miles.

4.4 Systems of Linear Inequalities

Student Learning Objective

After studying this section, you will be able to:

① **Graph a system of linear inequalities.**

① Graphing a System of Linear Inequalities

We learned how to graph a linear inequality in two variables in Section 3.4. We call two linear inequalities in two variables a **system of linear inequalities in two variables.** We now consider how to graph such a system. The solution to a system of inequalities is the intersection of the solution sets of the individual inequalities of the system.

EXAMPLE 1 Graph the solution of the system.

$$y \le -3x + 2$$
$$-2x + y \ge -1$$

Solution

In this example, we will first graph each inequality separately. The graph of $y \le -3x + 2$ is the region on and below the line $y = -3x + 2$.

The graph of $-2x + y \ge -1$ consists of the region on and above the line $-2x + y = -1$.

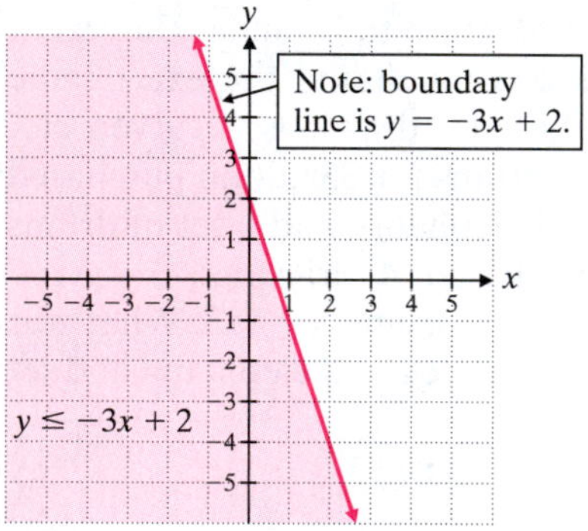

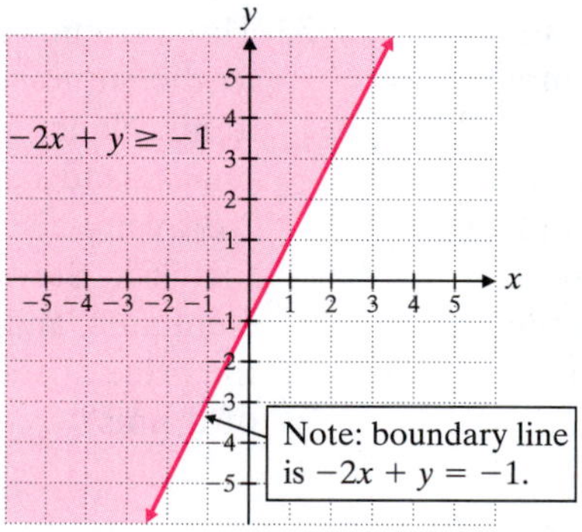

We will now place these graphs on one rectangular coordinate system. The darker shaded region is the intersection of the two graphs. Thus, the solution to the system of two inequalities is the darker shaded region and its boundary lines.

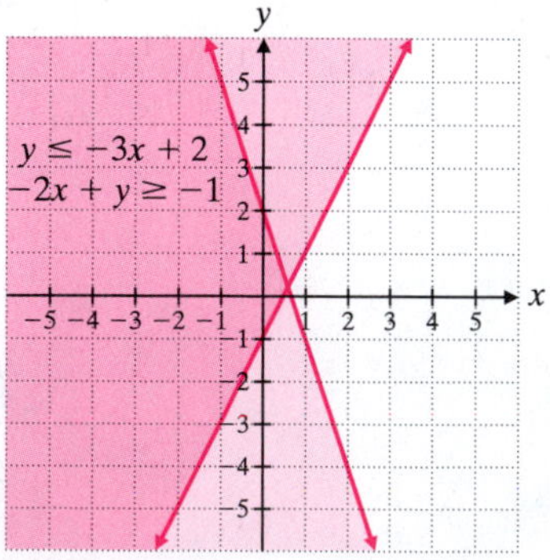

Student Practice 1 Graph the solution of the system.

$$-2x + y \le -3$$
$$x + 2y \ge 4$$

Student Practice 1

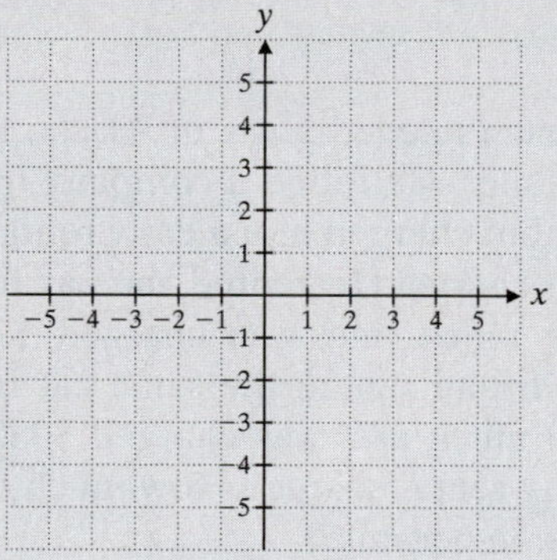

NOTE TO STUDENT: *Fully worked-out solutions to all of the Student Practice problems can be found at the back of the text starting at page SP-1.*

Usually we sketch the graphs of the individual inequalities on one set of axes. We will illustrate that concept with the following example.

EXAMPLE 2 Graph the solution of the system.

$$y < 4$$
$$y > \frac{3}{2}x - 2$$

Solution The graph of $y < 4$ is the region below the line $y = 4$. It does not include the line since we have the $<$ symbol. Thus, we use a dashed line to indicate that the boundary line is not part of the answer. The graph of $y > \frac{3}{2}x - 2$ is the region above the line $y = \frac{3}{2}x - 2$. Again, we use the dashed line to indicate that the boundary line is not part of the answer. The final solution is the darker shaded region. The solution does *not* include the dashed boundary lines.

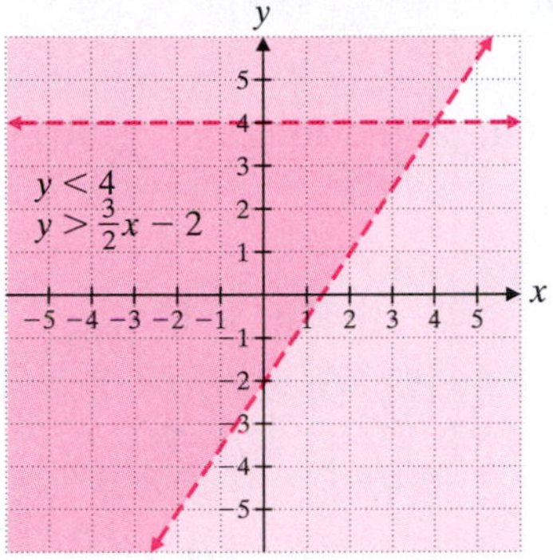

Student Practice 2 Graph the solution of the system.

$$y > -1$$
$$y < -\frac{3}{4}x + 2$$

Student Practice 2

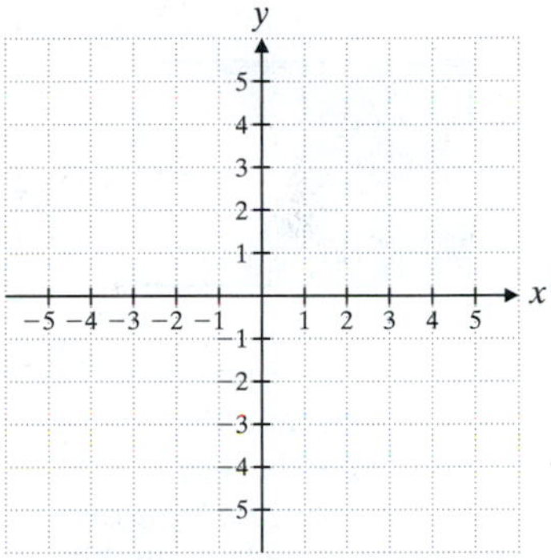

There are times when we require the exact location of the point where two boundary lines intersect. In these cases the boundary points are labeled on the final sketch of the solution.

EXAMPLE 3 Graph the solution of the following system of inequalities. Find the coordinates of any points where boundary lines intersect.

$$x + y \leq 5$$
$$x + 2y \leq 8$$
$$x \geq 0$$
$$y \geq 0$$

Solution The graph of $x + y \leq 5$ is the region on and below the line $x + y = 5$. The graph of $x + 2y \leq 8$ is the region on and below the line $x + 2y = 8$. We solve the system containing the equations $x + y = 5$ and $x + 2y = 8$ to find that their point of intersection is (2, 3). The graph of $x \geq 0$ is the y-axis and all the region to the right of the y-axis. The graph of $y \geq 0$ is the x-axis and all the region above the x-axis. Thus, the solution to the system is the shaded region and its boundary lines. There are four points where boundary lines intersect.

Continued on next page

Graphing Calculator

Graphing Systems of Linear Inequalities

On some graphing calculators, you can graph a system of linear inequalities. To graph the system in Example 1, first rewrite it as:

$$y \leq -3x + 2$$
$$y \geq 2x - 1.$$

Enter each expression into the Y = editor of your graphing calculator and then select the appropriate direction for shading. Display:

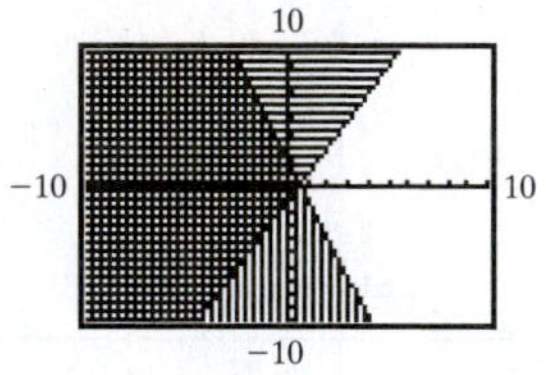

Notice that one inequality is shaded vertically and the other is shaded horizontally. The intersection is the solution. Note also that the graphing calculator will not indicate whether or not the boundary of the solution region is included in the solution.

These points are called the **vertices** of the solution. Thus, the vertices of the solution are (0, 0), (0, 4), (2, 3), and (5, 0).

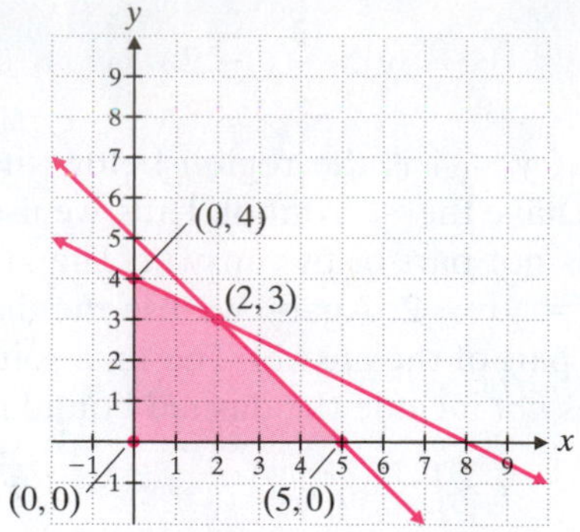

Student Practice 3 Graph the solution to the system of inequalities. Find the vertices of the solution.

$$x + y \leq 6$$
$$3x + y \leq 12$$
$$x \geq 0$$
$$y \geq 0$$

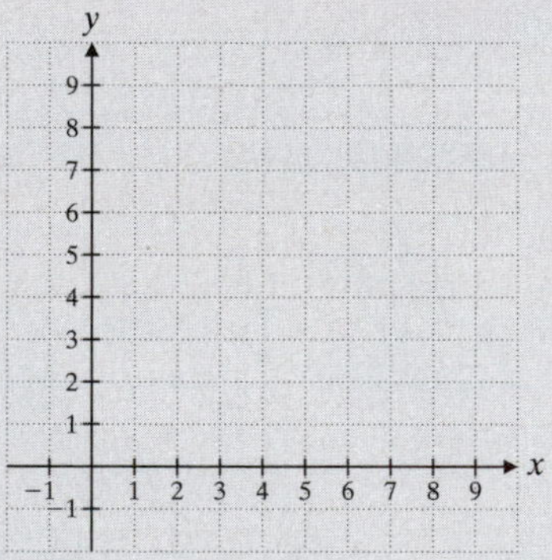

STEPS TO SUCCESS What Are the Absolute Essentials to Succeed in This Course?

If you are in a traditional class: Students who are successful in this course find there are six absolute essentials.

Here they are:

1. Attend every class session.
2. Read the textbook for every assigned section.
3. Take notes in class.
4. Do the assigned homework for every class.
5. Get help immediately when you need assistance.
6. Review what you are learning.

Making it personal: Which of these six suggestions is the one you have the greatest trouble actually doing? Will you make a personal commitment to doing that one thing faithfully for the next two weeks? You will be amazed at the results. ▼

If you are in an online class or a nontraditional class: Students in an online class or a self-paced class find it really helps to spread the homework and the reading over five different days each week.

Making it personal: Will you try to do your homework for five days each week over the next two weeks? Write out a study plan for the next two weeks. You will be amazed at the results. ▼

4.4 Exercises

MyMathLab® Watch the videos in MyMathLab Download the MyDashBoard App

Verbal and Writing Skills, Exercises 1–4

1. In the graph of the system $y > 3x + 1$ and $y < -2x + 5$, would the boundary lines be solid or dashed? Why?

2. In the graph of the system $y \geq -6x + 3$ and $y \leq -4x - 2$, would the boundary lines be solid or dashed? Why?

3. Stephanie wanted to know if the point $(3, -4)$ lies in the region that is a solution for $y < -2x + 3$ and $y > 5x - 3$. How could she determine if this is true?

4. John wanted to know if the point $(-5, 2)$ lies in the region that is a solution for $x + 2y < 3$ and $-4x + y > 2$. How could he determine if this is true?

Graph the solution of each of the following systems.

5. $y \geq 2x - 1$
$x + y \leq 6$

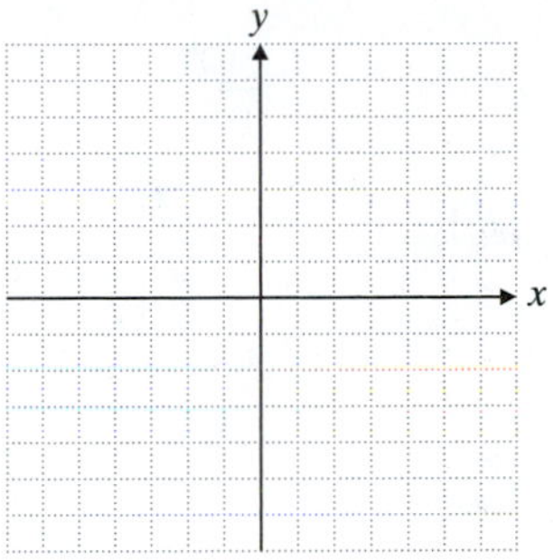

6. $y \geq x - 3$
$x + y \geq 2$

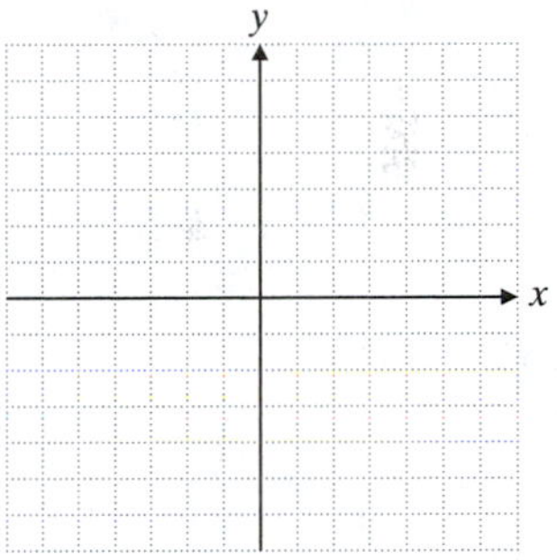

7. $y \geq -2x$
$y \geq 3x + 5$

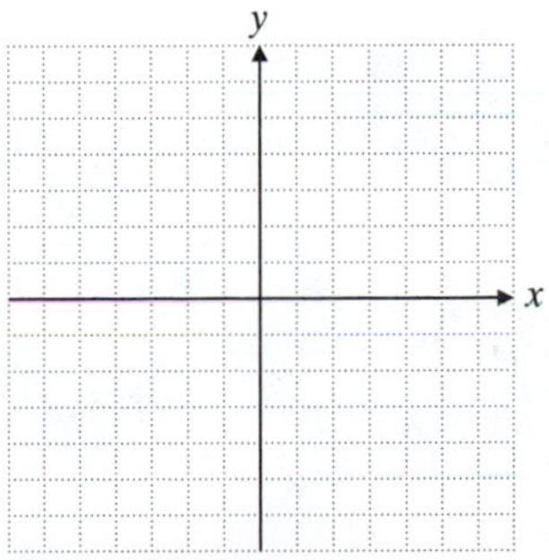

8. $y \geq x$
$y \leq -x + 2$

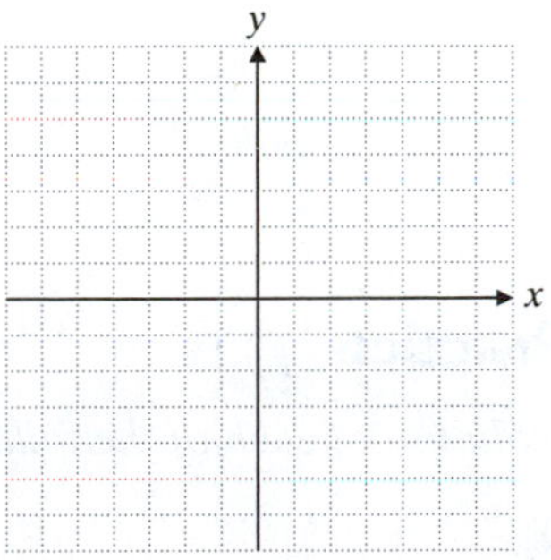

9. $y \geq 2x - 3$
$y \leq \frac{2}{3}x$

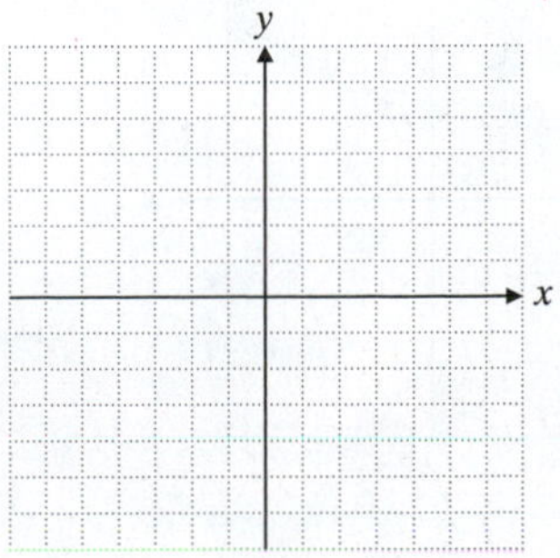

10. $y \leq \frac{1}{2}x - 3$
$y \geq -\frac{1}{2}x$

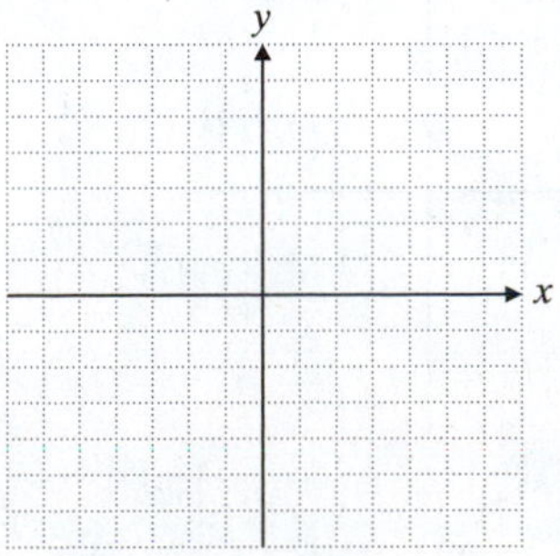

11. $x - y \geq -1$
$-3x - y \leq 4$

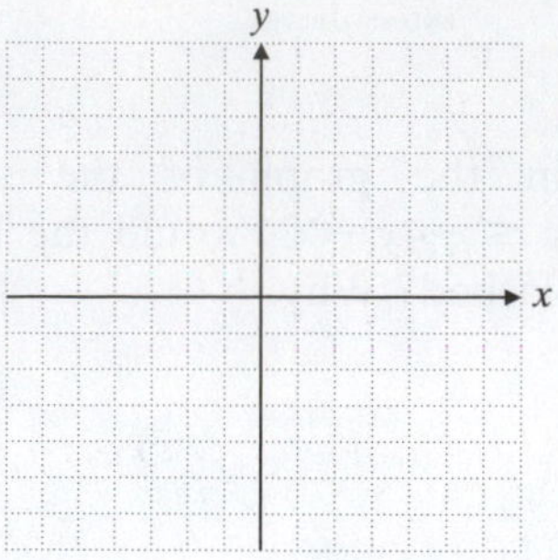

12. $x + y \leq 5$
$-3x + 4y \geq -8$

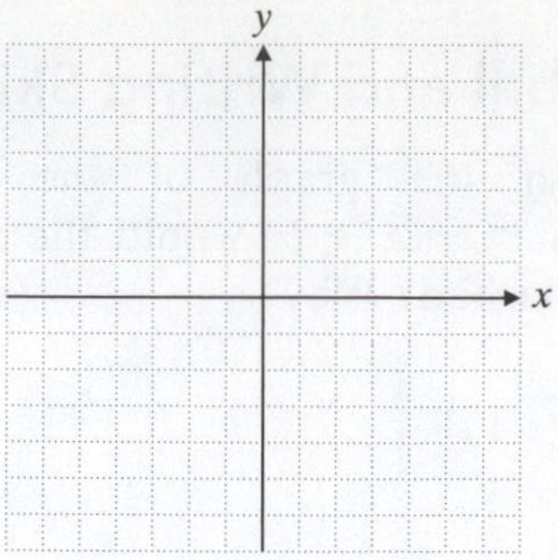

13. $x + 2y < 6$
$y < 3$

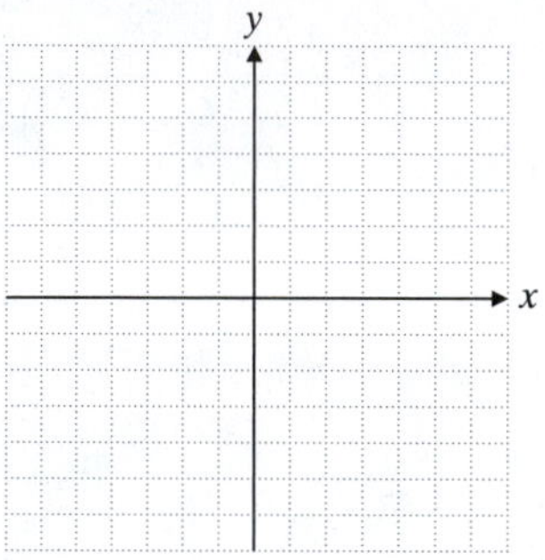

14. $-2x + y < -4$
$y > -4$

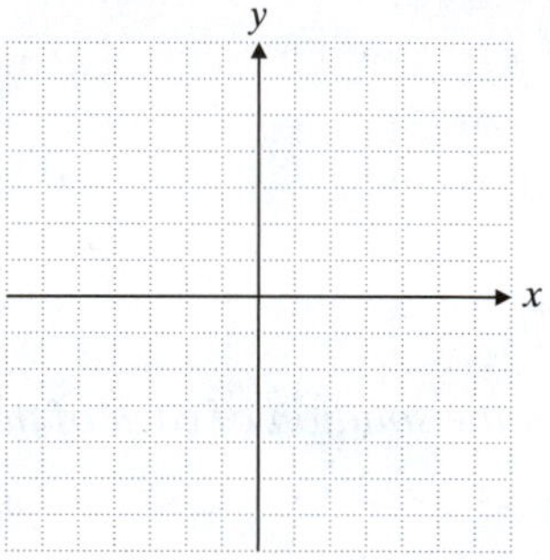

15. $y < 4$
$x > -2$

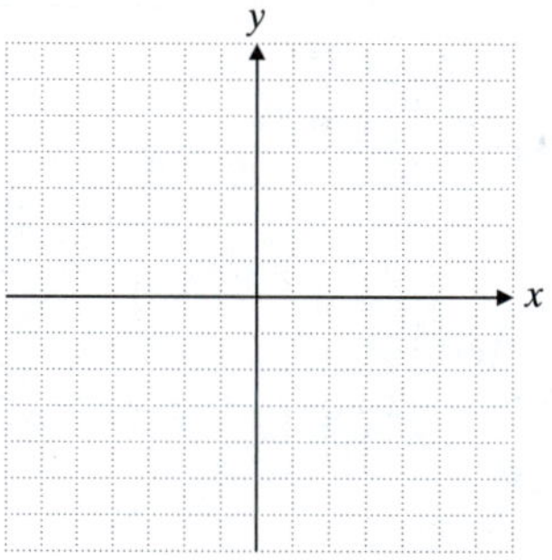

16. $y > -5$
$x < 3$

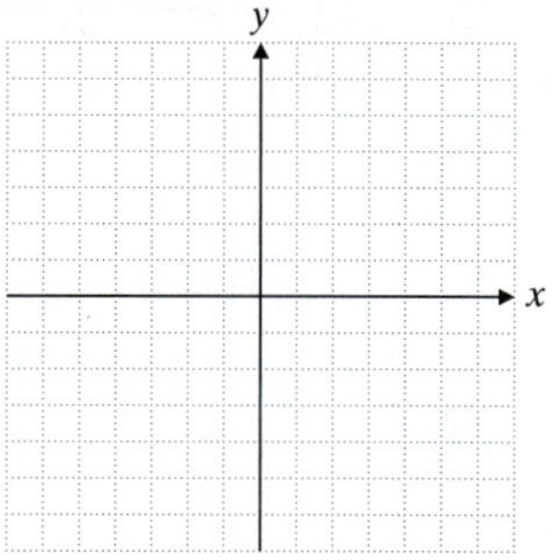

Mixed Practice

Graph the solution of each of the following systems.

17. $x - 4y \geq -4$
$3x + y \leq 3$

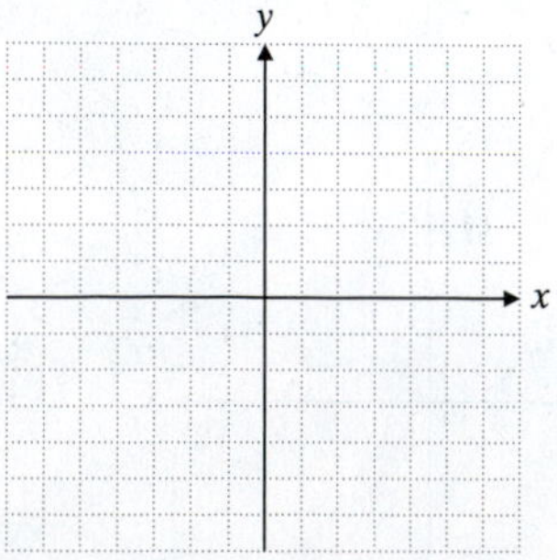

18. $5x - 2y \leq 10$
$x - y \geq -1$

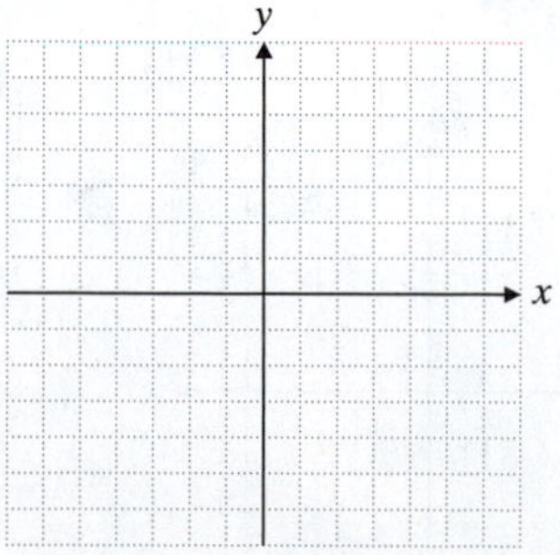

19. $3x + 2y < 6$
$3x + 2y > -6$

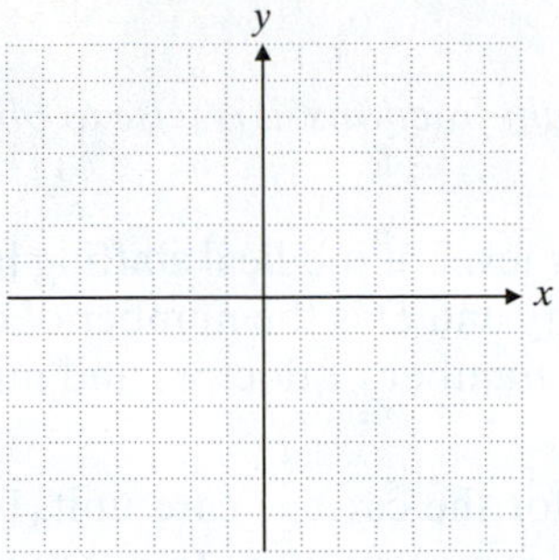

20. $2x - y < 2$
$2x - y > -2$

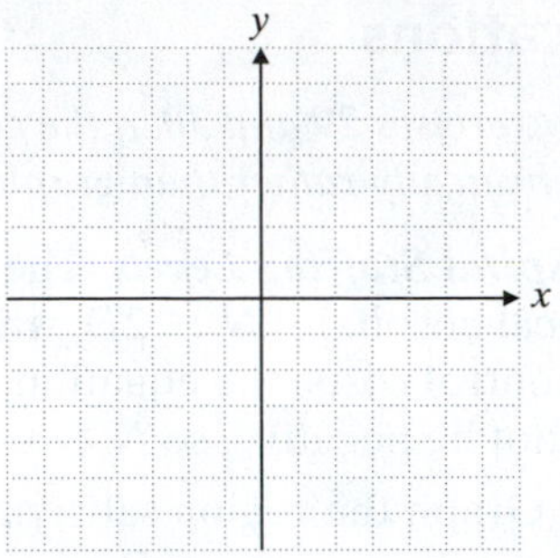

Graph the solution of the following systems of inequalities. Find the vertices of the solution.

21. $x + y \leq 5$
$2x - y \geq 1$

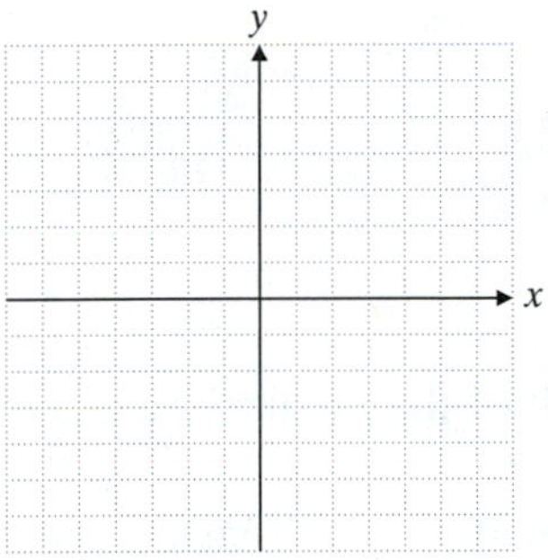

22. $x + y \geq 2$
$y + 4x \leq -1$

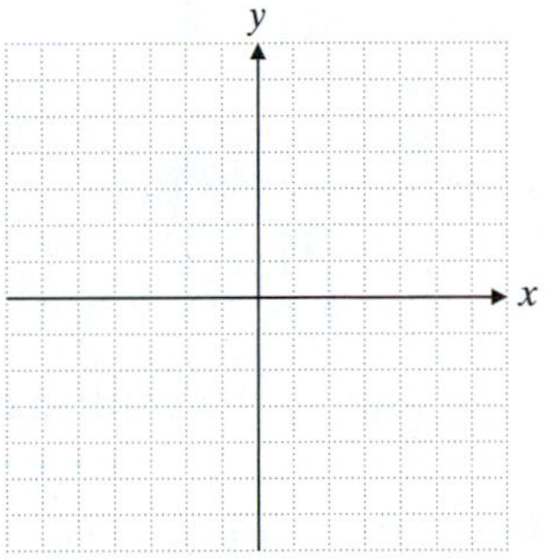

23. $x + 3y \leq 12$
$y < x$

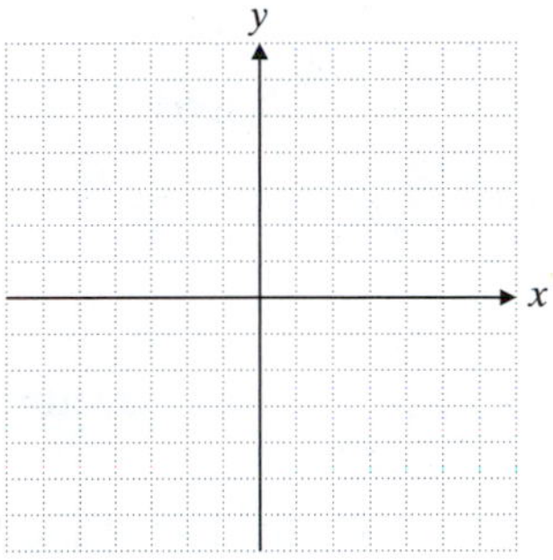

24. $x + 2y \leq 4$
$y < -x$

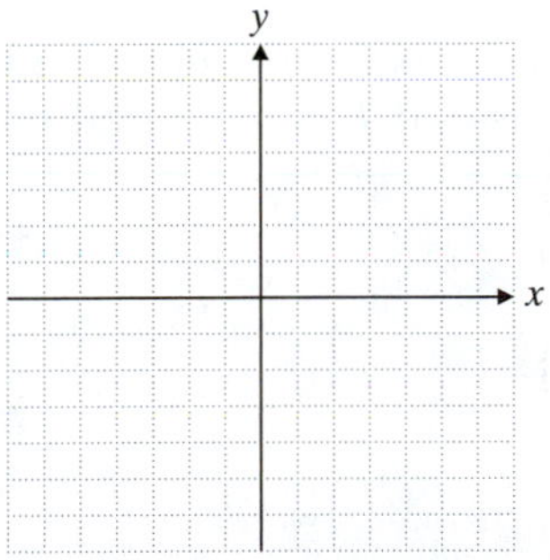

25. $y \leq x$
$x + y \geq 1$
$x \leq 5$

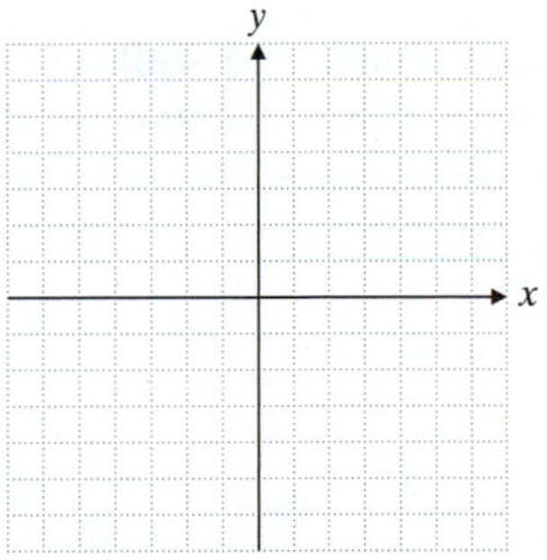

26. $x + y \leq 3$
$x - y \leq 1$
$x \geq -1$

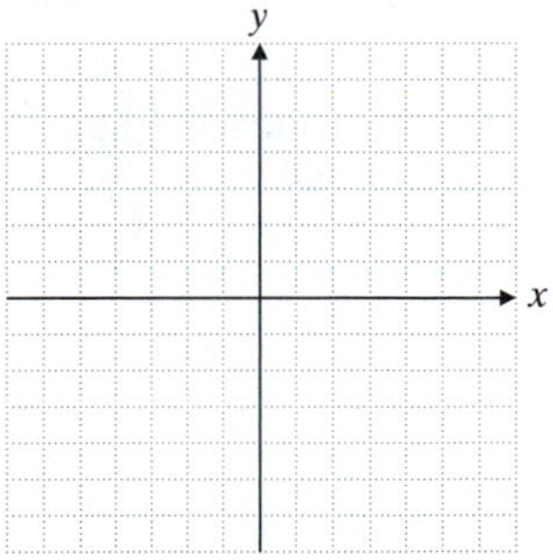

To Think About

Graph the region determined by each of the following systems.

27. $y \leq 3x + 6$
$4y + 3x \leq 3$
$x \geq -2$
$y \geq -3$

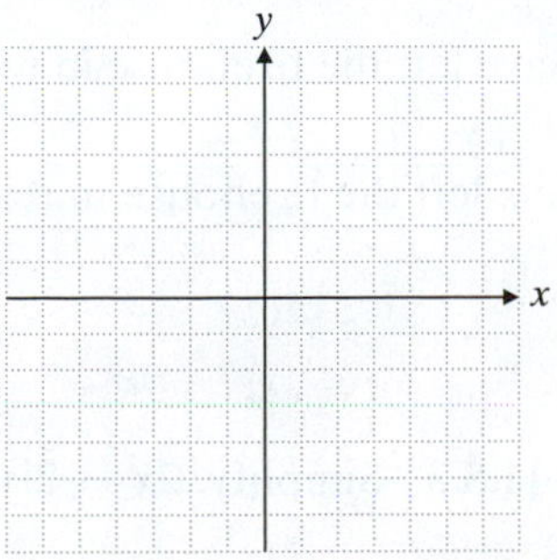

28. $-x + y \leq 100$
$x + 3y \leq 150$
$x \geq -80$
$y \geq 20$
(*Hint:* Use a scale of each square = 20 units on both axes.)

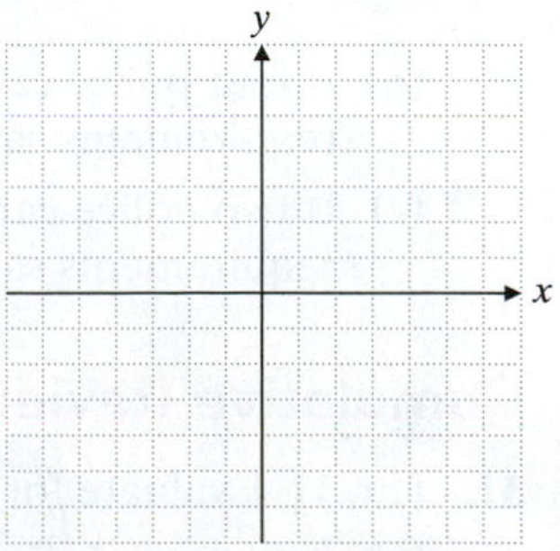

Applications

Hint: In exercises 29 and 30, if the coordinates of a boundary point contain fractions, it is wise to obtain the point of intersection algebraically rather than graphically.

29. ***Hospital Staffing Levels*** The equation that represents the proper level of medical staffing in the cardiac care unit of a local hospital is $N \leq 2D$, where N is the number of nurses on duty and D is the number of doctors on duty. In order to control costs, the equation $4N + 3D \leq 20$ is appropriate. The number of doctors and nurses on duty at any time cannot be negative, so $N \geq 0$ and $D \geq 0$.

(a) Graph the region satisfying all of the medical requirements for the cardiac care unit. Use the following special graph grid, where D is measured on the horizontal axis and N is measured on the vertical axis.

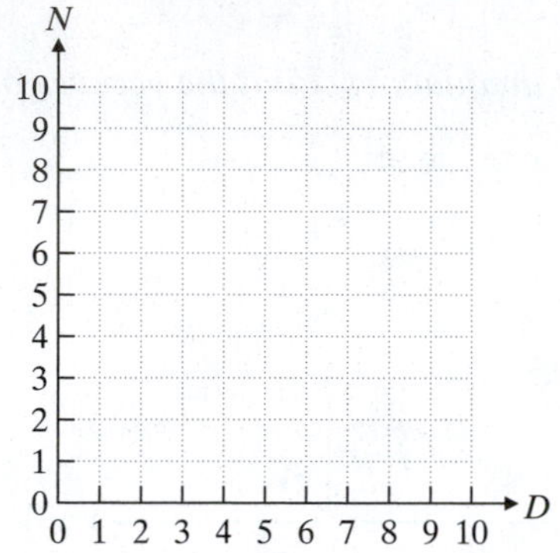

(b) If there are three doctors and two nurses on duty in the cardiac care unit, are all of the medical requirements satisfied?

(c) If there is one doctor and four nurses on duty in the cardiac care unit, are all of the medical requirements satisfied?

30. ***Traffic Control*** The equation that represents the proper traffic control and emergency vehicle response availability in the city of Salem is $P + 3F \leq 18$, where P is the number of police cars on active duty and F is the number of fire trucks that have left the firehouse and are involved in a response to a call. In order to comply with staffing limitations, the equation $4P + F \leq 28$ is appropriate. The number of police cars on active duty and the number of fire trucks that have left the firehouse cannot be negative, so $P \geq 0$ and $F \geq 0$.

(a) Graph the regions satisfying all of the availability and staffing limitation requirements for the city of Salem. Use the following special graph grid, where P is measured on the horizontal axis and F is measured on the vertical axis.

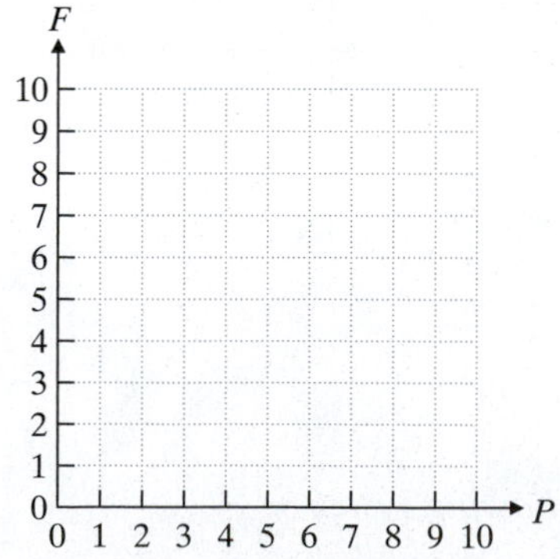

(b) If four police cars are on active duty and four fire trucks have left the firehouse in response to a call, are all of the requirements satisfied?

(c) If two police cars are on active duty and six fire trucks have left the firehouse in response to a call, are all of the requirements satisfied?

Cumulative Review

31. **[1.6.1]** Evaluate for $x = 2, y = -1$.
$-3x^2y - x^2 + 5y^2$

32. **[1.5.3]** Simplify. $2x - 2[y + 3(x - y)]$

33. **[4.3.1]** ***Driving Range Receipts*** Golf Galaxy Indoor Driving Range took in \$6050 on one rainy day and six sunny days. The next week the driving range took in \$7400 on four rainy days and three sunny days. What is the average amount of money taken in at Golf Galaxy on a rainy day? On a sunny day?

34. **[4.3.2]** ***Office Lunch Expenses*** Two weeks ago, Raquel took the office lunch order to Kramer's Deli and bought 3 chicken sandwiches, 2 side salads, and 3 sodas for \$17.30. One week ago it was Gerry's turn to take the trip to Kramer's. He bought 3 chicken sandwiches, 4 side salads, and 4 sodas for \$23.15. This week Sheryl made the trip. She purchased 4 chicken sandwiches, 3 side salads, and 5 sodas for \$24.75. What is the cost of one chicken sandwich? Of one side salad? Of one soda?

Quick Quiz 4.4

1. In graphing the inequality $3x + 5y < 15$, do you shade the region above the line or below the line?

2. In graphing the following region, should the boundary lines be dashed lines or solid lines?

$$4x + 5y \geq 20$$
$$2x - 3y \geq 6$$

3. Graph the solution of the following system of linear inequalities.

$$3x + 2y > 6$$
$$x - 2y < 2$$

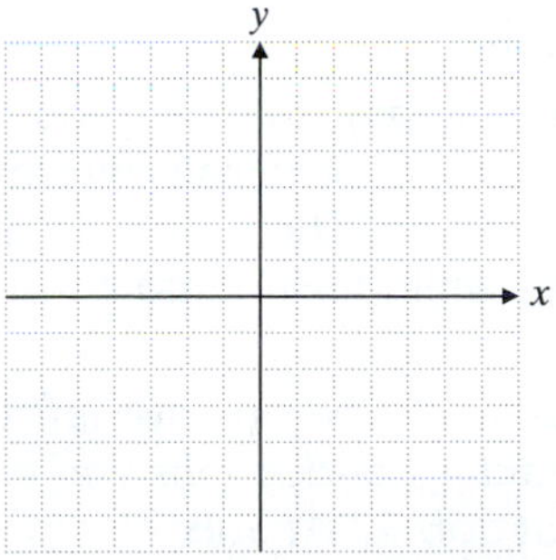

4. Concept Check Explain how you would graph the region described by the following.

$$y > x + 2$$
$$x < 3$$

Did You Know...

That How Much You Get Paid at Your Job Is Not How Much You Get to Spend Because of Expenses?

DETERMINING YOUR EARNINGS

Understanding the Problem:
The real hourly wage is the amount you make per hour after taxes and other work-related expenses. When calculating real hourly wage, we cannot just look at how much a job pays an hour or how much the paycheck will be. We need to consider work-related expenses as well as extra time spent for the job.

Sharon is not currently employed. She is considering two different positions and needs to determine which job she should take. If she takes a job, she estimates she'll have the following weekly expenses: daycare, \$175; after-school program, \$75; clothing, \$25; other work-related expenses, \$50, as well as \$0.35 per mile for transportation.

Making a Plan:
Sharon needs to determine what the income and expenses are for each position.

Step 1: First she calculates what her income would be for each position.

Job 1: Donivan Tech is offering her a full-time management position at a yearly salary of \$50,310. Sharon estimates that she will work an average of 45 hours per week when she considers the extra work commitment required for managers. The commute of 45 miles a day will be about five hours a week in driving. She estimates that payroll deductions will be 15% of her gross salary.

Job 2: J & R Financial Group is offering her a part-time position as an assistant manager. The work schedule would be 6 hours per day, 5 days a week at an hourly rate of \$20.50. The commute of 5 miles a day will be about one hour a week in driving. Due to the reduced work schedule, she calculates that payroll deductions will only be 12%, and daycare, after-school program, and other expenses will all be reduced by \$25 per week.

Task 1: *What is the weekly pay, before deductions or expenses, at Job 1? At Job 2?*

Step 2: Step 2: She needs to calculate what her expenses would be for each position.

Task 2: *How much are weekly payroll deductions at Job 1? At Job 2?*

Task 3: *What are the weekly job-related expenses at Job 1? At Job 2?*

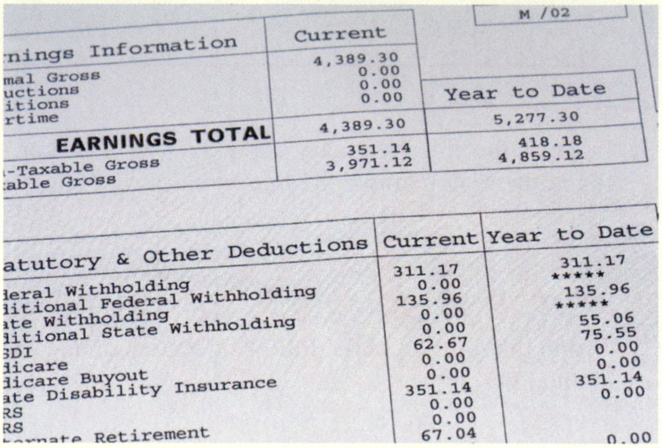

Finding a Solution

Step 3: Using weekly pay, payroll deductions and job-related expenses, Sharon can determine what her real hourly wage will be. Then she'll be able to compare the two jobs while taking into consideration all expenses.

Task 4: *How much will Sharon have left to spend every week after subtracting deductions and expenses from the weekly pay at Job 1? At Job 2?*

Task 5: *Considering the total time spent working and commuting, what is the real hourly wage at Job 1? At Job 2?*

Step 4: Sharon can now decide which job to take.

Task 6: *If Sharon is deciding on a job based only on real hourly wage, which job should she take?*

Task 7: *If Sharon is deciding on a job based on how much she will have left to spend each week after deductions and expenses, which job should she take?*

Applying the Situation to Your Life:
When comparing jobs, one needs to determine the real hourly wage because what you earn is not as important as what you actually get to spend. Work-related expenses and time spent commuting have an impact on the real hourly rate. Taking these into consideration will help you make an informed decision.

Chapter 4 Organizer

Topic and Procedure	Examples	You Try It

Finding a solution to a system of equations by the graphing method, p. 199

1. Graph the first equation.
2. Graph the second equation.
3. Approximate from your graph the coordinates of the point where the two lines intersect, if they intersect at one point.
4. If the lines are parallel, there is no solution. If the lines coincide, there is an infinite number of solutions.

Example: Solve by graphing. $x + y = 6$, $2x - y = 6$

Graph each line. The solution is (4, 2).

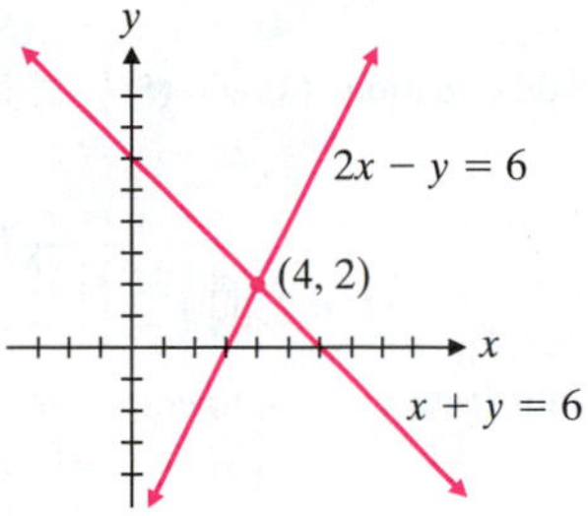

You Try It: 1. Solve by graphing. $-x + 3y = 6$, $2x - 3y = -9$

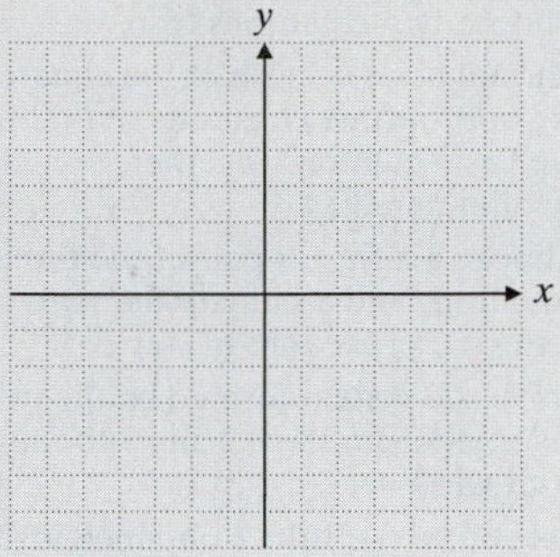

Solving a system of two linear equations by the substitution method, p. 200

The substitution method is most appropriate when *at least one variable has a coefficient of* 1 *or* −1.

1. Solve for one variable in one of the equations.
2. In the other equation, replace that variable with the expression you obtained in step 1.
3. Solve the resulting equation.
4. Substitute the numerical value you obtain for the variable into one of the original or equivalent equations and solve for the other variable.
5. Check the solution in both original equations.

Example: Solve.

$$2x + y = 11 \quad \textbf{(1)}$$
$$x + 3y = 18 \quad \textbf{(2)}$$

$y = 11 - 2x$ from equation **(1)**. Substitute into **(2)**.

$$x + 3(11 - 2x) = 18$$
$$x + 33 - 6x = 18$$
$$-5x = -15$$
$$x = 3$$

Substitute $x = 3$ into $2x + y = 11$.

$$2(3) + y = 11$$
$$y = 5$$

The solution is (3, 5).

You Try It: 2. Solve by the substitution method.

$$-3x + 4y = -10$$
$$x - 5y = 7$$

Solving a system of two linear equations by the addition (or elimination) method, p. 204

The addition method is most appropriate when the variables *all have coefficients other than* 1 *or* −1.

1. Arrange each equation in the form $ax + by = c$.
2. Multiply one or both equations by appropriate numerical values so that when the two resulting equations are added, one variable is eliminated.
3. Solve the resulting equation.
4. Substitute the numerical value you obtain for the variable into one of the original or equivalent equations.
5. Solve this equation to find the other variable.

Example: Solve.

$$2x + 3y = 5 \quad \textbf{(1)}$$
$$-3x - 4y = -2 \quad \textbf{(2)}$$

Multiply equation **(1)** by 3 and **(2)** by 2.

$$6x + 9y = 15$$
$$\underline{-6x - 8y = -4}$$
$$y = 11$$

Substitute $y = 11$ into equation **(1)**.

$$2x + 3(11) = 5$$
$$2x + 33 = 5$$
$$2x = -28$$
$$x = -14$$

The solution is (−14, 11).

You Try It: 3. Solve by the addition method.

$$4x - 5y = 5$$
$$-3x + 7y = 19$$

Inconsistent system of equations, p. 205

If there is *no solution* to a system of linear equations, the system of equations is inconsistent. When you try to solve an inconsistent system algebraically, you obtain an equation that is not true, such as $0 = 5$.

Example: Solve.

$$4x + 3y = 10 \quad \textbf{(1)}$$
$$-8x - 6y = 5 \quad \textbf{(2)}$$

Multiply equation **(1)** by 2 and add to **(2)**.

$$8x + 6y = 20$$
$$\underline{-8x - 6y = 5}$$
$$0 = 25$$

But $0 \neq 25$. Thus, there is no solution. The system of equations is inconsistent.

You Try It: 4. Solve.

$$2x + 12y = 3$$
$$x + 6y = 8$$

Dependent equations, p. 205

If there is an *infinite number of solutions* to a system of linear equations, at least one pair of equations is dependent. When you try to solve a system that contains dependent equations, you will obtain an equation that is always true (such as $0 = 0$ or $3 = 3$). These equations are called *identities*.

Example: Attempt to solve the system.

$$x - 2y = -5 \quad \textbf{(1)}$$
$$-3x + 6y = 15 \quad \textbf{(2)}$$

Multiply equation **(1)** by 3 and add to **(2)**.

$$3x - 6y = -15$$
$$\underline{-3x + 6y = 15}$$
$$0 = 0$$

There is an infinite number of solutions. The equations are dependent.

You Try It: 5. Solve.

$$x - y = 2$$
$$-4x + 4y = -8$$

Topic and Procedure	Examples	You Try It

Solving a system of three linear equations by algebraic methods, p. 213

If there is one solution to a system of three linear equations in three unknowns, it may be obtained in the following manner.

1. Choose two equations from the system.
2. Multiply one or both of the equations by the appropriate constants so that by adding the two equations together, one variable can be eliminated.
3. Choose a *different* pair of the three original equations and eliminate the *same* variable using the procedure of step 2.
4. Solve the system formed by the two equations resulting from steps 2 and 3 to find values for both variables.
5. Substitute the values obtained in step 4 into one of the original three equations to find the value of the third variable.

Examples

Solve.
$$2x - y - 2z = -1 \quad \textbf{(1)}$$
$$x - 2y - z = 1 \quad \textbf{(2)}$$
$$x + y + z = 4 \quad \textbf{(3)}$$

Add equations **(2)** and **(3)** to eliminate z.
$$2x - y = 5 \quad \textbf{(4)}$$

Multiply equation **(3)** by 2 and add to **(1)**.
$$4x + y = 7 \quad \textbf{(5)}$$

Add equations **(4)** and **(5)**.
$$\begin{aligned} 2x - y &= 5 \\ 4x + y &= 7 \\ \hline 6x &= 12 \\ x &= 2 \end{aligned}$$

Substitute $x = 2$ into equation **(5)**.
$$4(2) + y = 7$$
$$y = -1$$

Substitute $x = 2, y = -1$ into equation **(3)**.
$$2 + (-1) + z = 4$$
$$z = 3$$

The solution is $(2, -1, 3)$.

You Try It

6. Solve.
$$x + 2y - z = -13$$
$$-2x + 3y - 2z = -8$$
$$x - y + z = 3$$

Graphing the solution to a system of inequalities in two variables, p. 232

1. Determine the region that satisfies each individual inequality.
2. Shade the common region that satisfies all the inequalities.

Examples

Graph.
$$3x + 2y \le 10$$
$$-x + 2y \ge 2$$

1. $3x + 2y \le 10$ can be graphed more easily as $y \le -\frac{3}{2}x + 5$. We draw a solid line and shade the region below it. $-x + 2y \ge 2$ can be graphed more easily as $y \ge \frac{1}{2}x + 1$. We draw a solid line and shade the region above it.
2. The common region is shaded.

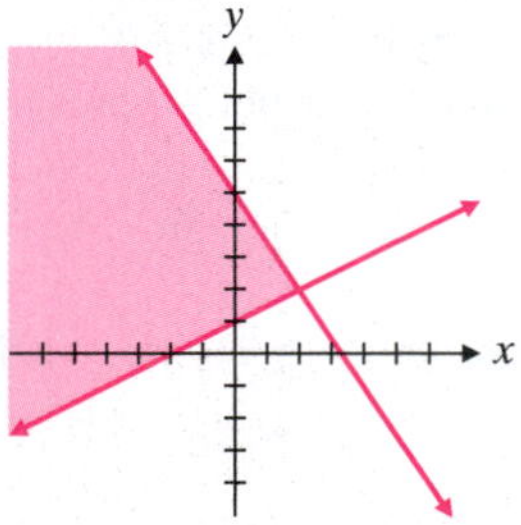

You Try It

7. Graph.
$$x - 3y < 6$$
$$2x + y > 5$$

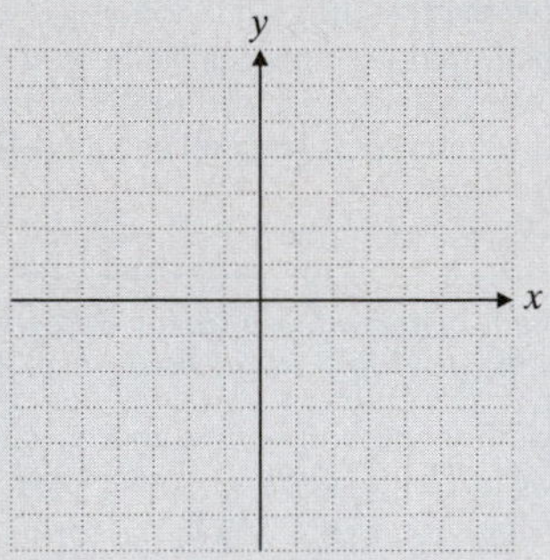

Chapter 4 Review Problems

Solve the following systems by graphing.

1. $x + 2y = 8$
$x - y = 2$

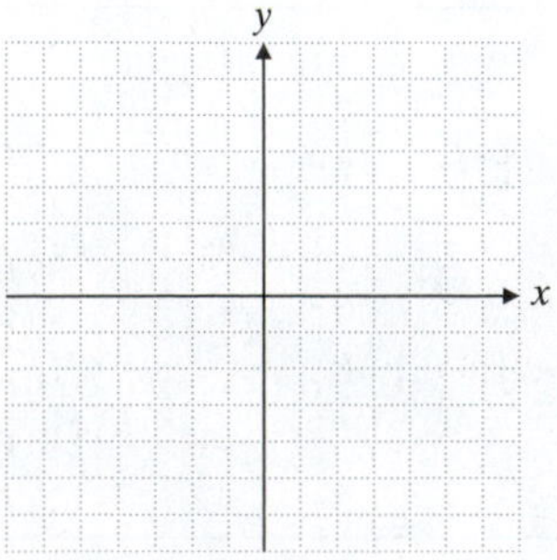

2. $2x + y = 6$
$3x + 4y = 4$

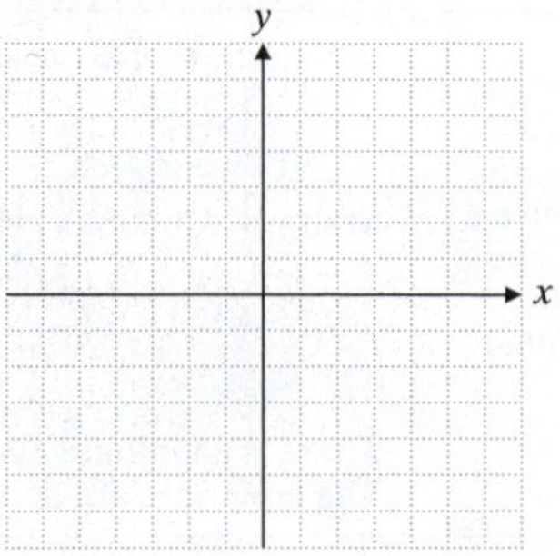

Solve the following systems by substitution.

3. $\begin{aligned} 3x - 2y &= -9 \\ 2x + y &= 1 \end{aligned}$

4. $\begin{aligned} 4x + 5y &= 2 \\ 3x - y &= 11 \end{aligned}$

5. $\begin{aligned} 3x - 4y &= -12 \\ x + 2y &= -4 \end{aligned}$

Solve the following systems by addition.

6. $\begin{aligned} -2x + 5y &= -12 \\ 3x + y &= 1 \end{aligned}$

7. $\begin{aligned} 7x - 4y &= 2 \\ 6x - 5y &= -3 \end{aligned}$

8. $\begin{aligned} 5x + 2y &= 3 \\ 7x + 5y &= -20 \end{aligned}$

Solve by any appropriate method. If there is no unique solution, state why.

9. $\begin{aligned} x &= 3 - 2y \\ 3x + 6y &= 8 \end{aligned}$

10. $\begin{aligned} x + 5y &= 10 \\ y &= 2 - \frac{1}{5}x \end{aligned}$

11. $\begin{aligned} 5x - 2y &= 15 \\ 3x + y &= -2 \end{aligned}$

12. $\begin{aligned} x + \frac{1}{3}y &= 1 \\ \frac{1}{4}x - \frac{3}{4}y &= -\frac{9}{4} \end{aligned}$

13. $\begin{aligned} 3a + 8b &= 0 \\ 9a + 2b &= 11 \end{aligned}$

14. $\begin{aligned} x + 3 &= 3y + 1 \\ 1 - 2(x - 2) &= 6y + 1 \end{aligned}$

15. $\begin{aligned} 10(x + 1) - 13 &= -8y \\ 4(2 - y) &= 5(x + 1) \end{aligned}$

16. $\begin{aligned} 0.2x - 0.1y &= 0.8 \\ 0.1x + 0.3y &= 1.1 \end{aligned}$

Solve by an appropriate method.

17. $\begin{aligned} 3x - 2y - z &= 3 \\ 2x + y + z &= 1 \\ -x - y + z &= -4 \end{aligned}$

18. $\begin{aligned} -2x + y - z &= -7 \\ x - 2y - z &= 2 \\ 6x + 4y + 2z &= 4 \end{aligned}$

19. $\begin{aligned} 2x + 5y + z &= 3 \\ x + y + 5z &= 42 \\ 2x + y &= 7 \end{aligned}$

20. $\begin{aligned} x + 2y + z &= 5 \\ 3x - 8y &= 17 \\ 2y + z &= -2 \end{aligned}$

21. $\begin{aligned} -3x - 4y + z &= -4 \\ x + 6y + 3z &= -8 \\ 5x + 3y - z &= 14 \end{aligned}$

22. $\begin{aligned} 3x + 2y &= 7 \\ 2x + 7z &= -26 \\ 5y + z &= 6 \end{aligned}$

Use a system of linear equations to solve each of the following exercises.

23. ***Commercial Airline*** A plane flies 720 miles against the wind in 3 hours. The return trip with the wind takes only $2\frac{1}{2}$ hours. Find the speed of the wind. Find the speed of the plane in still air.

24. ***Football*** During the 2010 Pro Bowl between the American Football Conference and the National Football Conference, 66 points were scored in touchdowns and field goals. The total number of touchdowns (6 points each) and field goals (3 points each) was 13. How many touchdowns and how many field goals were scored?

25. ***Temporary Help Expenses*** When the circus came to town last year, they hired general laborers at \$70 per day and mechanics at \$90 per day. They paid \$1950 for this temporary help for one day. This year they hired exactly the same number of people of each type, but they paid \$80 for general laborers and \$100 for mechanics for the one day. This year they paid \$2200 for temporary help. How many general laborers did they hire? How many mechanics did they hire?

26. ***Children's Theater Prices*** A total of 330 tickets were sold for the opening performance of *Frog and Toad*. Children's admission tickets were \$8, and adults' tickets were \$13. The ticket receipts for the performance totaled \$3215. How many children's tickets were sold? How many adults' tickets were sold?

27. ***Baseball Equipment*** A baseball coach bought two hats, five shirts, and four pairs of pants for \$129. His assistant purchased one hat, one shirt, and two pairs of pants for \$42. The next week the coach bought two hats, three shirts, and one pair of pants for \$63. What was the cost of each item?

28. ***Math Exam Scores*** Jess, Chris, and Nick scored a combined total of 249 points on their last math exam. Jess's score was 20 points higher than Chris's score. Twice Nick's score was 6 more than the sum of Chris's and Jess's scores. What did each of them score on the exam?

29. ***Food Costs*** Four jars of jelly, three jars of peanut butter, and five jars of honey cost \$23.90. Two jars of jelly, two jars of peanut butter, and one jar of honey cost \$10.50. Three jars of jelly, four jars of peanut butter, and two jars of honey cost \$19.50. Find the cost for one jar of jelly, one jar of peanut butter, and one jar of honey.

30. ***Transportation Logistics*** The church youth group is planning a trip to Mount Washington. A total of 127 people need rides. The church has available buses that hold forty passengers, and several parents have volunteered station wagons that hold eight passengers or sedans that hold five passengers. The youth leader is planning to use nine vehicles to transport the people. One parent said that if they didn't use any buses, tripled the number of station wagons, and doubled the number of sedans, they would be able to transport 126 people. How many buses, station wagons, and sedans are they planning to use if they use nine vehicles?

Mixed Practice, Exercises 31–40

Solve by any appropriate method.

31. $x - y = 1$
$5x + y = 7$

32. $2x + 5y = 4$
$5x - 7y = -29$

33. $\frac{x}{2} - 3y = -6$
$\frac{4}{3}x + 2y = 4$

34. $\frac{x}{2} - y = -12$
$x + \frac{3}{4}y = 9$

35. $\begin{aligned} 7(x + 3) &= 2y + 25 \\ 3(x - 6) &= -2(y + 1) \end{aligned}$

36. $\begin{aligned} 0.3x - 0.4y &= 0.9 \\ 0.2x - 0.3y &= 0.4 \end{aligned}$

37. $\begin{aligned} 1.2x - \quad y &= 1.6 \\ x + 1.5y &= 6 \end{aligned}$

38. $\begin{aligned} x - \frac{y}{2} + \frac{1}{2}z &= -1 \\ 2x \qquad + \frac{5}{2}z &= -1 \\ \frac{3}{2}y + 2z &= 1 \end{aligned}$

39. $\begin{aligned} x - 4y + 4z &= -1 \\ 2x - \quad y + 5z &= -3 \\ x - 3y + \quad z &= 4 \end{aligned}$

40. $\begin{aligned} x - 2y + z &= -5 \\ 2x + \qquad z &= -10 \\ y - z &= 15 \end{aligned}$

Graph the solution of each of the following systems of linear inequalities.

41. $\begin{aligned} y &\geq -\frac{1}{2}x - 1 \\ -x + y &\leq 5 \end{aligned}$

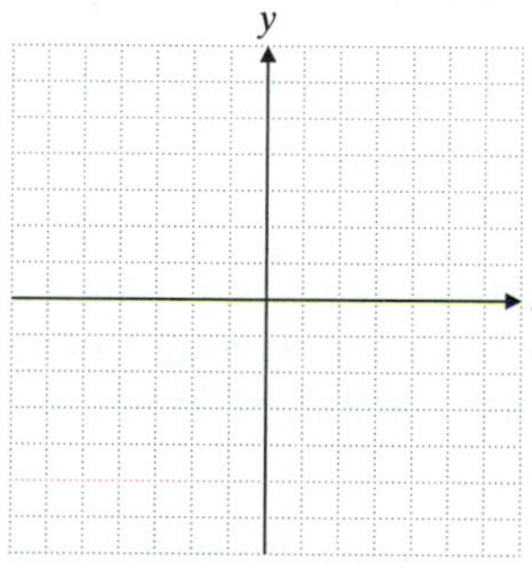

42. $\begin{aligned} -2x + 3y &< 6 \\ y &> -2 \end{aligned}$

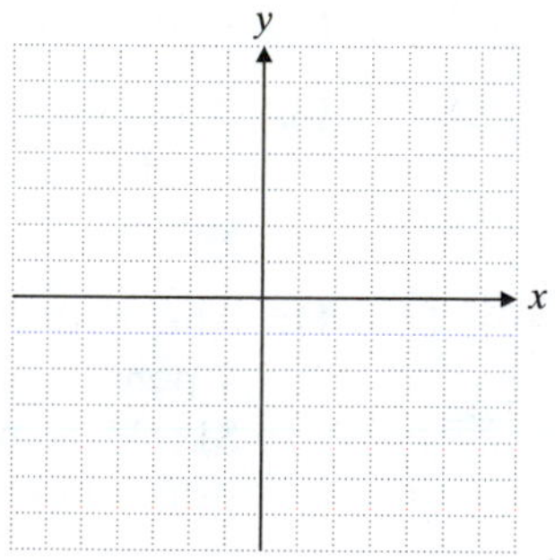

43. $\begin{aligned} x + y &> 1 \\ 2x - y &< 5 \end{aligned}$

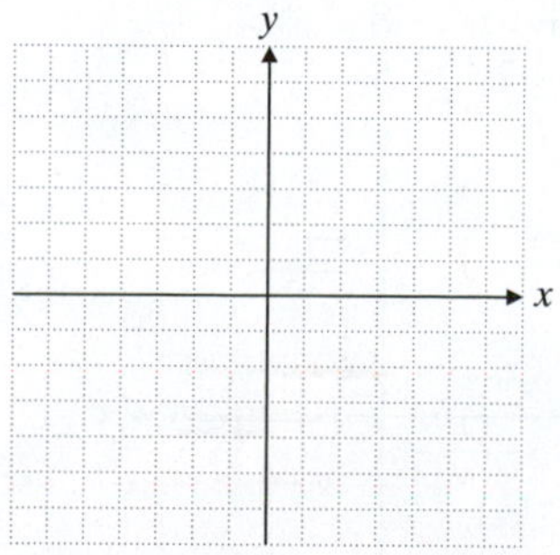

44. $\begin{aligned} x + y &\geq 4 \\ y &\leq x \\ x &\leq 6 \end{aligned}$

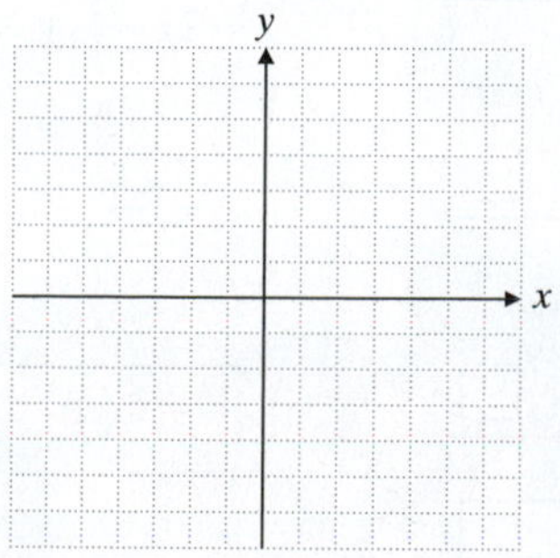

How Am I Doing? Chapter 4 Test

Chapter Test Prep Videos MATH COACH MyMathLab® YouTube™

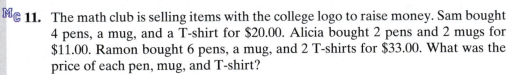

After you take this test read through the Math Coach on pages 247–248. Math Coach videos are available via MyMathLab and YouTube. Step-by-step test solutions in the Chapter Test Prep Videos are also available via MyMathLab and YouTube. (Search "TobeyInterAlg" and click on "Channels.")

Solve each system of equations. If there is no solution to the system, give a reason.

1. Solve using the substitution method.

$$\begin{aligned} x - y &= 3 \\ 2x - 3y &= -1 \end{aligned}$$

2. Solve using the addition method.

$$\begin{aligned} 3x + 2y &= 1 \\ 5x + 3y &= 3 \end{aligned}$$

In exercises 3–9, solve using any method.

3.
$$\begin{aligned} 5x - 3y &= 3 \\ 7x + y &= 25 \end{aligned}$$

4.
$$\begin{aligned} \frac{1}{4}a - \frac{3}{4}b &= -1 \\ \frac{1}{3}a + b &= \frac{5}{3} \end{aligned}$$

MC **5.**
$$\begin{aligned} \frac{1}{3}x + \frac{5}{6}y &= 2 \\ \frac{3}{5}x - y &= -\frac{7}{5} \end{aligned}$$

6.
$$\begin{aligned} 8x - 3y &= 5 \\ -16x + 6y &= 8 \end{aligned}$$

MC **7.**
$$\begin{aligned} 3x + 5y - 2z &= -5 \\ 2x + 3y - z &= -2 \\ 2x + 4y + 6z &= 18 \end{aligned}$$

8.
$$\begin{aligned} 3x + 2y \phantom{{}+3z} &= 0 \\ 2x - y + 3z &= 8 \\ 5x + 3y + z &= 4 \end{aligned}$$

9.
$$\begin{aligned} x + 5y + 4z &= -3 \\ x - y - 2z &= -3 \\ x + 2y + 3z &= -5 \end{aligned}$$

Use a system of linear equations to solve the following exercises.

10. A plane flew 1000 miles with a tailwind in 2 hours. The return trip against the wind took $2\frac{1}{2}$ hours. Find the speed of the wind and the speed of the plane in still air.

MC **11.** The math club is selling items with the college logo to raise money. Sam bought 4 pens, a mug, and a T-shirt for \$20.00. Alicia bought 2 pens and 2 mugs for \$11.00. Ramon bought 6 pens, a mug, and 2 T-shirts for \$33.00. What was the price of each pen, mug, and T-shirt?

12. Sue Miller had to move some supplies to Camp Cedarbrook for the summer camp program. She rented a Portland Rent-A-Truck in April for 5 days and drove 150 miles. She paid \$180 for the rental in April. Then in May she rented the same truck again for 7 days and drove 320 miles. She paid \$274 for the rental in May. How much does Portland Rent-A-Truck charge for a daily rental of the truck? How much does it charge per mile?

Solve the following systems of linear inequalities by graphing.

13.
$$\begin{aligned} x + 2y &\le 6 \\ -2x + y &\ge -2 \end{aligned}$$

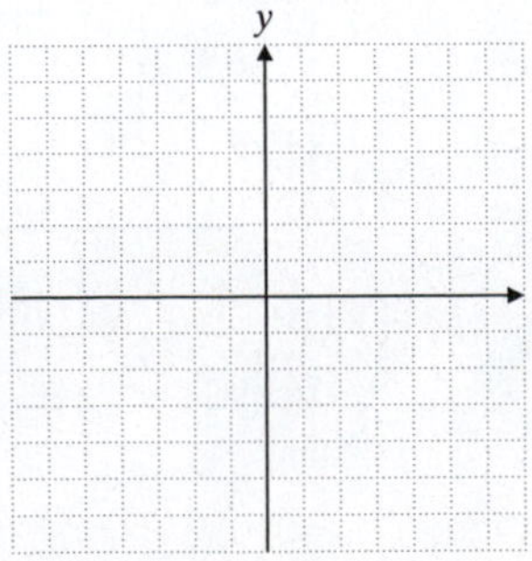

MC **14.**
$$\begin{aligned} 3x + y &> 8 \\ x - 2y &> 5 \end{aligned}$$

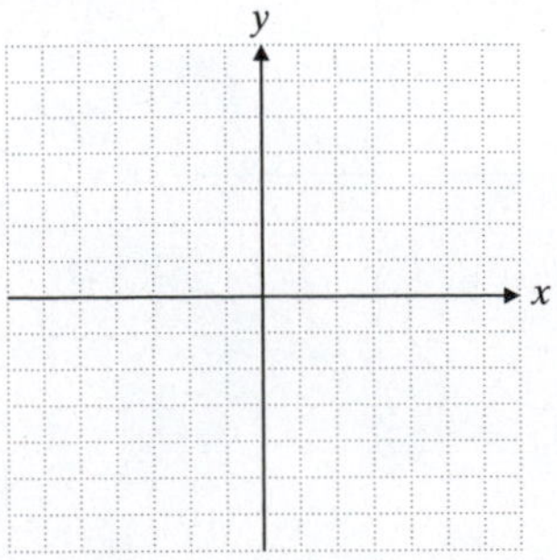

1. ______
2. ______
3. ______
4. ______
5. ______
6. ______
7. ______
8. ______
9. ______
10. ______
11. ______
12. ______
13. ______
14. ______

Total Correct: ______

MATH COACH

Mastering the skills you need to do well on the test.

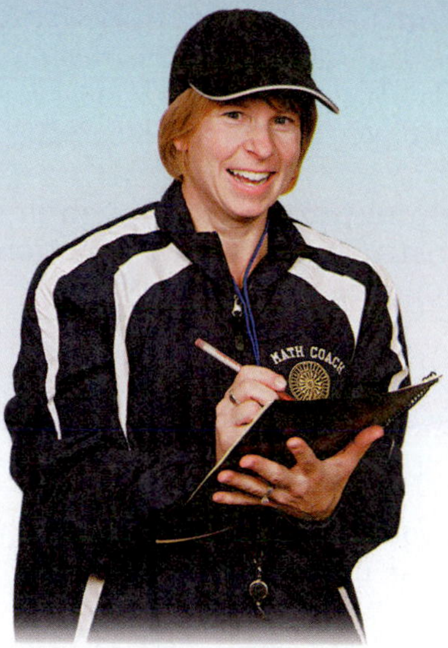

Students often make the same types of errors when they do the Chapter 4 Test. Here are some helpful hints to keep you from making these common errors on test problems.

Solving a System of Linear Equations with Fractions—Problem 5

Solve the system.

$$\begin{aligned} \frac{1}{3}x + \frac{5}{6}y &= 2 \\ \frac{3}{5}x - y &= -\frac{7}{5} \end{aligned}$$

Helpful Hint First clear fractions by multiplying the first equation by its LCD and multiplying the second equation by its LCD. Then choose the appropriate method to solve this system.

Did you multiply the first equation by the LCD, 6? Did you obtain the equation $2x + 5y = 12$?

Yes ____ No ____

If you answered No, consider why 6 is the LCD and perform the multiplication again. Remember to multiply each term of the equation by the LCD.

Did you multiply the second equation by 5? Did you obtain the equation $3x - 5y = -7$?

Yes ____ No ____

If you answered No, go back and perform those multiplications again. Be careful as you carry out the calculations.

Did you use the addition (elimination) method to complete the solution?

Yes ____ No ____

If you answered No, consider why this method might make the solution easiest to find.

If you answered Problem 5 incorrectly, go back and rework the problem using these suggestions.

Solving a System of Three Linear Equations in Three Variables—Problem 7

Solve the system.

(1) $3x + 5y - 2z = -5$

(2) $2x + 3y - z = -2$

(3) $2x + 4y + 6z = 18$

Helpful Hint Try to eliminate one of the variables from the original first and second equations. Then eliminate this same variable from the original second and third equations. The result should be a system of two equations in two variables.

Did you choose the variable z to eliminate?

Yes ____ No ____

If you answered No, consider why this might be the easiest variable to eliminate out of the three.

Did you multiply equation (2) by -2 and add the result to equation (1) to obtain the equation $-x - y = -1$?

Yes ____ No ____

If you answered No, stop and perform those operations using only original equations (2) and (1).

Next did you multiply the original equation (2) by 6 and add the result to equation (3) to obtain the equation $14x + 22y = 6$?

Yes ____ No ____

If you answered No, stop and perform those operations using only original equations (2) and (3).

Make sure that after performing these steps, your result is a system of two linear equations in two variables. Once you solve this system in two variables, remember to go back to one of the original equations and substitute in the resulting values for x and y to find the value of z.

Now go back and rework the problem using these suggestions.

Need help? Watch the MATH COACH videos in MyMathLab® or on YouTube™.

Solving an Application Using a System of Three Equations in Three Variables—Problem 11

The math club is selling items with the college logo to raise money. Sam bought 4 pens, a mug, and a T-shirt for \$20.00. Alicia bought 2 pens and 2 mugs for \$11.00. Ramon bought 6 pens, a mug, and 2 T-shirts for \$33.00. What was the price of each pen, mug, and T-shirt?

Helpful Hint Read through the word problem carefully. Specify what each variable represents. Then construct appropriate equations for Sam, Alicia, and Ramon. Solve the resulting system of three linear equations in three variables.

Did you realize that this problem requires a system of three equations in three variables?

Yes ____ No ____

Did you let x = the price of pens, y = the price of mugs, and z = the price of T shirts?

Yes ____ No ____

If you answered No to either question, go back and read the problem carefully again. Consider that you have three items with unknown prices. Consider that information is provided for three people—Sam, Alicia, and Ramon.

Did you write Sam's equation as $4x + y + z = 20$?

Yes ____ No ____

If you answered No, reread the second sentence in the word problem. Apply the information using the variables listed. Now translate this information into a linear equation in three variables.

Did you write Alicia's equation as $2x + 2y = 11$?

Yes ____ No ____

If you answered No, reread the third sentence in the word problem. Notice that we do not use the variable z because Alicia did not buy any T-shirts.

Now see if you can write a third equation for Ramon that totals to \$33. Then solve the resulting system.

If you answered Problem 11 incorrectly, go back and rework the problem using these suggestions.

Graphing a System of Linear Inequalities—Problem 14

Solve the system of linear inequalities by graphing.

$$3x + y > 8$$
$$x - 2y > 5$$

Helpful Hint Remember to use dashed lines when the inequality symbols are > or <. Take one inequality at a time and graph the border line. Then shade above or below each border line based on test points. The intersection of the two shaded regions is the solution to the system.

First graph the border line $3x + y = 8$. Did you see that this line passes through $(3, -1)$ and $(1, 5)$? Did you see that the line should be dashed?

Yes ____ No ____

If you answered No to any of these questions, stop and examine the first inequality again. Notice the inequality symbol used. Then carefully substitute $x = 3$ into the equation and solve for y. Substitute $x = 1$ into the equation and solve for y. Now you have two ordered pairs to use when graphing the border line.

Now you must decide to shade above or below the first border line. Did you substitute $(0, 0)$ into $3x + y > 8$ and decide to shade above the line?

Yes ____ No ____

If you answered No, remember that $3(0) + 0$ is not greater than 8. So we do not shade on the side of the line that contains $(0, 0)$. We must shade above the line.

Follow this procedure for the second border line. Remember that your solution is the intersection of the two shaded regions.

Now go back and rework the problem using these suggestions.

Need more help? Look for section examples marked with MC to review.

CHAPTER 5

Coral reefs around the world serve as the habitat for a variety of fish and other sea creatures. However, scientists are now studying coral reef diseases that are threatening the health of coral reefs worldwide. Scientists use polynomials to describe certain measures of health of a coral reef. In this chapter you will learn to perform a variety of mathematical operations using polynomials.

Polynomials

5.1 Introduction to Polynomials and Polynomial Functions: Adding, Subtracting, and Multiplying

Student Learning Objectives

After studying this section, you will be able to:

1. Identify types and degrees of polynomials.
2. Evaluate polynomial functions.
3. Add and subtract polynomials.
4. Multiply two binomials by FOIL.
5. Multiply two binomials $(a + b)(a - b)$.
6. Multiply two binomials $(a - b)^2$ or $(a + b)^2$.
7. Multiply polynomials with more than two terms.

1 Identifying Types and Degrees of Polynomials

A **polynomial** is an algebraic expression of one or more terms. A **term** is a number, a variable raised to a nonnegative integer power, or a product of numbers and variables raised to nonnegative integer powers. There must be no division by a variable. Three types of polynomials that you will see often are **monomials, binomials,** and **trinomials.**

1. A **monomial** has *one* term.
2. A **binomial** has *two* terms.
3. A **trinomial** has *three* terms.

Number of Variables	Monomials	Binomials	Trinomials	Other Polynomials
One Variable	$8x^3$	$2y^2 + 3y$	$5x^2 + 2x - 6$	$x^4 + 2x^3 - x^2 + 9$
Two Variables	$6x^2y$	$3x^2 - 5y^3$	$8x^2 + 5xy - 3y^2$	$x^3y + 5xy^2 + 3xy - 7y^5$
Three Variables	$12uvw^3$	$11a^2b + 5c^2$	$4a^2b^4 + 7c^4 - 2a^5$	$3c^2 + 4c - 8d + 2e - e^2$

The following are *not* polynomials.

$$2x^{-3} + 5x^2 - 3 \qquad 4ab^{\frac{1}{2}} \qquad \frac{2}{x} + \frac{3}{y}$$

TO THINK ABOUT: Understanding Polynomials Give a reason each expression above is not a polynomial.

Polynomials are also classified by degree. The **degree of a term** is the sum of the exponents of its variables. The **degree of a polynomial** is the degree of the highest-degree term in the polynomial. If the polynomial has no variable, then it has degree zero.

EXAMPLE 1 Name the type of polynomial and give its degree.

(a) $5x^6 + 3x^2 + 2$ **(b)** $7x + 6$
(c) $5x^2y + 3xy^3 + 6xy$ **(d)** $7x^4y^5$

Solution

(a) This is a trinomial of degree 6.
(b) This is a binomial of degree 1.
(c) This is a trinomial of degree 4.
(d) This is a monomial of degree 9.

Remember that if a variable has no exponent, the exponent is understood to be 1.

Student Practice 1 State the type of polynomial and give its degree.

(a) $3x^5 - 6x^4 + x^2$ **(b)** $5x^2 + 2$
(c) $3ab + 5a^2b^2 - 6a^4b$ **(d)** $16x^4y^6$

***NOTE TO STUDENT:** Fully worked-out solutions to all of the Student Practice problems can be found at the back of the text starting at page SP-1.*

Some polynomials contain only one variable. A **polynomial in** x is an expression of the form

$$a_n x^n + a_{n-1}x^{n-1} + a_{n-2}x^{n-2} + \cdots + a_0$$

where n is a nonnegative integer and the constants $a_n, a_{n-1}, a_{n-2}, \ldots, a_0$ are real numbers. We usually write polynomials in **descending order** of the variable. That is, the exponents on the variables decrease from left to right. For example, the polynomial $4x^5 - 2x^3 + 6x^2 + 5x - 8$ is written in descending order.

② Evaluating Polynomial Functions

A **polynomial function** is a function that is defined by a polynomial.
For example,

$$p(x) = 5x^2 - 3x + 6 \quad \text{and} \quad p(x) = 2x^5 - 3x^3 + 8x - 15$$

are both polynomial functions.

To evaluate a polynomial function, we use the skills developed in Section 3.5.

EXAMPLE 2 Evaluate the polynomial function $p(x) = -3x^3 + 2x^2 - 5x + 6$ to find **(a)** $p(-3)$ and **(b)** $p(6)$.

Solution

(a) $p(-3) = -3(-3)^3 + 2(-3)^2 - 5(-3) + 6$
$= -3(-27) + 2(9) - 5(-3) + 6$
$= 81 + 18 + 15 + 6$
$= 120$

(b) $p(6) = -3(6)^3 + 2(6)^2 - 5(6) + 6$
$= -3(216) + 2(36) - 5(6) + 6$
$= -648 + 72 - 30 + 6$
$= -600$

Student Practice 2 Evaluate the polynomial function $p(x) = 2x^4 - 3x^3 + 6x - 8$ to find

(a) $p(-2)$ **(b)** $p(5)$

③ Adding and Subtracting Polynomials

We can add and subtract polynomials by combining like terms as we learned in Section 1.5.

EXAMPLE 3 Add. $(5x^2 - 3x - 8) + (-3x^2 - 7x + 9)$

Solution

$5x^2 - 3x - 8 - 3x^2 - 7x + 9$ We remove the parentheses and combine like terms.
$= 2x^2 - 10x + 1$

Student Practice 3 Add. $(-7x^2 + 5x - 9) + (2x^2 - 3x + 5)$

To subtract real numbers, we add the opposite of the second number to the first. Thus, for real numbers a and b, we have $a - (b) = a + (-b)$ Similarly for polynomials, to subtract polynomials we add the opposite of the second polynomial to the first.

EXAMPLE 4 Subtract. $(-5x^2 - 19x + 15) - (3x^2 - 4x + 13)$

Solution

$(-5x^2 - 19x + 15) + (-3x^2 + 4x - 13)$ We add the opposite of the second polynomial to the first polynomial.

$= -8x^2 - 15x + 2$

Student Practice 4 Subtract. $(2x^2 - 14x + 9) - (-3x^2 + 10x + 7)$

④ Multiplying Two Binomials by FOIL

The FOIL method for multiplying two binomials has been developed to help you keep track of the order of the terms to be multiplied. The acronym FOIL means the following:

F	First
O	Outer
I	Inner
L	Last

That is, we multiply the first terms, then the outer terms, then the inner terms, and finally, the last terms.

EXAMPLE 5 Multiply. $(5x + 2)(7x - 3)$

Solution First Last First + Outer + Inner + Last

$(5x + 2)(7x - 3) = 35x^2 - 15x + 14x - 6$

$= 35x^2 - x - 6$

Inner

Outer

Student Practice 5 Multiply. $(7x + 3)(2x - 5)$

EXAMPLE 6 Multiply. $(7x^2 - 8)(2x - 3)$

Solution First Last

$(7x^2 - 8)(2x - 3) = 14x^3 - 21x^2 - 16x + 24$

Inner

Outer

Note that in this case we were not able to combine the inner and outer products.

Student Practice 6 Multiply. $(3x^2 - 2)(5x - 4)$

⑤ Multiplying $(a + b)(a - b)$

Products of the form $(a + b)(a - b)$ deserve special attention.

$$(a + b)(a - b) = a^2 - ab + ab - b^2 = a^2 - b^2$$

Notice that the middle terms, $-ab$ and $+ab$, equal zero when combined. The product is the difference of two squares, $a^2 - b^2$ This is always true when you multiply binomials of the form $(a + b)(a - b)$ You should memorize the following formula.

$$(a + b)(a - b) = a^2 - b^2$$

EXAMPLE 7 Multiply. $(2a - 9b)(2a + 9b)$

Solution $(2a - 9b)(2a + 9b) = (2a)^2 - (9b)^2 = 4a^2 - 81b^2$

Of course, we could have used the FOIL method, but recognizing the special product allowed us to save time.

Student Practice 7 Multiply. $(7x - 2y)(7x + 2y)$

⑥ Multiplying $(a - b)^2$ or $(a + b)^2$

Another special product is the square of a binomial.

$$(a - b)^2 = (a - b)(a - b) = a^2 - ab - ab + b^2 = a^2 - 2ab + b^2$$

Once you understand the pattern, you should memorize these two formulas.

$$(a - b)^2 = a^2 - 2ab + b^2 \qquad (a + b)^2 = a^2 + 2ab + b^2$$

This procedure is also called **expanding a binomial.** *Note*: $(a - b)^2 \neq a^2 - b^2$ and $(a + b)^2 \neq a^2 + b^2$

EXAMPLE 8 Multiply.

(a) $(5a - 8b)^2$ **(b)** $(3u + 11v^2)^2$

Solution

(a) $(5a - 8b)^2 = (5a)^2 - 2(5a)(8b) + (8b)^2 = 25a^2 - 80ab + 64b^2$

(b) Here $a = 3u$ and $b = 11v^2$.

$$\begin{aligned}(3u + 11v^2)^2 &= (3u)^2 + 2(3u)(11v^2) + (11v^2)^2 \\ &= 9u^2 + 66uv^2 + 121v^4\end{aligned}$$

Student Practice 8 Multiply.

(a) $(4u + 5v)^2$ **(b)** $(7x^2 - 3y^2)^2$

⑦ Multiplying Polynomials with More Than Two Terms

The distributive property is the basis for multiplying polynomials. Recall that

$$a(b + c) = ab + ac.$$

We can use this property to multiply a polynomial by a monomial.

$$\begin{aligned}3xy(5x^3 + 2x^2 - 4x + 1) &= 3xy(5x^3) + 3xy(2x^2) - 3xy(4x) + 3xy(1) \\ &= 15x^4y + 6x^3y - 12x^2y + 3xy\end{aligned}$$

A similar procedure can be used to multiply two binomials.

$$(3x + 5)(6x + 7) = (3x + 5)6x + (3x + 5)7$$ We use the distributive property again.

$$= (3x)(6x) + (5)(6x) + (3x)(7) + (5)(7)$$
$$= 18x^2 + 30x + 21x + 35$$
$$= 18x^2 + 51x + 35$$

The multiplication of a binomial and a trinomial is more involved. One way to multiply two polynomials is to write them vertically, as we do when multiplying two- and three-digit numbers. We then multiply them in the usual way.

EXAMPLE 9 Multiply. $(4x^2 - 2x + 3)(-3x + 4)$

Solution

$$\begin{array}{rrrrl} & 4x^2 & - 2x & + 3 & \\ & & -3x & + 4 & \\ \hline & 16x^2 & - 8x & + 12 & \text{Multiply } (4x^2 - 2x + 3)(+4). \\ -12x^3 & + 6x^2 & - 9x & & \text{Multiply } (4x^2 - 2x + 3)(-3x). \\ \hline -12x^3 & + 22x^2 & - 17x & + 12 & \text{Add the two products.} \end{array}$$

Student Practice 9 Multiply. $(2x^2 - 3x + 1)(x^2 - 5x)$

Another way to multiply polynomials is to multiply horizontally. We redo Example 9 in the following example.

EXAMPLE 10 Multiply horizontally. $(4x^2 - 2x + 3)(-3x + 4)$

Solution By the distributive law, we have the following:

$$(4x^2 - 2x + 3)(-3x + 4) = (4x^2 - 2x + 3)(-3x) + (4x^2 - 2x + 3)(4)$$
$$= -12x^3 + 6x^2 - 9x + 16x^2 - 8x + 12$$
$$= -12x^3 + 22x^2 - 17x + 12.$$

In actual practice you will find that you can do some of these steps mentally.

Student Practice 10 Multiply horizontally. $(2x^2 - 3x + 1)(x^2 - 5x)$

STEPS TO SUCCESS Look Ahead to See What Is Coming

You will find that learning new material is much easier if you know what is coming. Take a few minutes at the end of your study time to glance over the next section of the book. If you quickly look over the topics and ideas in this new section, it will help you get your bearings when the instructor presents new material. Students find that when they preview new material it enables them to see what is coming. It helps them to be able to grasp new ideas much more quickly.

Making it personal: Do this right now: Look ahead to the next section of the book. Glance over the ideas and concepts. Write down a couple of facts about the next section. ▼

5.1 Exercises

MyMathLab® Watch the videos in MyMathLab Download the MyDashBoard App

Name the type of polynomial and give its degree.

1. $2x^2 - 5x + 3$

2. $7x^3 + 6x^2 - 2$

3. $-3.2a^4bc^3$

4. $26.8a^3bc^2$

5. $\frac{3}{5}m^3n - \frac{2}{5}mn$

6. $\frac{2}{7}m^2n^2 + \frac{1}{2}mn^2$

For the polynomial function $p(x) = 5x^2 - 9x - 12$, evaluate the following:

7. $p(3)$

8. $p(-4)$

For the polynomial function $g(x) = -4x^3 - x^2 + 5x - 1$, evaluate the following:

9. $g(-2)$

10. $g(1)$

For the polynomial function $h(x) = 2x^4 - x^3 + 2x^2 - 4x - 3$, evaluate the following:

11. $h(-1)$

12. $h(3)$

Add or subtract the following polynomials as indicated.

13. $(x^2 + 3x - 2) + (-2x^2 - 5x + 1) + (x^2 - x - 5)$

14. $(3x^2 + 2x - 4) + (-x^2 - x + 1) + (-3x^2 + 5x + 7)$

15. $(7m^3 + 4m^2 - m + 2.5) - (-3m^3 + 5m + 3.8)$

16. $(5x^3 - x^2 + 6x - 3.8) - (x^3 + 2x^2 - 4x - 10.4)$

17. $(5a^3 - 2a^2 - 6a + 8) + (5a + 6) - (-a^2 - a + 2)$

18. $(a^5 + 3a^2) + (2a^4 - a^3 - 3a^2 + 2) - (a^4 + 3a^3 - 5)$

19. $\left(\frac{2}{3}x^2 + 5x\right) + \left(\frac{1}{2}x^2 - \frac{1}{3}x\right)$

20. $\left(\frac{5}{6}x^2 - \frac{1}{4}x\right) + \left(\frac{1}{6}x^2 + 3x\right)$

21. $(2.3x^3 - 5.6x^2 - 2) - (5.5x^3 - 7.4x^2 + 2)$

22. $(5.9x^3 + 3.4x^2 - 7) - (2.9x^3 - 9.6x^2 + 3)$

Multiply.

23. $(5x + 8)(2x + 9)$

24. $(6x + 7)(3x + 2)$

25. $(5w + 2d)(3a - 4b)$

26. $(7a + 8b)(5d - 8w)$

27. $(-6x + y)(2x - 5y)$

28. $(5a + y)(-2x - y)$

29. $(7r - s^2)(-4a - 11s^2)$

30. $(-3r - 2s^2)(5r - 6s^2)$

Multiply mentally. See Examples 7 and 8.

31. $(5x - 8y)(5x + 8y)$

32. $(2a - 7b)(2a + 7b)$

33. $(5a - 2b)^2$

34. $(4a + 3b)^2$

35. $(7m - 1)^2$

36. $(5r + 3)^2$

37. $(6 + 5x^2)(6 - 5x^2)$

38. $(8 - 3x^3)(8 + 3x^3)$

39. $(3m^3 + 1)^2$

40. $(4r^3 - 5)^2$

Multiply.

41. $2x(3x^2 - 5x + 1)$

42. $-5x(x^2 - 6x - 2)$

43. $-\frac{1}{2}ab(4a - 5b - 10)$

44. $\frac{3}{4}ab^2(a - 8b + 6)$

45. $(2x - 3)(x^2 - x + 1)$

46. $(4x + 1)(2x^2 + x + 1)$

47. $(3x^2 - 2xy - 6y^2)(2x - y)$

48. $(5x^2 + 3xy - 7y^2)(3x - 2y)$

49. $(3a^3 + 4a^2 - a - 1)(a - 5)$

50. $(4b^3 - b^2 + 2b - 6)(3b + 1)$

First multiply any two binomials in the exercise; then multiply the result by the third binomial.

51. $(x + 2)(x - 3)(2x - 5)$

52. $(x - 6)(x + 2)(3x + 2)$

53. $(a - 2)(5 + a)(3 - 2a)$

54. $(4 + 3a)(1 - 2a)(3 - a)$

Applications

▲ **55.** ***Geometry*** Ace Landscape Design makes large raised boxes for plantings. The height of the boxes they build is $(2x + 5)$ feet, and the area of the rectangular base measures $(2x^2 + x + 4)$ feet^2. If the box is filled with soil, what is the volume of soil needed?

▲ **56.** ***Geometry*** At Wilkinson Auditorium, $(3n + 8)$ rows of chairs can be arranged for large ceremonies. Each row can have $(3n^2 + 2n + 3)$ chairs. Find the number of chairs that can be arranged.

Cumulative Review

57. **[2.6.3]** Solve. $\frac{1}{2}x + 4 \leq \frac{2}{3}(x - 3) + 1$

58. **[2.1.1]** Solve. $2(x + 1) - 3 = 4 - (x + 5)$

Quick Quiz 5.1

1. Combine. $(7x - 4) + (5x - 6) - (3x^2 - 9x)$

2. Multiply. $(5x - 2y^2)^2$

3. Multiply. $(2x - 3)(4x^2 + 2x - 1)$

4. **Concept Check** Explain how you would multiply the following. $(3x + 1)(x - 4)(2x - 3)$

5.2 Dividing Polynomials

① Dividing a Polynomial by a Monomial

The easiest type of polynomial division occurs when the divisor is a monomial. We perform this type of division just as if we were dividing numbers. First we write the indicated division as the sum of separate fractions, and then we reduce each fraction (if possible).

EXAMPLE 1 Divide. $(15x^3 - 10x^2 + 40x) \div 5x$

Solution

$$\frac{15x^3 - 10x^2 + 40x}{5x} = \frac{15x^3}{5x} - \frac{10x^2}{5x} + \frac{40x}{5x}$$
$$= 3x^2 - 2x + 8$$

Student Practice 1 Divide. $(-16x^4 + 16x^3 + 8x^2 + 64x) \div 8x$

② Dividing a Polynomial by a Polynomial

When we divide polynomials by binomials or trinomials, we perform long division. This is much like the long division method for dividing numbers. Both polynomials must be written in descending order.

First we write the problem in the form of long division.

$$2x + 3\overline{)6x^2 + 17x + 12}$$

The divisor is $2x + 3$; the dividend is $6x^2 + 17x + 12$. Now we divide the first term of the dividend ($6x^2$) by the first term of the divisor ($2x$).

$$\begin{array}{r} 3x \\ 2x + 3\overline{)6x^2 + 17x + 12} \end{array} \qquad 6x^2 \div 2x = 3x$$

Now we multiply $3x$ (the first term of the quotient) by the divisor $2x + 3$.

$$\begin{array}{r} 3x \\ 2x + 3\overline{)6x^2 + 17x + 12} \\ 6x^2 + 9x \end{array} \qquad \leftarrow \text{The product } 3x(2x + 3).$$

Next, just as in long division with numbers, we subtract and bring down the next term.

$$\begin{array}{r} 3x \\ 2x + 3\overline{)6x^2 + 17x + 12} \\ \underline{6x^2 + 9x} \\ 8x + 12 \end{array}$$

Subtract $6x^2 + 9x$ from $6x^2 + 17x$.

Bring down the next term.

Now we divide the first term of this new dividend ($8x$) by the first term of the divisor ($2x$).

$$\begin{array}{r} 3x + 4 \\ 2x + 3\overline{)6x^2 + 17x + 12} \\ \underline{6x^2 + 9x} \\ 8x + 12 \\ \underline{8x + 12} \\ 0 \end{array}$$

$8x \div 2x = 4$

The product $4(2x + 3)$.

Note that we then multiplied $(2x + 3)(4)$ and subtracted, just as we did before. We continue this process until the remainder is zero. Thus, we find that

$$\frac{6x^2 + 17x + 12}{2x + 3} = 3x + 4.$$

The remainder is not always zero, as we shall see in Example 2.

Student Learning Objectives

After studying this section, you will be able to:

① **Divide a polynomial by a monomial.**

② **Divide a polynomial by a polynomial.**

***NOTE TO STUDENT:** Fully worked-out solutions to all of the Student Practice problems can be found at the back of the text starting at page SP-1.*

Graphing Calculator

Verifying Answers When Dividing Polynomials

One way to verify that the division was performed correctly is to graph

$$y_1 = 3x + 4$$

and $y_2 = \dfrac{6x^2 + 17x + 12}{2x + 3}$.

If the graphs appear to coincide, then we have an independent verification that

$$\frac{6x^2 + 17x + 12}{2x + 3} = 3x + 4.$$

DIVIDING A POLYNOMIAL BY A BINOMIAL OR TRINOMIAL

1. Write the division problem in long division form. Write both polynomials in descending order; write missing terms with a coefficient of zero.
2. Divide the *first* term of the divisor into the first term of the dividend. The result is the first term of the quotient.
3. Multiply the first term of the quotient by *every* term in the divisor.
4. Write the product under the dividend (align like terms) and subtract.
5. Treat this difference as a new dividend after bringing down the next term of the original dividend. Repeat steps 2 through 4. Continue until the remainder is zero or a polynomial of lower degree than the *first term* of the divisor.
6. If there is a remainder, write it as the numerator of a fraction with the divisor as the denominator. Add this fraction to the quotient.

EXAMPLE 2 Divide. $(6x^3 + 7x^2 + 3) \div (3x - 1)$

Solution There is no x-term in the dividend, so we write $0x$.

$$\begin{array}{r} 2x^2 + 3x + 1 \\ 3x - 1\overline{)6x^3 + 7x^2 + 0x + 3} \\ \underline{6x^3 - 2x^2} \\ 9x^2 + 0x \\ \underline{9x^2 - 3x} \\ 3x + 3 \\ \underline{3x - 1} \\ 4 \end{array}$$

Note that we calculate $7x^2 - (-2x^2)$ to obtain $9x^2$.

Note that we calculate $0x - (-3x)$ to obtain $3x$.

The quotient is $2x^2 + 3x + 1$ with a remainder of 4. We may write this as

$$2x^2 + 3x + 1 + \frac{4}{3x - 1}$$

Check.

$$(3x - 1)(2x^2 + 3x + 1) + 4 \stackrel{?}{=} 6x^3 + 7x^2 + 3$$
$$6x^3 + 7x^2 - 0x - 1 + 4 \stackrel{?}{=} 6x^3 + 7x^2 + 3$$
$$6x^3 + 7x^2 + 3 = 6x^3 + 7x^2 + 3 \quad \checkmark$$

Student Practice 2 Divide.
$(8x^3 - 10x^2 - 9x + 14) \div (4x - 3)$

EXAMPLE 3 Divide. $\dfrac{64x^3 - 125}{4x - 5}$

Solution This fraction is equivalent to the problem $(64x^3 - 125) \div (4x - 5)$.

Note that two terms are missing in the dividend. We write them with zero coefficients.

$$\begin{array}{r} 16x^2 + 20x + 25 \\ 4x - 5\overline{)64x^3 + 0x^2 + 0x - 125} \\ \underline{64x^3 - 80x^2} \\ 80x^2 + 0x \\ \underline{80x^2 - 100x} \\ 100x - 125 \\ \underline{100x - 125} \\ 0 \end{array}$$

Note that $0x^2 - (-80x^2) = 80x^2$.

Note that $0x - (-100x) = 100x$.

The quotient is $16x^2 + 20x + 25$.

Check. Verify that $(4x - 5)(16x^2 + 20x + 25) = 64x^3 - 125$.

Student Practice 3 Divide. $(8x^3 + 27) \div (2x + 3)$

EXAMPLE 4 Divide. $(7x^3 - 10x - 7x^2 + 2x^4 + 8) \div (2x^2 + x - 2)$

Solution Arrange the dividend in descending order before dividing.

$2x^4 \div 2x^2 = x^2$

$$
\begin{array}{r}
x^2 + 3x - 4 \\
2x^2 + x - 2\overline{)2x^4 + 7x^3 - 7x^2 - 10x + 8} \\
\underline{2x^4 + x^3 - 2x^2} \qquad\qquad \\
6x^3 - 5x^2 - 10x \qquad \\
\underline{6x^3 + 3x^2 - 6x} \qquad \\
-8x^2 - 4x + 8 \\
\underline{-8x^2 - 4x + 8} \\
0
\end{array}
$$

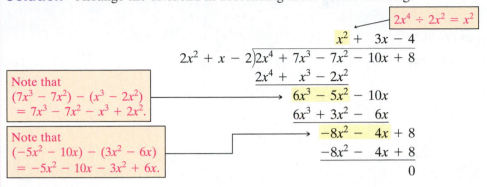

Note that $(7x^3 - 7x^2) - (x^3 - 2x^2) = 7x^3 - 7x^2 - x^3 + 2x^2$.

Note that $(-5x^2 - 10x) - (3x^2 - 6x) = -5x^2 - 10x - 3x^2 + 6x$.

The quotient is $x^2 + 3x - 4$.

Check. Verify that

$$(2x^2 + x - 2)(x^2 + 3x - 4) = 2x^4 + 7x^3 - 7x^2 - 10x + 8.$$

Student Practice 4 Divide. $(2x^4 + 3x^3 + 7x - 3 - x^2) \div (2x^2 - x + 3)$

STEPS TO SUCCESS Feeling Good While You Take the Test

Allow yourself plenty of time. Arrive a little early to collect your thoughts and ready yourself. This will help you feel more relaxed.

Successful students have found that the following tips have helped them. After you receive the test paper:

1. Take two or three moderately deep breaths. Inhale; then exhale slowly. You will feel your entire body begin to relax.
2. Immediately write down on the back of the exam any formulas or ideas that you need to remember.
3. Look over the entire test quickly. Notice how many points each exercise is worth. Spend more time on items of greater worth.
4. Work the exercises and answer the questions that are easiest for you first. Then go back and do the more difficult ones.
5. Don't get bogged down on one exercise for too long. Leave the tough exercise and go back to it when you have time later.
6. Check your work. This will help you catch minor errors.
7. Stay calm if others leave before you do. You are entitled to use the full amount of allotted time. The successful student is the one who takes advantage of every minute in order to do well on the test.

Making it personal: Select two or three of the suggestions given that you know are the ones you really need to follow in order to improve your test performance. Write down what you specifically need to do when you walk in to take your next test. ▼

5.2 Exercises

MyMathLab®

Watch the videos in MyMathLab

Download the MyDashBoard App

Divide.

1. $(24x^2 - 8x - 44) \div 4$

2. $(16x^2 - 56x + 40) \div 8$

3. $(27x^4 - 9x^3 + 63x^2) \div (-9x)$

4. $(22x^4 + 33x^3 - 121x^2) \div (-11x)$

5. $\dfrac{8b^4 - 6b^3 - b^2}{2b^2}$

6. $\dfrac{12w^4 - 6w^3 + w^2}{3w^2}$

7. $\dfrac{9a^3b^3 - 15a^3b^2 + 3a^2b}{3a^2b}$

8. $\dfrac{12m^2n^4 - 8mn^3 + 32mn^2}{4mn^2}$

Divide. Check your answers for exercises 11–16.

9. $(5x^2 - 17x + 6) \div (x - 3)$

10. $(6x^2 - 31x + 5) \div (x - 5)$

11. $(15x^2 + 23x + 4) \div (5x + 1)$

12. $(12x^2 + 11x + 2) \div (4x + 1)$

13. $(20x^2 - 17x + 3) \div (4x - 1)$

14. $(21x^2 - 2x - 8) \div (3x - 2)$

15. $(x^3 - x^2 + 11x - 1) \div (x + 1)$

16. $(x^3 + 2x^2 - 3x + 2) \div (x + 1)$

17. $(2x^3 - x^2 - 7) \div (x - 2)$

18. $(4x^3 - 6x - 11) \div (2x - 4)$

19. $\dfrac{4x^3 - 6x^2 - 3}{2x + 1}$

20. $\dfrac{9x^3 + 11x + 2}{3x + 1}$

21. $\dfrac{2x^4 - 3x^3 + 8x^2 - 18}{2x - 3}$

22. $\dfrac{18x^4 - 5x^2 - 11x + 8}{3x - 1}$

23. $\dfrac{6t^4 - 5t^3 - 8t^2 + 16t - 8}{3t^2 + 2t - 4}$

24. $\dfrac{2t^4 + 5t^3 - 11t^2 - 20t + 12}{t^2 + t - 6}$

Applications

▲ 25. ***Space Station*** For the space station an engineer has designed a new rectangular solar panel that has an area of $(18x^3 - 21x^2 + 11x - 2)$ square meters. The length of the solar panel is $(6x^2 - 5x + 2)$ meters. What is the width of the solar panel?

▲ 26. ***Space Station*** For the space station an engineer has designed a new rectangular solar panel that has an area of $(8x^3 + 22x^2 - 29x + 6)$ square meters. The length of the solar panel is $(2x^2 + 7x - 2)$ meters. What is the width of the solar panel?

Cumulative Review

27. **[3.2.1]** Find the slope of the line passing through $\left(-\frac{1}{3}, 0\right)$ and $\left(\frac{1}{2}, -1\right)$.

28. **[3.3.3]** Find the slope of a line that is perpendicular to $3y - 2x = 7$.

Solve for x.

29. **[2.2.1]** $5(x - 2) + 3y = 3x - (y - 1)$

30. **[2.2.1]** $\dfrac{2x + 4}{3} - \dfrac{y}{2} = x - 2y + 1$

Quick Quiz 5.2 *Divide.*

1. $(16x^3 - 20x^2 - 8x) \div (-4x)$

2. $(x^3 - 6x^2 + 7x - 2) \div (x - 1)$

3. $(2x^4 - x^3 - 12x^2 - 17x - 12) \div (2x + 3)$

4. **Concept Check** Explain how you would check your answer if you divided $(x^2 - 9x - 5) \div (x - 3)$ and obtained a result of $x - 6 - \dfrac{23}{x - 3}$.

5.3 Synthetic Division

Student Learning Objective

After studying this section, you will be able to:

① **Use synthetic division to divide polynomials.**

① Using Synthetic Division to Divide Polynomials

When dividing a polynomial by a binomial of the form $x + b$ you may find a procedure known as **synthetic division** quite efficient. Notice the following division exercises. The right-hand problem is the same as the left, but without the variables.

$$
\begin{array}{r r}
 & 3x^2 - 2x + 2 \\
x + 3\,\big) & 3x^3 + 7x^2 - 4x + 3 \\
 & 3x^3 + 9x^2 \qquad\qquad \\
\hline
 & -2x^2 - 4x \qquad \\
 & -2x^2 - 6x \qquad \\
\hline
 & 2x + 3 \\
 & 2x + 6 \\
\hline
 & -3
\end{array}
\qquad\qquad
\begin{array}{r r}
 & 3 \;\; -2 \;\; 2 \\
1 + 3\,\big) & 3 \;\; 7 \;\; -4 \;\; 3 \\
 & 3 \;\; 9 \qquad\quad \\
\hline
 & -2 \;\; -4 \quad \\
 & -2 \;\; -6 \quad \\
\hline
 & 2 \;\; 3 \\
 & 2 \;\; 6 \\
\hline
 & -3
\end{array}
$$

Eliminating the variables makes synthetic division efficient, and we can make the procedure simpler yet. Note that the highlighted numbers (3, -2, and 2) appear twice in the previous example, once in the quotient and again in the subtraction. Synthetic division makes it possible to write each number only once. Also, in synthetic division we change the subtraction that division otherwise requires to addition. We do this by dropping the 1, which is the coefficient of x in the divisor, and taking the opposite of the second number in the divisor. In our first example, this means dropping the 1 and changing 3 to -3. The following steps detail synthetic division.

Step 1

Divisor, without the 1 and with the opposite sign

$$
\begin{array}{r|rrrr}
-3 & 3 & 7 & -4 & 3 \\
 & & & & \\
\hline
 & 3 & & &
\end{array}
$$

Dividend, without variables

Step 2

$$
\begin{array}{r|rrrr}
-3 & 3 & 7 & -4 & 3 \\
 & & -9 & & \\
\hline
 & 3 & -2 & &
\end{array}
$$

Multiply $(-3)(3) = -9$ and add $7 + (-9) = -2$.

Step 3

$$
\begin{array}{r|rrrr}
-3 & 3 & 7 & -4 & 3 \\
 & & -9 & 6 & \\
\hline
 & 3 & -2 & 2 &
\end{array}
$$

Multiply $(-3)(-2) = 6$ and add $-4 + 6 = 2$.

Step 4

$$
\begin{array}{r|rrrr}
-3 & 3 & 7 & -4 & 3 \\
 & & -9 & 6 & -6 \\
\hline
 & 3 & -2 & 2 & \lfloor -3
\end{array}
$$

Multiply $(-3)(2) = -6$ and add $3 + (-6) = -3$.

$3x^2 - 2x + 2 +$ remainder of -3

Replace the variables in descending order. The degree of the quotient should be one less than the degree of the dividend.

The result is read from the bottom row. Our answer is $3x^2 - 2x + 2 + \dfrac{-3}{x + 3}$.

EXAMPLE 1 Divide using synthetic division. $(3x^3 - x^2 + 4x + 8) \div (x + 2)$

Solution

$\underline{-2}\rfloor$	3	−1	4	8
		−6	14	−36
	3	−7	18	$\lfloor -28$

The quotient is $3x^2 - 7x + 18 + \dfrac{-28}{x + 2}$.

Student Practice 1 Divide using synthetic division.
$(x^3 - 3x^2 + 4x - 5) \div (x + 3)$

***NOTE TO STUDENT:* Fully worked-out solutions to all of the Student Practice problems can be found at the back of the text starting at page SP-1.**

When a term is missing in the sequence of descending powers of x, we use a zero to indicate the coefficient of that term.

EXAMPLE 2 Divide using synthetic division.
$(3x^4 - 21x^3 + 31x^2 - 25) \div (x - 5)$

Solution Since $b = -5$, we use 5 as the divisor for synthetic division.

$\underline{5}\rfloor$	3	−21	31	0	−25
		15	−30	5	25
	3	−6	1	5	$\lfloor 0$

Note that the remainder is zero.

The quotient is $3x^3 - 6x^2 + x + 5$.

Student Practice 2 Divide using synthetic division.
$(2x^4 - x^2 - 39x - 36) \div (x - 3)$

EXAMPLE 3 Divide using synthetic division.
$(3x^4 - 4x^3 + 8x^2 - 5x - 5) \div (x - 2)$

Solution

$\underline{2}\rfloor$	3	−4	8	−5	−5
		6	4	24	38
	3	2	12	19	$\lfloor 33$

The quotient is $3x^3 + 2x^2 + 12x + 19 + \dfrac{33}{x - 2}$.

Student Practice 3 Divide using synthetic division.
$(2x^4 - 9x^3 + 5x^2 + 13x - 5) \div (x - 3)$

5.3 Exercises

MyMathLab®

Watch the videos in MyMathLab

Download the MyDashBoard App

Divide using synthetic division.

1. $(2x^2 - 11x - 8) \div (x - 6)$

2. $(2x^2 - 15x - 23) \div (x - 9)$

3. $(3x^3 + x^2 - x + 4) \div (x + 1)$

4. $(3x^3 + 10x^2 + 6x - 4) \div (x + 2)$

5. $(x^3 + 7x^2 + 17x + 15) \div (x + 3)$

6. $(3x^3 - x^2 + 4x + 8) \div (x + 2)$

7. $(4x^3 - 11x^2 - 20x + 5) \div (x - 4)$

8. $(5x^3 - 8x^2 - 4x + 10) \div (x - 2)$

9. $(x^3 - 2x^2 + 8) \div (x + 2)$

10. $(2x^3 + 7x^2 - 5) \div (x + 3)$

11. $(6x^4 + 13x^3 + 35x - 24) \div (x + 3)$

12. $(x^4 - 2x^3 - 11x^2 + 34) \div (x + 2)$

13. $(2x^4 + 3x^3 + x^2 + 2x + 5) \div (x + 1)$

14. $(x^4 - 4x^3 + 5x^2 - 6x + 1) \div (x - 3)$

15. $(2x^4 - 5x - 3) \div (x - 1)$

16. $(3x^4 + 8x - 10) \div (x + 2)$

17. $(2x^5 - 13x^3 + 10x^2 + 6) \div (x + 3)$

18. $(x^5 + 6x^4 - 28x^2 + 50) \div (x + 5)$

19. $(x^6 - 5x^3 + x^2 + 12) \div (x + 1)$

20. $(x^6 - 4) \div (x + 1)$

How do we use synthetic division when the divisor is in the form $ax + b$*? We divide the divisor by* a *to get* $x + \frac{b}{a}$*. Then, after performing the synthetic division, we divide each term of the result by* a*. The number that is the remainder does not change. To divide* $(2x^3 + 7x^2 - 5x - 4) \div (2x + 1)$*, we would use* $-\frac{1}{2}\big|\ \ 2 \quad 7 \quad -5 \quad -4$ *and then divide each term of the result by 2.*

In exercises 21 and 22, divide using synthetic division.

21. $(4x^3 - 6x^2 + 6) \div (2x + 3)$

22. $(2x^3 - 3x^2 + 6x + 4) \div (2x + 1)$

Cumulative Review

A total of 21 people were killed and 150 people injured in the Great Boston Molasses Flood in January 1919. A molasses storage tank burst and spilled 2.3 million gallons of molasses through the streets of Boston.

▲ **23.** **[1.2.2]** ***Boston Molasses Flood*** How many cubic feet of molasses were contained in the 2.3-million-gallon molasses tank? (Use 1 gallon $\approx$ 0.134 cubic feet.)

24. **[1.2.2]** ***Boston Molasses Flood*** The wave of molasses traveled through the streets at a speed of 56 kilometers per hour. What was the speed in miles per hour? (Use 1 km/hr $\approx$ 0.62 mi/hr.) Round to the nearest whole number.

25. **[3.5.3]** If $p(x) = 2x^4 - 3x^2 + 6x - 1$, find $p(-3)$.

Quick Quiz 5.3 *Divide using synthetic division.*

1. $(x^3 - 2x^2 - 5x - 2) \div (x + 1)$

2. $(x^3 - 8x^2 + 24) \div (x - 2)$

3. $(x^4 - 7x^2 + 2x - 12) \div (x + 3)$

4. **Concept Check** When doing a problem such as $(2x^4 - x + 3) \div (x - 2)$, why is it necessary to use zeros to represent $0x^3 + 0x^2$ when performing synthetic division?

5.4 Removing Common Factors; Factoring by Grouping

Student Learning Objectives

After studying this section, you will be able to:

① **Factor out the greatest common factor from a polynomial.**

② **Factor a polynomial by the grouping method.**

We learned to multiply polynomials in Section 5.1. When two or more algebraic expressions (monomials, binomials, and so on) are multiplied, each expression is called a **factor.**

In the rest of this chapter, we will learn how to find the factors of a polynomial. **Factoring** is multiplication in reverse and is an extremely important mathematical technique.

① Factoring Out the Greatest Common Factor

To factor out a common factor, we make use of the distributive property.

$$ab + ac = a(b + c)$$

The **greatest common factor** is simply the largest factor that is common to all terms of the expression. It must contain

1. The largest possible common factor of the numerical coefficients and
2. The largest possible common variable factor

EXAMPLE 1 Factor out the greatest common factor.

(a) $7x^2 - 14x$ **(b)** $40a^3 - 20a^2$

Solution

(a) $7x^2 - 14x = 7 \cdot x \cdot x - 7 \cdot 2 \cdot x = 7x(x - 2)$

Be careful. The greatest common factor is $7x$, not 7.

(b) $40a^3 - 20a^2 = 20a^2(2a - 1)$

The greatest common factor is $20a^2$.

Suppose we had written $10a(4a^2 - 2a)$ or $10a(2a)(2a - 1)$ as our answer. Although we have factored the expression, we have not found the *greatest* common factor.

Student Practice 1 Factor out the greatest common factor.

(a) $19x^3 - 38x^2$ **(b)** $100a^4 - 50a^2$

***NOTE TO STUDENT:** Fully worked-out solutions to all of the Student Practice problems can be found at the back of the text starting at page SP-1.*

EXAMPLE 2 Factor out the greatest common factor.

(a) $9x^2 - 18xy - 15y^2$ **(b)** $4a^3 - 12a^2b^2 - 8ab^3 + 6ab$

Solution

(a) $9x^2 - 18xy - 15y^2 = 3(3x^2 - 6xy - 5y^2)$

The greatest common factor is 3.

(b) $4a^3 - 12a^2b^2 - 8ab^3 + 6ab = 2a(2a^2 - 6ab^2 - 4b^3 + 3b)$

The greatest common factor is $2a$.

Student Practice 2 Factor out the greatest common factor.

(a) $21x^3 - 18x^2y + 24xy^2$ **(b)** $12xy^2 - 14x^2y + 20x^2y^2 + 36x^3y$

How do you know whether you have factored correctly? You should do two things to verify your answer.

1. Examine the polynomial in the parentheses. Its terms should not have any remaining common factors.
2. Multiply the two factors. You should obtain the original expression.

In the next two examples, you will be asked to **factor** a polynomial (i.e., to find the factors that, when multiplied, give the polynomial as a product). For each of these examples, this will require you to factor out the greatest common factor.

EXAMPLE 3 Factor and check your answer. $6x^3 - 9x^2y - 6x^2y^2$

Solution

$$6x^3 - 9x^2y - 6x^2y^2 = 3x^2(2x - 3y - 2y^2)$$

Check.

1. $(2x - 3y - 2y^2)$ has no common factors. If it did, we would know that we had not factored out the *greatest* common factor.
2. Multiply the two factors.

$$3x^2(2x - 3y - 2y^2) = 6x^3 - 9x^2y - 6x^2y^2$$

Observe that we do obtain the original polynomial.

Student Practice 3 Factor and check your answer. $9a^3 - 12a^2b^2 - 15a^4$

The greatest common factor need not be a monomial. It may be a binomial or even a trinomial. For example, note the following:

$$5a(x + 3) + 2(x + 3) = (x + 3)(5a + 2)$$
$$5a(x + 4y) + 2(x + 4y) = (x + 4y)(5a + 2)$$

The common factors are binomials.

EXAMPLE 4 Factor.

(a) $2x(x + 5) - 3(x + 5)$ **(b)** $5a(a + b) - 2b(a + b) - (a + b)$

Solution

(a) $2x(x + 5) - 3(x + 5) = (x + 5)(2x - 3)$

The common factor is $x + 5$.

(b) $5a(a + b) - 2b(a + b) - (a + b) = 5a(a + b) - 2b(a + b) - 1(a + b)$

$$= (a + b)(5a - 2b - 1)$$

The common factor is $a + b$.

Student Practice 4 Factor. $7x(x + 2y) - 8y(x + 2y) - (x + 2y)$

② Factoring by Grouping

Because the common factors in Example 4 were grouped inside parentheses, it was easy to pick them out. However, this rarely happens, so we have to learn how to manipulate expressions to find the greatest common factor.

Polynomials with four terms can often be factored by the method of Example 4(a). However, the parentheses are not always present in the original problem. When they are not present, we look for a way to remove a common factor from the first two terms. We then factor out a common factor from the first two terms and a common factor from the second two terms. Often, we can then find the greatest common factor of the original expression.

EXAMPLE 5 Factor. $ax + 2ay + 2bx + 4by$

Solution

Remove the greatest common factor (a) from the first two terms.

$$ax + 2ay + 2bx + 4by = a(x + 2y) + 2b(x + 2y)$$

Remove the greatest common factor ($2b$) from the last two terms.

Now we can see that $(x + 2y)$ is a common factor.

$$a(x + 2y) + 2b(x + 2y) = (x + 2y)(a + 2b)$$

Student Practice 5 Factor. $bx + 5by + 2wx + 10wy$

To factor an expression by this method, it may be necessary to rearrange the order of the four terms so that the first two terms do have a common factor.

EXAMPLE 6 Factor. $2x^2 - 18y - 12x + 3xy$

Solution First write the polynomial in this order: $2x^2 - 12x + 3xy - 18y$

Remove the greatest common factor ($2x$) from the first two terms.

$$2x^2 - 12x + 3xy - 18y = 2x(x - 6) + 3y(x - 6) = (x - 6)(2x + 3y)$$

Remove the greatest common factor ($3y$) from the last two terms.

Student Practice 6 Factor. $5x^2 - 12y - 15x + 4xy$

MC **EXAMPLE 7** Factor. $xy - 6 + 3x - 2y$

Solution

$xy + 3x - 2y - 6$ — Rearrange the terms so that the first two terms have a common factor and the last two terms have a common factor.

$= x(y + 3) - 2(y + 3)$ — Factor out the common factor x from the first two terms and -2 from the second two terms.

$= (y + 3)(x - 2)$ — Factor out the common binomial factor $y + 3$.

Student Practice 7 Factor.
$xy - 12 - 4x + 3y$

EXAMPLE 8 Factor and check your answer by multiplying.
$2x^3 + 21 - 7x^2 - 6x$

Solution

$2x^3 - 7x^2 - 6x + 21$ — Rearrange the terms.

$= x^2(2x - 7) - 3(2x - 7)$ — Factor out a common factor from each group of two terms.

$= (2x - 7)(x^2 - 3)$ — Factor out the common binomial factor $2x - 7$.

Check.

$$(2x - 7)(x^2 - 3) = 2x^3 - 6x - 7x^2 + 21 \quad \text{Multiply the two binomials.}$$
$$= 2x^3 + 21 - 7x^2 - 6x \quad \text{Rearrange the terms.}$$

The product is identical to the original expression.

Student Practice 8 Factor. $2x^3 - 15 - 10x + 3x^2$

STEPS TO SUCCESS I Don't Really Need to Actually Read This Book, Do I?

Successful students find that they can get a lot of benefit from reading the book. Here are some suggestions they have made after they completed the course:

1. This book was written by faculty who have learned how to help you succeed in doing math problems. So take the time to read over the assigned section of the book. You will be amazed by the understanding you will acquire.

2. Read your textbook with a paper and pen handy. As you come across a new definition or a new idea, underline it. As you come across a suggestion that helps you understand things better, be sure to underline it.

3. Interact with the book. When you see something you don't understand put a big question mark next to it. Try reading it again. Find a similar example in the book and go over the steps. If that doesn't help, ask your instructor or a classmate to explain the procedure or idea.

4. Use your pen to follow each step that is worked out for each example. Underline the steps that you think are especially challenging. This will help you when you do the homework.

Making it personal: Which of these four suggestions do you think you would benefit from the most? Put a check mark by that suggestion. Now write down what you need to do in order to take full advantage of your textbook. Make a commitment to doing this for your next homework assignment. ▼

5.4 Exercises

MyMathLab® Watch the videos in MyMathLab. Download the MyDashBoard App

Factor out the greatest common factor.

1. $80 - 10y$

2. $16x - 16$

3. $5a^2 - 25a$

4. $7a^2 - 14a$

5. $4a^2b^3 - 8ab + 32a$

6. $15c^2d^2 + 10c^2 - 60c$

7. $30y^4 + 24y^3 + 18y^2$

8. $16y^5 - 24y^4 - 40y^3$

9. $15ab^2 + 5ab - 10a^3b$

10. $-12x^2y - 18xy + 6x$

11. $10a^2b^3 - 30a^3b^3 + 10a^3b^2 - 40a^4b^2$

12. $28x^3y^2 - 12x^2y^4 + 4x^3y^4 - 32x^2y^2$

13. $3x(x + y) - 2(x + y)$

14. $7a(2a - b) + 3(2a - b)$

(*Hint:* Is the expression in the first parentheses equal to the expression in the second parentheses in exercises 15 and 16?)

15. $5b(a - 3b) + 8(-3b + a)$

16. $4y(x - 5y) - 3(-5y + x)$

17. $3x(a + 5b) + (a + 5b)$

18. $2w(s - 3t) - (s - 3t)$

19. $2a^2(3x - y) - 5b^3(3x - y)$

20. $7a^3(5a + 4) - 2(5a + 4)$

21. $3x(5x + y) - 8y(5x + y) - (5x + y)$

22. $2w(x - 4y) - 3z(x - 4y) + (x - 4y)$

23. $2a(a - 6b) - 3b(a - 6b) - 2(a - 6b)$

24. $3a(a + 4b) - 5b(a + 4b) - 9(a + 4b)$

Factor.

25. $x^3 + 5x^2 + 3x + 15$

26. $x^3 + 8x^2 + 2x + 16$

27. $2x + 6 - 3ax - 9a$

28. $2bc + 8b - 3c - 12$

29. $ab - 4a + 12 - 3b$

30. $2m^2 - 8mn - 5m + 20n$

31. $5x - 20 + 3xy - 12y$

32. $3x - 21 + 4xy - 28y$

33. $9y + 2x - 6 - 3xy$

34. $10y + 3x - 6 - 5xy$

35. $ab^3 + c + b^2 + abc$

36. $8x^2z - 6yz + 15y - 20x^2$

Applications

37. ***Stacked Oranges*** The total number of oranges stacked in a pile of x rows is given by the polynomial $\frac{1}{3}x^3 + \frac{1}{2}x^2 + \frac{1}{6}x$. Write this polynomial in factored form.

▲ 38. ***Geometry*** The volume of the box pictured below is given by the polynomial $4x^3 + 2x^2 - 6x$. Write this polynomial in factored form.

Cumulative Review

39. **[3.1.1]** Graph the line. $6x - 2y = -12$

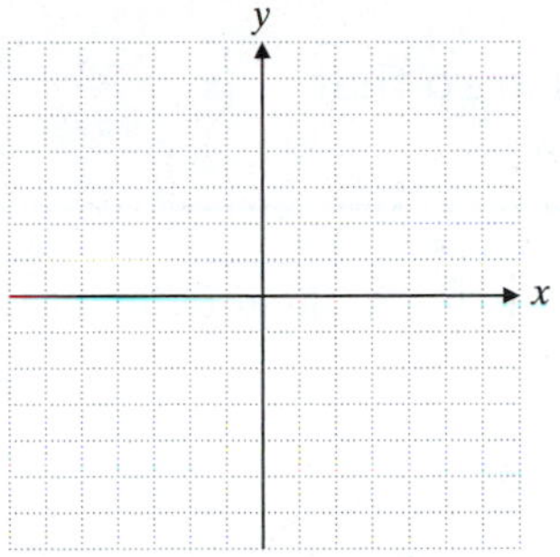

40. **[4.4.1]** Graph the region.
$x - y \geq -4$
$x + 2y \geq 2$

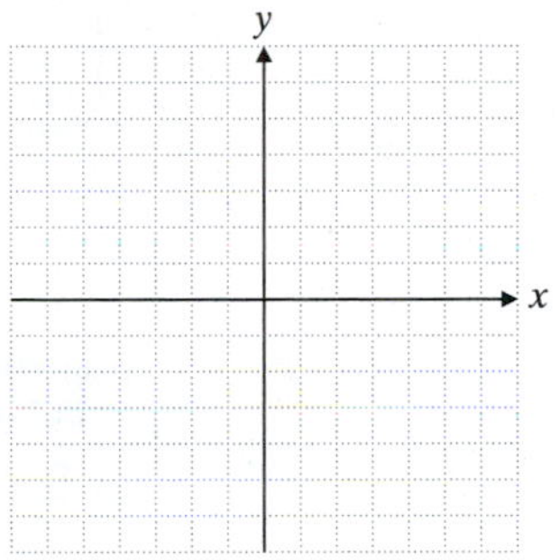

41. **[3.2.1]** Find the slope of the line that passes through $(5, -3)$ and $(2, 6)$.

42. **[3.3.1]** Find the slope and y-intercept of $2y + 6x = -3$.

Quick Quiz 5.4 *Factor completely.*

1. $15a^2b^3 - 10a^3b^3 - 25a^2b^4$

2. $2x(5x - 3y) - 5(5x - 3y)$

3. $3ac + 20b^2 - 4ab - 15bc$

4. **Concept Check** Explain how you would rearrange the order of $8xy - 15 - 10y + 12x$ in order to factor the polynomial.

5.5 Factoring Trinomials

Student Learning Objectives

After studying this section, you will be able to:

1. Factor trinomials of the form $x^2 + bx + c$.
2. Factor trinomials of the form $ax^2 + bx + c$.

① Factoring Trinomials of the Form $x^2 + bx + c$

If we multiply $(x + 4)(x + 5)$, we obtain $x^2 + 9x + 20$. But suppose that we already have the polynomial $x^2 + 9x + 20$ and need to factor it. In other words, suppose we need to find the expressions that, when multiplied, give us the polynomial. Let's use this example to find a general procedure.

The coefficient of x is the **sum** of these two numbers.

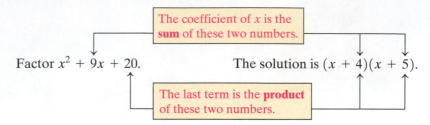

Factor $x^2 + 9x + 20$. The solution is $(x + 4)(x + 5)$.

The last term is the **product** of these two numbers.

FACTORING TRINOMIALS OF THE FORM $x^2 + bx + c$

1. The answer has the form $(x + m)(x + n)$, where m and n are real numbers.
2. The numbers m and n are chosen so that
   (a) $m \cdot n = c$ and **(b)** $m + n = b$.

If the last term of the trinomial is positive and the middle term is negative, the two numbers m and n will be negative numbers.

EXAMPLE 1 Factor. $x^2 - 14x + 24$

Solution We want to find two numbers whose product is 24 and whose sum is -14. They will both be negative numbers.

Factor Pairs of 24	*Sum of the Factors*	
$(-24)(-1)$	$-24 - 1 = -25$	
$(-12)(-2)$	$-12 - 2 = -14$	✓
$(-8)(-3)$	$-8 - 3 = -11$	
$(-6)(-4)$	$-6 - 4 = -10$	

The numbers whose product is 24 and whose sum is -14 are -12 and -2. Thus,

$$x^2 - 14x + 24 = (x - 12)(x - 2).$$

Student Practice 1 Factor. $x^2 - 10x + 21$

NOTE TO STUDENT: Fully worked-out solutions to all of the Student Practice problems can be found at the back of the text starting at page SP-1.

If the last term of the trinomial is negative, the two numbers m and n will be opposite in sign.

EXAMPLE 2 Factor. $x^2 + 11x - 26$

Solution We want to find two numbers whose product is -26 and whose sum is 11. One number will be positive and the other negative.

Factor Pairs of –26	*Sum of the Factors*	
$(-26)(1)$	$-26 + 1 = -25$	
$(26)(-1)$	$26 - 1 = 25$	
$(-13)(2)$	$-13 + 2 = -11$	
$(13)(-2)$	$13 - 2 = 11$	✓

The numbers whose product is -26 and whose sum is 11 are 13 and -2. Thus,

$$x^2 + 11x - 26 = (x + 13)(x - 2).$$

Student Practice 2 Factor. $x^2 - 13x - 48$

Sometimes we can make a substitution that makes a polynomial easier to factor, as shown in the following example.

EXAMPLE 3 Factor. $x^4 - 2x^2 - 24$

Solution We need to recognize that we can write this as $(x^2)^2 - 2(x^2) - 24$. We can make this polynomial easier to factor if we substitute y for x^2. Then we have $y^2 + (-2)y + (-24)$. So the factors will be $(y + m)(y + n)$. The two numbers whose product is -24 and whose sum is -2 are -6 and 4. Therefore, we have $(y - 6)(y + 4)$. But $y = x^2$, so our answer is

$$x^4 - 2x^2 - 24 = (x^2 - 6)(x^2 + 4).$$

Student Practice 3 Factor. $x^4 + 9x^2 + 8$

FACTS ABOUT SIGNS

Suppose $x^2 + bx + c = (x + m)(x + n)$. We know certain facts about m and n.

1. m and n have the same sign if c is positive. (*Note:* We did *not* say that they will have the same sign as c.)
   (a) They are positive if b is positive.
   (b) They are negative if b is negative.
2. m and n have opposite signs if c is negative. The larger number is positive if b is positive and negative if b is negative.

If you understand these sign facts, continue on to Example 4. If not, review Examples 1 through 3.

EXAMPLE 4 Factor.

(a) $y^2 + 5y - 36$ **(b)** $x^4 - 4x^2 - 12$

Solution

(a) $y^2 + 5y - 36 = (y + 9)(y - 4)$ The larger number (9) is positive because $b = 5$ is positive.

(b) $x^4 - 4x^2 - 12 = (x^2 - 6)(x^2 + 2)$ The larger number (6) is negative because $b = -4$ is negative.

Student Practice 4 Factor.

(a) $a^2 - 2a - 48$ **(b)** $x^4 + 2x^2 - 15$

Does the order in which we write the factors make any difference? In other words, is it true that $x^2 + bx + c = (x + n)(x + m)$? Since multiplication is commutative,

$$x^2 + bx + c = (x + n)(x + m) = (x + m)(x + n).$$

The order of the factors is not important.

We can also factor certain kinds of trinomials that have more than one variable.

EXAMPLE 5 Factor. **(a)** $x^2 - 21xy + 20y^2$ **(b)** $x^2 + 4xy - 21y^2$

Solution **(a)** $x^2 - 21xy + 20y^2 = (x - 20y)(x - y)$

The last terms in each factor contain the varible y.

(b) $x^2 + 4xy - 21y^2 = (x + 7y)(x - 3y)$

Student Practice 5 Factor.

(a) $x^2 - 16xy + 15y^2$ **(b)** $x^2 + xy - 42y^2$

If the terms of a trinomial have a common factor, you should remove the greatest common factor from the terms first. Then you will be able to follow the factoring procedure we have used in the previous examples.

EXAMPLE 6 Factor. $3x^2 - 30x + 48$

Solution The factor 3 is common to all three terms of the polynomial. Factoring out the 3 gives us the following:

$$3x^2 - 30x + 48 = 3(x^2 - 10x + 16)$$

Now we continue to factor the trinomial in the usual fashion.

$$3(x^2 - 10x + 16) = 3(x - 8)(x - 2)$$

Student Practice 6 Factor. $4x^2 - 44x + 72$

② Factoring Trinomials of the Form $ax^2 + bx + c$

Using the Grouping Number Method. One way to factor a trinomial $ax^2 + bx + c$ is to write it as four terms and factor it by grouping as we did in Section 5.4. For example, the trinomial $2x^2 + 11x + 12$ can be written as $2x^2 + 3x + 8x + 12$.

$$\begin{aligned} 2x^2 + 3x + 8x + 12 &= x(2x + 3) + 4(2x + 3) \\ &= (2x + 3)(x + 4) \end{aligned}$$

We can factor all factorable trinomials of the form $ax^2 + bx + c$ in this way. Use the following procedure.

GROUPING NUMBER METHOD FOR FACTORING TRINOMIALS OF THE FORM $ax^2 + bx + c$

1. Obtain the grouping number ac.
2. Find the factor pair of the grouping number whose sum is b.
3. Use those two factors to write bx as the sum of two terms.
4. Factor by grouping.

EXAMPLE 7 Factor. $2x^2 + 19x + 24$

Solution

1. The grouping number is $(a)(c) = (2)(24) = 48$.
2. The factor pairs of 48 are as follows:

$$48 \cdot 1 \quad 24 \cdot 2 \quad 16 \cdot 3 \quad 12 \cdot 4 \quad 8 \cdot 6$$

Now $b = 19$ so we want the factor pair of 48 whose sum is 19. Therefore, we select the factors 16 and 3.

3. We use the numbers 16 and 3 to write $19x$ as the sum of $16x$ and $3x$.

$$2x^2 + 19x + 24 = 2x^2 + 16x + 3x + 24$$

4. Factor by grouping.

$$\begin{aligned} 2x^2 + 16x + 3x + 24 &= 2x(x + 8) + 3(x + 8) \\ &= (x + 8)(2x + 3) \end{aligned}$$

Student Practice 7 Factor. $10x^2 - 9x + 2$

EXAMPLE 8 Factor. $6x^2 + 7x - 5$

Solution

1. The grouping number is -30.

2. We want the factor pair of -30 whose sum is 7.

$$\begin{aligned} -30 &= (-30)(1) & -30 &= (5)(-6) \\ &= (30)(-1) & &= (-5)(6) \\ &= (15)(-2) & &= (3)(-10) \\ &= (-15)(2) & &= (-3)(10) \end{aligned}$$

3. Since $-3 + 10 = 7$, use -3 and 10 to write $6x^2 + 7x - 5$ with four terms.

$$6x^2 + 7x - 5 = 6x^2 - 3x + 10x - 5$$

4. Factor by grouping.

$$\begin{aligned} 6x^2 - 3x + 10x - 5 &= 3x(2x - 1) + 5(2x - 1) \\ &= (2x - 1)(3x + 5) \end{aligned}$$

Student Practice 8 Factor. $3x^2 + 2x - 8$

If the three terms have a common factor, then prior to using the four-step grouping number procedure, we first factor out the greatest common factor from the terms of the trinomial.

MC **EXAMPLE 9** Factor. $6x^3 - 26x^2 + 24x$

Solution First we factor out the greatest common factor $2x$ from each term.

$$6x^3 - 26x^2 + 24x = 2x(3x^2 - 13x + 12)$$

Next we follow the four steps to factor $3x^2 - 13x + 12$.

1. The grouping number is 36.

2. We want the factor pair of 36 whose sum is -13. The two factors are -4 and -9.

3. We use -4 and -9 to write $3x^2 - 13x + 12$ with four terms.

$$3x^2 - 13x + 12 = 3x^2 - 4x - 9x + 12$$

4. Factor by grouping. Remember that we first factored out the factor $2x$. This factor must be part of the answer.

$$\begin{aligned} 2x(3x^2 - 4x - 9x + 12) &= 2x[x(3x - 4) - 3(3x - 4)] \\ &= 2x(3x - 4)(x - 3) \end{aligned}$$

Student Practice 9 Factor.

$9x^3 - 15x^2 - 6x$

Using the Trial-and-Error Method. Another way to factor trinomials of the form $ax^2 + bx + c$ is by trial and error. This method has an advantage if the grouping number is large and we would have to list many factors. In the trial-and-error method, we try different values and see which ones can be multiplied out to obtain the original expression.

If the last term is negative, there are many more sign possibilities.

EXAMPLE 10 Factor by trial and error. $10x^2 - 49x - 5$

Solution The first terms in the factors could be $10x$ and x or $5x$ and $2x$. The second terms could be $+1$ and -5 or -1 and $+5$. We list all the possibilities and look for one that will yield a middle term of $-49x$.

Possible Factors	*Middle Term of Product*
$(2x - 1)(5x + 5)$	$+5x$
$(2x + 1)(5x - 5)$	$-5x$
$(2x + 5)(5x - 1)$	$+23x$
$(2x - 5)(5x + 1)$	$-23x$
$(10x - 5)(x + 1)$	$+5x$
$(10x + 5)(x - 1)$	$-5x$
$(10x - 1)(x + 5)$	$+49x$
$(10x + 1)(x - 5)$	$-49x$

Thus,

$$10x^2 - 49x - 5 = (10x + 1)(x - 5)$$

As a check, it is always a good idea to multiply the two binomials to see whether you obtain the original expression.

$$\begin{aligned}(10x + 1)(x - 5) &= 10x^2 - 50x + 1x - 5 \\ &= 10x^2 - 49x - 5\end{aligned}$$

Student Practice 10 Factor by trial and error.

$$8x^2 - 6x - 5$$

EXAMPLE 11 Factor by trial and error. $6x^4 + x^2 - 12$

Solution The first term of each factor must contain x^2. Suppose that we try the following:

Possible Factors	*Middle Term of Product*
$(2x^2 - 3)(3x^2 + 4)$	$-x^2$

The middle term we get is $-x^2$, but we need its opposite, x^2. In this case, we just need to reverse the signs of -3 and 4. Do you see why? Therefore,

$$6x^4 + x^2 - 12 = (2x^2 + 3)(3x^2 - 4).$$

Student Practice 11 Factor by trial and error.

$$6x^4 + 13x^2 - 5$$

5.5 Exercises

MyMathLab® Watch the videos in MyMathLab Download the MyDashBoard App

Factor each polynomial.

1. $x^2 + 8x + 7$

2. $x^2 + 12x + 11$

3. $x^2 - 8x + 15$

4. $x^2 - 10x + 16$

5. $x^2 - 10x + 24$

6. $x^2 - 9x + 20$

7. $a^2 + 4a - 45$

8. $a^2 + 2a - 35$

9. $x^2 - xy - 42y^2$

10. $x^2 - xy - 56y^2$

11. $x^2 - 15xy + 14y^2$

12. $x^2 + 10xy + 9y^2$

13. $x^4 - 3x^2 - 40$

14. $x^4 + 8x^2 + 12$

15. $x^4 + 16x^2y^2 + 63y^4$

16. $x^4 - 6x^2 - 55$

Factor out the greatest common factor from the terms of the trinomial. Then factor the remaining trinomial.

17. $2x^2 + 26x + 44$

18. $2x^2 + 30x + 52$

19. $x^3 + x^2 - 20x$

20. $x^3 - 4x^2 - 45x$

Factor each polynomial. You may use the grouping number method or the trial-and-error method.

21. $2x^2 - x - 1$

22. $3x^2 + x - 2$

23. $6x^2 - 7x - 5$

24. $7x^2 - 19x - 6$

25. $3a^2 - 8a + 5$

26. $6a^2 + 11a + 3$

27. $4a^2 + a - 14$

28. $3a^2 - 20a + 12$

29. $2x^2 + 13x + 15$

30. $5x^2 - 8x - 4$

31. $3x^4 - 8x^2 - 3$

32. $3x^4 + 13x^2 - 10$

33. $6x^2 + 35xy + 11y^2$

34. $5x^2 + 12xy + 7y^2$

35. $7x^2 + 11xy - 6y^2$

36. $4x^2 - 13xy + 3y^2$

Factor out the greatest common factor from the terms of the trinomial. Then factor the remaining trinomial.

37. $8x^3 - 2x^2 - x$

38. $9x^3 - 9x^2 - 10x$

39. $10x^4 + 15x^3 + 5x^2$

40. $16x^4 + 48x^3 + 20x^2$

Mixed Practice *Factor each polynomial.*

41. $x^2 - 2x - 63$

42. $x^2 + 6x - 40$

43. $6x^2 + x - 2$

44. $5x^2 + 17x + 6$

45. $x^2 - 20x + 51$

46. $x^2 - 20x + 99$

47. $15x^2 + x - 2$

48. $12x^2 - 5x - 3$

49. $2x^2 + 4x - 96$

50. $3x^2 + 9x - 84$

51. $18x^2 + 21x + 6$

52. $24x^2 + 26x + 6$

53. $40ax^2 + 72ax - 16a$

54. $30bx^2 - 80bx - 30b$

55. $6x^3 + 26x^2 - 20x$

56. $24x^2 + 30x^2 + 9x$

57. $7x^4 + 13x^2 - 2$

58. $8x^4 + 2x^2 - 3$

59. $9a^2 - 18ab - 7b^2$

60. $15a^2 + ab - 2b^2$

61. $x^6 - 10x^3 - 39$

62. $x^6 - 3x^3 - 70$

63. $10x^3y - 15x^2y - 10xy$

64. $8x^3y + 20x^2y - 12xy$

Cumulative Review

▲ **65.** **[1.6.2]** Find the area of a circle of radius 3 inches. (Use $\pi \approx 3.14$.)

66. **[2.2.1]** Solve for b. $A = \frac{1}{3}(3b + 4a)$

67. **[4.3.1]** ***Airplane Seating*** A large commercial jetliner flies from Atlanta to San Francisco. The jetliner can carry a total of 184 passengers. There are two types of seats on the aircraft: first class and coach. The number of coach seats is sixteen more than six times the number of first-class seats. How many of each type of seat are there on the airplane?

68. **[3.1.1]** Graph. $6x + 4y = -12$

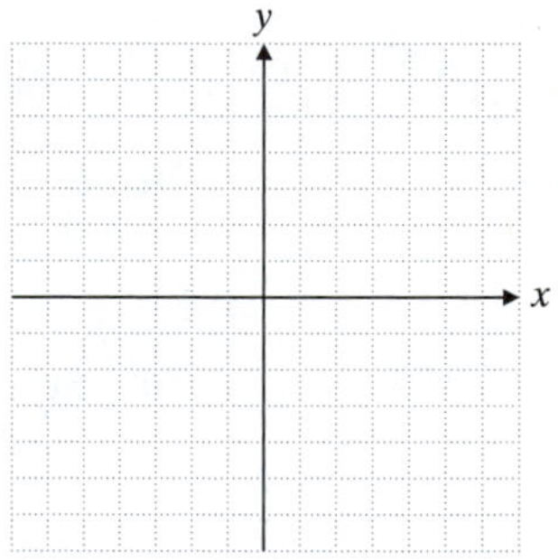

Quick Quiz 5.5 *Factor each polynomial.*

1. $x^2 + 5x - 84$

2. $6x^2 - 17x + 12$

3. $4x^3 + 6x^2 - 40x$

4. **Concept Check** Explain what the first step would be in factoring the following polynomial. $3x^2y^2 + 6xy^2 - 72y^2$

How Am I Doing? Sections 5.1–5.5

How are you doing with your homework assignments on Sections 5.1 to 5.5? Do you feel you have mastered the material so far? Do you understand the concepts you have covered? Before you go further in the textbook, take some time to do each of the following problems.

5.1

Simplify.

1. $(5x^2 - 3x + 2) + (-3x^2 - 5x - 8) - (x^2 + 3x - 10)$

2. $(x^2 - x - 6)(2x - 5)$

3. $(5a - 8)(a - 7)$

4. $(3y - 5)(3y + 5)$

5. $(3x^2 + 4)^2$

6. Evaluate the polynomial function $p(x) = 2x^3 - 5x^2 - 6x + 1$ for $p(-3)$.

5.2 and 5.3

Divide.

7. $(25x^3y^2 - 30x^2y^3 - 50x^2y^2) \div 5x^2y^2$

8. $(3y^3 - 5y^2 + 2y - 1) \div (y - 2)$

9. $(6x^4 - x^3 + 11x^2 - 2x - 2) \div (3x + 1)$

10. Use synthetic division to do the following:
$(2x^4 + 10x^3 + 11x^2 - 6x - 9) \div (x + 3)$

5.4

Factor completely.

11. $24a^3b^2 + 36a^4b^2 - 60a^3b^3$

12. $3x(4x - 3y) - 2(4x - 3y)$

13. $10wx + 6xz - 15yz - 25wy$

14. $18a^2 + 6ab - 15a - 5b$

5.5

Factor completely.

15. $x^2 - 7x + 10$

16. $4y^2 - 4y - 15$

17. $28x^2 - 19xy + 3y^2$

18. $2x^2 + 21x + 40$

19. $3x^2 - 6x - 72$

20. $24x^3 + 44x^2 + 16x$

Now turn to page SA-13 for the answers to each of these problems. Each answer also includes a reference to the objective in which the problem is first taught. If you missed any of these problems, you should stop and review the Examples and Student Practice problems in the referenced objective. A little review now will help you master the material in the upcoming sections of the text.

1. ____________
2. ____________
3. ____________
4. ____________
5. ____________
6. ____________
7. ____________
8. ____________
9. ____________
10. ____________
11. ____________
12. ____________
13. ____________
14. ____________
15. ____________
16. ____________
17. ____________
18. ____________
19. ____________
20. ____________

5.6 Special Cases of Factoring

Student Learning Objectives

After studying this section, you will be able to:

1. **Factor a binomial that is the difference of two squares.**
2. **Factor a perfect square trinomial.**
3. **Factor a binomial that is the sum or difference of two cubes.**

***NOTE TO STUDENT:** Fully worked-out solutions to all of the Student Practice problems can be found at the back of the text starting at page SP-1.*

① Factoring the Difference of Two Squares

Recall the special product formula: $(a + b)(a - b) = a^2 - b^2$. We can use it now as a factoring formula.

FACTORING THE DIFFERENCE OF TWO SQUARES

$$a^2 - b^2 = (a + b)(a - b)$$

EXAMPLE 1 Factor. $x^2 - 16$

Solution In this case $a = x$ and $b = 4$ in the formula.

$$\begin{array}{ccccccc} a^2 & - & b^2 & = & (a & + & b)(a & - & b) \\ \downarrow & & \downarrow & & \downarrow & & \downarrow \; \downarrow & & \downarrow \\ (x)^2 & - & (4)^2 & = & (x & + & 4)(x & - & 4) \end{array}$$

Student Practice 1 Factor. $x^2 - 9$

EXAMPLE 2 Factor. $25x^2 - 36$

Solution Here we will use the formula $a^2 - b^2 = (a + b)(a - b)$.

$$25x^2 - 36 = (5x)^2 - (6)^2 = (5x + 6)(5x - 6)$$

Student Practice 2 Factor. $64x^2 - 121$

EXAMPLE 3 Factor. $100w^4 - 9z^4$

Solution $100w^4 - 9z^4 = (10w^2)^2 - (3z^2)^2 = (10w^2 + 3z^2)(10w^2 - 3z^2)$

Student Practice 3 Factor. $49x^2 - 25y^4$

Whenever possible, a common factor should be factored out in the first step. Then the formula can be applied.

EXAMPLE 4 Factor. $75x^2 - 3$

Solution We factor out the common factor 3 from each term.

$$\begin{aligned} 75x^2 - 3 &= 3(25x^2 - 1) \\ &= 3(5x + 1)(5x - 1) \end{aligned}$$

Student Practice 4 Factor. $7x^2 - 28$

② Factoring Perfect Square Trinomials

Recall the formulas for squaring a binomial.

$$(a - b)^2 = a^2 - 2ab + b^2$$
$$(a + b)^2 = a^2 + 2ab + b^2$$

We can use these formulas to factor perfect square trinomials.

PERFECT SQUARE FACTORING FORMULAS

$$a^2 - 2ab + b^2 = (a - b)^2$$
$$a^2 + 2ab + b^2 = (a + b)^2$$

Recognizing these special cases will save you a lot of time when factoring. How can we recognize a perfect square trinomial?

1. The first and last terms are perfect squares. (The numerical values are 1, 4, 9, 16, 25, 36, . . . , and the variables have an exponent that is an even whole number.)
2. The middle term is twice the product of the values that, when squared, give the first and last terms.

EXAMPLE 5 Factor. $25x^2 - 20x + 4$

Solution Is this trinomial a perfect square? Yes.

1. The first and last terms are perfect squares.

$$25x^2 - 20x + 4 = (5x)^2 - 20x + (2)^2$$

2. The middle term is twice the product of the value $5x$ and the value 2. In other words, $2(5x)(2) = 20x$.

$$(5x)^2 - 2(5x)(2) + (2)^2 = (5x - 2)^2$$

Therefore, we can use the formula $a^2 - 2ab + b^2 = (a - b)^2$. Thus,

$$\downarrow \quad \downarrow \quad \downarrow \quad \downarrow \quad \downarrow$$

$$25x^2 - 20x + 4 = (5x - 2)^2.$$

Student Practice 5 Factor. $9x^2 - 30x + 25$

EXAMPLE 6 Factor. $16x^2 - 24x + 9$

Solution

1. The first and last terms are perfect squares: $16x^2 = (4x)^2$ and $9 = (3)^2$.
2. The middle term is twice the product $(4x)(3)$. Therefore, we have the following:

$$a^2 - 2ab + b^2 = (a - b)^2$$
$$16x^2 - 24x + 9 = (4x)^2 - 2(4x)(3) + (3)^2$$
$$= (4x - 3)^2.$$

Student Practice 6 Factor. $25x^2 - 70x + 49$

EXAMPLE 7 Factor. $200x^2 + 360x + 162$

Solution First we factor out the common factor 2.

$$200x^2 + 360x + 162 = 2(100x^2 + 180x + 81)$$
$$a^2 + 2ab + b^2 = (a + b)^2$$
$$2(100x^2 + 180x + 81) = 2[(10x)^2 + (2)(10x)(9) + (9)^2]$$
$$= 2(10x + 9)^2$$

Student Practice 7 Factor. $242x^2 + 88x + 8$

EXAMPLE 8 Factor.

(a) $x^4 + 14x^2 + 49$ **(b)** $9x^4 + 30x^2y^2 + 25y^4$

Solution

(a) $x^4 + 14x^2 + 49 = (x^2)^2 + 2(x^2)(7) + (7)^2$
$= (x^2 + 7)^2$

(b) $9x^4 + 30x^2y^2 + 25y^4 = (3x^2)^2 + 2(3x^2)(5y^2) + (5y^2)^2$
$= (3x^2 + 5y^2)^2$

Student Practice 8 Factor.

(a) $49x^4 + 28x^2 + 4$ **(b)** $36x^4 + 84x^2y^2 + 49y^4$

③ Factoring the Sum or Difference of Two Cubes

There are also special formulas for factoring cubic binomials. We see that the factors of $x^3 + 27$ are $(x + 3)(x^2 - 3x + 9)$, and that the factors of $x^3 - 64$ are $(x - 4)(x^2 + 4x + 16)$. We can generalize these patterns and derive the following factoring formulas.

SUM AND DIFFERENCE OF CUBES FACTORING FORMULAS

$$a^3 + b^3 = (a + b)(a^2 - ab + b^2)$$
$$a^3 - b^3 = (a - b)(a^2 + ab + b^2)$$

EXAMPLE 9 Factor. $125x^3 + y^3$

Solution Here $a = 5x$ and $b = y$.

$$a^3 + b^3 = (a + b)(a^2 - ab + b^2)$$
$$\downarrow \quad \downarrow \quad \downarrow \quad \downarrow \quad \downarrow$$
$$125x^3 + y^3 = (5x)^3 + (y)^3 = (5x + y)(25x^2 - 5xy + y^2)$$

Student Practice 9 Factor. $8x^3 + 125y^3$

EXAMPLE 10 Factor. $64x^3 + 27$

Solution Here $a = 4x$ and $b = 3$.

$$a^3 + b^3 = (a + b)(a^2 - ab + b^2)$$
$$\downarrow \quad \downarrow \quad \downarrow \quad \downarrow \quad \downarrow$$
$$64x^3 + 27 = (4x)^3 + (3)^3 = (4x + 3)(16x^2 - 12x + 9)$$

Student Practice 10 Factor. $64x^3 + 125y^3$

EXAMPLE 11 Factor. $125w^3 - 8z^6$

Solution Here $a = 5w$ and $b = 2z^2$.

$$a^3 - b^3 = (a - b)(a^2 + ab + b^2)$$
$$\downarrow \quad \downarrow \quad \downarrow \quad \downarrow \quad \downarrow$$
$$125w^3 - 8z^6 = (5w)^3 - (2z^2)^3 = (5w - 2z^2)(25w^2 + 10wz^2 + 4z^4)$$

Student Practice 11 Factor. $27w^3 - 125z^6$

EXAMPLE 12 Factor. $250x^3 - 2$

Solution First we factor out the common factor 2.

$$250x^3 - 2 = 2(125x^3 - 1)$$
$$= 2(5x - 1)\underbrace{(25x^2 + 5x + 1)}$$

Note that this trinomial cannot be factored.

Student Practice 12 Factor. $54x^3 - 16$

What should you do if a polynomial is the difference of two cubes *and* the difference of two squares? Usually, it's easier to use the difference of two squares formula first. Then apply the difference of two cubes formula.

EXAMPLE 13 Factor. $x^6 - y^6$

Solution We can write this binomial as $(x^2)^3 - (y^2)^3$ or as $(x^3)^2 - (y^3)^2$. Therefore, we can use either the difference of two cubes formula or the difference of two squares formula. It's usually better to use the difference of two squares formula first, so we'll do that.

$$x^6 - y^6 = (x^3)^2 - (y^3)^2$$

Here $a = x^3$ and $b = y^3$. Therefore,

$$(x^3)^2 - (y^3)^2 = (x^3 + y^3)(x^3 - y^3).$$

Now we use the sum of two cubes formula for the first factor and the difference of two cubes formula for the second factor.

$$x^3 + y^3 = (x + y)(x^2 - xy + y^2)$$
$$x^3 - y^3 = (x - y)(x^2 + xy + y^2)$$

Hence,

$$x^6 - y^6 = (x + y)(x^2 - xy + y^2)(x - y)(x^2 + xy + y^2).$$

Student Practice 13 Factor. $64a^6 - 1$

You'll see these special cases of factoring often. You should be very familiar with the following formulas. Be sure you understand how to use each one of them.

SPECIAL CASES OF FACTORING

Difference of Two Squares

$$a^2 - b^2 = (a + b)(a - b)$$

Perfect Square Trinomial

$$a^2 - 2ab + b^2 = (a - b)^2$$
$$a^2 + 2ab + b^2 = (a + b)^2$$

Sum and Difference of Cubes

$$a^3 + b^3 = (a + b)(a^2 - ab + b^2)$$
$$a^3 - b^3 = (a - b)(a^2 + ab + b^2)$$

5.6 Exercises

MyMathLab®

Watch the videos in MyMathLab

Download the MyDashBoard App

Verbal and Writing Skills, Exercise 1–4

1. How do you determine if a factoring problem will use the difference of two squares?

2. How do you determine if a factoring problem will use the perfect square trinomial formula?

3. How do you determine if a factoring problem will use the sum of two cubes formula?

4. How do you determine if a factoring problem will use the difference of two cubes formula?

Use the difference of two squares formula to factor. Be sure to factor out any common factors.

5. $a^2 - 64$

6. $y^2 - 49$

7. $16x^2 - 81$

8. $4x^2 - 25$

9. $64x^2 - 1$

10. $81x^2 - 1$

11. $49m^2 - 9n^2$

12. $36x^2 - 25y^2$

13. $100y^2 - 81$

14. $49y^2 - 144$

15. $1 - 36x^2y^2$

16. $1 - 64x^2y^2$

17. $27x^2 - 75$

18. $50x^2 - 8$

19. $3x - 27x^3$

20. $49x^3 - 36x$

Use the perfect square trinomial formulas to factor. Be sure to factor out any common factors.

21. $9x^2 - 6x + 1$

22. $16y^2 - 8y + 1$

23. $49x^2 - 14x + 1$

24. $100y^2 - 20y + 1$

25. $81w^2 + 36wt + 4t^2$

26. $25w^2 + 20wt + 4t^2$

27. $36x^2 + 60xy + 25y^2$

28. $64x^2 + 48xy + 9y^2$

29. $8x^2 + 40x + 50$

30. $64x^2 + 32x + 4$

31. $3x^3 - 24x^2 + 48x$

32. $72x^3 - 24x^2 + 2x$

Use the sum and difference of cubes formulas to factor. Be sure to factor out any common factors.

33. $x^3 - 27$

34. $x^3 - 8$

35. $x^3 + 125$

36. $x^3 + 64$

37. $64x^3 - 1$

38. $125x^3 - 1$

39. $8x^3 - 125$

40. $27x^3 - 64$

41. $1 - 27x^3$

42. $1 - 8x^3$

43. $64x^3 + 125$

44. $27x^3 + 125$

45. $64s^6 + t^6$

46. $125s^6 + t^6$

47. $5y^3 - 40$

48. $54y^3 - 2$

49. $250x^3 + 2$

50. $128y^3 + 2$

51. $x^5 - 8x^2y^3$

52. $x^5 - 27x^2y^3$

Mixed Practice *Factor by the methods of this section.*

53. $25w^4 - 1$

54. $16m^4 - 25$

55. $b^4 + 6b^2 + 9$

56. $a^4 + 12a^2 + 36$

57. $9m^6 - 64$

58. $144 - m^6$

59. $36y^6 - 60y^3 + 25$

60. $100n^6 - 140n^3 + 49$

61. $45z^8 - 5$

62. $2a^8 - 98$

63. $125m^3 + 8n^3$

64. $64z^3 - 27w^3$

65. $24a^3 - 3b^3$

66. $54w^3 + 250$

67. $4w^2 - 20wz + 25z^2$

68. $49x^4 - 42x^2 + 9$

69. $36a^2 - 81b^2$

70. $400x^4 - 36y^2$

71. $16x^4 - 81y^4$

72. $256x^4 - 1$

73. $125m^6 + 8$

74. $27n^6 + 125$

Try to factor the following four exercises by using the formulas for the perfect square trinomial. Why can't the formulas be used? Then factor each exercise correctly using an appropriate method.

75. $25x^2 + 25x + 4$

76. $16x^2 + 40x + 9$

77. $49x^2 - 35x + 4$

78. $4x^2 - 25x + 36$

Applications

▲ **79.** ***Carpentry*** Find the area of a maple cabinet surface that is constructed by a carpenter as a large square with sides of $4x$ feet and has a square cut-out region whose sides are y feet. Write your answer in factored form.

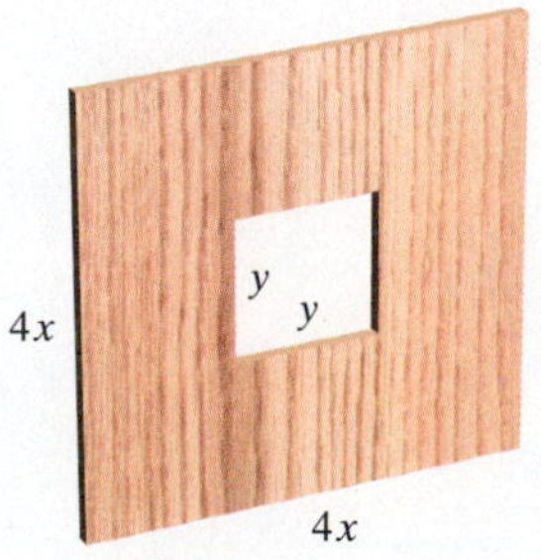

▲ **80.** ***Base of a Lamp*** A copper base for a lamp consists of a large circle of radius $2y$ inches with a cut-out area in the center of radius x inches. Write an expression for the area of the top of this copper base. Write your answer in factored form.

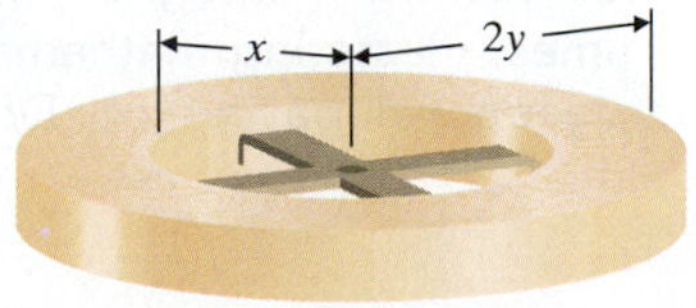

Cumulative Review

81. **[2.1.1]** ***Public School Salaries*** The average annual salary y, in thousands of dollars, paid to public school principals in the united states can be approximated by the equation $y = 1.57x + 84$, where x is the number of years since 2002. The average annual salary y, in thousands of dollars, paid to public high school classroom teachers can be approximated by the equation $y = 1.04x + 44$, where x is the number of years since 2002. If this trend continues, in what year will a public school principal be paid $50,000 more than a public school classroom teacher? (*Source:* www.census.gov)

82. **[2.4.1]** ***iPod Music Player Prices*** Three friends each bought a different iPod music player. The total for the three purchases was $365. Grant paid $30 more than Megan. Ryan paid $100 less than Megan. How much did each person pay?

Quick Quiz 5.6 *Factor completely.*

1. $25x^2 - 64y^2$

2. $49x^2 - 56xy + 16y^2$

3. $64x^3 - 27y^3$

4. **Concept Check** Explain why the formula $(a + b)^2 = a^2 + 2ab + b^2$ does not work when factoring the following expression.

$$36x^2 + 66xy + 121y^2$$

STEPS TO SUCCESS Keep Trying! Do Not Quit!

We live in a highly technical world. More and more it depends on mathematics. You cannot afford to give up on the study of mathematics. Dropping mathematics may prevent you from entering some career field that you find really interesting. Learning mathematics can open new doors for you.

Learning mathematics is a process. It takes time and effort. You will find that regular study and daily practice are necessary. This will help your level of academic success and lead you toward mastery of mathematics. In hard economic times, mastering mathematics will open a path for you to get a well-paying job. Do not quit! You can do it!

Making it personal: Which of these statements do you find the most helpful? What things can you do to help you master mathematics? Don't let yourself quit or let up on your studies. Your work now will help your financial success in the future! ▼

5.7 Factoring a Polynomial Completely

① Factoring Factorable Polynomials

Not all polynomials have the convenient form of one of the special formulas. Most do not. The following procedure will help you handle these common cases. You must practice this procedure until you can *recognize the various forms* and *determine which factoring method to use.*

Student Learning Objectives

After studying this section, you will be able to:

1. **Factor any factorable polynomial.**
2. **Recognize polynomials that are prime.**

COMPLETELY FACTORING A POLYNOMIAL

1. Check for a common factor. Factor out the greatest common factor (if there is one) before doing anything else.
2. (a) If the remaining polynomial has two terms, try to factor it as one of the following.
   (1) The difference of two squares: $a^2 - b^2 = (a + b)(a - b)$
   (2) The difference of two cubes: $a^3 - b^3 = (a - b)(a^2 + ab + b^2)$
   (3) The sum of two cubes: $a^3 + b^3 = (a + b)(a^2 - ab + b^2)$
   (b) If the remaining polynomial has three terms, try to factor it as one of the following.
   (1) A perfect square trinomial: $a^2 + 2ab + b^2 = (a + b)^2$ or $a^2 - 2ab + b^2 = (a - b)^2$
   (2) A general trinomial of the form $x^2 + bx + c$ or the form $ax^2 + bx + c$
   (c) If the remaining polynomial has four terms, try to factor by grouping.
3. Check to see whether the factors can be factored further.

EXAMPLE 1 Factor completely.

(a) $2x^2 - 18$ **(b)** $27x^4 - 8x$
(c) $27x^2 + 36xy + 12y^2$ **(d)** $2x^2 - 100x + 98$
(e) $6x^3 + 11x^2 - 10x$ **(f)** $5ax + 5ay - 20x - 20y$

Solution

(a) $2x^2 - 18 = 2(x^2 - 9)$ Factor out the common factor.
$= 2(x + 3)(x - 3)$ Use $a^2 - b^2 = (a + b)(a - b)$.

(b) $27x^4 - 8x = x(27x^3 - 8)$ Factor out the common factor.
$= x(3x - 2)(9x^2 + 6x + 4)$ Use $a^3 - b^3 = (a - b)(a^2 + ab + b^2)$.

(c) $27x^2 + 36xy + 12y^2 = 3(9x^2 + 12xy + 4y^2)$ Factor out the common factor.
$= 3(3x + 2y)^2$ Use $(a + b)^2 = a^2 + 2ab + b^2$.

(d) $2x^2 - 100x + 98 = 2(x^2 - 50x + 49)$ Factor out the common factor.
$= 2(x - 49)(x - 1)$ The trinomial has the form $x^2 + bx + c$.

(e) $6x^3 + 11x^2 - 10x = x(6x^2 + 11x - 10)$ Factor out the common factor.
$= x(3x - 2)(2x + 5)$ The trinomial has the form $ax^2 + bx + c$.

(f) $5ax + 5ay - 20x - 20y = 5(ax + ay - 4x - 4y)$ Factor out the common factor.
$= 5[a(x + y) - 4(x + y)]$ Factor by grouping.
$= 5(x + y)(a - 4)$

Continued on next page

***NOTE TO STUDENT:** Fully worked-out solutions to all of the Student Practice problems can be found at the back of the text starting at page SP-1.*

Student Practice 1 Factor completely.

(a) $7x^5 + 56x^2$

(b) $125x^2 + 50xy + 5y^2$

(c) $12x^2 - 75$

(d) $3x^2 - 39x + 126$

(e) $6ax + 6ay + 18bx + 18by$

(f) $6x^3 - x^2 - 12x$

② Recognizing Polynomials That Are Prime

Can all polynomials be factored? No. Many polynomials cannot be factored. If a polynomial cannot be factored using rational numbers, it is said to be **prime.**

EXAMPLE 2 If possible, factor $6x^2 + 10x + 3$.

Solution The trinomial has the form $ax^2 + bx + c$. The grouping number is 18. If the trinomial can be factored, we must find two numbers whose product is 18 and whose sum is 10.

Factor Pairs of 18	*Sum of the Factors*
(18)(1)	19
(9)(2)	11
(6)(3)	9

There are no numbers meeting the necessary conditions. Thus, the polynomial is prime. (If you use the trial-and-error method, try all the possible factors and show that none of them has a product with a middle term of $10x$.)

Student Practice 2 If possible, factor $3x^2 - 10x + 4$.

EXAMPLE 3 If possible, factor $25x^2 + 49$.

Solution Unless there is a common factor that can be factored out, binomials of the form $a^2 + b^2$ cannot be factored. Therefore, $25x^2 + 49$ is prime.

Student Practice 3 If possible, factor $16x^2 + 81$.

5.7 Exercises

MyMathLab® Watch the videos in MyMathLab Download the MyDashBoard App

Verbal and Writing Skills, Exercises 1–4

1. In any factoring problem the first step is ______________.

2. If $x^2 + bx + c = (x + e)(x + f)$ and c is positive and b is negative, what can you say about the signs of e and of f?

3. If you were asked to factor a polynomial such as $49x^2 + 9y^2$, how would you know immediately that this polynomial is prime?

4. If you were asked to factor a polynomial such as $x^2 + 9x + 12$, how would you know very quickly that this polynomial is prime?

Factor, if possible. These problems will require only one step.

5. $3xy - 6yz$

6. $50c - 40bc^2$

7. $y^2 + 7y - 18$

8. $b^2 - 7b + 12$

9. $3x^2 - 8x + 5$

10. $4x^2 - 11x - 3$

11. $4x^2 + 20x + 25$

12. $16x^2 - 24x + 9$

13. $8x^3 - 125y^3$

14. $27x^3 + 64y^3$

15. $a^3 - 3ab - ac$

16. $3a^3 + a^2b - 2a^2c$

17. $x^2 + 16$

18. $4x^4 + 25$

19. $64y^2 - 25z^2$

20. $100m^2 - n^2$

Mixed Practice *Factor if possible. Be sure to factor completely.*

21. $6x^2 - 23x - 4$

22. $5x^2 + x - 4$

23. $3x^2 - x - 1$

24. $5x^2 - x - 2$

25. $x^3 + 9x^2 + 14x$

26. $x^3 + x^2 - 20x$

27. $25x^2 - 40x + 16$

28. $4r^2 + 28r + 49$

29. $6a^2 - 6a - 36$

30. $9a^2 + 18a - 72$

31. $3x^2 - 3x - xy + y$

32. $ax - bx - 2a + 2b$

33. $81a^4 - 1$

34. $1 - 16x^4$

35. $2x^5 - 16x^3 - 18x$

36. $2x^4 - 2x^2 - 24$

37. $8a^3b - 50ab^3$

38. $50x^2y^2 - 32y^2$

39. $2ax - 8xy + aw - 4wy$

40. $3ax - xy + 6aw - 2wy$

Applications

Cattle Farming *A cattle pen is constructed with solid wood walls. The pen is divided into four rectangular compartments. Each compartment is x feet long and y feet wide. The walls are x − 10 feet high.*

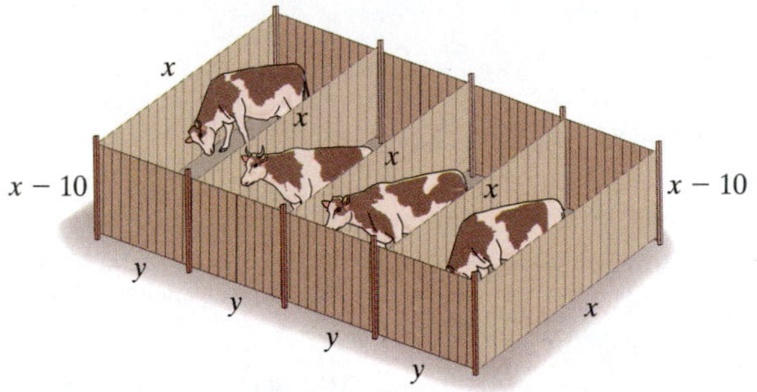

▲ **41.** Find the total surface area of the walls used in the cattle pen. Express the answer in factored form and in the form with the factors multiplied.

▲ **42.** The rancher who owns the cattle pen wants to increase the length x of each pen by 3 feet. Find the new total surface area of the walls that would be used in this enlarged cattle pen. Express the answer in factored form and in the form with the factors multiplied.

Cumulative Review *Solve the following inequalities.*

43. **[2.6.3]** $3x - 2 \leq -5 + 2(x - 3)$

44. **[2.8.1]** $|2 + 5x - 3| < 2$

45. **[2.8.2]** $\left|\frac{1}{3}(5 - 4x)\right| > 4$

46. **[2.7.2]** $x - 4 \geq 7$ *or* $4x + 1 \leq 17$

Quick Quiz 5.7 *Factor if possible.*

1. $49x^3 + 84x^2y + 36xy^2$

2. $6x^4 + x^2 - 12$

3. $9x^2 + 12x - 12$

4. **Concept Check** Explain why $4x^2 + 3x + 1$ is prime.

5.8 Solving Equations and Applications Using Polynomials

① Factoring to Find the Roots of a Quadratic Equation

Up until now, we have solved only first-degree equations. In this section we will solve quadratic, or second-degree, equations.

Student Learning Objectives

After studying this section, you will be able to:

① **Factor to find the roots of a quadratic equation.**

② **Solve applications that involve factorable quadratic equations.**

DEFINITION OF QUADRATIC EQUATION

A second-degree equation of the form $ax^2 + bx + c = 0$, where a, b, c are real numbers and $a \neq 0$, is a **quadratic equation.** $ax^2 + bx + c = 0$ is the **standard form** of a quadratic equation.

Before solving a quadratic equation, we will first write the equation in standard form. Although it is not necessary that *a*, *b*, and *c* be integers, the equation is usually written this way.

The key to solving quadratic equations by factoring is called the **zero factor property.** When we multiply two real numbers, the resulting product will be zero if one or both of the factors is zero. Thus, if the product of two real numbers is zero, at least one of the factors must be zero. We state this property of real numbers formally.

ZERO FACTOR PROPERTY

For all real numbers a and b,

$$\text{if } a \cdot b = 0, \text{ then } a = 0,\ b = 0, \text{ or both} = 0.$$

NOTE TO STUDENT: *Fully worked-out solutions to all of the Student Practice problems can be found at the back of the text starting at page SP-1.*

EXAMPLE 1 Solve the equation. $x^2 + 15x = 100$

Solution When we say "solve the equation" or "find the roots," we mean "find the values of x that satisfy the equation."

$x^2 + 15x - 100 = 0$ Subtract 100 from both sides so that one side is 0.

$(x + 20)(x - 5) = 0$ Factor the trinomial.

$x + 20 = 0$ or $x - 5 = 0$ Set each factor equal to 0.

$x = -20$ $\quad x = 5$ Solve each equation.

Check. Use the *original* equation $x^2 + 15x = 100$.

$$x = -20: \quad (-20)^2 + 15(-20) \stackrel{?}{=} 100$$
$$400 - 300 \stackrel{?}{=} 100$$
$$100 = 100 \checkmark$$
$$x = 5: \quad (5)^2 + 15(5) \stackrel{?}{=} 100$$
$$25 + 75 \stackrel{?}{=} 100$$
$$100 = 100 \checkmark$$

Thus, 5 and −20 are both roots of the quadratic equation $x^2 + 15x = 100$.

Student Practice 1 Find the roots.
$$x^2 + x = 56$$

For convenience, on the next page we list the steps we have employed to solve the quadratic equation.

Graphing Calculator

Finding Roots

Find the roots of

$10x(x + 1) = 83x + 12{,}012.$

This can be written in the form

$10x^2 - 73x - 12{,}012 = 0$

and factored to obtain the following:

$(5x + 156)(2x - 77) = 0$

$x = -\frac{156}{5}$ *and* $x = \frac{77}{2}$.

In decimal form the solutions are -31.2 and 38.5.

However, you can use the graphing calculator to graph

$y = 10x^2 - 73x - 12{,}012.$

By setting an appropriate viewing window and then using the Zoom and Trace features of your calculator, you can find the two places where $y = 0$.

Some calculators have a command that will find the zeros (roots) of a graph.

In a similar fashion graph each equation using the form $y = ax^2 + bx + c$ and use the graph to find the roots.

(a) $10x^2 - 189x - 12{,}834 = 0$

(b) $10x(x + 2) = 11{,}011 - 193x$

SOLVING A QUADRATIC EQUATION BY FACTORING

1. Rewrite the quadratic equation in standard form (so that one side of the equation is 0) and, if possible, *factor* the quadratic expression.
2. Set each factor equal to zero.
3. Solve the resulting equations to find both roots. (A quadratic equation has two roots.)
4. Check your solutions.

It is extremely important to remember that when you are placing the quadratic equation in standard form, one side of the equation must be zero. Several algebraic operations may be necessary to obtain that desired result before you can factor the polynomial.

EXAMPLE 2 Find the roots. $6x^2 + 4 = 7(x + 1)$

Solution

$6x^2 + 4 = 7x + 7$ Apply the distributive property.

$6x^2 - 7x - 3 = 0$ Rewrite the equation in standard form.

$(2x - 3)(3x + 1) = 0$ Factor the trinomial.

$2x - 3 = 0$ or $3x + 1 = 0$ Set each factor equal to 0.

$2x = 3$ $\quad 3x = -1$ Solve the equations.

$x = \frac{3}{2}$ $\quad x = -\frac{1}{3}$

Check. Use the *original* equation $6x^2 + 4 = 7(x + 1)$.

$$x = \frac{3}{2}: \quad 6\left(\frac{3}{2}\right)^2 + 4 \stackrel{?}{=} 7\left(\frac{3}{2} + 1\right)$$

$$6\left(\frac{9}{4}\right) + 4 \stackrel{?}{=} 7\left(\frac{5}{2}\right)$$

$$\frac{27}{2} + 4 \stackrel{?}{=} \frac{35}{2}$$

$$\frac{27}{2} + \frac{8}{2} \stackrel{?}{=} \frac{35}{2}$$

$$\frac{35}{2} = \frac{35}{2} \checkmark$$

It checks, so $\frac{3}{2}$ is a root. Verify that $-\frac{1}{3}$ is also a root.

If you are using a calculator to check your roots, you can complete the check more rapidly using the decimal values 1.5 and -0.33333333. The latter value is approximate, so some rounding error is expected.

Student Practice 2 Find the roots. $12x^2 - 9x + 2 = 2x$

EXAMPLE 3 Find the roots. $3x^2 - 5x = 0$

Solution

$3x^2 - 5x = 0$ The equation is already in standard form.

$x(3x - 5) = 0$ Factor.

$x = 0$ or $3x - 5 = 0$ Set each factor equal to 0.

$3x = 5$

$x = \frac{5}{3}$

Check. Verify that 0 and $\frac{5}{3}$ are roots of $3x^2 - 5x = 0$.

Student Practice 3 Find the roots. $7x^2 - 14x = 0$

EXAMPLE 4 Solve. $9x(x - 1) = 3x - 4$

Solution

$$9x^2 - 9x = 3x - 4 \quad \text{Remove parentheses.}$$
$$9x^2 - 9x - 3x + 4 = 0 \quad \text{Get 0 on one side.}$$
$$9x^2 - 12x + 4 = 0 \quad \text{Combine like terms.}$$
$$(3x - 2)^2 = 0 \quad \text{Factor.}$$
$$3x - 2 = 0 \quad \text{or} \quad 3x - 2 = 0$$
$$3x = 2 \qquad 3x = 2$$
$$x = \frac{2}{3} \qquad x = \frac{2}{3}$$

We obtain one solution twice. This value is called a **double root.**

Student Practice 4 Solve. $16x(x - 2) = 8x - 25$

The zero factor property can be extended to polynomial equations of degree greater than 2. In the following example, we will find the three roots of a third-degree polynomial equation.

EXAMPLE 5 Solve. $2x^3 = 24x - 8x^2$

Solution

$$2x^3 + 8x^2 - 24x = 0 \quad \text{Get 0 on one side of the equation.}$$
$$2x(x^2 + 4x - 12) = 0 \quad \text{Factor out the common factor } 2x.$$
$$2x(x + 6)(x - 2) = 0 \quad \text{Factor the trinomial.}$$
$$2x = 0 \quad \text{or} \quad x + 6 = 0 \quad \text{or} \quad x - 2 = 0 \quad \text{Zero factor property.}$$
$$x = 0 \qquad x = -6 \qquad x = 2 \quad \text{Solve for } x.$$

The solutions are 0, −6, and 2.

Student Practice 5 Solve. $3x^3 + 6x^2 = 45x$

② Solving Applications That Involve Factorable Quadratic Equations

Some applied exercises lead to factorable quadratic equations. Using the methods developed in this section, we can solve these types of exercises.

▲ **EXAMPLE 6** A racing sailboat has a triangular sail. Find the base and altitude of the triangular sail that has an area of 35 square meters and a base that is 3 meters shorter than the altitude.

Continued on next page

Solution

1. ***Understand the problem.*** We draw a sketch and recall the formula for the area of a triangle.

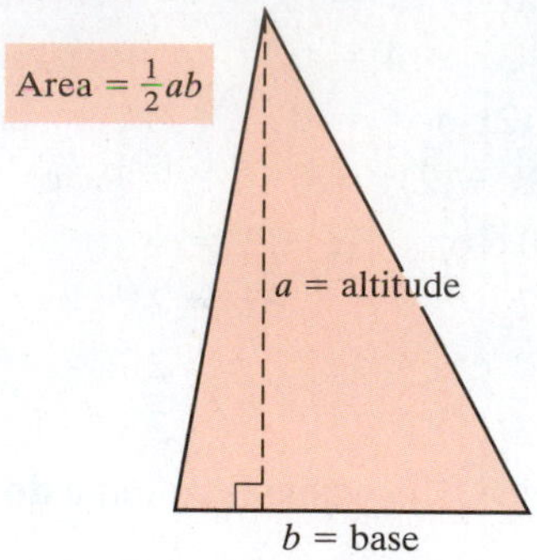

Let x = the length of the altitude in meters.
Then $x - 3$ = the length of the base in meters.

2. ***Write an equation.***

$$A = \frac{1}{2}ab$$

$$35 = \frac{1}{2}x(x - 3) \quad \text{Replace } A \text{ (area) by 35, } a \text{ (altitude) by } x, \text{ and } b \text{ (base) by } x - 3.$$

3. ***Solve the equation and state the answer.***

$$70 = x(x - 3) \quad \text{Multiply each side by 2.}$$
$$70 = x^2 - 3x \quad \text{Remove parentheses.}$$
$$0 = x^2 - 3x - 70 \quad \text{Subtract 70 from each side.}$$
$$0 = (x - 10)(x + 7)$$
$$x = 10 \quad \text{or} \quad x = -7$$

The altitude of a triangle must be a positive number, so we disregard -7. Thus,

$$\text{altitude} = x = 10 \text{ meters and}$$
$$\text{base} = x - 3 = 7 \text{ meters.}$$

The altitude of the triangular sail measures 10 meters, and the base of the sail measures 7 meters.

4. ***Check.*** Is the base 3 meters shorter than the altitude?

$$10 - 3 = 7 \ \checkmark$$

Is the area of the triangle 35 square meters?

$$A = \frac{1}{2}ab$$
$$A = \frac{1}{2}(10)(7) = 5(7) = 35 \ \checkmark$$

▲ **Student Practice 6** A racing sailboat has a triangular sail. Find the base and the altitude of the triangular sail if the area is 52 square feet and the altitude is 5 feet longer than the base.

▲ **EXAMPLE 7** A car manufacturer uses a square panel that holds fuses. The square panel that is used this year has an area that is 72 square centimeters greater than the area of the square panel used last year. The length of each side of the new panel is 3 centimeters more than double the length of last year's panel. Find the dimensions of each panel.

Solution

1. ***Understand the problem.*** We draw a sketch of each square panel and recall that the area of each panel is obtained by squaring its side.

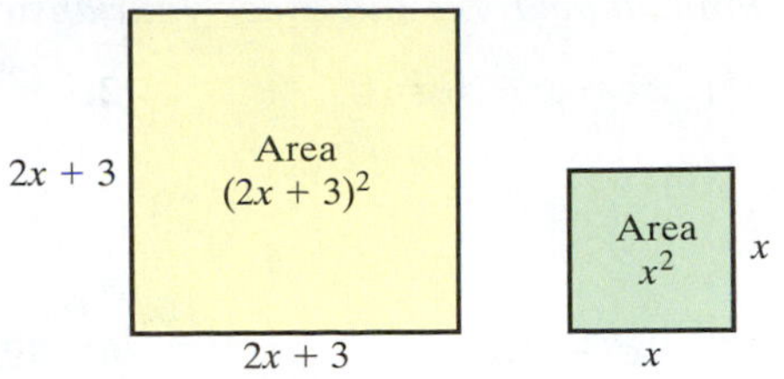

Let $x =$ the length in centimeters of last year's panel.

Then $2x + 3 =$ the length in centimeters of this year's panel.

The area of this year's panel is 72 square centimeters greater than the area of last year's panel.

2. ***Write an equation.***

Area of the larger square is 72 square centimeters greater than the area of the smaller square.

$$(2x + 3)^2 \quad = 72 \quad + \quad x^2$$

3. ***Solve the equation and state the answer.***

$4x^2 + 12x + 9 = 72 + x^2$	Remove parentheses.
$4x^2 + 12x + 9 - x^2 - 72 = 0$	Get 0 on one side of the equation.
$3x^2 + 12x - 63 = 0$	Simplify.
$3(x^2 + 4x - 21) = 0$	Factor out the common factor 3.
$3(x + 7)(x - 3) = 0$	Factor the trinomial.
$x + 7 = 0 \quad \text{or} \quad x - 3 = 0$	Use the zero factor property.
$x = -7 \qquad\qquad x = 3$	Solve for x.

A fuse box cannot measure -7 centimeters, so we reject the negative answer. We use $x = 3$; then $2x + 3 = 2(3) + 3 = 6 + 3 = 9$.

Thus, the old fuse panel measured 3 centimeters on a side. The new fuse panel measures 9 centimeters on a side.

4. ***Check.*** Verify that each answer is correct.

▲ **Student Practice 7** Last year Grandpa Jones had a small square garden. This year he has a square garden that is 112 square feet larger in area. Each side of the new garden is 2 feet longer than triple each side of the old garden. Find the dimensions of each garden.

STEPS TO SUCCESS Getting Help

Sometimes you study hard, listen in class, take good notes, and work diligently on the homework but you realize that is not enough. You need help! When you feel that way take immediate action.

Successful students find the following suggestions have given them significant help.

1. Make an appointment to talk to your instructor and explain your problem. Often this will clarify things and make it much easier to do the homework.
2. If your college has a tutoring center or a mathematics lab take advantage of it. People are on duty each day to help you with your problems.
3. Use MyMathLab or MathXL to have the computer take you through a problem step by step and find a similar example for you. Watch the video that explains that section of the book.
4. Call the Pearson Tutoring Center using the 1-800 number that is provided. The faculty member answering your call will talk you through the steps of any odd-numbered problem in the book.
5. Ask a friend in class to explain the concept to you. Often a classmate can relate to you in a simple way that helps your understanding.

Making it personal: Pick two of these suggestions that you think would be most helpful. Now make a plan and actually follow those two suggestions this week. Write down what specific action you will take. ▼

5.8 Exercises

MyMathLab® Watch the videos in MyMathLab Download the MyDashBoard App

Find all the roots and check your answers.

1. $x^2 - x - 6 = 0$

2. $x^2 + 5x - 14 = 0$

3. $5x^2 - 6x = 0$

4. $3x^2 + 5x = 0$

5. $9x^2 - 16 = 0$

6. $49x^2 - 25 = 0$

7. $3x^2 - 2x - 8 = 0$

8. $4x^2 - 13x + 3 = 0$

9. $8x^2 - 3 = 2x$

10. $4x^2 - 15 = 4x$

11. $8x^2 = 11x - 3$

12. $5x^2 = 11x - 2$

13. $x^2 + \frac{5}{3}x = \frac{2}{3}x$

14. $x^2 - \frac{5}{2}x = \frac{x}{2}$

15. $25x^2 + 10x + 1 = 0$

16. $36x^2 - 12x + 1 = 0$

Find all the roots and check your answers.

17. $x^3 + 7x^2 + 12x = 0$

18. $x^3 + 7x^2 + 10x = 0$

19. $\frac{x^3}{6} - 8x = \frac{x^2}{3}$

20. $\frac{x^3}{24} - \frac{x^2}{12} = x$

21. $3x^3 - 10x = 17x$

22. $5x^3 - 8x = 12x$

23. $2x^3 + 4x^2 = 30x$

24. $3x^3 + 12x^2 = 36x$

25. $\frac{7x^2 - 3}{2} = 2x$

26. $\frac{3x^2 + 3x}{2} = \frac{2}{3}$

27. $2(x + 3) = -3x + 2(x^2 - 3)$

28. $2(x^2 - 4) - 3x = 4x - 11$

29. $7x^2 + 6 = 2x^2 + 2(4x + 3)$

30. $11x^2 - 3x + 1 = 2(x^2 - 5x) + 1$

To Think About

31. The equation $2x^2 - 3x + c = 0$, has a solution of $-\frac{1}{2}$. What is the value of c? What is the other solution to the equation?

32. The equation $x^2 + bx - 12 = 0$ has a solution of -4. What is the value of b? What is the other solution to the equation?

Solve the following applied exercises.

▲ **33.** ***Warning Road Sign*** An orange triangular warning sign by the side of the road has an area of 180 square inches. The base of the sign is 2 inches longer than the altitude. Find the measurements of the base and altitude.

▲ **34.** ***Baseball Banner*** A triangular Boston Red Sox banner has an area of 150 square inches. The altitude of the triangle is 3 times the base. Find the measurements of the base and altitude.

▲ **35.** ***Neon Billboard*** The area of a triangular neon billboard advertising the local mall is 104 square feet. The base of the triangle is 2 feet longer than triple the length of the altitude.
(a) What are the dimensions of the triangular billboard in feet?
(b) What are the dimensions of the triangular billboard in yards?

▲ **36.** ***Entertainment Platform*** During halftime at the Super Bowl, one of the performers will sing on a triangular platform that measures 119 square yards. The base of the triangular stage is 6 yards longer than four times the length of the altitude. What are the dimensions of the triangular stage?

▲ **37.** ***Desk Telephone*** The area of the base of a rectangular desk telephone is 896 square centimeters. The length of the rectangular telephone is 4 centimeters longer than its width.
(a) What are the length and width, in centimeters, of the desk telephone?
(b) What are the length and width, in millimeters, of the desk telephone?

▲ **38.** ***Mouse Pad*** The area of a rectangular mouse pad is 480 square centimeters. Its length is 16 centimeters shorter than twice its width.
(a) What are the length and width of the mouse pad in centimeters?
(b) What are the length and width of the mouse pad in millimeters?

▲ **39.** ***Turkish Rug*** A rare square Turkish rug, belonging to the family of President John F. Kennedy, was auctioned off. The area of the Turkish rug in square feet is 96 more than its perimeter in feet. Find the length in feet of the side.

▲ **40.** ***Stadium Seating*** One of the "private luxury infield" suites in the new baseball stadium is in the shape of a square. The area of the square suite in square feet is 140 more than its perimeter in feet. Find the length of the side.

▲ **41.** ***Cereal Box*** The volume of a rectangular solid can be written as $V = LWH$, where L is the length of the solid, W is the width, and H is the height. A box of cereal has a width of 2 inches. Its height is 2 inches longer than its length. If the volume of the box is 198 cubic inches, what are the length and height of the box?

▲ **42.** ***College Catalog*** The volume of a rectangular solid can be written as $V = LWH$, where L is the length of the solid, W is the width, and H is the height. The North Shore Community College catalog is $\frac{1}{2}$ inch wide. Its height is 3 inches longer than its length. The volume of the catalog is 27 cubic inches. What are the height and length of the catalog?

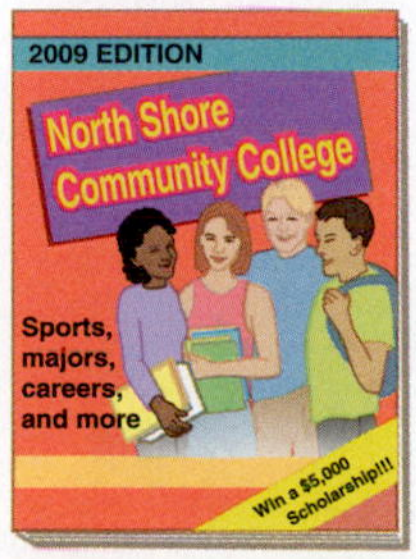

▲ **43.** ***Moon Exploration*** In planning for the first trip to the moon, NASA surveyed a rectangular area of 54 square miles. The length of the rectangle was 3 miles less than double the width. What were the dimensions of this potential landing area?

▲ **44.** ***Vegetable Garden*** The Patel family has a square vegetable garden. They increased the size of the garden so that the new, larger square garden is 144 square feet larger than the old garden. The new garden has a side that is 3 feet longer than double the length of the side of the old garden. What are the dimensions of the old and new gardens?

Debit Card Purchases *The transaction volume V (in billions of dollars) of purchases made on debit cards in the United States can be approximated by the equation* $V = 133.5x + 311$, *where x is the number of years since 2000. Use this equation to solve exercises 45–48. (Source: www.census.gov)*

45. Approximately how much was the transaction volume in the year 2000?

46. Approximately how much was the transaction volume in the year 2010?

47. In what year was the transaction volume $845,000,000,000?

48. In what year is the transaction volume expected to be $2447 billion?

Cumulative Review

Simplify. Do not leave negative exponents in your answers.

49. **[1.4.2]** $(2x^3y^2)^3(5xy^2)^2$

50. **[1.4.3]** $\dfrac{(2a^3b^2)^3}{16a^5b^8}$

51. **[4.2.2]** Solve the system.

$$\begin{aligned} x + 3y + z &= 5 \\ 2x - 3y - 2z &= 0 \\ x - 2y + 3z &= -9 \end{aligned}$$

52. **[3.3.2]** Find an equation of the line passing through (5, 2) and (6, 1).

Quick Quiz 5.8

Solve the following equations.

1. $x^2 = -4x + 32$

2. $15x^2 - 11x + 2 = 0$

3. The area of a triangular field is 160 square yards. The altitude of the triangle is 4 yards shorter than the base of the triangle. Find the altitude and the base of the triangle.

4. **Concept Check** Explain how you would solve the following equation. $\dfrac{3}{8}x^2 + \dfrac{1}{4}x = 1$

Did You Know...

That Saving a Few Dollars Every Day Can Add Up to Over a Thousand Dollars a Year?

EXPENSES THAT COST MORE THAN YOU THINK

Understanding the Problem:
Tracy and Max would like to go on an exciting vacation. They need to start saving so that they will be able to afford the trip, but they don't have a lot of extra cash left over at the end of the month. They wonder how much they could save if they stopped going out for lunch and instead made lunch themselves.

Making a Plan:
They decide to figure out how much they are spending now on lunches and what it will cost to make their own lunches. Then they can calculate how long it will take to save up enough money if they stop going out for lunch.

Step 1: They reviewed their weekly expenses, and they determined how much they are spending on average. Monday through Thursday Tracy goes to a sub shop and spends $8 a day. On Fridays she goes out with friends to a restaurant and spends $13 on lunch. Max orders out Monday through Friday and spends $10 a day. They total up their weekly spending.

Task 1: *Add up what Tracy and Max spend every week on lunch.*

Step 2: They calculate that it will cost $3 a day per person if they make their own lunches.

Task 2: *What is the weekly cost of bringing their own lunches?*

Step 3: Next, they figure out how much they will save per week by not going out for lunch.

Task 3: *How much will they save every week?*

Finding a Solution:

Step 4: Now that they know how much they can save per week, they want to know how long it will take to save up $2275, the cost of the trip.

Task 4: *After how many weeks of saving will they have enough money?*

Applying the Situation to Your Life:
There is probably money you spend on a daily basis that you could save by making changes. You could cut down on eating out by making more of your own meals. Instead of buying coffee, you could brew your own coffee. See where you could cut spending and tally up the weekly savings. Is there something you've thought of buying, but you haven't had the money? Use your predicted weekly savings to figure out how many weeks it will take you to save up enough money. Having a goal will help you stick to a savings plan.

Chapter 5 Organizer

Topic and Procedure	Examples	You Try It

Adding and subtracting polynomials, p. 251

Topic and Procedure

Combine like terms to add and subtract polynomials.

Examples

(a) $(5x^2 - 6x - 8) + (-2x^2 - 5x + 3)$
$= 3x^2 - 11x - 5$

(b) $(3a^2 - 2ab - 5b^2) - (-7a^2 + 6ab - b^2)$
$= (3a^2 - 2ab - 5b^2) + (7a^2 - 6ab + b^2)$
$= 10a^2 - 8ab - 4b^2$

You Try It

1. Add or subtract.
   (a) $(2a^2 - 9a + 8) + (3a^2 - a + 4)$
   (b) $(a^2 + 3ab - 10b^2) - (2a^2 + ab + 2b^2)$

Multiplying polynomials, p. 253

Topic and Procedure

1. Multiply each term of the first polynomial by each term of the second polynomial. If two binomials are being multiplied, use the FOIL (First, Outer, Inner, Last) method.
2. Combine like terms.

Examples

(a) $2x^2(3x^2 - 5x - 6) = 6x^4 - 10x^3 - 12x^2$

(b) $(3x + 4)(2x - 7) = 6x^2 - 21x + 8x - 28$
$= 6x^2 - 13x - 28$

(c) $(x - 3)(x^2 + 5x + 8)$
$= x^3 + 5x^2 + 8x - 3x^2 - 15x - 24$
$= x^3 + 2x^2 - 7x - 24$

You Try It

2. Multiply.
   (a) $3x(5x^3 + x^2 - 2x - 4)$
   (b) $(3x - 1)(3x - 6)$
   (c) $(x + 2)(2x^2 + 3x - 1)$

Division of a polynomial by a monomial, p. 257

Topic and Procedure

1. Write the division as the sum of separate fractions.
2. If possible, reduce the separate fractions.

Examples

$$(16x^3 - 24x^2 + 56x) \div (-8x)$$
$$= \frac{16x^3}{-8x} - \frac{24x^2}{-8x} + \frac{56x}{-8x}$$
$$= -2x^2 + 3x - 7$$

You Try It

3. Divide.
   $(-30a^4 + 5a^3 + 15a^2 - 15a) \div (5a)$

Dividing a polynomial by a binomial or a trinomial, p. 257

Topic and Procedure

1. Write the division exercise in long division form. Write both polynomials in descending order; write any missing terms with a coefficient of zero.
2. Divide the *first* term of the divisor into the first term of the dividend. The result is the first term of the quotient.
3. Multiply the first term of the quotient by *every* term in the divisor.
4. Write this product under the dividend (align like terms) and subtract. Bring down the next term.
5. Treat this difference as a new dividend. Repeat steps 2 to 4. Continue until the remainder is zero or is a polynomial of lower degree than the divisor.
6. If there is a remainder, write it as the numerator of a fraction with the divisor as the denominator. Add this fraction to the quotient.

Examples

Divide. $(6x^3 + 5x^2 - 2x + 1) \div (3x + 1)$

$$\begin{array}{r}
2x^2 + x - 1 \\
3x + 1\overline{)6x^3 + 5x^2 - 2x + 1} \\
\underline{6x^3 + 2x^2} \\
3x^2 - 2x \\
\underline{3x^2 + x} \\
-3x + 1 \\
\underline{-3x - 1} \\
2
\end{array}$$

The quotient is $2x^2 + x - 1 + \frac{2}{3x + 1}$.

You Try It

4. Divide.
   $(8x^3 - 6x^2 + 7x + 2) \div (2x - 1)$

Synthetic division, p. 262

Topic and Procedure

Synthetic division can be used if the divisor is in the form $x + b$.

1. Write the coefficients of the terms of the dividend in descending order. Write any missing terms with a coefficient of zero.
2. The divisor must be of the form $x + b$. Write down the opposite of b to the left.
3. Bring down the first coefficient to the bottom row.
4. Multiply the coefficient in the bottom row by the opposite of b. Write the result in the second row under the next coefficient in the top row.
5. Add the values in the top and second rows and place the result in the bottom row.
6. Repeat steps 4 and 5 until the bottom row is filled.

Examples

Divide.

$$(3x^5 - 2x^3 + x^2 - x + 7) \div (x + 2)$$

$$\begin{array}{r|rrrrrr}
-2 & 3 & 0 & -2 & 1 & -1 & 7 \\
 & & -6 & 12 & -20 & 38 & -74 \\
\hline
 & 3 & -6 & 10 & -19 & 37 & -67
\end{array}$$

The quotient is

$$3x^4 - 6x^3 + 10x^2 - 19x + 37 + \frac{-67}{x + 2}.$$

You Try It

5. Use synthetic division to divide.
   $(2x^4 - 3x^3 - 8x^2 - 1) \div (x - 3)$

Topic and Procedure	Examples	You Try It
Factoring out a common factor, p. 266 Remove the greatest common factor from each term. Many factoring problems are two steps, of which this is the first.	**(a)** $5x^3 - 25x^2 - 10x = 5x(x^2 - 5x - 2)$ **(b)** $20a^3b^2 - 40a^4b^3 + 30a^3b^3$ $= 10a^3b^2(2 - 4ab + 3b)$	**6.** Factor out the greatest common factor. **(a)** $12a^4 - 18a^2 - 24a$ **(b)** $14x^2y^4 + 21x^3y^5 + 28xy^3$
Factoring by grouping, p. 267 **1.** Make sure that the first two terms have a common factor and the last two terms have a common factor; otherwise, rearrange the terms. **2.** Factor out the common factor in each group. **3.** Factor out the common binomial factor.	$6xy - 8y + 3xw - 4w$ $= 2y(3x - 4) + w(3x - 4)$ $= (3x - 4)(2y + w)$	**7.** Factor by grouping. $20ab - 15b - 4ac + 3c$
Factoring trinomials of the form $x^2 + bx + c$, p. 272 The factors will be of the form $(x + m)(x + n)$, where $m \cdot n = c$ and $m + n = b$.	**(a)** $x^2 - 7x + 12 = (x - 4)(x - 3)$ **(b)** $3x^2 - 36x + 60 = 3(x^2 - 12x + 20)$ $= 3(x - 2)(x - 10)$ **(c)** $x^2 + 2x - 15 = (x + 5)(x - 3)$ **(d)** $2x^2 - 44x - 96 = 2(x^2 - 22x - 48)$ $= 2(x - 24)(x + 2)$	**8.** Factor. **(a)** $x^2 + 9x + 18$ **(b)** $3x^2 - 21x + 36$ **(c)** $x^2 + 2x - 80$ **(d)** $2x^2 - 2x - 40$
Factoring trinomials of the form $ax^2 + bx + c$, p. 274 Use the trial-and-error method or the grouping number method.	**(a)** $2x^2 + 7x + 3 = (2x + 1)(x + 3)$ **(b)** $8x^2 - 26x + 6 = 2(4x^2 - 13x + 3)$ $= 2(4x - 1)(x - 3)$ **(c)** $7x^2 + 20x - 3 = (7x - 1)(x + 3)$ **(d)** $5x^3 - 18x^2 - 8x = x(5x^2 - 18x - 8)$ $= x(5x + 2)(x - 4)$	**9.** Factor. **(a)** $3x^2 + 5x + 2$ **(b)** $6a^2 - 33a + 15$ **(c)** $6x^2 + x - 5$ **(d)** $4y^3 - 15y^2 - 4y$
Factoring the difference of two squares, p. 280 $a^2 - b^2 = (a + b)(a - b)$	**(a)** $9x^2 - 1 = (3x + 1)(3x - 1)$ **(b)** $8x^2 - 50 = 2(4x^2 - 25) = 2(2x + 5)(2x - 5)$	**10.** Factor. **(a)** $25 - 4a^2$ **(b)** $48c^2 - 3$
Factoring a perfect square trinomial, p. 280 $a^2 + 2ab + b^2 = (a + b)^2$ $a^2 - 2ab + b^2 = (a - b)^2$	**(a)** $16x^2 + 40x + 25 = (4x + 5)^2$ **(b)** $18x^2 + 120xy + 200y^2 = 2(9x^2 + 60xy + 100y^2)$ $= 2(3x + 10y)^2$ **(c)** $4x^2 - 36x + 81 = (2x - 9)^2$ **(d)** $25a^3 - 10a^2b + ab^2 = a(25a^2 - 10ab + b^2)$ $= a(5a - b)^2$	**11.** Factor. **(a)** $4x^2 + 12x + 9$ **(b)** $75a^2 + 60ab + 12b^2$ **(c)** $9a^2 - 48a + 64$ **(d)** $36x^3 - 12x^2y + xy^2$
Factoring the sum and difference of two cubes, p. 282 $a^3 + b^3 = (a + b)(a^2 - ab + b^2)$ $a^3 - b^3 = (a - b)(a^2 + ab + b^2)$	**(a)** $8x^3 + 27 = (2x + 3)(4x^2 - 6x + 9)$ **(b)** $250x^3 + 2y^3 = 2(125x^3 + y^3)$ $= 2(5x + y)(25x^2 - 5xy + y^2)$ **(c)** $27x^3 - 64 = (3x - 4)(9x^2 + 12x + 16)$ **(d)** $125y^4 - 8y = y(125y^3 - 8)$ $= y(5y - 2)(25y^2 + 10y + 4)$	**12.** Factor. **(a)** $27x^3 + 1$ **(b)** $2y^3 + 128x^3$ **(c)** $a^3 - 125$ **(d)** $24a^3 - 81b^3$

Topic and Procedure	Examples	You Try It
Solving a quadratic equation by factoring, p. 291 **1.** Write the equation in standard form. **2.** Factor, if possible. **3.** Set each factor equal to 0. **4.** Solve each of the resulting equations.	Solve. $(x+3)(x-2)=5(x+3)$ $x^2-2x+3x-6=5x+15$ $x^2+x-6=5x+15$ $x^2+x-6-5x-15=0$ $x^2-4x-21=0$ $(x-7)(x+3)=0$ $x-7=0$ or $x+3=0$ $x=7$ $\quad x=-3$	**13.** Solve. $(x-1)(x+4)=6(x+1)$

Chapter 5 Review Problems

Perform the operations indicated.

1. $(x^2-3x+5)+(-2x^2-7x+8)$

2. $(-4x^2y-7xy+y)+(5x^2y+2xy-9y)$

3. $(-6x^2+7xy-3y^2)-(5x^2-3xy-9y^2)$

4. $(-13x^2+9x-14)-(-2x^2-6x+1)$

5. $(5x+2)-(6-x)+(2x+3)$

6. $(4x-5)-(3x^2+x)+(x^2-2)$

For the polynomial function $p(x)=3x^3-2x^2-6x+1$ *find the following.*

7. $p(-4)$

8. $p(-1)$

9. $p(3)$

For the polynomial function $g(x)=-x^4+2x^3-x+5$ *find the following.*

10. $g(-1)$

11. $g(3)$

12. $g(0)$

Multiply.

13. $3xy(x^2-xy+y^2)$

14. $(3x^2+1)(2x-1)$

15. $(5x^2+3)^2$

16. $(x-3)(2x-5)(x+2)$

17. $(x^2-3x+1)(-2x^2+x-2)$

18. $(3x-5)(3x^2+2x-4)$

19. $(5ab-2)(5ab+2)$

20. $(3a-b^2)(6a-5b^2)$

Divide.

21. $(25x^3y - 15x^2y - 100xy) \div (-5xy)$

22. $(16a^4 - 28a^3 - 2a^2) \div (4a)$

23. $(12x^2 - 5x - 2) \div (3x - 2)$

24. $(2x^3 + x^2 - x + 1) \div (2x + 3)$

25. $(2x^4 - x^2 + 6x + 3) \div (x - 1)$

26. $(2x^4 - 13x^3 + 16x^2 - 9x + 20) \div (x - 5)$

Use synthetic division to perform the division.

27. $(3x^4 + 5x^3 - x^2 + x - 2) \div (x + 2)$

28. $(3y^3 - 2y + 5) \div (y - 3)$

Remove the greatest common factor.

29. $15a^2b + 5ab^2 - 10ab$

30. $x^5 - 3x^4 + 2x^2$

31. $12mn - 8m$

Factor using the grouping method.

32. $xy - 6y + 3x - 18$

33. $8x^2y + x^2b + 8y + b$

34. $3ab - 15a - 2b + 10$

Factor the trinomials.

35. $x^2 - 9x - 22$

36. $4x^2 - 5x - 6$

37. $6x^2 + 5x - 21$

Factor using one of the formulas from Section 5.6.

38. $100x^2 - 49$

39. $4x^2 - 28x + 49$

40. $8a^3 - 27$

Mixed Practice

In exercises 41–61, factor, if possible. Be sure to factor completely.

41. $9x^2 - 121$

42. $5x^2 - 11x + 2$

43. $x^3 + 8x^2 + 12x$

44. $x^2 - 8wy + 4wx - 2xy$

45. $36x^2 + 25$

46. $2x^2 - 7x - 3$

47. $27x^4 - x$

48. $-3a^3b^3 + 2a^2b^4 - a^2b^3$

49. $3x^4 - 5x^2 - 2$

50. $9a^2b + 15ab - 14b$

51. $2x^2 + 7x - 6$

52. $4y^4 - 13y^3 + 9y^2$

53. $10x^2y^2 - 20x^2y + 5x^2$

54. $a^2 + 5ab^3 + 4b^6$

55. $3x^2 - 12 - 8x + 2x^3$

56. $2x^4 - 12x^2 - 54$

57. $8a + 8b - 4bx - 4ax$

58. $4x^3 + 10x^2 - 6x$

59. $9x^4y^2 - 30x^2y + 25$

60. $27x^3y - 3xy$

61. $5bx - 28y + 4by - 35x$

Solve the following equations.

62. $5x^2 - 9x - 2 = 0$

63. $2x^2 - 11x + 12 = 0$

64. $(6x + 5)(x + 2) = -2$

65. $6x^2 = 24x$

66. $3x^2 + 14x + 3 = -1 + 4(x + 1)$

67. $x^3 + 7x^2 = -12x$

Use a quadratic equation to solve each of the following exercises.

▲ **68.** ***Geometry*** The area of a triangle is 75 square meters. The altitude of the triangle is 5 meters longer than the base of the triangle. Find the base and the altitude of the triangle.

▲ **69.** ***Ping-Pong Table*** A regulation ping-pong table has an area of 45 square feet. The length of the table is one foot less than double the width. Find the dimensions of a regulation ping-pong table.

70. ***Manufacturing Profit*** The hourly profit in dollars made by a scientific calculator manufacturing plant is given by the equation $P = 3x^2 - 7x - 10$, where x is the number of calculators assembled in 1 hour. Find the number of calculators that should be made in 1 hour if the hourly profit is to be \$30.

▲ **71.** ***Sound Insulators*** A square sound insulator is constructed for a restaurant. It does not provide enough insulation, so a larger square is constructed. The larger square has 24 square yards more insulation. The side of the larger square is 3 yards longer than double the side of the smaller square. Find the dimensions of each square.

How Am I Doing? Chapter 5 Test

CHAPTER Test Prep VIDEOS MATH COACH MyMathLab® YouTube™

After you take this test read through the Math Coach on pages 307–308. Math Coach videos are available via MyMathLab and YouTube. Step-by-step test solutions in the Chapter Test Prep Videos are also available via MyMathLab and YouTube. (Search "TobeyInterAlg" and click on "Channels.")

Combine.

1. $(3x^2y - 2xy^2 - 6) + (5 + 2xy^2 - 7x^2y)$

2. $(5a^2 - 3) - (2 + 5a) - (4a - 3)$

Multiply.

3. $-2x(x + 3y - 4)$

4. $(2x - 3y^2)^2$

5. $(x - 2)(2x^2 + x - 1)$

Divide.

6. $(-15x^3 - 12x^2 + 21x) \div (-3x)$

MC **7.** $(2x^4 - 7x^3 + 7x^2 - 9x + 10) \div (2x - 5)$

8. $(x^3 - x^2 - 5x + 2) \div (x + 2)$

Use synthetic division to perform this division.

9. $(x^4 + x^3 - x - 3) \div (x + 1)$

Factor, if possible.

10. $121x^2 - 25y^2$

11. $9x^2 + 30xy + 25y^2$

12. $x^3 - 26x^2 + 48x$

13. $4x^3y + 8x^2y^2 + 4x^2y$

14. $x^2 - 6wy + 3xy - 2wx$

15. $2x^2 - 3x + 2$

1. ______ ☐

2. ______ ☐

3. ______ ☐

4. ______ ☐

5. ______ ☐

6. ______ ☐

7. ______ ☐

8. ______ ☐

9. ______ ☐

10. ______ ☐

11. ______ ☐

12. ______ ☐

13. ______ ☐

14. ______ ☐

15. ______ ☐

16. ☐

17. ☐

18. ☐

19. ☐

20. ☐

21. ☐

22. ☐

23. ☐

24. ☐

25. ☐

26. ☐

Total Correct: ☐

MC **16.** $18x^2 + 3x - 15$

17. $54a^4 - 16a$

18. $9x^5 - 6x^3y + xy^2$

19. $3x^4 + 17x^2 + 10$

MC **20.** $3x - 10ay + 6y - 5ax$

Find the following if $p(x) = -2x^3 - x^2 + 6x - 10$.

21. $p(2)$

22. $p(-3)$

Solve the following equations.

MC **23.** $x^2 = 5x + 14$

24. $3x^2 - 11x - 4 = 0$

25. $7x^2 + 6x = 8x$

Use a quadratic equation to solve the following exercise.

▲ **26.** The area of a triangular road sign is 70 square inches. The altitude of the triangle is 4 inches less than the base of the triangle. Find the altitude and the base of the triangle.

MATH COACH

Mastering the skills you need to do well on the test.

Students often make the same types of errors when they do the Chapter 5 Test. Here are some helpful hints to keep you from making those common errors on test problems.

Performing Long Division with Polynomials—Problem 7

Divide. $(2x^4 - 7x^3 + 7x^2 - 9x + 10) \div (2x - 5)$

Helpful Hint Be careful when multiplying the first term of the quotient by the divisor. Make sure to multiply by both terms of the divisor. Watch out for + and − signs during the subtraction process. This is where most errors occur.

After the first step of division, did you obtain x^3 for the first term of the quotient?

Yes ____ No ____

When you multiplied x^3 by $2x - 5$, did you get $2x^4 - 5x^3$?

Yes ____ No ____

When you subtracted for the first time, did you get $-2x^3$?

Yes ____ No ____

If you answered No to any of these questions, go back and complete these steps again. Remember that $(2x^4 - 7x^3) - (2x^4 - 5x^3) = 2x^4 - 7x^3 - 2x^4 + 5x^3$.

After the second step of division, did you obtain $-x^2$ for the second term of the quotient?

Yes ____ No ____

When you multiplied $-x^2$ by $2x - 5$, did you get $-2x^3 + 5x^2$?

Yes ____ No ____

When you subtracted for the second time, did you get $2x^2$?

Yes ____ No ____

If you answered No to any of these questions, go back and complete these steps again. Remember that $(-2x^3 + 7x^2) - (-2x^3 + 5x^2) = -2x^3 + 7x^2 + 2x^3 - 5x^2$.

If you answered Problem 7 incorrectly, go back and rework the problem using these suggestions.

Factoring a Trinomial with a Greatest Common Factor—Problem 16

Factor, if possible. $18x^2 + 3x - 15$

Helpful Hint Always look for a common factor before performing any other factoring step. Factor out the greatest common factor first. Then try to factor the remaining trinomial.

Did you notice that the greatest common factor of the trinomial is 3?

Yes ____ No ____

After factoring out the GCF, did you obtain $3(6x^2 + x - 5)$?

Yes ____ No ____

If you answered No to either question, carefully perform that step again. Remember to factor out the GCF of 3 from each term.

Did you notice that $6x^2 + x - 5$ can be factored into the form $(?\,x \quad 1)(?\,x \quad 5)$?

Yes ____ No ____

Did you notice that the parentheses must end with one positive number and with one negative number since the product must be -5?

Yes ____ No ____

If you answered No, stop and complete that step again.

Now go back and rework the problem using these suggestions.

Need help? Watch the MATH COACH videos in MyMathLab® or on YouTube™.

Factoring a Polynomial with Four Terms—Problem 20 Factor, if possible. $3x - 10ay + 6y - 5ax$

Helpful Hint Check to see if the first two terms have a common factor. if they do not, rearrange the order of the terms of the polynomial. Group the terms so that the first two terms have a common factor and the last two terms have a common factor.

Did you rearrange the order of the terms of the polynomial to $3x + 6y - 10ay - 5ax$?

Yes ____ No ____

Did you factor out a common factor of 3 from the first two terms to obtain $3(x + 2y)$?

Yes ____ No ____

If you answered No to these questions, go back and interchange the second term with the third term. Then factor out a common factor of 3 from the first two terms of the rearranged polynomial.

Did you factor out a common factor of $-5a$ from the second two terms of the rearranged polynomial to obtain $-5a(2y + x)$?

Yes ____ No ____

If you answered No, go back to the rearranged polynomial and factor out the GCF of the last two terms. Note that by the commutative property of addition, the expression $(2y + x)$ is equivalent to $(x + 2y)$. Simplify your result so that the common factor appears only once.

If you answered Problem 20 incorrectly, go back and rework the problem using these suggestions.

Solving a Quadratic Equation by Factoring—Problem 23 Solve. $x^2 = 5x + 14$

Helpful Hint Before factoring, write the equation in standard form, which is $ax^2 + bx + c = 0$.

Did you first add $-5x - 14$ to each side of the equation to obtain $x^2 - 5x - 14 = 0$?

Yes ____ No ____

If you answered No, stop and perform that step so that your equation is written in standard form.

Did you factor the rewritten equation to obtain $(x - 7)(x + 2)$ or $(x + 2)(x - 7)$?

Yes ____ No ____

If you answered No, go back and complete the factoring step again. Be careful of signs. Remember that once you finish factoring, you must set each factor equal to 0 to find the two solutions to the equation.

Now go back and rework the problem using these suggestions.

Need more help? Look for section examples marked with MC to review.

CHAPTER 6

In some Southern states it is very important to keep track of the number of alligators that are in the wild. One method that is used is called the capture-release-recapture method. Alligators are captured and tagged and then released. At a later time they are recaptured. Using the mathematics of this chapter, an accurate estimate is made of the size of the alligator population.

Rational Expressions and Equations

6.1 Rational Expressions and Functions: Simplifying, Multiplying, and Dividing

Student Learning Objectives

After studying this section, you will be able to:

1. **Simplify a rational expression or function.**
2. **Simplify the product of two or more rational expressions.**
3. **Simplify the quotient of two or more rational expressions.**

1 Simplifying a Rational Expression or Function

You may recall from Chapter 1 that a *rational number* is an exact quotient $\frac{a}{b}$ of two integers a and b with $b \neq 0$. A **rational expression** is an expression of the form $\frac{P}{Q}$, where P and Q are polynomials and Q is not zero. For example, $\frac{7}{x+2}$ and $\frac{x+5}{x-3}$ are rational expressions. Since the denominator cannot equal zero, the first expression is undefined if $x + 2 = 0$. This occurs when $x = -2$. We say that x can be any real number except -2. Look at the second expression. When is this expression undefined? Why? For what values of x is the expression defined? Note that the numerator can be zero because any fraction $\frac{0}{a}$ $(a \neq 0)$ is just 0.

A function defined by a rational expression is a **rational function.** The domain of a rational function is the set of values that can be used to replace the variable. Thus, the domain of $f(x) = \frac{7}{x+2}$ is all real numbers except -2. The domain of $g(x) = \frac{x+5}{x-3}$ is all real numbers except 3.

Graphing Calculator

Finding Domains

Find the domain in Example 1.

Since the domain will be values other than those that make the denominator equal to zero, we can graph $y = x^2 + 8x - 20$ in order to identify the values that are not in the domain.

Display:

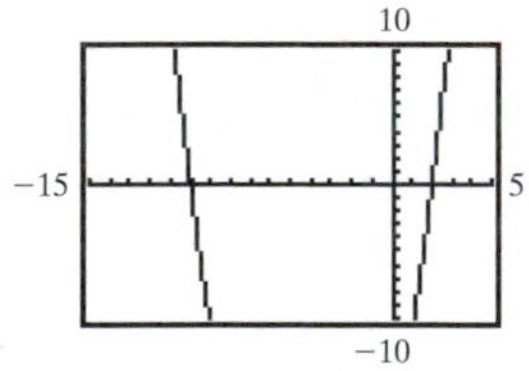

Use the Zoom and Trace features or the zero command to find where $y = 0$.

Try to find the domain of the following function. Round to the nearest tenth any values where the function is not defined.

$$y = \frac{x+9}{2.1x^2 + 5.2x - 3.1}$$

EXAMPLE 1 Find the domain of the function $f(x) = \frac{x-7}{x^2 + 8x - 20}$.

Solution The domain will be all real numbers except those that make the denominator equal to zero. What value(s) will make the denominator equal to zero? We determine this by solving the equation $x^2 + 8x - 20 = 0$.

$$(x + 10)(x - 2) = 0 \quad \text{Factor.}$$

$$x + 10 = 0 \quad \text{or} \quad x - 2 = 0 \quad \text{Use the zero factor property.}$$

$$x = -10 \qquad\qquad x = 2 \quad \text{Solve for } x.$$

The domain of $y = f(x)$ is all real numbers except -10 and 2.

Student Practice 1 Find the domain of $f(x) = \frac{4x+3}{x^2 - 9x - 22}$.

NOTE TO STUDENT: Fully worked-out solutions to all of the Student Practice problems can be found at the back of the text starting at page SP-1.

We have learned that we can simplify fractions (or reduce them to lowest terms) by factoring the numerator and denominator into prime factors and dividing out the common factors. For example,

$$\frac{15}{25} = \frac{3 \cdot \cancel{5}}{5 \cdot \cancel{5}} = \frac{3}{5}.$$

We can do this by using the **basic rule of fractions.**

BASIC RULE OF FRACTIONS

For any polynomials a, b, and c,

$$\frac{ac}{bc} = \frac{a}{b}, \quad \text{where } b \text{ and } c \neq 0.$$

What we are doing is factoring out the common factor c, and $(\frac{c}{c} = 1)$. We have

$$\frac{ac}{bc} = \frac{a}{b} \cdot \frac{c}{c} = \frac{a}{b} \cdot 1 = \frac{a}{b}.$$

Note that c must be a factor of the numerator *and* the denominator. Thus, the basic rule of fractions simply says that we may divide out a *common* factor from the numerator and the denominator. This nonzero factor can be a number or an algebraic expression.

EXAMPLE 2 Simplify. $\dfrac{2a^2 - ab - b^2}{a^2 - b^2}$

Solution

$$\frac{(2a + b)(a - b)}{(a + b)(a - b)} = \frac{2a + b}{a + b} \cdot 1 = \frac{2a + b}{a + b}$$

As you become more familiar with this basic rule, you won't have to write out every step. We did so here to show the application of the rule. We cannot simplify this fraction any further.

Student Practice 2 Simplify.

$$\frac{x^2 - 36y^2}{x^2 - 3xy - 18y^2}$$

EXAMPLE 3 Simplify. $\dfrac{2x - 3y}{2x^2 - 7xy + 6y^2}$

Solution

$$\frac{(2x - 3y)1}{(2x - 3y)(x - 2y)} = \frac{1}{x - 2y}$$

Note: Do you see why it is necessary to have a 1 in the numerator of the answer?

Student Practice 3 Simplify.

$$\frac{3xy}{3xy^2 + 6x^2y}$$

EXAMPLE 4 Simplify. $\dfrac{2x^2 + 2x - 12}{x^3 + 7x^2 + 12x}$

Solution

$$\frac{2x^2 + 2x - 12}{x^3 + 7x^2 + 12x} = \frac{2(x^2 + x - 6)}{x(x^2 + 7x + 12)} = \frac{2(x + 3)(x - 2)}{x(x + 3)(x + 4)} = \frac{2(x - 2)}{x(x + 4)}$$

We usually leave the answer in factored form.

Student Practice 4 Simplify. $\dfrac{x^3 - 4x^2 - 5x}{3x^2 - 30x + 75}$

Be alert for situations in which one factor in the numerator is the opposite of another factor in the denominator. In such cases you should factor -1 from one of the factors.

EXAMPLE 5 Simplify. $\dfrac{25y^2 - 16x^2}{8x^2 - 14xy + 5y^2}$

Solution

$$\frac{(5y + 4x)(5y - 4x)}{(4x - 5y)(2x - y)} = \frac{(5y + 4x)(5y - 4x)}{-1(-4x + 5y)(2x - y)} = \frac{5y + 4x}{-1(2x - y)} = -\frac{5y + 4x}{2x - y}$$

Observe that $4x - 5y = -1(-4x + 5y)$.

Student Practice 5 Simplify. $\dfrac{7a^2 - 23ab + 6b^2}{4b^2 - 49a^2}$

Sometimes you will need to factor a different negative number (other than -1) from the numerator, denominator, or one of the factors.

EXAMPLE 6 Simplify. $\dfrac{-2x + 14y}{x^2 - 5xy - 14y^2}$

Solution

$$\frac{-2x + 14y}{x^2 - 5xy - 14y^2} = \frac{-2(x - 7y)}{(x + 2y)(x - 7y)}$$ Factor -2 from each term of the numerator and factor the denominator.

$$= \frac{-2}{x + 2y}$$ Use the basic rule of fractions.

Student Practice 6 Simplify. $\dfrac{-3x + 6y}{x^2 - 7xy + 10y^2}$

② Simplifying the Product of Two or More Rational Expressions

Multiplication of rational expressions follows the same rule as multiplication of integer fractions. However, it is particularly helpful to use the basic rule of fractions to simplify whenever possible.

MULTIPLYING RATIONAL EXPRESSIONS

For any polynomials a, b, c, and d,

$$\frac{a}{b} \cdot \frac{c}{d} = \frac{ac}{bd}, \quad \text{where } b \text{ and } d \neq 0.$$

EXAMPLE 7 Multiply. $\dfrac{2x^2 - 4x}{x^2 - 5x + 6} \cdot \dfrac{x^2 - 9}{2x^4 + 14x^3 + 24x^2}$

Solution We first use the basic rule of fractions; that is, we factor (if possible) the numerator and denominator and divide out common factors.

$$\frac{2x(x - 2)}{(x - 2)(x - 3)} \cdot \frac{(x + 3)(x - 3)}{2x^2(x^2 + 7x + 12)} = \frac{2x(x - 2)(x + 3)(x - 3)}{(2x)x(x - 2)(x - 3)(x + 3)(x + 4)}$$

$$= \frac{2x}{2x} \cdot \frac{1}{x} \cdot \frac{x - 2}{x - 2} \cdot \frac{x + 3}{x + 3} \cdot \frac{x - 3}{x - 3} \cdot \frac{1}{x + 4}$$

$$= 1 \cdot \frac{1}{x} \cdot 1 \cdot 1 \cdot 1 \cdot \frac{1}{x + 4}$$

$$= \frac{1}{x(x + 4)} \quad \text{or} \quad \frac{1}{x^2 + 4x}$$

Although either form of the answer is correct, we usually use the factored form.

Student Practice 7 Multiply. $\dfrac{2x^2 + 5x + 2}{4x^2 - 1} \cdot \dfrac{10x^2 + 5x - 5}{x^2 + x - 2}$

EXAMPLE 8 Multiply. $\dfrac{7x + 7y}{4ax + 4ay} \cdot \dfrac{8a^2x^2 - 8b^2x^2}{35ax^3 - 35bx^3}$

Solution

$$\frac{7(x + y)}{4a(x + y)} \cdot \frac{8x^2(a^2 - b^2)}{35x^3(a - b)} = \frac{7(x + y)}{4a(x + y)} \cdot \frac{8x^2(a + b)(a - b)}{35x^3(a - b)}$$

$$= \frac{\cancel{7}\cancel{(x + y)}}{\cancel{4}a\cancel{(x + y)}} \cdot \frac{\overset{2}{\cancel{8}}\cancel{x^2}(a + b)\cancel{(a - b)}}{\underset{5}{\cancel{35}}\underset{x}{\cancel{x^3}}\cancel{(a - b)}}$$

$$= \frac{2(a + b)}{5ax} \quad \text{or} \quad \frac{2a + 2b}{5ax}$$

Note that we shortened our steps by not writing out every factor 1 as we did in Example 7. Either way is correct.

Student Practice 8 Multiply. $\dfrac{9x + 9y}{5ax + 5ay} \cdot \dfrac{10a^2x^2 - 40b^2x^2}{27ax^2 - 54bx^2}$

3 Simplifying the Quotient of Two or More Rational Expressions

When we divide fractions, we take the **reciprocal** of the second fraction and then multiply the fractions. (Remember that the reciprocal of a fraction $\dfrac{m}{n}$ is $\dfrac{n}{m}$. Thus, the reciprocal of $\dfrac{2}{3}$ is $\dfrac{3}{2}$, and the reciprocal of $\dfrac{3x}{11y^2}$ is $\dfrac{11y^2}{3x}$.) We divide rational expressions in the same way.

DIVIDING RATIONAL EXPRESSIONS

For any polynomials a, b, c, and d,

$$\frac{a}{b} \div \frac{c}{d} = \frac{a}{b} \cdot \frac{d}{c}, \quad \text{where } b, c, \text{ and } d \neq 0.$$

EXAMPLE 9 Divide. $\dfrac{4x^2 - y^2}{x^2 + 4xy + 4y^2} \div \dfrac{4x - 2y}{3x + 6y}$

Solution We take the reciprocal of the second fraction and multiply the fractions.

$$\frac{4x^2 - y^2}{x^2 + 4xy + 4y^2} \cdot \frac{3x + 6y}{4x - 2y} = \frac{(2x + y)\cancel{(2x - y)}}{\cancel{(x + 2y)}(x + 2y)} \cdot \frac{3\cancel{(x + 2y)}}{2\cancel{(2x - y)}}$$

$$= \frac{3(2x + y)}{2(x + 2y)} \quad \text{or} \quad \frac{6x + 3y}{2x + 4y}$$

Student Practice 9 Divide.

$$\frac{8x^3 + 27y^3}{64x^3 - y^3} \div \frac{4x^2 - 9y^2}{16x^2 + 4xy + y^2}$$

EXAMPLE 10 Divide. $\dfrac{24 + 10x - 4x^2}{2x^2 + 13x + 15} \div (2x - 8)$

Solution We take the reciprocal of the second fraction and multiply the fractions.

The reciprocal of $(2x - 8)$ is $\dfrac{1}{2x - 8}$.

$$\frac{-4x^2 + 10x + 24}{2x^2 + 13x + 15} \cdot \frac{1}{2x - 8} = \frac{-2(2x^2 - 5x - 12)}{(2x + 3)(x + 5)} \cdot \frac{1}{2(x - 4)}$$

$$= \frac{\overset{-1}{\cancel{-2}}\,\cancel{(x - 4)}\,\cancel{(2x + 3)}}{\cancel{(2x + 3)}(x + 5)} \cdot \frac{1}{\underset{1}{\cancel{2}}\,\cancel{(x - 4)}}$$

$$= \frac{-1}{x + 5} \text{ or } -\frac{1}{x + 5}$$

Student Practice 10 Divide.

$$\frac{4x^2 - 9}{2x^2 + 11x + 12} \div (-6x + 9)$$

STEPS TO SUCCESS How Important Is the Quick Quiz?

At the end of each homework exercises section is a Quick Quiz. Please be sure to complete that section. You will be amazed at how it will help you prove to yourself that you have mastered the concepts discussed in that section. Make sure you do the Quick Quiz each time as a necessary part of your homework. It will really help you gain confidence in what you have learned. The Quick Quiz is essential.

Here is a fun way you can use the Quick Quiz. Ask a friend in the class if he or she will "quiz" you about five minutes before class by asking you to work out (without using your book) the solution to one of the problems on the Quick Quiz. Tell your friend you can do the same thing for him or her. This will force you to "be ready for anything" when you come to class. This little trick will keep you sharp and ready for class.

Making It personal: Which of these suggestions do you find most helpful? Use those suggestions as you do each section of Chapter 6. ▼

6.1 Exercises

MyMathLab® Watch the videos in MyMathLab Download the MyDashBoard App

Find the domain of each of the following rational functions.

1. $f(x) = \dfrac{5x + 6}{2x - 6}$

2. $f(x) = \dfrac{3x - 8}{4x + 20}$

3. $g(x) = \dfrac{-3x + 5}{x^2 + 5x - 24}$

4. $g(x) = \dfrac{3x}{x^2 - 5x - 6}$

Simplify completely.

5. $\dfrac{-18x^4y}{12x^2y^6}$

6. $\dfrac{-7xy^2}{28x^5y}$

7. $\dfrac{3x^3 - 24x^2}{6x - 48}$

8. $\dfrac{10x^2 + 15x}{35x^2 - 5x}$

9. $\dfrac{9x^2}{12x^2 - 15x}$

10. $\dfrac{24x^2}{8x^2 - 4x}$

11. $\dfrac{5x^2y^2 - 15xy^2}{10x^3y - 20x^3y^2}$

12. $\dfrac{6m^2n^4 + 2mn^3}{4mn^2 - 8n^2}$

13. $\dfrac{2x + 10}{2x^2 - 50}$

14. $\dfrac{6x^2 - 15x}{3x}$

15. $\dfrac{3y^2 - 27}{3y + 9}$

16. $\dfrac{x + 2}{7x^2 - 28}$

17. $\dfrac{2x^2 - x^3 - x^4}{x^4 - x^3}$

18. $\dfrac{30x - x^2 - x^3}{x^3 - x^2 - 20x}$

19. $\dfrac{2y^2 + y - 10}{4 - y^2}$

20. $\dfrac{36 - b^2}{3b^2 - 16b - 12}$

Multiply.

21. $\dfrac{-8mn^5}{3m^4n^3} \cdot \dfrac{9m^3n^3}{6mn}$

22. $\dfrac{24x^4y^2}{-12x^5y} \cdot \dfrac{8x^3}{30x^2y}$

23. $\dfrac{3a^2}{a^2 + 4a + 4} \cdot \dfrac{a^2 - 4}{3a}$

24. $\dfrac{8x^2}{x^2 - 9} \cdot \dfrac{x^2 + 6x + 9}{16x^3}$

25. $\dfrac{x^2 + 5x + 7}{x^2 - 5x + 6} \cdot \dfrac{3x - 6}{x^2 + 5x + 7}$

26. $\dfrac{x - 5}{10x - 2} \cdot \dfrac{25x^2 - 1}{x^2 - 10x + 25}$

27. $\dfrac{x^2 - 3xy - 10y^2}{x + y} \cdot \dfrac{x^2 + 7xy + 6y^2}{x + 2y}$

28. $\dfrac{x + 8y}{x^2 + 5xy - 24y^2} \cdot \dfrac{x^2 + 2xy - 15y^2}{x + 5y}$

29. $\dfrac{y^2 - y - 12}{2y^2 + y - 1} \cdot \dfrac{2y^2 + 7y - 4}{2y^2 - 32}$

30. $\dfrac{6a^2 - a - 1}{a^2 - 3a - 4} \cdot \dfrac{5a^2 - 5}{3a^2 - 2a - 1}$

31. $\dfrac{x^3 - 125}{x^5y} \cdot \dfrac{x^3y^2}{x^2 + 5x + 25}$

32. $\dfrac{3a^3b^2}{8a^3 - b^3} \cdot \dfrac{4a^2 + 2ab + b^2}{12ab^4}$

Divide.

33. $\dfrac{2mn - m}{15m^3} \div \dfrac{2n - 1}{3m^2}$

34. $\dfrac{3y + 12}{8y^3} \div \dfrac{9y + 36}{16y^3}$

35. $\dfrac{b^2 - 6b + 9}{5b^2 - 16b + 3} \div \dfrac{6b - 3}{15b - 3}$

36. $\dfrac{a^2 - 3a}{a^2 - 9} \div \dfrac{a^3 - a^2}{4a^2 - a - 3}$

37. $\dfrac{x^2 - xy - 6y^2}{x^2 + 2} \div (x^2 + 2xy)$

38. $\dfrac{x^2 - 5x + 4}{2x - 8} \div (3x^2 - 3x)$

Mixed Practice *Perform the operation indicated. When no operation is indicated, simplify the rational expression completely.*

39. $\dfrac{7x}{y^2} \div 21x^3$

40. $\dfrac{10m^4}{9a^2} \cdot 3a^2b$

41. $\dfrac{5x^2 - 2x}{15x - 6}$

42. $\dfrac{3x^2 - 5x}{18x - 30}$

43. $\dfrac{x^2y - 49y}{x^2y^3} \cdot \dfrac{3x^2y - 21xy}{x^2 - 14x + 49}$

44. $\dfrac{x^2 + 6x + 9}{2x^2y - 18y} \div \dfrac{6xy + 18y}{3x^2y - 27y}$

45. $\dfrac{x^2 - 9x + 14}{x^3y^4} \div \dfrac{x^2 - 6x - 7}{x^2y^2}$

46. $\dfrac{x^2 - 10x + 16}{x^5y} \div \dfrac{x^2 - 6x - 16}{x^2y^5}$

47. $\dfrac{a^2 - a - 12}{2a^2 + 5a - 12}$

48. $\dfrac{3y^2 - 3y - 36}{2y^2 - y - 3}$

Optional Graphing Calculator Problems *Find the domain of the following functions. Round any values that you exclude to the nearest tenth.*

49. $f(x) = \dfrac{2x + 5}{3.6x^2 + 1.8x - 4.3}$

50. $f(x) = \dfrac{5x - 4}{1.6x^2 - 1.3x - 5.9}$

Applications

Tropical Fish *The total number of fish will depend on the size of a pond or aquarium. At the Reading Mandarin Restaurant a total of 90 tropical fish were placed in an aquarium. For the next 12 months the total number of fish could be predicted by the equation*

$$P(x) = \frac{90(1 + 1.5x)}{1 + 0.5x},$$

where P is the number of fish and x is the number of months since the fish were placed in the aquarium.

51. One month after the fish are first placed in the aquarium, what is the total number of fish?

52. Three months after the fish are first placed in the aquarium, what is the total number of fish?

53. Six months after the fish are first placed in the aquarium, what is the total number of fish?

54. One year after the fish are first placed in the aquarium, what is the total number of fish? Round your answer to the nearest whole number.

55. Using your answers for exercises 51 to 54, graph the function $P(x)$.

56. Based on your graph, what would you estimate to be the total number of fish eleven months after the fish are first placed in the aquarium? Round your answer to the nearest ten.

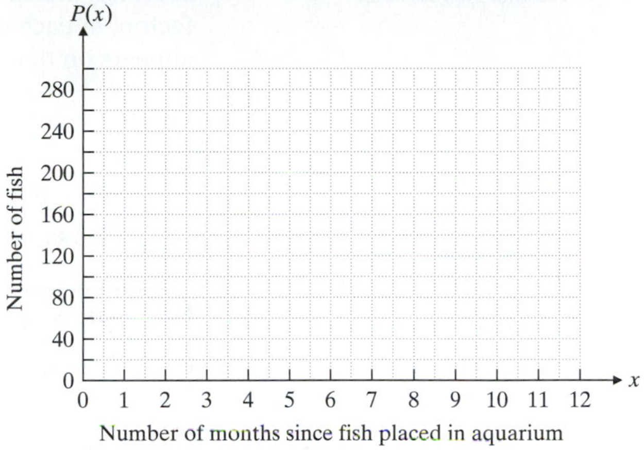

Cumulative Review

57. **[2.6.3]** Solve. $3x - (x + 5) < 7x + 10$

58. **[5.1.6]** Multiply. $(4x - 1)^2$

59. **[1.2.2]** ***Motorcyclist Deaths*** Between 1996 and 2008, the number of annual deaths from motorcycle accidents went from 2160 to 5290. (This may have been the result of many states loosening laws requiring riders to wear helmets.) What was the percent increase in the number of deaths between 1996 and 2008? In 2008, 59% of riders killed were not wearing helmets. Of the number of people killed in 2008 in motorcycle accidents, how many were not wearing helmets? (*Source:* www.dot.gov)

Quick Quiz 6.1 *Simplify.*

1. $\dfrac{x^3 + 11x^2 + 30x}{3x^3 + 17x^2 - 6x}$

2. $\dfrac{2x^2 - 128}{x^2 + 16x + 64} \cdot \dfrac{3x^2 + 30x + 48}{x^2 - 6x - 16}$

3. $\dfrac{x^2 + x - 6}{x - 5} \div \dfrac{x^2 + 8x + 15}{15 - 3x}$

4. **Concept Check** Explain how you would simplify the following.

$$\frac{9 - x^2}{x^2 - 7x + 12}$$

6.2 Adding and Subtracting Rational Expressions

Student Learning Objectives

After studying this section, you will be able to:

1. **Find the LCD of two or more rational expressions.**
2. **Add or subtract two or more rational expressions.**

① Finding the LCD of Two or More Rational Expressions

Recall that if we wish to add or subtract fractions, the denominators must be the same. If the denominators are not the same, we use the basic rule of fractions to rewrite one or both fractions so that the denominators are the same. For example,

$$\frac{3}{7}+\frac{2}{7}=\frac{5}{7} \quad \text{and} \quad \frac{2}{3}+\frac{3}{4}=\frac{2}{3}\cdot\frac{4}{4}+\frac{3}{4}\cdot\frac{3}{3}=\frac{8}{12}+\frac{9}{12}=\frac{17}{12}.$$

How did we know that 12 was the least common denominator (LCD)? The least common denominator of two or more fractions is the product of the different prime factors in each denominator. If a factor is repeated, we use the highest power that appears on that factor in the denominators.

$$3 = 3$$
$$4 = 2\cdot 2$$
$$\text{LCD} = 3\cdot 2\cdot 2 = 12$$

This same technique is used to add or subtract rational expressions.

HOW TO FIND THE LCD

1. Factor each denominator completely into prime factors.
2. List all the different prime factors.
3. The LCD is the product of these factors, each of which is raised to the highest power that appears on that factor in the denominators.

Now we'll do some sample problems.

EXAMPLE 1 Find the LCD of the rational expressions $\frac{7}{x^2-4}$ and $\frac{2}{x-2}$.

Solution

Step 1 We factor each denominator completely (into prime factors).

$$x^2-4=(x+2)(x-2)$$
$x-2$ cannot be factored.

Step 2 We list all the *different* prime factors. The different factors are $x+2$ and $x-2$.

Step 3 Since no factor occurs more than once in the denominators, the LCD is the product $(x+2)(x-2)$.

Student Practice 1 Find the LCD of the rational expressions

$$\frac{8}{x^2-x-12} \text{ and } \frac{3}{x-4}.$$

***NOTE TO STUDENT:** Fully worked-out solutions to all of the Student Practice problems can be found at the back of the text starting at page SP-1.*

EXAMPLE 2 Find the LCD of the rational expressions $\frac{7}{12xy^2}$ and $\frac{4}{15x^3y}$.

Solution

Step 1 We factor each denominator.

$$12xy^2 = 2\cdot 2\cdot 3\cdot x\cdot y\cdot y$$
$$15x^3y = 3\cdot 5\cdot x\cdot x\cdot x\cdot y$$

Step 2 Our LCD will require each of the different factors that appear in either denominator. They are 2, 3, 5, x, and y.

Step 3 The factor 2 and the factor y each occur twice in $12xy^2$.
The factor x occurs three times in $15x^3y$.
Thus, the LCD $= 2 \cdot 2 \cdot 3 \cdot 5 \cdot x \cdot x \cdot x \cdot y \cdot y$
$= 60x^3y^2$

Student Practice 2 Find the LCD of $\dfrac{2}{15x^3y^2}$ and $\dfrac{13}{25xy^3}$.

② Adding or Subtracting Two or More Rational Expressions

We can add and subtract rational expressions with the same denominator just as we do in arithmetic: We simply add or subtract the numerators.

ADDITION AND SUBTRACTION OF RATIONAL EXPRESSIONS

For any polynomials a, b, and c we have the following:

$$\frac{a}{b} + \frac{c}{b} = \frac{a + c}{b}, \qquad b \neq 0$$

$$\frac{a}{b} - \frac{c}{b} = \frac{a - c}{b}, \qquad b \neq 0.$$

EXAMPLE 3 Subtract. $\dfrac{5x + 2}{(x + 3)(x - 4)} - \dfrac{6x}{(x + 3)(x - 4)}$

Solution $\dfrac{5x + 2}{(x + 3)(x - 4)} - \dfrac{6x}{(x + 3)(x - 4)} = \dfrac{-x + 2}{(x + 3)(x - 4)}$

Student Practice 3 Subtract.

$$\frac{4x}{(x + 6)(2x - 1)} - \frac{3x + 1}{(x + 6)(2x - 1)}$$

If the two rational expressions have different denominators, we first need to find the LCD. Then we must rewrite each fraction as an equivalent fraction that has the LCD as the denominator.

EXAMPLE 4 Add the rational expressions. $\dfrac{7}{(x + 2)(x - 2)} + \dfrac{2}{x - 2}$

Solution The LCD $= (x + 2)(x - 2)$.

Before we can add the fractions, we must rewrite our fractions as fractions with the LCD. The first fraction needs no change, but the second fraction does.

$$\frac{7}{(x + 2)(x - 2)} + \frac{2}{x - 2} \cdot \frac{x + 2}{x + 2}$$

Since $\dfrac{x + 2}{x + 2} = 1$, we have not changed the *value* of the fraction. We are simply writing it in an equivalent form. Thus, we now have the following:

$$\frac{7}{(x + 2)(x - 2)} + \frac{2(x + 2)}{(x + 2)(x - 2)} = \frac{7 + 2(x + 2)}{(x + 2)(x - 2)}$$

$$= \frac{7 + 2x + 4}{(x + 2)(x - 2)} = \frac{2x + 11}{(x + 2)(x - 2)}$$

Student Practice 4 Add the rational expressions.

$$\frac{8}{(x - 4)(x + 3)} + \frac{3}{x - 4}$$

Graphing Calculator

Exploration

You can verify the answer to Example 5 with a graphing calculator.

Graph $y_1 = \frac{4}{x+6} + \frac{5}{6x}$ and $y_2 = \frac{29x+30}{6x(x+6)}$ on the same set of axes. Since y_1 is equivalent to y_2, you will obtain exactly the same curve.

Verify on your graphing calculator whether y_1 *is* or *is not* equivalent to y_2 in the following:

$$y_1 = \frac{x}{x+3} - \frac{3-x}{x^2-9}$$

$$\text{and } y_2 = \frac{x+1}{x+3}$$

EXAMPLE 5 Add. $\frac{4}{x+6} + \frac{5}{6x}$

Solution The LCD $= 6x(x+6)$. We need to multiply the fractions by 1 to obtain two equivalent fractions that have the LCD for the denominator. We multiply the first fraction by $1 = \frac{6x}{6x}$. We multiply the second fraction by $1 = \frac{x+6}{x+6}$.

$$\frac{4(6x)}{(x+6)(6x)} + \frac{5(x+6)}{6x(x+6)}$$

$$= \frac{24x}{6x(x+6)} + \frac{5x+30}{6x(x+6)} = \frac{29x+30}{6x(x+6)}$$

Student Practice 5 Add. $\frac{5}{x+4} + \frac{3}{4x}$

EXAMPLE 6 Add. $\frac{7}{2x^2y} + \frac{3}{xy^2}$

Solution You should be able to see that the LCD of these fractions is $2x^2y^2$.

$$\frac{7}{2x^2y} \cdot \frac{y}{y} + \frac{3}{xy^2} \cdot \frac{2x}{2x} = \frac{7y}{2x^2y^2} + \frac{6x}{2x^2y^2} = \frac{6x+7y}{2x^2y^2}$$

Student Practice 6 Add. $\frac{7}{4ab^3} + \frac{1}{3a^3b^2}$

When two rational expressions are subtracted, we must be very careful with the signs in the numerator of the second fraction.

EXAMPLE 7 Subtract. $\frac{2}{x^2+3x+2} - \frac{4}{x^2+4x+3}$

Solution

$$x^2 + 3x + 2 = (x+1)(x+2)$$

$$x^2 + 4x + 3 = (x+1)(x+3)$$

Therefore, the LCD is $(x+1)(x+2)(x+3)$. We now have the following:

$$\frac{2}{(x+1)(x+2)} \cdot \frac{x+3}{x+3} - \frac{4}{(x+1)(x+3)} \cdot \frac{x+2}{x+2}$$

$$= \frac{2x+6}{(x+1)(x+2)(x+3)} - \frac{4x+8}{(x+1)(x+2)(x+3)}$$

$$= \frac{2x+6-4x-8}{(x+1)(x+2)(x+3)}$$

$$= \frac{-2x-2}{(x+1)(x+2)(x+3)} = \frac{-2\cancel{(x+1)}}{\cancel{(x+1)}(x+2)(x+3)}$$

$$= \frac{-2}{(x+2)(x+3)}$$

Study this problem carefully. Be sure you understand the reason for each step. You'll see this type of problem often.

Student Practice 7 Subtract.

$$\frac{4x+2}{x^2+x-12} - \frac{3x+8}{x^2+6x+8}$$

The following example involves repeated factors in the denominator. Read it through carefully to be sure you understand each step.

EXAMPLE 8 Subtract. $\dfrac{2x+1}{25x^2+10x+1} - \dfrac{6x}{25x+5}$

Solution

Step 1 Factor each denominator into prime factors.

$$25x^2 + 10x + 1 = (5x+1)^2$$
$$25x + 5 = 5(5x+1)$$

Step 2 The different factors are 5 and $5x + 1$. However, $5x + 1$ appears to the first power *and* to the second power. So we need step 3.

Step 3 We must use the *highest* power of each factor. In this example the highest power of the factor 5 is 1. The highest power of the factor $5x + 1$ is 2, so we use $(5x + 1)^2$. Thus, the LCD is $5(5x + 1)^2$.

Now after finding the LCD, we first we write our problem in factored form.

$$\frac{2x+1}{(5x+1)^2} - \frac{6x}{5(5x+1)}$$

Next, we must multiply our fractions by the appropriate factor to change them to equivalent fractions with the LCD.

$$\frac{2x+1}{(5x+1)^2}\cdot\frac{5}{5} - \frac{6x}{5(5x+1)}\cdot\frac{5x+1}{5x+1} = \frac{5(2x+1) - 6x(5x+1)}{5(5x+1)^2}$$

$$= \frac{10x + 5 - 30x^2 - 6x}{5(5x+1)^2}$$

$$= \frac{-30x^2 + 4x + 5}{5(5x+1)^2}$$

Student Practice 8 Subtract.

$$\frac{7x-3}{4x^2+20x+25} - \frac{3x}{4x+10}$$

CAUTION: Adding and subtracting rational expressions is somewhat difficult. You should take great care in finding the LCD. Students sometimes make careless errors when finding the LCD.

Likewise, great care should be taken to copy correctly all + and − signs. It is very easy to make a sign error when combining the equivalent fractions. Try to work very neatly and very carefully. A little extra diligence will result in greater accuracy.

6.2 Exercises

MyMathLab® Watch the videos in MyMathLab Download the MyDashBoard App

Verbal and Writing Skills, Exercises 1 and 2

1. Explain how to find the LCD of the fractions $\frac{3}{5xy}$ and $\frac{11}{y^3}$.

2. Explain how to find the LCD of the fractions $\frac{8}{7xy}$ and $\frac{3}{x^2}$.

Find the LCD.

3. $\frac{3}{x-1}, \frac{4}{x^2-2x+1}$

4. $\frac{5}{x-6}, \frac{7}{x^2-12x+36}$

5. $\frac{7}{2m^3n}, \frac{3}{2mn^2}$

6. $\frac{5}{7ab^5}, \frac{2}{7a^3b^3}$

7. $\frac{3x}{(2x+5)^3}, \frac{x-2}{(x+1)(2x+5)^2}$

8. $\frac{10x}{(x+5)(x-3)^2}, \frac{12xy}{(x-3)^3}$

9. $\frac{15xy}{3x^2+2x}, \frac{17y}{18x^2+9x-2}$

10. $\frac{8x}{3x^2-4x}, \frac{10xy}{3x^2+5x-12}$

Add or subtract and simplify your answers.

11. $\frac{3}{x+4}+\frac{2}{x^2-16}$

12. $\frac{6}{x^2-7x+10}+\frac{3}{x-5}$

13. $\frac{9}{4xy}+\frac{3}{4y^2}$

14. $\frac{5}{6a^2}+\frac{1}{6ab}$

15. $\frac{3}{x^2-7x+12}+\frac{5}{x^2-4x}$

16. $\frac{7}{x^2-1}+\frac{5}{3x^2+3x}$

17. $\frac{6x}{2x-5}+4$

18. $\frac{14}{8a+1}+3$

19. $\frac{-5y}{y^2-1}+\frac{6}{y^2-2y+1}$

20. $\frac{7y}{y^2+6y+9}+\frac{5}{y^2-9}$

Mixed Practice *Add or subtract and simplify your answers.*

21. $\frac{a + 4}{3a - 6} + \frac{a - 1}{a^2 - 4}$

22. $\frac{4b}{b^2 - 2b - 15} + \frac{b + 2}{2b - 10}$

23. $\frac{5}{x - 4} - \frac{3}{x + 1}$

24. $\frac{7}{x + 2} - \frac{4}{2x - 3}$

25. $\frac{1}{x^2 - x - 2} - \frac{3}{x^2 + 2x + 1}$

26. $\frac{3x}{x^2 + 3x - 10} - \frac{2x}{x^2 + x - 6}$

27. $\frac{4y}{y^2 + 3y + 2} - \frac{y - 3}{y + 2}$

28. $\frac{3y^2}{y^2 - 1} - \frac{y + 2}{y + 1}$

29. $a + 2 + \frac{3}{2a - 5}$

30. $a - 6 + \frac{2}{4a + 1}$

Applications

31. ***Artificial Lung*** If an artificial lung company manufactures more than five machines per day, the revenue function in thousands of dollars to manufacture and sell x machines is given by

$$R(x) = \frac{80 - 24x}{2 - x}.$$

The cost function in thousands of dollars to manufacture x machines is given by

$$C(x) = \frac{60 - 12x}{3 - x}.$$

Determine the profit function in thousands of dollars for this company when more than five machines per day are manufactured by obtaining $P(x) = R(x) - C(x)$.

32. ***Artificial Heart*** If an artificial heart company manufactures more than five machines per day, the revenue function in thousands of dollars to manufacture and sell x machines is given by

$$R(x) = \frac{150 - 38x}{3 - x}.$$

The cost function in thousands of dollars to manufacture x machines is given by

$$C(x) = \frac{120 - 25x}{2 - x}.$$

Determine the profit function in thousands of dollars for this company when more than five machines per day are manufactured by obtaining $P(x) = R(x) - C(x)$.

33. ***Artificial Lung*** Determine the daily profit of the artificial lung company in exercise 31 if ten machines per day are manufactured. Round your answer to the nearest dollar.

34. ***Artificial Heart*** Determine the daily profit of the artificial heart company in exercise 32 if twenty machines per day are manufactured. Round your answer to the nearest dollar.

Cumulative Review

35. **[1.3.3]** Evaluate. $\dfrac{-6 \cdot 9 \div |1 - 10|}{4^2 - 5 \cdot 2 + 3\sqrt{25}}$

36. **[1.4.5]** Write using scientific notation. 0.000351

37. **[2.5.1]** ***Purchase Price of Automobiles*** Jamie, Denise, and Amanda each purchased a car. The total cost of the cars was $24,000. Jamie purchased a car that cost twice as much as the one Amanda purchased. Denise purchased a car that cost $2000 more than the cost of Amanda's car. How much did each car cost?

38. **[2.5.1]** ***Chemical Mixtures*** A chemist at Argonne Laboratories must combine a mixture that is 15% acid with a mixture that is 30% acid to obtain 60 liters of a mixture that is 20% acid. How much of each kind should he use?

Quick Quiz 6.2 *Add or subtract and simplify your answers.*

1. $\dfrac{5}{x} - \dfrac{4}{x + 3}$

2. $\dfrac{4}{x^2 + 3x + 2} + \dfrac{3}{x^2 + 6x + 8}$

3. $\dfrac{6x}{3x - 7} + 5$

4. **Concept Check** Explain how you would find the LCD of $\dfrac{2}{x^2 - 16}$ and $\dfrac{3}{x^2 - 6x + 8}$.

6.3 Complex Rational Expressions

① Simplifying Complex Rational Expressions

A **complex rational expression** is a large fraction that has at least one rational expression in the numerator, in the denominator, or in both the numerator and the denominator. The following are three examples of complex rational expressions.

$$\frac{7 + \frac{1}{x}}{x + 2}, \quad \frac{2}{\frac{x}{y} + 3}, \quad \frac{\frac{a + b}{7}}{\frac{1}{x} + \frac{1}{x + a}}$$

There are two ways to simplify complex rational expressions. You can use whichever method you like.

Student Learning Objective

After studying this section, you will be able to:

① **Simplify complex rational expressions.**

Graphing Calculator

Exploration

You can verify the answer for Example 1 on a graphing calculator. Graph y_1 and y_2 on the same set of axes.

$$y_1 = \frac{x + \frac{1}{x}}{\frac{1}{x} + \frac{3}{x^2}}$$

$$y_2 = \frac{x(x^2 + 1)}{x + 3}$$

The domain of y_1 is more restrictive than that of y_2. If we use the domain of y_1, then the graphs of y_1 and y_2 should be identical.

Show on your graphing calculator whether y_1 is or is not equivalent to y_2 in the following:

$$y_1 = \frac{1 + \frac{3}{x + 2}}{1 + \frac{6}{x - 1}}$$

$$y_2 = \frac{x - 1}{x + 2}$$

EXAMPLE 1 Simplify. $\dfrac{x + \frac{1}{x}}{\frac{1}{x} + \frac{3}{x^2}}$

Solution

Method 1

1. Simplify numerator and denominator.

$$x + \frac{1}{x} = \frac{x^2 + 1}{x}$$

$$\frac{1}{x} + \frac{3}{x^2} = \frac{x + 3}{x^2}$$

2. Divide the numerator by the denominator.

$$\frac{\frac{x^2 + 1}{x}}{\frac{x + 3}{x^2}} = \frac{x^2 + 1}{x} \div \frac{x + 3}{x^2}$$

$$= \frac{x^2 + 1}{x} \cdot \frac{x^2}{x + 3}$$

$$= \frac{x^2 + 1}{\cancel{x}} \cdot \frac{\overset{x}{\cancel{x^2}}}{x + 3}$$

$$= \frac{x(x^2 + 1)}{x + 3}$$

3. The result is already simplified.

Method 2

1. Find the LCD of all the fractions in the numerator and denominator. The LCD is x^2.

2. Multiply the numerator and denominator by the LCD. Use the distributive property.

$$\frac{x + \frac{1}{x}}{\frac{1}{x} + \frac{3}{x^2}} \cdot \frac{x^2}{x^2} = \frac{x^3 + x}{x + 3}$$

3. The result is already simplified, but we will write it in factored form.

$$\frac{x^3 + x}{x + 3} = \frac{x(x^2 + 1)}{x + 3}$$

Student Practice 1 Simplify.

$$\frac{y + \frac{3}{y}}{\frac{2}{y^2} + \frac{5}{y}}$$

NOTE TO STUDENT: Fully worked-out solutions to all of the Student Practice problems can be found at the back of the text starting at page SP-1.

NOTE TO STUDENT reference: page SP-1

METHOD 1: COMBINING FRACTIONS IN BOTH NUMERATOR AND DENOMINATOR

1. Simplify the numerator and denominator, if possible, by combining quantities to obtain one fraction in the numerator and one fraction in the denominator.
2. Divide the numerator by the denominator (that is, multiply the numerator by the reciprocal of the denominator).
3. Simplify the expression.

EXAMPLE 2 Simplify. $\dfrac{\dfrac{1}{2x+6}+\dfrac{3}{2}}{\dfrac{3}{x^2-9}+\dfrac{x}{x-3}}$

Solution

Method 1

1. Simplify the numerator.

$$\frac{1}{2x+6}+\frac{3}{2}=\frac{1}{2(x+3)}+\frac{3}{2}$$
$$=\frac{1}{2(x+3)}+\frac{3(x+3)}{2(x+3)}$$
$$=\frac{1+3x+9}{2(x+3)}$$
$$=\frac{3x+10}{2(x+3)}$$

Simplify the denominator.

$$\frac{3}{x^2-9}+\frac{x}{x-3}=\frac{3}{(x+3)(x-3)}+\frac{x}{x-3}$$
$$=\frac{3}{(x+3)(x-3)}+\frac{x(x+3)}{(x+3)(x-3)}$$
$$=\frac{x^2+3x+3}{(x+3)(x-3)}$$

2. Divide the numerator by the denominator.

$$\frac{3x+10}{2(x+3)}\div\frac{x^2+3x+3}{(x+3)(x-3)}=\frac{3x+10}{2\cancel{(x+3)}}\cdot\frac{\cancel{(x+3)}(x-3)}{x^2+3x+3}$$
$$=\frac{(3x+10)(x-3)}{2(x^2+3x+3)}$$

3. Simplify. The answer is already simplified.

Student Practice 2 Simplify the following fraction by Method 1. If your instructor prefers that you use Method 2, described on the next page, then use this space to do so.

$$\frac{\dfrac{4}{16x^2-1}+\dfrac{3}{4x+1}}{\dfrac{x}{4x-1}+\dfrac{5}{4x+1}}$$

Before we continue the example, we state Method 2 in the following box.

METHOD 2: MULTIPLYING EACH TERM OF THE NUMERATOR AND DENOMINATOR BY THE LCD OF ALL INDIVIDUAL FRACTIONS

1. Find the LCD of all the rational expressions in the numerator and denominator.
2. Multiply the numerator and denominator of the complex fraction by the LCD.
3. Simplify the result.

We will now proceed to do Example 2 by Method 2.

Method 2

1. To find the LCD, we factor.

$$\frac{\frac{1}{2x+6}+\frac{3}{2}}{\frac{3}{x^2-9}+\frac{x}{x-3}} = \frac{\frac{1}{2(x+3)}+\frac{3}{2}}{\frac{3}{(x+3)(x-3)}+\frac{x}{x-3}}$$

The LCD of the two fractions in the numerator and the two fractions in the denominator is

$$2(x+3)(x-3).$$

2. Multiply the numerator and the denominator by the LCD.

$$\frac{\frac{1}{2(x+3)}+\frac{3}{2}}{\frac{3}{(x+3)(x-3)}+\frac{x}{x-3}} \cdot \frac{2(x+3)(x-3)}{2(x+3)(x-3)}$$

$$= \frac{\frac{1}{\cancel{2}\cancel{(x+3)}} \cdot \cancel{2}\cancel{(x+3)}(x-3) + \frac{3}{\cancel{2}} \cdot \cancel{2}(x+3)(x-3)}{\frac{3}{\cancel{(x+3)}\cancel{(x-3)}} \cdot 2\cancel{(x+3)}\cancel{(x-3)} + \frac{x}{\cancel{x-3}} \cdot 2(x+3)\cancel{(x-3)}}$$

$$= \frac{x-3+3(x+3)(x-3)}{6+2x(x+3)}$$

$$= \frac{3x^2+x-30}{2x^2+6x+6} = \frac{(3x+10)(x-3)}{2(x^2+3x+3)}$$

3. Simplify. The answer is already simplified.

Whether we use Method 1 or Method 2, we can leave the answer in factored form, or we can multiply it out to obtain

$$\frac{3x^2+x-30}{2x^2+6x+6}.$$

EXAMPLE 3 Simplify by Method 1. $\dfrac{x+3}{\dfrac{9}{x}-x}$

Solution

$$\frac{x+3}{\dfrac{9}{x}-\dfrac{x}{1}\cdot\dfrac{x}{x}}=\frac{x+3}{\dfrac{9}{x}-\dfrac{x^2}{x}}=\frac{\dfrac{x+3}{1}}{\dfrac{9-x^2}{x}}=\frac{x+3}{1}\div\frac{9-x^2}{x}=\frac{x+3}{1}\cdot\frac{x}{9-x^2}$$

$$=\frac{\cancel{(x+3)}}{1}\cdot\frac{x}{\cancel{(3+x)}(3-x)}=\frac{x}{3-x}$$

Student Practice 3 Simplify by Method 1.

$$\frac{4+x}{x-\dfrac{16}{x}}$$

EXAMPLE 4 Simplify by Method 2. $\dfrac{\dfrac{3}{x+2}+\dfrac{1}{x}}{\dfrac{3}{y}-\dfrac{2}{x}}$

Solution The LCD of the numerator is $x(x+2)$. The LCD of the denominator is xy. Thus, the LCD of the complex fraction is $xy(x+2)$.

$$\frac{\dfrac{3}{x+2}+\dfrac{1}{x}}{\dfrac{3}{y}-\dfrac{2}{x}}\cdot\frac{xy(x+2)}{xy(x+2)}=\frac{3xy+xy+2y}{3x(x+2)-2y(x+2)}$$

$$=\frac{4xy+2y}{(x+2)(3x-2y)}$$

$$=\frac{2y(2x+1)}{(x+2)(3x-2y)}$$

Student Practice 4 Simplify by Method 2.

$$\frac{\dfrac{7}{y+3}-\dfrac{3}{y}}{\dfrac{2}{y}+\dfrac{5}{y+3}}$$

6.3 Exercises

MyMathLab®

Watch the videos in MyMathLab

Download the MyDashBoard App

Simplify the complex fractions by any method.

1. $\dfrac{\frac{7}{x}}{\frac{3}{xy}}$

2. $\dfrac{-\frac{2}{ab^2}}{\frac{5}{a}}$

3. $\dfrac{\frac{2x}{x+5}}{\frac{x^2}{x-1}}$

4. $\dfrac{\frac{4y+1}{9y^4}}{\frac{y+2}{y^3}}$

5. $\dfrac{1-\frac{4}{3y}}{\frac{2}{y}+1}$

6. $\dfrac{4-\frac{1}{3y}}{\frac{5}{6y}+1}$

7. $\dfrac{\frac{y}{6}-\frac{1}{2y}}{\frac{3}{2y}-\frac{1}{y}}$

8. $\dfrac{\frac{1}{3y}+\frac{1}{6y}}{\frac{1}{2y}+\frac{3}{4y}}$

9. $\dfrac{\frac{2}{y^2-9}}{\frac{3}{y+3}+1}$

10. $\dfrac{\frac{2}{y+4}}{\frac{3}{y-4}-\frac{1}{y^2-16}}$

11. $\dfrac{\frac{3}{2x+4}+\frac{1}{2}}{\frac{2}{x^2-4}+\frac{x}{x+2}}$

12. $\dfrac{\frac{5}{2x+8}+\frac{3}{2}}{\frac{3}{x^2-16}+\frac{1}{x+4}}$

13. $\dfrac{-8}{\frac{6x}{x-1}-4}$

14. $\dfrac{10}{5x-\frac{5}{x+2}}$

15. $\dfrac{\frac{1}{2x+1}+\frac{4}{4x^2+4x+1}}{\frac{6}{2x^2+x}}$

16. $\dfrac{\frac{4}{3x-2}-\frac{1}{9x^2-4}}{\frac{5x}{3x^2-2x}}$

17. $\dfrac{\frac{4}{x+y}+\frac{1}{3}}{\frac{x}{x+y}-1}$

18. $\dfrac{\frac{2}{y+3}+1}{\frac{1}{3}-\frac{2}{y+3}}$

19. $\dfrac{\dfrac{1}{x-a}-\dfrac{1}{x}}{a}$

20. $\dfrac{\dfrac{1}{x+a}-\dfrac{1}{x}}{a}$

Cumulative Review *Solve for x.*

21. **[2.3.1]** $|2-3x|=4$

22. **[2.8.1]** $|7x-3-2x|<6$

23. **[2.5.1]** Anthony won $20,000 in the lottery and decided to invest part at 5% simple interest and part at 9% simple interest. After one year, he had earned $1480 in interest. How much was invested at each rate?

24. **[3.5.3]** If $f(x)=2x^2-4x+1$, find $f(-1)$.

Quick Quiz 6.3 *Simplify.*

1. $\dfrac{\dfrac{1}{4x}+\dfrac{1}{2x}}{\dfrac{1}{3y}+\dfrac{5}{6y}}$

2. $\dfrac{\dfrac{3}{x+4}-2}{4-\dfrac{3}{x+4}}$

3. $\dfrac{\dfrac{3}{x+3}-\dfrac{1}{x}}{\dfrac{5}{x^2+3x}}$

4. **Concept Check** Explain how you would simplify the following.

$$\dfrac{\dfrac{1}{x}+\dfrac{2}{y}}{\dfrac{3}{x}-\dfrac{4}{y}}$$

How Am I Doing? Sections 6.1–6.3

How are you doing with your homework assignments in Sections 6.1 to 6.3? Do you feel you have mastered the material so far? Do you understand the concepts you have covered? Before you go further in the textbook, take some time to do each of the following problems.

6.1

Simplify.

1. $\dfrac{49x^2 - 9}{7x^2 + 4x - 3}$

2. $\dfrac{x^2 + 2x - 24}{x^2 - 6x + 8}$

3. $\dfrac{2x^3 - 5x^2 - 3x}{x^3 - 8x^2 + 15x}$

4. $\dfrac{6a - 30}{3a + 3} \cdot \dfrac{9a^2 + a - 8}{2a^2 - 15a + 25}$

5. $\dfrac{5x^3y^2}{x^2y + 10xy^2 + 25y^3} \div \dfrac{2x^4y^5}{3x^3 - 75xy^2}$

6. $\dfrac{8x^3 + 1}{4x^2 + 4x + 1} \cdot \dfrac{6x + 3}{4x^2 - 2x + 1}$

6.2

Add or subtract. Simplify your answers.

7. $\dfrac{x}{2x - 4} - \dfrac{5}{2x}$

8. $\dfrac{2}{x + 5} + \dfrac{3}{x - 5} + \dfrac{7x}{x^2 - 25}$

9. $\dfrac{y - 1}{y^2 - 2y - 8} - \dfrac{y + 2}{y^2 + 6y + 8}$

10. $\dfrac{x + 1}{x + 4} + \dfrac{4 - x^2}{x^2 - 16}$

6.3

Simplify.

11. $\dfrac{\dfrac{1}{8x} + \dfrac{7}{16x}}{\dfrac{3}{8x^2}}$

12. $\dfrac{\dfrac{x}{4x^2 - 1}}{3 - \dfrac{2}{2x + 1}}$

13. $\dfrac{\dfrac{5}{x} + 3}{\dfrac{6}{x} - 2}$

14. $\dfrac{\dfrac{1}{x} + \dfrac{x}{x + 3}}{\dfrac{x + 1}{x + 3} + \dfrac{4}{x}}$

1. ____________

2. ____________

3. ____________

4. ____________

5. ____________

6. ____________

7. ____________

8. ____________

9. ____________

10. ____________

11. ____________

12. ____________

13. ____________

14. ____________

Now turn to page SA-16 for the answers to each of these problems. Each answer also includes a reference to the objective in which the problem is first taught. If you missed any of these problems, you should stop and review the Examples and Student Practice problems in the referenced objective. A little review now will help you master the material in the upcoming sections of the text.

6.4 Rational Equations

Student Learning Objectives

After studying this section, you will be able to:

1. **Solve a rational equation that has a solution and be able to check the solution.**
2. **Identify those rational equations that have no solution.**

Graphing Calculator

Solving Rational Equations

To solve Example 2 on your graphing calculator, find the point of intersection of

$$y_1 = \frac{2}{3x+6}$$

and $y_2 = \dfrac{1}{6} - \dfrac{1}{2x+4}$.

Use the Zoom and Trace features or the intersection command to find that the solution is 5.00 (to the nearest hundredth).

***NOTE TO STUDENT:* Fully worked-out solutions to all of the Student Practice problems can be found at the back of the text starting at page SP-1.**

① Solving a Rational Equation

A **rational equation** is an equation that has one or more rational expressions as terms. To solve a rational equation, we find the LCD of all fractions in the equation and multiply each side of the equation by the LCD. We then solve the resulting equation.

EXAMPLE 1 Solve and check your solution. $\dfrac{9}{4} - \dfrac{1}{2x} = \dfrac{4}{x}$

Solution First we multiply each side of the equation by the LCD, which is $4x$.

$$4x\left(\frac{9}{4} - \frac{1}{2x}\right) = 4x\left(\frac{4}{x}\right)$$

$$4x\left(\frac{9}{4}\right) - 4x\left(\frac{1}{2x}\right) = 4x\left(\frac{4}{x}\right) \quad \text{Use the distributive property.}$$

$$9x - 2 = 16 \quad \text{Simplify.}$$

$$9x = 18 \quad \text{Combine like terms.}$$

$$x = 2 \quad \text{Divide each side by the coefficient of } x.$$

Check.

$$\frac{9}{4} - \frac{1}{2(2)} \stackrel{?}{=} \frac{4}{2}$$

$$\frac{9}{4} - \frac{1}{4} \stackrel{?}{=} 2$$

$$\frac{8}{4} \stackrel{?}{=} 2$$

$$2 = 2 \checkmark$$

Student Practice 1 Solve and check. $\dfrac{4}{3x} + \dfrac{x+1}{x} = \dfrac{1}{2}$

Usually, we combine the first two steps of a problem like Example 1 and show only the step of multiplying each term of the equation by the LCD. The next few problems require an additional step of factoring as we begin the problem. Thus, it is helpful to make this change in the remaining examples in this section.

This is another illustration of the need to understand a mathematical principle rather than merely copying down a step without understanding it. Because we understand the distributive property, we can move directly to simplifying a rational equation by multiplying each term of the equation by the LCD.

MC **EXAMPLE 2** Solve and check. $\dfrac{2}{3x+6} = \dfrac{1}{6} - \dfrac{1}{2x+4}$

Solution

$$\frac{2}{3(x+2)} = \frac{1}{6} - \frac{1}{2(x+2)} \quad \text{Factor each denominator.}$$

$$6(x+2)\left[\frac{2}{3(x+2)}\right] = 6(x+2)\left[\frac{1}{6}\right] - 6(x+2)\left[\frac{1}{2(x+2)}\right]$$

Multiply each term by the LCD $6(x+2)$.

$$4 = x + 2 - 3 \quad \text{Simplify.}$$

$$4 = x - 1 \quad \text{Combine like terms.}$$

$$5 = x \quad \text{Solve for } x.$$

Check. Verify that 5 is the solution.

Student Practice 2 Solve and check.

$$\frac{1}{3x-9} = \frac{1}{2x-6} - \frac{5}{6}$$

EXAMPLE 3 Solve. $\dfrac{y^2 - 10}{y^2 - y - 20} = 1 + \dfrac{7}{y - 5}$

Solution

$$\frac{y^2 - 10}{(y - 5)(y + 4)} = 1 + \frac{7}{y - 5}$$ Factor each denominator. Multiply each term by the LCD $(y - 5)(y + 4)$.

$$(y-5)(y+4)\left[\frac{y^2 - 10}{(y-5)(y+4)}\right] = (y - 5)(y + 4)(1) + (y-5)(y + 4)\left[\frac{7}{(y-5)}\right]$$

$$y^2 - 10 = (y - 5)(y + 4)(1) + 7(y + 4)$$ Divide out common factors.

$$y^2 - 10 = y^2 - y - 20 + 7y + 28$$ Simplify.

$$y^2 - 10 = y^2 + 6y + 8$$ Combine like terms.

$$-10 = 6y + 8$$ Subtract y^2 from each side.

$$-18 = 6y$$ Add -8 to each side.

$$-3 = y$$ Divide each side by the coefficient of y.

Check.

$$\frac{(-3)^2 - 10}{(-3)^2 - (-3) - 20} \stackrel{?}{=} 1 + \frac{7}{-3 - 5}$$

$$\frac{9 - 10}{9 + 3 - 20} \stackrel{?}{=} 1 + \frac{7}{-8}$$

$$\frac{-1}{-8} \stackrel{?}{=} 1 - \frac{7}{8}$$

$$\frac{1}{8} = \frac{1}{8} \quad ✓$$

Student Practice 3 Solve. $\dfrac{y^2 + 4y - 2}{y^2 - 2y - 8} = 1 + \dfrac{4}{y - 4}$

② Identifying Equations with No Solution

Some rational equations have no solution. This can happen in two distinct ways. In the first case, when you attempt to solve the equation, you obtain a contradiction, such as $0 = 1$. This occurs because the variable "drops out" of the equation. No solution can be obtained. In the second case, we may solve an equation to get an *apparent* solution, but it may not satisfy the original equation. We call the apparent solution an **extraneous solution.** An equation that yields only an extraneous solution has no solution.

Case 1: The Variable Drops Out. In this case, when you attempt to solve the equation, the coefficient of the variable term becomes zero. Thus, you are left with a statement such as $0 = 1$, which is false. In such a case you know that there is no value for the variable that could make $0 = 1$; hence, there cannot be any solution.

EXAMPLE 4 Solve. $\dfrac{z + 1}{z^2 - 3z + 2} + \dfrac{3}{z - 1} = \dfrac{4}{z - 2}$

Solution

$$\frac{z + 1}{(z - 2)(z - 1)} + \frac{3}{z - 1} = \frac{4}{z - 2}$$ Factor to find the LCD $(z - 2)(z - 1)$. Then multiply each term by the LCD.

$$(z-2)(z-1)\left[\frac{z + 1}{(z-2)(z-1)}\right] + (z - 2)(z-1)\left[\frac{3}{z-1}\right] = (z-2)(z - 1)\left[\frac{4}{(z-2)}\right]$$

Continued on next page

$$z + 1 + 3(z - 2) = 4(z - 1)$$ Divide out common factors.

$$z + 1 + 3z - 6 = 4z - 4$$ Simplify.

$$4z - 5 = 4z - 4$$ Combine like terms.

$$4z - 4z = -4 + 5$$ Move variable terms to one side and constant values to the other.

$$0 = 1$$

Of course, $0 \neq 1$. Therefore, no value of z makes the original equation true. Hence, the equation has **no solution.**

Student Practice 4 Solve. $\dfrac{2x - 1}{x^2 - 7x + 10} + \dfrac{3}{x - 5} = \dfrac{5}{x - 2}$

Case 2: The Obtained Value of the Variable Leads to a Denominator of Zero

EXAMPLE 5 Solve. $\dfrac{4y}{y + 3} - \dfrac{12}{y - 3} = \dfrac{4y^2 + 36}{y^2 - 9}$

Solution

$$\frac{4y}{y + 3} - \frac{12}{y - 3} = \frac{4y^2 + 36}{(y + 3)(y - 3)}$$ Factor each denominator to find the LCD $(y + 3)(y - 3)$. Multiply each term by the LCD.

$$\cancel{(y + 3)}(y - 3)\left[\frac{4y}{\cancel{y + 3}}\right] - (y + 3)\cancel{(y - 3)}\left[\frac{12}{\cancel{y - 3}}\right] = \cancel{(y + 3)}\,\cancel{(y - 3)}\left[\frac{4y^2 + 36}{\cancel{(y + 3)}\,\cancel{(y - 3)}}\right]$$

$$4y(y - 3) - 12(y + 3) = 4y^2 + 36$$ Divide out common factors.

$$4y^2 - 12y - 12y - 36 = 4y^2 + 36$$ Remove parentheses.

$$4y^2 - 24y - 36 = 4y^2 + 36$$ Combine like terms.

$$-24y - 36 = 36$$ Subtract $4y^2$ from each side.

$$-24y = 72$$ Add 36 to each side.

$$y = \frac{72}{-24}$$ Divide each side by -24.

$$y = -3$$

Check.

$$\frac{4(-3)}{-3 + 3} - \frac{12}{-3 - 3} \stackrel{?}{=} \frac{4(-3)^2 + 36}{(-3)^2 - 9}$$

$$\frac{-12}{0} - \frac{12}{-6} \stackrel{?}{=} \frac{36 + 36}{0}$$

You cannot divide by zero. Division by zero is not defined. A value of a variable that makes a denominator in the original equation zero is not a solution to the equation. Thus, this equation has **no solution.** Sometimes we refer to $y = -3$ as an extraneous solution.

Student Practice 5 Solve and check. $\dfrac{y}{y - 2} - 3 = 1 + \dfrac{2}{y - 2}$

TO THINK ABOUT: Quick Solution Checks Sometimes you may find that you do not have sufficient time for a complete check, but you still wish to make sure that you do not have a "no solution" situation. In those instances you can do a quick analysis to be sure that your obtained value for the variable does not make a denominator zero. If you were solving the equation $\dfrac{4x - 1}{x^2 + 5x - 14} = \dfrac{1}{x - 2} - \dfrac{2}{x + 7}$, you would know immediately that you could not have 2 or -7 as a solution. Do you see why?

6.4 Exercises

MyMathLab® Watch the videos in MyMathLab Download the MyDashBoard App

Solve the equations and check your solutions. If there is no solution, say so.

1. $\frac{2}{x} + \frac{3}{2x} = \frac{7}{6}$

2. $\frac{1}{x} + \frac{2}{3x} = \frac{1}{3}$

3. $2 - \frac{1}{x} = \frac{1}{5x}$

4. $\frac{5}{4x} + 2 = \frac{1}{x}$

5. $\frac{5}{2x + 3} + \frac{1}{x} = \frac{3}{x}$

6. $\frac{4}{3x - 1} + \frac{1}{2x} = \frac{3}{2x}$

7. $\frac{2}{y} = \frac{5}{y - 3}$

8. $\frac{9}{2x + 1} = \frac{4}{x}$

9. $\frac{y + 6}{y + 3} - 2 = \frac{3}{y + 3}$

10. $4 - \frac{8x}{x + 1} = \frac{8}{x + 1}$

11. $\frac{1}{3x} - \frac{2}{x} = \frac{-5}{x + 4}$

12. $\frac{2}{x} - \frac{1}{5x} = \frac{3}{2x - 3}$

13. $\frac{2x + 3}{x + 3} = \frac{2x}{x + 1}$

14. $\frac{4x}{x + 1} = \frac{4x - 3}{x - 2}$

15. $\frac{3}{x^2 - 4} = \frac{2}{2x^2 + 4x}$

16. $\frac{4}{y^2 + y} = \frac{2}{y^2 - 1}$

Mixed Practice

Solve the equations and check your solutions. If there is no solution, say so.

17. $\frac{1}{x^2 + x} + \frac{5}{x} = \frac{3}{x + 1}$

18. $\frac{2}{3x - 1} + \frac{2}{3x + 1} = \frac{4x}{9x^2 - 1}$

19. $\frac{5}{y - 3} + 2 = \frac{3}{3y - 9}$

20. $\frac{2}{5} + \frac{1}{y - 5} = \frac{y - 7}{5y - 25}$

21. $1 - \frac{10}{z - 3} = \frac{-5}{3z - 9}$

22. $\frac{3}{2} + \frac{2}{2z - 8} = \frac{1}{z - 4}$

23. $\frac{8}{3x + 2} - \frac{7x + 4}{3x^2 + 5x + 2} = \frac{2}{x + 1}$

24. $\frac{1}{x - 1} - \frac{2x + 1}{2x^2 + 5x - 7} = \frac{6}{2x + 7}$

25. $\frac{4}{z^2 - 9} = \frac{2}{z^2 - 3z}$

26. $\frac{z^2 + 16}{z^2 - 16} = \frac{z}{z + 4} - \frac{4}{z - 4}$

27. $\frac{3x + 1}{3} + \frac{1}{x + 2} = x$

28. $\frac{4x - 3}{4} - x = -\frac{3}{x + 4}$

Verbal and Writing Skills

29. In what situations will a rational equation have no solution?

30. What does "extraneous solution" mean? What must we do to determine whether a solution is an extraneous solution?

Cumulative Review *Factor completely.*

31. **[5.7.1]** $7x^2 - 63$

32. **[5.7.1]** $2x^2 + 20x + 50$

33. **[5.7.1]** $64x^3 - 27y^3$

34. **[5.7.1]** $3x^2 - 13x + 14$

35. **[5.1.6]** Multiply. $(x + 3y)^2$

36. **[5.2.2]** Divide. $(2x^3 - 3x^2 + 3x - 4) \div (x - 2)$

Quick Quiz 6.4 *Solve.*

1. $2 + \frac{x}{x + 3} = \frac{3x}{x - 3}$

2. $\frac{1}{x + 2} - \frac{1}{3} = \frac{-2}{3x + 6}$

3. $\frac{1}{3x - 2} + \frac{2x}{x + 1} = 2$

4. **Concept Check** Why does the equation $\frac{1}{x - 4} - \frac{2}{2x - 8} = \frac{3}{2}$ have no solution? Explain how this can be determined.

6.5 Applications: Formulas and Advanced Ratio Exercises

Student Learning Objectives

After studying this section, you will be able to:

① Solve a formula for a particular variable.

② Solve advanced exercises involving ratio and rate.

① Solving a Formula for a Particular Variable

In science, economics, business, and mathematics, we use formulas that contain rational expressions. We often have to solve these formulas for a specific variable in terms of the other variables.

EXAMPLE 1 Solve for a. $\frac{1}{f} = \frac{1}{a} + \frac{1}{b}$

Solution This formula is used in optics in the study of light passing through a lens. It relates the focal length f of the lens to the distance a of an object from the lens and the distance b of the image from the lens.

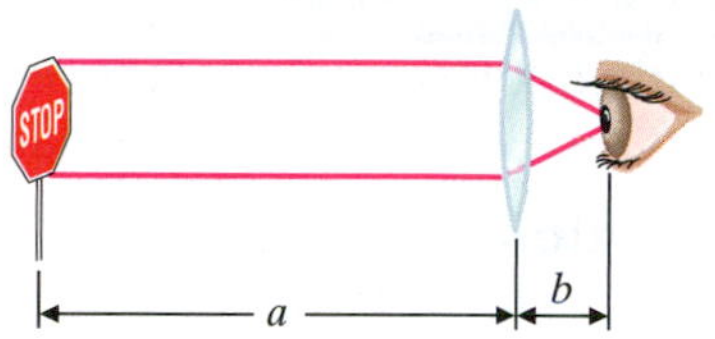

$$ab\cancel{f}\left[\frac{1}{\cancel{f}}\right] = \cancel{a}bf\left[\frac{1}{\cancel{a}}\right] + a\cancel{b}f\left[\frac{1}{\cancel{b}}\right]$$ Multiply each term by the LCD abf.

$$ab = bf + af$$ Simplify.

$$ab - af = bf$$ Collect all the terms containing the variable a on one side of the equation.

$$a(b - f) = bf$$ Factor.

$$a = \frac{bf}{b - f}$$ Divide each side by $b - f$.

Student Practice 1 Solve for t. $\frac{1}{t} = \frac{1}{c} + \frac{1}{d}$

This formula relates the total amount of time t in hours that is required for two workers to complete a job working together if one worker can complete it alone in c hours and the other worker in d hours.

NOTE TO STUDENT: Fully worked-out solutions to all of the Student Practice problems can be found at the back of the text starting at page SP-1.

EXAMPLE 2 The gravitational force F between two masses m_1 and m_2 a distance d apart is represented by the following formula. Solve for m_2.

$$F = \frac{Gm_1m_2}{d^2}$$

Solution The subscripts on the variable m mean that m_1 and m_2 are *different*. (The m stands for "mass." G stands for "gravitational constant.")

$$F = \frac{Gm_1m_2}{d^2}$$

$$d^2[F] = \cancel{d^2}\left[\frac{Gm_1m_2}{\cancel{d^2}}\right]$$ Multiply each side by the LCD d^2.

$$d^2F = Gm_1m_2$$ Simplify.

$$\frac{d^2F}{Gm_1} = \frac{Gm_1m_2}{Gm_1}$$ Divide each side by the coefficient of m_2, which is Gm_1.

$$\frac{d^2F}{Gm_1} = m_2$$

Student Practice 2 The number of telephone calls C between two cities of populations p_1 and p_2 that are a distance d apart may be represented by the formula

$$C = \frac{Bp_1p_2}{d^2}.$$

Solve this equation for p_1.

② Solving Advanced Exercises Involving Ratio and Rate

You have already encountered the idea of proportions in a previous course. In that course you learned that a **proportion** is an equation that says that two ratios are equal.

For example, if Wendy's car traveled 180 miles on 7 gallons of gas, how many miles can the car travel on 11 gallons of gas? You probably remember that you can solve this type of exercise quickly if you let x be the number of miles the car can travel. Then to find x you solve the proportion $\frac{7}{180} = \frac{11}{x}$. Rounded to the nearest mile, the answer is 283 miles. Can you obtain that answer? So far in Chapters 1–5 in this book in the Cumulative Review Exercises, there have been several ratio and proportion exercises for you to solve at this elementary level.

Now we proceed with more advanced exercises in which the ratio of two quantities is more difficult to establish. Study the next two examples carefully. See whether you can follow the reasoning in each case.

EXAMPLE 3 A company plans to employ 910 people with a ratio of two managers for every eleven workers. How many managers should be hired? How many workers?

Solution If we let $x =$ the number of managers for the company, then $910 - x =$ the number of workers. We are given the ratio of managers to workers, so let's set up our proportion in that way.

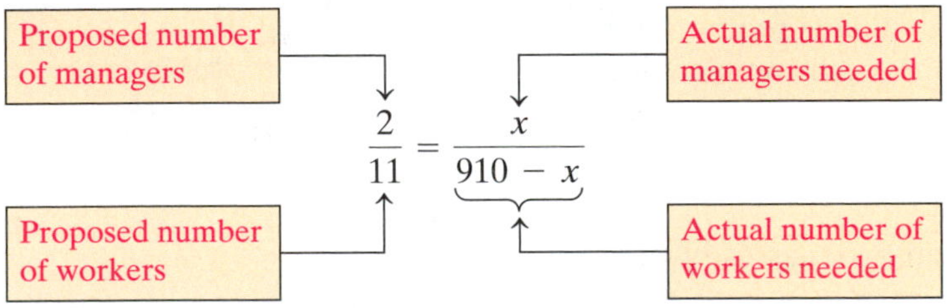

$$\frac{2}{11} = \frac{x}{910 - x}$$

The LCD is $11(910 - x)$. Multiplying by the LCD, we get the following:

$$\cancel{11}(910 - x)\left[\frac{2}{\cancel{11}}\right] = 11\cancel{(910 - x)}\left[\frac{x}{\cancel{910 - x}}\right]$$

$$2(910 - x) = 11x$$
$$1820 - 2x = 11x$$
$$1820 = 13x$$
$$140 = x$$

$$910 - x = 910 - 140 = 770$$

The number of managers needed is 140. The number of workers needed is 770.

Student Practice 3 Western University has 1932 faculty and students. The university always maintains a student-to-faculty ratio of 21 : 2. How many students are enrolled? How many faculty members are there?

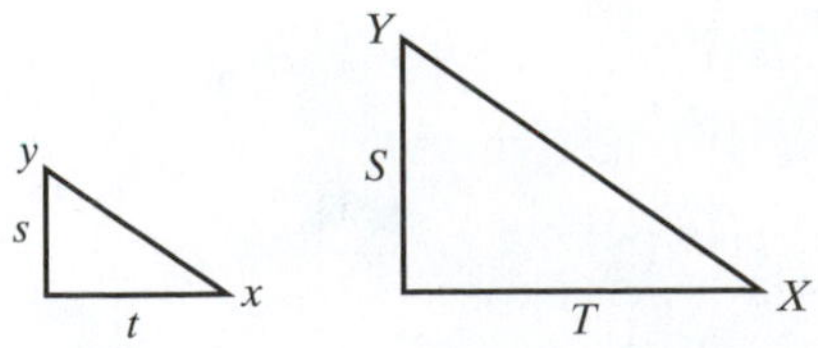

The next example concerns **similar triangles.** These triangles have corresponding angles that are equal and corresponding sides that are *proportional* (not equal). Similar triangles are frequently used to determine distances that cannot be conveniently measured. For example, in the following sketch, x and X are corresponding angles, y and Y are corresponding angles, and s and S are corresponding sides.

Hence, angle x = angle X, angle y = angle Y, and side s is proportional to side S. (Again, note that we did not say that side s is equal to side S.) Also, side t is proportional to side T. So, really, one triangle is just a magnification of the other triangle. If the sides are corresponding, they are in the same ratio.

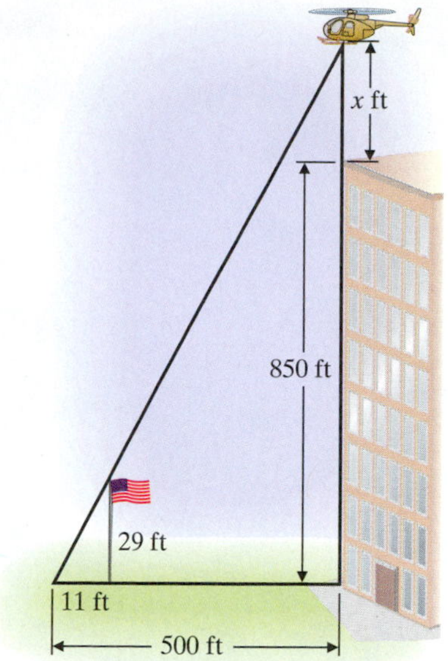

▲ **EXAMPLE 4** A helicopter is hovering an unknown distance above an 850-foot building. A man watching the helicopter is 500 feet from the base of the building and 11 feet from a flagpole that is 29 feet tall. The man's line of sight to the helicopter is directly above the flagpole, as you can see in the sketch. How far above the building is the helicopter? Round your answer to the nearest foot.

Solution

1. ***Understand the problem.***

Can you see the two triangles in the diagram in the margin? For convenience, we separate them out in the sketch on the right. We want to find the distance x. Are the triangles similar? The angles at the bases of the triangles are equal. (Why?) It follows, then, that the top angles must also be equal. (Remember that the angles of any triangle add up to 180°.) Since the angles are equal, the triangles are similar and the sides are proportional.

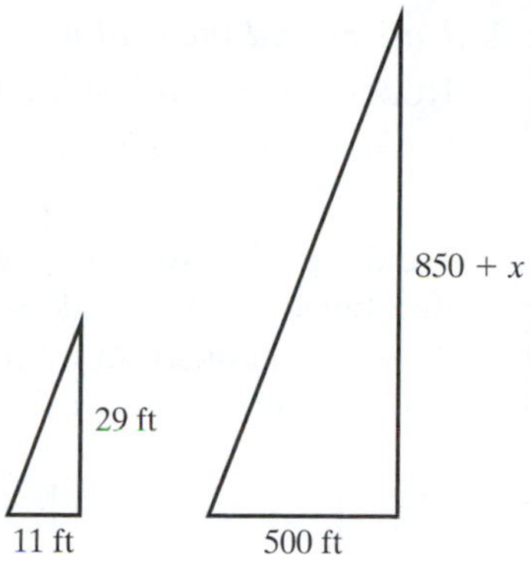

2. ***Write an equation.***

We can set up our proportion like this:

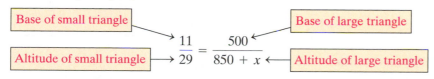

Base of small triangle → 11; Base of large triangle → 500

Altitude of small triangle → 29; Altitude of large triangle → 850 + x

$$\frac{11}{29} = \frac{500}{850 + x}$$

3. ***Solve the equation and state the answer.***

The LCD is $29(850 + x)$.

$$29(850 + x)\left[\frac{11}{29}\right] = 29(850 + x)\left[\frac{500}{850 + x}\right]$$

$$11(850 + x) = 29(500)$$

$$9350 + 11x = 14{,}500$$

$$11x = 5150$$

$$x = \frac{5150}{11} \approx 468.18$$

So the helicopter is about 468 feet above the building.

▲ **Student Practice 4** Solve the exercise in Example 4 for a man watching 450 feet from the base of a 900-foot building as shown in the figure in the margin. The flagpole is 35 feet tall, and the man is 10 feet from the flagpole.

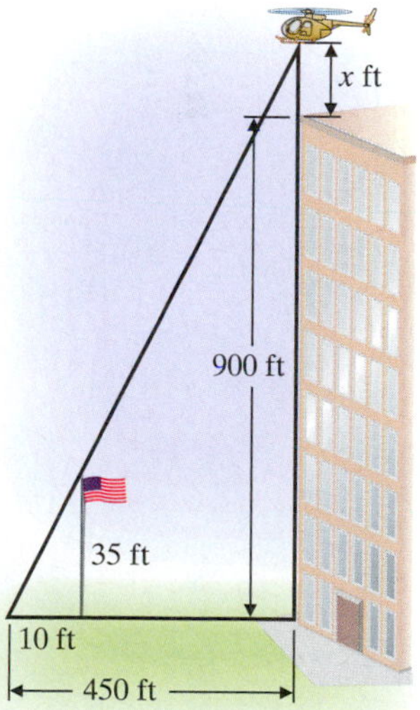

We will see some challenging exercises involving similar triangles in exercises 51 and 52.

We will sometimes encounter exercises in which two or more people or machines are working together to complete a certain task. These types of exercises are sometimes called *work problems*. In general, these types of exercises can be analyzed by using the following concept.

Part of task done by first person + Part of task done by second person = 1 (one complete task finished)

We will also use a general idea about the rate at which something is done. If Robert can do a task in 3 hours, then he can do $\frac{1}{3}$ of the task in 1 hour. If Susan can do the same task in 2 hours, then she can do $\frac{1}{2}$ of the task in 1 hour. In general, if a person can do a task in t hours, then that person can do $\frac{1}{t}$ of the task in 1 hour.

EXAMPLE 5 Robert can paint the kitchen in 3 hours. Susan can paint the kitchen in 2 hours. How long will it take Robert and Susan to paint the kitchen if they work together?

Solution

1. *Understand the problem.*

Robert can paint $\frac{1}{3}$ of the kitchen in 1 hour.
Susan can paint $\frac{1}{2}$ of the kitchen in 1 hour.

We do not know how long it will take them working together, so we let $x =$ the number of hours it takes them to paint the kitchen working together. To assist us, we will construct a table that relates the data. We will use the concept that (rate)(time) = fraction of task done.

	Rate of Work per Hour	Time Worked in Hours	Fraction of Task Done
Robert	$\frac{1}{3}$	x	$\frac{x}{3}$
Susan	$\frac{1}{2}$	x	$\frac{x}{2}$

2. *Write an equation.*

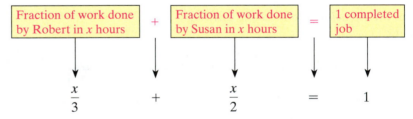

Fraction of work done by Robert in x hours + Fraction of work done by Susan in x hours = 1 completed job

$$\frac{x}{3} + \frac{x}{2} = 1$$

3. *Solve the equation and state the answer.*

Multiply each side of the equation by the LCD and use the distributive property.

$$6\left(\frac{x}{3}\right) + 6\left(\frac{x}{2}\right) = 6(1)$$
$$2x + 3x = 6$$
$$5x = 6$$
$$x = \frac{6}{5} = 1\frac{1}{5} = 1.2 \text{ hours}$$

Working together, Robert and Susan can paint the kitchen in 1.2 hours.

Student Practice 5 Alfred can mow the huge lawn at his Vermont farm with his new lawn mower in 4 hours. His young son can mow the lawn with the old lawn mower in 5 hours. If they work together, how long will it take them to mow the lawn?

6.5 Exercises

MyMathLab® Watch the videos in MyMathLab Download the MyDashBoard App

Solve for the indicated variable.

1. $x = \frac{y - b}{m}$; for m

2. $m = \frac{y - b}{x}$; for x

3. $\frac{1}{f} = \frac{1}{a} + \frac{1}{b}$; for b

4. $\frac{1}{f} = \frac{1}{a} + \frac{1}{b}$; for f

5. $\frac{V}{lh} = w$; for h

6. $r = \frac{I}{pt}$; for p

7. $G = \frac{ab + ac}{3}$; for a

8. $H = \frac{xy + xz}{4}$; for x

9. $\frac{r^3}{V} = \frac{3}{4\pi}$; for V

10. $\frac{V}{\pi r^2} = h$; for r^2

11. $\frac{E}{e} = \frac{R + r}{r}$; for e

12. $\frac{E}{e} = \frac{R + r}{r}$; for r

13. $\frac{QR_1}{S_1} = \frac{QR_2}{S_2}$; for S_1

14. $\frac{QR_1}{S_1} = \frac{QR_2}{S_2}$; for S_2

15. $\frac{S - 2lw}{2w + 2l} = h$; for w

16. $\frac{S - 2lw}{2w + 2l} = h$; for l

Mixed Practice *Solve for the indicated variable.*

17. $E = T_1 - \frac{T_1}{T_2}$; for T_1

18. $E = T_1 - \frac{T_1}{T_2}$; for T_2

19. $m = \frac{y_2 - y_1}{x_2 - x_1}$; for x_1

20. $m = \frac{y_2 - y_1}{x_2 - x_1}$; for x_2

21. $\frac{2D - at^2}{2t} = V$; for D

22. $\frac{2D - at^2}{2t} = V$; for a

23. $A = \frac{h(B + b)}{2}$; for b

24. $A = \frac{h(B + b)}{2}$; for B

25. $\frac{T_2W}{T_2 - T_1} = q$; for T_2

26. $d = \frac{LR_2}{R_2 + R_1}$; for R_1

27. $\frac{s - s_0}{v_0 + gt} = t$; for v_0

28. $\frac{A - P}{Pr} = t$; for P

Round your answers to four decimal places.

29. Solve for T. $\frac{1.98V}{1.96V_0} = 0.983 + 5.936(T - T_0)$

30. Solve for r_1. $\frac{1}{R} = \frac{1}{r_1} + \frac{1}{0.368} + \frac{1}{0.736}$

Applications

31. ***Map Scale*** On a map of Nigeria, the cities of Benin City and Onitsha are 6.5 centimeters apart. The map scale shows that 3 centimeters = 55 kilometers on the map. How far apart are the two cities? Round to the nearest tenth.

32. ***Scale Drawing*** The lobby of the Fine Arts building at Normandale Community College was recently renovated. The scale drawing is 12 inches wide and 18 inches long. The blueprint has a scale of 3 : 50. Find the length and width of the lobby measured in inches. Find the length and width of the lobby measured in feet.

33. ***Speed Conversion*** A speed of 60 miles per hour (mph) is equivalent to 88 kilometers per hour (kph). On a trip through Nova Scotia, Lora passes a sign stating that the speed limit is 80 kph. What is the speed limit in miles per hour? Round to the nearest hundredth.

34. ***Currency Conversion*** Miguel traveled to Italy on business for a week and will now travel to Australia for a week. He wants to convert the 80 euros he now has to Australian dollars. 0.5 euro is currently equivalent to 0.7 Australian dollar. How many Australian dollars will he receive for his 80 euros?

In exercises 35–50, round your answers to the nearest hundredth, unless otherwise directed.

35. ***Grizzly Bear Population*** Thirty-five grizzly bears were captured and tagged by wildlife personnel in the Yukon and then released back into the wild. Later that same year, fifty were captured, and twenty-two had tags. Estimate the number of grizzly bears in that part of the Yukon. (Round this answer to the nearest whole number.)

36. ***Alligator Population*** Thirty alligators were captured in the Louisiana bayou (swamp) and tagged by National Park officials and then put back into the bayou. One month later, fifty alligators were captured from the same part of the bayou, eighteen of which had been tagged. Estimate to the nearest whole number how many alligators are in that part of Louisiana.

37. ***Composition of Russian Navy*** A ship in the Russian navy has a ratio of two officers for every seven seamen. The crew of the ship totals 117 people. How many officers and how many seamen are on this ship?

38. ***Camp Counselors*** A YMCA summer camp employs a total of 180 counselors each summer. The ratio of employees is three new counselors for every seven experienced counselors. How many experienced counselors are employed at the camp? How many new counselors are employed at the camp?

39. ***Cell Phone Employees*** Southwestern Cell Phones has a total of 187 employees in the marketing and sales offices in Dallas. For every four people employed in marketing, there are thirteen people employed in sales. How many people in the Dallas office work in marketing? How many people in the Dallas office work in sales?

40. ***Education*** At Elmwood University there are nine men for every five women on the faculty. If Elmwood University employs 182 faculty members, how many are men and how many are women?

▲ **41.** ***Picture Framing*** Rose is carefully framing an 8-inch × 10-inch family photo so that the length and width of the frame have the same ratio as the photo. She is using a piece of frame 54 inches in length. What are the length and width of the frame?

▲ **42.** ***Enlarging Photographs*** Becky DeWitt is a photographer at Photographics. She wants to enlarge a photograph that measures 3 inches wide and 5 inches long to an oversize photograph with the same width-to-length ratio. The perimeter of the new photograph is 115.2 inches. What are the length and the width of the new oversize photograph?

43. ***E-Mail Software*** A college recently installed new e-mail software on all of its computers. A poll was taken to find out how many of the faculty and staff preferred the new software. The ratio of those preferring the new software to the old software was 3 : 11. If there are 280 faculty and staff at the college, how many prefer the new software?

44. ***Hospital Staff*** The ratio of doctors to nurses at Carson City Hospital is 2 : 5. The hospital has a total of 119 doctors and nurses. How many are doctors? How many are nurses?

▲ **45.** ***Shadow Length*** A 12-foot marble statue in Italy casts a shadow that is 15 feet long. At the same time of day, a wall casts a shadow that is 8 feet long. How high is the wall?

▲ **46.** ***Shadow Length*** A 3.5-foot-tall child casts a shadow that is 6 feet long. At the same time of day, a tree casts a shadow that is 42 feet long. How tall is the tree?

47. ***Snow Plowing*** Matt can plow out all the driveways on his street in Duluth, Minnesota, with his new four-wheel-drive truck in 6 hours. Using a snow blade on a lawn tractor, his neighbor can plow out the same number of driveways in 9 hours. How long would it take them to do the work together?

48. ***Mowing Lawns*** The new employee at Pete's Yard Service can mow four yards in 5 hours. Pete, the owner, can do the job in 3 hours. How long would it take them to do the work together?

49. ***Swimming Pool*** Houghton College has just built a new Olympic-size swimming pool. One large pipe with cold water and one small pipe with hot water are used to fill the pool. When both pipes are used, the pool can be filled in 2 hours. If only the cold water pipe is used, the pool can be filled in 3 hours. How long would it take to fill the pool with just the hot water pipe?

50. ***Splitting Firewood*** A lumberjack and his cousin Fred can split a cord of seasoned oak firewood in 3 hours. The lumberjack can split the wood alone without any help from Fred in 4 hours. How long would it take Fred if he worked without the lumberjack?

To Think About

River Width *To find the width of a river, a hiking club laid out a triangular pattern of measurements. See the figure. Use your knowledge of similar triangles to solve exercises 51 and 52.*

▲ **51.** If any observer stands at the point shown in the figure, then $a = 8$ feet, $b = 5$ feet, and $c = 116$ feet. What is the width of the river?

▲ **52.** What is the width of the river if $a = 5$ feet, $b = 3$ feet, and $c = 295$ feet?

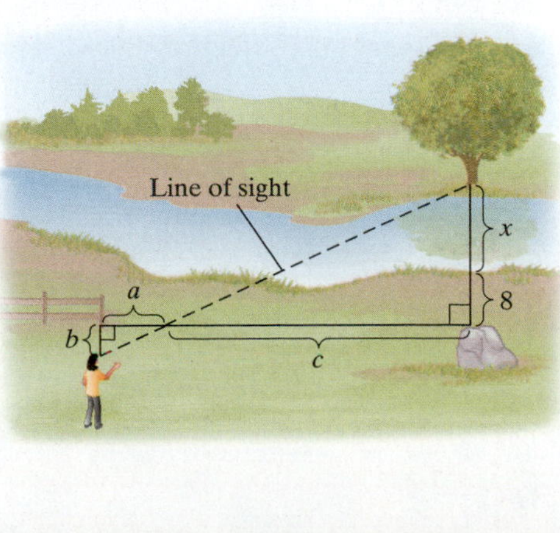

Cumulative Review

53. **[3.3.1]** Find the slope and the y-intercept of the line: $7x - 3y = 8$

54. **[3.2.1]** Find the slope of the line passing through $(-1, 2.5)$ and $(1.5, 2)$

55. **[4.1.6]** Solve the system.

$$\begin{aligned} x - 5y &= -9 \\ 3x + 2y &= -10 \end{aligned}$$

56. **[4.2.2]** Solve the system.

$$\begin{aligned} x + 4y - z &= 10 \\ 3x + 2y + z &= 4 \\ 2x - 3y + 2z &= -7 \end{aligned}$$

Quick Quiz 6.5

1. Solve for H. $\dfrac{3W + 2AH}{4W - 8H} = A$

2. For every 3 students who were accepted for admission to Springfield University, there were 22 students who were not accepted. If 2400 students applied, how many were accepted for admission?

3. A window at Don Work's house is 22 inches wide and 35 inches tall. He took that window out and replaced it with a larger window that has a perimeter of 342 inches. The new window has the same proportional ratio of width to length as the old window. What is the width and what is the length of the new window?

4. **Concept Check** Explain how you would solve for H in the following equation.

$$S = \frac{2A + 3F}{B - H}$$

Did You Know...

That You Can Save in Fuel Costs with a Hybrid Vehicle Compared to a Less Efficient Vehicle?

AFFORDABLE TRANSPORTATION

Understanding the Problem:
Michele wants to reduce her auto expenses. Her current vehicle is a 2003 SUV that only gets 16 miles per gallon. Different options for a replacement vehicle are

- **Option A:** A small SUV that gets 20 miles per gallon (MPG) and sells for $21,540.
- **Option B:** A hybrid model of the same SUV that gets 32 MPG and sells for $28,150.
- **Option C:** The most fuel-efficient hybrid car that gets 60 MPG and sells for $23,770.

She wants to determine the cost and potential savings she would enjoy from the three different types of cars.

Making a Plan:
We need to calculate the cost to own the different vehicles.

Step 1: Michele drives about 15,000 miles a year. She wants to know what fuel cost would be for the three options.

Task 1: *Determine the number of gallons of gasoline that each car (Options A, B, and C) would use if it is driven 15,000 miles in one year.*

Task 2: *If the average price of gasoline is $3.60 per gallon (Feb 2011), how much would it cost to fuel each car for 15,000 miles?*

Analyzing the Options

Step 2: The SUV in option B would cost Michele $6610 more to purchase then the SUV in Option A. Compare the different costs of ownership.

Task 3: *How much gas money would Michele save in one year with Option B compared to Option A?*

Task 4: *How many years would it take for Michele to save $6610 in gas money with Option B compared to Option A?*

Task 5: *How much gas money would Michele save after five years with Option B compared to Option A? After 10 years?*

Step 3: The car in Option C would cost Michele $2230 more to purchase than the SUV in Option A. Compare the different costs of ownership.

Task 6: *How much gas money would Michele save in one year with Option C compared to Option A?*

Task 7: *How many years would it take for Michele to save $2230 in gas money with Option C compared to Option A?*

Task 8: *How much gas money would Michele save after five years with Option C compared to Option A? After 10 years?*

Making a Decision:

Step 4: Even though Michele enjoyed the comfort of her SUV, she could not ignore the potential savings in gas money she saw with Option C. She decided it was time to make a smart financial decision instead of one based on comfort. She expects to have her new car for at least five years. At that point, Option C will have saved her $9000 in gas money!

Applying the Situation to Your Life:

The number of miles you drive per year as well as the price you pay for a gallon of gas is going to affect this analysis when applied to your own situation. The more miles you drive, the greater the savings from driving a fuel-efficient vehicle. Also, the higher the price of gas, the more you will save by switching.

Chapter 6 Organizer

Topic and Procedure	Examples	You Try It
Simplifying rational expressions, p. 310 1. Factor the numerator and denominator, if possible. 2. Any factor that is common to both numerator and denominator can be divided out. This is an application of the basic rule of fractions. *Basic rule of fractions:* For any polynomials *a*, *b*, and *c* (where $b \neq 0$ and $c \neq 0$), $\frac{ac}{bc} = \frac{a}{b}.$	Simplify. $\frac{6x^2 - 14x + 4}{6x^2 - 20x + 6}$ $\frac{2(3x^2 - 7x + 2)}{2(3x^2 - 10x + 3)} = \frac{\cancel{2}\cancel{(3x - 1)}(x - 2)}{\cancel{2}\cancel{(3x - 1)}(x - 3)}$ $= \frac{x - 2}{x - 3}$	1. Simplify. $\frac{3x^2 - 48}{6x^2 - 42x + 72}$
Multiplying rational expressions, p. 312 1. Factor all numerators and denominators, if possible. 2. Any factor that is common to a numerator of one fraction and the denominator of the same fraction or any fraction that is multiplied by it can be divided out. 3. Write the product of the remaining factors in the numerator. Write the product of the remaining factors in the denominator. $\frac{a}{b} \cdot \frac{c}{d} = \frac{ac}{bd}$	Multiply. $\frac{x^2 - 4x}{6x - 12} \cdot \frac{3x^2 - 6x}{x^3 + 3x^2}$ $\frac{\cancel{x}(x - 4)}{\underset{2}{\cancel{6}}\cancel{(x - 2)}} \cdot \frac{\overset{1}{\cancel{3}}\cancel{x}\cancel{(x - 2)}}{\cancel{x^2}(x + 3)} = \frac{x - 4}{2(x + 3)}$ or $\frac{x - 4}{2x + 6}$	2. Multiply. $\frac{x^3 + 3x^2}{x^2 - 9} \cdot \frac{x^2 - 6x + 9}{2x^2 - 6x}$
Dividing rational expressions, p. 313 1. Invert the second fraction and multiply it by the first fraction. $\frac{a}{b} \div \frac{c}{d} = \frac{a}{b} \cdot \frac{d}{c}$ 2. Apply the steps for multiplying rational expressions.	Divide. $\frac{6x^2 - 5x - 6}{24x^2 + 13x - 2} \div \frac{4x^2 + x - 3}{8x^2 + 7x - 1}$ $\frac{6x^2 - 5x - 6}{24x^2 + 13x - 2} \cdot \frac{8x^2 + 7x - 1}{4x^2 + x - 3}$ $= \frac{\cancel{(3x + 2)}(2x - 3)}{\cancel{(3x + 2)}\cancel{(8x - 1)}} \cdot \frac{\cancel{(8x - 1)}\cancel{(x + 1)}}{\cancel{(x + 1)}(4x - 3)} = \frac{2x - 3}{4x - 3}$	3. Divide. $\frac{2x^2 - x - 1}{6x^2 + 5x + 1} \div \frac{5x^2 - 6x + 1}{25x^2 - 1}$
Adding rational expressions, p. 319 1. If all fractions have a common denominator, add the numerators and place the result over the common denominator. $\frac{a}{c} + \frac{b}{c} = \frac{a + b}{c}$ 2. If the fractions do not have a common denominator, factor the denominators (if necessary) and determine the least common denominator (LCD). 3. Rewrite each fraction as an equivalent fraction with the LCD as the denominator. 4. Add the numerators and place the result over the common denominator. 5. Simplify, if possible.	Add. $\frac{7x}{x^2 - 9} + \frac{x + 2}{x + 3}$ $\frac{7x}{(x + 3)(x - 3)} + \frac{x + 2}{x + 3}$ LCD $= (x + 3)(x - 3)$ $\frac{7x}{(x + 3)(x - 3)} + \frac{x + 2}{x + 3} \cdot \frac{x - 3}{x - 3}$ $= \frac{7x}{(x + 3)(x - 3)} + \frac{x^2 - x - 6}{(x + 3)(x - 3)}$ $= \frac{x^2 + 6x - 6}{(x + 3)(x - 3)}$	4. Add. $\frac{2}{2x + 1} + \frac{13}{10x^2 - 3x - 4}$
Subtracting rational expressions, p. 319 Follow the procedure for adding rational expressions through step 3. Then subtract the second numerator from the first and place the result over the common denominator. $\frac{a}{c} - \frac{b}{c} = \frac{a - b}{c}$ Simplify, if possible.	Subtract. $\frac{4x}{3x - 2} - \frac{5x}{x + 4}$ LCD $= (3x - 2)(x + 4)$ $\frac{4x}{3x - 2} \cdot \frac{x + 4}{x + 4} - \frac{5x}{x + 4} \cdot \frac{3x - 2}{3x - 2}$ $= \frac{4x^2 + 16x}{(3x - 2)(x + 4)} - \frac{15x^2 - 10x}{(x + 4)(3x - 2)}$ $= \frac{-11x^2 + 26x}{(3x - 2)(x + 4)}$ $= \frac{x(-11x + 26)}{(3x - 2)(x + 4)}$	5. Subtract. $\frac{2x}{x - 4} - \frac{x + 1}{2x + 3}$

Topic and Procedure	Examples	You Try It
Simplifying a complex rational expression by Method 1, p. 326 1. Simplify the numerator and denominator, if possible, by combining quantities to obtain one fraction in the numerator and one fraction in the denominator. 2. Divide the numerator by the denominator. (That is, multiply the numerator by the reciprocal of the denominator.) 3. Simplify the result.	Simplify by Method 1. $\dfrac{4 - \frac{1}{x^2}}{\frac{2}{x} + \frac{1}{x^2}}$ **Step 1:** $\dfrac{\frac{4x^2}{x^2} - \frac{1}{x^2}}{\frac{2x}{x^2} + \frac{1}{x^2}} = \dfrac{\frac{4x^2 - 1}{x^2}}{\frac{2x + 1}{x^2}}$ **Step 2:** $\dfrac{4x^2 - 1}{x^2} \cdot \dfrac{x^2}{2x + 1}$ **Step 3:** $\dfrac{\cancel{(2x + 1)}(2x - 1)}{\cancel{x^2}} \cdot \dfrac{\cancel{x^2}}{\cancel{2x + 1}} = 2x - 1$	6. Simplify by Method 1. $\dfrac{\frac{2}{x} - \frac{5}{3x}}{3 + \frac{4}{x^2}}$
Simplifying a complex rational expression by Method 2, p. 327 1. Find the LCD of all the rational expressions in the numerator and the denominator. 2. Multiply the numerator and the denominator of the complex fraction by the LCD. 3. Simplify the result.	Simplify by Method 2. $\dfrac{4 - \frac{1}{x^2}}{\frac{2}{x} - \frac{1}{x^2}}$ **Step 1:** The LCD of the rational expressions is x^2. **Step 2:** $\dfrac{\left[4 - \frac{1}{x^2}\right]x^2}{\left[\frac{2}{x} - \frac{1}{x^2}\right]x^2} = \dfrac{4(x^2) - \left(\frac{1}{x^2}\right)(x^2)}{\left(\frac{2}{x}\right)(x^2) - \left(\frac{1}{x^2}\right)(x^2)}$ $= \dfrac{4x^2 - 1}{2x - 1}$ **Step 3:** $\dfrac{(2x + 1)\cancel{(2x - 1)}}{\cancel{(2x - 1)}} = 2x + 1$	7. Simplify by Method 2. $\dfrac{\frac{2}{x} - \frac{5}{3x}}{3 + \frac{4}{x^2}}$
Solving rational equations, p. 332 1. Determine the LCD of all denominators in the equation. 2. Multiply each term in the equation by the LCD. 3. Simplify and remove parentheses. 4. Combine any like terms. 5. Solve for the variable. If the variable term drops out and a false statement results, there is no solution. 6. Check your answer. Be sure that the value you obtained does not make any denominator in the original equation 0. If so, there is no solution.	Solve. $\dfrac{4}{y - 1} + \dfrac{-y + 5}{3y^2 - 4y + 1} = \dfrac{9}{3y - 1}$ LCD $= (y - 1)(3y - 1)$ $\cancel{(y - 1)}(3y - 1)\left[\dfrac{4}{\cancel{y - 1}}\right]$ $+ \cancel{(y - 1)}\cancel{(3y - 1)}\left[\dfrac{-y + 5}{\cancel{(y - 1)}\cancel{(3y - 1)}}\right]$ $= (y - 1)\cancel{(3y - 1)}\left[\dfrac{9}{\cancel{3y - 1}}\right]$ $4(3y - 1) + (-y + 5) = 9(y - 1)$ $12y - 4 - y + 5 = 9y - 9$ $11y + 1 = 9y - 9$ $11y - 9y = -9 - 1$ $2y = -10$ $y = -5$ *Check.* $\dfrac{4}{-5 - 1} + \dfrac{-(-5) + 5}{3(-5)^2 - 4(-5) + 1} \stackrel{?}{=} \dfrac{9}{3(-5) - 1}$ $\dfrac{4}{-6} + \dfrac{10}{96} \stackrel{?}{=} \dfrac{9}{-16}$ $-\dfrac{9}{16} = -\dfrac{9}{16}$ ✓	8. Solve. $\dfrac{3}{4x + 1} - \dfrac{2}{x - 2} = \dfrac{x - 20}{4x^2 - 7x - 2}$

Topic and Procedure	Examples	You Try It
Solving formulas containing rational expressions for a specified variable, p. 337 **1.** Multiply each term of the equation by the LCD. **2.** Add a quantity to or subtract a quantity from each side of the equation so that only terms containing the desired variable are on one side of the equation while all other terms are on the other side. **3.** If there are two or more unlike terms containing the desired variable, remove that variable as a common factor. **4.** Divide each side of the equation by the coefficient of the desired variable. **5.** Simplify, if possible.	Solve for n. $v = c - \frac{ct}{n}$ $n(v) = n(c) - n\left(\frac{ct}{n}\right)$ $nv = nc - ct$ $nv - nc = -ct$ $n(v - c) = -ct$ $n = \frac{-ct}{v - c}$ or $\frac{ct}{c - v}$	**9.** Solve for d. $\frac{dr}{a} + w = a$
Using advanced ratios to solve applied problems, p. 338 **1.** Determine a given ratio in the problem for which both values are known. **2.** Use variables to describe each of the quantities in the other ratio. **3.** Set the two ratios equal to each other to form an equation. **4.** Solve the resulting equation. Determine both quantities in the second ratio.	A new navy cruiser has a crew of 304 people. For every thirteen seamen there are three officers. How many officers and how many seamen are in the crew? The ratio of seamen to officers is: $\frac{13}{3}$ Let x be the number of officers. Because the crew totals 304, it follows that $304 - x$ represents the number of seamen. $\frac{\text{seamen} \rightarrow 13}{\text{officers} \rightarrow 3} = \frac{304 - x \leftarrow \text{seamen}}{x \leftarrow \text{officers}}$ $13x = 3(304 - x)$ $13x = 912 - 3x$ $16x = 912$ $x = 57$ $304 - x = 247$ There are 57 officers and 247 seamen.	**10.** At Stonehill College there are a total of 228 faculty and staff. For every eleven staff, there are eight faculty. How many staff and how many faculty are there at the college?

Chapter 6 Review Problems

Simplify.

1. $\frac{6x^3 - 9x^2}{12x^2 - 18x}$

2. $\frac{26x^3y^2}{39xy^4}$

3. $\frac{2x^2 - 5x + 3}{3x^2 + 2x - 5}$

4. $\frac{ax + 2a - bx - 2b}{3x^2 - 12}$

5. $\frac{4x^2 - 1}{x^2 - 4} \cdot \frac{2x^2 + 4x}{4x + 2}$

6. $\frac{3y}{4xy - 6y^2} \cdot \frac{2x - 3y}{12xy}$

7. $\frac{y^2 + 8y - 20}{y^2 + 6y - 16} \cdot \frac{y^2 + 3y - 40}{y^2 + 6y - 40}$

8. $\frac{3x + 9}{5x - 20} \div \frac{3x^2 - 3x - 36}{x^2 - 8x + 16}$

9. $\frac{4a^3b^2}{4x^2 - 4x - 3} \div \frac{8ab^4}{6x^2 - 11x + 3}$

10. $\frac{4a^2 + 12a + 5}{2a^2 - 7a - 13} \div (4a^2 + 2a)$

Add or subtract the rational expressions and simplify your answers.

11. $\frac{x - 5}{2x + 1} - \frac{x + 1}{x - 2}$

12. $\frac{5}{4x} + \frac{-3}{x + 4}$

13. $\dfrac{2y-1}{12y} - \dfrac{3y+2}{9y}$

14. $\dfrac{2}{y^2-4} + \dfrac{y}{y^2+4y+4}$

15. $\dfrac{y^2+2y+20}{y^2+8y+12} - \dfrac{2y+1}{y+6}$

16. $\dfrac{a}{5-a} - \dfrac{2}{a+3} + \dfrac{2a^2-2a}{a^2-2a-15}$

17. $\dfrac{2}{x^2+8x+16} - \dfrac{x}{2x^2+9x+4}$

18. $\dfrac{1}{x} + \dfrac{3}{2x} + 3 + 2x$

Simplify the complex rational expressions.

19. $\dfrac{\dfrac{5}{x}+1}{1-\dfrac{25}{x^2}}$

20. $\dfrac{\dfrac{4}{x+3}}{\dfrac{2}{x-2} - \dfrac{1}{x^2+x-6}}$

21. $\dfrac{\dfrac{8}{y+2}-4}{\dfrac{2}{y+2}-1}$

22. $\dfrac{\dfrac{1}{a}+\dfrac{a}{a+1}}{\dfrac{a}{a+1}-\dfrac{1}{a}}$

23. $\dfrac{\dfrac{y^2}{y^2-x^2}-1}{x+\dfrac{xy}{x-y}}$

24. $\dfrac{\dfrac{3}{x}-\dfrac{2}{x+1}}{\dfrac{5}{x^2+5x+4}-\dfrac{1}{x+4}}$

Solve for the variable and check your solutions. If there is no solution, say so.

25. $\dfrac{3}{7} + \dfrac{4}{x+1} = 1$

26. $\dfrac{7}{2x-3} + \dfrac{3}{x+2} = \dfrac{6}{2x-3}$

27. $\dfrac{9}{5x-2} - \dfrac{2}{x} = \dfrac{1}{x}$

28. $\dfrac{5}{2a} = \dfrac{2}{a} - \dfrac{1}{12}$

29. $\dfrac{1}{y} + \dfrac{1}{2y} = 2$

30. $\dfrac{3}{a+4} + \dfrac{a+1}{3a+12} = \dfrac{5}{3}$

31. $\dfrac{3x-23}{2x^2-5x-3} + \dfrac{2}{x-3} = \dfrac{5}{2x+1}$

32. $\dfrac{2x-10}{2x^2-5x-3} + \dfrac{1}{x-3} = \dfrac{3}{2x+1}$

Solve for the variable indicated.

33. $\dfrac{N}{V} = \dfrac{m}{M+N}$; for M

34. $m = \dfrac{y-y_0}{x-x_0}$; for x

35. $\dfrac{1}{f} = \dfrac{1}{a} + \dfrac{1}{b}$; for a

36. $S = \dfrac{V_1 t + V_2 t}{2}$; for t

37. $d = \dfrac{LR_2}{R_2+R_1}$; for R_2

38. $\dfrac{S-P}{Pr} = t$; for r

Mixed Practice, Exercises 39–43

Simplify.

39. $\dfrac{x^2-x-42}{x^2-2x-35}$

40. $\dfrac{2x^2-5x-3}{x^2-9} \cdot \dfrac{2x^2+5x-3}{2x^2+5x+2}$

41. $\dfrac{-2x-1}{x+4} + 4x + 3$

42. $\dfrac{\dfrac{1}{x^2-3x+2}}{\dfrac{3}{x-2}-\dfrac{2}{x-1}}$

43. Solve for x. $\dfrac{5}{2} - \dfrac{3}{x+1} = \dfrac{2-x}{x+1}$

Solve the following exercises. If necessary, round your answers to the nearest hundredth.

44. ***Calculator Supply*** At the beginning of the fall semester, the campus bookstore at Boston University ordered 320 scientific and graphing calculators. For every seven graphing calculators, they ordered three scientific calculators. How many graphing calculators did they order? How many scientific calculators did they order?

▲ **45.** ***Photograph Enlargement*** Jill VanderWoude decided to enlarge a photograph that measures 5 inches wide and 7 inches long into a poster-size photograph with a perimeter of 168 inches. The new photograph will maintain the same width-to-length ratio. How wide will the enlarged photograph be? How long will the enlarged photograph be?

46. ***Swimming Pool*** How long will it take a pump to empty a 4900-gallon swimming pool if the same pump can empty a 3500-gallon swimming pool in 4 hours?

47. ***Rabbit Population*** In a sanctuary, a sample of one hundred wild rabbits is tagged and released by the wildlife management team. In a few weeks, after they have mixed with the general rabbit population, a sample of forty rabbits is caught, and eight have a tag. Estimate the population of rabbits in the sanctuary.

48. ***Police Statistics*** The ratio of officers to state troopers is 2 : 9. If there are 154 men and women on the force, how many are officers?

▲ **49.** ***Shadow Length*** A 7-foot-tall tree casts a shadow that is 6 feet long. At the same time of day, a building casts a shadow that is 156 feet long. How tall is the building?

50. ***Planting Time*** Each spring it takes Marianne 10 hours to plant flowers in her yard. She estimates her son would take one and a half times as long, or 15 hours, to do the job. How many hours would it take if they both worked together?

51. ***Whirlpool Tub*** If the hot water faucet at Mike's house is left on, it takes 15 minutes to fill the whirlpool tub. If the cold water faucet is left on, it takes 10 minutes to fill the whirlpool tub. How many minutes would it take if both faucets are left on?

Online Retail Spending *The bar graph below shows the growth in online spending in the United States. Use this bar graph to answer exercises 52–55. Round all answers to the nearest billion.*

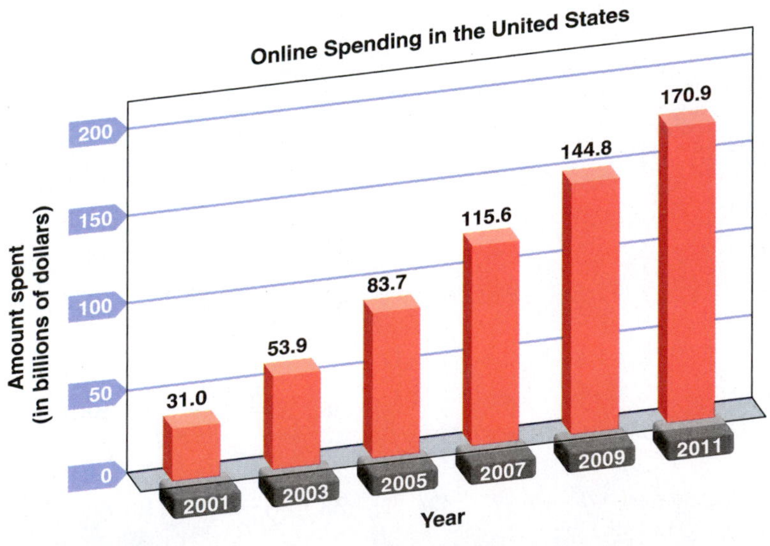

Source: www.census.gov

52. If the increase per year in online spending that occurred from 2007 to 2011 continues at the same increase per year, what would be the expected amount spent online in 2013?

53. If the ratio of the amount spent online in 2011 to the amount spent online in 2009 is proportional to the ratio of the amount spent online in 2013 to the amount spent online in 2011, what would be the expected amount spent online in 2013?

54. If the increase per year in online spending that occurred from 2001 to 2005 is double the increase in spending from 2011 to 2013, what would be the expected amount spent online in 2013?

55. If the increase per year in online spending that occurred from 2005 to 2007 is one-half as great as the increase in spending from 2011 to 2013, what would be the expected amount spent online in 2013?

How Am I Doing? Chapter 6 Test

CHAPTER Test Prep VIDEOS MATH COACH MyMathLab® YouTube™

After you take this test read through the Math Coach on pages 352–353. Math Coach videos are available via MyMathLab and YouTube. Step-by-step test solutions in the Chapter Test Prep Videos are also available via MyMathLab and YouTube. (Search "TobeyInterAlg" and click on "Channels.")

Simplify.

1. $\dfrac{x^3 + 3x^2 + 2x}{x^3 - 2x^2 - 3x}$

2. $\dfrac{-25p^4qr^3}{45pqr^6}$

3. $\dfrac{2y^2 + 7y - 4}{y^2 + 2y - 8} \cdot \dfrac{2y^2 - 8}{3y^2 + 11y + 10}$

4. $\dfrac{4 - 2x}{3x^2 - 2x - 8} \div \dfrac{2x^2 + x - 1}{9x + 12}$

5. $\dfrac{3}{x} - \dfrac{2}{x + 1}$

MC **6.** $\dfrac{2}{x^2 + 5x + 6} + \dfrac{3x}{x^2 + 6x + 9}$

7. $\dfrac{\dfrac{4}{y + 2} - 2}{5 - \dfrac{10}{y + 2}}$

MC **8.** $\dfrac{\dfrac{1}{x} - \dfrac{3}{x + 2}}{\dfrac{2}{x^2 + 2x}}$

Solve for the variable and check your answers. If no solution exists, say so.

9. $\dfrac{7}{4} = \dfrac{x + 4}{x}$

10. $2 + \dfrac{x}{x + 4} = \dfrac{3x}{x - 4}$

MC **11.** $\dfrac{1}{2y + 4} - \dfrac{1}{6} = \dfrac{-2}{3y + 6}$

12. $2 + \dfrac{3}{x} = \dfrac{2x}{x - 1}$

13. Solve for W. $h = \dfrac{S - 2WL}{2W + 2L}$

14. Solve for h. $\dfrac{3V}{\pi h} = r^2$

MC **15.** A total of 286 employees at Kaiser Telecommunication Systems were eligible this year for a high-performance bonus. The company president announced that for every three employees who got the bonus, nineteen employees did not. If the president was correct, how many employees got the high-performance bonus? How many did not get the high-performance bonus?

▲ **16.** The Newbury Elementary School had a rectangular playground that measured 500 feet wide and 850 feet long. When the school had a major addition to its property, it was decided to increase the playground to a large rectangular shape with the same proportional ratio that had a perimeter of 8100 feet. What are the width and the length of the new playground?

1. ______ ☐

2. ______ ☐

3. ______ ☐

4. ______ ☐

5. ______ ☐

6. ______ ☐

7. ______ ☐

8. ______ ☐

9. ______ ☐

10. ______ ☐

11. ______ ☐

12. ______ ☐

13. ______ ☐

14. ______ ☐

15. ______ ☐

16. ______ ☐

Total Correct: ☐

MATH COACH

Mastering the skills you need to do well on the test.

Students often make the same types of errors when they do the Chapter 6 Test. Here are some helpful hints to keep you from making those common errors on test problems.

Adding or Subtracting Two or More Rational Expressions—

Problem 6 Simplify. $\dfrac{2}{x^2 + 5x + 6} + \dfrac{3x}{x^2 + 6x + 9}$

> **Helpful Hint** First, factor each denominator. Form the LCD by taking the product of all the different factors you have obtained. Then multiply the numerator and denominator of each fraction by any factors that do not appear in the denominator.

Did you factor the first denominator into $(x + 2)(x + 3)$?

Yes ____ No ____

Did you factor the second denominator into $(x + 3)(x + 3)$?

Yes ____ No ____

If you answered No to either question, stop and carefully factor each denominator.

Did you determine that the LCD is $(x + 2)(x + 3)(x + 3)$?

Yes ____ No ____

Did you multiply the first fraction by $\dfrac{(x + 3)}{(x + 3)}$?

Yes ____ No ____

Did you multiply the second fraction by $\dfrac{(x + 2)}{(x + 2)}$?

Yes ____ No ____

If you answered No to any of these questions, review how to find the LCD and then multiply each fraction by the correct expression.

If you answered Problem 6 incorrectly, go back and rework the problem using these suggestions.

Simplifying a Complex Rational Expression—Problem 8

Simplify. $\dfrac{\dfrac{1}{x} - \dfrac{3}{x + 2}}{\dfrac{2}{x^2 + 2x}}$

> **Helpful Hint** You may use either Method 1 or Method 2. We follow Method 1 in the solution to this problem because many students find Method 1 easier to use. For this problem, the two fractions in the numerator need to be combined, but the denominator does not involve combining fractions.

Did you find the LCD of the two fractions in the numerator to be $x(x + 2)$?

Yes ____ No ____

Did you multiply the first fraction in the numerator by $\dfrac{(x + 2)}{(x + 2)}$?

Yes ____ No ____

Did you multiply the second fraction in the numerator by $\dfrac{x}{x}$?

Yes ____ No ____

If you answered No to any of these questions, stop and review how to find the LCD. Consider that when combining two fractions, we multiply the numerator and denominator of each fraction by any factor in the LCD that is not presently in the denominator. Please go back and complete these steps again.

Did you combine the two fractions in the numerator to obtain $\dfrac{-2x + 2}{x(x + 2)}$?

Yes ____ No ____

If you answered No, remember that you need to combine $x + 2 - 3x$ to obtain $-2x + 2$. Now complete the division process just as you would divide two fractions.

Now go back and rework the problem using these suggestions.

Solving a Rational Equation—Problem 11

Solve for the variable and check your answers.

$$\frac{1}{2y + 4} - \frac{1}{6} = \frac{-2}{3y + 6}$$

Helpful Hint Factor the denominators. Find the LCD. Multiply each fraction on both sides of the equation by the LCD. The resulting equation should not contain any fractions. Simplify and solve for the variable.

Did you find that the LCD is $(2)(3)(y + 2)$ or $6(y + 2)$?

Yes _____ No _____

If you answered No, remember that the denominator of the first fraction factors to $2(y + 2)$, the denominator of the second fraction factors to $2 \cdot 3$, and the denominator of the third fraction factors to $3(y + 2)$. Work the problem again to find the correct LCD.

When you multiplied the LCD by each fraction, did you obtain the unsimplified equation $3 - (y + 2) = 2(-2)$?

If you answered No, carefully multiply $6(y + 2)$ by each of the three fractions. Make sure that your resulting equation is cleared of all fractions. Then simplify the equation and solve for y. Check your solution for y by substituting the value in the original equation.

If you answered Problem 11 incorrectly, go back and rework the problem using these suggestions.

Solving an Applied Problem Involving Ratios—Problem 15

A total of 286 employees at Kaiser Telecommunication Systems were eligible this year for a high-performance bonus. The company president announced that for every three employees who got the bonus, nineteen employees did not. If the president was correct, how many employees got the high-performance bonus? How many did not get the high-performance bonus?

Helpful Hint Read through the problem carefully. Make sure you can identify a variable expression for the number of employees who got a bonus and the number of employees who did not get a bonus. Setting up an equation requires that you first find these two expressions.

If you let x be the number of employees who got a bonus, can you determine that $286 - x$ is the number of employees who did not get a bonus?

Yes _____ No _____

If you answered No, remember that out of a total of 286 employees, some received bonuses and the remaining number did not. So if x people got a bonus, then $286 - x$ did not get a bonus.

Did you set up a rational equation with one fraction as the ratio of $\frac{3}{19}$ and the other fraction as $\frac{x}{286 - x}$?

Yes _____ No _____

If you answered No, remember that if you choose x as one of the numerators, then 3 and x should be the two numerators. This means that 19 and $286 - x$ are the two denominators, respectively. Another possibility is to set up the equation $\frac{19}{3} = \frac{286 - x}{x}$. The key is to have the same value represented for the numerator in each fraction and the same value represented for the denominator of each fraction.

After multiplying by the LCD, did you obtain the equation $3(286 - x) = 19x$ or $19x = 3(286 - x)$?

Yes _____ No _____

If you answered No, remember that you must set the two ratios equal to each other and then multiply by the LCD to get an equation without fractions. Now solve the equation for x. Remember to find the values for both x and $286 - x$ in your final answer.

If you answered Problem 15 incorrectly, go back and rework the problem using these suggestions.

Cumulative Test for Chapters 1–6

This test provides a comprehensive review of the key objectives for Chapters 1–6.

1. Simplify. $(3x^{-2}y)^2(x^4y^{-3})$

2. Evaluate. $\sqrt{100} + 4(3 - 5)^3 - (-20)$

3. Write in scientific notation. 14,200,000

4. Evaluate $2x^2 - 3x - 4y^2$ when $x = -2$ and $y = 3$.

5. Simplify. $2x - [6x - 3(x + 5y)]$

6. Solve for x. $\frac{3}{4}(x + 2) = \frac{1}{3}x - 2$

7. Solve for x. $0.4x - 0.5(2x + 3) = -0.7(x + 3)$

8. Brenda invested \$7000 in two accounts at the bank. One account earns 5% simple interest. The other earns 8% simple interest. She earned \$539 interest in 1 year. How much was invested at each rate?

9. Solve for x and graph. $3(2 - 6x) > 4(x + 1) + 24$

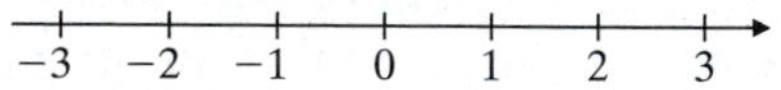

10. Solve for x. $|2x + 1| \leq 8$

11. Graph the line. $-6x + 2y = -12$

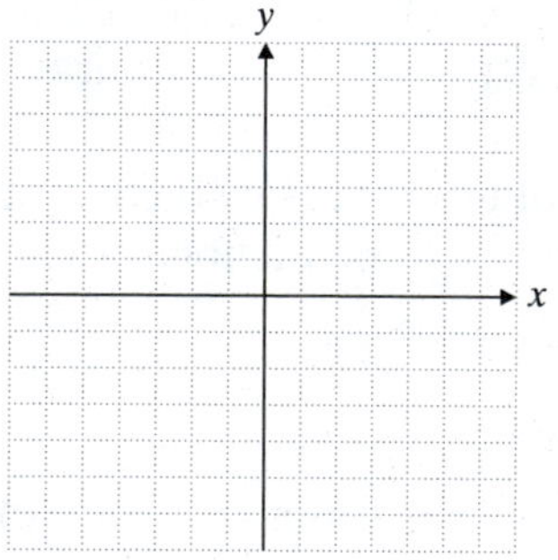

12. Find the standard form of the equation of the line parallel to $5x - 6y = 8$ that passes through $(-1, -3)$.

13. Find the standard form of the equation of the line that passes through the points $(3, 0)$ and $(-1, 4)$.

If $f(x) = \frac{1}{2}x^2 - 5x + 3$, *find*

14. $f(2)$

15. $f(-4)$

Solve the system.

16. $7x - 6y = 17$
$3x + y = 18$

1. ______
2. ______
3. ______
4. ______
5. ______
6. ______
7. ______
8. ______
9. ______
10. ______
11. ______
12. ______
13. ______
14. ______
15. ______
16. ______

17. $3x - 2y - 9z = 9$

$x - y + z = 8$

$2x + 3y - z = -2$

18. Jane went shopping at Old Navy. She bought six T-shirts and nine sweatshirts for her children. She spent a total of $156. Her sister, Lisa, spent a total of $124 on her children. She bought eight T-shirts and five sweatshirts. How much did a T-shirt cost? How much did a sweatshirt cost?

19. Graph the region.

$y \geq \frac{1}{2}x + 3$

$y \leq -x + 1$

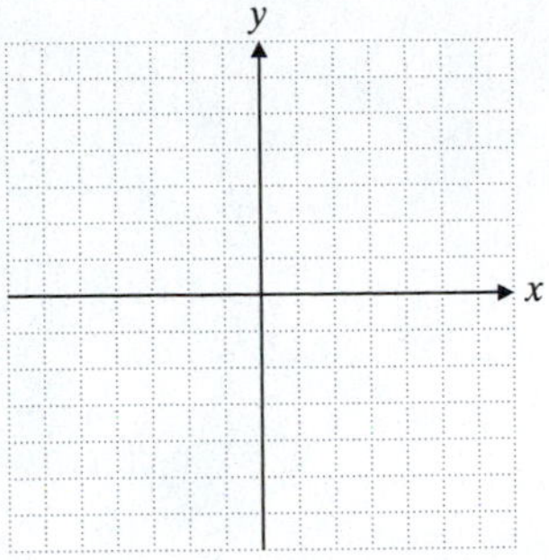

20. Multiply. $(2x + 5)^2$

21. Divide. $(9x^3 + 11x - 4) \div (3x - 1)$

Factor.

22. $8x^3 - 125y^3$

23. $81x^3 - 90x^2y + 25xy^2$

24. Solve for x. $x^2 + 20x + 36 = 0$

Simplify.

25. $\dfrac{2x^2 + x - 1}{2x^2 - 9x + 4} \cdot \dfrac{3x^2 - 12x}{6x + 15}$

26. $\dfrac{\dfrac{1}{2x + 1} + 1}{4 - \dfrac{3}{4x^2 - 1}}$

27. $\dfrac{3}{x - 6} + \dfrac{4}{x + 4}$

28. Solve for x. $\dfrac{1}{2x + 3} - \dfrac{4}{4x^2 - 9} = \dfrac{3}{2x - 3}$

29. Solve for b. $H = \dfrac{3b + 2x}{5 - 4b}$

30. The mayor of Chicago decided that in the coming year the city police force should be deployed in a particular ratio. He decided that for every three police officers who patrol on foot, there should be eleven police officers patrolling in squad cars. If 3234 police officers are normally on duty during the day, how many are patrolling on foot, and how many are patrolling in squad cars?

17. ______

18. ______

19. ______

20. ______

21. ______

22. ______

23. ______

24. ______

25. ______

26. ______

27. ______

28. ______

29. ______

30. ______

CHAPTER 7

At what speed should a satellite travel to maintain a circular orbit around Earth at a distance of 100 miles above Earth? By how much should the speed be increased if the orbiting distance is changed to 125 miles above Earth? These and other questions about satellite travel require calculations using radicals. The material contained in this chapter will allow you to explore questions related to satellite travel.

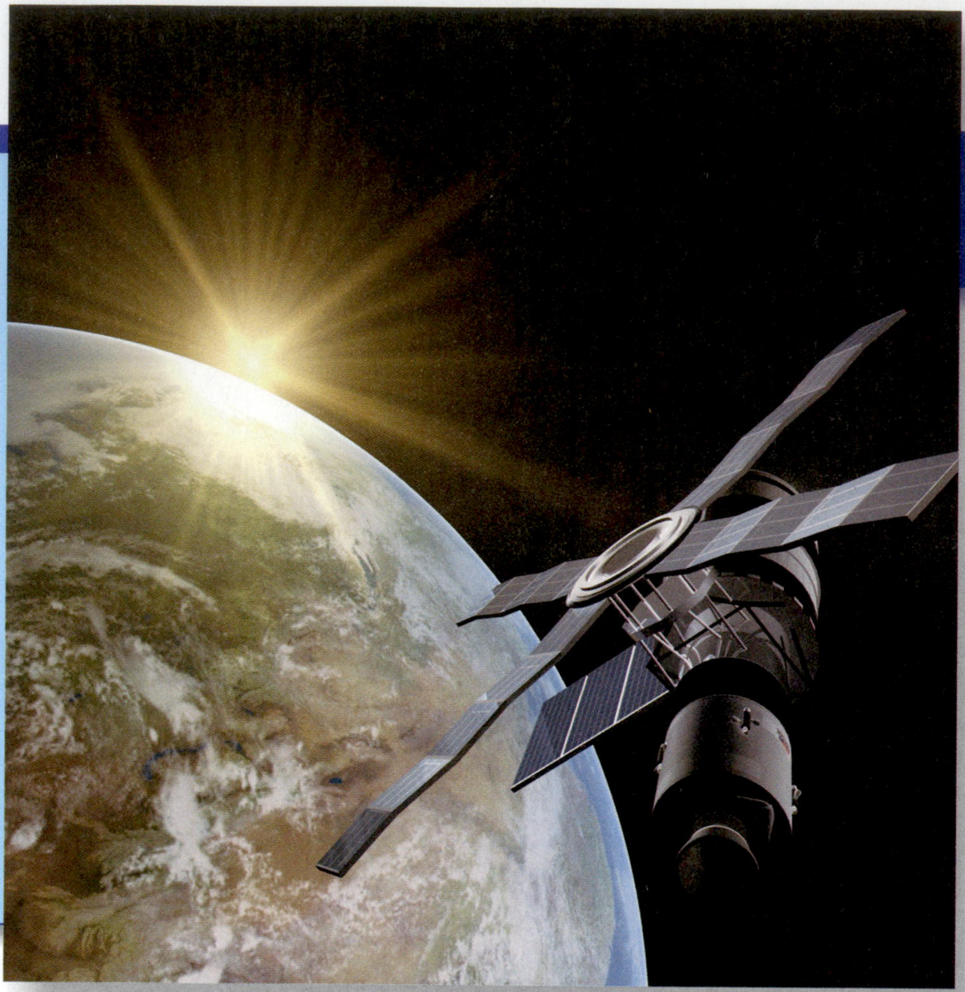

Rational Exponents and Radicals

7.1 Rational Exponents

① Simplifying Expressions with Rational Exponents

Student Learning Objectives

After studying this section, you will be able to apply the laws of exponents to:

① **Simplify expressions with rational exponents.**

② **Add expressions with rational exponents.**

③ **Factor expressions with rational exponents.**

Before studying this section, you may need to review Section 1.4. For convenience, we list the rules of exponents that we learned there.

$$x^m x^n = x^{m+n} \qquad x^0 = 1$$

$$\frac{x^m}{x^n} = x^{m-n} \qquad (x^m)^n = x^{mn}$$

$$x^{-n} = \frac{1}{x^n} \qquad (xy)^n = x^n y^n$$

$$\frac{x^{-n}}{y^{-m}} = \frac{y^m}{x^n} \qquad \left(\frac{x}{y}\right)^n = \frac{x^n}{y^n}$$

To ensure that you understand these rules, study Example 1 carefully and work Student Practice 1.

EXAMPLE 1 Simplify. $\left(\dfrac{5xy^{-3}}{2x^{-4}y}\right)^{-2}$

Solution

$$\left(\frac{5xy^{-3}}{2x^{-4}y}\right)^{-2} = \frac{(5xy^{-3})^{-2}}{(2x^{-4}y)^{-2}} \qquad \left(\frac{x}{y}\right)^n = \frac{x^n}{y^n}.$$

$$= \frac{5^{-2}x^{-2}(y^{-3})^{-2}}{2^{-2}(x^{-4})^{-2}y^{-2}} \qquad (xy)^n = x^n y^n.$$

$$= \frac{5^{-2}x^{-2}y^{6}}{2^{-2}x^{8}y^{-2}} \qquad (x^m)^n = x^{mn}.$$

$$= \frac{5^{-2}}{2^{-2}} \cdot \frac{x^{-2}}{x^{8}} \cdot \frac{y^{6}}{y^{-2}}$$

$$= \frac{2^2}{5^2} \cdot x^{-2-8} \cdot y^{6+2} \qquad \frac{x^{-n}}{y^{-m}} = \frac{y^m}{x^n}; \frac{x^m}{x^n} = x^{m-n}.$$

$$= \frac{4}{25}x^{-10}y^{8}$$

The answer can also be written as $\dfrac{4y^8}{25x^{10}}$. Explain why.

Student Practice 1 Simplify.

$$\left(\frac{3x^{-2}y^{4}}{2x^{-5}y^{2}}\right)^{-3}$$

***NOTE TO STUDENT:** Fully worked-out solutions to all of the Student Practice problems can be found at the back of the text starting at page SP-1.*

SIDELIGHT: Example 1 Follow-Up Deciding when to use the rule $\dfrac{x^{-n}}{y^{-m}} = \dfrac{y^m}{x^n}$ is entirely up to you. In Example 1, we could have begun by writing

$$\left(\frac{5xy^{-3}}{2x^{-4}y}\right)^{-2} = \left(\frac{5x \cdot x^4}{2y \cdot y^3}\right)^{-2} = \left(\frac{5x^5}{2y^4}\right)^{-2}.$$

Complete the steps to simplify this expression.

Likewise, in the fourth step in Example 1, we could have written

$$\frac{5^{-2}x^{-2}y^{6}}{2^{-2}x^{8}y^{-2}} = \frac{2^2y^6y^2}{5^2x^8x^2}.$$

Complete the steps to simplify this expression. Are the two answers the same as the answer in Example 1? Why or why not?

We generally begin to simplify a rational expression with exponents by raising a power to a power because sometimes negative powers become positive. The order in which you use the rules of exponents is up to you. Work carefully. Keep track of your exponents and where you are as you simplify the rational expression.

These rules for exponents can also be extended to include rational exponents—that is, exponents that are fractions. As you recall, rational numbers are of the form $\frac{a}{b}$, where a and b are integers and b does not equal zero. We will write fractional exponents using diagonal lines. Thus, we will write $\frac{5}{6}$ as $5/6$, and if a fractional exponent contains a variable such as $\frac{a}{b}$, we will write it as a/b. For now we restrict the base to *positive* real numbers. Later we will talk about negative bases.

EXAMPLE 2 Simplify.

(a) $(x^{2/3})^4$ **(b)** $\dfrac{x^{5/6}}{x^{1/6}}$ **(c)** $x^{2/3} \cdot x^{-1/3}$ **(d)** $5^{3/7} \cdot 5^{2/7}$

Solution We will not write out every step or every rule of exponents that we use. You should be able to follow the solutions.

(a) $(x^{2/3})^4 = x^{(2/3)(4/1)} = x^{8/3}$

(b) $\dfrac{x^{5/6}}{x^{1/6}} = x^{5/6-1/6} = x^{4/6} = x^{2/3}$

(c) $x^{2/3} \cdot x^{-1/3} = x^{2/3-1/3} = x^{1/3}$

(d) $5^{3/7} \cdot 5^{2/7} = 5^{3/7+2/7} = 5^{5/7}$

Student Practice 2 Simplify.

(a) $(x^4)^{3/8}$ **(b)** $\dfrac{x^{3/7}}{x^{2/7}}$ **(c)** $x^{-7/5} \cdot x^{4/5}$

Sometimes fractional exponents will not have the same denominator. Remember that you need to change the fractions to equivalent fractions with the same denominator when the rules of exponents require you to add or to subtract them.

EXAMPLE 3 Simplify. Express your answers with positive exponents only.

(a) $(2x^{1/2})(3x^{1/3})$ **(b)** $\dfrac{18x^{1/4}y^{-1/3}}{-6x^{-1/2}y^{1/6}}$

Solution

(a) $(2x^{1/2})(3x^{1/3}) = 6x^{1/2+1/3} = 6x^{3/6+2/6} = 6x^{5/6}$

(b)
$$\begin{aligned}\frac{18x^{1/4}y^{-1/3}}{-6x^{-1/2}y^{1/6}} &= -3x^{1/4-(-1/2)}y^{-1/3-1/6} \\ &= -3x^{1/4+2/4}y^{-2/6-1/6} \\ &= -3x^{3/4}y^{-3/6} \\ &= -3x^{3/4}y^{-1/2} \\ &= -\frac{3x^{3/4}}{y^{1/2}}\end{aligned}$$

Student Practice 3 Simplify. Express your answers with positive exponents only.

(a) $(-3x^{1/4})(2x^{1/2})$ **(b)** $\dfrac{13x^{1/12}y^{-1/4}}{26x^{-1/3}y^{1/2}}$

EXAMPLE 4 Multiply and simplify. $-2x^{5/6}(3x^{1/2} - 4x^{-1/3})$

Solution We will need to be very careful when we add the exponents for x as we use the distributive property. Study each step of the following example. Be sure you understand each operation.

$$-2x^{5/6}(3x^{1/2} - 4x^{-1/3}) = -6x^{5/6+1/2} + 8x^{5/6-1/3}$$
$$= -6x^{5/6+3/6} + 8x^{5/6-2/6}$$
$$= -6x^{8/6} + 8x^{3/6}$$
$$= -6x^{4/3} + 8x^{1/2}$$

Student Practice 4 Multiply and simplify. $-3x^{1/2}(2x^{1/4} + 3x^{-1/2})$

Sometimes we can use the rules of exponents to simplify numerical values raised to rational powers.

EXAMPLE 5 Evaluate.

(a) $(25)^{3/2}$ **(b)** $(27)^{2/3}$

Solution

(a) $(25)^{3/2} = (5^2)^{3/2} = 5^{2/1 \cdot 3/2} = 5^3 = 125$

(b) $(27)^{2/3} = (3^3)^{2/3} = 3^{3/1 \cdot 2/3} = 3^2 = 9$

Student Practice 5 Evaluate.

(a) $(4)^{5/2}$ **(b)** $(27)^{4/3}$

② Adding Expressions with Rational Exponents

Adding expressions with rational exponents may require several steps. Sometimes this involves removing negative exponents. For example, to add $2x^{-1/2} + x^{1/2}$, we begin by writing $2x^{-1/2}$ as $\frac{2}{x^{1/2}}$. This is a rational expression. Recall that to add rational expressions we need to have a common denominator. Take time to look at the steps needed to write $2x^{-1/2} + x^{1/2}$ as one term.

EXAMPLE 6 Write as one fraction with positive exponents. $2x^{-1/2} + x^{1/2}$

Solution

$$2x^{-1/2} + x^{1/2} = \frac{2}{x^{1/2}} + \frac{x^{1/2} \cdot x^{1/2}}{x^{1/2}} = \frac{2}{x^{1/2}} + \frac{x^1}{x^{1/2}} = \frac{2 + x}{x^{1/2}}$$

Student Practice 6 Write as one fraction with only positive exponents.
$3x^{1/3} + x^{-1/3}$

③ Factoring Expressions with Rational Exponents

To factor expressions, we need to be able to recognize common factors. If the terms of the expression contain exponents, we look for the same exponential factor in each term. For example, in the expression $6x^5 + 4x^3 - 8x^2$, the common factor of each term is $2x^2$. Thus, we can factor out the common factor $2x^2$ from each term. The expression then becomes $2x^2(3x^3 + 2x - 4)$.

We do exactly the same thing when we factor expressions with rational exponents. The key is to identify the exponent of the common factor. In the expression $6x^{3/4} + 4x^{1/2} - 8x^{1/4}$, the common factor is $2x^{1/4}$. Thus, we factor the expression $6x^{3/4} + 4x^{1/2} - 8x^{1/4}$ as $2x^{1/4}(3x^{1/2} + 2x^{1/4} - 4)$. We do not always need to factor out the greatest common factor. In the following examples we simply factor out a common factor.

EXAMPLE 7 Factor out the common factor $2x$. $2x^{3/2} + 4x^{5/2}$

Solution We rewrite the exponent of each term so that we can see that each term contains the factor $2x$ or $2x^{2/2}$.

$$\begin{aligned} 2x^{3/2} + 4x^{5/2} &= 2x^{2/2+1/2} + 4x^{2/2+3/2} \\ &= 2(x^{2/2})(x^{1/2}) + 4(x^{2/2})(x^{3/2}) \\ &= 2x(x^{1/2} + 2x^{3/2}) \end{aligned}$$

Student Practice 7 Factor out the common factor $4y$. $4y^{3/2} - 8y^{5/2}$

For convenience we list here the properties of exponents that we have discussed in this section, as well as the property $x^0 = 1$.

When x and y are **positive real numbers** and a and b are **rational numbers:**

$$x^a x^b = x^{a+b} \qquad \frac{x^a}{x^b} = x^{a-b} \qquad x^0 = 1$$

$$x^{-a} = \frac{1}{x^a} \qquad \frac{x^{-a}}{y^{-b}} = \frac{y^b}{x^a}$$

$$(x^a)^b = x^{ab} \qquad (xy)^a = x^a y^a \qquad \left(\frac{x}{y}\right)^a = \frac{x^a}{y^a}$$

7.1 Exercises

MyMathLab® Watch the videos in MyMathLab Download the MyDashBoard App

Simplify. Express your answer with positive exponents.

1. $\left(\frac{3xy^{-1}}{z^2}\right)^4$

2. $\left(\frac{3xy^{-2}}{x^3}\right)^2$

3. $\left(\frac{4ab^{-2}}{3b}\right)^2$

4. $\left(\frac{-a^2b}{2b^{-1}}\right)^3$

5. $\left(\frac{2x^2}{y}\right)^{-3}$

6. $\left(\frac{4x}{y^3}\right)^{-3}$

7. $\left(\frac{3xy^{-2}}{y^3}\right)^{-2}$

8. $\left(\frac{5x^{-2}y}{x^4}\right)^{-2}$

9. $(x^{3/4})^2$

10. $(x^{5/6})^3$

11. $(y^{12})^{2/3}$

12. $(y^2)^{5/2}$

13. $\frac{x^{3/5}}{x^{1/5}}$

14. $\frac{y^{6/7}}{y^{3/7}}$

15. $\frac{x^{8/9}}{x^{2/9}}$

16. $\frac{x^{11/12}}{x^{5/12}}$

17. $\frac{a^2}{a^{1/4}}$

18. $\frac{x^4}{x^{1/3}}$

19. $x^{1/7} \cdot x^{3/7}$

20. $a^{3/4} \cdot a^{1/4}$

21. $a^{3/8} \cdot a^{1/2}$

22. $b^{2/5} \cdot b^{2/15}$

23. $y^{3/5} \cdot y^{-1/10}$

24. $y^{7/10} \cdot y^{-1/5}$

Write each expression with positive exponents.

25. $x^{-3/4}$

26. $x^{-5/6}$

27. $a^{-5/6}b^{1/3}$

28. $4a^{-3/8}b^{1/2}$

29. $6^{-1/2}$

30. $4^{-1/3}$

31. $3a^{-1/3}$

32. $-3y^{-5/4}$

Evaluate or simplify the numerical expressions.

33. $(27)^{5/3}$

34. $(16)^{3/4}$

35. $(4)^{3/2}$

36. $(9)^{3/2}$

37. $(-8)^{5/3}$

38. $(-27)^{5/3}$

39. $(-27)^{2/3}$

40. $(-64)^{2/3}$

Mixed Practice, Exercises 41–64

Simplify and express your answers with positive exponents. Evaluate or simplify the numerical expressions.

41. $(x^{1/4}y^{-1/3})(x^{3/4}y^{1/2})$

42. $(x^{1/4}y^{1/2})(x^{3/8}y^{-1/2})$

43. $(7x^{1/3}y^{1/4})(-2x^{1/4}y^{-1/6})$

44. $(8x^{-1/5}y^{1/3})(-3x^{-1/4}y^{1/6})$

45. $6^2 \cdot 6^{-2/3}$

46. $11^{1/2} \cdot 11^3$

47. $\dfrac{2x^{1/5}}{x^{-1/2}}$

48. $\dfrac{3y^{2/3}}{y^{-1/4}}$

49. $\dfrac{-20x^2y^{-1/5}}{5x^{-1/2}y}$

50. $\dfrac{12x^{-2/3}y}{-6xy^{-3/4}}$

51. $\left(\dfrac{8a^2b^6}{a^{-1}b^3}\right)^{1/3}$

52. $\left(\dfrac{25a^4b^{-3}}{a^{-2}b^{-9}}\right)^{1/2}$

53. $(-4x^{1/4}y^{5/2}z^{1/2})^2$

54. $(4x^{2/3}y^{-1/2}z^{3/5})^3$

55. $x^{2/3}(x^{4/3} - x^{1/5})$

56. $y^{-2/3}(y^{2/3} + y^{3/2})$

57. $m^{7/8}(m^{-1/2} + 2m)$

58. $n^{1/3}(n^{1/9} + 2n)$

59. $(8)^{-1/3}$

60. $(100)^{-1/2}$

61. $(25)^{-3/2}$

62. $(16)^{-5/4}$

63. $(81)^{3/4} + (25)^{1/2}$

64. $(49)^{1/2} + 8^{2/3}$

Write each expression as one fraction with positive exponents.

65. $3y^{1/2} + y^{-1/2}$

66. $2y^{1/3} + y^{-2/3}$

67. $x^{-1/3} + 6^{4/3}$

68. $5^{-1/4} + x^{-1/2}$

Factor out the common factor 2a.

69. $10a^{5/4} - 4a^{8/5}$

70. $6a^{4/3} - 8a^{3/2}$

Factor out the common factor 3x.

71. $12x^{4/3} - 3x^{5/2}$

72. $18x^{5/3} - 6x^{11/8}$

To Think About

73. What is the value of a if $x^a \cdot x^{1/4} = x^{-1/8}$?

74. What is the value of b if $x^b \div x^{1/3} = x^{-1/12}$?

Applications

Radius and Volume of a Sphere *The radius needed to create a sphere with a given volume V can be approximated by the equation* $r = 0.62(V)^{1/3}$. *Find the radius of the spheres with the following volumes.*

▲**75.** 27 cubic meters

▲**76.** 64 cubic meters

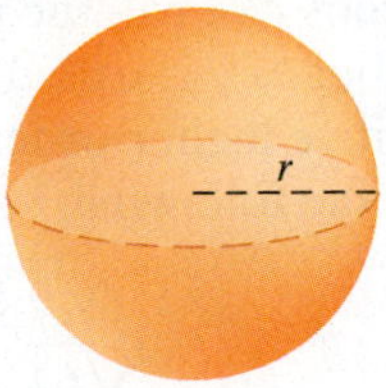

Radius and Volume of a Cone *The radius required for a cone to have a volume V and a height h is given by the equation* $r = \left(\frac{3V}{\pi h}\right)^{1/2}$. *Find the necessary radius to have a cone with the properties below. Use* $\pi \approx 3.14$.

▲**77.** $V = 314$ cubic feet and $h = 12$ feet.

▲**78.** $V = 3140$ cubic feet and $h = 30$ feet.

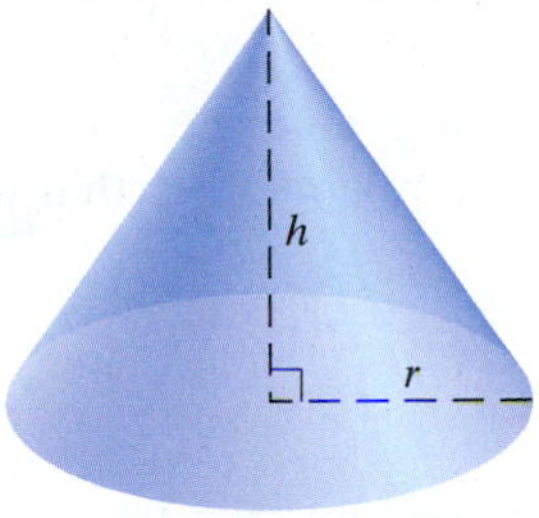

Cumulative Review

79. **[2.1.1]** Solve for x. $-4(x + 1) = \frac{1}{3}(3 - 2x)$

80. **[2.2.1]** Solve for b. $A = \frac{h}{2}(a + b)$

Quick Quiz 7.1 *Simplify.*

1. $(-4x^{2/3}y^{1/4})(3x^{1/6}y^{1/2})$

2. $\dfrac{16x^4}{8x^{2/3}}$

3. $(25x^{1/4})^{3/2}$

4. **Concept Check** Explain how you would simplify the following.

$$x^{9/10} \cdot x^{-1/5}$$

7.2 Radical Expressions and Functions

Student Learning Objectives

After studying this section, you will be able to:

1. **Evaluate radical expressions and functions.**
2. **Change radical expressions to expressions with rational exponents.**
3. **Change expressions with rational exponents to radical expressions.**
4. **Evaluate higher-order radicals containing variable radicands that represent any real number (including negative real numbers).**

① Evaluating Radical Expressions and Functions

In Section 1.3 we studied simple radical expressions called square roots. The **square root** of a number is a value that when multiplied by itself is equal to the original number. That is, since $3 \cdot 3 = 9$, 3 is a square root of 9. But $(-3) \cdot (-3) = 9$, so -3 is also a square root. We call the positive square root the **principal square root.**

The symbol $\sqrt{}$ is called a **radical sign.** We use it to denote positive square roots (and positive higher-order roots also). A negative square root is written $-\sqrt{}$. Thus, we have the following:

$$\sqrt{9} = 3 \qquad -\sqrt{9} = -3$$
$$\sqrt{64} = 8 \quad \text{(because } 8 \cdot 8 = 64\text{)}$$
$$\sqrt{121} = 11 \quad \text{(because } 11 \cdot 11 = 121\text{)}$$

Because $\sqrt{9} = \sqrt{3 \cdot 3} = \sqrt{3^2} = 3$, we can say the following:

DEFINITION OF SQUARE ROOT

If x is a nonnegative real number, then $\sqrt{x}$ is the *nonnegative* (or principal) *square root* of x; in other words, $(\sqrt{x})^2 = x$.

Note that x must be *nonnegative*. Why? Suppose we want to find $\sqrt{-36}$. We must find a real number that when multiplied by itself gives -36. Is there one? No, because

$$6 \cdot 6 = 36 \quad \text{and}$$
$$(-6)(-6) = 36.$$

So there is no real number that we can square to get -36.

We call $\sqrt[n]{x}$ a **radical expression.** The $\sqrt{}$ symbol is the radical sign, the x is the **radicand,** and the n is the **index** of the radical. When no number for n appears in the radical expression, it is understood that 2 is the index, which means that we are looking for the square root. For example, in the radical expression $\sqrt{25}$, with no number given for the index n, we take the index to be 2. Thus, $\sqrt{25}$ is the principal square root of 25.

We can extend the notion of square root to **higher-order roots,** such as cube roots, fourth roots, and so on. A **cube root** of a number is a value that when cubed is equal to the original number. The index n of the radical is 3, and the radical used is $\sqrt[3]{}$. Similarly, a **fourth root** of a number is a value that when raised to the fourth power is equal to the original number. The index n of the radical is 4, and the radical used is $\sqrt[4]{}$. Thus, we have the following:

$$\sqrt[3]{27} = 3 \quad \text{because } 3 \cdot 3 \cdot 3 = 3^3 = 27.$$
$$\sqrt[4]{81} = 3 \quad \text{because } 3 \cdot 3 \cdot 3 \cdot 3 = 3^4 = 81.$$
$$\sqrt[5]{32} = 2 \quad \text{because } 2 \cdot 2 \cdot 2 \cdot 2 \cdot 2 = 2^5 = 32.$$
$$\sqrt[3]{-64} = -4 \quad \text{because } (-4)(-4)(-4) = (-4)^3 = -64.$$
$$\sqrt[6]{729} = 3 \quad \text{because } 3 \cdot 3 \cdot 3 \cdot 3 \cdot 3 \cdot 3 = 3^6 = 729.$$

You should be able to see a pattern here.

$$\sqrt[3]{27} = \sqrt[3]{3^3} = 3$$
$$\sqrt[4]{81} = \sqrt[4]{3^4} = 3$$
$$\sqrt[5]{32} = \sqrt[5]{2^5} = 2$$
$$\sqrt[3]{-64} = \sqrt[3]{(-4)^3} = -4$$
$$\sqrt[6]{729} = \sqrt[6]{3^6} = 3$$

In these cases, we see that $\sqrt[n]{x^n} = x$. We now give the following definition.

DEFINITION OF HIGHER-ORDER ROOTS

1. If x is a *nonnegative* real number, then $\sqrt[n]{x}$ is a nonnegative nth root and has the property that

$$(\sqrt[n]{x})^n = x.$$

2. If x is a *negative* real number, then
   (a) When n is an *odd integer*, $\sqrt[n]{x}$ has the property that $(\sqrt[n]{x})^n = x$.
   (b) When n is an *even integer*, $\sqrt[n]{x}$ is *not* a real number.

EXAMPLE 1 If possible, find the root of each negative number. If there is no real number root, say so.

(a) $\sqrt[3]{-216}$ **(b)** $\sqrt[5]{-32}$ **(c)** $\sqrt[4]{-16}$ **(d)** $\sqrt[6]{-64}$

Solution

(a) $\sqrt[3]{-216} = \sqrt[3]{(-6)^3} = -6$ because $(-6)^3 = -216$.

(b) $\sqrt[5]{-32} = \sqrt[5]{(-2)^5} = -2$ because $(-2)^5 = -32$.

(c) $\sqrt[4]{-16}$ is not a real number because n is even and x is negative.

(d) $\sqrt[6]{-64}$ is not a real number because n is even and x is negative.

Student Practice 1 If possible, find the roots. If there is no real number root, say so.

(a) $\sqrt[3]{216}$ **(b)** $\sqrt[5]{32}$ **(c)** $\sqrt[3]{-8}$ **(d)** $\sqrt[4]{-81}$

***NOTE TO STUDENT:** Fully worked-out solutions to all of the Student Practice problems can be found at the back of the text starting at page SP-1.*

Because the symbol $\sqrt{x}$ represents exactly one real number for all real numbers x that are nonnegative, we can use it to define the **square root function** $f(x) = \sqrt{x}$.

This function has a domain of all real numbers x that are greater than or equal to zero.

EXAMPLE 2 Find the indicated function values of the function $f(x) = \sqrt{2x + 4}$. Round your answers to the nearest tenth when necessary.

(a) $f(-2)$ **(b)** $f(6)$ **(c)** $f(3)$

Solution

(a) $f(-2) = \sqrt{2(-2) + 4} = \sqrt{-4 + 4} = \sqrt{0} = 0$ The square root of zero is zero.

(b) $f(6) = \sqrt{2(6) + 4} = \sqrt{12 + 4} = \sqrt{16} = 4$

(c) $f(3) = \sqrt{2(3) + 4} = \sqrt{6 + 4} = \sqrt{10} \approx 3.2$ We use a calculator or a square root table to approximate $\sqrt{10}$.

Student Practice 2 Find the indicated values of the function $f(x) = \sqrt{4x - 3}$. Round your answers to the nearest tenth when necessary.

(a) $f(3)$ **(b)** $f(4)$ **(c)** $f(7)$

EXAMPLE 3 Find the domain of the function. $f(x) = \sqrt{3x - 6}$

Solution We know that the expression $3x - 6$ must be nonnegative. That is, $3x - 6 \geq 0$.

$$3x - 6 \geq 0$$
$$3x \geq 6$$
$$x \geq 2$$

Thus, the domain is all real numbers x where $x \geq 2$.

Student Practice 3 Find the domain of the function. $f(x) = \sqrt{0.5x + 2}$

EXAMPLE 4 Graph the function $f(x) = \sqrt{x + 2}$. Use the values $f(-2)$, $f(-1)$, $f(0)$, $f(1)$, $f(2)$, and $f(7)$. Round your answers to the nearest tenth when necessary.

Solution We show the table of values here.

x	f(x)
−2	0
−1	1
0	1.4
1	1.7
2	2
7	3

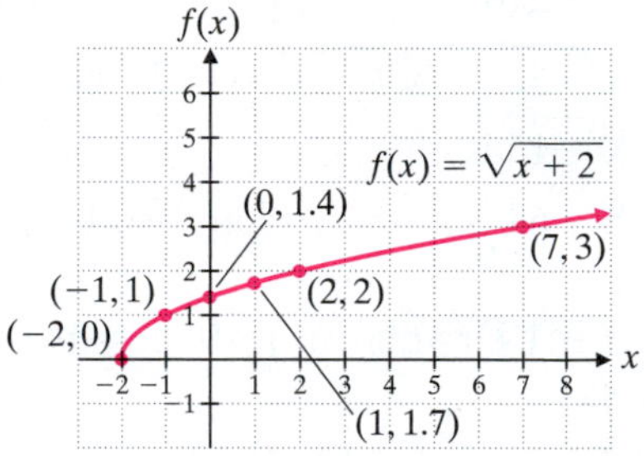

Student Practice 4

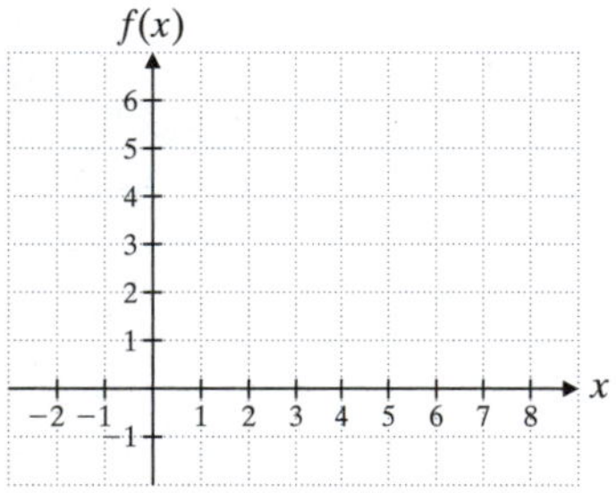

Student Practice 4 Graph the function $f(x) = \sqrt{3x - 9}$. Use the values $f(3)$, $f(4)$, $f(5)$, and $f(6)$.

② Changing Radical Expressions to Expressions with Rational Exponents

Now we want to extend our definition of roots to rational exponents. By the laws of exponents we know that

$$x^{1/2} \cdot x^{1/2} = x^{1/2+1/2} = x^1 = x.$$

Since $x^{1/2}x^{1/2} = x$, $x^{1/2}$ must be a square root of x. That is, $x^{1/2} = \sqrt{x}$. Is this true? By the definition of square root, $(\sqrt{x})^2 = x$. Does $(x^{1/2})^2 = x$? Using the law of exponents we have

$$(x^{1/2})^2 = x^{(1/2)(2)} = x^1 = x.$$

We conclude that

$$x^{1/2} = \sqrt{x}.$$

In the same way we can write the following:

$$x^{1/3} \cdot x^{1/3} \cdot x^{1/3} = x \qquad x^{1/3} = \sqrt[3]{x}$$
$$x^{1/4} \cdot x^{1/4} \cdot x^{1/4} \cdot x^{1/4} = x \qquad x^{1/4} = \sqrt[4]{x}$$
$$\vdots \qquad\qquad \vdots$$
$$\underbrace{x^{1/n} \cdot x^{1/n} \cdot \cdots \cdot x^{1/n}}_{n \text{ factors}} = x \qquad x^{1/n} = \sqrt[n]{x}$$

Therefore, we are ready to define fractional exponents in general.

DEFINITION
If n is a positive integer and x is a nonnegative real number, then

$$x^{1/n} = \sqrt[n]{x}.$$

EXAMPLE 5 Change to rational exponents and simplify. Assume that all variables are nonnegative real numbers.

(a) $\sqrt[4]{x^4}$ **(b)** $\sqrt[5]{(32)^5}$

Solution

(a) $\sqrt[4]{x^4} = (x^4)^{1/4} = x^{4/4} = x^1 = x$

(b) $\sqrt[5]{(32)^5} = (32^5)^{1/5} = 32^{5/5} = 32^1 = 32$

Student Practice 5 Change to rational exponents and simplify. Assume that all variables are nonnegative real numbers.

(a) $\sqrt[3]{x^3}$ **(b)** $\sqrt[4]{y^4}$

EXAMPLE 6 Replace all radicals with rational exponents.

(a) $\sqrt[3]{x^2}$ **(b)** $(\sqrt[5]{w})^7$

Solution

(a) $\sqrt[3]{x^2} = (x^2)^{1/3} = x^{2/3}$

(b) $(\sqrt[5]{w})^7 = (w^{1/5})^7 = w^{7/5}$

Student Practice 6 Replace all radicals with rational exponents.

(a) $\sqrt[4]{x^3}$ **(b)** $\sqrt[5]{(xy)^7}$

EXAMPLE 7 Evaluate or simplify. Assume that all variables are nonnegative.

(a) $\sqrt[5]{32x^{10}}$ **(b)** $\sqrt[3]{125x^9}$ **(c)** $(16x^4)^{3/4}$

Solution

(a) $\sqrt[5]{32x^{10}} = (2^5x^{10})^{1/5} = 2x^2$

(b) $\sqrt[3]{125x^9} = [5^3x^9]^{1/3} = 5x^3$

(c) $(16x^4)^{3/4} = (2^4x^4)^{3/4} = 2^3x^3 = 8x^3$

Student Practice 7 Evaluate or simplify. Assume that all variables are nonnegative.

(a) $\sqrt[4]{81x^{12}}$ **(b)** $\sqrt[3]{27x^6}$ **(c)** $(32x^5)^{3/5}$

③ Changing Expressions with Rational Exponents to Radical Expressions

Sometimes we need to change an expression with rational exponents to a radical expression. This is especially helpful because the value of the radical form of an expression is sometimes more recognizable. For example, because of our experience with radicals, we know that $\sqrt{25} = 5$. It is not as easy to see that $25^{1/2} = 5$. Therefore, we simplify expressions with rational exponents by first rewriting them as radical expressions. Recall that

$$x^{1/n} = \sqrt[n]{x}.$$

Again, using the laws of exponents, we know that

$$x^{m/n} = (x^m)^{1/n} = (x^{1/n})^m,$$

when x is nonnegative. We can make the following general definition.

DEFINITION

For positive integers m and n and any real number x for which $x^{1/n}$ is defined,

$$x^{m/n} = \left(\sqrt[n]{x}\right)^m = \sqrt[n]{x^m}.$$

If it is also true that $x \neq 0$, then

$$x^{-m/n} = \frac{1}{x^{m/n}} = \frac{1}{\left(\sqrt[n]{x}\right)^m} = \frac{1}{\sqrt[n]{x^m}}.$$

EXAMPLE 8 Change to radical form.

(a) $(xy)^{5/7}$ **(b)** $w^{-2/3}$ **(c)** $3x^{3/4}$ **(d)** $(3x)^{3/4}$

Solution

(a) $(xy)^{5/7} = \sqrt[7]{(xy)^5} = \sqrt[7]{x^5y^5}$ or $(xy)^{5/7} = \left(\sqrt[7]{xy}\right)^5$

(b) $w^{-2/3} = \dfrac{1}{w^{2/3}} = \dfrac{1}{\sqrt[3]{w^2}}$ or $w^{-2/3} = \dfrac{1}{w^{2/3}} = \dfrac{1}{\left(\sqrt[3]{w}\right)^2}$

(c) $3x^{3/4} = 3\sqrt[4]{x^3}$ or $3x^{3/4} = 3\left(\sqrt[4]{x}\right)^3$

(d) $(3x)^{3/4} = \sqrt[4]{(3x)^3} = \sqrt[4]{27x^3}$ or $(3x)^{3/4} = \left(\sqrt[4]{3x}\right)^3$

Student Practice 8 Change to radical form.

(a) $(xy)^{3/4}$ **(b)** $y^{-1/3}$ **(c)** $(2x)^{4/5}$ **(d)** $2x^{4/5}$

Graphing Calculator

Rational Exponents

To evaluate Example 9(a) on a graphing calculator we use

125 [^] [(] 2 [÷] 3 [)] [ENTER]

Note the need to include the parentheses around the quantity $\frac{2}{3}$. Try each part of Example 9 on your graphing calculator.

EXAMPLE 9 Change to radical form and evaluate.

(a) $125^{2/3}$ **(b)** $(-16)^{5/2}$ **(c)** $144^{-1/2}$

Solution

(a) $125^{2/3} = \left(\sqrt[3]{125}\right)^2 = (5)^2 = 25$

(b) $(-16)^{5/2} = \left(\sqrt{-16}\right)^5$; however, $\sqrt{-16}$ is not a real number. Thus, $(-16)^{5/2}$ is not a real number.

(c) $144^{-1/2} = \dfrac{1}{144^{1/2}} = \dfrac{1}{\sqrt{144}} = \dfrac{1}{12}$

Student Practice 9 Change to radical form and evaluate.

(a) $8^{2/3}$ **(b)** $(-8)^{4/3}$ **(c)** $100^{-3/2}$

④ Evaluating Higher-Order Radicals Containing Variable Radicands That Represent Any Real Number (Including Negative Real Numbers)

We now give a definition of higher-order radicals that works for all radicals, no matter what their signs are.

> **DEFINITION**
>
> For all real numbers x (including negative real numbers),
>
> $\sqrt[n]{x^n} = |x|$ when n is an *even* positive integer, and
>
> $\sqrt[n]{x^n} = x$ when n is an *odd* positive integer.

EXAMPLE 10 Evaluate; x may be any real number.

(a) $\sqrt[3]{(-2)^3}$ **(b)** $\sqrt[4]{(-2)^4}$ **(c)** $\sqrt[5]{x^5}$ **(d)** $\sqrt[6]{x^6}$

Solution

(a) $\sqrt[3]{(-2)^3} = -2$ because the index is odd.

(b) $\sqrt[4]{(-2)^4} = |-2| = 2$ because the index is even.

(c) $\sqrt[5]{x^5} = x$ because the index is odd.

(d) $\sqrt[6]{x^6} = |x|$ because the index is even.

Student Practice 10 Evaluate; y and w may be any real numbers.

(a) $\sqrt[5]{(-3)^5}$ **(b)** $\sqrt[4]{(-5)^4}$ **(c)** $\sqrt[4]{w^4}$ **(d)** $\sqrt[7]{y^7}$

EXAMPLE 11 Simplify. Assume that x and y may be any real numbers.

(a) $\sqrt{49x^2}$ **(b)** $\sqrt[4]{81y^{16}}$ **(c)** $\sqrt[3]{27x^6y^9}$

Solution

(a) We observe that the index is an even positive number. We will need the absolute value. $\sqrt{49x^2} = 7|x|$

(b) Again, we need the absolute value. $\sqrt[4]{81y^{16}} = 3|y^4|$
Since we know that $3y^4$ is positive (anything to the fourth power will be positive), we can write $3|y^4|$ without the absolute value symbol. Thus, $\sqrt[4]{81y^{16}} = 3y^4$.

(c) The index is an odd integer. The absolute value is never needed in such a case.
$\sqrt[3]{27x^6y^9} = \sqrt[3]{3^3(x^2)^3(y^3)^3} = 3x^2y^3$

Student Practice 11 Simplify. Assume that x and y may be any real numbers.

(a) $\sqrt{36x^2}$ **(b)** $\sqrt[4]{16y^8}$ **(c)** $\sqrt[3]{125x^3y^6}$

7.2 Exercises

MyMathLab® — Watch the videos in MyMathLab — Download the MyDashBoard App

Verbal and Writing Skills, Exercises 1–4

1. In a simple sentence, explain what a square root is.

2. In a simple sentence, explain what a cube root is.

3. Give an example to show why the cube root of a negative number is a negative number.

4. Give an example to show why it is not possible to find a real number that is the square root of a negative number.

Evaluate if possible.

5. $\sqrt{100}$ **6.** $\sqrt{64}$ **7.** $\sqrt{16} + \sqrt{81}$ **8.** $\sqrt{36} + \sqrt{121}$ **9.** $-\sqrt{\frac{1}{9}}$

10. $-\sqrt{\frac{4}{25}}$ **11.** $\sqrt{-36}$ **12.** $\sqrt{-49}$ **13.** $\sqrt{0.04}$ **14.** $\sqrt{0.81}$

For the given function, find the indicated function values. Find the domain of each function. Round your answers to the nearest tenth when necessary. You may use a table of square roots or a calculator as needed.

15. $f(x) = \sqrt{3x + 21}$; $f(0), f(1), f(5), f(-4)$

16. $f(x) = \sqrt{10x - 4}$; $f(1), f(2), f(3), f(4)$

17. $f(x) = \sqrt{0.5x - 5}$; $f(10), f(12), f(18), f(20)$

18. $f(x) = \sqrt{1.5x + 3}$; $f(2), f(4), f(6), f(8)$

Graph each of the following functions. Plot at least four points for each function.

19. $f(x) = \sqrt{x - 1}$

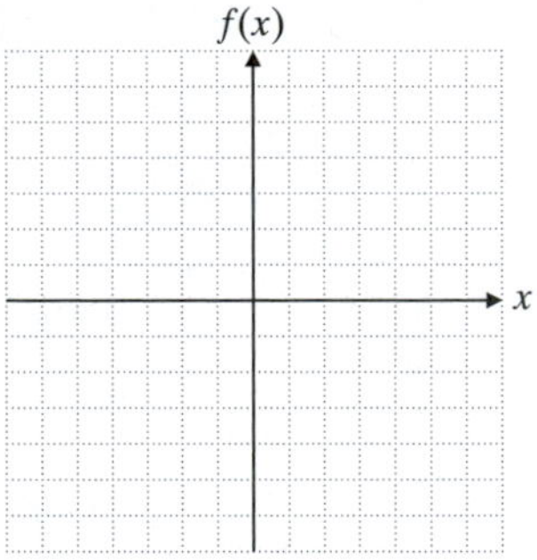

20. $f(x) = \sqrt{x - 3}$

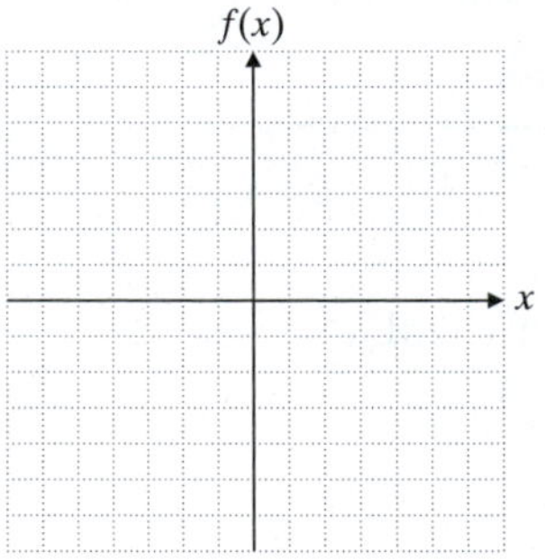

21. $f(x) = \sqrt{3x + 9}$

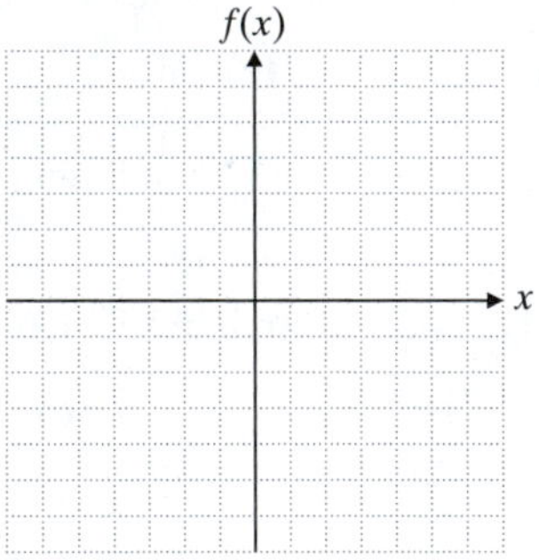

22. $f(x) = \sqrt{2x + 4}$

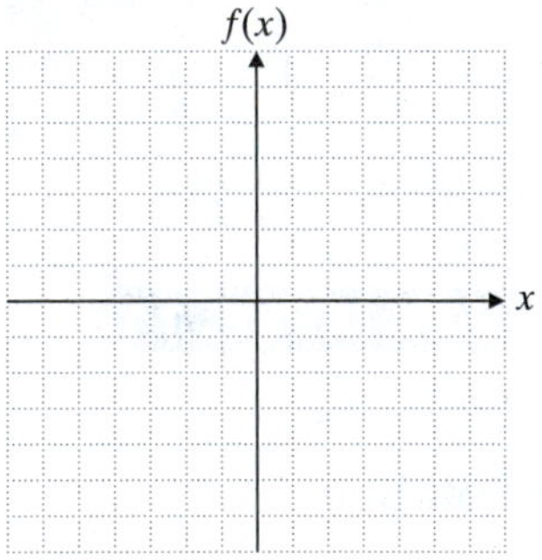

Evaluate if possible.

23. $\sqrt[3]{64}$ **24.** $\sqrt[3]{27}$ **25.** $\sqrt[3]{-1000}$ **26.** $\sqrt[3]{-125}$ **27.** $\sqrt[4]{16}$

28. $\sqrt[4]{625}$ **29.** $\sqrt[4]{81}$ **30.** $-\sqrt[6]{64}$ **31.** $\sqrt[5]{(8)^5}$ **32.** $\sqrt[6]{(9)^6}$

33. $\sqrt[8]{(5)^8}$ **34.** $\sqrt[7]{(11)^7}$ **35.** $\sqrt[3]{-\frac{1}{8}}$ **36.** $\sqrt[3]{-\frac{8}{27}}$

For exercises 37–78, assume that variables represent positive real numbers. Replace all radicals with rational exponents.

37. $\sqrt[3]{y}$ **38.** $\sqrt{a}$ **39.** $\sqrt[5]{m^3}$ **40.** $\sqrt[6]{b^5}$

41. $\sqrt[4]{2a}$ **42.** $\sqrt[5]{4b}$ **43.** $\sqrt[7]{(a + b)^3}$ **44.** $\sqrt[9]{(a - b)^5}$

45. $\sqrt{\sqrt[3]{x}}$ **46.** $\sqrt[5]{\sqrt{y}}$ **47.** $(\sqrt[6]{3x})^5$ **48.** $(\sqrt[5]{2x})^3$

Simplify.

49. $\sqrt[6]{(12)^6}$ **50.** $\sqrt[5]{(-11)^5}$ **51.** $\sqrt[3]{x^{12}y^3}$ **52.** $\sqrt[3]{x^6y^{15}}$

53. $\sqrt{36x^8y^4}$ **54.** $\sqrt{64x^6y^{10}}$ **55.** $\sqrt[4]{16a^8b^4}$ **56.** $\sqrt[4]{81a^{12}b^{20}}$

Change to radical form.

57. $y^{4/7}$ **58.** $x^{5/6}$ **59.** $7^{-2/3}$ **60.** $5^{-3/5}$

61. $(a + 5b)^{3/4}$ **62.** $(3a + b)^{2/7}$ **63.** $(-x)^{3/5}$ **64.** $(-y)^{5/7}$

65. $(2xy)^{3/5}$ **66.** $(3ab)^{2/7}$

Mixed Practice, Exercises 67–78

Evaluate or simplify. Assume that variables represent positive real numbers.

67. $9^{3/2}$ **68.** $8^{2/3}$ **69.** $\left(\frac{4}{25}\right)^{1/2}$ **70.** $\left(\frac{1}{16}\right)^{3/4}$

71. $\left(\frac{1}{8}\right)^{-1/3}$ **72.** $\left(\frac{25}{9}\right)^{-1/2}$ **73.** $(64x^4)^{-1/2}$ **74.** $(81y^{10})^{-1/2}$

75. $\sqrt{121x^4}$ **76.** $\sqrt{49x^8}$ **77.** $\sqrt{144a^6b^{24}}$ **78.** $\sqrt{36a^{20}b^{10}}$

Simplify. Assume that the variables represent any real number.

79. $\sqrt{25x^2}$ **80.** $\sqrt{100x^2}$ **81.** $\sqrt[3]{-8x^6}$ **82.** $\sqrt[3]{-27x^9}$ **83.** $\sqrt[4]{x^8y^{16}}$

84. $\sqrt[4]{x^{16}y^{40}}$ **85.** $\sqrt[4]{a^{12}b^4}$ **86.** $\sqrt[4]{a^4b^{20}}$ **87.** $\sqrt{25x^{12}y^4}$ **88.** $\sqrt{49x^8y^8}$

Cumulative Review

89. **[2.2.1]** Solve for x. $-5x + 2y = 6$

90. **[2.2.1]** Solve for y. $x = \frac{2}{3}y + 4$

Quick Quiz 7.2

Simplify. Assume that variables represent positive real numbers.

1. $\left(\frac{4}{25}\right)^{3/2}$

2. $\sqrt[3]{-64}$

3. $\sqrt{121x^{10}y^{12}}$

4. **Concept Check** Explain how you would simplify the following.

$$\sqrt[4]{81x^8y^{16}}$$

7.3 Simplifying, Adding, and Subtracting Radicals

Student Learning Objectives

After studying this section, you will be able to:

① **Simplify a radical by using the product rule.**

② **Add and subtract like radical terms.**

NOTE TO STUDENT: Fully worked-out solutions to all of the Student Practice problems can be found at the back of the text starting at page SP-1.

① Simplifying a Radical by Using the Product Rule

When we simplify a radical, we want to get an equivalent expression with the smallest possible quantity in the radicand. We can use the product rule for radicals to simplify radicals.

PRODUCT RULE FOR RADICALS

For all nonnegative real numbers a and b and positive integers n,

$$\sqrt[n]{a}\,\sqrt[n]{b} = \sqrt[n]{ab}.$$

You should be able to derive the product rule from your knowledge of the laws of exponents. We have

$$\sqrt[n]{a}\,\sqrt[n]{b} = a^{1/n}b^{1/n} = (ab)^{1/n} = \sqrt[n]{ab}.$$

Throughout the remainder of this chapter, assume that all variables in any radicand represent nonnegative numbers, unless a specific statement is made to the contrary.

EXAMPLE 1 Simplify. $\sqrt{32}$

Solution 1: $\sqrt{32} = \sqrt{16 \cdot 2} = \sqrt{16}\sqrt{2} = 4\sqrt{2}$

Solution 2: $\sqrt{32} = \sqrt{4 \cdot 8} = \sqrt{4}\sqrt{8} = 2\sqrt{8} = 2\sqrt{4 \cdot 2} = 2\sqrt{4}\sqrt{2} = 4\sqrt{2}$

Although we obtained the same answer both times, the first solution is much shorter. You should try to use the largest factor that is a perfect square when you use the product rule.

Student Practice 1 Simplify. $\sqrt{20}$

EXAMPLE 2 Simplify. $\sqrt{48}$

Solution

$$\sqrt{48} = \sqrt{16}\sqrt{3} = 4\sqrt{3}$$

Student Practice 2 Simplify. $\sqrt{27}$

EXAMPLE 3 Simplify.

(a) $\sqrt[3]{16}$ **(b)** $\sqrt[3]{-81}$

Solution

(a) $\sqrt[3]{16} = \sqrt[3]{8}\sqrt[3]{2} = 2\sqrt[3]{2}$

(b) $\sqrt[3]{-81} = \sqrt[3]{-27}\sqrt[3]{3} = -3\sqrt[3]{3}$

Student Practice 3 Simplify.

(a) $\sqrt[3]{24}$ **(b)** $\sqrt[3]{-108}$

EXAMPLE 4 Simplify. $\sqrt[4]{48}$

Solution $\sqrt[4]{48} = \sqrt[4]{16}\sqrt[4]{3} = 2\sqrt[4]{3}$

Student Practice 4 Simplify. $\sqrt[4]{64}$

EXAMPLE 5 Simplify.

(a) $\sqrt{27x^3y^4}$ **(b)** $\sqrt[3]{16x^4y^3z^6}$

Solution

(a) $\sqrt{27x^3y^4} = \sqrt{9 \cdot 3 \cdot x^2 \cdot x \cdot y^4} = \sqrt{9x^2y^4}\sqrt{3x}$ Factor out the perfect squares.

$= 3xy^2\sqrt{3x}$

(b) $\sqrt[3]{16x^4y^3z^6} =$

$\sqrt[3]{8 \cdot 2 \cdot x^3 \cdot x \cdot y^3 \cdot z^6} = \sqrt[3]{8x^3y^3z^6}\sqrt[3]{2x}$ Factor out the perfect cubes. Why is z^6 a perfect cube?

$= 2xyz^2\sqrt[3]{2x}$

Student Practice 5 Simplify.

(a) $\sqrt{45x^6y^7}$ **(b)** $\sqrt{27a^7b^8c^9}$

② Adding and Subtracting Like Radical Terms

Only like radicals can be added or subtracted. Two radicals are **like radicals** if they have the same radicand and the same index. $2\sqrt{5}$ and $3\sqrt{5}$ are like radicals. $2\sqrt{5}$ and $2\sqrt{3}$ are not like radicals; $2\sqrt{5}$ and $2\sqrt[3]{5}$ are not like radicals. When we combine radicals, we combine like terms by using the distributive property.

EXAMPLE 6 Combine. $2\sqrt{5} + 3\sqrt{5} - 4\sqrt{5}$

Solution $2\sqrt{5} + 3\sqrt{5} - 4\sqrt{5} = (2 + 3 - 4)\sqrt{5} = 1\sqrt{5} = \sqrt{5}$

Student Practice 6 Combine. $19\sqrt{xy} + 5\sqrt{xy} - 10\sqrt{xy}$

Sometimes when you simplify radicands, you may find you have like radicals.

EXAMPLE 7 Combine. $5\sqrt{3} - \sqrt{27} + 2\sqrt{48}$

Solution

$$\begin{aligned} 5\sqrt{3} - \sqrt{27} + 2\sqrt{48} &= 5\sqrt{3} - \sqrt{9}\sqrt{3} + 2\sqrt{16}\sqrt{3} \\ &= 5\sqrt{3} - 3\sqrt{3} + 2(4)\sqrt{3} \\ &= 5\sqrt{3} - 3\sqrt{3} + 8\sqrt{3} \\ &= 10\sqrt{3} \end{aligned}$$

Student Practice 7 Combine.

$4\sqrt{2} - 5\sqrt{50} - 3\sqrt{98}$

EXAMPLE 8 Combine. $6\sqrt{x} + 4\sqrt{12x} - \sqrt{75x} + 3\sqrt{x}$

Solution

$$\begin{aligned}6\sqrt{x} + 4\sqrt{12x} - \sqrt{75x} + 3\sqrt{x} &= 6\sqrt{x} + 4\sqrt{4}\sqrt{3x} - \sqrt{25}\sqrt{3x} + 3\sqrt{x}\\ &= 6\sqrt{x} + 8\sqrt{3x} - 5\sqrt{3x} + 3\sqrt{x}\\ &= 6\sqrt{x} + 3\sqrt{x} + 8\sqrt{3x} - 5\sqrt{3x}\\ &= 9\sqrt{x} + 3\sqrt{3x}\end{aligned}$$

Student Practice 8 Combine.

$4\sqrt{2x} + \sqrt{18x} - 2\sqrt{125x} - 6\sqrt{20x}$

EXAMPLE 9 Combine. $2\sqrt[3]{81x^3y^4} + 3xy\sqrt[3]{24y}$

Solution

$$\begin{aligned}2\sqrt[3]{81x^3y^4} + 3xy\sqrt[3]{24y} &= 2\sqrt[3]{27x^3y^3}\sqrt[3]{3y} + 3xy\sqrt[3]{8}\sqrt[3]{3y}\\ &= 2(3xy)\sqrt[3]{3y} + 3xy(2)\sqrt[3]{3y}\\ &= 6xy\sqrt[3]{3y} + 6xy\sqrt[3]{3y}\\ &= 12xy\sqrt[3]{3y}\end{aligned}$$

Student Practice 9 Combine.

$3x\sqrt[3]{54x^4} - 3\sqrt[3]{16x^7}$

7.3 Exercises

MyMathLab® — Watch the videos in MyMathLab — Download the MyDashBoard App

Simplify. Assume that all variables are nonnegative real numbers.

1. $\sqrt{8}$

2. $\sqrt{12}$

3. $\sqrt{18}$

4. $\sqrt{75}$

5. $\sqrt{28}$

6. $\sqrt{54}$

7. $\sqrt{50}$

8. $\sqrt{80}$

9. $\sqrt{9x^3}$

10. $\sqrt{16x^5}$

11. $\sqrt{40a^6b^7}$

12. $\sqrt{48a^5b^{10}}$

13. $\sqrt{90x^3yz^4}$

14. $\sqrt{24xy^8z^3}$

15. $\sqrt[3]{8}$

16. $\sqrt[3]{125}$

17. $\sqrt[3]{40}$

18. $\sqrt[3]{128}$

19. $\sqrt[3]{54a^2}$

20. $\sqrt[3]{32n}$

21. $\sqrt[3]{27a^5b^9}$

22. $\sqrt[3]{125a^6b^2}$

23. $\sqrt[3]{40x^{12}y^{13}}$

24. $\sqrt[3]{108x^5y^{17}}$

25. $\sqrt[4]{81kp^{23}}$

26. $\sqrt[4]{16k^{12}p^{18}}$

27. $\sqrt[5]{-32x^5y^6}$

28. $\sqrt[5]{-243x^4y^{10}}$

To Think About, Exercises 29 and 30

29. $\sqrt[4]{1792} = a\sqrt[4]{7}$. What is the value of a?

30. $\sqrt[3]{3072} = b\sqrt[3]{6}$. What is the value of b?

Combine.

31. $4\sqrt{5} + 8\sqrt{5}$

32. $3\sqrt{13} + 7\sqrt{13}$

33. $4\sqrt{3} + \sqrt{7} - 5\sqrt{7}$

34. $2\sqrt{6} + \sqrt{2} - 5\sqrt{6}$

35. $3\sqrt{32} - \sqrt{2}$

36. $\sqrt{90} - \sqrt{10}$

37. $4\sqrt{12} + \sqrt{27}$

38. $5\sqrt{75} + \sqrt{48}$

39. $\sqrt{8} + \sqrt{50} - 2\sqrt{72}$

40. $\sqrt{45} + \sqrt{80} - 3\sqrt{20}$

41. $\sqrt{48} - 2\sqrt{27} + \sqrt{12}$

42. $-4\sqrt{18} + \sqrt{98} - \sqrt{32}$

43. $-5\sqrt{45} + 6\sqrt{20} + 3\sqrt{5}$

44. $-7\sqrt{10} + 4\sqrt{40} - 8\sqrt{90}$

Combine. Assume that all variables represent nonnegative real numbers.

45. $3\sqrt{48x} - 2\sqrt{12x}$

46. $5\sqrt{27x} - 4\sqrt{75x}$

47. $5\sqrt{2x} + 2\sqrt{18x} + 2\sqrt{32x}$

48. $4\sqrt{50x} + 3\sqrt{2x} + \sqrt{72x}$

49. $\sqrt{44} - 3\sqrt{63x} + 4\sqrt{28x}$

50. $\sqrt{63x} - \sqrt{54x} + \sqrt{24x}$

51. $\sqrt{200x^3} - x\sqrt{32x}$

52. $\sqrt{75a^3} + a\sqrt{12a}$

53. $\sqrt[3]{16} + 3\sqrt[3]{54}$

54. $\sqrt[3]{128} - 4\sqrt[3]{16}$

55. $4\sqrt[3]{x^4y^3} - 3\sqrt[3]{xy^5}$

56. $2y\sqrt[3]{27x^3} - 3\sqrt[3]{125x^3y^3}$

To Think About

57. Use a calculator to show that $\sqrt{48} + \sqrt{27} + \sqrt{75} = 12\sqrt{3}$.

58. Use a calculator to show that $\sqrt{98} + \sqrt{50} + \sqrt{128} = 20\sqrt{2}$.

Applications ***Electric Current*** *We can approximate the amount of current in amps I (amperes) drawn by an appliance in the home using the formula*

$$I = \sqrt{\frac{P}{R}},$$

where P is the power measured in watts and R is the resistance measured in ohms. In exercises 59 and 60, round your answers to three decimal places.

59. What is the amount of current drawn by a refrigerator if $P = 825$ watts and $R = 15$ ohms?

60. What is the amount of current drawn by a clothes dryer if $P = 3600$ watts and $R = 12$ ohms?

Period of a Pendulum *The* ***period*** *of a pendulum is the amount of time it takes the pendulum to make one complete swing back and forth. If the length of the pendulum L is measured in feet, then its period T measured in seconds is given by the formula*

$$T = 2\pi\sqrt{\frac{L}{32}}.$$

Use $\pi \approx 3.14$ *for exercises 61–62.*

61. Find the period of a pendulum if its length is 8 feet.

62. How much longer is the period of a pendulum consisting of a person swinging on a 30-foot rope than that of a person swinging on a 10-foot rope? Round your answer to the nearest hundredth.

Cumulative Review *Factor completely.*

63. [5.7.1] $16x^3 - 56x^2y + 49xy^2$

64. [5.7.1] $81x^2y - 25y$

FDA Recommendations for Phosphorus *The FDA recommends that an adult's minimum daily intake of the mineral phosphorus be 1 gram. A small serving of scallops (six average-size scallops) has 0.2 gram of phosphorus, while one small serving of skim milk (1 cup) has 0.25 gram of phosphorus.*

65. [4.3.1] If the number of servings of scallops and the number of servings of skim milk totals 4.5 servings, how many of each would you need to meet the minimum daily requirement of phosphorus?

66. [1.2.2] If you eat only scallops, how many servings would you need to obtain the minimum daily requirement of phosphorus? If you drink only skim milk, how many servings would you need to obtain the minimum daily requirement of phosphorus?

Quick Quiz 7.3 *Simplify.*

1. $\sqrt{120x^7y^8}$

2. $\sqrt[3]{16x^{15}y^{10}}$

3. Combine.
$2\sqrt{75} + 3\sqrt{48} - 4\sqrt{27}$

4. Concept Check Explain how you would simplify the following.

$$\sqrt[4]{16x^{13}y^{16}}$$

7.4 Multiplying and Dividing Radicals

Student Learning Objectives

After studying this section, you will be able to:

1. **Multiply radical expressions.**
2. **Divide radical expressions.**
3. **Simplify radical expressions by rationalizing the denominator.**

***NOTE TO STUDENT:** Fully worked-out solutions to all of the Student Practice problems can be found at the back of the text starting at page SP-1.*

① Multiplying Radical Expressions

We use the product rule for radicals to multiply radical expressions. Recall that $\sqrt[n]{a}\sqrt[n]{b} = \sqrt[n]{ab}$.

EXAMPLE 1 Multiply. $(3\sqrt{2})(5\sqrt{11x})$

Solution $(3\sqrt{2})(5\sqrt{11x}) = (3)(5)\sqrt{2 \cdot 11x} = 15\sqrt{22x}$

Student Practice 1 Multiply. $(-4\sqrt{2})(-3\sqrt{13x})$

EXAMPLE 2 Multiply. $\sqrt{6x}\,(\sqrt{3} + \sqrt{2x} + \sqrt{5})$

Solution

$$\begin{aligned}\sqrt{6x}(\sqrt{3} + \sqrt{2x} + \sqrt{5}) &= (\sqrt{6x})(\sqrt{3}) + (\sqrt{6x})(\sqrt{2x}) + (\sqrt{6x})(\sqrt{5})\\ &= \sqrt{18x} + \sqrt{12x^2} + \sqrt{30x}\\ &= \sqrt{9}\sqrt{2x} + \sqrt{4x^2}\sqrt{3} + \sqrt{30x}\\ &= 3\sqrt{2x} + 2x\sqrt{3} + \sqrt{30x}\end{aligned}$$

Student Practice 2 Multiply. $\sqrt{2x}\,(\sqrt{5} + 2\sqrt{3x} + \sqrt{8})$

To multiply two binomials containing radicals, we can use the distributive property. Most students find that the FOIL method is helpful in remembering how to find the four products.

EXAMPLE 3 Multiply. $(\sqrt{2} + 3\sqrt{5})(2\sqrt{2} - \sqrt{5})$

Solution By FOIL:

$$\begin{aligned}(\sqrt{2} + 3\sqrt{5})(2\sqrt{2} - \sqrt{5}) &= 2\sqrt{4} - \sqrt{10} + 6\sqrt{10} - 3\sqrt{25}\\ &= 4 + 5\sqrt{10} - 15\\ &= -11 + 5\sqrt{10}\end{aligned}$$

By the distributive property:

$$\begin{aligned}(\sqrt{2} + 3\sqrt{5})(2\sqrt{2} - \sqrt{5}) &= (\sqrt{2} + 3\sqrt{5})(2\sqrt{2}) - (\sqrt{2} + 3\sqrt{5})(\sqrt{5})\\ &= (\sqrt{2})(2\sqrt{2}) + (3\sqrt{5})(2\sqrt{2}) - (\sqrt{2})(\sqrt{5}) - (3\sqrt{5})(\sqrt{5})\\ &= 2\sqrt{4} + 6\sqrt{10} - \sqrt{10} - 3\sqrt{25}\\ &= 4 + 5\sqrt{10} - 15\\ &= -11 + 5\sqrt{10}\end{aligned}$$

Student Practice 3 Multiply. $(\sqrt{7} + 4\sqrt{2})(2\sqrt{7} - 3\sqrt{2})$

EXAMPLE 4 Multiply. $(7 - 3\sqrt{2})(4 - \sqrt{3})$

Solution

$$(7 - 3\sqrt{2})(4 - \sqrt{3}) = 28 - 7\sqrt{3} - 12\sqrt{2} + 3\sqrt{6}$$

Student Practice 4 Multiply. $(2 - 5\sqrt{5})(3 - 2\sqrt{2})$

EXAMPLE 5 Multiply. $(\sqrt{7} + \sqrt{3x})^2$

Solution

Method 1: We can use the FOIL method or the distributive property.

$$(\sqrt{7} + \sqrt{3x})(\sqrt{7} + \sqrt{3x}) = \sqrt{49} + \sqrt{21x} + \sqrt{21x} + \sqrt{9x^2}$$
$$= 7 + 2\sqrt{21x} + 3x$$

Method 2: We could also use the Chapter 5 formula.

$$(a + b)^2 = a^2 + 2ab + b^2,$$

where $a = \sqrt{7}$ and $b = \sqrt{3x}$. Then

$$(\sqrt{7} + \sqrt{3x})^2 = (\sqrt{7})^2 + 2\sqrt{7}\sqrt{3x} + (\sqrt{3x})^2$$
$$= 7 + 2\sqrt{21x} + 3x$$

Student Practice 5 Multiply, using the approach that seems easiest to you.

$$(\sqrt{5x} + \sqrt{10})^2$$

EXAMPLE 6 Multiply.

(a) $\sqrt[3]{3x}(\sqrt[3]{x^2} + 3\sqrt[3]{4y})$ **(b)** $(\sqrt[3]{2y} + \sqrt[3]{4})(2\sqrt[3]{4y^2} - 3\sqrt[3]{2})$

Solution

(a) $\sqrt[3]{3x}(\sqrt[3]{x^2} + 3\sqrt[3]{4y}) = (\sqrt[3]{3x})(\sqrt[3]{x^2}) + 3(\sqrt[3]{3x})(\sqrt[3]{4y})$
$= \sqrt[3]{3x^3} + 3\sqrt[3]{12xy}$
$= x\sqrt[3]{3} + 3\sqrt[3]{12xy}$

(b) $(\sqrt[3]{2y} + \sqrt[3]{4})(2\sqrt[3]{4y^2} - 3\sqrt[3]{2}) = 2\sqrt[3]{8y^3} - 3\sqrt[3]{4y} + 2\sqrt[3]{16y^2} - 3\sqrt[3]{8}$
$= 2(2y) - 3\sqrt[3]{4y} + 2\sqrt[3]{8}\sqrt[3]{2y^2} - 3(2)$
$= 4y - 3\sqrt[3]{4y} + 4\sqrt[3]{2y^2} - 6$

Student Practice 6 Multiply.

(a) $\sqrt[3]{2x}(\sqrt[3]{4x^2} + 3\sqrt[3]{y})$

(b) $(\sqrt[3]{7} + \sqrt[3]{x^2})(2\sqrt[3]{49} - \sqrt[3]{x})$

② Dividing Radical Expressions

We can use the laws of exponents to develop a rule for dividing two radicals.

$$\sqrt[n]{\frac{a}{b}} = \left(\frac{a}{b}\right)^{1/n} = \frac{a^{1/n}}{b^{1/n}} = \frac{\sqrt[n]{a}}{\sqrt[n]{b}}$$

This quotient rule is very useful. We now state it more formally.

QUOTIENT RULE FOR RADICALS

For all nonnegative real numbers a, all positive real numbers b, and positive integers n,

$$\frac{\sqrt[n]{a}}{\sqrt[n]{b}} = \sqrt[n]{\frac{a}{b}}.$$

Sometimes it will be best to change $\sqrt[n]{\frac{a}{b}}$ to $\frac{\sqrt[n]{a}}{\sqrt[n]{b}}$, whereas at other times it will be best to change $\frac{\sqrt[n]{a}}{\sqrt[n]{b}}$ to $\sqrt[n]{\frac{a}{b}}$. To use the quotient rule for radicals, you need to have good number sense. You should know your squares up to 15^2 and your cubes up to 5^3.

EXAMPLE 7 Divide.

(a) $\frac{\sqrt{48}}{\sqrt{3}}$ **(b)** $\sqrt[3]{\frac{125}{8}}$ **(c)** $\frac{\sqrt{28x^5y^3}}{\sqrt{7x}}$

Solution

(a) $\frac{\sqrt{48}}{\sqrt{3}} = \frac{\sqrt{48}}{\sqrt{3}} = \sqrt{16} = 4$

(b) $\sqrt[3]{\frac{125}{8}} = \frac{\sqrt[3]{125}}{\sqrt[3]{8}} = \frac{5}{2}$

(c) $\frac{\sqrt{28x^5y^3}}{\sqrt{7x}} = \sqrt{\frac{28x^5y^3}{7x}} = \sqrt{4x^4y^3} = 2x^2y\sqrt{y}$

Student Practice 7 Divide.

(a) $\frac{\sqrt{75}}{\sqrt{3}}$ **(b)** $\sqrt[3]{\frac{27}{64}}$ **(c)** $\frac{\sqrt{54a^3b^7}}{\sqrt{6b^5}}$

3 Simplifying Radical Expressions by Rationalizing the Denominator

Recall that to simplify a radical we want to get the smallest possible quantity in the radicand. Whenever possible, we find the square root of a perfect square. Thus, to simplify $\sqrt{\frac{7}{16}}$ we have

$$\sqrt{\frac{7}{16}} = \frac{\sqrt{7}}{\sqrt{16}} = \frac{\sqrt{7}}{4}.$$

Notice that the denominator does not contain a square root. The expression $\frac{\sqrt{7}}{4}$ is in simplest form.

Let's look at $\sqrt{\frac{16}{7}}$. We have

$$\sqrt{\frac{16}{7}} = \frac{\sqrt{16}}{\sqrt{7}} = \frac{4}{\sqrt{7}}.$$

Notice that the denominator contains a square root. If an expression contains a square root in the denominator, it is not considered to be simplified. How can we rewrite $\frac{4}{\sqrt{7}}$ as an equivalent expression that does not contain the $\sqrt{7}$ in the denominator? Since $\sqrt{7}\sqrt{7} = 7$, we can multiply the numerator and the denominator by the radical in the denominator.

$$\frac{4}{\sqrt{7}} \cdot \frac{\sqrt{7}}{\sqrt{7}} = \frac{4\sqrt{7}}{\sqrt{49}} = \frac{4\sqrt{7}}{7}$$

This expression is considered to be in simplest form. We call this process rationalizing the denominator.

Rationalizing the denominator is the process of transforming a fraction with one or more radicals in the denominator into an equivalent fraction without a radical in the denominator.

EXAMPLE 8 Simplify by rationalizing the denominator. $\frac{3}{\sqrt{2}}$

Solution

$$\frac{3}{\sqrt{2}} = \frac{3}{\sqrt{2}} \cdot \frac{\sqrt{2}}{\sqrt{2}} \quad \text{Since } \frac{\sqrt{2}}{\sqrt{2}} = 1.$$

$$= \frac{3\sqrt{2}}{\sqrt{4}} \quad \text{Product rule for radicals.}$$

$$= \frac{3\sqrt{2}}{2}$$

Student Practice 8 Simplify by rationalizing the denominator.

$$\frac{7}{\sqrt{3}}$$

We can rationalize the denominator either before or after we simplify the denominator.

EXAMPLE 9 Simplify. $\frac{3}{\sqrt{12x}}$

Solution

Method 1: First we simplify the radical in the denominator, and then we multiply in order to rationalize the denominator.

$$\frac{3}{\sqrt{12x}} = \frac{3}{\sqrt{4}\sqrt{3x}} = \frac{3}{2\sqrt{3x}} \cdot \frac{\sqrt{3x}}{\sqrt{3x}} = \frac{\cancel{3}\sqrt{3x}}{2(\cancel{3}x)} = \frac{\sqrt{3x}}{2x}$$

Method 2: We can multiply numerator and denominator by a value that will make the radicand in the denominator a perfect square (i.e., rationalize the denominator).

$$\frac{3}{\sqrt{12x}} = \frac{3}{\sqrt{12x}} \cdot \frac{\sqrt{3x}}{\sqrt{3x}}$$

$$= \frac{3\sqrt{3x}}{\sqrt{36x^2}} \quad \text{Since } \sqrt{12x}\sqrt{3x} = \sqrt{36x^2}.$$

$$= \frac{3\sqrt{3x}}{6x} = \frac{\sqrt{3x}}{2x}$$

Student Practice 9 Simplify.

$$\frac{8}{\sqrt{20x}}$$

If a radicand contains a fraction, it is not considered to be simplified. We can use the quotient rule for radicals and then rationalize the denominator to simplify the radical. We have already rationalized denominators when they contain square roots. Now we will rationalize denominators when they contain radical expressions that are cube roots or higher-order roots.

EXAMPLE 10 Simplify. $\sqrt[3]{\frac{2}{3x^2}}$

Solution

Method 1: $\sqrt[3]{\frac{2}{3x^2}} = \frac{\sqrt[3]{2}}{\sqrt[3]{3x^2}}$ Quotient rule for radicals.

$= \frac{\sqrt[3]{2}}{\sqrt[3]{3x^2}} \cdot \frac{\sqrt[3]{9x}}{\sqrt[3]{9x}}$ Multiply the numerator and denominator by an appropriate value so that the new radicand in the denominator will be a perfect cube.

$= \frac{\sqrt[3]{18x}}{\sqrt[3]{27x^3}}$ Observe that we can evaluate the cube root in the denominator.

$= \frac{\sqrt[3]{18x}}{3x}$

Method 2: $\sqrt[3]{\frac{2}{3x^2}} = \sqrt[3]{\frac{2}{3x^2} \cdot \frac{9x}{9x}}$

$= \sqrt[3]{\frac{18x}{27x^3}}$

$= \frac{\sqrt[3]{18x}}{\sqrt[3]{27x^3}}$

$= \frac{\sqrt[3]{18x}}{3x}$

Student Practice 10 Simplify.

$$\sqrt[3]{\frac{6}{5x}}$$

If the denominator of a radical expression contains a sum or difference with radicals, we multiply the numerator and denominator by the *conjugate* of the denominator. For example, the conjugate of $x + \sqrt{y}$ is $x - \sqrt{y}$; similarly, the conjugate of $x - \sqrt{y}$ is $x + \sqrt{y}$. What is the conjugate of $3 + \sqrt{2}$? It is $3 - \sqrt{2}$. How about $\sqrt{11} + \sqrt{xyz}$? It is $\sqrt{11} - \sqrt{xyz}$.

CONJUGATES

The expressions $a + b$ and $a - b$, where a and b represent any algebraic term, are called **conjugates**. Each expression is the conjugate of the other expression.

Multiplying by conjugates is simply an application of the formula

$$(a + b)(a - b) = a^2 - b^2.$$

For example,

$$(\sqrt{x} + \sqrt{y})(\sqrt{x} - \sqrt{y}) = (\sqrt{x})^2 - (\sqrt{y})^2 = x - y.$$

EXAMPLE 11 Simplify. $\dfrac{5}{3 + \sqrt{2}}$

Solution

$$\frac{5}{3 + \sqrt{2}} = \frac{5}{3 + \sqrt{2}} \cdot \frac{3 - \sqrt{2}}{3 - \sqrt{2}}$$

Multiply the numerator and denominator by the conjugate of $3 + \sqrt{2}$.

$$= \frac{15 - 5\sqrt{2}}{3^2 - (\sqrt{2})^2}$$

$$= \frac{15 - 5\sqrt{2}}{9 - 2} = \frac{15 - 5\sqrt{2}}{7}$$

Student Practice 11 Simplify.

$$\frac{4}{2 + \sqrt{5}}$$

EXAMPLE 12 Simplify. $\dfrac{\sqrt{7} + \sqrt{3}}{\sqrt{7} - \sqrt{3}}$

Solution The conjugate of $\sqrt{7} - \sqrt{3}$ is $\sqrt{7} + \sqrt{3}$.

$$\frac{\sqrt{7} + \sqrt{3}}{\sqrt{7} - \sqrt{3}} \cdot \frac{\sqrt{7} + \sqrt{3}}{\sqrt{7} + \sqrt{3}} = \frac{\sqrt{49} + 2\sqrt{21} + \sqrt{9}}{(\sqrt{7})^2 - (\sqrt{3})^2}$$

$$= \frac{7 + 2\sqrt{21} + 3}{7 - 3}$$

$$= \frac{10 + 2\sqrt{21}}{4}$$

$$= \frac{2(5 + \sqrt{21})}{2 \cdot 2}$$

$$= \frac{\cancel{2}(5 + \sqrt{21})}{\cancel{2} \cdot 2}$$

$$= \frac{5 + \sqrt{21}}{2}$$

Student Practice 12 Simplify.

$$\frac{\sqrt{11} + \sqrt{5}}{\sqrt{11} - \sqrt{5}}$$

7.4 Exercises

MyMathLab® Watch the videos in MyMathLab Download the MyDashBoard App

Multiply and simplify. Assume that all variables represent nonnegative numbers.

1. $\sqrt{5}\sqrt{7}$

2. $\sqrt{11}\sqrt{6}$

3. $(5\sqrt{2})(-6\sqrt{5})$

4. $(-4\sqrt{5})(-2\sqrt{3})$

5. $(3\sqrt{10})(-4\sqrt{2})$

6. $(-5\sqrt{6})(2\sqrt{3})$

7. $(-3\sqrt{y})(\sqrt{5x})$

8. $(\sqrt{2x})(-7\sqrt{3y})$

9. $(3x\sqrt{2x})(-2\sqrt{10xy})$

10. $(4\sqrt{3a})(a\sqrt{6ab})$

11. $5\sqrt{a}(3\sqrt{b} - 5)$

12. $-\sqrt{x}(5\sqrt{y} + 3)$

13. $-3\sqrt{a}(\sqrt{2b} + 2\sqrt{5})$

14. $4\sqrt{x}(\sqrt{3y} - 3\sqrt{6})$

15. $-\sqrt{a}(\sqrt{a} - 2\sqrt{b})$

16. $-2\sqrt{ab}(5\sqrt{a} - \sqrt{ab})$

17. $7\sqrt{x}(2\sqrt{3} - 5\sqrt{x})$

18. $3\sqrt{y}(4\sqrt{6} + 11\sqrt{y})$

19. $(3 - \sqrt{2})(8 + \sqrt{2})$

20. $(\sqrt{6} + 3)(\sqrt{6} - 1)$

21. $(2\sqrt{3} + \sqrt{2})(2\sqrt{3} - 4\sqrt{2})$

22. $(3\sqrt{3} + \sqrt{5})(\sqrt{3} - 2\sqrt{5})$

23. $(\sqrt{7} + 4\sqrt{5x})(2\sqrt{7} + 3\sqrt{5x})$

24. $(\sqrt{6} + 3\sqrt{3y})(5\sqrt{6} + 2\sqrt{3y})$

25. $(\sqrt{3} + 2\sqrt{2})(\sqrt{5} + \sqrt{3})$

26. $(3\sqrt{5} + \sqrt{3})(\sqrt{2} + 2\sqrt{5})$

27. $(\sqrt{5} - 2\sqrt{6})^2$

28. $(\sqrt{3} + 4\sqrt{7})^2$

29. $(9 - 2\sqrt{b})^2$

30. $(3\sqrt{a} + 4)^2$

31. $(\sqrt{3x + 4} + 3)^2$

32. $(\sqrt{2x + 1} - 2)^2$

33. $(\sqrt[3]{x^2})(3\sqrt[3]{4x} - 4\sqrt[3]{x^5})$

34. $(2\sqrt[3]{x})(\sqrt[3]{4x^2} - \sqrt[3]{14x})$

35. $(\sqrt[3]{3} + \sqrt[3]{2})(\sqrt[3]{9} - \sqrt[3]{4})$

36. $(\sqrt[3]{4} - \sqrt[3]{6})(\sqrt[3]{2} + \sqrt[3]{36})$

Divide and simplify. Assume that all variables represent positive numbers.

37. $\sqrt{\dfrac{49}{25}}$

38. $\sqrt{\dfrac{16}{36}}$

39. $\sqrt{\dfrac{12x}{49y^6}}$

40. $\sqrt{\dfrac{27a^4}{64x^2}}$

41. $\sqrt[3]{\dfrac{8x^5y^6}{27}}$

42. $\sqrt[3]{\dfrac{125a^3b^4}{64}}$

43. $\dfrac{\sqrt[3]{5y^8}}{\sqrt[3]{27x^3}}$

44. $\dfrac{\sqrt[3]{5y^{10}}}{\sqrt[3]{216x^3}}$

Simplify by rationalizing the denominator.

45. $\dfrac{3}{\sqrt{2}}$

46. $\dfrac{5}{\sqrt{7}}$

47. $\sqrt{\dfrac{4}{3}}$

48. $\sqrt{\dfrac{25}{2}}$

49. $\dfrac{1}{\sqrt{5y}}$

50. $\dfrac{1}{\sqrt{3x}}$

51. $\dfrac{\sqrt{14a}}{\sqrt{2y}}$

52. $\dfrac{\sqrt{24x}}{\sqrt{3y}}$

53. $\dfrac{\sqrt{2}}{\sqrt{6x}}$

54. $\dfrac{\sqrt{5y}}{\sqrt{10x}}$

55. $\dfrac{x}{\sqrt{5} - \sqrt{2}}$

56. $\dfrac{a}{\sqrt{10} + \sqrt{2}}$

57. $\dfrac{2y}{\sqrt{6} + \sqrt{5}}$

58. $\dfrac{3x}{\sqrt{10} - \sqrt{2}}$

59. $\dfrac{\sqrt{y}}{\sqrt{6} + \sqrt{2y}}$

60. $\dfrac{\sqrt{y}}{\sqrt{3y} + \sqrt{5}}$

61. $\dfrac{\sqrt{5} + \sqrt{3}}{\sqrt{5} - \sqrt{3}}$

62. $\dfrac{\sqrt{11} - \sqrt{5}}{\sqrt{11} + \sqrt{5}}$

63. $\dfrac{\sqrt{3x} - 2\sqrt{y}}{\sqrt{3x} + \sqrt{y}}$

64. $\dfrac{\sqrt{x} + \sqrt{y}}{\sqrt{x} - 2\sqrt{y}}$

Mixed Practice *Simplify each of the following.*

65. $2\sqrt{32} - \sqrt{72} + 3\sqrt{18}$

66. $\sqrt{45} - 2\sqrt{125} - 3\sqrt{20}$

67. $(3\sqrt{2} - 5\sqrt{3})(\sqrt{2} + 2\sqrt{3})$

68. $(5\sqrt{6} - 3\sqrt{2})(\sqrt{6} + 2\sqrt{2})$

69. $\dfrac{9}{\sqrt{8x}}$

70. $\dfrac{5}{\sqrt{12x}}$

71. $\dfrac{\sqrt{5}+1}{\sqrt{5}+2}$

72. $\dfrac{\sqrt{2}-1}{2\sqrt{2}+1}$

To Think About

73. A student rationalized the denominator of $\dfrac{\sqrt{6}}{2\sqrt{3}-\sqrt{2}}$ and obtained $\dfrac{\sqrt{3}+3\sqrt{2}}{5}$. Find a decimal approximation of each expression. Are the decimals equal? Did the student do the work correctly?

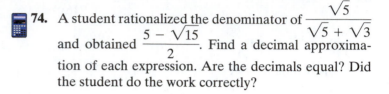

74. A student rationalized the denominator of $\dfrac{\sqrt{5}}{\sqrt{5}+\sqrt{3}}$ and obtained $\dfrac{5-\sqrt{15}}{2}$. Find a decimal approximation of each expression. Are the decimals equal? Did the student do the work correctly?

In calculus, students are sometimes required to rationalize the numerator of an expression. In this case the numerator will not have a radical in the answer. Rationalize the numerator in each of the following:

75. $\dfrac{\sqrt{2}+3\sqrt{5}}{4}$

76. $\dfrac{\sqrt{6}-3\sqrt{3}}{7}$

Applications

Fertilizer Costs *The cost of fertilizing a lawn is \$1.25 per square yard. Find the cost to fertilize each of the triangular lawns in exercises 77 and 78. Round your answers to the nearest cent.*

▲ **77.** The base of the triangle is $(12+\sqrt{5})$ yards, and the altitude is $\sqrt{80}$ yards.

▲ **78.** The base of the triangle is $(14+\sqrt{6})$ yards, and the altitude is $\sqrt{96}$ yards.

▲ **79.** ***Pacemaker Control Panel*** A medical doctor has designed a pacemaker that has a rectangular control panel. This rectangle has a width of $(\sqrt{x}+3)$ millimeters and a length of $(\sqrt{x}+5)$ millimeters. Find the area in square millimeters of this rectangle.

▲ **80.** ***FBI Listening Device*** An FBI agent has designed a secret listening device that has a rectangular base. The rectangle has a width of $(\sqrt{x}+7)$ centimeters and a length of $(\sqrt{x}+11)$ centimeters. Find the area in square centimeters of this rectangle.

Cumulative Review

81. **[4.1.6]** Solve the system.

$$-2x+3y=21$$
$$3x+2y=1$$

82. **[4.2.2]** Solve the system.

$$2x+3y-z=8$$
$$-x+2y+3z=-14$$
$$3x-y-z=10$$

Price of Engagement Rings *Tiffany & Co., one of the most famous diamond merchants, offers engagement rings in a variety of price ranges. These data are displayed in the bar graph below. The data indicate the price of all diamond rings sold during a recent year.*

83. **[1.2.2]** What percent of the rings sold for $23,000 or less?

84. **[1.2.2]** If Tiffany & Co. sold 85,000 diamond engagement rings last year, what is the number of rings that cost more than $5000?

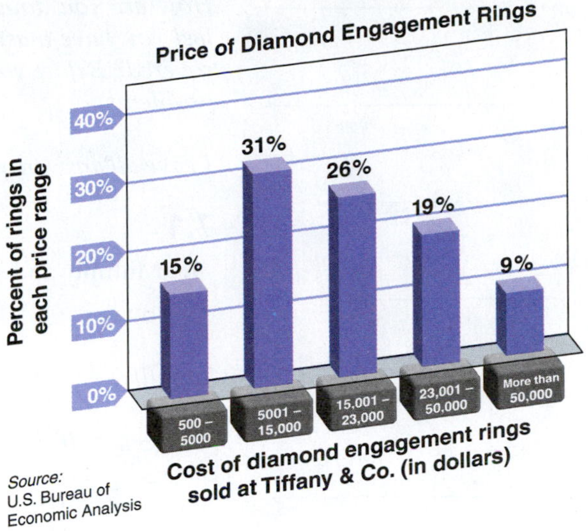

Quick Quiz 7.4

1. Multiply and simplify. $(2\sqrt{3} - \sqrt{5})(3\sqrt{3} + 2\sqrt{5})$

Rationalize the denominator.

2. $\dfrac{9}{\sqrt{3x}}$

3. $\dfrac{1 + 2\sqrt{5}}{4 - \sqrt{5}}$

4. **Concept Check** Explain how you would rationalize the denominator of the following.

$$\frac{5\sqrt{3} - 3\sqrt{2}}{3\sqrt{2} - 2\sqrt{3}}$$

1. ______

2. ______

3. ______

4. ______

5. ______

6. ______

7. ______

8. ______

9. ______

10. ______

11. ______

12. ______

13. ______

14. ______

15. ______

16. ______

17. ______

How Am I Doing? Sections 7.1–7.4

How are you doing with your homework assignments in Sections 7.1 to 7.4? Do you feel you have mastered the material so far? Do you understand the concepts you have covered? Before you go further in the textbook, take some time to do each of the following problems.

Leave all answers with positive exponents. All variables represent positive real numbers.

7.1

1. Multiply and simplify your answer. $(-3x^{1/4}y^{1/2})(-2x^{-1/2}y^{1/3})$

Simplify.

2. $(-a^{-2/3}b^{1/2})^6$

3. $\dfrac{-24x^{-1}y}{3x^{-4}y^{2/3}}$

4. $\left(\dfrac{27x^2y^{-5}}{x^{-4}y^4}\right)^{2/3}$

7.2

Evaluate.

5. $27^{-4/3}$

6. $\sqrt[5]{-243}$

7. $\sqrt{169} + \sqrt[3]{-64}$

8. $\sqrt{64a^8y^{16}}$

9. $\sqrt[3]{27a^{12}b^6c^{15}}$

10. Replace the radical with a rational exponent. $\left(\sqrt[6]{4x}\right)^5$

7.3

11. Simplify. $\sqrt[4]{16x^{20}y^{28}}$

12. Simplify. $\sqrt[3]{32x^8y^{15}}$

13. Combine. $\sqrt{44} - 2\sqrt{99} + 7\sqrt{11}$

14. Combine where possible. $3\sqrt{48y^3} - 2\sqrt[3]{16} + 3\sqrt[3]{54} - 5y\sqrt{12y}$

7.4

15. Multiply and simplify. $\left(5\sqrt{2} - 3\sqrt{5}\right)\left(\sqrt{8} + 3\sqrt{5}\right)$

16. Rationalize the denominator and simplify your answer. $\dfrac{5}{\sqrt{18x}}$

17. Rationalize the denominator and simplify your answer. $\dfrac{\sqrt{2} + \sqrt{3}}{\sqrt{2} - \sqrt{3}}$

Now turn to page SA-18 for the answers to each of these problems. Each answer also includes a reference to the objective in which the problem is first taught. If you missed any of these problems, you should stop and review the Examples and Student Practice problems in the referenced objective. A little review now will help you master the material in the upcoming sections of the text.

7.5 Radical Equations

① Solving a Radical Equation by Squaring Each Side Once

A **radical equation** is an equation with a variable in one or more of the radicals. $3\sqrt{x} = 8$ and $\sqrt{3x-1} = 5$ are radical equations. We solve radical equations by raising each side of the equation to the appropriate power. In other words, we square both sides if the radicals are square roots, cube both sides if the radicals are cube roots, and so on. Once we have done this, solving for the unknown becomes routine.

Sometimes after we square each side, we obtain a quadratic equation. In this case we collect all terms on one side and use the zero factor method that we developed in Section 5.8. After solving the equation, *always* check your answers to see whether extraneous solutions have been introduced.

We will now generalize this rule because it is very useful in higher-level mathematics courses.

RAISING EACH SIDE OF AN EQUATION TO A POWER

If $y = x$, then $y^n = x^n$ for all natural numbers n.

EXAMPLE 1 Solve. $\sqrt{2x+9} = x + 3$

Solution

$(\sqrt{2x+9})^2 = (x+3)^2$	Square each side.
$2x + 9 = x^2 + 6x + 9$	Simplify.
$0 = x^2 + 4x$	Collect all terms on one side.
$0 = x(x+4)$	Factor.
$x = 0$ or $x + 4 = 0$	Set each factor equal to zero.
$x = 0$ $x = -4$	Solve for x.

Check.

For $x = 0$: $\sqrt{2(0)+9} \stackrel{?}{=} 0 + 3$, $\sqrt{9} \stackrel{?}{=} 3$, $3 = 3$ ✓

For $x = -4$: $\sqrt{2(-4)+9} \stackrel{?}{=} -4 + 3$, $\sqrt{1} \stackrel{?}{=} -1$, $1 \neq -1$

Therefore, 0 is the only solution to this equation.

Student Practice 1 Solve and check your solution(s). $\sqrt{3x-8} = x - 2$

As you begin to solve more complicated radical equations, it is important to make sure that one radical expression is alone on one side of the equation. This is often referred to as **isolating the radical term.**

MC **EXAMPLE 2** Solve. $\sqrt{10x+5} - 1 = 2x$

Solution

$\sqrt{10x+5} = 2x + 1$	Isolate the radical term.
$(\sqrt{10x+5})^2 = (2x+1)^2$	Square each side.
$10x + 5 = 4x^2 + 4x + 1$	Simplify.
$0 = 4x^2 - 6x - 4$	Collect all terms on one side.
$0 = 2(2x^2 - 3x - 2)$	Factor out the common factor.
$0 = 2(2x+1)(x-2)$	Factor completely.
$2x + 1 = 0$ or $x - 2 = 0$	Set each factor equal to zero.
$2x = -1$ $x = 2$	Solve for x.
$x = -\frac{1}{2}$	

Continued on next page

Student Practice 2 Solve and check your solution(s). $-4 + \sqrt{x+4} = x$

Student Learning Objectives

After studying this section, you will be able to:

① **Solve a radical equation that requires squaring each side once.**

② **Solve a radical equation that requires squaring each side twice.**

Graphing Calculator

Solving Radical Equations

On a graphing calculator Example 1 can be solved in two ways. Let $y_1 = \sqrt{2x+9}$ and let $y_2 = x + 3$. Use your graphing calculator to determine where y_1 intersects y_2. What value of x do you obtain? Now let $y = \sqrt{2x+9} - x - 3$ and find the value of x when $y = 0$. What value of x do you obtain? Which method seems more efficient?

Use the method above that you found most efficient to solve the following equations, and round your answers to the nearest tenth.

$$\sqrt{x+9.5} = x - 2.3$$

$$\sqrt{6x+1.3} = 2x - 1.5$$

***NOTE TO STUDENT:** Fully worked-out solutions to all of the Student Practice problems can be found at the back of the text starting at page SP-1.*

Check.

$x = -\frac{1}{2}$: $\sqrt{10\left(-\frac{1}{2}\right) + 5} - 1 \stackrel{?}{=} 2\left(-\frac{1}{2}\right)$ $\qquad$ $x = 2$: $\sqrt{10(2) + 5} - 1 \stackrel{?}{=} 2(2)$

$\sqrt{-5 + 5} - 1 \stackrel{?}{=} -1$ $\qquad$ $\sqrt{25} - 1 \stackrel{?}{=} 4$

$\sqrt{0} - 1 \stackrel{?}{=} -1$ $\qquad$ $5 - 1 \stackrel{?}{=} 4$

$-1 = -1$ ✓ $\qquad$ $4 = 4$ ✓

Both answers check, so $-\frac{1}{2}$ and 2 are roots of the equation.

② Solving a Radical Equation by Squaring Each Side Twice

In some exercises, we must square each side twice in order to remove all the radicals. It is important to isolate at least one radical before squaring each side.

EXAMPLE 3 Solve. $\sqrt{5x + 1} - \sqrt{3x} = 1$

Solution

$$\sqrt{5x + 1} = 1 + \sqrt{3x} \qquad \text{Isolate one of the radicals.}$$

$$(\sqrt{5x + 1})^2 = (1 + \sqrt{3x})^2 \qquad \text{Square each side.}$$

$$5x + 1 = (1 + \sqrt{3x})(1 + \sqrt{3x})$$

$$5x + 1 = 1 + 2\sqrt{3x} + 3x$$

$$2x = 2\sqrt{3x} \qquad \text{Isolate the remaining radical.}$$

$$x = \sqrt{3x} \qquad \text{Divide each side by 2.}$$

$$(x)^2 = (\sqrt{3x})^2 \qquad \text{Square each side.}$$

$$x^2 = 3x$$

$$x^2 - 3x = 0 \qquad \text{Collect all terms on one side.}$$

$$x(x - 3) = 0 \qquad \text{Factor.}$$

$$x = 0 \quad \text{or} \quad x - 3 = 0 \qquad \text{Solve for } x.$$

$$x = 3$$

Check.

$x = 0$: $\sqrt{5(0) + 1} - \sqrt{3(0)} \stackrel{?}{=} 1$ $\qquad$ $x = 3$: $\sqrt{5(3) + 1} - \sqrt{3(3)} \stackrel{?}{=} 1$

$\sqrt{1} - \sqrt{0} \stackrel{?}{=} 1$ $\qquad$ $\sqrt{16} - \sqrt{9} \stackrel{?}{=} 1$

$1 = 1$ ✓ $\qquad$ $1 = 1$ ✓

Both answers check. The solutions are 0 and 3.

Student Practice 3 Solve and check your solution(s).

$$\sqrt{2x + 5} - 2\sqrt{2x} = 1$$

We will now formalize the procedure for solving radical equations.

PROCEDURE FOR SOLVING RADICAL EQUATIONS

1. Perform algebraic operations to obtain one radical by itself on one side of the equation.
2. If the equation contains square roots, square each side of the equation. Otherwise, raise each side to the appropriate power for third- and higher-order roots.
3. Simplify, if possible.
4. If the equation still contains a radical, repeat steps 1 to 3.
5. Collect all terms on one side of the equation.
6. Solve the resulting equation.
7. Check all apparent solutions. Solutions to radical equations must be verified.

EXAMPLE 4 Solve. $\sqrt{2y+5} - \sqrt{y-1} = \sqrt{y+2}$

Solution

$$\left(\sqrt{2y+5} - \sqrt{y-1}\right)^2 = \left(\sqrt{y+2}\right)^2$$

$$\left(\sqrt{2y+5} - \sqrt{y-1}\right)\left(\sqrt{2y+5} - \sqrt{y-1}\right) = y+2$$

$$2y+5-2\sqrt{(y-1)(2y+5)} + y - 1 = y+2$$

$$-2\sqrt{(y-1)(2y+5)} = -2y-2$$

$$\sqrt{(y-1)(2y+5)} = y+1$$ Divide each side by −2.

$$\left(\sqrt{2y^2+3y-5}\right)^2 = (y+1)^2$$ Square each side.

$$2y^2+3y-5 = y^2+2y+1$$

$$y^2+y-6=0$$ Collect all terms on one side.

$$(y+3)(y-2)=0$$

$$y=-3 \quad \text{or} \quad y=2$$

Check. Verify that 2 is a valid solution but −3 is not a valid solution.

Student Practice 4 Solve and check your solution(s).

$$\sqrt{y-1} + \sqrt{y-4} = \sqrt{4y-11}$$

7.5 Exercises

MyMathLab® Watch the videos in MyMathLab Download the MyDashBoard App

Verbal and Writing Skills, Exercises 1 and 2

1. Before squaring each side of a radical equation, what step should be taken first?

2. Why do we have to check the solutions when we solve radical equations?

Solve each radical equation. Check your solution(s).

3. $\sqrt{8x + 1} = 5$

4. $\sqrt{5x - 4} = 6$

5. $\sqrt{7x - 3} - 2 = 0$

6. $1 = 5 - \sqrt{9x - 2}$

7. $y + 1 = \sqrt{5y - 1}$

8. $\sqrt{y + 10} = y - 2$

9. $2x = \sqrt{11x + 3}$

10. $3x = \sqrt{15x - 4}$

11. $2 = 5 + \sqrt{2x + 1}$

12. $12 + \sqrt{4x + 5} = 7$

13. $y - \sqrt{y - 3} = 5$

14. $\sqrt{2y - 4} + 2 = y$

15. $y = \sqrt{y + 3} - 3$

16. $y = \sqrt{2y + 9} + 3$

17. $x - 2\sqrt{x - 3} = 3$

18. $2\sqrt{4x + 1} + 5 = x + 9$

19. $\sqrt{3x^2 - x} = x$

20. $\sqrt{5x^2 - 3x} = 2x$

21. $\sqrt[3]{2x + 3} = 2$

22. $\sqrt[3]{3x + 7} = 4$

23. $\sqrt[3]{4x - 1} = 3$

24. $\sqrt[3]{3 - 5x} = 2$

Solve each radical equation. This will usually involve squaring each side twice. Check your solutions.

25. $\sqrt{x + 4} = 1 + \sqrt{x - 3}$

26. $\sqrt{x + 6} + 1 = \sqrt{x + 15}$

27. $\sqrt{7x + 1} = 1 + \sqrt{5x}$

28. $2\sqrt{x + 4} = 1 + \sqrt{2x + 9}$

29. $\sqrt{x + 6} = 1 + \sqrt{x + 2}$

30. $\sqrt{3x + 1} - \sqrt{x - 4} = 3$

31. $\sqrt{3x + 13} = 1 + \sqrt{x + 4}$

32. $\sqrt{8x + 17} = \sqrt{9x + 7} + 1$

33. $\sqrt{2x + 9} - \sqrt{x + 1} = 2$

34. $\sqrt{2x + 6} = \sqrt{7 - 2x} + 1$

35. $\sqrt{4x + 6} = \sqrt{x + 1} - \sqrt{x + 5}$

36. $\sqrt{3x + 4} + \sqrt{x + 5} = \sqrt{7 - 2x}$

37. $2\sqrt{x} - \sqrt{x - 5} = \sqrt{2x - 2}$

38. $\sqrt{3 - 2\sqrt{x}} = \sqrt{x}$

Optional Graphing Calculator Problems

Solve for x. Round your answer to four decimal places.

39. $x = \sqrt{4.28x - 3.15}$

40. $\sqrt[3]{5.62x + 9.93} = 1.47$

Applications

41. ***Length of Skid Marks*** When a car traveling on wet pavement at a speed V in miles per hour stops suddenly, it will produce skid marks of length S feet according to the formula $V = 2\sqrt{3S}$.

(a) Solve the equation for S.

(b) Use your result from **(a)** to find the length of the skid mark S if the car is traveling at 30 miles per hour

42. ***Flight Data Recorder*** The volume V of a steel container inside a flight data recorder is defined by the equation

$$x = \sqrt{\frac{V}{5}},$$

where x is the sum of the length and the width of the container in inches and the height of the container is 5 inches.

(a) Solve the equation for V.

(b) Use the result from **(a)** to find the volume of the container whose length and width total 3.5 inches.

Stopping Distance *Recently an experiment was conducted relating the speed a car is traveling and the stopping distance. In this experiment, a car is traveling on dry pavement at a constant rate of speed. From the instant that a driver recognizes the need to stop, the number of feet it takes for him to stop the car is recorded. For example, for a driver traveling at 50 miles per hour, it requires a stopping distance of 190 feet. In general, the stopping distance x in feet is related to the speed of the car y in miles per hour by the equation*

$$0.11y + 1.25 = \sqrt{3.7625 + 0.22x}.$$

(*Source:* National Highway Traffic Safety Administration)

43. Solve this equation for x.

44. Use your answer from exercise 43 to find what the stopping distance x would have been for a car traveling at $y = 60$ miles per hour.

To Think About

45. The solution to the equation

$$\sqrt{x^2 - 3x + c} = x - 2$$

is $x = 3$. What is the value of c?

46. The solution to the equation

$$\sqrt{x + a} - \sqrt{x} = -4$$

is $x = 25$. What is the value of a?

Cumulative Review *Simplify.*

47. [7.1.1] $(4^3x^6)^{2/3}$

48. [7.1.1] $(2^{-3}x^{-6})^{1/3}$

49. [7.2.2] $\sqrt[3]{-216x^6y^9}$

50. [7.2.2] $\sqrt[5]{-32x^{15}y^5}$

51. [4.3.1] ***Mississippi Queen*** The Mississippi Queen is a steamboat that travels up and down the Mississippi River at a cruising speed of 12 miles per hour in still water. After traveling 4 hours downstream with the current, it takes 5 hours to get upstream against the current and return to its original starting point. What is the speed of the Mississippi River's current?

52. [4.3.1] ***Veterinarian Costs*** The Concord Veterinary Clinic saw a total of 28 dogs and cats in one day for a standard annual checkup. The checkup cost for a cat is \$55, and the checkup cost for a dog is \$68. The total amount charged that day for checkups was \$1748. How many dogs were examined? How many cats?

Quick Quiz 7.5 *Solve and check your solutions.*

1. $\sqrt{5x - 4} = x$

2. $x = 3 - \sqrt{2x - 3}$

3. $4 - \sqrt{x - 4} = \sqrt{2x - 1}$

4. Concept Check When you try to solve the equation $2 + \sqrt{x + 10} = x$, you obtain the values $x = -1$ and $x = 6$. Explain how you would determine if either of these values is a solution of the radical equation.

7.6 Complex Numbers

① Simplifying Expressions Involving Complex Numbers

Until now we have not been able to solve an equation such as $x^2 = -4$ because there is no *real* number that satisfies this equation. However, this equation *does* have a nonreal solution. This solution is an *imaginary number*.

We define a new number:

$$i = \sqrt{-1} \text{ or } i^2 = -1$$

Now let us use the product rule

$$\sqrt{-a} = \sqrt{-1}\sqrt{a} \quad \text{and see if it is valid.}$$

If $x^2 = -4$, $x = \sqrt{-4}$. Then $\sqrt{-4} = \sqrt{4(-1)} = \sqrt{4}\sqrt{-1} = \sqrt{4} \cdot i = 2i$.

Thus, one solution to the equation $x^2 = -4$ is $2i$. Let's check it.

$$\begin{aligned} x^2 &= -4 \\ (2i)^2 &\stackrel{?}{=} -4 \\ 4i^2 &\stackrel{?}{=} -4 \\ 4(-1) &\stackrel{?}{=} -4 \\ -4 &= -4 \quad \checkmark \end{aligned}$$

The value $-2i$ is also a solution. You should verify this.

Now we formalize our definitions and give some examples of imaginary numbers.

Student Learning Objectives

After studying this section, you will be able to:

1. **Simplify expressions involving complex numbers.**
2. **Add and subtract complex numbers.**
3. **Multiply complex numbers.**
4. **Evaluate complex numbers of the form i^n.**
5. **Divide two complex numbers.**

DEFINITION OF IMAGINARY NUMBERS

The **imaginary number i** is defined as follows:

$$i = \sqrt{-1} \quad \text{and} \quad i^2 = -1.$$

The set of imaginary numbers consists of numbers of the form bi, where b is a real number and $b \neq 0$.

DEFINITION

For all positive real numbers a,

$$\sqrt{-a} = \sqrt{-1}\sqrt{a} = i\sqrt{a}.$$

EXAMPLE 1 Simplify.

(a) $\sqrt{-36}$ **(b)** $\sqrt{-17}$

Solution

(a) $\sqrt{-36} = \sqrt{-1}\sqrt{36} = (i)(6) = 6i$

(b) $\sqrt{-17} = \sqrt{-1}\sqrt{17} = i\sqrt{17}$

Student Practice 1 Simplify. **(a)** $\sqrt{-49}$ **(b)** $\sqrt{-31}$

***NOTE TO STUDENT:** Fully worked-out solutions to all of the Student Practice problems can be found at the back of the text starting at page SP-1.*

To avoid confusing $\sqrt{17i}$ with $\sqrt{17}i$, we write the i before the radical. That is, we write $i\sqrt{17}$.

EXAMPLE 2 Simplify. $\sqrt{-45}$

Solution

$$\sqrt{-45} = \sqrt{-1}\sqrt{45} = i\sqrt{45} = i\sqrt{9}\sqrt{5} = 3i\sqrt{5}$$

Student Practice 2 Simplify. $\sqrt{-98}$

The rule $\sqrt{a}\sqrt{b} = \sqrt{ab}$ requires that $a \geq 0$ and $b \geq 0$. Therefore, we cannot use our product rule when the radicands are negative unless we first use the definition of $\sqrt{-1}$. Notice that

$$\sqrt{-1} \cdot \sqrt{-1} = i \cdot i = i^2 = -1.$$

EXAMPLE 3 Multiply. $\sqrt{-16} \cdot \sqrt{-25}$

Solution First we must use the definition $\sqrt{-1} = i$. Thus, we have the following:

$$\begin{aligned}(\sqrt{-16})(\sqrt{-25}) &= (i\sqrt{16})(i\sqrt{25}) \\ &= i^2(4)(5) \\ &= -1(20) \qquad i^2 = -1. \\ &= -20\end{aligned}$$

Student Practice 3 Multiply. $\sqrt{-8} \cdot \sqrt{-2}$

Now we formally define a complex number.

DEFINITION OF COMPLEX NUMBER

A number that can be written in the form $a + bi$, where a and b are real numbers, is a **complex number.** We say that a is the **real part** and bi is the **imaginary part.**

Under this definition, every real number is also a complex number. For example, the real number 5 can be written as $5 + 0i$. Therefore, 5 is a complex number. In a similar fashion, the imaginary number $2i$ can be written as $0 + 2i$. So $2i$ is a complex number. Thus, the set of complex numbers includes the set of real numbers and the set of imaginary numbers.

DEFINITION

Two complex numbers $a + bi$ and $c + di$ are equal if and only if $a = c$ and $b = d$.

This definition means that two complex numbers are equal if and only if their real parts are equal *and* their imaginary parts are equal.

Graphing Calculator

Complex Numbers

Some graphing calculators, such as the TI-84, have a complex number mode. If your graphing calculator has this capability, you will be able to use it to do complex number operations. First you must use the Mode command to transfer selection from "Real" to "Complex" or "$a + bi$." To verify your status, try to find $\sqrt{-7}$ on your graphing calculator. If you obtain an approximate answer of "2.645751311 i," then your calculator is operating in the complex number mode. If you obtain "ERROR: NONREAL ANSWER," then your calculator is not operating in the complex number mode.

EXAMPLE 4 Find the real numbers x and y if $x + 3i\sqrt{7} = -2 + yi$.

Solution By our definition, the real parts must be equal, so x must be -2; the imaginary parts must also be equal, so y must be $3\sqrt{7}$.

Student Practice 4 Find the real numbers x and y if

$$-7 + 2yi\sqrt{3} = x + 6i\sqrt{3}.$$

② Adding and Subtracting Complex Numbers

ADDING AND SUBTRACTING COMPLEX NUMBERS

For all real numbers a, b, c, and d,

$$(a + bi) + (c + di) = (a + c) + (b + d)i \quad \text{and}$$

$$(a + bi) - (c + di) = (a - c) + (b - d)i.$$

In other words, to combine complex numbers we add (or subtract) the real parts, and we add (or subtract) the imaginary parts.

EXAMPLE 5 Subtract. $(6 - 2i) - (3 - 5i)$

Solution

$$(6 - 2i) - (3 - 5i) = (6 - 3) + [-2 - (-5)]i = 3 + (-2 + 5)i = 3 + 3i$$

Student Practice 5 Subtract. $(3 - 4i) - (-2 - 18i)$

③ Multiplying Complex Numbers

As we might expect, the procedure for multiplying complex numbers is similar to the procedure for multiplying polynomials. We will see that the complex numbers obey the associative, commutative, and distributive properties.

EXAMPLE 6 Multiply. $(7 - 6i)(2 + 3i)$

Solution Use FOIL.

$$\begin{aligned}(7 - 6i)(2 + 3i) &= (7)(2) + (7)(3i) + (-6i)(2) + (-6i)(3i)\\ &= 14 + 21i - 12i - 18i^2\\ &= 14 + 21i - 12i - 18(-1)\\ &= 14 + 21i - 12i + 18\\ &= 32 + 9i\end{aligned}$$

Student Practice 6 Multiply. $(4 - 2i)(3 - 7i)$

EXAMPLE 7 Multiply. $3i(4 - 5i)$

Solution Use the distributive property.

$$\begin{aligned} 3i(4 - 5i) &= (3)(4)i + (3)(-5)i^2 \\ &= 12i - 15i^2 \\ &= 12i - 15(-1) \\ &= 15 + 12i \end{aligned}$$

Student Practice 7 Multiply. $-2i(5 + 6i)$

④ Evaluating Complex Numbers of the Form in i^n

How would you evaluate i^n, where n is any positive integer? We look for a pattern. We have defined

$$i^2 = -1.$$

We could write

$$i^3 = i^2 \cdot i = (-1)i = -i.$$

We also have the following:

$$\begin{aligned} i^4 &= i^2 \cdot i^2 = (-1)(-1) = +1 \\ i^5 &= i^4 \cdot i = (+1)i = +i \end{aligned}$$

We notice that $i^5 = i$. Let's look at i^6.

$$i^6 = i^4 \cdot i^2 = (+1)(-1) = -1$$

We begin to see a pattern that starts with i and repeats itself for i^5. Will $i^7 = -i$? Why or why not?

VALUES OF i^n

$i = i$	$i^5 = i$	$i^9 = i$
$i^2 = -1$	$i^6 = -1$	$i^{10} = -1$
$i^3 = -i$	$i^7 = -i$	$i^{11} = -i$
$i^4 = +1$	$i^8 = +1$	$i^{12} = +1$

We can use this pattern to evaluate powers of i.

EXAMPLE 8 Evaluate. **(a)** i^{36} **(b)** i^{27}

Solution

(a) $i^{36} = (i^4)^9 = (1)^9 = 1$

(b) $i^{27} = (i^{24+3}) = (i^{24})(i^3) = (i^4)^6(i^3) = (1)^6(-i) = -i$

This suggests a quick method for evaluating powers of i. Divide the exponent by 4. i^4 raised to any power will be 1. Then use the first column of the values of i^n chart above to evaluate the remainder.

Student Practice 8 Evaluate. **(a)** i^{42} **(b)** i^{53}

⑤ Dividing Two Complex Numbers

The complex numbers $a + bi$ and $a - bi$ are called **conjugates.** The product of two complex conjugates is always a real number.

$$\begin{aligned}(a + bi)(a - bi) &= a^2 - abi + abi - b^2i^2\\ &= a^2 - b^2(-1)\\ &= a^2 + b^2\end{aligned}$$

When dividing two complex numbers, we want to remove any expression involving i from the denominator. So we multiply the numerator and denominator by the conjugate of the denominator. This is just what we did when we rationalized the denominators of radical expressions.

EXAMPLE 9 Divide. $\dfrac{7 + i}{3 - 2i}$

Solution

$$\begin{aligned}\frac{7 + i}{3 - 2i}\cdot\frac{3 + 2i}{3 + 2i} &= \frac{21 + 14i + 3i + 2i^2}{9 - 4i^2} = \frac{21 + 17i + 2(-1)}{9 - 4(-1)}\\ &= \frac{21 + 17i - 2}{9 + 4}\\ &= \frac{19 + 17i}{13} \quad\text{or}\quad \frac{19}{13} + \frac{17}{13}i\end{aligned}$$

Student Practice 9 Divide.

$$\frac{4 + 2i}{3 + 4i}$$

EXAMPLE 10 Divide. $\dfrac{3 - 2i}{4i}$

Solution The conjugate of $0 + 4i$ is $0 - 4i$ or simply $-4i$.

$$\begin{aligned}\frac{3 - 2i}{4i}\cdot\frac{-4i}{-4i} &= \frac{-12i + 8i^2}{-16i^2} = \frac{-12i + 8(-1)}{-16(-1)}\\ &= \frac{-8 - 12i}{16} = \frac{\not{4}(-2 - 3i)}{\not{4}\cdot 4}\\ &= \frac{-2 - 3i}{4} \quad\text{or}\quad -\frac{1}{2} - \frac{3}{4}i\end{aligned}$$

Student Practice 10 Divide.

$$\frac{5 - 6i}{-2i}$$

Graphing Calculator

Complex Operations

When we perform complex number operations on a graphing calculator, the answer will usually be displayed as an approximate value in decimal form. Try Example 9 on your graphing calculator by entering $(7 + i) \div (3 - 2i)$. You should obtain an approximate answer of $1.461538462 + 1.307692308i$.

7.6 Exercises

MyMathLab® Watch the videos in MyMathLab Download the MyDashBoard App

Verbal and Writing Skills, Exercises 1–4

1. Does $x^2 = -9$ have a real number solution? Why or why not?

2. Describe a complex number and give an example.

3. Are the complex numbers $2 + 3i$ and $3 + 2i$ equal? Why or why not?

4. Describe in your own words how to add or subtract complex numbers.

Simplify. Express in terms of i.

5. $\sqrt{-25}$

6. $\sqrt{-100}$

7. $\sqrt{-50}$

8. $\sqrt{-48}$

9. $\sqrt{-\frac{25}{4}}$

10. $\sqrt{-\frac{49}{64}}$

11. $-\sqrt{-81}$

12. $-\sqrt{-36}$

13. $2 + \sqrt{-3}$

14. $5 + \sqrt{-7}$

15. $-2.8 + \sqrt{-16}$

16. $\frac{3}{2} + \sqrt{-121}$

17. $-3 + \sqrt{-24}$

18. $-6 - \sqrt{-32}$

19. $(\sqrt{-5})(\sqrt{-2})$

20. $(\sqrt{-7})(\sqrt{-3})$

21. $(\sqrt{-36})(\sqrt{-4})$

22. $(\sqrt{-25})(\sqrt{-9})$

Find the real numbers x and y.

23. $x - 3i = 5 + yi$

24. $x - 9i = 8 + yi$

25. $1.3 - 2.5yi = x - 5i$

26. $3.4 - 0.8i = 2x - yi$

27. $23 + yi = 17 - x + 3i$

28. $2 + x - 11i = 19 + yi$

Perform the addition or subtraction.

29. $(1 + 8i) + (-6 + 3i)$

30. $(-10 - 4i) + (7 + i)$

31. $\left(-\frac{3}{2} + \frac{1}{2}i\right) + \left(\frac{5}{2} - \frac{3}{2}i\right)$

32. $\left(\frac{2}{3} - \frac{1}{3}i\right) + \left(\frac{10}{3} + \frac{4}{3}i\right)$

33. $(2.8 - 0.7i) - (1.6 - 2.8i)$

34. $(6.5 + 7.2i) - (2.3 + 4.9i)$

Multiply and simplify your answers. Place in i notation before doing any other operations.

35. $(2i)(7i)$

36. $(6i)(3i)$

37. $(-7i)(6i)$

38. $(i)(-3i)$

39. $(2 + 3i)(2 - i)$

40. $(5 - 2i)(1 + 4i)$

41. $9i - 3(-2 + i)$

42. $15i - 5(4 + i)$

43. $2i(5i - 6)$

44. $4i(7 - 2i)$

45. $\left(\frac{1}{2} + i\right)^2$

46. $\left(\frac{1}{5} - i\right)^2$

47. $\left(i\sqrt{3}\right)\left(i\sqrt{7}\right)$

48. $\left(i\sqrt{2}\right)\left(i\sqrt{6}\right)$

49. $\left(3 + \sqrt{-2}\right)\left(4 + \sqrt{-5}\right)$

50. $\left(2 + \sqrt{-3}\right)\left(6 + \sqrt{-2}\right)$

Evaluate.

51. i^{17}

52. i^{21}

53. i^{24}

54. i^{16}

55. i^{46}

56. i^{83}

57. i^{47}

58. i^{10}

59. $i^{30} + i^{28}$

60. $i^{32} + i^{42}$

61. $i^{100} - i^{7}$

62. $3i^{64} - 2i^{11}$

Divide.

63. $\frac{2 + i}{3 - i}$

64. $\frac{4 + 2i}{2 - i}$

65. $\frac{2i}{3 + 3i}$

66. $\frac{-3i}{2 + 5i}$

67. $\frac{5 - 2i}{6i}$

68. $\frac{7 + 10i}{3i}$

69. $\frac{2}{i}$

70. $\frac{-5}{i}$

71. $\frac{7}{5 - 6i}$

72. $\frac{3}{4 + 2i}$

73. $\frac{5 - 2i}{3 + 2i}$

74. $\frac{6 + 3i}{6 - 3i}$

Mixed Practice *Simplify.*

75. $\sqrt{-98}$

76. $\sqrt{-72}$

77. $(8 - 5i) - (-1 + 3i)$

78. $(-4 + 6i) - (3 - 4i)$

79. $(3i - 1)(5i - 3)$

80. $(4i + 5)(2i - 5)$

81. $\dfrac{2 - 3i}{2 + i}$

82. $\dfrac{4 - 3i}{5 + 2i}$

Applications *The impedance Z in an alternating current circuit (like the one used in your home and in your classroom) is given by the formula* $Z = \frac{V}{I}$, *where V is the voltage and I is the current.*

83. Find the value of Z if $V = 3 + 2i$ and $I = 3i$.

84. Find the value of Z if $V = 4 + 2i$ and $I = -3i$.

Cumulative Review

85. [2.4.1] ***Factory Production*** A grape juice factory produces juice in three different types of containers. $x + 3$ hours per week are spent on producing juice in glass bottles. $2x - 5$ hours per week are spent on producing juice in cans. $4x + 2$ hours per week are spent on producing juice in plastic bottles. If the factory operates 105 hours per week, how much time is spent producing juice in each type of container?

86. [1.2.2] ***Donation of Computers*** Citizens Bank has decided to donate its older personal computers to the Boston Public Schools. Each computer donated is worth \$120 in tax-deductible dollars to the bank. In addition, the computer company supplying the bank with its new computers gives a 7% rebate to any customer donating used computers to schools. If sixty new computers are purchased at a list price of \$1850 each and sixty older computers are donated to the Boston Public Schools, what is the net cost to the bank for this purchase?

Quick Quiz 7.6 *Simplify.*

1. $(6 - 7i)(3 + 2i)$

2. $\dfrac{4 + 3i}{1 - 2i}$

3. i^{33}

4. Concept Check Explain how you would simplify the following.

$$(3 + 5i)^2$$

7.7 Variation

① Solving Problems Using Direct Variation

Many times in daily life we observe how a change in one quantity produces a change in another. If we order one large pepperoni pizza, we pay \$8.95. If we order two large pepperoni pizzas, we pay \$17.90. For three large pepperoni pizzas, it is \$26.85. The change in the number of pizzas we order results in a corresponding change in the price we pay.

Notice that the price we pay for each pizza stays the same. That is, each pizza costs \$8.95. The number of pizzas changes, and the corresponding price of the order changes. From our experience with functions and with equations, we see that the cost of the order is $y = \$8.95x$, where the price y depends on the number of pizzas x. We see that the variable y is a constant multiple of x. The two variables are said to *vary directly*. That is, y varies directly with x. We write a general equation that represents this idea as follows: $y = kx$.

When we solve problems using direct variation, we usually are not given the value of the constant of variation k. This is something that we must find. Usually all we are given is a point of reference. That is, we are given the value of y for a specific value of x. Using this information, we can find k.

Student Learning Objectives

After studying this section, you will be able to:

① **Solve problems requiring the use of direct variation.**

② **Solve problems requiring the use of inverse variation.**

③ **Solve problems requiring the use of joint or combined variation.**

EXAMPLE 1 The time of a pendulum's period varies directly with the square root of its length. If the pendulum is 1 foot long when the time is 0.2 second, find the time if the length is 4 feet.

Solution Let $t =$ the time and $L =$ the length.

We then have the equation

$$t = k\sqrt{L}.$$

We can evaluate k by substituting $L = 1$ and $t = 0.2$ into the equation.

$$t = k\sqrt{L}$$
$$0.2 = k(\sqrt{1})$$
$$0.2 = k \quad \text{Because } \sqrt{1} = 1.$$

Now we know the value of k and can write the equation more completely.

$$t = 0.2\sqrt{L}$$

When $L = 4$, we have the following:

$$t = 0.2\sqrt{4}$$
$$t = (0.2)(2)$$
$$t = 0.4 \text{ second}$$

Student Practice 1 The maximum speed of a racing car varies directly with the square root of the horsepower of the engine. If the maximum speed of a car with 256 horsepower is 128 miles per hour, what is the maximum speed of a car with 225 horsepower?

***NOTE TO STUDENT:** Fully worked-out solutions to all of the Student Practice problems can be found at the back of the text starting at page SP-1.*

② Solving Problems Using Inverse Variation

In some cases when one variable increases, another variable decreases. For example, as the amount of money you earn each year increases, the percentage of your income that you get to keep after taxes decreases. If one variable is a constant multiple of the reciprocal of the other, the two variables are said to *vary inversely*.

EXAMPLE 2 If y varies inversely with x and $y = 12$ when $x = 5$, find the value of y when $x = 14$.

Solution If y varies inversely with x, we can write the equation $y = \frac{k}{x}$. We can find the value of k by substituting the values $y = 12$ and $x = 5$.

$$12 = \frac{k}{5}$$

$$60 = k$$

We can now write the equation

$$y = \frac{60}{x}.$$

To find the value of y when $x = 14$, we substitute 14 for x in the equation.

$$y = \frac{60}{14}$$

$$y = \frac{30}{7}$$

Student Practice 2 If y varies inversely with x and $y = 45$ when $x = 16$, find the value of y when $x = 36$.

EXAMPLE 3 The amount of light from a light source varies inversely with the square of the distance to the light source. If an object receives 6.25 lumens when the light source is 8 meters away, how much light will the object receive if the light source is 4 meters away?

Solution Let $L =$ the amount of light and $d =$ the distance to the light source.

Since the amount of light varies inversely with the *square of the distance* to the light source, we have

$$L = \frac{k}{d^2}.$$

Substituting the known values of $L = 6.25$ and $d = 8$, we can find the value of k.

$$6.25 = \frac{k}{8^2}$$

$$6.25 = \frac{k}{64}$$

$$400 = k$$

We are now able to write a more specific equation,

$$L = \frac{400}{d^2}.$$

We will use this to find L when $d = 4$ meters.

$$L = \frac{400}{4^2}$$

$$L = \frac{400}{16}$$

$$L = 25 \text{ lumens}$$

Check. Does this answer seem reasonable? Would we expect to have more light if we move closer to the light source? ✓

Student Practice 3 The weight that can safely be supported on top of a cylindrical column varies inversely with the square of its height. If a 7.5-ft column can support 2 tons, how much weight can a 3-ft column support?

③ Solving Problems Using Joint or Combined Variation

Sometimes a quantity depends on the variation of two or more variables. This is called joint or **combined variation.**

EXAMPLE 4 y varies directly with x and z and inversely with d^2. When $x = 7$, $z = 3$, and $d = 4$, the value of y is 20. Find the value of y when $x = 5$, $z = 6$, and $d = 2$.

Solution We can write the equation

$$y = \frac{kxz}{d^2}.$$

To find the value of k, we substitute into the equation $y = 20$, $x = 7$, $z = 3$, and $d = 4$.

$$20 = \frac{k(7)(3)}{4^2}$$

$$20 = \frac{21k}{16}$$

$$320 = 21k$$

$$\frac{320}{21} = k$$

Now we substitute $\frac{320}{21}$ for k into our original equation.

$$y = \frac{\frac{320}{21}xz}{d^2} \quad \text{or} \quad y = \frac{320xz}{21d^2}$$

Continued on next page

We use this equation to find y for the known values of x, z, and d. We want to find y when $x = 5$, $z = 6$, and $d = 2$.

$$y = \frac{320(5)(6)}{21(2)^2} = \frac{9600}{84}$$

$$y = \frac{800}{7}$$

Student Practice 4 y varies directly with z and w^2 and inversely with x. $y = 20$ when $z = 3$, $w = 5$, and $x = 4$. Find y when $z = 4$, $w = 6$, and $x = 2$.

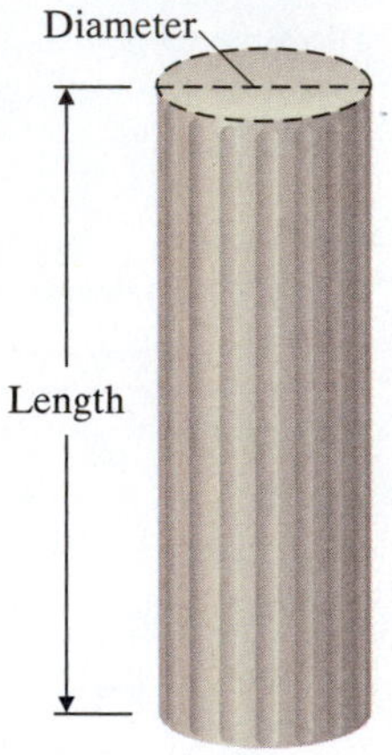

Many applied problems involve joint variation. For example, a cylindrical concrete column has a safe load capacity that varies directly with the diameter raised to the fourth power and inversely with the square of its length.

Therefore, if d = diameter and l = length, the equation would be of the form

$$y = \frac{kd^4}{l^2}.$$

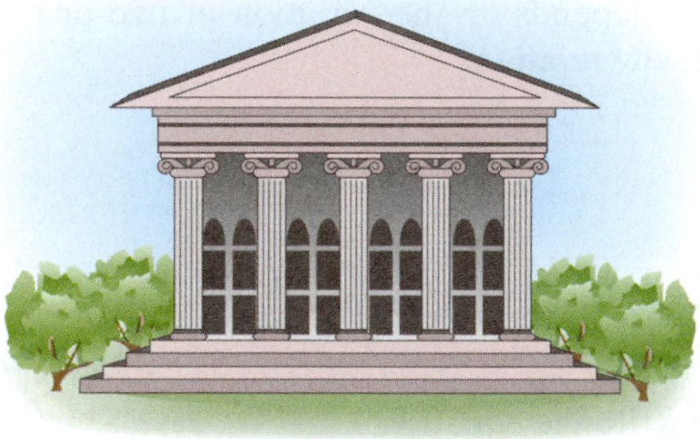

7.7 Exercises

MyMathLab®

Watch the videos in MyMathLab

Download the MyDashBoard App

Verbal and Writing Skills, Exercises 1–4

1. Give an example in everyday life of direct variation and write an equation as a mathematical model.

2. The general equation $y = kx$ means that y varies ________ with x. k is called the ________ of variation.

3. If y varies inversely with x, we write the equation ________.

4. Write a mathematical model for the following situation: The strength of a rectangular beam varies directly with its width and the square of its depth.

Round all answers to the nearest tenth unless otherwise directed.

5. If y varies directly with x and $y = 15$ when $x = 40$, find y when $x = 64$.

6. If y varies directly with x and $y = 30$ when $x = 18$, find y when $x = 120$.

7. ***Pressure on a Submarine*** A marine biology submarine was searching the waters for blue whales at 50 feet below the surface, where it experienced a pressure of 21 pounds per square inch (psi). If the pressure of water on a submerged object varies directly with its distance beneath the surface, how much pressure would the submarine experience if it had to dive to 170 feet?

8. ***Spring Stretching*** The distance a spring stretches varies directly with the weight of the object hung on the spring. If a 10-pound weight stretches a spring 6 inches, how far will a 35-pound weight stretch this spring?

9. ***Stopping Distance*** A car's stopping distance varies directly with the square of its speed. A car that is traveling 30 miles per hour can stop in 40 feet. What distance will it take to stop if it is traveling 60 miles per hour?

10. ***Time of Fall in Gravitation*** When an object is dropped, the distance it falls in feet varies directly with the square of the duration of the fall in seconds. An apple that falls from a tree falls 1 foot in $\frac{1}{4}$ second. How far will it fall in 1 second? How far will it fall in 2 seconds?

11. If y varies inversely with the square of x, and $y = 10$ when $x = 2$, find y when $x = 0.5$.

12. If y varies inversely with the square root of x, and $y = 2.4$ when $x = 0.09$, find y when $x = 0.2$.

13. ***Gasoline Prices*** Last summer the price of gasoline changed frequently. One station owner noticed that the number of gallons he sold each day seemed to vary inversely with the price per gallon. If he sold 2800 gallons when the price was \$3.10, how many gallons could he expect to sell if the price fell to \$2.90? Round your answer to the nearest gallon.

14. ***Weight of an Object*** The weight of an object on Earth's surface varies inversely with the square of its distance from the center of Earth. An object weighs 1000 pounds on Earth's surface. This is approximately 4000 miles from the center of Earth. How much would this object weigh 4500 miles from the center of Earth?

15. ***Beach Cleanup*** Every year on Earth Day, a group of volunteers pick up garbage at Hidden Falls Park. The time it takes to clean the park varies inversely with the number of people picking up garbage. Last year, 39 volunteers took 6 hours to clean the park. If 60 volunteers come to pick up garbage this year, how long will it take to clean the park?

16. ***Volume of a Gas*** If temperature remains constant, the volume of a gas varies inversely as the pressure of the gas on its container. If a pressure of 60 pounds per square inch corresponds to a volume of 45 cubic inches, what pressure corresponds to a volume of 70 cubic inches?

17. ***Support Beam*** The weight that can be safely supported by a 2- by 6-inch support beam varies inversely with its length. A builder finds that a support beam that is 8 feet long will support 900 pounds. Find the weight that can be safely supported by a beam that is 18 feet long.

18. ***Satellite Orbit Speed*** The speed that is required to maintain a satellite in a circular orbit around Earth varies directly with the square root of the distance of the satellite from the center of Earth. We will assume that the radius of Earth is approximately 4000 miles. A satellite that is 100 miles above the surface of Earth is orbiting at approximately 18,000 miles per hour. What speed would be necessary for the satellite to orbit 500 miles above the surface of Earth? Round to the nearest mile per hour.

19. ***Whirlpool Tub*** The amount of time it takes to fill a whirlpool tub is inversely proportional to the square of the radius of the pipe used to fill it. If a pipe of radius 2.5 inches can fill the tub in 6 minutes, how long will it take the tub to fill if a pipe of radius 3.5 inches is used?

20. ***Wind Generator*** The force on a blade of a wind generator varies jointly with the product of the blade's area and the square of the wind velocity. The force of the wind is 20 pounds when the area is 3 square feet and the velocity is 30 feet per second. Find the force when the area is increased to 5 square feet and the velocity is reduced to 25 feet per second.

Cumulative Review *Solve each of the following equations or word problems.*

21. **[5.8.1]** $3x^2 - 8x + 4 = 0$

22. **[5.8.1]** $4x^2 = -28x + 32$

23. **[1.2.2]** ***Sales Tax*** In Champaign, Illinois, the sales tax is 6.25%. Donny bought an amplifier for his stereo that cost \$488.75 after tax. What was the original price of the amplifier?

24. **[1.2.2]** ***Tennis Courts*** It takes 7.5 gallons of white paint to properly paint lines on three tennis courts. How much paint is needed to paint twenty-two tennis courts?

Quick Quiz 7.7

1. If y varies inversely with x and $y = 9$ when $x = 3$, find the value of y when $x = 6$.

2. Suppose y varies directly with x and inversely with the square of z. $y = 6$ when $x = 3$ and $z = 5$. Find y when $x = 6$ and $z = 10$.

3. The distance a pickup truck requires to stop varies directly with the square of the speed. A new Ford pickup truck traveling on dry pavement can stop in 80 feet when traveling at 50 miles per hour. What distance will the truck take to stop if it is traveling at 65 miles per hour?

4. **Concept Check** If y varies directly with the square root of x and $y = 50$ when $x = 5$, explain how you would find the constant of variation k.

Did You Know...

That Balancing Your Checkbook Can Help You Keep Track of What You Are Spending?

BALANCING YOUR FINANCES

Understanding the Problem:
One of the first steps in saving money is to determine your current spending trends. The first step in that process is learning to balance your finances.

Terry balanced his checkbook once a month when he received his bank statement. Below is a table that records the deposits Terry made for the month of May. The beginning balance for May was $300.50.

Date	Deposit
May 1	$200.00
May 3	$150.50
May 10	$120.25
May 25	$50.00
May 28	$25.00

Keeping a Record of Checks:
Terry needs to know if he is depositing enough money to cover his monthly expenses. Below is a table that records each check Terry wrote for the month of May.

Date	Check Number	Checks
May 2	102	$238.50
May 6	103	$75.00
May 12	104	$200.00
May 28	105	$28.56
May 30	106	$36.00

Finding the Facts:

Step 1: Terry needs to know how much he deposits into the bank every month.

Task 1: *Determine how much Terry deposited in the bank in May.*

Step 2: Terry needs to know how much he spends each month.

Task 2: *Determine the total amount of the checks Terry wrote in May.*

Task 3: *Based on the given information, will Terry be able to cover all his expenses for the month of May?*

Making a Decision

Step 3: Terry needs to know if he can continue to spend money at the same rate, or if he needs to cut back on his spending.

Task 4: *Assuming all the checks cleared for May, what would Terry's balance be at the beginning of June?*

Task 5: *If Terry continues these spending habits, what will happen?*

Applying the Situation to Your Life:
Knowing your monthly income and spending habits can help you to save. Balance your checkbook and monitor your spending habits each month, and try to cut out unnecessary expenses. If you are charged ATM fees for making withdrawals from your checking account with an ATM card be sure to subtract those costs from your checkbook balance. If you pay monthly fees to the bank for the cost of your checking account be sure to subtract those costs from your checkbook balance.

Task 6: *How often do you balance your checkbook?*

Task 7: *Do you have any unnecessary expenses that can be cut out?*

Chapter 7 Organizer

Topic and Procedure	Examples	You Try It
Raising a variable with an exponent to a power, p. 357 $(x^m)^n = x^{mn}$ $(xy)^n = x^n y^n$ $\left(\frac{x}{y}\right)^n = \frac{x^n}{y^n}, \quad y \neq 0$	Simplify. **(a)** $(x^{-1/2})^{-2/3} = x^{1/3}$ **(b)** $(3x^{-2}y^{-1/2})^{2/3} = 3^{2/3}x^{-4/3}y^{-1/3}$ **(c)** $\left(\frac{4x^{-2}}{3^{-1}y^{-1/2}}\right)^{1/4} = \frac{4^{1/4}x^{-1/2}}{3^{-1/4}y^{-1/8}}$	**1.** Simplify. **(a)** $(x^{-2/5})^{-1/2}$ **(b)** $(2a^{-3}b^{-1/4})^{1/3}$ **(c)** $\left(\frac{5x^{-3}}{4^{-2}y^2}\right)^{1/4}$
Multiplication of variables with rational exponents, p. 358 $x^m x^n = x^{m+n}$	Multiply. $(3x^{1/5})(-2x^{3/5}) = -6x^{4/5}$	**2.** Multiply. $(-a^{1/2})(4a^{1/3})$
Division of variables with rational exponents, p. 358 $\frac{x^m}{x^n} = x^{m-n}, \quad x \neq 0$	Divide. $\frac{-16x^{3/20}}{24x^{5/20}} = -\frac{2x^{-1/10}}{3}$	**3.** Divide. $\frac{12x^{7/12}}{-2x^{1/3}}$
Removing negative exponents, p. 358 $x^{-n} = \frac{1}{x^n}, \quad x \text{ and } y \neq 0$ $\frac{x^{-n}}{y^{-m}} = \frac{y^m}{x^n}$	Write with positive exponents. **(a)** $3x^{-4} = \frac{3}{x^4}$ **(b)** $\frac{2x^{-6}}{5y^{-8}} = \frac{2y^8}{5x^6}$ **(c)** $4^{-2} = \frac{1}{4^2} = \frac{1}{16}$	**4.** Write with positive exponents. **(a)** $5a^{-3}$ **(b)** $\frac{3a^{-2}}{6a^{-6}}$ **(c)** 2^{-5}
Multiplication of expressions with rational exponents, p. 359 Add exponents whenever expressions with the same base are multiplied.	Multiply. $x^{2/3}(x^{1/3} - x^{1/4}) = x^{3/3} - x^{2/3+1/4} = x - x^{11/12}$	**5.** Multiply. $x^{1/2}(x^{2/3} - x^{1/2})$
Zero exponent, p. 360 $x^0 = 1 \quad (\text{if } x \neq 0)$	Simplify. $(3x^{1/2})^0 = 1$	**6.** Simplify. $(-5x^{2/3})^0$
Higher-order roots, p. 365 If x is a nonnegative real number, $\sqrt[n]{x}$ is a nonnegative nth root and has the property that $(\sqrt[n]{x})^n = x.$ If x is a negative real number, and n is an odd integer, then $\sqrt[n]{x}$ has the property that $(\sqrt[n]{x})^n = x$. If x is a negative real number, and n is an even integer, $\sqrt[n]{x}$ is not a real number.	Simplify. **(a)** $\sqrt[3]{27} = 3$ because $3^3 = 27$. **(b)** $\sqrt[5]{-32} = -2$ because $(-2)^5 = -32$. **(c)** $\sqrt[4]{-16}$ is *not* a real number.	**7.** Simplify. **(a)** $\sqrt[4]{16}$ **(b)** $\sqrt[5]{-1}$ **(c)** $\sqrt[6]{-64}$
Rational exponents and radicals, p. 368 For positive integers m and n and any real number x for which $x^{1/n}$ is defined, $x^{m/n} = (\sqrt[n]{x})^m = \sqrt[n]{x^m}.$ If it is also true that $x \neq 0$, then $x^{-m/n} = \frac{1}{x^{m/n}} = \frac{1}{(\sqrt[n]{x})^m} = \frac{1}{\sqrt[n]{x^m}}.$	**(a)** Write as a radical. $x^{3/7} = \sqrt[7]{x^3}$ or $(\sqrt[7]{x})^3$, $3^{1/5} = \sqrt[5]{3}$ **(b)** Write as an expression with a fractional exponent. $\sqrt[3]{w^4} = w^{4/3}$ **(c)** Evaluate. $25^{3/2} = (\sqrt{25})^3 = (5)^3 = 125$	**8. (a)** Write as a radical. $x^{4/5}$ **(b)** Write as an expression with a fractional exponent. $\sqrt[4]{v^9}$ **(c)** Evaluate. $27^{4/3}$
Higher-order roots and absolute value, p. 369 $\sqrt[n]{x^n} = \|x\|$ when n is an even positive integer. $\sqrt[n]{x^n} = x$ when n is an odd positive integer.	Simplify. Assume x can be any real number. **(a)** $\sqrt[6]{x^6} = \|x\|$ **(b)** $\sqrt[5]{x^5} = x$	**9.** Simplify. Assume x can be any real number. **(a)** $\sqrt[4]{x^4}$ **(b)** $\sqrt[7]{y^7}$
Evaluation of higher-order roots, p. 369 Use exponent notation.	Simplify. $\sqrt[5]{-32x^{15}} = \sqrt[5]{(-2)^5x^{15}}$ $= [(-2)^5x^{15}]^{1/5} = (-2)^1x^3 = -2x^3$	**10.** Simplify. $\sqrt[3]{-64m^{18}}$

Topic and Procedure	Examples	You Try It
Simplification of radicals with the product rule, p. 372 For nonnegative real numbers a and b and positive integers n, $\sqrt[n]{a}\,\sqrt[n]{b} = \sqrt[n]{ab}$.	Simplify when $x \geq 0, y \geq 0$. **(a)** $\sqrt{75x^3} = \sqrt{25x^2}\sqrt{3x}$ $= 5x\sqrt{3x}$ **(b)** $\sqrt[3]{16x^5y^6} = \sqrt[3]{8x^3y^6}\,\sqrt[3]{2x^2}$ $= 2xy^2\sqrt[3]{2x^2}$	**11.** Simplify when $x \geq 0, y \geq 0$. **(a)** $\sqrt{24x^5}$ **(b)** $\sqrt[3]{54r^4s^9}$
Combining radicals, p. 373 Simplify radicals and combine them if they have the same index and the same radicand.	Combine. $2\sqrt{50} - 3\sqrt{98} = 2\sqrt{25}\sqrt{2} - 3\sqrt{49}\sqrt{2}$ $= 2(5)\sqrt{2} - 3(7)\sqrt{2}$ $= 10\sqrt{2} - 21\sqrt{2} = -11\sqrt{2}$	**12.** Combine. $3\sqrt{72} + 4\sqrt{18}$
Multiplying radicals, p. 378 **1.** Multiply coefficients outside the radical and then multiply the radicands. **2.** Simplify your answer.	Multiply. **(a)** $(2\sqrt{3})(4\sqrt{5}) = 8\sqrt{15}$ **(b)** $2\sqrt{6}(\sqrt{2} - 3\sqrt{12}) = 2\sqrt{12} - 6\sqrt{72}$ $= 2\sqrt{4}\sqrt{3} - 6\sqrt{36}\sqrt{2}$ $= 4\sqrt{3} - 36\sqrt{2}$ **(c)** $(\sqrt{2} + \sqrt{3})(2\sqrt{2} - \sqrt{3})$ Use the FOIL method. $= 2\sqrt{4} - \sqrt{6} + 2\sqrt{6} - \sqrt{9}$ $= 4 + \sqrt{6} - 3$ $= 1 + \sqrt{6}$	**13.** Multiply. **(a)** $(\sqrt{6})(3\sqrt{5})$ **(b)** $3\sqrt{3}(2\sqrt{6} - \sqrt{15})$ **(c)** $(\sqrt{3} - \sqrt{5})(2\sqrt{3} + \sqrt{5})$
Simplifying quotients of radicals with the quotient rule, p. 379 For nonnegative real numbers a, positive real numbers b, and positive integers n, $\sqrt[n]{\frac{a}{b}} = \frac{\sqrt[n]{a}}{\sqrt[n]{b}}$.	Simplify. $\sqrt[3]{\frac{5}{27}} = \frac{\sqrt[3]{5}}{\sqrt[3]{27}} = \frac{\sqrt[3]{5}}{3}$	**14.** Simplify. $\sqrt[4]{\frac{3}{16}}$
Rationalizing denominators, p. 380 Multiply numerator and denominator by a value that eliminates the radical in the denominator.	Rationalize the denominator. **(a)** $\frac{2}{\sqrt{7}} = \frac{2}{\sqrt{7}} \cdot \frac{\sqrt{7}}{\sqrt{7}} = \frac{2\sqrt{7}}{7}$ **(b)** $\frac{3}{\sqrt{5} + \sqrt{2}} = \frac{3}{\sqrt{5} + \sqrt{2}} \cdot \frac{\sqrt{5} - \sqrt{2}}{\sqrt{5} - \sqrt{2}}$ $= \frac{3\sqrt{5} - 3\sqrt{2}}{(\sqrt{5})^2 - (\sqrt{2})^2} = \frac{3\sqrt{5} - 3\sqrt{2}}{5 - 2}$ $= \frac{3\sqrt{5} - 3\sqrt{2}}{3} = \sqrt{5} - \sqrt{2}$	**15.** Rationalize the denominator. **(a)** $\frac{3}{\sqrt{6}}$ **(b)** $\frac{4}{\sqrt{2} - \sqrt{3}}$
Solving radical equations, p. 389 **1.** Perform algebraic operations to obtain one radical by itself on one side of the equation. **2.** If the equation contains square roots, square each side of the equation. Otherwise, raise each side to the appropriate power for third- and higher-order roots. **3.** Simplify, if possible. **4.** If the equation still contains a radical, repeat steps 1 to 3. **5.** Collect all terms on one side of the equation. **6.** Solve the resulting equation. **7.** Check all apparent solutions. Solutions to radical equations must be verified.	Solve. $x = \sqrt{2x + 9} - 3$ $x + 3 = \sqrt{2x + 9}$ $(x + 3)^2 = (\sqrt{2x + 9})^2$ $x^2 + 6x + 9 = 2x + 9$ $x^2 + 6x - 2x + 9 - 9 = 0$ $x^2 + 4x = 0$ $x(x + 4) = 0$ $x = 0$ or $x = -4$ *Check.* $x = 0$: $0 \stackrel{?}{=} \sqrt{2(0) + 9} - 3$ $0 \stackrel{?}{=} \sqrt{9} - 3$ $0 = 3 - 3$ ✓ $x = -4$: $-4 \stackrel{?}{=} \sqrt{2(-4) + 9} - 3$ $-4 \stackrel{?}{=} \sqrt{1} - 3$ $-4 \neq -2$ The only solution is 0.	**16.** Solve. $5 + \sqrt{3x - 11} = x$

Topic and Procedure	Examples	You Try It
Simplifying imaginary numbers, p. 395 Use $i = \sqrt{-1}$, $i^2 = -1$, and $\sqrt{-a} = \sqrt{-1}\sqrt{a}$.	Simplify. **(a)** $\sqrt{-16} = \sqrt{-1}\sqrt{16} = 4i$ **(b)** $\sqrt{-18} = \sqrt{-1}\sqrt{18} = i\sqrt{9}\sqrt{2} = 3i\sqrt{2}$	**17.** Simplify. **(a)** $\sqrt{-100}$ **(b)** $\sqrt{-24}$
Adding and subtracting complex numbers, p. 397 Combine real parts and imaginary parts separately.	Add or subtract. **(a)** $(5 + 6i) + (2 - 4i) = 7 + 2i$ **(b)** $(-8 + 3i) - (4 - 2i) = -8 + 3i - 4 + 2i$ $= -12 + 5i$	**18.** Add or subtract. **(a)** $(8 + i) + (3 - 7i)$ **(b)** $(4 + 3i) - (5 + i)$
Multiplying complex numbers, p. 397 Use the FOIL method and $i^2 = -1$.	Multiply. $(5 - 6i)(2 - 4i) = 10 - 20i - 12i + 24i^2$ $= 10 - 32i + 24(-1)$ $= 10 - 32i - 24$ $= -14 - 32i$	**19.** Multiply. $(1 - 3i)(3 + 2i)$
Raising i to a power, p. 398 $i^1 = i$ $i^2 = -1$ $i^3 = -i$ $i^4 = 1$	Evaluate. $i^{27} = i^{24} \cdot i^3$ $= (i^4)^6 \cdot i^3$ $= (1)^6(-i)$ $= -i$	**20.** Evaluate. i^{36}
Dividing complex numbers, p. 399 Multiply the numerator and denominator by the conjugate of the denominator.	Divide. $\frac{5 + 2i}{4 - i} = \frac{5 + 2i}{4 - i} \cdot \frac{4 + i}{4 + i} = \frac{20 + 5i + 8i + 2i^2}{4^2 - i^2}$ $= \frac{20 + 13i + 2(-1)}{16 - (-1)}$ $= \frac{20 + 13i - 2}{16 + 1}$ $= \frac{18 + 13i}{17}$ or $\frac{18}{17} + \frac{13}{17}i$	**21.** Divide. $\frac{4 - 5i}{2 + i}$
Direct variation, p. 403 If y varies directly with x, there is a constant of variation k such that $y = kx$. After k is determined, other values of y or x can easily be computed.	y varies directly with x. When $x = 2$, $y = 7$. $y = kx$ $7 = k(2)$ Substitute. $k = \frac{7}{2}$ Solve. $y = \frac{7}{2}x$ What is y when $x = 18$? $y = \frac{7}{2}x = \frac{7}{2} \cdot 18 = 63$	**22.** If y varies directly with the square of x, and $y = 16$ when $x = 2$, what is y when $x = -3$?
Inverse variation, p. 403 If y varies inversely with x, the constant k is such that $y = \frac{k}{x}$.	y varies inversely with x. When x is 5, y is 12. What is y when x is 30? $y = \frac{k}{x}$ $12 = \frac{k}{5}$ Substitute. $k = 60$ Solve. $y = \frac{60}{x}$ Substitute. When $x = 30$, $y = \frac{60}{30} = 2$.	**23.** If w varies inversely with z, and $w = 5$ when $z = 0.5$, what is w when $z = 5$?

Chapter 7 Review Problems

In all exercises assume that the variables represent positive real numbers unless otherwise stated. Simplify using only positive exponents in your answers.

1. $(3xy^{1/2})(5x^2y^{-3})$

2. $(16a^6b^5)^{1/2}$

3. $3^{1/2} \cdot 3^{1/6}$

4. $\dfrac{6x^{2/3}y^{1/10}}{12x^{1/6}y^{-1/5}}$

5. $(2x^{-1/5}y^{1/10}z^{4/5})^{-5}$

6. $\left(\dfrac{49a^3b^6}{a^{-7}b^4}\right)^{1/2}$

7. $\left(\dfrac{8a^4}{a^{-2}}\right)^{1/3}$

8. $(4^{5/3})^{6/5}$

9. Combine as one fraction containing only positive exponents. $2x^{1/3} + x^{-2/3}$

10. Factor out the common factor $3x$ from $6x^{3/2} - 9x^{1/2}$.

In exercises 11–36, assume that all variables represent nonnegative real numbers. Evaluate, if possible.

11. $-\sqrt{16}$

12. $\sqrt[5]{-32}$

13. $-\sqrt{\dfrac{1}{25}}$

14. $\sqrt{0.04}$

15. $\sqrt[4]{-256}$

16. $\sqrt[3]{-\dfrac{1}{8}}$

17. $64^{2/3}$

18. $125^{4/3}$

Simplify.

19. $\sqrt{49x^4y^{10}z^2}$

20. $\sqrt[3]{64a^{12}b^{30}}$

21. $\sqrt[3]{-8a^{12}b^{15}c^{21}}$

22. $\sqrt{49x^{22}y^2}$

Replace radicals with rational exponents.

23. $\sqrt[5]{a^2}$

24. $\sqrt{2b}$

25. $\sqrt[3]{5a}$

26. $(\sqrt[5]{xy})^7$

Change to radical form.

27. $m^{1/2}$

28. $y^{3/5}$

29. $(3z)^{2/3}$

30. $(2x)^{3/7}$

Evaluate or simplify.

31. $16^{3/4}$

32. $(-27)^{2/3}$

33. $\left(\dfrac{1}{9}\right)^{1/2}$

34. $(0.49)^{1/2}$

35. $\left(\dfrac{1}{36}\right)^{-1/2}$

36. $(25a^2b^4)^{3/2}$

Combine where possible.

37. $\sqrt{50} + 2\sqrt{32} - \sqrt{8}$

38. $\sqrt{28} - 4\sqrt{7} + 5\sqrt{63}$

39. $2\sqrt{12} - \sqrt{48} + 5\sqrt{75}$

40. $\sqrt{125x^3} + x\sqrt{45x}$

41. $2\sqrt{32x} - 5x\sqrt{2} + \sqrt{18x}$

42. $3\sqrt[3]{16} - 4\sqrt[3]{54}$

Multiply and simplify.

43. $(5\sqrt{12})(3\sqrt{6})$

44. $(-2\sqrt{15})(4x\sqrt{3})$

45. $3\sqrt{x}\,(2\sqrt{8x} - 3\sqrt{48})$

46. $\sqrt{3a}\,(4 - \sqrt{21a})$

47. $2\sqrt{7b}\,(\sqrt{ab} - b\sqrt{3bc})$

48. $(5\sqrt{2} + \sqrt{3})(\sqrt{2} - 2\sqrt{3})$

49. $(2\sqrt{5} - 3\sqrt{6})^2$

50. $(\sqrt[3]{2x} + \sqrt[3]{6})(\sqrt[3]{4x^2} - \sqrt[3]{y})$

51. Let $f(x) = \sqrt{4x + 16}$.

(a) Find $f(12)$.

(b) What is the domain of $f(x)$?

52. Let $f(x) = \sqrt{36 - 3x}$.

(a) Find $f(9)$.

(b) What is the domain of $f(x)$?

Rationalize the denominator and simplify the expression.

53. $\sqrt{\dfrac{6y^2}{x}}$

54. $\dfrac{3}{\sqrt{5y}}$

55. $\dfrac{3\sqrt{7x}}{\sqrt{21x}}$

56. $\dfrac{2}{\sqrt{6} - \sqrt{5}}$

57. $\dfrac{\sqrt{x}}{3\sqrt{x} + \sqrt{y}}$

58. $\dfrac{\sqrt{5}}{\sqrt{7} - 3}$

59. $\dfrac{2\sqrt{3} + \sqrt{6}}{\sqrt{3} + 2\sqrt{6}}$

60. $\dfrac{2xy}{\sqrt[3]{16xy^5}}$

61. Simplify. $\sqrt{-16} + \sqrt{-45}$

62. Find x and y. $2x - 3i + 5 = yi - 2 + \sqrt{6}$

Simplify by performing the operation indicated.

63. $(-12 - 6i) + (3 - 5i)$

64. $(2 - i) - (12 - 3i)$

65. $(5 - 2i)(3 + 3i)$

66. $(6 - 2i)^2$

67. $2i(3 + 4i)$

68. $3 - 4(2 + i)$

69. Evaluate. i^{34}

70. Evaluate. i^{65}

Divide.

71. $\dfrac{7 - 2i}{3 + 4i}$

72. $\dfrac{5 - 2i}{1 - 3i}$

73. $\dfrac{4 - 3i}{5i}$

74. $\dfrac{12}{3 - 5i}$

Solve and check your solution(s).

75. $\sqrt{3x - 2} = 5$

76. $\sqrt[3]{3x - 1} = 2$

77. $\sqrt{2x + 1} = 2x - 5$

78. $1 + \sqrt{3x + 1} = x$

79. $\sqrt{3x + 1} - \sqrt{2x - 1} = 1$

80. $\sqrt{7x + 2} = \sqrt{x + 3} + \sqrt{2x - 1}$

Round all answers to the nearest tenth.

81. If y varies directly with x and $y = 11$ when $x = 4$, find the value of y when $x = 6$.

82. ***Saturated Fat Intake*** The maximum amount of saturated fat a person should consume varies directly with the number of calories consumed. A person who consumes 2000 calories per day should have a maximum of 18 grams of saturated fat per day. For a person who consumes 2500 calories, what is the maximum amount of saturated fat he or she should consume?

83. ***Time of Falling Object*** The time it takes a falling object to drop a given distance varies directly with the square root of the distance traveled. A steel ball takes 2 seconds to drop a distance of 64 feet. How many seconds will it take to drop a distance of 196 feet?

84. If y varies inversely with x and $y = 8$ when $x = 3$, find the value of y when $x = 48$.

85. Suppose that y varies directly with x and inversely with the square of z. $y = 20$ when $x = 10$ and $z = 5$. Find y when $x = 8$ and $z = 2$.

▲**86.** ***Capacity of a Cylinder*** The capacity of a cylinder varies directly with the height and the square of the radius. A cylinder with a radius of 3 centimeters and a height of 5 centimeters has a capacity of approximately 135 cubic centimeters. What is the capacity of a cylinder with a height of 9 centimeters and a radius of 4 centimeters?

How Am I Doing? Chapter 7 Test

CHAPTER Test Prep VIDEOS   MATH COACH   MyMathLab®   YouTube™

After you take this test read through the Math Coach on pages 419–420. Math Coach videos are available via MyMathLab and YouTube. Step-by-step test solutions in the Chapter Test Prep Videos are also available via MyMathLab and YouTube. (Search "TobeyInterAlg" and click on "Channels.")

Simplify.

1. $(2x^{1/2}y^{1/3})(-3x^{1/3}y^{1/6})$

2. $\dfrac{7x^3}{4x^{3/4}}$

3. $(8x^{1/3})^{3/2}$

4. Evaluate. $\left(\dfrac{4}{9}\right)^{3/2}$

5. Evaluate. $\sqrt[5]{-32}$

Evaluate.

6. $8^{-2/3}$

MC **7.** $16^{5/4}$

Simplify. Assume that all variables are nonnegative.

8. $\sqrt{75a^4b^9}$

9. $\sqrt{49a^4b^{10}}$

10. $\sqrt[3]{54m^3n^5}$

Combine where possible.

11. $3\sqrt{24} - \sqrt{18} + \sqrt{50}$

MC **12.** $\sqrt{40x} - \sqrt{27x} + 2\sqrt{12x}$

Multiply and simplify.

13. $(-3\sqrt{2y})(5\sqrt{10xy})$

14. $2\sqrt{3}\,(3\sqrt{6} - 5\sqrt{2})$

15. $(5\sqrt{3} - \sqrt{6})(2\sqrt{3} + 3\sqrt{6})$

1. ______
2. ______
3. ______
4. ______
5. ______
6. ______
7. ______
8. ______
9. ______
10. ______
11. ______
12. ______
13. ______
14. ______
15. ______

16.

17.

18.

19.

20.

21.

22.

23.

24.

25.

26.

27.

28.

29.

30.

Total Correct:

Rationalize the denominator.

16. $\dfrac{30}{\sqrt{5x}}$

17. $\sqrt{\dfrac{xy}{3}}$

MC **18.** $\dfrac{1 + 2\sqrt{3}}{3 - \sqrt{3}}$

Solve and check your solution(s).

19. $\sqrt{3x - 2} = x$

MC **20.** $5 + \sqrt{x + 15} = x$

21. $5 - \sqrt{x - 2} = \sqrt{x + 3}$

Simplify by using the properties of complex numbers.

22. $(8 + 2i) - 3(2 - 4i)$

23. $i^{18} + \sqrt{-16}$

24. $(3 - 2i)(4 + 3i)$

25. $\dfrac{2 + 5i}{1 - 3i}$

26. $(6 + 3i)^2$

27. i^{43}

28. If y varies inversely with x and $y = 9$ when $x = 2$, find the value of y when $x = 6$.

29. Suppose y varies directly with x and inversely with the square of z. When $x = 8$ and $z = 4$, then $y = 3$. Find y when $x = 5$ and $z = 6$.

30. A car's stopping distance varies directly with the square of its speed. A car traveling on pavement can stop in 30 feet when traveling at 30 miles per hour. What distance will the car take to stop if it is traveling at 50 miles per hour?

MATH COACH

Mastering the skills you need to do well on the test.

Students often make the same types of errors when they do the Chapter 7 Test. Here are some helpful hints to keep you from making those common errors on test problems.

Simplifying Expressions with Rational Exponents—

Problem 7 Evaluate. $16^{5/4}$

Helpful Hint First see if the numerical base can be written in exponent form. If there are two alternate ways to rewrite this number, use the form that involves the smallest possible number as the base and has the largest exponent.

Did you write 16 as 2^4?

Yes ____ No ____

If you answered No, think of the different ways that 16 can be written in exponent form: 16^1, 4^2, or 2^4.

Your choice is most likely between 4^2 and 2^4. The problem works out more easily with the smallest possible base, which is 2. See if you can redo this step correctly.

Do you see that if we use the rule $(x^m)^n = x^{mn}$, we obtain the expression $(2^4)^{5/4} = 2^{4/1 \cdot 5/4}$?

Yes ____ No ____

If you answered No, please review the rule about raising a power to a power and see if you can obtain the right result. Now simplify this expression further and evaluate the resulting exponential expression.

If you answered Problem 7 incorrectly, go back and rework the problem using these suggestions.

Adding and Subtracting Radical Expressions—Problem 12

Combine where possible. $\sqrt{40x} - \sqrt{27x} + 2\sqrt{12x}$

Helpful Hint Remember to simplify each radical expression first. For this problem, remember to keep the variable x inside the radical.

Did you simplify $\sqrt{40x}$ to $2\sqrt{10x}$?

Yes ____ No ____

Did you simplify $\sqrt{27x}$ to $3\sqrt{3x}$?

Yes ____ No ____

If you answered No to either question, remember that $\sqrt{40x} = \sqrt{4} \cdot \sqrt{10x}$ and $\sqrt{27x} = \sqrt{9} \cdot \sqrt{3x}$. You can take the square roots of both 4 and 9. Try to do that part of the problem again and then simplify to see if you obtain the same results.

Did you simplify $2\sqrt{12x}$ to $4\sqrt{3x}$?

Yes ____ No ____

If you answered No, remember that $2\sqrt{12x} = 2 \cdot \sqrt{4} \cdot \sqrt{3x}$. Go back and see if you can obtain the correct answer to that step. Remember that in your final step, you can only add radical expressions if they have the same radicand.

Now go back and rework the problem using these suggestions.

Need help? Watch the MATH COACH videos in MyMathLab® or on YouTube®.

Rationalizing the Denominator—Problem 18 Rationalize the denominator. $\frac{1 + 2\sqrt{3}}{3 - \sqrt{3}}$

Helpful Hint You will need to multiply both the numerator and the denominator by the conjugate of the denominator. Do this very carefully using the FOIL method.

Did you multiply the numerator and the denominator by $(3 + \sqrt{3})$?

Yes ____ No ____

If you answered No, review the definition of conjugate. Note that the conjugate of $3 - \sqrt{3}$ is $3 + \sqrt{3}$.

Did you multiply the binomials in the denominator to get $9 + 3\sqrt{3} - 3\sqrt{3} - \sqrt{9}$?

Yes ____ No ____

Did you multiply the binomials in the numerator to get $3 + \sqrt{3} + 6\sqrt{3} + 2\sqrt{9}$?

Yes ____ No ____

If you answered No to either question, go back over each step of multiplying the binomial expressions using the FOIL method. Be careful in writing which numbers go outside the radical and which ones go inside the radical. Now simplify each expression.

If you answered Problem 18 incorrectly, go back and rework the problem using these suggestions.

Solving a Radical Equation—Problem 20 Solve and check your solution(s). $5 + \sqrt{x + 15} = x$

Helpful Hint Always isolate a radical term on one side of the equation before squaring each side.

Did you isolate the radical term first to obtain the equation $\sqrt{x + 15} = x - 5$?

Yes ____ No ____

After squaring each side, did you obtain $x + 15 = x^2 - 10x + 25$?

Yes ____ No ____

If you answered No to either question, review the process for solving a radical equation. Remember that $(\sqrt{x + 15})^2 = x + 15$ and $(x - 5)^2 = x^2 - 10x + 25$. Go back and try to complete these steps again.

Did you collect all like terms on one side to obtain the equation $0 = x^2 - 11x + 10$?

Yes ____ No ____

If you answered No, remember to add $-x - 15$ to both sides of the equation.

Now factor the quadratic equation and set each factor equal to 0 to find the solution(s). Make sure to check your possible solutions and discard any solution that does not check when substituted back into the original equation.

If you answered Problem 20 incorrectly, go back and rework the problem using these suggestions.

Need more help? Look for section examples marked with MC to review.

CHAPTER 8

One of the most exciting events in baseball is the home run. The path of the ball can be predicted by a quadratic function. The mathematics of this chapter provides players, coaches, and managers information that allows players to improve in their ability to hit home runs.

Quadratic Equations and Inequalities

8.1 Quadratic Equations

Student Learning Objectives

After studying this section, you will be able to:

① **Solve quadratic equations by the square root property.**

② **Solve quadratic equations by completing the square.**

① Solving Quadratic Equations by the Square Root Property

Recall that an equation written in the form $ax^2 + bx + c = 0$, where a, b, and c are real numbers and $a \neq 0$, is called a **quadratic equation.** Recall also that we call this the **standard form** of a quadratic equation. We have previously solved quadratic equations using the zero factor property. This has allowed us to factor the left side of an equation such as $x^2 - 7x + 12 = 0$ and obtain $(x - 3)(x - 4) = 0$ and then solve to find that $x = 3$ and $x = 4$. In this chapter we develop new methods of solving quadratic equations. The first method is often called the **square root property.**

THE SQUARE ROOT PROPERTY

If $x^2 = a$, then $x = \pm\sqrt{a}$ for all real numbers a.

The notation $\pm\sqrt{a}$ is a shorthand way of writing "$+\sqrt{a}$ or $-\sqrt{a}$." The symbol $\pm$ is read "plus or minus." We can justify this property by using the zero factor property. If we write $x^2 = a$ in the form $x^2 - a = 0$, we can factor it to obtain $(x + \sqrt{a})(x - \sqrt{a}) = 0$ and thus, $x = -\sqrt{a}$ or $x = +\sqrt{a}$. This can be written more compactly as $x = \pm\sqrt{a}$.

EXAMPLE 1 Solve and check. $x^2 - 36 = 0$

Solution If we add 36 to each side, we have $x^2 = 36$.

$$x = \pm\sqrt{36}$$
$$x = \pm 6$$

Thus, the two roots are 6 and -6.

Check.

$$(6)^2 - 36 \stackrel{?}{=} 0 \qquad (-6)^2 - 36 \stackrel{?}{=} 0$$
$$36 - 36 \stackrel{?}{=} 0 \qquad 36 - 36 \stackrel{?}{=} 0$$
$$0 = 0 \ \checkmark \qquad 0 = 0 \ \checkmark$$

Student Practice 1 Solve and check. $x^2 - 121 = 0$

NOTE TO STUDENT: *Fully worked-out solutions to all of the Student Practice problems can be found at the back of the text starting at page SP-1.*

EXAMPLE 2 Solve. $x^2 = 48$

Solution

$$x = \pm\sqrt{48} = \pm\sqrt{16 \cdot 3}$$
$$x = \pm 4\sqrt{3}$$

The roots are $4\sqrt{3}$ and $-4\sqrt{3}$.

Student Practice 2 Solve. $x^2 = 18$

EXAMPLE 3 Solve and check. $3x^2 + 2 = 77$

Solution

$$3x^2 + 2 = 77$$
$$3x^2 = 75$$
$$x^2 = 25$$
$$x = \pm\sqrt{25}$$
$$x = \pm 5$$

The roots are 5 and -5.

Check.

$$3(5)^2 + 2 \stackrel{?}{=} 77 \qquad 3(-5)^2 + 2 \stackrel{?}{=} 77$$
$$3(25) + 2 \stackrel{?}{=} 77 \qquad 3(25) + 2 \stackrel{?}{=} 77$$
$$75 + 2 \stackrel{?}{=} 77 \qquad 75 + 2 \stackrel{?}{=} 77$$
$$77 = 77 \checkmark \qquad 77 = 77 \checkmark$$

Student Practice 3 Solve. $5x^2 + 1 = 46$

Sometimes we obtain roots that are complex numbers.

EXAMPLE 4 Solve and check. $4x^2 = -16$

Solution

$$x^2 = -4$$
$$x = \pm\sqrt{-4}$$
$$x = \pm 2i \qquad \text{Simplify using } \sqrt{-1} = i.$$

The roots are $2i$ and $-2i$.

Check.

$$4(2i)^2 \stackrel{?}{=} -16 \qquad 4(-2i)^2 \stackrel{?}{=} -16$$
$$4(4i^2) \stackrel{?}{=} -16 \qquad 4(4i^2) \stackrel{?}{=} -16$$
$$4(-4) \stackrel{?}{=} -16 \qquad 4(-4) \stackrel{?}{=} -16$$
$$-16 = -16 \checkmark \qquad -16 = -16 \checkmark$$

Student Practice 4 Solve and check. $3x^2 = -27$

EXAMPLE 5 Solve. $(4x - 1)^2 = 5$

Solution

$$4x - 1 = \pm\sqrt{5}$$
$$4x = 1 \pm \sqrt{5}$$
$$x = \frac{1 \pm \sqrt{5}}{4}$$

The roots are $\frac{1 + \sqrt{5}}{4}$ and $\frac{1 - \sqrt{5}}{4}$.

Student Practice 5 Solve. $(2x + 3)^2 = 7$

② Solving Quadratic Equations by Completing the Square

Often, a quadratic equation cannot be factored (or it may be difficult to factor). So we use another method of solving the equation, called **completing the square.** When we complete the square, we are changing the polynomial to a perfect square trinomial. The form of the equation then becomes $(x + d)^2 = e$.

We already know that

$$(x + d)^2 = x^2 + 2dx + d^2.$$

Notice three things about the quadratic equation on the right-hand side.

1. The coefficient of the quadratic term (x^2) is 1.
2. The coefficient of the linear (x) term is $2d$.
3. The constant term (d^2) is the square of *half* the coefficient of the linear term.

For example, in the trinomial $x^2 + 6x + 9$, the coefficient of the linear term is 6 and the constant term is $\left(\frac{6}{2}\right)^2 = (3)^2 = 9$.

For the trinomial $x^2 - 10x + 25$, the coefficient of the linear term is -10 and the constant term is $\left(\frac{-10}{2}\right)^2 = (-5)^2 = 25$.

What number n makes the trinomial $x^2 + 12x + n$ a perfect square?

$$n = \left(\frac{12}{2}\right)^2 = 6^2 = 36$$

Hence, the trinomial $x^2 + 12x + 36$ is a perfect square trinomial and can be written as $(x + 6)^2$.

Now let's solve some equations.

EXAMPLE 6 Solve by completing the square and check. $x^2 + 6x + 1 = 0$

Solution

Step 1 First we rewrite the equation in the form $ax^2 + bx = c$ by subtracting 1 from each side of the equation. Thus, we obtain

$$x^2 + 6x = -1.$$

Step 2 We verify that the coefficient of the quadratic term (x^2) is 1. If the coefficient of the x^2 term were not 1, you would divide each term of the equation by the coefficient.

Step 3 We want to complete the square of $x^2 + 6x$. That is, we want to add a constant term to $x^2 + 6x$ so that we get a perfect square trinomial. We do this by taking half the coefficient of x and squaring it.

$$\left(\frac{6}{2}\right)^2 = 3^2 = 9$$

Adding 9 to $x^2 + 6x$ gives the perfect square trinomial $x^2 + 6x + 9$, which we factor as $(x + 3)^2$. But we cannot add 9 to the left side of our equation unless we also add 9 to the right side. (Why?) We now have

$$x^2 + 6x + 9 = -1 + 9$$

Step 4 Now we factor the left side.

$$(x + 3)^2 = 8$$

Then we use the square root property.

$$x + 3 = \pm\sqrt{8}$$
$$x + 3 = \pm 2\sqrt{2}$$

Step 5 Next we solve for x by subtracting 3 from each side of the equation.

$$x = -3 \pm 2\sqrt{2}$$

The roots are $-3 + 2\sqrt{2}$ and $-3 - 2\sqrt{2}$.

Step 6 We *must* check our solution in the *original* equation (not the perfect square trinomial we constructed).

$$x^2 + 6x + 1 = 0$$
$$(-3 + 2\sqrt{2})^2 + 6(-3 + 2\sqrt{2}) + 1 \stackrel{?}{=} 0$$
$$9 - 12\sqrt{2} + 8 - 18 + 12\sqrt{2} + 1 \stackrel{?}{=} 0$$
$$18 - 18 - 12\sqrt{2} + 12\sqrt{2} \stackrel{?}{=} 0$$
$$0 = 0 \quad ✓$$

$$x^2 + 6x + 1 = 0$$
$$(-3 - 2\sqrt{2})^2 + 6(-3 - 2\sqrt{2}) + 1 \stackrel{?}{=} 0$$
$$9 + 12\sqrt{2} + 8 - 18 - 12\sqrt{2} + 1 \stackrel{?}{=} 0$$
$$18 - 18 + 12\sqrt{2} - 12\sqrt{2} \stackrel{?}{=} 0$$
$$0 = 0 \quad ✓$$

Student Practice 6 Solve by completing the square. $x^2 + 8x + 3 = 0$

Let us summarize for future reference the six steps we use to solve a quadratic equation by completing the square.

COMPLETING THE SQUARE

1. Put the equation in the form $ax^2 + bx = c$.
2. If $a \neq 1$, divide each term of the equation by a.
3. Square half of the numerical coefficient of the linear term. Add the result to both sides of the equation.
4. Factor the left side; then take the square root of both sides of the equation.
5. Solve the resulting equation for x.
6. Check the solutions in the original equation.

Notice that step 2 is used when the coefficient of x^2 is not 1, as in Example 7.

EXAMPLE 7 Solve by completing the square. $3x^2 - 8x + 1 = 0$

Solution

$$3x^2 - 8x = -1 \qquad \text{Subtract 1 from each side.}$$

$$\frac{3x^2}{3} - \frac{8x}{3} = -\frac{1}{3} \qquad \text{Divide each term by 3. (Remember that the coefficient of the quadratic term must be 1.)}$$

$$x^2 - \frac{8}{3}x + \frac{16}{9} = -\frac{1}{3} + \frac{16}{9}$$

$$\left(x - \frac{4}{3}\right)^2 = \frac{13}{9}$$

$$x - \frac{4}{3} = \pm\sqrt{\frac{13}{9}}$$

$$x - \frac{4}{3} = \pm\frac{\sqrt{13}}{3}$$

$$x = \frac{4}{3} \pm \frac{\sqrt{13}}{3}$$

$$x = \frac{4 \pm \sqrt{13}}{3}$$

Check. For $x = \dfrac{4 + \sqrt{13}}{3}$,

$$3\left(\frac{4 + \sqrt{13}}{3}\right)^2 - 8\left(\frac{4 + \sqrt{13}}{3}\right) + 1 \stackrel{?}{=} 0$$

$$\frac{16 + 8\sqrt{13} + 13}{3} - \frac{32 + 8\sqrt{13}}{3} + 1 \stackrel{?}{=} 0$$

$$\frac{16 + 8\sqrt{13} + 13 - 32 - 8\sqrt{13}}{3} + 1 \stackrel{?}{=} 0$$

$$\frac{29 - 32}{3} + 1 \stackrel{?}{=} 0$$

$$-\frac{3}{3} + 1 \stackrel{?}{=} 0$$

$$-1 + 1 = 0 \checkmark$$

See whether you can check the solution $\dfrac{4 - \sqrt{13}}{3}$.

Student Practice 7 Solve by completing the square. $2x^2 + 4x + 1 = 0$

8.1 Exercises

MyMathLab®

Watch the videos in MyMathLab

Download the MyDashBoard App

Solve the equations by using the square root property. Express any complex numbers using i notation.

1. $x^2 = 100$

2. $x^2 = 49$

3. $3x^2 - 45 = 0$

4. $4x^2 - 68 = 0$

5. $2x^2 - 80 = 0$

6. $5x^2 - 40 = 0$

7. $x^2 = -81$

8. $x^2 = -64$

9. $x^2 + 16 = 0$

10. $x^2 + 121 = 0$

11. $(x - 3)^2 = 12$

12. $(x + 5)^2 = 24$

13. $(x + 9)^2 = 21$

14. $(x - 8)^2 = 23$

15. $(2x + 1)^2 = 7$

16. $(3x + 2)^2 = 5$

17. $(4x - 3)^2 = 36$

18. $(5x - 2)^2 = 25$

19. $(2x + 5)^2 = 49$

20. $(7x + 1)^2 = 64$

21. $2x^2 - 9 = 0$

22. $4x^2 - 7 = 0$

Solve the equations by completing the square. Simplify your answers.

23. $x^2 + 10x + 5 = 0$

24. $x^2 + 6x + 2 = 0$

25. $x^2 - 8x = 17$

26. $x^2 - 12x = 4$

27. $x^2 - 14x = -48$

28. $x^2 - 16x = -48$

29. $\frac{x^2}{2} + \frac{3}{2}x = 4$

30. $\frac{x^2}{3} - \frac{5}{3}x = 1$

31. $2y^2 + 10y = -11$

32. $7x^2 + 4x - 5 = 0$

33. $3x^2 + 10x - 2 = 0$

34. $5x^2 + 4x - 3 = 0$

Mixed Practice, Exercises 35–44

Solve the equations by any method. Simplify your answers. Express any complex numbers using i notation.

35. $x^2 + 4x - 6 = 0$

36. $x^2 + 8x + 10 = 0$

37. $\frac{x^2}{2} - x = 4$

38. $\frac{x^2}{4} - 3x = -9$

39. $3x^2 + 1 = x$

40. $2x^2 + 5 = -3x$

41. $x^2 + 2 = x$

42. $x^2 + 5 = 4x$

43. $2x^2 + 2 = 3x$

44. $3x^2 + 8x + 3 = 2$

45. Check the solution $x = -1 + \sqrt{6}$ in the equation $x^2 + 2x - 5 = 0$.

46. Check the solution $x = 2 + \sqrt{3}$ in the equation $x^2 - 4x + 1 = 0$.

Applications *Volume of a Box The sides of the box shown are labeled with the dimensions in feet.*

▲ **47.** What is the value of x if the volume of the box is 648 cubic feet?

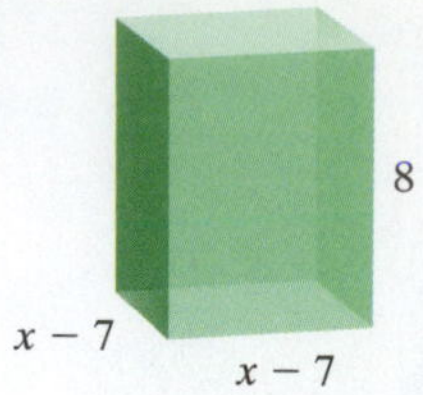

▲ **48.** What is the value of x if the volume of the box is 1800 cubic feet?

Basketball *The time a basketball player spends in the air when shooting a basket is called the "hang time." The vertical leap L measured in feet is related to the hang time t measured in seconds by the equation $L = 4t^2$.*

49. Kadour Ziani is a member of Slam Nation, a group that performs slam dunk shows. Ziani holds the vertical leap world record, with a leap of 5 feet. Find the hang time for this leap. Round to the nearest hundredth.

50. Spudd Webb, one of the shortest NBA players at 5 feet 7 inches, was known for his amazing 3.8-foot vertical leap. Find the hang time for this leap. Round to the nearest hundredth.

Cumulative Review *Evaluate the expressions for the given values.*

51. **[1.6.1]** $\sqrt{b^2 - 4ac}$; $b = 4, a = 3, c = -4$

52. **[1.6.1]** $\sqrt{b^2 - 4ac}$; $b = -5, a = 2, c = -3$

53. **[1.6.1]** $5x^2 - 6x + 8$; $x = -2$

54. **[1.6.1]** $2x^2 + 3x - 5$; $x = -3$

Quick Quiz 8.1

1. Solve by using the square root property.
$(4x - 3)^2 = 12$

2. Solve by completing the square. $x^2 - 8x = 28$

3. Solve by completing the square. $2x^2 + 10x = -11$

4. **Concept Check** Explain how you would decide what to add to each side of the equation to complete the square for the following equation.

$$x^2 + x = 1$$

8.2 The Quadratic Formula and Solutions to Quadratic Equations

Student Learning Objectives

After studying this section, you will be able to:

1. **Solve a quadratic equation by using the quadratic formula.**
2. **Use the discriminant to determine the nature of the roots of a quadratic equation.**
3. **Write a quadratic equation given the solutions of the equation.**

① Solving a Quadratic Equation by Using the Quadratic Formula

The last method we'll study for solving quadratic equations is the **quadratic formula.** This method works for *any* quadratic equation.

The quadratic formula is developed from completing the square. We begin with the **standard form** of the quadratic equation.

$$ax^2 + bx + c = 0$$

To complete the square, we want the equation to be in the form $x^2 + dx = e$. Thus, we divide by a and subtract the constant term $\frac{c}{a}$ from each side.

$$\frac{ax^2}{a} + \frac{b}{a}x + \frac{c}{a} = 0$$

$$x^2 + \frac{b}{a}x = -\frac{c}{a}$$

Now we complete the square by adding $\left(\frac{b}{2a}\right)^2$ to each side.

$$x^2 + \frac{b}{a}x + \left(\frac{b}{2a}\right)^2 = -\frac{c}{a} + \left(\frac{b}{2a}\right)^2$$

We factor the left side and write the right side as one fraction.

$$\left(x + \frac{b}{2a}\right)^2 = \frac{b^2 - 4ac}{4a^2}$$

Now we use the square root property.

$$x + \frac{b}{2a} = \pm\sqrt{\frac{b^2 - 4ac}{4a^2}}$$

We solve for x and simplify.

$$x = -\frac{b}{2a} \pm \sqrt{\frac{b^2 - 4ac}{4a^2}}$$

$$x = \frac{-b \pm \sqrt{b^2 - 4ac}}{2a}$$

This is the quadratic formula.

QUADRATIC FORMULA

For all equations $ax^2 + bx + c = 0$ where $a \neq 0$

$$x = \frac{-b \pm \sqrt{b^2 - 4ac}}{2a}$$

EXAMPLE 1 Solve by using the quadratic formula. $x^2 + 8x = -3$

Solution The standard form is $x^2 + 8x + 3 = 0$. We substitute $a = 1, b = 8$, and $c = 3$.

$$x = \frac{-b \pm \sqrt{b^2 - 4ac}}{2a}$$

$$x = \frac{-8 \pm \sqrt{8^2 - 4(1)(3)}}{2(1)}$$

$$x = \frac{-8 \pm \sqrt{64 - 12}}{2} = \frac{-8 \pm \sqrt{52}}{2} = \frac{-8 \pm \sqrt{4}\sqrt{13}}{2}$$

$$x = \frac{-8 \pm 2\sqrt{13}}{2} = \frac{\cancel{2}(-4 \pm \sqrt{13})}{\cancel{2}}$$

$$x = -4 \pm \sqrt{13}$$

Student Practice 1 Solve by using the quadratic formula.

$$x^2 + 5x = -1 + 2x$$

***NOTE TO STUDENT:** Fully worked-out solutions to all of the Student Practice problems can be found at the back of the text starting at page SP-1.*

EXAMPLE 2 Solve by using the quadratic formula. $3x^2 - x - 2 = 0$

Solution Here $a = 3, b = -1$, and $c = -2$.

$$x = \frac{-b \pm \sqrt{b^2 - 4ac}}{2a}$$

$$x = \frac{-(-1) \pm \sqrt{(-1)^2 - 4(3)(-2)}}{2(3)}$$

$$x = \frac{1 \pm \sqrt{1 + 24}}{6} = \frac{1 \pm \sqrt{25}}{6}$$

$$x = \frac{1 + 5}{6} = \frac{6}{6} \quad \text{or} \quad x = \frac{1 - 5}{6} = -\frac{4}{6}$$

$$x = 1 \qquad\qquad x = -\frac{2}{3}$$

Student Practice 2 Solve by using the quadratic formula.

$$2x^2 + 7x + 6 = 0$$

EXAMPLE 3 Solve by using the quadratic formula. $2x^2 - 48 = 0$

Solution This equation is equivalent to $2x^2 - 0x - 48 = 0$. Therefore, we know that $a = 2, b = 0$, and $c = -48$.

$$x = \frac{-b \pm \sqrt{b^2 - 4ac}}{2a}$$

$$x = \frac{-0 \pm \sqrt{(0)^2 - 4(2)(-48)}}{2(2)}$$

$$x = \frac{\pm\sqrt{384}}{4}$$ but this is not simplified

$$x = \frac{\pm\sqrt{64}\sqrt{6}}{4} = \frac{\pm 8\sqrt{6}}{4}$$

$$x = \pm 2\sqrt{6}$$

Student Practice 3 Solve by using the quadratic formula.

$$2x^2 - 26 = 0$$

EXAMPLE 4 A small company that manufactures canoes makes a daily profit p according to the equation $p = -100x^2 + 3400x - 26{,}196$, where p is measured in dollars and x is the number of canoes made per day. Find the number of canoes that must be made each day to produce a zero profit for the company. Round your answer to the nearest whole number.

Solution Since $p = 0$, we are solving the equation $0 = -100x^2 + 3400x - 26{,}196$.

In this case we have $a = -100$, $b = 3400$, and $c = -26{,}196$. Now we substitute these into the quadratic formula.

$$x = \frac{-b \pm \sqrt{b^2 - 4ac}}{2a}$$

$$x = \frac{-3400 \pm \sqrt{(3400)^2 - 4(-100)(-26{,}196)}}{2(-100)}$$

We will use a calculator to assist us with computation in this problem.

$$x = \frac{-3400 \pm \sqrt{11{,}560{,}000 - 10{,}478{,}400}}{-200}$$

$$x = \frac{-3400 \pm \sqrt{1{,}081{,}600}}{-200}$$

$$x = \frac{-3400 \pm 1040}{-200}$$

We now obtain two answers.

$$x = \frac{-3400 + 1040}{-200} = \frac{-2360}{-200} = 11.8 \approx 12$$

$$x = \frac{-3400 - 1040}{-200} = \frac{-4440}{-200} = 22.2 \approx 22$$

A zero profit is obtained when approximately 12 canoes are produced or when approximately 22 canoes are produced. Actually a slight profit of \$204 is made when these numbers of canoes are produced. The discrepancy is due to the rounding error that occurs when we approximate. By methods that we will learn later in this chapter, the maximum profit is produced when 17 canoes are made at the factory. We will investigate exercises of this kind later.

Student Practice 4 A company that manufactures modems makes a daily profit p according to the equation $p = -100x^2 + 4800x - 52{,}559$, where p is measured in dollars and x is the number of modems made per day. Find the number of modems that must be made each day to produce a zero profit for the company. Round your answer to the nearest whole number.

When an equation contains fractions or rational expressions, eliminate them by multiplying each term by the LCD. Then rewrite the equation in standard form before using the quadratic formula.

EXAMPLE 5 Solve by using the quadratic formula.

$$\frac{2x}{x+2} = 1 - \frac{3}{x+4}$$

Solution The LCD is $(x + 2)(x + 4)$.

$$\frac{2x}{x+2} = 1 - \frac{3}{x+4}$$

$$\frac{2x}{\cancel{x+2}}\cancel{(x+2)}(x+4) = 1(x+2)(x+4) - \frac{3}{\cancel{x+4}}(x+2)\cancel{(x+4)}$$

$$2x(x+4) = (x+2)(x+4) - 3(x+2)$$

$$2x^2 + 8x = x^2 + 6x + 8 - 3x - 6$$

$$2x^2 + 8x = x^2 + 3x + 2$$

$$x^2 + 5x - 2 = 0 \quad \text{Now we have an equation that is quadratic.}$$

Now the equation is in standard form, and we can use the quadratic formula with $a = 1, b = 5$, and $c = -2$.

$$x = \frac{-5 \pm \sqrt{5^2 - 4(1)(-2)}}{2(1)} = \frac{-5 \pm \sqrt{25 + 8}}{2}$$

$$x = \frac{-5 \pm \sqrt{33}}{2}$$

Student Practice 5 Solve by using the quadratic formula.

$$\frac{1}{x} + \frac{1}{x-1} = \frac{5}{6}$$

Some quadratic equations will have solutions that are not real numbers. You should use i notation to simplify the solutions that are nonreal complex numbers.

EXAMPLE 6 Solve and simplify your answer. $8x^2 - 4x + 1 = 0$

Solution $a = 8, b = -4$, and $c = 1$.

$$x = \frac{-(-4) \pm \sqrt{(-4)^2 - 4(8)(1)}}{2(8)}$$

$$x = \frac{4 \pm \sqrt{16 - 32}}{16} = \frac{4 \pm \sqrt{-16}}{16}$$

$$x = \frac{4 \pm 4i}{16} = \frac{4(1 \pm i)}{16} = \frac{1 \pm i}{4}$$

Student Practice 6 Solve by using the quadratic formula.

$$2x^2 - 4x + 5 = 0$$

You may have noticed that complex roots come in pairs. In other words, if $a + bi$ is a solution of a quadratic equation, its conjugate $a - bi$ is also a solution.

② Using the Discriminant to Determine the Nature of the Roots of a Quadratic Equation

So far we have used the quadratic formula to solve quadratic equations that had two real roots. Sometimes the roots were rational, and sometimes they were irrational. We have also solved equations like Example 6 with nonreal complex numbers. Such solutions occur when the expression $b^2 - 4ac$, the radicand in the quadratic formula, is negative.

$$x = \frac{-b \pm \sqrt{b^2 - 4ac}}{2a}$$

The expression $b^2 - 4ac$ is called the **discriminant.** Depending on the value of the discriminant and whether the discriminant is positive, zero, or negative, the roots of the quadratic equation will be rational, irrational, or complex. We summarize the types of solutions in the following table.

If the Discriminant $b^2 - 4ac$ is:	Then the Quadratic Equation $ax^2 + bx + c = 0$, where a, b, and c are integers, $a \neq 0$, will have:
A positive number that is also a perfect square	Two different rational solutions (Such an equation can always be factored.)
A positive number that is not a perfect square	Two different irrational solutions
Zero	One rational solution
Negative	Two complex solutions containing i (They will be complex conjugates.)

EXAMPLE 7 What type of solutions does the equation $2x^2 - 9x - 35 = 0$ have? Do not solve the equation.

Solution $a = 2, b = -9$, and $c = -35$. Thus,

$$b^2 - 4ac = (-9)^2 - 4(2)(-35) = 361.$$

Since the discriminant is positive, the equation has two real roots.

Since $(19)^2 = 361$, 361 is a perfect square. Thus, the equation has two different rational solutions. This type of quadratic equation can always be factored.

Student Practice 7 Use the discriminant to find what type of solutions the equation $9x^2 + 12x + 4 = 0$ has. Do not solve the equation.

EXAMPLE 8 Use the discriminant to determine the type of solutions each of the following equations has.

(a) $3x^2 - 4x + 2 = 0$ **(b)** $5x^2 - 3x - 5 = 0$

Solution

(a) Here $a = 3, b = -4$, and $c = 2$. Thus,

$$\begin{aligned} b^2 - 4ac &= (-4)^2 - 4(3)(2) \\ &= 16 - 24 = -8 \end{aligned}$$

Since the discriminant is negative, the equation will have two complex solutions containing i.

(b) Here $a = 5, b = -3$, and $c = -5$. Thus,

$$\begin{aligned} b^2 - 4ac &= (-3)^2 - 4(5)(-5) \\ &= 9 + 100 = 109 \end{aligned}$$

Since this positive number is not a perfect square, the equation will have two different irrational solutions.

Student Practice 8 Use the discriminant to determine the type of solutions each of the following equations has.

(a) $x^2 - 4x + 13 = 0$

(b) $9x^2 + 6x - 7 = 0$

③ Writing a Quadratic Equation Given the Solutions of the Equation

By using the zero factor property in reverse, we can find a quadratic equation that contains two given solutions. To illustrate, if 3 and 7 are the two solutions, then we could write the equation $(x - 3)(x - 7) = 0$, and therefore, a quadratic equation that has these two solutions is $x^2 - 10x + 21 = 0$.

This answer is not unique. Any constant multiple of $x^2 - 10x + 21 = 0$ would also have roots of 3 and 7. Thus, $2x^2 - 20x + 42 = 0$ also has roots of 3 and 7.

EXAMPLE 9 Find a quadratic equation whose roots are 5 and -2.

Solution

$$x = 5 \qquad x = -2$$
$$x - 5 = 0 \qquad x + 2 = 0$$
$$(x - 5)(x + 2) = 0$$
$$x^2 - 3x - 10 = 0$$

Student Practice 9 Find a quadratic equation whose roots are -10 and -6.

EXAMPLE 10 Find a quadratic equation whose solutions are $3i$ and $-3i$.

Solution First we write the two equations.

$$x - 3i = 0 \quad \text{and} \quad x + 3i = 0$$
$$(x - 3i)(x + 3i) = 0$$
$$x^2 + 3ix - 3ix - 9i^2 = 0$$
$$x^2 - 9(-1) = 0 \quad \text{Use } i^2 = -1.$$
$$x^2 + 9 = 0$$

Student Practice 10 Find a quadratic equation whose solutions are $2i\sqrt{3}$ and $-2i\sqrt{3}$.

8.2 Exercises

MyMathLab®

Watch the videos in MyMathLab

Download the MyDashBoard App

Verbal and Writing Skills, Exercises 1–4

1. How is the quadratic formula used to solve a quadratic equation?

2. The discriminant in the quadratic formula is the expression _______________.

3. If the discriminant in the quadratic formula is zero, then the quadratic equation will have _______________ solution(s).

4. If the discriminant in the quadratic formula is a perfect square, then the quadratic equation will have _______________ solution(s).

Solve by the quadratic formula. Simplify your answers.

5. $x^2 + x - 5 = 0$

6. $x^2 - 3x - 1 = 0$

7. $3x^2 - x - 1 = 0$

8. $4x^2 + x - 1 = 0$

9. $x^2 = \frac{2}{3}x$

10. $\frac{4}{5}x^2 = x$

11. $3x^2 - x - 2 = 0$

12. $7x^2 + 4x - 3 = 0$

13. $4x^2 + 3x - 2 = 0$

14. $6x^2 - 2x - 1 = 0$

15. $4x^2 + 1 = 7$

16. $6x^2 - 1 = 9$

Simplify each equation. Then solve by the quadratic formula. Simplify your answers and use i notation for nonreal complex numbers.

17. $2x(x + 3) - 3 = 4x - 2$

18. $5 + 3x(x - 2) = 4$

19. $x(x + 3) - 2 = 3x + 7$

20. $3(x^2 - 12) - 2x = 2x(x - 1)$

21. $(x - 2)(x + 1) = \dfrac{2x + 3}{2}$

22. $3x(x + 1) = \dfrac{7x + 1}{3}$

23. $\dfrac{1}{x + 2} + \dfrac{1}{x} = \dfrac{1}{3}$

24. $2y = \dfrac{1}{y} + \dfrac{5}{2}$

25. $\dfrac{1}{12} + \dfrac{1}{y} = \dfrac{2}{y + 2}$

26. $\dfrac{1}{4} + \dfrac{6}{y + 2} = \dfrac{6}{y}$

27. $x(x + 4) = -12$

28. $x^2 = 3(x - 3)$

29. $2x^2 + 11 = 0$

30. $3x^2 = -7$

31. $3x^2 - 8x + 7 = 0$

32. $3x^2 - 4x + 6 = 0$

Use the discriminant to find what type of solutions (two rational, two irrational, one rational, or two nonreal complex) each of the following equations has. Do not solve the equation.

33. $3x^2 + 4x = 2$

34. $5x^2 - 10x + 5 = 0$

35. $2x^2 + 10x + 8 = 0$

36. $2x^2 - 9x + 4 = 0$

37. $9x^2 + 4 = 12x$

38. $6x^2 + 7x - 2 = 0$

39. $7x(x - 1) + 15 = 10$

40. $x^2 - 3(x - 8) = 2x$

Write a quadratic equation having the given solutions.

41. $13, 2$

42. $5, 11$

43. $-3, -9$

44. $-8, -5$

45. $4i, -4i$

46. $6i, -6i$

47. $3, -\frac{5}{2}$

48. $-2, \frac{5}{6}$

Solve for x by using the quadratic formula. Approximate your answers to four decimal places.

49. $0.162x^2 + 0.094x - 0.485 = 0$

50. $20.6x^2 - 73.4x + 41.8 = 0$

Applications

51. ***Business Profit*** A company that manufactures mountain bikes makes a daily profit p according to the equation $p = -100x^2 + 4800x - 54{,}351$, where p is measured in dollars and x is the number of mountain bikes made per day. Find the number of mountain bikes that must be made each day to produce a zero profit for the company. Round your answer to the nearest whole number.

52. ***Business Profit*** A company that manufactures sport parachutes makes a daily profit p according to the equation $p = -100x^2 + 4200x - 39{,}476$, where p is measured in dollars and x is the number of parachutes made per day. Find the number of parachutes that must be made each day to produce a zero profit for the company. Round your answer to the nearest whole number.

Cumulative Review *Simplify.*

53. **[1.5.1]** $9x^2 - 6x + 3 - 4x - 12x^2 + 8$

54. **[1.5.3]** $3y(2 - y) + \frac{1}{5}(10y^2 - 15y)$

Quick Quiz 8.2 *Solve each of the following equations. Use i notation for any complex numbers.*

1. $11x^2 - 9x - 1 = 0$

2. $\frac{3}{4} + \frac{5}{4x} = \frac{2}{x^2}$

3. $(x + 2)(x + 1) + (x - 4)^2 = 9$

4. **Concept Check** Explain how you would determine if $2x^2 - 6x + 3 = 3$ has two rational, two irrational, one rational, or two nonreal complex solutions.

8.3 Equations That Can Be Transformed into Quadratic Form

Student Learning Objectives

After studying this section, you will be able to:

① **Solve equations of degree greater than 2 that can be transformed into quadratic form.**

② **Solve equations with fractional exponents that can be transformed into quadratic form.**

① Solving Equations of Degree Greater than 2

Some higher-order equations can be solved by writing them in the form of quadratic equations. An equation is **quadratic in form** if we can substitute a linear term for the variable raised to the lowest power and get an equation of the form $ay^2 + by + c = 0$.

NOTE TO STUDENT: Fully worked-out solutions to all of the Student Practice problems can be found at the back of the text starting at page SP-1.

EXAMPLE 1 Solve. $x^4 - 13x^2 + 36 = 0$

Solution Let $y = x^2$. Then $y^2 = x^4$. Thus, we obtain a new equation and solve it as follows:

$y^2 - 13y + 36 = 0$ Replace x^2 by y and x^4 by y^2.

$(y - 4)(y - 9) = 0$ Factor.

$y - 4 = 0$ or $y - 9 = 0$ Solve for y.

$y = 4$ $y = 9$ These are *not* the roots to the original equation. We must replace y by x^2.

$x^2 = 4$ $x^2 = 9$

$x = \pm\sqrt{4}$ $x = \pm\sqrt{9}$

$x = \pm 2$ $x = \pm 3$

Thus, there are *four* solutions to the original equation: $x = +2$, $x = -2$, $x = +3$, and $x = -3$. Check these values to verify that they are solutions.

Student Practice 1 Solve.

$x^4 - 5x^2 - 36 = 0$

EXAMPLE 2 Solve for all real roots. $2x^6 - x^3 - 6 = 0$

Solution Let $y = x^3$. Then $y^2 = x^6$. Thus, we have the following:

$2y^2 - y - 6 = 0$ Replace x^3 by y and x^6 by y^2.

$(2y + 3)(y - 2) = 0$ Factor.

$2y + 3 = 0$ or $y - 2 = 0$ Solve for y.

$y = -\frac{3}{2}$ $y = 2$

$x^3 = -\frac{3}{2}$ $x^3 = 2$ Replace y by x^3.

$x = \sqrt[3]{-\frac{3}{2}}$ $x = \sqrt[3]{2}$ Take the cube root of each side of the equation.

$x = \frac{\sqrt[3]{-12}}{2}$ Simplify $\sqrt[3]{-\frac{3}{2}}$ by rationalizing the denominator.

Check these solutions.

Student Practice 2 Solve for all real roots. $x^6 - 5x^3 + 4 = 0$

② Solving Equations with Fractional Exponents

EXAMPLE 3 Solve and check your solutions. $x^{2/3} - 3x^{1/3} + 2 = 0$

Solution Let $y = x^{1/3}$. Then $y^2 = x^{2/3}$.

$y^2 - 3y + 2 = 0$ Replace $x^{1/3}$ by y and $x^{2/3}$ by y^2.

$(y - 2)(y - 1) = 0$ Factor.

$$y - 2 = 0 \quad \text{or} \quad y - 1 = 0$$
$$y = 2 \qquad y = 1 \qquad \text{Solve for } y.$$
$$x^{1/3} = 2 \qquad x^{1/3} = 1 \qquad \text{Replace } y \text{ by } x^{1/3}.$$
$$(x^{1/3})^3 = (2)^3 \qquad (x^{1/3})^3 = (1)^3 \qquad \text{Cube each side of the equation.}$$
$$x = 8 \qquad x = 1$$

Check.

$$x = 8: \quad (8)^{2/3} - 3(8)^{1/3} + 2 \stackrel{?}{=} 0$$
$$(\sqrt[3]{8})^2 - 3(\sqrt[3]{8}) + 2 \stackrel{?}{=} 0$$
$$(2)^2 - 3(2) + 2 \stackrel{?}{=} 0$$
$$4 - 6 + 2 \stackrel{?}{=} 0$$
$$0 = 0 \checkmark$$

$$x = 1: \quad (1)^{2/3} - 3(1)^{1/3} + 2 \stackrel{?}{=} 0$$
$$(\sqrt[3]{1})^2 - 3(\sqrt[3]{1}) + 2 \stackrel{?}{=} 0$$
$$1 - 3 + 2 \stackrel{?}{=} 0$$
$$0 = 0 \checkmark$$

Student Practice 3 Solve and check your solutions.

$$3x^{4/3} - 5x^{2/3} + 2 = 0$$

The exercises that appear in this section are somewhat difficult to solve. Part of the difficulty lies in the fact that the equations have different numbers of solutions. A fourth-degree equation like the one in Example 1 has four different solutions. Whereas a sixth-degree equation such as the one in Example 2 has only two solutions, some sixth-degree equations will have as many as six solutions. Although the equation that we examined in Example 3 has only two solutions, other equations with fractional exponents may have one solution or even no solution at all. It is good to take some time to carefully examine your work to determine that you have obtained the correct number of solutions.

A graphing program on a computer such as TI Interactive, Derive, or Maple can be very helpful in determining or verifying the solutions to these types of problems. Of course a graphing calculator can be most helpful, particularly in verifying the value of a solution and the number of solutions.

Optional Graphing Calculator Exploration: If you have a graphing calculator, verify the solutions for Example 3 by graphing the equation

$$y = x^{2/3} - 3x^{1/3} + 2.$$

Determine from your graph whether the curve does in fact cross the x-axis (that is, $y = 0$) when $x = 1$ and $x = 8$. You will have to carefully select the window so that you can see the behavior of the curve clearly. For this equation a useful window is $[-1, 12]$ by $[-2, 2]$. Remember that with most graphing calculators, you will need to surround the exponents with parentheses.

EXAMPLE 4 Solve and check your solutions. $2x^{1/2} = 5x^{1/4} + 12$

Solution

$$2x^{1/2} - 5x^{1/4} - 12 = 0 \qquad \text{Place in standard form.}$$
$$2y^2 - 5y - 12 = 0 \qquad \text{Replace } x^{1/4} \text{ by } y \text{ and } x^{1/2} \text{ by } y^2.$$
$$(2y + 3)(y - 4) = 0 \qquad \text{Factor.}$$
$$2y + 3 = 0 \quad \text{or} \quad y - 4 = 0$$
$$y = -\frac{3}{2} \qquad y = 4 \qquad \text{Solve for } y.$$
$$x^{1/4} = -\frac{3}{2} \qquad x^{1/4} = 4 \qquad \text{Replace } y \text{ by } x^{1/4}.$$
$$(x^{1/4})^4 = \left(-\frac{3}{2}\right)^4 \qquad (x^{1/4})^4 = (4)^4 \qquad \text{Solve for } x.$$
$$x = \frac{81}{16} \qquad x = 256$$

Continued on next page

Check.

$$x = \frac{81}{16}: \quad 2\left(\frac{81}{16}\right)^{1/2} - 5\left(\frac{81}{16}\right)^{1/4} - 12 \stackrel{?}{=} 0 \qquad x = 256: \quad 2(256)^{1/2} - 5(256)^{1/4} - 12 \stackrel{?}{=} 0$$

$$2\left(\frac{9}{4}\right) - 5\left(\frac{3}{2}\right) - 12 \stackrel{?}{=} 0 \qquad 2(16) - 5(4) - 12 \stackrel{?}{=} 0$$

$$\frac{9}{2} - \frac{15}{2} - 12 \stackrel{?}{=} 0 \qquad 32 - 20 - 12 \stackrel{?}{=} 0$$

$$-15 \neq 0 \qquad 0 = 0 \checkmark$$

$\frac{81}{16}$ is extraneous and not a valid solution. The only valid solution is 256.

Student Practice 4 Solve and check your solutions.

$$3x^{1/2} = 8x^{1/4} - 4$$

EXAMPLE 5 Solve and check your solutions. $x^{-2} = 5x^{-1} + 14$

Solution

$$x^{-2} - 5x^{-1} - 14 = 0 \qquad \text{Place in standard form.}$$

$$y^2 - 5y - 14 = 0 \qquad \text{Replace } x^{-1} \text{ by } y \text{ and } x^{-2} \text{ by } y^2.$$

$$(y - 7)(y + 2) = 0 \qquad \text{Factor.}$$

$$y - 7 = 0 \quad \text{or} \quad y + 2 = 0$$

$$y = 7 \qquad y = -2 \qquad \text{Solve for } y.$$

$$x^{-1} = 7 \qquad x^{-1} = -2 \qquad \text{Replace } y \text{ by } x^{-1}.$$

$$(x^{-1})^{-1} = (7)^{-1} \qquad (x^{-1})^{-1} = (-2)^{-1} \qquad \text{Solve for } x.$$

$$x = \frac{1}{7} \qquad x = -\frac{1}{2}$$

Check.

$$x = \frac{1}{7}: \left(\frac{1}{7}\right)^{-2} - 5\left(\frac{1}{7}\right)^{-1} - 14 \stackrel{?}{=} 0 \qquad x = -\frac{1}{2}: \left(-\frac{1}{2}\right)^{-2} - 5\left(-\frac{1}{2}\right)^{-1} - 14 \stackrel{?}{=} 0$$

$$49 - 5(7) - 14 \stackrel{?}{=} 0 \qquad 4 - 5(-2) - 14 \stackrel{?}{=} 0$$

$$0 = 0 \checkmark \qquad 0 = 0 \checkmark$$

Student Practice 5 Solve and check your solutions.

$$2x^{-2} + x^{-1} = 6$$

Although we have covered just five basic examples here, this substitution technique can be extended to other types of equations. In each case we substitute y for an appropriate expression in order to obtain a quadratic equation. The following table lists some substitutions that would be appropriate.

If You Want to Solve:	Then You Would Use the Substitution:	Resulting Equation
$x^4 - 13x^2 + 36 = 0$	$y = x^2$	$y^2 - 13y + 36 = 0$
$2x^6 - x^3 - 6 = 0$	$y = x^3$	$2y^2 - y - 6 = 0$
$x^{2/3} - 3x^{1/3} + 2 = 0$	$y = x^{1/3}$	$y^2 - 3y + 2 = 0$
$6(x - 1)^{-2} + (x - 1)^{-1} - 2 = 0$	$y = (x - 1)^{-1}$	$6y^2 + y - 2 = 0$
$(2x^2 + x)^2 + 4(2x^2 + x) + 3 = 0$	$y = 2x^2 + x$	$y^2 + 4y + 3 = 0$
$\left(\frac{1}{x - 1}\right)^2 + \frac{1}{x - 1} - 6 = 0$	$y = \frac{1}{x - 1}$	$y^2 + y - 6 = 0$
$2x - 5x^{1/2} + 2 = 0$	$y = x^{1/2}$	$2y^2 - 5y + 2 = 0$

Exercises like these appear in the exercise set.

8.3 Exercises

MyMathLab®

Watch the videos in MyMathLab

Download the MyDashBoard App

Solve. Express any nonreal complex numbers with i notation.

1. $x^4 - 9x^2 + 20 = 0$

2. $x^4 - 11x^2 + 18 = 0$

3. $x^4 + x^2 - 12 = 0$

4. $x^4 - 2x^2 - 8 = 0$

5. $4x^4 - x^2 - 3 = 0$

6. $3x^4 - 11x^2 - 4 = 0$

In exercises 7–10, find all valid real roots for each equation.

7. $x^6 - 7x^3 - 8 = 0$

8. $x^6 - 3x^3 - 4 = 0$

9. $x^6 - 3x^3 = 0$

10. $x^6 + 8x^3 = 0$

Solve for real roots.

11. $x^8 = 17x^4 - 16$

12. $x^8 = 3x^4$

13. $3x^8 + 13x^4 = 10$

14. $3x^8 - 10x^4 = 8$

Solve for real roots.

15. $x^{2/3} + x^{1/3} - 12 = 0$ **16.** $x^{2/3} + 2x^{1/3} - 8 = 0$ **17.** $12x^{2/3} + 5x^{1/3} - 2 = 0$ **18.** $2x^{2/3} - 7x^{1/3} - 4 = 0$

19. $2x^{1/2} - 5x^{1/4} - 3 = 0$ **20.** $3x^{1/2} + 13x^{1/4} - 10 = 0$ **21.** $4x^{1/2} + 11x^{1/4} - 3 = 0$

22. $2x^{1/2} - x^{1/4} - 1 = 0$ **23.** $x^{2/5} - x^{1/5} - 2 = 0$ **24.** $2x^{2/5} - 7x^{1/5} + 6 = 0$

Mixed Practice *In each exercise make an appropriate substitution in order to obtain a quadratic equation. Find the complex values for x.*

25. $x^6 - 5x^3 = 14$ **26.** $x^6 + 2x^3 = 15$

27. $(x^2 - x)^2 - 10(x^2 - x) + 24 = 0$

28. $(x^2 + 2x)^2 - (x^2 + 2x) - 12 = 0$

29. $x - 5x^{1/2} + 6 = 0$

30. $x - 5x^{1/2} - 36 = 0$

31. $x^{-2} + 3x^{-1} = 0$

32. $2x^{-2} - x^{-1} = 0$

33. $x^{-2} = 3x^{-1} + 10$

34. $x^{-2} = -7x^{-1} - 6$

To Think About

Solve. Find all valid real roots for each equation.

35. $15 - \dfrac{2x}{x - 1} = \dfrac{x^2}{x^2 - 2x + 1}$

36. $4 - \dfrac{x^3 + 1}{x^3 + 6} = \dfrac{x^3 - 3}{x^3 + 2}$

Cumulative Review

37. **[4.1.6]** Solve the system.

$$2x + 3y = 5$$
$$-5x - 2y = 4$$

38. **[6.3.1]** Simplify. $\dfrac{5 + \frac{2}{x}}{\frac{7}{3x} - 1}$

Multiply and simplify.

39. [7.4.1] $3\sqrt{2}(\sqrt{5} - 2\sqrt{6})$

40. [7.4.1] $(\sqrt{2} + \sqrt{6})(3\sqrt{2} - 2\sqrt{5})$

41. [1.2.2] ***Salary and Educational Attainment*** How much greater is the average annual salary of a man than a woman at each of the four levels of educational attainment shown in the graph? Express your answers to the nearest tenth of a percent.

42. [1.2.2] ***Salary and Educational Attainment*** Approximately how many extra years would an average woman with an associate's degree have to work to earn the same lifetime earnings as a male counterpart who worked for 30 years? Round your answer to the nearest tenth of a year.

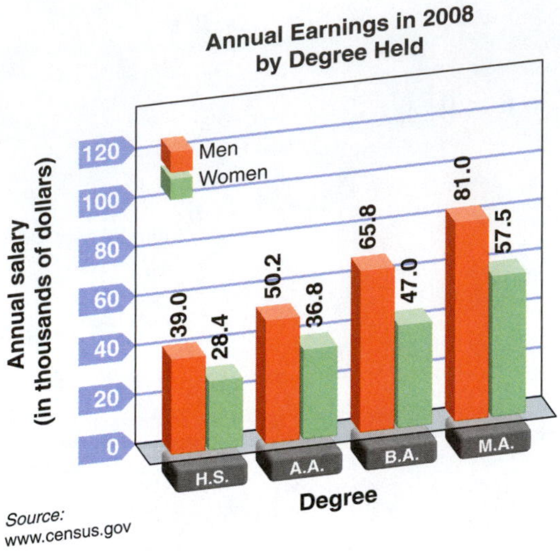

Quick Quiz 8.3 *Solve for any valid real roots.*

1. $x^4 - 18x^2 + 32 = 0$

2. $2x^{-2} - 3x^{-1} - 20 = 0$

3. $x^{2/3} - 4x^{1/3} - 5 = 0$

4. Concept Check Explain how you would solve the following. $x^8 - 6x^4 = 0$

How Am I Doing? Sections 8.1–8.3

How are you doing with your homework assignments in Sections 8.1 to 8.3? Do you feel you have mastered the material so far? Do you understand the concepts you have covered? Before you go further in the textbook take some time to do each of the following problems.

8.1

Simplify your answer to all problems.

1. Solve by the square root property. $2x^2 + 3 = 39$

2. Solve by the square root property. $(3x - 1)^2 = 28$

3. Solve by completing the square. $x^2 - 8x = -12$

4. Solve by completing the square. $2x^2 - 4x - 3 = 0$

8.2

Solve by the quadratic formula. Express any complex roots using i notation.

5. $2x^2 - 8x + 3 = 0$

6. $(x - 1)(x + 5) = 2$

7. $4x^2 = -12x - 17$

8. $5x^2 + 4x - 12 = 0$

9. $8x^2 + 3x = 3x^2 + 4x$

10. $4x^2 - 3x = -6$

11. $\dfrac{18}{x} + \dfrac{12}{x + 1} = 9$

8.3

Solve for any real roots and check your answers.

12. $x^6 - 7x^3 - 8 = 0$

13. $w^{4/3} - 6w^{2/3} + 8 = 0$

14. $x^8 = 5x^4 - 6$

15. $3x^{-2} = 11x^{-1} + 4$

1. __________
2. __________
3. __________
4. __________
5. __________
6. __________
7. __________
8. __________
9. __________
10. __________
11. __________
12. __________
13. __________
14. __________
15. __________

Now turn to page SA-20 for the answers to each of these problems. Each answer also includes a reference to the objective in which the problem is first taught. If you missed any of these problems, you should stop and review the Examples and Student Practice problems in the referenced objective. A little review now will help you master the material in the upcoming sections of the text.

8.4 Formulas and Applications

Student Learning Objectives

After studying this section, you will be able to:

1. Solve a quadratic equation containing several variables.
2. Solve problems requiring the use of the Pythagorean Theorem.
3. Solve applied problems requiring the use of a quadratic equation.

① Solving a Quadratic Equation Containing Several Variables

In mathematics, physics, and engineering, we must often solve an equation for a variable in terms of other variables. You recall we solved linear equations in several variables in Section 2.2. We now examine several cases where the variable that we are solving for is squared. If the variable we are solving for is squared, and there is no other term containing that variable, then the equation can be solved using the square root property.

EXAMPLE 1 The surface area of a sphere is given by $A = 4\pi r^2$. Solve this equation for r. (You do not need to rationalize the denominator.)

Solution

$$A = 4\pi r^2$$

$$\frac{A}{4\pi} = r^2$$

$$\pm\sqrt{\frac{A}{4\pi}} = r \quad \text{Use the square root property.}$$

$$\pm\frac{1}{2}\sqrt{\frac{A}{\pi}} = r \quad \text{Simplify.}$$

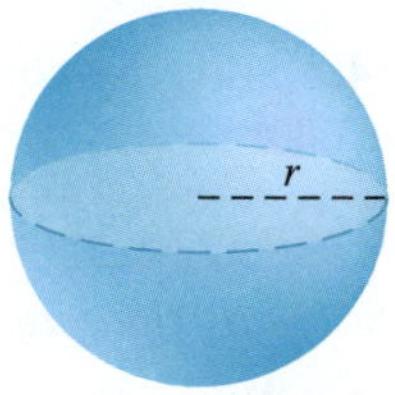

Since the radius of a sphere must be a positive value, we use only the principal root.

$$r = \frac{1}{2}\sqrt{\frac{A}{\pi}}$$

Student Practice 1 The volume of a cylindrical cone is $V = \frac{1}{3}\pi r^2 h$. Solve this equation for r. (You do not need to rationalize the denominator.)

NOTE TO STUDENT: Fully worked-out solutions to all of the Student Practice problems can be found at the back of the text starting at page SP-1.

Some quadratic equations containing many variables can be solved for one variable by factoring.

EXAMPLE 2 Solve for y. $y^2 - 2yz - 15z^2 = 0$

Solution

$$(y + 3z)(y - 5z) = 0 \quad \text{Factor.}$$

$$y + 3z = 0 \quad \text{or} \quad y - 5z = 0 \quad \text{Set each factor equal to 0.}$$

$$y = -3z \qquad y = 5z \quad \text{Solve for } y.$$

Student Practice 2 Solve for y.

$$2y^2 + 9wy + 7w^2 = 0$$

Sometimes the quadratic formula is required in order to solve the equation.

EXAMPLE 3 Solve for x. $2x^2 + 3wx - 4z = 0$

Solution We use the quadratic formula where the variable is considered to be x and the letters w and z are considered constants. Thus, $a = 2$, $b = 3w$, and $c = -4z$.

$$x = \frac{-b \pm \sqrt{b^2 - 4ac}}{2a}$$

$$x = \frac{-3w \pm \sqrt{(3w)^2 - 4(2)(-4z)}}{2(2)} = \frac{-3w \pm \sqrt{9w^2 + 32z}}{4}$$

Note that this answer cannot be simplified any further.

Student Practice 3 Solve for y.

$$3y^2 + 2fy - 7g = 0$$

EXAMPLE 4 The formula for the curved surface area S of a right circular cone of altitude h and with base of radius r is $S = \pi r\sqrt{r^2 + h^2}$. Solve for r^2.

Solution

$$S = \pi r\sqrt{r^2 + h^2}$$

$$\frac{S}{\pi r} = \sqrt{r^2 + h^2} \quad \text{Isolate the radical.}$$

$$\frac{S^2}{\pi^2 r^2} = r^2 + h^2 \quad \text{Square both sides.}$$

$$\frac{S^2}{\pi^2} = r^4 + h^2 r^2 \quad \text{Multiply each term by } r^2.$$

$$0 = r^4 + h^2 r^2 - \frac{S^2}{\pi^2} \quad \text{Subtract } \frac{S^2}{\pi^2} \text{ from both sides.}$$

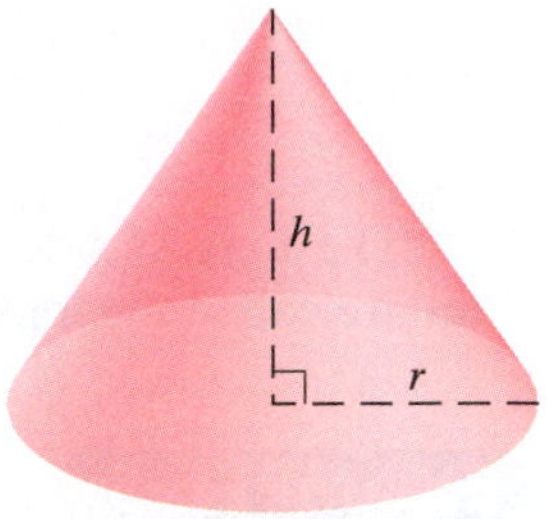

This equation is quadratic in form. If we let $y = r^2$, then we have

$$0 = y^2 + h^2 y - \frac{S^2}{\pi^2}.$$

By the quadratic formula we have the following:

$$y = \frac{-h^2 \pm \sqrt{(h^2)^2 - 4(1)\left(-\frac{S^2}{\pi^2}\right)}}{2}$$

$$y = \frac{-h^2 \pm \sqrt{\frac{\pi^2 h^4}{\pi^2} + \frac{4S^2}{\pi^2}}}{2} \quad \text{Obtain a common denominator under the radical.}$$

$$y = \frac{-h^2 \pm \frac{1}{\pi}\sqrt{\pi^2 h^4 + 4S^2}}{2} \quad \text{Factor out } \frac{1}{\pi^2} \text{ from both terms under the radical. } \sqrt{\frac{1}{\pi^2}} = \frac{1}{\pi} \text{ is written outside the radical.}$$

$$y = \frac{-\pi h^2 \pm \sqrt{\pi^2 h^4 + 4S^2}}{2\pi} \quad \text{Multiply the rational expression by } \frac{\pi}{\pi}.$$

Since $y = r^2$, we have

$$r^2 = \frac{-\pi h^2 \pm \sqrt{\pi^2 h^4 + 4S^2}}{2\pi}.$$

Student Practice 4 The formula for the number of diagonals d in a polygon with n sides is $d = \dfrac{n^2 - 3n}{2}$. Solve for n.

② Solving Problems Requiring the Use of the Pythagorean Theorem

A very useful formula is the Pythagorean Theorem for right triangles.

PYTHAGOREAN THEOREM

If c is the length of the longest side of a right triangle and a and b are the lengths of the other two sides, then $a^2 + b^2 = c^2$.

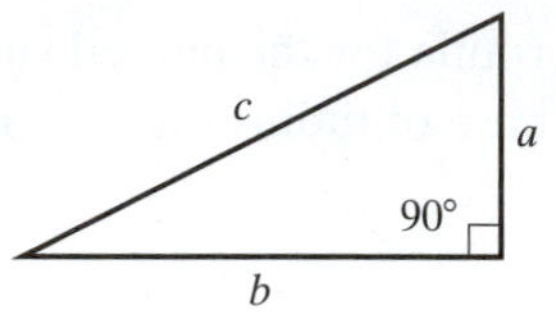

The longest side of a right triangle is called the **hypotenuse.** The other two sides are called the **legs** of the triangle.

EXAMPLE 5

(a) Solve the Pythagorean Theorem $a^2 + b^2 = c^2$ for a.

(b) Find the value of a if $c = 13$ and $b = 5$.

Solution

(a) $a^2 = c^2 - b^2$ Subtract b^2 from each side.

$a = \pm\sqrt{c^2 - b^2}$ Use the square root property.

a, b, and c must be positive numbers because they represent lengths, so we use only the positive root, $a = \sqrt{c^2 - b^2}$.

(b) $a = \sqrt{c^2 - b^2}$

$$a = \sqrt{(13)^2 - (5)^2} = \sqrt{169 - 25} = \sqrt{144} = 12$$

Thus, $a = 12$.

Student Practice 5

(a) Solve the Pythagorean Theorem for b.

(b) Find the value of b if $c = 26$ and $a = 24$.

There are many practical uses for the Pythagorean Theorem. For hundreds of years, it was used in land surveying and in ship navigation. So, for hundreds of years students learned it in public schools and were shown applications in surveying and navigation. In almost any situation where you have a right triangle and you know two sides and need to find the third side, you can use the Pythagorean Theorem.

See if you can see how it is used in Example 6. Notice that it is helpful right at the beginning of the problem to draw a picture of the right triangle and label information that you know.

▲ **EXAMPLE 6** The perimeter of a triangular piece of land is 12 miles. One leg of the triangle is 1 mile longer than the other leg. Find the length of each boundary of the land if the triangle is a right triangle.

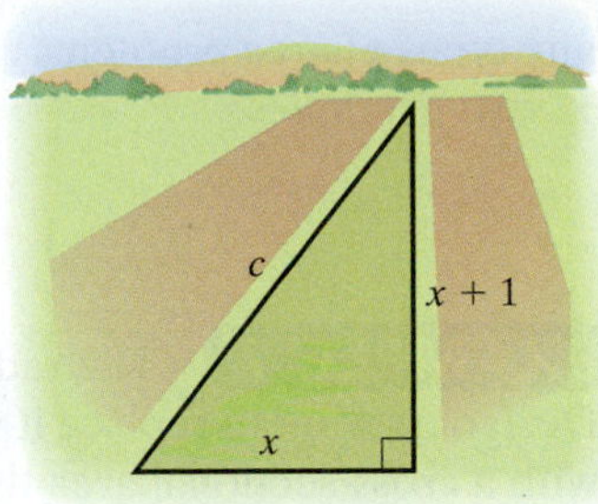

Solution

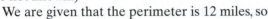

1. ***Understand the problem.*** Draw a picture of the piece of land and label the sides of the triangle.
2. ***Write an equation.*** We can use the Pythagorean Theorem. First, we want only one variable in our equation. (Right now, both c and x are not known.)

We are given that the perimeter is 12 miles, so

$$x + (x + 1) + c = 12.$$

Thus,

$$c = -2x + 11.$$

By the Pythagorean Theorem,

$$x^2 + (x + 1)^2 = (-2x + 11)^2.$$

3. ***Solve the equation and state the answer.***

$$x^2 + (x + 1)^2 = (-2x + 11)^2$$
$$x^2 + x^2 + 2x + 1 = 4x^2 - 44x + 121$$
$$0 = 2x^2 - 46x + 120$$
$$0 = x^2 - 23x + 60$$

By the quadratic formula, we have the following:

$$x = \frac{-(-23) \pm \sqrt{(-23)^2 - 4(1)(60)}}{2(1)}$$
$$x = \frac{23 \pm \sqrt{289}}{2}$$
$$x = \frac{23 \pm 17}{2}$$
$$x = \frac{40}{2} = 20 \quad \text{or} \quad x = \frac{6}{2} = 3.$$

The answer $x = 20$ cannot be right because the perimeter (the sum of *all* the sides) is only 12. The only answer that makes sense is $x = 3$. Thus, the sides of the triangle are $x = 3$, $x + 1 = 3 + 1 = 4$, and $-2x + 11 = -2(3) + 11 = 5$. The longest boundary of this triangular piece of land is 5 miles. The other two boundaries are 4 miles and 3 miles.

Notice that we could have factored the quadratic equation instead of using the quadratic formula. $x^2 - 23x + 60 = 0$ can be written as $(x - 20)(x - 3) = 0$.

4. ***Check.***

Is the perimeter 12 miles?
$5 + 4 + 3 = 12$ ✓

Is one leg 1 mile longer than the other?
$4 = 3 + 1$ ✓

▲ **Student Practice 6** The perimeter of a triangular piece of land is 30 miles. One leg of the triangle is 7 miles shorter than the other leg. Find the length of each boundary of the land if the triangle is a right triangle.

③ Solving Applied Problems Requiring the Use of a Quadratic Equation

Many types of area problems can be solved with quadratic equations as shown in the next two examples.

EXAMPLE 7 The radius of an old circular pipe under a roadbed is 10 feet. Designers want to replace it with a smaller pipe and have decided they can use one with a cross-sectional area that is 36π square feet smaller. What should the radius of the new pipe be?

Solution First we need the formula for the area of a circle,

$$A = \pi r^2,$$

where A is the area and r is the radius. The area of the cross section of the old pipe is as follows:

$$A_{\text{old}} = \pi(10)^2 = 100\pi$$

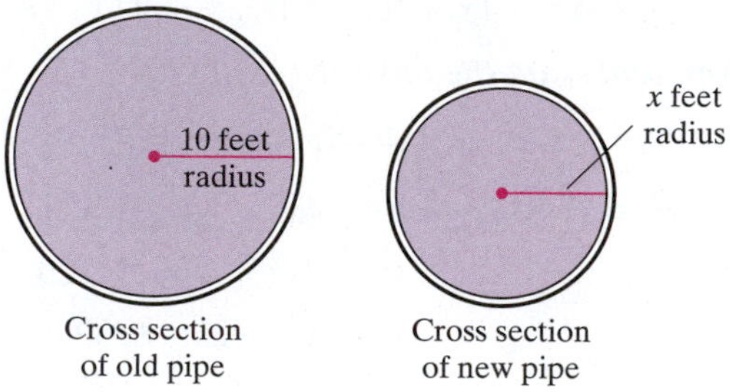

Let x = the radius of the new pipe.

$$(\text{area of old pipe}) - (\text{area of new pipe}) = 36\pi$$
$$100\pi - \pi x^2 = 36\pi$$

$64\pi = \pi x^2$ Add πx^2 to each side and subtract 36π from each side.

$\dfrac{64\pi}{\pi} = \dfrac{\pi x^2}{\pi}$ Divide each side by π.

$64 = x^2$

$\pm 8 = x$ Use the square root property.

Since the radius must be positive, we select $x = 8$. The radius of the new pipe is 8 feet. Check to verify this solution.

Student Practice 7 Redo Example 7 when the radius of the pipe under the roadbed is 6 feet and the designers want to replace it with a pipe that has a cross-sectional area that is 45π square feet larger. What should the radius of the new pipe be?

▲ **EXAMPLE 8** A triangular sign marks the edge of the rocks in Rockport Harbor. The sign has an area of 35 square meters. Find the base and altitude of this triangular sign if the base is 3 meters shorter than the altitude.

Solution The area of a triangle is given by

$$A = \frac{1}{2}ab.$$

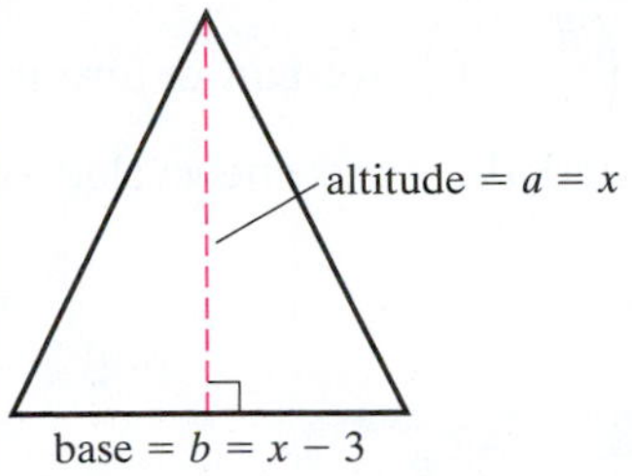

Let x = the length in meters of the altitude. Then $x - 3$ = the length in meters of the base.

$35 = \frac{1}{2}x(x - 3)$	Replace A (area) by 35, a (altitude) by x, and b (base) by $x - 3$.
$70 = x(x - 3)$	Multiply each side by 2.
$70 = x^2 - 3x$	Use the distributive property.
$0 = x^2 - 3x - 70$	Subtract 70 from each side.
$0 = (x - 10)(x + 7)$	
$x = 10$ or $x = -7$	

The length of a side of a triangle must be a positive number, so we disregard -7. Thus,

$$\text{altitude} = x = 10 \text{ meters and}$$
$$\text{base} = x - 3 = 7 \text{ meters.}$$

The check is left to the student.

▲ **Student Practice 8** The length of a rectangle is 3 feet shorter than twice the width. The area of the rectangle is 54 square feet. Find the dimensions of the rectangle.

We will now examine a few word problems that require the use of the formula distance = (rate)(time) or $d = rt$.

EXAMPLE 9 When Barbara was training for a bicycle race, she rode a total of 135 miles on Monday and Tuesday. On Monday she rode for 75 miles in the rain. On Tuesday she rode 5 miles per hour faster because the weather was better. Her total cycling time for the 2 days was 8 hours. Find her speed for each day.

Solution We can find each distance. If Barbara rode 75 miles on Monday and a total of 135 miles during the 2 days, then she rode $135 - 75 = 60$ miles on Tuesday.

Let $x =$ the cycling rate in miles per hour on Monday. Since Barbara rode 5 miles per hour faster on Tuesday, $x + 5 =$ the cycling rate in miles per hour on Tuesday.

Since distance divided by rate is equal to time $\left(\frac{d}{r} = t\right)$, we can determine that the time Barbara cycled on Monday was $\frac{75}{x}$ and the time she cycled on Tuesday was $\frac{60}{x+5}$.

Day	Distance	Rate	Time
Monday	75	x	$\frac{75}{x}$
Tuesday	60	$x + 5$	$\frac{60}{x+5}$
Total	135	(not used)	8

Since the total cycling time was 8 hours, we have the following:

$$\text{time cycling Monday} + \text{time cycling Tuesday} = 8 \text{ hours}$$
$$\frac{75}{x} + \frac{60}{x+5} = 8$$

The LCD of this equation is $x(x + 5)$. Multiply each term by the LCD.

$$\cancel{x}(x+5)\left(\frac{75}{\cancel{x}}\right) + x\cancel{(x+5)}\left(\frac{60}{\cancel{x+5}}\right) = x(x+5)(8)$$
$$75(x + 5) + 60x = 8x(x + 5)$$
$$75x + 375 + 60x = 8x^2 + 40x$$
$$0 = 8x^2 - 95x - 375$$
$$0 = (x - 15)(8x + 25)$$
$$x - 15 = 0 \quad \text{or} \quad 8x + 25 = 0$$
$$x = 15 \qquad\qquad x = -\frac{25}{8}$$

We disregard the negative answer. The cyclist did not have a negative rate of speed—unless she was moving backward! Thus, $x = 15$. So Barbara's rate of speed on Monday was 15 mph, and her rate of speed on Tuesday was $x + 5 = 15 + 5 = 20$ mph.

Student Practice 9 Carlos traveled in his car at a constant speed on a secondary road for 150 miles. Then he traveled 10 mph faster on a better road for 240 miles. If Carlos drove for 7 hours, find the car's speed for each part of the trip.

8.4 Exercises

MyMathLab®

Watch the videos in MyMathLab

Download the MyDashBoard App

Solve for the variable specified. Assume that all other variables are nonzero.

1. $S = 16t^2$; for t

2. $E = mc^2$; for c

3. $S = 9\pi r^2$; for r

4. $A = \frac{1}{2}r^2\theta$; for r

5. $5M = \frac{1}{2}by^2$; for y

6. $3G = \frac{5}{2}hx^2$; for x

7. $4(y^2 + w) - 5 = 7R$; for y

8. $9x^2 - 2 = 3B$; for x

9. $Q = \frac{3mwM^2}{2c}$; for M

10. $H = \frac{5abT^2}{7k}$; for T

11. $V = \pi(r^2 + R^2)h$; for r

12. $H = b(a^2 + w^2)$; for w

13. $7bx^2 - 3ax = 0$; for x

14. $2x^2 - 5ax = 0$; for x

15. $P = EI - RI^2$; for I

16. $A = P(1 + r)^2$; for r

17. $9w^2 + 5tw - 2 = 0$; for w

18. $6w^2 - sw + 4 = 0$; for w

19. $S = 2\pi rh + \pi r^2$; for r

20. $B = 3abx^2 - 5x$; for x

21. $(a + 1)x^2 + 5x + 2w = 0$; for x

22. $(b - 2)x^2 - 3x + 5y = 0$; for x

In exercises 23–28, use the Pythagorean Theorem to find the unknown side(s).

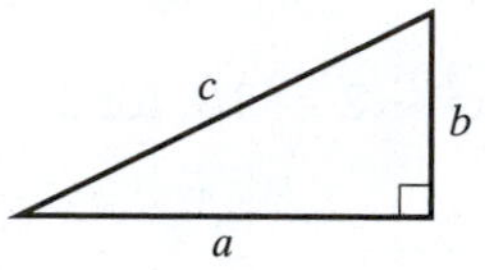

▲ **23.** $c = 6, a = 4$; find b

▲ **24.** $b = 5, c = 10$; find a

▲ **25.** $c = \sqrt{34}, b = \sqrt{19}$; find a

▲ **26.** $c = \sqrt{21}, a = \sqrt{5}$; find b

▲ **27.** $c = 12, b = 2a$; find b and a

▲ **28.** $c = 15, a = 2b$; find b and a

Applications

▲ **29.** ***Brace for a Shelf*** A brace for a shelf has the shape of a right triangle. Its hypotenuse is 10 inches long and the two legs are equal in length. How long are the legs of the triangle?

▲ **30.** ***Tent*** The two sides of a tent are equal in length and meet to form a right triangle. If the width of the floor of the tent is 8 feet, what is the length of the two sides?

▲ **31.** ***College Parking Lot*** Knox College is creating a new rectangular parking lot. The length is 0.07 mile longer than the width and the area of the parking lot is 0.026 square mile. Find the length and width of the parking lot.

▲ **32.** ***Computer Chip*** A high-tech company is producing a new rectangular computer chip. The length is 0.3 centimeter less than twice the width. The chip has an area of 1.04 square centimeters. Find the length and width of the chip.

▲**33.** ***Barn*** The area of a rectangular wall of a barn is 126 square feet. Its length is 4 feet longer than twice its width. Find the length and width of the wall of the barn.

▲**34.** ***Tennis Court*** The area of a doubles tennis court is 312 square yards. Its length is 2 yards longer than twice its width. Find the length and width of the tennis court.

▲**35.** ***Triangular Flag*** The area of a triangular flag is 72 square centimeters. Its altitude is 2 centimeters longer than twice its base. Find the lengths of the altitude and the base.

▲**36.** ***Children's Playground*** A children's playground is triangular in shape. Its altitude is 2 yards shorter than its base. The area of the playground is 60 square yards. Find the base and altitude of the playground.

37. ***Driving Speed*** Roberto drove at a constant speed in a rainstorm for 225 miles. He took a break, and the rain stopped. He then drove 150 miles at a speed that was 5 miles per hour faster than his previous speed. If he drove for 8 hours, find the car's speed for each part of the trip.

38. ***Driving Speed*** Benita traveled at a constant speed on a gravel road for 40 miles. She then traveled 10 miles per hour faster on a paved road for 150 miles. If she drove for 4 hours, find the car's speed for each part of the trip.

39. ***Commuter Traffic*** Bob drove from home to work at 50 mph. After work the traffic was heavier, and he drove home at 45 mph. His driving time to and from work was 1 hour and 16 minutes. How far does he live from his job?

40. ***Commercial Trucking*** A driver drove his heavily loaded truck from the company warehouse to a delivery point at 35 mph. He unloaded the truck and drove back to the warehouse at 45 mph. The total trip took 5 hours and 20 minutes. How far is the delivery point from the warehouse?

Incarcerated Adults *The number of incarcerated adults N (measured in thousands) in the United States can be approximated by the quadratic equation* $N = -3.9x^2 + 76.6x + 1927$, *where x is the number of years since 2000. For example, when* $x = 1$, $N = 1999.7$. *This tells us that in 2001 there were approximately 1,999,700 incarcerated adults in the United States. Use this equation to answer the following questions.* (*Source:* www.bjs.ojp.usdoj.gov)

41. How many incarcerated adults does the equation indicate there were in the year 2003?

42. How many incarcerated adults does the equation predict there will be in the year 2012?

43. In what year was the number of incarcerated adults 2,171,000?

44. Determine the year in which the number of incarcerated adults is expected to be 2,263,700.

Cumulative Review *Rationalize the denominators.*

45. **[7.4.3]** $\dfrac{4}{\sqrt{3x}}$

46. **[7.4.3]** $\dfrac{5\sqrt{6}}{2\sqrt{5}}$

47. **[7.4.3]** $\dfrac{3}{\sqrt{x} + \sqrt{y}}$

48. **[7.4.3]** $\dfrac{2\sqrt{3}}{\sqrt{3} - \sqrt{6}}$

Quick Quiz 8.4

1. Solve for y.

$$H = \frac{5ab}{y^2}$$

2. Solve for z.

$$6z^2 + 7yz - 5w = 0$$

3. The area of a rectangular field is 175 square yards. The length of the field is 4 yards longer than triple the width of the field. What is the width of the field? What is the length of the field?

4. **Concept Check** In a right triangle the hypotenuse measures 12 meters. One of the legs of the triangle is three times as long as the other. Explain how you would find the length of each leg.

8.5 Quadratic Functions

① Finding the Vertex and Intercepts of a Quadratic Function

In Section 3.6 we graphed functions such as $p(x) = x^2$ and $g(x) = (x + 2)^2$. We will now study quadratic functions in more detail.

DEFINITION OF A QUADRATIC FUNCTION

A **quadratic function** is a function of the form

$f(x) = ax^2 + bx + c$, where a, b, and c are real numbers and $a \neq 0$.

Graphs of quadratic functions written in this form will be parabolas opening upward if $a > 0$ or downward if $a < 0$. The **vertex** of a parabola is the lowest point on a parabola opening upward or the highest point on a parabola opening downward. The vertex will occur at $x = \frac{-b}{2a}$. To find the y-value, or $f(x)$, when $x = \frac{-b}{2a}$, we find $f\left(\frac{-b}{2a}\right)$. Therefore, we can say that a quadratic function has its vertex at $\left(\frac{-b}{2a}, f\left(\frac{-b}{2a}\right)\right)$.

It is helpful to know the x-intercepts and the y-intercept when graphing a quadratic function.

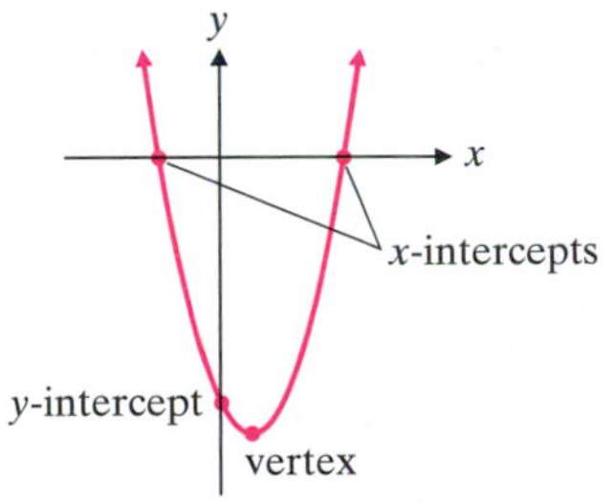

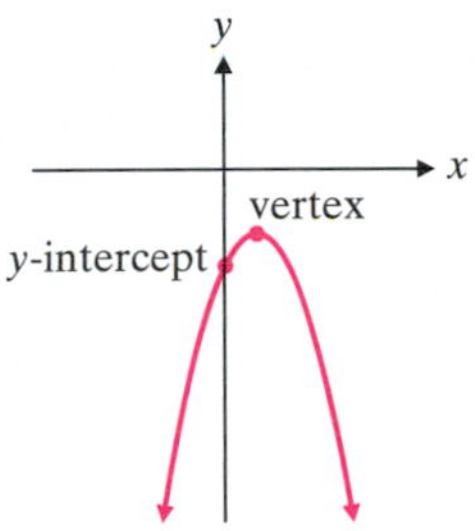

A quadratic function will always have exactly one y-intercept. However, it may have zero, one, or two x-intercepts. Why?

WHEN GRAPHING QUADRATIC FUNCTIONS OF THE FORM $f(x) = ax^2 + bx + c$, $a \neq 0$

1. The coordinates of the vertex are $\left(\frac{-b}{2a}, f\left(\frac{-b}{2a}\right)\right)$.
2. The y-intercept is at $f(0)$.
3. The x-intercepts (if they exist) occur where $f(x) = 0$. They can always be found with the quadratic formula and can sometimes be found by factoring.

Since we may replace y by $f(x)$, the graph is equivalent to the graph of $y = ax^2 + bx + c$.

Student Learning Objectives

After studying this section, you will be able to:

① **Find the vertex and intercepts of a quadratic function.**

② **Graph a quadratic function.**

Graphing Calculator

Finding the x-intercepts and the Vertex

To find the intercepts of the quadratic function

$$f(x) = x^2 - 4x + 3,$$

graph $y = x^2 - 4x + 3$ on a graphing calculator using an appropriate window.

Display:

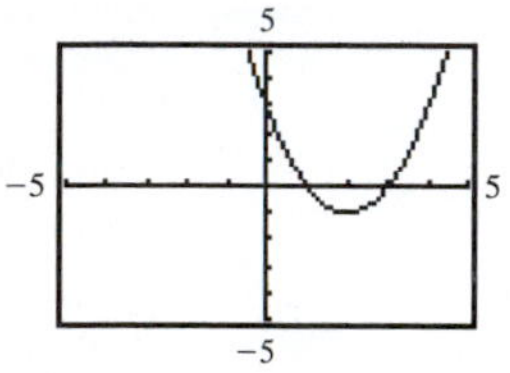

Next you can use the Trace and Zoom features or zero command of your calculator to find the x-intercepts.

You can also use the Trace and Zoom features to determine the vertex.

Some calculators have a feature that will calculate the maximum or minimum point on the graph. Use the feature that calculates the minimum point to find the vertex of $f(x) = x^2 - 4x + 3$.

Display:

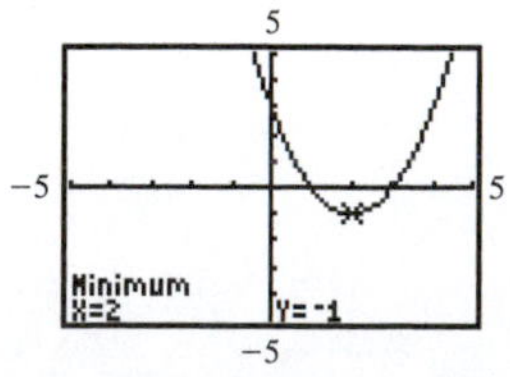

Thus, the vertex is (2, −1).

***NOTE TO STUDENT:** Fully worked-out solutions to all of the Student Practice problems can be found at the back of the text starting at page SP-1.*

EXAMPLE 1 Find the coordinates of the vertex and the intercepts of the quadratic function $f(x) = x^2 - 8x + 15$.

Solution For this function $a = 1$, $b = -8$, and $c = 15$.

Step 1 The vertex occurs at $x = \frac{-b}{2a}$. Thus,

$$x = \frac{-(-8)}{2(1)} = \frac{8}{2} = 4.$$

The vertex has an x-coordinate of 4. To find the y-coordinate, we evaluate $f(4)$.

$$f(4) = 4^2 - 8(4) + 15 = 16 - 32 + 15 = -1$$

Thus, the vertex is $(4, -1)$.

Step 2 The y-intercept is at $f(0)$. We evaluate $f(0)$ to find the y-coordinate when x is 0.

$$f(0) = 0^2 - 8(0) + 15 = 15$$

The y-intercept is $(0, 15)$.

Step 3 If there are x-intercepts, they will occur when $f(x) = 0$—that is, when $x^2 - 8x + 15 = 0$. We solve for x.

$$(x - 5)(x - 3) = 0$$

$$x - 5 = 0 \qquad x - 3 = 0$$

$$x = 5 \qquad x = 3$$

Thus, we conclude that the x-intercepts are $(5, 0)$ and $(3, 0)$.

We list these four important points of the function in table form.

Name	x	f(x)
Vertex	4	−1
y-intercept	0	15
x-intercept	5	0
x-intercept	3	0

Student Practice 1 Find the coordinates of the vertex and the intercepts of the quadratic function $f(x) = x^2 - 6x + 5$.

② Graphing a Quadratic Function

It is helpful to find the vertex and the intercepts of a quadratic function before graphing it.

EXAMPLE 2 Find the vertex and the intercepts, and then graph the function $f(x) = x^2 + 2x - 4$.

Solution Here $a = 1, b = 2$, and $c = -4$. Since $a > 0$, the parabola opens *upward*.

Step 1 We find the vertex.

$$x = \frac{-b}{2a} = \frac{-2}{2(1)} = \frac{-2}{2} = -1$$

$$f(-1) = (-1)^2 + 2(-1) - 4 = 1 + (-2) - 4 = -5$$

The vertex is $(-1, -5)$.

Step 2 We find the y-intercept. The y-intercept is at $f(0)$.

$$f(0) = (0)^2 + 2(0) - 4 = -4$$

The y-intercept is $(0, -4)$.

Step 3 We find the x-intercepts. The x-intercepts occur when $f(x) = 0$. Thus, we solve $x^2 + 2x - 4 = 0$ for x. We cannot factor this equation, so we use the quadratic formula.

$$x = \frac{-b \pm \sqrt{b^2 - 4ac}}{2a} = \frac{-2 \pm \sqrt{2^2 - 4(1)(-4)}}{2(1)} = \frac{-2 \pm \sqrt{20}}{2} = -1 \pm \sqrt{5}$$

To aid our graphing, we will approximate the value of x to the nearest tenth by using a square root table or a scientific calculator.

1 [+/−] [+] 5 [√] [=] 1.236068

$x \approx 1.2$

1 [+/−] [−] 5 [√] [=] −3.236068

$x \approx -3.2$

The x-intercepts are approximately $(-3.2, 0)$ and $(1.2, 0)$.

We have found that the vertex is $(-1, -5)$; the y-intercept is $(0, -4)$; and the x-intercepts are approximately $(-3.2, 0)$ and $(1.2, 0)$. We connect these points by a smooth curve to graph the parabola.

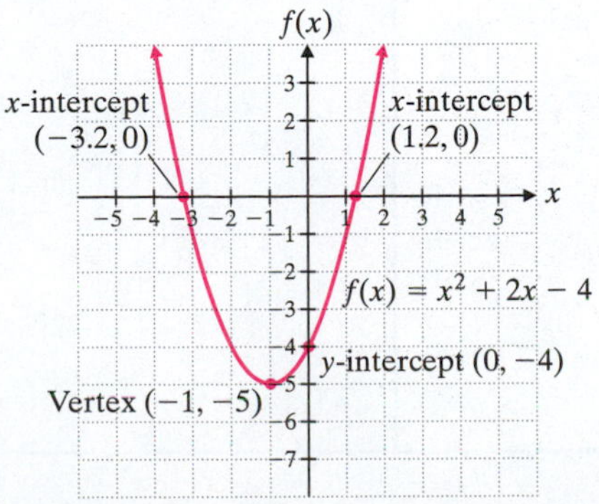

Student Practice 2 Find the vertex and the intercepts, and then graph the function $g(x) = x^2 - 2x - 2$.

Student Practice 2

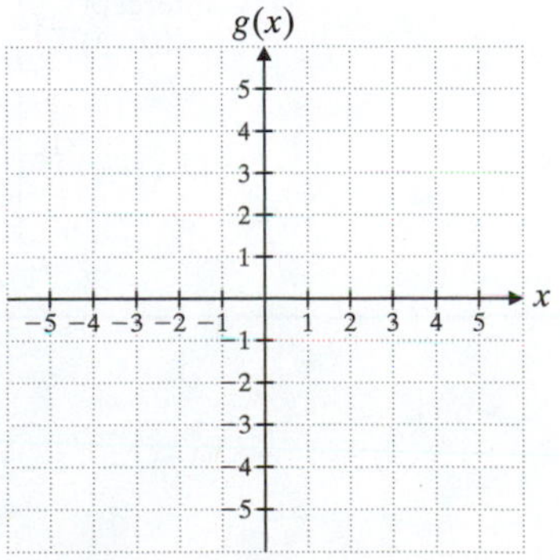

EXAMPLE 3 Find the vertex and the intercepts, and then graph the function $f(x) = -2x^2 + 4x - 3$.

Solution Here $a = -2, b = 4$, and $c = -3$. Since $a < 0$, the parabola opens *downward*.

The vertex occurs at $x = \frac{-b}{2a}$.

$$x = \frac{-4}{2(-2)} = \frac{-4}{-4} = 1$$

$$f(1) = -2(1)^2 + 4(1) - 3 = -2 + 4 - 3 = -1$$

The vertex is $(1, -1)$.

The y-intercept is at $f(0)$.

$$f(0) = -2(0)^2 + 4(0) - 3 = -3$$

The y-intercept is $(0, -3)$.

If there are any x-intercepts, they will occur when $f(x) = 0$. We use the quadratic formula to solve $-2x^2 + 4x - 3 = 0$ for x.

$$x = \frac{-4 \pm \sqrt{4^2 - 4(-2)(-3)}}{2(-2)} = \frac{-4 \pm \sqrt{-8}}{-4}$$

Because $\sqrt{-8}$ yields an imaginary number, there are no real roots. Thus, there are no x-intercepts for the graph of the function. That is, the graph does not intersect the x-axis.

We know that the parabola opens *downward*. Thus, the vertex is a maximum value at $(1, -1)$. Since this graph has no x-intercepts, we will look for three additional points to help us in drawing the graph. We try $f(2)$, $f(3)$, and $f(-1)$.

$$f(2) = -2(2)^2 + 4(2) - 3 = -8 + 8 - 3 = -3$$
$$f(3) = -2(3)^2 + 4(3) - 3 = -18 + 12 - 3 = -9$$
$$f(-1) = -2(-1)^2 + 4(-1) - 3 = -2 - 4 - 3 = -9$$

We plot the vertex, the y-intercept, and the points $(2, -3)$, $(3, -9)$, and $(-1, -9)$.

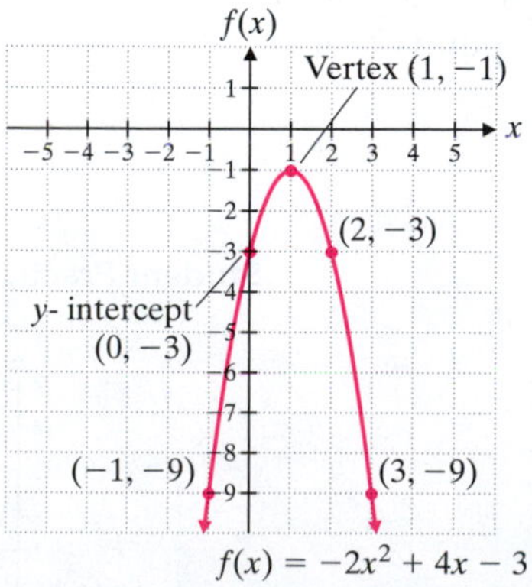

Student Practice 3 Find the vertex and the intercepts, and then graph the function $g(x) = -2x^2 - 8x - 6$.

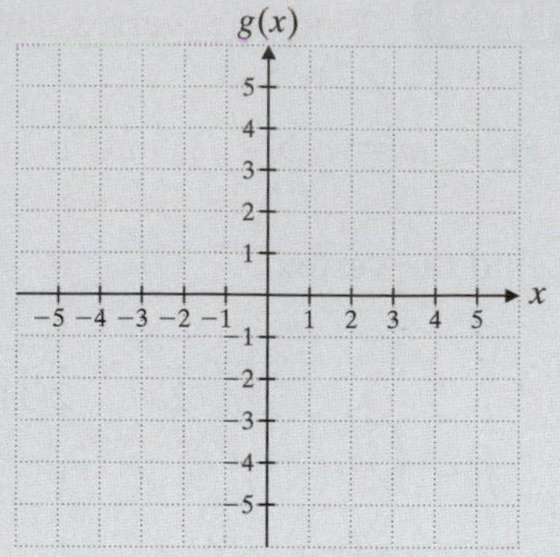

8.5 Exercises

MyMathLab® Watch the videos in MyMathLab Download the MyDashBoard App

Find the coordinates of the vertex, the y-intercept, and the x-intercepts (if any exist) of each of the following quadratic functions. When necessary, approximate the x-intercepts to the nearest tenth.

1. $f(x) = x^2 - 2x - 8$

2. $f(x) = x^2 - 4x - 5$

3. $g(x) = -x^2 - 8x + 9$

4. $g(x) = -x^2 + 4x + 21$

5. $p(x) = 3x^2 + 12x + 3$

6. $p(x) = 2x^2 + 4x + 1$

7. $r(x) = -3x^2 - 2x - 6$

8. $f(x) = -2x^2 + 3x - 2$

9. $f(x) = 2x^2 + 2x - 4$

10. $f(x) = 5x^2 + 2x - 3$

In each of the following exercises, find the vertex, the y-intercept, and the x-intercepts (if any exist), and then graph the function.

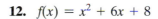

11. $f(x) = x^2 - 6x + 8$

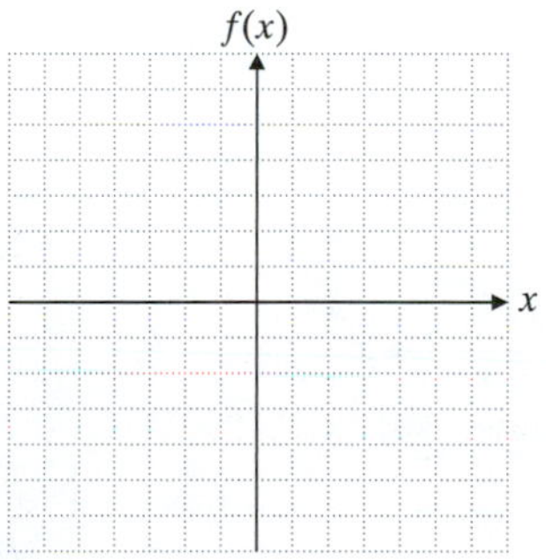

12. $f(x) = x^2 + 6x + 8$

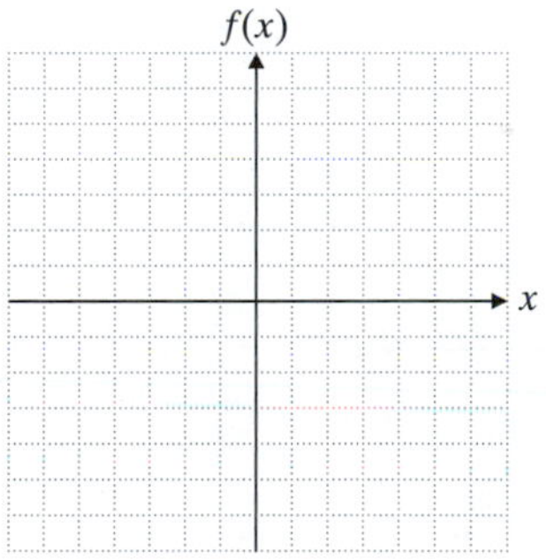

13. $g(x) = x^2 + 2x - 8$

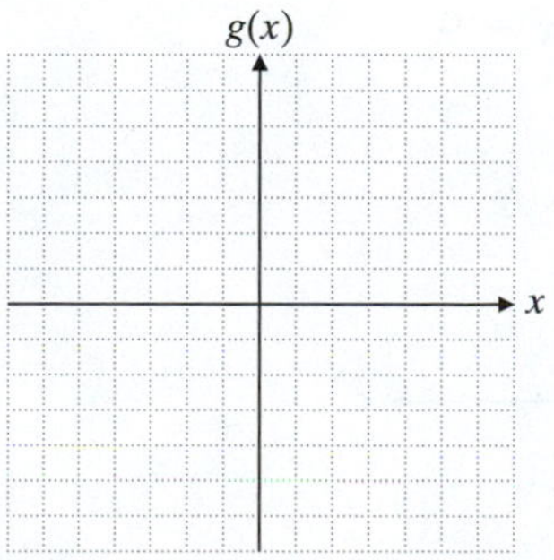

14. $g(x) = x^2 - 2x - 8$

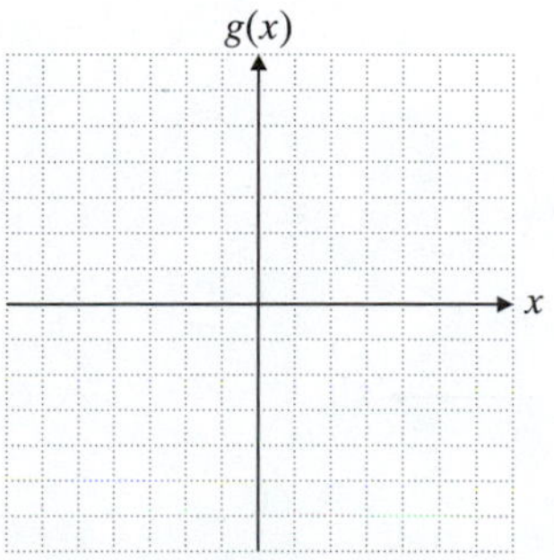

15. $p(x) = -x^2 + 8x - 12$

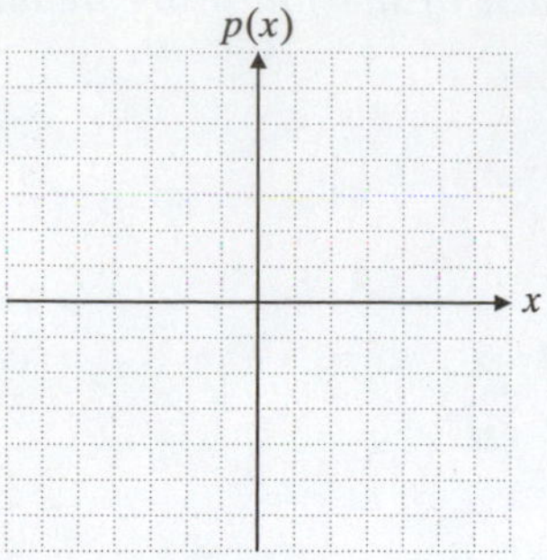

16. $p(x) = -x^2 - 8x - 12$

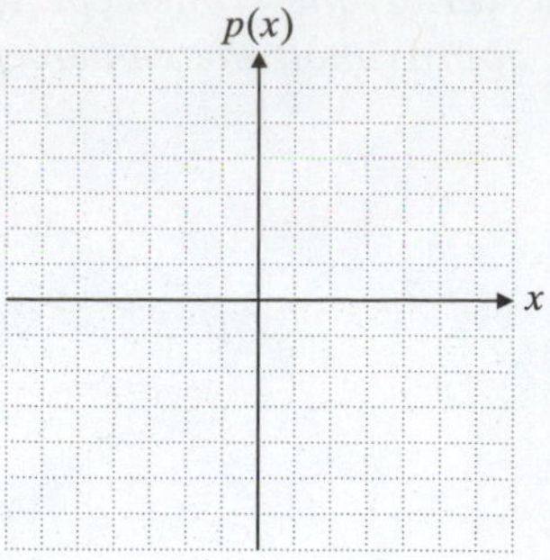

17. $r(x) = 3x^2 + 6x + 4$

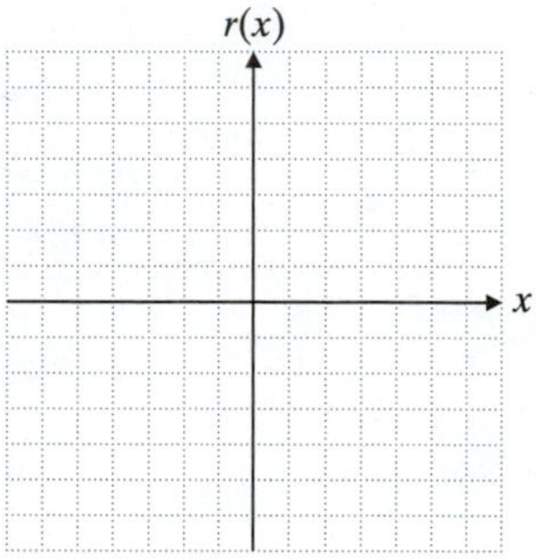

18. $r(x) = -3x^2 + 6x - 4$

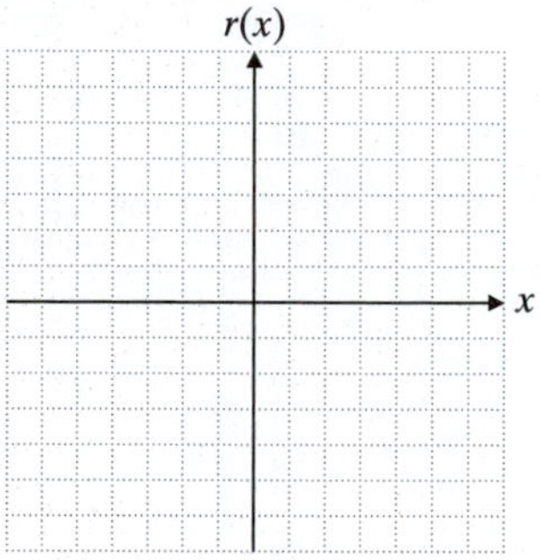

19. $f(x) = x^2 - 6x + 5$

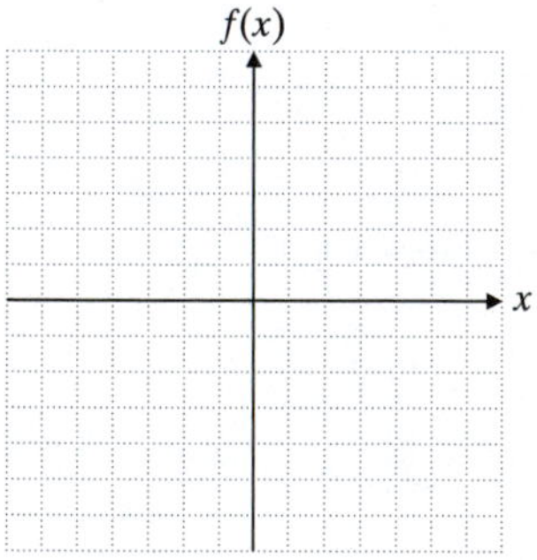

20. $g(x) = 2x^2 - 2x + 1$

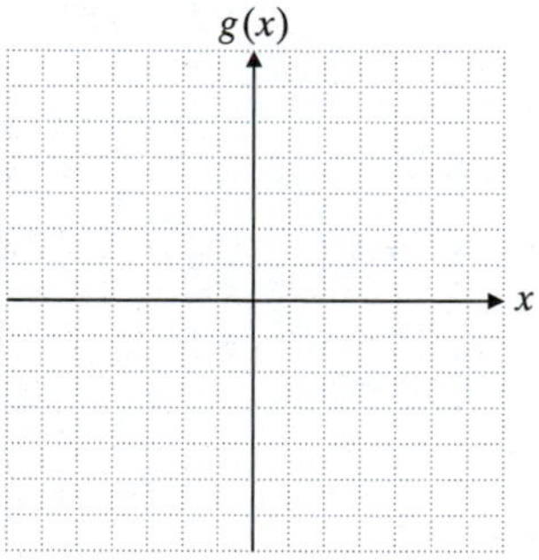

21. $f(x) = x^2 - 4x + 4$

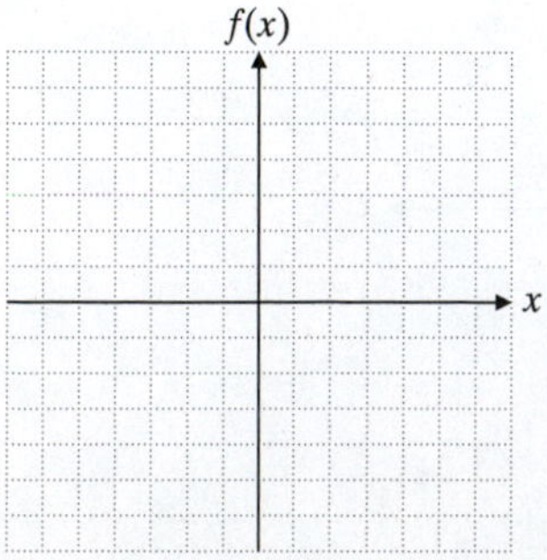

22. $g(x) = -x^2 + 6x - 9$

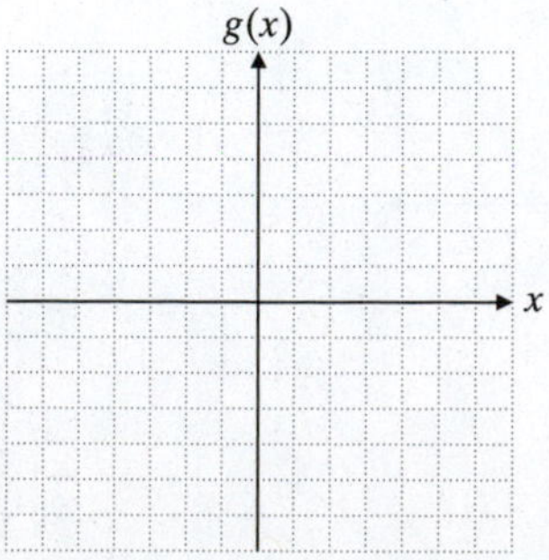

23. $f(x) = x^2 - 4$

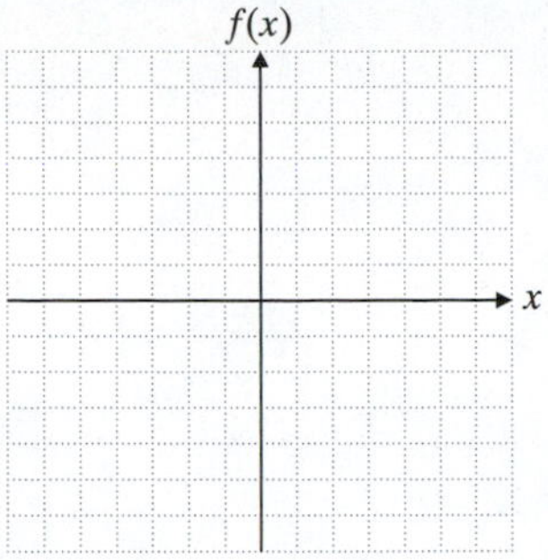

24. $r(x) = -x^2 + 1$

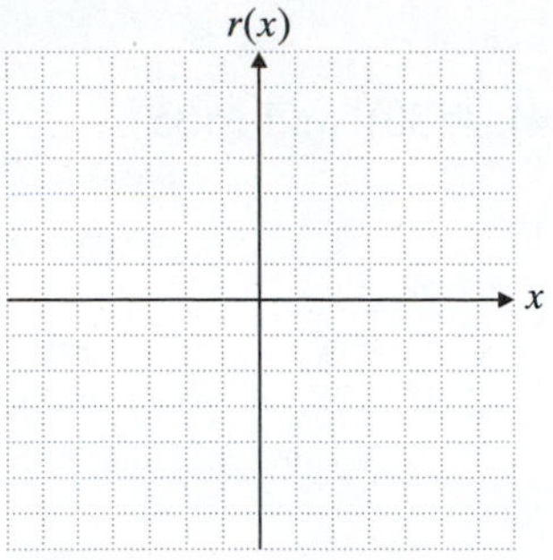

Applications

Boating *Some leisure activities such as boating are expensive. People with larger incomes are more likely to participate in such an activity. The number of people N (measured in thousands) who participate in boating (motor and power boating) can be described by the function* $N(x) = 1.99x^2 - 6.76x + 1098.4$, *where x is the mean income (measured in thousands) and* $x \geq 10$. *Use this information to answer exercises 25–30.* (*Source:* www.census.gov)

25. Find $N(10)$, $N(30)$, $N(50)$, $N(70)$, and $N(90)$.

26. Use the results of exercise 25 to graph the function from $x = 10$ to $x = 90$. You may use the graph grid provided at the bottom of the page.

27. Find $N(60)$ from your graph. Explain what $N(60)$ means.

28. Find $N(60)$ from the equation for $N(x)$. Compare your answers for exercises 27 and 28.

29. Use your graph to find the value of x when $N(x)$ is equal to 10,000. Explain what this means.

30. Use your graph to find the value of x when $N(x)$ is equal to 4000. Explain what this means.

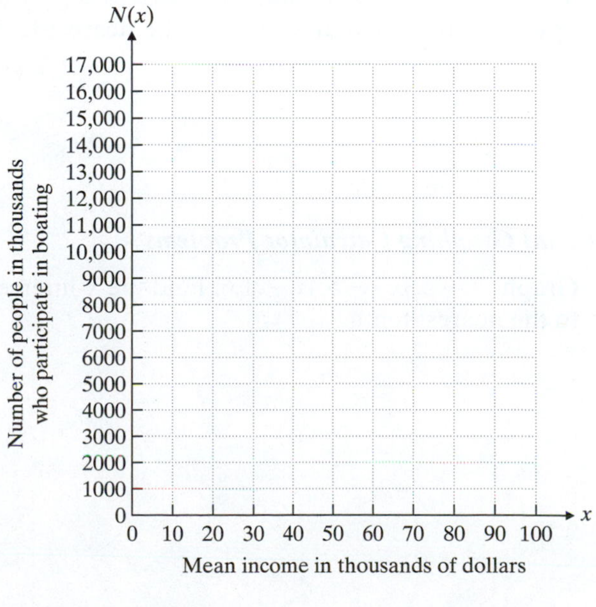

Manufacturing Profit *The daily profit P in dollars of the Pine Tree Table Company is described by the function* $P(x) = -6x^2 + 312x - 3672$, *where x is the number of tables that are manufactured in 1 day. Use this information to answer exercises 31–36.*

31. Find $P(16)$, $P(20)$, $P(24)$, $P(30)$, and $P(35)$.

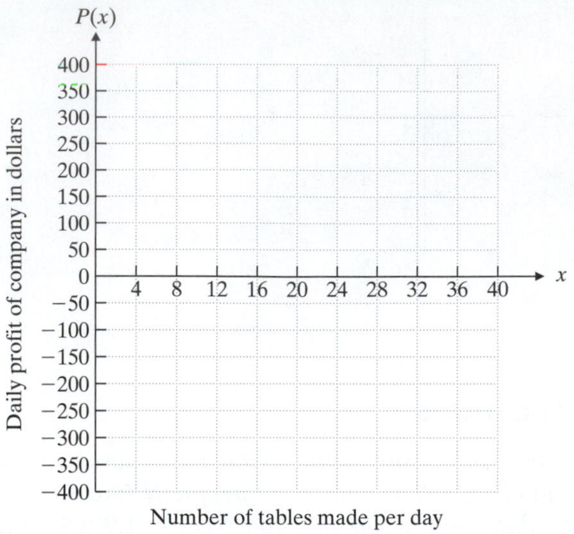

32. Use the results of exercise 31 to graph the function from $x = 16$ to $x = 35$.

33. The maximum profit of the company occurs at the vertex of the parabola. How many tables should be made per day in order to obtain the maximum profit for the company? What is the maximum profit?

34. How many tables per day should be made in order to obtain a daily profit of $330? Why are there two answers to this question?

35. How many tables are made per day if the company has a daily profit of zero dollars?

36. How many tables are made per day if the company has a daily *loss* of $102?

37. ***Softball*** Susan throws a softball upward into the air at a speed of 32 feet per second from a 40-foot platform. The height of the ball after t seconds is given by the function $d(t) = -16t^2 + 32t + 40$. What is the maximum height of the softball? How many seconds does it take to reach the ground after first being thrown upward? (Round your answer to the nearest tenth.)

38. ***Baseball*** Henry is standing on a platform overlooking a baseball stadium. It is 160 feet above the playing field. When he throws a baseball upward at 64 feet per second, the distance d from the baseball to the ground after t seconds is given by the function $d(t) = -16t^2 + 64t + 160$. What is the maximum height of the baseball if he throws it upward? How many seconds does it take until the ball finally hits the ground? (Round your answer to the nearest tenth.)

Optional Graphing Calculator Problems

39. Graph $y = 2.3x^2 - 5.4x - 1.6$. Find the x-intercepts to the nearest tenth.

40. Graph $y = -4.6x^2 + 7.2x - 2.3$. Find the x-intercepts to the nearest tenth.

To Think About

41. A graph of a quadratic equation of the form $y = ax^2 + bx + c$ passes through the points $(0, -10)$, $(3, 41)$, and $(-1, -15)$. What are the values of a, b, and c?

42. A graph of a quadratic equation of the form $y = ax^2 + bx + c$ has a vertex of $(1, -25)$ and passes through the point $(0, -24)$. What are the values of a, b, and c?

Cumulative Review *Solve each system.*

43. **[4.2.2]**
$$3x - y + 2z = 12$$
$$2x - 3y + z = 5$$
$$x + 3y + 8z = 22$$

44. **[4.2.2]**
$$7x + 3y - z = -2$$
$$x + 5y + 3z = 2$$
$$x + 2y + z = 1$$

Quick Quiz 8.5 *Consider the function $f(x) = -2x^2 - 4x + 6$.*

1. Find the vertex.

2. Find the y-intercept and the x-intercepts.

3. Using the above points, graph the function.

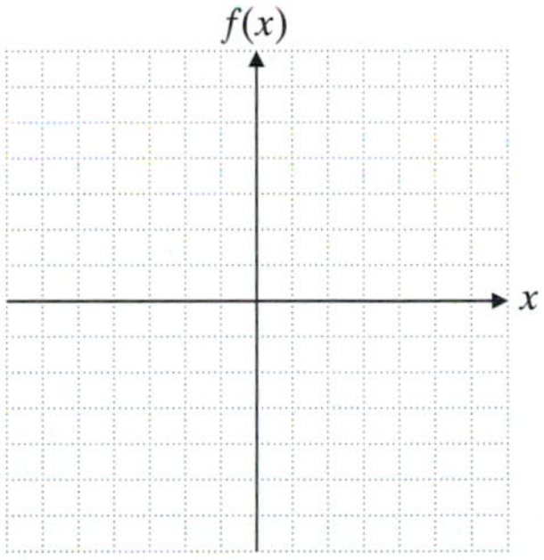

4. Concept Check Explain how you would find the vertex of $f(x) = 4x^2 - 9x - 5$.

8.6 Quadratic Inequalities in One Variable

Student Learning Objectives

After studying this section, you will be able to:

1. **Solve a factorable quadratic inequality in one variable.**
2. **Solve a nonfactorable quadratic inequality in one variable.**

1 Solving a Factorable Quadratic Inequality in One Variable

We will now solve quadratic inequalities such as $x^2 - 2x - 3 > 0$ and $2x^2 + x - 15 < 0$. A **quadratic inequality** has the form $ax^2 + bx + c < 0$ (or replace $<$ by $>$, $\leq$, or $\geq$), where a, b, and c are real numbers and $a \neq 0$. We use our knowledge of solving quadratic equations to solve quadratic inequalities.

Let's solve the inequality $x^2 - 2x - 3 > 0$. We want to find the two points where the expression on the left side is equal to zero. We call these the **boundary points.** To do this, we replace the inequality symbol by an equals sign and solve the resulting equation.

$$x^2 - 2x - 3 = 0$$

$$(x + 1)(x - 3) = 0 \quad \text{Factor.}$$

$$x + 1 = 0 \quad \text{or} \quad x - 3 = 0 \quad \text{Zero factor property}$$

$$x = -1 \qquad\qquad x = 3$$

These two solutions form boundary points that divide the number line into three segments.

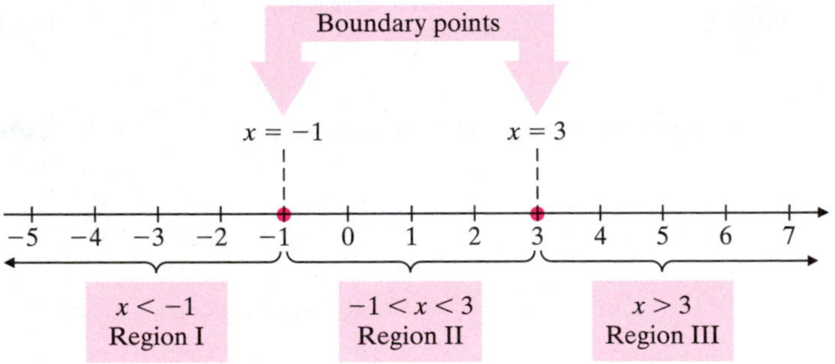

All values of x in a given segment produce results that are greater than zero, or all values of x in a given segment produce results that are less than zero.

To solve the quadratic inequality, we pick an arbitrary test point in each region and then substitute it into the inequality to determine whether it satisfies the inequality. If one point in a region satisfies the inequality, then *all* points in the region satisfy the inequality. We will test three values of x in the expression $x^2 - 2x - 3$.

$x < -1$, *Region I*: A sample point is $x = -2$.

$$(-2)^2 - 2(-2) - 3 = 4 + 4 - 3 = 5 > 0$$

$-1 < x < 3$, *Region II*: A sample point is $x = 0$.

$$(0)^2 - 2(0) - 3 = 0 + 0 - 3 = -3 < 0$$

$x > 3$, *Region III*: A sample point is $x = 4$.

$$(4)^2 - 2(4) - 3 = 16 - 8 - 3 = 5 > 0$$

Thus, we see that $x^2 - 2x - 3 > 0$ when $x < -1$ or $x > 3$. No points in Region II satisfy the inequality. The graph of the solution is shown next.

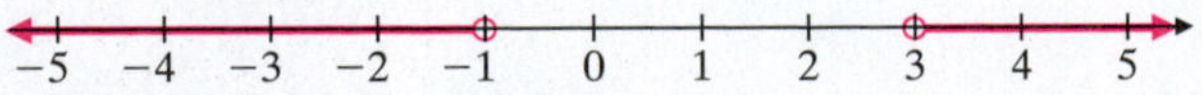

We summarize our method.

SOLVING A QUADRATIC INEQUALITY

1. Replace the inequality symbol by an equals sign. Solve the resulting equation to find the boundary points.
2. Use the boundary points to separate the number line into three distinct regions.
3. Evaluate the quadratic expression at a test point in each region.
4. Determine which regions satisfy the original conditions of the quadratic inequality.

EXAMPLE 1 Solve and graph $x^2 - 10x + 24 > 0$.

Solution

1. We replace the inequality symbol by an equals sign and solve the resulting equation.

$$x^2 - 10x + 24 = 0$$
$$(x - 4)(x - 6) = 0$$
$$x - 4 = 0 \quad \text{or} \quad x - 6 = 0$$
$$x = 4 \qquad\qquad x = 6$$

2. We use the boundary points to separate the number line into distinct regions.

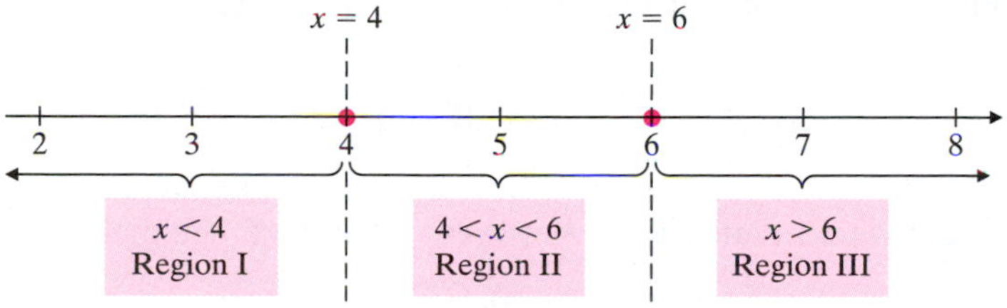

3. We evaluate the quadratic expression at a test point in each of the regions.

$$x^2 - 10x + 24$$

$x < 4$, Region I: We pick the test point $x = 1$.

$$(1)^2 - 10(1) + 24 = 1 - 10 + 24 = 15 > 0$$

$4 < x < 6$, Region II: We pick the test point $x = 5$.

$$(5)^2 - 10(5) + 24 = 25 - 50 + 24 = -1 < 0$$

$x > 6$, Region III: We pick the test point $x = 7$.

$$(7)^2 - 10(7) + 24 = 49 - 70 + 24 = 3 > 0$$

4. We determine which regions satisfy the original conditions of the quadratic inequality.

$$x^2 - 10x + 24 > 0 \text{ when } x < 4 \text{ or when } x > 6.$$

The graph of the solution is shown next.

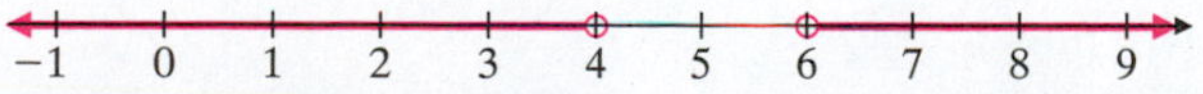

Student Practice 1 Solve and graph $x^2 - 2x - 8 < 0$.

***NOTE TO STUDENT:** Fully worked-out solutions to all of the Student Practice problems can be found at the back of the text starting at page SP-1.*

Student Practice 1

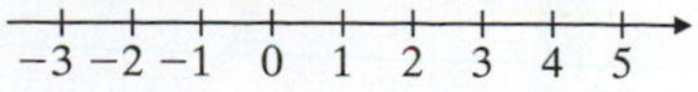

EXAMPLE 2 Solve and graph $2x^2 + x - 6 \leq 0$.

Solution We replace the inequality symbol by an equals sign and solve the resulting equation.

$$2x^2 + x - 6 = 0$$
$$(2x - 3)(x + 2) = 0$$
$$2x - 3 = 0 \quad \text{or} \quad x + 2 = 0$$
$$2x = 3 \qquad x = -2$$
$$x = \frac{3}{2} = 1.5$$

We use the boundary points to separate the number line into distinct regions. The boundary points are $x = -2$ and $x = 1.5$. Now we arbitrarily pick a test point in each region.

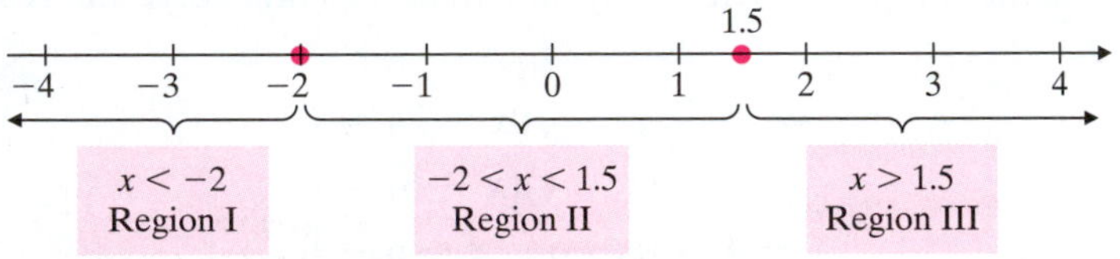

We will pick 3 values of x for the polynomial $2x^2 + x - 6$.

Region I: First we need an x-value less than -2. We pick $x = -3$.

$$2(-3)^2 + (-3) - 6 = 18 - 3 - 6 = 9 > 0$$

Region II: Next we need an x-value between -2 and 1.5. We pick $x = 0$.

$$2(0)^2 + (0) - 6 = 0 + 0 - 6 = -6 < 0$$

Region III: Finally we need an x-value greater than 1.5. We pick $x = 2$.

$$2(2)^2 + (2) - 6 = 8 + 2 - 6 = 4 > 0$$

Since our inequality is $\leq$ and not just $<$, we need to include the boundary points. Thus, $2x^2 + x - 6 \leq 0$ when $-2 \leq x \leq 1.5$. The graph of our solution is shown next.

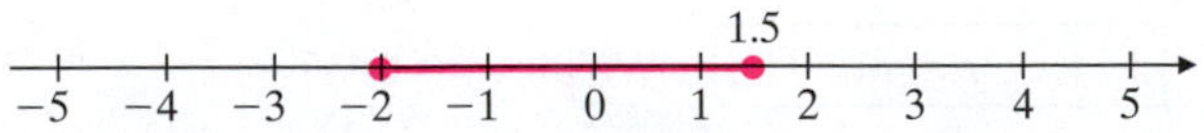

Student Practice 2 Solve and graph $3x^2 - x - 2 \geq 0$.

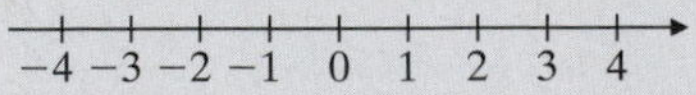

Graphing Calculator

Solving Quadratic Inequalities

To solve Example 2 on a graphing calculator, graph $y = 2x^2 + x - 6$, zoom in on the two x-intercepts, and use the Trace feature to find the value of x where y changes from a negative value to a positive value or zero. You can use the Zero command if your calculator has it. Can you verify from your graph that the solution to Example 2 is $-2 \leq x \leq 1.5$? Use your graphing calculator and this method to solve the exercises below. Round your answers to the nearest hundredth.

1. $3x^2 - 3x - 60 \geq 0$
2. $2.1x^2 + 4.3x - 29.7 > 0$
3. $15.3x^2 - 20.4x + 6.8 \geq 0$
4. $16.8x^2 > -16.8x - 35.7$

If you are using a graphing calculator and you think there is no solution to a quadratic inequality, how can you *verify* this from the graph? Why is this possible?

② Solving a Nonfactorable Quadratic Inequality in One Variable

If the quadratic expression in a quadratic inequality cannot be factored, then we will use the quadratic formula to obtain the boundary points.

EXAMPLE 3 Solve and graph $x^2 + 4x > 6$. Round your answer to the nearest tenth.

Solution First we write $x^2 + 4x - 6 > 0$. Because we cannot factor $x^2 + 4x - 6$, we use the quadratic formula to find the boundary points.

$$x = \frac{-4 \pm \sqrt{4^2 - 4(1)(-6)}}{2(1)} = \frac{-4 \pm \sqrt{16 + 24}}{2}$$

$$= \frac{-4 \pm \sqrt{40}}{2} = \frac{-4 \pm 2\sqrt{10}}{2} = -2 \pm \sqrt{10}$$

Using a calculator or our table of square roots, we find the following:

$$-2 + \sqrt{10} \approx -2 + 3.162 \approx 1.162 \text{ or about } 1.2$$

$$-2 - \sqrt{10} \approx -2 - 3.162 \approx -5.162 \text{ or about } -5.2$$

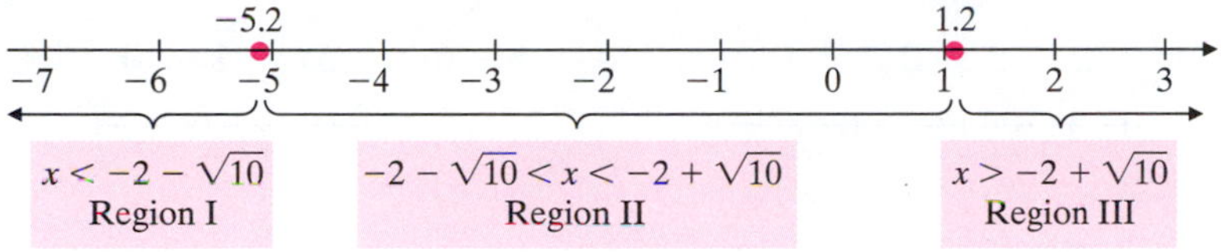

We will see where $x^2 + 4x - 6 > 0$.

Region I: Test $x = -6$.

$$(-6)^2 + 4(-6) - 6 = 36 - 24 - 6 = 6 > 0$$

Region II: Test $x = 0$.

$$(0)^2 + 4(0) - 6 = 0 + 0 - 6 = -6 < 0$$

Region III: Test $x = 2$.

$$(2)^2 + 4(2) - 6 = 4 + 8 - 6 = 6 > 0$$

Thus, $x^2 + 4x > 6$ when $x^2 + 4x - 6 > 0$, and this occurs when $x < -2 - \sqrt{10}$ or $x > -2 + \sqrt{10}$. Rounding to the nearest tenth, our answer is

$$x < -5.2 \quad \text{or} \quad x > 1.2.$$

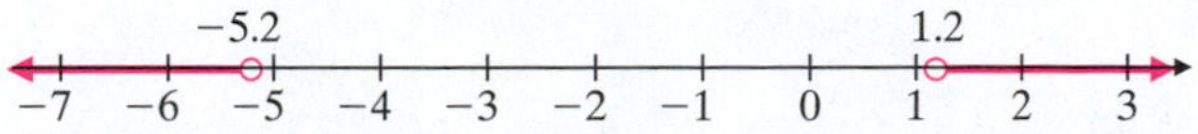

Student Practice 3 Solve and graph $x^2 + 2x < 7$. Round your answer to the nearest tenth.

Student Practice 3

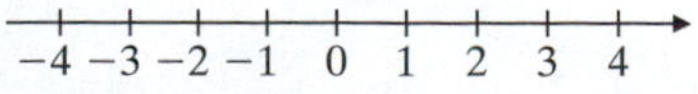

8.6 Exercises

MyMathLab® Watch the videos in MyMathLab Download the MyDashBoard App

Verbal and Writing Skills, Exercises 1 and 2

1. When solving a quadratic inequality, why is it necessary to find the boundary points?

2. What is the difference between solving an exercise like $ax^2 + bx + c > 0$ and an exercise like $ax^2 + bx + c \geq 0$?

Solve and graph.

3. $x^2 + x - 12 < 0$

4. $x^2 - x - 6 > 0$

5. $x^2 \geq 4$

6. $x^2 - 9 \leq 0$

7. $2x^2 + x - 3 < 0$

8. $6x^2 - 5x + 1 < 0$

Solve.

9. $x^2 + x - 20 > 0$

10. $x^2 + 6x - 27 > 0$

11. $4x^2 \leq 11x + 3$

12. $4x^2 - 5 \leq -8x$

13. $6x^2 - 5x > 6$

14. $3x^2 + 13x > -4$

15. $-2x + 30 \geq x(x + 5)$
Hint: Put variables on the right and zero on the left in your first step.

16. $55 - x^2 \geq 6x$
Hint: Put variables on the right and zero on the left in your first step.

17. $x^2 - 2x \geq -1$

18. $x^2 + 25 \geq 10x$

19. $x^2 - 4x \leq -4$

20. $x^2 - 6x \leq -9$

Solve each of the following quadratic inequalities if possible. Round your answer to the nearest tenth.

21. $x^2 - 2x > 4$

22. $x^2 + 2x > 5$

23. $2x^2 - 2x < 3$

24. $2x^2 < 4x + 1$

25. $2x^2 \geq x^2 - 4$

26. $4x^2 \geq 3x^2 - 9$

27. $5x^2 \leq 4x^2 - 1$

28. $x^2 - 1 \leq -17$

Applications

Projectile Flight *In exercises 29 and 30, a projectile is fired vertically with an initial velocity of 640 feet per second. The distance s in feet above the ground after t seconds is given by the equation* $s = -16t^2 + 640t$.

29. For what range of time t (measured in seconds) will the height s be greater than 6000 feet?

30. For what range of time t (measured in seconds) will the height s be less than 4800 feet?

Manufacturing Profit *In exercises 31 and 32, the profit of a manufacturing company is determined by the number of units x manufactured each day according to the given equation.* ***(a)*** *Find when the profit is greater than zero.* ***(b)*** *Find the daily profit when 50 units are manufactured.* ***(c)*** *Find the daily profit when 60 units are manufactured.*

31. Profit $= -10(x^2 - 200x + 1800)$

32. Profit $= -20(x^2 - 320x + 3600)$

Cumulative Review

33. **[2.6.3]** ***Test Scores*** The university's synchronized swimming team will not let Mona participate unless she passes biology with a C (70 or better) average. There are six tests in the semester, and she failed the first one (with a score of 0). She decided to find a tutor. Since then, she has scored 81, 92, and 80 on the next three tests. What must her minimum scores be on the last two tests to pass the course with a minimum grade of 70 and participate in synchronized swimming?

34. **[2.4.1]** ***Gourmet Foods*** The Java Express Assortment sold on www.igourmet.com contains Dutch coffee, chocolate banana bites, and Belgian chocolate. The total weight of the assortment is 16 ounces. There are twice as many ounces of Dutch coffee as chocolate banana bites. There are 1.6 ounces less of Belgian chocolate as chocolate banana bites. How many ounces of each are in the Java Express Assortment? (*Source:* www.igourmet.com)

Cost of a Cruise *The Circle Line Cruise is a 2-hour, 24-mile cruise around southern Manhattan (New York City). The charge for adults is $18; children 12 and under, $10; and seniors over 62, $16. For the 3-hour, 35-mile cruise around all of Manhattan, the charge for adults is $22; children 12 and under, $12; and seniors over 62, $19. The Yoffa family has come to New York for their family reunion and is planning family activities. The family has ten adults, fourteen children under 12, and five senior members.*

35. **[1.2.2]** What would it cost for all of the family to take the 2-hour trip? The 3-hour trip?

36. **[4.3.2]** Six people do not take the cruise. If the rest of the family takes a 2-hour cruise, it will cost $314. If the rest of the family takes a 3-hour cruise, it will cost $380. How many adults, how many children, and how many senior members plan to take a cruise?

Quick Quiz 8.6 *Solve.*

1. $x^2 - 7x + 6 > 0$

2. $6x^2 - x - 2 < 0$

3. Use a calculator or a square root table to approximate to the nearest tenth the solution for $x^2 + 4x - 8 \geq 0$.

4. **Concept Check** Explain what happens when you solve the inequality $x^2 + 2x + 8 > 0$.

Did You Know...

That You Can Save Money in a Semester by Choosing the Right Meal Plan?

CHOOSING A MEAL PLAN

Understanding the Problem:
Jake is a college student who lives in a college dormitory. He is going over the meal plans and calculating his food budget for the semester. He is interested in choosing a meal plan that will cost him the least while still meeting his needs.

Making a Plan:
First he needs to compare the prices and options for the meal plans. Then he needs to determine how many meals he will eat in the dining halls so he can determine which plan is right for him.

Step 1: There are three dining plans that cover the 15-week semester.

- The Gold plan costs $2205 a semester and covers 21 meals a week.
- The Silver plan costs $1764 a semester and covers 16 meals a week, with extra meals costing $10.
- The Bronze plan costs $1386 a semester and covers 12 meals a week, with extra meals costing $10.

Which plan is best for the following situations:

Task 1: *Jake plans on eating 20 meals per week in the dining halls?*

Task 2: *Jake plans on eating 18 meals per week in the dining halls?*

Task 3: *Jake plans on eating 16 meals per week in the dining halls?*

Task 4: *Jake plans on eating 14 meals per week in the dining halls?*

Step 2: Jake also has the option of buying "dining dollars" to use at other eating establishments on campus. With dining dollars Jake will get 10% off all food purchases.

How much will Jake save over the course of the semester by using dining dollars in the following situations:

Task 5: *if he spends $20 a week at these establishments?*

Task 6: *if he spends $40 a week at these establishments?*

Task 7: *if he spends $60 a week at these establishments?*

Finding a Solution:

Step 3: Jake plans on eating 17 meals a week in the dining halls and spending $30 a week at other establishments.

Task 8: *Which plan should he buy?*

Task 9: *How much will Jake spend on food for the semester?*

Applying the Situation to Your Life:

You can do these same calculations for your own circumstances.

- See what plans are offered.
- Determine how often you will eat in the dining halls.
- Decide which plan would be right for you.
- Calculate how much you will need to spend on food for the semester.

Chapter 8 Organizer

Topic and Procedure	Examples	You Try It
Solving a quadratic equation by using the square root property, p. 422 If $x^2 = a$, then $x = \pm\sqrt{a}$.	Solve. $2x^2 - 50 = 0$ $2x^2 = 50$ $x^2 = 25$ $x = \pm\sqrt{25}$ $x = \pm 5$	**1.** Solve. $3x^2 - 60 = 0$
Solving a quadratic equation by completing the square, p. 425 **1.** Rewrite the equation in the form $ax^2 + bx = c$. **2.** If $a \neq 1$, divide each term of the equation by a. **3.** Square half of the numerical coefficient of the linear term. Add the result to both sides of the equation. **4.** Factor the left side; then take the square root of both sides of the equation. **5.** Solve the resulting equation for x. **6.** Check the solutions in the original equation.	Solve. $2x^2 - 4x - 1 = 0$ $2x^2 - 4x = 1$ $\frac{2x^2}{2} - \frac{4x}{2} = \frac{1}{2}$ $x^2 - 2x + \_\_\_\_ = \frac{1}{2} + \_\_\_\_$ $x^2 - 2x + 1 = \frac{1}{2} + 1$ $(x - 1)^2 = \frac{3}{2}$ $x - 1 = \pm\sqrt{\frac{3}{2}}$ $x - 1 = \pm\frac{\sqrt{6}}{2}$ $x = 1 \pm \frac{\sqrt{6}}{2} = \frac{2 \pm \sqrt{6}}{2}$	**2.** Solve. $2x^2 + 6x - 3 = 0$
Placing a quadratic equation in standard form, p. 429 A quadratic equation in standard form is an equation of the form $ax^2 + bx + c = 0$, where a, b, and c are real numbers and $a \neq 0$. It is often necessary to remove parentheses and clear away fractions by multiplying each term of the equation by the LCD to obtain the standard form.	Rewrite in quadratic form. $\frac{2}{x-3} + \frac{x}{x+3} = \frac{5}{x^2 - 9}$ $(x+3)(x-3)\left[\frac{2}{x-3}\right] + (x+3)(x-3)\left[\frac{x}{x+3}\right]$ $= (x+3)(x-3)\left[\frac{5}{(x+3)(x-3)}\right]$ $2(x+3) + x(x-3) = 5$ $2x + 6 + x^2 - 3x = 5$ $x^2 - x + 1 = 0$	**3.** Rewrite in quadratic form. $\frac{5}{x-1} + \frac{3}{x+2} = 2$
Solve a quadratic equation by using the quadratic formula, p. 429 If $ax^2 + bx + c = 0$, where $a \neq 0$, $x = \frac{-b \pm \sqrt{b^2 - 4ac}}{2a}$. **1.** Rewrite the equation in standard form. **2.** Determine the values of a, b, and c. **3.** Substitute the values of a, b, and c into the formula. **4.** Simplify the result to obtain the values of x. **5.** Any imaginary solutions to the quadratic equation should be simplified by using the definition $\sqrt{-a} = i\sqrt{a}$, where $a > 0$.	Solve. $2x^2 = 3x - 2$ $2x^2 - 3x + 2 = 0$ $a = 2, b = -3, c = 2$ $x = \frac{-(-3) \pm \sqrt{(-3)^2 - 4(2)(2)}}{2(2)}$ $x = \frac{3 \pm \sqrt{9 - 16}}{4}$ $x = \frac{3 \pm \sqrt{-7}}{4}$ $x = \frac{3 \pm i\sqrt{7}}{4}$	**4.** Solve. $x^2 - 8x = 5$

Topic and Procedure | Examples | You Try It

Equations that can be transformed into quadratic form, p. 438

If the exponents are positive:

1. Find the variable with the smallest exponent. Let this quantity be replaced by y.
2. Continue to make substitutions for the remaining variable terms based on the first substitution. (You should be able to replace the variable with the largest exponent by y^2.)
3. Solve the resulting equation for y.
4. Reverse the substitution used in step 1.
5. Solve the resulting equation for x.
6. Check your solution in the *original* equation.

Examples

Solve. $x^{2/3} - x^{1/3} - 2 = 0$

Let $y = x^{1/3}$. Then $y^2 = x^{2/3}$.

$$y^2 - y - 2 = 0$$
$$(y - 2)(y + 1) = 0$$
$$y = 2 \quad \text{or} \quad y = -1$$
$$x^{1/3} = 2 \qquad x^{1/3} = -1$$
$$(x^{1/3})^3 = 2^3 \qquad (x^{1/3})^3 = (-1)^3$$
$$x = 8 \qquad x = -1$$

Check.

$$x = 8: \quad (8)^{2/3} - (8)^{1/3} - 2 \stackrel{?}{=} 0$$
$$2^2 - 2 - 2 \stackrel{?}{=} 0$$
$$4 - 4 = 0 \ \checkmark$$
$$x = -1: \quad (-1)^{2/3} - (-1)^{1/3} - 2 \stackrel{?}{=} 0$$
$$(-1)^2 - (-1) - 2 \stackrel{?}{=} 0$$
$$1 + 1 - 2 = 0 \ \checkmark$$

Both 8 and -1 are solutions.

You Try It

5. Solve. $x + 3x^{1/2} - 4 = 0$

Solving quadratic equations containing two or more variables, p. 446

Treat the letter to be solved for as a variable, but treat all other letters as constants. Solve the equation by factoring, by using the square root property, or by using the quadratic formula.

Examples

Solve for x.

(a) $6x^2 - 11xw + 4w^2 = 0$

(b) $4x^2 + 5b = 2w^2$

(c) $2x^2 + 3xz - 10z = 0$

(a) By factoring:

$$(3x - 4w)(2x - w) = 0$$
$$3x - 4w = 0 \quad \text{or} \quad 2x - w = 0$$
$$x = \frac{4w}{3} \qquad x = \frac{w}{2}$$

(b) Using the square root property:

$$4x^2 = 2w^2 - 5b$$
$$x^2 = \frac{2w^2 - 5b}{4}$$
$$x = \pm\sqrt{\frac{2w^2 - 5b}{4}} = \pm\frac{1}{2}\sqrt{2w^2 - 5b}$$

(c) By the quadratic formula, with $a = 2, b = 3z, c = -10z$:

$$x = \frac{-3z \pm \sqrt{9z^2 + 80z}}{4}$$

You Try It

6. Solve for x.

(a) $2x^2 - 5xy - 3y^2 = 0$

(b) $x^2 + 4y^2 = 4a$

(c) $4x^2 - 2xw = 9w$

The Pythagorean Theorem, p. 448

In any right triangle, if c is the length of the hypotenuse and a and b are the lengths of the two legs, then

$$c^2 = a^2 + b^2.$$

Examples

Find a if $c = 7$ and $b = 5$.

$$49 = a^2 + 25$$
$$49 - 25 = a^2$$
$$24 = a^2$$
$$\sqrt{24} = a$$
$$2\sqrt{6} = a$$

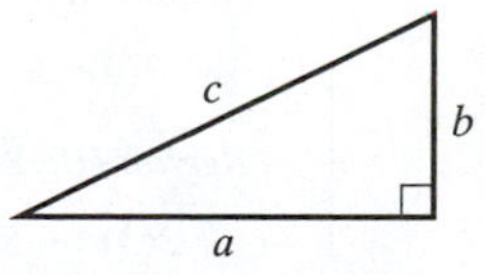

You Try It

7. Find b if $c = 10$ and $a = 5$.

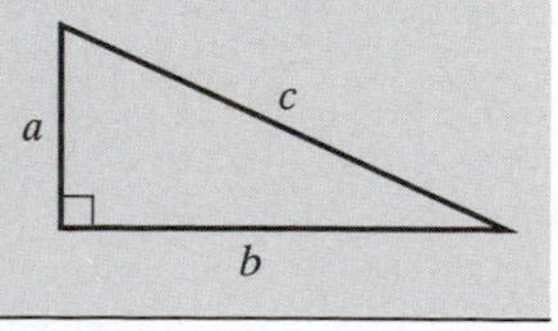

Topic and Procedure | Examples | You Try It

Graphing quadratic functions, p. 457

Graph quadratic functions of the form $f(x) = ax^2 + bx + c$ with $a \neq 0$ as follows:

1. Find the vertex at $\left(\frac{-b}{2a}, f\left(\frac{-b}{2a}\right)\right)$.
2. Find the y-intercept, which occurs at $f(0)$.
3. Find the x-intercepts if they exist. Solve $f(x) = 0$ for x.
4. Connect the points with a smooth curve.

Examples

Graph. $f(x) = x^2 + 6x + 8$

Vertex:

$$x = \frac{-6}{2} = -3$$

$$f(-3) = (-3)^2 + 6(-3) + 8 = -1$$

The vertex is $(-3, -1)$.

Intercepts: $f(0) = (0)^2 + 6(0) + 8 = 8$
The y-intercept is $(0, 8)$.

$$x^2 + 6x + 8 = 0$$
$$(x + 2)(x + 4) = 0$$
$$x = -2, x = -4$$

The x-intercepts are $(-2, 0)$ and $(-4, 0)$.

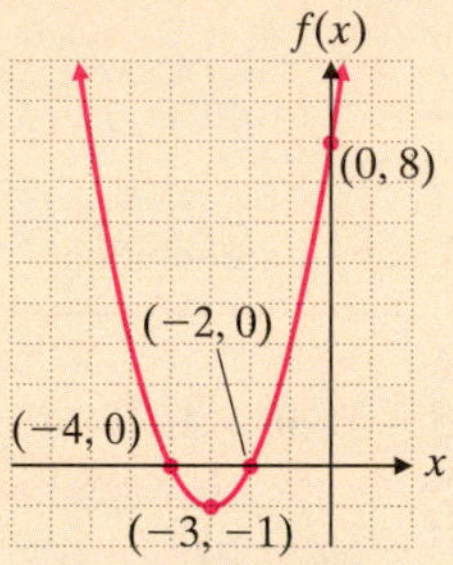

$f(x) = x^2 + 6x + 8$

You Try It

8. Graph. $f(x) = x^2 + 3x - 4$

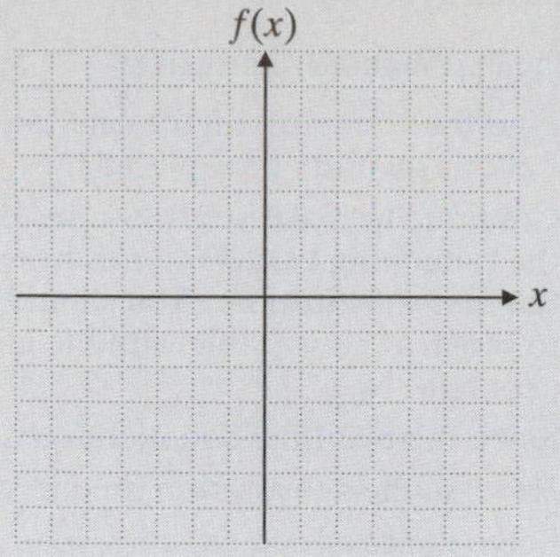

Solving quadratic inequalities in one variable, p. 467

1. Replace the inequality symbol by an equals sign. Solve the resulting equation to find the boundary points.
2. Use the boundary points to separate the number line into distinct regions.
3. Evaluate the quadratic expression at a test point in each region.
4. Determine which regions satisfy the original conditions of the quadratic inequality.

Examples

Solve and graph. $3x^2 + 5x - 2 > 0$

1. $3x^2 + 5x - 2 = 0$

$$(3x - 1)(x + 2) = 0$$

$$3x - 1 = 0 \quad \text{or} \quad x + 2 = 0$$

$$x = \frac{1}{3} \qquad\qquad x = -2$$

Boundary points are -2 and $\frac{1}{3}$.

2.

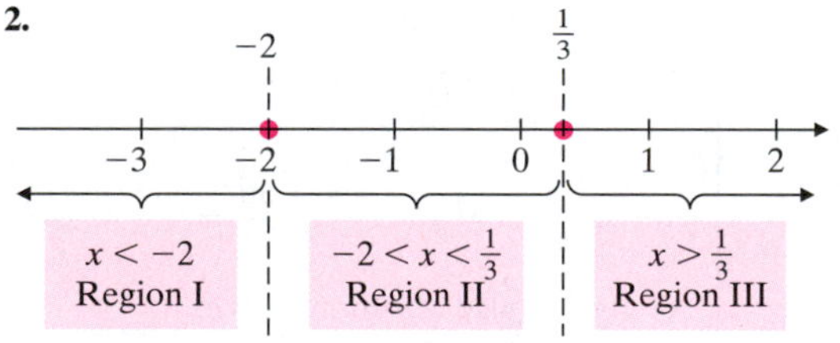

3. $3x^2 + 5x - 2$

Region I: Pick $x = -3$.

$$3(-3)^2 + 5(-3) - 2 = 27 - 15 - 2 = 10 > 0$$

Region II: Pick $x = 0$.

$$3(0)^2 + 5(0) - 2 = 0 + 0 - 2 = -2 < 0$$

Region III: Pick $x = 3$.

$$3(3)^2 + 5(3) - 2 = 27 + 15 - 2 = 40 > 0$$

4. We know that the expression is greater than zero (that is, $3x^2 + 5x - 2 > 0$) when

$$x < -2 \text{ or } x > \frac{1}{3}.$$

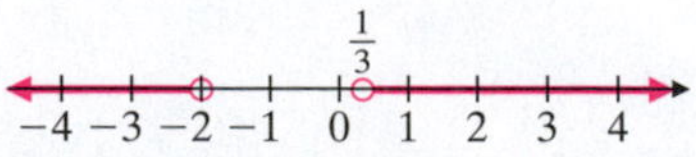

You Try It

9. Solve and graph.
$2x^2 - 3x - 9 < 0$

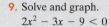

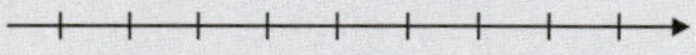

Chapter 8 Review Problems

Solve each of the following exercises by the specified method. Simplify all answers.

Solve by the square root property.

1. $6x^2 = 24$

2. $(x + 8)^2 = 81$

Solve by completing the square.

3. $x^2 + 8x + 13 = 0$

4. $4x^2 - 8x + 1 = 0$

Solve by the quadratic formula.

5. $x^2 - 4x - 2 = 0$

6. $3x^2 - 8x + 4 = 0$

Solve by any appropriate method and simplify your answers. Express any nonreal complex solutions using i notation.

7. $4x^2 - 12x + 9 = 0$

8. $x^2 - 14 = 5x$

9. $6x^2 - 23x = 4x$

10. $2x^2 = 5x - 1$

11. $5x^2 - 10 = 0$

12. $3x^2 - 2x = 15x - 10$

13. $6x^2 + 12x - 24 = 0$

14. $7x^2 + 24 = 5x^2$

15. $3x^2 + 5x + 1 = 0$

16. $2x(x - 4) - 4 = -x$

17. $9x(x + 2) + 2 = 12x$

18. $\frac{4}{5}x^2 + x + \frac{1}{5} = 0$

19. $y + \frac{5}{3y} + \frac{17}{6} = 0$

20. $\frac{15}{y^2} - \frac{2}{y} = 1$

21. $y(y + 1) + (y + 2)^2 = 4$

22. $\frac{2x}{x + 3} + \frac{3x - 1}{x + 1} = 3$

Determine the nature of each of the following quadratic equations. Do not solve the equation. Find the discriminant in each case and determine whether the equation has one rational solution, two rational solutions, two irrational solutions, or two nonreal complex solutions.

23. $4x^2 - 5x - 3 = 0$

24. $2x^2 - 7x + 6 = 0$

25. $25x^2 - 20x + 4 = 0$

Write a quadratic equation having the given numbers as solutions.

26. $5, -5$

27. $3i, -3i$

28. $-\frac{1}{4}, -\frac{3}{2}$

Solve for any valid real roots.

29. $x^4 - 6x^2 + 8 = 0$

30. $2x^6 - 5x^3 - 3 = 0$

31. $x^{2/3} - 3 = 2x^{1/3}$

32. $1 + 4x^{-8} = 5x^{-4}$

Solve for the variable specified. Assume that all radical expressions obtained have a positive radicand.

33. $3M = \frac{2A^2}{N}$; for A

34. $3t^2 + 4b = t^2 + 6ay$; for t

35. $yx^2 - 3x - 7 = 0$; for x

36. $20d^2 - xd - x^2 = 0$; for d

37. $2y^2 + 4ay - 3a = 0$; for y

38. $AB = 3x^2 + 2y^2 - 4x$; for x

Use the Pythagorean Theorem to find the unknown side. Assume that c is the length of the hypotenuse of a right triangle and that a and b are the lengths of the legs. Leave your answer as a radical in simplified form.

▲ **39.** $a = 3\sqrt{2}, b = 2$; find c

▲ **40.** $c = 16, b = 4$; find a

▲ **41.** ***Airplane Flight*** A plane is 6 miles away from an observer and exactly 5 miles above the ground. The plane is directly above a car. How far is the car from the observer? Round your answer to the nearest tenth of a mile.

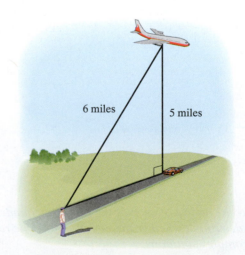

▲ **42.** ***Geometry*** The area of a triangle is 70 square centimeters. Its altitude is 6 meters longer than twice the length of the base. Find the lengths of the altitude and base.

▲ **43.** ***Geometry*** The area of a rectangle is 203 square meters. Its length is 1 meter longer than four times its width. Find the length and width of the rectangle.

44. ***Cruise Ship*** A cruise ship left port and traveled 80 miles at a constant speed. Then for 10 miles it traveled 10 miles per hour slower while circling an island before stopping. The trip took 5 hours. Find the ship's speed for each part of the trip.

45. ***Car Travel in the Rain*** Jessica drove at a constant speed for 200 miles. Then it started to rain. So for the next 90 miles she traveled 5 miles per hour slower. The entire trip took 6 hours of driving time. Find her speed for each part of the trip.

▲ **46.** ***Garden Walkway*** Mr. and Mrs. Gomez are building a rectangular garden that is 10 feet by 6 feet. Around the outside of the garden, they will build a brick walkway. They have 100 square feet of brick. How wide should they make the brick walkway? Round your answer to the nearest tenth of a foot.

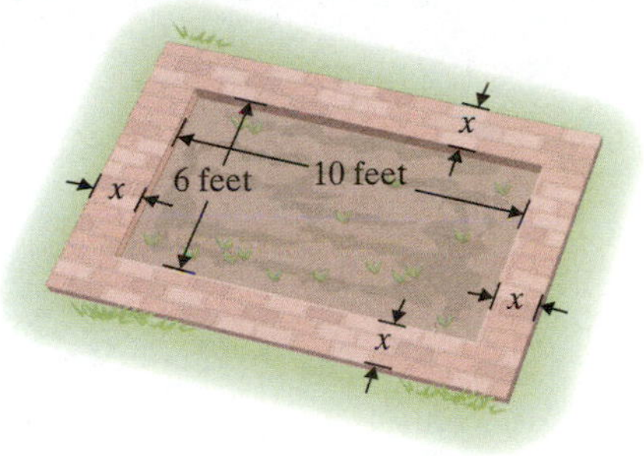

▲ **47.** ***Swimming Pool*** The local YMCA is building a new Olympic-sized pool of 50 meters by 25 meters. The builders want to make a walkway around the pool with a nonslip concrete surface. They have enough material to make 76 square meters of nonslip concrete surface. How wide should the walkway be?

Find the vertex and the intercepts of the following quadratic functions.

48. $g(x) = -x^2 + 6x - 11$

49. $f(x) = x^2 + 10x + 25$

In each of the following exercises, find the vertex, the y-intercept, and the x-intercepts (if any exist) and then graph the function.

50. $f(x) = x^2 + 6x + 5$

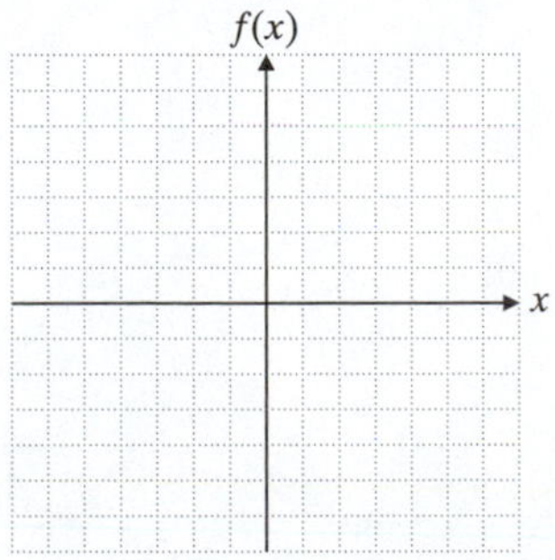

51. $f(x) = -x^2 + 6x - 5$

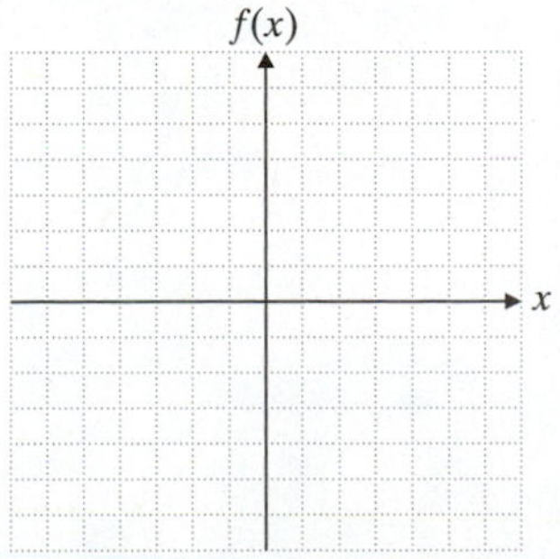

52. ***Rocket Flight*** A model rocket is launched upward from a platform 40 feet above the ground. The height of the rocket h is given at any time t in seconds by the function $h(t) = -16t^2 + 400t + 40$. Find the maximum height of the rocket. How long will it take the rocket to go through its complete flight and then hit the ground? (Assume that the rocket does *not* have a parachute.) Round your answer to the nearest tenth.

53. ***Revenue for a Store*** A salesman for an electronics store finds that in 1 month he can sell $(1200 - x)$ compact disc players that each sell for x dollars. Write a function for the revenue. What is the price x that will result in the maximum revenue for the store?

Solve and graph your solutions.

54. $x^2 + 7x - 18 < 0$

55. $x^2 - 9x + 20 > 0$

Solve each of the following if possible. Approximate, if necessary, any irrational solutions to the nearest tenth.

56. $2x^2 - x - 6 \leq 0$

57. $3x^2 - 13x + 12 \leq 0$

58. $9x^2 - 4 > 0$

59. $4x^2 - 8x \leq 12 + 5x^2$

60. $x^2 + 13x > 16 + 7x$

61. $3x^2 - 12x > -11$

62. $4x^2 + 12x + 9 < 0$

To Think About

Solve.

63. $(x + 4)(x - 2)(3 - x) > 0$

64. $(x + 1)(x + 4)(2 - x) < 0$

How Am I Doing? Chapter 8 Test

CHAPTER Test Prep VIDEOS   MATH COACH   MyMathLab®   YouTube™

After you take this test read through the Math Coach on pages 483–484. Math Coach videos are available via MyMathLab and YouTube. Step-by-step test solutions in the Chapter Test Prep Videos are also available via MyMathLab and YouTube. (Search "TobeyInterAlg" and click on "Channels.")

Solve the quadratic equations and simplify your answers. Use i notation for any complex numbers.

1. $8x^2 + 9x = 0$

2. $6x^2 - 3x = 1$

3. $\frac{3x}{2} - \frac{8}{3} = \frac{2}{3x}$

4. $x(x - 3) - 30 = 5(x - 2)$

5. $7x^2 - 4 = 52$

MC **6.** $\frac{2x}{2x + 1} - \frac{6}{4x^2 - 1} = \frac{x + 1}{2x - 1}$

7. $2x^2 - 6x + 5 = 0$

8. $2x(x - 3) = -3$

Solve for any valid real roots.

MC **9.** $x^4 - 11x^2 + 18 = 0$

10. $3x^{-2} - 11x^{-1} - 20 = 0$

11. $x^{2/3} - 3x^{1/3} - 4 = 0$

1. ______ ☐

2. ______ ☐

3. ______ ☐

4. ______ ☐

5. ______ ☐

6. ______ ☐

7. ______ ☐

8. ______ ☐

9. ______ ☐

10. ______ ☐

11. ______ ☐

12.

13.

14.

15.

16.

17.

18.

19.

20.

Total Correct:

Solve for the variable specified.

12. $B = \dfrac{xyw}{z^2}$; for z

MC 13. $5y^2 + 2by + 6w = 0$; for y

▲ 14. The area of a rectangle is 80 square miles. Its length is 1 mile longer than three times its width. Find its length and width.

▲ 15. Find the hypotenuse of a right triangle if the lengths of its legs are 6 and $2\sqrt{3}$.

16. Shirley and Bill paddled a canoe at a constant speed for 6 miles. They rested, had lunch, and then paddled 1 mile per hour faster for an additional 3 miles. The travel time for the entire trip was 4 hours. How fast did they paddle during each part of the trip?

MC 17. Find the vertex and the intercepts of $f(x) = -x^2 - 6x - 5$. Then graph the function.

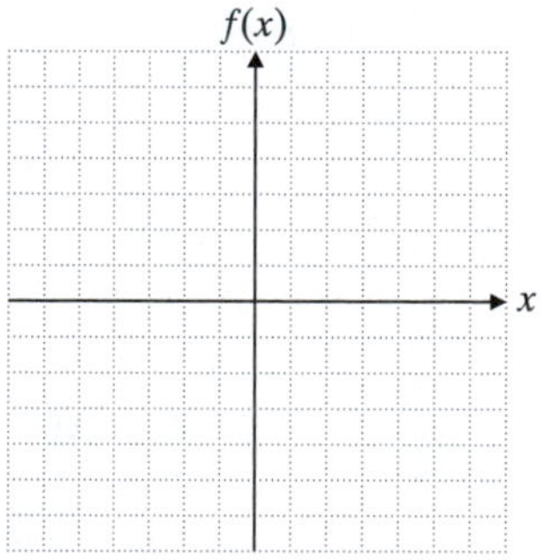

Solve.

18. $2x^2 + 3x \geq 27$

19. $x^2 - 5x - 14 < 0$

20. Use a calculator or square root table to approximate to the nearest tenth the solution to $x^2 + 3x - 7 > 0$.

MATH COACH

Mastering the skills you need to do well on the test.

Students often make the same types of errors when they do the Chapter 8 Test. Here are some helpful hints to keep you from making those common errors on test problems.

Solving Quadratic Equations Involving Fractions—Problem 6

Solve the quadratic equation. $\dfrac{2x}{2x+1} - \dfrac{6}{4x^2-1} = \dfrac{x+1}{2x-1}$

Helpful Hint If any denominators need to be factored, do that first. Then determine the LCD of all the denominators in the equation. Multiply each term of the equation by the LCD before solving for x.

Did you factor $4x^2 - 1$ as $(2x + 1)(2x - 1)$?

Yes ____ No ____

Did you identify the LCD to be $(2x + 1)(2x - 1)$?

Yes ____ No ____

If you answered No to these questions, review how to factor the difference of two squares and how to find the LCD of polynominal denominators.

Did you multiply the LCD by each term of the equation and remove parentheses to obtain $2x^2 - 5x - 7 = 0$?

Yes ____ No ____

Did you then use the quadratic formula and substitute for a, b, and c to get $x = \dfrac{-(-5) \pm \sqrt{(-5)^2 - 4(2)(-7)}}{2(2)}$?

Yes ____ No ____

If you answered No to these questions, remember that your final equation should be in the form $ax^2 + bx + c = 0$. Then use $a = 2$, $b = -5$, and $c = -7$ in the quadratic formula and simplify your result.

If you answered Problem 6 incorrectly, go back and rework the problem using these suggestions.

Solving Equations That Are Quadratic in Form—Problem 9

Solve for any valid real roots.
$x^4 - 11x^2 + 18 = 0$

Helpful Hint Let $y = x^2$ and then $y^2 = x^4$. Write the new quadratic equation after these replacements have been made.

After making the necessary replacements, did you obtain the equation $y^2 - 11y + 18 = 0$?

Yes ____ No ____

Did you solve the quadratic equation using any method to result in $y = 9$ and $y = 2$?

Yes ____ No ____

If you answered No to these questions, stop and complete these steps again.

If $y = 9$ and $y = 2$, can you conclude that $x^2 = 9$ and $x^2 = 2$?

Yes ____ No ____

If you take the square root of each side of each equation, can you obtain $x = \pm 3$ and $x = \pm\sqrt{2}$?

Yes ____ No ____

If you answered No to these equations, remember that when you take the square root of each side of the equation there are two sign possibilities. Note that your final solution should consist of four values for x.

Now go back and rework the problem using these suggestions.

Need help? Watch the MATH COACH videos in MyMathLab® or on YouTube™.

Solving a Quadratic Equation with Several Variables—Problem 13

Solve for the variable specified. $5y^2 + 2by + 6w = 0$; for y

Helpful Hint Think of the equation being written as $ay^2 + by + c = 0$. The quantities for a, b, or c may contain variables. Use the quadratic formula to solve for y.

Did you determine that $a = 5$, $b = 2b$, and $c = 6w$?

Yes ____ No ____

Did you substitute these values into the quadratic formula and simplify to obtain $y = \dfrac{-2b \pm \sqrt{4b^2 - 120w}}{10}$?

Yes ____ No ____

If you answered No to these questions, review the Helpful Hint again to make sure that you find the correct values for a, b, and c. Carefully substitute these values into the quadratic formula and simplify.

Were you able to simplify the radical expression by factoring out a 4 to obtain $\sqrt{4(b^2 - 30w)}$?

Yes ____ No ____

Did you simplify this expression further to get $2\sqrt{b^2 - 30w}$?

Yes ____ No ____

If you answered No to these questions, remember to always simplify radicals whenever possible. Make sure to write your solution as a simplified rational expression.

Now go back and rework the problem using these suggestions.

Find the Vertex, Intercepts, and Graph of a Quadratic Function—Problem 17

Find the vertex and the intercepts of $f(x) = -x^2 - 6x - 5$. Then graph the function.

Helpful Hint When the function is written in $f(x) = ax^2 + bx + c$ form, we can find the vertex by using $x = \dfrac{-b}{2a}$ to find the x value of the vertex. We can solve for the intercepts using the substitutions $x = 0$ and $f(x) = 0$ to find the unknown coordinates. And, if $a < 0$, the graph is a parabola opening downward.

Do you see that $a = -1$, $b = -6$, and $c = -5$?

Yes ____ No ____

Did you use $x = \dfrac{-b}{2a}$ to discover that the vertex point has an x-coordinate of -3?

Yes ____ No ____

If you answered No to these questions, notice that the function is written in $f(x) = ax^2 + bx + c$ form and review the formula: $x = \dfrac{-b}{2a}$. Substitute the resulting value for x into the original function to find the y-coordinate of the vertex point.

To find the y-intercept, did you substitute 0 for x into the original function to find the value for y?

Yes ____ No ____

After letting $f(x) = 0$ and substituting the values for a, b, and c into the quadratic formula, did you get the expression

$x = \dfrac{-(-6) \pm \sqrt{(-6)^2 - 4(-1)(-5)}}{2(-1)}$?

Yes ____ No ____

If you answered No to these questions, remember that the y-intercept will be an ordered pair in the form $(0, y)$, or in this case, $(0, f(x))$, and the x-intercept will be an ordered pair in the form $(x, 0)$. Be careful when substituting values for a, b, and c into the quadratic formula and remember to evaluate $\sqrt{16}$ as both 4 and -4. Simplify the expression to find the possible x-values.

Since $a < 0$, the parabola will open downward. Plot the vertex, x-intercept, and y-intercept points and connect these points with a curve to find the graph of the function.

If you answered Problem 17 incorrectly, go back and rework the problem using these suggestions.

Need more help? Look for section examples marked with MC to review.

CHAPTER 9

Comets often seem to be heading toward Earth but then veer off and head back into space. Some of them follow the path of a hyperbola during the time they are close to our planet. Thus the study of hyperbolas and other conic sections is very important to astronomers as they make predictions about how close a comet will come to Earth.

The Conic Sections

9.1 The Distance Formula and the Circle

Student Learning Objectives

After studying this section, you will be able to:

1. Find the distance between two points.
2. Find the center and radius of a circle and graph the circle if the equation is in standard form.
3. Write the equation of a circle in standard form given its center and radius.
4. Rewrite the equation of a circle in standard form.

In this chapter we'll talk about the equations and graphs of four special geometric figures—the circle, the parabola, the ellipse, and the hyperbola. These shapes are called **conic sections** because they can be formed by slicing a cone with a plane. The equation of any conic section is of degree 2.

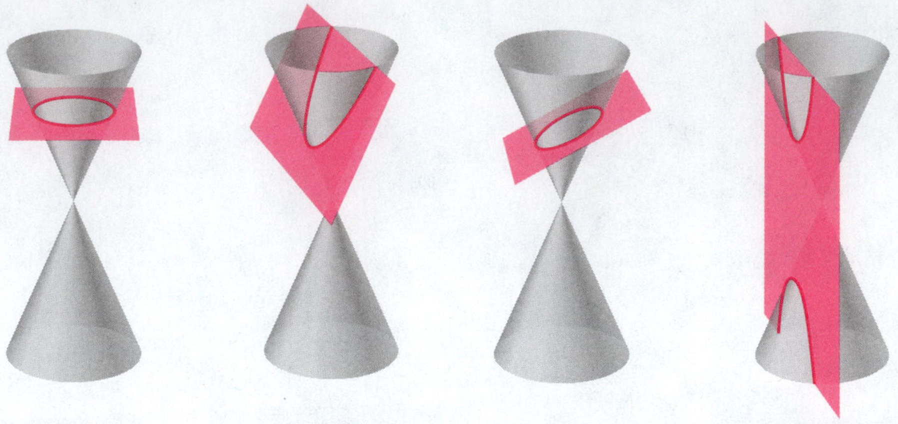

(a) Circle **(b)** Parabola **(c)** Ellipse **(d)** Hyperbola

Conic sections are an important and interesting subject. They are studied along with many other things in a branch of mathematics called *analytic geometry.* Conic sections can be found in applications of physics and engineering. Satellite transmission dishes have parabolic shapes; the orbits of planets are ellipses; the orbits of some comets are hyperbolas; the path of a ball, rocket, or bullet is a parabola (if we neglect air resistance).

1 Finding the Distance Between Two Points

Before we investigate the conic sections, we need to know how to find the distance between two points in the *xy*-plane. We will derive a *distance formula* and use it to find the equations for the conic sections.

Recall from Chapter 1 that to find the distance between two points on the real number line, we simply find the absolute value of the difference of the values of the points. For example, the distance from -3 to 5 on the *x*-axis is

$$|5 - (-3)| = |5 + 3| = 8.$$

Remember that absolute value is another name for distance. We could have written

$$|-3 - 5| = |-8| = 8.$$

Similarly, the distance from -3 to 5 on the *y*-axis is

$$|5 - (-3)| = 8.$$

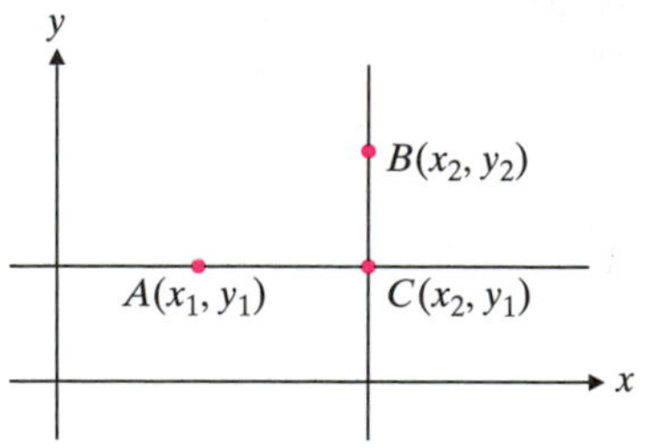

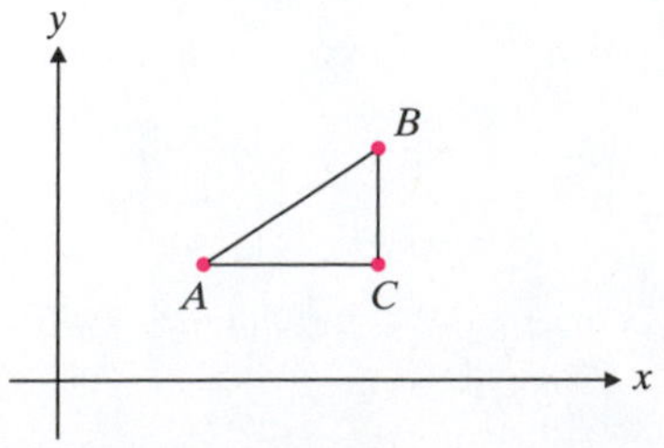

We use this simple fact to find the distance between two points in the *xy*-plane. Let $A(x_1, y_1)$ and $B(x_2, y_2)$ be points in a plane. First we draw a horizontal line through A, and then we draw a vertical line through B. (We could have drawn a horizontal line through B and a vertical line through A.) The lines intersect at point $C(x_2, y_1)$. Why are the coordinates (x_2, y_1)? The distance from A to C is $|x_2 - x_1|$ and from B to C, $|y_2 - y_1|$.

Now, if we draw a line from A to B, we have the right triangle ABC. We can use the Pythagorean Theorem to find the length (distance) of the line from A to B. By the Pythagorean Theorem,

$$(AB)^2 = (AC)^2 + (BC)^2.$$

Let's rename the distance AB as d. Then

$$d^2 = (|x_2 - x_1|)^2 + (|y_2 - y_1|)^2$$
$$d = \sqrt{(x_2 - x_1)^2 + (y_2 - y_1)^2}.$$

(Can you give a reason why we can drop the absolute value bars here?)

This is the **distance formula.**

DISTANCE FORMULA

The distance between two points (x_1, y_1) and (x_2, y_2) is

$$d = \sqrt{(x_2 - x_1)^2 + (y_2 - y_1)^2}.$$

***NOTE TO STUDENT:** Fully worked-out solutions to all of the Student Practice problems can be found at the back of the text starting at page SP-1.*

EXAMPLE 1 Find the distance between $(3, -4)$ and $(-2, -5)$.

Solution To use the formula, we arbitrarily let $(x_1, y_1) = (3, -4)$ and $(x_2, y_2) = (-2, -5)$.

$$\begin{aligned} d &= \sqrt{(x_2 - x_1)^2 + (y_2 - y_1)^2} \\ &= \sqrt{(-2 - 3)^2 + [-5 - (-4)]^2} \\ &= \sqrt{(-5)^2 + (-5 + 4)^2} \\ &= \sqrt{(-5)^2 + (-1)^2} \\ &= \sqrt{25 + 1} = \sqrt{26} \end{aligned}$$

Student Practice 1 Find the distance between $(-6, -2)$ and $(3, 1)$.

The choice of which point is (x_1, y_1) and which point is (x_2, y_2) is up to you. We would obtain exactly the same answer in Example 1 if $(x_1, y_1) = (-2, -5)$ and if $(x_2, y_2) = (3, -4)$. Try it for yourself and see whether you obtain the same result.

② Finding the Center and Radius of a Circle and Graphing the Circle

A **circle** is defined as the set of all points in a plane that are at a fixed distance from a point in that plane. The fixed distance is called the **radius,** and the point is called the **center** of the circle.

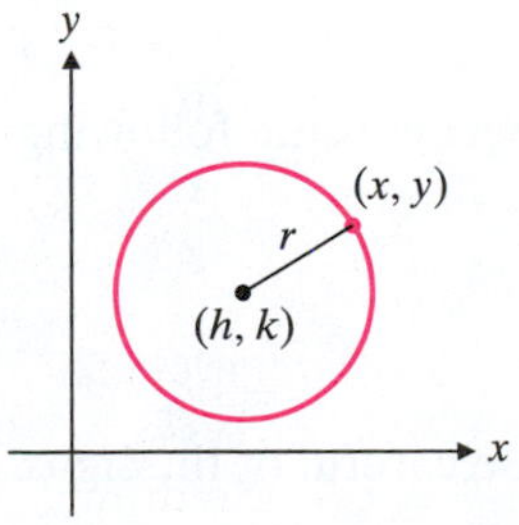

We can use the distance formula to find the equation of a circle. Let a circle of radius r have its center at (h, k). For any point (x, y) on the circle, the distance formula tells us that

$$\sqrt{(x - h)^2 + (y - k)^2} = r.$$

Squaring each side gives

$$(x - h)^2 + (y - k)^2 = r^2.$$

This is the equation of a circle with center at (h, k) and radius r.

Graphing Calculator

Graphing Circles

A graphing calculator is designed to graph *functions.* In order to graph a circle, you need to separate it into two halves, each of which is a function. Thus, in order to graph the circle in Example 2 on the next page, first solve for y.

$$(y - 3)^2 = 25 - (x - 2)^2$$
$$y - 3 = \pm\sqrt{25 - (x - 2)^2}$$
$$y = 3 \pm \sqrt{25 - (x - 2)^2}$$

Now graph the two functions

$$y_1 = 3 + \sqrt{25 - (x - 2)^2}$$

(the upper half of the circle) and

$$y_2 = 3 - \sqrt{25 - (x - 2)^2}$$

(the lower half of the circle).

To get a proper-looking circle, use a "square" window setting. Window settings will vary depending on the calculator. Display:

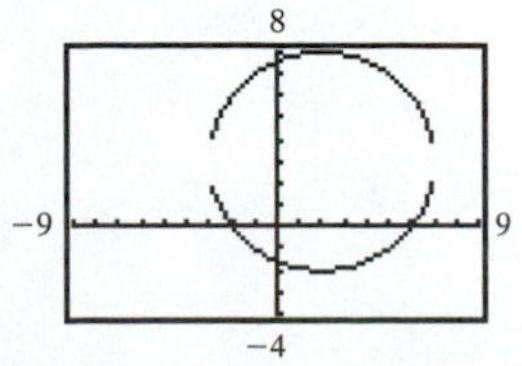

Notice that due to limitations in the calculator, it is not a perfect circle and two gaps appear.

STANDARD FORM OF THE EQUATION OF A CIRCLE

The standard form of the equation of a circle with center at (h, k) and radius r is

$$(x - h)^2 + (y - k)^2 = r^2.$$

EXAMPLE 2 Find the center and radius of the circle $(x - 2)^2 + (y - 3)^2 = 25$. Then sketch its graph.

Solution From the equation of a circle,

$$(x - h)^2 + (y - k)^2 = r^2,$$

we see that $(h, k) = (2, 3)$. Thus, the center of the circle is at $(2, 3)$. Since $r^2 = 25$, the radius of the circle is $r = 5$.

The graph of this circle is shown on the right.

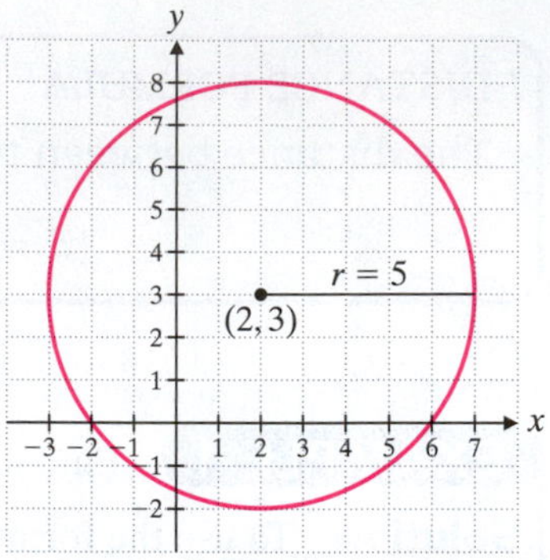

$(x - 2)^2 + (y - 3)^2 = 25$

Student Practice 2 Find the center and radius of the circle

$$(x + 1)^2 + (y + 2)^2 = 9.$$

Then sketch its graph.

Student Practice 2

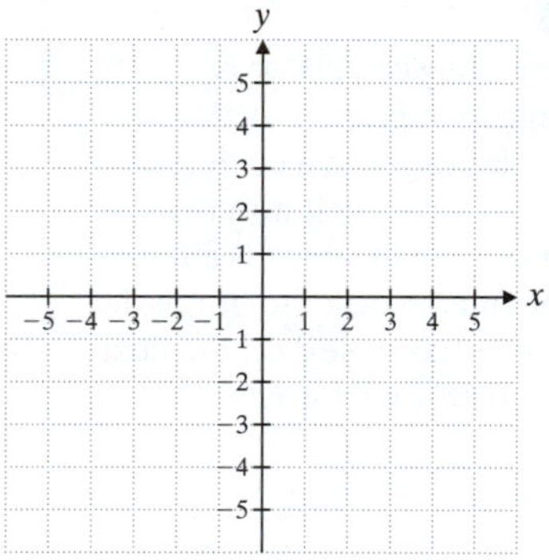

③ Writing the Equation of a Circle in Standard Form Given the Center and Radius

We can write the standard form of the equation of a specific circle if we are given the center and the radius. We use the definition of the standard form of the equation of a circle to write the equation we want.

EXAMPLE 3 Write the equation of the circle with center $(-1, 3)$ and radius $\sqrt{5}$. Put your answer in standard form.

Solution We are given that $(h, k) = (-1, 3)$ and $r = \sqrt{5}$. Thus,

$$(x - h)^2 + (y - k)^2 = r^2$$

becomes the following:

$$[x - (-1)]^2 + (y - 3)^2 = \left(\sqrt{5}\right)^2$$
$$(x + 1)^2 + (y - 3)^2 = 5$$

Be careful of the signs. It is easy to make a sign error in these steps.

Student Practice 3 Write the equation of the circle with center $(-5, 0)$ and radius $\sqrt{3}$. Put your answer in standard form.

④ Rewriting the Equation of a Circle in Standard Form

The standard form of the equation of a circle helps us sketch the graph of the circle. Sometimes the equation of a circle is not given in standard form, and we need to rewrite the equation.

EXAMPLE 4 Write the equation of the circle $x^2 + 2x + y^2 + 6y + 6 = 0$ in standard form. Find the radius and center of the circle and sketch its graph.

Solution The standard form of the equation of a circle is

$$(x - h)^2 + (y - k)^2 = r^2.$$

If we multiply out the terms in the equation, we get

$$(x^2 - 2hx + h^2) + (y^2 - 2ky + k^2) = r^2.$$

Comparing this with the equation we were given,

$$(x^2 + 2x) + (y^2 + 6y) = -6,$$

suggests that we can complete the squares to put the equation in standard form.

$$x^2 + 2x + \_\_\_\_ + y^2 + 6y + \_\_\_\_ = -6$$

$$x^2 + 2x + 1 + y^2 + 6y + 9 = -6 + 1 + 9$$

$$x^2 + 2x + 1 + y^2 + 6y + 9 = 4$$

$$(x + 1)^2 + (y + 3)^2 = 4$$

Thus, the center is at $(-1, -3)$, and the radius is 2. The sketch of the circle is shown below.

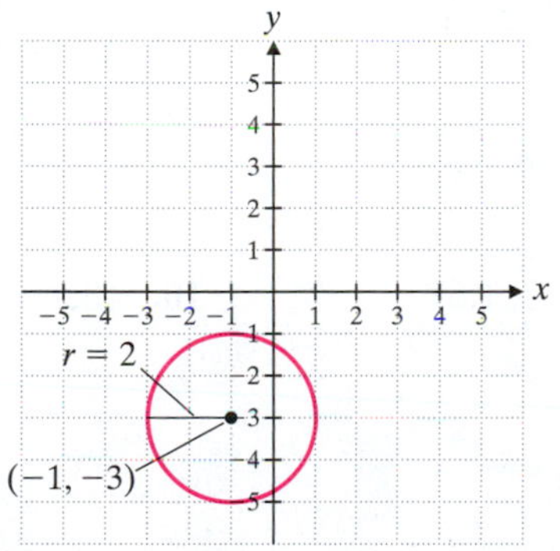

$(x + 1)^2 + (y + 3)^2 = 4$

Student Practice 4 Write the equation of the circle $x^2 + 4x + y^2 + 2y - 20 = 0$ in standard form. Find the radius and center of the circle and sketch its graph.

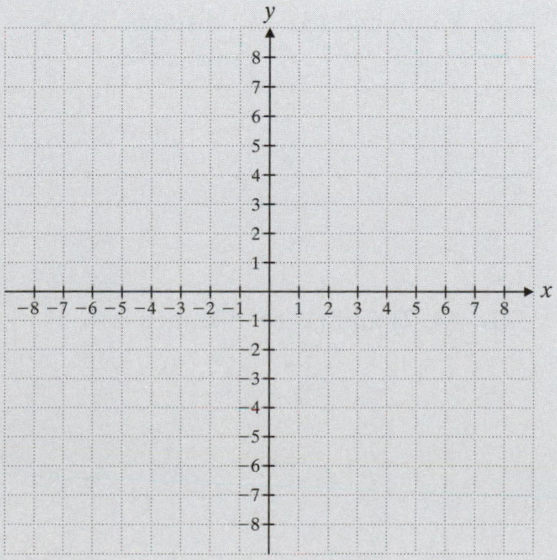

Graphing Calculator

Exploration

One way to graph the equation in Example 4 is to write it as a quadratic equation in y and then employ the quadratic formula.

$$y^2 + 6y + (x^2 + 2x + 6) = 0$$
$$ay^2 + by + c = 0$$

$a = 1, b = 6$, and
$c = x^2 + 2x + 6$

$$y = \frac{-6 \pm \sqrt{36 - 4(1)(x^2 + 2x + 6)}}{2(1)}$$

$$y = \frac{-6 \pm \sqrt{12 - 8x - 4x^2}}{2}$$

Thus, we have the two halves of the circle.

$$y_1 = \frac{-6 + \sqrt{12 - 8x - 4x^2}}{2}$$

$$y_2 = \frac{-6 - \sqrt{12 - 8x - 4x^2}}{2}$$

We can graph these on one coordinate system to obtain the graph. Some graphing calculators have a feature for getting a background grid for your graph. If you have this feature, use it to find the coordinates of the center.

9.1 Exercises

MyMathLab®

Watch the videos in MyMathLab

Download the MyDashBoard App

Verbal and Writing Skills, Exercises 1–4

1. Explain how you would find the distance from -2 to 4 on the y-axis.

2. Explain how you would find the distance between $(3, -1)$ and $(-4, 0)$ in the xy-plane.

3. $(x - 1)^2 + (y + 2)^2 = 9$ is the equation of a circle. Explain how to determine the center and the radius of the circle.

4. $x^2 - 6x + y^2 - 2y = 6$ is the equation of a circle. Explain how you would rewrite the equation in standard form.

Find the distance between each pair of points. Simplify your answers.

5. $(1, 6)$ and $(2, 4)$

6. $(4, 6)$ and $(7, 5)$

7. $(-5, 2)$ and $(1, 5)$

8. $(0, -2)$ and $(5, 3)$

9. $(4, -5)$ and $(-2, -13)$

10. $(-7, 13)$ and $(-12, 1)$

11. $\left(\frac{5}{4}, -\frac{1}{3}\right)$ and $\left(\frac{1}{4}, -\frac{2}{3}\right)$

12. $\left(-1, \frac{1}{5}\right)$ and $\left(-\frac{1}{2}, \frac{11}{5}\right)$

13. $\left(\frac{1}{3}, \frac{3}{5}\right)$ and $\left(\frac{7}{3}, \frac{1}{5}\right)$

14. $\left(-\frac{1}{4}, \frac{1}{7}\right)$ and $\left(\frac{3}{4}, \frac{6}{7}\right)$

15. $(1.3, 2.6)$ and $(-5.7, 1.6)$

16. $(8.2, 3.5)$ and $(6.2, -0.5)$

Find the two values of the unknown coordinate so that the distance between the points is as given.

17. $(7, 2)$ and $(1, y)$; distance is 10

18. $(3, y)$ and $(3, -5)$; distance is 9

19. $(1.5, 2)$ and $(0, y)$; distance is 2.5

20. $\left(1, \frac{15}{2}\right)$ and $\left(x, -\frac{1}{2}\right)$; distance is 10

21. $(4, 7)$ and $(x, 5)$; distance is $\sqrt{5}$

22. $(1, 6)$ and $(3, y)$; distance is $2\sqrt{2}$

Applications, Exercises 23 and 24

Radar Detection *Use the following information to solve exercises 23 and 24. An airport is located at point O. A short-range radar tower is located at point R. The maximum range at which the radar can detect a plane is 4 miles from point R.*

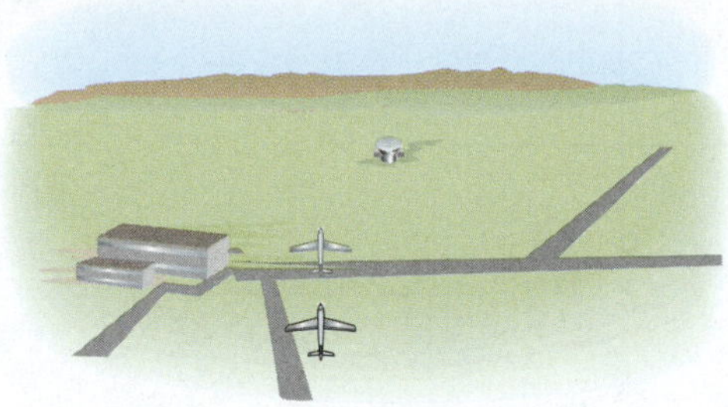

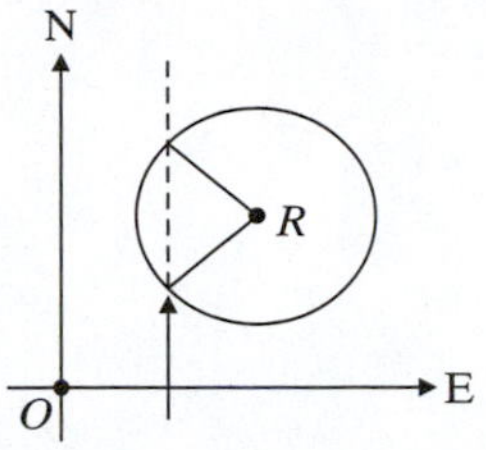

23. Assume that R is 5 miles east of O and 7 miles north of O. In other words, R is located at the point $(5, 7)$. An airplane is flying parallel to and 2 miles east of the north axis. (In other words, the plane is flying along the path $x = 2$.) What is the *shortest distance* north of the airport at which the plane can be detected by the radar tower at R? Round your answer to the nearest tenth of a mile.

24. Assume that R is 6 miles east of O and 6 miles north of O. In other words, R is located at the point $(6, 6)$. An airplane is flying parallel to and 4 miles east of the north axis. (In other words, the plane is flying along the path $x = 4$.) What is the *greatest distance* north of the airport at which the plane can still be detected by the radar tower at R? Round your answer to the nearest tenth of a mile.

Write in standard form the equation of the circle with the given center and radius.

25. Center $(-3, 7)$; $r = 6$

26. Center $(9, -4)$; $r = 5$

27. Center $(-2.4, 0)$; $r = \frac{3}{4}$

28. Center $\left(0, \frac{1}{2}\right)$; $r = \frac{1}{3}$

29. Center $\left(0, \frac{3}{8}\right)$; $r = \sqrt{3}$

30. Center $(-2.5, 0)$; $r = \sqrt{6}$

Give the center and radius of each circle. Then sketch its graph.

31. $x^2 + y^2 = 25$

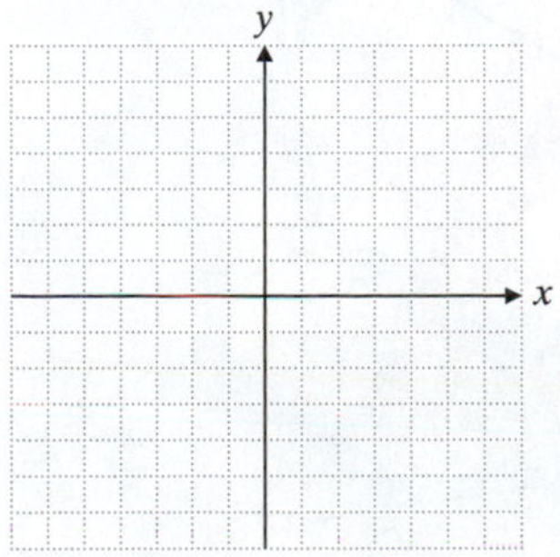

32. $x^2 + y^2 = 9$

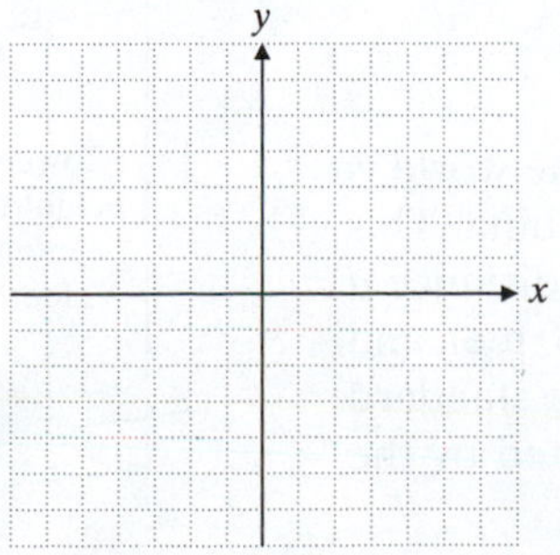

33. $(x - 5)^2 + (y - 3)^2 = 16$

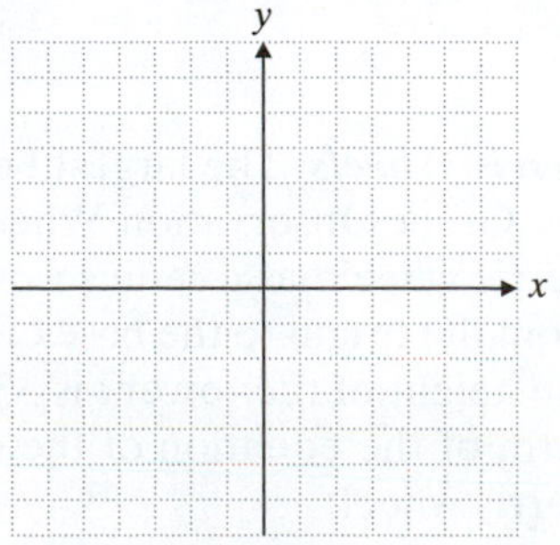

34. $(x - 3)^2 + (y - 2)^2 = 4$

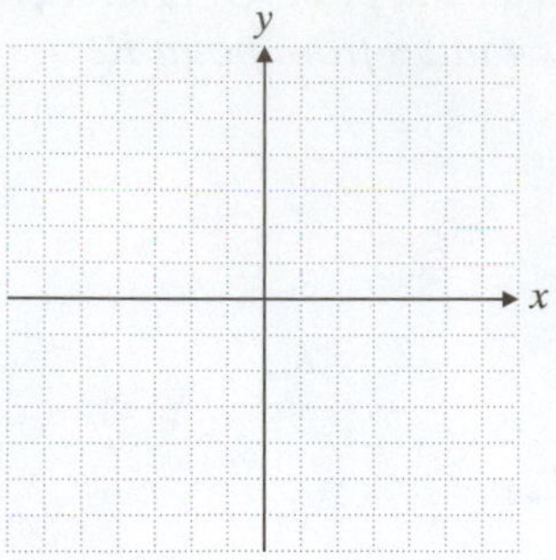

35. $(x + 2)^2 + (y - 3)^2 = 25$

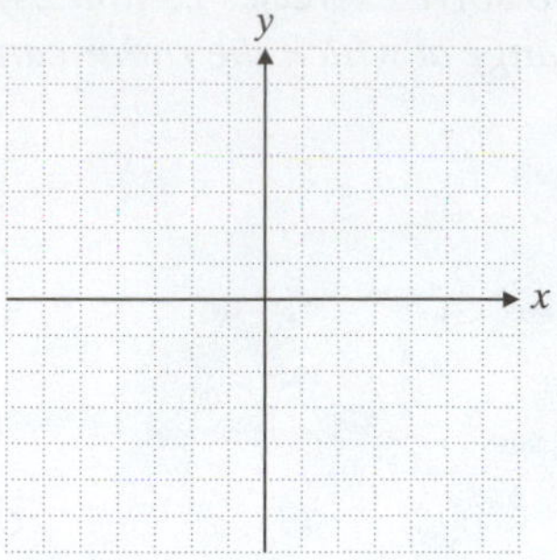

36. $\left(x - \frac{3}{2}\right)^2 + (y + 2)^2 = 9$

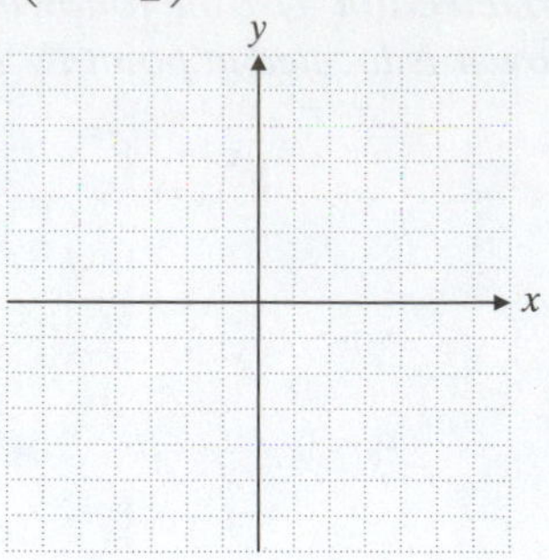

Rewrite each equation in standard form, using the approach of Example 4. Find the center and radius of each circle.

37. $x^2 + y^2 + 8x - 6y - 24 = 0$

38. $x^2 + y^2 + 6x - 4y - 3 = 0$

39. $x^2 + y^2 - 10x + 6y - 2 = 0$

40. $x^2 + y^2 - 4x + 8y + 16 = 0$

41. $x^2 + y^2 + 3x - 2 = 0$

42. $x^2 + y^2 - 5x - 1 = 0$

43. ***Ferris Wheels*** A Ferris wheel has a radius r of 44.2 feet. The height of the tower t is 55.8 feet. The distance d from the origin to the base of the tower is 61.5 feet. Find the standard form of the equation of the circle represented by the Ferris wheel.

44. ***Ferris Wheels*** The largest Ferris wheel in the world is the Great Observation Wheel in Beijing, China. This Ferris wheel has a radius r of 325 feet. The distance d from the origin to the base of the tower is 340 feet, and the height of the tower t is 357 feet. Find the standard form of the equation of the circle represented by the Ferris wheel.

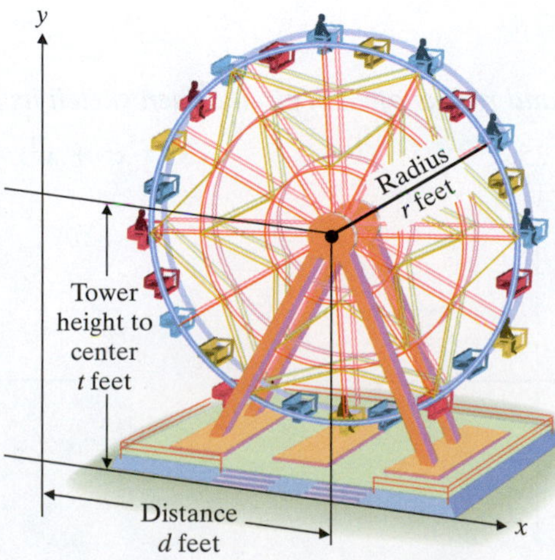

Optional Graphing Calculator Problems *Graph each circle with your graphing calculator.*

45. $(x - 5.32)^2 + (y + 6.54)^2 = 47.28$

46. $x^2 + 9.56x + y^2 - 7.12y + 8.9995 = 0$

Cumulative Review *Solve the quadratic equations by using the quadratic formula.*

47. **[8.2.1]** $4x^2 + 2x = 1$

48. **[8.2.1]** $5x^2 - 6x - 7 = 0$

▲ **49.** **[1.6.2]** ***Volcano Eruptions*** The 1980 eruptions of Mt. Saint Helens blew down or scorched 230 square miles of forest. A deposit of rock and sediments soon filled up a 20-square-mile area to an average depth of 150 feet. How many cubic feet of rock and sediments settled in this region?

50. **[1.2.2]** ***Volcano Eruptions*** Within a 15-mile radius north of Mt. Saint Helens, the blast of its 1980 eruption traveled at up to 670 miles per hour. If an observer 15 miles north of the volcano saw the blast and attempted to run for cover, how many seconds did he have to run before the blast reached his original location?

Quick Quiz 9.1

1. Find the distance between $(3, -4)$ and $(-2, -6)$.

2. Write in standard form the equation of the circle with center $(5, -6)$ and radius 7.

3. Rewrite the equation in standard form. Find the center and radius of the circle.

$$x^2 + 4x + y^2 - 6y + 4 = 0$$

4. **Concept Check** Explain how you would find the values of the unknown coordinate x if the distance between $(-6, 8)$ and $(x, 12)$ is 4.

9.2 The Parabola

Student Learning Objectives

After studying this section, you will be able to:

1. **Graph vertical parabolas.**
2. **Graph horizontal parabolas.**
3. **Rewrite the equation of a parabola in standard form.**

If we pass a plane through a cone so that the plane is parallel to but not touching a side of the cone, we form a parabola. A **parabola** is defined as the set of points that are the same distance from some fixed line (called the **directrix**) and some fixed point (called the **focus**) that is *not* on the line.

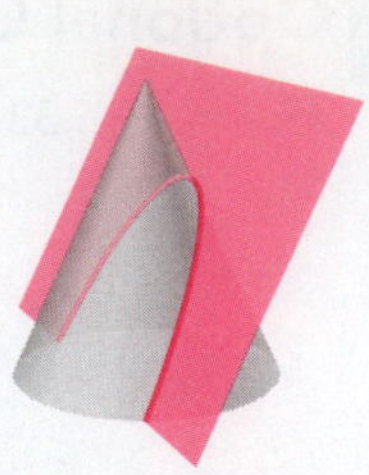

The shape of a parabola is a common one. For example, the cables that are used to support the weight of a bridge are often in the shape of parabolas.

The simplest form for the equation is one variable = (another variable)^2. That is, $y = x^2$ or $x = y^2$. We will make a table of values for each equation, plot the points, and draw a graph. For the first equation we choose values for x and find y. For the second equation we choose values for y and find x.

$y = x^2$

x	y
−2	4
−1	1
0	0
1	1
2	4

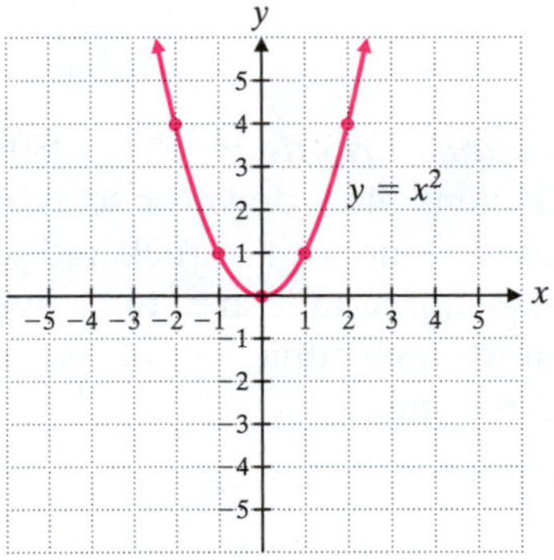

Vertical Parabola

$x = y^2$

x	y
4	−2
1	−1
0	0
1	1
4	2

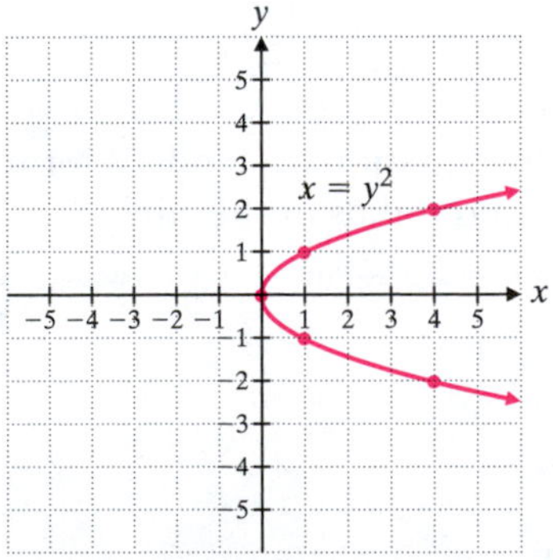

Horizontal Parabola

Notice that the graph of $y = x^2$ is symmetric about the y-axis. That is, if you folded the graph along the y-axis, the two parts of the curve would coincide. For this parabola, the y-axis is the **axis of symmetry.**

What is the axis of symmetry for the parabola $x = y^2$? Every parabola has an axis of symmetry. This axis can be *any* line; it depends on the location and orientation of the parabola in the rectangular coordinate system. The point at which the parabola crosses the axis of symmetry is the **vertex.** What are the coordinates of the vertex for $y = x^2$? For $x = y^2$?

① Graphing Vertical Parabolas

EXAMPLE 1 Graph $y = (x - 2)^2$. Identify the vertex and the axis of symmetry.

Student Practice 1

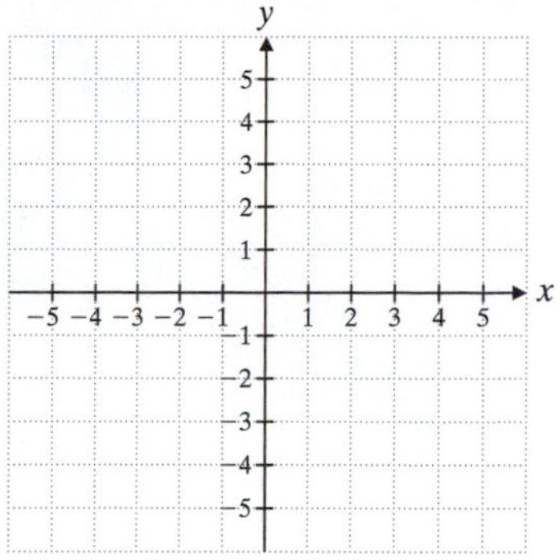

Solution We make a table of values. We begin with $x = 2$ in the middle of the table of values because $(2 - 2)^2 = 0$. That is, when $x = 2$, $y = 0$. We then fill in the x- and y-values above and below $x = 2$. We plot the points and draw the graph.

$y = (x - 2)^2$

x	y
4	4
3	1
2	0
1	1
0	4

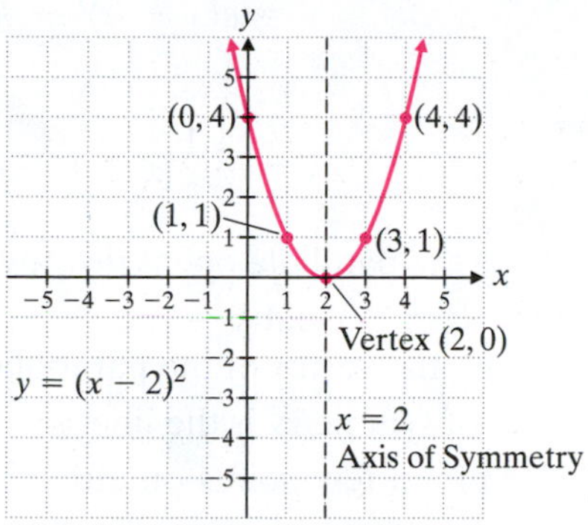

The vertex is $(2, 0)$, and the axis of symmetry is the line $x = 2$.

***NOTE TO STUDENT:** Fully worked-out solutions to all of the Student Practice problems can be found at the back of the text starting at page SP-1.*

Student Practice 1 Graph $y = -(x + 3)^2$. Identify the vertex and the axis of symmetry.

EXAMPLE 2 Graph $y = (x - 2)^2 + 3$. Find the vertex, the axis of symmetry, and the y-intercept.

Solution This graph looks just like the graph of $y = x^2$, except that it is shifted 2 units to the right and 3 units up. The vertex is $(2, 3)$. The axis of symmetry is $x = 2$. We can find the y-intercept by letting $x = 0$ in the equation. We get

$$y = (0 - 2)^2 + 3 = 4 + 3 = 7.$$

Thus, the y-intercept is $(0, 7)$.

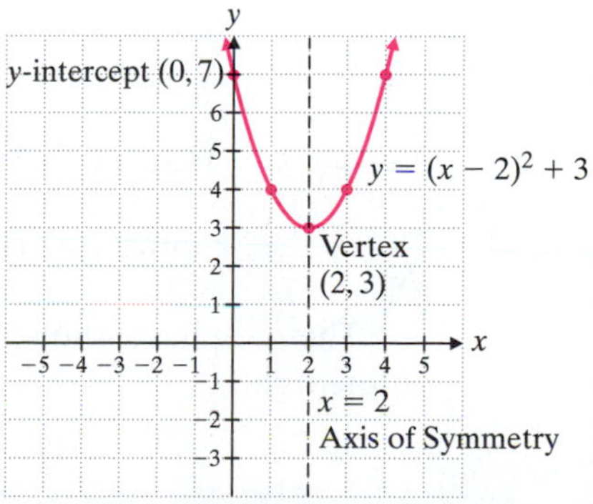

Student Practice 2 Graph the parabola $y = (x - 6)^2 + 4$.

Student Practice 2

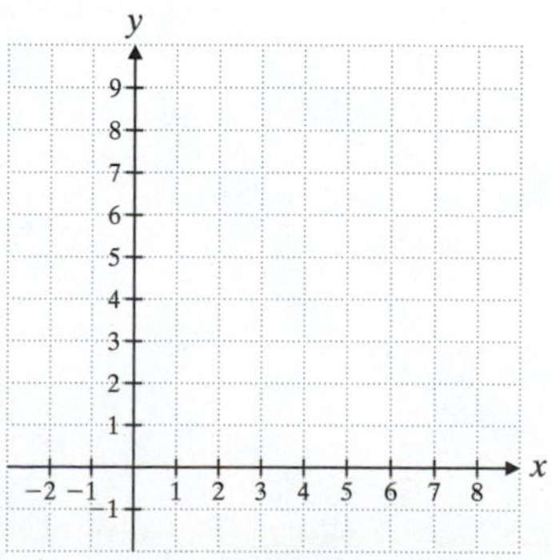

The examples we have studied illustrate the following properties of the standard form of the equation of a vertical parabola.

STANDARD FORM OF THE EQUATION OF A VERTICAL PARABOLA

1. The graph of $y = a(x - h)^2 + k$, where $a \neq 0$, is a vertical parabola.
2. The parabola opens upward ◡ if $a > 0$ and downward ◠ if $a < 0$.
3. The vertex of the parabola is (h, k).
4. The axis of symmetry is the line $x = h$.
5. The y-intercept is the point where the parabola crosses the y-axis (i.e., where $x = 0$).

We can use these properties as steps to graph a parabola. If we want greater accuracy, we should also plot a few other points.

EXAMPLE 3 Graph $y = -\frac{1}{2}(x + 3)^2 - 1$.

Solution

Step 1 The equation has the form $y = a(x - h)^2 + k$, where $a = -\frac{1}{2}$, $h = -3$, and $k = -1$, so it is a vertical parabola.

$$y = a(x - h)^2 + k$$

$$y = -\frac{1}{2}[x - (-3)]^2 + (-1)$$

Step 2 $a < 0$; so the parabola opens downward.

Step 3 We have $h = -3$ and $k = -1$.
Therefore, the vertex of the parabola is $(-3, -1)$.

Step 4 The axis of symmetry is the line $x = -3$.

We plot a few points on either side of the axis of symmetry. We try $x = -1$ because $(-1 + 3)^2$ is 4 and $-\frac{1}{2}(4)$ is an integer. We avoid fractions. When $x = -1$, $y = -\frac{1}{2}(-1 + 3)^2 - 1 = -3$. Thus, the point is $(-1, -3)$. The image of this point on the other side of the axis of symmetry is $(-5, -3)$. We now try $x = 1$. When $x = 1$, $y = -\frac{1}{2}(1 + 3)^2 - 1 = -9$. Thus, the point is $(1, -9)$. The image of this point on the other side of the axis of symmetry is $(-7, -9)$.

Step 5 When $x = 0$, we have the following:

$$y = -\frac{1}{2}(0 + 3)^2 - 1$$
$$= -\frac{1}{2}(9) - 1$$
$$= -4.5 - 1 = -5.5$$

Thus, the y-intercept is $(0, -5.5)$.

The graph is shown on the right.

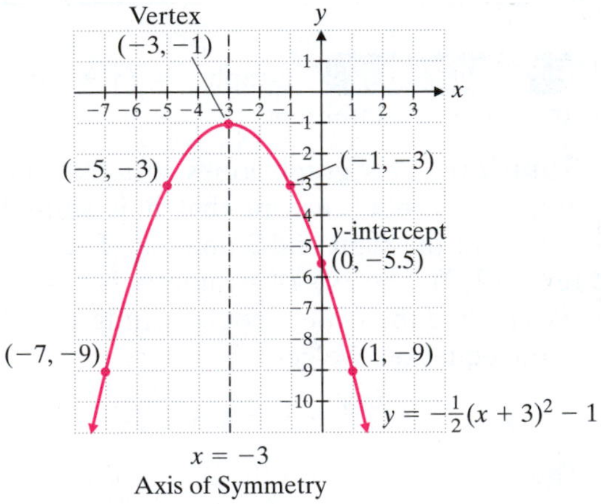

Student Practice 3

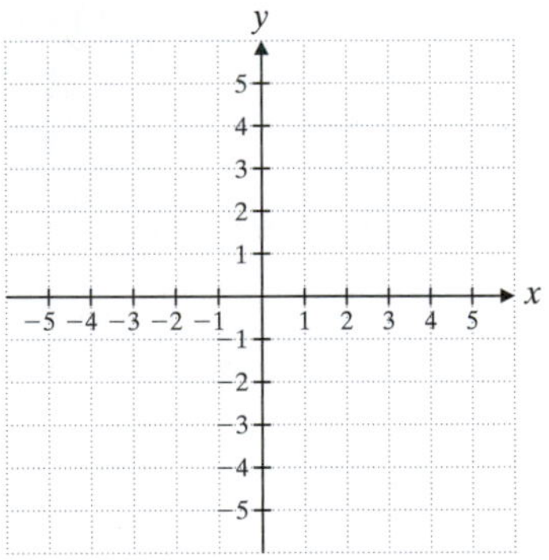

Student Practice 3 Graph $y = \frac{1}{4}(x - 2)^2 + 3$.

② Graphing Horizontal Parabolas

Recall that the equation $x = y^2$, in which the squared term is the y-variable, describes a horizontal parabola. Horizontal parabolas open to the left or right. They are symmetric about the x-axis or about a line parallel to the x-axis. We now look at examples of horizontal parabolas.

EXAMPLE 4 Graph $x = -2y^2$.

Solution Notice that the y-term is squared. This means that the parabola is horizontal. We make a table of values, plot points, and draw the graph. To make the table of values, we choose values for y and find x. We begin with $y = 0$.

$x = -2y^2$

x	y
-8	-2
-2	-1
0	0
-2	1
-8	2

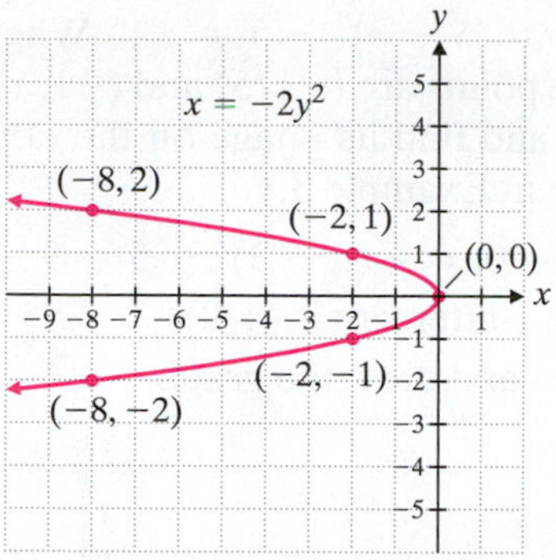

The parabola $x = -2y^2$ has its vertex at $(0, 0)$. The axis of symmetry is the x-axis.

Student Practice 4 Graph the parabola $x = -2y^2 + 4$.

Student Practice 4

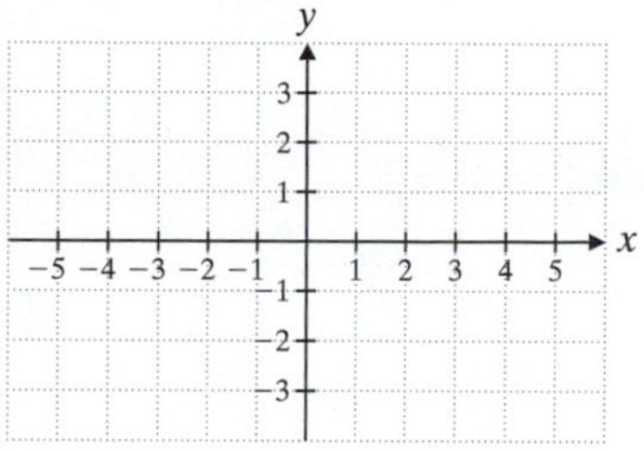

TO THINK ABOUT: Example 4 Follow-up Compare the graphs in Example 4 and Student Practice 4 to the graph of $x = y^2$. How are they different? How are they the same? What does the coefficient -2 in the equation $x = -2y^2$ do to the graph of the equation $x = y^2$? What does the constant 4 in $x = -2y^2 + 4$ do to the graph of $x = -2y^2$?

Now we can make the same type of observations for horizontal parabolas as we did for vertical ones.

STANDARD FORM OF THE EQUATION OF A HORIZONTAL PARABOLA

1. The graph of $x = a(y - k)^2 + h$, where $a \neq 0$, is a horizontal parabola.
2. The parabola opens to the right (if $a > 0$ and opens to the left) if $a < 0$.
3. The vertex of the parabola is (h, k).
4. The axis of symmetry is the line $y = k$.
5. The x-intercept is the point where the parabola crosses the x-axis (i.e., where $y = 0$).

EXAMPLE 5 Graph $x = (y - 3)^2 - 5$. Find the vertex, the axis of symmetry, and the x-intercept.

Solution

Step 1 The equation has the form $x = a(y - k)^2 + h$, where $a = 1$, $k = 3$, and $h = -5$, so it is a horizontal parabola.

$$x = a(y - k)^2 + h$$

$$\downarrow \quad \downarrow \quad \downarrow$$

$$x = 1(y - 3)^2 + (-5)$$

Step 2 $a > 0$; so the parabola opens to the right.

Step 3 We have $k = 3$ and $h = -5$. Therefore, the vertex is $(-5, 3)$.

Step 4 The line $y = 3$ is the axis of symmetry.

We look for a few points on either side of the axis of symmetry. We will try y-values close to the vertex $(-5, 3)$. We try $y = 4$ and $y = 2$. When $y = 4$, $x = (4 - 3)^2 - 5 = -4$. When $y = 2$, $x = (2 - 3)^2 - 5 = -4$. Thus, the points are $(-4, 4)$ and $(-4, 2)$. (Remember to list the x-value first in a coordinate pair.) We try $y = 5$ and $y = 1$. When

Continued on next page

$y = 5$, $x = (5 - 3)^2 - 5 = -1$. When $y = 1$, $x = (1 - 3)^2 - 5 = -1$. Thus, the points are $(-1, 5)$ and $(-1, 1)$. You may prefer to find one point, graph it, and find its image on the other side of the axis of symmetry, as was done in Example 3.

Step 5 When $y = 0$, $x = (0 - 3)^2 - 5 = 9 - 5 = 4$.
Thus, the x-intercept is $(4, 0)$.

We plot the points and draw the graph.

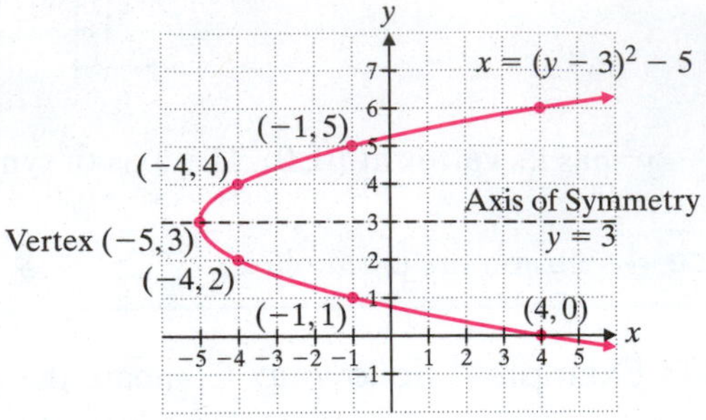

Notice that the graph also crosses the y-axis. You can find the y-intercepts by setting x equal to 0 and solving the resulting quadratic equation. Try it.

Student Practice 5 Graph the parabola $x = -(y + 1)^2 - 3$. Find the vertex, the axis of symmetry, and the x-intercept.

Student Practice 5

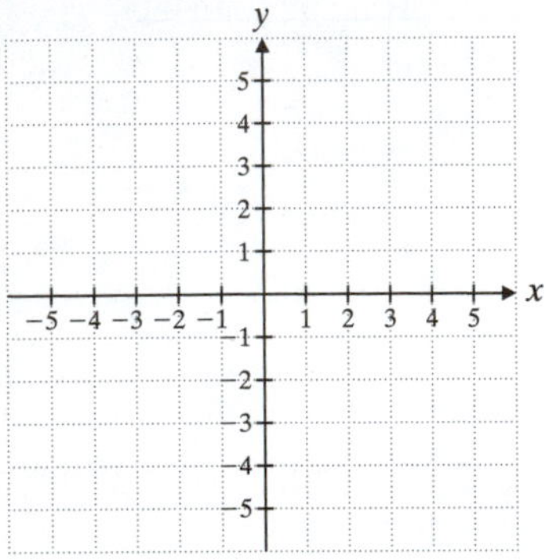

③ Rewriting the Equation of a Parabola in Standard Form

EXAMPLE 6 Place the equation $x = y^2 + 4y + 1$ in standard form. Then graph it.

Solution Since the y-term is squared, we have a horizontal parabola. So the standard form is

$$x = a(y - k)^2 + h.$$

Now we have the following:

$x = y^2 + 4y + \_\_\_\_ - \_\_\_\_ + 1$ Whatever we add to the right side we must also subtract from the right side.

$x = y^2 + 4y + \left(\frac{4}{2}\right)^2 - \left(\frac{4}{2}\right)^2 + 1$ Complete the square.

$x = (y^2 + 4y + 4) - 3$ Simplify.

$x = (y + 2)^2 - 3$ Standard form.

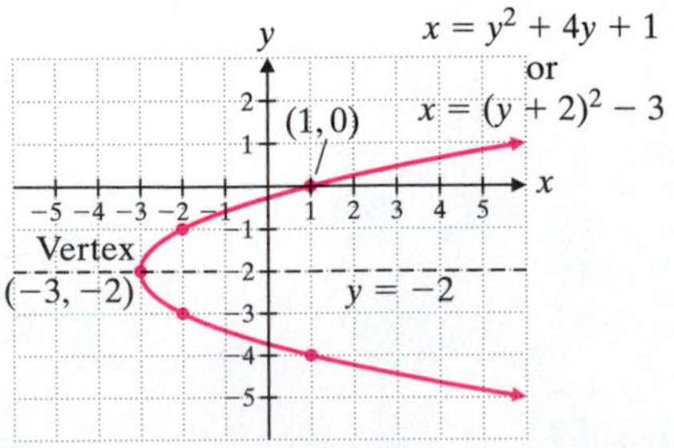

We see that $a = 1$, $k = -2$, and $h = -3$. Since a is positive, the parabola opens to the right. The vertex is $(-3, -2)$. The axis of symmetry is $y = -2$. If we let $y = 0$, we find that the x-intercept is $(1, 0)$.

Student Practice 6 Place the equation $x = y^2 - 6y + 13$ in standard form and graph it.

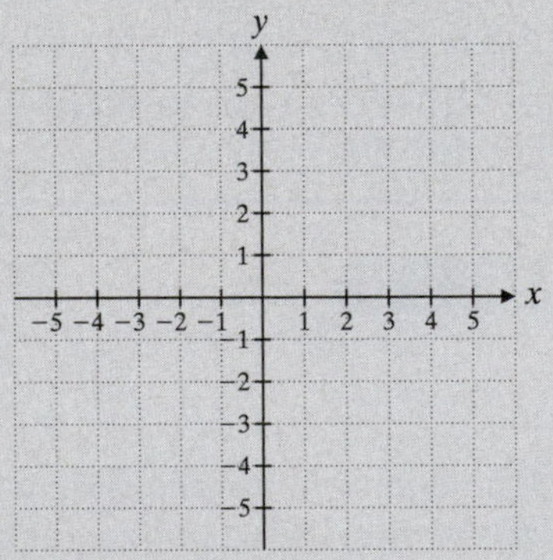

EXAMPLE 7 Place the equation $y = 2x^2 - 4x - 1$ in standard form. Then graph it.

Solution This time the x-term is squared, so we have a vertical parabola. The standard form is

$$y = a(x - h)^2 + k.$$

We need to complete the square.

$$\begin{aligned} y &= 2(x^2 - 2x + \underline{\quad}) - \underline{\quad} - 1 \\ &= 2[x^2 - 2x + (1)^2] - 2(1)^2 - 1 \\ &= 2(x - 1)^2 - 3 \end{aligned}$$

The parabola opens upward ($a > 0$), the vertex is $(1, -3)$, the axis of symmetry is $x = 1$, and the y-intercept is $(0, -1)$.

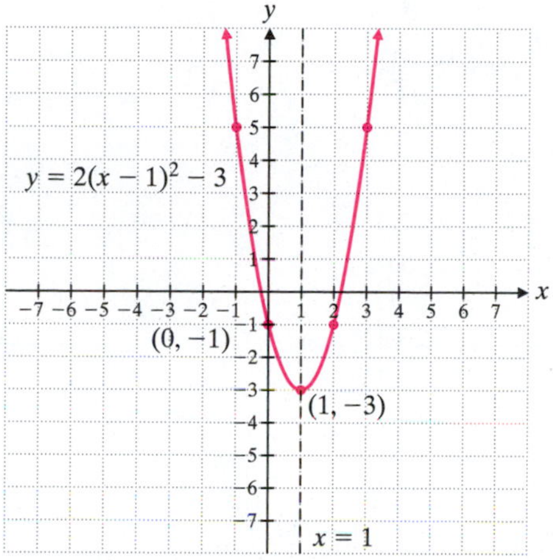

Student Practice 7 Place $y = 2x^2 + 8x + 9$ in standard form and graph it.

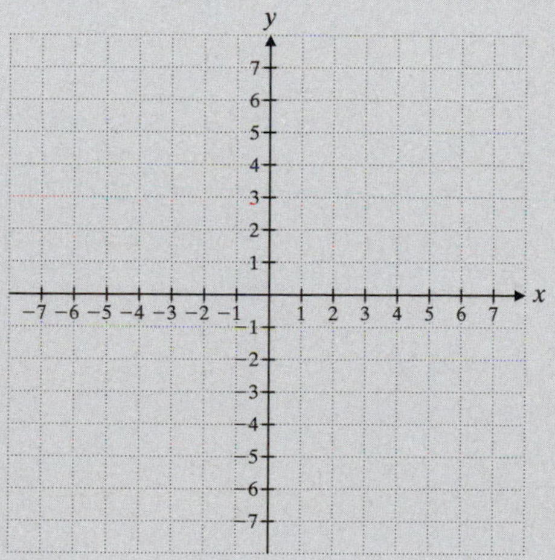

Graphing Calculator

Graphing Parabolas

Graphing horizontal parabolas such as the one in Example 6 on a graphing calculator requires dividing the curve into two halves. In this case the halves would be

$$y_1 = -2 + \sqrt{x + 3}$$

and

$$y_2 = -2 - \sqrt{x + 3}$$

Vertical parabolas can be graphed immediately on a graphing calculator. Why is this? How can you tell whether it is necessary to divide a curve into two halves?

9.2 Exercises

MyMathLab®

Watch the videos in MyMathLab

Download the MyDashBoard App

Verbal and Writing Skills, Exercises 1–4

1. The graph of $y = x^2$ is symmetric about the ________. The graph of $x = y^2$ is symmetric about the ________.

2. Explain how to determine the axis of symmetry of the parabola $x = \frac{1}{2}(y + 5)^2 - 1$.

3. Explain how to determine the vertex of the parabola $y = 2(x - 3)^2 + 4$.

4. How does the coefficient -6 affect the graph of the parabola $y = -6x^2$?

Graph each parabola and label the vertex. Find the y-intercept.

5. $y = -4x^2$

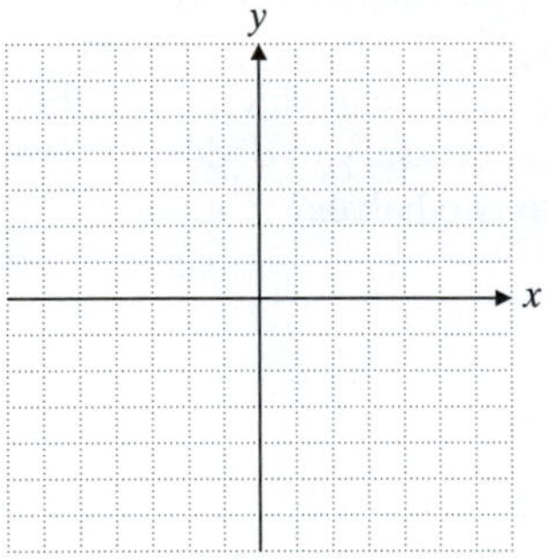

6. $y = -3x^2$

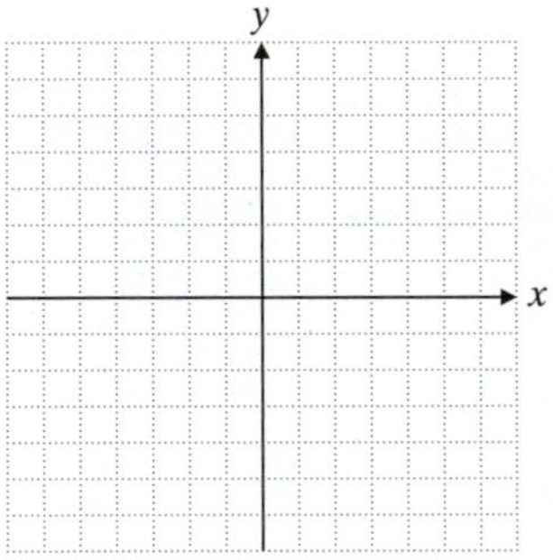

7. $y = x^2 - 6$

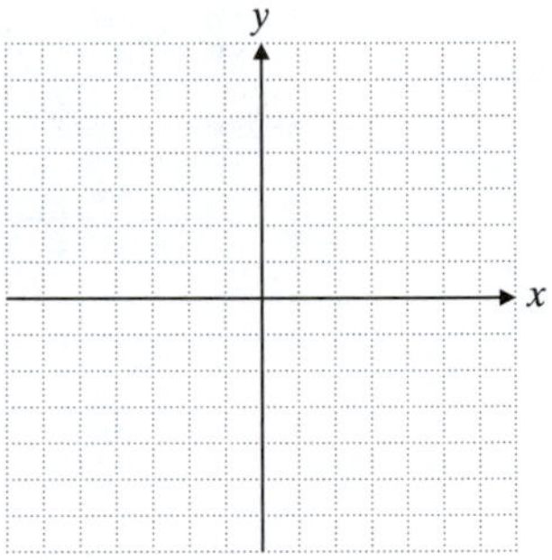

8. $y = x^2 + 2$

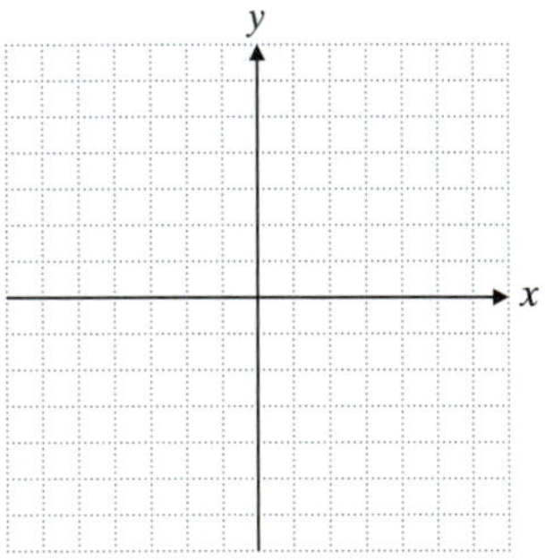

9. $y = \frac{1}{2}x^2 - 2$

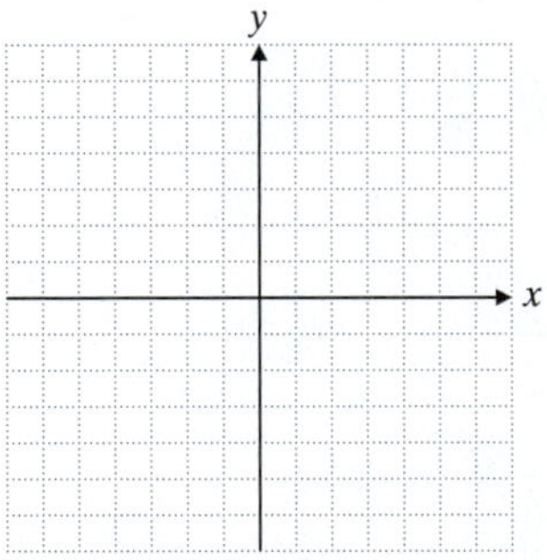

10. $y = \frac{1}{4}x^2 + 1$

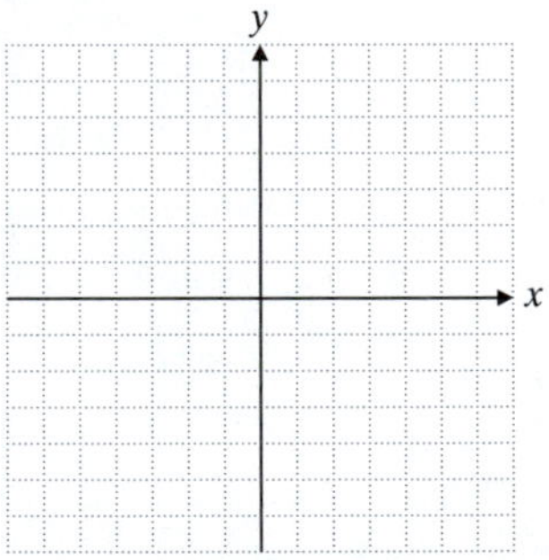

11. $y = (x - 3)^2 - 2$

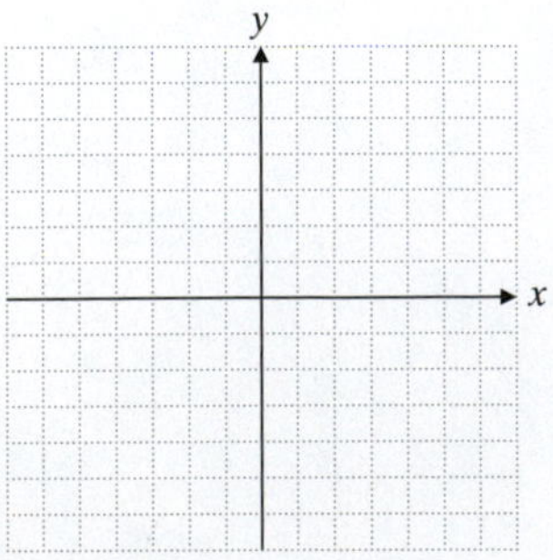

12. $y = (x - 2)^2 - 4$

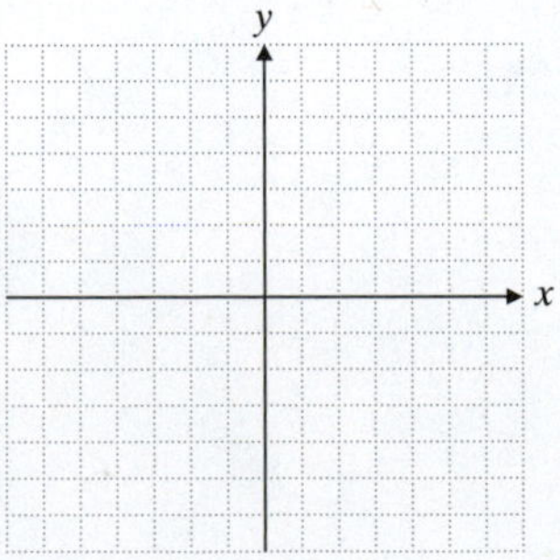

13. $y = 2(x - 1)^2 + \frac{3}{2}$

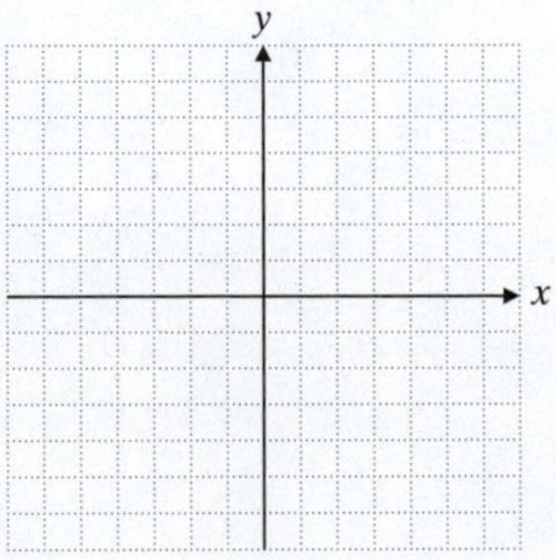

14. $y = 2(x - 2)^2 + \frac{5}{2}$

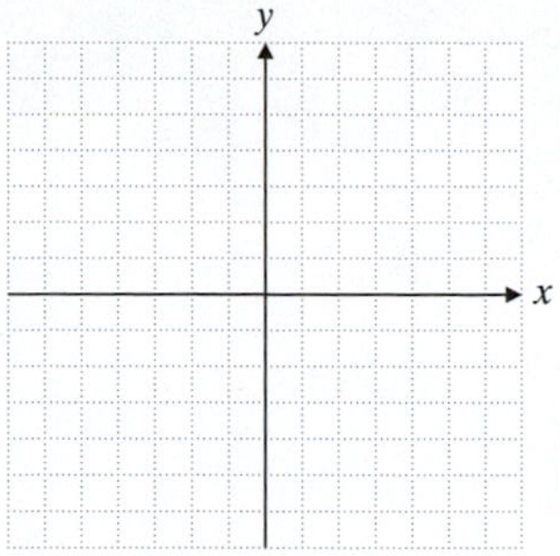

15. $y = -4\left(x + \frac{3}{2}\right)^2 + 5$

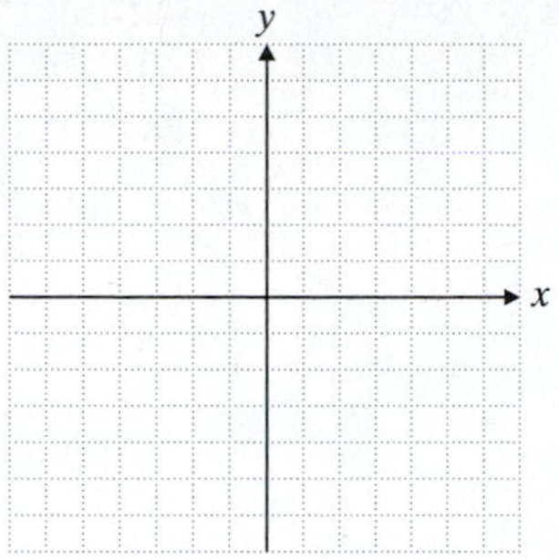

16. $y = -2\left(x + \frac{1}{2}\right)^2 - 1$

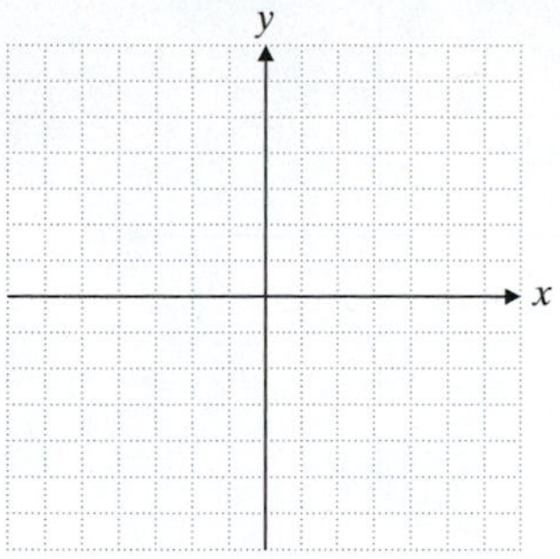

Graph each parabola and label the vertex. Find the x-intercept.

17. $x = \frac{1}{2}y^2$

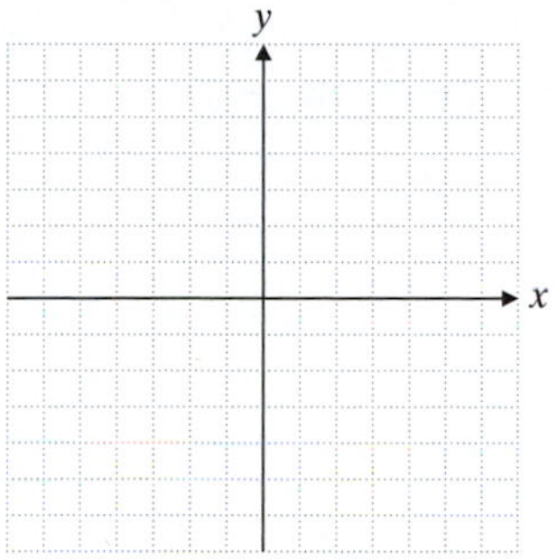

18. $x = \frac{2}{3}y^2$

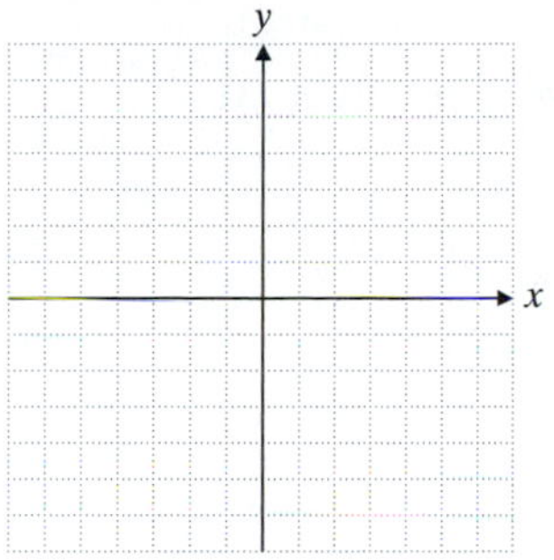

19. $x = \frac{1}{4}y^2 - 2$

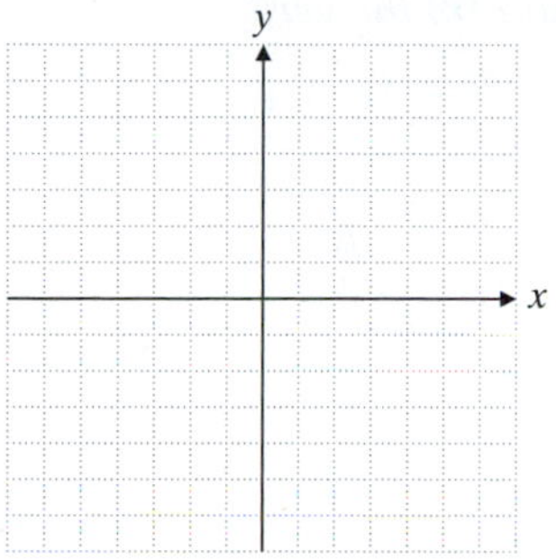

20. $x = \frac{1}{3}y^2 + 1$

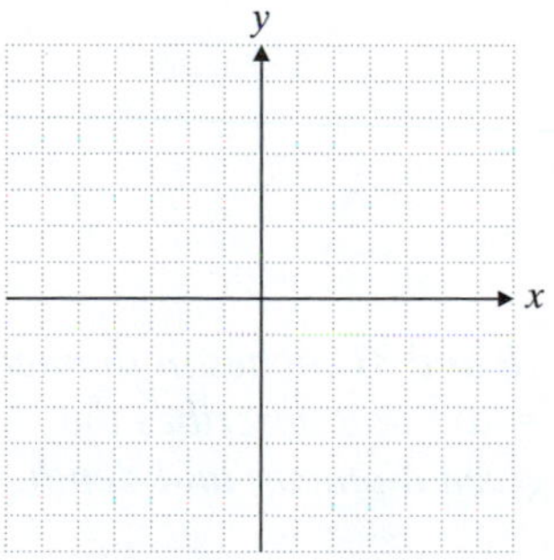

21. $x = -y^2 + 2$

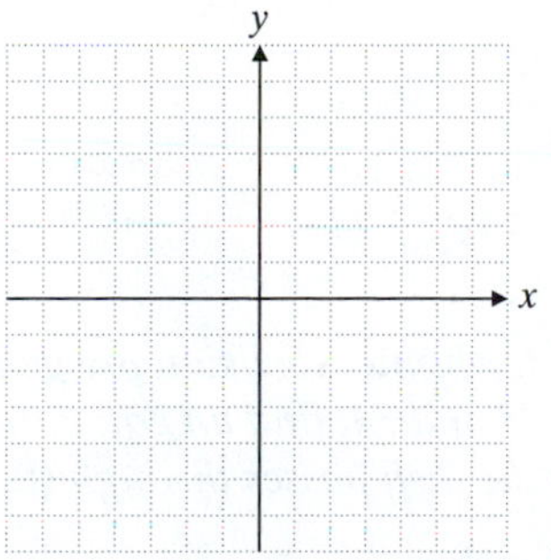

22. $x = -y^2 - 1$

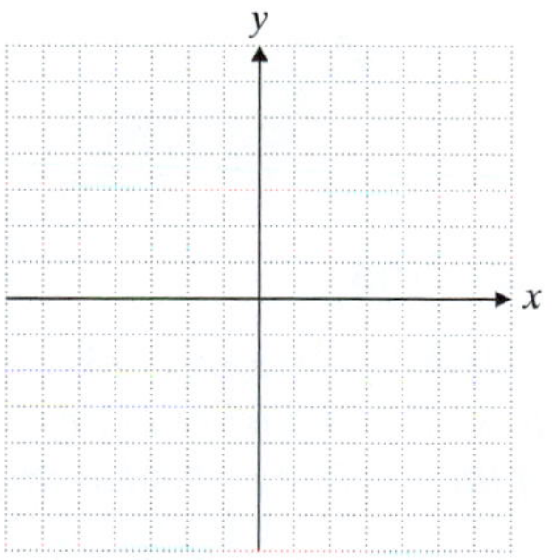

23. $x = (y - 2)^2 + 3$

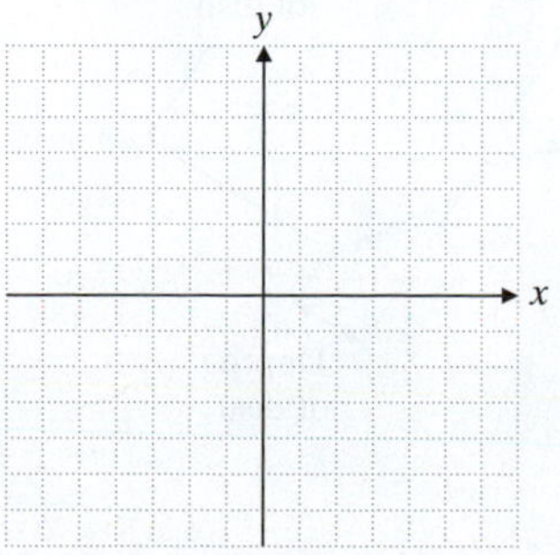

24. $x = (y - 4)^2 + 1$

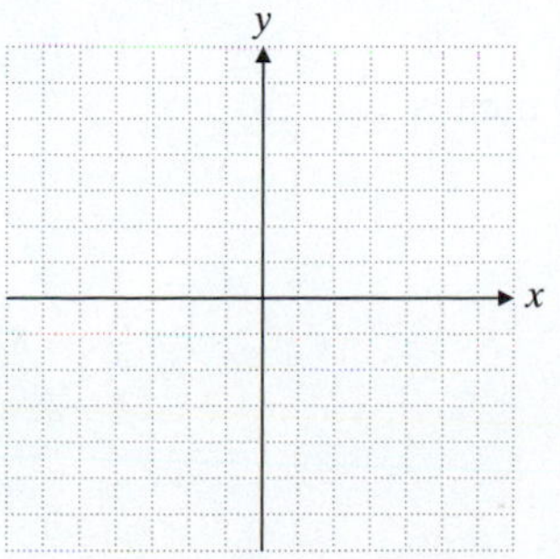

25. $x = -3(y + 1)^2 - 2$

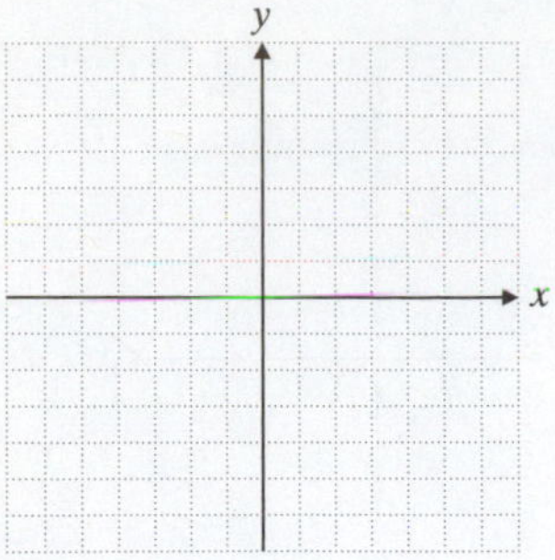

26. $x = -2(y + 3)^2 - 1$

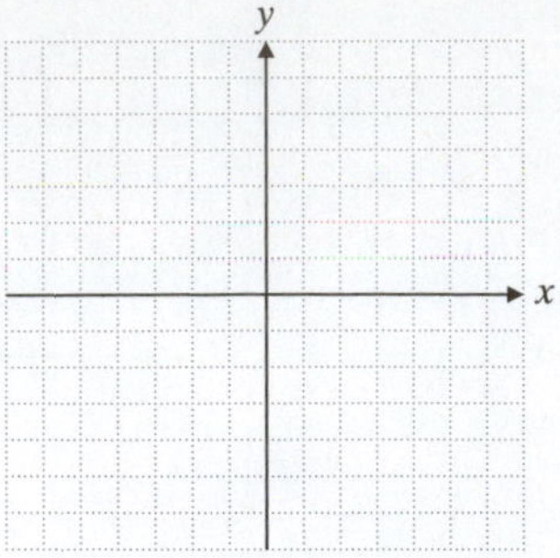

Rewrite each equation in standard form. Determine ***(a)*** *whether the parabola is horizontal or vertical,* ***(b)*** *the direction it opens, and* ***(c)*** *the vertex.*

27. $y = x^2 - 4x - 1$

28. $y = x^2 + 10x + 31$

29. $y = -2x^2 + 8x - 1$

30. $y = -3x^2 + 6x + 2$

31. $x = y^2 + 8y + 9$

32. $x = y^2 + 10y + 23$

Applications

Satellite Dishes *Modern satellite dishes intended for home television are generally between 18 and 31 inches in diameter and are less than 10 inches in depth. For exercises 33 and 34, find an equation of the form* $y = ax^2$ *that describes the outline of the satellite dish such that the bottom of the dish passes through* (0, 0) *and has the given diameter and depth.*

33. The diameter is 20 inches, and the depth is 5 inches.

34. The diameter is 30 inches, and the depth is 9 inches.

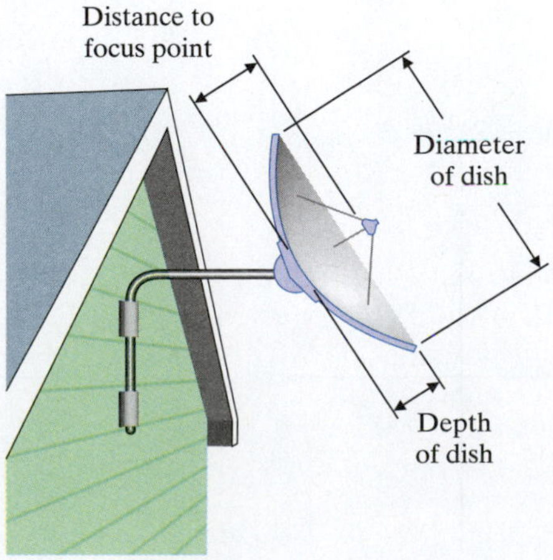

35. ***Satellite Dishes*** If the outline of a satellite dish is described by the equation $y = ax^2$, then the distance p from the center of the dish to the focus point of the dish is given by the equation $a = \frac{1}{4p}$. Find the distance p for the dish in exercise 33.

36. ***Satellite Dishes*** If the outline of a satellite dish is described by the equation $y = ax^2$, then the distance p from the center of the dish to the focus point of the dish is given by the equation $a = \frac{1}{4p}$. Find the distance p for the dish in exercise 34.

Optional Graphing Calculator Problems

Find the vertex and y-intercept of each parabola. Find the two x-intercepts.

37. $y = 2x^2 + 6.48x - 0.1312$

38. $y = -3x^2 + 33.66x - 73.5063$

Applications

By writing a quadratic equation in the form $y = a(x - h)^2 + k$, we can find the maximum or minimum value of the equation and the value of x at which it occurs. Remember that the equation $y = a(x - h)^2 + k$ is a vertical parabola. For $a > 0$, the parabola opens upward. Thus, the y-coordinate of the vertex is the smallest (or minimum) value of the equation. Similarly, when $a < 0$, the parabola opens downward, so the y-coordinate of the vertex is the maximum value of the equation. Since the vertex occurs at (h, k), the maximum value of the equation occurs when $x = h$. Then

$$y = -a(x - h)^2 + k = -a(0) + k = k.$$

For example, suppose the weekly profit of a manufacturing company in dollars is $P = -2(x - 45)^2 + 2300$ for x units manufactured. By looking at the equation, we see that the maximum profit per week is \$2300 and is attained when 45 units are manufactured. Use this approach for exercises 39–42.

39. ***Business*** A specialty watch company's monthly profit equation is

$$P = -2x^2 + 400x + 35{,}000,$$

where x is the number of watches manufactured. Find the maximum monthly profit and the number of watches that must be produced each month to attain the maximum profit.

40. ***Business*** A small bicycle manufacturing company's monthly profit equation is

$$P = -4x^2 + 400x + 10{,}000,$$

where x is the number of bicycles manufactured. Find the maximum monthly profit and the number of bicycles that must be produced each month to attain the maximum profit.

41. ***Orange Grove Yield*** The effective yield from a grove of miniature orange trees is described by the equation $E = x(900 - x)$, where x is the number of orange trees per acre. What is the maximum effective yield? How many orange trees per acre should be planted to achieve the maximum yield?

42. ***Drug Sensitivity*** A research pharmacologist has determined that sensitivity S to a drug depends on the dosage d in milligrams, according to the equation $S = 650d - 2d^2$. What is the maximum sensitivity that will occur? What dosage will produce that maximum sensitivity?

Cumulative Review

43. [2.5.1] ***Produce Delivery*** A driver delivered eggs from a farm to two supermarkets. He drove at a steady rate to the first supermarket for $1\frac{1}{2}$ hours. He then drove 20 mph faster to the second supermarket than he drove to the first. The total trip took 2.5 hours and covered 90 miles. What was his speed on the way to the second supermarket?

44. [2.5.1] ***Commuting*** Matthew drives from work to his home at 40 mph. One morning, an accident on the road delayed him for 15 minutes. The driving time including the delay was 56 minutes. How far does Matthew live from his job?

45. [1.2.2] ***Rose Bushes*** Sir George Tipkin of Sussex has a collection of eight large English rose bushes, each having approximately 1050 buds. In normal years this type of bush produces blooms from 73% of its buds. During years of drought this figure drops to 44%. During years of heavy rainfall the figure rises to 88%. How many blooms can Sir George expect on these bushes if there is heavy rainfall this year?

46. [1.2.2] ***Rose Bushes*** Last year Sir George had only six of the type of bushes described in exercise 45. It was a drought year, and he counted 2900 blooms. Using the bloom rates given in exercise 45, determine approximately how many buds appeared on each of these six bushes. (Round your answer to the nearest whole number.)

Quick Quiz 9.2

1. Find the vertex and the y-intercept of the parabola whose equation is $y = -3(x + 2)^2 + 5$.

2. Find the standard form of the parabola that opens to the right, has a vertex at the point $(3, 2)$, and crosses the x-axis at $(7, 0)$. Place your answer in the form $x = (y - k)^2 + h$.

3. Place the equation $y = 2x^2 - 12x + 12$ in standard form. Then graph it.

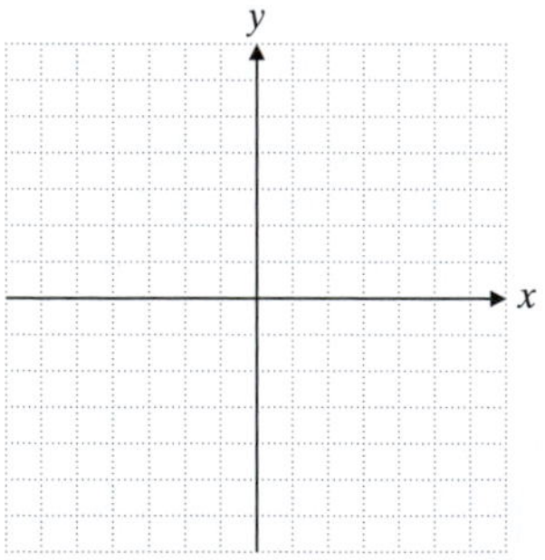

4. Concept Check Explain how you can tell which way the parabola opens for these equations: $y = 2x^2$, $y = -2x^2$, $x = 2y^2$, and $x = -2y^2$.

9.3 The Ellipse

Suppose a plane cuts a cone at an angle so that the plane intersects all sides of the cone. If the plane is not perpendicular to the axis of the cone, the conic section that is formed is called an ellipse.

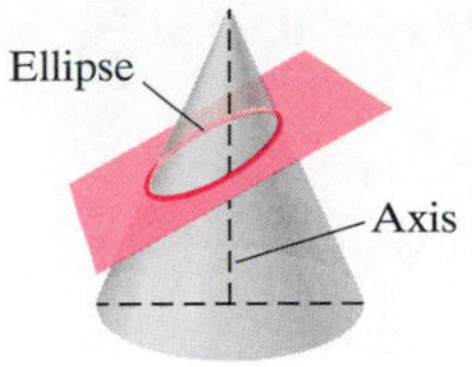

Student Learning Objectives

After studying this section, you will be able to:

1. Graph an ellipse whose center is at the origin.
2. Graph an ellipse whose center is at (h, k).

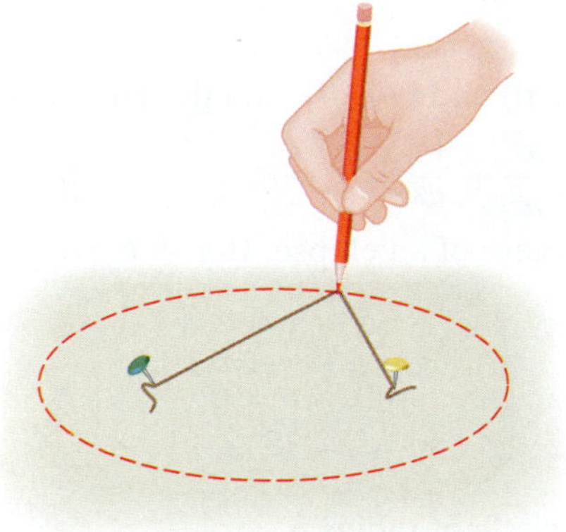

We define an **ellipse** as the set of points in a plane such that for each point in the set, the *sum* of its distances to two fixed points is constant. The fixed points are called **foci** (plural of *focus*).

We can use this definition to draw an ellipse using a piece of string tied at each end to a thumbtack. Place a pencil as shown in the drawing and draw the curve, keeping the pencil pushed tightly against the string. The two thumbtacks are the foci of the ellipse that results.

Examples of the ellipse can be found in the real world. The orbit of Earth (and each of the other planets) is approximately an ellipse with the Sun at one focus. In the sketch at the right the Sun is located at F and the other focus of the orbit of Earth is at F'.

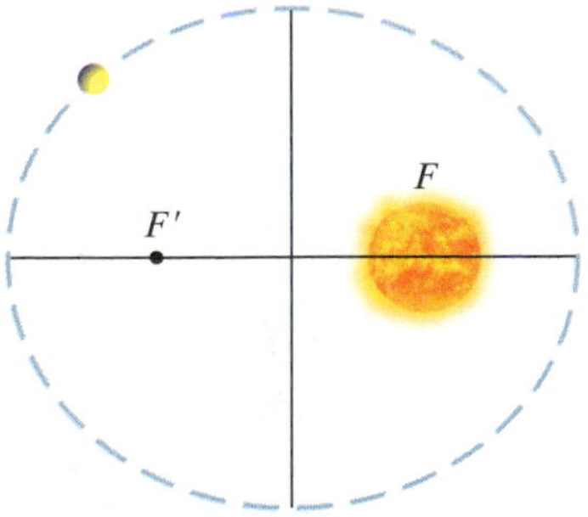

An elliptical surface has a special reflecting property. When sound, light, or some other object originating at one focus reaches the ellipse, it is reflected in such a way that it passes through the other focus. This property can be found in the United States Capitol in a famous room known as the Statuary Hall. If a person whispers at the focus of one end of this elliptically shaped room, a person at the other focus can easily hear him or her.

① Graphing an Ellipse Whose Center Is at the Origin

The equation of an ellipse is similar to the equation of a circle. The standard form of the equation of an ellipse centered at the origin is given next.

STANDARD FORM OF THE EQUATION OF AN ELLIPSE

An ellipse with center at the origin has the equation

$$\frac{x^2}{a^2} + \frac{y^2}{b^2} = 1, \qquad \text{where } a \text{ and } b > 0.$$

The **vertices** of this ellipse are at $(a, 0)$, $(-a, 0)$, $(0, b)$, and $(0, -b)$.

To plot the ellipse, we need the x- and y-intercepts.

$$\frac{x^2}{a^2} + \frac{y^2}{b^2} = 1$$

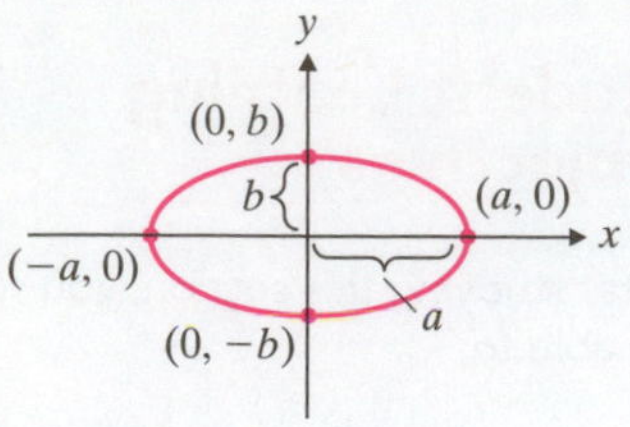

If $x = 0$***, then*** $\frac{y^2}{b^2} = 1.$

$$y^2 = b^2$$
$$\pm\sqrt{y^2} = \pm\sqrt{b^2}$$
$$\pm y = \pm b \quad \text{or} \quad y = \pm b$$

If $y = 0$***, then*** $\frac{x^2}{a^2} = 1.$

$$x^2 = a^2$$
$$\pm\sqrt{x^2} = \pm\sqrt{a^2}$$
$$\pm x = \pm a \quad \text{or} \quad x = \pm a$$

So the x-intercepts are $(a, 0)$ and $(-a, 0)$, and the y-intercepts are $(0, b)$ and $(0, -b)$ for an ellipse of the form $\frac{x^2}{a^2} + \frac{y^2}{b^2} = 1$.

A circle is a special case of an ellipse. If $a = b$, we get the following:

$$\frac{x^2}{a^2} + \frac{y^2}{a^2} = 1$$
$$x^2 + y^2 = a^2$$

This is the equation of a circle of radius a.

EXAMPLE 1 Graph $x^2 + 3y^2 = 12$. Label the intercepts.

Solution Before we can graph this ellipse, we need to rewrite the equation in standard form.

$$\frac{x^2}{12} + \frac{3y^2}{12} = \frac{12}{12} \quad \text{Divide each side by 12.}$$

$$\frac{x^2}{12} + \frac{y^2}{4} = 1 \quad \text{Simplify.}$$

Thus, we have the following:

$$a^2 = 12 \quad \text{so} \quad a = 2\sqrt{3}$$
$$b^2 = 4 \quad \text{so} \quad b = 2$$

The x-intercepts are $(-2\sqrt{3}, 0)$ and $(2\sqrt{3}, 0)$, and the y-intercepts are $(0, 2)$ and $(0, -2)$. We plot these points and draw the ellipse.

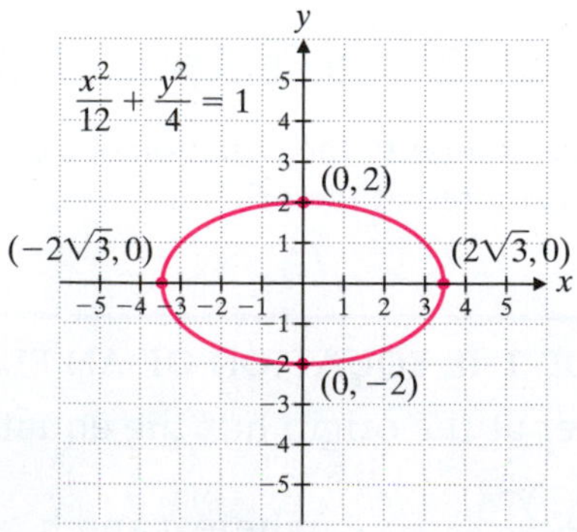

Student Practice 1 Graph $4x^2 + y^2 = 16$. Label the intercepts.

Student Practice 1

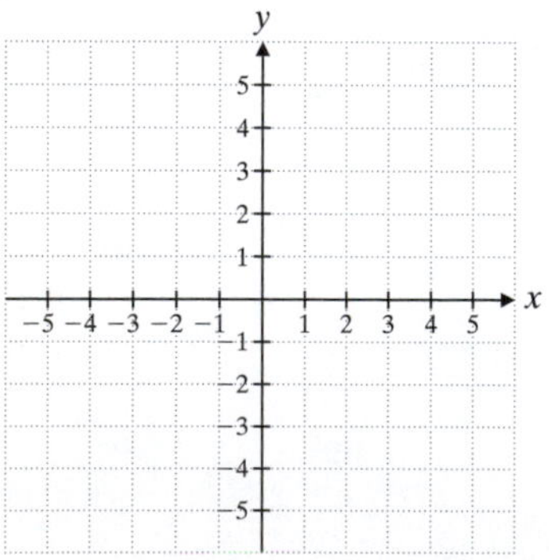

NOTE TO STUDENT: Fully worked-out solutions to all of the Student Practice problems can be found at the back of the text starting at page SP-1.

② Graphing an Ellipse Whose Center Is at (h, k)

If the center of the ellipse is not at the origin but at some point whose coordinates are (h, k), then the standard form of the equation is changed.

An ellipse with center at (h, k) has the equation

$$\frac{(x-h)^2}{a^2} + \frac{(y-k)^2}{b^2} = 1,$$

where a and $b > 0$.

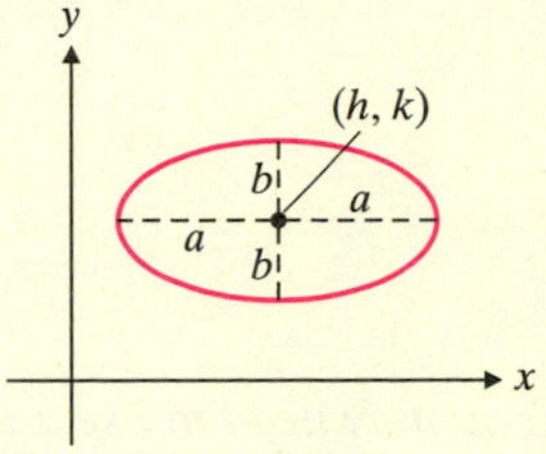

Note that a and b are *not* the x-intercepts and y-intercepts now. Why is this? Look at the sketch. You'll see that a is the horizontal distance from the center of the ellipse to a point on the ellipse. Similarly, b is the vertical distance. Hence, when the center of the ellipse is not at the origin, the ellipse may not cross either axis.

Graphing Calculator

Graphing Ellipses

In order to graph the ellipse in Example 2 on a graphing calculator, we first need to solve for y.

$$\frac{(y-6)^2}{4} = 1 - \frac{(x-5)^2}{9}$$

$$(y-6)^2 = 4\left[1 - \frac{(x-5)^2}{9}\right]$$

$$y = 6 \pm 2\sqrt{1 - \frac{(x-5)^2}{9}}$$

Is it necessary to break up the curve into two halves in order to graph the ellipse? Why or why not?

EXAMPLE 2 Graph $\dfrac{(x-5)^2}{9} + \dfrac{(y-6)^2}{4} = 1$.

Solution The center of the ellipse is $(5, 6)$, $a = 3$, and $b = 2$. Therefore, we begin at $(5, 6)$. We plot points 3 units to the left, 3 units to the right, 2 units up, and 2 units down from $(5, 6)$. The points we plot are the *vertices* of the ellipse.

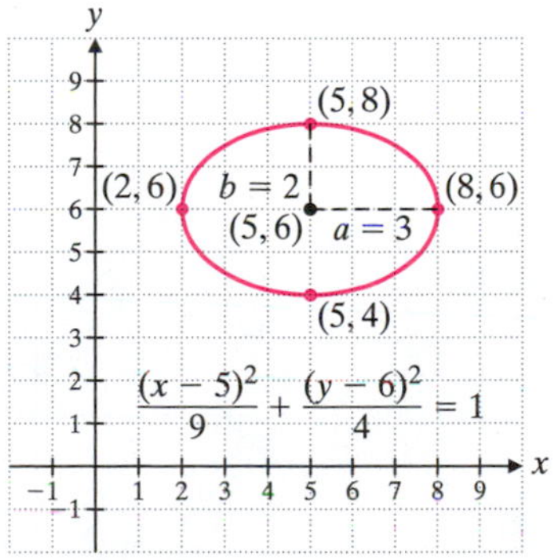

Student Practice 2 Graph $\dfrac{(x-2)^2}{16} + \dfrac{(y+3)^2}{9} = 1$.

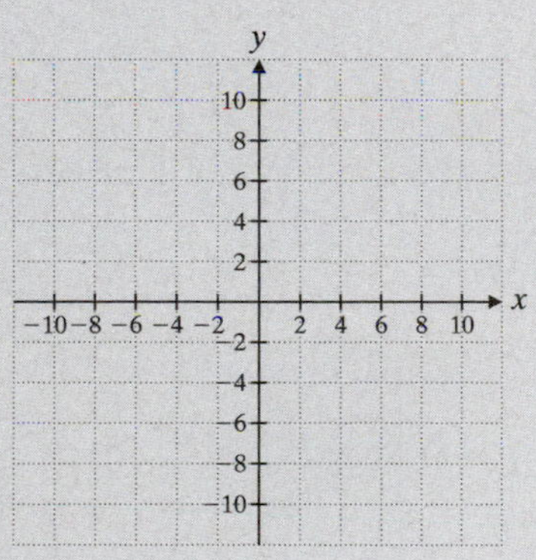

9.3 Exercises

MyMathLab®

Watch the videos in MyMathLab

Download the MyDashBoard App

Verbal and Writing Skills, Exercises 1 and 2

1. Explain how to determine the center of the ellipse $\frac{(x+2)^2}{4} + \frac{(y-3)^2}{9} = 1$.

2. Explain how to determine the x- and y-intercepts of the ellipse $\frac{x^2}{9} + \frac{y^2}{16} = 1$.

Graph each ellipse. Label the intercepts. You may need to use a scale other than 1 square = 1 unit.

3. $\frac{x^2}{36} + \frac{y^2}{4} = 1$

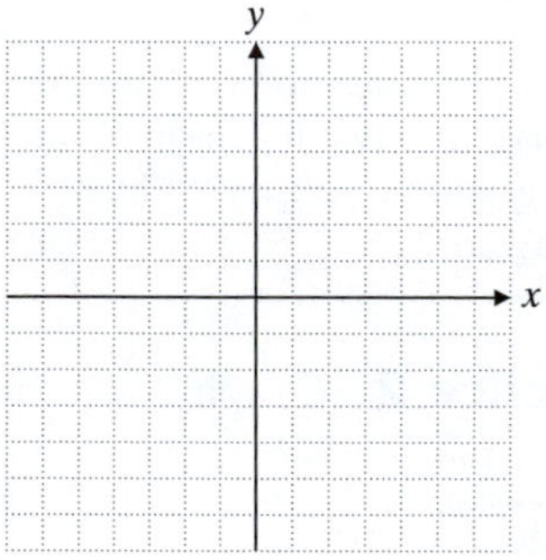

4. $\frac{x^2}{49} + \frac{y^2}{25} = 1$

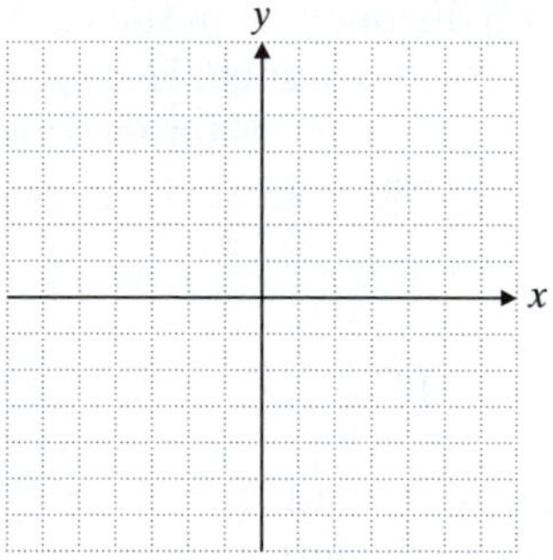

5. $\frac{x^2}{81} + \frac{y^2}{100} = 1$

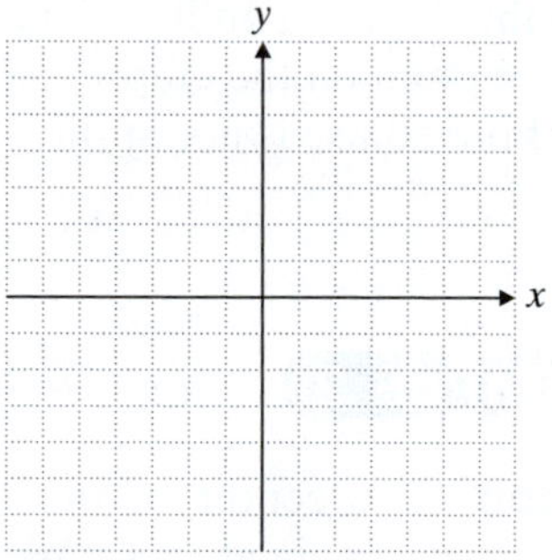

6. $\frac{x^2}{121} + \frac{y^2}{144} = 1$

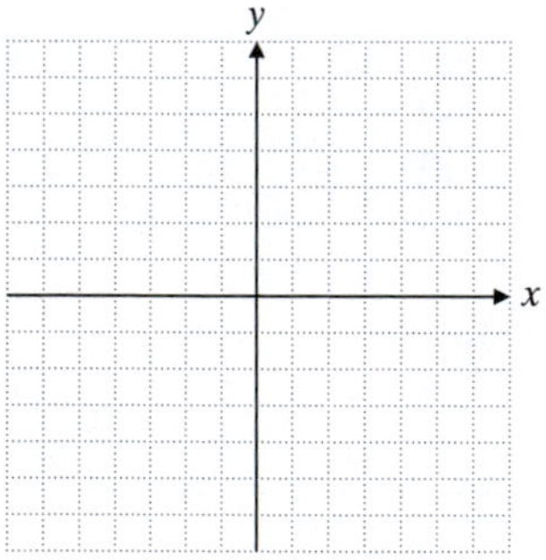

7. $4x^2 + y^2 - 36 = 0$

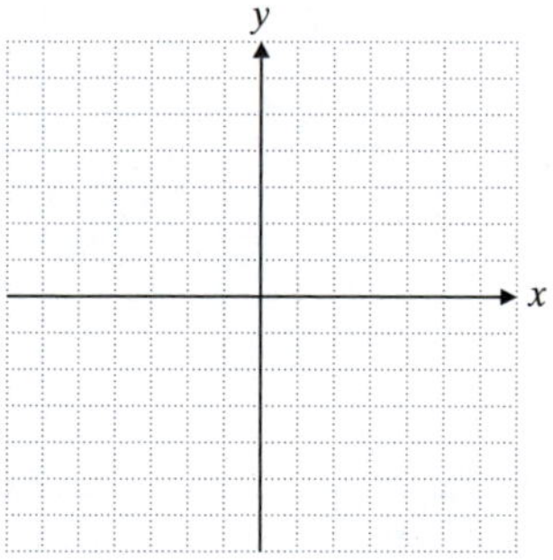

8. $9x^2 + y^2 - 9 = 0$

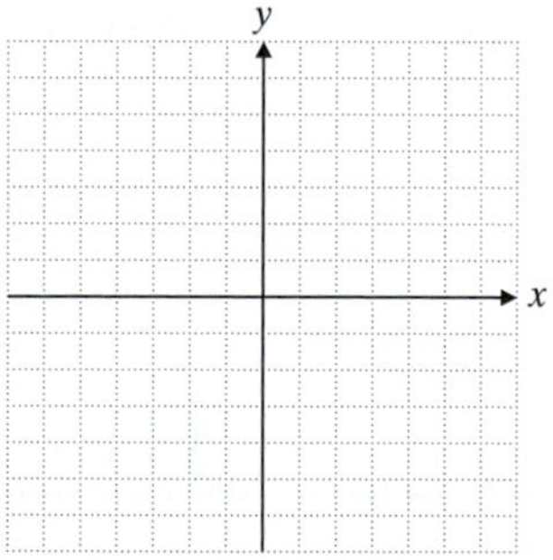

9. $x^2 + 9y^2 = 81$

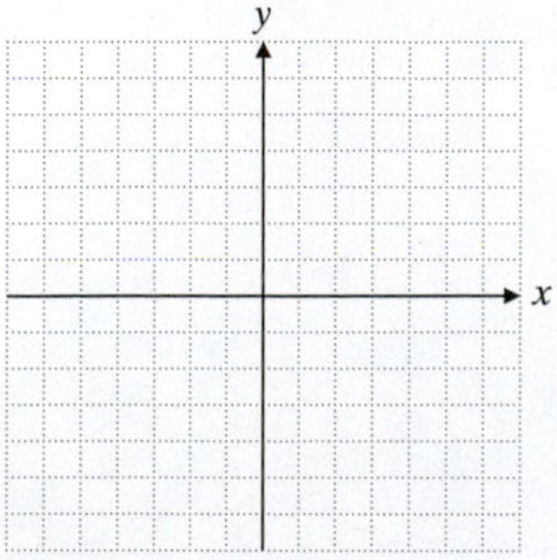

10. $4x^2 + 25y^2 = 100$

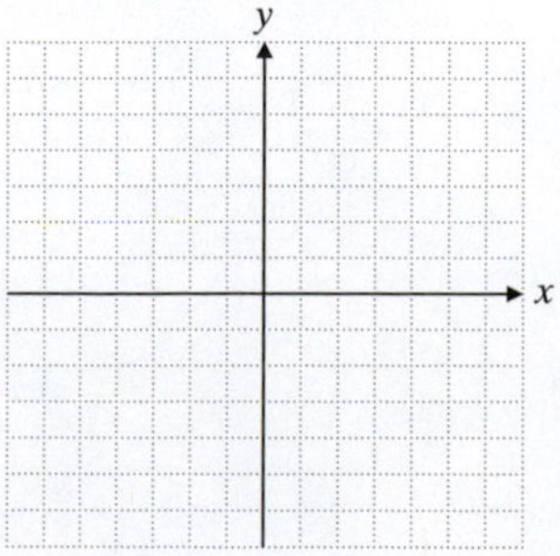

11. $x^2 + 12y^2 = 36$

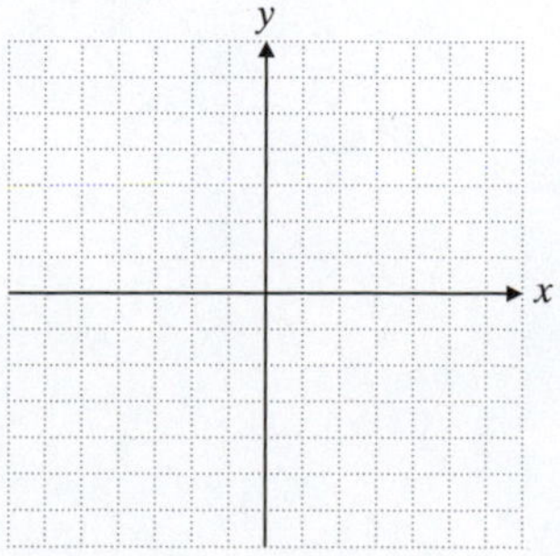

12. $2x^2 + 3y^2 = 18$

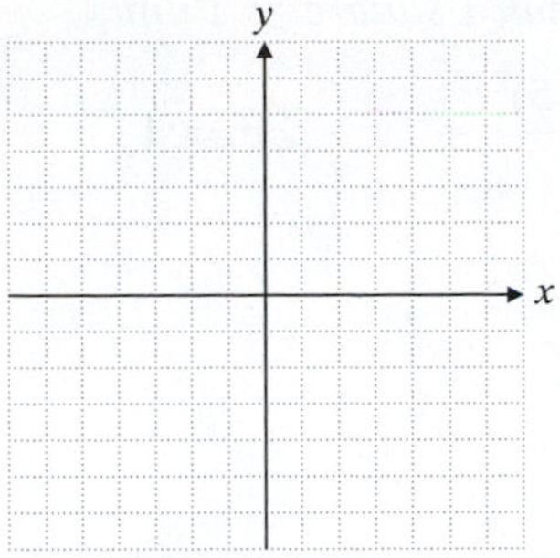

13. $\dfrac{x^2}{\frac{25}{4}} + \dfrac{y^2}{\frac{16}{9}} = 1$

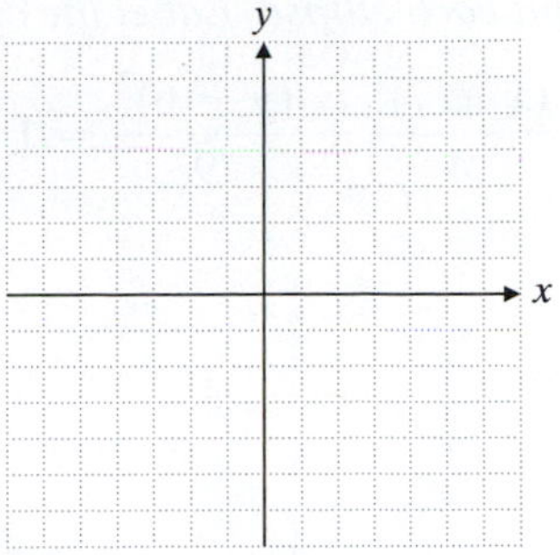

14. $\dfrac{x^2}{\frac{81}{4}} + \dfrac{y^2}{\frac{25}{16}} = 1$

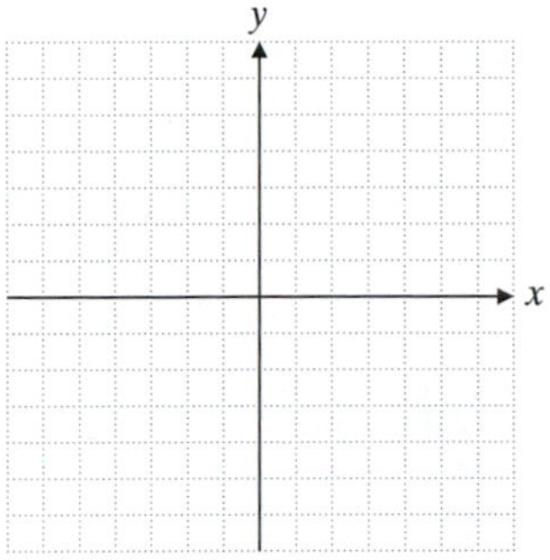

15. $121x^2 + 64y^2 = 7744$

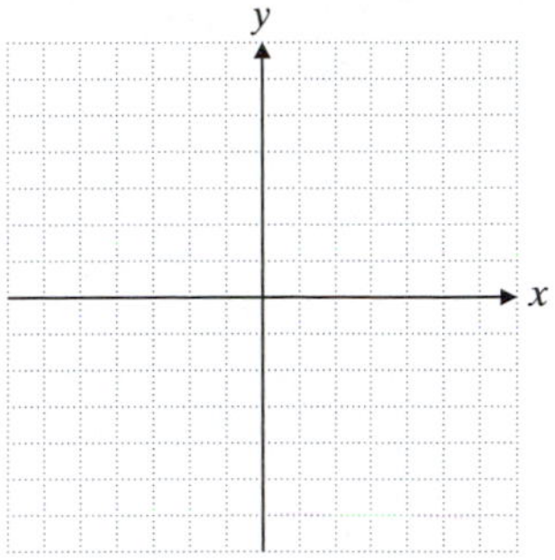

16. Write in standard form the equation of the ellipse with center at the origin, an x-intercept at $(-7, 0)$, and a y-intercept at $(0, 11)$.

17. Write in standard form the equation of the ellipse with center at the origin, an x-intercept at $(13, 0)$, and a y-intercept at $(0, -12)$.

18. Write in standard form the equation of the ellipse with center at the origin, an x-intercept at $(5\sqrt{2}, 0)$, and a y-intercept at $(0, -1)$.

19. Write in standard form the equation of the ellipse with center at the origin, an x-intercept at $(6, 0)$, and a y-intercept at $(0, 4\sqrt{3})$

Applications

20. ***Window Design*** The window shown in the sketch is in the shape of half of an ellipse. Find the equation for the ellipse if the center of the ellipse is at point $A = (0, 0)$.

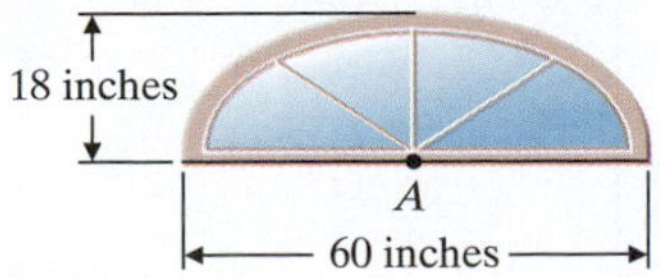

21. ***Orbit of Venus*** The orbit of Venus is an ellipse with the Sun as one focus. If we say that the center of the ellipse is at the origin, an approximate equation for the orbit is

$$\frac{x^2}{5013} + \frac{y^2}{4970} = 1,$$

where x and y are measured in millions of miles. Find the largest possible distance across the ellipse. Round your answer to the nearest million miles.

Graph each ellipse. Label the center. You may need to use a scale other than 1 *square* = 1 *unit.*

22. $\dfrac{(x-7)^2}{4}+\dfrac{(y-6)^2}{9}=1$

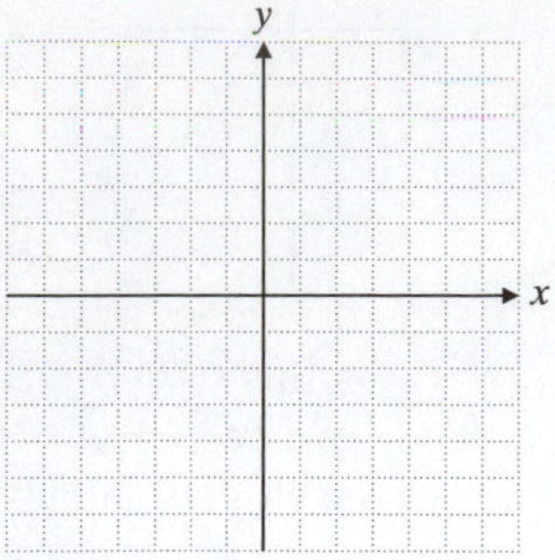

23. $\dfrac{(x-5)^2}{9}+(y-2)^2=1$

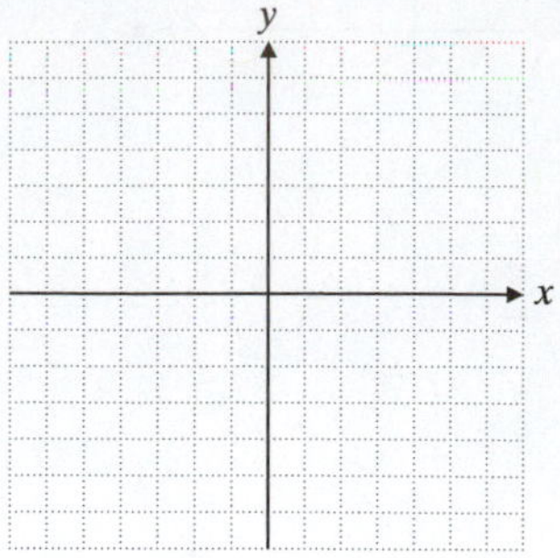

24. $\dfrac{(x+2)^2}{49}+\dfrac{y^2}{25}=1$

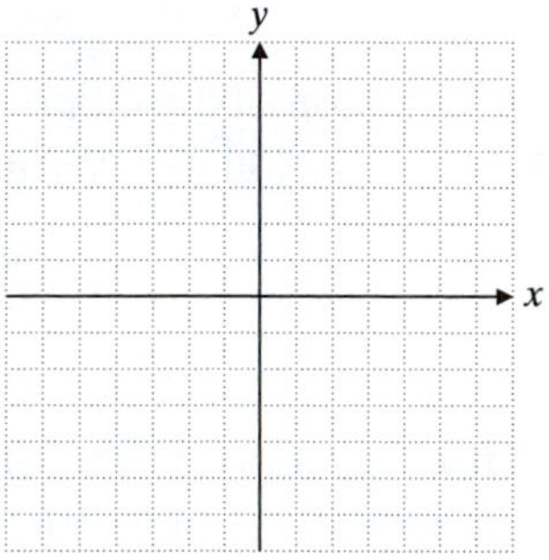

25. $\dfrac{x^2}{25}+\dfrac{(y-4)^2}{16}=1$

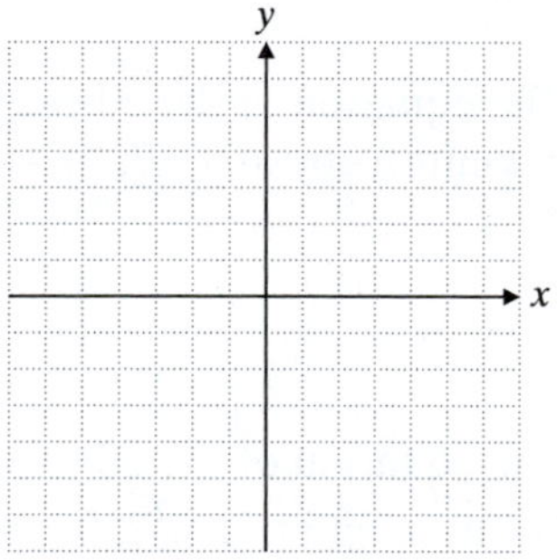

26. $\dfrac{(x+1)^2}{36}+\dfrac{(y+4)^2}{16}=1$

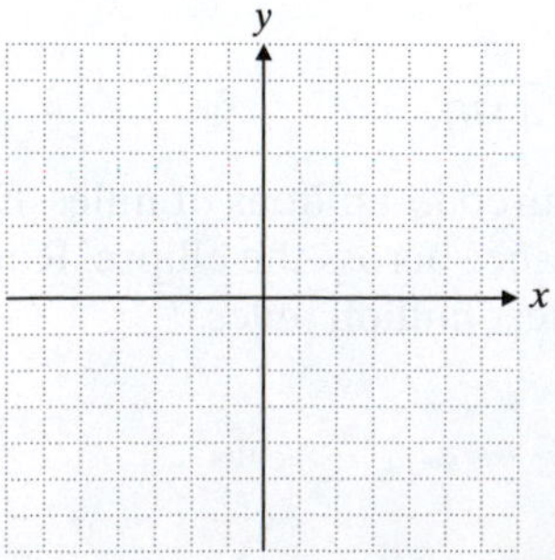

27. $\dfrac{(x+5)^2}{16}+\dfrac{(y+2)^2}{36}=1$

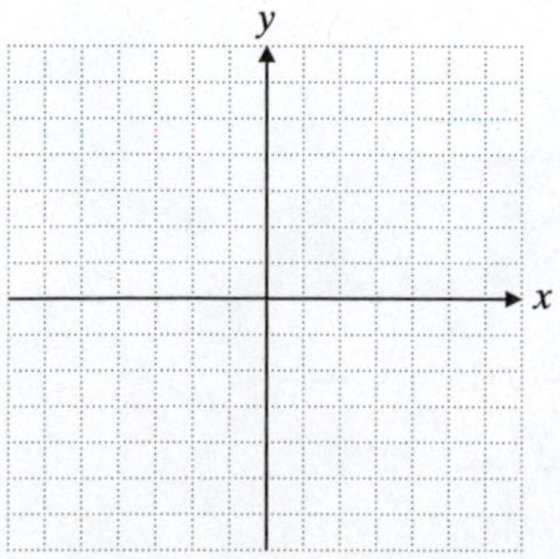

28. Write in standard form the equation of the ellipse whose vertices are $(-7, 1)$, $(3, 1)$, $(-2, 4)$, and $(-2, -2)$.

29. Write in standard form the equation of the ellipse whose vertices are $(2, 3)$, $(6, 3)$, $(4, 7)$, and $(4, -1)$.

30. For what values of a does the ellipse

$$\frac{(x+5)^2}{4} + \frac{(y+a)^2}{9} = 1$$

pass through the point $(-5, 4)$?

31. ***Pet Exercise Area*** Bob's backyard is a rectangle 40 meters by 60 meters. He uses this backyard for an exercise area for his dog. He drove two posts into the ground and fastened a rope to each post, passing the rope through the metal ring on his dog's collar. When the dog pulls on the rope while running, its path is an ellipse. (See the figure.) If the dog can just reach all four sides of the rectangle, find the equation of the elliptical path.

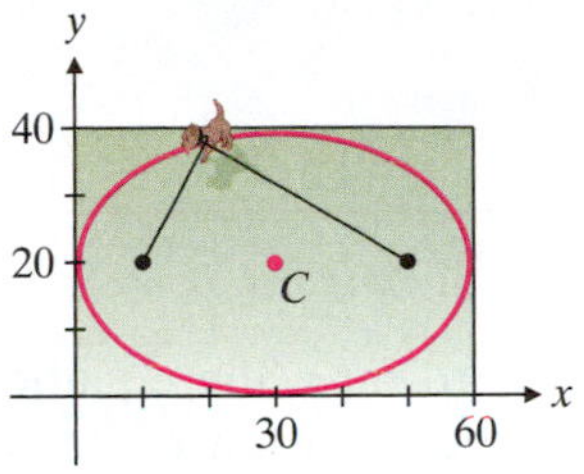

Optional Graphing Calculator Problems

Find the four intercepts, rounded to four decimal places, for each ellipse.

32. $\dfrac{x^2}{12} + \dfrac{y^2}{19} = 1$

33. $\dfrac{(x-3.6)^2}{14.98} + \dfrac{(y-5.3)^2}{28.98} = 1$

To Think About

The area enclosed by the ellipse $\dfrac{x^2}{a^2} + \dfrac{y^2}{b^2} = 1$ *is given by the equation* $A = \pi ab$. *Use the value* $\pi \approx 3.1416$ *to find an approximate value for the following answer.*

34. ***Mirror Design*** An oval mirror has an outer boundary in the shape of an ellipse. The width of the mirror is 20 inches, and the length of the mirror is 45 inches. Find the area of the mirror. Round your answer to the nearest tenth.

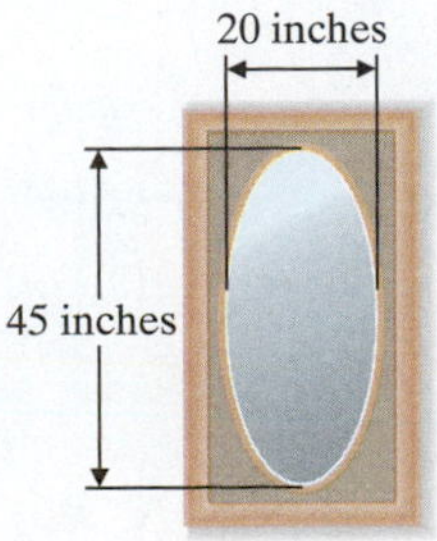

Cumulative Review

35. **[7.4.3]** Rationalize the denominator. $\dfrac{5}{\sqrt{2x} - \sqrt{y}}$

36. **[7.4.1]** Multiply and simplify.
$(2\sqrt{3} + 4\sqrt{2})(5\sqrt{6} - \sqrt{2})$

37. **[1.2.2]** ***Empire State Building*** The Empire State Building was the tallest building in the world for many years. Construction began on March 17, 1930, and the framework rose at a rate of 4.5 stories per week. How many weeks did it take to complete the framework for all 102 stories?

Quick Quiz 9.3

1. Write in standard form the equation of the ellipse with center at the origin, an x-intercept at $(5, 0)$, and a y-intercept at $(0, -6)$.

2. Write in standard form the equation of the ellipse with vertices at $(-7, -2)$, $(1, -2)$, $(-3, 1)$, and $(-3, -5)$.

3. Graph the ellipse. Label the center and the four vertices.

$$\frac{(x-2)^2}{9} + \frac{(y+1)^2}{25} = 1$$

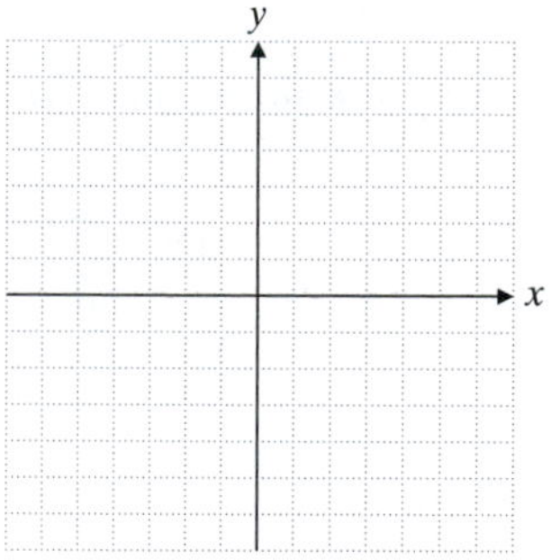

4. **Concept Check** Explain how you would find the four vertices of the ellipse $12x^2 + y^2 - 36 = 0$.

How Am I Doing? Sections 9.1–9.3

How are you doing with your homework assignments in Sections 9.1 to 9.3? Before you go further in the textbook, take some time to do each of the following problems. Simplify all answers.

9.1

1. Write the standard form of the equation of the circle with center at $(-3, 7)$ and a radius of $\sqrt{2}$.

2. Find the distance between $(-6, -2)$ and $(-3, 4)$.

3. Rewrite the equation $x^2 + y^2 - 2x - 4y + 1 = 0$ in standard form. Find the center and radius of the circle and sketch its graph.

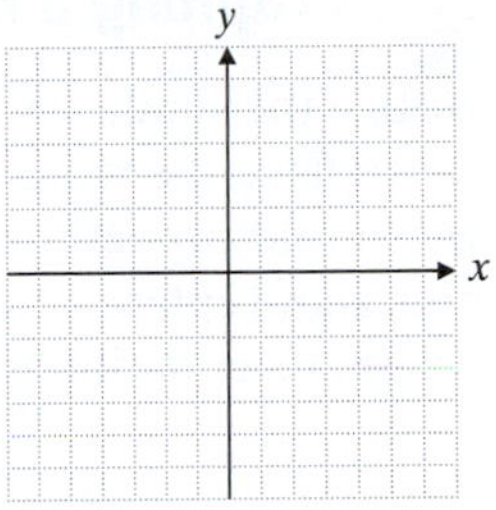

9.2

4. What is the axis of symmetry of the parabola $y = 4(x - 3)^2 + 5$?

5. Find the vertex of the parabola $y = \frac{1}{3}(x + 4)^2 + 6$.

Graph each parabola. Write the equation in standard form.

6. $x = (y + 1)^2 + 2$

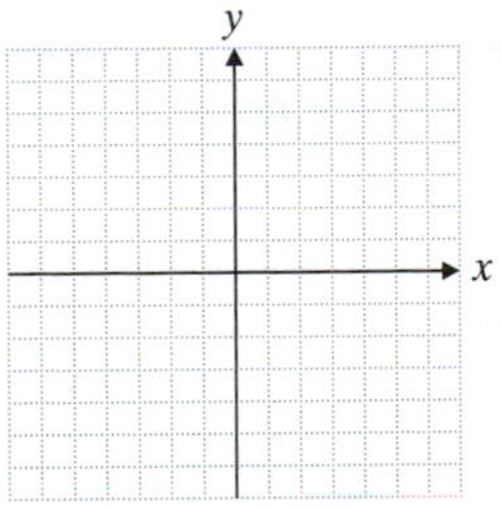

7. $x^2 = y - 4x - 1$

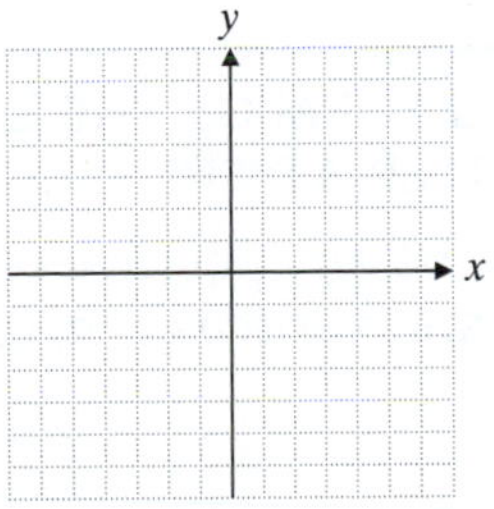

9.3

8. Write in standard form the equation of the ellipse with center at the origin, an x-intercept at $(-10, 0)$, and a y-intercept at $(0, 7)$.

Graph each ellipse. Write the equations in standard form.

9. $4x^2 + y^2 - 36 = 0$

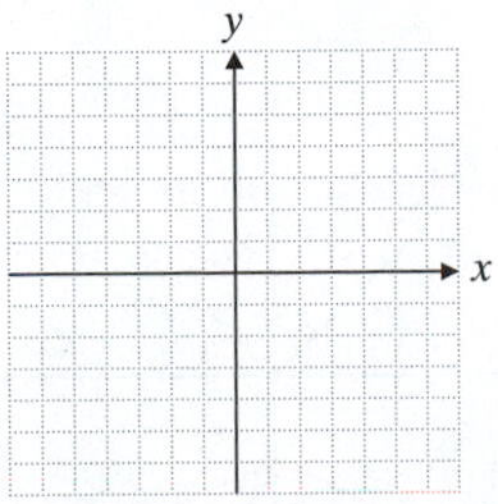

10. $\frac{(x + 3)^2}{25} + \frac{(y - 1)^2}{16} = 1$

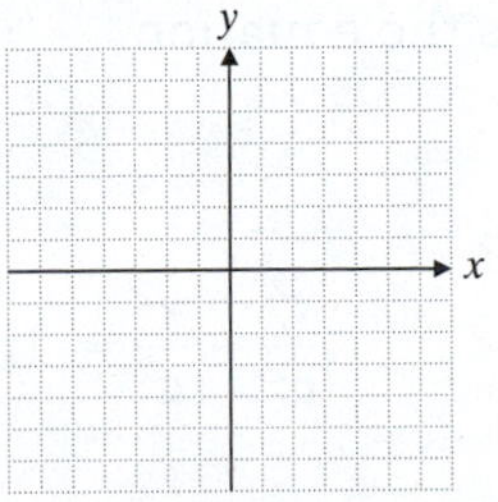

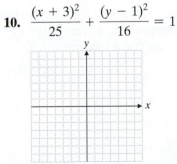

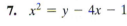

Now turn to page SA-24 for the answers to each of these problems. If you missed any of these problems, you should stop and review the Examples and Student Practice problems in the referenced objective.

1. ______________

2. ______________

3. ______________

4. ______________

5. ______________

6. ______________

7. ______________

8. ______________

9. ______________

10. ______________

9.4 The Hyperbola

Student Learning Objectives

After studying this section, you will be able to:

① **Graph a hyperbola whose center is at the origin.**

② **Graph a hyperbola whose center is at (h, k).**

By cutting two branches of a cone by a plane as shown in the sketch, we obtain the two branches of a hyperbola. A comet moving with more than enough kinetic energy to escape the Sun's gravitational pull will travel in a hyperbolic path. Similarly, a rocket traveling with more than enough velocity to escape Earth's gravitational field will follow a hyperbolic path.

We define a **hyperbola** as the set of points in a plane such that for each point in the set, the absolute value of the *difference* of its distances to two fixed points (called **foci**) is constant.

① Graphing a Hyperbola Whose Center Is at the Origin

Notice the similarity of the definition of a hyperbola to the definition of an ellipse. If we replace the word *difference* by *sum,* we have the definition of an ellipse. Hence, we should expect that the equation of a hyperbola will be that of an ellipse with the plus sign replaced by a minus sign. And it is. If the hyperbola has its center at the origin, its equation is

$$\frac{x^2}{a^2} - \frac{y^2}{b^2} = 1 \quad \text{or} \quad \frac{y^2}{b^2} - \frac{x^2}{a^2} = 1.$$

A hyperbola has two branches. If the center of the hyperbola is at the origin and each branch has one x-intercept but no y-intercepts, the hyperbola is a horizontal hyperbola, and its **axis** is the x-axis. If the center of the hyperbola is at the origin and each branch has one y-intercept but no x-intercepts, the hyperbola is a vertical hyperbola, and its axis is the y-axis.

The points where the hyperbola intersects its axis are called the **vertices** of the hyperbola.

For hyperbolas centered at the origin, the vertices are also the intercepts.

STANDARD FORM OF THE EQUATION OF A HYPERBOLA WITH CENTER AT THE ORIGIN

Let a and b be any positive real numbers. A hyperbola with center at the origin and vertices $(-a, 0)$ and $(a, 0)$ has the equation

$$\frac{x^2}{a^2} - \frac{y^2}{b^2} = 1.$$

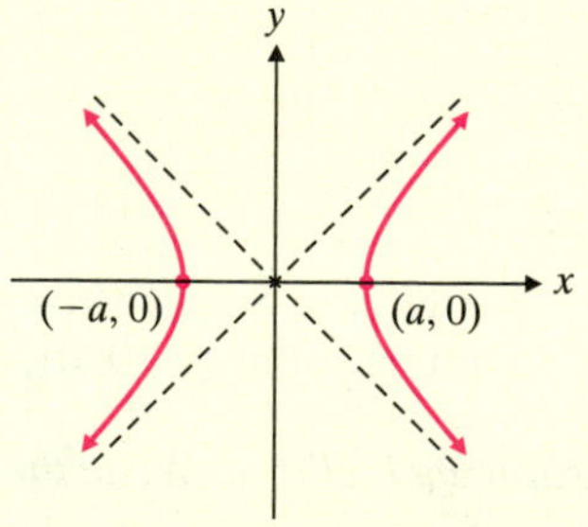

This is called a *horizontal hyperbola*.

A hyperbola with center at the origin and vertices $(0, b)$ and $(0, -b)$ has the equation

$$\frac{y^2}{b^2} - \frac{x^2}{a^2} = 1.$$

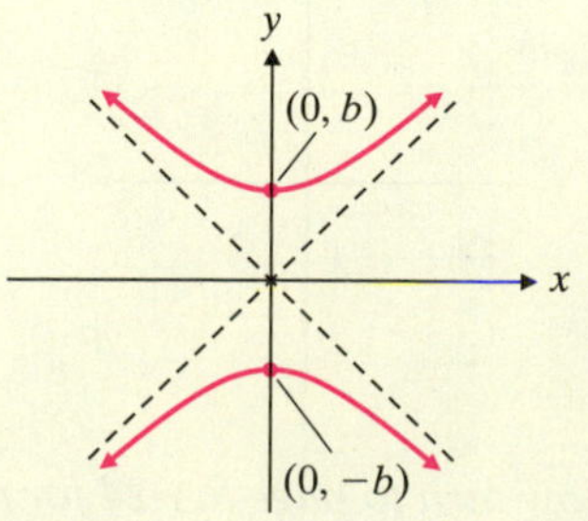

This is called a *vertical hyperbola*.

Notice that the two equations are slightly different. Be aware of this difference so that when you look at an equation you will be able to tell whether the hyperbola is horizontal or vertical.

Notice also the diagonal lines that we've drawn on the graphs of the hyperbolas. These lines are called **asymptotes.** The two branches of the hyperbola come increasingly closer to the asymptotes as the value of $|x|$ gets very large. By drawing the asymptotes and plotting the vertices, we can easily graph a hyperbola.

ASYMPTOTES OF HYPERBOLAS

The asymptotes of the hyperbolas $\frac{x^2}{a^2} - \frac{y^2}{b^2} = 1$ and $\frac{y^2}{b^2} - \frac{x^2}{a^2} = 1$ are

$$y = \frac{b}{a}x \quad \text{and} \quad y = -\frac{b}{a}x.$$

Note that $\frac{b}{a}$ and $-\frac{b}{a}$ are the slopes of the asymptotes.

An easy way to find the asymptotes is to draw extended diagonals of the rectangle whose center is at the origin and whose corners are at (a, b), $(a, -b)$, $(-a, b)$, and $(-a, -b)$. (This rectangle is sometimes called the **fundamental rectangle.**) We draw the fundamental rectangle and the asymptotes with dashed lines because they are not part of the curve.

***NOTE TO STUDENT:** Fully worked-out solutions to all of the Student Practice problems can be found at the back of the text starting at page SP-1.*

EXAMPLE 1 Graph $\frac{x^2}{25} - \frac{y^2}{16} = 1$.

Solution The equation has the form $\frac{x^2}{a^2} - \frac{y^2}{b^2} = 1$, so it is a horizontal hyperbola. $a^2 = 25$, so $a = 5$; $b^2 = 16$, so $b = 4$. Since the hyperbola is horizontal, it has vertices at $(a, 0)$ and $(-a, 0)$ or $(5, 0)$ and $(-5, 0)$.

To draw the asymptotes, we construct a fundamental rectangle with corners at $(5, 4)$, $(5, -4)$, $(-5, 4)$, and $(-5, -4)$. We draw extended diagonals of the rectangle as the asymptotes. We construct each branch of the curve so that it passes through a vertex and gets closer to the asymptotes as it moves away from the origin.

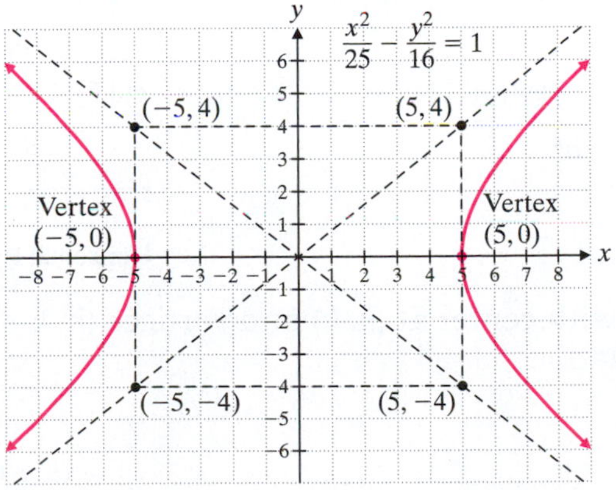

Student Practice 1

Graph $\frac{x^2}{16} - \frac{y^2}{25} = 1$.

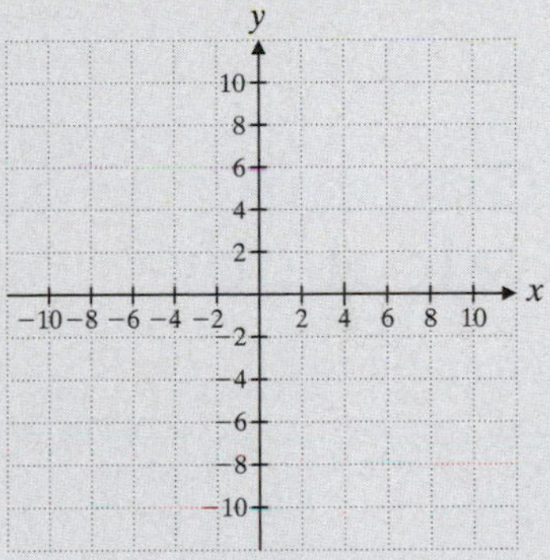

EXAMPLE 2 Graph $4y^2 - 7x^2 = 28$.

Solution To find the vertices and asymptotes, we must rewrite the equation in standard form. Divide each term by 28.

$$\frac{4y^2}{28} - \frac{7x^2}{28} = \frac{28}{28}$$

$$\frac{y^2}{7} - \frac{x^2}{4} = 1$$

Thus, we have the standard form of a vertical hyperbola with center at the origin. Here $b^2 = 7$, so $b = \sqrt{7}$; $a^2 = 4$, so $a = 2$. The hyperbola has vertices at $(0, \sqrt{7})$ and $(0, -\sqrt{7})$. The fundamental rectangle has corners at $(2, \sqrt{7})$, $(2, -\sqrt{7})$, $(-2, \sqrt{7})$, and $(-2, -\sqrt{7})$. To aid us in graphing, we measure the distance $\sqrt{7}$ as approximately 2.6.

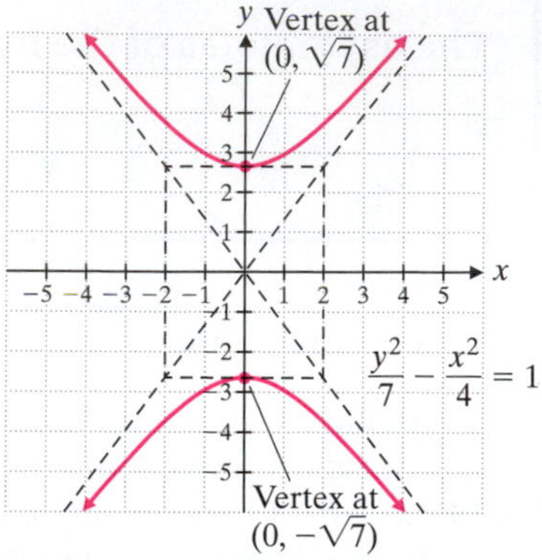

Student Practice 2 Graph $y^2 - 4x^2 = 4$.

Student Practice 2

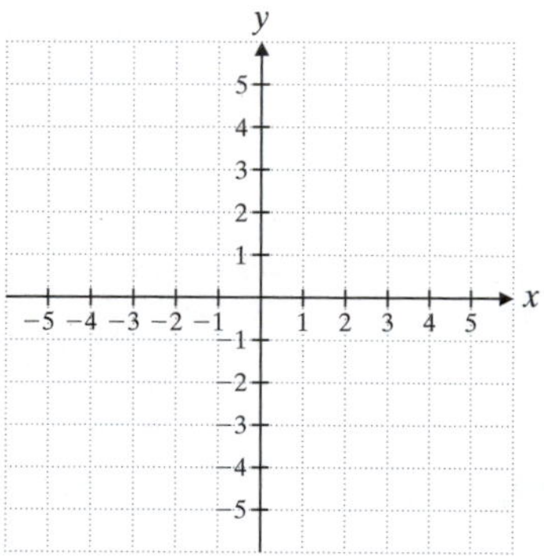

② Graphing a Hyperbola Whose Center Is at (*h*, *k*)

If a hyperbola does not have its center at the origin but is shifted h units to the right or left and k units up or down, its equation is one of the following:

STANDARD FORM OF THE EQUATION OF A HYPERBOLA WITH CENTER AT (h, k)

Let a and b be any positive real numbers. A horizontal hyperbola with center at (h, k) and vertices $(h - a, k)$ and $(h + a, k)$ has the equation

$$\frac{(x - h)^2}{a^2} - \frac{(y - k)^2}{b^2} = 1.$$

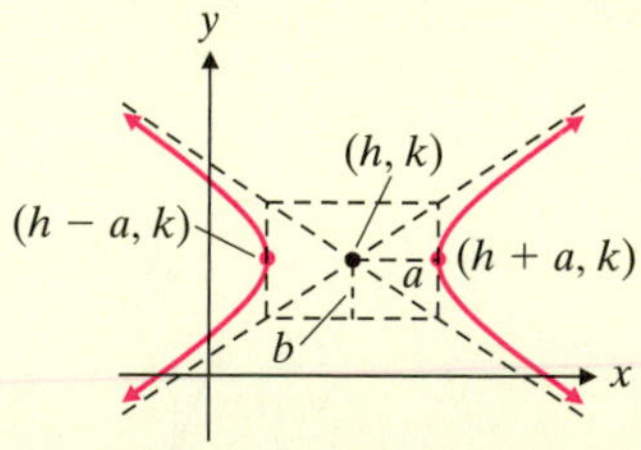

Horizontal Hyperbola

A vertical hyperbola with center at (h, k) and vertices $(h, k + b)$ and $(h, k - b)$ has the equation

$$\frac{(y - k)^2}{b^2} - \frac{(x - h)^2}{a^2} = 1.$$

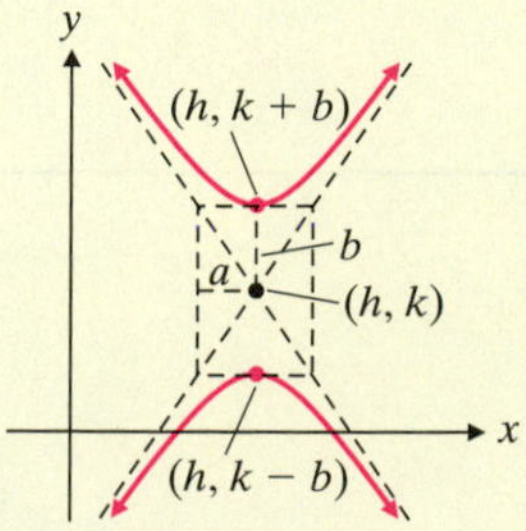

Vertical Hyperbola

Graphing Calculator

Graphing Hyperbolas

Graph on your graphing calculator the hyperbola in Example 2 using

$$y_1 = \frac{\sqrt{28 + 7x^2}}{2}$$

and

$$y_2 = -\frac{\sqrt{28 + 7x^2}}{2}.$$

Do you see how we obtained y_1 and y_2?

EXAMPLE 3 Graph $\frac{(x-4)^2}{9} - \frac{(y-5)^2}{4} = 1$.

Solution The center is at (4, 5), and the hyperbola is horizontal. We have $a = 3$ and $b = 2$, so the vertices are $(4 \pm 3, 5)$, or (7, 5) and (1, 5). We can sketch the hyperbola more readily if we can draw a fundamental rectangle. Using (4, 5) as the center, we construct a rectangle $2a$ units wide and $2b$ units high. We then draw and extend the diagonals of the rectangle. The extended diagonals are the asymptotes for the branches of the hyperbola.

In this example, since $a = 3$ and $b = 2$, we draw a rectangle $2a = 6$ units wide and $2b = 4$ units high with a center at (4, 5). We draw extended diagonals through the rectangle. From the vertex at (7, 5), we draw a branch of the hyperbola opening to the right. From the vertex at (1, 5), we draw a branch of the hyperbola opening to the left. The graph of the hyperbola is shown.

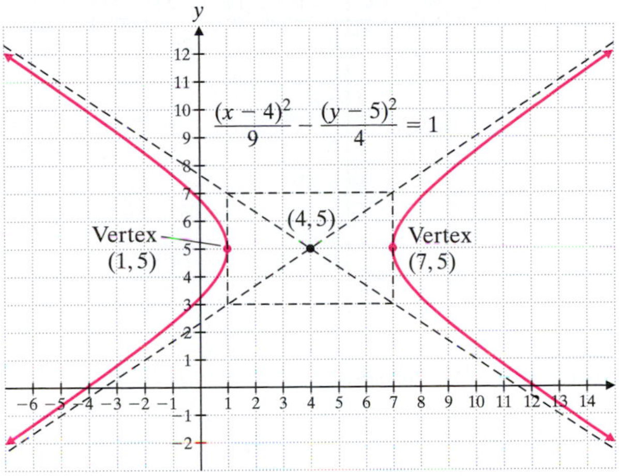

Student Practice 3 Graph $\frac{(y+2)^2}{9} - \frac{(x-3)^2}{16} = 1$.

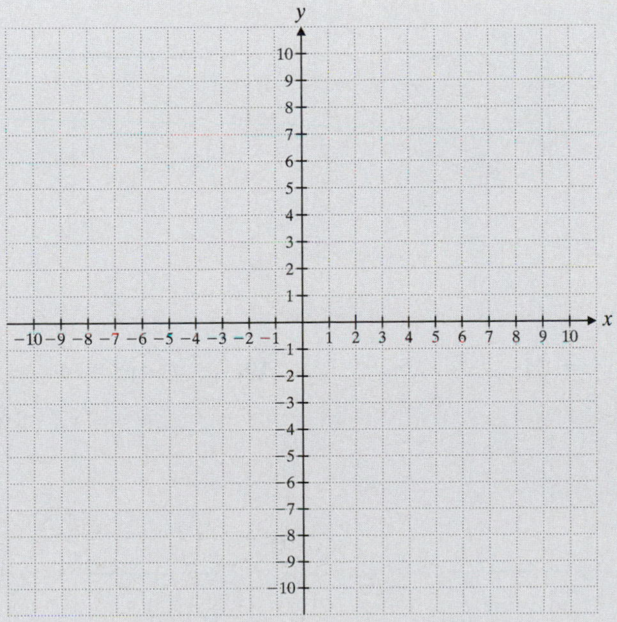

Graphing Calculator

Exploration

Graph the hyperbola in Example 3 using

$$y_1 = 5 + \sqrt{\frac{4(x-4)^2}{9} - 4}$$

and

$$y_2 = 5 - \sqrt{\frac{4(x-4)^2}{9} - 4}.$$

Do you see how we obtained y_1 and y_2?

9.4 Exercises

Verbal and Writing Skills, Exercises 1–4

1. What is the standard form of the equation of a horizontal hyperbola centered at the origin?

2. What are the vertices of the hyperbola $\frac{y^2}{9} - \frac{x^2}{4} = 1$? Is this a horizontal hyperbola or a vertical hyperbola? Why?

3. Explain in your own words how you would draw the graph of the hyperbola $\frac{x^2}{16} - \frac{y^2}{4} = 1$.

4. Explain how you determine the center of the hyperbola $\frac{(x - 2)^2}{4} - \frac{(y + 3)^2}{25} = 1$.

Find the vertices and graph each hyperbola. If the equation is not in standard form, write it as such.

5. $\frac{x^2}{4} - \frac{y^2}{25} = 1$

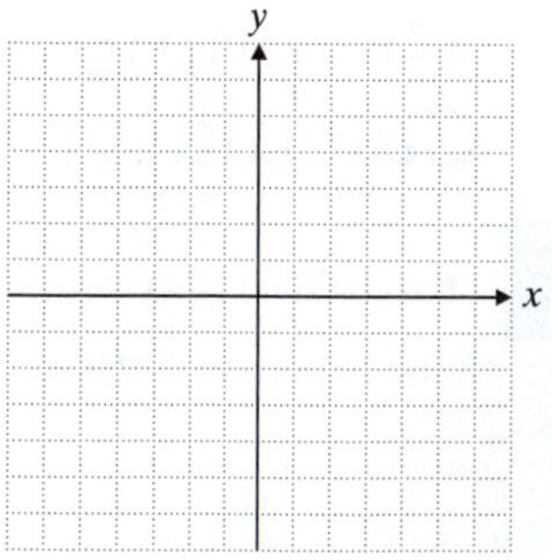

6. $\frac{x^2}{9} - \frac{y^2}{36} = 1$

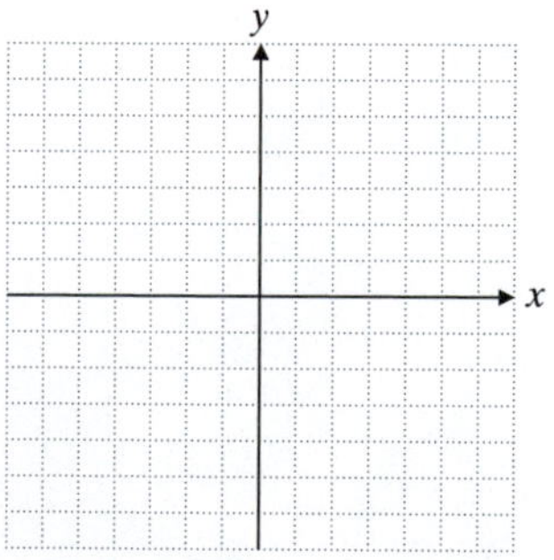

7. $\frac{y^2}{25} - \frac{x^2}{16} = 1$

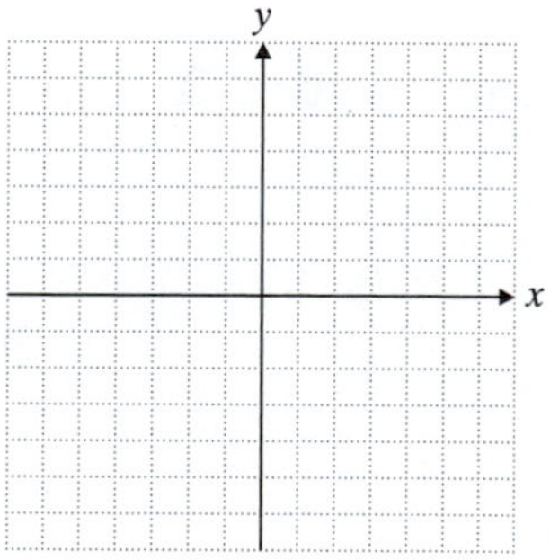

8. $\frac{y^2}{9} - \frac{x^2}{4} = 1$

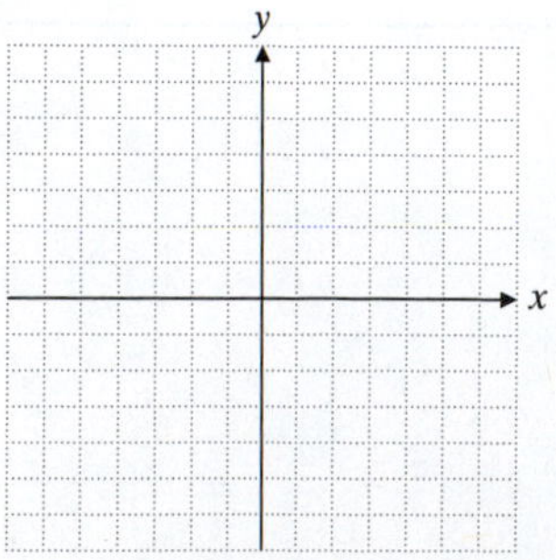

9. $4x^2 - y^2 = 64$

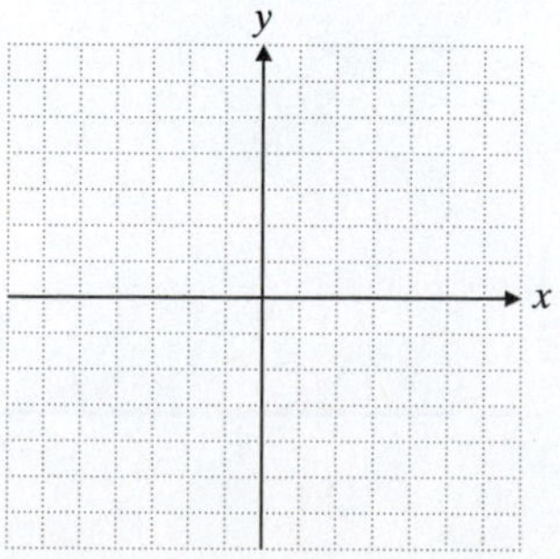

10. $x^2 - 4y^2 = 4$

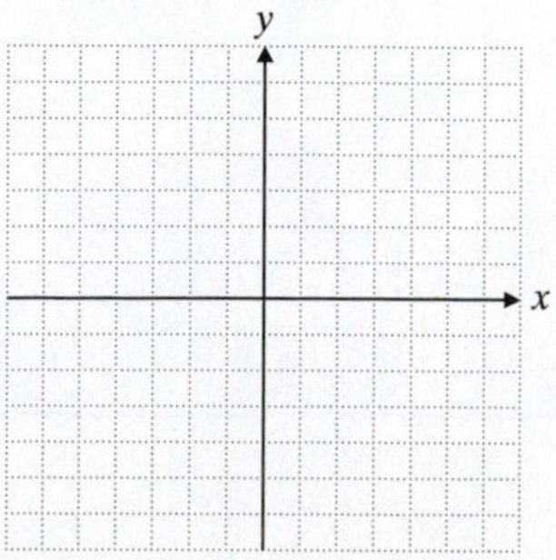

11. $8x^2 - y^2 = 16$

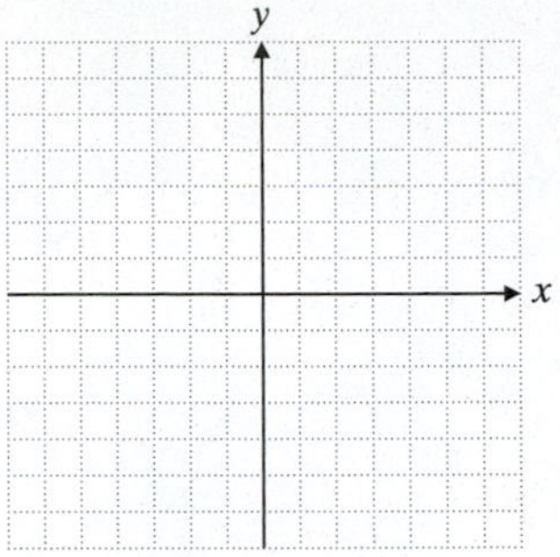

12. $12x^2 - y^2 = 36$

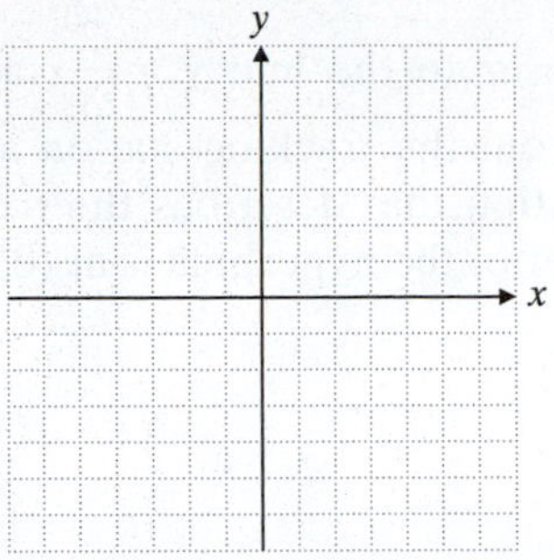

13. $4y^2 - 3x^2 = 48$

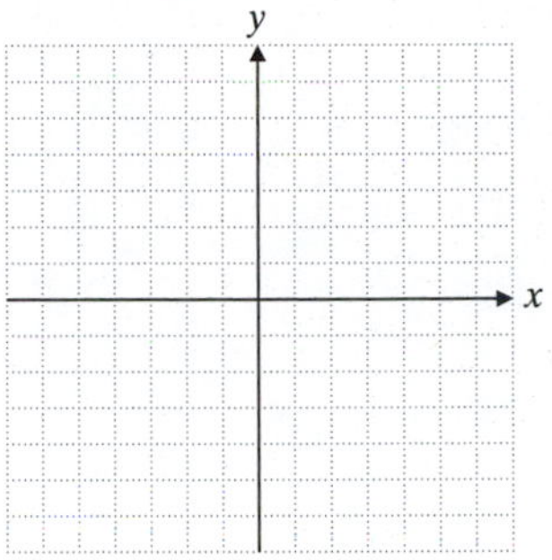

14. $8x^2 - 3y^2 = 24$

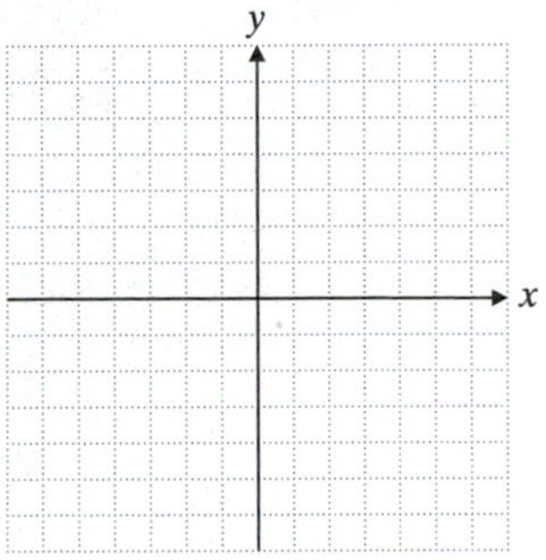

Find the equation of the hyperbola with center at the origin and with the following vertices and asymptotes.

15. Vertices at $(3, 0)$ and $(-3, 0)$; asymptotes $y = \frac{4}{3}x$, $y = -\frac{4}{3}x$

16. Vertices at $(2, 0)$ and $(-2, 0)$; asymptotes $y = \frac{3}{2}x$, $y = -\frac{3}{2}x$

17. Vertices $(0, 11)$ and $(0, -11)$; asymptotes $y = \frac{11}{13}x$, $y = -\frac{11}{13}x$

18. Vertices $(0, \sqrt{15})$ and $(0, -\sqrt{15})$; asymptotes $y = \frac{\sqrt{15}}{4}x$, $y = -\frac{\sqrt{15}}{4}x$

Applications

19. ***Comet Orbits*** Some comets have orbits that are hyperbolic in shape with the Sun at the focus of the hyperbola. A comet is heading toward Earth but then veers off as shown in the graph. It comes within 120 million miles of Earth. As it travels into the distance, it moves closer and closer to the line $y = 3x$ with Earth at the origin. Find the equation that describes the path of the comet.

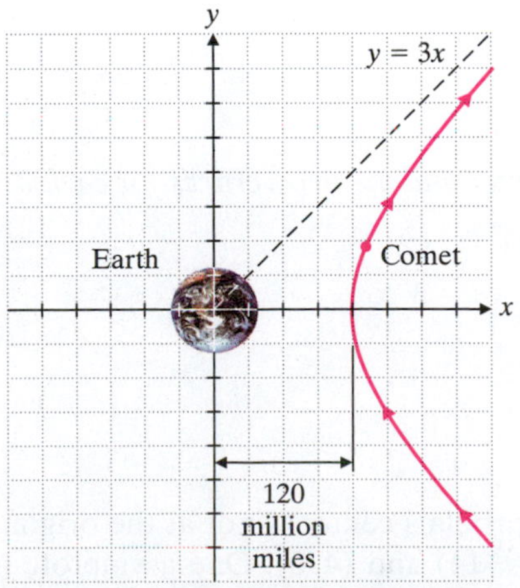

Scale on x-axis: each square is 30 million miles.
Scale on y-axis: each square is 90 million miles.

20. ***Rocket Path*** A rocket following the hyperbolic path shown in the graph turns rapidly at $(4, 0)$ and then moves closer and closer to the line $y = \frac{2}{3}x$ as the rocket gets farther from the tracking station at the origin. Find the equation that describes the path of the rocket if the center of the hyperbola is at $(0, 0)$.

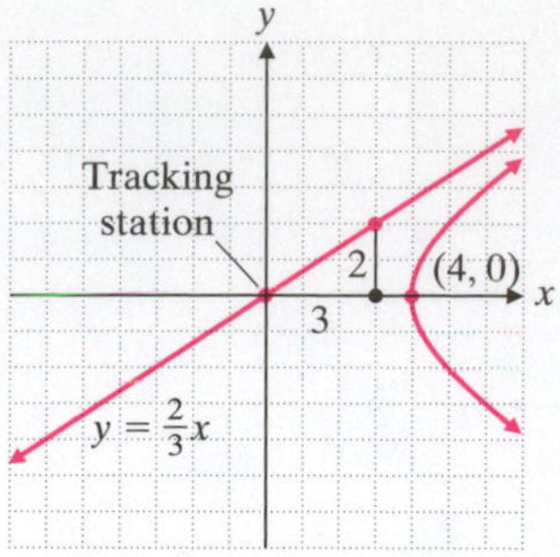

Find the center and then graph each hyperbola. You may want to use a scale other than 1 square = 1 unit.

21. $\dfrac{(x-1)^2}{4} - \dfrac{(y+2)^2}{9} = 1$

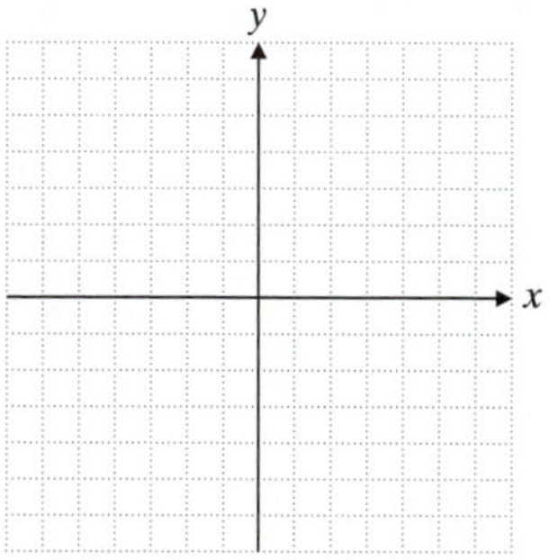

22. $\dfrac{(x+3)^2}{16} - \dfrac{(y-1)^2}{4} = 1$

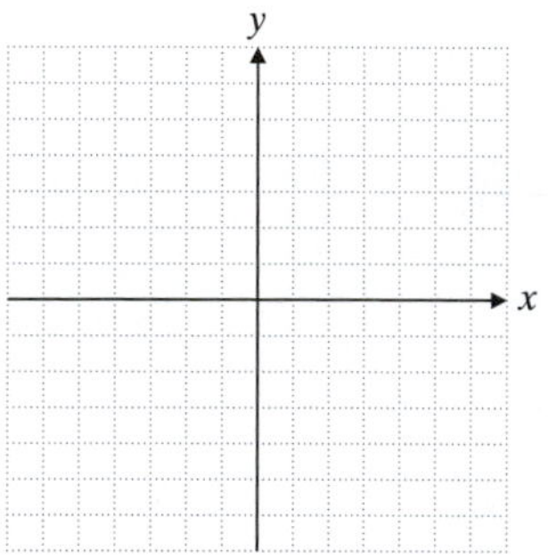

23. $\dfrac{(y+2)^2}{36} - \dfrac{(x+1)^2}{81} = 1$

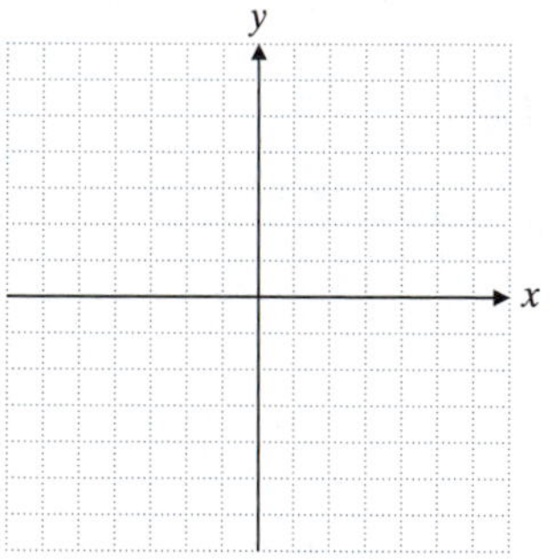

24. $\dfrac{(y+1)^2}{49} - \dfrac{(x+3)^2}{81} = 1$

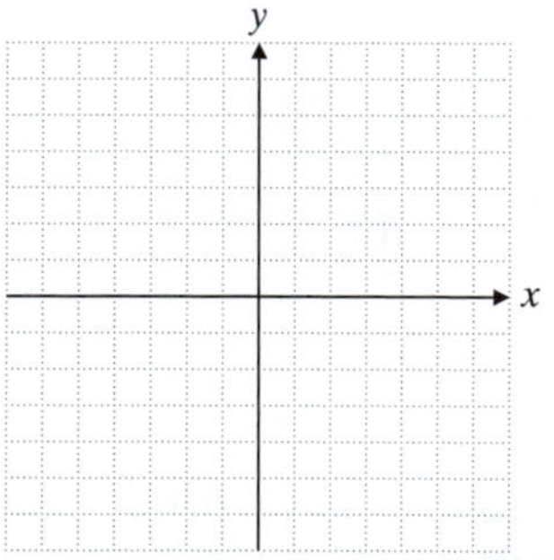

Find the center and the two vertices for each of the following hyperbolas.

25. $\dfrac{(x+6)^2}{7} - \dfrac{y^2}{3} = 1$

26. $\dfrac{x^2}{6} - \dfrac{(y-4)^2}{12} = 1$

27. A hyperbola's center is not at the origin. Its vertices are $(4, -14)$ and $(4, 0)$. One asymptote is $y = -\frac{7}{4}x$. Find the equation of the hyperbola.

28. A hyperbola's center is not at the origin. Its vertices are $(5, 0)$ and $(5, 14)$; one asymptote is $y = \frac{7}{5}x$. Find the equation of the hyperbola.

Optional Graphing Calculator Problems

29. For the hyperbola $8x^2 - y^2 = 16$, if $x = 3.5$, what are the two values of y?

30. For the hyperbola $x^2 - 12y^2 = 36$, if $x = 8.2$, what are the two values of y?

Cumulative Review *Combine.*

31. **[6.2.2]** $\dfrac{3}{x^2 - 5x + 6} + \dfrac{2}{x^2 - 4}$

32. **[6.2.2]** $\dfrac{2x}{5x^2 + 9x - 2} - \dfrac{3}{5x - 1}$

33. **[2.5.1]** ***Box Office*** For the weekend of January 7–9, 2011, two of the top ten movies in U.S. theaters were *True Grit* and *Season of the Witch*. Together, these two movies grossed \$25.7 million, which accounted for 28.3% of the total amount grossed by the top ten movies. What was the amount grossed by the top ten movies during that weekend? Round to the nearest tenth. (*Source:* www.boxofficemojo.com)

34. **[1.2.2]** ***Oil Consumption*** At the beginning of 2011, it was predicted that global oil use would reach a record 87.8 million barrels per day. This was a 1.7% increase in oil use from 2010. How many barrels of oil were used daily in 2010? Round your answer to the nearest tenth. (*Source:* www.energy.gov)

Quick Quiz 9.4

1. What are the two vertices of the hyperbola $36y^2 - 9x^2 = 36$?

2. Find the equation in standard form of the hyperbola with center at the origin and vertices at $(4, 0)$ and $(-4, 0)$. One asymptote is $y = \frac{5}{4}x$.

3. Graph the hyperbola. Label the vertices.

$$\frac{y^2}{9} - \frac{x^2}{4} = 1$$

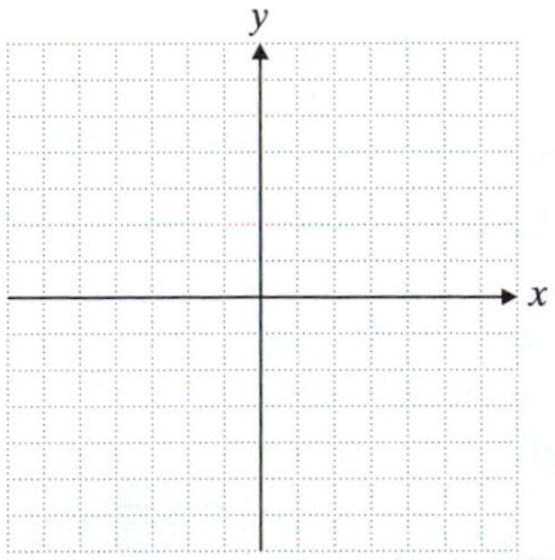

4. **Concept Check** Explain how you would find the equation of one of the asymptotes for the hyperbola $49x^2 - 4y^2 = 196$.

9.5 Nonlinear Systems of Equations

Student Learning Objectives

After studying this section, you will be able to:

① **Solve a nonlinear system by the substitution method.**

② **Solve a nonlinear system by the addition method.**

① Solving a Nonlinear System by the Substitution Method

Any equation that is of second degree or higher is a **nonlinear equation.** In other words, the graph of the equation is not a straight line (which is what the word *nonlinear* means), and the equation can't be written in the form $y = mx + b$. A **nonlinear system of equations** includes at least one nonlinear equation.

The most frequently used method for solving a nonlinear system is the method of substitution. This method works especially well when one equation of the system is linear. A sketch can often be used to verify the solution(s).

EXAMPLE 1 Solve the following nonlinear system and verify your answer with a sketch.

$$x + y - 1 = 0 \quad \textbf{(1)}$$
$$y - 1 = x^2 + 2x \quad \textbf{(2)}$$

Solution We'll use the substitution method.

$$y = -x + 1 \quad \textbf{(3)} \qquad \text{Solve for } y \text{ in equation } \textbf{(1)}.$$
$$(-x + 1) - 1 = x^2 + 2x \qquad \text{Substitute } \textbf{(3)} \text{ into equation } \textbf{(2)}.$$
$$-x + 1 - 1 = x^2 + 2x$$
$$0 = x^2 + 3x \qquad \text{Solve the resulting quadratic equation.}$$
$$0 = x(x + 3)$$
$$x = 0 \quad \text{or} \quad x = -3$$

Now substitute the values for x in the equation $y = -x + 1$.

For $x = -3$: $\quad y = -(-3) + 1 = +3 + 1 = 4$

For $x = 0$: $\quad y = -(0) + 1 = +1 = 1$

Thus, the solutions of the system are $(-3, 4)$ and $(0, 1)$.

To sketch the system, we see that equation **(2)** describes a parabola. We can rewrite it in the form

$$y = x^2 + 2x + 1 = (x + 1)^2.$$

This is a parabola opening upward with its vertex at $(-1, 0)$. Equation **(1)** can be written as $y = -x + 1$, which is a straight line with slope $= -1$ and y-intercept $(0, 1)$.

A sketch shows the two graphs intersecting at $(0, 1)$ and $(-3, 4)$. Thus, the solutions are verified.

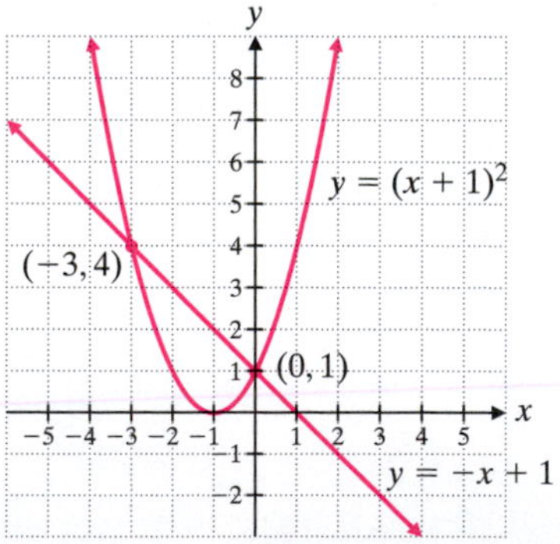

Student Practice 1 Solve the system.

$$\frac{x^2}{4} - \frac{y^2}{4} = 1$$
$$x + y + 1 = 0$$

***NOTE TO STUDENT:** Fully worked-out solutions to all of the Student Practice problems can be found at the back of the text starting at page SP-1.*

EXAMPLE 2 Solve the following nonlinear system and verify your answer with a sketch.

$$y - 2x = 0 \quad \textbf{(1)}$$

$$\frac{x^2}{4} + \frac{y^2}{9} = 1 \quad \textbf{(2)}$$

Solution

$$y = 2x \quad \textbf{(3)}$$ Solve equation **(1)** for y.

$$\frac{x^2}{4} + \frac{(2x)^2}{9} = 1$$ Substitute **(3)** into equation **(2)**.

$$\frac{x^2}{4} + \frac{4x^2}{9} = 1$$ Simplify.

$$36\left(\frac{x^2}{4}\right) + 36\left(\frac{4x^2}{9}\right) = 36(1)$$ Clear the fractions.

$$9x^2 + 16x^2 = 36$$

$$25x^2 = 36$$

$$x^2 = \frac{36}{25}$$

$$x = \pm\sqrt{\frac{36}{25}}$$

$$x = \pm\frac{6}{5} = \pm 1.2$$

For $x = +1.2$: $\quad y = 2(1.2) = 2.4.$

For $x = -1.2$: $\quad y = 2(-1.2) = -2.4.$

Thus, the solutions are $(1.2, 2.4)$ and $(-1.2, -2.4)$.

We recognize $\frac{x^2}{4} + \frac{y^2}{9} = 1$ as an ellipse with center at the origin and vertices $(0, 3)$, $(0, -3)$, $(2, 0)$, and $(-2, 0)$. When we rewrite $y - 2x = 0$ as $y = 2x$, we recognize it as a straight line with slope 2 passing through the origin. The sketch shows that the points of intersection at $(1.2, 2.4)$ and $(-1.2, -2.4)$ seem reasonable.

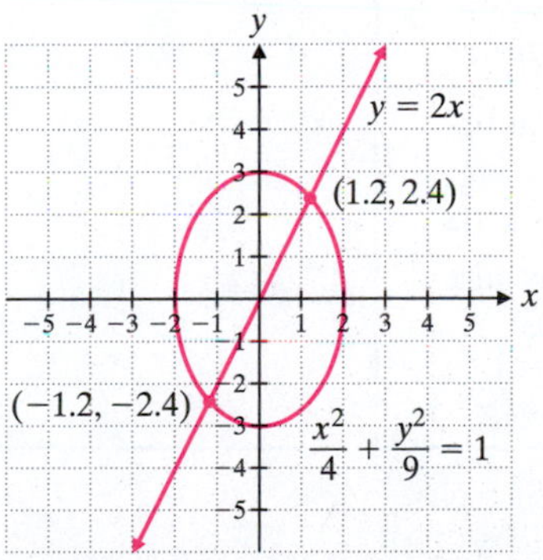

Student Practice 2 Solve the system. Verify your answer with a sketch.

$$2x - 9 = y$$
$$xy = -4$$

Graphing Calculator

Solving Nonlinear Systems

Use a graphing calculator to solve the following system. Round your answers to the nearest tenth.

$$30x^2 + 256y^2 = 7680$$
$$3x + y - 40 = 0$$

First we will need to obtain the equations

$$y_1 = \frac{\sqrt{7680 - 30x^2}}{16},$$

$$y_2 = -\frac{\sqrt{7680 - 30x^2}}{16},$$

and

$$y_3 = 40 - 3x$$

and graph them to approximate the solutions. Be sure that your window includes enough of the graphs to find the points of intersection.

Student Practice 2

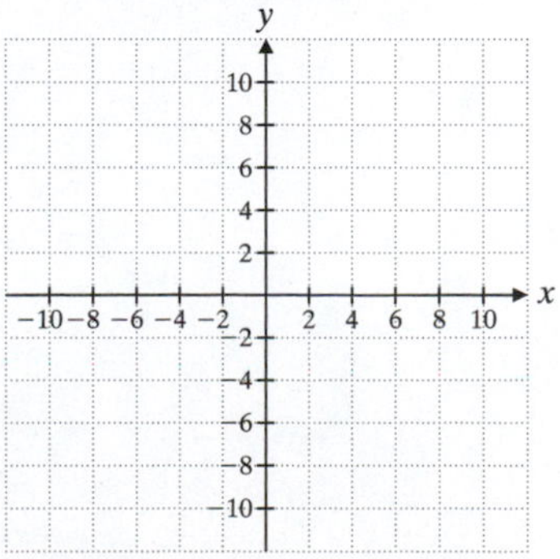

② Solving a Nonlinear System by the Addition Method

Sometimes a system may be solved more readily by adding the equations together. It should be noted that some systems have no solution.

EXAMPLE 3 Solve the system.

$$4x^2 + y^2 = 1 \quad \textbf{(1)}$$
$$x^2 + 4y^2 = 1 \quad \textbf{(2)}$$

Solution Although we could use the substitution method, it is easier to use the addition method because neither equation is linear.

$$\begin{aligned} -16x^2 - 4y^2 &= -4 \\ x^2 + 4y^2 &= 1 \\ \hline -15x^2 &= -3 \end{aligned}$$

Multiply equation **(1)** by -4 and add to equation **(2)**.

$$x^2 = \frac{-3}{-15}$$
$$x^2 = \frac{1}{5}$$
$$x = \pm\sqrt{\frac{1}{5}}$$

If $x = +\sqrt{\frac{1}{5}}$, then $x^2 = \frac{1}{5}$. Substituting this value into equation **(2)** gives

$$\frac{1}{5} + 4y^2 = 1$$
$$4y^2 = \frac{4}{5}$$
$$y^2 = \frac{1}{5}$$
$$y = \pm\sqrt{\frac{1}{5}}$$

Similarly, if $x = -\sqrt{\frac{1}{5}}$, then $y = \pm\sqrt{\frac{1}{5}}$. It is important to determine exactly how many solutions a nonlinear system of equations actually has. In this case, we have four solutions. When x is negative, there are two values for y. When x is positive, there are two values for y. If we rationalize each expression, the four solutions are $\left(\frac{\sqrt{5}}{5}, \frac{\sqrt{5}}{5}\right), \left(\frac{\sqrt{5}}{5}, -\frac{\sqrt{5}}{5}\right), \left(-\frac{\sqrt{5}}{5}, \frac{\sqrt{5}}{5}\right)$, and $\left(-\frac{\sqrt{5}}{5}, -\frac{\sqrt{5}}{5}\right)$.

Student Practice 3 Solve the system.

$$x^2 + y^2 = 12$$
$$3x^2 - 4y^2 = 8$$

9.5 Exercises

MyMathLab®

Watch the videos in MyMathLab

Download the MyDashBoard App

Solve each of the following systems by the substitution method. Graph each equation to verify that the answer seems reasonable.

1. $y^2 = 2x$
$y = -2x + 2$

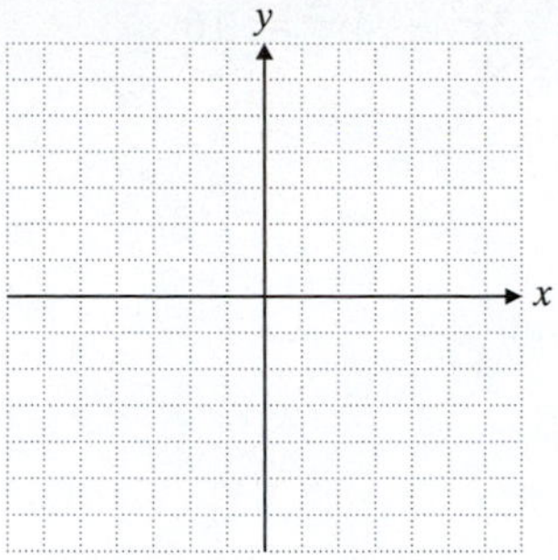

2. $y^2 = 4x$
$y = x + 1$

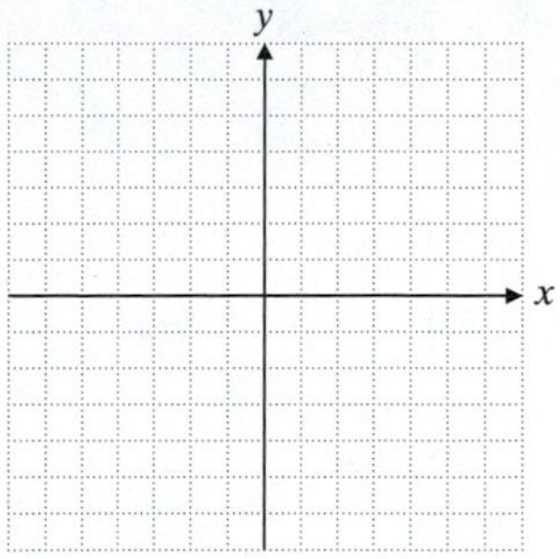

3. $x + 2y = 0$
$x^2 + 4y^2 = 32$

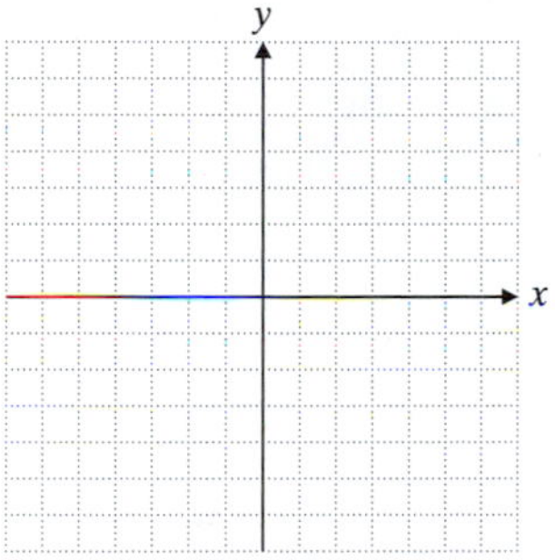

4. $y - 4x = 0$
$4x^2 + y^2 = 20$

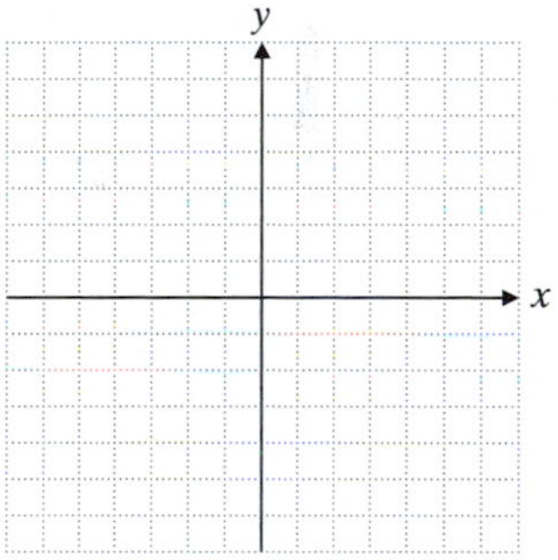

Solve each of the following systems by the substitution method.

5. $\frac{x^2}{1} - \frac{y^2}{3} = 1$
$x + y = 1$

6. $y = (x + 3)^2 - 3$
$2x - y + 2 = 0$

7. $x^2 + y^2 - 25 = 0$
$3y = x + 5$

8. $x^2 + y^2 - 9 = 0$
$2y = 3 - x$

9. $x^2 + 2y^2 = 4$
$y = -x + 2$

10. $2x^2 + 3y^2 = 27$
$y = x + 3$

11. $\frac{x^2}{4} - \frac{y^2}{4} = 1$
$x + y - 4 = 0$

12. $y^2 - x^2 = 8$
$y = 3x$

Solve each of the following systems by the addition method. Graph each equation to verify that the answer seems reasonable.

13. $2x^2 - 5y^2 = -2$
$3x^2 + 2y^2 = 35$

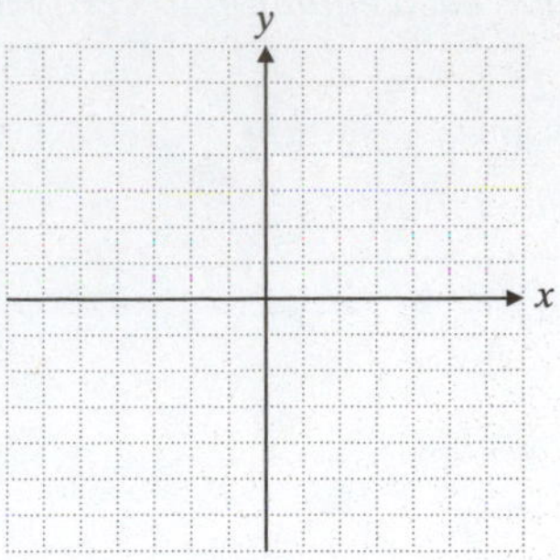

14. $2x^2 - 3y^2 = 5$
$3x^2 + 4y^2 = 16$

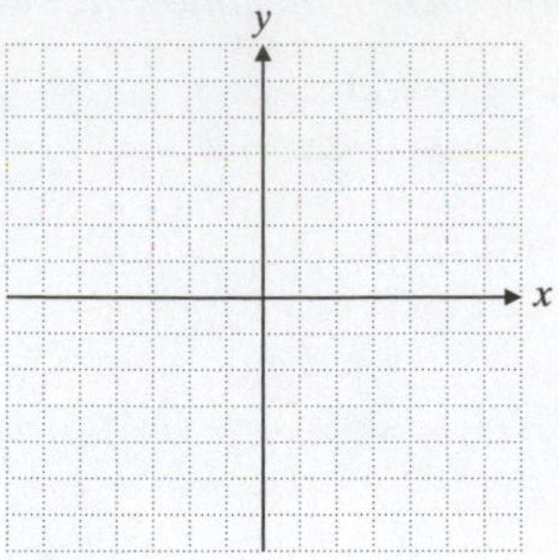

Solve each of the following systems by the addition method.

15. $x^2 + y^2 = 9$
$2x^2 - y^2 = 3$

16. $5x^2 - y^2 = 4$
$x^2 + 3y^2 = 4$

17. $x^2 + 2y^2 = 8$
$x^2 - y^2 = 1$

18. $x^2 + 6y^2 = 21$
$x^2 - 2y^2 = 5$

Mixed Practice

Solve each of the following systems by any appropriate method. If there is no real number solution, so state.

19. $x^2 + y^2 = 7$
$\frac{x^2}{3} - \frac{y^2}{9} = 1$

20. $x^2 + 2y^2 = 4$
$x^2 + y^2 = 4$

21. $2xy = 5$
$x - 4y = 3$

22. $4xy = -1$
$x + 8y = 1$

23. $xy = -6$
$2x + y = -4$

24. $xy = 1$
$3x - y + 2 = 0$

25. $x + y = 5$
$x^2 + y^2 = 4$

26. $x^2 + y^2 = 9$
$x - y = 5$

Applications

27. ***Path of a Meteor*** The outline of Earth can be considered a circle with a radius of approximately 4000 miles. Thus, if we say that the center of Earth is located at (0, 0), then an equation for this circle is $x^2 + y^2 = 16{,}000{,}000$. Suppose an incoming meteor is approaching Earth in a hyperbolic path. The equation for the meteor's path is $25{,}000{,}000x^2 - 9{,}000{,}000y^2 = 2.25 \times 10^{14}$. Will the meteor strike Earth? Why or why not? If so, locate the point (x, y) where the meteor will strike Earth. Assume that x and y are both positive. Round your answer to three significant digits.

28. ***Path of a Meteor*** Suppose that a second incoming meteor is approaching Earth in a hyperbolic path. The equation for this second meteor's path is $16{,}000{,}000x^2 - 25{,}000{,}000y^2 = 4.0 \times 10^{14}$. Will the second meteor strike Earth? Why or why not? If so, locate the point (x, y) where the meteor will strike Earth. Assume that x and y are both positive. Round your answer to three significant digits.

Cumulative Review

29. **[5.2.2]** Divide. $(3x^3 - 8x^2 - 33x - 10) \div (3x + 1)$

30. **[6.1.1]** Simplify. $\dfrac{6x^4 - 24x^3 - 30x^2}{3x^3 - 21x^2 + 30x}$

Quick Quiz 9.5 *Solve each nonlinear system.*

1. $2x - y = 4$
$y^2 - 4x = 0$

2. $y - x^2 = -4$
$x^2 + y^2 = 16$

3. $(x + 2)^2 + (y - 1)^2 = 9$
$x = 2 - y$

4. **Concept Check** Explain how you would solve the following system.

$$y^2 + 2x^2 = 18$$
$$xy = 4$$

USE MATH TO SAVE MONEY

Did You Know...

That the Government Will Pay Interest on Certain Student Loans While You Are in College?

CHOOSING A STUDENT LOAN

Understanding the Problem:
Alicia needs a $10,000 college loan. A credit union offers a 20-year private loan with a 4.65% fixed rate. A subsidized federal loan provides a 20-year loan at a 6% fixed rate with the government paying interest during school (4.5 years). Which option is the better deal?

Making a Plan:

Step 1: Notice that the two interest rates are different for each loan. Also, the total number of payments is different because the government will pay for Alicia's interest for 4.5 years out of the 20-year life of the subsidized loan.

Task 1: *What is the total amount that Alicia must pay back for each loan?*

Task 2: *What is surprising about the results of Task 1?*

Step 2: The private loan requires that Alicia make more payments overall, and she must start making payments immediately. The subsidized loan allows for fewer payments, and she can wait until 6 months after graduation to begin making payments.

Task 3: *What is the monthly payment for each loan?*

Task 4: *Which loan offers the lowest monthly payment?*

Finding the Solution:

Step 3: Let's compare the loans.

Task 5: *What are the values for the missing places in this table?*

Type of Loan	Total Amount	Number of Monthly Payments	Monthly Payment Amount	When Payments Begin
Private loan				Immediately
Subsidized loan				6 months after graduation

Task 6: *Which option provides the best deal? Explain your reasoning.*

Applying the Solution to Your Life:
When choosing between the two loans, students must consider the total amount that they must repay, whether the government will pay interest while they are in school, and when they start making their monthly payments. To apply for a subsidized loan, students must fill out a Free Application for Federal Student Aid (FAFSA) form. Approval for subsidized loans is based on financial need, but you do not need a high credit score. On the other hand, a private loan from a bank or credit union will require a good credit score. Credit unions sometimes offer lower cost private loans than banks. Make sure to check into all of your options before making a decision.

Chapter 9 Organizer

Topic and Procedure	Examples	You Try It

Distance between two points, p. 487

The distance d between points (x_1, y_1) and (x_2, y_2) is

$$d = \sqrt{(x_2 - x_1)^2 + (y_2 - y_1)^2}.$$

Find the distance between $(-6, -3)$ and $(5, -2)$.

$$\begin{aligned} d &= \sqrt{[5 - (-6)]^2 + [-2 - (-3)]^2} \\ &= \sqrt{(5 + 6)^2 + (-2 + 3)^2} \\ &= \sqrt{121 + 1} \\ &= \sqrt{122} \end{aligned}$$

1. Find the distance between $(7, -4)$ and $(3, -1)$.

Standard form of the equation of a circle, p. 488

The standard form of the equation of a circle with center at (h, k) and radius r is

$$(x - h)^2 + (y - k)^2 = r^2.$$

Graph $(x - 3)^2 + (y + 4)^2 = 16$.

Center at $(h, k) = (3, -4)$. Radius $= 4$.

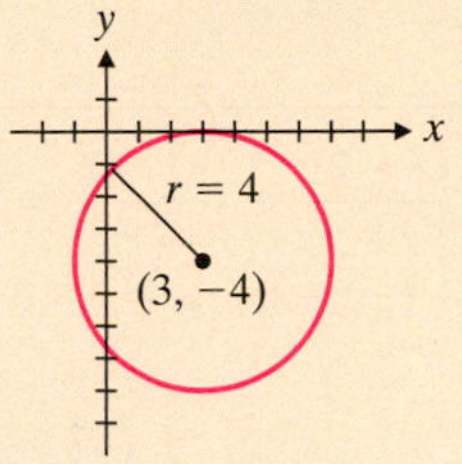

2. Graph $(x + 2)^2 + (y - 4)^2 = 9$.

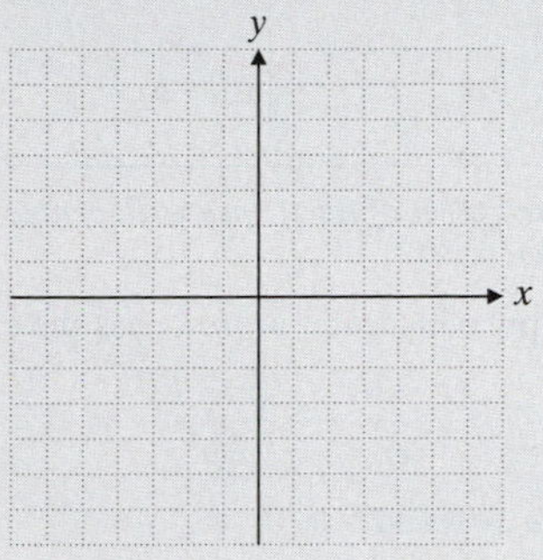

Standard form of the equation of a vertical parabola, p. 495

The equation of a vertical parabola with its vertex at (h, k) can be written in the form $y = a(x - h)^2 + k$. It opens upward if $a > 0$ and downward if $a < 0$.

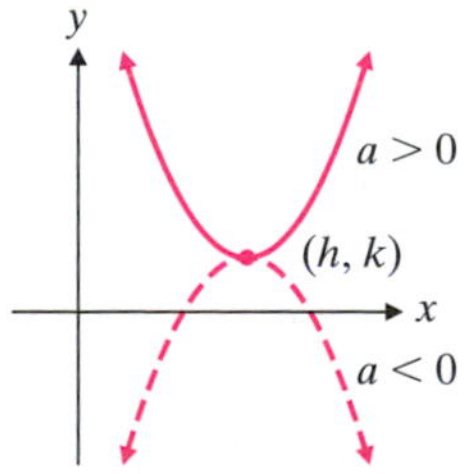

Graph $y = \frac{1}{2}(x - 3)^2 + 5$.

$a = \frac{1}{2}$, so parabola opens upward.

Vertex at $(h, k) = (3, 5)$. If $x = 0$, $y = 9.5$.

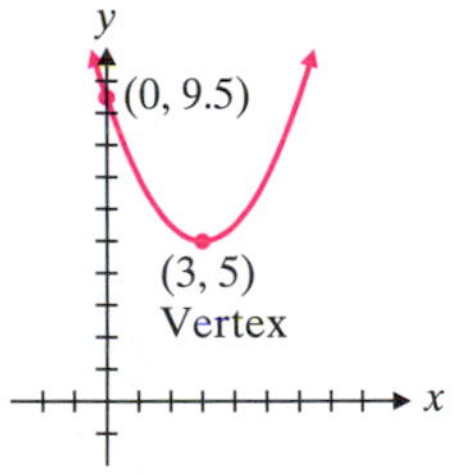

3. Graph $y = -2(x + 1)^2 + 3$.

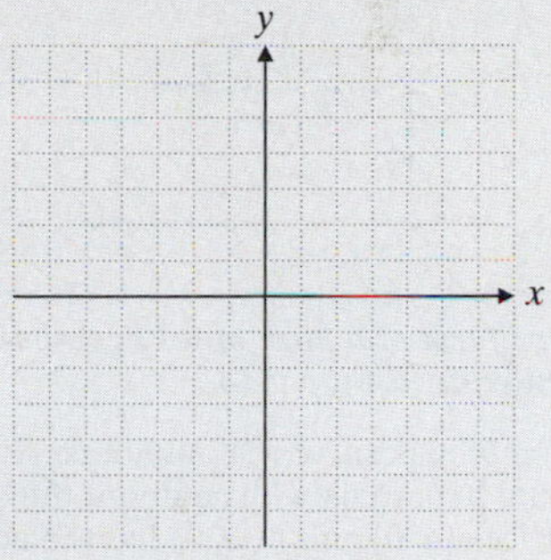

Standard form of the equation of a horizontal parabola, p. 497

The equation of a horizontal parabola with its vertex at (h, k) can be written in the form $x = a(y - k)^2 + h$. It opens to the right if $a > 0$ and to the left if $a < 0$.

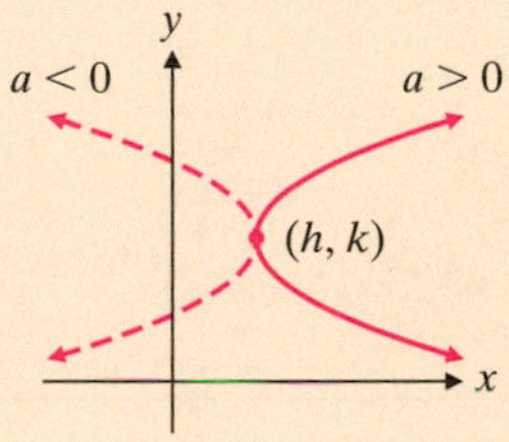

Graph $x = \frac{1}{3}(y + 2)^2 - 4$.

$a = \frac{1}{3}$, so parabola opens to the right.

Vertex at $(h, k) = (-4, -2)$.

If $y = 0$, $x = -\frac{8}{3}$.

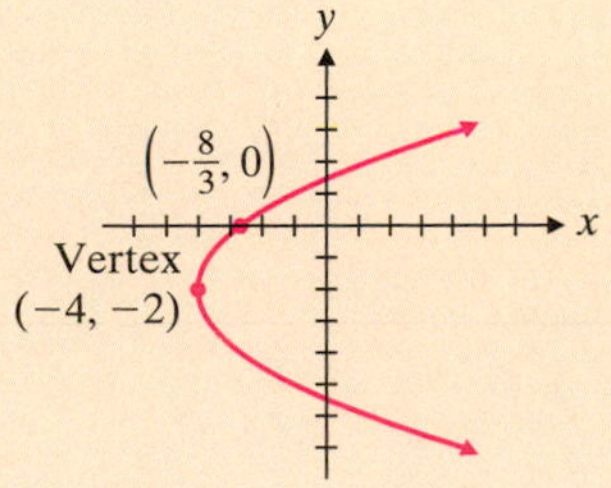

4. Graph $x = (y - 4)^2 - 1$.

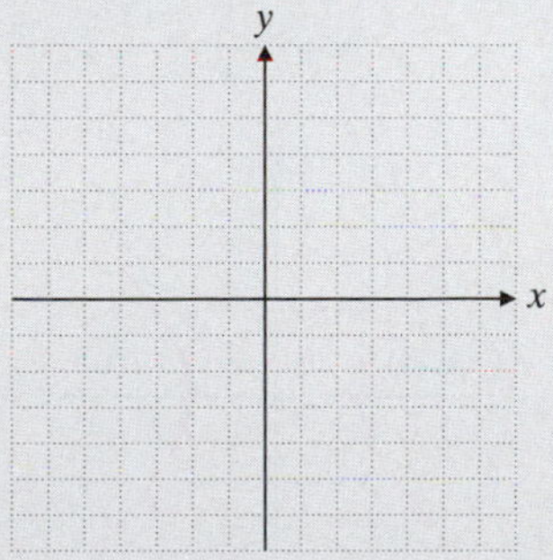

Topic and Procedure	Examples	You Try It

Standard form of the equation of an ellipse with center at* (0, 0)*, p. 505

An ellipse with center at the origin has the equation

$$\frac{x^2}{a^2} + \frac{y^2}{b^2} = 1,$$

where $a > 0$ and $b > 0$.

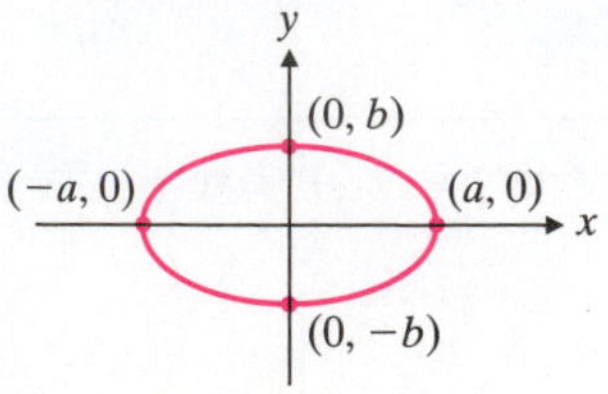

Graph $\frac{x^2}{16} + \frac{y^2}{4} = 1.$

$a^2 = 16, \quad a = 4; \quad b^2 = 4, \quad b = 2$

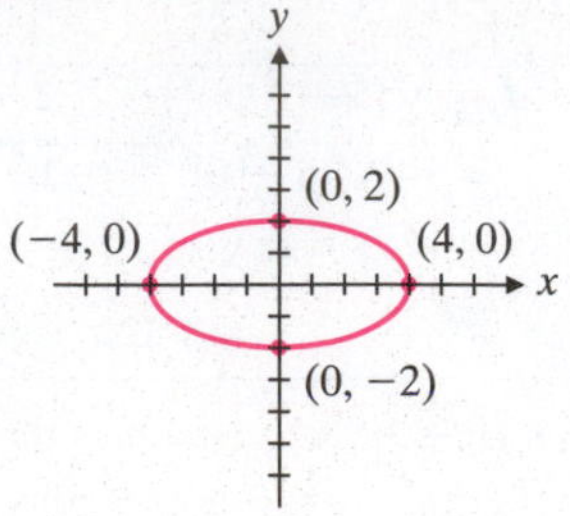

5. Graph $\frac{x^2}{36} + y^2 = 1.$

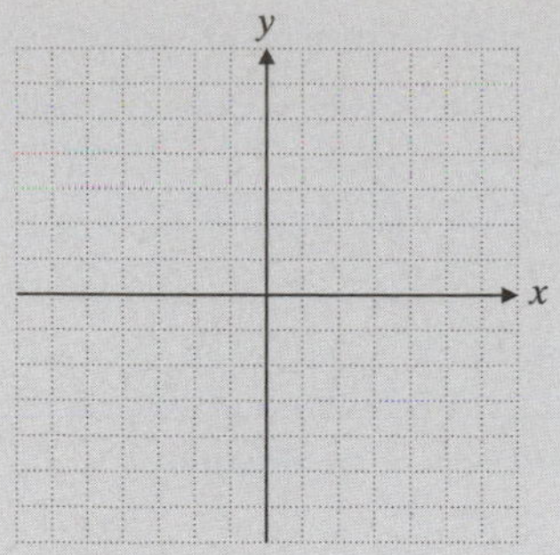

Standard form of an ellipse with center at* (h, k)*, p. 507

An ellipse with center at (h, k) has the equation

$$\frac{(x-h)^2}{a^2} + \frac{(y-k)^2}{b^2} = 1,$$

where $a > 0$ and $b > 0$.

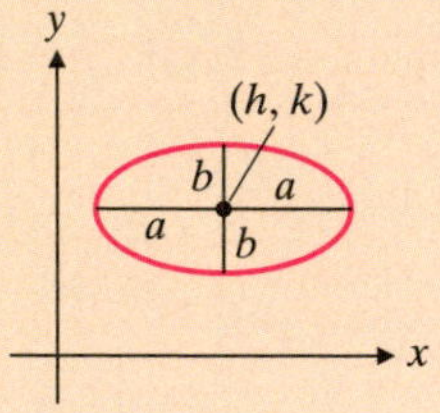

Graph $\frac{(x+2)^2}{9} + \frac{(y+4)^2}{25} = 1.$

$(h, k) = (-2, -4); \quad a = 3, \quad b = 5$

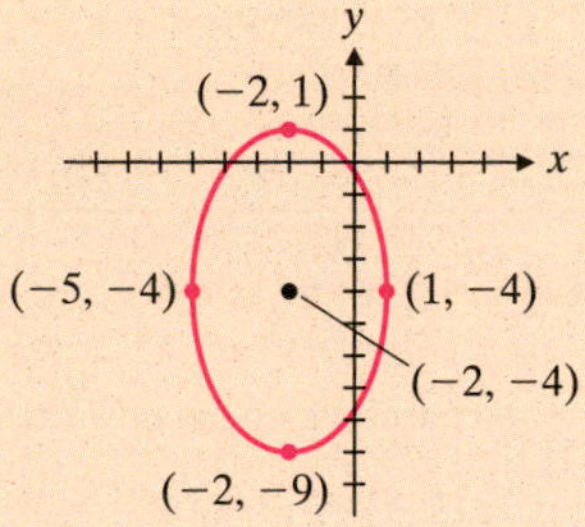

6. Graph $\frac{(x-1)^2}{9} + \frac{(y+3)^2}{4} = 1.$

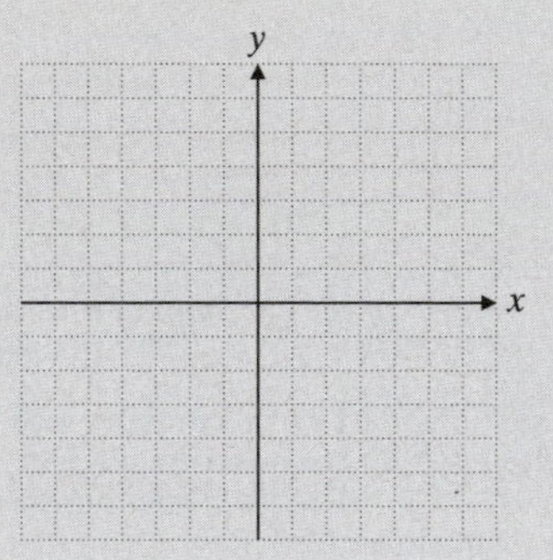

Standard form of a horizontal hyperbola with center at* (0, 0)*, p. 514

Let a and b be positive real numbers. A horizontal hyperbola with center at the origin and vertices $(a, 0)$ and $(-a, 0)$ has the equation

$$\frac{x^2}{a^2} - \frac{y^2}{b^2} = 1$$

and asymptotes

$$y = \pm\frac{b}{a}x.$$

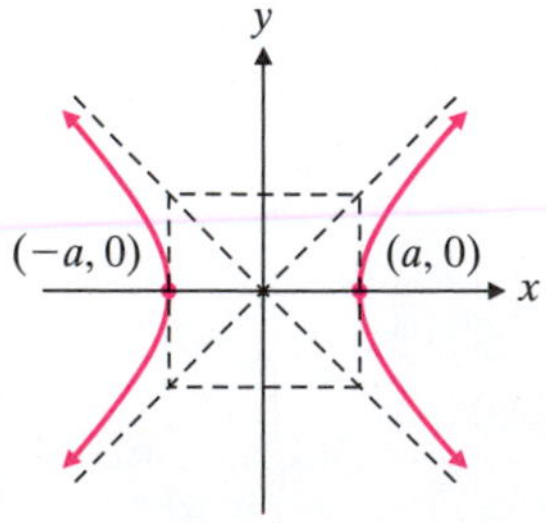

Graph $\frac{x^2}{25} - \frac{y^2}{9} = 1.$

$a = 5, \quad b = 3$

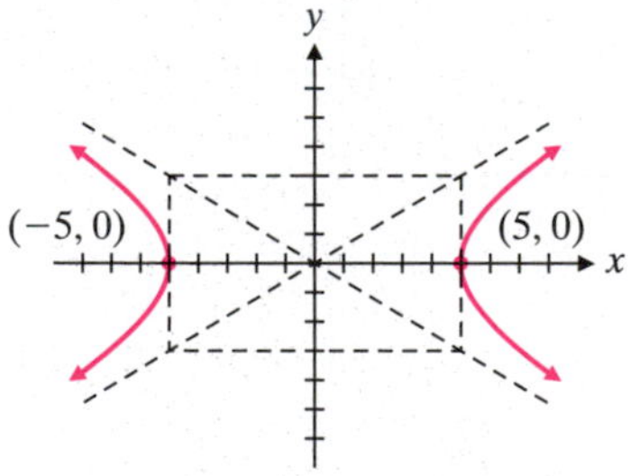

7. Graph $\frac{x^2}{36} - \frac{y^2}{4} = 1.$

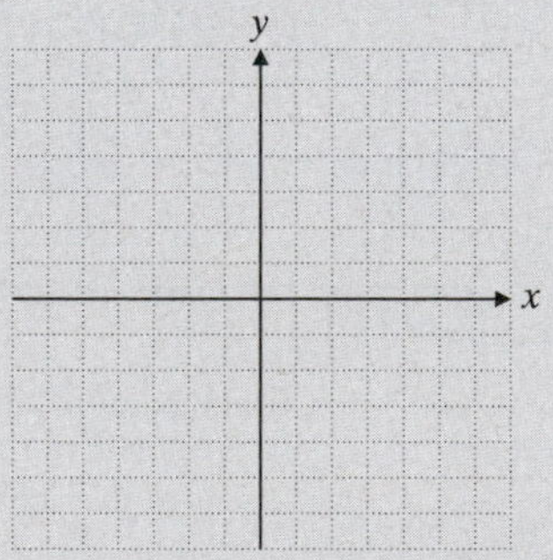

Topic and Procedure	Examples	You Try It

Standard form of a vertical hyperbola with center at (0, 0), p. 514

Let a and b be positive real numbers. A vertical hyperbola with center at the origin and vertices $(0, b)$ and $(0, -b)$ has the equation

$$\frac{y^2}{b^2} - \frac{x^2}{a^2} = 1$$

and asymptotes

$$y = \pm\frac{b}{a}x.$$

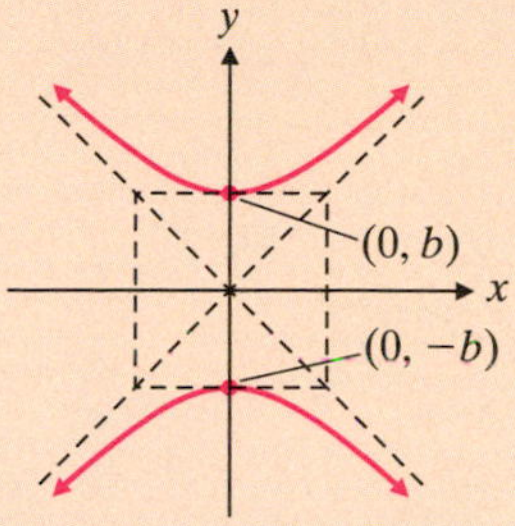

Graph $\frac{y^2}{9} - \frac{x^2}{4} = 1.$

$b = 3, \quad a = 2$

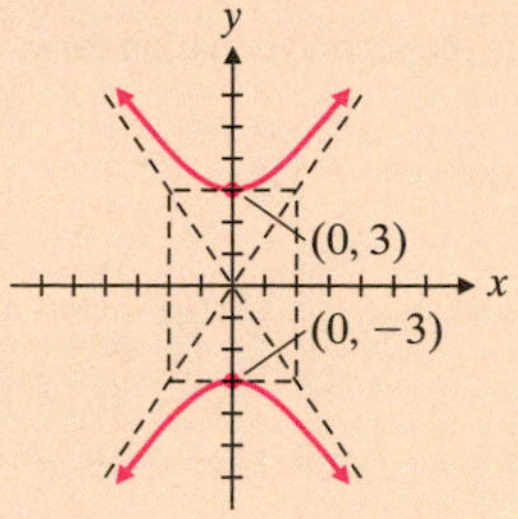

8. Graph $\frac{y^2}{16} - x^2 = 1.$

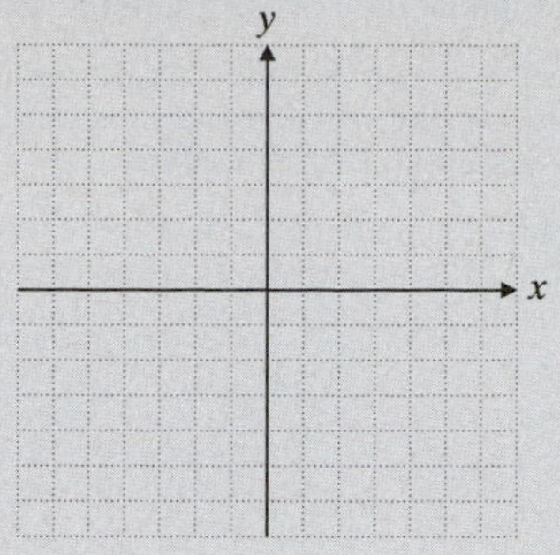

Standard form of a horizontal hyperbola with center at (h, k), p. 516

Let a and b be positive real numbers. A horizontal hyperbola with center at (h, k) and vertices $(h - a, k)$ and $(h + a, k)$ has the equation

$$\frac{(x-h)^2}{a^2} - \frac{(y-k)^2}{b^2} = 1.$$

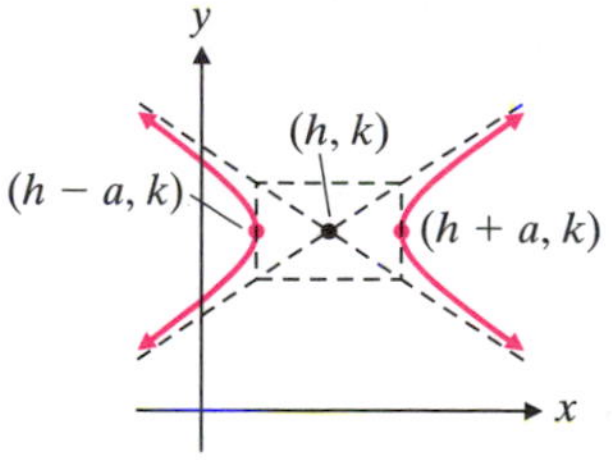

Graph $\frac{(x-2)^2}{4} - \frac{(y-3)^2}{25} = 1.$

Center at $(2, 3)$; $a = 2, b = 5$

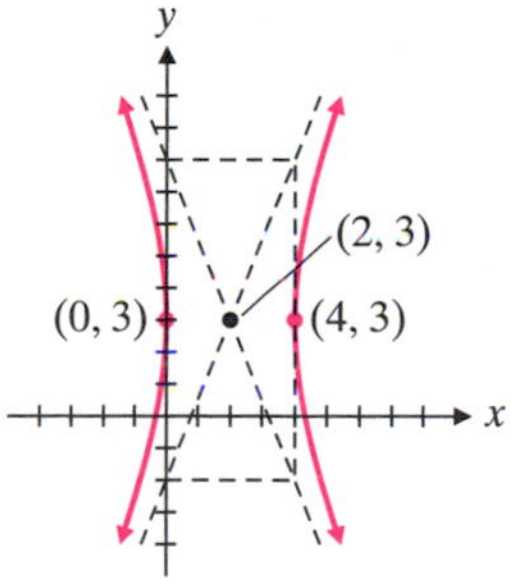

9. Graph $\frac{(x-1)^2}{9} - \frac{(y+2)^2}{16} = 1.$

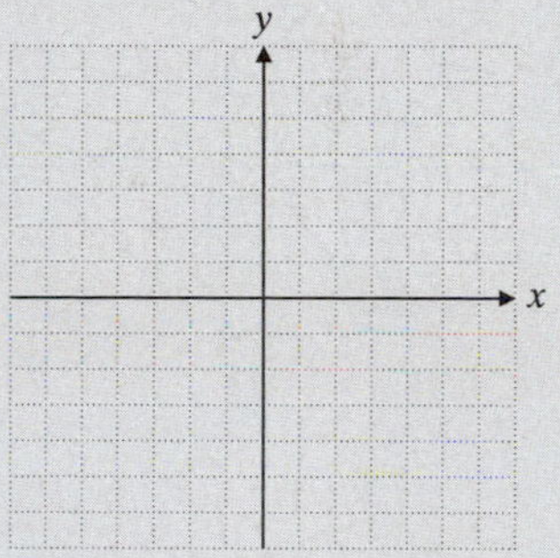

Standard form of a vertical hyperbola with center at (h, k), p. 516

Let a and b be positive real numbers. A vertical hyperbola with center at (h, k) and vertices $(h, k + b)$ and $(h, k - b)$ has the equation

$$\frac{(y-k)^2}{b^2} - \frac{(x-h)^2}{a^2} = 1.$$

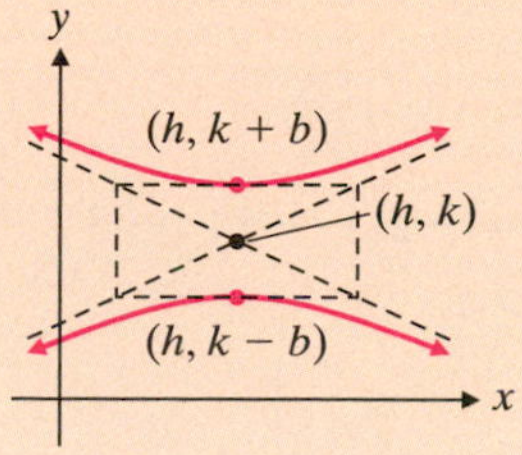

Graph $\frac{(y-5)^2}{9} - \frac{(x-4)^2}{4} = 1.$

Center at $(4, 5)$; $b = 3, a = 2$

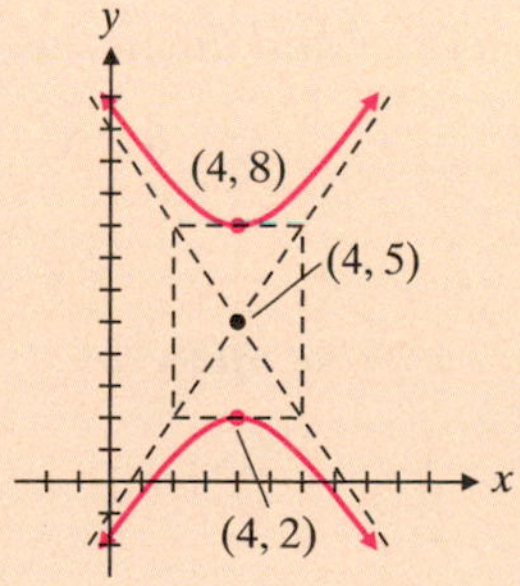

10. Graph $\frac{(y+3)^2}{4} - \frac{(x-2)^2}{4} = 1.$

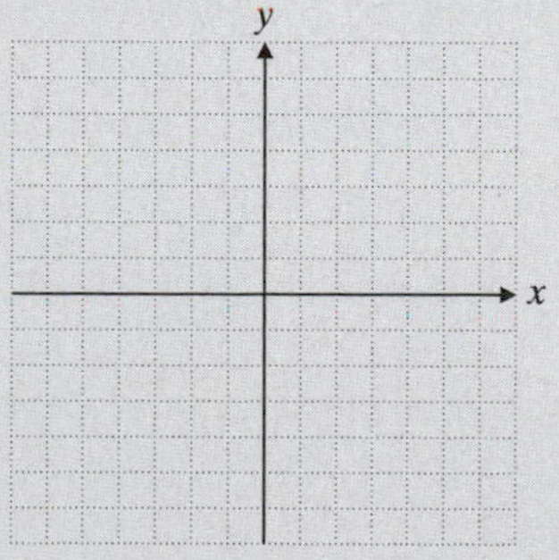

Topic and Procedure	Examples	You Try It
Nonlinear systems of equations, pp. 522–524	Solve by substitution.	**11.** Solve by substitution.

Topic and Procedure

We can solve a nonlinear system by the substitution method or the addition method.

In the substitution method we solve one equation for one variable and substitute that expression into the other equation.

In the addition method, we multiply one or more equations by a numerical value and then add them together so that one variable is eliminated.

Examples

Solve by substitution.

$$2x^2 + y^2 = 18$$
$$xy = 4$$

Solving the second equation for y, we have $y = \frac{4}{x}$.

$$2x^2 + \left(\frac{4}{x}\right)^2 = 18$$
$$2x^2 + \frac{16}{x^2} = 18$$
$$2x^4 + 16 = 18x^2$$
$$2x^4 - 18x^2 + 16 = 0$$
$$x^4 - 9x^2 + 8 = 0$$
$$(x^2 - 1)(x^2 - 8) = 0$$
$$x^2 - 1 = 0 \qquad x^2 - 8 = 0$$
$$x^2 = 1 \qquad x^2 = 8$$
$$x = \pm 1 \qquad x = \pm 2\sqrt{2}$$

Since $xy = 4$,

if $x = 1$, then $y = 4$.
if $x = -1$, then $y = -4$.
if $x = 2\sqrt{2}$, then $y = \sqrt{2}$.
if $x = -2\sqrt{2}$, then $y = -\sqrt{2}$.

The solutions are $(1, 4)$, $(-1, -4)$, $(2\sqrt{2}, \sqrt{2})$, and $(-2\sqrt{2}, -\sqrt{2})$.

You Try It

11. Solve by substitution.

$$x^2 + 3y^2 = 12$$
$$y - 3x = -2$$

Chapter 9 Review Problems

In exercises 1 and 2, find the distance between the points.

1. $(0, -6)$ and $(-3, 2)$

2. $(-7, 3)$ and $(-2, -1)$

3. Write in standard form the equation of the circle with center at $(-6, 3)$ and radius $\sqrt{15}$.

4. Write in standard form the equation of the circle with center at $(0, -7)$ and radius 5.

Rewrite each equation in standard form. Find the center and the radius of each circle.

5. $x^2 + y^2 + 2x - 6y + 5 = 0$

6. $x^2 + y^2 - 10x + 12y + 52 = 0$

Graph each parabola. Label its vertex and plot at least one intercept.

7. $x = \frac{1}{3}y^2$

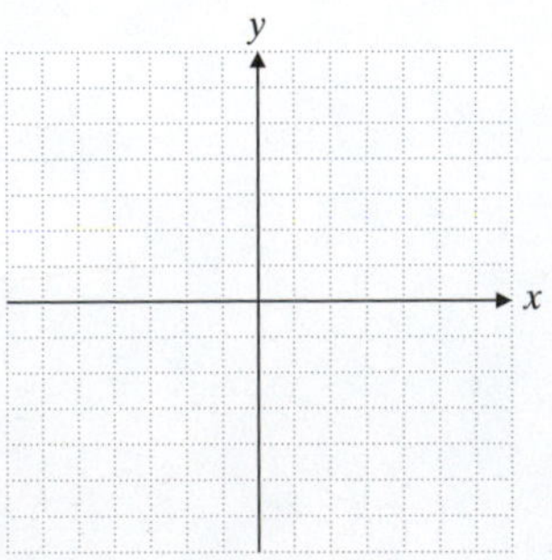

8. $x = \frac{1}{2}(y - 2)^2 + 4$

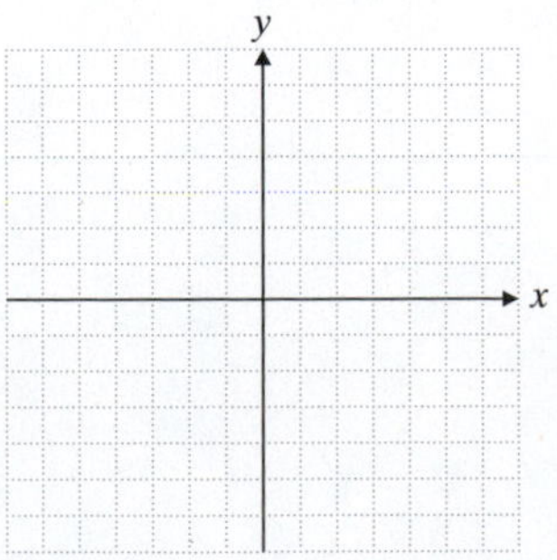

9. $y = -2(x + 1)^2 - 3$

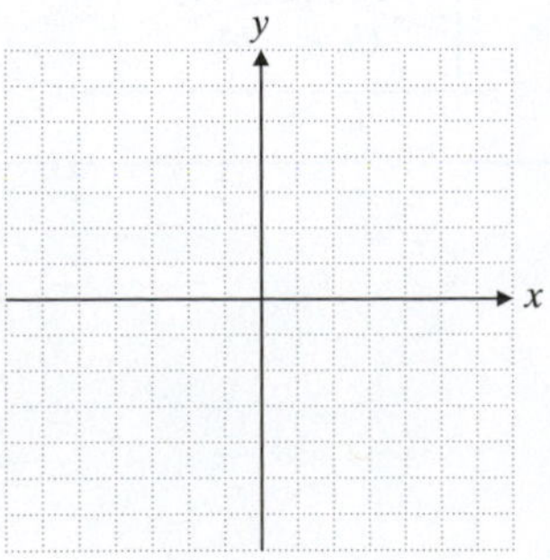

Rewrite each equation in standard form. Find the vertex and determine in which direction the parabola opens.

10. $x^2 + 6x = y - 4$

11. $x + 8y = y^2 + 10$

Graph each ellipse. Label its center and four other points.

12. $\dfrac{x^2}{4} + y^2 = 1$

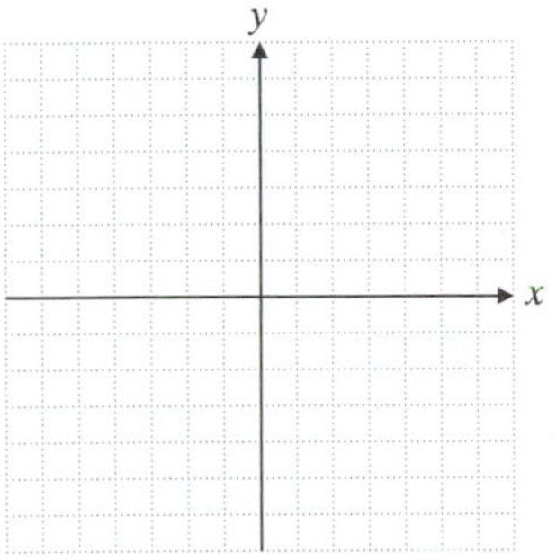

13. $16x^2 + y^2 - 32 = 0$

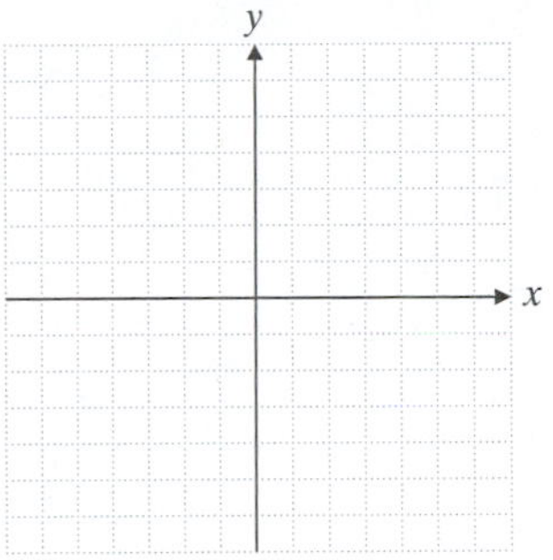

14. Determine the vertices and the center of the ellipse. $\dfrac{(x + 5)^2}{4} + \dfrac{(y + 3)^2}{25} = 1$

Find the center and vertices of each hyperbola and graph it.

15. $x^2 - 4y^2 - 16 = 0$

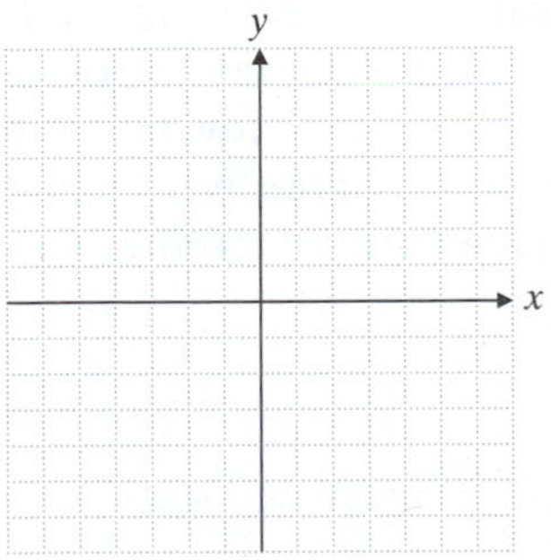

16. $3y^2 - x^2 = 27$

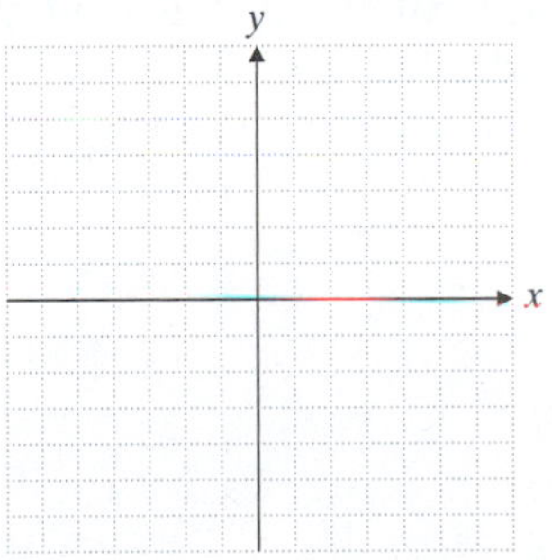

17. Determine the vertices and the center of the hyperbola. $\dfrac{(x - 2)^2}{4} - \dfrac{(y + 3)^2}{25} = 1$

Solve each nonlinear system. If there is no real number solution, so state.

18. $x^2 + y = 9$
$y - x = 3$

19. $x^2 + y^2 = 4$
$x + y = 2$

20. $2x^2 + y^2 = 17$
$x^2 + 2y^2 = 22$

21. $3x^2 - 4y^2 = 12$
$7x^2 - y^2 = 8$

22. $y = x^2 + 1$
$x^2 + y^2 - 8y + 7 = 0$

23. $2x^2 + y^2 = 18$
$xy = 4$

24. $y^2 - 2x^2 = 2$
$2y^2 - 3x^2 = 5$

25. $y^2 = 2x$
$y = \frac{1}{2}x + 1$

Applications

26. ***Searchlights*** The side view of an airport searchlight is shaped like a parabola. The center of the light source of the searchlight is located 2 feet from the base along the axis of symmetry, and the opening is 5 feet across. How deep should the searchlight be? Round your answer to the nearest hundredth.

27. ***Satellite Dishes*** The side view of a satellite dish on Jason and Wendy's house is shaped like a parabola. The signals that come from the satellite hit the surface of the dish and are then reflected to the point where the signal receiver is located. This point is the focus of the parabolic dish. The dish is 10 feet across at its opening and 4 feet deep at its center. How far from the center of the dish should the signal receiver be placed? Round your answer to the nearest hundredth.

How Am I Doing? Chapter 9 Test

CHAPTER Test Prep VIDEOS MATH COACH MyMathLab® YouTube™

After you take this test read through the Math Coach on pages 537–538. Math Coach videos are available via MyMathLab and YouTube. Step-by-step test solutions in the Chapter Test Prep Videos are also available via MyMathLab and YouTube. (Search "TobeyInterAlg" and click on "Channels.")

1. Find the distance between $(-6, -8)$ and $(-2, 5)$.

Rewrite the equation in standard form. Identify the conic, find the center or vertex, plot at least one other point, and sketch the curve.

MC **2.** $y^2 - 6y - x + 13 = 0.$

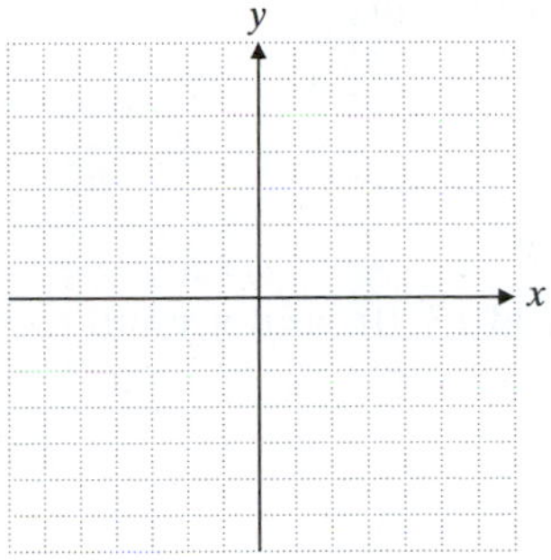

3. $x^2 + y^2 + 6x - 4y + 9 = 0$

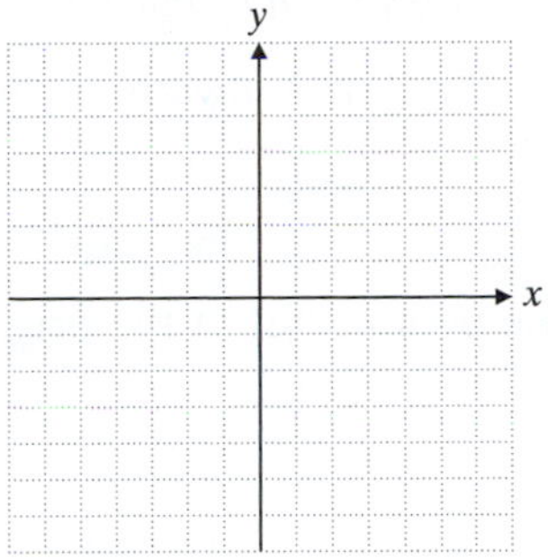

Identify and graph each conic section. Label the center and/or vertex as appropriate.

4. $\dfrac{x^2}{25} + \dfrac{y^2}{1} = 1$

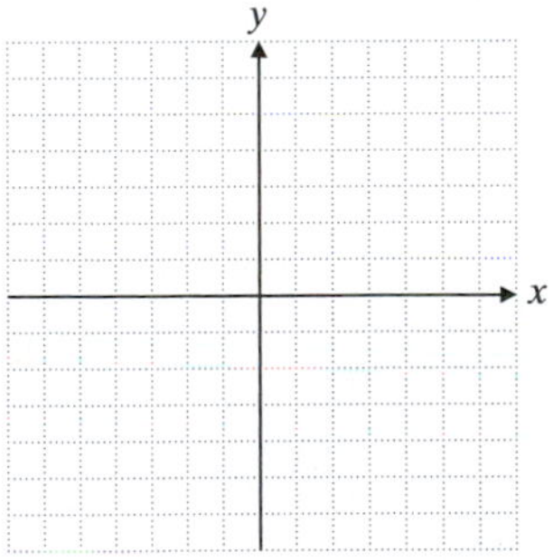

MC **5.** $\dfrac{x^2}{10} - \dfrac{y^2}{9} = 1$

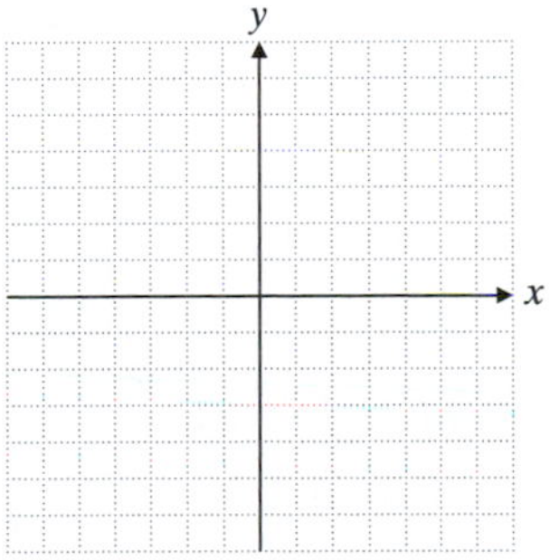

6. $y = -2(x + 3)^2 + 4$

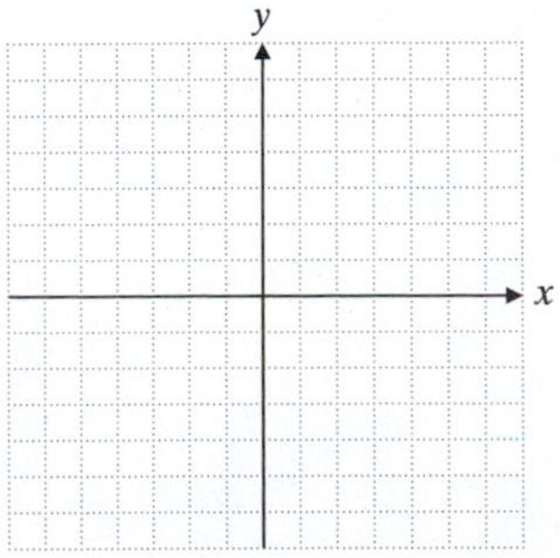

MC **7.** $\dfrac{(x + 2)^2}{16} + \dfrac{(y - 5)^2}{4} = 1$

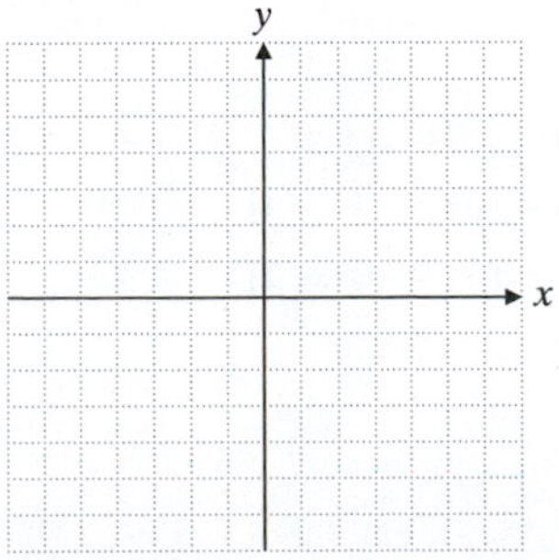

1. ______ ☐

2. ______ ☐

3. ______ ☐

4. ______ ☐

5. ______ ☐

6. ______ ☐

7. ______ ☐

8.
9.
10.
11.
12.
13.
14.
15.
16.

Total correct:

8. $7y^2 - 7x^2 = 28$

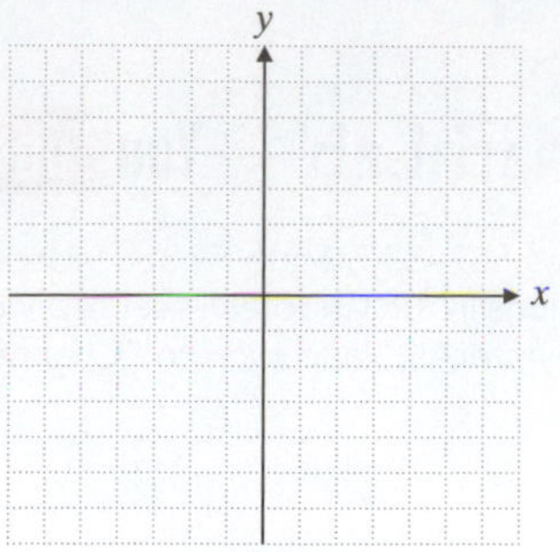

Find the standard form of the equation of each of the following:

9. Circle of radius $\sqrt{8}$ with its center at $(3, -5)$.

10. Ellipse with center at the origin, an x-intercept at $(3, 0)$, and a y-intercept at $(0, 5)$.

11. Parabola with its vertex at $(-7, 3)$ and that opens to the right. This parabola crosses the x-axis at $(2, 0)$. It is of the form $x = (y - k)^2 + h$.

12. Hyperbola with center at the origin, vertices at $(3, 0)$ and $(-3, 0)$. One asymptote is $y = \frac{5}{3}x$.

Solve each nonlinear system.

13.
$$-2x + y = 5$$
$$x^2 + y^2 - 25 = 0$$

14.
$$x^2 + y^2 = 9$$
$$y = x - 3$$

15.
$$4x^2 + y^2 - 4 = 0$$
$$9x^2 - 4y^2 - 9 = 0$$

MC **16.**
$$x^2 + 2y^2 = 15$$
$$x^2 - y^2 = 6$$

MATH COACH

Mastering the skills you need to do well on the test.

Students often make the same types of errors when they do the Chapter 9 Test. Here are some helpful hints to keep you from making those common errors on test problems.

Rewriting and Graphing a Parabola—Problem 2

Rewrite the equation in standard form. Find the center or vertex, plot at least one other point, identify the conic, and sketch the curve.

$$y^2 - 6y - x + 13 = 0$$

Helpful Hint If the equation of the parabola has a y^2 term, then it can be written in the standard form $x = a(y - k)^2 + h$. The vertex is (h, k), and the graph is a horizontal parabola.

Did you first rewrite the equation as $x = y^2 - 6y + 13$?

Yes ____ No ____

Next did you complete the square to obtain $x = (y - 3)^2 + 4$?

Yes ____ No ____

If you answered No to these questions, remember that when completing the square, we take half of -6 and square it, which results in 9. We must add both a positive and a negative 9 to the right side of the equation. Go back and try to complete these steps again.

Did you see that since $a = 1$ and $a > 0$, the parabola opens to the right?

Yes ____ No ____

Did you identify the vertex as $(4, 3)$?

Yes ____ No ____

If you answered No to these questions, review the helpful hint and other information about horizontal parabolas. Stop and perform this step again.

In your final answer, be sure to give the equation in standard form, list the vertex, identify the conic, and graph the equation.

If you answered Problem 2 incorrectly, go back and rework the problem using these suggestions.

Identifying and Graphing a Hyperbola with a Center at the Origin—Problem 5

Identify and graph the conic section. Label the center and/or vertex as appropriate. $\dfrac{x^2}{10} - \dfrac{y^2}{9} = 1$

Helpful Hint The standard form of a hyperbola with the center at the origin and vertices $(-a, 0)$ and $(a, 0)$ has the equation $\dfrac{x^2}{a^2} - \dfrac{y^2}{b^2} = 1$.

Did you realize that this equation represents a horizontal hyperbola with vertices at $(-\sqrt{10}, 0)$ and $(\sqrt{10}, 0)$?

Yes ____ No ____

If you answered No, review the rules for the standard form of an equation of a hyperbola with its center at the origin and read the Helpful Hint. Then remember that since $a^2 = 10$ and $a > 0$, $a = \sqrt{10}$. Likewise, if $b^2 = 9$ and $b > 0$, then $b = 3$.

Did you obtain a fundamental rectangle with corners at $(\sqrt{10}, 3)$, $(\sqrt{10}, -3)$, $(-\sqrt{10}, 3)$ and $(-\sqrt{10}, -3)$?

Yes ____ No ____

If you answered No, note that drawing a fundamental rectangle with corners at (a, b), $(a, -b)$, $(-a, b)$, and $(-a, -b)$ can help in creating the graph. The extended diagonals of the rectangle are the asymptotes.

In your final answer, remember to identify the conic section, create its graph, and label the center and the vertices.

Now go back and rework the problem using these suggestions.

Identifying and Graphing an Ellipse with a Center Not at the Origin—Problem 7

Identify and graph the conic section. Label the center and/or vertex as appropriate. $\frac{(x+2)^2}{16} + \frac{(y-5)^2}{4} = 1$

Helpful Hint When an ellipse has a center that is not at the origin, the center has the coordinates (h, k) and the equation in standard form is $\frac{(x-h)^2}{a^2} + \frac{(y-k)^2}{b^2} = 1$, where both a and b are greater than zero.

Did you determine that the center of the ellipse is at $(-2, 5)$ with $a = 4$ and $b = 2$?

Yes ____ No ____

If you answered No, remember that the standard form of the equation involves $x - h$ and $y - k$, so you must be careful in determining the signs of h and k.

Did you start at the center and find points a units to the left, a units to the right, b units up, and b units down to plot the points $(-6, 5)$, $(2, 5)$, $(-2, 7)$, and $(-2, 3)$?

Yes ____ No ____

If you answered No, remember that to find these four points you need to find the following: $(h - a, k)$, $(h + a, k)$, $(h, k + b)$, and $(h, k - b)$.

Plot all four points and the center and label these on your graph. Then use the four points to make a sketch of the ellipse. Remember to identify the conic as an ellipse in your final answer.

If you answered Problem 7 incorrectly, go back and rework the problem using these suggestions.

Solving a System of Nonlinear Equations—Problem 16

Solve. $x^2 + 2y^2 = 15$
$x^2 - y^2 = 6$

Helpful Hint When two equations in a system have the form $ax^2 + by^2 = c$, where a, b, and c are real numbers, then it may be easiest to solve the system by the addition method.

If you multiplied the second equation by 2 and added the result to the first equation, do you get the equation $3x^2 = 27$?

Yes ____ No ____

Can you solve this equation for x to get $x = 3$ and $x = -3$?

Yes ____ No ____

If you answered No to these questions, remember that when you add the two equations together, the y^2 term adds to 0.

If you substitute $x = 3$ into the first equation, do you get the equation $9 + 2y^2 = 15$?

Yes ____ No ____

Can you solve this equation for y to get $y = \sqrt{3}$ and $y = -\sqrt{3}$?

Yes ____ No ____

If you answered No to these questions, try substituting $x = 3$ into the equation again and be careful to avoid calculation errors. Remember that you must also perform this same step with $x = -3$. Since x is squared, your results for y should be the same.

Your final answer should have a total of four possible ordered pair solutions to this system.

Now go back and rework this problem using these suggestions.

Need more help? Look for section examples marked with MC to review.

CHAPTER 10

Scientists are currently experimenting with solar cars. These vehicles use solar panels to collect the sun's rays and convert them into electric current. The types of functions studied in this chapter are used to describe the amount of electric current derived from the sunlight and the amount of current required to propel these vehicles at highway speeds.

Additional Properties of Functions

10.1 Function Notation

Student Learning Objectives

After studying this section, you will be able to:

1. **Use function notation to evaluate expressions.**
2. **Use function notation to solve application exercises.**

① Using Function Notation to Evaluate Expressions

Function notation is useful in solving a number of interesting exercises. Suppose you wanted to skydive from an airplane. Your instructor tells you that you must wait 20 seconds before you pull the cord to open the parachute. How far will you fall in that time?

The approximate distance an object in free-fall travels when there is no initial downward velocity is given by the distance function $d(t) = 16t^2$, where time t is measured in seconds and distance $d(t)$ is measured in feet. (Neglect air resistance.)

How far will a person in free-fall travel in 20 seconds if he or she leaves the airplane with no downward velocity? We want to find the distance $d(20)$ where $t = 20$ seconds. To find the distance, we substitute 20 for t in the function $d(t) = 16t^2$.

$$d(20) = 16(20)^2 = 16(400) = 6400$$

Thus, a person in free-fall will travel approximately 6400 feet in 20 seconds.

Now suppose that person waits longer than he or she should to pull the parachute cord. How will this affect the distance he or she falls? Suppose the person falls for a seconds beyond the 20-second mark before pulling the parachute cord.

$$\begin{aligned} d(20 + a) &= 16(20 + a)^2 \\ &= 16(20 + a)(20 + a) \\ &= 16(400 + 40a + a^2) \\ &= 6400 + 640a + 16a^2 \end{aligned}$$

Thus, if the person waited 5 seconds too long before pulling the cord, he or she would fall the following distance.

$$\begin{aligned} d(20 + 5) &= 6400 + 640(5) + 16(5)^2 \\ &= 6400 + 3200 + 16(25) \\ &= 6400 + 3200 + 400 \\ &= 10{,}000 \text{ feet} \end{aligned}$$

This is 3600 feet farther than the person would have fallen in 20 seconds. Obviously, a delay of 5 seconds could have life or death consequences.

We now revisit a topic that we first discussed in Chapter 3, evaluating a function for particular values of the variable.

EXAMPLE 1 If $g(x) = 5 - 3x$, find the following:

(a) $g(a)$ **(b)** $g(a + 3)$ **(c)** $g(a) + g(3)$

Solution

(a) $g(a) = 5 - 3a$

(b) $g(a + 3) = 5 - 3(a + 3) = 5 - 3a - 9 = -4 - 3a$

(c) This exercise requires us to find each addend separately. Then we add them together.

$$g(a) = 5 - 3a$$
$$g(3) = 5 - 3(3) = 5 - 9 = -4$$

Thus,
$$\begin{aligned} g(a) + g(3) &= (5 - 3a) + (-4) \\ &= 5 - 3a - 4 \\ &= 1 - 3a \end{aligned}$$

Notice that $g(a + 3) \neq g(a) + g(3)$.

Student Practice 1 If $g(x) = \frac{1}{2}x - 3$, find the following:

(a) $g(a)$ **(b)** $g(a + 4)$ **(c)** $g(a) + g(4)$

NOTE TO STUDENT: Fully worked-out solutions to all of the Student Practice problems can be found at the back of the text starting at page SP-1.

TO THINK ABOUT: Understanding Function Notation
Is $g(a + 4) = g(a) + g(4)$ in Example 1? Why or why not?

EXAMPLE 2 If $p(x) = 2x^2 - 3x + 5$, find the following:

(a) $p(-2)$ **(b)** $p(a)$
(c) $p(3a)$ **(d)** $p(a - 2)$

Solution

(a)
$$\begin{aligned} p(-2) &= 2(-2)^2 - 3(-2) + 5 \\ &= 2(4) - 3(-2) + 5 \\ &= 8 + 6 + 5 \\ &= 19 \end{aligned}$$

(b) $p(a) = 2(a)^2 - 3(a) + 5 = 2a^2 - 3a + 5$

(c)
$$\begin{aligned} p(3a) &= 2(3a)^2 - 3(3a) + 5 \\ &= 2(9a^2) - 3(3a) + 5 \\ &= 18a^2 - 9a + 5 \end{aligned}$$

(d)
$$\begin{aligned} p(a - 2) &= 2(a - 2)^2 - 3(a - 2) + 5 \\ &= 2(a - 2)(a - 2) - 3(a - 2) + 5 \\ &= 2(a^2 - 4a + 4) - 3(a - 2) + 5 \\ &= 2a^2 - 8a + 8 - 3a + 6 + 5 \\ &= 2a^2 - 11a + 19 \end{aligned}$$

Student Practice 2
If $p(x) = -3x^2 + 2x + 4$, find the following:

(a) $p(-3)$
(b) $p(a)$
(c) $p(2a)$
(d) $p(a - 3)$

EXAMPLE 3 If $r(x) = \frac{4}{x + 2}$, find

(a) $r(a + 3)$ **(b)** $r(a)$
(c) $r(a + 3) - r(a)$. Express this result as one fraction.

Solution

(a) $r(a + 3) = \frac{4}{a + 3 + 2} = \frac{4}{a + 5}$

(b) $r(a) = \frac{4}{a + 2}$

(c) $r(a + 3) - r(a) = \frac{4}{a + 5} - \frac{4}{a + 2}$

To express this as one fraction, we note that the LCD $= (a + 5)(a + 2)$.

$$\begin{aligned} r(a + 3) - r(a) &= \frac{4(a + 2)}{(a + 5)(a + 2)} - \frac{4(a + 5)}{(a + 2)(a + 5)} \\ &= \frac{4a + 8}{(a + 5)(a + 2)} - \frac{4a + 20}{(a + 5)(a + 2)} \\ &= \frac{4a + 8 - 4a - 20}{(a + 5)(a + 2)} \\ &= \frac{-12}{(a + 5)(a + 2)} \end{aligned}$$

Continued on next page

Student Practice 3 If $r(x) = \frac{-3}{x + 1}$, find

(a) $r(a + 2)$ **(b)** $r(a)$

(c) $r(a + 2) - r(a)$. Express this result as one fraction.

EXAMPLE 4 Let $f(x) = 3x - 7$. Find $\frac{f(x + h) - f(x)}{h}$.

Solution First

$$f(x + h) = 3(x + h) - 7 = 3x + 3h - 7$$

and

$$f(x) = 3x - 7.$$

So

$$\begin{aligned} f(x + h) - f(x) &= (3x + 3h - 7) - (3x - 7) \\ &= 3x + 3h - 7 - 3x + 7 \\ &= 3h. \end{aligned}$$

Therefore,

$$\frac{f(x + h) - f(x)}{h} = \frac{3h}{h} = 3.$$

Student Practice 4

Suppose that $g(x) = 2 - 5x$. Find $\frac{g(x + h) - g(x)}{h}$.

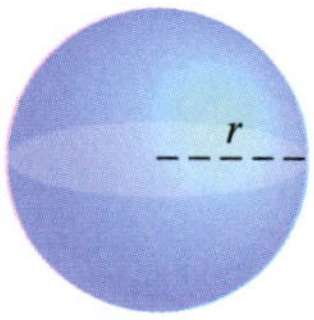

② Using Function Notation to Solve Application Exercises

▲ **EXAMPLE 5** The surface area of a sphere is given by $S = 4\pi r^2$ where r is the radius. If we use $\pi \approx 3.14$ as an approximation, this becomes $S = 4(3.14)r^2$, or $S = 12.56r^2$.

(a) Write the surface area of a sphere as a function of radius r.

(b) Find the surface area of a sphere with a radius of 3 centimeters.

(c) Suppose that an error is made and the radius of the sphere is calculated to be $(3 + e)$ centimeters. Find an expression for the surface area as a function of the error e.

(d) Evaluate the surface area for $r = (3 + e)$ centimeters when $e = 0.2$. Round your answer to the nearest hundredth of a centimeter. What is the difference in the surface area due to the error in measurement?

Solution

(a) $S(r) = 12.56r^2$

(b) $S(3) = 12.56(3)^2 = 12.56(9) = 113.04$ square centimeters

(c)
$$\begin{aligned} S(e) &= 12.56(3 + e)^2 \\ &= 12.56(3 + e)(3 + e) \\ &= 12.56(9 + 6e + e^2) \\ &= 113.04 + 75.36e + 12.56e^2 \end{aligned}$$

(d) If an error in measurement is made so that the radius is calculated to be $r = (3 + e)$ centimeters, where $e = 0.2$, we can use the function generated in part **(c)**.

$$S(e) = 113.04 + 75.36e + 12.56e^2$$

$$S(0.2) = 113.04 + 75.36(0.2) + 12.56(0.2)^2$$

$$= 113.04 + 15.072 + 0.5024$$

$$= 128.6144$$

Rounding, we have $S = 128.61$ square centimeters.

Thus, if the radius of 3 centimeters was incorrectly calculated as 3.2 centimeters, the surface area would be approximately $128.61 - 113.04 = 15.57$ square centimeters too large.

▲ **Student Practice 5** The surface area of a cylinder of height 8 meters and radius r is given by $S = 16\pi r + 2\pi r^2$.

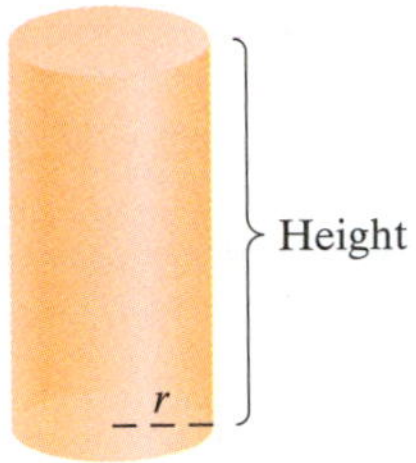

(a) Write the surface area of a cylinder of height 8 meters (using $\pi \approx 3.14$) and radius r as a function of r.

(b) Find the surface area if the radius is 2 meters.

(c) Suppose that an error is made and the radius is calculated to be $(2 + e)$ meters. Find an expression for the surface area as a function of the error e.

(d) Evaluate the surface area for $r = (2 + e)$ meters when $e = 0.3$. Round your answer to the nearest hundredth of a meter. What is the difference in the surface area due to the error in measurement?

10.1 Exercises

MyMathLab® Watch the videos in MyMathLab Download the MyDashBoard App

For the function $f(x) = 3x - 5$, *find the following.*

1. $f\left(-\frac{2}{3}\right)$

2. $f(1.5)$

3. $f(a - 4)$

4. $f(b + 3)$

For the function $g(x) = \frac{1}{2}x - 3$, *find the following.*

5. $g(4) + g(a)$

6. $g(6) + g(b)$

7. $g(4a) - g(a)$

8. $g(6b) + g(-2b)$

9. $g(2a - 4)$

10. $g(3a + 1)$

11. $g(a^2) - g\left(\frac{2}{5}\right)$

12. $g(b^2) - g\left(\frac{4}{3}\right)$

If $p(x) = 3x^2 + 4x - 2$, *find the following.*

13. $p(-2)$

14. $p(-5)$

15. $p\left(\frac{1}{2}\right)$

16. $p(2.5)$

17. $p(a + 1)$

18. $p(b - 1)$

19. $p\left(-\frac{2a}{3}\right)$

20. $p\left(-\frac{3a}{2}\right)$

If $h(x) = \sqrt{x + 5}$, *find the following.*

21. $h(4)$

22. $h(-5)$

23. $h(7)$

24. $h(19)$

25. $h(a^2 - 1)$

26. $h(a^2 + 4)$

27. $h(-2b)$

28. $h(3b)$

29. $h(4a - 1)$

30. $h(8a - 1)$

31. $h(b^2 + b)$

32. $h(b^2 + b - 5)$

If $r(x) = \frac{7}{x - 3}$, *find the following and write your answers as one fraction.*

33. $r(7)$

34. $r(-4)$

35. $r(3.5)$

36. $r(-0.5)$

37. $r(a^2)$

38. $r(3b^2)$

39. $r(a + 2)$

40. $r(a - 3)$

41. $r\left(\frac{1}{2}\right) + r(8)$

42. $r\left(\frac{3}{2}\right) + r(5)$

Find $\dfrac{f(x+h)-f(x)}{h}$ *for the following functions.*

43. $f(x) = 2x - 3$

44. $f(x) = 5 - 2x$

45. $f(x) = x^2 - x$

46. $f(x) = 2x^2$

Applications

47. ***Wind Generators*** A turbine wind generator produces P kilowatts of power for wind speed w (measured in miles per hour) according to the equation $P = 2.5w^2$.

(a) Write the number of kilowatts P as a function of w.

(b) Find the power in kilowatts when the wind speed is $w = 20$ miles per hour.

(c) Suppose that an error is made and the speed of the wind is calculated to be $(20 + e)$ miles per hour. Find an expression for the power as a function of error e.

(d) Evaluate the power for $w = (20 + e)$ miles per hour when $e = 2$.

▲ **48.** ***Geometry*** The area of a circle is $A = \pi r^2$.

(a) Write the area of a circle as the function of the radius r. Use $\pi \approx 3.14$.

(b) Find the area of a circle with a radius of 4.0 feet.

(c) Suppose that an error is made and the radius is calculated to be $(4 + e)$ feet. Find an expression for the area as a function of error e.

(d) Evaluate the area for $r = (4 + e)$ feet when $e = 0.4$. Round your answer to the nearest hundredth.

Lead Levels in Air *Because of the elimination of lead in automobile gasoline and increased use of emission controls in automobiles and industrial operations, the amount of lead in the air in the United States has shown a marked decrease. The percent of lead in the air $p(x)$ expressed in terms of 1984 levels is given in the line graph. The variable x indicates the number of years since 1984. The function value $p(x)$ indicates the amount of lead that remains in the air in selected regions of the United States, expressed as a percent of the amount of lead in the air in 1984.*

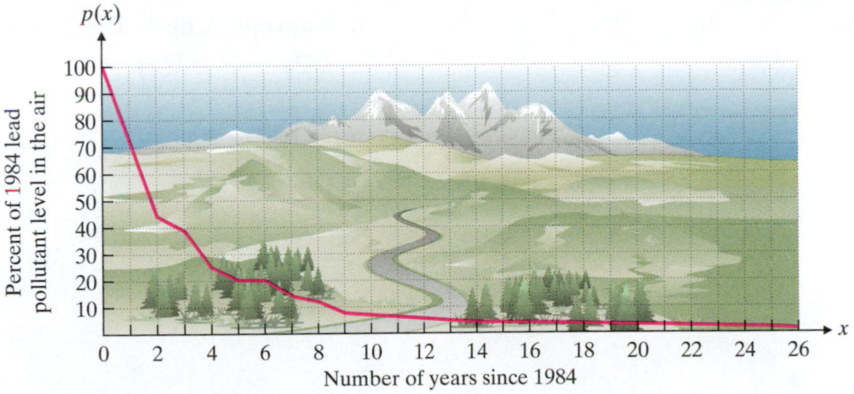

Source: Environmental Protection Agency.

49. If a new function were defined as $p(x) - 13$, what would happen to the function values associated with x? Find $p(3) - 13$.

50. If a new function were defined as $p(x + 2)$, what would happen to the function values associated with x? Find $p(x + 2)$ when $x = 4$.

If $f(x) = 3x^2 - 4.6x + 1.23$, *find each of the following function values to the nearest thousandth.*

51. $f(0.026a)$

52. $f(a + 2.23)$

Cumulative Review *Solve for x.*

53. [6.4.1] $\frac{7}{6} + \frac{5}{x} = \frac{3}{2x}$

54. [6.4.1] $\frac{1}{6} - \frac{2}{3x + 6} = \frac{1}{2x + 4}$

▲ **55. [1.6.2]** ***Planet Volume*** The diameter of the planet Mars is 4211 miles, while that of Earth is almost twice as large with a diameter of 7927 miles. How many times greater is the volume of Earth compared to the volume of Mars?

▲ **56. [1.6.2]** ***Planet Volume*** The radius of the planet Neptune is 15,345 miles. The radius of Saturn (excluding its rings) is 37,366 miles. How many times greater is the volume of Saturn compared to the volume of Neptune?

Quick Quiz 10.1

1. If $f(x) = \frac{3}{5}x - 4$, find $f(a) - f(-3)$.

2. If $g(x) = 2x^2 - 3x + 4$, find $g\left(\frac{2}{3}a\right)$.

3. If $h(x) = \frac{3}{x + 4}$, find $h(a - 6)$.

4. Concept Check Explain how you would find $k(2a - 1)$ if $k(x) = \sqrt{3x + 1}$.

10.2 General Graphing Procedures for Functions

Student Learning Objectives

After studying this section, you will be able to:

1. **Use the vertical line test to determine whether a graph represents a function.**
2. **Graph a function of the form $f(x + h) + k$ by means of horizontal and vertical shifts of the graph of $f(x)$.**

① Using the Vertical Line Test to Determine Whether a Graph Represents a Function

Not every graph we observe is that of a function. By definition, a function must have no ordered pairs that have the same first coordinates and different second coordinates. A graph that includes the points (4, 2) and (4, −2), for example, would not be the graph of a function. Thus, the graph of $x = y^2$ would not be the graph of a function.

If any vertical line crosses a graph of a relation in more than one place, the relation is not a function. If no such line exists, the relation is a function.

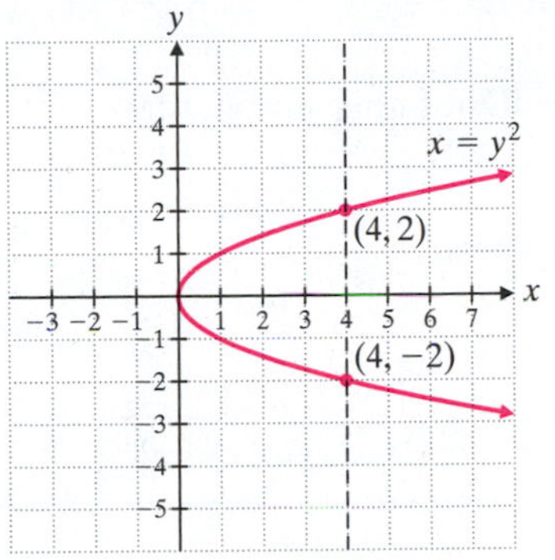

VERTICAL LINE TEST

If any vertical line intersects the graph of a relation more than once, the relation is not a function. If no such line exists, the relation is a function.

In the following sketches, we observe that the dashed vertical line crosses the curve of a function no more than once. The dashed vertical line crosses the curve of a relation that is *not* a function more than once.

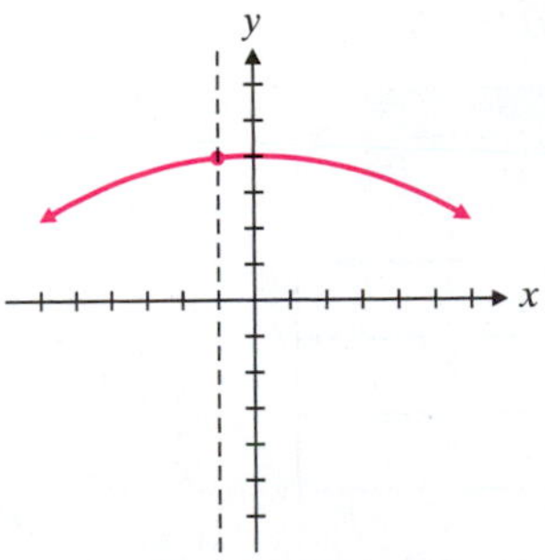

A Function

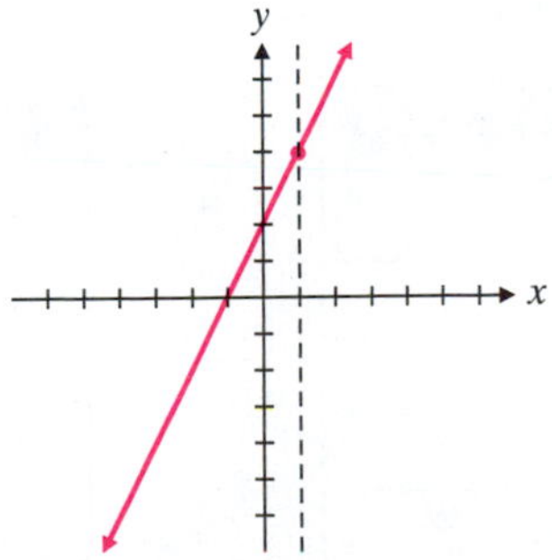

A Function

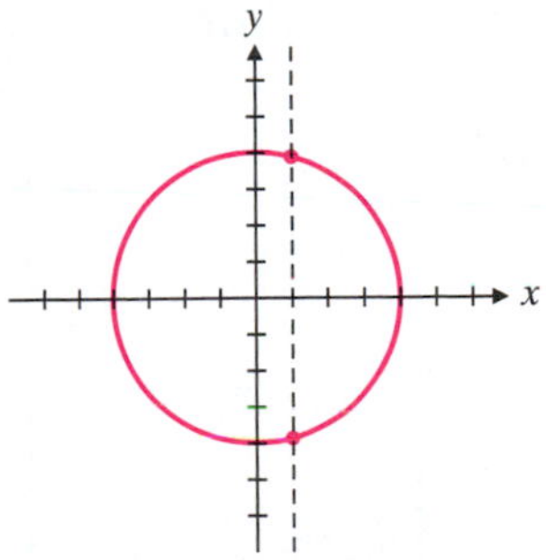

Not a Function

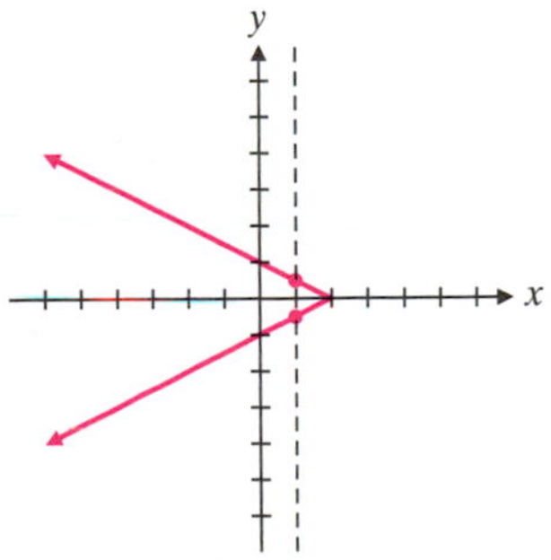

Not a Function

EXAMPLE 1 Determine whether each of the following is the graph of a function.

(a)

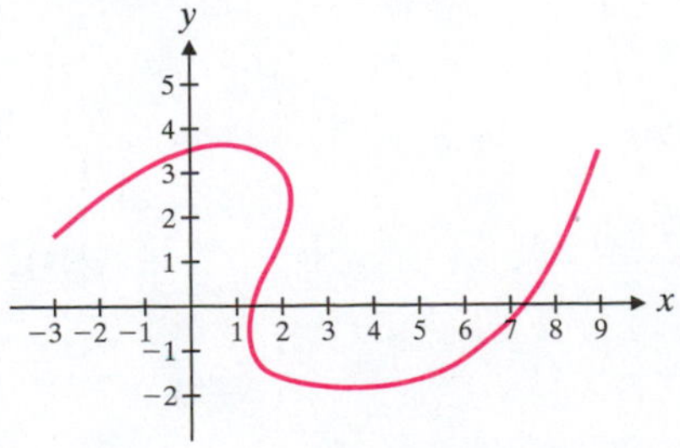

(b)

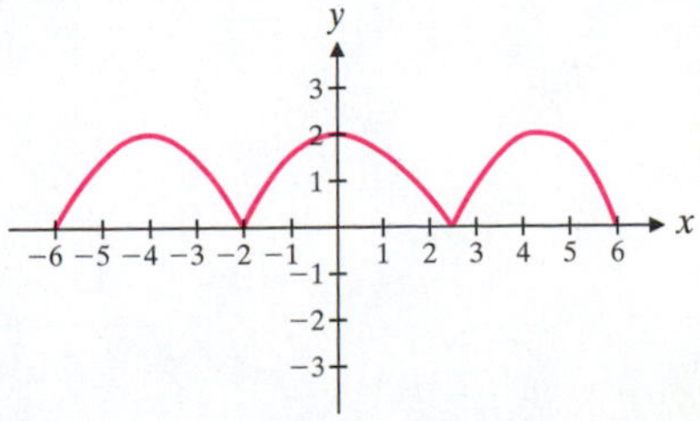

Continued on next page

Solution

(a) By the vertical line test, this relation is not a function.

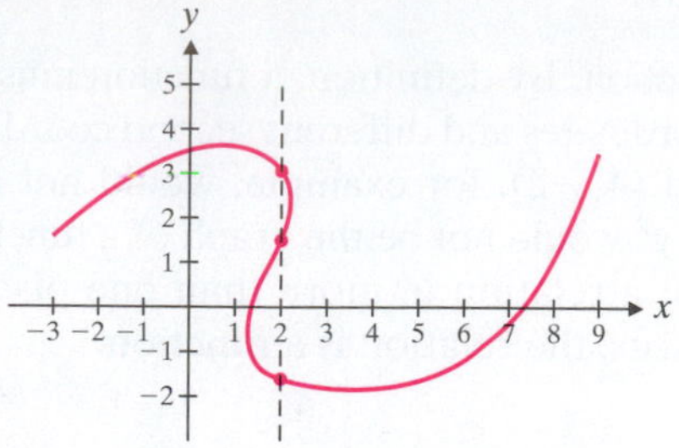

(b) By the vertical line test, this relation is a function.

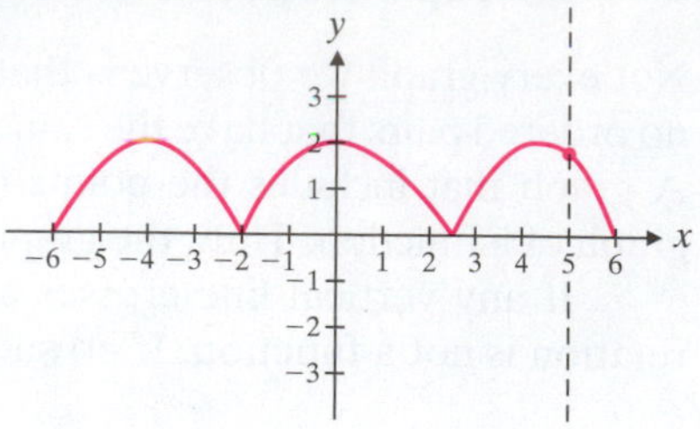

***NOTE TO STUDENT:** Fully worked-out solutions to all of the Student Practice problems can be found at the back of the text starting at page SP-1.*

Student Practice 1 Does this graph represent a function? Why or why not?

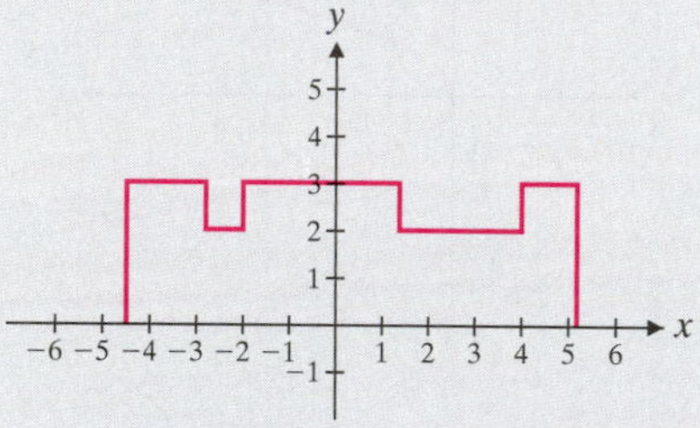

② Graphing a Function of the Form $f(x + h) + k$ by Means of Horizontal and Vertical Shifts of the Graph of $f(x)$

The graphs of some functions are simple vertical shifts of the graphs of similar functions.

EXAMPLE 2 Graph the functions on one coordinate plane.

$$f(x) = x^2 \quad \text{and} \quad h(x) = x^2 + 2$$

Solution First we make a table of values for $f(x)$ and for $h(x)$.

x	$f(x) = x^2$
−2	4
−1	1
0	0
1	1
2	4

x	$h(x) = x^2 + 2$
−2	6
−1	3
0	2
1	3
2	6

Now we graph each function on the same coordinate plane.

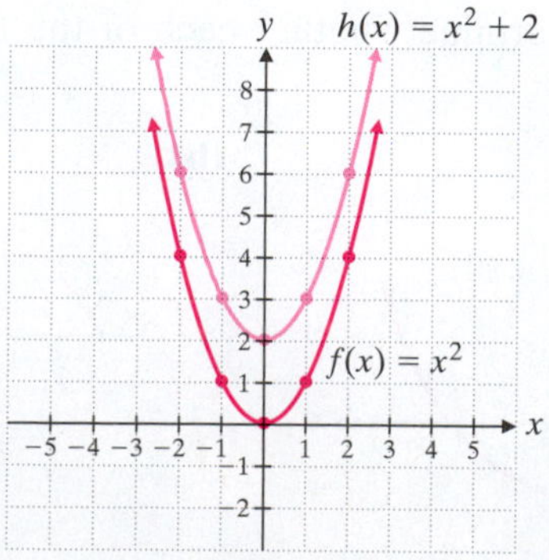

Notice that the graph of $h(x)$ is the graph of $f(x)$ moved 2 units upward.

Student Practice 2 Graph the functions on one coordinate plane.

$$f(x) = x^2 \quad \text{and} \quad h(x) = x^2 - 5$$

Student Practice 2

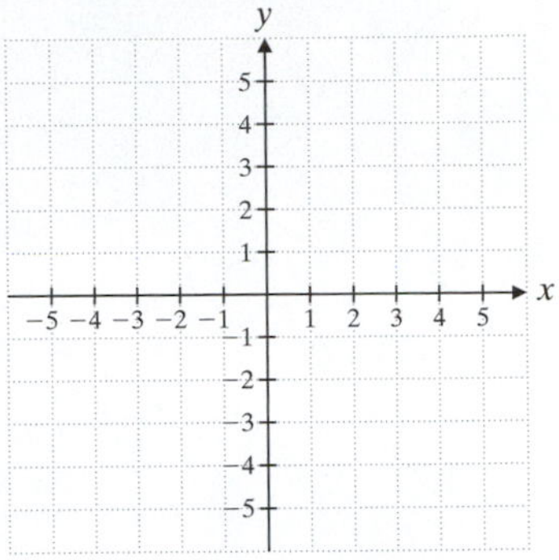

We have the following general summary.

VERTICAL SHIFTS

Suppose that k is a positive number.

1. To obtain the graph of $f(x) + k$, shift the graph of $f(x)$ up k units.
2. To obtain the graph of $f(x) - k$, shift the graph of $f(x)$ down k units.

Now we turn to the topic of horizontal shifts.

EXAMPLE 3 Graph the functions on one coordinate plane.

$$f(x) = |x| \quad \text{and} \quad p(x) = |x - 3|$$

Solution First we make a table of values for $f(x)$ and for $p(x)$.

x	$f(x) = \lvert x \rvert$
−2	2
−1	1
0	0
1	1
2	2
3	3
4	4

x	$p(x) = \lvert x - 3 \rvert$
−2	5
−1	4
0	3
1	2
2	1
3	0
4	1

Now we graph each function on the same coordinate plane.

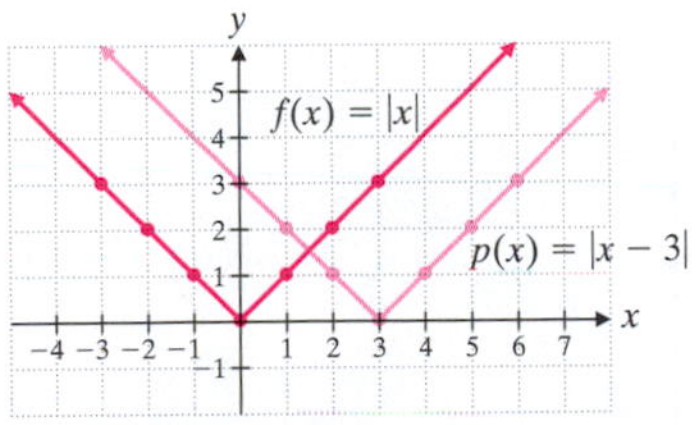

Notice that the graph of $p(x)$ is the graph of $f(x)$ shifted 3 units to the right.

Student Practice 3 Graph the functions on one coordinate plane.

$$f(x) = |x| \quad \text{and} \quad p(x) = |x + 2|$$

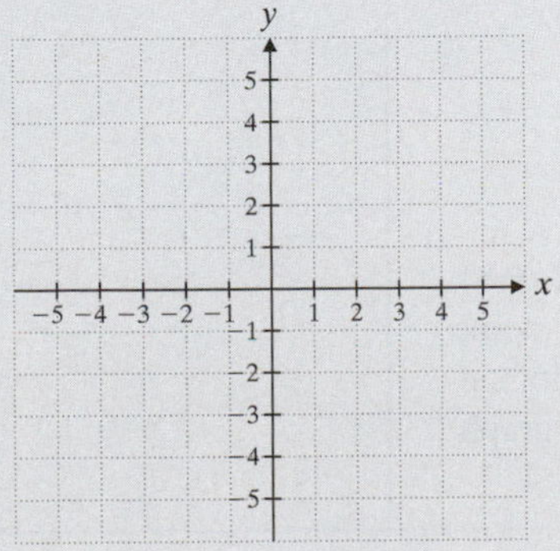

Graphing Calculator

Exploration

Most graphing calculators have an absolute value function (abs). Use this function to graph $f(x)$ and $p(x)$ from Example 3 on one coordinate plane.

Now we can write the following general summary.

HORIZONTAL SHIFTS

Suppose that h is a positive number.

1. To obtain the graph of $f(x - h)$, shift the graph of $f(x)$ to the right h units.
2. To obtain the graph of $f(x + h)$, shift the graph of $f(x)$ to the left h units.

Some graphs will involve both horizontal and vertical shifts.

EXAMPLE 4 Graph the functions on one coordinate plane.

$$f(x) = x^3 \quad \text{and} \quad h(x) = (x - 3)^3 - 2$$

Solution First we make a table of values for $f(x)$ and graph the function.

x	f(x)
−2	−8
−1	−1
0	0
1	1
2	8

Next we recognize that $h(x)$ will have a similar shape, but the curve will be shifted 3 units to the *right* and 2 units *downward*. We draw the graph of $h(x)$ using these shifts.

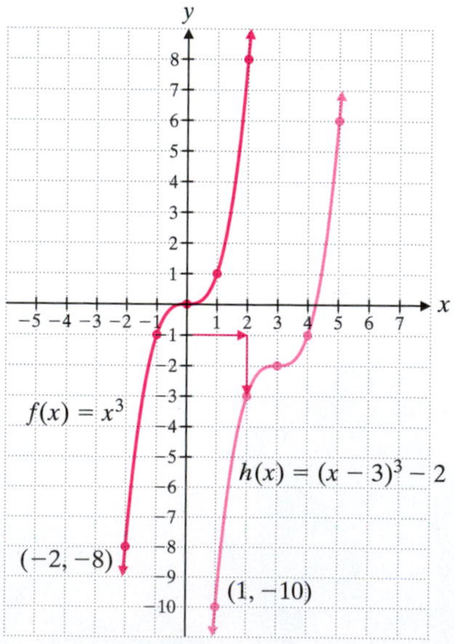

TO THINK ABOUT: Shifting Points The point $(-2, -8)$ has been shifted 3 units to the right and 2 units down to the point $(-2 + 3, -8 + (-2))$ or $(1, -10)$. The point $(-1, -1)$ is a point on $f(x)$. Use the same reasoning to find the image of $(-1, -1)$ on the graph of $h(x)$. Verify by checking the graphs.

Student Practice 4 Graph the functions on one coordinate plane.

$$f(x) = x^3 \quad \text{and} \quad h(x) = (x + 4)^3 + 3$$

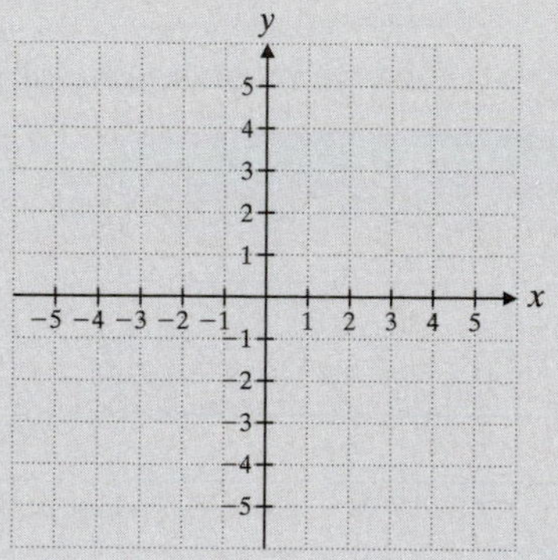

All the functions that we have sketched in this section so far have had a domain of all real numbers. Some functions have restricted domains.

EXAMPLE 5 Graph the functions on one coordinate plane. State the domain of each function.

$$f(x) = \frac{4}{x} \quad \text{and} \quad g(x) = \frac{4}{x + 3} + 1$$

Solution First we make a table of values for $f(x)$. The domain of $f(x)$ is all real numbers, where $x \neq 0$. Note that $f(x)$ is not defined when $x = 0$ since we cannot divide by 0.

x	f(x)
−4	−1
−2	−2
−1	−4
$-\frac{1}{2}$	−8
$\frac{1}{2}$	8
1	4
2	2
4	1

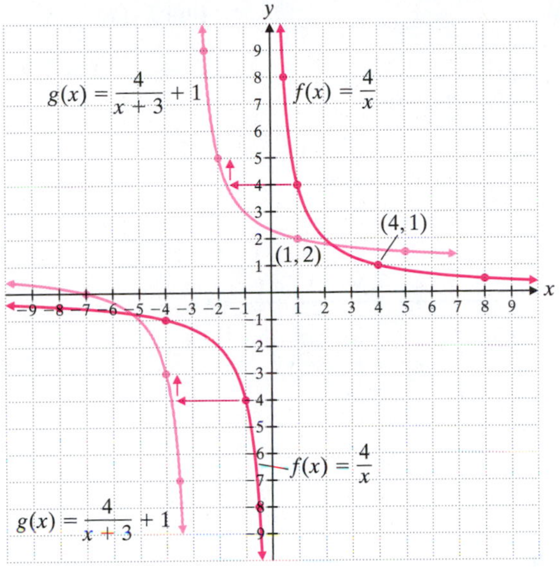

We draw $f(x)$ and note key points. From the equation, we see that the graph of $g(x)$ is 3 units to the left of and 1 unit above $f(x)$. We can find the image of each of the key points on $f(x)$ as a guide in graphing $g(x)$. For example, the image of (4, 1) is $(4 - 3, 1 + 1)$ or (1, 2).

Each point on $f(x)$ is shifted

$\Leftarrow$ 3 units left and
$\Uparrow$ 1 unit up

to form the graph of $g(x)$.

What is the domain of $g(x)$? Why? $g(x)$ contains the denominator $x + 3$. But $x + 3 \neq 0$. Therefore, $x \neq -3$. The domain of $g(x)$ is all real numbers, where $x \neq -3$.

Student Practice 5 Graph the functions on one coordinate plane.

$$f(x) = \frac{2}{x} \quad \text{and} \quad g(x) = \frac{2}{x + 1} - 2$$

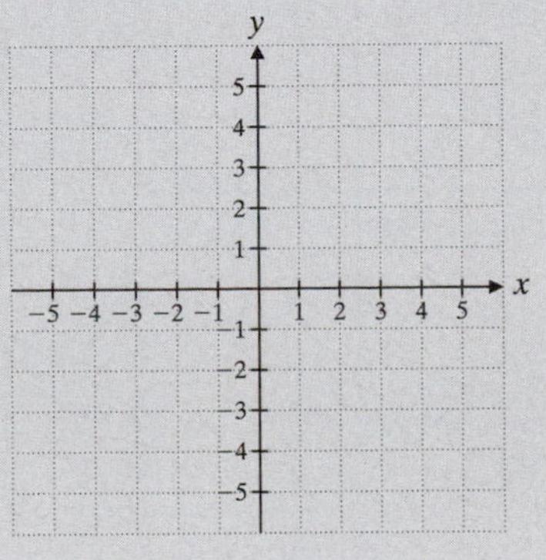

10.2 Exercises

MyMathLab®

Watch the videos in MyMathLab

Download the MyDashBoard App

Verbal and Writing Skills, Exercises 1–4

1. Does $f(x + 2) = f(x) + f(2)$? Why or why not? Give an example.

2. Explain what the vertical line test is and why it works.

3. To obtain the graph of $f(x) + k$ for $k > 0$, shift the graph of $f(x)$ __________ k units.

4. To obtain the graph of $f(x - h)$ for $h > 0$, shift the graph of $f(x)$ __________ h units.

Determine whether or not each graph represents a function.

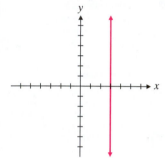

5.

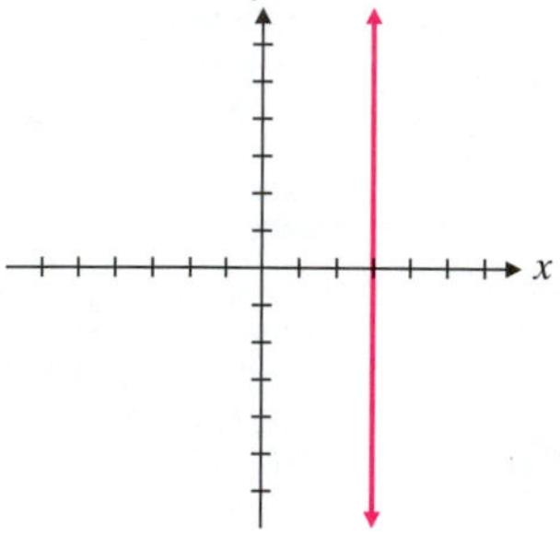

6.

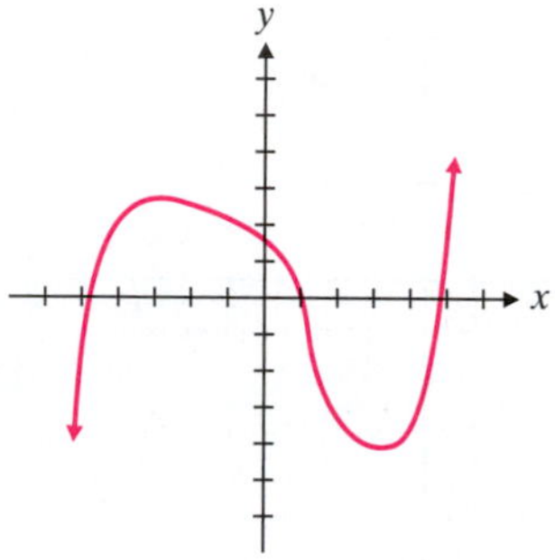

7.

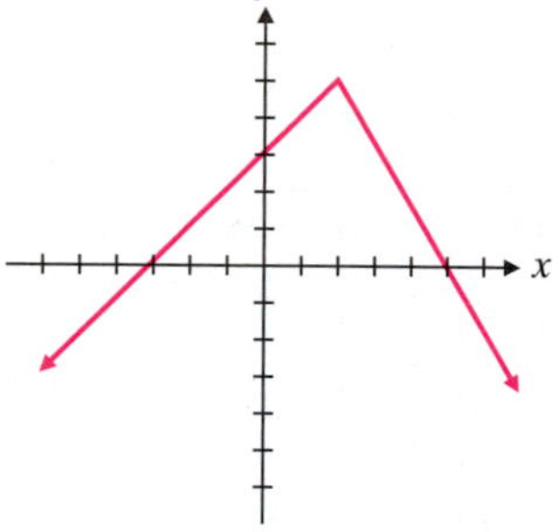

8.

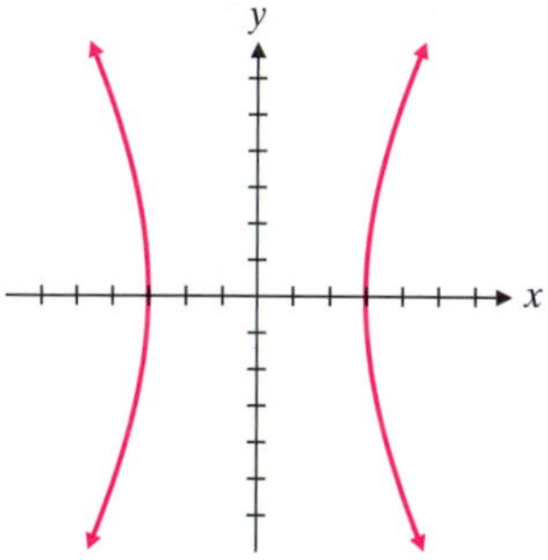

9.

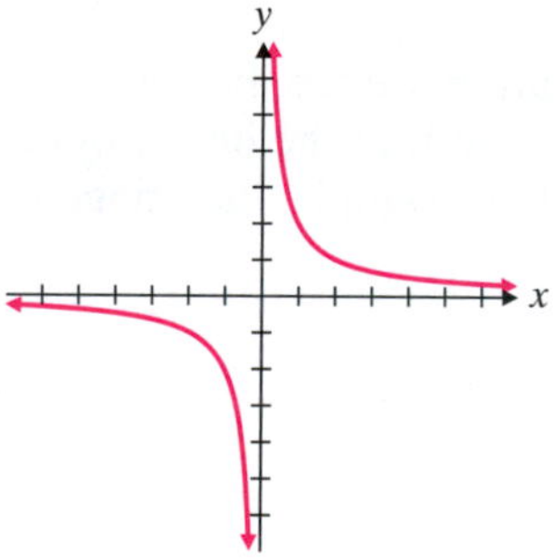

10.

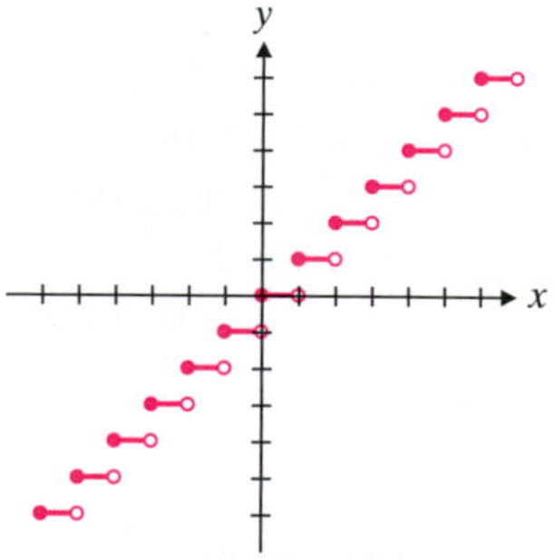

Hint: The open circle means that the function value does not exist at that point.

11.

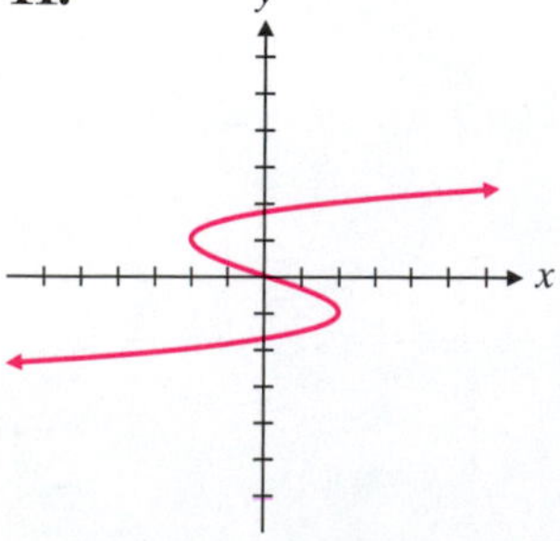

12.

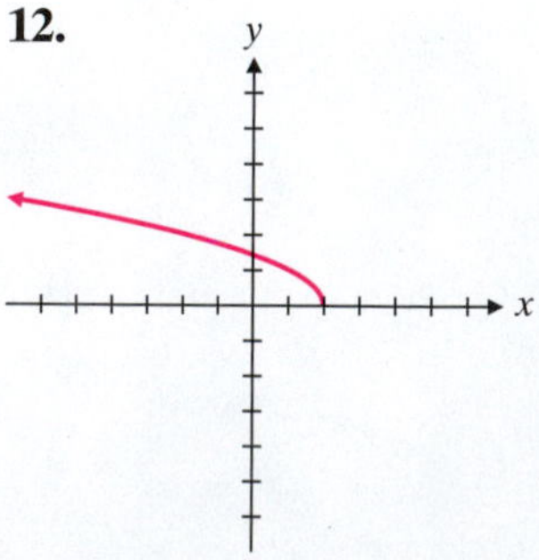

13.

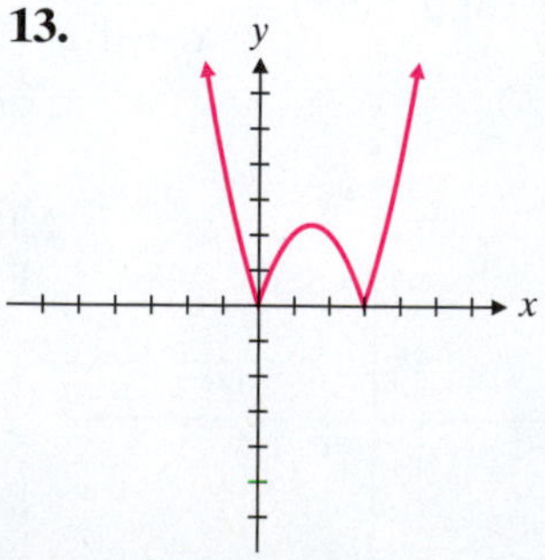

14.

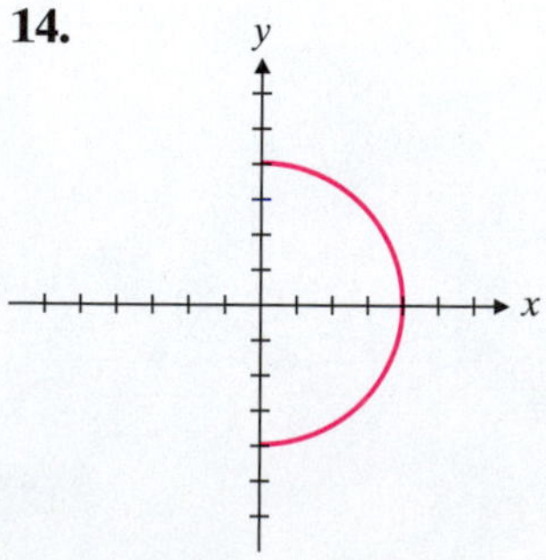

For each of exercises 15–28, graph the two functions on one coordinate plane.

15. $f(x) = x^2$
$h(x) = x^2 - 3$

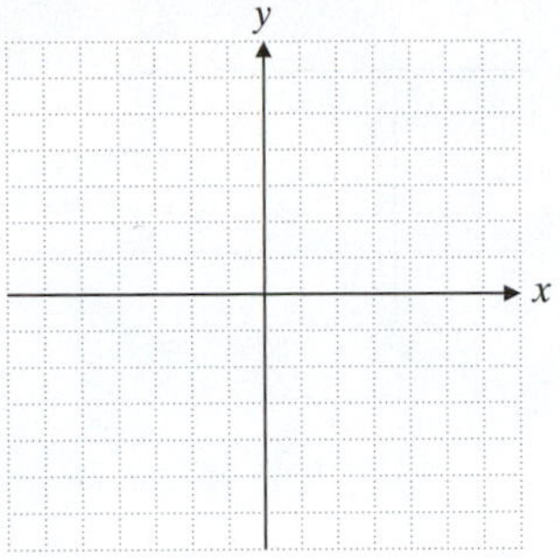

16. $f(x) = x^2$
$h(x) = x^2 + 4$

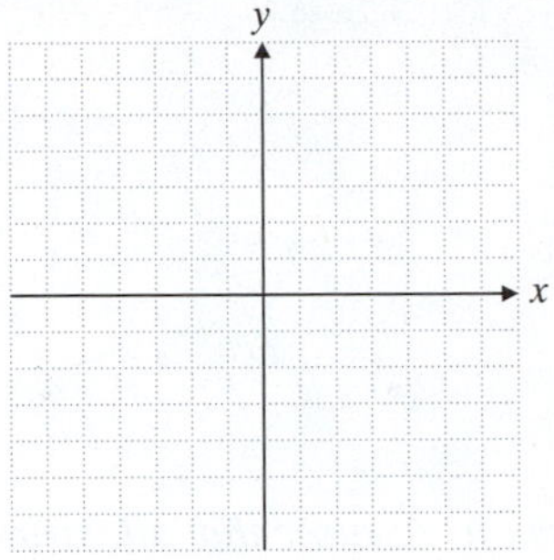

17. $f(x) = x^2$
$p(x) = (x + 1)^2$

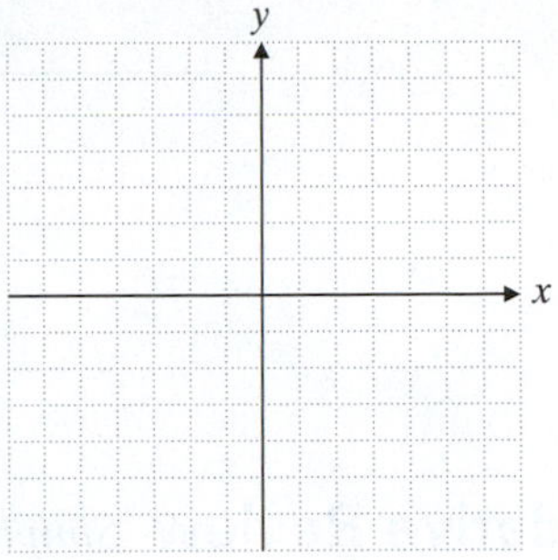

18. $f(x) = x^2$
$p(x) = (x - 2)^2$

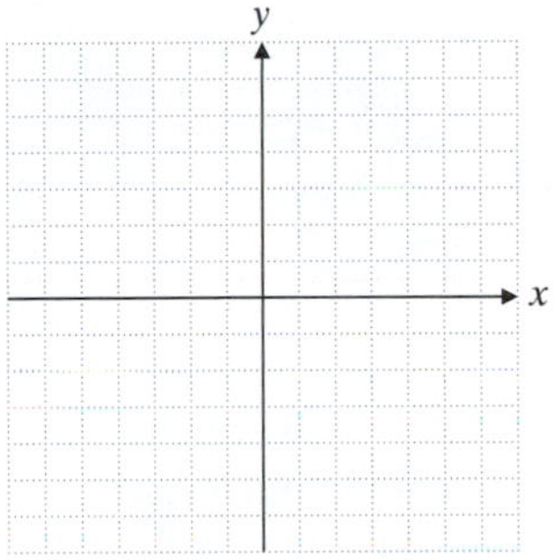

19. $f(x) = x^2$
$g(x) = (x - 2)^2 + 1$

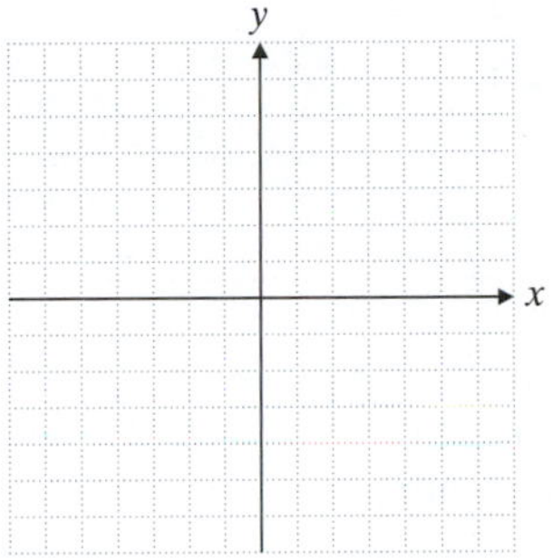

20. $f(x) = x^2$
$g(x) = (x + 1)^2 - 2$

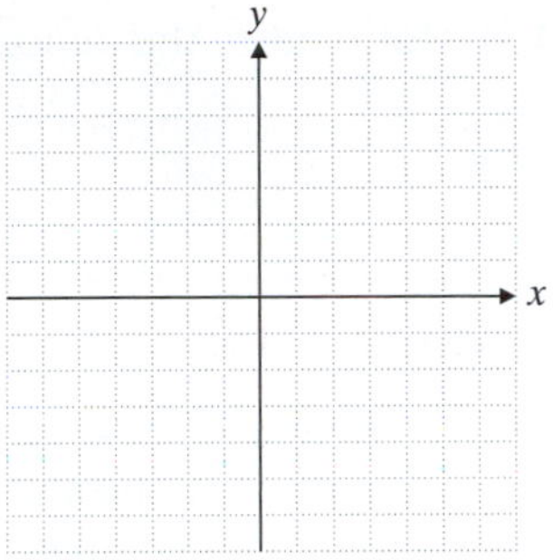

21. $f(x) = x^3$
$r(x) = x^3 - 1$

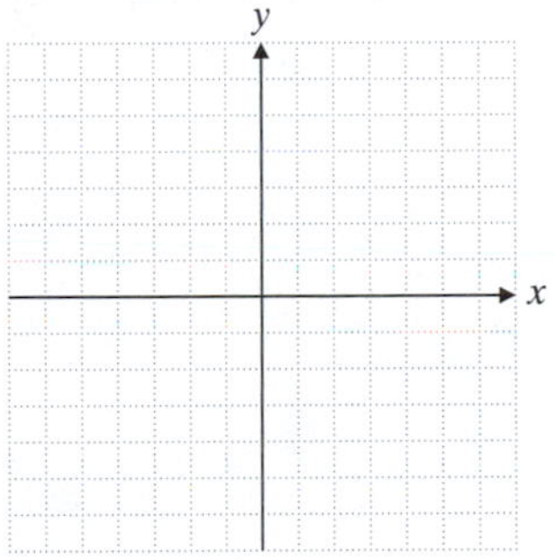

22. $f(x) = x^3$
$r(x) = x^3 + 2$

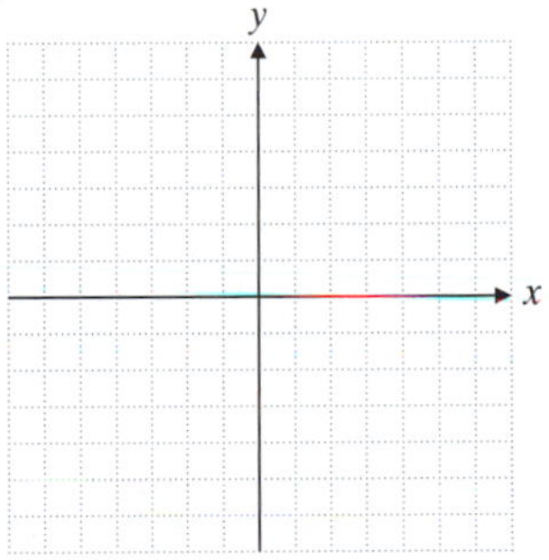

23. $f(x) = |x|$
$s(x) = |x + 4|$

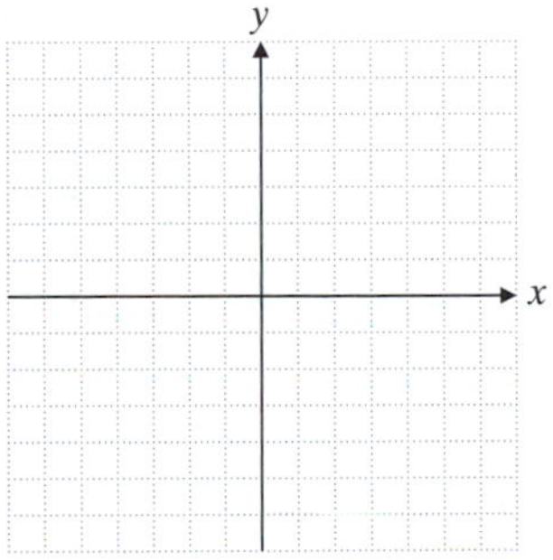

24. $f(x) = |x|$
$s(x) = |x - 5|$

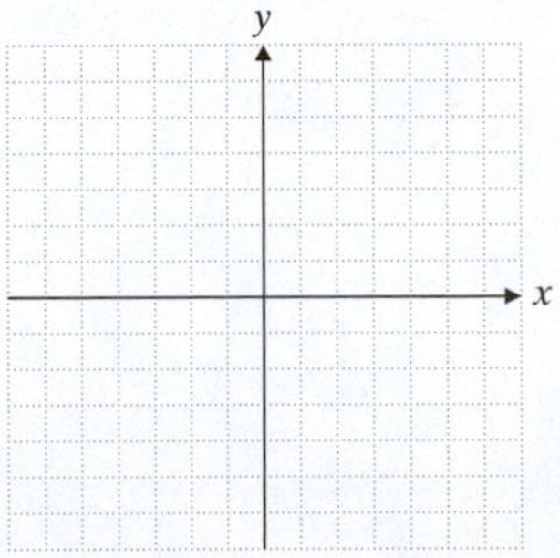

25. $f(x) = |x|$
$t(x) = |x - 3| - 4$

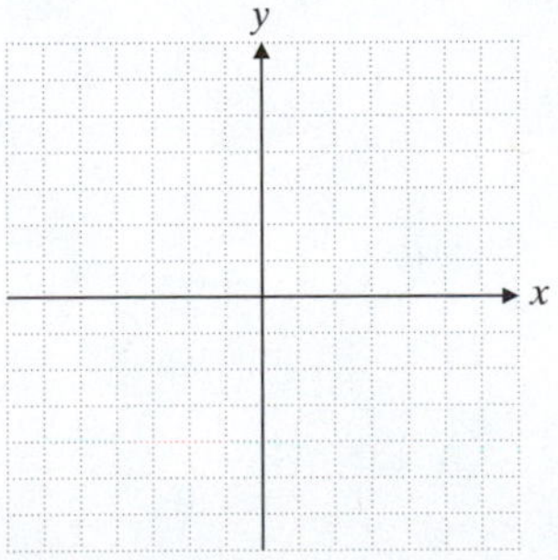

26. $f(x) = |x|$
$t(x) = |x + 2| + 4$

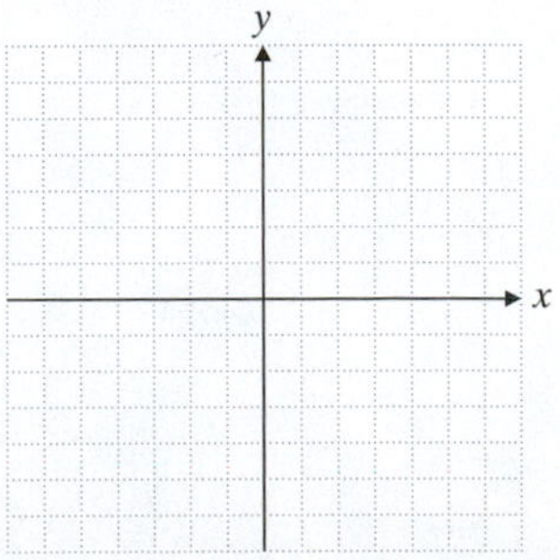

27. $f(x) = \dfrac{3}{x}$

$g(x) = \dfrac{3}{x} - 2$

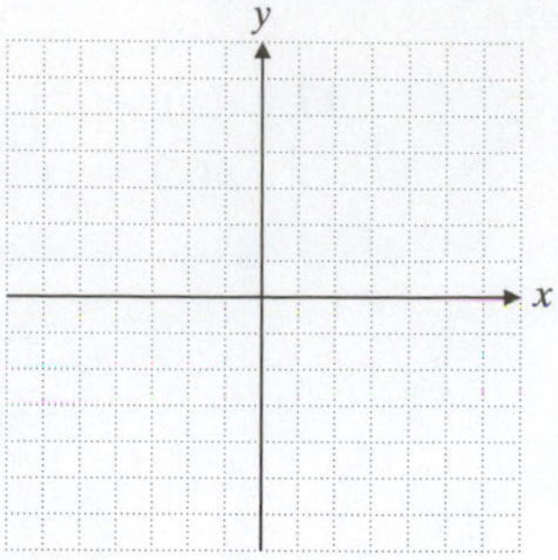

28. $f(x) = \dfrac{2}{x}$

$g(x) = \dfrac{2}{x} + 3$

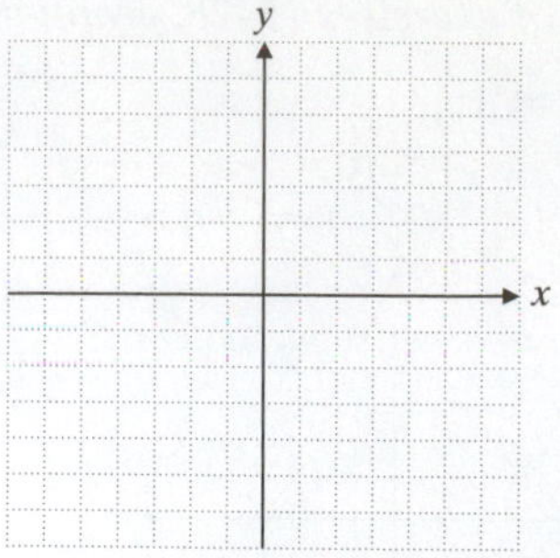

Cumulative Review *Simplify each expression. Assume that all variables are positive.*

29. [7.3.2] $\sqrt{12} + 3\sqrt{50} - 4\sqrt{27}$

30. [7.4.1] $(\sqrt{3x} - 1)^2$

31. [7.4.3] Rationalize the denominator. $\dfrac{\sqrt{5} - 2}{\sqrt{5} + 1}$

Quick Quiz 10.2

1. If $f(x) = x^2$ and $h(x) = (x - 3)^2$ were graphed on one coordinate plane, how would you describe where the graph of $h(x)$ is compared to the graph of $f(x)$?

2. If $f(x) = x^3$ and $g(x) = x^3 - 4$ were graphed on one coordinate plane, how would you describe where the graph of $g(x)$ is compared to the graph of $f(x)$?

3. Graph each pair of functions on one coordinate plane.

$$f(x) = |x|, \quad k(x) = |x - 1| - 4$$

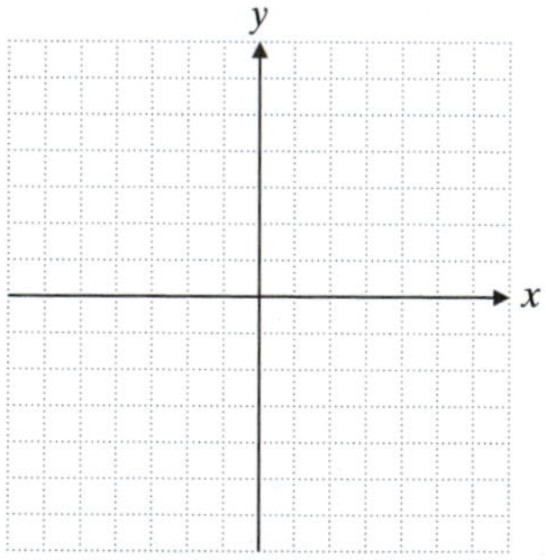

4. Concept Check Explain how you can use a vertical line to determine whether a graph represents a function.

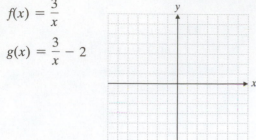

How Am I Doing? Sections 10.1–10.2

How are you doing with your homework assignments in Sections 10.1 and 10.2? Do you feel you have mastered the material so far? Do you understand the concepts you have covered? Before you go further in the textbook, take some time to do each of the following problems.

10.1 *For the function* $f(x) = 2x - 6$, *find the following:*

1. $f(-3)$ **2.** $f(a)$ **3.** $f(2a)$ **4.** $f(a + 2)$

For $f(x) = 5x^2 + 2x - 3$, *find the following:*

5. $f(-2)$ **6.** $f(a)$ **7.** $f(a - 1)$ **8.** $f(-2a)$

For $f(x) = \dfrac{3x}{x + 2}$, *find the following:*

9. $f(a) + f(a - 2)$. Express your answer as one fraction.

10. $f(3a) - f(3)$. Express your answer as one fraction.

10.2 *Determine whether or not each graph represents a function.*

11.

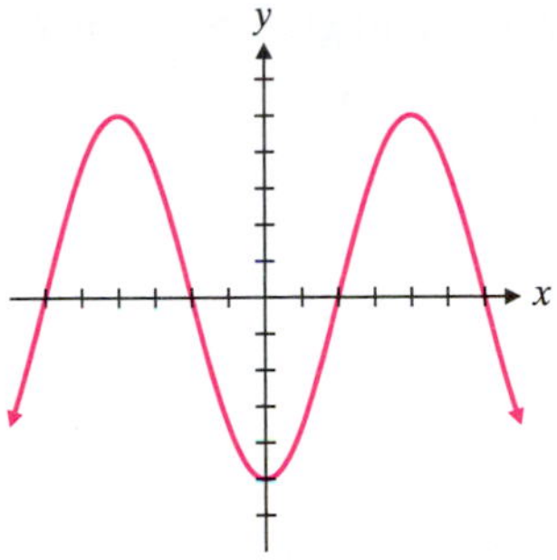

12.

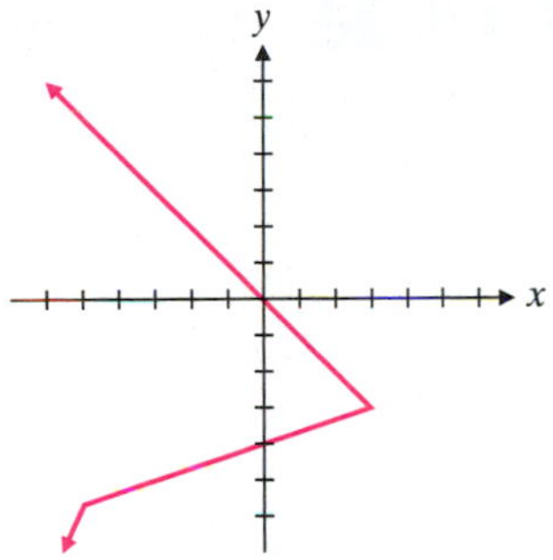

Graph the given functions on one coordinate plane.

13. $f(x) = |x|$ and $s(x) = |x - 3|$.

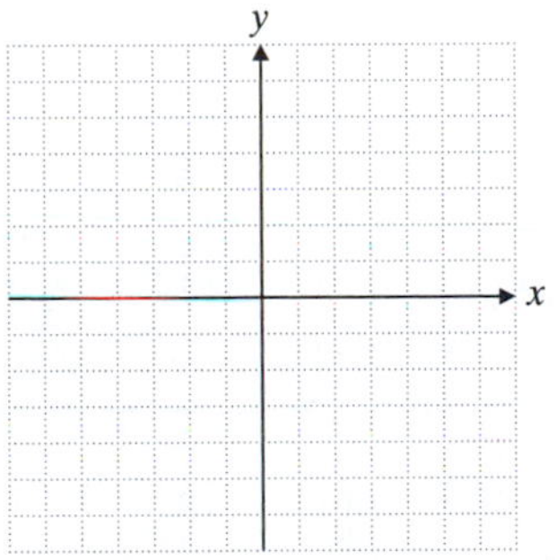

14. $f(x) = x^2$ and $h(x) = (x + 2)^2 + 3$.

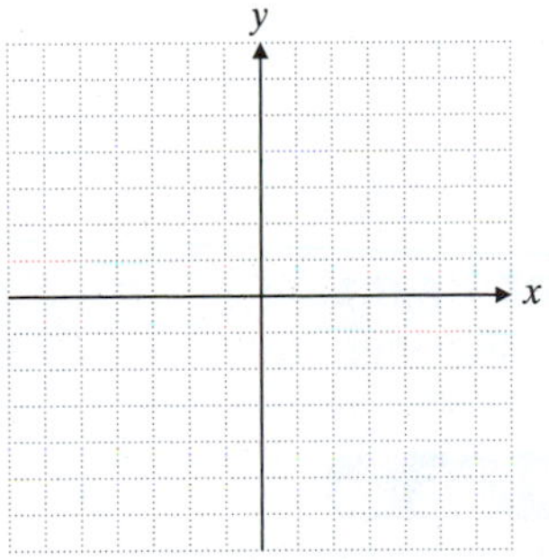

15. Graph $f(x) = \dfrac{4}{x + 2}$.

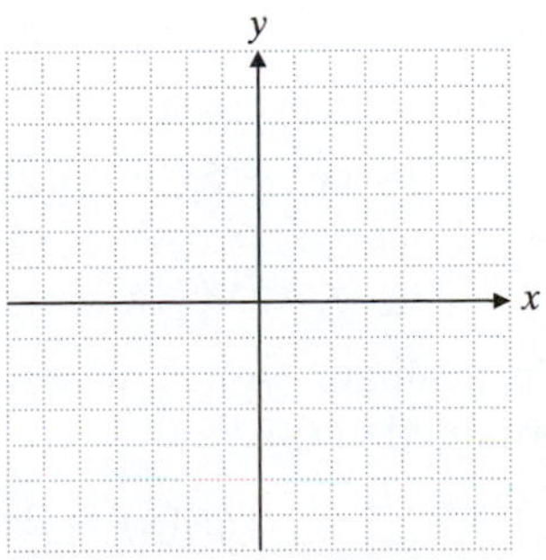

16. If you were to graph $g(x) = \dfrac{4}{x + 2} - 2$, how would it compare to the graph of $f(x)$ in problem 15?

Now turn to page SA-27 for the answers to each of these problems. Each answer also includes a reference to the objective in which the problem is first taught. If you missed any of these problems, you should stop and review the Examples and Student Practice problems in the referenced objective. A little review now will help you master the material in the upcoming sections of the text.

1. ______

2. ______

3. ______

4. ______

5. ______

6. ______

7. ______

8. ______

9. ______

10. ______

11. ______

12. ______

13. ______

14. ______

15. ______

16. ______

10.3 Algebraic Operations on Functions

Student Learning Objectives

After studying this section, you will be able to:

① **Find the sum, difference, product, and quotient of two functions.**

② **Find the composition of two functions.**

① Finding the Sum, Difference, Product, and Quotient of Two Functions

If f represents one function and g represents a second function, we can define new functions as follows:

NEW FUNCTIONS CAN BE FORMED BY COMBINING TWO GIVEN FUNCTIONS USING ALGEBRAIC OPERATIONS.

The Sum, Difference, Product, and Quotient of Two Functions

Sum of Functions	$(f+g)(x)=f(x)+g(x)$
Difference of Functions	$(f-g)(x)=f(x)-g(x)$
Product of Functions	$(fg)(x)=f(x)\cdot g(x)$
Quotient of Functions	$\left(\dfrac{f}{g}\right)(x)=\dfrac{f(x)}{g(x)},\ g(x)\neq 0$

EXAMPLE 1 Suppose that $f(x)=3x^2-3x+5$ and $g(x)=5x-2$.

(a) Find $(f+g)(x)$. **(b)** Evaluate $(f+g)(x)$ when $x=3$.

Solution

(a)
$$\begin{aligned}(f+g)(x)&=f(x)+g(x)\\&=(3x^2-3x+5)+(5x-2)\\&=3x^2-3x+5+5x-2\\&=3x^2+2x+3\end{aligned}$$

(b) To evaluate $(f+g)(x)$ when $x=3$, we write $(f+g)(3)$ and use the formula obtained in **(a)**.

$$\begin{aligned}(f+g)(x)&=3x^2+2x+3\\(f+g)(3)&=3(3)^2+2(3)+3\\&=3(9)+2(3)+3\\&=27+6+3=36\end{aligned}$$

Student Practice 1 Given $f(x)=4x+5$ and $g(x)=2x^2+7x-8$, find the following: **(a)** $(f+g)(x)$ **(b)** $(f+g)(4)$

***NOTE TO STUDENT:** Fully worked-out solutions to all of the Student Practice problems can be found at the back of the text starting at page SP-1.*

EXAMPLE 2 Given $f(x)=x^2-5x+6$ and $g(x)=2x-1$, find the following: **(a)** $(fg)(x)$ **(b)** $(fg)(-4)$

Solution

(a)
$$\begin{aligned}(fg)(x)&=f(x)\cdot g(x)\\&=(x^2-5x+6)(2x-1)\\&=2x^3-10x^2+12x-x^2+5x-6\\&=2x^3-11x^2+17x-6\end{aligned}$$

(b) To evaluate $(fg)(x)$ when $x=-4$, we write $(fg)(-4)$ and use the formula obtained in **(a)**.

$$\begin{aligned}(fg)(x)&=2x^3-11x^2+17x-6\\(fg)(-4)&=2(-4)^3-11(-4)^2+17(-4)-6\\&=2(-64)-11(16)+17(-4)-6\\&=-128-176-68-6\\&=-378\end{aligned}$$

Student Practice 2 Given $f(x) = 3x + 2$ and $g(x) = x^2 - 3x - 4$, find the following:

(a) $(fg)(x)$ **(b)** $(fg)(2)$

When finding the quotient of a function, we must be careful to avoid division by zero. Thus, we always specify any values of x that must be eliminated from the domain.

EXAMPLE 3 Given $f(x) = 3x + 1$, $g(x) = 2x - 1$, and $h(x) = 9x^2 + 6x + 1$, find the following:

(a) $\left(\frac{f}{g}\right)(x)$ **(b)** $\left(\frac{f}{h}\right)(x)$ **(c)** $\left(\frac{f}{h}\right)(-2)$

Solution

(a) $\left(\frac{f}{g}\right)(x) = \frac{3x + 1}{2x - 1}$

The denominator of the quotient can never be zero. Since $2x - 1 \neq 0$, we know that $x \neq \frac{1}{2}$.

(b) $\left(\frac{f}{h}\right)(x) = \frac{3x + 1}{9x^2 + 6x + 1} = \frac{3x + 1}{(3x + 1)(3x + 1)} = \frac{1}{3x + 1}$

Since $3x + 1 \neq 0$, we know that $x \neq -\frac{1}{3}$.

(c) To find $\left(\frac{f}{h}\right)(-2)$, we must evaluate $\left(\frac{f}{h}\right)(x)$ when $x = -2$.

$$\left(\frac{f}{h}\right)(x) = \frac{1}{3x + 1}$$

$$\left(\frac{f}{h}\right)(-2) = \frac{1}{(3)(-2) + 1} = \frac{1}{-6 + 1} = -\frac{1}{5}$$

Student Practice 3 Given $p(x) = 5x^2 + 6x + 1$, $h(x) = 3x - 2$, and $g(x) = 5x + 1$, find the following. (Be careful to find the two values for x in part b that must be eliminated from the domain.)

(a) $\left(\frac{g}{h}\right)(x)$ **(b)** $\left(\frac{g}{p}\right)(x)$ **(c)** $\left(\frac{g}{h}\right)(3)$

② Finding the Composition of Two Functions

Suppose that a new online music vendor finds that the number of sales of digital albums on a given day is generally equal to 25% of the number of people who visit the site on that day. Thus, if x = the number of people who visit the site, then the sales S can be modeled by the equation $S(x) = 0.25x$.

Suppose that the average album on the site sells for \$15. Then if S = the number of albums sold on a given day, the income for that day can be modeled by the equation $P(S) = 15S$. Suppose that eight thousand people visit the site.

$$S(x) = 0.25x$$
$$S(8000) = 0.25(8000) = 2000$$

Thus, two thousand albums would be sold.

If two thousand albums were sold and the average price of an album is \$15, then we would have the following:

$$P(S) = 15S$$
$$P(2000) = 15(2000) = 30{,}000$$

That is, the income from the sales of albums would be \$30,000.

Let us analyze the functions we have described and record a few values of x, $S(x)$, and $P(S)$.

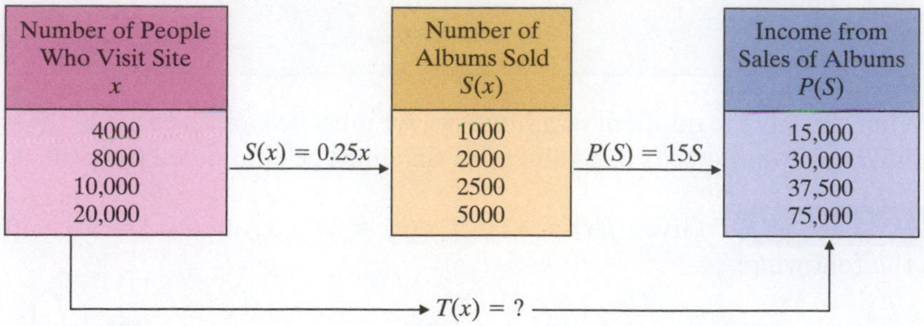

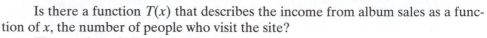

Is there a function $T(x)$ that describes the income from album sales as a function of x, the number of people who visit the site?

The number of albums sold is

$$S(x) = 0.25x.$$

Thus, $0.25x$ is the number of albums sold.

If we replace S in $P(S) = 15S$ by $S(x)$, we have

$$P[S(x)] = P(0.25x) = 15(0.25x) = 3.75x.$$

Thus, the formula $T(x)$ that describes the income in terms of the number of visitors is

$$T(x) = 3.75x.$$

Is this correct? Let us check by finding $T(20{,}000)$. From our table the result should be 75,000.

$$\text{If} \qquad T(x) = 3.75x,$$
$$\text{then} \quad T(20{,}000) = 3.75(20{,}000) = 75{,}000.$$

Thus, we have found a function T that is the composition of the functions P and S: $T(x) = P[S(x)]$.

We now state a definition of the composition of one function with another.

The **composition** of the functions f and g, denoted $f \circ g$, is defined as follows: $(f \circ g)(x) = f[g(x)]$. The domain of $f \circ g$ is the set of all x-values in the domain of g such that $g(x)$ is in the domain of f.

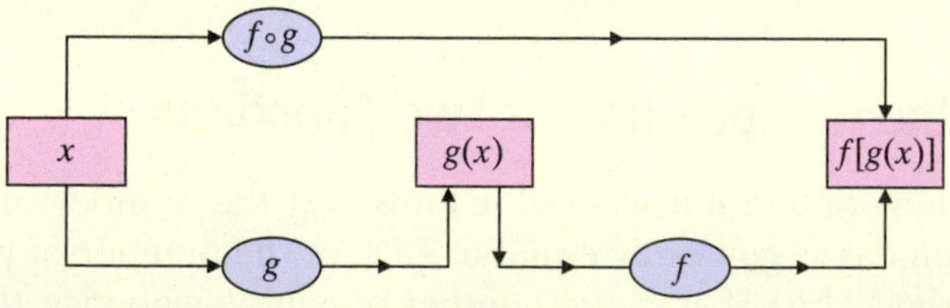

EXAMPLE 4 Given $f(x) = 3x - 2$ and $g(x) = 2x + 5$, find $f[g(x)]$.

Solution

$f[g(x)]$	$= f(2x + 5)$	Substitute $g(x) = 2x + 5$.
	$= 3(2x + 5) - 2$	Apply the formula for $f(x)$.
	$= 6x + 15 - 2$	Remove parentheses.
	$= 6x + 13$	Simplify.

Student Practice 4 Given $f(x) = 2x - 1$ and $g(x) = 3x - 4$, find $f[g(x)]$.

In most situations $f[g(x)]$ and $g[f(x)]$ are not the same.

EXAMPLE 5 Given $f(x) = \sqrt{x - 4}$ with $x \geq 4$ and $g(x) = 3x + 1$, find the following:

(a) $f[g(x)]$ **(b)** $g[f(x)]$

Solution

(a) $f[g(x)] = f(3x + 1)$ Substitute $g(x) = 3x + 1$.

$= \sqrt{(3x + 1) - 4}$ Apply the formula for $f(x)$.

$= \sqrt{3x + 1 - 4}$ Remove parentheses.

$= \sqrt{3x - 3}$ Simplify.

(b) $g[f(x)] = g(\sqrt{x - 4})$ Substitute $f(x) = \sqrt{x - 4}$.

$= 3(\sqrt{x - 4}) + 1$ Apply the formula for $g(x)$.

$= 3\sqrt{x - 4} + 1$ Remove parentheses.

We note that $g[f(x)] \neq f[g(x)]$.

Student Practice 5 Given $f(x) = 2x^2 - 3x + 1$ and $g(x) = x + 2$, find the following:

(a) $f[g(x)]$ **(b)** $g[f(x)]$

Graphing Calculator

Composition of Functions

You can evaluate the composition of functions on most graphing calculators by using the y-variable function (Y-VARS). To do Example 6 on most graphing calculators, you would use the following equations.

$$y_1 = \frac{1}{3x - 4}$$

$$y_2 = 2(y_1)$$

To find the function value, you can use the TableSet command to let $x = 2$. Then enter Table and you will see displayed $y_1 = 0.5$, which represents $g(2) = 0.5$, and $y_2 = 1$, which represents $f[g(2)] = 1$.

MC **EXAMPLE 6** Given $f(x) = 2x$ and $g(x) = \dfrac{1}{3x - 4}, x \neq \dfrac{4}{3}$, find the following:

(a) $(f \circ g)(x)$

(b) $(f \circ g)(2)$

Solution

(a) $(f \circ g)(x) = f[g(x)] = f\left(\dfrac{1}{3x - 4}\right)$ Substitute $g(x) = \dfrac{1}{3x - 4}$.

$= 2\left(\dfrac{1}{3x - 4}\right)$ Apply the formula for $f(x)$.

Simplify.

$= \dfrac{2}{3x - 4}$

(b) $(f \circ g)(2) = \dfrac{2}{3(2) - 4} = \dfrac{2}{6 - 4} = \dfrac{2}{2} = 1$

Student Practice 6 Given $f(x) = 3x + 1$ and $g(x) = \dfrac{2}{x - 3}, x \neq 3$, find the following:

(a) $(g \circ f)(x)$

(b) $(g \circ f)(-3)$

10.3 Exercises

MyMathLab®

Watch the videos in MyMathLab

Download the MyDashBoard App

For the following functions, find **(a)** $(f + g)(x)$, **(b)** $(f - g)(x)$, **(c)** $(f + g)(2)$, *and* **(d)** $(f - g)(-1)$.

1. $f(x) = -2x + 3, g(x) = 2 + 4x$

2. $f(x) = 2x + 1, g(x) = 2 - 3x$

3. $f(x) = 3x^2 - x, g(x) = 5x + 2$

4. $f(x) = 6 - 5x, g(x) = x^2 - 7x - 3$

5. $f(x) = x^3 - \frac{1}{2}x^2 + x, g(x) = x^2 - \frac{x}{4} - 5$

6. $f(x) = 2.4x^2 + x - 3.5, g(x) = 1.1x^3 - 2.2x$

7. $f(x) = -5\sqrt{x + 6}, x \geq -6;\ g(x) = 8\sqrt{x + 6}, x \geq -6$

8. $f(x) = 3\sqrt{3 - x},\ x \leq 3; g(x) = -5\sqrt{3 - x}, x \leq 3$

For the following functions, find **(a)** $(fg)(x)$ *and* **(b)** $(fg)(-3)$.

9. $f(x) = 2x - 3, g(x) = -2x^2 - 3x + 1$

10. $f(x) = x^2 - 3x + 2, g(x) = 1 - x$

11. $f(x) = \frac{2}{x^2},\ x \neq 0; g(x) = x^2 - x$

12. $f(x) = \frac{3x}{x + 4}, x \neq -4; g(x) = \frac{x}{3}$

13. $f(x) = \sqrt{-2x + 1},\ x \leq \frac{1}{2}; g(x) = -3x$

14. $f(x) = 4x, g(x) = \sqrt{3x + 10}, x \geq -\frac{10}{3}$

For the following functions, find **(a)** $\left(\frac{f}{g}\right)(x)$ *and* **(b)** $\left(\frac{f}{g}\right)(2)$.

15. $f(x) = x - 6, g(x) = 3x$

16. $f(x) = 3x, g(x) = 4x - 1$

17. $f(x) = x^2 - 1, g(x) = x - 1$

18. $f(x) = x, g(x) = x^2 - 5x$

19. $f(x) = x^2 + 10x + 25, g(x) = x + 5$

20. $f(x) = 2x^2 + 5x - 12, g(x) = x + 4$

21. $f(x) = 4x - 1, g(x) = 4x^2 + 7x - 2$

22. $f(x) = 3x + 2, g(x) = 3x^2 - x - 2$

Let $f(x) = 3x + 2, g(x) = x^2 - 2x$, *and* $h(x) = \frac{x - 2}{3}$. *Find the following:*

23. $(f - g)(x)$

24. $(g - f)(x)$

25. $\left(\frac{g}{h}\right)(x)$

26. $\left(\frac{g}{h}\right)(-2)$

27. $(fg)(-1)$ **28.** $(gh)(3)$ **29.** $\left(\frac{g}{f}\right)(-1)$ **30.** $\left(\frac{g}{f}\right)(x)$

Find $f[g(x)]$ for each of the following:

31. $f(x) = 2 - 3x, g(x) = 2x + 5$

32. $f(x) = 3x + 2, g(x) = 4x - 1$

33. $f(x) = 2x^2 + 5, g(x) = x - 1$

34. $f(x) = 9 + x^2, g(x) = x - 3$

35. $f(x) = 8 - 5x, g(x) = x^2 + 3$

36. $f(x) = 4 - x, g(x) = 1 - 3x + x^2$

37. $f(x) = \frac{7}{2x - 3}, g(x) = x + 2$

38. $f(x) = \frac{11}{6 - x}, g(x) = x + 3$

39. $f(x) = |x + 3|, g(x) = 2x - 1$

40. $f(x) = \left|\frac{1}{2}x - 5\right|, g(x) = 4x + 6$

Let $f(x) = x^2 + 2, g(x) = 3x + 5, h(x) = \frac{1}{x}$, and $p(x) = \sqrt{x - 1}$. Find each of the following:

41. $f[g(x)]$

42. $g[h(x)]$

43. $g[f(x)]$

44. $h[g(x)]$

45. $g[f(0)]$

46. $h[g(0)]$

47. $(p \circ f)(x)$

48. $(f \circ h)(x)$

49. $(g \circ h)(\sqrt{2})$

50. $(f \circ p)(x)$

51. $(p \circ f)(-5)$

52. $(f \circ p)(7)$

Applications

53. ***Temperature Scales*** Consider the Celsius function $C(F) = \frac{5F - 160}{9}$, which converts degrees Fahrenheit to degrees Celsius. A different temperature scale, called the Kelvin scale, is used by many scientists in their research. The Kelvin scale is similar to the Celsius scale, but it begins at absolute zero (the coldest possible temperature, which is around $-273°C$). To convert a Celsius temperature to a temperature on the Kelvin scale, we use the function $K(C) = C + 273$. Find $K[C(F)]$, which is the composite function that defines the temperature in Kelvins in terms of the temperature in degrees Fahrenheit.

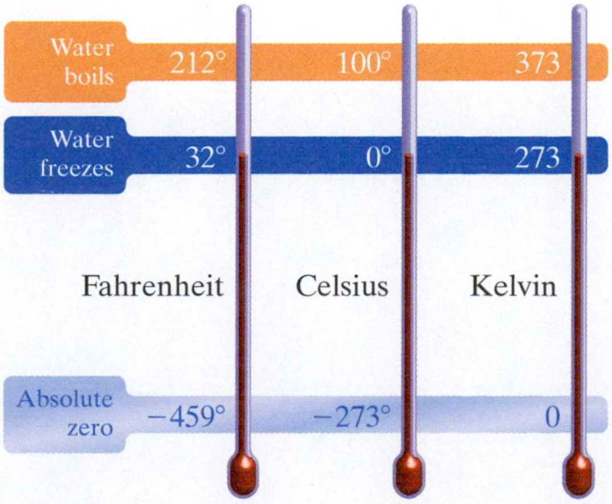

54. ***Business*** Suppose the dollar cost to produce n items in a factory is $c(n) = 5n + 4$. Furthermore, the number of items n produced in x hours is $n(x) = 3x$. Find $c[n(x)]$, which is the composite function that defines the dollar cost in terms of the number of hours of production x.

▲ **55.** ***Oil Slicks*** An oil tanker with a ruptured hull is leaking oil off the coast of Africa. There is no wind or significant current, so the oil slick is spreading in a circle whose radius is defined by the function $r(t) = 3t$, where t is the time in minutes since the tanker began to leak. The area of the slick for any given radius is approximately determined by the function $a(r) = 3.14r^2$, where r is the radius of the circle measured in feet. Find $a[r(t)]$, which is the composite function that defines the area of the oil slick in terms of the minutes t since the beginning of the leak. How large is the area after 20 minutes?

Cumulative Review *Factor each of the following:*

56. **[5.6.2]** $36x^2 - 12x + 1$

57. **[5.6.1]** $25x^4 - 1$

58. **[5.6.1]** $x^4 - 10x^2 + 9$

59. **[5.5.2]** $3x^2 - 7x + 2$

Quick Quiz 10.3

1. If $f(x) = 2x^2 - 4x - 8$ and $g(x) = -3x^2 + 5x - 2$, find $(f - g)(x)$.

2. If $f(x) = x^2 - 3$ and $g(x) = \frac{x - 4}{2}$, find $f[g(x)]$.

3. If $f(x) = x - 7$ and $g(x) = 2x - 5$, find $\left(\frac{g}{f}\right)(2)$.

4. **Concept Check** If $f(x) = 5 - 2x^2$ and $g(x) = 3x^2 - 5x + 1$, explain how you would find $(f - g)(-4)$.

10.4 Inverse of a Function

Americans driving in Canada or Mexico need to be able to convert miles per hour to kilometers per hour and vice versa.

If someone is driving at 55 miles per hour, how fast is he or she going in kilometers per hour?

Approximate Value in Miles per Hour	Approximate Value in Kilometers per Hour
35	56
40	64
45	72
50	80
55	88
60	96
65	104

A function f that converts from miles per hour to an approximate value in kilometers per hour is $f(x) = 1.6x$.

For example, $f(40) = 1.6(40) = 64$.

This tells us that 40 miles per hour is approximately equivalent to 64 kilometers per hour.

We can come up with a function that does just the opposite—that is, that converts kilometers per hour to an approximate value in miles per hour. This function is $f^{-1}(x) = 0.625x$.

For example, $f^{-1}(64) = 0.625(64) = 40$.

This tells us that 64 kilometers per hour is approximately equivalent to 40 miles per hour.

Miles per hour **Kilometers per hour**

40 ⟶ $f(x) = 1.6x$ ⟶ 64

40 ⟵ $f^{-1}(x) = 0.625x$ ⟵ 64

We call a function f^{-1} that reverses the domain and range of a function f the **inverse function *f*.**

Most American cars have numbers showing kilometers per hour in smaller print on the car speedometer. Unfortunately, these numbers are usually hard to read. If we made a list of several function values of f and several inverse function values of f^{-1}, we could create a conversion scale like the one below that we could use if we should travel to Mexico or Canada with an American car.

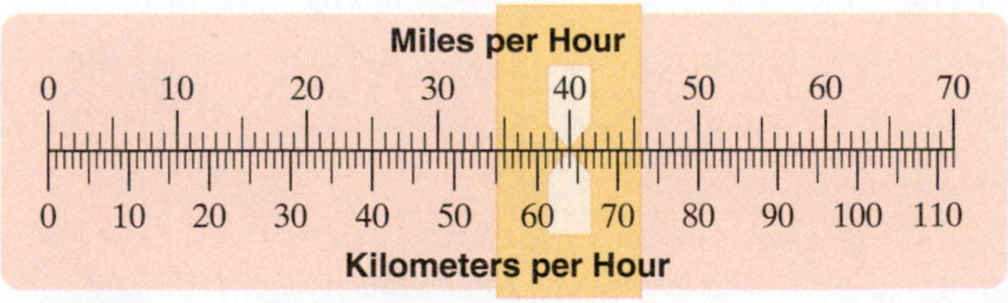

Student Learning Objectives

After studying this section, you will be able to:

1. **Determine whether a function is a one-to-one function.**
2. **Find the inverse function for a given function.**
3. **Graph a function and its inverse function.**

The original function that we studied converts miles per hour to kilometers per hour. The corresponding inverse function converts kilometers per hour to miles per hour. How do we find inverse functions? Do all functions have inverse functions? These are questions we will explore in this section.

① Determining Whether a Function Is a One-to-One Function

First we state that not all functions have inverse functions. To have an inverse that is a function, a function must be one-to-one. This means that for every value of y, there is only one value of x. Or, in the language of ordered pairs, no ordered pairs have the same second coordinate.

DEFINITION OF A ONE-TO-ONE FUNCTION

A **one-to-one function** is a function in which no ordered pairs have the same second coordinate.

TO THINK ABOUT: Relationship of One-to-One and Inverses Why must a function be one-to-one in order to have an inverse that is a function?

EXAMPLE 1 Indicate whether the following are one-to-one functions.

(a) $M = \{(1, 3), (2, 7), (5, 8), (6, 12)\}$ **(b)** $P = \{(1, 4), (2, 9), (3, 4), (4, 18)\}$

Solution

(a) M is a function because no ordered pairs have the same first coordinate. M is also a one-to-one function because no ordered pairs have the same second coordinate.

(b) P is a function, but it is not one-to-one because the ordered pairs $(1, 4)$ and $(3, 4)$ have the same second coordinate.

Student Practice 1

(a) Is the function $A = \{(-2, -6), (-3, -5), (-1, 2), (3, 5)\}$ one-to-one?

(b) Is the function $B = \{(0, 0), (1, 1), (2, 4), (3, 9), (-1, 1)\}$ one-to-one?

***NOTE TO STUDENT:** Fully worked-out solutions to all of the Student Practice problems can be found at the back of the text starting at page SP-1.*

By examining the graph of a function, we can quickly tell whether it is one-to-one. If any horizontal line crosses the graph of a function in more than one place, the function is not one-to-one. If no such line exists, then the function is one-to-one.

HORIZONTAL LINE TEST

If any horizontal line intersects the graph of a function more than once, the function is not one-to-one. If no such line exists, the function is one-to-one.

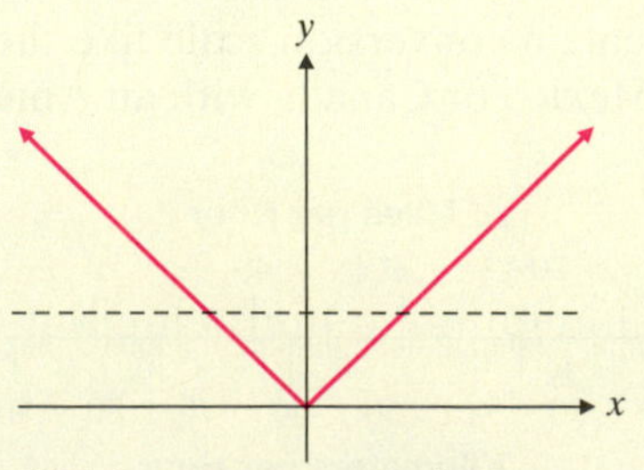

This is *not* a one-to-one function.

EXAMPLE 2 Determine whether the functions graphed are one-to-one functions.

(a)

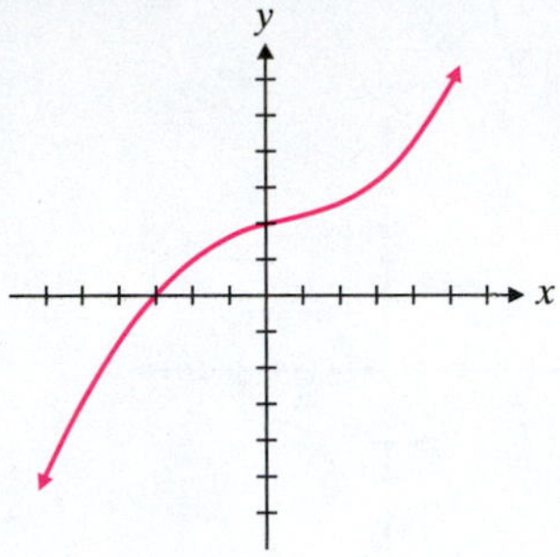

(b)

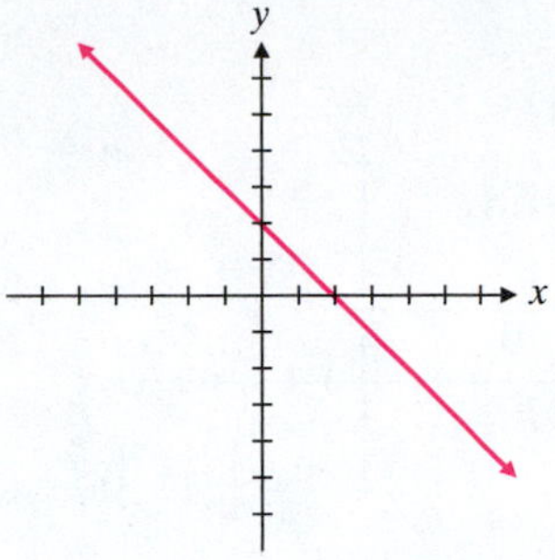

(c)

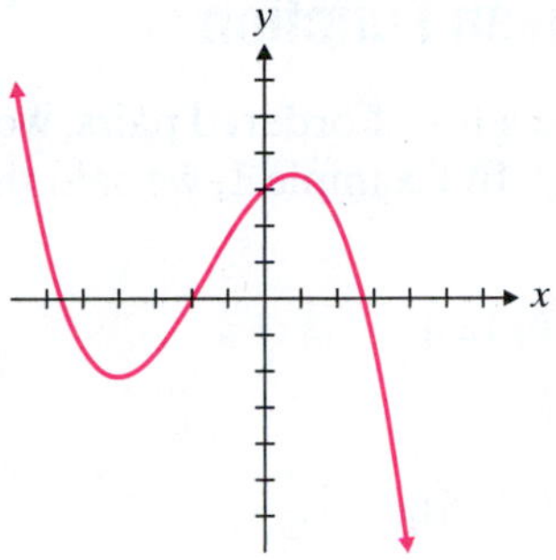

(d)

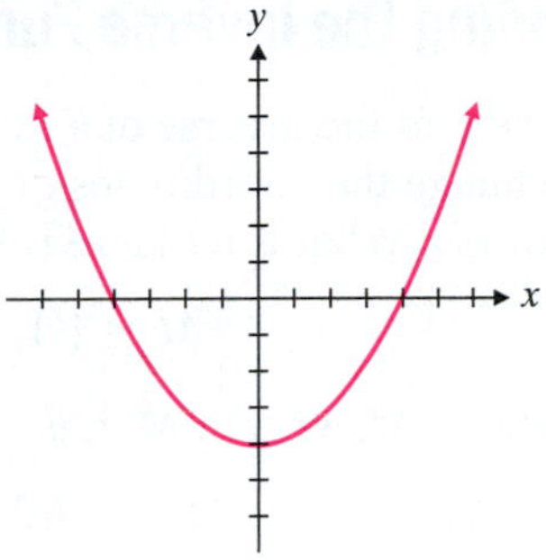

Solution The graphs of **(a)** and **(b)** represent one-to-one functions. Horizontal lines cross the graphs at most once.

(a)

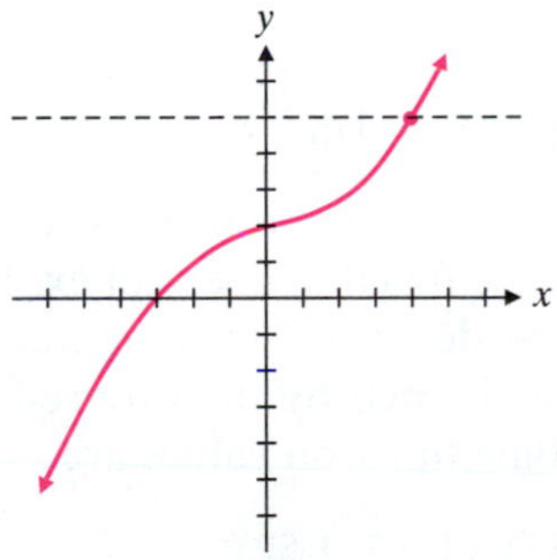

(b)

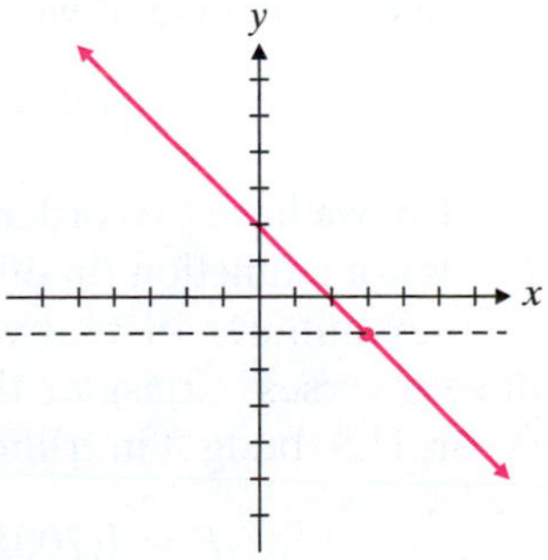

The graphs of **(c)** and **(d)** do not represent one-to-one functions. A horizontal line exists that crosses the graphs more than once.

(a)

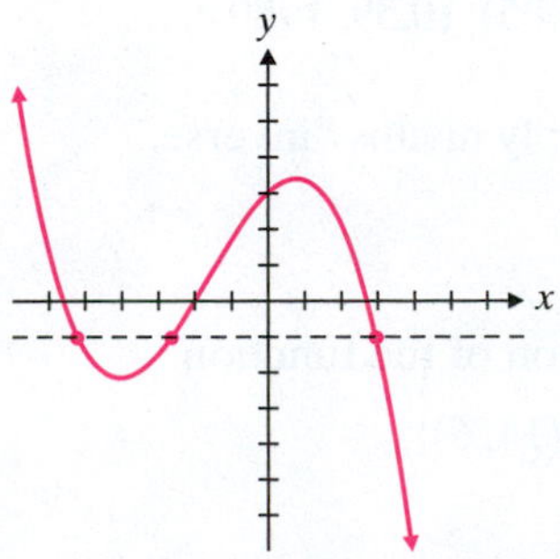

(b)

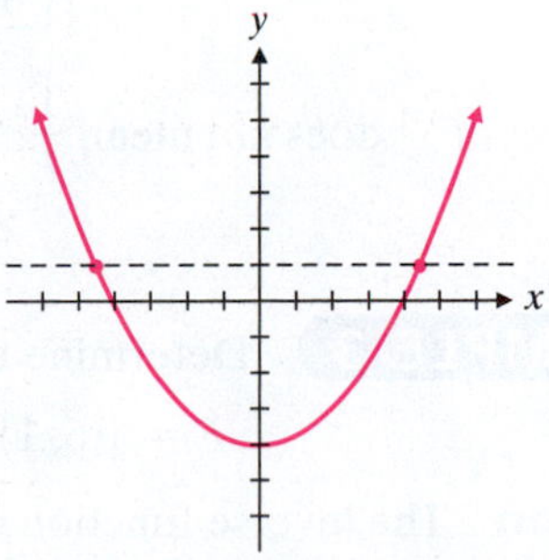

Continued on next page

Student Practice 2 Do each of the following graphs of a function represent a one-to-one function? Why or why not?

(a)

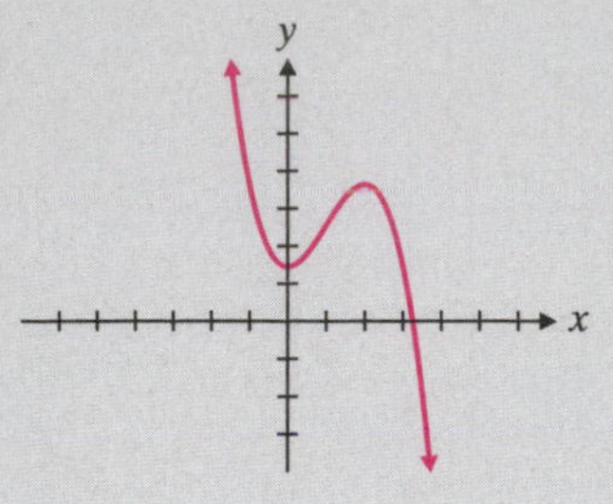

(b)

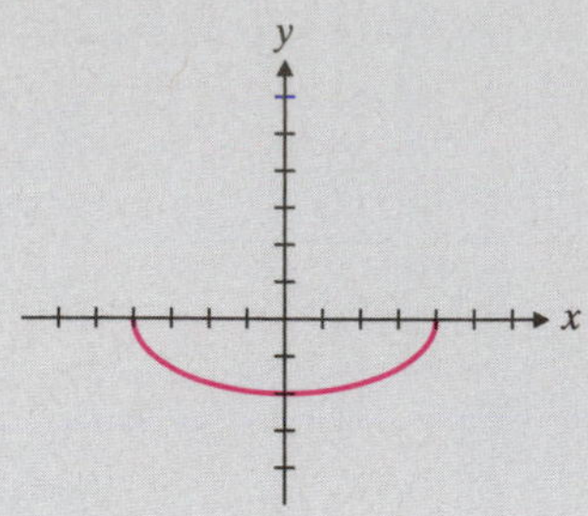

② Finding the Inverse Function for a Given Function

How do we find the inverse of a function? If we have a list of ordered pairs, we simply interchange the coordinates of each ordered pair. In Example 1, we said that M has an inverse. What is it? Here is M again.

$$M = \{(1, 3), (2, 7), (5, 8), (6, 12)\}$$

The inverse of M, written M^{-1}, is

$$M^{-1} = \{(3, 1), (7, 2), (8, 5), (12, 6)\}.$$

Now do you see why a function must be one-to-one in order to have an inverse that is a function? Let's look at the function P from Example 1.

$$P = \{(1, 4), (2, 9), (3, 4), (4, 18)\}$$

If P had an inverse, it would be

$$P^{-1} = \{(4, 1), (9, 2), (4, 3), (18, 4)\}.$$

But we have two ordered pairs with the same first coordinate. Therefore, P^{-1} is not a function (in other words, the inverse function does not exist).

A number of real-world situations are described by functions that have inverses. Consider the function that is defined by the ordered pairs (year, U.S. budget in trillions of dollars). Some function values are

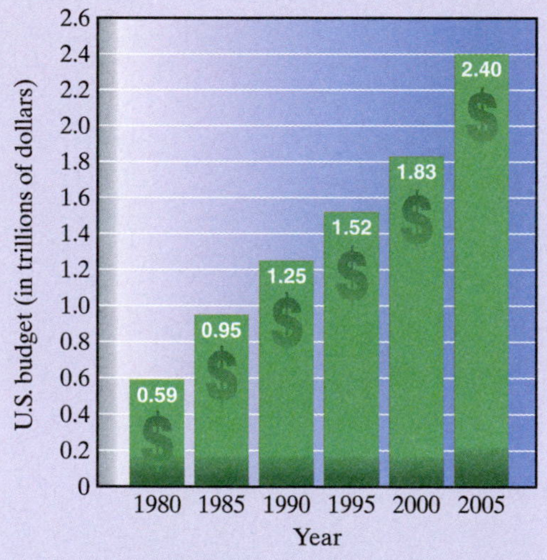

Source: U.S. Budget Office

$$F = \{(2005, 2.40), (2000, 1.83), (1995, 1.52), (1990, 1.25), (1985, 0.95), (1980, 0.59)\}.$$

In this case the inverse of the function is

$$F^{-1} = \{(2.40, 2005), (1.83, 2000), (1.52, 1995), (1.25, 1990), (0.95, 1985), (0.59, 1980)\}.$$

By the way, F^{-1} does *not* mean $\dfrac{1}{F}$. Here the -1 simply means "inverse."

EXAMPLE 3 Determine the inverse function of the function

$$F = \{(6, 1), (12, 2), (13, 5), (14, 6)\}.$$

Solution The inverse function of F is

$$F^{-1} = \{(1, 6), (2, 12), (5, 13), (6, 14)\}.$$

Student Practice 3 Find the inverse of the one-to-one function.
$B = \{(1, 2), (7, 8), (8, 7), (10, 12)\}$

Suppose that a function is given in the form of an equation. How do we find the inverse? Since, by definition, we interchange the ordered pairs to find the inverse of a function, this means that the x-values of the function become the y-values of the inverse function and vice versa.

Four steps will help us find the inverse of a one-to-one function when we are given its equation.

FINDING THE INVERSE OF A ONE-TO-ONE FUNCTION

1. Replace $f(x)$ with y.
2. Interchange x and y.
3. Solve for y in terms of x.
4. Replace y with $f^{-1}(x)$.

EXAMPLE 4 Find the inverse of $f(x) = 7x - 4$.

Solution

Step 1 $y = 7x - 4$ Replace $f(x)$ with y.

Step 2 $x = 7y - 4$ Interchange the variables x and y.

Step 3 $x + 4 = 7y$ Solve for y in terms of x.

$$\frac{x+4}{7} = y$$

Step 4 $f^{-1}(x) = \dfrac{x+4}{7}$ Replace y with $f^{-1}(x)$.

Student Practice 4 Find the inverse of the function $g(x) = 4 - 6x$.

Let's see whether this technique works on a situation similar to the opening example in which we converted miles per hour to approximate values in kilometers per hour.

EXAMPLE 5 Find the inverse function of $f(x) = \frac{9}{5}x + 32$, which converts Celsius temperature (x) into equivalent Fahrenheit temperature.

Solution

Step 1 $y = \frac{9}{5}x + 32$ Replace $f(x)$ with y.

Step 2 $x = \frac{9}{5}y + 32$ Interchange x and y.

Step 3 $5(x) = 5\left(\frac{9}{5}\right)y + 5(32)$ Solve for y in terms of x.

$$5x = 9y + 160$$

$$5x - 160 = 9y$$

$$\frac{5x - 160}{9} = \frac{9y}{9}$$

$$\frac{5x - 160}{9} = y$$

Step 4 $f^{-1}(x) = \dfrac{5x - 160}{9}$ Replace y with $f^{-1}(x)$.

Continued on next page

Graphing Calculator

Inverse Functions

If f and g are inverse functions, then their graphs will be symmetric about the line $y = x$. Use a square window setting when graphing. For Example 6, graph

$$y_1 = 3x - 2,\ y_2 = \frac{x+2}{3},$$

and $y_3 = x$.

Display:

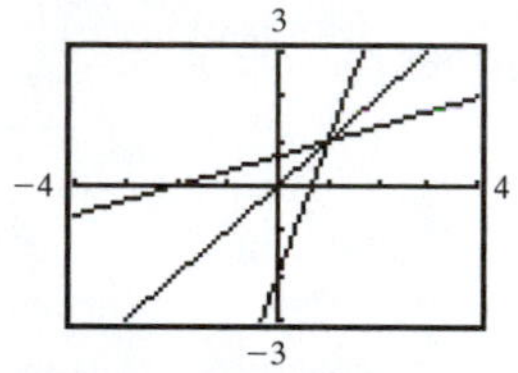

Note: Our inverse function $f^{-1}(x)$ will now convert Fahrenheit temperature to Celsius temperature. For example, suppose we wanted to know the Celsius temperature that corresponds to 86°F.

$$f^{-1}(86) = \frac{5(86) - 160}{9} = \frac{270}{9} = 30$$

This tells us that a temperature of 86°F corresponds to a temperature of 30°C.

Student Practice 5 Find the inverse function of $f(x) = 0.75 + 0.55(x - 1)$, which gives the cost of a telephone call for any call over 1 minute if the telephone company charges 75 cents for the first minute and 55 cents for each minute thereafter. Here $x =$ the number of minutes.

③ Graphing a Function and Its Inverse Function

The graph of a function and its inverse are symmetric about the line $y = x$. Why do you think that this is so?

EXAMPLE 6 If $f(x) = 3x - 2$, find $f^{-1}(x)$. Graph f and f^{-1} on the same set of axes. Draw the line $y = x$ as a dashed line for reference.

Solution

$$f(x) = 3x - 2$$
$$y = 3x - 2$$
$$x = 3y - 2$$
$$x + 2 = 3y$$
$$\frac{x+2}{3} = y$$
$$f^{-1}(x) = \frac{x+2}{3}$$

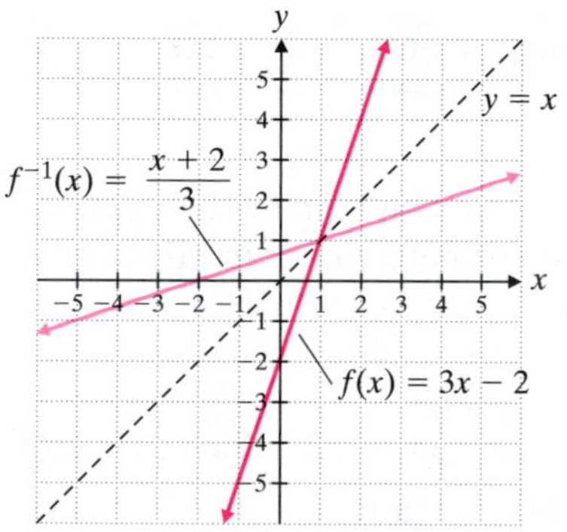

Now we graph each line.

We see that the graphs of f and f^{-1} are symmetric about the line $y = x$. If we folded the graph paper along the line $y = x$, the graph of f would touch the graph of f^{-1}. Try it. Redraw the functions on a separate piece of graph paper. Fold the graph paper on the line $y = x$.

Student Practice 6

If $f(x) = -\frac{1}{4}x + 1$, find $f^{-1}(x)$. Graph f and f^{-1} on the same coordinate plane. Draw the line $y = x$ as a dashed line for reference.

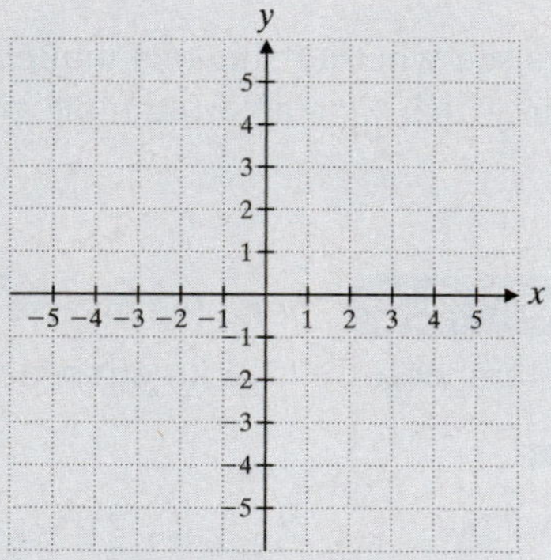

10.4 Exercises

MyMathLab®

Watch the videos in MyMathLab

Download the MyDashBoard App

Verbal and Writing Skills, Exercises 1–6

Complete the following:

1. A one-to-one function is a function in which no ordered pairs ______________________________.

2. If any horizontal line intersects the graph of a function more than once, the function ______________.

3. The graphs of a function f and its inverse f^{-1} are symmetric about the line ______.

4. Do all functions have inverse functions? Why or why not?

5. Does the graph of a horizontal line represent a function? Why or why not? Does it represent a one-to-one function? Explain.

6. Does the graph of a vertical line represent a function? Why or why not? Does it represent a one-to-one function? Explain.

Indicate whether each function is one-to-one.

7. $B = \{(0, 1), (1, 0), (10, 0)\}$

8. $A = \{(-6, -2), (6, 2), (3, 4)\}$

9. $F = \{(\frac{2}{3}, 2), (3, -\frac{4}{5}), (-\frac{2}{3}, -2), (-3, \frac{4}{5})\}$

10. $C = \{(12, 3), (-6, 1), (6, 3)\}$

11. $E = \{(1, 2.8), (3, 6), (-1, -2.8), (2.8, 1)\}$

12. $F = \{(6, 5), (-6, -5), (5, 6), (-5, -6)\}$

Indicate whether each graph represents a one-to-one function.

13.

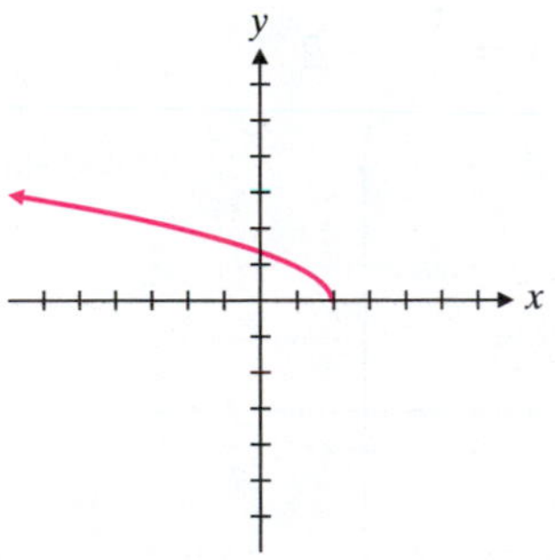

14.

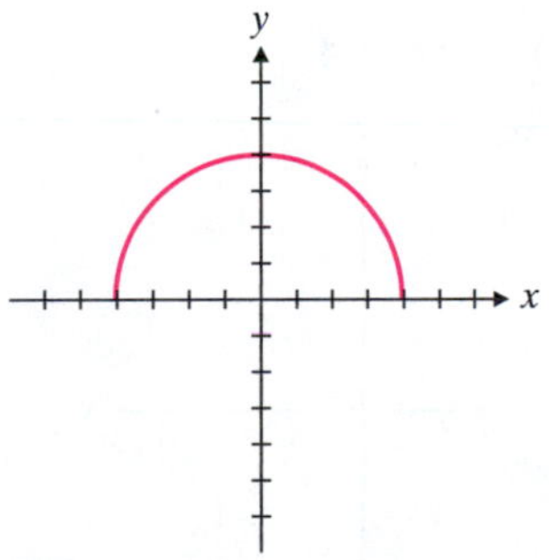

15.

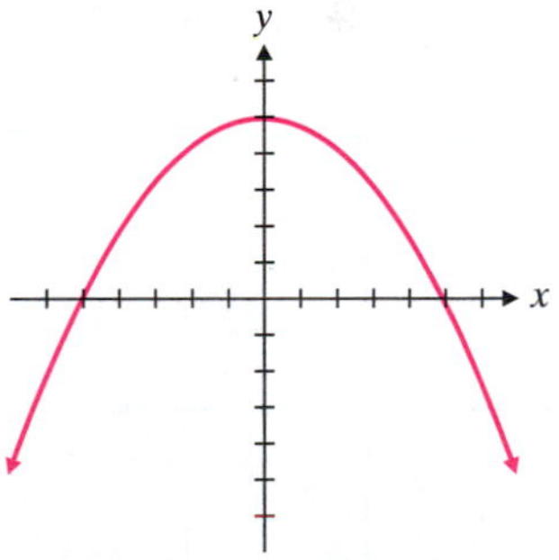

16.

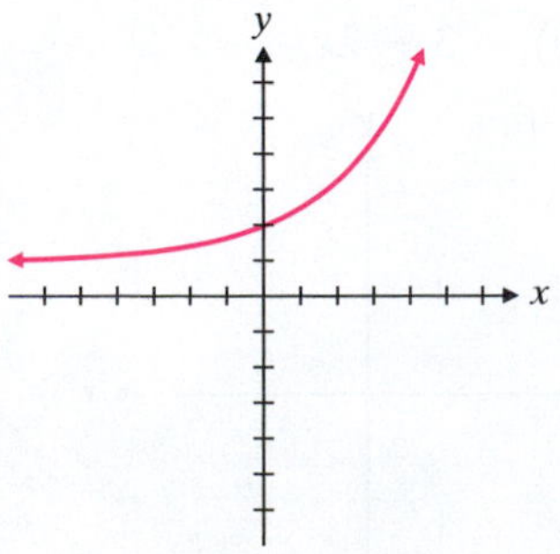

17.

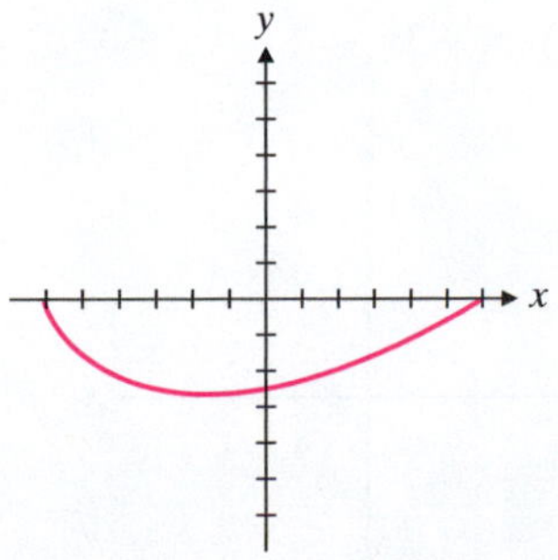

18.

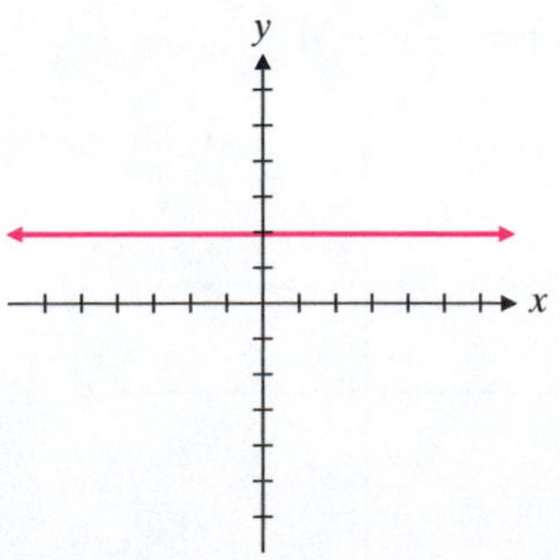

Find the inverse of each one-to-one function. Graph the function and its inverse on one coordinate plane.

19. $J = \{(8, 2), (1, 1), (0, 0), (-8, -2)\}$

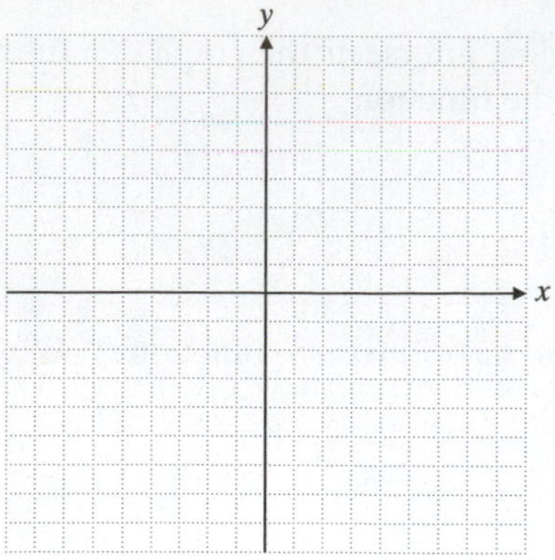

20. $K = \{(-7, 1), (6, 2), (3, -1), (2, 5)\}$

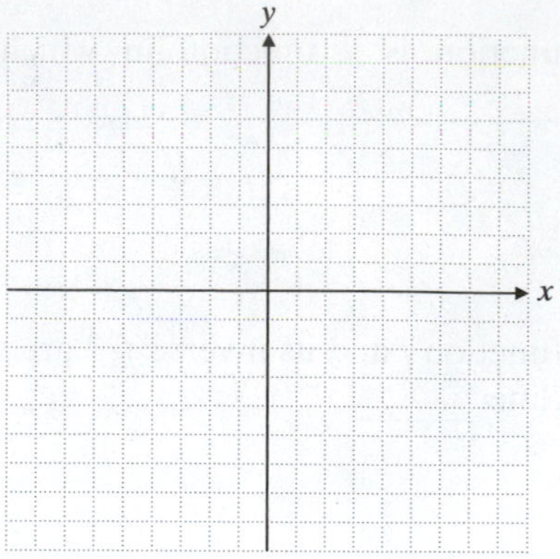

Find the inverse of each function.

21. $f(x) = 4x - 5$

22. $f(x) = 8 - 3x$

23. $f(x) = x^3 + 7$

24. $f(x) = 3 - x^3$

25. $f(x) = -\dfrac{4}{x}$

26. $f(x) = \dfrac{3}{x}$

27. $f(x) = \dfrac{4}{x - 5}$

28. $f(x) = \dfrac{2}{3 + x}$

For the given function f and its inverse, find $f[f^{-1}(x)]$.

29. $f(x) = \dfrac{x - 3}{5}; f^{-1}(x) = 5x + 3$

30. $f(x) = \dfrac{x}{2} + 1; f^{-1}(x) = 2x - 2$

Find the inverse of each function. Graph the function and its inverse on one coordinate plane. Graph the line $y = x$ *as a dashed line.*

31. $g(x) = 2x + 5$

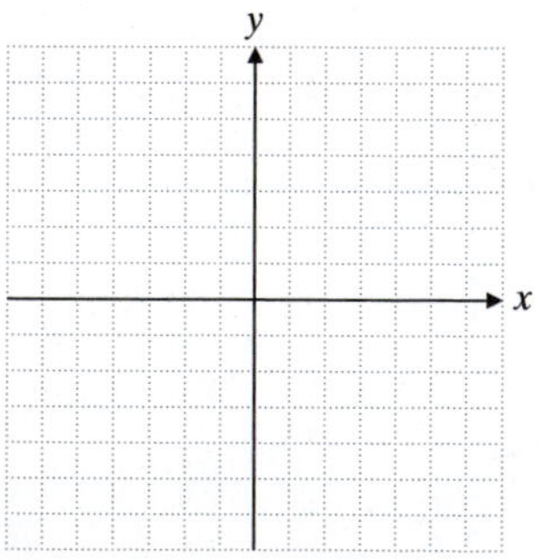

32. $f(x) = 3x + 4$

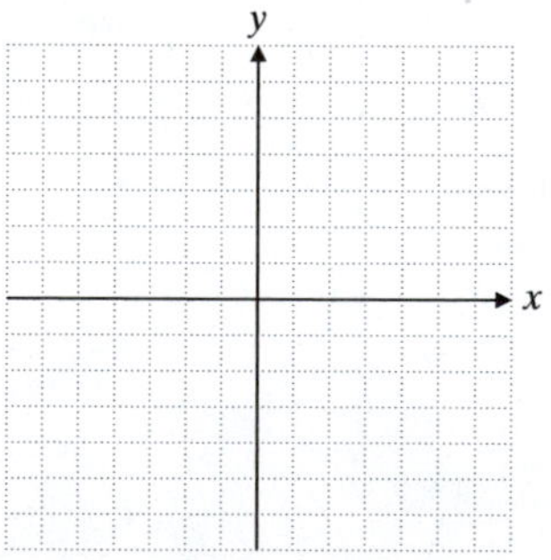

33. $h(x) = \dfrac{1}{2}x - 2$

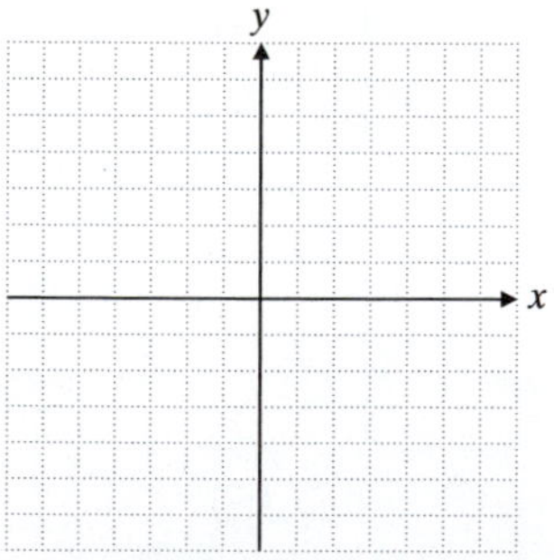

34. $p(x) = \dfrac{2}{3}x - 4$

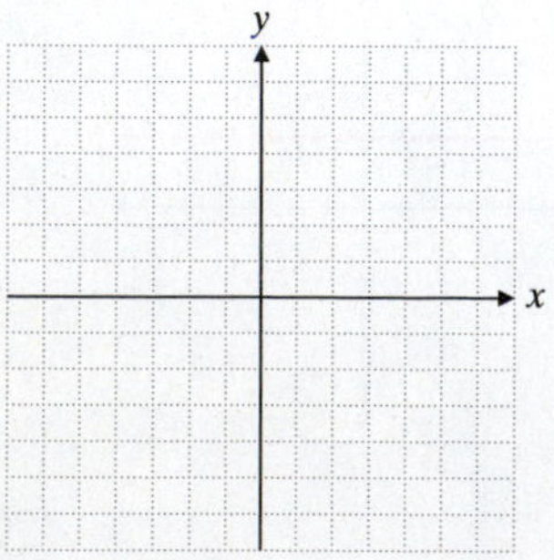

35. $r(x) = -3x - 1$

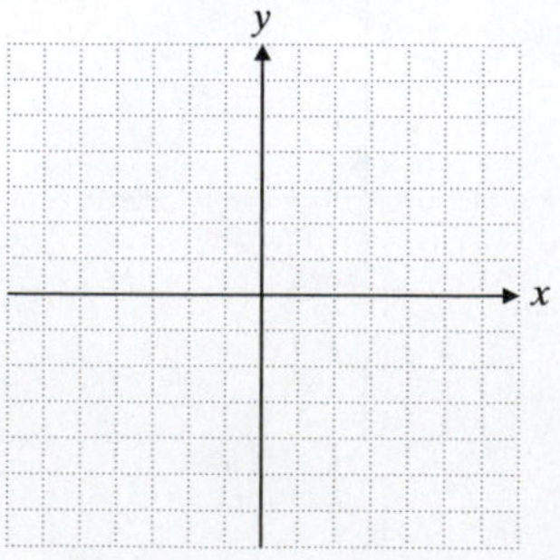

36. $k(x) = 3 - 2x$

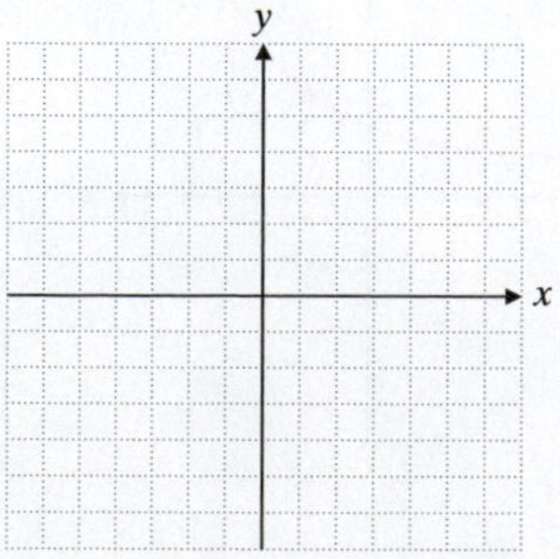

To Think About

37. Can you find an inverse function for the function $f(x) = 2x^2 + 3$? Why or why not?

38. Can you find an inverse function for the function $f(x) = |3x + 4|$? Why or why not?

For every function f and its inverse, f^{-1}, it is true that $f[f^{-1}(x)] = x$ and $f^{-1}[f(x)] = x$. Show that this is true for each pair of inverse functions.

39. $f(x) = 2x + \frac{3}{2}, f^{-1}(x) = \frac{1}{2}x - \frac{3}{4}$

40. $f(x) = -3x - 10, f^{-1}(x) = \frac{-x - 10}{3}$

Cumulative Review *Solve for x.*

41. **[7.5.1]** $\sqrt{20 - x} = x$

42. **[8.3.2]** $x^{2/3} + 7x^{1/3} + 12 = 0$

43. **[1.2.2]** ***Canadian Forests*** Forests are a dominant feature of Canada. Ten percent of all the world's forests lie in Canada. One out of every fifteen people in the labor force in Canada works in a job that relates to forests. If the labor force in Canada in 2005 was 17,300,000 people, how many people worked in a job related to forests in that year? Round to the nearest whole number. (*Source:* www.encarta.msn.com)

44. **[1.2.2]** Bulgaria is one of the top ten countries with the biggest decline in population. The population of Bulgaria in 2010 was 7,564,000. By 2050, the population is expected to be 4,992,240. What is Bulgaria's expected percent of decrease in population? (*Source:* www.geography.about.com)

Quick Quiz 10.4

1. $A = \{(3, -4), (2, -6), (5, 6), (-3, 4)\}$

(a) Is A a function?

(b) Is A a one-to-one function?

(c) If the answers to **(a)** and **(b)** are yes, find A^{-1}.

2. Find the inverse function for $f(x) = 5 - 2x$. Graph f and its inverse. Graph the line $y = x$ as a dashed line.

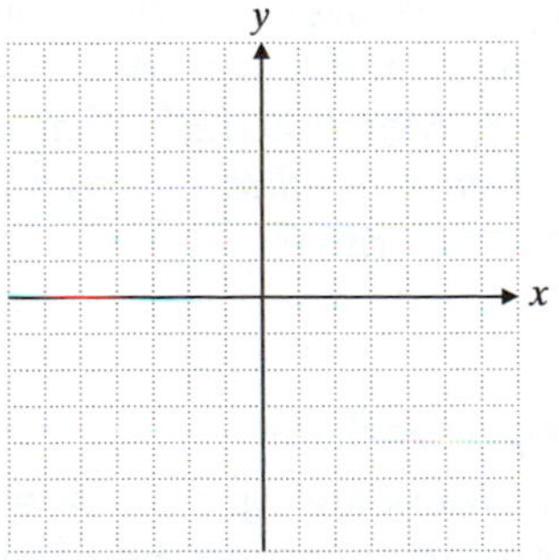

3. Find the inverse function for $f(x) = 2 - x^3$.

4. **Concept Check** If $f(x) = \frac{x - 5}{3}$, explain how you would find the inverse function.

Did You Know...

That You Can Save Money by Turning Down Your Thermostat This Winter?

ADJUST THE THERMOSTAT

Understanding the Problem:
Mark heats his home with oil. He is interested in how much he needs to budget for the heating season and how changing the settings on his thermostat will affect that amount.

Making a Plan:
Mark needs to calculate the cost of his current oil usage and potential savings.

Step 1: During the winter of 2009 Mark paid $2.95 per gallon for home heating oil. The price has since increased to $3.55 per gallon during the winter of 2011.

Task 1: *The oil company will only deliver 100 gallons of heating oil or more. Determine the cost of 100 gallons in December 2009 and November 2011.*

Task 2: *Mark knows he will need a 100-gallon heating oil delivery every month during the winter (November and December 2011 and January, February, March 2012). What can Mark expect to pay in home heating oil costs for the upcoming five months of winter?*

Making a Decision

Step 2: Mark knows that for every 1 degree change in his thermostat setting, he can see a 2% change in his utility bills.

Task 3: *Mark typically keeps his thermostat at 72°. How much will he save on his heating costs if he turns the thermostat down to 68° for the entire winter?*

Task 4: *How low would Mark have to set his thermostat to save one month's worth of heating costs ($355)?*

Applying the Situation to Your Life:
If you lower your thermostat a few degrees, you too can realize savings on your heating bill. If you are not comfortable with the temperature set lower, you can still turn the heat down when you are away at work or school, as well as when you are sleeping. A programmable thermostat that will adjust the temperature depending on the time of day would make it easy to do that.

Task 5: *At what temperature do you keep your thermostat?*

Task 6: *Calculate your savings if you lower the temperature of your thermostat during the winter.*

Chapter 10 Organizer

Topic and Procedure	Examples	You Try It

Topic and Procedure

Finding function values, p. 540

Replace the variable by the quantity inside the parentheses. Simplify the result.

Examples

If $f(x) = 2x^2 + 3x - 4$, then we have the following:

(a) $f(-2) = 2(-2)^2 + 3(-2) - 4$
$= 8 - 6 - 4 = -2$

(b) $f(a) = 2a^2 + 3a - 4$

(c) $f(a + 2) = 2(a + 2)^2 + 3(a + 2) - 4$
$= 2(a^2 + 4a + 4) + 3(a + 2) - 4$
$= 2a^2 + 8a + 8 + 3a + 6 - 4$
$= 2a^2 + 11a + 10$

(d) $f(3a) = 2(3a)^2 + 3(3a) - 4$
$= 2(9a^2) + 3(3a) - 4$
$= 18a^2 + 9a - 4$

You Try It

1. If $f(x) = -x^2 + x - 5$, find the following.
   (a) $f(1)$
   (b) $f(a)$
   (c) $f(a - 1)$
   (d) $f(4a)$

Topic and Procedure

Vertical line test, p. 547

If any vertical line intersects the graph of a relation more than once, the relation is not a function. If no such line exists, the relation is a function.

Examples

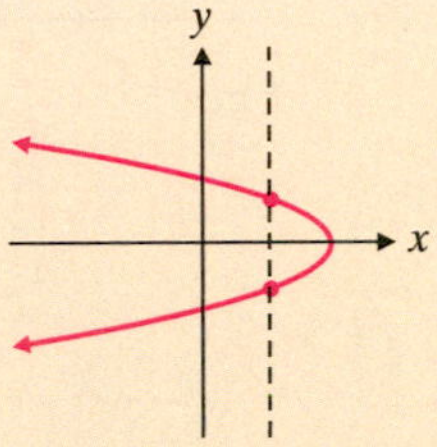

Does this graph represent a function?

No, because a vertical line intersects the curve more than once.

You Try It

2. Does this graph represent a function?

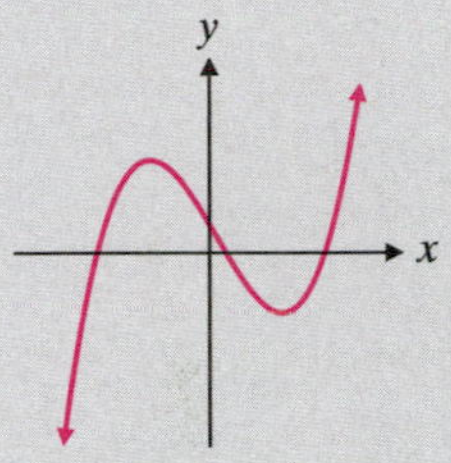

Topic and Procedure

Vertical shifts of the graph of a function, p. 549

If $k > 0$:

1. The graph of $f(x) + k$ is shifted k units *upward* from the graph of $f(x)$.
2. The graph of $f(x) - k$ is shifted k units *downward* from the graph of $f(x)$.

Examples

(a) Graph $f(x) = x^2$ and $g(x) = x^2 + 3$.

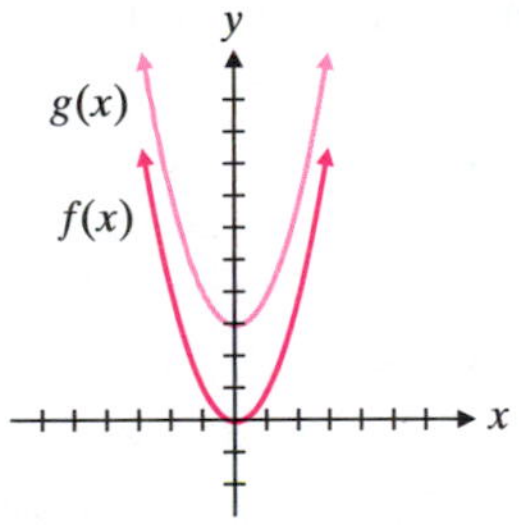

(b) Graph $f(x) = |x|$ and $g(x) = |x| - 2$.

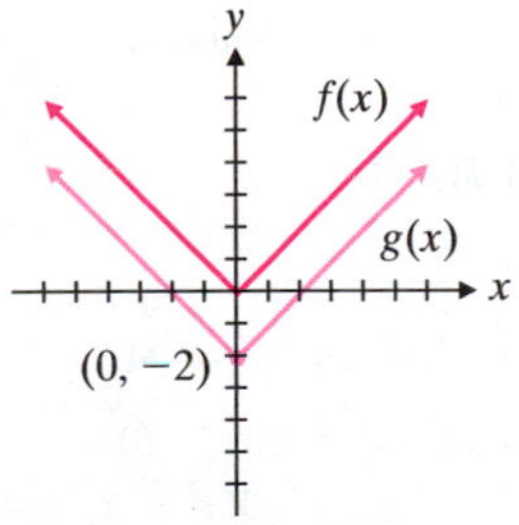

You Try It

3. **(a)** Graph $f(x) = x^2$ and $g(x) = x^2 - 2$.

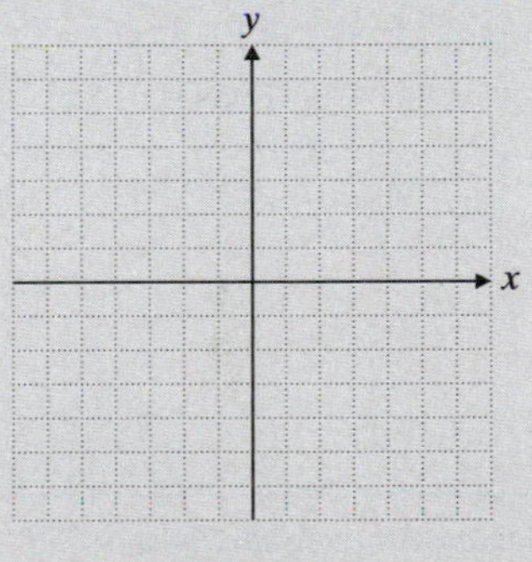

(b) Graph $f(x) = |x|$ and $g(x) = |x| + 4$.

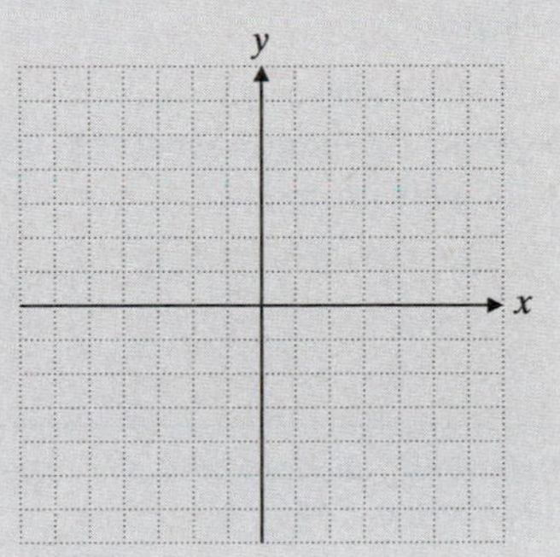

Topic and Procedure	Examples	You Try It

Horizontal shifts of the graph of a function, p. 550

If $h > 0$:

1. The graph of $f(x - h)$ is shifted h units to the *right* of the graph of $f(x)$.
2. The graph of $f(x + h)$ is shifted h units to the *left* of the graph of $f(x)$.

(a) Graph $f(x) = x^2$ and $g(x) = (x - 3)^2$.

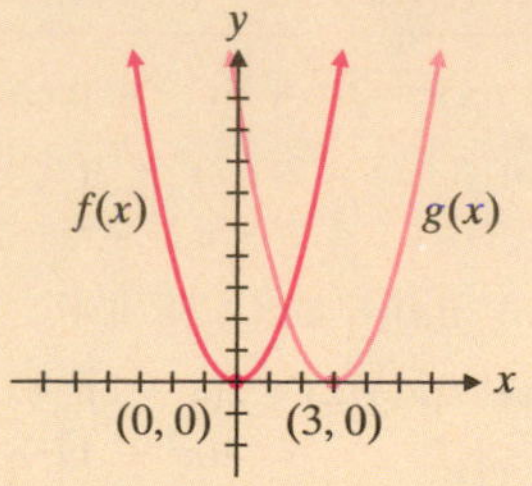

(b) Graph $f(x) = x^3$ and $g(x) = (x + 4)^3$.

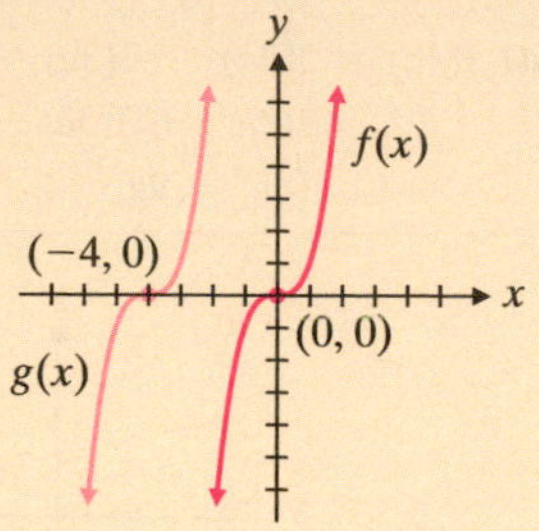

4. (a) Graph $f(x) = x^2$ and $g(x) = (x + 5)^2$.

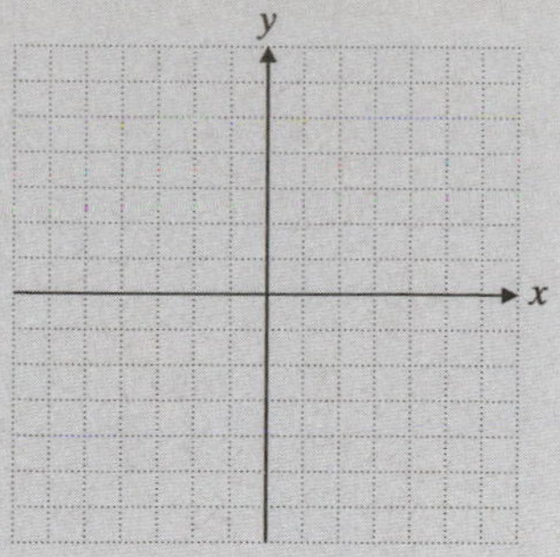

(b) Graph $f(x) = x^3$ and $g(x) = (x - 3)^3$.

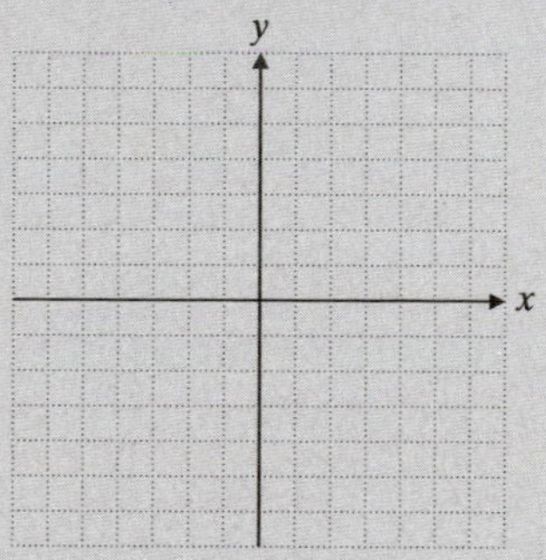

Sum, difference, product, and quotient of functions, p. 556

1. $(f + g)(x) = f(x) + g(x)$
2. $(f - g)(x) = f(x) - g(x)$
3. $(f \cdot g)(x) = f(x) \cdot g(x)$
4. $\left(\frac{f}{g}\right)(x) = \frac{f(x)}{g(x)}, g(x) \neq 0$

If $f(x) = 2x + 3$ and $g(x) = 3x - 4$, then we have the following:

(a) $(f + g)(x) = (2x + 3) + (3x - 4)$
$= 5x - 1$

(b) $(f - g)(x) = (2x + 3) - (3x - 4)$
$= 2x + 3 - 3x + 4$
$= -x + 7$

(c) $(f \cdot g)(x) = (2x + 3)(3x - 4)$
$= 6x^2 + x - 12$

(d) $\left(\frac{f}{g}\right)(x) = \frac{2x + 3}{3x - 4}, x \neq \frac{4}{3}$

5. If $f(x) = x^2 + 3x$ and $g(x) = x - 6$, find the following.

(a) $(f + g)(x)$

(b) $(f - g)(x)$

(c) $(f \cdot g)(x)$

(d) $\left(\frac{f}{g}\right)(x)$

Composition of functions, p. 558

The composition of functions f and g is written as $(f \circ g)(x) = f[g(x)]$. To find $f[g(x)]$ do the following:

1. Replace $g(x)$ by its equation.
2. Apply the formula for $f(x)$ to this expression.
3. Simplify the results.

Usually, $f[g(x)] \neq g[f(x)]$.

If $f(x) = x^2 - 5$ and $g(x) = -3x + 4$, find $f[g(x)]$ and $g[f(x)]$.

(a) $f[g(x)] = f(-3x + 4)$
$= (-3x + 4)^2 - 5$
$= 9x^2 - 24x + 16 - 5$
$= 9x^2 - 24x + 11$

(b) $g[f(x)] = g(x^2 - 5)$
$= -3(x^2 - 5) + 4$
$= -3x^2 + 15 + 4$
$= -3x^2 + 19$

6. If $f(x) = -x + 2$ and $g(x) = x^2 + 3$, find the following.

(a) $f[g(x)]$

(b) $g[f(x)]$

Relations, functions, and one-to-one functions, p. 564

A relation is any set of ordered pairs.

A function is a relation in which no ordered pairs have the same first coordinate.

A one-to-one function is a function in which no ordered pairs have the same second coordinate.

Is $\{(3, 6), (2, 8), (9, 1), (4, 6)\}$ a one-to-one function?
No, since (3, 6) and (4, 6) have the same second coordinate.

7. Is $\{(-1, 0), (2, 3), (2, 5), (0, 4)\}$ a one-to-one function?

Topic and Procedure	Examples	You Try It
Horizontal line test, p. 564 If any horizontal line intersects the graph of a function more than once, the function is not one-to-one. If no such line exists, the function is one-to-one.	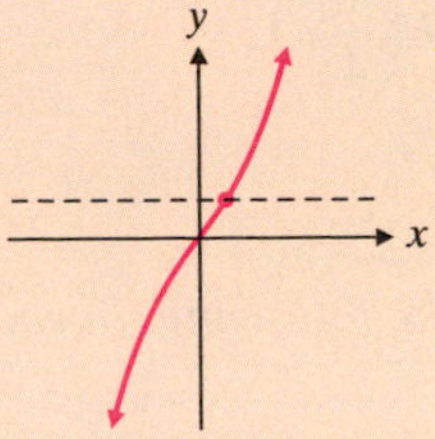 Does this graph represent a one-to-one function? Yes, any horizontal line will cross this curve at most once.	8. Does this graph represent a one-to-one function? 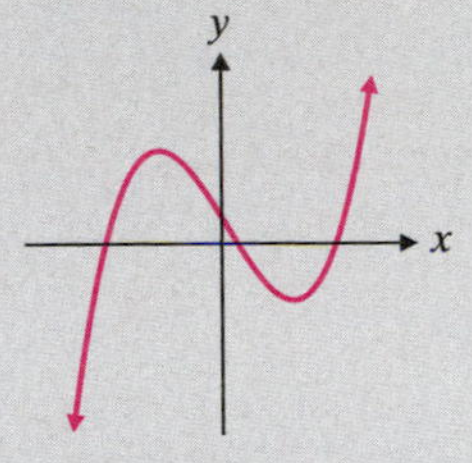
Finding the inverse of a function defined by a set of ordered pairs, p. 566 Reverse the order of the coordinates of each ordered pair from (a, b) to (b, a).	Find the inverse of $A = \{(5, 6), (7, 8), (9, 10)\}$. $A^{-1} = \{(6, 5), (8, 7), (10, 9)\}$	9. Find the inverse of $C = \{(3, -2), (0, 4), (5, 1)\}$.
Finding the inverse of a function defined by an equation, p. 567 Any one-to-one function has an inverse function. To find the inverse f^{-1} of a one-to-one function f, do the following: **1.** Replace $f(x)$ with y. **2.** Interchange x and y. **3.** Solve for y in terms of x. **4.** Replace y with $f^{-1}(x)$.	Find the inverse of $f(x) = -\frac{2}{3}x + 4$. $y = -\frac{2}{3}x + 4$ $x = -\frac{2}{3}y + 4$ $3x = -2y + 12$ $3x - 12 = -2y$ $\frac{3x - 12}{-2} = y$ $-\frac{3}{2}x + 6 = y$ $f^{-1}(x) = -\frac{3}{2}x + 6$	10. Find the inverse of $f(x) = \frac{x}{2} - 5$.
Graphing the inverse of a function, p. 568 Graph the line $y = x$ as a dashed line for reference. **1.** Graph $f(x)$. **2.** Graph $f^{-1}(x)$. The graphs of f and f^{-1} are symmetric about the line $y = x$.	$f(x) = 2x + 3$ $f^{-1}(x) = \frac{x - 3}{2}$ Graph f and f^{-1} on the same set of axes. 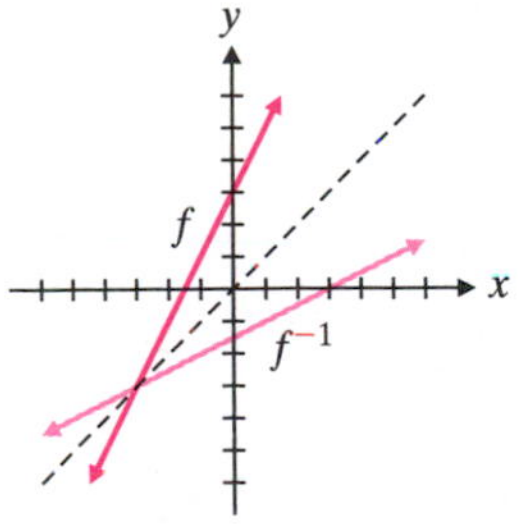	11. Graph $f(x) = \frac{x}{3} + 2$ and $f^{-1}(x) = 3x - 6$ on the same set of axes. 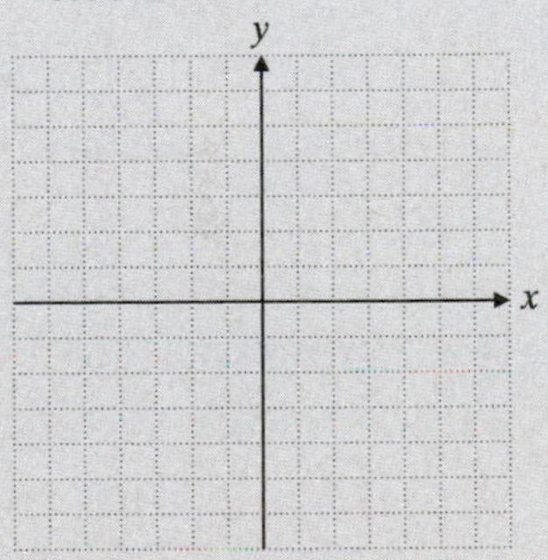

Chapter 10 Review Problems

For the function $f(x) = \frac{1}{2}x + 3$, *find the following:*

1. $f(a - 1)$

2. $f(a - 1) - f(a)$

3. $f(b^2 - 3)$

For the function $p(x) = -2x^2 + 3x - 1$, *find the following:*

4. $p(-3)$

5. $p(2a) + p(-2)$

6. $p(a + 2)$

For the function $h(x) = |2x - 1|$, *find the following:*

7. $h(0)$

8. $h\left(\frac{1}{4}a\right)$

9. $h(2a^2 - 3a)$

For the function $r(x) = \frac{3x}{x + 4}$, $x \neq -4$, *find the following. In each case, write your answers as one fraction, if possible.*

10. $r(5)$

11. $r(2a - 5)$

12. $r(3) + r(a)$

Find $\frac{f(x + h) - f(x)}{h}$ *for the following:*

13. $f(x) = 7x - 4$

14. $f(x) = 6x - 5$

15. $f(x) = 2x^2 - 5x$

Examine each of the following graphs. ***(a)*** *Does the graph represent a function?* ***(b)*** *Does the graph represent a one-to-one function?*

16.

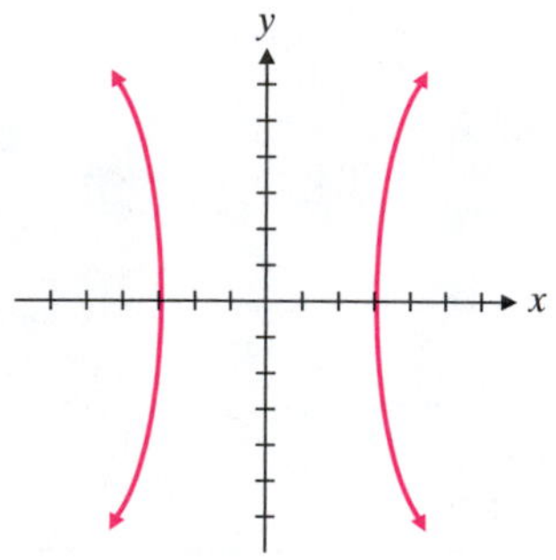

17.

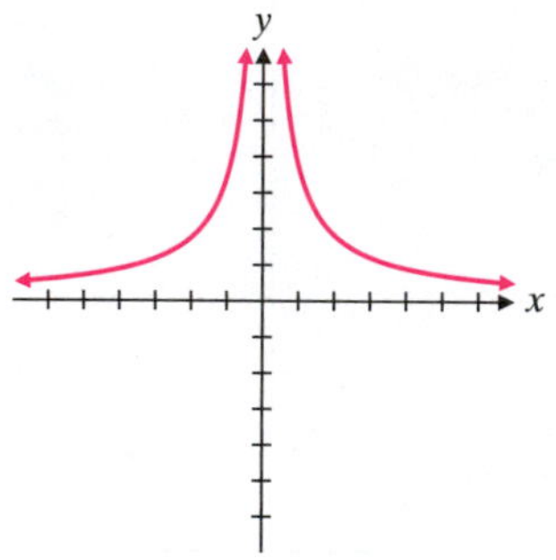

18.

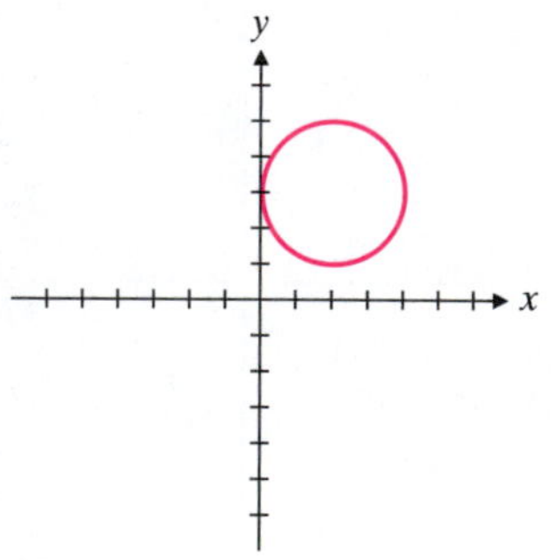

19.

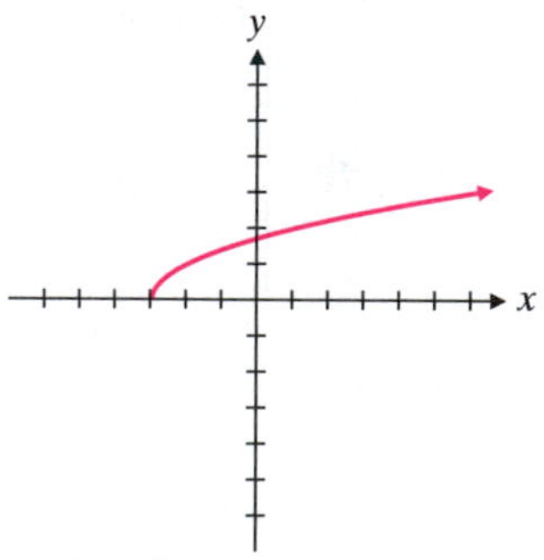

Graph each pair of functions on one set of axes.

20. $f(x) = x^2$
$g(x) = (x + 2)^2 + 4$

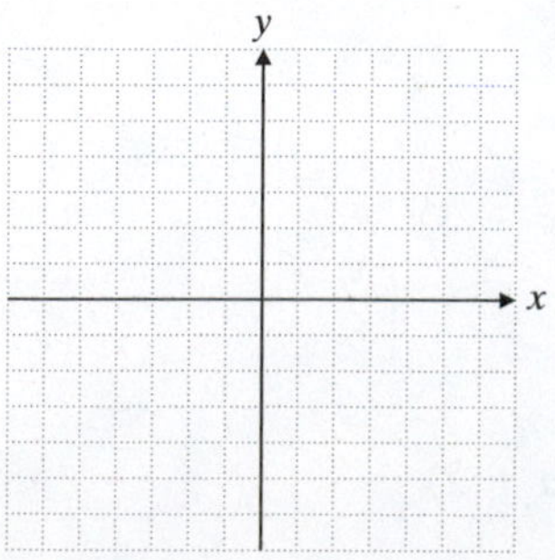

21. $f(x) = |x|$
$g(x) = |x - 4|$

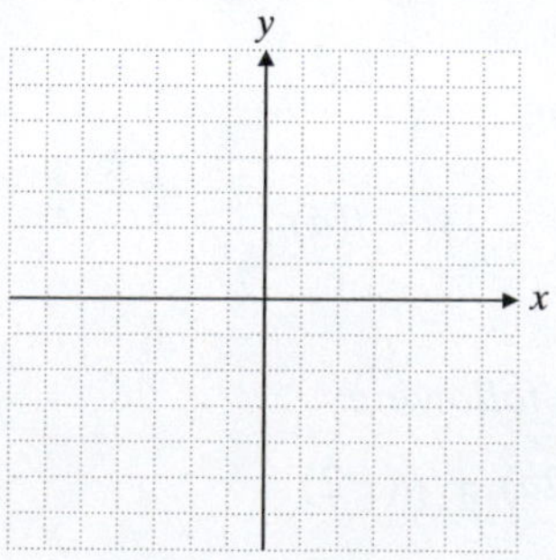

22. $f(x) = |x|$
$h(x) = |x| + 3$

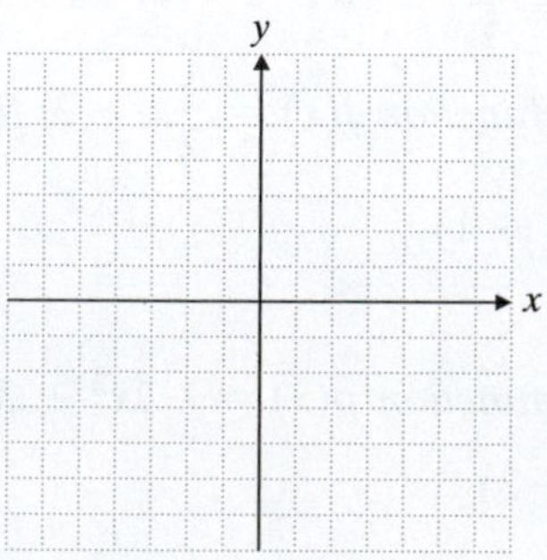

23. $f(x) = x^3$
$r(x) = (x + 3)^3 + 1$

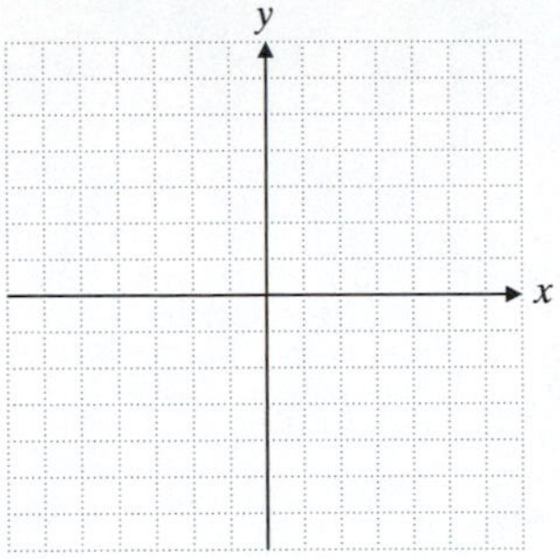

24. $f(x) = x^3$
$r(x) = (x - 1)^3 + 5$

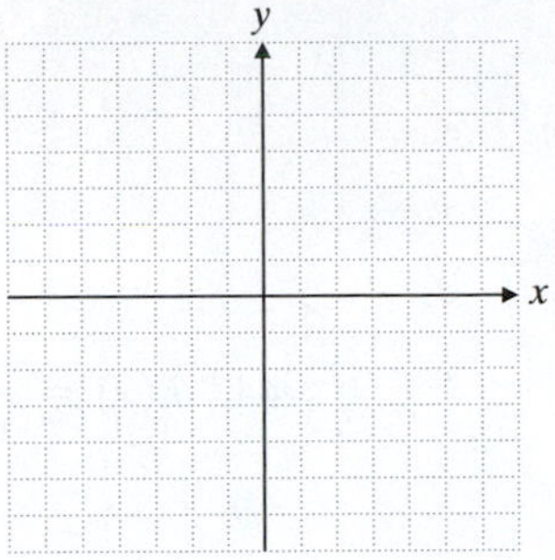

25. $f(x) = \frac{2}{x}, x \neq 0$
$r(x) = \frac{2}{x + 3} - 2, x \neq -3$

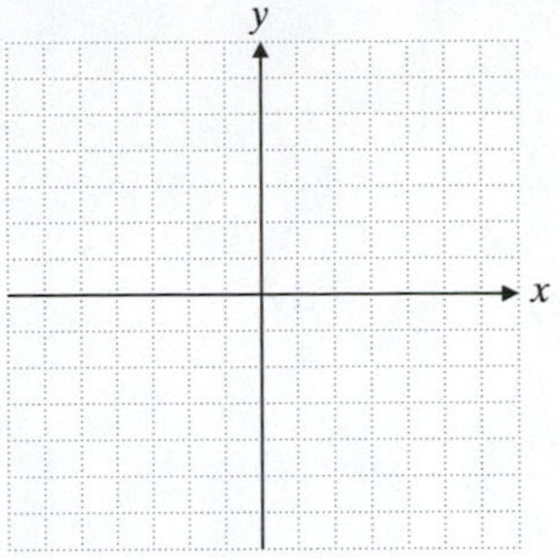

Use these functions for problems 26–35:

$$f(x) = 3x + 5; \quad g(x) = \frac{2}{x}, x \neq 0; \quad s(x) = \sqrt{x - 2}, x \geq 2;$$

$$h(x) = \frac{x + 1}{x - 4}, x \neq 4; \quad p(x) = 2x^2 - 3x + 4; \quad \text{and} \quad t(x) = -\frac{1}{2}x - 3,$$

find each of the following:

26. $(f + t)(x)$

27. $(p - f)(x)$

28. $(p - f)(2)$

29. $(fg)(x)$

30. $\left(\frac{g}{h}\right)(x)$

31. $\left(\frac{g}{h}\right)(-2)$

32. $p[f(x)]$

33. $s[p(x)]$

34. $s[p(2)]$

35. Show that $f[g(x)] \neq g[f(x)]$.

36. Given that $f(x) = \frac{2}{3}x + \frac{1}{2}$ and $f^{-1}(x) = \frac{6x - 3}{4}$, find $f^{-1}[f(x)]$.

For each set, determine **(a)** *the domain,* **(b)** *the range,* **(c)** *whether the set defines a function, and* **(d)** *whether the set defines a one-to-one function. (Hint: You can review the concept of domain and range in Section 3.5 if necessary.)*

37. $B = \{(3, 7), (7, 3), (0, 8), (0, -8)\}$

38. $A = \{(100, 10), (200, 20), (300, 30), (400, 10)\}$

39. $D = \left\{\left(\frac{1}{2}, 2\right), \left(\frac{1}{4}, 4\right), \left(-\frac{1}{3}, -3\right), \left(4, \frac{1}{4}\right)\right\}$

40. $F = \{(3, 7), (2, 1), (0, -3), (1, 1)\}$

Find the inverse of each of the following one-to-one functions.

41. $A = \left\{\left(3, \frac{1}{3}\right), \left(-2, -\frac{1}{2}\right), \left(-4, -\frac{1}{4}\right), \left(5, \frac{1}{5}\right)\right\}$

42. $f(x) = -\frac{3}{4}x + 2$

43. $g(x) = -8 - 4x$

44. $h(x) = \frac{6}{x + 5}$

45. $p(x) = \sqrt[3]{x + 1}$

46. $r(x) = x^3 + 2$

Find the inverse of each function. Graph the function and its inverse on one coordinate plane. Then on that same set of axes, graph the line $y = x$ as a dashed line.

47. $f(x) = \frac{-x - 2}{3}$

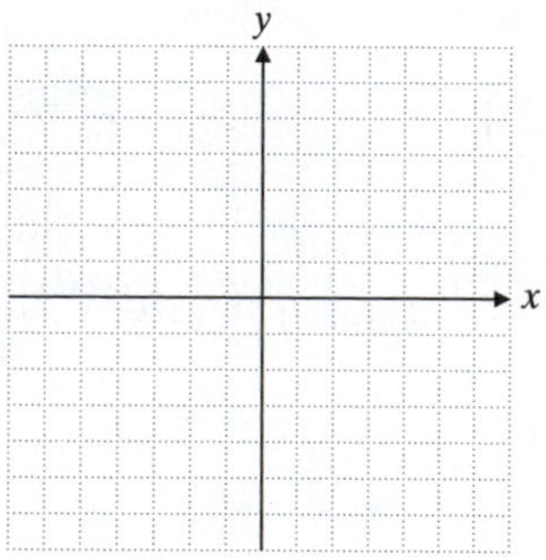

48. $f(x) = -\frac{3}{4}x + 1$

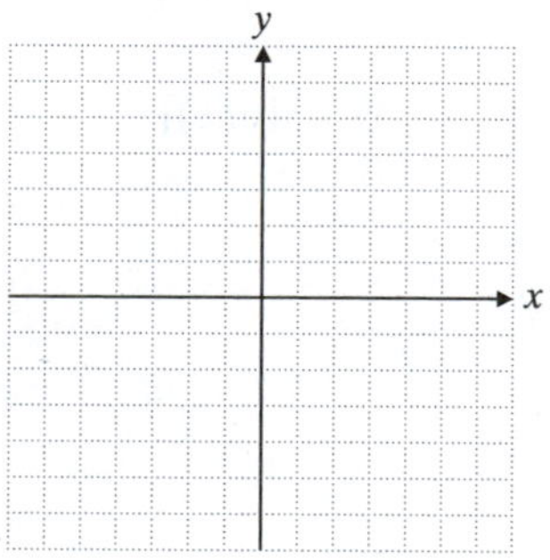

How Am I Doing? Chapter 10 Test

CHAPTER Test Prep VIDEOS MATH COACH MyMathLab® YouTube™

After you take this test read through the Math Coach on pages 581–582. Math Coach videos are available via MyMathLab and YouTube. Step-by-step test solutions in the Chapter Test Prep Videos are also available via MyMathLab and YouTube. (Search "TobeyInterAlg" and click on "Channels.")

For the function $f(x) = \frac{3}{4}x - 2$, *find the following:*

1. $f(-8)$ **2.** $f(3a)$ **3.** $f(a) - f(2)$

For the function $f(x) = 3x^2 - 2x + 4$, *find the following:*

4. $f(-6)$ MC **5.** $f(a + 1)$

6. $f(a) + f(1)$ **7.** $f(-2a) - 2$

Look at each graph below. **(a)** *Does the graph represent a function?* **(b)** *Does the graph represent a one-to-one function?*

8.

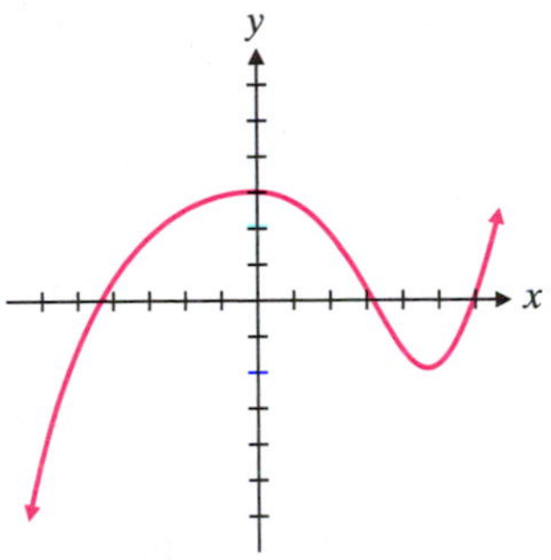

9.

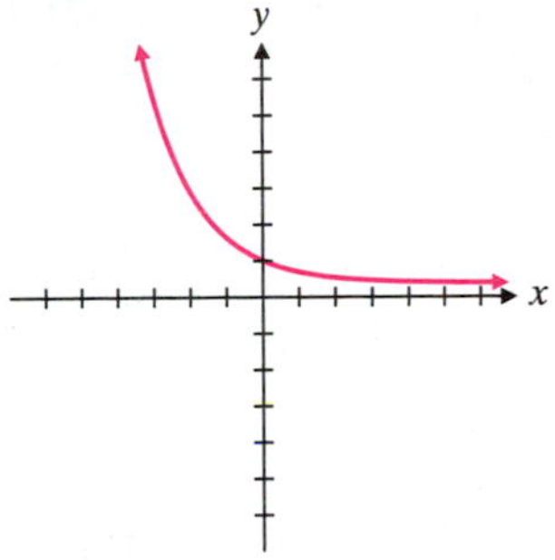

Graph each pair of functions on one coordinate plane.

MC **10.** $f(x) = x^2$
$g(x) = (x - 1)^2 + 3$

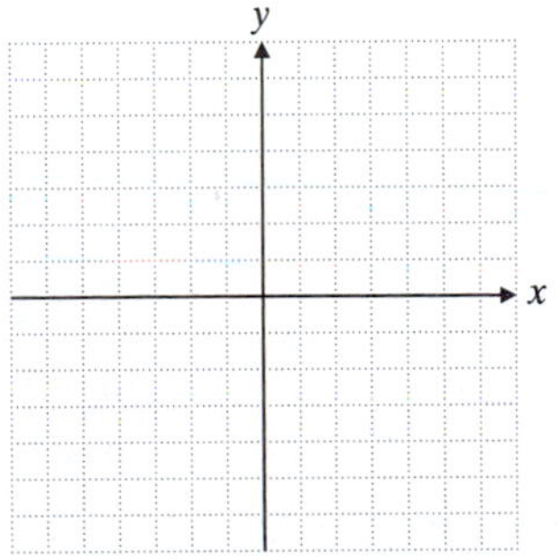

11. $f(x) = |x|$
$g(x) = |x + 1| + 2$

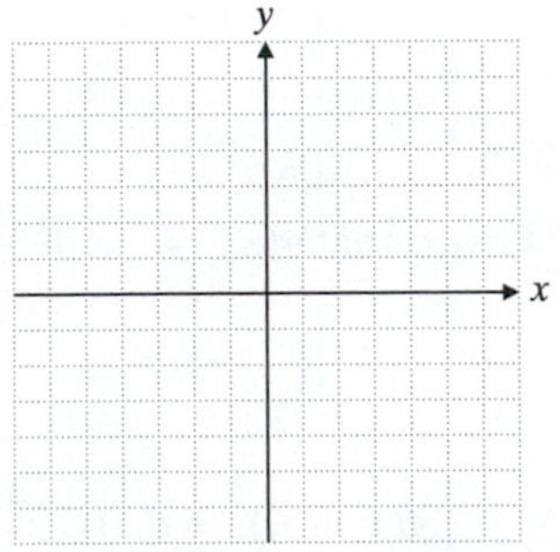

12. If $f(x) = 3x^2 - x - 6$ and $g(x) = -2x^2 + 5x + 7$, find the following:
(a) $(f + g)(x)$ **(b)** $(f - g)(x)$ **(c)** $(f - g)(-2)$

1. ____
2. ____
3. ____
4. ____
5. ____
6. ____
7. ____
8. (a) ____
(b) ____
9. (a) ____
(b) ____
10. ____
11. ____
12. (a) ____
(b) ____
(c) ____

13. (a)

(b)

(c)

14. (a)

(b)

(c)

15. (a)

(b)

16. (a)

(b)

17.

18.

19.

20.

Total Correct:

13. If $f(x) = \frac{3}{x}, x \neq 0$ and $g(x) = 2x - 1$, find the following:

(a) $(fg)(x)$ **(b)** $\left(\frac{f}{g}\right)(x)$ **(c)** $f[g(x)]$

14. If $f(x) = \frac{1}{2}x - 3$ and $g(x) = 4x + 5$, find the following:

(a) $(f \circ g)(x)$ **(b)** $(g \circ f)(x)$ **(c)** $(f \circ g)\left(\frac{1}{4}\right)$

Look at the following functions. **(a)** *Is the function one-to-one?* **(b)** *If so, find the inverse of the function.*

15. $B = \{(1, 8), (8, 1), (9, 10), (-10, 9)\}$

16. $A = \{(1, 5), (2, 1), (4, -7), (0, 7), (-1, 5)\}$

17. Determine the inverse of $f(x) = \sqrt[3]{2x - 1}$.

18. Find f^{-1}. Graph f and its inverse f^{-1} on one coordinate plane. Graph $y = x$ as a dashed line for a reference.

$$f(x) = -3x + 2$$

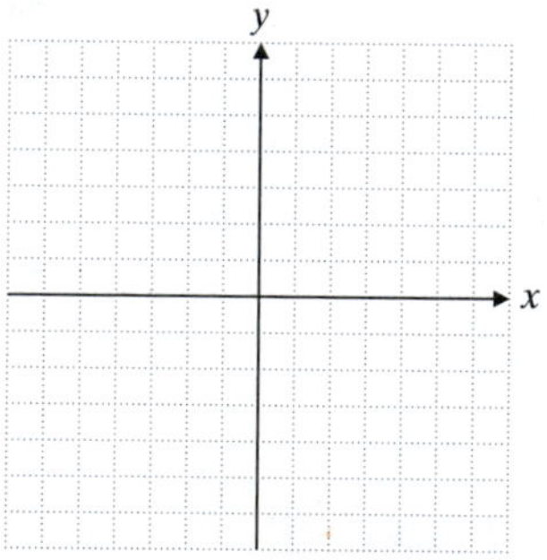

19. Given that $f(x) = \frac{3}{7}x + \frac{1}{2}$ and that $f^{-1}(x) = \frac{14x - 7}{6}$, find $f^{-1}[f(x)]$.

20. Find $\frac{f(x + h) - f(x)}{h}$ for $f(x) = 7 - 8x$.

MATH COACH

Mastering the skills you need to do well on the test.

Students often make the same types of errors when they do the Chapter 10 Test. Here are some helpful hints to keep you from making those common errors on test problems.

Using Function Notation to Evaluate Expressions—Problem 5

For the function $f(x) = 3x^2 - 2x + 4$, find $f(a + 1)$.

Helpful Hint The key idea is to replace every x by the expression $a + 1$ and then simplify the result. Use parentheses around the substitutions to avoid calculation errors.

Did you substitute $a + 1$ into the function to get $3(a + 1)^2 - 2(a + 1) + 4$?

Yes _____ No _____

Did you evaluate $(a + 1)^2$ to get $a^2 + 2a + 1$?

Yes _____ No _____

If you answered No to these questions, remember to replace every x with $a + 1$. Use parentheses to avoid calculation errors. Note that when you square a binomial, you must be sure to write down all the terms.

Next, did you simplify further to obtain $3a^2 + 6a + 3 - 2a - 2 + 4$?

Yes _____ No _____

If you answered No, remember to multiply all three terms of $a^2 + 2a + 1$ by 3. Multiply both terms of $a + 1$ by -2.

If your final step, you can collect like terms to write your answer in simplest form.

If you answered Problem 5 incorrectly, go back and rework the problem using these suggestions.

Graphing a Function with a Horizontal and Vertical Shift—Problem 10

Graph each pair of functions on one coordinate plane.

$$f(x) = x^2$$
$$g(x) = (x - 1)^2 + 3$$

Helpful Hint First graph $f(x)$. The graph of $f(x - h) + k$ is the graph of $f(x)$ shifted h units to the right and k units upward (assuming that $h > 0$ and $k > 0$).

Can you determine that the graph of $f(x) = x^2$ passes through the points $(0, 0)$, $(1, 1)$, $(-1, 1)$, $(2, 4)$, and $(-2, 4)$?

Yes _____ No _____

If you answered No, try building a table of values in which you replace x with 0, 1, -1, 2, and -2 and find the corresponding values of $f(x)$. Plot those points and connect the points with a curve to form the graph of $f(x) = x^2$.

Do you see that the function $g(x) = (x - 1)^2 + 3$ has the values of $h = 1$ and $k = 3$ when you apply the Helpful Hint?

Yes _____ No _____

Did you find that the graph of $g(x)$ is the graph of $f(x)$ shifted one unit to the right and 3 units up?

Yes _____ No _____

If you answered No to these questions, reread the Helpful Hint carefully. Notice that the values of h and k are both greater than zero.

Finish by graphing $g(x)$ on the same coordinate plane as your graph of $f(x)$.

If you answered Problem 10 incorrectly, go back and rework the problem using these suggestions.

Finding the Composition of Two Functions—Problem 14(a)

If $f(x) = \frac{1}{2}x - 3$ and $g(x) = 4x + 5$, find $(f \circ g)(x)$

Helpful Hint First rewrite $(f \circ g)(x)$ as $f[g(x)]$. Most students find this expression more logical. Then substitute $g(x)$ for the value of x in $f(x)$.

First, did you rewrite the problem as $f[g(x)] = \frac{1}{2}(4x + 5) - 3$?

Yes ____ No ____

If you answered No, substitute the expression for $g(x)$ for the value of x in the expression for $f(x)$.

Next did you simplify the resulting expression to $f[g(x)] = 2x + \frac{5}{2} - 3$?

Yes ____ No ____

If you answered No, remember that $\frac{1}{2}(4x) = 2x$ and $\frac{1}{2}(5) = \frac{5}{2}$. As your final step, combine like terms to write your expression in simplest form.

If you answered Problem 14(a) incorrectly, go back and rework the problem using these suggestions.

Finding and Graphing the Inverse of a Function—Problem 18

Given $f(x) = -3x + 2$, find f^{-1}. Graph f and its inverse f^{-1} on one coordinate plane. Graph $y = x$ as a dashed line for reference.

Helpful Hint Use the following four steps to find the inverse of a function:

1. Replace $f(x)$ with y.
2. Interchange x and y.
3. Solve for y in terms of x.
4. Replace y with $f^{-1}(x)$.

Did you substitute y for $f(x)$ to get $y = -3x + 2$ and then interchange x and y to get $x = -3y + 2$?

Yes ____ No ____

If you answered No, review the first two steps in the Helpful Hint and perform these steps again.

Did you solve the equation for y to get $y = -\frac{1}{3}x + \frac{2}{3}$?

Yes ____ No ____

If you answered No, remember to add $3y$ to each side. Next add $-x$ to each side and then divide each side of the equation by 3. As your last step in finding the inverse, replace y with $f^{-1}(x)$.

Remember to graph $f(x)$ and $f^{-1}(x)$ on the same coordinate plane. Add the graph of $y = x$ as a dashed line for reference.

If you answered Problem 18 incorrectly, go back and rework the problem using these suggestions.

Need more help? Look for section examples marked with MC to review.

CHAPTER 11

We all enjoy a beautiful lake in a lovely natural setting. However, today many lakes are contaminated by acid rain. Scientists studying this problem measure the level of acidity using a pH scale, which is based on logarithms. The material you study in this chapter will enable you to better understand some of the scientific results and recommendations about reducing the effects of acid rain.

Logarithmic and Exponential Functions

11.1 The Exponential Function

Student Learning Objectives

After studying this section, you will be able to:

1. **Graph an exponential function.**
2. **Solve elementary exponential equations.**
3. **Solve applications requiring the use of an exponential equation.**

① Graphing an Exponential Function

We have defined exponential equations $a^x = b$ for any rational number x. For example,

$$2^{-2} = \frac{1}{4},$$
$$2^{1/2} = \sqrt{2}, \text{ and}$$
$$2^{1.7} = 2^{17/10} = \sqrt[10]{2^{17}}.$$

We can also define such equations when x is an irrational number, such as π or $\sqrt{2}$. However, we will leave this definition for a more-advanced course.

We define an **exponential function** for all real values of x as follows:

DEFINITION OF EXPONENTIAL FUNCTION

The function $f(x) = b^x$, where $b > 0$, $b \neq 1$, and x is a real number, is called an **exponential function.** The number b is called the **base** of the function.

Now let's look at some graphs of exponential functions.

EXAMPLE 1 Graph $f(x) = 2^x$.

Solution We make a table of values for x and $f(x)$.

$$f(-1) = 2^{-1} = \frac{1}{2}, \quad f(0) = 2^0 = 1, \quad f(1) = 2^1 = 2$$

Verify the other values in the table below. We then draw the graph.

x	$f(x)$
-2	$\frac{1}{4}$
-1	$\frac{1}{2}$
0	1
1	2
2	4
3	8

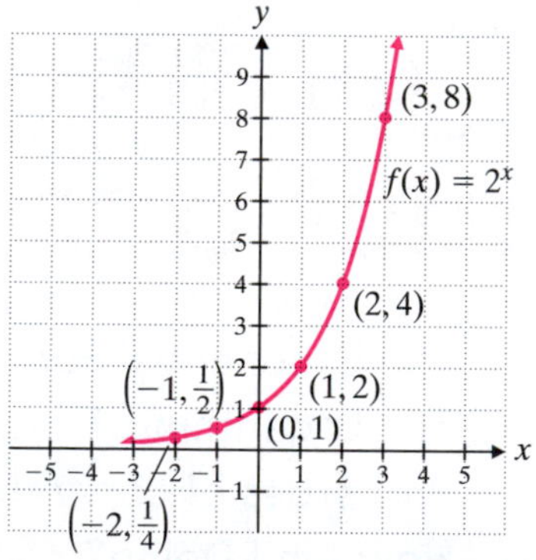

Notice how the curve of $f(x) = 2^x$ comes *very close to* the x-axis but *never touches* it. The x-axis is an **asymptote** for every exponential function. You should also notice that $f(x)$ is always positive, so the range of f is the set of all positive real numbers (whereas the domain is the set of all real numbers). When the base is greater than one, as x increases, $f(x)$ increases faster and faster (that is, the curve gets steeper).

Student Practice 1 Graph $f(x) = 3^x$.

Student Practice 1

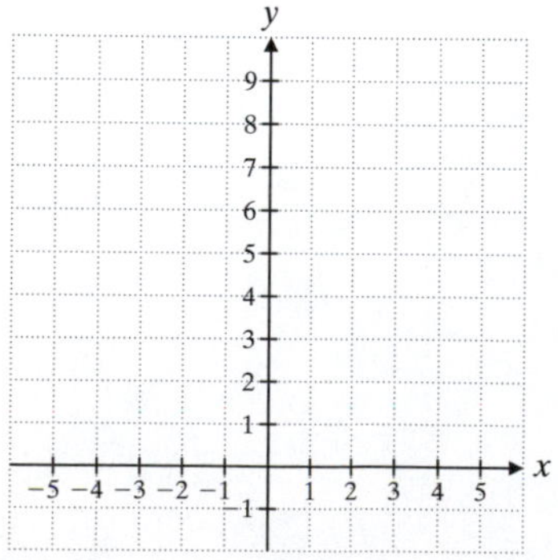

***NOTE TO STUDENT:** Fully worked-out solutions to all of the Student Practice problems can be found at the back of the text starting at page SP-1.*

EXAMPLE 2 Graph $f(x) = \left(\frac{1}{2}\right)^x$.

Solution We can write $f(x) = \left(\frac{1}{2}\right)^x$ as $f(x) = \left(\frac{1}{2}\right)^x = (2^{-1})^x = 2^{-x}$ and evaluate it for a few values of x. We then draw the graph.

x	$f(x)$
-3	8
-2	4
-1	2
0	1
1	$\frac{1}{2}$
2	$\frac{1}{4}$

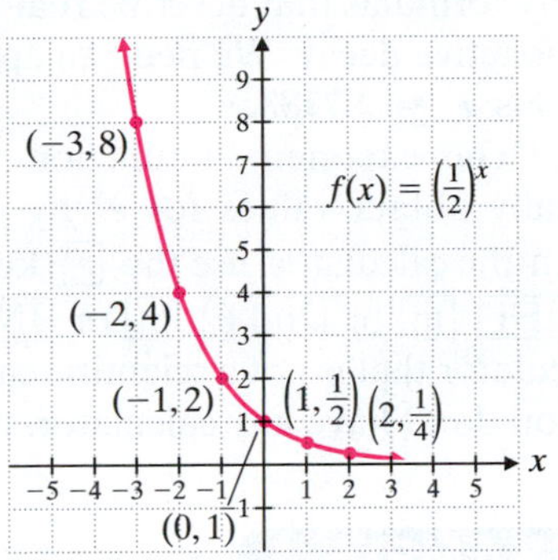

Note that as x increases, $f(x)$ decreases.

Student Practice 2 Graph $f(x) = \left(\frac{1}{3}\right)^x$.

TO THINK ABOUT: Comparing Graphs Look at the graph of $f(x) = 2^x$ in Example 1 and the graph of $f(x) = \left(\frac{1}{2}\right)^x = 2^{-x}$ in Example 2. How are the two graphs related?

EXAMPLE 3 Graph $f(x) = 3^{x-2}$.

Solution We will make a table of values for a few values of x. Then we will graph the function.

$$f(0) = 3^{0-2} = 3^{-2} = \frac{1}{3^2} = \frac{1}{9}$$
$$f(1) = 3^{1-2} = 3^{-1} = \frac{1}{3}$$
$$f(2) = 3^{2-2} = 3^0 = 1$$
$$f(3) = 3^{3-2} = 3^1 = 3$$
$$f(4) = 3^{4-2} = 3^2 = 9$$

x	$f(x)$
0	$\frac{1}{9}$
1	$\frac{1}{3}$
2	1
3	3
4	9

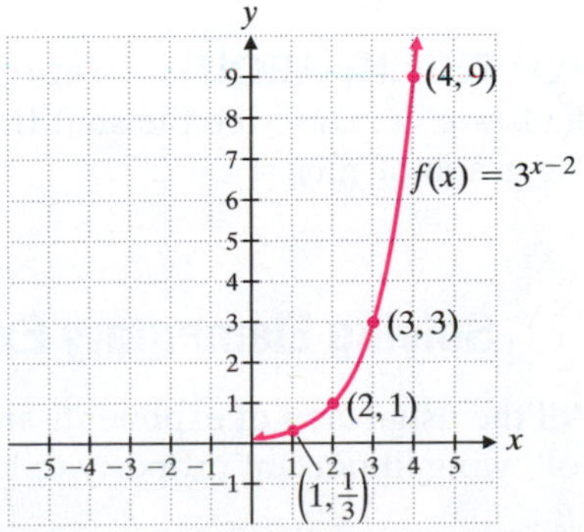

We observe that the curve is that of $f(x) = 3^x$ except that it has been shifted 2 units to the right.

Student Practice 3 Graph $f(x) = 3^{x+2}$ using the grid on p. 586.

Student Practice 2

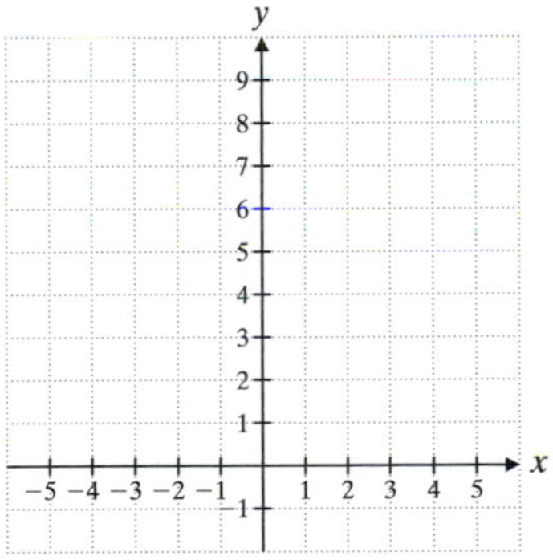

Student Practice 3

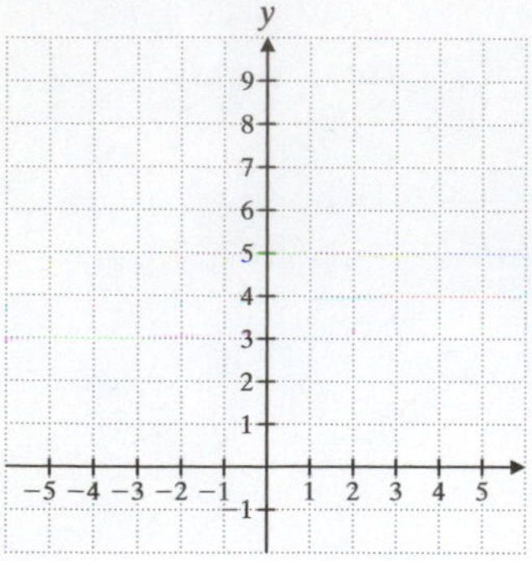

TO THINK ABOUT: Graph Shifts How is the graph of $f(x) = 3^{x+2}$ related to the graph of $f(x) = 3^x$? Without making a table of values, draw the graph of $f(x) = 3^{x+3}$. Draw the graph of $f(x) = 3^{x-3}$.

For the next example we need to discuss a special number that is denoted by the letter e. The letter e is a number like π. It is an irrational number. It occurs in many formulas that describe real-world phenomena, such as the growth of cells and radioactive decay. We need an approximate value for e to use this number in calculations: $e \approx$ **2.7183.**

The exponential function $f(x) = e^x$ is an extremely useful function. We usually obtain values for e^x by using a calculator or a computer. If you have a scientific calculator, use the $\boxed{e^x}$ key. (Many scientific calculators require you to press $\boxed{\text{SHIFT}}$ $\boxed{\ln}$ or $\boxed{\text{2nd F}}$ $\boxed{\ln}$ or $\boxed{\text{INV}}$ $\boxed{\ln}$ to obtain the operation e^x.) If you have a calculator that is not a scientific calculator, use $e \approx 2.7183$ as an approximate value. If you don't have any calculator, use Table A-2 in the appendix.

EXAMPLE 4 Graph $f(x) = e^x$.

Solution We evaluate $f(x)$ for some negative and some positive values of x. We begin with $f(-2)$. Notice that the x-column in Table A-2 has only positive values. To find the value of $f(-2) = e^{-2}$, we must locate 2 in the x-column and then read across to the value in the column under e^{-x}. Thus, we see that $f(-2) \approx 0.1353$, or 0.14 rounded to the nearest hundredth.

To find $f(2) = e^2$ on a scientific calculator, we enter 2 $\boxed{e^x}$ and obtain 7.389056099 as an approximation. (On some scientific calculators you will need to use the keystrokes 2 $\boxed{\text{2nd F}}$ $\boxed{\ln}$ or 2 $\boxed{\text{SHIFT}}$ $\boxed{\ln}$ or 2 $\boxed{\text{INV}}$ $\boxed{\ln}$.) Thus, $f(2) = e^2 \approx 7.39$ to the nearest hundredth.

x	$f(x)$
−2	0.14
−1	0.37
0	1
1	2.72
2	7.39

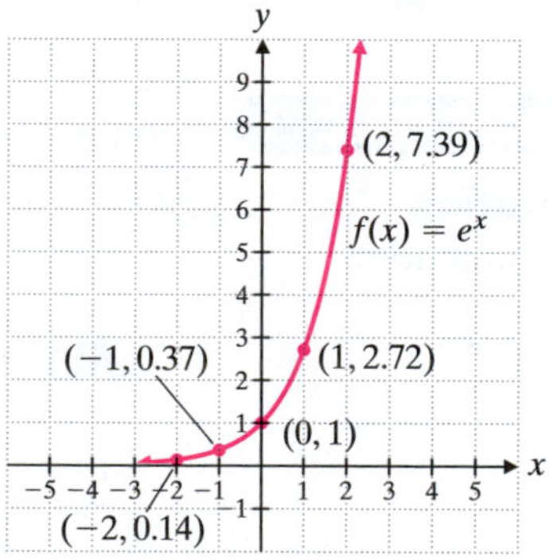

Student Practice 4 Graph $f(x) = e^{x-2}$.

Student Practice 4

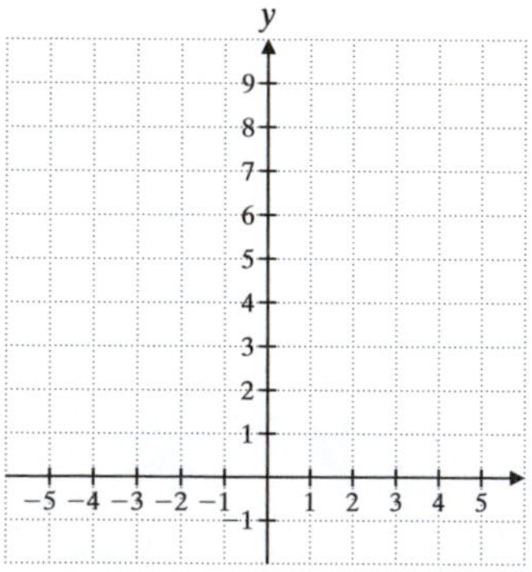

TO THINK ABOUT: Graph Shifts Look at the graphs of $f(x) = e^x$ and $f(x) = e^{x-2}$. Describe the shift that occurs. Without making a table of values, draw the graph of $f(x) = e^{x+3}$.

② Solving Elementary Exponential Equations

All the usual laws of exponents are true for exponential functions. We also have the following important property to help us solve exponential equations.

PROPERTY OF EXPONENTIAL EQUATIONS

If $b^x = b^y$, then $x = y$ for $b > 0$ and $b \neq 1$.

EXAMPLE 5 Solve. $2^x = \frac{1}{16}$

Solution To use the property of exponential equations, we must have the same base on both sides of the equation.

$$2^x = \frac{1}{16}$$

$$2^x = \frac{1}{2^4} \quad \text{Because } 2^4 = 16.$$

$$2^x = 2^{-4} \quad \text{Because } \frac{1}{2^4} = 2^{-4}.$$

$$x = -4 \quad \text{Property of exponential equations.}$$

Student Practice 5 Solve. $2^x = \frac{1}{32}$

3 Solving Applications Requiring the Use of an Exponential Equation

An exponential function can be used to solve compound interest exercises. If a principal amount P is invested at an interest rate r compounded annually, the amount of money A accumulated after t years is $A = P(1 + r)^t$.

EXAMPLE 6 If a young married couple invests \$5000 in a mutual fund that pays 16% interest compounded annually, how much will they have in 3 years?

Solution Here $P = 5000$, $r = 0.16$, and $t = 3$.

$$\begin{aligned} A &= P(1 + r)^t \\ &= 5000(1 + 0.16)^3 \\ &= 5000(1.16)^3 \\ &= 5000(1.560896) \\ &= 7804.48 \end{aligned}$$

The couple will have \$7804.48.

If you have a scientific calculator, you can find the value of $5000(1.16)^3$ immediately by using the $\boxed{\times}$ key and the $\boxed{y^x}$ key. On most scientific calculators you can use the following keystrokes.

$$5000 \ \boxed{\times}\ 1.16 \ \boxed{y^x}\ 3 \ \boxed{=} \quad 7804.48$$

Student Practice 6 If Uncle Jose invests \$4000 in a mutual fund that pays 11% interest compounded annually, how much will he have in 2 years?

Interest is often compounded quarterly or monthly or even daily. Therefore, a more useful form of the interest formula that allows for variable compounding is needed. If a principal P is invested at an annual interest rate r that is compounded n times a year, then the amount of money A accumulated after t years is

$$A = P\left(1 + \frac{r}{n}\right)^{nt}.$$

EXAMPLE 7 If we invest \$8000 in a fund that pays 15% annual interest compounded monthly, how much will we have after 6 years?

Solution In this situation $P = 8000$, $r = 15\% = 0.15$, and $n = 12$. The interest is compounded monthly or twelve times per year. Finally, $t = 6$ since the interest will be compounded for 6 years.

Continued on next page

$$\begin{aligned} A &= 8000\left(1 + \frac{0.15}{12}\right)^{(12)(6)} \\ &= 8000(1 + 0.0125)^{72} \\ &= 8000(1.0125)^{72} \\ &\approx 8000(2.445920268) \\ &\approx 19{,}567.36214 \end{aligned}$$

Rounding to the nearest cent, we obtain the answer \$19,567.36. Using a scientific calculator, we could have found the answer directly by using the following keystrokes.

8000 [×] 1.0125 [y^x] 72 [=] 19,567.36215

Depending on your calculator, your answer may contain fewer or more digits.

Student Practice 7 How much money would Collette have if she invested \$1500 for 8 years at 8% annual interest if the interest is compounded quarterly? Round your answer to the nearest cent.

Exponential functions are used to describe radioactive decay. The equation $A = Ce^{kt}$ tells us how much of a radioactive chemical element is left in a sample after a specified time.

EXAMPLE 8 The radioactive decay of the element americium 241 can be described by the equation

$$A = Ce^{-0.0016008t},$$

where C is the original amount of the element in the sample, A is the amount of the element remaining after t years, and $k = -0.0016008$, the decay constant for americium 241. If 10 milligrams (mg) of americium 241 is sealed in a laboratory container today, how much will theoretically be present in 2000 years? Round your answer to the nearest hundredth.

Solution Here $C = 10$ and $t = 2000$.

$$A = 10e^{-0.0016008(2000)} = 10e^{-3.2016}$$

Using a calculator, we have

$$A \approx 10(0.040697) = 0.40697 \approx 0.41 \text{ mg}.$$

The expression $10e^{-3.2016}$ can be found directly on some scientific calculators as follows:

10 [×] 3.2016 [+/−] [e^x] [=] 0.406970366

(Scientific calculators with no [e^x] key will require the keystrokes [INV] [ln] or [2nd F] [ln] or [SHIFT] [ln] in place of the [e^x].)

Thus, 0.41 milligrams of americium 241 would be present in 2000 years. (If you do not have a scientific calculator, use Table A-2 to find $e^{-3.2}$.)

Student Practice 8 If 20 milligrams of americium 241 is present in a sample now, how much will theoretically be present in 5000 years? Round your answer to the nearest thousandth.

Graphing Calculator

Exploration

Graph the function $f(t) = 10e^{-0.0016008t}$ from Example 8 for $t = 0$ to $t = 100$ years. Now graph the function for $t = 0$ to $t = 2000$ years. What significant change is there in the two graphs? From the graphs, estimate a value of t for which $f(t) = 5.0$. (Round your value of t to the nearest hundredth.)

11.1 Exercises

MyMathLab®

Watch the videos in MyMathLab

Download the MyDashBoard App

Verbal and Writing Skills, Exercises 1 and 2

1. An exponential function is a function of the form ____________________________________.

2. The irrational number e is a number that is approximately equal to ________. (Give your answer with four decimal places.)

Graph each function.

3. $f(x) = 3^x$

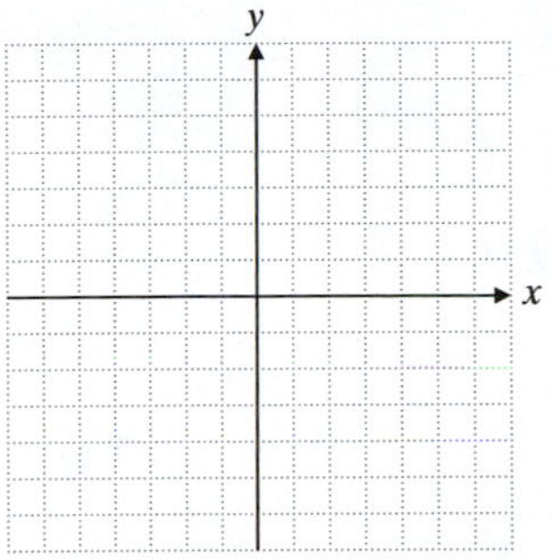

4. $f(x) = 2^x$

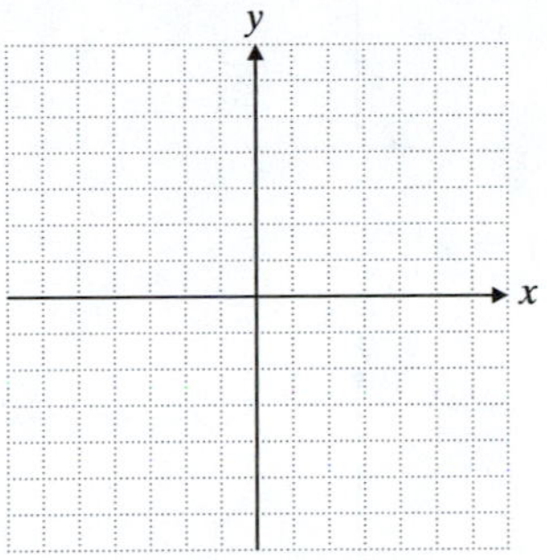

5. $f(x) = 2^{-x}$

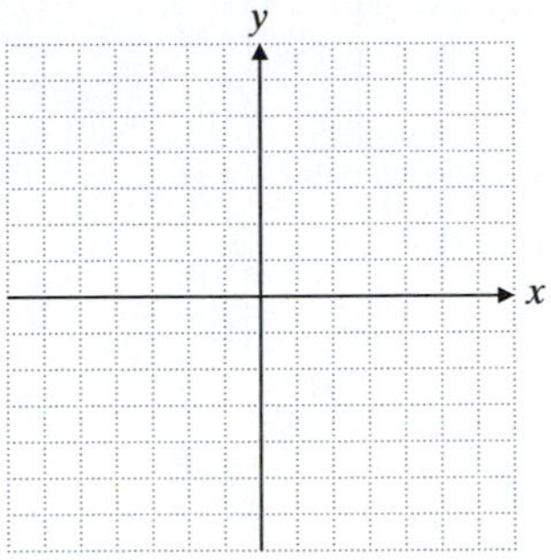

6. $f(x) = 5^{-x}$

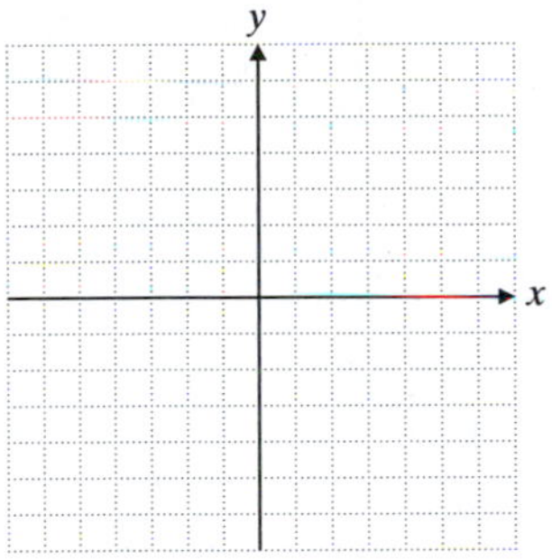

7. $f(x) = 3^{-x}$

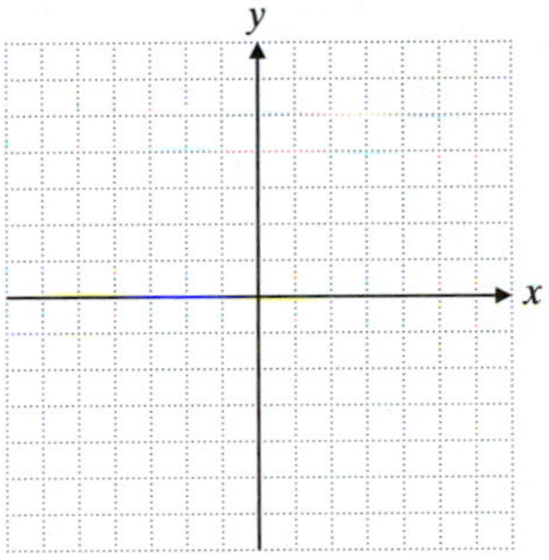

8. $f(x) = 4^{-x}$

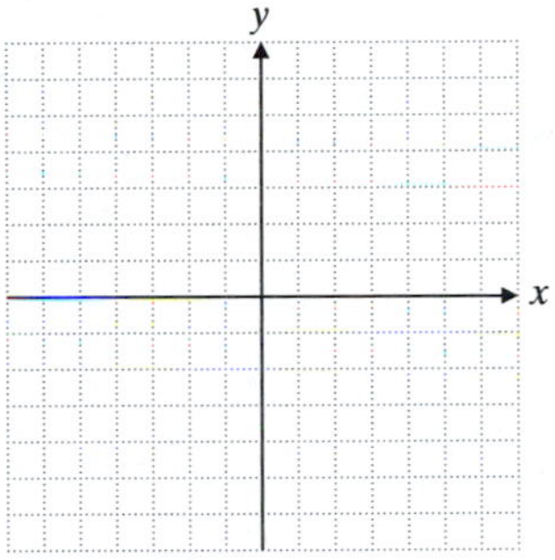

9. $f(x) = 2^{x+3}$

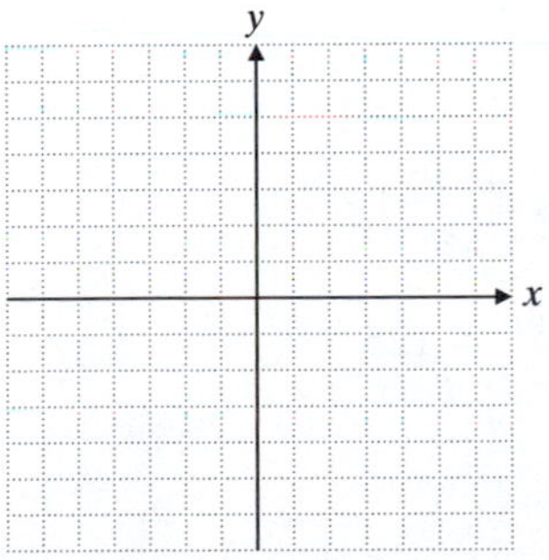

10. $f(x) = 2^{x+2}$

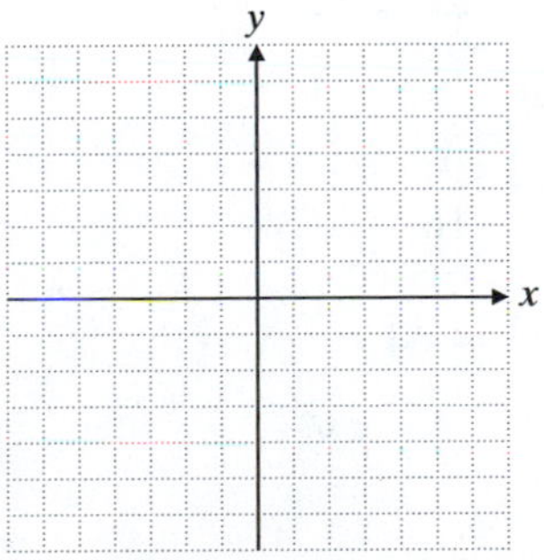

11. $f(x) = 3^{x-3}$

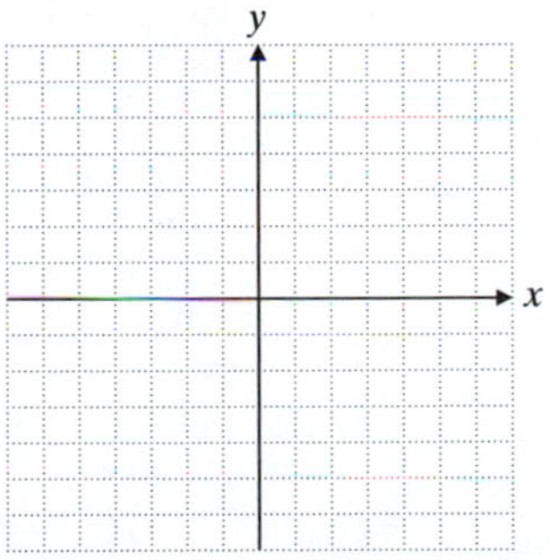

12. $f(x) = 3^{x-1}$

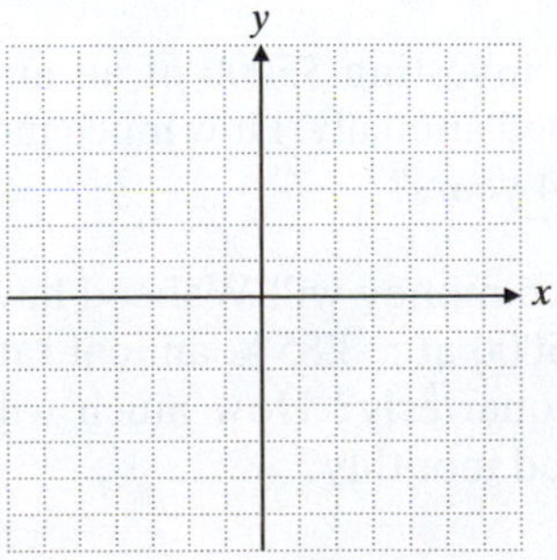

13. $f(x) = 2^x + 2$

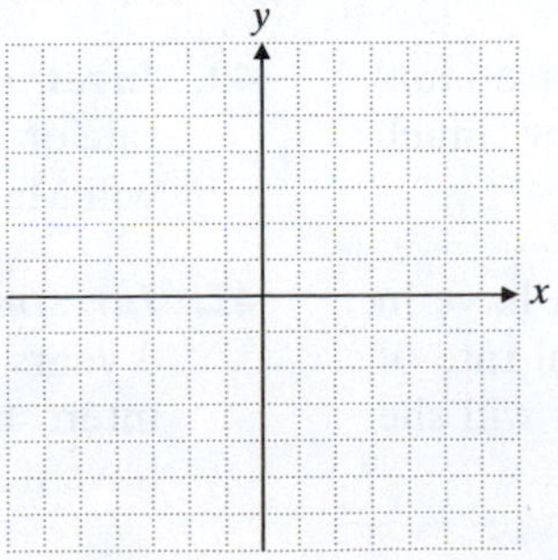

14. $f(x) = 2^x - 3$

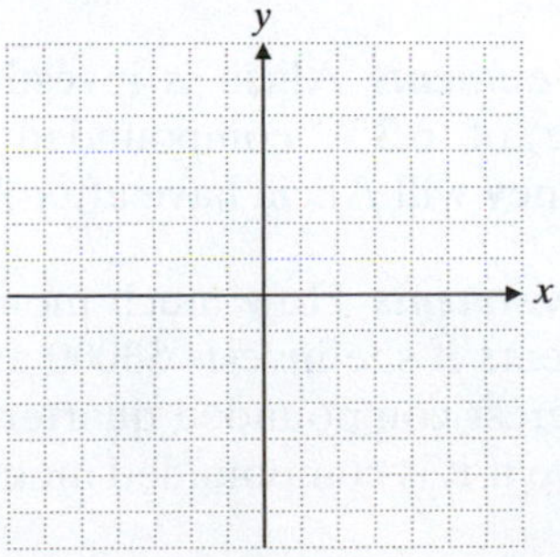

Graph each function. Use a calculator or Table A-2.

15. $f(x) = e^{x-1}$

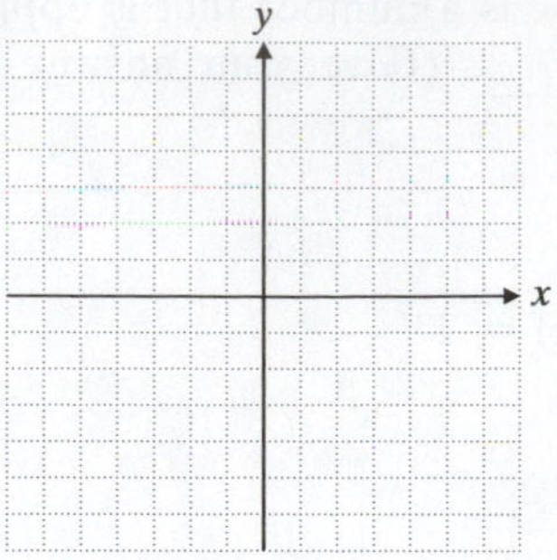

16. $f(x) = e^{x+1}$

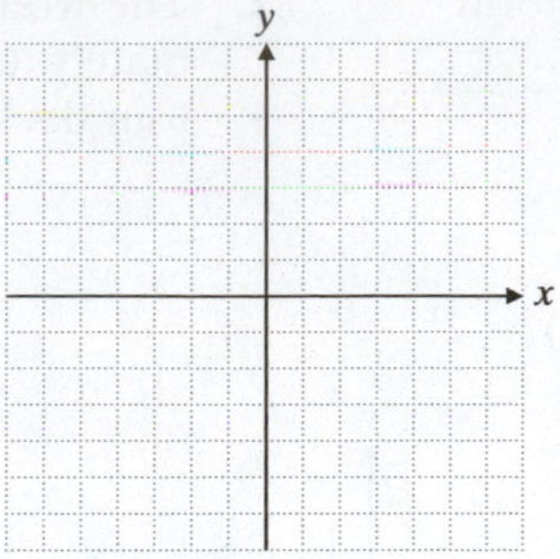

17. $f(x) = 2e^{x}$

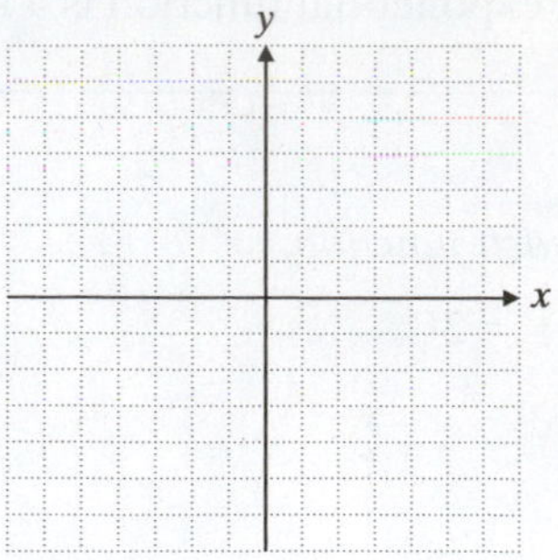

18. $f(x) = 3e^{x}$

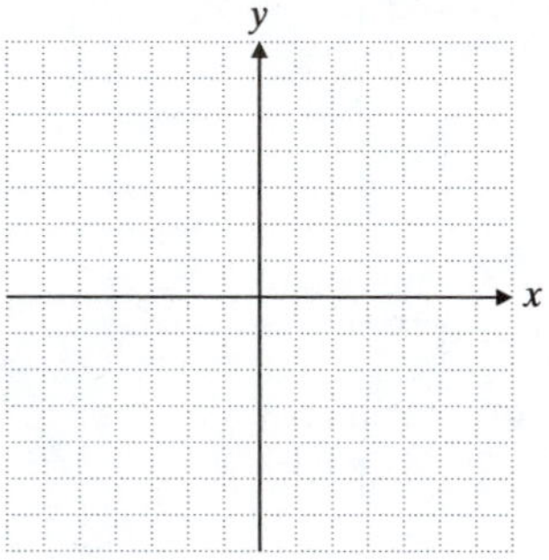

19. $f(x) = e^{1-x}$

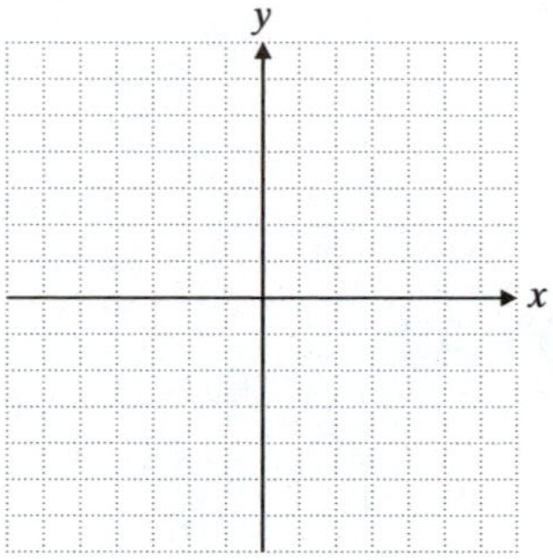

20. $f(x) = e^{2-x}$

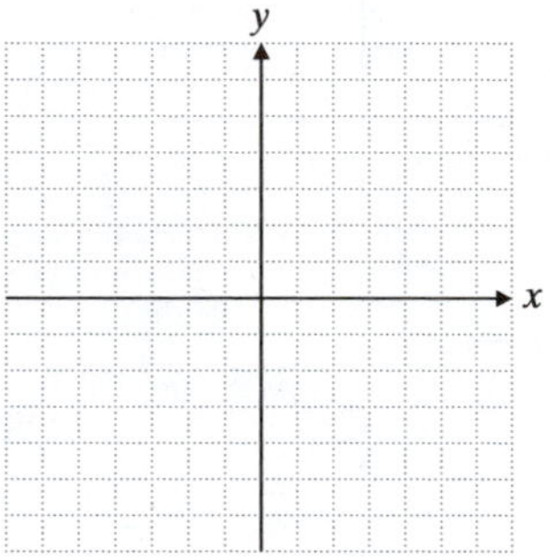

Solve for x.

21. $2^x = 4$

22. $2^x = 8$

23. $2^x = 1$

24. $2^x = 2$

25. $2^x = \frac{1}{2}$

26. $2^x = \frac{1}{64}$

27. $3^x = 81$

28. $3^x = 27$

29. $3^x = 1$

30. $3^x = 243$

31. $3^{-x} = \frac{1}{9}$

32. $3^{-x} = \frac{1}{3}$

33. $4^x = 256$

34. $4^{-x} = \frac{1}{16}$

35. $5^{x+1} = 125$

36. $2^{x-3} = 128$

37. $8^{3x-1} = 64$

38. $5^{4x+1} = 125$

To solve exercises 39–42, use the interest formula $A = P\left(1 + \frac{r}{n}\right)^{nt}$. *Round your answers to the nearest cent.*

39. ***Investments*** Alicia is investing \$2000 at an annual rate of 6.3% compounded annually. How much money will Alicia have after 3 years?

40. ***Investments*** Manza is investing \$5000 at an annual rate of 7.1% compounded annually. How much money will Manza have after 4 years?

41. ***Investments*** How much money will Isabela have in 6 years if she invests \$3000 at a 3.2% annual rate of interest compounded quarterly? How much will she have if it is compounded monthly?

42. ***Investments*** How much money will Waheed have in 3 years if he invests \$5000 at a 3.85% annual rate of interest compounded quarterly? How much will he have if it is compounded monthly?

43. ***Bacteria Culture*** The number of bacteria in a culture is given by $B(t) = 4000(2^t)$, where t is the time in hours. How many bacteria will grow in the culture in the first 3 hours? In the first 9 hours?

44. ***College Tuition*** Suppose that the cost of a college education is increasing 4% per year. The equation $C(t) = P(1.04)^t$ forecasts the tuition cost t years from now and is based on the present cost P in dollars. How much will a college now charging \$3000 for tuition charge in 10 years? How much will a college now charging \$12,000 for tuition charge in 15 years?

45. ***Diving Depth*** U.S. Navy divers off the coast of Nantucket are searching for the wreckage of an old World War II–era submarine. They have found that if the water is relatively clear and the surface is calm, the ocean filters out 18% of the sunlight for each 4 feet they descend. How much sunlight is available at a depth of 20 feet? The divers need to use underwater spotlights when the amount of sunlight is less than 10%. Will they need spotlights when working at a depth of 48 feet?

46. ***Sewer Systems*** The city of Manchester just put in a municipal sewer to solve an underground water contamination problem, and many homeowners would like to have sewer lines connected to their homes. It is expected that each year the number of homeowners who use their own private septic tanks rather than the public sewer system will decrease by 8%. What percentage of people will still be using their private septic tanks in 5 years? The city feels that the underground water contamination problem will be solved when the number of homeowners still using septic tanks is less than 10%. Will that goal be achieved in the next 25 years?

Use an exponential equation to solve each problem. Round your answers to the nearest hundredth.

47. ***Radium Decay*** The radioactive decay of radium 226 can be described by the equation $A = Ce^{-0.0004279t}$, where C is the original amount of radium and A is the amount of radium remaining after t years. If 6 milligrams of radium are sealed in a container now, how much radium will be in the container after 1000 years?

48. ***Radon Decay*** The radioactive decay of radon 222 can be described by the equation $A = Ce^{-0.1813t}$, where C is the original amount of radon and A is the amount of radon after t days. If 1.5 milligrams are in a laboratory container today, how much was there in the container 10 days ago?

Atmospheric Pressure *Use the following information for exercises 49 and 50. The atmospheric pressure measured in pounds per square inch is given by the equation* $P = 14.7e^{-0.21d}$, *where d is the distance in miles above sea level. Round your answers to the nearest hundredth.*

49. What is the pressure in pounds per square inch on an American Airlines jet plane flying 10 miles above sea level?

50. What is the pressure in pounds per square inch experienced by a man on a Colorado mountain at 2 miles above sea level?

Major League Baseball Salaries *The average salary of an MLB player increased exponentially between 1995 and 2010. The average annual salary A (in millions) can be approximated by the equation* $A = 1.191e^{0.072t}$, *where t is the number of years since 1995. Round your answers to the nearest whole number.* (*Source:* www.baseball-almanac.com)

51. Using the given equation, determine the average salary in 1995. In 2003. What was the percent increase from 1995 to 2003?

52. Using the given equation, determine the average salary in 2000. In 2010. What was the percent increase from 2000 to 2010?

World Population *Since the year 1750, the population of the world has been growing exponentially. The following table and graph contain population data for selected years and show a pattern of significant increases.*

Year	1750	1800	1850	1900	1950	2000	2050
Approximate World Population in Billions	0.8	0.9	1.2	1.61	2.5	6.07	9

*estimated
Source: www.census.gov

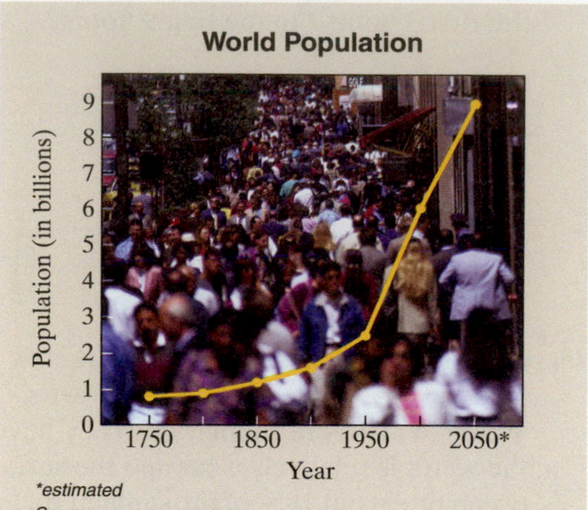

*estimated
Source: www.census.gov

53. Based on the graph, in approximately what year did the world population reach three billion people?

54. Based on the graph, what was the approximate world population in 1900?

55. The growth rate of the world population peaked in 1963 at 2.1% per year, and growth continued at that rate through 1970. If that rate would have continued from 2000 to 2025, what would have been the world population in 2025? According to the graph, what will the population be in 2025?

56. Some leaders have reported that the growth rate of the world population in 2010 was 1.14% but continues to decrease. If this rate were to continue from 2000 to 2050, what would be the population in 2050? According to the graph, what will the population actually be in 2050?

Cumulative Review *Solve for x.*

57. **[2.1.1]** $5 - 2(3 - x) = 2(2x + 5) + 1$

58. **[2.1.1]** $\frac{7}{12} + \frac{3}{4}x + \frac{5}{4} = -\frac{1}{6}x$

Quick Quiz 11.1

1. Graph $f(x) = 2^{x+4}$.

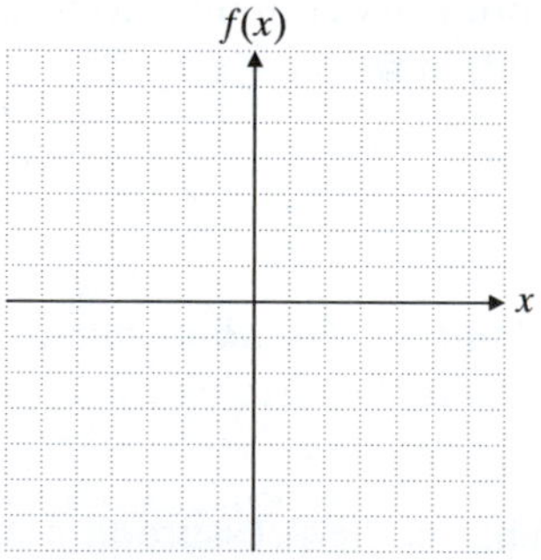

2. Dwayne invested $5500 at an annual rate of 4% compounded semiannually. How much money will Dwayne have after 6 years?

3. Solve for x. $3^{x+2} = 81$

4. **Concept Check** Explain how you would solve $4^{-x} = \frac{1}{64}$.

11.2 The Logarithmic Function

Student Learning Objectives

After studying this section, you will be able to:

1. **Write exponential equations in logarithmic form.**
2. **Write logarithmic equations in exponential form.**
3. **Solve elementary logarithmic equations.**
4. **Graph a logarithmic function.**

Logarithms were invented about 400 years ago by the Scottish mathematician John Napier. Napier's amazing invention reduced complicated exercises to simple subtraction and addition. Astronomers quickly saw the immense value of logarithms and began using them. The work of Johannes Kepler, Isaac Newton, and others would have been much more difficult without logarithms.

The most important thing to know for this chapter is that a logarithm is an exponent. In Section 11.1 we solved the equation $2^x = 8$. We found that $x = 3$. The question we faced was, "To what power do we raise 2 to get 8?" The answer was 3. Mathematicians have to solve this type of problem so often that we have invented a short-hand notation for asking the question. Instead of asking, "To what power do we raise 2 to get 8?" we say instead, "What is $\log_2 8$?" Both questions mean the same thing.

Now suppose we had a general equation $x = b^y$ and someone asked, "To what power do we raise b to get x?" We would abbreviate this question by asking, "What is $\log_b x$?" Thus, we see that $y = \log_b x$ is an equivalent form of the equation $x = b^y$.

The key concept you must remember is that a logarithm is an exponent. We write $\log_b x = y$ to mean that b to the power y is x. y is the exponent.

DEFINITION OF LOGARITHM

The **logarithm,** base b, of a *positive* number x is the power (exponent) to which the base b must be raised to produce x. That is, $y = \log_b x$ is the same as $x = b^y$, where $b > 0$ and $b \neq 1$.

Often you will need to convert logarithmic statements to exponential statements, and vice versa, to solve equations.

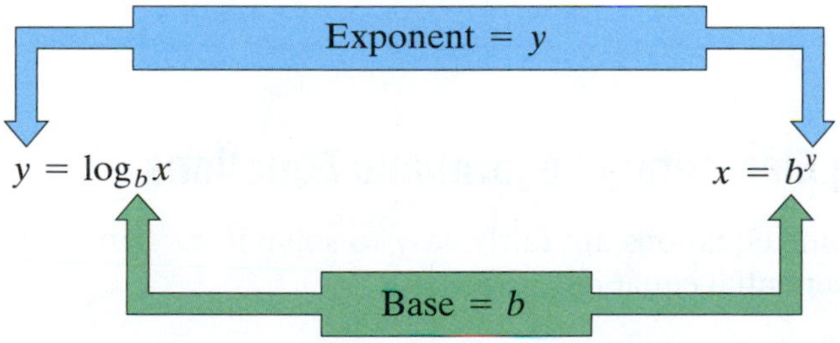

① Writing Exponential Equations in Logarithmic Form

We begin by converting exponential statements to logarithmic statements.

EXAMPLE 1 Write in logarithmic form.

(a) $81 = 3^4$ **(b)** $\frac{1}{100} = 10^{-2}$

Solution We use the fact that $x = b^y$ is equivalent to $\log_b x = y$.

(a) $81 = 3^4$

Here $x = 81$, $b = 3$, and $y = 4$. So $4 = \log_3 81$.

(b) $\frac{1}{100} = 10^{-2}$

Here $x = \frac{1}{100}$, $b = 10$, and $y = -2$. So $-2 = \log_{10}\left(\frac{1}{100}\right)$.

Continued on next page

NOTE TO STUDENT: ***Fully worked-out solutions to all of the Student Practice problems can be found at the back of the text starting at page SP-1.***

Student Practice 1 Write in logarithmic form.

(a) $49 = 7^2$ **(b)** $\frac{1}{64} = 4^{-3}$

② Writing Logarithmic Equations in Exponential Form

If we have an equation with a logarithm in it, we can write it in the form of an exponential equation. This is a very important skill. Carefully study the following example.

EXAMPLE 2 Write in exponential form.

(a) $2 = \log_5 25$ **(b)** $-4 = \log_{10}\left(\frac{1}{10{,}000}\right)$

Solution

(a) $2 = \log_5 25$

Here $y = 2$, $b = 5$, and $x = 25$. Thus, since $x = b^y$, $25 = 5^2$.

(b) $-4 = \log_{10}\left(\frac{1}{10{,}000}\right)$

Here $y = -4$, $b = 10$, and $x = \frac{1}{10{,}000}$. So $\frac{1}{10{,}000} = 10^{-4}$.

Student Practice 2 Write in exponential form.

(a) $3 = \log_5 125$ **(b)** $-2 = \log_6\left(\frac{1}{36}\right)$

③ Solving Elementary Logarithmic Equations

Many logarithmic equations are fairly easy to solve if we first convert them to an equivalent exponential equation.

EXAMPLE 3 Solve for the variable.

(a) $\log_5 x = -3$ **(b)** $\log_a 16 = 4$

Solution

(a) $5^{-3} = x$

$\frac{1}{5^3} = x$

$\frac{1}{125} = x$

(b) $a^4 = 16$

$a^4 = 2^4$

$a = 2$

Student Practice 3 Solve for the variable.

(a) $\log_b 125 = 3$ **(b)** $\log_{1/2} 32 = x$

With this knowledge we have the ability to solve an additional type of exercise.

EXAMPLE 4 Evaluate. $\log_3 81$

Solution Now, what exactly is the exercise asking for? It is asking, "To what power must we raise 3 to get 81?" Since we do not know the power, we call it x. We have

$\log_3 81 = x$

$81 = 3^x$ Write an equivalent exponential equation.

$3^4 = 3^x$ Write 81 as 3^4.

$x = 4$ If $b^x = b^y$, then $x = y$ for $b > 0$ and $b \neq 1$.

Thus, $\log_3 81 = 4$.

Student Practice 4 Evaluate. $\log_{10} 0.1$

④ Graphing a Logarithmic Function

We found in Chapter 10 that the graphs of a function and its inverse have an interesting property. They are symmetric to one another with respect to the line $y = x$. We also found in Chapter 10 that the procedure for finding the inverse of a function is to interchange the x and y variables. For example, $y = 2x + 3$ and $x = 2y + 3$ are inverse functions. In similar fashion, $y = 2^x$ and $x = 2^y$ are inverse functions. Another way to write $x = 2^y$ is the logarithmic equation $y = \log_2 x$. Thus, the logarithmic function $y = \log_2 x$ is the **inverse** of the exponential function $y = 2^x$. If we graph the function $y = 2^x$ and $y = \log_2 x$ on the same set of axes, the graph of one is the reflection of the other about the line $y = x$.

EXAMPLE 5 Graph $y = \log_2 x$.

Solution If we write $y = \log_2 x$ in exponential form, we have $x = 2^y$. We make a table of values and graph the function $x = 2^y$.

In each case, we pick a value of y as a first step.

If $y = -2$, $\quad x = 2^y = 2^{-2} = \dfrac{1}{2^2} = \dfrac{1}{4}$.

If $y = -1$, $\quad x = 2^{-1} = \dfrac{1}{2}$.

If $y = 0$, $\quad x = 2^0 = 1$.

If $y = 1$, $\quad x = 2^1 = 2$.

If $y = 2$, $\quad x = 2^2 = 4$.

x	y
$\frac{1}{4}$	-2
$\frac{1}{2}$	-1
1	0
2	1
4	2

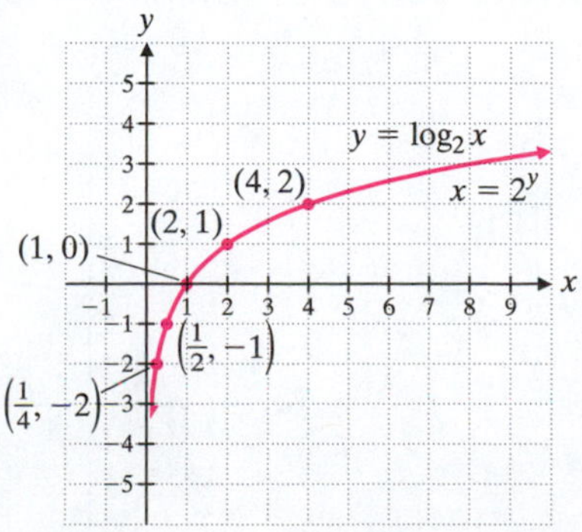

Continued on next page

Student Practice 5 Graph $y = \log_{1/2} x$.

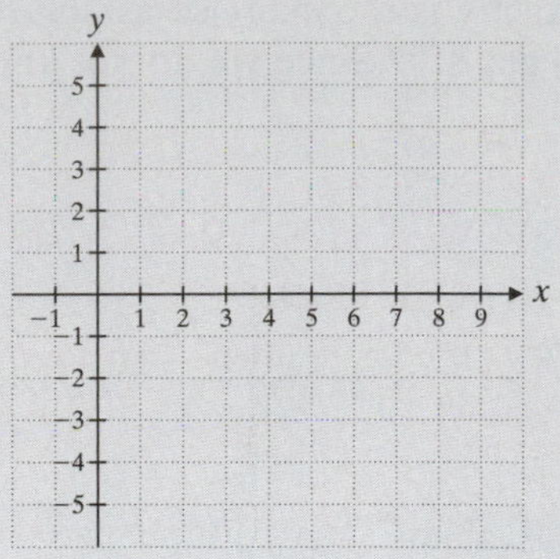

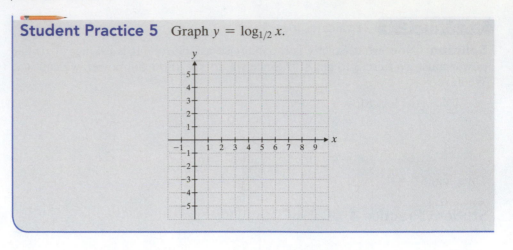

$f(x) = a^x$ and $f(x) = \log_a x$ are inverse functions. As such they have all the properties of inverse functions. We will review a few of these properties as we study the graphs of two inverse functions, $y = 2^x$ and $y = \log_2 x$.

EXAMPLE 6 Graph $y = \log_2 x$ and $y = 2^x$ on the same set of axes.

Solution Make a table of values (ordered pairs) for each equation. Then draw each graph.

$y = 2^x$

x	y
-1	$\frac{1}{2}$
0	1
1	2
2	4

$y = \log_2 x$

x	y
$\frac{1}{2}$	-1
1	0
2	1
4	2

Coordinates of ordered pairs are reversed

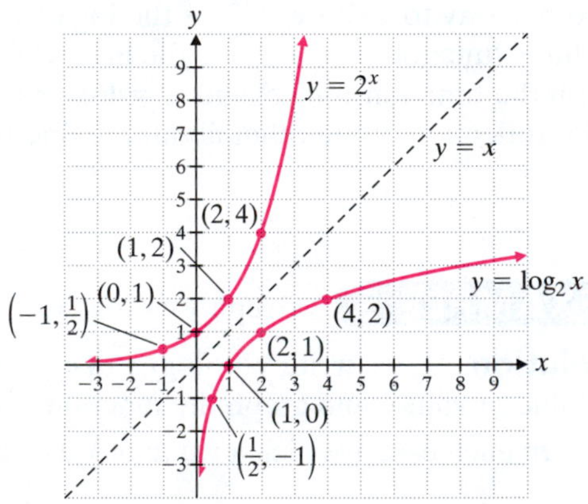

Note that $y = \log_2 x$ is the inverse of $y = 2^x$ because the ordered pairs (x, y) are reversed. The sketch of the two equations shows that they are inverses. If we reflect the graph of $y = 2^x$ about the line $y = x$, it will coincide with the graph of $y = \log_2 x$.

Recall that in function notation, f^{-1} means the inverse function of f. Thus, if we write $f(x) = \log_2 x$, then $f^{-1}(x) = 2^x$.

Student Practice 6 Graph $y = \log_6 x$ and $y = 6^x$ on the same set of axes.

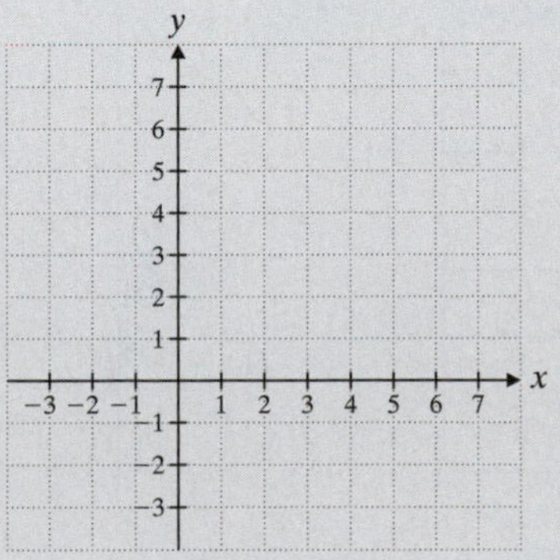

11.2 Exercises

MyMathLab®

Watch the videos in MyMathLab

Download the MyDashBoard App

Verbal and Writing Skills, Exercises 1–4

1. A logarithm is an ______________.

2. In the equation $y = \log_b x$, the value b is called the ______________.

3. In the equation $y = \log_b x$, the domain (the set of permitted values of x) is ______________.

4. In the equation $y = \log_b x$, the permitted values of b are ______________.

Write in logarithmic form.

5. $49 = 7^2$

6. $512 = 8^3$

7. $16 = 2^4$

8. $100 = 10^2$

9. $0.001 = 10^{-3}$

10. $0.01 = 10^{-2}$

11. $\frac{1}{32} = 2^{-5}$

12. $\frac{1}{64} = 2^{-6}$

13. $y = e^5$

14. $y = e^{-12}$

Write in exponential form.

15. $2 = \log_3 9$

16. $2 = \log_2 4$

17. $0 = \log_{17} 1$

18. $0 = \log_{13} 1$

19. $\frac{1}{2} = \log_{16} 4$

20. $\frac{1}{2} = \log_{81} 9$

21. $-2 = \log_{10}(0.01)$

22. $-1 = \log_{10}(0.1)$

23. $-4 = \log_3\left(\frac{1}{81}\right)$

24. $-7 = \log_2\left(\frac{1}{128}\right)$

25. $-\frac{3}{2} = \log_e x$

26. $-\frac{5}{4} = \log_e x$

Solve.

27. $\log_2 x = 4$

28. $3 = \log_3 x$

29. $\log_{10} x = -3$

30. $\log_{10} x = -2$

31. $\log_4 64 = y$

32. $\log_7 343 = y$

33. $\log_8\left(\frac{1}{64}\right) = y$

34. $\log_3\left(\frac{1}{243}\right) = y$

35. $\log_a 121 = 2$

36. $\log_a 81 = 4$

37. $\log_a 1000 = 3$

38. $\log_a 100 = 2$

39. $\log_{25} 5 = w$

40. $\log_8 2 = w$

41. $\log_3\left(\frac{1}{3}\right) = w$

42. $\log_{37} 1 = w$

43. $\log_{15} w = 0$

44. $\log_{10} w = -4$

45. $\log_w 3 = \frac{1}{2}$

46. $\log_w 2 = \frac{1}{3}$

Evaluate.

47. $\log_{10}(0.001)$

48. $\log_{10}(0.0001)$

49. $\log_2 128$

50. $\log_5 125$

51. $\log_{23} 1$

52. $\log_{18} \frac{1}{18}$

53. $\log_6 \sqrt{6}$

54. $\log_3 \sqrt{3}$

55. $\log_{57} 1$

56. $\log_3 \frac{1}{27}$

Graph.

57. $\log_3 x = y$

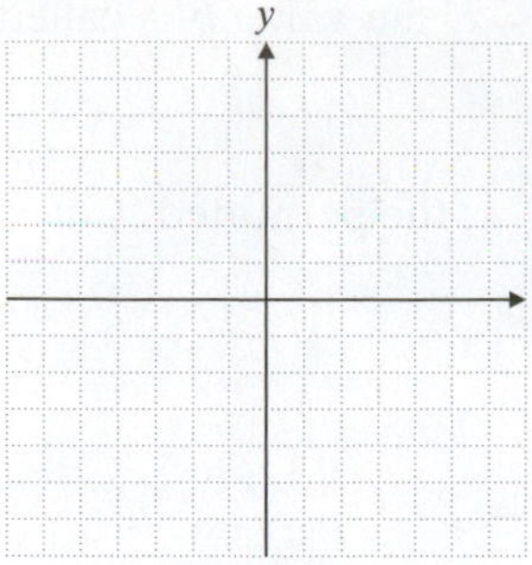

58. $\log_4 x = y$

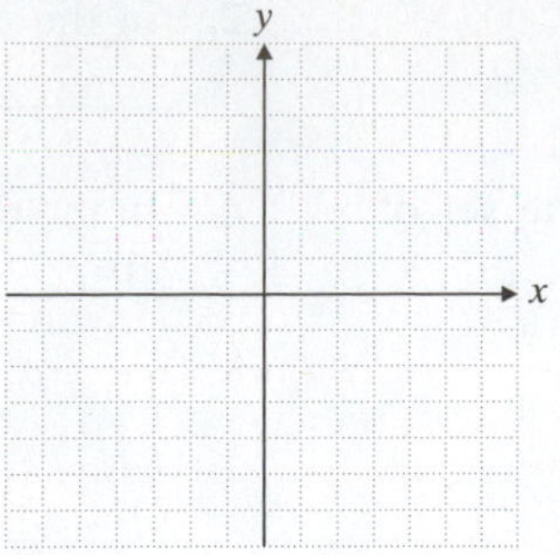

59. $\log_{1/4} x = y$

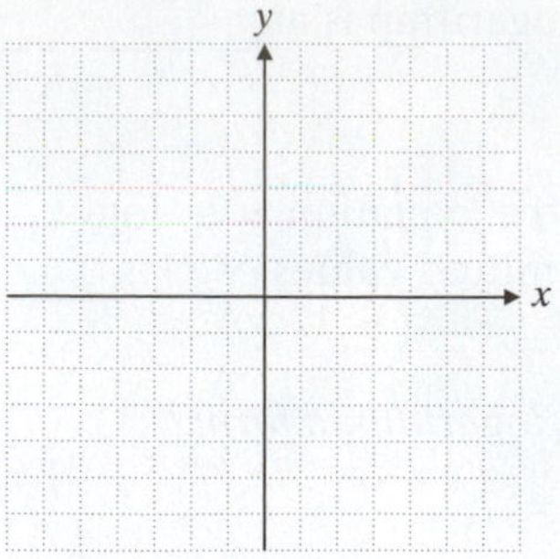

60. $\log_{1/3} x = y$

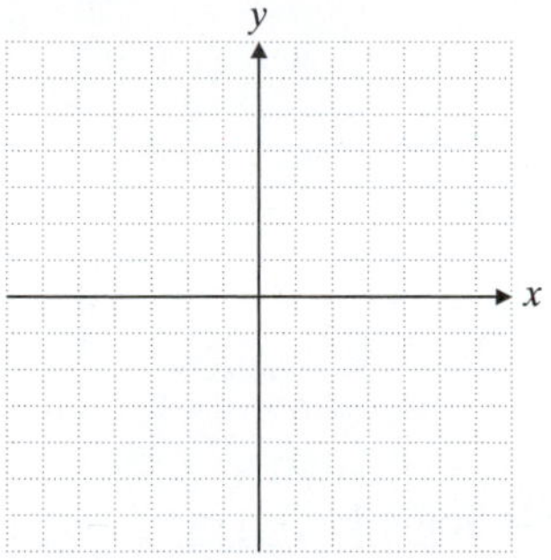

61. $\log_{10} x = y$

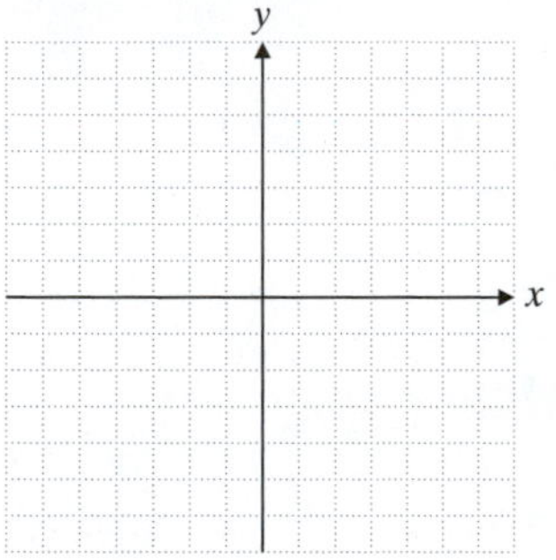

62. $\log_8 x = y$

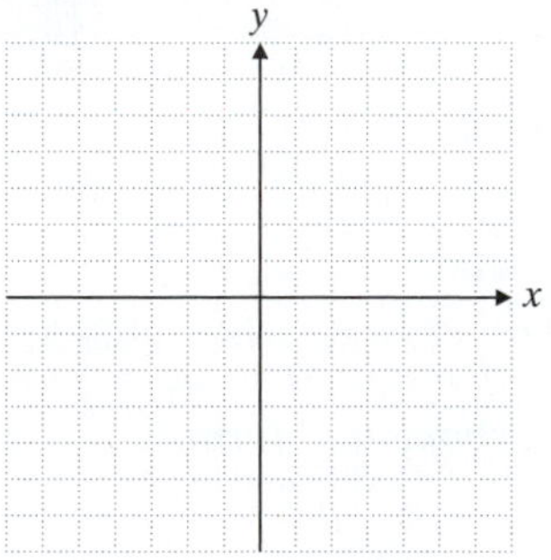

On one coordinate plane, graph the function f and the function f^{-1}. *Then graph a dashed line for the equation* $y = x$.

63. $f(x) = \log_3 x, f^{-1}(x) = 3^x$

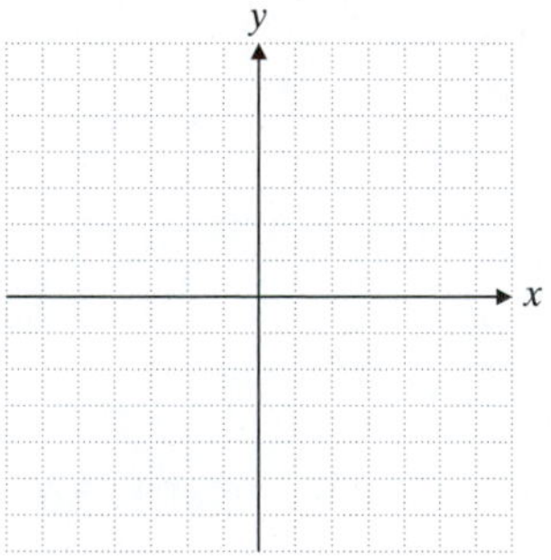

64. $f(x) = \log_4 x, f^{-1}(x) = 4^x$

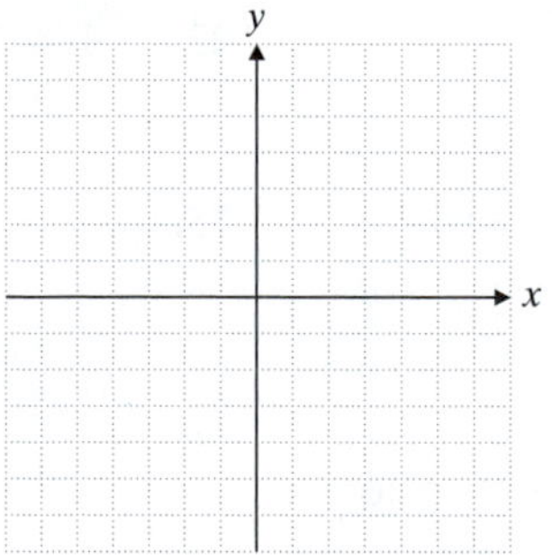

Applications

pH *Solutions* *To determine whether a solution is an acid or a base, chemists check the solution's* pH. *A solution is an acid if its* pH *is less than 7 and a base if its* pH *is greater than 7. The* pH *is defined by* $\text{pH} = -\log_{10}[H^+]$, *where* $[H^+]$ *is the concentration of hydrogen ions in the solution.*

65. The concentration of hydrogen ions in toothpaste is approximately 10^{-10}. What is the pH of toothpaste?

66. The concentration of hydrogen ions in tomatoes is approximately $10^{-4.3}$. What is the pH of tomatoes?

67. A food scientist knows that for jelly to gel properly, its pH needs to be close to 3.5. Find the concentration of hydrogen ions in the jelly.

68. A lawn care employee tested a customer's soil and found that it had a pH of 7. Find the concentration of hydrogen ions in the soil.

69. The chef of The Depot Restaurant has prepared a special balsamic vinaigrette salad dressing. What is the pH of the dressing if the concentration of hydrogen ions is 1.103×10^{-3}? The logarithm, base 10, of 0.001103 is approximately -2.957424488. Round your answer to three decimal places.

70. The EPA is testing a batch of experimental industrial solvent with a pH of 9.25. Find the concentration of hydrogen ions in the solution. Give your answer in scientific notation rounded to three decimal places.

Tax Software *Easy Tax produces tax software for people who want to use their home computers to file their income tax returns. In the first two years of sales of the software, they have found that for N sets of software to be sold, they need to invest d dollars for advertising according to the equation*

$$N = 1200 + (2500)(\log_{10} d),$$

where d is always a positive number not less than 1.

71. How many sets of software were sold when they spent \$10,000 on advertising?

72. How many sets of software were sold when they spent \$100,000 on advertising?

73. They have a goal for two years from now of selling 18,700 sets of software. How much should they spend on advertising two years from now?

74. They have a goal for next year of selling 16,200 sets of software. How much should they spend on advertising next year?

Cumulative Review

75. **[3.3.3]** Find the equation of the line perpendicular to $y = -\frac{2}{3}x + 4$ that contains $(-4, 1)$.

76. **[3.2.1]** Find the slope of the line containing $(-6, 3)$ and $(-1, 2)$.

77. **[11.1.3]** ***Viral Culture*** The number of viral cells in a laboratory culture in a biology research project is given by $A(t) = 9000(2^t)$, where t is measured in hours.

(a) How many viral cells will be in the culture after the first 2 hours?

(b) How many viral cells will be in the culture after the first 12 hours?

78. **[11.1.3]** ***College Tuition*** Assume the cost of a college education is increasing at 4% per year. The equation $C(t) = P(1.04)^t$ forecasts the tuition cost t years from now, based on the present cost P in dollars.

(a) How much will a college now charging \$4400 for tuition charge in 5 years?

(b) How much will a college now charging \$16,500 for tuition charge in 10 years?

Quick Quiz 11.2

1. Evaluate $\log_3 81$.

2. Solve for x.

$$\log_2 x = 6$$

3. Solve for w.

$$\log_{27} 3 = w$$

4. **Concept Check** Explain how you would solve for x.

$$-\frac{1}{2} = \log_e x$$

11.3 Properties of Logarithms

Student Learning Objectives

After studying this section, you will be able to:

1. Use the property $\log_b MN = \log_b M + \log_b N$.
2. Use the property $\log_b\left(\frac{M}{N}\right) = \log_b M - \log_b N$.
3. Use the property $\log_b M^p = p \log_b M$.
4. Solve simple logarithmic equations.

① Using the Property $\log_b MN = \log_b M + \log_b N$

We have already said that logarithms reduce complex expressions to addition and subtraction. The following properties show us how to use logarithms in this way.

PROPERTY 1: THE LOGARITHM OF A PRODUCT

For any positive real numbers M and N and any positive base $b \neq 1$,

$$\log_b MN = \log_b M + \log_b N.$$

To see that this property is true, we let

$$\log_b M = x \quad \text{and} \quad \log_b N = y,$$

where x and y are any values. Now we write the statements in exponential notation:

$$b^x = M \quad \text{and} \quad b^y = N.$$

Then

$$MN = b^x b^y = b^{x+y} \quad \text{Laws of exponents.}$$

If we convert this equation to logarithmic form, then we have the following:

$$\log_b MN = x + y \quad \text{Definition of logarithm.}$$

$$\log_b MN = \log_b M + \log_b N \quad \text{By substitution.}$$

Note that the logarithms must have the same base.

EXAMPLE 1 Write $\log_3 XZ$ as a sum of logarithms.

Solution By property 1, $\log_3 XZ = \log_3 X + \log_3 Z$.

Student Practice 1 Write $\log_4 WXY$ as a sum of logarithms.

NOTE TO STUDENT: Fully worked-out solutions to all of the Student Practice problems can be found at the back of the text starting at page SP-1.

EXAMPLE 2 Write $\log_3 16 + \log_3 x + \log_3 y$ as a single logarithm.

Solution If we extend our rule, we have $\log_b MNP = \log_b M + \log_b N + \log_b P$. Thus,

$$\log_3 16 + \log_3 x + \log_3 y = \log_3 16xy.$$

Student Practice 2 Write $\log_7 w + \log_7 8 + \log_7 x$ as a single logarithm.

② Using the Property $\log_b\left(\frac{M}{N}\right) = \log_b M - \log_b N$

Property 2 is similar to property 1 except that it involves two expressions that are divided, not multiplied.

PROPERTY 2: THE LOGARITHM OF A QUOTIENT

For any positive real numbers M and N and any positive base $b \neq 1$,

$$\log_b\left(\frac{M}{N}\right) = \log_b M - \log_b N.$$

Property 2 can be proved using a similar approach to the one used to prove property 1. We encourage you to try to prove property 2.

EXAMPLE 3 Write $\log_3\left(\frac{29}{7}\right)$ as the difference of two logarithms.

Solution

$$\log_3\left(\frac{29}{7}\right) = \log_3 29 - \log_3 7$$

Student Practice 3 Write $\log_3\left(\frac{17}{5}\right)$ as the difference of two logarithms.

EXAMPLE 4 Express $\log_b 36 - \log_b 9$ as a single logarithm.

Solution $\log_b 36 - \log_b 9 = \log_b\left(\frac{36}{9}\right) = \log_b 4$

Student Practice 4 Express $\log_b 132 - \log_b 4$ as a single logarithm.

CAUTION: Be sure you understand property 2!

$$\frac{\log_b M}{\log_b N} \neq \log_b M - \log_b N$$

Do you see why?

③ Using the Property $\log_b M^p = p \log_b M$

We now introduce the third property. We encourage you to try to prove Property 3.

PROPERTY 3: THE LOGARITHM OF A NUMBER RAISED TO A POWER

For any positive real number M, any real number p, and any positive base $b \neq 1$,

$$\log_b M^p = p \log_b M.$$

MC **EXAMPLE 5** Write $\frac{1}{3}\log_b x + 2\log_b w - 3\log_b z$ as a single logarithm.

Solution First, we must eliminate the coefficients of the logarithmic terms.

$$\log_b x^{1/3} + \log_b w^2 - \log_b z^3 \quad \text{By property 3.}$$

Now we can combine either the sum or difference of the logarithms. We'll do the sum.

$$\log_b x^{1/3}w^2 - \log_b z^3 \quad \text{By property 1.}$$

Now we combine the difference.

$$\log_b\left(\frac{x^{1/3}w^2}{z^3}\right) \quad \text{By property 2.}$$

Student Practice 5 Write $\frac{1}{3}\log_7 x - 5\log_7 y$ as one logarithm.

EXAMPLE 6 Write $\log_b\left(\frac{x^4y^3}{z^2}\right)$ as a sum or difference of logarithms.

Solution

$$\log_b\left(\frac{x^4y^3}{z^2}\right) = \log_b x^4y^3 - \log_b z^2 \quad \text{By property 2.}$$
$$= \log_b x^4 + \log_b y^3 - \log_b z^2 \quad \text{By property 1.}$$
$$= 4\log_b x + 3\log_b y - 2\log_b z \quad \text{By property 3.}$$

Student Practice 6 Write $\log_3\left(\frac{x^4y^5}{z}\right)$ as a sum or difference of logarithms.

4 Solving Simple Logarithmic Equations

A major goal in solving many logarithmic equations is to obtain a logarithm on one side of the equation and no logarithm on the other side. In Example 7 we will use property 1 to combine two separate logarithms that are added.

EXAMPLE 7 Find x if $\log_2 x + \log_2 5 = 3$.

Solution

$$\log_2 5x = 3 \quad \text{Use property 1.}$$
$$5x = 2^3 \quad \text{Convert to exponential form.}$$
$$5x = 8 \quad \text{Simplify.}$$
$$x = \frac{8}{5} \quad \text{Divide both sides by 5.}$$

Student Practice 7 Find x if $\log_4 x + \log_4 5 = 2$.

In Example 8, two logarithms are subtracted on the left side of the equation. We can use property 2 to combine these two logarithms. This will allow us to obtain the form of one logarithm on one side of the equation and no logarithm on the other side.

MC **EXAMPLE 8** Find x if $\log_3(x + 4) - \log_3(x - 4) = 2$.

Solution

$$\log_3\left(\frac{x+4}{x-4}\right) = 2 \quad \text{Use property 2.}$$
$$\frac{x+4}{x-4} = 3^2 \quad \text{Convert to exponential form.}$$
$$x + 4 = 9(x - 4) \quad \text{Multiply each side by } (x - 4).$$
$$x + 4 = 9x - 36 \quad \text{Simplify.}$$
$$40 = 8x$$
$$5 = x$$

Student Practice 8 Find x if $\log_{10} x - \log_{10}(x + 3) = -1$.

To solve some logarithmic equations, we need a few additional properties of logarithms. We state these properties now.

The following properties are true for all positive values of $b \neq 1$ and all positive values of x and y.

Property 4 $\log_b b = 1$

Property 5 $\log_b 1 = 0$

Property 6 If $\log_b x = \log_b y$, then $x = y$.

We now illustrate each property in Example 9.

EXAMPLE 9

(a) Evaluate $\log_7 7$.

(b) Evaluate $\log_5 1$.

(c) Find x if $\log_3 x = \log_3 17$.

Solution

(a) $\log_7 7 = 1$ because $\log_b b = 1$. Property 4.

(b) $\log_5 1 = 0$ because $\log_b 1 = 0$. Property 5.

(c) If $\log_3 x = \log_3 17$, then $x = 17$. Property 6.

Student Practice 9 Evaluate.

(a) $\log_7 1$ **(b)** $\log_8 8$ **(c)** Find y if $\log_{12} 13 = \log_{12}(y + 2)$.

We now have the mathematical tools needed to solve a variety of logarithmic equations.

EXAMPLE 10 Find x if $2 \log_7 3 - 4 \log_7 2 = \log_7 x$.

Solution We can use property 3 in two cases.

$$2 \log_7 3 = \log_7 3^2 = \log_7 9$$
$$4 \log_7 2 = \log_7 2^4 = \log_7 16$$

By substituting these results, we have the following.

$$\log_7 9 - \log_7 16 = \log_7 x$$
$$\log_7\left(\frac{9}{16}\right) = \log_7 x \quad \text{Property 2.}$$
$$\frac{9}{16} = x \quad \text{Property 6.}$$

Student Practice 10 Find x if $\log_3 2 - \log_3 5 = \log_3 6 + \log_3 x$.

11.3 Exercises

MyMathLab®

Watch the videos in MyMathLab

Download the MyDashBoard App

Express as a sum of logarithms.

1. $\log_3 AB$

2. $\log_{12} CD$

3. $\log_5(7 \cdot 11)$

4. $\log_6(13 \cdot 5)$

5. $\log_b 9f$

6. $\log_b 5d$

Express as a difference of logarithms.

7. $\log_9\left(\frac{2}{7}\right)$

8. $\log_{11}\left(\frac{23}{17}\right)$

9. $\log_b\left(\frac{H}{10}\right)$

10. $\log_a\left(\frac{G}{7}\right)$

11. $\log_a\left(\frac{E}{F}\right)$

12. $\log_m\left(\frac{3}{A}\right)$

Express as a product.

13. $\log_8 a^7$

14. $\log_3 c^8$

15. $\log_b A^{-2}$

16. $\log_a B^{-5}$

17. $\log_5 \sqrt{w}$

18. $\log_6 \sqrt{z}$

Mixed Practice

Write each expression as a sum or difference of logarithms of a single variable.

19. $\log_8 x^2 y$

20. $\log_6 xy^4$

21. $\log_{11}\left(\frac{6M}{N}\right)$

22. $\log_5\left(\frac{2C}{7}\right)$

23. $\log_2\left(\frac{5xy^4}{\sqrt{z}}\right)$

24. $\log_4\left(\frac{2x^3\sqrt[4]{y}}{z^5}\right)$

25. $\log_a\left(\sqrt[3]{\frac{x^4}{y}}\right)$

26. $\log_a\left(\sqrt[5]{\frac{y}{z^3}}\right)$

Write as a single logarithm.

27. $\log_4 13 + \log_4 y + \log_4 3$

28. $\log_8 15 + \log_8 a + \log_8 b$

29. $5 \log_3 x - \log_3 7$

30. $3 \log_8 5 - \log_8 z$

31. $2 \log_b 7 + 3 \log_b y - \frac{1}{2} \log_b z$

32. $4 \log_b 2 + \frac{1}{3} \log_b z - 5 \log_b y$

Use the properties of logarithms to simplify each of the following.

33. $\log_3 3$

34. $\log_7 7$

35. $\log_e e$

36. $\log_{10} 10$

37. $\log_9 1$

38. $\log_e 1$

39. $3\log_7 7 + 4\log_7 1$

40. $\frac{1}{2}\log_5 5 - 8\log_5 1$

Find x in each of the following.

41. $\log_8 x = \log_8 7$

42. $\log_9 x = \log_9 5$

43. $\log_5(2x + 7) = \log_5 29$

44. $\log_{15} 26 = \log_{15}(3x - 1)$

45. $\log_3 1 = x$

46. $\log_8 1 = x$

47. $\log_7 7 = x$

48. $\log_5 5 = x$

49. $\log_{10} x + \log_{10} 25 = 2$

50. $\log_{10} x + \log_{10} 5 = 1$

51. $\log_2 7 = \log_2 x - \log_2 3$

52. $\log_5 1 = \log_5 x - \log_5 8$

53. $3\log_5 x = \log_5 8$

54. $\frac{1}{2}\log_3 x = \log_3 4$

55. $\log_e x = \log_e 5 + 1$

56. $\log_e x + \log_e 7 = 2$

57. $\log_6(5x + 21) - \log_6(x + 3) = 1$

58. $\log_3(4x + 6) - \log_3(x - 1) = 2$

59. It can be shown that $y = b^{\log_b y}$. Use this property to evaluate $5^{\log_5 4} + 3^{\log_3 2}$.

60. It can be shown that $x = \log_b b^x$. Use this property to evaluate $\log_7\sqrt[4]{7} + \log_6\sqrt[12]{6}$.

Cumulative Review

▲ **61.** **[1.6.2]** Find the volume of a cylinder with a radius of 2 meters and a height of 5 meters. Round your answer to the nearest tenth.

▲ **62.** **[1.6.2]** Find the area of a circle whose radius is 4 meters. Round your answer to the nearest tenth.

63. **[4.1.4]** Solve the system.

$$5x + 3y = 9$$
$$7x - 2y = 25$$

64. **[4.2.2]** Solve the system.

$$2x - y + z = 3$$
$$x + 2y + 2z = 1$$
$$4x + y + 2z = 0$$

At the end of 2010, a study ranked the top ten cities where households owed the highest percent of their average yearly income to credit card companies. Below is the information for Wilmington, North Carolina (ranked #1), and Toledo, Ohio (ranked #3). For each city, find the average credit card debt per household. Then find the percent of the average yearly household income that was owed to credit card companies. Round the debt to the nearest dollar and the percent to the nearest hundredth of a percent.

65. [1.2.2; 1.4.5] ***Wilmington, North Carolina***
Number of households: 1.53×10^5
Total credit card debt for all households: $\$1.117 \times 10^9$
Average yearly household income: $42,392

66. [1.2.2; 1.4.5] ***Toledo, Ohio***
Number of households: 2.56×10^5
Total credit card debt for all households: $\$1.9 \times 10^9$
Average yearly household income: $44,349

Quick Quiz 11.3

1. Express $\log_5\left(\frac{\sqrt[3]{x}}{y^4}\right)$ as a sum or difference of single logarithms.

2. Express $3\log_6 x + \log_6 y - \log_6 5$ as a single logarithm.

3. Find x if $\frac{1}{2}\log_4 x = \log_4 25$.

4. Concept Check Explain how you would simplify $\log_{10}(0.001)$.

How Am I Doing? Sections 11.1–11.3

How are you doing with your homework assignments in Sections 11.1 to 11.3? Do you feel you have mastered the material so far? Do you understand the concepts you have covered? Before you go further in the textbook, take some time to do each of the following problems.

11.1

1. Sketch the graph of $f(x) = 2^{-x}$. Plot at least four points.

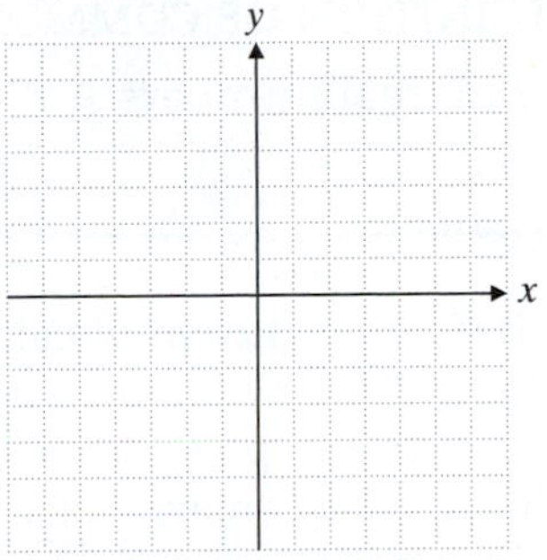

2. Solve for x. $3^{x+2} = 81$

3. Solve for x. $2^x = \frac{1}{32}$

4. Solve for x. $125 = 5^{3x+4}$

5. When a principal amount P is invested at interest rate r compounded annually, the amount of money A earned after t years is given by the equation $A = P(1 + r)^t$. How much money will Nancy have in 4 years if she invests \$10,000 in a mutual fund that pays 12% interest compounded annually?

11.2

6. Write in logarithmic form. $\frac{1}{49} = 7^{-2}$

7. Write in exponential form. $-3 = \log_{10}(0.001)$

8. Solve for x. $\log_5 x = 3$

9. Solve for x. $\log_x 49 = -2$

10. Evaluate $\log_{10}(10{,}000)$.

11.3

11. Write $\log_5\left(\frac{x^2y^5}{z^3}\right)$ as a sum or difference of logarithms.

12. Express $\frac{1}{2}\log_4 x - 3\log_4 w$ as a single logarithm.

13. Find x if $\log_3 x + \log_3 2 = 4$.

14. Find x if $\log_7 x = \log_7 8$.

15. Find x if $\log_9 1 = x$.

16. Find x if $\log_3 2x = 2$.

17. Find x if $1 = \log_4 3x$.

18. Find x if $\log_e x + \log_e 3 = 1$.

Now turn to page SA-30 for the answers to each of these problems. Each answer also includes a reference to the objective in which the problem is first taught. If you missed any of these problems, you should stop and review the Examples and Student Practice problems in the referenced objective. A little review now will help you master the material in the upcoming sections of the text.

1. ______________
2. ______________
3. ______________
4. ______________
5. ______________
6. ______________
7. ______________
8. ______________
9. ______________
10. ______________
11. ______________
12. ______________
13. ______________
14. ______________
15. ______________
16. ______________
17. ______________
18. ______________

11.4 Common Logarithms, Natural Logarithms, and Change of Base of Logarithms

Student Learning Objectives

After studying this section, you will be able to:

1. Find common logarithms.
2. Find the antilogarithm of a common logarithm.
3. Find natural logarithms.
4. Find the antilogarithm of a natural logarithm.
5. Evaluate a logarithm to a base other than 10 or e.

1 Finding Common Logarithms on a Scientific Calculator

Although we can find a logarithm of a number for any positive base except 1, the most frequently used bases are 10 and e. Base 10 logarithms are called *common logarithms* and are usually written with no subscript.

DEFINITION OF COMMON LOGARITHM

For all real numbers $x > 0$, the **common logarithm** of x is

$$\log x = \log_{10} x.$$

Before the advent of calculators and computers, people used tables of common logarithms. Now most work with logarithms is done with the aid of a scientific calculator or a graphing calculator. We will take that approach in this section of the text. To find the common logarithm of a number on a scientific calculator, enter the number and then press the [log x] or [log] key.

EXAMPLE 1 On a scientific calculator or a graphing calculator, find a decimal approximation for each of the following.

(a) $\log 7.32$ **(b)** $\log 73.2$ **(c)** $\log 0.314$

Solution

(a) 7.32 [log] ≈ 0.864511081

(b) 73.2 [log] ≈ 1.864511081

(c) 0.314 [log] ≈ -0.503070352

Note that the only difference in the two answers is the 1 before the decimal point.

Note: Your calculator may display fewer or more digits in the answer.

Student Practice 1 On a scientific calculator or a graphing calculator, find a decimal approximation for each of the following.

(a) $\log 4.36$ **(b)** $\log 436$ **(c)** $\log 0.2418$

***NOTE TO STUDENT:** Fully worked-out solutions to all of the Student Practice problems can be found at the back of the text starting at page SP-1.*

TO THINK ABOUT: Decimal Point Placement Why is the difference in the answers to Example 1 **(a)** and **(b)** equal to 1.00? Consider the following.

$$\log 73.2 = \log(7.32 \times 10^1) \quad \text{Use scientific notation.}$$
$$= \log 7.32 + \log 10^1 \quad \text{By property 1.}$$
$$= \log 7.32 + 1 \quad \text{Because } \log_b b = 1.$$
$$\approx 0.864511081 + 1 \quad \text{Use a calculator.}$$
$$\approx 1.864511081 \quad \text{Add the decimals.}$$

2 Finding the Antilogarithm of a Common Logarithm on a Scientific Calculator

We have previously discussed the function $f(x) = \log x$ and the corresponding inverse function $f^{-1}(x) = 10^x$. The inverse of a logarithmic function is an exponential function. There is another name for this function. It is called an **antilogarithm.**

If $f(x) = \log x$ (here the base is understood to be 10), then $f^{-1}(x) = \text{antilog } x = 10^x$.

EXAMPLE 2 Find an approximate value for x if $\log x = 4.326$.

Solution Here we are given the value of the logarithm, and we want to find the number that has that logarithm. In other words, we want the antilogarithm. We know that $\log_{10} x = 4.326$ is equivalent to $10^{4.326} = x$. So to solve this problem, we want to find the value of 10 raised to the 4.326 power. Using a calculator, we have the following.

$$4.326 \boxed{10^x} \approx 21{,}183.61135$$

Thus, $x \approx 21{,}183.61135$. (If your scientific calculator does not have a $\boxed{10^x}$ key, you can usually use $\boxed{\text{2nd F}}$ $\boxed{\log}$ or $\boxed{\text{INV}}$ $\boxed{\log}$ or $\boxed{\text{SHIFT}}$ $\boxed{\log}$ to perform the operation.)

Student Practice 2 Using a scientific calculator, find an approximate value for x if $\log x = 2.913$.

EXAMPLE 3 Evaluate antilog(-1.6784).

Solution Asking what is antilog(-1.6784) is equivalent to asking what the value is of $10^{-1.6784}$. To determine this on most scientific calculators, it will be necessary to enter the number 1.6784 followed by the $\boxed{+/-}$ key. You may need slightly different steps on a graphing calculator.

$$1.6784 \boxed{+/-} \boxed{10^x} \approx 0.020970076$$

Thus, antilog$(-1.6784) \approx 0.020970076$.

Student Practice 3 Evaluate antilog(-3.0705).

EXAMPLE 4 Using a scientific calculator, find an approximate value for x.

(a) $\log x = 0.07318$ **(b)** $\log x = -3.1621$

Solution

(a) $\log x = 0.07318$ is equivalent to $10^{0.07318} = x$.

$$0.07318 \boxed{10^x} \approx 1.183531987$$

Thus, $x \approx 1.183531987$.

(b) $\log x = -3.1621$ is equivalent to $10^{-3.1621} = x$.

$$3.1621 \boxed{+/-} \boxed{10^x} \approx 0.0006884937465$$

Thus, $x \approx 0.0006884937465$.

(Some calculators may give the answer in scientific notation as $6.884937465 \times 10^{-4}$. This is often displayed on a calculator screen as 6.884937465 −4.)

Student Practice 4 Using a scientific calculator, find an approximate value for x.

(a) $\log x = 0.06134$ **(b)** $\log x = -4.6218$

③ Finding Natural Logarithms on a Scientific Calculator

For most theoretical work in mathematics and other sciences, the most useful base for logarithms is e. Logarithms with base e are known as *natural logarithms* and are usually written $\ln x$.

DEFINITION OF NATURAL LOGARITHM

For all real numbers $x > 0$, the **natural logarithm** of x is

$$\ln x = \log_e x.$$

On a scientific calculator we can usually approximate natural logarithms with the $\boxed{\ln x}$ or $\boxed{\ln}$ key.

EXAMPLE 5 On a scientific calculator, approximate the following values.

(a) $\ln 7.21$ **(b)** $\ln 72.1$ **(c)** $\ln 0.0356$

Solution

(a) $7.21 \boxed{\ln} \approx 1.975468951$

(b) $72.1 \boxed{\ln} \approx 4.278054044$

(c) $0.0356 \boxed{\ln} \approx -3.335409641$

Note that there is no simple relationship between the answers to parts **(a)** and **(b).** Do you see why these are different from common logarithms?

Student Practice 5 On a scientific calculator, approximate the following values.

(a) $\ln 4.82$ **(b)** $\ln 48.2$ **(c)** $\ln 0.0793$

4 Finding the Antilogarithm of a Natural Logarithm on a Scientific Calculator

EXAMPLE 6 On a scientific calculator, find an approximate value for x for each equation.

(a) $\ln x = 2.9836$ **(b)** $\ln x = -1.5619$

Solution

(a) If $\ln x = 2.9836$, then $e^{2.9836} = x$.

$$2.9836 \boxed{e^x} \approx 19.75882051$$

(b) If $\ln x = -1.5619$, then $e^{-1.5619} = x$.

$$1.5619 \boxed{+/-} \boxed{e^x} \approx 0.209737192$$

Student Practice 6 On a scientific calculator, find an approximate value for x for each equation.

(a) $\ln x = 3.1628$ **(b)** $\ln x = -2.0573$

An alternative notation is sometimes used. This is $\text{antilog}_e(x)$.

5 Evaluating a Logarithm to a Base Other Than 10 or *e*

Although a scientific calculator or a graphing calculator has specific keys for finding common logarithms (base 10) and natural logarithms (base e), there are no keys for finding logarithms with other bases. What do we do in such cases? The logarithm of a number for a base other than 10 or e can be found with the following formula.

CHANGE OF BASE FORMULA

$$\log_b x = \frac{\log_a x}{\log_a b},$$

where a, b, and $x > 0$, $a \neq 1$, and $b \neq 1$.

Let's see how this formula works. If we want to use common logarithms to find $\log_3 56$, we must first note that the value of b in the formula is 3. We then write

$$\log_3 56 = \frac{\log_{10} 56}{\log_{10} 3} = \frac{\log 56}{\log 3}.$$

Do you see why?

EXAMPLE 7 Evaluate using common logarithms. $\log_3 5.12$

Solution $\log_3 5.12 = \dfrac{\log 5.12}{\log 3}$

On a calculator, we find the following.

5.12 [log] [÷] 3 [log] [=] 1.486561234

Our answer is an approximate value with nine decimal places. Your answer may have more or fewer digits depending on your calculator.

Student Practice 7 Evaluate using common logarithms. $\log_9 3.76$

If we desire to use base e, then the change of base formula is used with natural logarithms.

EXAMPLE 8 Obtain an approximate value for $\log_4 0.005739$ using natural logarithms.

Solution Using the change of base formula, with $a = e$, $b = 4$, and $x = 0.005739$, we have the following.

$$\log_4 0.005739 = \frac{\log_e 0.005739}{\log_e 4} = \frac{\ln 0.005739}{\ln 4}$$

This is done on some scientific calculators as follows.

0.005739 [ln] [÷] 4 [ln] [=] −3.722492455

Thus, we have $\log_4 0.005739 \approx -3.722492455$.

Check. To check our answer we want to know the following.

$$4^{-3.722492455} \stackrel{?}{=} 0.005739$$

Using a calculator, we can verify this with the [y^x] key.

4 [y^x] 3.722492455 [+/−] [=] 0.005739 ✓

Student Practice 8 Obtain an approximate value for $\log_8 0.009312$ using natural logarithms.

Graphing Calculator

Graphing Logarithmic Functions

You can use the change of base formula to graph logarithmic functions on a graphing calculator.

To graph $y = \log_2 x$ in Example 9 on a graphing calculator, enter the function $y = \dfrac{\log x}{\log 2}$ into the Y = editor of your calculator.

Display:

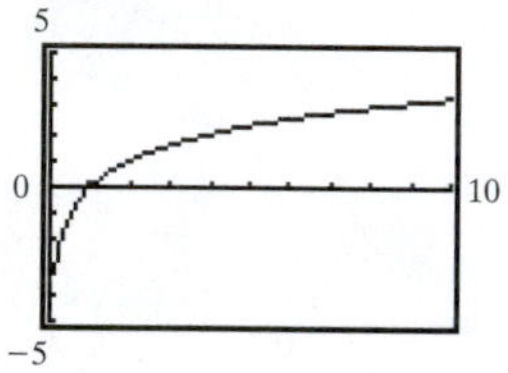

EXAMPLE 9 Using a scientific calculator, graph $y = \log_2 x$.

Solution If we use common logarithms ($\log_{10} x$), then for each value of x, we will need to calculate $\dfrac{\log x}{\log 2}$. Therefore, to find y when $x = 3$, we need to calculate $\dfrac{\log 3}{\log 2}$. On most scientific calculators, we would enter 3 [log] [÷] 2 [log] [=] and obtain 1.584962501. Rounded to the nearest tenth, we have $x = 3$ and $y = 1.6$. In a similar fashion we find other table values and then graph them.

x	$y = \log_2 x$
0.5	−1
1	0
2	1
3	1.6
4	2
6	2.6
8	3

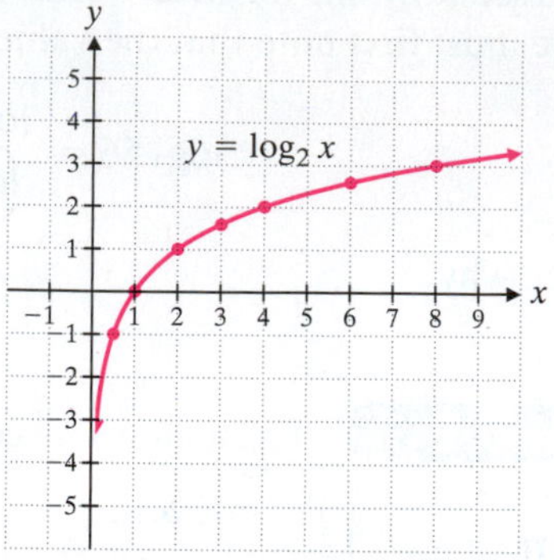

Student Practice 9 Using a scientific calculator, graph $y = \log_5 x$.

x	$y = \log_5 x$

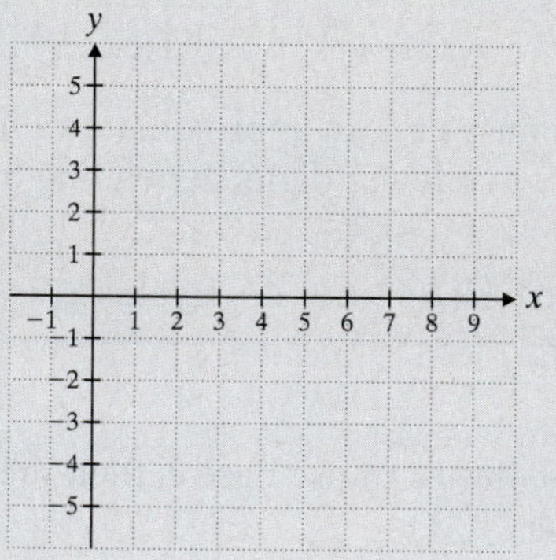

11.4 Exercises

MyMathLab®

Watch the videos in MyMathLab

Download the MyDashBoard App

Verbal and Writing Skills, Exercises 1 and 2

1. Try to find $\log(-5.08)$ on a scientific calculator. What happens? Why?

2. Try to find $\log(-6.63)$ on a scientific calculator or a graphing calculator. What happens? Why?

Use a scientific calculator to approximate the following.

3. $\log 12.3$

4. $\log 11.7$

5. $\log 25.6$

6. $\log 65.9$

7. $\log 8$

8. $\log 6$

9. $\log 125{,}000$

10. $\log 52{,}300$

11. $\log 0.0123$

12. $\log 0.567$

Find an approximate value for x using a scientific calculator or a graphing calculator.

13. $\log x = 2.016$

14. $\log x = 2.754$

15. $\log x = -2$

16. $\log x = -3$

17. $\log x = 3.9304$

18. $\log x = 3.9576$

19. $\log x = 6.4683$

20. $\log x = 5.6274$

21. $\log x = -3.3893$

22. $\log x = -4.0458$

23. $\log x = -1.5672$

24. $\log x = -1.0075$

Approximate the following with a scientific calculator or a graphing calculator.

25. $\text{antilog}(7.6215)$

26. $\text{antilog}(4.3894)$

27. $\text{antilog}(-1.0826)$

28. $\text{antilog}(-3.1145)$

29. $\ln 5.62$

30. $\ln 8.81$

31. $\ln 1.53$

32. $\ln 7.35$

33. $\ln 136{,}000$

34. $\ln 129{,}000$

35. $\ln 0.00579$

36. $\ln 0.00134$

Find an approximate value for x using a scientific calculator or a graphing calculator.

37. $\ln x = 0.95$

38. $\ln x = 0.55$

39. $\ln x = 2.4$

40. $\ln x = 4.4$

41. $\ln x = -0.05$

42. $\ln x = -0.03$

43. $\ln x = -2.7$

44. $\ln x = -3.8$

Approximate the following with a scientific calculator or a graphing calculator.

45. $\text{antilog}_e(6.1582)$

46. $\text{antilog}_e(1.9047)$

47. $\text{antilog}_e(-2.1298)$

48. $\text{antilog}_e(-3.3712)$

Use a scientific calculator or a graphing calculator and **common logarithms** *to evaluate the following.*

49. $\log_3 9.2$

50. $\log_2 6.13$

51. $\log_7 7.35$

52. $\log_9 9.85$

53. $\log_6 0.127$

54. $\log_5 0.173$

55. $\log_{15} 12$

56. $\log_{17} 18$

Use a scientific calculator or a graphing calculator and **natural logarithms** *to evaluate the following.*

57. $\log_4 0.07733$

58. $\log_7 0.004462$

59. $\log_{21} 436$

60. $\log_{30} 913$

Mixed Practice

Use a scientific calculator or a graphing calculator to find an approximate value for each of the following.

61. $\ln 1537$

62. $\log 92.81$

63. $\text{antilog}_e(-1.874)$

64. $\log_6 0.5437$

Find an approximate value for x for each of the following.

65. $\log x = 8.5634$

66. $\ln x = 7.9631$

67. $\log_4 x = 0.8645$

68. $\log_3 x = 0.5649$

Use a graphing calculator to graph the following.

69. $y = \log_6 x$

70. $y = \log_4 x$

71. $y = \log_{0.4} x$

72. $y = \log_{0.2} x$

Applications

Median Age *The median age of people in the United States is slowly increasing as the population becomes older. In 1980 the median age of the population was 30.0 years. This means that approximately half the population of the country was under 30.0 years old and approximately half the population of the country was over 30.0 years old. By 2005 the median age had increased to 36.4 years. An equation that can be used to predict the median age N (in years) of the population of the United States is* $N = 33.62 + 1.0249 \ln x$, *where x is the number of years since 1990 and* $x \geq 1$. (*Source:* www.census.gov)

73. Use the equation to find the median age of the U.S. population in 1995 and in 2005. Round to the nearest hundredth. If this model is correct, by what percent did the median age increase from 1995 to 2005? Round to the nearest tenth of a percent.

74. Use the equation to find the median age of the U.S. population in 2003 and in 2013. Round to the nearest hundredth. If this model is correct, by what percent will the median age increase from 2003 to 2013? Round to the nearest tenth of a percent.

Earthquakes *Suppose that we want to measure the magnitude of an earthquake. If an earthquake has a shock wave x times greater than the smallest shock wave that can be measured by a seismograph, then its magnitude R on the Richter scale is given by the equation $R = \log x$. An earthquake that has a shock wave 25,000 times greater than the smallest shock wave that can be detected will have a magnitude of $R = \log 25{,}000 \approx 4.40$. (Usually we round the magnitude of an earthquake to the nearest hundredth.)*

75. What is the magnitude of an earthquake that has a shock wave that is 56,000 times greater than the smallest shock wave that can be detected?

76. What is the magnitude of an earthquake that has a shock wave that is 184,000 times greater than the smallest shock wave that can be detected?

77. In October 2010, an earthquake of magnitude $R = 6.7$ occurred in the Gulf of California. What can you say about the size of the earthquake's shock wave?

78. In November 2010, an earthquake of magnitude $R = 5.3$ occurred in Serbia. What can you say about the size of the earthquake's shock wave?

Cumulative Review *Solve the quadratic equations. Simplify your answers.*

79. **[8.2.1]** $3x^2 - 11x - 5 = 0$

80. **[8.2.1]** $2y^2 + 4y - 3 = 0$

Highway Exits *On a specific portion of Interstate 91, there are six exits. There is a distance of 12 miles between odd-numbered exits. There is a distance of 15 miles between even-numbered exits. The total distance between Exit 1 and Exit 6 is 36 miles.*

81. **[1.2.2]** Find the distance between Exit 1 and Exit 2. Find the distance between Exit 1 and Exit 3.

82. **[1.2.2]** Find the distance between Exit 1 and Exit 4. Find the distance between Exit 1 and Exit 5.

Quick Quiz 11.4 *Use a scientific calculator to evaluate the following.*

1. Find $\log 9.36$. Round to the nearest ten-thousandth.

2. Find x if $\log x = 0.2253$. Round to the nearest hundredth.

3. Find $\log_5 8.26$. Round to the nearest ten-thousandth.

4. **Concept Check** Explain how you would find x using a scientific calculator if $\ln x = 1.7821$.

11.5 Exponential and Logarithmic Equations

Student Learning Objectives

After studying this section, you will be able to:

1. **Solve logarithmic equations.**
2. **Solve exponential equations.**
3. **Solve applications using logarithmic or exponential equations.**

① Solving Logarithmic Equations

In general, when solving logarithmic equations we try to collect all the logarithms on one side of the equation and all the numerical values on the other side. Then we seek to use the properties of logarithms to obtain a single logarithmic expression on one side.

We can describe a general procedure for solving logarithmic equations.

Step 1 If an equation contains some logarithms and some terms without logarithms, try to get one logarithm alone on one side and one numerical value on the other.

Step 2 Then convert to an exponential equation using the definition of a logarithm.

Step 3 Solve the equation.

EXAMPLE 1 Solve. $\log 5 = 2 - \log(x + 3)$

Solution

$\log 5 + \log(x + 3) = 2$	Add $\log(x + 3)$ to each side.
$\log[5(x + 3)] = 2$	Property 1.
$\log(5x + 15) = 2$	Simplify.
$5x + 15 = 10^2$	Write the equation in exponential form.
$5x + 15 = 100$	Simplify.
$5x = 85$	Subtract 15 from each side.
$x = 17$	Divide each side by 5.

Check. $\log 5 \stackrel{?}{=} 2 - \log(17 + 3)$

$\log 5 \stackrel{?}{=} 2 - \log 20$

Since these are common logarithms (base 10), the easiest way to check the answer is to find decimal approximations for each logarithm on a calculator.

$$0.698970004 \stackrel{?}{=} 2 - 1.301029996$$

$$0.698970004 = 0.698970004 \quad ✓$$

Student Practice 1 Solve. $\log(x + 5) = 2 - \log 5$

Graphing Calculator

Solving Logarithmic Equations

Example 1 could be solved with a graphing calculator in the following way. First write the equation as

$\log 5 + \log(x + 3) - 2 = 0$

and then graph the function

$y = \log 5 + \log(x + 3) - 2$

to find an approximate value for x when $y = 0$.

***NOTE TO STUDENT:** Fully worked-out solutions to all of the Student Practice problems can be found at the back of the text starting at page SP-1.*

EXAMPLE 2 Solve. $\log_3(x + 6) - \log_3(x - 2) = 2$

Solution

$\log_3\left(\dfrac{x + 6}{x - 2}\right) = 2$	Property 2.
$\dfrac{x + 6}{x - 2} = 3^2$	Write the equation in exponential form.
$\dfrac{x + 6}{x - 2} = 9$	Evaluate 3^2.
$x + 6 = 9(x - 2)$	Multiply each side by $(x - 2)$.
$x + 6 = 9x - 18$	Simplify.
$24 = 8x$	Add $18 - x$ to each side.
$3 = x$	Divide each side by 8.

Check.

$$\log_3(3+6) - \log_3(3-2) \stackrel{?}{=} 2$$
$$\log_3 9 - \log_3 1 \stackrel{?}{=} 2$$
$$2 - 0 \stackrel{?}{=} 2$$
$$2 = 2 \ \checkmark$$

Student Practice 2 Solve. $\log(x+3) - \log x = 1$

Some equations consist of logarithmic terms only. In such cases we may be able to use property 6 to solve them. Recall that this rule states that if $b > 0$, $b \neq 1, x > 0, y > 0$, and $\log_b x = \log_b y$, then $x = y$.

What if one of our possible solutions is the logarithm of a negative number? Can we evaluate the logarithm of a negative number? Look again at the graph of $y = \log_2 x$ on page 595. Note that the domain of this function is $x > 0$. (The curve is located on the positive side of the x-axis.) Therefore, the logarithm of a negative number is *not defined*.

You should be able to see this by using the definition of logarithms. If $\log(-2)$ were valid, we could write the following.

$$y = \log_{10}(-2)$$
$$10^y = -2$$

Obviously, no value of y can make this equation true. Thus, we see that **it is not possible to take the logarithm of a negative number.**

Sometimes when we attempt to solve a logarithmic equation, we obtain a possible solution that leads to the logarithm of a negative number. We can immediately discard such a solution.

EXAMPLE 3 Solve. $\log(x+6) + \log(x+2) = \log(x+20)$

Solution

$$\log[(x+6)(x+2)] = \log(x+20)$$
$$\log(x^2 + 8x + 12) = \log(x+20)$$
$$x^2 + 8x + 12 = x + 20$$
$$x^2 + 7x - 8 = 0$$
$$(x+8)(x-1) = 0$$
$$x + 8 = 0 \qquad x - 1 = 0$$
$$x = -8 \qquad x = 1$$

Check. $\log(x+6) + \log(x+2) = \log(x+20)$

$x = 1$: $\log(1+6) + \log(1+2) \stackrel{?}{=} \log(1+20)$

$$\log(7) + \log(3) \stackrel{?}{=} \log(21)$$
$$\log(7 \cdot 3) \stackrel{?}{=} \log 21$$
$$\log 21 = \log 21 \ \checkmark$$

$x = -8$: $\log(-8+6) + \log(-8+2) \stackrel{?}{=} \log(-8+20)$

$$\log(-2) + \log(-6) \neq \log(12)$$

We can discard -8 because it leads to taking the logarithm of a negative number, which is not allowed. Only $x = 1$ is a solution. The only solution is 1.

Student Practice 3 Solve $\log 5 - \log x = \log(6x - 7)$ and check your solution.

② Solving Exponential Equations

You might expect that property 6 can be used in the reverse direction. It seems logical, for example, that if $x = 3$, we should be able to state that $\log_4 x = \log_4 3$. This is exactly the case, and we will formally state it as a property.

PROPERTY 7

If x and $y > 0$ and $x = y$, then $\log_b x = \log_b y$, where $b > 0$ and $b \neq 1$.

Property 7 is often referred to as "taking the logarithm of each side of the equation." Usually we will take the common logarithm of each side of the equation, but any base can be used.

EXAMPLE 4 Solve $2^x = 7$. Leave your answer in exact form.

Solution

$$\log 2^x = \log 7 \quad \text{Take the logarithm of each side (property 7).}$$

$$x \log 2 = \log 7 \quad \text{Property 3.}$$

$$x = \frac{\log 7}{\log 2} \quad \text{Divide each side by log 2.}$$

Student Practice 4 Solve $3^x = 5$. Leave your answer in exact form.

When we solve exponential equations, it will often be useful to find an approximate value for the answer.

EXAMPLE 5 Solve $3^x = 7^{x-1}$. Approximate your answer to the nearest thousandth.

Solution

$$\log 3^x = \log 7^{x-1}$$

$$x \log 3 = (x - 1) \log 7$$

$$x \log 3 = x \log 7 - \log 7$$

$$x \log 3 - x \log 7 = -\log 7$$

$$x(\log 3 - \log 7) = -\log 7$$

$$x = \frac{-\log 7}{\log 3 - \log 7}$$

We can approximate the value for x on most scientific calculators by using the following keystrokes.

7 [log] [+/−] [÷] [(] 3 [log] [−] 7 [log] [)] [=] 2.296606943

Rounding to the nearest thousandth, we have $x \approx 2.297$.

Student Practice 5 Solve $2^{3x+1} = 9^{x+1}$. Approximate your answer to the nearest thousandth.

If the exponential equation involves e raised to a power, it is best to take the natural logarithm of each side of the equation.

MC **EXAMPLE 6** Solve $e^{2.5x} = 8.42$. Round your answer to the nearest ten-thousandth.

Solution

$$\ln e^{2.5x} = \ln 8.42 \quad \text{Take the natural logarithm of each side.}$$

$$(2.5x)(\ln e) = \ln 8.42 \quad \text{Property 3.}$$

$$2.5x = \ln 8.42 \quad \ln e = 1.\ \text{Property 4.}$$

$$x = \frac{\ln 8.42}{2.5} \quad \text{Divide each side by 2.5.}$$

On most scientific calculators, the value of x can be approximated with the following keystrokes.

8.42 [ln] [÷] 2.5 [=] 0.85224393

Rounding to the nearest ten-thousandth, we have $x \approx 0.8522$.

Student Practice 6 Solve $20.98 = e^{3.6x}$. Round your answer to the nearest ten-thousandth.

③ Solving Applications Using Logarithmic or Exponential Equations

We now return to the compound interest formula and consider some other exercises that can be solved with it. For example, perhaps we would like to know how long it will take for a deposit to grow to a specified goal.

EXAMPLE 7 If P dollars are invested in an account that earns interest at 12% compounded annually, the amount available after t years is $A = P(1 + 0.12)^t$. How many years will it take for \$300 in this account to grow to \$1500? Round your answer to the nearest whole year.

Solution

$$1500 = 300(1 + 0.12)^t \quad \text{Substitute } A = 1500 \text{ and } P = 300.$$

$$1500 = 300(1.12)^t \quad \text{Simplify.}$$

$$\frac{1500}{300} = (1.12)^t \quad \text{Divide each side by 300.}$$

$$5 = (1.12)^t \quad \text{Simplify.}$$

$$\log 5 = \log(1.12)^t \quad \text{Take the common logarithm of each side.}$$

$$\log 5 = t(\log 1.12) \quad \text{Property 3.}$$

$$\frac{\log 5}{\log 1.12} = t \quad \text{Divide each side by log 1.12.}$$

On a scientific calculator we have the following.

5 [log] [÷] 1.12 [log] [=] 14.20150519

Thus, it would take approximately 14 years. Look at the graph on the next page to get a visual image of how the investment increases.

Continued on next page

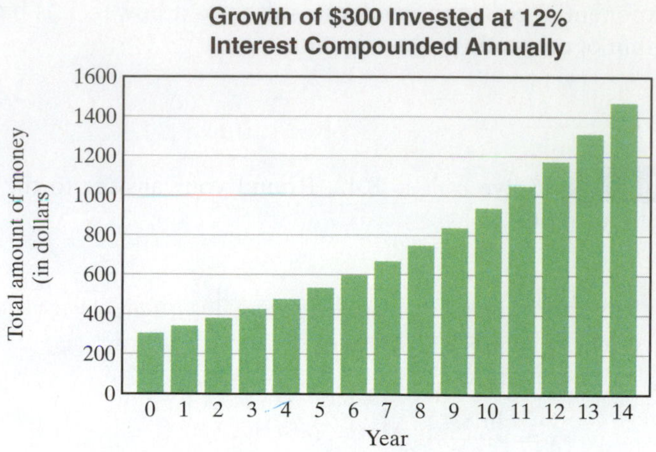

Student Practice 7 Mon Ling's father has an investment account that earns 8% interest compounded annually. How many years would it take for \$4000 to grow to \$10,000 in that account? Round your answer to the nearest whole year.

The growth equation for things that appear to be growing continuously is $A = A_0e^{rt}$, where A is the final amount, A_0 is the original amount, r is the rate at which things are growing in a unit of time, and t is the total number of units of time.

For example, if a laboratory starts with 5000 cells and they reproduce at a rate of 35% per hour, the number of cells in 18 hours can be described by the following equation.

$$A = 5000e^{(0.35)(18)} = 5000e^{6.3} \approx 5000(544.57) = 2{,}722{,}850 \text{ cells}$$

EXAMPLE 8 At the beginning of 2011, the world population was seven billion people and the growth rate was 1.2% per year. If this growth rate continues, how many years will it take for the population to double to fourteen billion people?

Solution

$$A = A_0e^{rt}$$

To make our calculation easier, we write the values for the population in terms of billions. This will allow us to avoid writing numbers like 14,000,000,000 and 7,000,000,000. Do you see why we can do this?

$$14 = 7e^{0.012t} \qquad \text{Substitute known values.}$$

$$\frac{14}{7} = e^{0.012t} \qquad \text{Divide each side by 7.}$$

$$\ln\left(\frac{14}{7}\right) = \ln e^{0.012t} \qquad \text{Take the natural logarithm of each side.}$$

$$\ln(2) = (0.012t)\ln e \qquad \text{Property 3.}$$

$$\ln 2 = 0.012t \qquad \text{Since } \ln e = 1.\text{ Property 4.}$$

$$\frac{\ln 2}{0.012} = t \qquad \text{Divide each side by 0.012}$$

Using our calculator, we obtain the following.

2 [ln] [÷] 0.012 [=] 57.76226505

Rounding to the nearest whole year, we find that the population will grow from seven billion to fourteen billion in about 58 years if the growth continues at the same rate.

Student Practice 8 The wildlife management team in one region of Alaska has determined that the black bear population is growing at the rate of 4.3% per year. This region currently has approximately 1300 black bears. A food shortage will develop if the population reaches 5000. If the growth rate remains unchanged, how many years will it take for this food shortage problem to occur?

EXAMPLE 9 The magnitude of an earthquake (amount of energy released) is measured by the formula $R = \log\left(\frac{I}{I_0}\right)$, where I is the intensity of the earthquake and I_0 is the minimum measurable intensity. A 1964 earthquake in Anchorage, Alaska, had a magnitude of 8.4. A 1906 earthquake in Taiwan had a magnitude of 7.1. How many times more intense was the Anchorage earthquake than the Taiwan earthquake? (*Source:* National Oceanic and Atmospheric Administration)

Solution

Let I_A = intensity of the Alaska earthquake. Then

$$8.4 = \log\left(\frac{I_A}{I_0}\right) = \log I_A - \log I_0.$$

Solving for $\log I_0$ gives

$\log I_0 = \log I_A - 8.4.$

Let I_T = intensity of the Taiwan earthquake. Then

$$7.1 = \log\left(\frac{I_T}{I_0}\right) = \log I_T - \log I_0.$$

Solving for $\log I_0$ gives

$\log I_0 = \log I_T - 7.1.$

Therefore,

$$\log I_A - 8.4 = \log I_T - 7.1.$$
$$\log I_A - \log I_T = 8.4 - 7.1$$
$$\log \frac{I_A}{I_T} = 1.3$$
$$10^{1.3} = \frac{I_A}{I_T}$$
$$19.95262315 \approx \frac{I_A}{I_T} \quad \text{Use a calculator.}$$
$$20 \approx \frac{I_A}{I_T} \quad \text{Round to the nearest whole number.}$$
$$20I_T \approx I_A$$

The Alaska earthquake was approximately twenty times more intense than the Taiwan earthquake.

Student Practice 9 A 1933 earthquake in Japan had a magnitude of 8.9. A 1989 earthquake in San Francisco had a magnitude of 7.1. How many times more intense was the Japan earthquake than the San Francisco earthquake? (*Source:* National Oceanic and Atmospheric Administration)

11.5 Exercises

MyMathLab®

Watch the videos in MyMathLab

Download the MyDashBoard App

Solve each logarithmic equation and check your solutions.

1. $\log_7\left(\frac{2}{3}x + 3\right) + \log_7 3 = 2$

2. $\log_5\left(\frac{1}{2}x - 2\right) + \log_5 2 = 1$

3. $\log_6(x + 3) + \log_6 4 = 2$

4. $\log_2 4 + \log_2(x - 1) = 5$

5. $\log_2\left(x + \frac{4}{3}\right) = 5 - \log_2 6$

6. $\log_5(3x + 1) = 1 - \log_5 2$

7. $\log(30x + 40) = 2 + \log(x - 1)$

8. $1 + \log x = \log(9x + 1)$

9. $2 + \log_6(x - 1) = \log_6(12x)$

10. $\log_2 x = \log_2(x + 5) - 1$

11. $\log(75x + 50) - \log x = 2$

12. $\log_{11}(x + 7) = \log_{11}(x - 3) + 1$

13. $\log_3(x + 6) + \log_3 x = 3$

14. $\log_8 x + \log_8(x - 2) = 1$

15. $1 + \log(x - 2) = \log(6x)$

16. $\log_5(2x) - \log_5(x - 3) = 3\log_5 2$

17. $\log_2(x + 5) - 2 = \log_2 x$

18. $\log_5(5x + 10) - 2 = \log_5(x - 2)$

19. $2\log_7 x = \log_7(x + 4) + \log_7 2$

20. $\log x + \log(x - 1) = \log 12$

21. $\ln 10 - \ln x = \ln(x - 3)$

22. $\ln(2 + 2x) = 2\ln(x + 1)$

Solve each exponential equation. Leave your answers in exact form. Do not approximate.

23. $7^{x+3} = 12$

24. $4^{x-2} = 6$

25. $2^{3x+4} = 17$

26. $5^{2x-1} = 11$

Solve each exponential equation. Use your calculator to approximate your solutions to the nearest thousandth.

27. $8^{2x-1} = 90$

28. $15^{3x-2} = 230$

29. $5^x = 4^{x+1}$

30. $3^x = 2^{x+3}$

31. $28 = e^{x-2}$

32. $e^{x+2} = 88$

33. $88 = e^{2x+1}$

34. $3 = e^{1-x}$

Applications

Compound Interest *When a principal P earns an annual interest rate r compounded yearly, the amount A after t years is* $A = P(1 + r)^t$. *Use this information to solve exercises 35–40. Round all answers to the nearest whole year.*

35. How long will it take $1500 to grow to $5000 at 8% compounded annually?

36. How long will it take $1000 to grow to $4500 at 7% compounded annually?

37. How long will it take for a principal to triple at 6% compounded annually?

38. How long will it take for a principal to double at 5% compounded annually?

39. What interest rate would be necessary to obtain \$6500 in 6 years if \$5000 is the amount of the original investment and the interest is compounded yearly? (Express the interest rate as a percent rounded to the nearest tenth.)

40. If \$3000 is invested for 3 years with annual interest compounded yearly, what interest rate is needed to achieve an amount of \$3600? (Express the interest rate as a percent rounded to the nearest tenth.)

World Population *The growth of the world population can be described by the equation* $A = A_0e^{rt}$, *where time* t *is measured in years,* A_0 *is the population of the world at time* $t = 0$, r *is the annual growth rate, and* A *is the population at time* t. *Use this information to solve exercises 41–44. Round your answers to the nearest whole year.*

41. In 1963, the world population was about three billion people and the growth rate was 2.1% per year. At that rate, how long would it have taken for the population to increase to seven billion?

42. At a growth rate of 2.1% per year, how long would it take a population to double?

43. By 2009, the world population was about six billion and the growth rate had slowed to 1.3%. At that rate, how long would it take for the population to increase to nine billion?

44. At a growth rate of 1.3% per year, how long would it take a population to double?

Biomedical Engineers *According to the Bureau of Labor Statistics, the biomedical engineer field has been the fastest growing occupation since 2008. The number* N *of biomedical engineers in the United States can be approximated by the equation* $N = 16{,}000(1.057)^x$, *where* x *is the number of years since 2008. Use this equation to answer the following questions.* (*Source:* www.bls.gov)

45. Approximately how many biomedical engineers were there in 2010? Round to the nearest whole number.

46. Approximately how many biomedical engineers will there be in 2013? Round to the nearest whole number.

47. In what year will the number of biomedical engineers reach 23,600?

48. In what year will the number of biomedical engineers reach 29,000?

Use the equation $A = A_0e^{rt}$ to solve exercises 49–54. Round your answers to the nearest whole number.

49. ***Population*** The population of Bethel is 80,000 people, and it is growing at the rate of 1.5% per year. How many years will it take for the population to grow to 120,000 people?

50. ***Population*** The 2010 population of Melbourne, Australia, was approximately four million people. The growth rate is 2% per year. In how many years will there be 4.5 million people?

51. ***Skin Grafts*** The number of new skin cells on a revolutionary skin graft is growing at a rate of 4% per hour. How many hours will it take for 200 cells to become 1800 cells?

52. ***Workforce*** The workforce in a state is increasing at the rate of 1.5% per year. During the last measured year, the workforce was 3.5 million. If this rate continues, how many years will it be before the workforce reaches 4.5 million?

53. ***Lyme Disease*** Unfortunately, some U.S. deer carry ticks that spread Lyme disease. The number of people who are infected by the virus is increasing by 6% every year. If 29,959 people were confirmed to have Lyme disease in 2009, how many are expected to be infected by the end of the year 2015? (*Source:* www.cdc.gov)

54. ***DVD Rentals*** In the city of Scranton, the number of DVD rentals is increasing by 7.5% per year. For the last year that data are available, 1.3 million DVDs were rented. How many years will it be before 2.0 million DVDs are rented per year?

To Think About

Earthquakes *The magnitude of an earthquake (amount of energy released) is described by the formula* $R = \log\left(\frac{I}{I_0}\right)$, *where I is the intensity of the earthquake and I_0 is the minimum measurable intensity. Use this formula to solve exercises 55–58. Round answers to the nearest tenth.*

55. October 17, 1989, brought tragedy to the San Francisco/Oakland area. An earthquake measuring 7.1 on the Richter scale and centered in the Loma Prieta area (Santa Cruz Mountains) collapsed huge sections of freeway, killing sixty-three people. Almost 6 years later, an earthquake measuring 8.2 on the Richter scale killed 190 people in the Kurile Islands of Japan and Russia. How many times more intense was the Kurile earthquake than the Loma Prieta earthquake? (*Source:* National Oceanic and Atmospheric Administration)

56. On January 17, 1993, in Northridge, California, residents experienced an earthquake that measured 6.8 on the Richter scale, killed sixty-one people, and undermined supposedly earthquake-proof steel-framed buildings. Exactly one year later near Kobe, Japan, an earthquake measuring 7.2 on the Richter scale killed more than 5300 people, injured more than 35,000, and destroyed nearly 200,000 homes, in spite of construction codes reputed to be the best in the world. How many times more intense was the Kobe earthquake than the Northridge earthquake? (*Source:* National Oceanic and Atmospheric Administration)

57. The 1906 earthquake in San Francisco had a magnitude of 8.3. In 1971 an earthquake in Japan measured 6.8. How many times more intense was the San Francisco earthquake than the Japan earthquake? (*Source:* National Oceanic and Atmospheric Administration)

58. The 1933 Japan earthquake had a magnitude of 8.9. In Turkey a 1975 earthquake had a magnitude of 6.7. How many times more intense was the Japan earthquake than the Turkey earthquake? (*Source:* National Oceanic and Atmospheric Administration)

Cumulative Review *Simplify. Assume that x and y are positive real numbers.*

59. **[7.4.1]** $(\sqrt{3} + 2\sqrt{2})(\sqrt{6} - \sqrt{2})$

60. **[7.3.1]** $\sqrt{98x^3y^2}$

61. **[1.2.2]** ***Spelling Bee*** 273 students competed in the 2010 Scripps National Spelling Bee. A total of thirty-nine students were 12 years old. Twenty-two more students were 13 years old than were 12 years old. One hundred forty-seven students were 14 or 15 years old. All the remaining students were younger. Only two students, the youngest in the competition, were 9 years old. There was one 10-year-old student. Twenty-two more students were 11 years old than were 10 years old. How many students were there in each age category (ages 9, 10, 11, 12, 13, and 14–15)?

62. **[1.2.2]** ***London Subway*** The London subway system has a staff of 19,000, and an average of 3.4 million passengers use the subway each weekday. Between 1993 and 1999, an expansion of the system was built that extended the system 16 kilometers and cost 3.5 billion British pounds. How many dollars per mile did this extension cost? (Use 1 kilometer = 0.62 miles and 1 U.S. dollar = 0.64 British pounds.) (*Source:* en.wikipedia.org)

Quick Quiz 11.5 *Solve the equation.*

1. $2 - \log_6 x = \log_6(x + 5)$

2. $\log_3(x - 5) + \log_3(x + 1) = \log_3 7$

3. $6^{x+2} = 9$ (Round your answer to the nearest thousandth.)

4. **Concept Check** Explain how you would solve $26 = 52\, e^{3x}$.

Did You Know...

That You Can Save Money by Purchasing Store Brand Products?

FOOD AND RICE PRICES

Understanding the Problem:
Many large stores and chains have their own brands of products. Store brand products often cost much less than the equivalent name brand products. For example, a one-pound bag of rice from a national name brand can cost $2.50. A one-pound bag of the same rice from a store brand might cost only $1.00.

Making a Plan:
Lucy and her family enjoy rice as part of their dinner four times a week. Lucy has a family of four that consumes approximately 2/3 cup of rice with each meal. Lucy wants to calculate the cost of feeding her family and see if they can save money.

Step 1: Lucy needs to know how much rice her family consumes in a week.

Task 1: *How many cups of rice does Lucy's family eat in a week?*

Task 2: *If each cup of raw weighs approximately 21 ounces, what is the weight in ounces of the rice Lucy's family eats in a week?*

Task 3: *There are 16 ounces in a pound. Find the number of pounds of raw rice Lucy's family eats each week?*

Step 2: Lucy notices that her supermarket has its own brand of rice that is much less expensive than the brand she normally buys. She usually buys name brand rice for $2.66 per pound. The store brand is only 88 cents per pound.

Task 4: *Find out how much Lucy spends per week on the name brand rice.*

Task 5: *Find out how much Lucy could save per week by buying the same amount of rice from the store brand.*

Task 6: *Find the percent savings that Lucy gets by buying the store brand rice instead of the name brand.*

Finding a Solution:

Step 3: Lucy realizes that she can have this kind of savings each week on all the food she buys by choosing the store brand products over the name brand.

Task 7: *If Lucy typically spends $162 per week on groceries by buying the name brand products, how much could she save in a year by purchasing all store brand products?*

Task 8: *Lucy finds that her family still prefers some of the name brand products. She continues to buy some of the name brand products and some of the store brands. She finds that her weekly grocery bill is reduced to around $130 per week. It this continues, how much will she save in the course of a year?*

Applying the Situation to Your Life:
You should calculate how much you spend on groceries every week. Then see how you can reduce that by purchasing store brands. You may not change to the store brand for all your shopping, but for every case where you do, you can often save money.

Chapter 11 Organizer

Topic and Procedure	Examples	You Try It

Exponential function, p. 584

$f(x) = b^x$, where $b > 0$, $b \neq 1$, and x is a real number.

Examples

Graph $f(x) = \left(\frac{2}{3}\right)^x$.

x	f(x)
−2	2.25
−1	1.5
0	1
1	$0.\overline{6}$
2	$0.\overline{4}$

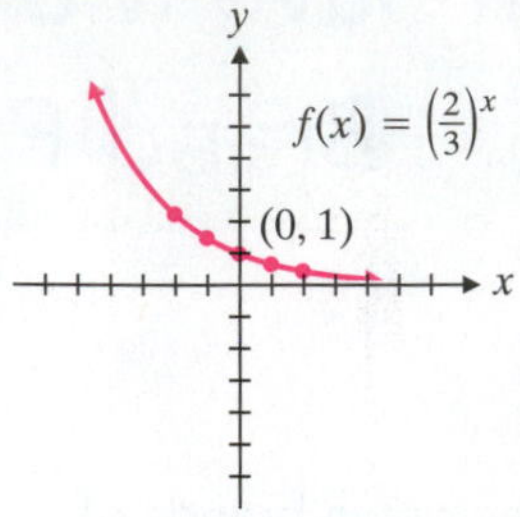

You Try It

1. Graph $f(x) = \left(\frac{1}{4}\right)^x$.

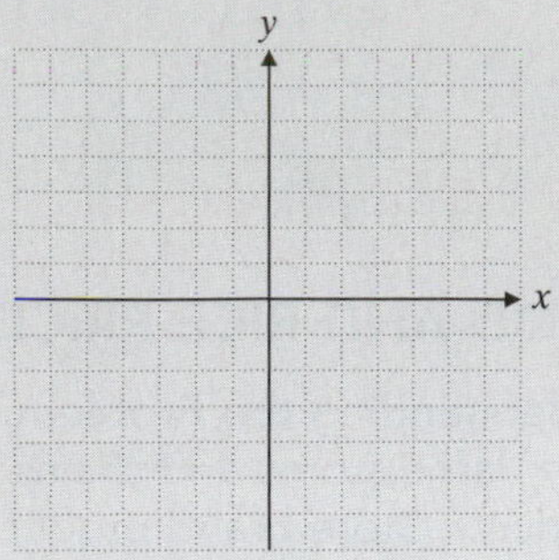

Property of exponential equations, p. 586

When $b > 0$ and $b \neq 1$, if $b^x = b^y$, then $x = y$.

Examples

Solve for x. $2^x = \frac{1}{32}$

$$2^x = \frac{1}{2^5}$$
$$2^x = 2^{-5}$$
$$x = -5$$

You Try It

2. Solve for x. $3^x = \frac{1}{81}$

Definition of logarithm, p. 593

$y = \log_b x$ is the same as $x = b^y$, where $x > 0$, $b > 0$, and $b \neq 1$.

Examples

(a) Write in exponential form. $\log_3 17 = 2x$

$$3^{2x} = 17$$

(b) Write in logarithmic form. $18 = 3^x$

$$\log_3 18 = x$$

(c) Solve for x. $\log_6\left(\frac{1}{36}\right) = x$

$$6^x = \frac{1}{36}$$
$$6^x = 6^{-2}$$
$$x = -2$$

You Try It

3. **(a)** Write in exponential form. $\log_4 9 = 0.5x$

(b) Write in logarithmic form. $28 = 5^x$

(c) Solve for x. $\log_4\left(\frac{1}{64}\right) = x$

Properties of logarithms, pp. 600–603

Suppose that $M > 0$, $N > 0$, $b > 0$, and $b \neq 1$.

$\log_b MN = \log_b M + \log_b N$

$\log_b\left(\frac{M}{N}\right) = \log_b M - \log_b N$

$\log_b M^p = p \log_b M$

$\log_b b = 1$

$\log_b 1 = 0$

If $\log_b x = \log_b y$, then $x = y$.

If $x = y$, then $\log_b x = \log_b y$.

Examples

(a) Write as separate logarithms of x, y, and w.

$$\log_3\left(\frac{x^2\sqrt[3]{y}}{w}\right)$$
$$= 2\log_3 x + \frac{1}{3}\log_3 y - \log_3 w$$

(b) Write as one logarithm.

$$5\log_6 x - 2\log_6 w - \frac{1}{4}\log_6 z$$
$$= \log_6\left(\frac{x^5}{w^2\sqrt[4]{z}}\right)$$

(c) Simplify.

$$\log 10^5 + \log_3 3 + \log_5 1$$
$$= 5\log 10 + \log_3 3 + \log_5 1$$
$$= 5 + 1 + 0$$
$$= 6$$

You Try It

4. **(a)** Write as separate logarithms of a, b, and c. $\log_2\left(\frac{a\sqrt{b}}{c^2}\right)$

(b) Write as one logarithm.

$3\log_5 a + \frac{2}{3}\log_5 b - 2\log_5 c$

(c) Simplify.

$\log_8 1 - \log 10^3 + \log_4 4$

Finding logarithms, pp. 608, 610

On a scientific calculator:

$\log x = \log_{10} x$, for all $x > 0$

$\ln x = \log_e x$, for all $x > 0$

Examples

(a) Find log 3.82.

3.82 [log]

$\log 3.82 \approx 0.5820634$

(b) Find ln 52.8.

52.8 [ln]

$\ln 52.8 \approx 3.9665112$

You Try It

5. Find each logarithm.

(a) log 16.5

(b) ln 32.7

Topic and Procedure	Examples	You Try It

Finding antilogarithms, pp. 608, 610

If $\log x = b$, then $10^b = x$.
If $\ln x = b$, then $e^b = x$.
Use a calculator or a table to solve.

(a) Find x if $\log x = 2.1416$.

$$10^{2.1416} = x$$

2.1416 [10^x] $\approx$ 138.54792

(b) Find x if $\ln x = 0.6218$.

$$e^{0.6218} = x$$

0.6218 [e^x] $\approx$ 1.8622771

6. Find x.

(a) $\log x = 2.075$

(b) $\ln x = 1.528$

Finding a logarithm to a different base, p. 611

Change of base formula:

$$\log_b x = \frac{\log_a x}{\log_a b},$$ where a, b, and $x > 0$, $a \neq 1$, and $b \neq 1$.

Evaluate $\log_7 1.86$.

$$\log_7 1.86 = \frac{\log 1.86}{\log 7}$$

1.86 [log] [÷] 7 [log] [=] 0.3189132

7. Evaluate $\log_8 2.5$.

Solving logarithmic equations, p. 616

1. If some but not all of the terms of an equation have logarithms, try to rewrite the equation with one single logarithm on one side and one numerical value on the other. Then convert the equation to exponential form.
2. If an equation contains logarithmic terms only, try to get only one logarithm on each side of the equation. Then use the property that if $\log_b x = \log_b y$, $x = y$.

Note: Always check your solutions when solving logarithmic equations.

Solve for x. $\log_5 3x - \log_5(x^2 - 1) = \log_5 2$

$$\log_5 3x = \log_5 2 + \log_5(x^2 - 1)$$
$$\log_5 3x = \log_5[2(x^2 - 1)]$$
$$3x = 2x^2 - 2$$
$$0 = 2x^2 - 3x - 2$$
$$0 = (2x + 1)(x - 2)$$
$$2x + 1 = 0 \qquad x - 2 = 0$$
$$x = -\frac{1}{2} \qquad x = 2$$

Check.

$x = 2$: $\log_5 3(2) - \log_5(2^2 - 1) \stackrel{?}{=} \log_5 2$

$$\log_5 6 - \log_5 3 \stackrel{?}{=} \log_5 2$$
$$\log_5\left(\frac{6}{3}\right) \stackrel{?}{=} \log_5 2$$
$$\log_5 2 = \log_5 2 \quad ✓$$

$x = -\frac{1}{2}$: For the expression $\log_5(3x)$, we would obtain $\log_5(-1.5)$. You cannot take the logarithm of a negative number. $x = -\frac{1}{2}$ is not a solution. The solution is 2.

8. Solve for x.
$-\log_6 3 + \log_6 8x = \log_6(x^2 - 1)$

Solving exponential equations, p. 618

1. See whether each expression can be written so that only one base appears on one side of the equation and the same base appears on the other side. Then use the property that if $b^x = b^y$, $x = y$.
2. If you can't do step 1, take the logarithm of each side of the equation and use the properties of logarithms to solve for the variable.

Solve for x. $2^{x-1} = 7$

$$\log 2^{x-1} = \log 7$$
$$(x - 1) \log 2 = \log 7$$
$$x \log 2 - \log 2 = \log 7$$
$$x \log 2 = \log 7 + \log 2$$
$$x = \frac{\log 7 + \log 2}{\log 2}$$

(We can approximate the answer as $x \approx 3.8073549$.)

9. Solve for x. $3^{x+2} = 11$

Chapter 11 Review Problems

Graph the functions in exercises 1 and 2.

1. $f(x) = 4^{3+x}$

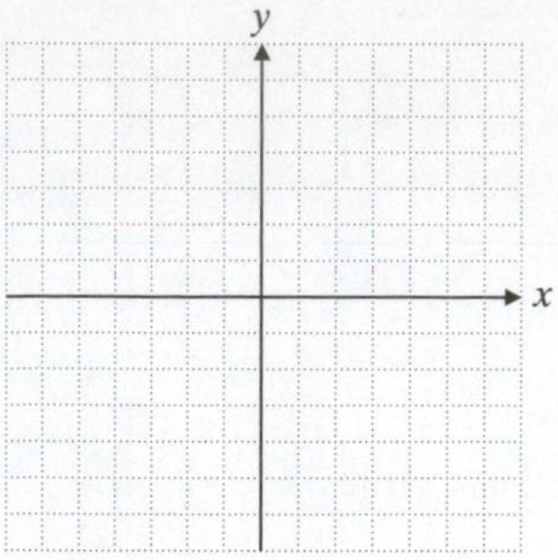

2. $f(x) = e^{x-3}$

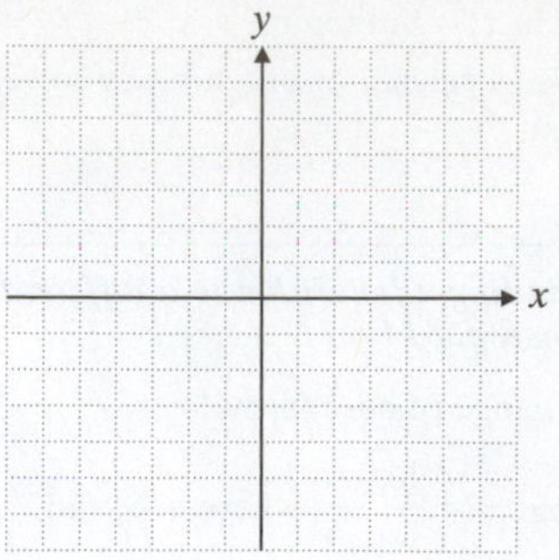

3. Solve. $3^{3x+1} = 81$

4. $2^{x+9} = 128$

5. Write in exponential form. $-2 = \log(0.01)$

6. Change to logarithmic form. $8 = 4^{3/2}$

Solve.

7. $\log_w 16 = 4$

8. $\log_8 x = 0$

9. $\log_7 w = -1$

10. $\log_w 64 = 3$

11. $\log_{10} w = -1$

12. $\log_{10} 1000 = x$

13. $\log_2 64 = x$

14. $\log_2\left(\frac{1}{4}\right) = x$

15. Graph the equation $y = \log_3 x$.

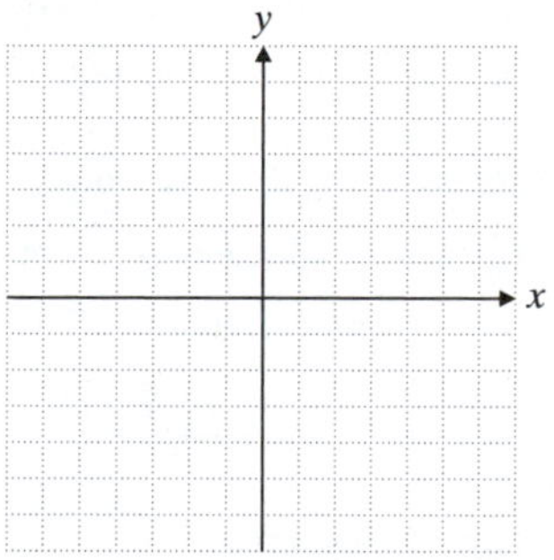

Write each expression as the sum or difference of $\log_2 w$, $\log_2 x$, $\log_2 y$, *and* $\log_2 z$.

16. $\log_2\left(\frac{5x}{\sqrt{w}}\right)$

17. $\log_2 x^3\sqrt{y}$

Write as a single logarithm.

18. $\log_3 x + \log_3 w^{1/2} - \log_3 2$

19. $4 \log_8 w - \frac{1}{3}\log_8 z$

20. Evaluate $\log_e e^6$.

Find the value with a scientific calculator.

21. $\log 23.8$

22. $\log 0.0817$

23. $\ln 3.92$

24. $\ln 803$

25. Find n if $\log n = 1.1367$. **26.** Find n if $\ln n = 1.7$. **27.** Evaluate. $\log_8 2.81$ **28.** Evaluate. $\log_4 72$

Solve each equation and check your solutions.

29. $\log_5 100 - \log_5 x = \log_5 4$

30. $\log_8 x + \log_8 3 = \log_8 75$

31. $\log_{11}\left(\frac{4}{3}x + 7\right) + \log_{11} 3 = 2$

32. $\log_8(x - 3) = -1 + \log_8 6x$

33. $\log(2t + 3) + \log(4t - 1) = 2 \log 3$

34. $\log(2t + 4) - \log(3t + 1) = \log 6$

Solve each equation. Leave your answers in exact form. Do not approximate.

35. $3^x = 14$ **36.** $5^{x+3} = 130$ **37.** $e^{2x-1} = 100$

Solve each equation. Round your answers to the nearest ten-thousandth.

38. $2^{3x+1} = 5^x$ **39.** $e^{3x-4} = 20$ **40.** $(1.03)^x = 20$

Compound Interest *For exercises 41 and 42, use $A = P(1 + r)^t$, the formula for interest that is compounded annually.*

41. How long will it take Frances to double the money in her account if the interest rate is 8% compounded annually? (Round your answer to the nearest year.)

42. How much money would Chou Lou have after 4 years if he invested \$5000 at 6% compounded annually?

Populations *The growth of many populations can be described by the equation $A = A_0e^{rt}$, where time t is measured in years, A_0 is the population at time $t = 0$, r is the annual growth rate, and A is the population at time t. Use this information to solve exercises 43 and 44. Round your answers to the nearest whole year.*

43. How long will it take a population of seven billion to increase to sixteen billion if $r = 2\%$ per year?

44. The number of moose in northern Maine is increasing at a rate of 3% per year. It is estimated in one county that there are now 2000 moose. If the growth rate remains unchanged, how many years will it be until there are 2600 moose in that county?

45. ***Gas Volume*** The work W done by a volume of gas expanding at a constant temperature from volume V_0 to volume V_1 is given by $W = p_0V_0 \ln\left(\frac{V_1}{V_0}\right)$, where p_0 is the pressure at volume V_0.

(a) Find W when $p_0 = 40$ pounds per cubic inch, $V_0 = 15$ cubic inches, and $V_1 = 24$ cubic inches.

(b) If the amount of work is 100 pounds per cubic inch, $V_0 = 8$ cubic inches, and $V_1 = 40$ cubic inches, find p_0.

46. ***Earthquakes*** An earthquake's magnitude is given by $M = \log\left(\frac{I}{I_0}\right)$, where I is the intensity of the earthquake and I_0 is the minimum measurable intensity. A 1964 earthquake in Anchorage, Alaska, had a magnitude of 8.4. A 1975 earthquake in Turkey had a magnitude of 6.7. How many times more intense was the Alaska earthquake than the Turkey earthquake? (*Source:* National Oceanic and Atmospheric Administration)

How Am I Doing? Chapter 11 Test

Chapter Test Prep Videos MATH COACH MyMathLab® YouTube™

After you take this test read through the Math Coach on pages 633–634. Math Coach videos are available via MyMathLab and YouTube. Step-by-step test solutions in the Chapter Test Prep Videos are also available via MyMathLab and YouTube. (Search "TobeyInterAlg" and click on "Channels.")

1. Graph $f(x) = 3^{x+1}$.

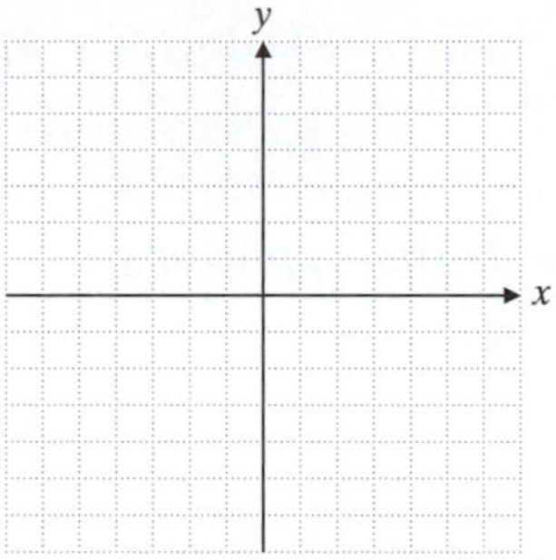

2. Graph $f(x) = \log_2 x$.

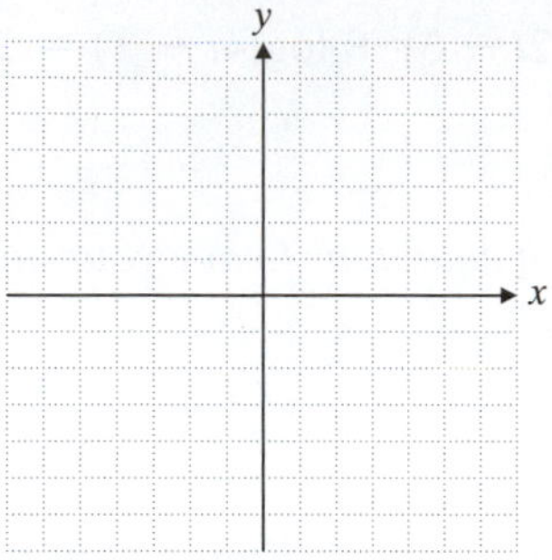

MC **3.** Solve. $4^{x+3} = 64$

In exercises 4 and 5, solve for the variable.

4. $\log_w 125 = 3$

5. $\log_8 x = -2$

MC **6.** Write as a single logarithm. $2 \log_7 x + \log_7 y - \log_7 4$

Evaluate using a calculator. Round your answers to the nearest ten-thousandth.

7. $\ln 5.99$

8. $\log 23.6$

9. $\log_3 1.62$

Use a scientific calculator to approximate x.

10. $\log x = 3.7284$

11. $\ln x = 0.14$

Solve the equation and check your solutions for problems 12 and 13.

MC **12.** $\log_8(x + 3) - \log_8 2x = \log_8 4$

13. $\log_8 2x + \log_8 6 = 2$

MC **14.** Solve the equation. Leave your answer in exact form. Do not approximate. $e^{5x-3} = 57$

15. Solve. $5^{3x+6} = 17$ (Approximate your answer to the nearest ten-thousandth.)

16. How much money will Henry have if he invests $2000 for 5 years at 8% annual interest compounded annually?

17. How long will it take for Barb to double her money if she invests it at 5% compounded annually? Round to the nearest whole year.

1. ______
2. ______
3. ______
4. ______
5. ______
6. ______
7. ______
8. ______
9. ______
10. ______
11. ______
12. ______
13. ______
14. ______
15. ______
16. ______
17. ______

Total correct: ______

MATH COACH

Mastering the skills you need to do well on the test.

Students often make the same types of errors when they do the Chapter 11 Test. Here are some helpful hints to keep you from making those common errors on test problems.

Solving an Exponential Equation—Problem 3

Solve. $4^{x+3} = 64$

Helpful Hint If one side of the equation is in exponential form, it is best to try to write the other side of the equation in exponential form. Then the procedure will be easier to complete.

Did you rewrite the equation as $4^{x+3} = 4^3$?

Yes ____ No ____

If you answered No, remember to write your equation in the form $b^x = b^y$. Note that $64 = 4^3$, and you want the base of the exponent on each side of the equation to be the same number, b, such that $b > 0$ and $b \neq 1$.

Did you use the property of exponential equations to write the equation $x + 3 = 3$?

Yes ____ No ____

If you answered No, remember that if $b^x = b^y$, then $x = y$ for any $b > 0$ and $b \neq 1$. Now solve the equation for x.

If you answered Problem 3 incorrectly, go back and rework the problem using these suggestions.

Using the Properties of Logarithms to Write Sums and Differences of Logarithms as a Single Logarithm—Problem 6

Write as a single logarithm. $2 \log_7 x + \log_7 y - \log_7 4$

Helpful Hint Try your best to memorize the three properties of logarithms in Objectives 11.3.1, 11.3.2, and 11.3.3. They are essential to know when working with logarithmic expressions.

Did you use property 3 to eliminate the coefficient of the first logarithmic term, $2 \log_7 x$, and rewrite that term as $\log_7 x^2$?

Yes ____ No ____

If you answered No, review property 3 in Objective 11.3.3 and complete this step again.

Did you use property 1 to combine the sum of the first two logarithmic terms and obtain the expression, $\log_7 x^2 y - \log_7 4$?

Yes ____ No ____

If you answered No, review property 1 in Objective 11.3.1 and complete this step again.

In your last step, you will need to use property 2 in Objective 11.3.2 to combine the difference of two logarithms.

Now go back and rework the problem using these suggestions.

Need help? Watch the MATH COACH videos in MyMathLab® or on YouTube™.

Solving a Logarithmic Equation—Problem 12

Solve the equation and check your solution. $\log_8(x + 3) - \log_8 2x = \log_8 4$

Helpful Hint Use the properties of logarithms to rewrite the equation such that one logarithmic term appears on each side of the equation.

First, did you combine the two logarithms on the left side of the equation and rewrite the equation as $\log_8\left(\frac{x + 3}{2x}\right) = \log_8 4$?

Yes ____ No ____

If you answered No, use property 2 from Objective 11.3.2 to complete this first step again.

Next, did you rewrite the equation as $\frac{x + 3}{2x} = 4$?

Yes ____ No ____

If you answered No, notice that you now have one logarithmic term on each side of the equation, which is the goal mentioned in the Helpful Hint. You can use property 6 from Objective 11.3.4 to evaluate these two logarithms.

In your final step, solve the equation for x. Check your solution by substituting for x in the original equation, evaluating the logarithms and then simplifying the resulting equation.

If you answered Problem 12 incorrectly, go back and rework the problem using these suggestions.

Solving an Exponential Equation Involving e Raised to a Power—Problem 14

Solve the equation. Leave your answer in exact form. Do not approximate. $e^{5x-3} = 57$

Helpful Hint The first step is to take the natural logarithm of each side of the equation. Then you can simplify the equation further using the properties of logarithms.

Did you take the natural logarithm of each side of the equation and rewrite as $\ln e^{5x-3} = \ln 57$?

Yes ____ No ____

Next did you rewrite this equation as $(5x - 3)(\ln e) = \ln 57$?

Yes ____ No ____

If you answered No to these questions, review property 3 of logarithms from Objective 11.3.3 and complete this step again.

Did you rewrite the equation as $5x - 3 = \ln 57$?

Yes ____ No ____

If you answered No, remember that $\ln e = \log_e e = 1$. This combines the definition of natural logarithms or $\ln e = \log_e e$ and property 4 of logarithms from Objective 11.3.4, which indicates that $\log_e e = 1$.

In your final step, solve the equation for x without evaluating ln 57.

Now go back and rework the problem using these suggestions.

Need more help? Look for section examples marked with MC to review.

Practice Final Examination

Review Chapters 1–11 and then try to solve the problems in this Practice Final Examination.

Chapter 1

1. Evaluate.
$(4 - 3)^2 + \sqrt{9} \div (-3) + 4$

2. Write using scientific notation.
36,250,000

3. Simplify.
$3a + 6b - a + 5ab + 3a^2 + b$

4. Simplify. $3[2x - 5(x + y)]$

5. $F = \frac{9}{5}C + 32$. Find F when $C = -35$.

Chapter 2

6. Solve for y. $\frac{1}{3}y - 4 = \frac{1}{2}y + 1$

7. Solve for b. $A = \frac{1}{2}a(b + c)$

8. Solve for x. $\left|\frac{2}{3}x - 4\right| = 2$

9. Solve for x.
$2x - 3 < x - 2(3x - 2)$

▲ **10.** A piece of land is rectangular and has a perimeter of 1760 meters. The length is 200 meters less than twice the width. Find the dimensions of the land.

11. A man invested $4000, part at 12% interest and part at 14% interest. After 1 year he had earned $508 in interest. How much was invested at each interest rate?

12. Find the values of x that satisfy the given conditions.
$x + 5 \leq -4$ *or* $2 - 7x \leq 16$

13. Solve the inequality.
$|2x - 5| < 10$

Chapter 3

14. Find the intercepts and then graph the line. $7x - 2y = -14$

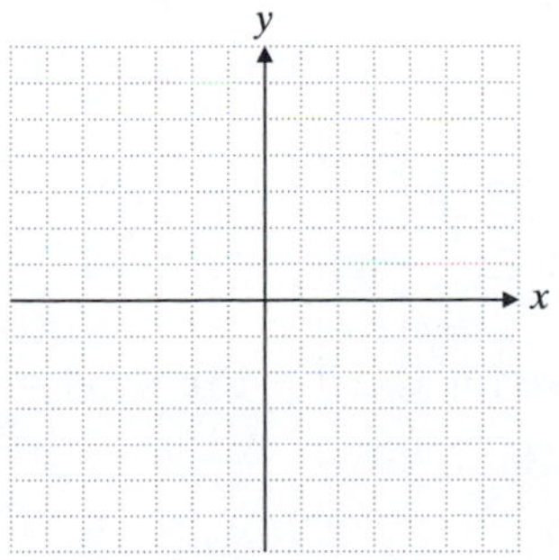

15. Graph the region. $3x - 4y \leq 6$

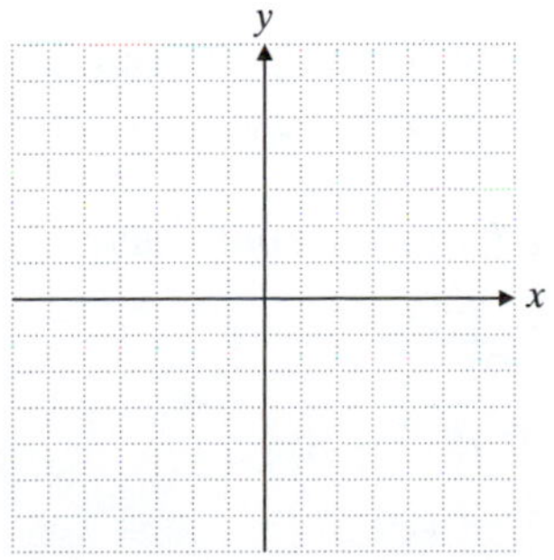

16. Find the slope of the line passing through $(1, 5)$ and $(-2, -3)$.

17. Write the equation in standard form of the line that is parallel to $3x + 2y = 8$ and passes through $(-1, 4)$.

Given the function defined by $f(x) = 3x^2 - 4x - 3$, find the following.

18. $f(3)$

19. $f(-2)$

1. ____
2. ____
3. ____
4. ____
5. ____
6. ____
7. ____
8. ____
9. ____
10. ____
11. ____
12. ____
13. ____
14. ____
15. ____
16. ____
17. ____
18. ____
19. ____

20. ______

21. ______

22. ______

23. ______

24. ______

25. ______

26. ______

27. ______

28. ______

29. ______

30. ______

31. ______

32. ______

33. ______

34. ______

35. ______

36. ______

20. Graph the function. $f(x) = |2x - 4|$

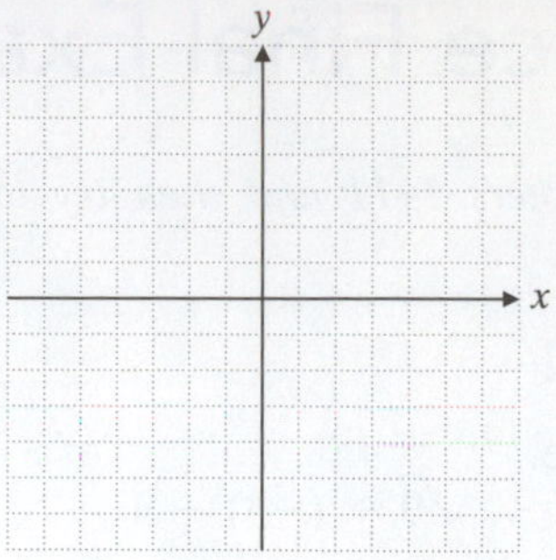

Chapter 4

21. Solve for x and y.

$$2x + y = 15$$
$$3x - 2y = 5$$

22. Solve for x and y.

$$4x - 3y = 12$$
$$3x - 4y = 2$$

23. Solve for x, y, and z.

$$2x + 3y - z = 16$$
$$x - y + 3z = -9$$
$$5x + 2y - z = 15$$

24. Solve for x and y.

$$3x - 2y = 7$$
$$-9x + 6y = 2$$

25. Reynoso sold 15 tickets for the drama club's play. Adults' tickets cost \$8 and children's tickets cost \$3. Reynoso collected \$100. How many of each kind of ticket did he sell?

26. Graph the region.

$$3y \geq 8x - 12$$
$$2x + 3y \leq -6$$

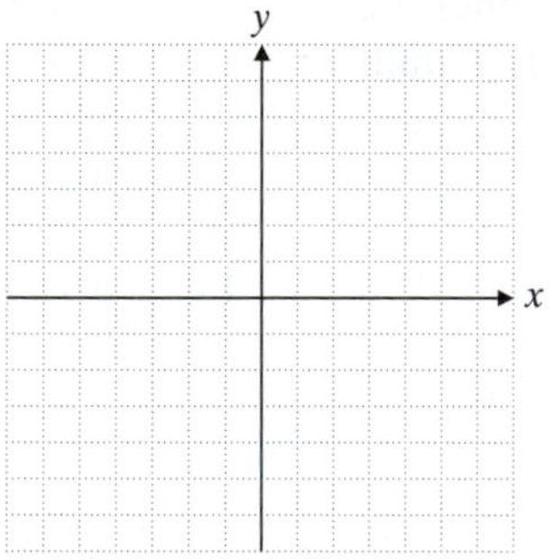

Chapter 5

27. Multiply and simplify your answer.

$(3x - 2)(2x^2 - 4x + 3)$

28. Divide.

$(25x^3 + 9x + 2) \div (5x + 1)$

Factor the following completely.

29. $8x^3 - 27$

30. $x^3 + 2x^2 - 4x - 8$

31. $2x^3 + 15x^2 - 8x$

32. Solve for x. $x^2 + 15x + 54 = 0$

Chapter 6

Simplify the following.

33. $\dfrac{9x^3 - x}{3x^2 - 8x - 3}$

34. $\dfrac{x^2 - 9}{2x^2 + 7x + 3} \div \dfrac{x^2 - 3x}{2x^2 + 11x + 5}$

35. $\dfrac{3x}{x + 5} - \dfrac{2}{x^2 + 7x + 10}$

36. $\dfrac{\dfrac{3}{2x} - 1}{\dfrac{5}{2} + \dfrac{1}{x}}$

37. Solve for x. $\dfrac{x-1}{x^2-4} = \dfrac{2}{x+2} + \dfrac{4}{x-2}$

Chapter 7

38. Evaluate. $16^{3/2}$

39. Simplify. $\sqrt{44a^4b^7c}$

40. Combine like terms.
$5\sqrt{2} - 3\sqrt{50} + 4\sqrt{98}$

41. Rationalize the denominator.
$\dfrac{5}{\sqrt{7}-2}$

42. Simplify and add together.
$i^3 + \sqrt{-25} + \sqrt{-16}$

43. Solve for x and check your solutions. $\sqrt{x+7} = x + 5$

44. If y varies directly with the square of x and $y = 15$ when $x = 2$, what will y be when $x = 3$?

Chapter 8

45. Solve for x. $5x(x+1) = 1 + 6x$

46. Solve for x. $5x^2 - 9x = -12x$

47. Solve for x. $x^{2/3} + 5x^{1/3} - 14 = 0$

48. Solve. $3x^2 - 11x - 4 \geq 0$

▲ **49.** Graph the quadratic function $f(x) = -x^2 - 4x + 5$. Label the vertex and the intercepts.

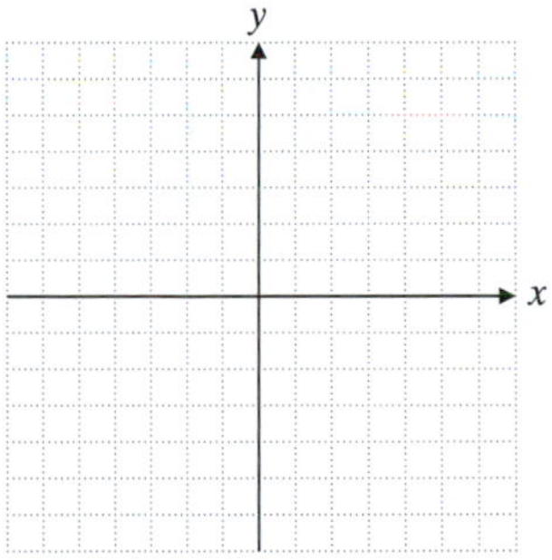

▲ **50.** The area of a rectangle is 52 square centimeters. The length of the rectangle is 1 centimeter longer than 3 times its width. Find the dimensions of the rectangle.

Chapter 9

51. Place the equation of the circle in standard form. Find its center and radius.

$$x^2 + y^2 + 6x - 4y = -9$$

Identify and graph.

52. $\dfrac{x^2}{16} + \dfrac{y^2}{25} = 1$

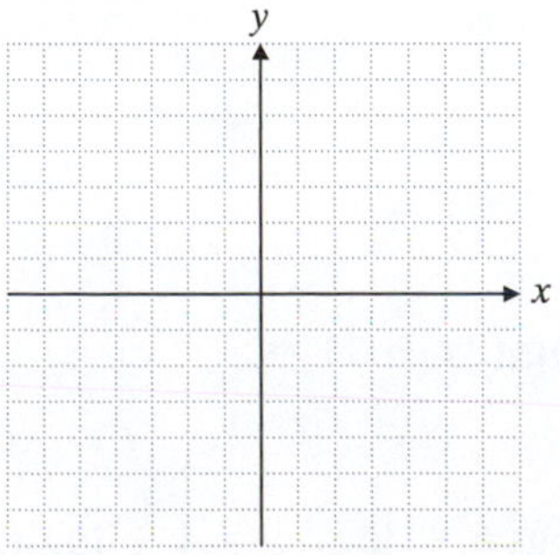

53. $\dfrac{x^2}{4} - \dfrac{y^2}{9} = 1$

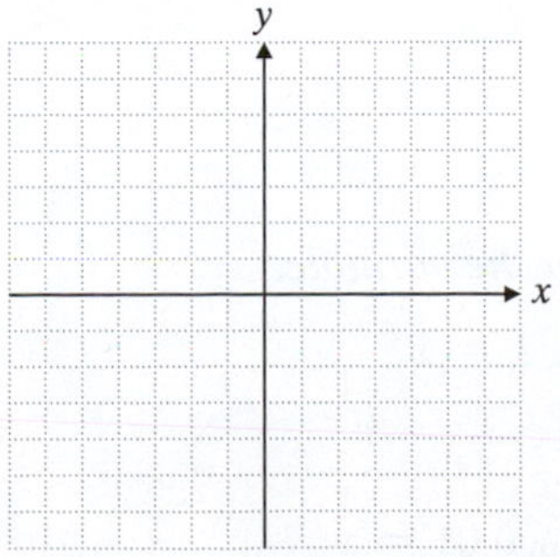

37. ____________

38. ____________

39. ____________

40. ____________

41. ____________

42. ____________

43. ____________

44. ____________

45. ____________

46. ____________

47. ____________

48. ____________

49. ____________

50. ____________

51. ____________

52. ____________

53. ____________

54. ______

55. ______

56. (a) ______

(b) ______

(c) ______

57. ______

58. ______

59. ______

60. ______

61. ______

62. ______

63. ______

64. ______

54. $x = (y - 3)^2 + 5$

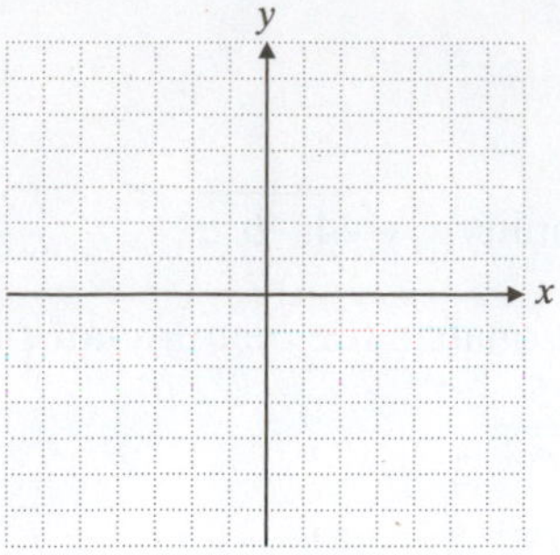

55. Solve the system of equations.

$$x^2 + y^2 = 16$$
$$x^2 - y = 4$$

Chapter 10

56. Let $f(x) = 3x^2 - 2x + 5$.

(a) Find $f(-1)$.

(b) Find $f(a)$.

(c) Find $f(a + 2)$.

57. If $f(x) = 5x^2 - 3$ and $g(x) = -4x - 2$, find $f[g(x)]$.

58. If $f(x) = \frac{1}{2}x - 7$, find $f^{-1}(x)$.

59. Is M a one-to-one function? $M = \left\{(3, 7), (2, 8), \left(7, \frac{1}{2}\right), (-3, 7)\right\}$

Chapter 11

60. Graph $f(x) = 2^{1-x}$. Plot three points.

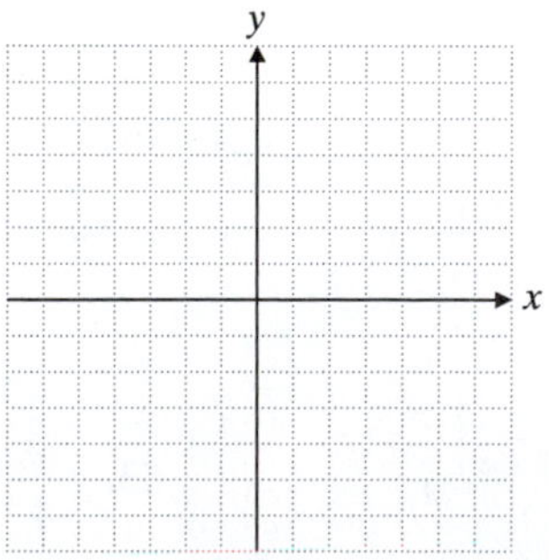

Solve for the variable.

61. $\log_6 1 = x$

62. $\log_4(3x + 1) = 3$

63. $\log_{10} 0.01 = y$

64. $\log_2 6 + \log_2 x = 4 + \log_2(x - 5)$

Appendix A Tables

Table A-1 Table of Square Roots

x	$\sqrt{x}$	x	$\sqrt{x}$	x	$\sqrt{x}$	x	$\sqrt{x}$	x	$\sqrt{x}$
1	1.000	**41**	6.403	**81**	9.000	**121**	11.000	**161**	12.689
2	1.414	**42**	6.481	**82**	9.055	**122**	11.045	**162**	12.728
3	1.732	**43**	6.557	**83**	9.110	**123**	11.091	**163**	12.767
4	2.000	**44**	6.633	**84**	9.165	**124**	11.136	**164**	12.806
5	2.236	**45**	6.708	**85**	9.220	**125**	11.180	**165**	12.845
6	2.449	**46**	6.782	**86**	9.274	**126**	11.225	**166**	12.884
7	2.646	**47**	6.856	**87**	9.327	**127**	11.269	**167**	12.923
8	2.828	**48**	6.928	**88**	9.381	**128**	11.314	**168**	12.961
9	3.000	**49**	7.000	**89**	9.434	**129**	11.358	**169**	13.000
10	3.162	**50**	7.071	**90**	9.487	**130**	11.402	**170**	13.038
11	3.317	**51**	7.141	**91**	9.539	**131**	11.446	**171**	13.077
12	3.464	**52**	7.211	**92**	9.592	**132**	11.489	**172**	13.115
13	3.606	**53**	7.280	**93**	9.644	**133**	11.533	**173**	13.153
14	3.742	**54**	7.348	**94**	9.695	**134**	11.576	**174**	13.191
15	3.873	**55**	7.416	**95**	9.747	**135**	11.619	**175**	13.229
16	4.000	**56**	7.483	**96**	9.798	**136**	11.662	**176**	13.266
17	4.123	**57**	7.550	**97**	9.849	**137**	11.705	**177**	13.304
18	4.243	**58**	7.616	**98**	9.899	**138**	11.747	**178**	13.342
19	4.359	**59**	7.681	**99**	9.950	**139**	11.790	**179**	13.379
20	4.472	**60**	7.746	**100**	10.000	**140**	11.832	**180**	13.416
21	4.583	**61**	7.810	**101**	10.050	**141**	11.874	**181**	13.454
22	4.690	**62**	7.874	**102**	10.100	**142**	11.916	**182**	13.491
23	4.796	**63**	7.937	**103**	10.149	**143**	11.958	**183**	13.528
24	4.899	**64**	8.000	**104**	10.198	**144**	12.000	**184**	13.565
25	5.000	**65**	8.062	**105**	10.247	**145**	12.042	**185**	13.601
26	5.099	**66**	8.124	**106**	10.296	**146**	12.083	**186**	13.638
27	5.196	**67**	8.185	**107**	10.344	**147**	12.124	**187**	13.675
28	5.292	**68**	8.246	**108**	10.392	**148**	12.166	**188**	13.711
29	5.385	**69**	8.307	**109**	10.440	**149**	12.207	**189**	13.748
30	5.477	**70**	8.367	**110**	10.488	**150**	12.247	**190**	13.784
31	5.568	**71**	8.426	**111**	10.536	**151**	12.288	**191**	13.820
32	5.657	**72**	8.485	**112**	10.583	**152**	12.329	**192**	13.856
33	5.745	**73**	8.544	**113**	10.630	**153**	12.369	**193**	13.892
34	5.831	**74**	8.602	**114**	10.677	**154**	12.410	**194**	13.928
35	5.916	**75**	8.660	**115**	10.724	**155**	12.450	**195**	13.964
36	6.000	**76**	8.718	**116**	10.770	**156**	12.490	**196**	14.000
37	6.083	**77**	8.775	**117**	10.817	**157**	12.530	**197**	14.036
38	6.164	**78**	8.832	**118**	10.863	**158**	12.570	**198**	14.071
39	6.245	**79**	8.888	**119**	10.909	**159**	12.610	**199**	14.107
40	6.325	**80**	8.944	**120**	10.954	**160**	12.649	**200**	14.142

Square root values ending in 000 are exact. All other values are approximate and are rounded to the nearest thousandth.

Table A-2 Exponential Values

x	e^x	e^{-x}	x	e^x	e^{-x}
0.00	1.0000	1.0000	**0.95**	2.5857	0.3867
0.01	1.0101	0.9900	**1.0**	2.7183	0.3679
0.02	1.0202	0.9802	**1.1**	3.0042	0.3329
0.03	1.0305	0.9704	**1.2**	3.3201	0.3012
0.04	1.0408	0.9608	**1.3**	3.6693	0.2725
0.05	1.0513	0.9512	**1.4**	4.0552	0.2466
0.06	1.0618	0.9418	**1.5**	4.4817	0.2231
0.07	1.0725	0.9324	**1.6**	4.9530	0.2019
0.08	1.0833	0.9231	**1.7**	5.4739	0.1827
0.09	1.0942	0.9139	**1.8**	6.0496	0.1653
0.10	1.1052	0.9048	**1.9**	6.6859	0.1496
0.11	1.1163	0.8958	**2.0**	7.3891	0.1353
0.12	1.1275	0.8869	**2.1**	8.1662	0.1225
0.13	1.1388	0.8781	**2.2**	9.0250	0.1108
0.14	1.1503	0.8694	**2.3**	9.9742	0.1003
0.15	1.1618	0.8607	**2.4**	11.023	0.0907
0.16	1.1735	0.8521	**2.5**	12.182	0.0821
0.17	1.1853	0.8437	**2.6**	13.464	0.0743
0.18	1.1972	0.8353	**2.7**	14.880	0.0672
0.19	1.2092	0.8270	**2.8**	16.445	0.0608
0.20	1.2214	0.8187	**2.9**	18.174	0.0550
0.21	1.2337	0.8106	**3.0**	20.086	0.0498
0.22	1.2461	0.8025	**3.1**	22.198	0.0450
0.23	1.2586	0.7945	**3.2**	24.533	0.0408
0.24	1.2712	0.7866	**3.3**	27.113	0.0369
0.25	1.2840	0.7788	**3.4**	29.964	0.0334
0.26	1.2969	0.7711	**3.5**	33.115	0.0302
0.27	1.3100	0.7634	**3.6**	36.598	0.0273
0.28	1.3231	0.7558	**3.7**	40.447	0.0247
0.29	1.3364	0.7483	**3.8**	44.701	0.0224
0.30	1.3499	0.7408	**3.9**	49.402	0.0202
0.35	1.4191	0.7047	**4.0**	54.598	0.0183
0.40	1.4918	0.6703	**4.1**	60.340	0.0166
0.45	1.5683	0.6376	**4.2**	66.686	0.0150
0.50	1.6487	0.6065	**4.3**	73.700	0.0136
0.55	1.7333	0.5769	**4.4**	81.451	0.0123
0.60	1.8221	0.5488	**4.5**	90.017	0.0111
0.65	1.9155	0.5220	**4.6**	99.484	0.0101
0.70	2.0138	0.4966	**4.7**	109.95	0.0091
0.75	2.1170	0.4724	**4.8**	121.51	0.0082
0.80	2.2255	0.4493	**4.9**	134.29	0.0074
0.85	2.3396	0.4274	**5.0**	148.41	0.0067
0.90	2.4596	0.4066	**5.5**	244.69	0.0041

x	e^x	e^{-x}	x	e^x	e^{-x}
6.0	403.43	0.0025	**9.5**	13,360	0.000075
6.5	665.14	0.0015	**10**	22,026	0.000045
7.0	1096.6	0.00091	**11**	59,874	0.000017
7.5	1808.0	0.00055	**12**	162,755	0.0000061
8.0	2981.0	0.00034	**13**	442,413	0.0000023
8.5	4914.8	0.00020	**14**	1,202,604	0.0000008
9.0	8103.1	0.00012	**15**	3,269,017	0.0000003

Appendix B Determinants and Cramer's Rule

Student Learning Objectives

After studying this section, you will be able to:

1. Evaluate a second-order determinant.
2. Evaluate a third-order determinant.
3. Solve a system of two linear equations with two unknowns using Cramer's rule.
4. Solve a system of three linear equations with three unknowns using Cramer's rule.

① Evaluating a Second-Order Determinant

Mathematicians have developed techniques to solve systems of linear equations by focusing on the coefficients of the variables and the constants in the equations. The computational techniques can be easily carried out by computers or calculators. We will learn to do them by hand so that you will have a better understanding of what is involved.

To begin, we need to define a matrix and a determinant. A **matrix** is any rectangular array of numbers that is arranged in rows and columns. We use the symbol [ ] to indicate a matrix.

$$\begin{bmatrix} 3 & 2 & 4 \\ -1 & 4 & 0 \end{bmatrix}, \quad \begin{bmatrix} 4 & -3 \\ 2 & \frac{1}{2} \\ 1 & 5 \end{bmatrix}, \quad [-4 \quad 1 \quad 6], \quad \text{and} \quad \begin{bmatrix} \frac{1}{4} \\ 3 \\ -2 \end{bmatrix}$$

are matrices. If you have a graphing calculator, you can enter the elements of a matrix and store them for future use. Let's examine two systems of equations.

$$\begin{aligned} 3x + 2y &= 16 \\ x + 4y &= 22 \end{aligned} \quad \text{and} \quad \begin{aligned} -6x \qquad &= 18 \\ x + 3y &= 9 \end{aligned}$$

We could write the coefficients of the variables in each of these systems as a matrix.

$$\begin{aligned} 3x + 2y \\ x + 4y \end{aligned} \Rightarrow \begin{bmatrix} 3 & 2 \\ 1 & 4 \end{bmatrix} \quad \text{and} \quad \begin{aligned} -6x \qquad \\ x + 3y \end{aligned} \Rightarrow \begin{bmatrix} -6 & 0 \\ 1 & 3 \end{bmatrix}$$

Now we define a determinant. A **determinant** is a *square* arrangement of numbers. We use the symbol | | to indicate a determinant.

$$\begin{vmatrix} 3 & 2 \\ 1 & 4 \end{vmatrix} \quad \text{and} \quad \begin{vmatrix} -6 & 0 \\ 1 & 3 \end{vmatrix}$$

are determinants. The value of a determinant is a *real number* and is defined as follows:

VALUE OF A SECOND-ORDER DETERMINANT

The value of the second-order determinant $\begin{vmatrix} a & c \\ b & d \end{vmatrix}$ is $ad - bc$.

EXAMPLE 1 Find the value of each determinant.

(a) $\begin{vmatrix} -6 & 2 \\ -1 & 4 \end{vmatrix}$ **(b)** $\begin{vmatrix} 0 & -3 \\ -2 & 6 \end{vmatrix}$

Solution

(a) $\begin{vmatrix} -6 & 2 \\ -1 & 4 \end{vmatrix} = (-6)(4) - (-1)(2) = -24 - (-2) = -24 + 2 = -22$

(b) $\begin{vmatrix} 0 & -3 \\ -2 & 6 \end{vmatrix} = (0)(6) - (-2)(-3) = 0 - (+6) = -6$

Student Practice 1 Find the value of each determinant.

(a) $\begin{vmatrix} -7 & 3 \\ -4 & -2 \end{vmatrix}$ **(b)** $\begin{vmatrix} 5 & 6 \\ 0 & -5 \end{vmatrix}$

***NOTE TO STUDENT:** Fully worked-out solutions to all of the Student Practice problems can be found at the back of the text starting at page SP-1.*

② Evaluating a Third-Order Determinant

Third-order determinants have three rows and three columns. Again, each determinant has exactly one value.

VALUE OF A THIRD-ORDER DETERMINANT

The value of the third-order determinant

$$\begin{vmatrix} a_1 & b_1 & c_1 \\ a_2 & b_2 & c_2 \\ a_3 & b_3 & c_3 \end{vmatrix}$$

is

$$a_1b_2c_3 + b_1c_2a_3 + c_1a_2b_3 - a_3b_2c_1 - b_3c_2a_1 - c_3a_2b_1.$$

Because this definition is difficult to memorize and cumbersome to use, we evaluate third-order determinants by a simpler method called **expansion by minors.** The **minor** of an element (number or variable) of a third-order determinant is the second-order determinant that remains after we delete the row and column in which the element appears.

EXAMPLE 2 Find **(a)** the minor of 6 and **(b)** the minor of -3 in the determinant.

$$\begin{vmatrix} 6 & 1 & 2 \\ -3 & 4 & 5 \\ -2 & 7 & 8 \end{vmatrix}$$

Solution

(a) Since the element 6 appears in the first row and the first column, we delete them.

$$\begin{vmatrix} \not{6} & \not{1} & \not{2} \\ -\not{3} & 4 & 5 \\ -\not{2} & 7 & 8 \end{vmatrix}$$

Therefore, the minor of 6 is

$$\begin{vmatrix} 4 & 5 \\ 7 & 8 \end{vmatrix}.$$

(b) Since -3 appears in the first column and the second row, we delete them.

$$\begin{vmatrix} \not{6} & 1 & 2 \\ -\not{3} & \not{4} & \not{5} \\ -\not{2} & 7 & 8 \end{vmatrix}$$

The minor of -3 is

$$\begin{vmatrix} 1 & 2 \\ 7 & 8 \end{vmatrix}.$$

Student Practice 2 Find **(a)** the minor of 3 and **(b)** the minor of -6 in the determinant.

$$\begin{vmatrix} 1 & 2 & 7 \\ -4 & -5 & -6 \\ 3 & 4 & -9 \end{vmatrix}$$

To evaluate a third-order determinant, we use expansion by minors of elements in the first column; for example, we have

$$\begin{vmatrix} a_1 & b_1 & c_1 \\ a_2 & b_2 & c_2 \\ a_3 & b_3 & c_3 \end{vmatrix} = a_1\begin{vmatrix} b_2 & c_2 \\ b_3 & c_3 \end{vmatrix} - a_2\begin{vmatrix} b_1 & c_1 \\ b_3 & c_3 \end{vmatrix} + a_3\begin{vmatrix} b_1 & c_1 \\ b_2 & c_2 \end{vmatrix}$$

Note that the signs alternate. We then evaluate the second-order determinants according to our definition.

EXAMPLE 3 Evaluate the determinant $\begin{vmatrix} 2 & 3 & 6 \\ 4 & -2 & 0 \\ 1 & -5 & -3 \end{vmatrix}$ by expanding it by minors of elements in the first column.

Solution

$$\begin{vmatrix} 2 & 3 & 6 \\ 4 & -2 & 0 \\ 1 & -5 & -3 \end{vmatrix} = 2\begin{vmatrix} -2 & 0 \\ -5 & -3 \end{vmatrix} - 4\begin{vmatrix} 3 & 6 \\ -5 & -3 \end{vmatrix} + 1\begin{vmatrix} 3 & 6 \\ -2 & 0 \end{vmatrix}$$

$$\begin{aligned} &= 2[(-2)(-3) - (-5)(0)] - 4[(3)(-3) - (-5)(6)] + 1[(3)(0) - (-2)(6)] \\ &= 2(6 - 0) - 4[-9 - (-30)] + 1[0 - (-12)] \\ &= 2(6) - 4(21) + 1(12) \\ &= 12 - 84 + 12 \\ &= -60 \end{aligned}$$

Student Practice 3

Evaluate the determinant. $\begin{vmatrix} 1 & 2 & -3 \\ 2 & -1 & 2 \\ 3 & 1 & 4 \end{vmatrix}$

③ Solving a System of Two Linear Equations with Two Unknowns Using Cramer's Rule

We can solve a linear system of two equations with two unknowns by Cramer's rule. The rule is named for Gabriel Cramer, a Swiss mathematician who lived from 1704 to 1752. Cramer's rule expresses the solution for each variable of a linear system as the quotient of two determinants. Computer programs are available to solve systems of equations by Cramer's rule.

CRAMER'S RULE

The solution to

$$a_1x + b_1y = c_1$$
$$a_2x + b_2y = c_2$$

is

$$x = \frac{D_x}{D} \quad \text{and} \quad y = \frac{D_y}{D}, \quad D \neq 0,$$

where

$$D_x = \begin{vmatrix} c_1 & b_1 \\ c_2 & b_2 \end{vmatrix}, \quad D_y = \begin{vmatrix} a_1 & c_1 \\ a_2 & c_2 \end{vmatrix}, \quad \text{and} \quad D = \begin{vmatrix} a_1 & b_1 \\ a_2 & b_2 \end{vmatrix}.$$

EXAMPLE 4 Solve by Cramer's rule.

$$\begin{aligned} -3x + y &= 7 \\ -4x - 3y &= 5 \end{aligned}$$

Solution

$$D = \begin{vmatrix} -3 & 1 \\ -4 & -3 \end{vmatrix} = (-3)(-3) - (-4)(1) = 9 - (-4) = 9 + 4 = 13$$

$$D_x = \begin{vmatrix} 7 & 1 \\ 5 & -3 \end{vmatrix} = (7)(-3) - (5)(1) = -21 - 5 = -26$$

$$D_y = \begin{vmatrix} -3 & 7 \\ -4 & 5 \end{vmatrix} = (-3)(5) - (-4)(7) = -15 - (-28) = -15 + 28 = 13$$

Hence,

$$x = \frac{D_x}{D} = \frac{-26}{13} = -2$$

$$y = \frac{D_y}{D} = \frac{13}{13} = 1.$$

The solution to the system is $x = -2$ and $y = 1$. Verify this.

Student Practice 4 Solve by Cramer's rule.

$$\begin{aligned} 5x + 3y &= 17 \\ 2x - 5y &= 13 \end{aligned}$$

④ Solving a System of Three Linear Equations with Three Unknowns Using Cramer's Rule

It is quite easy to extend Cramer's rule to three linear equations.

CRAMER'S RULE

The solution to the system

$$\begin{aligned} a_1x + b_1y + c_1z &= d_1 \\ a_2x + b_2y + c_2z &= d_2 \\ a_3x + b_3y + c_3z &= d_3 \end{aligned}$$

is

$$x = \frac{D_x}{D}, \quad y = \frac{D_y}{D}, \quad \text{and} \quad z = \frac{D_z}{D}, \quad D \neq 0,$$

where

$$D = \begin{vmatrix} a_1 & b_1 & c_1 \\ a_2 & b_2 & c_2 \\ a_3 & b_3 & c_3 \end{vmatrix}, \quad D_x = \begin{vmatrix} d_1 & b_1 & c_1 \\ d_2 & b_2 & c_2 \\ d_3 & b_3 & c_3 \end{vmatrix},$$

$$D_y = \begin{vmatrix} a_1 & d_1 & c_1 \\ a_2 & d_2 & c_2 \\ a_3 & d_3 & c_3 \end{vmatrix}, \quad \text{and} \quad D_z = \begin{vmatrix} a_1 & b_1 & d_1 \\ a_2 & b_2 & d_2 \\ a_3 & b_3 & d_3 \end{vmatrix}.$$

EXAMPLE 5 Use Cramer's rule to solve the system.

$$\begin{aligned} 2x - y + z &= 6 \\ 3x + 2y - z &= 5 \\ 2x + 3y - 2z &= 1 \end{aligned}$$

Solution We will expand each determinant by the first column.

$$D = \begin{vmatrix} 2 & -1 & 1 \\ 3 & 2 & -1 \\ 2 & 3 & -2 \end{vmatrix}$$

$$= 2\begin{vmatrix} 2 & -1 \\ 3 & -2 \end{vmatrix} - 3\begin{vmatrix} -1 & 1 \\ 3 & -2 \end{vmatrix} + 2\begin{vmatrix} -1 & 1 \\ 2 & -1 \end{vmatrix}$$

$$= 2[-4 - (-3)] - 3(2 - 3) + 2(1 - 2)$$

$$= 2(-1) - 3(-1) + 2(-1)$$

$$= -2 + 3 - 2$$

$$= -1$$

$$D_x = \begin{vmatrix} 6 & -1 & 1 \\ 5 & 2 & -1 \\ 1 & 3 & -2 \end{vmatrix}$$

$$= 6\begin{vmatrix} 2 & -1 \\ 3 & -2 \end{vmatrix} - 5\begin{vmatrix} -1 & 1 \\ 3 & -2 \end{vmatrix} + 1\begin{vmatrix} -1 & 1 \\ 2 & -1 \end{vmatrix}$$

$$= 6[-4 - (-3)] - 5(2 - 3) + 1(1 - 2)$$

$$= 6(-1) - 5(-1) + 1(-1)$$

$$= -6 + 5 - 1$$

$$= -2$$

$$D_y = \begin{vmatrix} 2 & 6 & 1 \\ 3 & 5 & -1 \\ 2 & 1 & -2 \end{vmatrix}$$

$$= 2\begin{vmatrix} 5 & -1 \\ 1 & -2 \end{vmatrix} - 3\begin{vmatrix} 6 & 1 \\ 1 & -2 \end{vmatrix} + 2\begin{vmatrix} 6 & 1 \\ 5 & -1 \end{vmatrix}$$

$$= 2[-10 - (-1)] - 3(-12 - 1) + 2(-6 - 5)$$

$$= 2(-9) - 3(-13) + 2(-11)$$

$$= -18 + 39 - 22$$

$$= -1$$

$$D_z = \begin{vmatrix} 2 & -1 & 6 \\ 3 & 2 & 5 \\ 2 & 3 & 1 \end{vmatrix}$$

$$= 2\begin{vmatrix} 2 & 5 \\ 3 & 1 \end{vmatrix} - 3\begin{vmatrix} -1 & 6 \\ 3 & 1 \end{vmatrix} + 2\begin{vmatrix} -1 & 6 \\ 2 & 5 \end{vmatrix}$$

$$= 2(2 - 15) - 3(-1 - 18) + 2(-5 - 12)$$

$$= 2(-13) - 3(-19) + 2(-17)$$

$$= -26 + 57 - 34$$

$$= -3$$

$$x = \frac{D_x}{D} = \frac{-2}{-1} = 2; \qquad y = \frac{D_y}{D} = \frac{-1}{-1} = 1; \qquad z = \frac{D_z}{D} = \frac{-3}{-1} = 3$$

Graphing Calculator

Copying Matrices

If you are using a graphing calculator to evaluate the four determinants in Example 5 or similar exercises, first enter matrix D into the calculator. Then copy the matrix using the copy function to three additional locations. Usually we store matrix D as matrix A. Then store a copy of it as matrices B, C, and D. Finally, use the Edit function and modify one column of each of matrices B, C, and D so that they become D_x, D_y, and D_z. This allows you to evaluate all four determinants in a minimum amount of time.

Student Practice 5 Find the solution to the system by Cramer's rule.

$$\begin{aligned} 2x + 3y - z &= -1 \\ 3x + 5y - 2z &= -3 \\ x + 2y + 3z &= 2 \end{aligned}$$

Cramer's rule cannot be used for every system of linear equations. If the equations are dependent or if the system of equations is inconsistent, the determinant of coefficients (D) will be zero. Division by zero is not defined. In such a situation the system will not have a unique answer.

If $D = 0$, then the following are true:

1. If $D_x = 0$ and $D_y = 0$ (and $D_z = 0$, if there are three equations), then the equations are *dependent*. Such a system will have an infinite number of solutions.
2. If at least one of D_x or D_y (or D_z if there are three equations) is nonzero, then the system of equations is *inconsistent*. Such a system will have no solution.

Appendix B Exercises

MyMathLab®

Watch the videos in MyMathLab

Download the MyDashBoard App

Evaluate each determinant.

1. $\begin{vmatrix} 5 & 6 \\ 2 & 1 \end{vmatrix}$ **2.** $\begin{vmatrix} 3 & 4 \\ 1 & 8 \end{vmatrix}$ **3.** $\begin{vmatrix} 2 & -1 \\ 3 & 6 \end{vmatrix}$ **4.** $\begin{vmatrix} -4 & 2 \\ 1 & 5 \end{vmatrix}$ **5.** $\begin{vmatrix} -\frac{1}{2} & -\frac{2}{3} \\ 9 & 8 \end{vmatrix}$

6. $\begin{vmatrix} 10 & 4 \\ -\frac{3}{2} & -\frac{2}{5} \end{vmatrix}$ **7.** $\begin{vmatrix} -5 & 3 \\ -4 & -7 \end{vmatrix}$ **8.** $\begin{vmatrix} 2 & -3 \\ -4 & -6 \end{vmatrix}$ **9.** $\begin{vmatrix} 0 & -6 \\ 3 & -4 \end{vmatrix}$ **10.** $\begin{vmatrix} -5 & 0 \\ 2 & -7 \end{vmatrix}$

11. $\begin{vmatrix} 2 & -5 \\ -4 & 10 \end{vmatrix}$ **12.** $\begin{vmatrix} -3 & 6 \\ 7 & -14 \end{vmatrix}$ **13.** $\begin{vmatrix} 0 & 0 \\ -2 & 6 \end{vmatrix}$ **14.** $\begin{vmatrix} -4 & 0 \\ -3 & 0 \end{vmatrix}$ **15.** $\begin{vmatrix} 0.3 & 0.6 \\ 1.2 & 0.4 \end{vmatrix}$

16. $\begin{vmatrix} 0.1 & 0.7 \\ 0.5 & 0.8 \end{vmatrix}$ **17.** $\begin{vmatrix} 7 & 4 \\ b & -a \end{vmatrix}$ **18.** $\begin{vmatrix} \frac{1}{4} & \frac{3}{5} \\ \frac{2}{3} & \frac{1}{5} \end{vmatrix}$ **19.** $\begin{vmatrix} \frac{3}{7} & -\frac{1}{3} \\ -\frac{1}{4} & \frac{1}{2} \end{vmatrix}$ **20.** $\begin{vmatrix} -3 & y \\ -2 & x \end{vmatrix}$

In the determinant $\begin{vmatrix} 3 & -4 & 7 \\ -2 & 6 & 10 \\ 1 & -5 & 9 \end{vmatrix}$,

21. Find the minor of 3.

22. Find the minor of -2.

23. Find the minor of 10.

24. Find the minor of 9.

Evaluate each of the following determinants.

25. $\begin{vmatrix} 4 & 1 & 2 \\ 3 & -1 & 0 \\ 1 & 2 & 3 \end{vmatrix}$ **26.** $\begin{vmatrix} 2 & 3 & 1 \\ -3 & 1 & 0 \\ 2 & 1 & 4 \end{vmatrix}$ **27.** $\begin{vmatrix} -4 & 0 & -1 \\ 2 & 1 & -1 \\ 0 & 3 & 2 \end{vmatrix}$ **28.** $\begin{vmatrix} 3 & -4 & -1 \\ -2 & 1 & 3 \\ 0 & 1 & 4 \end{vmatrix}$

29. $\begin{vmatrix} \frac{1}{2} & 1 & -1 \\ \frac{3}{2} & 1 & 2 \\ 3 & 0 & -2 \end{vmatrix}$ **30.** $\begin{vmatrix} 1 & 2 & 3 \\ 4 & -2 & -1 \\ 5 & -3 & 2 \end{vmatrix}$ **31.** $\begin{vmatrix} 4 & 1 & 2 \\ -1 & -2 & -3 \\ 4 & -1 & 3 \end{vmatrix}$ **32.** $\begin{vmatrix} -\frac{1}{2} & 2 & 3 \\ \frac{5}{2} & -2 & -1 \\ \frac{3}{4} & -3 & 2 \end{vmatrix}$

33. $\begin{vmatrix} 2 & 0 & -2 \\ -1 & 0 & 2 \\ 3 & 4 & 3 \end{vmatrix}$ **34.** $\begin{vmatrix} 7 & 0 & 2 \\ 1 & 0 & -5 \\ 3 & 0 & 6 \end{vmatrix}$ **35.** $\begin{vmatrix} 6 & -4 & 3 \\ 1 & 2 & 4 \\ 0 & 0 & 0 \end{vmatrix}$ **36.** $\begin{vmatrix} 7 & 0 & 3 \\ 1 & 2 & 4 \\ 3 & 0 & -7 \end{vmatrix}$

Optional Graphing Calculator Problems

If you have a graphing calculator, use the determinant function to evaluate the following:

37. $\begin{vmatrix} 1.3 & 1.8 & 2.5 \\ 7.9 & 5.3 & 6.0 \\ 1.7 & 1.8 & 2.8 \end{vmatrix}$ **38.** $\begin{vmatrix} 0.7 & 5.3 & 0.4 \\ 1.6 & 0.3 & 3.7 \\ 0.8 & 6.7 & 4.2 \end{vmatrix}$ **39.** $\begin{vmatrix} -55 & 17 & 19 \\ -62 & 23 & 31 \\ 81 & 51 & 74 \end{vmatrix}$ **40.** $\begin{vmatrix} 82 & -20 & 56 \\ 93 & -18 & 39 \\ 65 & -27 & 72 \end{vmatrix}$

Solve each system by Cramer's rule.

41. $$\begin{aligned} x + 2y &= 8 \\ 2x + y &= 7 \end{aligned}$$

42. $$\begin{aligned} x + 3y &= 6 \\ 2x + y &= 7 \end{aligned}$$

43. $$\begin{aligned} 5x + 4y &= 10 \\ -x + 2y &= 12 \end{aligned}$$

44. $$\begin{aligned} 3x + 5y &= 11 \\ 2x + y &= -2 \end{aligned}$$

45. $$\begin{aligned} x - 5y &= 0 \\ x + 6y &= 22 \end{aligned}$$

46. $$\begin{aligned} x - 3y &= 4 \\ -3x + 4y &= -12 \end{aligned}$$

47. $$\begin{aligned} 0.3x + 0.5y &= 0.2 \\ 0.1x + 0.2y &= 0.0 \end{aligned}$$

48. $$\begin{aligned} 0.5x + 0.3y &= -0.7 \\ 0.4x + 0.5y &= -0.3 \end{aligned}$$

Solve by Cramer's rule. Round your answers to four decimal places.

49. $$\begin{aligned} 52.9634x - 27.3715y &= 86.1239 \\ 31.9872x + 61.4598y &= 44.9812 \end{aligned}$$

50. $$\begin{aligned} 0.0076x + 0.0092y &= 0.01237 \\ -0.5628x - 0.2374y &= -0.7635 \end{aligned}$$

Solve each system by Cramer's rule.

51. $$\begin{aligned} 2x + y + z &= 4 \\ x - y - 2z &= -2 \\ x + y - z &= 1 \end{aligned}$$

52. $$\begin{aligned} x + 2y - z &= -4 \\ x + 4y - 2z &= -6 \\ 2x + 3y + z &= 3 \end{aligned}$$

53. $$\begin{aligned} 2x + 2y + 3z &= 6 \\ x - y + z &= 1 \\ 3x + y + z &= 1 \end{aligned}$$

54. $$\begin{aligned} 4x + y + 2z &= 6 \\ x + y + z &= 1 \\ -x + 3y - z &= -5 \end{aligned}$$

55. $$\begin{aligned} x + 2y + z &= 1 \\ 3x - 4z &= 8 \\ 3y + 5z &= -1 \end{aligned}$$

56. $$\begin{aligned} 3x + y + z &= 2 \\ 2y + 3z &= -6 \\ 2x - y &= -1 \end{aligned}$$

Optional Graphing Calculator Problems *Round your answers to the nearest thousandth.*

57. $$\begin{aligned} 10x + 20y + 10z &= -2 \\ -24x - 31y - 11z &= -12 \\ 61x + 39y + 28z &= -45 \end{aligned}$$

58. $$\begin{aligned} 121x + 134y + 101z &= 146 \\ 315x - 112y - 108z &= 426 \\ 148x + 503y + 516z &= -127 \end{aligned}$$

59. $$\begin{aligned} 28w + 35x - 18y + 40z &= 60 \\ 60w + 32x + 28y &= 400 \\ 30w + 15x + 18y + 66z &= 720 \\ 26w - 18x - 15y + 75z &= 125 \end{aligned}$$

Appendix C Solving Systems of Linear Equations Using Matrices

Student Learning Objective

After studying this section, you will be able to:

① **Solve a system of linear equations using matrices.**

① Solving a System of Linear Equations Using Matrices

In Appendix B we defined a matrix as any rectangular array of numbers that is arranged in rows and columns.

$$\begin{bmatrix} 2 & 3 \\ 5 & 6 \end{bmatrix}$$ This is a 2×2 matrix with two rows and two columns.

$$\begin{bmatrix} 1 & -5 & -6 & 2 \\ 3 & 4 & -8 & -2 \\ 2 & 7 & 9 & -4 \end{bmatrix}$$ This is a 3×4 matrix with three rows and four columns.

A matrix that is derived from a system of linear equations is called the **augmented matrix** of the system. This augmented matrix is made up of two smaller matrices separated by a vertical line. The coefficients of each variable in the linear system are placed to the left of the vertical line. The constants are placed to the right of the vertical line.

The augmented matrix for the system of equations

$$\begin{aligned} -3x + 5y &= -22 \\ 2x - y &= 10 \end{aligned}$$

is the 2×3 matrix

$$\left[\begin{array}{cc|c} -3 & 5 & -22 \\ 2 & -1 & 10 \end{array}\right].$$

The augmented matrix for the system of equations

$$\begin{aligned} 3x - 5y + 2z &= 8 \\ x + y + z &= 3 \\ 3x - 2y + 4z &= 10 \end{aligned}$$

is the 3×4 matrix

$$\left[\begin{array}{ccc|c} 3 & -5 & 2 & 8 \\ 1 & 1 & 1 & 3 \\ 3 & -2 & 4 & 10 \end{array}\right].$$

EXAMPLE 1 Write the solution to the system of linear equations represented by the following matrix.

$$\left[\begin{array}{cc|c} 1 & -3 & -7 \\ 0 & 1 & 4 \end{array}\right]$$

Solution The matrix represents the system of equations

$$\begin{aligned} x - 3y &= -7 \quad \text{and} \\ 0x + y &= 4. \end{aligned}$$

Since we know that $y = 4$, we can find x by substitution.

$$\begin{aligned} x - 3y &= -7 \\ x - 3(4) &= -7 \\ x - 12 &= -7 \\ x &= 5 \end{aligned}$$

Thus, the solution to the system is $x = 5; y = 4$. We can also write the solution as $(5, 4)$.

Student Practice 1 Write the solution to the system of linear equations represented by the following matrix.

$$\left[\begin{array}{cc|c} 1 & 9 & 33 \\ 0 & 1 & 3 \end{array}\right]$$

***NOTE TO STUDENT:** Fully worked-out solutions to all of the Student Practice problems can be found at the back of the text starting at page SP-1.*

To solve a system of linear equations in matrix form, we use three row operations of the matrix.

MATRIX ROW OPERATIONS

1. Any two rows of a matrix may be interchanged.
2. All the numbers in a row may be multiplied or divided by any nonzero number.
3. All the numbers in any row or any multiple of a row may be added to the corresponding numbers of any other row.

To obtain the values for x and y in a system of two linear equations, we use row operations to obtain an augmented matrix in a form similar to the form of the matrix in Example 1.

The desired form is

$$\left[\begin{array}{cc|c} 1 & a & b \\ 0 & 1 & c \end{array}\right] \quad \text{or} \quad \left[\begin{array}{ccc|c} 1 & a & b & d \\ 0 & 1 & c & e \\ 0 & 0 & 1 & f \end{array}\right].$$

The last row of the matrix will allow us to find the value of one of the variables. We can then use substitution to find the other variables.

EXAMPLE 2 Use matrices to solve the system.

$$\begin{aligned} 4x - 3y &= -13 \\ x + 2y &= 5 \end{aligned}$$

Solution The augmented matrix for this system of linear equations is

$$\left[\begin{array}{cc|c} 4 & -3 & -13 \\ 1 & 2 & 5 \end{array}\right].$$

First we want to obtain a 1 as the first element in the first row. We can obtain this by interchanging rows one and two.

$$\left[\begin{array}{cc|c} 1 & 2 & 5 \\ 4 & -3 & -13 \end{array}\right] \quad R_1 \longleftrightarrow R_2$$

Next we wish to obtain a 0 as the first element of the second row. To obtain this we multiply -4 by all the elements of row one and add this to row two.

$$\left[\begin{array}{cc|c} 1 & 2 & 5 \\ 0 & -11 & -33 \end{array}\right] \quad -4R_1 + R_2$$

Next, to obtain a 1 as the second element of the second row, we multiply each element of row two by $-\frac{1}{11}$.

$$\left[\begin{array}{cc|c} 1 & 2 & 5 \\ 0 & 1 & 3 \end{array}\right] \quad -\frac{1}{11}R_2$$

This final matrix is in the desired form. It represents the linear system

$$\begin{aligned} x + 2y &= 5 \\ y &= 3. \end{aligned}$$

Continued on next page

Since we know that $y = 3$, we substitute this value into the first equation.

$$\begin{aligned} x + 2(3) &= 5 \\ x + 6 &= 5 \\ x &= -1 \end{aligned}$$

Thus, the solution to the system is $(-1, 3)$.

Student Practice 2 Use matrices to solve the system.

$$\begin{aligned} 3x - 2y &= -6 \\ x - 3y &= 5 \end{aligned}$$

Now we continue with a similar example involving three equations and three unknowns.

EXAMPLE 3 Use matrices to solve the system.

$$\begin{aligned} 2x + 3y - z &= 11 \\ x + 2y + z &= 12 \\ 3x - y + 2z &= 5 \end{aligned}$$

Solution The augmented matrix that represents this system of linear equations is

$$\left[\begin{array}{rrr|r} 2 & 3 & -1 & 11 \\ 1 & 2 & 1 & 12 \\ 3 & -1 & 2 & 5 \end{array}\right].$$

To obtain a 1 as the first element of the first row, we interchange the first and second rows.

$$\left[\begin{array}{rrr|r} 1 & 2 & 1 & 12 \\ 2 & 3 & -1 & 11 \\ 3 & -1 & 2 & 5 \end{array}\right] \quad R_1 \longleftrightarrow R_2$$

Now, in order to obtain a 0 as the first element of the second row, we multiply row one by -2 and add the result to row two. In order to obtain a 0 as the first element of the third row, we multiply row one by -3 and add the result to row three.

$$\left[\begin{array}{rrr|r} 1 & 2 & 1 & 12 \\ 0 & -1 & -3 & -13 \\ 0 & -7 & -1 & -31 \end{array}\right] \quad \begin{array}{l} -2R_1 + R_2 \\ -3R_1 + R_3 \end{array}$$

To obtain a 1 as the second element of row two, we multiply all the elements of row two by -1.

$$\left[\begin{array}{rrr|r} 1 & 2 & 1 & 12 \\ 0 & 1 & 3 & 13 \\ 0 & -7 & -1 & -31 \end{array}\right] \quad -1R_2$$

Next, in order to obtain a 0 as the second element of row three, we add 7 times row two to row three.

$$\left[\begin{array}{rrr|r} 1 & 2 & 1 & 12 \\ 0 & 1 & 3 & 13 \\ 0 & 0 & 20 & 60 \end{array}\right] \quad 7R_2 + R_3$$

Finally, we multiply all the elements of row three by $\frac{1}{20}$. Thus, we have the following:

$$\left[\begin{array}{rrr|r} 1 & 2 & 1 & 12 \\ 0 & 1 & 3 & 13 \\ 0 & 0 & 1 & 3 \end{array}\right] \quad \frac{1}{20}R_3$$

From the final line of the matrix, we see that $z = 3$. If we substitute this value into the equation represented by the second row, we have

$$\begin{aligned} y + 3z &= 13 \\ y + 3(3) &= 13 \\ y + 9 &= 13 \\ y &= 4. \end{aligned}$$

Now we substitute the values obtained for y and z into the equation represented by the first row of the matrix.

$$\begin{aligned} x + 2y + z &= 12 \\ x + 2(4) + 3 &= 12 \\ x + 8 + 3 &= 12 \\ x + 11 &= 12 \\ x &= 1 \end{aligned}$$

Thus, the solution to this linear system of three equations is $(1, 4, 3)$.

Student Practice 3 Use matrices to solve the system.

$$\begin{aligned} 2x + y - 2z &= -15 \\ 4x - 2y + z &= 15 \\ x + 3y + 2z &= -5 \end{aligned}$$

We could continue to use these row operations to obtain an augmented matrix of the form

$$\left[\begin{array}{cc|c} 1 & 0 & a \\ 0 & 1 & b \end{array}\right] \quad \text{or} \quad \left[\begin{array}{ccc|c} 1 & 0 & 0 & a \\ 0 & 1 & 0 & b \\ 0 & 0 & 1 & c \end{array}\right].$$

This form of the augmented matrix is given a special name. It is known as the **reduced row echelon form**. If the augmented matrix of a system of linear equations is placed in this form, we would immediately know the solution to the system. Thus, if a system of linear equations in the variables x, y, and z has an augmented matrix that could be placed in the form

$$\left[\begin{array}{ccc|c} 1 & 0 & 0 & 7 \\ 0 & 1 & 0 & 32 \\ 0 & 0 & 1 & 18 \end{array}\right],$$

we could determine directly that $x = 7$, $y = 32$, and $z = 18$. A similar pattern is obtained for a system of four equations in four unknowns, and so on. Thus, if a system of linear equations in the variables w, x, y, and z has an augmented matrix that could be placed in the form

$$\left[\begin{array}{cccc|c} 1 & 0 & 0 & 0 & 23.4 \\ 0 & 1 & 0 & 0 & 48.6 \\ 0 & 0 & 1 & 0 & 0.73 \\ 0 & 0 & 0 & 1 & 5.97 \end{array}\right],$$

we could directly conclude that $w = 23.4$, $x = 48.6$, $y = 0.73$, and $z = 5.97$. Reducing a matrix to reduced row echelon form is readily done on computers. Many mathematical software packages contain matrix operations that will obtain the reduced row echelon form of an augmented matrix. A number of graphing calculators such as the TI-84 can be used to obtain the reduced row echelon form by using the **rref** command on a given matrix.

Graphing Calculator

Obtaining a Reduced Row Echelon Form of an Augmented Matrix

If your graphing calculator has a routine to obtain the **reduced row echelon form** of a matrix **(rref)**, then this routine will allow you to quickly obtain the solution of a system of linear equations if one exists. If your calculator has this capability, solve the following system.

$$\begin{aligned} 5w + 2x + 3y + 4z &= -8.3 \\ -4w + 3x + 2y + 7z &= -70.1 \\ 6w + x + 4y + 5z &= -13.3 \\ 7w + 4x + y + 2z &= 14.1 \end{aligned}$$

Ans:

$w = 3.1$, $x = 2.2$,

$y = 4.6$, $z = -10.5$

Appendix C Exercises

MyMathLab® Watch the videos in MyMathLab Download the MyDashBoard App

Solve each system of equations by the matrix method. Round your answers to the nearest tenth when necessary.

1. $$\begin{aligned} 2x + 3y &= 5 \\ 5x + y &= 19 \end{aligned}$$

2. $$\begin{aligned} 3x + 5y &= -15 \\ 2x + 7y &= -10 \end{aligned}$$

3. $$\begin{aligned} 2x + y &= -3 \\ 5x - y &= 24 \end{aligned}$$

4. $$\begin{aligned} x + 5y &= -9 \\ 4x - 3y &= -13 \end{aligned}$$

5. $$\begin{aligned} 5x + 2y &= 6 \\ 3x + 4y &= 12 \end{aligned}$$

6. $$\begin{aligned} -5x + y &= 24 \\ x + 5y &= 10 \end{aligned}$$

7. $$\begin{aligned} 3x - 2y + 3 &= 5 \\ x + 4y - 1 &= 9 \end{aligned}$$

8. $$\begin{aligned} 3x + y - 4 &= 12 \\ -2x + 3y + 2 &= -5 \end{aligned}$$

9. $$\begin{aligned} -7x + 3y &= 2.7 \\ 6x + 5y &= 25.7 \end{aligned}$$

10. $$\begin{aligned} x - 2y - 3z &= 4 \\ 2x + 3y + z &= 1 \\ -3x + y - 2z &= 5 \end{aligned}$$

11. $$\begin{aligned} x + y - z &= -2 \\ 2x - y + 3z &= 19 \\ 4x + 3y - z &= 5 \end{aligned}$$

12. $$\begin{aligned} 5x - y + 4z &= 5 \\ 6x + y - 5z &= 17 \\ 2x - 3y + z &= -11 \end{aligned}$$

13. $$\begin{aligned} x + y - z &= -3 \\ x + y + z &= 3 \\ 3x - y + z &= 7 \end{aligned}$$

14. $$\begin{aligned} 2x - y + z &= 5 \\ x + 2y - z &= -2 \\ x + y - 2z &= -5 \end{aligned}$$

15. $$\begin{aligned} 2x - 3y + z &= 11 \\ x + y + 2z &= 8 \\ x + 3y - z &= -11 \end{aligned}$$

16. $$\begin{aligned} 4x + 3y + 5z &= 2 \\ 2y + 7z &= 16 \\ 2x - y &= 6 \end{aligned}$$

17. $$\begin{aligned} 6x - y + z &= 9 \\ 2x + 3z &= 16 \\ 4x + 7y + 5z &= 20 \end{aligned}$$

18. $$\begin{aligned} 3x + 2y &= 44 \\ 4y + 3z &= 19 \\ 2x + 3z &= -5 \end{aligned}$$

Optional Graphing Calculator Problems

If your graphing calculator has the necessary capability, solve the following exercises.

19. $$\begin{aligned} 5x + 6y + 7z &= 45.6 \\ 1.4x - 3.2y + 1.6z &= 3.12 \\ 9x - 8y + 22z &= 70.8 \end{aligned}$$

20. $$\begin{aligned} 2x + 12y + 9z &= 37.9 \\ 1.6x + 1.8y - 2.5z &= -20.53 \\ 7x + 8y + 4z &= 39.6 \end{aligned}$$

21. $$\begin{aligned} 6w + 5x + 3y + 1.5z &= 41.7 \\ 2w + 6.7x - 5y + 7z &= -21.92 \\ 12w + x + 5y - 6z &= 58.4 \\ 3w + 8x - 15y + z &= -142.8 \end{aligned}$$

22. $$\begin{aligned} 2w + 3x + 11y - 14z &= 6.7 \\ 5w + 8x + 7y + 3z &= 25.3 \\ -4w + x + 1.5y - 9z &= -53.4 \\ 9w + 7x - 2.5y + 6z &= 22.9 \end{aligned}$$

Solutions to Student Practice Problems

Chapter 1

1.1 Student Practice

1. (a) 1.26 is a rational number and a real number.
   (b) 3 is a natural number, a whole number, an integer, a rational number, and a real number.
   (c) $\frac{3}{7}$ is a rational number and a real number.
   (d) -2 is an integer, a rational number, and a real number.
   (e) 5.182671 ... has no repeating pattern. Therefore, it is an irrational number and a real number.
2. (a) Commutative property of addition
   (b) Identity property of addition
3. (a) Associative property of multiplication
   (b) Inverse property of multiplication
   (c) Distributive property of multiplication over addition

1.2 Student Practice

1. (a) $|-4| = 4$
   (b) $|3.16| = 3.16$
   (c) $|8 - 8| = |0| = 0$
   (d) $|12 - 7| = |5| = 5$
   (e) $\left|-2\frac{1}{3}\right| = 2\frac{1}{3}$
2. (a) $3.4 + 2.6 = 6.0$
   (b) $-\frac{3}{4} + \left(-\frac{1}{6}\right) = -\frac{9}{12} + \left(-\frac{2}{12}\right) = -\frac{11}{12}$
   (c) $-5 + (-37) = -42$
3. (a) $24 + (-30) = -6$
   (b) $-\frac{1}{5} + \frac{2}{3} = -\frac{3}{15} + \frac{10}{15} = \frac{7}{15}$
4. (a) $-8 - (-3) = -8 + 3 = -5$
   (b) $\frac{1}{2} - \left(-\frac{1}{4}\right) = \frac{2}{4} + \frac{1}{4} = \frac{3}{4}$
   (c) $-0.35 - 0.67 = -0.35 + (-0.67) = -1.02$
5. (a) $\left(-\frac{2}{5}\right)\left(\frac{3}{4}\right) = -\frac{6}{20} = -\frac{3}{10}$
   (b) $\frac{150}{-30} = -5$
   (c) $\frac{0.27}{-0.003} = -90$
   (d) $\frac{-12}{24} = -\frac{1}{2}$
6. (a) $-\frac{2}{7}\cdot\left(-\frac{3}{5}\right) = \frac{6}{35}$
   (b) $-60 \div (-5) = 12$
   (c) $\frac{3}{7} \div \frac{1}{5} = \frac{3}{7}\cdot\frac{5}{1} = \frac{15}{7}$
7. $5 + 7(-2) - (-3) + 50 \div (-2)$
   $= 5 + (-14) + 3 + 50 \div (-2)$
   $= 5 + (-14) + 3 + (-25)$
   $= -9 + 3 + (-25)$
   $= -6 + (-25)$
   $= -31$
8. (a) $6(-2) + (-20) \div (2)(3) = -12 + (-10)(3)$
   $= -12 + (-30)$
   $= -42$
   (b) $\frac{7 + 2 - 12 - (-1)}{(-5)(-6) + 4(-8)} = \frac{7 + 2 - 12 + 1}{30 + (-32)} = \frac{-2}{-2} = 1$

1.3 Student Practice

1. (a) $(-4)(-4)(-4)(-4) = (-4)^4$
   (b) $z \cdot z \cdot z \cdot z \cdot z \cdot z \cdot z = z^7$
2. (a) $(-3)^5 = (-3)(-3)(-3)(-3)(-3) = -243$
   (b) $(-3)^6 = (-3)(-3)(-3)(-3)(-3)(-3) = 729$
   (c) $(-4)^4 = (-4)(-4)(-4)(-4) = 256$
   (d) $-4^4 = -(4 \cdot 4 \cdot 4 \cdot 4) = -256$
   (e) $\left(\frac{1}{5}\right)^2 = \left(\frac{1}{5}\right)\left(\frac{1}{5}\right) = \frac{1}{25}$
3. Since $(-7)^2 = 49$ and $7^2 = 49$, the square roots of 49 are -7 and 7. The principal square root is 7.
4. (a) $\sqrt{100} = 10$ because $10^2 = 100$.
   (b) $\sqrt{1} = 1$ because $1^2 = 1$.
   (c) $-\sqrt{36} = -(\sqrt{36}) = -(6) = -6$
5. (a) $(0.3)^2 = (0.3)(0.3) = 0.09$, therefore, $\sqrt{0.09} = 0.3$.
   (b) $\sqrt{\frac{4}{81}} = \frac{\sqrt{4}}{\sqrt{81}} = \frac{2}{9}$
   (c) This is not a real number.
6. (a) $6(12 - 8) + 4 = 6(4) + 4 = 24 + 4 = 28$
   (b) $5[6 - 3(7 - 9)] - 8 = 5[6 - 3(-2)] - 8$
   $= 5[6 + 6] - 8$
   $= 5[12] - 8$
   $= 60 - 8$
   $= 52$
7. $\frac{(-6)(3)(2)}{5 - 12 + 3} = \frac{-36}{-4} = 9$
8. (a) $\sqrt{(-5)^2 + 12^2} = \sqrt{25 + 144} = \sqrt{169} = 13$
   (b) $|-3 - 7 + 2 - (-4)| = |-3 - 7 + 2 + 4| = |-4| = 4$
9. $-7 - 2(-3) + (4 - 5)^3$
   $= -7 - 2(-3) + (-1)^3$
   $= -7 - 2(-3) + (-1)$
   $= -7 + 6 + (-1)$
   $= -2$
10. $5 + 6 \cdot 2 - 12 \div (-2) + 3\sqrt{4}$
    $= 5 + 6 \cdot 2 - 12 \div (-2) + 3(2)$
    $= 5 + 12 + 6 + 6$
    $= 29$
11. $\frac{2(3) + 5(-2)}{1 + 2 \cdot 3^2 + 5(-3)} = \frac{6 - 10}{1 + 2 \cdot 9 + 5(-3)}$
    $= \frac{-4}{1 + 18 + (-15)}$
    $= \frac{-4}{4}$
    $= -1$

1.4 Student Practice

1. (a) $3^{-2} = \frac{1}{3^2} = \frac{1}{9}$
   (b) $z^{-8} = \frac{1}{z^8}$
2. $\left(\frac{3}{4}\right)^{-2} = \frac{1}{\left(\frac{3}{4}\right)^2} = \frac{1}{\frac{9}{16}} = 1 \cdot \frac{16}{9} = \frac{16}{9}$
3. (a) $2^8 \cdot 2^{15} = 2^{8+15} = 2^{23}$
   (b) $x^2 \cdot x^8 \cdot x^6 = x^{2+8+6} = x^{16}$
   (c) $(x + 2y)^4(x + 2y)^{10} = (x + 2y)^{4+10} = (x + 2y)^{14}$
4. (a) $(7w^3)(2w) = (7 \cdot 2)(w^3 \cdot w^1) = 14w^4$
   (b) $(-5xy)(-2x^2y^3) = (-5)(-2)(x^1 \cdot x^2)(y^1 \cdot y^3) = 10x^3y^4$

5. $(7xy^{-2})(2x^{-5}y^{-6}) = 14x^{1-5}y^{-2-6} = 14x^{-4}y^{-8} = \frac{14}{x^4y^8}$

6. (a) $\frac{w^8}{w^6} = w^{8-6} = w^2$
   (b) $\frac{3^7}{3^3} = 3^{7-3} = 3^4$
   (c) $\frac{x^5}{x^{16}} = x^{5-16} = x^{-11} = \frac{1}{x^{11}}$
   (d) $\frac{4^5}{4^8} = 4^{5-8} = 4^{-3} = \frac{1}{4^3}$

7. (a) $6y^0 = 6(1) = 6$
   (b) $(3xy)^0 = 1$. Note that the entire expression is raised to the zero power.
   (c) $(5^{-3})(2a)^0 = (5^{-3})(1) = \frac{1}{5^3} = \frac{1}{125}$

8. (a) $\frac{30x^6y^5}{20x^3y^2} = \frac{30}{20} \cdot \frac{x^6}{x^3} \cdot \frac{y^5}{y^2} = \frac{3}{2} \cdot x^3 \cdot y^3 = \frac{3x^3y^3}{2}$
   (b) $\frac{-15a^3b^4c^4}{3a^5b^4c^2} = \frac{-15}{3} \cdot \frac{a^3}{a^5} \cdot \frac{b^4}{b^4} \cdot \frac{c^4}{c^2} = -5 \cdot a^{-2} \cdot b^0 \cdot c^2 = -\frac{5c^2}{a^2}$

9. $\frac{2x^{-3}y}{4x^{-2}y^5} = \frac{1}{2}x^{-3-(-2)}y^{1-5} = \frac{1}{2}x^{-1}y^{-4} = \frac{1}{2xy^4}$

10. (a) $(w^3)^8 = w^{3 \cdot 8} = w^{24}$
    (b) $(5^2)^5 = 5^{2 \cdot 5} = 5^{10}$
    (c) $[(x-2y)^3]^3 = (x-2y)^{3 \cdot 3} = (x-2y)^9$

11. (a) $(4x^3y^4)^2 = 4^2x^6y^8 = 16x^6y^8$
    (b) $\left(\frac{4xy}{3x^5y^6}\right)^3 = \frac{4^3x^3y^3}{3^3x^{15}y^{18}} = \frac{64x^3y^3}{27x^{15}y^{18}} = \frac{64}{27x^{12}y^{15}}$
    (c) $(3xy^2)^{-2} = 3^{-2}x^{-2}y^{-4} = \frac{1}{3^2x^2y^4} = \frac{1}{9x^2y^4}$

12. (a) $\frac{7x^2y^{-4}z^{-3}}{8x^{-5}y^{-6}z^2} = \frac{7x^2x^5y^6}{8y^4z^3z^2} = \frac{7x^7y^2}{8z^5}$
    (b) $\left(\frac{4x^2y^{-2}}{x^{-4}y^{-3}}\right)^{-3} = \frac{4^{-3}x^{-6}y^6}{x^{12}y^9} = \frac{y^6}{4^3x^{12}x^6y^9} = \frac{1}{64x^{18}y^3}$

13. $(2x^{-3})^2(-3xy^{-2})^{-3} = 2^2x^{-6} \cdot (-3)^{-3}x^{-3}y^6$
    $= \frac{2^2y^6}{(-3)^3x^6x^3} = -\frac{4y^6}{27x^9}$

14. (a) $128{,}320 = 1.2832 \times 10^5$ (b) $476 = 4.76 \times 10^2$
    (c) $0.0786 = 7.86 \times 10^{-2}$ (d) $0.007 = 7 \times 10^{-3}$

15. (a) $4.62 \times 10^6 = 4{,}620{,}000$ (b) $1.973 \times 10^{-3} = 0.001973$
    (c) $4.931 \times 10^{-1} = 0.4931$

16. $\frac{(55{,}000)(3{,}000{,}000)}{5{,}500{,}000} = \frac{(5.5 \times 10^4)(3.0 \times 10^6)}{5.5 \times 10^6}$
    $= \frac{3.0}{1} \times \frac{10^{10}}{10^6}$
    $= 3.0 \times 10^{10-6}$
    $= 3.0 \times 10^4$

17. $(6.0 \times 10^{24})(3.4 \times 10^5)$
    $= 6.0 \times 3.4 \times 10^{24} \times 10^5$
    $= 20.4 \times 10^{29}$
    $= 2.04 \times 10^{30}$
    The mass of the star is 2.04×10^{30} kilograms.

1.5 Student Practice

1. (a) $7x$ is the product of a real number (7) and a variable (x), so $7x$ is a term.
   $-2w^3$ is the product of a real number (-2) and a variable (w^3), so $-2w^3$ is a term.
   (b) 5 is a real number, so it is a term. $6x$ is the product of a real number (6) and a variable (x), so $6x$ is a term. $2y$ is the product of a real number (2) and a variable (y), so $2y$ is a term.

2. (a) The coefficient of the x^2y term is 5.
   The coefficient of the w term is -3.5.
   (b) The coefficient of the x^3 term is $\frac{3}{4}$.
   The coefficient of the x^2y term is $-\frac{5}{7}$.
   (c) The coefficient of the abc term is -5.6, of the ab term is -0.34, and of the bc term is 8.56.

3. (a) $9x - 12x = (9 - 12)x = -3x$
   (b) $4ab^2c + 15ab^2c = (4 + 15)ab^2c = 19ab^2c$

4. (a) $12x^3 - 5x^2 + 7x - 3x^3 - 8x^2 + x$
   $= 9x^3 - 13x^2 + 8x$
   (b) $\frac{1}{3}a^2 - \frac{1}{5}a - \frac{4}{15}a^2 + \frac{1}{2}a + 5$
   $= \frac{5}{15}a^2 - \frac{4}{15}a^2 - \frac{2}{10}a + \frac{5}{10}a + 5$
   $= \frac{1}{15}a^2 + \frac{3}{10}a + 5$
   (c) $4.5x^3 - 0.6x - 9.3x^3 + 0.8x$
   $= -4.8x^3 + 0.2x$

5. $-3x^2(2x - 5) = (-3x^2)(2x) - (-3x^2)(5)$
   $= -6x^3 + 15x^2$

6. (a) $-5x(2x^2 - 3x - 1) = -10x^3 + 15x^2 + 5x$
   (b) $3ab(4a^3 + 2b^2 - 6) = 12a^4b + 6ab^3 - 18ab$

7. (a) $(7x^2 - 8) = 1(7x^2 - 8) = 7x^2 - 8$
   (b) $-(3x + 2y - 6) = -1(3x + 2y - 6) = -3x - 2y + 6$
   (c) $-5x^2(x + 2xy) = -5x^3 - 10x^3y$
   (d) $\frac{3}{4}(8x^2 + 12x - 3) = \frac{3}{4}(8x^2) + \frac{3}{4}(12x) - \frac{3}{4}(3)$
   $= 6x^2 + 9x - \frac{9}{4}$

8. $-7(a + b) - 8a(2 - 3b) + 5a$
   $= -7a - 7b - 16a + 24ab + 5a$
   $= -18a + 24ab - 7b$

9. $-2\{4x - 3[x - 2x(1 + x)]\}$
   $= -2\{4x - 3[x - 2x - 2x^2]\}$
   $= -2\{4x - 3[-x - 2x^2]\}$
   $= -2\{4x + 3x + 6x^2\}$
   $= -2\{7x + 6x^2\}$
   $= -14x - 12x^2$

1.6 Student Practice

1. $2x^2 + 3x - 8 = 2(-3)^2 + 3(-3) - 8$
   $= 2(9) + 3(-3) - 8$
   $= 18 - 9 - 8$
   $= 1$

2. $(x-3)^2 - 2xy = [(-3)-3]^2 - 2(-3)(4)$
   $= [-6]^2 - 2(-3)(4)$
   $= 36 - 2(-3)(4)$
   $= 36 + 24$
   $= 60$

3. (a) $(-3x)^2 = [-3(-4)]^2 = (12)^2 = 144$
   (b) $-3x^2 = -3(-4)^2 = -3(16) = -48$

4. Let $C = 70$.
   $F = \frac{9}{5}C + 32$
   $= \frac{9}{5}(70) + 32$
   $= 9(14) + 32$
   $= 126 + 32$
   $= 158$
   The equivalent Fahrenheit temperature is 158°F.

5. When $L = 128$ and $g = 32$,

$$T = 2\pi\sqrt{\frac{L}{g}}$$
$$= 2\pi\sqrt{\frac{128}{32}}$$
$$= 2\pi\sqrt{4}$$
$$= 2(3.14)(2)$$
$$= 12.56$$

It takes Tarzan about 12.6 seconds.

6. When $p = 600$, $r = 9\%$, and $t = 3$,

$$A = p(1 + rt)$$
$$= 600[1 + (0.09)(3)]$$
$$= 600(1 + 0.27)$$
$$= 600(1.27)$$
$$= 762$$

The amount is \$762.

7. When $l = 0.76$ and $w = 0.38$,

$$P = 2l + 2w$$
$$= 2(0.76) + 2(0.38)$$
$$= 1.52 + 0.76$$
$$= 2.28$$

The perimeter is 2.28 centimeters.

8. When $a = 12$ and $b = 14$,

$$A = \frac{1}{2}ab$$
$$= \frac{1}{2}(12)(14)$$
$$= 84$$

The area is 84 square meters.

9. When $r = 6$ and $h = 10$,

$$V = \pi r^2 h$$
$$= \pi(6)^2(10)$$
$$\approx 3.14(36)(10)$$
$$= 1130.4$$

The volume is approximately 1130.4 cubic meters.

Chapter 2

2.1 Student Practice

1. Replace a by the value $-\frac{3}{2}$ in the equation $6a - 5 = -4a + 10$.

$$6\left(-\frac{3}{2}\right) - 5 \stackrel{?}{=} -4\left(-\frac{3}{2}\right) + 10$$
$$-9 - 5 \stackrel{?}{=} 6 + 10$$
$$-14 \neq 16$$

This last statement is not true. Thus $-\frac{3}{2}$ is not a solution of $6a - 5 = -4a + 10$.

2.

$$x + 5.2 = -2.8$$
$$x + 5.2 - 5.2 = -2.8 - 5.2$$
$$x = -8$$

Check.

$$x + 5.2 = -2.8$$
$$-8 + 5.2 \stackrel{?}{=} -2.8$$
$$-2.8 = -2.8 \checkmark$$

3.

$$\frac{1}{5}w = -6$$
$$5\left(\frac{1}{5}\right)w = 5(-6)$$
$$w = -30$$

Check.

$$\frac{1}{5}w = -6$$
$$\frac{1}{5}(-30) \stackrel{?}{=} -6$$
$$-6 = -6 \checkmark$$

4.

$$8w - 3 = 2w - 7w + 4$$
$$8w - 3 = -5w + 4$$
$$8w + 5w - 3 = -5w + 5w + 4$$
$$13w - 3 = 4$$
$$13w - 3 + 3 = 4 + 3$$
$$13w = 7$$
$$\frac{13w}{13} = \frac{7}{13}$$
$$w = \frac{7}{13}$$

Check.

$$8\left(\frac{7}{13}\right) - 3 \stackrel{?}{=} 2\left(\frac{7}{13}\right) - 7\left(\frac{7}{13}\right) + 4$$
$$\frac{56}{13} - \frac{39}{13} \stackrel{?}{=} \frac{14}{13} - \frac{49}{13} + \frac{52}{13}$$
$$\frac{17}{13} = \frac{17}{13} \checkmark$$

Thus, $\frac{7}{13}$ is the solution.

5.

$$a - 4(2a - 7) = 3(a + 6)$$
$$a - 8a + 28 = 3a + 18$$
$$-7a + 28 = 3a + 18$$
$$-7a - 3a + 28 = 3a - 3a + 18$$
$$-10a + 28 = 18$$
$$-10a + 28 - 28 = 18 - 28$$
$$-10a = -10$$
$$\frac{-10a}{-10} = \frac{-10}{-10}$$
$$a = 1$$

Check.

$$1 - 4[2(1) - 7] \stackrel{?}{=} 3(1 + 6)$$
$$1 - 4(2 - 7) \stackrel{?}{=} 3(7)$$
$$1 - 4(-5) \stackrel{?}{=} 21$$
$$1 + 20 \stackrel{?}{=} 21$$
$$21 = 21 \checkmark$$

Thus, the solution is 1.

6. The LCD is 12.

$$\frac{y}{3} + \frac{1}{2} = 5 + \frac{y - 9}{4}$$
$$12\left(\frac{y}{3} + \frac{1}{2}\right) = 12\left(5 + \frac{y}{4} - \frac{9}{4}\right)$$
$$12\left(\frac{y}{3}\right) + 12\left(\frac{1}{2}\right) = 12(5) + 12\left(\frac{y}{4}\right) - 12\left(\frac{9}{4}\right)$$
$$4y + 6 = 60 + 3y - 27$$
$$4y + 6 = 3y + 33$$
$$4y - 3y + 6 = 3y - 3y + 33$$
$$y + 6 = 33$$
$$y + 6 - 6 = 33 - 6$$
$$y = 27$$

Check.

$$\frac{27}{3} + \frac{1}{2} \stackrel{?}{=} 5 + \frac{27 - 9}{4}$$
$$9 + \frac{1}{2} \stackrel{?}{=} 5 + \frac{18}{4}$$
$$\frac{18}{2} + \frac{1}{2} \stackrel{?}{=} \frac{10}{2} + \frac{9}{2}$$
$$\frac{19}{2} = \frac{19}{2} \checkmark$$

Thus, the solution is 27.

7.
$$4(0.01x + 0.09) - 0.07(x - 8) = 0.83$$
$$0.04x + 0.36 - 0.07x + 0.56 = 0.83$$
$$100(0.04x) + 100(0.36) - 100(0.07x) + 100(0.56) = 100(0.83)$$
$$4x + 36 - 7x + 56 = 83$$
$$-3x + 92 = 83$$
$$-3x + 92 - 92 = 83 - 92$$
$$-3x = -9$$
$$x = 3$$

Check.
$$4[0.01(3) + 0.09] - 0.07(3 - 8) \stackrel{?}{=} 0.83$$
$$4(0.03 + 0.09) - 0.07(-5) \stackrel{?}{=} 0.83$$
$$4(0.12) + 0.35 \stackrel{?}{=} 0.83$$
$$0.48 + 0.35 \stackrel{?}{=} 0.83$$
$$0.83 = 0.83 \checkmark$$

Thus, the solution is 3.

8.
$$7 + 14x - 3 = 2(x - 4) + 12x$$
$$4 + 14x = 2x - 8 + 12x$$
$$4 + 14x = 14x - 8$$
$$4 + 14x - 14x = 14x - 8 - 14x$$
$$4 = -8$$

Clearly, no value of x will make this equation true. No solution.

9.
$$13x - 7(x + 5) = 4x - 35 + 2x$$
$$13x - 7x - 35 = 6x - 35$$
$$6x - 35 = 6x - 35$$
$$6x - 35 - 6x = 6x - 35 - 6x$$
$$-35 = -35$$

Any real number is a solution.

2.2 Student Practice

1.
$$P = 2L + 2W$$
$$P - 2L = 2W$$
$$\frac{P - 2L}{2} = \frac{2W}{2}$$
$$\frac{P - 2L}{2} = W$$

2.
$$H = \frac{3}{4}(a + 2b - 4)$$
$$H = \frac{3}{4}a + \frac{3}{2}b - 3$$
$$4H = 4\left(\frac{3}{4}a\right) + 4\left(\frac{3}{2}b\right) - 4(3)$$
$$4H = 3a + 6b - 12$$
$$4H - 6b + 12 = 3a$$
$$\frac{4H - 6b + 12}{3} = \frac{3a}{3}$$
$$\frac{4H - 6b + 12}{3} = a$$

3.
$$-2(ab - 3x) + 2(8 - ab) = 5x + 4ab$$
$$-2ab + 6x + 16 - 2ab = 5x + 4ab$$
$$-4ab + 6x + 16 = 5x + 4ab$$
$$6x + 16 = 5x + 4ab + 4ab$$
$$6x + 16 = 5x + 8ab$$
$$6x - 5x + 16 = 8ab$$
$$x + 16 = 8ab$$
$$\frac{x + 16}{8a} = \frac{8ab}{8a}$$
$$\frac{x + 16}{8a} = b$$

4.
$$t = -0.4x + 81$$
$$10t = -4x + 810$$
$$4x = 810 - 10t$$
$$x = \frac{810 - 10t}{4}$$
$$x = \frac{405 - 5t}{2}$$

Solve for x when $t = 71$.
$$x = \frac{405 - 5(71)}{2}$$
$$x = \frac{405 - 355}{2}$$
$$x = \frac{50}{2} = 25$$

We estimate that this winning time will occur in the year $1990 + 25 = 2015$.

5. (a)
$$A = 2\pi rh + 2\pi r^2$$
$$A - 2\pi r^2 = 2\pi rh$$
$$\frac{A - 2\pi r^2}{2\pi r} = h$$

(b)
$$h = \frac{A - 2\pi r^2}{2\pi r}$$
$$= \frac{100 - 2(3.14)(2.0)^2}{2(3.14)(2.0)}$$
$$= \frac{100 - (6.28)(4)}{12.56}$$
$$= \frac{100 - 25.12}{12.56}$$
$$= \frac{74.88}{12.56}$$
$$\approx 5.96$$

Thus, $h \approx 5.96$.

2.3 Student Practice

1. $|3x - 4| = 23$

$3x - 4 = 23$ or $3x - 4 = -23$
$3x = 27$ $\quad$ $3x = -19$
$x = 9$ $\quad$ $x = -\frac{19}{3}$

Check.

if $x = 9$
$$|3x - 4| = 23$$
$$|3(9) - 4| \stackrel{?}{=} 23$$
$$|27 - 4| \stackrel{?}{=} 23$$
$$|23| \stackrel{?}{=} 23$$
$$23 = 23 \checkmark$$

if $x = -\frac{19}{3}$
$$|3x - 4| = 23$$
$$\left|3\left(-\frac{19}{3}\right) - 4\right| \stackrel{?}{=} 23$$
$$|-19 - 4| \stackrel{?}{=} 23$$
$$|-23| \stackrel{?}{=} 23$$
$$23 = 23 \checkmark$$

2. $\left|\frac{2}{3}x + 4\right| = 2$

$\frac{2}{3}x + 4 = 2$ or $\frac{2}{3}x + 4 = -2$
$2x + 12 = 6$ $\quad$ $2x + 12 = -6$
$2x = -6$ $\quad$ $2x = -18$
$x = -3$ $\quad$ $x = -9$

Check.

if $x = -3$
$$\left|\frac{2}{3}(-3) + 4\right| \stackrel{?}{=} 2$$
$$|-2 + 4| \stackrel{?}{=} 2$$
$$|2| \stackrel{?}{=} 2$$
$$2 = 2 \checkmark$$

if $x = -9$
$$\left|\frac{2}{3}(-9) + 4\right| \stackrel{?}{=} 2$$
$$|-6 + 4| \stackrel{?}{=} 2$$
$$|-2| \stackrel{?}{=} 2$$
$$2 = 2 \checkmark$$

3.
$$|2x + 1| + 3 = 8$$
$$|2x + 1| + 3 - 3 = 8 - 3$$
$$|2x + 1| = 5$$

$2x + 1 = 5$ or $2x + 1 = -5$
$2x = 4$ $\quad$ $2x = -6$
$x = 2$ $\quad$ $x = -3$

Check.

if $x = 2$
$$|2(2) + 1| + 3 \stackrel{?}{=} 8$$
$$|4 + 1| + 3 \stackrel{?}{=} 8$$
$$|5| + 3 \stackrel{?}{=} 8$$
$$5 + 3 \stackrel{?}{=} 8$$
$$8 = 8 \checkmark$$

if $x = -3$
$$|2(-3) + 1| + 3 \stackrel{?}{=} 8$$
$$|-6 + 1| + 3 \stackrel{?}{=} 8$$
$$|-5| + 3 \stackrel{?}{=} 8$$
$$5 + 3 \stackrel{?}{=} 8$$
$$8 = 8 \checkmark$$

4. $|x - 6| = |5x + 8|$

$$x - 6 = 5x + 8 \quad \text{or} \quad x - 6 = -(5x + 8)$$
$$x - 5x = 6 + 8 \qquad x - 6 = -5x - 8$$
$$-4x = 14 \qquad x + 5x = 6 - 8$$
$$x = \frac{14}{-4} = -\frac{7}{2} \qquad 6x = -2$$
$$x = \frac{-2}{6} = -\frac{1}{3}$$

Check.

if $x = -\frac{7}{2}$

$$\left|-\frac{7}{2} - 6\right| \stackrel{?}{=} \left|5\left(-\frac{7}{2}\right) + 8\right|$$
$$\left|-\frac{7}{2} - 6\right| \stackrel{?}{=} \left|-\frac{35}{2} + 8\right|$$
$$\left|-\frac{7}{2} - \frac{12}{2}\right| \stackrel{?}{=} \left|-\frac{35}{2} + \frac{16}{2}\right|$$
$$\left|-\frac{19}{2}\right| \stackrel{?}{=} \left|-\frac{19}{2}\right|$$
$$\frac{19}{2} = \frac{19}{2} \checkmark$$

if $x = -\frac{1}{3}$

$$\left|-\frac{1}{3} - 6\right| \stackrel{?}{=} \left|5\left(-\frac{1}{3}\right) + 8\right|$$
$$\left|-\frac{1}{3} - 6\right| \stackrel{?}{=} \left|-\frac{5}{3} + 8\right|$$
$$\left|-\frac{1}{3} - \frac{18}{3}\right| \stackrel{?}{=} \left|-\frac{5}{3} + \frac{24}{3}\right|$$
$$\left|-\frac{19}{3}\right| \stackrel{?}{=} \left|\frac{19}{3}\right|$$
$$\frac{19}{3} = \frac{19}{3} \checkmark$$

2.4 Student Practice

1. Let n = the hours of computer use time.
Then $8n$ = the charge for n hours of computer use time at \$8 per hour.
The fixed cost is 400(12).

$$400(12) + 8n = 7680$$
$$4800 + 8n = 7680$$
$$8n = 2880$$
$$n = 360$$

They used the computer for 360 hours.

2. Let n = the number of weeks on tour.
weekly income = 5(6000)(14) = 420,000
weekly expenses = 5(48,000) + 150,000 = 390,000
earnings goal = 5(60,000) = 300,000

$$420{,}000n - 390{,}000n = 300{,}000$$
$$30{,}000n = 300{,}000$$
$$n = 10$$

The group needs to tour for 10 weeks.

3. Let x = the length of the second side.
Then $2x$ = the length of the first side and $3x - 6$ = the length of the third side.

$$P = a + b + c$$
$$162 = x + 2x + 3x - 6$$
$$162 = 6x - 6$$
$$168 = 6x$$
$$x = 28$$

The 1st side is $2x = 2(28) = 56$ meters.
The 2nd side is $x = 28$ meters.
The 3rd side is $3x - 6 = 3(28) - 6 = 78$ meters.

2.5 Student Practice

1. Let x = amount of sales last year, then $0.15x$ = the increase in sales over last year.

$$x + 0.15x = 6900$$
$$1.15x = 6900$$
$$x = 6000$$

6000 computer workstations were sold last year.

2. Let a = the number of cars Alicia sold this month. Then $43 - a$ = the number of cars Heather sold. If Alicia doubles her sales next month, then she will sell $2a$ cars. If Heather triples hers, she will sell $3(43 - a)$ cars.

$$2a + 3(43 - a) = 108$$
$$2a + 129 - 3a = 108$$
$$-a + 129 = 108$$
$$-a = -21$$
$$a = 21$$

Therefore, Alicia sold 21 cars and Heather sold $43 - 21 = 22$ cars.

3. Let x = the amount invested at 8%, then $5500 - x$ = the amount invested at 12%.

$$0.08x + 0.12(5500 - x) = 540$$
$$0.08x + 660 - 0.12x = 540$$
$$660 - 0.04x = 540$$
$$-0.04x = -120$$
$$x = 3000$$

Therefore, she invested \$3000 at 8% and \$5500 − \$3000 = \$2500 at 12%.

4.

	(A) Number of Grams of the Mixture	(B) % Pure Gold	(C) Number of Grams of Pure Gold
68% pure gold source	x	68%	$0.68x$
83% pure gold source	$200 - x$	83%	$0.83(200 - x)$
Final mixture of 80% pure gold	200	80%	0.80(200)

Form an equation from the entries in Column C.

$$0.68x + 0.83(200 - x) = 0.80(200)$$
$$0.68x + 166 - 0.83x = 160$$
$$166 - 0.15x = 160$$
$$-0.15x = -6$$
$$\frac{-0.15x}{-0.15} = \frac{-6}{-0.15}$$
$$x = 40$$

If $x = 40$, then $200 - x = 200 - 40 = 160$.
Thus, he needs to use 40 grams of 68% pure gold and 160 grams of 83% pure gold.

5.

	r	t	d
Steady speed	x	4	$4x$
Slower speed	$x - 10$	2	$2(x - 10)$
Entire trip	Not appropriate	6	352

$$4x + 2(x - 10) = 352$$
$$4x + 2x - 20 = 352$$
$$6x - 20 = 352$$
$$6x = 372$$
$$x = 62$$

Thus, Wally drove 62 miles per hour for 4 hours and $x - 10 = 52$ miles per hour for 2 hours.

2.6 Student Practice

1. (a) $-1 > -2$ since -1 is to the right of -2 on a number line.

(b) $\frac{2}{3} = \frac{8}{12}, \frac{3}{4} = \frac{9}{12}$

Since $8 < 9$, then $\frac{8}{12} < \frac{9}{12}$ so $\frac{2}{3} < \frac{3}{4}$.

(c) $-0.561 < -0.5555$ because $-0.5610 < -0.5555$.
Notice that since both numbers are negative, -0.5610 is to the left of -0.5555 on a number line.

2. (a) $(-8 - 2) \; ? \; (-3 - 12)$

$-10 > -15$

Thus, $(-8 - 2) > (-3 - 12)$.

(b) $|-15 + 8| \; ? \; |7 - 13|$

$|-7| \; ? \; |-6|$

$7 > 6$

Thus, $|-15 + 8| > |7 - 13|$.

3. (a) $x > 3.5$

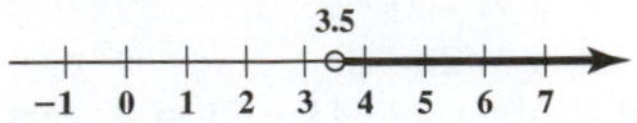

(b) $x \leq -10$

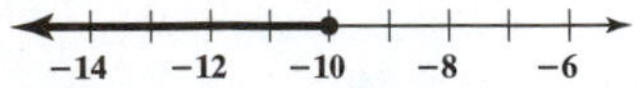

(c) $x \geq -3$

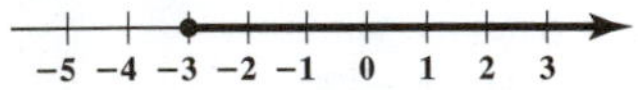

(d) $-4 > x$

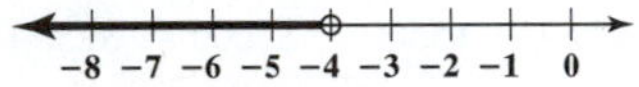

4.
$$x + 2 > -12$$
$$x + 2 - 2 > -12 - 2$$
$$x > -14$$

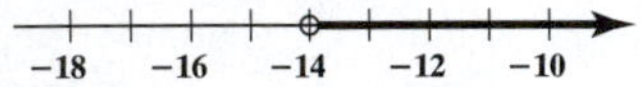

To check, we choose -13.
$$-13 + 2 \overset{?}{>} -12$$
$$-11 > -12 \quad ✓ \text{ True}$$

5.
$$8x - 8 \geq 5x + 1$$
$$8x - 8 + 8 \geq 5x + 1 + 8$$
$$8x \geq 5x + 9$$
$$8x - 5x \geq 5x - 5x + 9$$
$$3x \geq 9$$
$$\frac{3x}{3} \geq \frac{9}{3}$$
$$x \geq 3$$

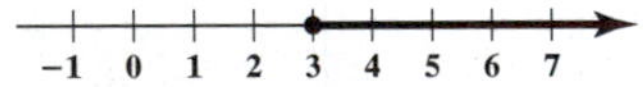

6.
$$2 - 12x > 7(1 - x)$$
$$2 - 12x > 7 - 7x$$
$$2 - 12x + 7x > 7 - 7x + 7x$$
$$2 - 5x > 7$$
$$2 - 2 - 5x > 7 - 2$$
$$-5x > 5$$
$$\frac{-5x}{-5} < \frac{5}{-5}$$
$$x < -1$$

7.
$$-0.8x + 0.9 \geq 0.5x - 0.4$$
$$10(-0.8x + 0.9) \geq 10(0.5x - 0.4)$$
$$-8x + 9 \geq 5x - 4$$
$$-8x - 5x + 9 \geq 5x - 5x - 4$$
$$-13x + 9 \geq -4$$
$$-13x + 9 - 9 \geq -4 - 9$$
$$-13x \geq -13$$
$$\frac{-13x}{-13} \leq \frac{-13}{-13}$$
$$x \leq 1$$

8.
$$\frac{1}{5}(x - 6) < \frac{1}{3}(x - 2)$$
$$\frac{x}{5} - \frac{6}{5} < \frac{x}{3} - \frac{2}{3}$$
$$15\left(\frac{x}{5}\right) - 15\left(\frac{6}{5}\right) < 15\left(\frac{x}{3}\right) - 15\left(\frac{2}{3}\right)$$
$$3x - 18 < 5x - 10$$
$$3x - 18 + 18 < 5x - 10 + 18$$
$$3x < 5x + 8$$
$$3x - 5x < 5x - 5x + 8$$
$$-2x < 8$$
$$\frac{-2x}{-2} > \frac{8}{-2}$$
$$x > -4$$

9. Let $x =$ the number of minutes they talk after the first minute. The cost must be less than or equal to \$13.90
$$3.50 + 0.65x \leq 13.90$$
$$0.65x \leq 10.40$$
$$x \leq 16$$
Add the 16 minutes to the one minute that cost \$3.50. This gives 17 minutes. Thus the maximum amount of time they can talk is 17 minutes.

2.7 Student Practice

1. $-8 < x < -2$

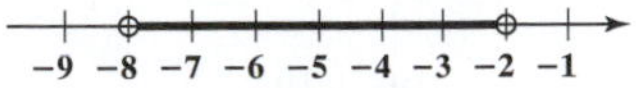

2. $-1 \leq x \leq 5$

3. $-10 < x \leq -5.5$

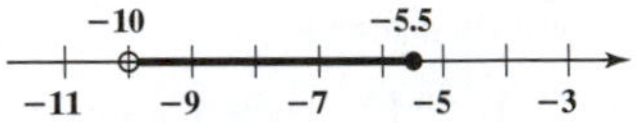

4. "At least \$200" means $s \geq \$200$ and "never more than \$950" means $s \leq \$950$.

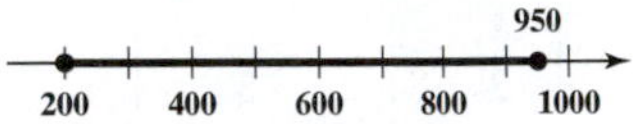

5. $x < 8$ *or* $x > 12$

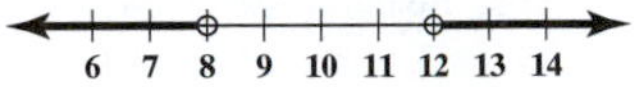

6. $x \leq -6$ *or* $x > 3$

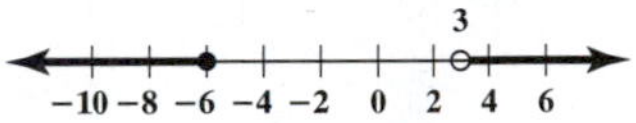

7. Rejected applicants' heights will be less than 56 inches ($h < 56$) or will be greater than 70 inches ($h > 70$).

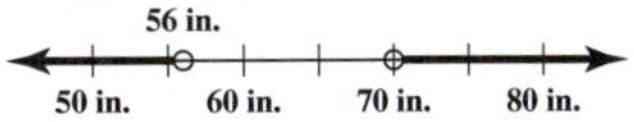

8. $3x - 4 < -1$ or $2x + 3 > 13$

$3x < 3$ $\quad 2x > 10$

$x < 1$ or $x > 5$

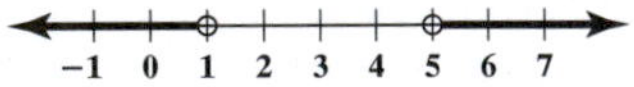

9. $3x + 6 > -6$ and $4x + 5 < 1$

$3x > -12$ $\quad 4x < -4$

$x > -4$ and $x < -1$

$-4 < x < -1$

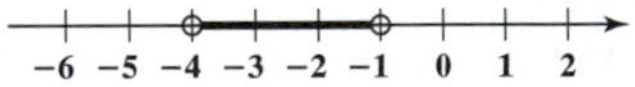

10. $-2x + 3 < -7$ *and* $7x - 1 > -15$

$-2x < -10$ $\quad 7x > -14$

$x > 5$ *and* $x > -2$

$x > 5$ *and* at the same time $x > -2$. Thus $x > 5$ is the solution to the compound inequality.

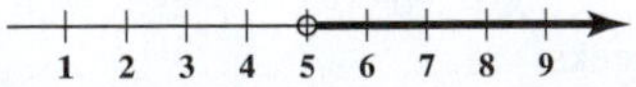

11. $-3x - 11 < -26$ *and* $5x + 4 < 14$

$-3x < -15$ $\quad 5x < 10$

$x > 5$ *and* $x < 2$

Now clearly it is impossible for one number to be greater than 5 *and* at the same time less than 2. There is no solution.

2.8 Student Practice

1. $|x| < 2$

$-2 < x < 2$

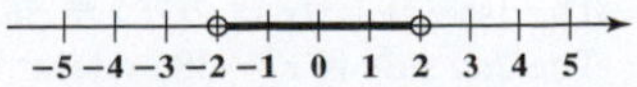

2. $|x - 6| < 15$
$$-15 < x - 6 < 15$$
$$-15 + 6 < x - 6 + 6 < 15 + 6$$
$$-9 < x < 21$$

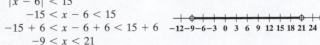

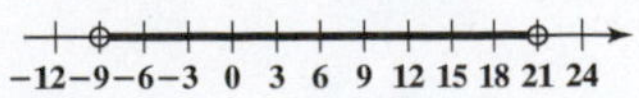

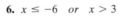

3. $\left|x + \frac{3}{4}\right| \le \frac{7}{6}$

$$-\frac{7}{6} \le x + \frac{3}{4} \le \frac{7}{6}$$
$$12\left(\frac{-7}{6}\right) \le 12(x) + 12\left(\frac{3}{4}\right) \le 12\left(\frac{7}{6}\right)$$
$$-14 \le 12x + 9 \le 14$$
$$-14 - 9 \le 12x + 9 - 9 \le 14 - 9$$
$$-23 \le 12x \le 5$$
$$\frac{-23}{12} \le \frac{12x}{12} \le \frac{5}{12}$$
$$-1\frac{11}{12} \le x \le \frac{5}{12}$$

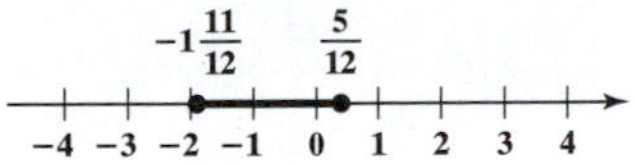

4. $|2 + 3(x - 1)| < 20$

$$|2 + 3x - 3| < 20$$
$$|-1 + 3x| < 20$$
$$-20 < -1 + 3x < 20$$
$$-20 + 1 < 1 - 1 + 3x < 20 + 1$$
$$-19 < 3x < 21$$
$$\frac{-19}{3} < \frac{3x}{3} < \frac{21}{3}$$
$$-\frac{19}{3} < x < 7$$
$$-6\frac{1}{3} < x < 7$$

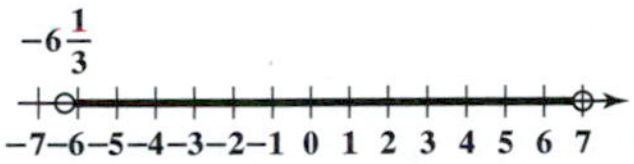

5. $|x| > 2.5$

$x < -2.5$ *or* $x > 2.5$

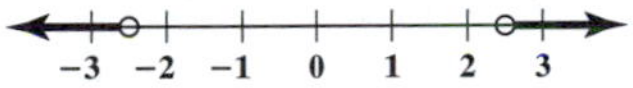

6. $|x + 6| > 2$

$x + 6 < -2$ *or* $x + 6 > 2$

$x + 6 - 6 < -2 - 6$ $\quad$ $x + 6 - 6 > 2 - 6$

$x < -8$ *or* $x > -4$

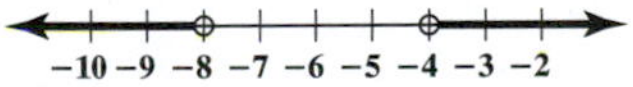

7. $|-5x - 2| > 13$

$-5x - 2 > 13$ *or* $-5x - 2 < -13$

$-5x > 15$ $\quad$ $-5x < -11$

$\frac{-5x}{-5} < \frac{15}{-5}$ $\quad$ $\frac{-5x}{-5} > \frac{-11}{-5}$

$x < -3$ $\quad$ $x > \frac{11}{5}$

$x > 2\frac{1}{5}$

$x < -3$ *or* $x > 2\frac{1}{5}$

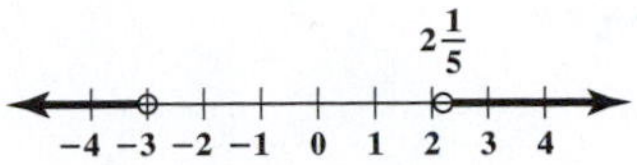

8. $\left|4 - \frac{3}{4}x\right| \ge 5$

$4 - \frac{3}{4}x \ge 5$ *or* $4 - \frac{3}{4}x \le -5$

$4(4) - 4\left(\frac{3}{4}x\right) \ge 4(5)$ $\quad$ $4(4) - 4\left(\frac{3}{4}x\right) \le 4(-5)$

$16 - 3x \ge 20$ $\quad$ $16 - 3x \le -20$

$-3x \ge 4$ $\quad$ $-3x \le -36$

$\frac{-3x}{-3} \le \frac{4}{-3}$ $\quad$ $\frac{-3x}{-3} \ge \frac{-36}{-3}$

$x \le -\frac{4}{3}$ $\quad$ $x \ge 12$

$x \le -1\frac{1}{3}$

$x \le -1\frac{1}{3}$ *or* $x \ge 12$

9.
$$|d - s| \le 0.37$$
$$|d - 276.53| \le 0.37$$
$$-0.37 \le d - 276.53 \le 0.37$$
$$-0.37 + 276.53 \le d - 276.53 + 276.53 \le 0.37 + 276.53$$
$$276.16 \le d \le 276.90$$

Thus the diameter of the transmission must be at least 276.16 millimeters, but not greater than 276.90 millimeters.

Chapter 3 3.1 Student Practice

1. Use $x = -1, x = 0$, and $x = 1$.
For $x = -1, y = -4(-1) + 2 = 6$.
For $x = 0, y = -4(0) + 2 = 2$.
For $x = 1, y = -4(1) + 2 = -2$.

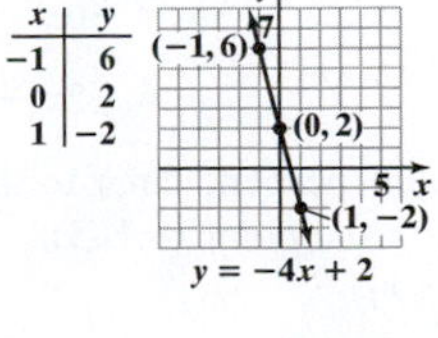

2. Let $y = 0$.

$$3x - 2(0) = -6$$
$$3x = -6$$
$$x = -2$$

The x-intercept is $(-2, 0)$.
Let $x = 0$.

$$3(0) - 2y = -6$$
$$-2y = -6$$
$$y = 3$$

The y-intercept is $(0, 3)$.
Use $x = -1$ for the third point.

$$3(-1) - 2y = -6$$
$$-3 - 2y = -6$$
$$-2y = -3$$
$$y = \frac{3}{2}$$

The third point is $\left(-1, \frac{3}{2}\right)$.

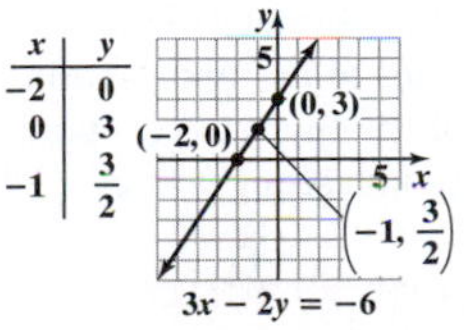

3. **(a)** The equation $x = 4$ means that for any value of y, x is 4. The graph of $x = 4$ is a vertical line 4 units to the right of the origin.
(b) The equation $3y + 12 = 0$ can be simplified.

$$3y + 12 = 0$$
$$3y = -12$$
$$y = -4$$

The equation $y = -4$ means that for any value of x, y is -4.
The graph of $y = -4$ is a horizontal line 4 units below the x-axis.

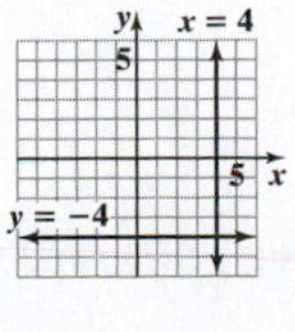

4. When $n = 0$, then $C = 300 + 0.15(0) = 300 + 0 = 300$.
When $n = 1000$, then $C = 300 + 0.15(1000) = 300 + 150 = 450$.
When $n = 2000$, then $C = 300 + 0.15(2000) = 300 + 300 = 600$.
Since n varies from 0 to 2000 and C varies from 300 to 600, we let each square on the horizontal scale represent 200 products and each square on the vertical scale represent \$100.

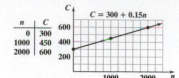

n	C
0	300
1000	450
2000	600

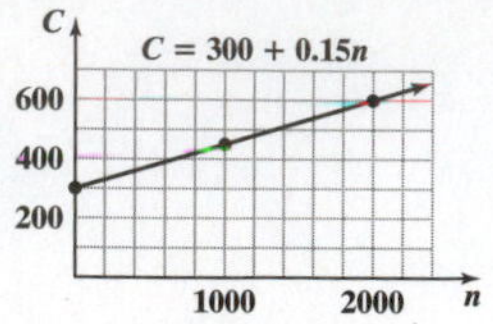

3.2 Student Practice

1. Identify the y-coordinates and the x-coordinates for the points $(-6, 1)$ and $(-5, -2)$.

$$y_2 - y_1$$
$$(-6, 1) \qquad (-5, -2)$$
$$x_2 - x_1$$

Use the formula.

$$\text{Slope} = m = \frac{y_2 - y_1}{x_2 - x_1} = \frac{-2 - 1}{-5 - (-6)} = \frac{-3}{1} = -3$$

2. (a) $m = \dfrac{-3.4 - (-6.2)}{-2.2 - 1.8} = \dfrac{2.8}{-4} = -0.7$

(b) $m = \dfrac{-\frac{3}{4} - \left(-\frac{1}{2}\right)}{\frac{4}{15} - \frac{1}{5}} = \dfrac{-\frac{1}{4}}{\frac{1}{15}} = -\dfrac{15}{4}$

3. slope $= \dfrac{\text{rise}}{\text{run}} = \dfrac{25.92}{1296} = 0.02$ or $\dfrac{1}{50}$

4. $m_l = \dfrac{-2 - 0}{6 - 5} = \dfrac{-2}{1} = -2$

$m_h = \dfrac{1}{2}$

5. $m_{AB} = \dfrac{1 - 5}{-1 - 1} = \dfrac{-4}{-2} = 2$

$m_{BC} = \dfrac{-1 - 1}{-2 - (-1)} = \dfrac{-2}{-2 + 1} = \dfrac{-2}{-1} = 2$

Since the slopes are the same and the line segments have a point in common, the points lie on the same line.

6. From the 4.5-mile point to the 3.5-mile point, the jet traveled 1200 feet downward over 1 horizontal mile. From the 3.5-mile point to the 1.3-mile point, the jet traveled 2640 feet downward over 2.2 horizontal miles. Thus, the first slope is 1200 feet per mile, and the second slope is 1200 feet per mile.

The slopes of the two portions of the flight are the same. The two portions of the flight have a point in common. Therefore the jet is descending in a straight line.

3.3 Student Practice

1. $y = mx + b$

Substitute 4 for m and $-\dfrac{3}{2}$ for b.

$$y = (4)x + \left(-\frac{3}{2}\right)$$
$$y = 4x - \frac{3}{2}$$

2. The y-intercept is $(0, -3)$, thus $b = -3$. Another point on the line is $(5, 0)$. Thus

$(x_1, y_1) = (0, -3)$
$(x_2, y_2) = (5, 0)$

$$m = \frac{y_2 - y_1}{x_2 - x_1} = \frac{0 - (-3)}{5 - 0} = \frac{3}{5}$$

In $y = mx + b$, substitute $\dfrac{3}{5}$ for m and -3 for b.

$$y = \frac{3}{5}x - 3$$

3.
$$3x - 4y = -8$$
$$-4y = -3x - 8$$
$$\frac{-4y}{-4} = \frac{-3x}{-4} + \frac{-8}{-4}$$
$$y = \frac{3}{4}x + 2$$

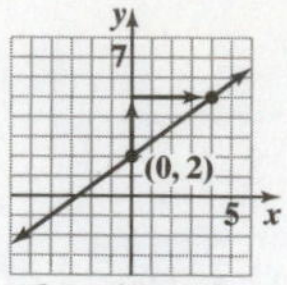

$3x - 4y = -8$

The slope is $\dfrac{3}{4}$ and the y-intercept is $(0, 2)$.

Plot the point $(0, 2)$. From this point go up 3 units and to the right 4 units to locate a second point. Draw the straight line that contains these two points.

4.
$$y - y_1 = m(x - x_1)$$
$$y - (-2) = \frac{3}{4}(x - 5)$$
$$y + 2 = \frac{3}{4}x - \frac{15}{4}$$
$$4y + 4(2) = 4\left(\frac{3}{4}x\right) - 4\left(\frac{15}{4}\right)$$
$$4y + 8 = 3x - 15$$
$$-3x + 4y = -15 - 8$$
$$3x - 4y = 23$$

5. $(-4, 1)$ and $(-2, -3)$

$$m = \frac{y_2 - y_1}{x_2 - x_1} = \frac{-3 - 1}{-2 - (-4)} = \frac{-4}{-2 + 4} = \frac{-4}{2} = -2$$

Substitute $m = -2$ and $(x_1, y_1) = (-4, 1)$ into the point–slope equation.

$$y - y_1 = m(x - x_1)$$
$$y - 1 = -2[x - (-4)]$$
$$y - 1 = -2(x + 4)$$
$$y - 1 = -2x - 8$$
$$y = -2x - 7$$

6. First, we need to find the slope of the line $5x - 3y = 10$. We do this by writing the equation in slope–intercept form.

$$5x - 3y = 10$$
$$-3y = -5x + 10$$
$$y = \frac{5}{3}x - \frac{10}{3}$$

The slope is $\frac{5}{3}$. Use the point–slope form with the point $(4, -5)$.

$$y - y_1 = m(x - x_1)$$
$$y - (-5) = \frac{5}{3}(x - 4)$$
$$y + 5 = \frac{5}{3}x - \frac{20}{3}$$
$$3y + 3(5) = 3\left(\frac{5}{3}x\right) - 3\left(\frac{20}{3}\right)$$
$$3y + 15 = 5x - 20$$
$$-5x + 3y = -35$$
$$5x - 3y = 35$$

7. Find the slope of the line $6x + 3y = 7$ by rewriting it in slope–intercept form.

$$6x + 3y = 7$$
$$3y = -6x + 7$$
$$y = -2x + \frac{7}{3}$$

The slope is -2. A line perpendicular to this has slope $\frac{1}{2}$. Use the point–slope form with the point $(-4, 3)$.

$$y - y_1 = m(x - x_1)$$
$$y - 3 = \frac{1}{2}[x - (-4)]$$
$$y - 3 = \frac{1}{2}(x + 4)$$
$$y - 3 = \frac{1}{2}x + 2$$
$$2y - 2(3) = 2\left(\frac{1}{2}x\right) + 2(2)$$
$$2y - 6 = x + 4$$
$$-x + 2y = 10$$
$$x - 2y = -10$$

3.4 Student Practice

1. The boundary line is $y = 3x + 1$. Graph the boundary line with a dashed line since the inequality contains $>$. Substituting $(0, 0)$ into the inequality $y > 3x + 1$ gives $0 > 1$, which is false. Thus we shade on the side of the boundary opposite $(0, 0)$. The solution is the shaded region not including the dashed line.

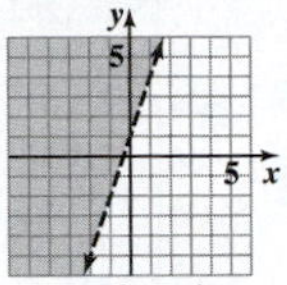

$y > 3x + 1$

2. The boundary line is $-4x + 5y = -10$. Graph the boundary line with a solid line because the inequality contains $\leq$. Substituting $(0, 0)$ into the inequality $-4x + 5y \leq -10$ gives $0 \leq -10$, which is false. Therefore, shade the region on the side of the boundary opposite $(0, 0)$. The solution is the shaded region including the solid line.

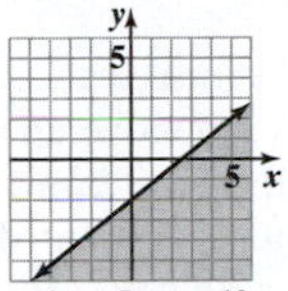

$-4x + 5y \leq -10$

3. The boundary line is $3y + x = 0$. We use a dashed line since the inequality contains $<$. We cannot use $(0, 0)$ to test the inequality. Instead use $(-2, -3)$. Substituting $(-2, -3)$ into $3y + x < 0$ gives $-11 < 0$, which is true. Therefore, shade the side of the boundary line that contains $(-2, -3)$. The solution is the shaded region not including the dashed line.

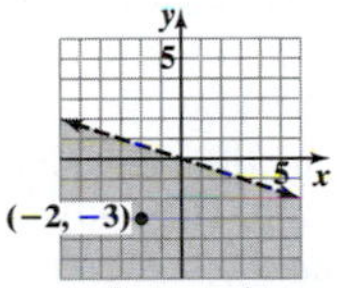

$3y + x < 0$

4. Simplify the original inequality by dividing each side by 6, which gives $y \leq 3$. We draw a solid horizontal line at $y = 3$. We use a solid line since the inequality contains $\leq$. The region we want to shade is the region below the horizontal line. The solution is the line $y = 3$ and the shaded region below that line.

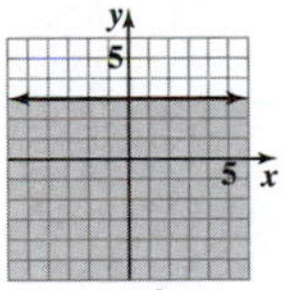

$y \leq 3$

3.5 Student Practice

1. *Table*

Year	1936	1948	1960	1972	1984	1996	2008
Time	11.5	11.9	11.0	11.07	10.97	10.94	10.78

Ordered pairs
$\{(1936, 11.5), (1948, 11.9), (1960, 11.0), (1972, 11.07), (1984, 10.97), (1996, 10.94), (2008, 10.78)\}$

Graph To save space, use the time values from 10 seconds to 12 seconds on the vertical axis. Plot the ordered pairs.

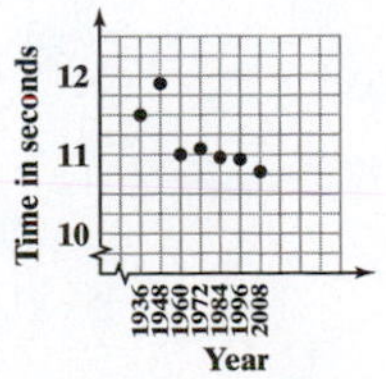

2. The domain is the set of all the first items in each ordered pair. The range is the set of all the second items in each ordered pair. Thus,

$$\text{Domain} = \{158, 161, 163, 160\}$$
$$\text{Range} = \{98.6, 89.2, 101.4, 102.3, 94.9\}$$

$(161, 89.2)$ and $(161, 94.9)$ have the same first coordinate. This relation is not a function.

3. (a) This is a function. No vertical line will pass through more than one ordered pair on the curve.

(b) This is not a function. A vertical line could pass through two points on the curve.

4. (a) $f(-3) = 2(-3)^2 - 8 = 2(9) - 8 = 18 - 8 = 10$

(b) $f(4) = 2(4)^2 - 8 = 2(16) - 8 = 32 - 8 = 24$

(c) $f\left(\frac{1}{2}\right) = 2\left(\frac{1}{2}\right)^2 - 8 = 2\left(\frac{1}{4}\right) - 8 = \frac{1}{2} - \frac{16}{2} = -\frac{15}{2}$

3.6 Student Practice

1. Income = 10,000 + 15% of total sales
Let x = amount of total sales in dollars.

$$d(x) = 10{,}000 + 0.15x$$
$$d(0) = 10{,}000 + 0.15(0) = 10{,}000$$
$$d(40{,}000) = 10{,}000 + 0.15(40{,}000) = 16{,}000$$
$$d(80{,}000) = 10{,}000 + 0.15(80{,}000) = 22{,}000$$

Modify the table by recording x and $d(x)$ in thousands to make the task of graphing easier.

x	$d(x)$
0	10
40	16
80	22

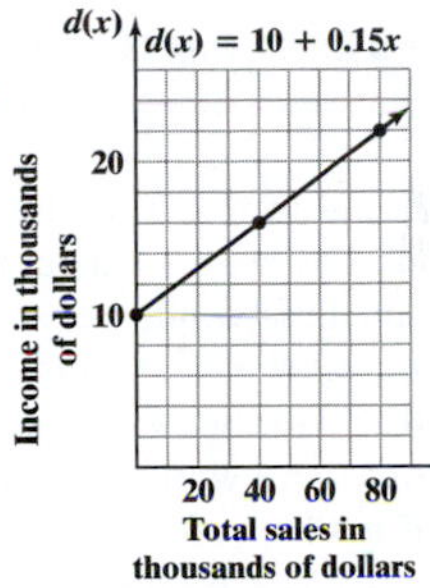

2. The table of values and the resulting graph are shown.

$h(x) = |x + 2|$

x	$h(x)$
−4	2
−3	1
−2	0
−1	1
0	2

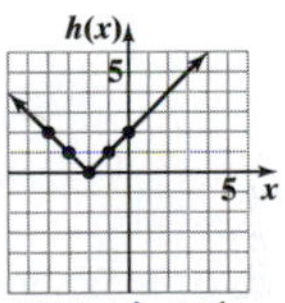

$h(x) = |x + 2|$

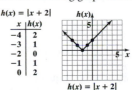

3. For each function we will choose five values of x. Since these are not linear functions we use a curved line to connect the points. The tables of values and the resulting graphs are shown.

(a) $r(x) = (x - 2)^2$

x	$r(x)$
0	4
1	1
2	0
3	1
4	4

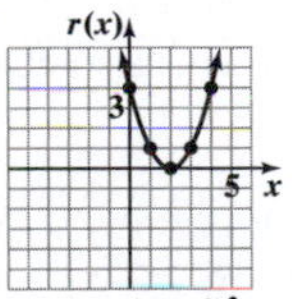

$r(x) = (x - 2)^2$

(b) $s(x) = x^2 + 2$

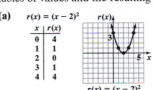

x	$s(x)$
−2	6
−1	3
0	2
1	3
2	6

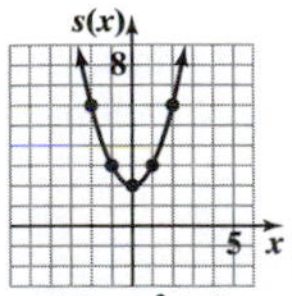

$s(x) = x^2 + 2$

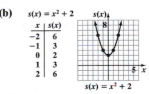

4. Pick five values of x, find the corresponding function values, and plot the four best points to assist in sketching the graph.

$f(x) = x^3 - 2$

x	$f(x)$
−2	−10
−1	−3
0	−2
1	−1
2	6

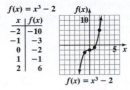

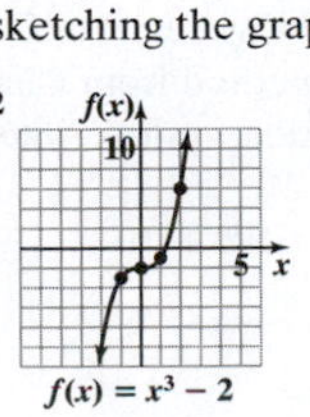

$f(x) = x^3 - 2$

5. We cannot choose x to be zero. Therefore, choose five values for x greater than zero and five values for x less than zero.

$q(x) = -\frac{4}{x}$

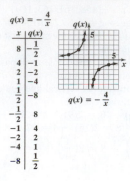

x	$q(x)$
8	$-\frac{1}{2}$
4	-1
2	-2
1	-4
$\frac{1}{2}$	-8
$-\frac{1}{2}$	8
-1	4
-2	2
-4	1
-8	$\frac{1}{2}$

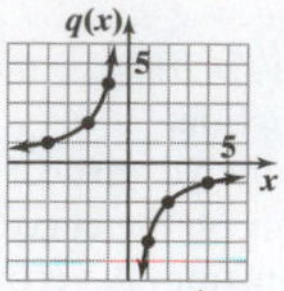

6. (a)

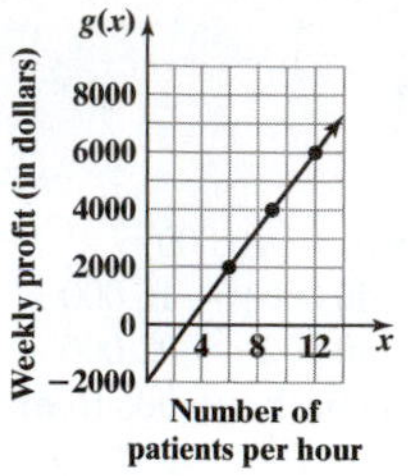

(b) The function is a linear function. As the number of patients seen per hour increases at a fixed rate, so does the weekly profit.

(c) We find that the value $x = 3$ on the graph corresponds to the value $y = 0$. Thus we would expect that if the doctor saw 3 patients per hour there would be \$0 profit.

(d) We find that the value $x = 0$ on the graph corresponds to the value $y = -2000$. Thus we would expect that if the doctor saw 0 patients per hour there would be a loss of \$2000.

(e) No. In terms of mathematics, the function will increase indefinitely; however, is it reasonable to think that a doctor could see an infinite number of patients?

7. **(a)** Since windchill depends on wind speed, wind speed is the independent variable and windchill is the dependent variable. Label the vertical axis "Windchill" and the horizontal axis "Wind speed." Plot the points. Use a smooth curve to connect the points.

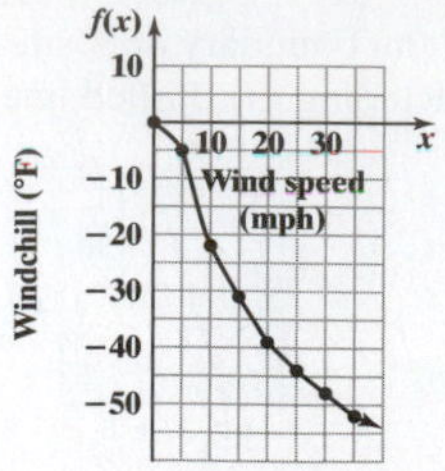

(b) Find 32 along the horizontal axis. Move down to the curve, then move left until you intersect the vertical axis. The function value on the vertical axis for 32 mph is about −49°F.

(c) Since the last windchill number in the table is for a wind speed of 35 mph, we need to extend the curve. At 40 mph the windchill is about −56°F.

(d) Find −24 along the vertical scale. Move to the right until you intersect the curve. Then move up until you interect the horizontal axis. This is slightly more than 11 mph.

(e) In theory based on our graph, the domain is all nonnegative real numbers. The range is all nonpositive real numbers.

8.

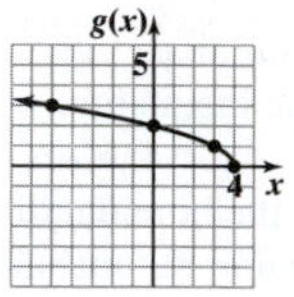

(a) Find $x = -3$ along the horizontal axis. Move up until you intersect the graph. Then move to the right until you intersect the vertical axis. Read the scale. Thus, $g(-3)$ is approximately 2.5.

(b) Find $g(x) = 1.5$ along the vertical axis. Move to the right until you intersect the graph. Then move down until you intersect the horizontal axis. Read the scale. Thus, x is approximately 1.6.

(c) As x increases, the curve is moving downward more quickly. Thus, the slope is decreasing as x increases.

Chapter 4 4.1 Student Practice

1. Substitute $(-3, 4)$ into the first equation to see if the ordered pair is a solution.

$$2x + 3y = 6$$
$$2(-3) + 3(4) \stackrel{?}{=} 6$$
$$-6 + 12 \stackrel{?}{=} 6$$
$$6 = 6 \ \checkmark$$

Likewise, we will determine if $(-3, 4)$ is a solution to the second equation.

$$3x - 4y = 7$$
$$3(-3) - 4(4) \stackrel{?}{=} 7$$
$$-9 - 16 \stackrel{?}{=} 7$$
$$-25 \neq 7$$

Since $(-3, 4)$ is not a solution to both equations in the system, it is not a solution to the system itself.

2. You can use any method from Chapter 3 to graph each line. We will change each equation to slope-intercept form to graph.

$$3x + 2y = 10$$
$$2y = -3x + 10$$
$$y = -\frac{3}{2}x + 5$$
$$x - y = 5$$
$$y = x - 5$$

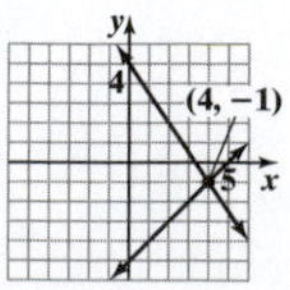

The solution is $(4, -1)$.

The lines intersect at the point $(4, -1)$. Thus $(4, -1)$ is the solution. We verify this by substituting $x = 4$ and $y = -1$ into the system of equations.

$$3x + 2y = 10 \qquad\qquad x - y = 5$$
$$3(4) + 2(-1) \stackrel{?}{=} 10 \qquad\qquad 4 - (-1) \stackrel{?}{=} 5$$
$$12 - 2 \stackrel{?}{=} 10 \qquad\qquad 5 = 5 \ \checkmark$$
$$10 = 10 \ \checkmark$$

3. $2x - y = 7$ [1]
$3x + 4y = -6$ [2]

Solve equation [1] for y.

$$-y = 7 - 2x$$
$$y = -7 + 2x \quad [3]$$

Substitute $-7 + 2x$ for y in equation [2].

$$3x + 4(-7 + 2x) = -6$$
$$3x - 28 + 8x = -6$$
$$11x - 28 = -6$$
$$11x = 22$$
$$x = 2$$

Substitute $x = 2$ into equation [1].

$$2(2) - y = 7$$
$$4 - y = 7$$
$$-y = 3$$
$$y = -3$$

The solution is $(2, -3)$.

4. $\frac{1}{2}x + \frac{2}{3}y = 1$ [1]

$\frac{1}{3}x + y = -1$ [2]

Clear both equations of fractions.

$$6\left(\frac{1}{2}x\right) + 6\left(\frac{2}{3}y\right) = 6(1)$$
$$3x + 4y = 6 \quad [3]$$
$$3\left(\frac{1}{3}x\right) + 3(y) = 3(-1)$$
$$x + 3y = -3 \quad [4]$$

Solve equation [4] for x.

$$x = -3 - 3y$$

Substitute $-3 - 3y$ for x in equation [3].

$$3(-3 - 3y) + 4y = 6$$
$$-9 - 9y + 4y = 6$$
$$-9 - 5y = 6$$
$$-5y = 15$$
$$y = -3$$

Substitute $y = -3$ into equation [2].

$$\frac{x}{3} - 3 = -1$$
$$\frac{x}{3} = 2$$
$$x = 6$$

The solution is $(6, -3)$.

5. $-3x + y = 5$ [1]

$2x + 3y = 4$ [2]

Multiply equation [1] by -3.

$$-3(-3x) + (-3)(y) = -3(5)$$
$$9x - 3y = -15 \quad [3]$$

Add equations [3] and [2].

$$\begin{aligned} 9x - 3y &= -15 \\ 2x + 3y &= 4 \\ \hline 11x \quad &= -11 \\ x &= -1 \end{aligned}$$

Now substitute $x = -1$ into equation [1].

$$-3(-1) + y = 5$$
$$3 + y = 5$$
$$y = 2$$

The solution is $(-1, 2)$.

6. $\frac{x}{4} + \frac{y}{5} = \frac{23}{20}$ [1]

$\frac{7}{15}x - \frac{y}{5} = 1$ [2]

Clear both equations of fractions.

$$20\left(\frac{x}{4}\right) + 20\left(\frac{y}{5}\right) = 20\left(\frac{23}{20}\right)$$
$$5x + 4y = 23 \quad [3]$$
$$15\left(\frac{7}{15}x\right) - 15\left(\frac{y}{5}\right) = 15(1)$$
$$7x - 3y = 15 \quad [4]$$

We now have an equivalent system.

$$5x + 4y = 23 \quad [3]$$
$$7x - 3y = 15 \quad [4]$$

Multiply equation [3] by 3 and equation [4] by 4.

$$\begin{aligned} 15x + 12y &= 69 \\ 28x - 12y &= 60 \\ \hline 43x \quad &= 129 \\ x &= 3 \end{aligned}$$

Now substitute $x = 3$ into equation [3].

$$5(3) + 4y = 23$$
$$15 + 4y = 23$$
$$4y = 8$$
$$y = 2$$

The solution is $(3, 2)$.

7. $4x - 2y = 6$ [1]

$-6x + 3y = 9$ [2]

Multiply equation [1] by 3 and equation [2] by 2.

$$3(4x) - 3(2y) = 3(6)$$
$$12x - 6y = 18 \quad [3]$$
$$2(-6x) + 2(3y) = 2(9)$$
$$-12x + 6y = 18 \quad [4]$$

When we add equations [3] and [4] we get $0 = 36$.
This statement is of course false. Thus, we conclude that this system of equations is inconsistent, so there is no solution.

8. $0.3x - 0.9y = 1.8$ [1]

$-0.4x + 1.2y = -2.4$ [2]

Multiply both equations by 10 to obtain a more convenient form.

$$3x - 9y = 18 \quad [3]$$
$$-4x + 12y = -24 \quad [4]$$

Multiply equation [3] by 4 and equation [4] by 3.

$$\begin{aligned} 12x - 36y &= 72 \\ -12x + 36y &= -72 \\ \hline 0 &= 0 \end{aligned}$$

This statement is always true. Hence these are dependent equations. There is an infinite number of solutions.

9. **(a)** $3x + 5y = 1485$

$x + 2y = 564$

Solve for x in the second equation and solve using the substitution method.

$$x = -2y + 564$$
$$3(-2y + 564) + 5y = 1485$$
$$-6y + 1692 + 5y = 1485$$
$$-y = -207$$
$$y = 207$$

Substitute $y = 207$ into the second equation and solve for x.

$$x + 2(207) = 564$$
$$x + 414 = 564$$
$$x = 150$$

The solution is $(150, 207)$.

(b) $7x + 6y = 45$

$6x - 5y = -2$

Use the addition method. Multiply the first equation by -6 and the second equation by 7.

$$-6(7x) + (-6)(6y) = (-6)(45)$$
$$7(6x) - 7(5y) = 7(-2)$$

$$\begin{aligned} -42x - 36y &= -270 \\ 42x - 35y &= -14 \\ \hline -71y &= -284 \\ y &= 4 \end{aligned}$$

Substitute $y = 4$ into the first equation and solve for x.

$$7x + 6(4) = 45$$
$$7x + 24 = 45$$
$$7x = 21$$
$$x = 3$$

The solution is $(3, 4)$.

4.2 Student Practice

1. Substitute $x = 3$, $y = -2$, $z = 2$ into each equation.

$$2(3) + 4(-2) + 2 \stackrel{?}{=} 0$$
$$6 - 8 + 2 \stackrel{?}{=} 0$$
$$0 = 0 \quad ✓$$
$$3 - 2(-2) + 5(2) \stackrel{?}{=} 17$$
$$3 + 4 + 10 \stackrel{?}{=} 17$$
$$17 = 17 \quad ✓$$
$$3(3) - 4(-2) + 2 \stackrel{?}{=} 19$$
$$9 + 8 + 2 \stackrel{?}{=} 19$$
$$19 = 19 \quad ✓$$

Since we obtained three true statements, the ordered triple $(3, -2, 2)$ is a solution to the system.

2. $x + 2y + 3z = 4$ [1]
$2x + y - 2z = 3$ [2]
$3x + 3y + 4z = 10$ [3]
Eliminate x by multiplying equation [1] by -2 (call it equation [4]) and adding it to equation [2].

$$\begin{array}{rl} -2x - 4y - 6z = -8 & [4] \\ 2x + y - 2z = 3 & [2] \\ \hline -3y - 8z = -5 & [5] \end{array}$$

Now eliminate x by multiplying equation [1] by -3 (call it equation [6]) and adding it to equation [3].

$$\begin{array}{rl} -3x - 6y - 9z = -12 & [6] \\ 3x + 3y + 4z = 10 & [3] \\ \hline -3y - 5z = -2 & [7] \end{array}$$

Now eliminate y and solve for z in the system formed by equation [5] and equation [7].

$$\begin{array}{rl} -3y - 8z = -5 & [5] \\ -3y - 5z = -2 & [7] \end{array}$$

To do this, multiply equation [5] by -1 (call it equation [8]) and add it to equation [7].

$$\begin{array}{rl} 3y + 8z = 5 & [8] \\ -3y - 5z = -2 & [7] \\ \hline 3z = 3 & \\ z = 1 & \end{array}$$

Substitute $z = 1$ into equation [8] and solve for y.

$$\begin{aligned} 3y + 8(1) &= 5 \\ 3y &= -3 \\ y &= -1 \end{aligned}$$

Substitute $z = 1$ and $y = -1$ into equation [1] and solve for x.

$$\begin{aligned} x + 2y + 3z &= 4 \\ x + 2(-1) + 3(1) &= 4 \\ x - 2 + 3 &= 4 \\ x &= 3 \end{aligned}$$

The solution is $(3, -1, 1)$.

3. $2x + y + z = 11$ [1]
$4y + 3z = -8$ [2]
$x - 5y = 2$ [3]
Multiply equation [1] by -3 and add the result to equation [2], thus eliminating the z terms.

$$\begin{array}{rl} -6x - 3y - 3z = -33 & [4] \\ 4y + 3z = -8 & [2] \\ \hline -6x + y = -41 & [5] \end{array}$$

We can solve the system formed by equation [3] and equation [5].

$$\begin{array}{rl} x - 5y = 2 & [3] \\ -6x + y = -41 & [5] \end{array}$$

Multiply equation [3] by 6 and add the result to equation [5].

$$\begin{array}{rl} 6x - 30y = 12 & [6] \\ -6x + y = -41 & [5] \\ \hline -29y = -29 & \\ y = 1 & \end{array}$$

Substitute $y = 1$ into equation [3] and solve for x.

$$\begin{aligned} x - 5(1) &= 2 \\ x &= 7 \end{aligned}$$

Now substitute $y = 1$ into equation [2] and solve for z.

$$\begin{aligned} 4y + 3z &= -8 \\ 4(1) + 3z &= -8 \\ 3z &= -12 \\ z &= -4 \end{aligned}$$

The solution is $(7, 1, -4)$.

4.3 Student Practice

1. Let $x =$ the number of baseballs purchased and $y =$ the number of bats purchased.
Last week: $6x + 21y = 318$ [1]
This week: $5x + 17y = 259$ [2]
Multiply equation [1] by 5 and equation [2] by -6.

$$\begin{array}{rl} 30x + 105y = & 1590 \\ -30x - 102y = & -1554 \\ \hline 3y = & 36 \\ y = & 12 \end{array}$$

Substitute $y = 12$ into equation [2].

$$\begin{aligned} 5x + 17(12) &= 259 \\ 5x &= 55 \\ x &= 11 \end{aligned}$$

Thus 11 baseballs and 12 bats were purchased.

2. Let $x =$ the number of small chairs and $y =$ the number of large chairs. (*Note*: All hours have been changed to minutes.)

$$\begin{array}{rl} 30x + 40y = 1560 & [1] \\ 75x + 80y = 3420 & [2] \end{array}$$

Multiply equation [1] by -2 and add the results to equation [2].

$$\begin{array}{rl} -60x - 80y = & -3120 \\ 75x + 80y = & 3420 \\ \hline 15x = & 300 \\ x = & 20 \end{array}$$

Substitute $x = 20$ in equation [1] and solve for y.

$$\begin{aligned} 600 + 40y &= 1560 \\ 40y &= 960 \\ y &= 24 \end{aligned}$$

Therefore, the company can make 20 small chairs and 24 large chairs each day.

3. Let $a =$ the speed of the airplane in still air in kilometers per hour and $w =$ the speed of the wind in kilometers per hour.

	R	· T	= D
Against the wind	$a - w$	3	1950
With the wind	$a + w$	2	1600

From the chart, we have the following equations.

$$\begin{aligned} (a - w)(3) &= 1950 \\ (a + w)(2) &= 1600 \end{aligned}$$

Remove the parentheses.

$$\begin{array}{rl} 3a - 3w = 1950 & [1] \\ 2a + 2w = 1600 & [2] \end{array}$$

Multiply equation [1] by 2 and equation [2] by 3 and add the resulting equations.

$$\begin{array}{r} 6a - 6w = 3900 \\ 6a + 6w = 4800 \\ \hline 12a = 8700 \\ a = 725 \end{array}$$

Substitute $a = 725$ into equation [2].

$$\begin{aligned} 2(725) + 2w &= 1600 \\ 1450 + 2w &= 1600 \\ 2w &= 150 \\ w &= 75 \end{aligned}$$

Thus, the speed of the plane in still air is 725 kilometers per hour and the speed of the wind is 75 kilometers per hour.

4. Let $A =$ the number of boxes that machine A can wrap in 1 hour, $B =$ the number that machine B can wrap in 1 hour, and $C =$ the number that machine C can wrap in one hour.

$$\begin{array}{rl} A + B + C = 260 & [1] \\ 3A + 2B = 390 & [2] \\ 3B + 4C = 655 & [3] \end{array}$$

Multiply equation [1] by -3 and add it to equation [2].

$$\begin{array}{rl} -3A - 3B - 3C = -780 & [4] \\ 3A + 2B = 390 & [2] \\ \hline -B - 3C = -390 & [5] \end{array}$$

Now multiply equation [5] by 3 and add it to equation [3].

$$\begin{array}{rl} -3B - 9C = -1170 & [6] \\ 3B + 4C = 655 & [3] \\ \hline -5C = -515 & \\ C = 103 & \end{array}$$

Substitute $C = 103$ into equation [3] and solve for B.

$$3B + 412 = 655$$
$$3B = 243$$
$$B = 81$$

Now substitute $B = 81$ into equation [2] and solve for A.

$$3A + 162 = 390$$
$$3A = 228$$
$$A = 76$$

Machine A wraps 76 boxes per hour, machine B wraps 81 boxes per hour, and machine C wraps 103 boxes per hour.

4.4 Student Practice

1. The graph of $-2x + y \leq -3$ is the region on and below the line $-2x + y = -3$. The graph of $x + 2y \geq 4$ is the region on and above the line $x + 2y = 4$.

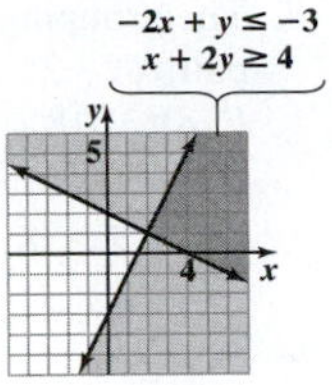

2. The graph of $y > -1$ is the region above the line $y = -1$, not including the line. The graph of $y < -\frac{3}{4}x + 2$ is the region below the line $y = -\frac{3}{4}x + 2$, not including the line.

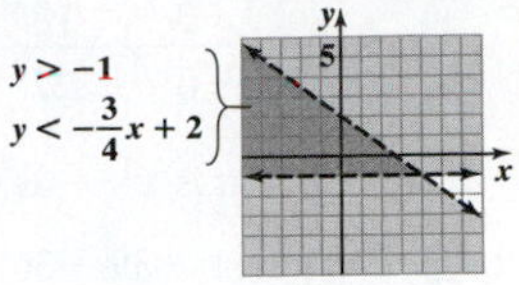

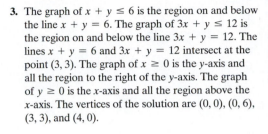

3. The graph of $x + y \leq 6$ is the region on and below the line $x + y = 6$. The graph of $3x + y \leq 12$ is the region on and below the line $3x + y = 12$. The lines $x + y = 6$ and $3x + y = 12$ intersect at the point (3, 3). The graph of $x \geq 0$ is the y-axis and all the region to the right of the y-axis. The graph of $y \geq 0$ is the x-axis and all the region above the x-axis. The vertices of the solution are (0, 0), (0, 6), (3, 3), and (4, 0).

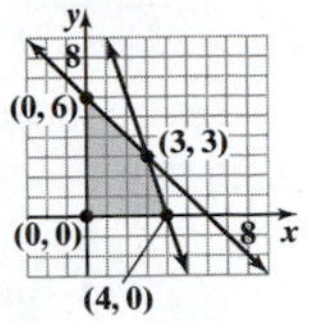

Chapter 5

To Think About page 250

$2x^3 + 5x^2 - 3$ is not a polynomial because it contains a negative exponent. $4ab^{1/2}$ is not a polynomial because it contains an exponent that is not an integer. $\frac{2}{x} + \frac{3}{y}$ is not a polynomial because it contains quotients, not products, of numbers and variables.

5.1 Student Practice

1. **(a)** $3x^5 - 6x^4 + x^2$ This is a trinomial of degree 5.
   (b) $5x^2 + 2$ This is a binomial of degree 2.
   (c) $3ab + 5a^2b^2 - 6a^4b$ This is a trinomial of degree 5.
   (d) $16x^4y^6$ This is a monomial of degree 10.

2. $p(x) = 2x^4 - 3x^3 + 6x - 8$

   (a)
   $$\begin{aligned} p(-2) &= 2(-2)^4 - 3(-2)^3 + 6(-2) - 8 \\ &= 2(16) - 3(-8) + 6(-2) - 8 \\ &= 32 + 24 - 12 - 8 \\ &= 36 \end{aligned}$$

   (b)
   $$\begin{aligned} p(5) &= 2(5)^4 - 3(5)^3 + 6(5) - 8 \\ &= 2(625) - 3(125) + 6(5) - 8 \\ &= 1250 - 375 + 30 - 8 \\ &= 897 \end{aligned}$$

3.
$$\begin{aligned} &(-7x^2 + 5x - 9) + (2x^2 - 3x + 5) \\ &= -7x^2 + 5x - 9 + 2x^2 - 3x + 5 \\ &= -5x^2 + 2x - 4 \end{aligned}$$

4.
$$\begin{aligned} &(2x^2 - 14x + 9) - (-3x^2 + 10x + 7) \\ &= (2x^2 - 14x + 9) + (3x^2 - 10x - 7) \\ &= 5x^2 - 24x + 2 \end{aligned}$$

5.
$$(7x + 3)(2x - 5)$$
$$= 14x^2 - 35x + 6x - 15 = 14x^2 - 29x - 15$$

6.
$$(3x^2 - 2)(5x - 4)$$
$$= 15x^3 - 12x^2 - 10x + 8$$

7.
$$\begin{aligned} (7x - 2y)(7x + 2y) &= (7x)^2 - (2y)^2 \\ &= 49x^2 - 4y^2 \end{aligned}$$

8. **(a)**
$$\begin{aligned} (4u + 5v)^2 &= (4u)^2 + 2(4u)(5v) + (5v)^2 \\ &= 16u^2 + 40uv + 25v^2 \end{aligned}$$

   (b)
$$\begin{aligned} (7x^2 - 3y^2)^2 &= (7x^2)^2 - 2(7x^2)(3y^2) + (3y^2)^2 \\ &= 49x^4 - 42x^2y^2 + 9y^4 \end{aligned}$$

9.
$$\begin{array}{r} 2x^2 - 3x + 1 \\ x^2 - 5x \\ \hline -10x^3 + 15x^2 - 5x \\ 2x^4 - 3x^3 + x^2 \\ \hline 2x^4 - 13x^3 + 16x^2 - 5x \end{array}$$

10.
$$\begin{aligned} &(2x^2 - 3x + 1)(x^2 - 5x) \\ &= (2x^2 - 3x + 1)(x^2) + (2x^2 - 3x + 1)(-5x) \\ &= 2x^4 - 3x^3 + x^2 - 10x^3 + 15x^2 - 5x \\ &= 2x^4 - 13x^3 + 16x^2 - 5x \end{aligned}$$

5.2 Student Practice

1.
$$\begin{aligned} &\frac{-16x^4 + 16x^3 + 8x^2 + 64x}{8x} \\ &= \frac{-16x^4}{8x} + \frac{16x^3}{8x} + \frac{8x^2}{8x} + \frac{64x}{8x} \\ &= -2x^3 + 2x^2 + x + 8 \end{aligned}$$

2.
$$\begin{array}{r} 2x^2 - x - 3 \\ 4x - 3 \overline{) 8x^3 - 10x^2 - 9x + 14} \\ \underline{8x^3 - 6x^2} \\ -4x^2 - 9x \\ \underline{-4x^2 + 3x} \\ -12x + 14 \\ \underline{-12x + 9} \\ 5 \end{array}$$

The answer is $2x^2 - x - 3$ with a remainder of 5 or $2x^2 - x - 3 + \frac{5}{4x - 3}$.

3.
$$\begin{array}{r} 4x^2 - 6x + 9 \\ 2x + 3 \overline{) 8x^3 + 0x^2 + 0x + 27} \\ \underline{8x^3 + 12x^2} \\ -12x^2 + 0x \\ \underline{-12x^2 - 18x} \\ 18x + 27 \\ \underline{18x + 27} \\ 0 \end{array}$$

The answer is $4x^2 - 6x + 9$.

4. Write $2x^4 + 3x^3 + 7x - 3 - x^2$ as $2x^4 + 3x^3 - x^2 + 7x - 3$.
$$\begin{array}{r} x^2 + 2x - 1 \\ 2x^2 - x + 3 \overline{) 2x^4 + 3x^3 - x^2 + 7x - 3} \\ \underline{2x^4 - x^3 + 3x^2} \\ 4x^3 - 4x^2 + 7x \\ \underline{4x^3 - 2x^2 + 6x} \\ -2x^2 + x - 3 \\ \underline{-2x^2 + x - 3} \\ 0 \end{array}$$

The answer is $x^2 + 2x - 1$.

5.3 Student Practice

1.

-3	1	-3	4	-5
		-3	18	-66
	1	-6	22	-71

The quotient is $x^2 - 6x + 22 + \frac{-71}{x+3}$.

2.

3	2	0	-1	-39	-36
		6	18	51	-36
	2	6	17	12	0

The quotient is $2x^3 + 6x^2 + 17x + 12$.

3.

3	2	-9	5	13	-5
		6	-9	-12	3
	2	-3	-4	1	-2

The quotient is $2x^3 - 3x^2 - 4x + 1 + \frac{-2}{x-3}$.

5.4 Student Practice

1. **(a)** $19x^3 - 38x^2 = 19 \cdot x \cdot x \cdot x - 19 \cdot 2 \cdot x \cdot x = 19x^2(x-2)$
   (b) $100a^4 - 50a^2 = 50a^2(2a^2 - 1)$
2. **(a)** $21x^3 - 18x^2y + 24xy^2 = 3x(7x^2 - 6xy + 8y^2)$
   (b) $12xy^2 - 14x^2y + 20x^2y^2 + 36x^3y$
   $= 2xy(6y - 7x + 10xy + 18x^2)$
3. $9a^3 - 12a^2b^2 - 15a^4 = 3a^2(3a - 4b^2 - 5a^2)$
   Check: $3a - 4b^2 - 5a^2$ has no common factors.
   $3a^2(3a - 4b^2 - 5a^2) = 9a^3 - 12a^2b^2 - 15a^4$
   This is the original polynomial. It checks.
4. $7x(x+2y) - 8y(x+2y) - (x+2y)$
   $= 7x(x+2y) - 8y(x+2y) - 1(x+2y)$
   $= (x+2y)(7x - 8y - 1)$
5. $bx + 5by + 2wx + 10wy = b(x+5y) + 2w(x+5y)$
   $= (x+5y)(b+2w)$
6. To factor $5x^2 - 12y - 15x + 4xy$, rearrange the terms. Then factor.
   $5x^2 - 15x + 4xy - 12y = 5x(x-3) + 4y(x-3)$
   $= (x-3)(5x+4y)$
7. To factor $xy - 12 - 4x + 3y$, rearrange the terms. Then factor.
   $xy - 4x + 3y - 12 = x(y-4) + 3(y-4)$
   $= (y-4)(x+3)$
8. To factor $2x^3 - 15 - 10x + 3x^2$, rearrange the terms. Then factor.
   $2x^3 - 10x + 3x^2 - 15 = 2x(x^2-5) + 3(x^2-5)$
   $= (x^2-5)(2x+3)$

5.5 Student Practice

1.

Factor Pairs of 21	Sum of the Factors
$(-21)(-1)$	$-21 - 1 = -22$
$(-7)(-3)$	$-7 - 3 = -10$ ✓

The numbers whose product is 21 and whose sum is -10 are -7 and -3. Thus, $x^2 - 10x + 21 = (x-7)(x-3)$.

2.

Factor Pairs of −48	Sum of the Factors
$(-48)(1)$	$-48 + 1 = -47$
$(48)(-1)$	$48 - 1 = 47$
$(-24)(2)$	$-24 + 2 = -22$
$(24)(-2)$	$24 - 2 = 22$
$(-16)(3)$	$-16 + 3 = -13$ ✓
$(16)(-3)$	$16 - 3 = 13$
$(-12)(4)$	$-12 + 4 = -8$
$(12)(-4)$	$12 - 4 = 8$
$(-8)(6)$	$-8 + 6 = -2$
$(8)(-6)$	$8 - 6 = 2$

The numbers whose product is -48 and whose sum is -13 are -16 and 3. Thus, $x^2 - 13x - 48 = (x-16)(x+3)$.

3. $x^4 + 9x^2 + 8 = (x^2)^2 + 9(x^2) + 8$
   Let $y = x^2$. Then $x^4 + 9x^2 + 8 = y^2 + 9y + 8$. The two numbers whose product is 8 and whose sum is 9 are 8 and 1, so $y^2 + 9y + 8 = (y+8)(y+1)$.
   Thus, $x^4 + 9x^2 + 8 = (x^2+8)(x^2+1)$.
4. **(a)** $a^2 - 2a - 48 = (a+6)(a-8)$
   (b) $x^4 + 2x^2 - 15 = (x^2+5)(x^2-3)$
5. **(a)** $x^2 - 16xy + 15y^2 = (x-15y)(x-y)$
   (b) $x^2 + xy - 42y^2 = (x+7y)(x-6y)$
6. $4x^2 - 44x + 72 = 4(x^2 - 11x + 18) = 4(x-9)(x-2)$
7. The grouping number is $(a)(c) = (10)(2) = 20$. The factor pairs of 20 are $(-20)(-1)$, $(-10)(-2)$, and $(-5)(-4)$. Since $-5 + (-4) = -9$, use -5 and -4.
   $10x^2 - 9x + 2 = 10x^2 - 5x - 4x + 2$
   $= 5x(2x-1) - 2(2x-1)$
   $= (2x-1)(5x-2)$
8. The grouping number is $(a)(c) = (3)(-8) = -24$. The factor pairs of -24 are $(-24)(1)$, $(24)(-1)$, $(-12)(2)$, $(12)(-2)$, $(-8)(3)$, $(8)(-3)$, $(-6)(4)$, $(6)(-4)$. Since $6 + (-4) = 2$, use 6 and -4.
   $3x^2 + 2x - 8 = 3x^2 + 6x - 4x - 8$
   $= 3x(x+2) - 4(x+2)$
   $= (x+2)(3x-4)$
9. $9x^3 - 15x^2 - 6x = 3x(3x^2 - 5x - 2)$
   $= 3x(3x^2 - 6x + x - 2)$
   $= 3x[3x(x-2) + 1(x-2)]$
   $= 3x(x-2)(3x+1)$
10. The first terms of the factors could be $8x$ and x or $4x$ and $2x$. The second terms could be $+1$ and -5 or -1 and $+5$.

Possible Factors	Middle Term of Product
$(8x+1)(x-5)$	$-39x$
$(8x-1)(x+5)$	$+39x$
$(8x+5)(x-1)$	$-3x$
$(8x-5)(x+1)$	$+3x$
$(4x+1)(2x-5)$	$-18x$
$(4x-1)(2x+5)$	$+18x$
$(4x+5)(2x-1)$	$+6x$
$(4x-5)(2x+1)$	$-6x$

Thus, $8x^2 - 6x - 5 = (4x-5)(2x+1)$.

11. $6x^4 + 13x^2 - 5 = (2x^2+5)(3x^2-1)$

5.6 Student Practice

1. $x^2 - 9 = (x)^2 - (3)^2 = (x+3)(x-3)$
2. $64x^2 - 121 = (8x)^2 - (11)^2 = (8x+11)(8x-11)$
3. $49x^2 - 25y^4 = (7x)^2 - (5y^2)^2 = (7x+5y^2)(7x-5y^2)$
4. $7x^2 - 28 = 7(x^2-4) = 7(x+2)(x-2)$
5. $9x^2 - 30x + 25 = (3x)^2 - 2(3x)(5) + (5)^2 = (3x-5)^2$
6. $25x^2 - 70x + 49 = (5x)^2 - 2(5x)(7) + (7)^2 = (5x-7)^2$
7. $242x^2 + 88x + 8 = 2(121x^2 + 44x + 4)$
   $= 2[(11x)^2 + 2(11x)(2) + (2)^2]$
   $= 2(11x+2)^2$
8. **(a)** $49x^4 + 28x^2 + 4 = (7x^2)^2 + 2(7x^2)(2) + (2)^2 = (7x^2+2)^2$
   (b) $36x^4 + 84x^2y^2 + 49y^4 = (6x^2)^2 + 2(6x^2)(7y^2) + (7y^2)^2$
   $= (6x^2 + 7y^2)^2$
9. $8x^3 + 125y^3 = (2x)^3 + (5y)^3 = (2x+5y)(4x^2 - 10xy + 25y^2)$
10. $64x^3 + 125y^3 = (4x)^3 + (5y)^3 = (4x+5y)(16x^2 - 20xy + 25y^2)$
11. $27w^3 - 125z^6 = (3w)^3 - (5z^2)^3 = (3w - 5z^2)(9w^2 + 15wz^2 + 25z^4)$
12. $54x^3 - 16 = 2(27x^3 - 8)$
    $= 2(3x-2)(9x^2 + 6x + 4)$
13. Use the difference of two squares formula first.
    $64a^6 - 1 = (8a^3)^2 - (1)^2$
    $= (8a^3+1)(8a^3-1)$
    $= [(2a)^3 + (1)^3][(2a)^3 - (1)^3]$
    $= (2a+1)(4a^2 - 2a + 1)(2a-1)(4a^2 + 2a + 1)$

5.7 Student Practice

1. **(a)** $7x^5 + 56x^2 = 7x^2(x^3+8)$
   $= 7x^2(x+2)(x^2 - 2x + 4)$
   (b) $125x^2 + 50xy + 5y^2 = 5(25x^2 + 10xy + y^2)$
   $= 5(5x+y)^2$
   (c) $12x^2 - 75 = 3(4x^2 - 25)$
   $= 3(2x+5)(2x-5)$

(d) $3x^2 - 39x + 126 = 3(x^2 - 13x + 42)$
$= 3(x - 7)(x - 6)$

(e) $6ax + 6ay + 18bx + 18by$
$= 6(ax + ay + 3bx + 3by)$
$= 6[a(x + y) + 3b(x + y)]$
$= 6(x + y)(a + 3b)$

(f) $6x^3 - x^2 - 12x = x(6x^2 - x - 12)$
$= x(2x - 3)(3x + 4)$

2. $3x^2 - 10x + 4$
Prime. There are no factors of 12 whose sum is -10.

3. $16x^2 + 81$
Prime. Binomials of the form $a^2 + b^2$ cannot be factored.

5.8 Student Practice

1. $x^2 + x = 56$
$x^2 + x - 56 = 0$
$(x + 8)(x - 7) = 0$
$x + 8 = 0$ or $x - 7 = 0$
$x = -8$ $\quad x = 7$

2. $12x^2 - 9x + 2 = 2x$
$12x^2 - 11x + 2 = 0$
$(4x - 1)(3x - 2) = 0$
$4x - 1 = 0$ or $3x - 2 = 0$
$x = \frac{1}{4}$ $\quad x = \frac{2}{3}$

3. $7x^2 - 14x = 0$
$7x(x - 2) = 0$
$7x = 0$ or $x - 2 = 0$
$x = 0$ $\quad x = 2$

4. $16x(x - 2) = 8x - 25$
$16x^2 - 32x = 8x - 25$
$16x^2 - 32x - 8x + 25 = 0$
$16x^2 - 40x + 25 = 0$
$(4x - 5)^2 = 0$
$4x - 5 = 0$ or $4x - 5 = 0$
$4x = 5$ $\quad 4x = 5$
$x = \frac{5}{4}$ $\quad x = \frac{5}{4}$
$x = \frac{5}{4}$ is a double root.

5. $3x^3 + 6x^2 = 45x$
$3x^3 + 6x^2 - 45x = 0$
$3x(x^2 + 2x - 15) = 0$
$3x(x + 5)(x - 3) = 0$
$3x = 0$ or $x + 5 = 0$ or $x - 3 = 0$
$x = 0$ $\quad x = -5$ $\quad x = 3$

6. $A = \frac{1}{2}ab$
Let $x =$ the length of the base, then
$x + 5 =$ the length of the altitude.
$52 = \frac{1}{2}(x + 5)(x)$
$104 = (x + 5)(x)$
$104 = x^2 + 5x$
$0 = x^2 + 5x - 104$
$0 = (x - 8)(x + 13)$
$x - 8 = 0$ or $x + 13 = 0$
$x = 8$ $\quad x = -13$
The base of a triangle must be a positive number, so we disregard -13. Thus,
base $= x = 8$ feet
altitude $= x + 5 = 13$ feet

7. Let $x =$ the length in feet of last year's garden, then $3x + 2 =$ the length in feet of this year's garden.
$(3x + 2)^2 = 112 + x^2$
$9x^2 + 12x + 4 = 112 + x^2$
$8x^2 + 12x - 108 = 0$
$4(2x^2 + 3x - 27) = 0$
$4(x - 3)(2x + 9) = 0$
$x - 3 = 0$ or $2x + 9 = 0$
$x = 3$ $\quad x = -\frac{9}{2}$
Length cannot be negative, so we reject the negative answer. We use $x = 3$. Last year's garden was a square with each side measuring 3 feet. This year's garden measures $3x + 2 = 3(3) + 2 = 11$ feet on each side.

Chapter 6

6.1 Student Practice

1. Solve the equation $x^2 - 9x - 22 = 0$.
$(x + 2)(x - 11) = 0$
$x + 2 = 0$ or $x - 11 = 0$
$x = -2$ $\quad x = 11$
The domain of $y = f(x)$ is all real numbers except -2 and 11.

2. $$\frac{x^2 - 36y^2}{x^2 - 3xy - 18y^2} = \frac{(x + 6y)(x - 6y)}{(x + 3y)(x - 6y)} = \frac{x + 6y}{x + 3y}$$

3. $$\frac{3xy}{3xy^2 + 6x^2y} = \frac{3xy \cdot 1}{3xy(y + 2x)} = \frac{1}{y + 2x}$$

4. $$\frac{x^3 - 4x^2 - 5x}{3x^2 - 30x + 75} = \frac{x(x + 1)(x - 5)}{3(x - 5)(x - 5)} = \frac{x(x + 1)}{3(x - 5)}$$

5. $$\frac{7a^2 - 23ab + 6b^2}{4b^2 - 49a^2} = \frac{(7a - 2b)(a - 3b)}{(2b + 7a)(2b - 7a)}$$
$$= \frac{(7a - 2b)(a - 3b)}{(2b + 7a)(-1)(-2b + 7a)}$$
$$= \frac{a - 3b}{-1(2b + 7a)}$$
$$= -\frac{a - 3b}{2b + 7a}$$

6. $$\frac{-3x + 6y}{x^2 - 7xy + 10y^2} = \frac{-3(x - 2y)}{(x - 5y)(x - 2y)} = \frac{-3}{x - 5y}$$

7. $$\frac{2x^2 + 5x + 2}{4x^2 - 1} \cdot \frac{10x^2 + 5x - 5}{x^2 + x - 2}$$
$$= \frac{(2x + 1)(x + 2)(5)(2x - 1)(x + 1)}{(2x + 1)(2x - 1)(x + 2)(x - 1)}$$
$$= \frac{2x + 1}{2x + 1} \cdot \frac{x + 2}{x + 2} \cdot \frac{5}{1} \cdot \frac{2x - 1}{2x - 1} \cdot \frac{1}{x - 1} \cdot \frac{x + 1}{1}$$
$$= 1 \cdot 1 \cdot \frac{5}{1} \cdot 1 \cdot \frac{1}{x - 1} \cdot \frac{x + 1}{1}$$
$$= \frac{5(x + 1)}{x - 1}$$

8. $$\frac{9x + 9y}{5ax + 5ay} \cdot \frac{10a^2x^2 - 40b^2x^2}{27ax^2 - 54bx^2}$$
$$= \frac{\cancel{9}\cancel{(x + y)}}{\cancel{5}a\cancel{(x + y)}} \cdot \frac{\overset{2}{\cancel{10}}\,\cancel{x^2}(a^2 - 4b^2)}{\underset{3}{\cancel{27}}\,\cancel{x^2}(a - 2b)}$$
$$= \frac{2(a + 2b)(a - 2b)}{3a(a - 2b)} = \frac{2(a + 2b)}{3a}$$

9. $$\frac{8x^3 + 27y^3}{64x^3 - y^3} \div \frac{4x^2 - 9y^2}{16x^2 + 4xy + y^2}$$
$$= \frac{8x^3 + 27y^3}{64x^3 - y^3} \cdot \frac{16x^2 + 4xy + y^2}{4x^2 - 9y^2}$$
$$= \frac{\cancel{(2x + 3y)}(4x^2 - 6xy + 9y^2)}{(4x - y)\cancel{(16x^2 + 4xy + y^2)}} \cdot \frac{\cancel{16x^2 + 4xy + y^2}}{\cancel{(2x + 3y)}(2x - 3y)}$$
$$= \frac{4x^2 - 6xy + 9y^2}{(4x - y)(2x - 3y)}$$

10. $\dfrac{4x^2 - 9}{2x^2 + 11x + 12} \div (-6x + 9)$

$$= \frac{4x^2 - 9}{2x^2 + 11x + 12} \cdot \frac{1}{-6x + 9}$$

$$= \frac{\cancel{(2x + 3)}\cancel{(2x - 3)}}{\cancel{(2x + 3)}(x + 4)} \cdot \frac{1}{-3\cancel{(2x - 3)}}$$

$$= -\frac{1}{3(x + 4)}$$

6.2 Student Practice

1. Find the LCD. $\dfrac{8}{x^2 - x - 12}, \dfrac{3}{x - 4}$

Factor each denominator completely.

$x^2 - x - 12 = (x - 4)(x + 3)$

$x - 4$ cannot be factored.

The LCD is the product of all the different prime factors.

LCD $= (x - 4)(x + 3)$

2. Find the LCD. $\dfrac{2}{15x^3y^2}, \dfrac{13}{25xy^3}$

Factor each denominator.

$$15x^3y^2 = 3 \cdot 5 \cdot x \cdot x \cdot x \cdot y \cdot y$$
$$25xy^3 = 5 \cdot 5 \cdot x \cdot y \cdot y \cdot y$$
$$\text{LCD} = 3 \cdot 5 \cdot 5 \cdot x \cdot x \cdot x \cdot y \cdot y \cdot y$$
$$= 75x^3y^3$$

3. $\dfrac{4x}{(x + 6)(2x - 1)} - \dfrac{3x + 1}{(x + 6)(2x - 1)} = \dfrac{x - 1}{(x + 6)(2x - 1)}$

4. LCD $= (x - 4)(x + 3)$

$$\frac{8}{(x - 4)(x + 3)} + \frac{3}{x - 4}$$

$$= \frac{8}{(x - 4)(x + 3)} + \frac{3}{x - 4} \cdot \frac{x + 3}{x + 3}$$

$$= \frac{8 + 3(x + 3)}{(x - 4)(x + 3)}$$

$$= \frac{8 + 3x + 9}{(x - 4)(x + 3)} = \frac{3x + 17}{(x - 4)(x + 3)}$$

5. LCD $= 4x(x + 4)$

$$\frac{5}{x + 4} + \frac{3}{4x}$$

$$= \frac{5}{x + 4} \cdot \frac{4x}{4x} + \frac{3}{4x} \cdot \frac{x + 4}{x + 4}$$

$$= \frac{20x}{4x(x + 4)} + \frac{3x + 12}{4x(x + 4)} = \frac{23x + 12}{4x(x + 4)}$$

6. LCD $= 12a^3b^3$

$$\frac{7}{4ab^3} + \frac{1}{3a^3b^2}$$

$$= \frac{7}{4ab^3} \cdot \frac{3a^2}{3a^2} + \frac{1}{3a^3b^2} \cdot \frac{4b}{4b}$$

$$= \frac{21a^2}{12a^3b^3} + \frac{4b}{12a^3b^3}$$

$$= \frac{21a^2 + 4b}{12a^3b^3}$$

7. $x^2 + x - 12 = (x + 4)(x - 3)$

$x^2 + 6x + 8 = (x + 4)(x + 2)$

LCD $= (x + 4)(x - 3)(x + 2)$

$$\frac{4x + 2}{x^2 + x - 12} - \frac{3x + 8}{x^2 + 6x + 8}$$

$$= \frac{4x + 2}{(x + 4)(x - 3)} \cdot \frac{x + 2}{x + 2} - \frac{3x + 8}{(x + 4)(x + 2)} \cdot \frac{x - 3}{x - 3}$$

$$= \frac{4x^2 + 10x + 4}{(x + 4)(x - 3)(x + 2)} - \frac{3x^2 - x - 24}{(x + 4)(x - 3)(x + 2)}$$

$$= \frac{4x^2 + 10x + 4 - 3x^2 + x + 24}{(x + 4)(x - 3)(x + 2)}$$

$$= \frac{x^2 + 11x + 28}{(x + 4)(x - 3)(x + 2)}$$

$$= \frac{(x + 4)(x + 7)}{(x + 4)(x - 3)(x + 2)}$$

$$= \frac{x + 7}{(x - 3)(x + 2)}$$

8. $4x^2 + 20x + 25 = (2x + 5)^2$

$4x + 10 = 2(2x + 5)$

LCD $= 2(2x + 5)^2$

$$\frac{7x - 3}{(2x + 5)^2} - \frac{3x}{2(2x + 5)}$$

$$= \frac{7x - 3}{(2x + 5)^2} \cdot \frac{2}{2} - \frac{3x}{2(2x + 5)} \cdot \frac{2x + 5}{2x + 5}$$

$$= \frac{2(7x - 3) - 3x(2x + 5)}{2(2x + 5)^2}$$

$$= \frac{14x - 6 - 6x^2 - 15x}{2(2x + 5)^2} = \frac{-6x^2 - x - 6}{2(2x + 5)^2}$$

6.3 Student Practice

1. $\dfrac{y + \dfrac{3}{y}}{\dfrac{2}{y^2} + \dfrac{5}{y}}$

METHOD 1

$$y + \frac{3}{y} = \frac{y^2 + 3}{y}$$

$$\frac{2}{y^2} + \frac{5}{y} = \frac{2 + 5y}{y^2}$$

Divide the numerator by the denominator.

$$\frac{\dfrac{y^2 + 3}{y}}{\dfrac{2 + 5y}{y^2}} = \frac{y^2 + 3}{y} \div \frac{2 + 5y}{y^2}$$

$$= \frac{y^2 + 3}{y} \cdot \frac{y^2}{2 + 5y}$$

$$= \frac{y(y^2 + 3)}{2 + 5y}$$

METHOD 2

LCD $= y^2$

$$\frac{y + \dfrac{3}{y}}{\dfrac{2}{y^2} + \dfrac{5}{y}} \cdot \frac{y^2}{y^2} = \frac{y^3 + 3y}{2 + 5y} = \frac{y(y^2 + 3)}{2 + 5y}$$

2. Simplify.

$$\frac{\dfrac{4}{16x^2 - 1} + \dfrac{3}{4x + 1}}{\dfrac{x}{4x - 1} + \dfrac{5}{4x + 1}}$$

METHOD 1

Simplify the numerator.

$$\frac{4}{(4x+1)(4x-1)} + \frac{3(4x-1)}{(4x+1)(4x-1)} = \frac{12x+1}{(4x+1)(4x-1)}$$

Simplify the denominator.

$$\frac{x(4x+1)}{(4x-1)(4x+1)} + \frac{5(4x-1)}{(4x+1)(4x-1)} = \frac{4x^2+21x-5}{(4x-1)(4x+1)}$$

To divide the numerator by the denominator, we multiply the numerator by the reciprocal of the denominator.

$$\frac{12x+1}{(4x+1)(4x-1)} \cdot \frac{(4x+1)(4x-1)}{4x^2+21x-5} = \frac{12x+1}{4x^2+21x-5}$$

METHOD 2

Multiply the numerator and denominator by the LCD of all the rational expressions in the numerator and denominator.

LCD $= (4x+1)(4x-1)$.

$$\frac{\left[\frac{4}{(4x+1)(4x-1)} + \frac{3}{(4x+1)}\right](4x+1)(4x-1)}{\left[\frac{x}{4x-1} + \frac{5}{4x+1}\right](4x+1)(4x-1)}$$

$$= \frac{\frac{4(4x+1)(4x-1)}{(4x+1)(4x-1)} + \frac{3(4x+1)(4x-1)}{4x+1}}{\frac{x}{4x-1}\cdot(4x+1)(4x-1) + \frac{5}{4x+1}\cdot(4x+1)(4x-1)}$$

$$= \frac{4+3(4x-1)}{x(4x+1)+5(4x-1)} = \frac{4+12x-3}{4x^2+x+20x-5} = \frac{12x+1}{4x^2+21x-5}$$

3. Simplify by **METHOD 1.**

$$\frac{4+x}{x-\frac{16}{x}} = \frac{4+x}{\frac{x}{1}\cdot\frac{x}{x} - \frac{16}{x}}$$

$$= \frac{4+x}{\frac{x^2-16}{x}} = \frac{4+x}{1} \div \frac{x^2-16}{x}$$

$$= \frac{4+x}{1}\cdot\frac{x}{(x-4)(x+4)} = \frac{x}{x-4}$$

4. Simplify by **METHOD 2.**

LCD $= y(y+3)$

$$\frac{\frac{7}{y+3} - \frac{3}{y}}{\frac{2}{y} + \frac{5}{y+3}} = \frac{\frac{7}{y+3} - \frac{3}{y}}{\frac{2}{y} + \frac{5}{y+3}} \cdot \frac{y(y+3)}{y(y+3)}$$

$$= \frac{7y-3(y+3)}{2(y+3)+5y} = \frac{7y-3y-9}{2y+6+5y} = \frac{4y-9}{7y+6}$$

6.4 Student Practice

1.

$$\frac{4}{3x} + \frac{x+1}{x} = \frac{1}{2}$$

$$6x\left(\frac{4}{3x} + \frac{x+1}{x}\right) = 6x\left(\frac{1}{2}\right)$$

$$6x\left(\frac{4}{3x}\right) + 6x\left(\frac{x+1}{x}\right) = 6x\left(\frac{1}{2}\right)$$

$$8 + 6(x+1) = 3x$$

$$8 + 6x + 6 = 3x$$

$$3x = -14$$

$$x = -\frac{14}{3}$$

Check.

$$\frac{4}{3\left(\frac{-14}{3}\right)} + \frac{\frac{-14}{3}+1}{-\frac{14}{3}} \stackrel{?}{=} \frac{1}{2}$$

$$\frac{4}{-14} + \frac{\frac{-11}{3}}{\frac{-14}{3}} \stackrel{?}{=} \frac{1}{2}$$

$$-\frac{4}{14} + \frac{11}{14} \stackrel{?}{=} \frac{1}{2}$$

$$\frac{7}{14} \stackrel{?}{=} \frac{1}{2}$$

$$\frac{1}{2} = \frac{1}{2} \checkmark$$

2.

$$\frac{1}{3x-9} = \frac{1}{2x-6} - \frac{5}{6}$$

$$\frac{1}{3(x-3)} = \frac{1}{2(x-3)} - \frac{5}{6}$$

$$6(x-3)\left[\frac{1}{3(x-3)}\right] = 6(x-3)\left[\frac{1}{2(x-3)}\right] - 6(x-3)\left(\frac{5}{6}\right)$$

$$2(1) = 3(1) - (x-3)(5)$$

$$2 = 3 - 5x + 15$$

$$2 = 18 - 5x$$

$$-16 = -5x$$

$$\frac{16}{5} = x$$

Check.

$$\frac{1}{3\left(\frac{16}{5}\right) - 9} \stackrel{?}{=} \frac{1}{2\left(\frac{16}{5}\right) - 6} - \frac{5}{6}$$

$$\frac{1}{\frac{48}{5} - 9} \stackrel{?}{=} \frac{1}{\frac{32}{5} - 6} - \frac{5}{6}$$

$$\frac{1}{\frac{3}{5}} \stackrel{?}{=} \frac{1}{\frac{2}{5}} - \frac{5}{6}$$

$$\frac{5}{3} \stackrel{?}{=} \frac{5}{2} - \frac{5}{6}$$

$$\frac{5}{3} = \frac{5}{3} \checkmark$$

3.

$$\frac{y^2+4y-2}{y^2-2y-8} = 1 + \frac{4}{y-4}$$

$$\frac{y^2+4y-2}{(y+2)(y-4)} = 1 + \frac{4}{y-4}$$

$$(y+2)(y-4)\left[\frac{y^2+4y-2}{(y+2)(y-4)}\right]$$

$$= (y+2)(y-4)(1) + (y+2)(y-4)\left(\frac{4}{y-4}\right)$$

$$y^2 + 4y - 2 = y^2 - 2y - 8 + 4y + 8$$

$$y^2 + 4y - 2 = y^2 + 2y$$

$$4y - 2 = 2y$$

$$2y = 2$$

$$y = 1$$

4. $\dfrac{2x-1}{x^2-7x+10}+\dfrac{3}{x-5}=\dfrac{5}{x-2}$

$$\frac{2x-1}{(x-2)(x-5)}+\frac{3}{x-5}=\frac{5}{x-2}$$

$$(x-2)(x-5)\left[\frac{2x-1}{(x-2)(x-5)}\right]+(x-2)(x-5)\left(\frac{3}{x-5}\right)$$
$$=(x-2)(x-5)\left(\frac{5}{x-2}\right)$$

$$2x-1+3(x-2)=5(x-5)$$
$$2x-1+3x-6=5x-25$$
$$5x-7=5x-25$$
$$0=-18$$

Of course $0 \neq -18$. Therefore, no value of x makes the original equation true. Hence the equation has **no solution.**

5. $\dfrac{y}{y-2}-3=1+\dfrac{2}{y-2}$

$$(y-2)\left(\frac{y}{y-2}\right)-(y-2)(3)=(y-2)(1)+(y-2)\left(\frac{2}{y-2}\right)$$
$$y-3(y-2)=1(y-2)+2$$
$$y-3y+6=y-2+2$$
$$-2y+6=y$$
$$6=3y$$
$$2=y$$

Check.

$$\frac{2}{2-2}-3\stackrel{?}{=}1+\frac{2}{2-2}$$
$$\frac{2}{0}-3\stackrel{?}{=}1+\frac{2}{0}$$

Division by zero is not defined. The value $y=2$ is therefore not a solution to the original equation. There is **no solution.**

6.5 Student Practice

1. Solve for t.

$$\frac{1}{t}=\frac{1}{c}+\frac{1}{d}$$
$$cdt\left[\frac{1}{t}\right]=cdt\left[\frac{1}{c}\right]+cdt\left[\frac{1}{d}\right]$$
$$cd=dt+ct$$
$$cd=t(d+c)$$
$$\frac{cd}{d+c}=t$$

2. Solve for p_1.

$$C=\frac{Bp_1p_2}{d^2}$$
$$d^2[C]=d^2\left[\frac{Bp_1p_2}{d^2}\right]$$
$$Cd^2=Bp_1p_2$$
$$\frac{Cd^2}{Bp_2}=\frac{Bp_1p_2}{Bp_2}$$
$$\frac{Cd^2}{Bp_2}=p_1$$

3. Let $x =$ the number of faculty. Then $1932-x =$ the number of students.

$$\frac{2}{21}=\frac{x}{1932-x}$$
$$21(1932-x)\left[\frac{2}{21}\right]=21(1932-x)\left[\frac{x}{1932-x}\right]$$
$$2(1932-x)=21x$$
$$3864-2x=21x$$
$$3864=23x$$
$$168=x$$
$$1932-x=1932-168=1764$$

There are 168 faculty members and 1764 students.

4. We will draw the picture as two similar triangles.

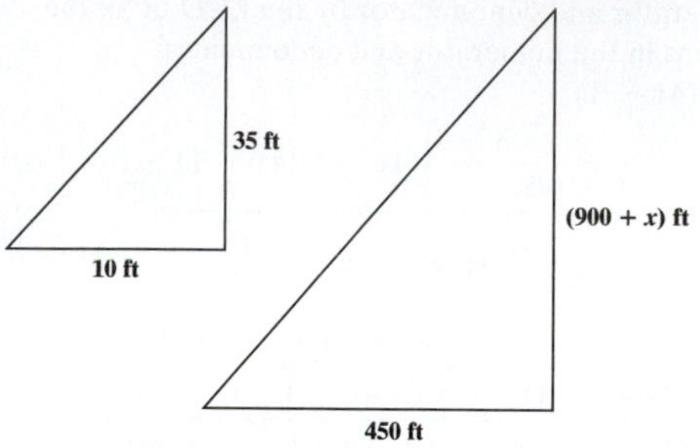

We write a proportion and solve.

$$\frac{10}{35}=\frac{450}{900+x}$$
$$\text{LCD}=35(900+x)$$
$$35(900+x)\left(\frac{10}{35}\right)=35(900+x)\left(\frac{450}{900+x}\right)$$
$$10(900+x)=35(450)$$
$$9000+10x=15{,}750$$
$$10x=6750$$
$$x=675$$

The helicopter is 675 feet over the building.

5. We construct a table for the data.

	Rate of Work per Hour	Time Worked in Hours	Fraction of Task Done
Alfred	$\frac{1}{4}$	x	$\frac{x}{4}$
Son	$\frac{1}{5}$	x	$\frac{x}{5}$

The fraction of work done by Alfred plus the fraction of work done by his son equals 1 completed job.
Write an equation and solve.

$$\frac{x}{4}+\frac{x}{5}=1$$
$$20\left(\frac{x}{4}\right)+20\left(\frac{x}{5}\right)=20(1)$$
$$5x+4x=20$$
$$9x=20$$
$$x=2.2\overline{2}$$

Alfred and his son can mow the lawn working together in approximately 2.2 hours.

Chapter 7

7.1 Student Practice

1. $\left(\dfrac{3x^{-2}y^4}{2x^{-5}y^2}\right)^{-3}=\dfrac{(3x^{-2}y^4)^{-3}}{(2x^{-5}y^2)^{-3}}$

$$=\frac{3^{-3}(x^{-2})^{-3}(y^4)^{-3}}{2^{-3}(x^{-5})^{-3}(y^2)^{-3}}$$
$$=\frac{3^{-3}x^6y^{-12}}{2^{-3}x^{15}y^{-6}}$$
$$=\frac{3^{-3}}{2^{-3}}\cdot\frac{x^6}{x^{15}}\cdot\frac{y^{-12}}{y^{-6}}$$
$$=\frac{2^3}{3^3}\cdot x^{6-15}\cdot y^{-12+6}$$
$$=\frac{8}{27}x^{-9}y^{-6}\text{ or }\frac{8}{27x^9y^6}$$

2. (a) $(x^4)^{3/8} = x^{(4/1)(3/8)} = x^{3/2}$

(b) $\dfrac{x^{3/7}}{x^{2/7}} = x^{3/7-2/7} = x^{1/7}$

(c) $x^{-7/5} \cdot x^{4/5} = x^{-7/5+4/5} = x^{-3/5}$

3. (a) $(-3x^{1/4})(2x^{1/2}) = -6x^{1/4+1/2} = -6x^{1/4+2/4} = -6x^{3/4}$

(b) $\dfrac{13x^{1/12}y^{-1/4}}{26x^{-1/3}y^{1/2}} = \dfrac{x^{1/12-(-1/3)}y^{-1/4-1/2}}{2}$

$= \dfrac{x^{1/12+4/12}y^{-1/4-2/4}}{2}$

$= \dfrac{x^{5/12}y^{-3/4}}{2}$

$= \dfrac{x^{5/12}}{2y^{3/4}}$

4. $-3x^{1/2}(2x^{1/4} + 3x^{-1/2}) = -6x^{1/2+1/4} - 9x^{1/2-1/2}$

$= -6x^{2/4+1/4} - 9x^0$

$= -6x^{3/4} - 9$

5. (a) $(4)^{5/2} = (2^2)^{5/2} = 2^{2/1 \cdot 5/2} = 2^5 = 32$

(b) $(27)^{4/3} = (3^3)^{4/3} = 3^{3/1 \cdot 4/3} = 3^4 = 81$

6. $3x^{1/3} + x^{-1/3} = 3x^{1/3} + \dfrac{1}{x^{1/3}}$

$= \dfrac{x^{1/3} \cdot 3x^{1/3}}{x^{1/3}} + \dfrac{1}{x^{1/3}}$

$= \dfrac{3x^{2/3} + 1}{x^{1/3}}$

7. $4y^{3/2} - 8y^{5/2} = 4y^{2/2+1/2} - 8y^{2/2+3/2}$

$= 4(y^{2/2})(y^{1/2}) - 8(y^{2/2})(y^{3/2})$

$= 4y(y^{1/2} - 2y^{3/2})$

7.2 Student Practice

1. (a) $\sqrt[3]{216} = \sqrt[3]{6^3} = 6$

(b) $\sqrt[5]{32} = \sqrt[5]{2^5} = 2$

(c) $\sqrt[3]{-8} = \sqrt[3]{(-2)^3} = -2$

(d) $\sqrt[4]{-81}$ is not a real number.

2. (a) $f(3) = \sqrt{4(3) - 3} = \sqrt{12 - 3} = \sqrt{9} = 3$

(b) $f(4) = \sqrt{4(4) - 3} = \sqrt{16 - 3} = \sqrt{13} \approx 3.6$

(c) $f(7) = \sqrt{4(7) - 3} = \sqrt{28 - 3} = \sqrt{25} = 5$

3. $0.5x + 2 \geq 0$

$0.5x \geq -2$

$x \geq -4$

The domain is all real numbers x where $x \geq -4$.

4.

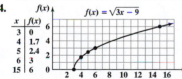

x	$f(x)$
3	0
4	1.7
5	2.4
6	3
15	6

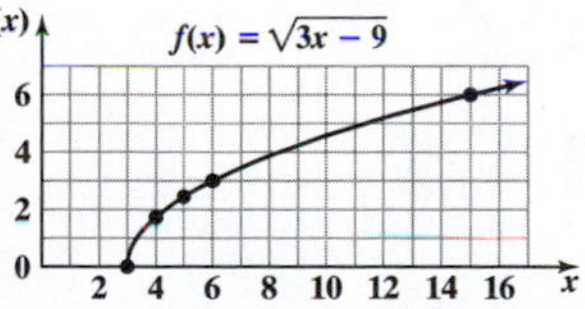

5. (a) $\sqrt[3]{x^3} = (x^3)^{1/3} = x^{3/3} = x^1 = x$

(b) $\sqrt[4]{y^4} = (y^4)^{1/4} = y^{4/4} = y^1 = y$

6. (a) $\sqrt[4]{x^3} = (x^3)^{1/4} = x^{3/4}$

(b) $\sqrt[5]{(xy)^7} = [(xy)^7]^{1/5} = (xy)^{7/5}$

7. (a) $\sqrt[4]{81x^{12}} = (3^4x^{12})^{1/4} = 3x^3$

(b) $\sqrt[3]{27x^6} = (3^3x^6)^{1/3} = 3x^2$

(c) $(32x^5)^{3/5} = (2^5x^5)^{3/5} = 2^3x^3 = 8x^3$

8. (a) $(xy)^{3/4} = \sqrt[4]{(xy)^3} = \sqrt[4]{x^3y^3}$ or $(xy)^{3/4} = (\sqrt[4]{xy})^3$

(b) $y^{-1/3} = \dfrac{1}{y^{1/3}} = \dfrac{1}{\sqrt[3]{y}}$

(c) $(2x)^{4/5} = \sqrt[5]{(2x)^4} = \sqrt[5]{16x^4}$ or $(2x)^{4/5} = (\sqrt[5]{2x})^4$

(d) $2x^{4/5} = 2\sqrt[5]{x^4}$ or $2x^{4/5} = 2(\sqrt[5]{x})^4$

9. (a) $8^{2/3} = (\sqrt[3]{8})^2 = 2^2 = 4$

(b) $(-8)^{4/3} = (\sqrt[3]{-8})^4 = (-2)^4 = 16$

(c) $100^{-3/2} = \dfrac{1}{100^{3/2}} = \dfrac{1}{(\sqrt{100})^3} = \dfrac{1}{10^3} = \dfrac{1}{1000}$

10. (a) $\sqrt[5]{(-3)^5} = -3$

(b) $\sqrt[4]{(-5)^4} = |-5| = 5$

(c) $\sqrt[4]{w^4} = |w|$

(d) $\sqrt[7]{y^7} = y$

11. (a) $\sqrt{36x^2} = 6|x|$

(b) $\sqrt[4]{16y^8} = 2|y^2| = 2y^2$

(c) $\sqrt[3]{125x^3y^6} = \sqrt[3]{5^3x^3(y^2)^3} = 5xy^2$

7.3 Student Practice

1. $\sqrt{20} = \sqrt{4 \cdot 5} = \sqrt{4}\sqrt{5} = 2\sqrt{5}$

2. $\sqrt{27} = \sqrt{9}\sqrt{3} = 3\sqrt{3}$

3. (a) $\sqrt[3]{24} = \sqrt[3]{8}\sqrt[3]{3} = 2\sqrt[3]{3}$

(b) $\sqrt[3]{-108} = \sqrt[3]{-27}\sqrt[3]{4} = -3\sqrt[3]{4}$

4. $\sqrt[4]{64} = \sqrt[4]{16}\sqrt[4]{4} = 2\sqrt[4]{4}$

5. (a) $\sqrt{45x^6y^7} = \sqrt{9 \cdot 5 \cdot x^6 \cdot y^6 \cdot y} = \sqrt{9x^6y^6}\sqrt{5y} = 3x^3y^3\sqrt{5y}$

(b) $\sqrt{27a^7b^8c^9} = \sqrt{9 \cdot 3 \cdot a^6 \cdot a \cdot b^8 \cdot c^8 \cdot c}$

$= \sqrt{9a^6b^8c^8}\sqrt{3ac}$

$= 3a^3b^4c^4\sqrt{3ac}$

6. $19\sqrt{xy} + 5\sqrt{xy} - 10\sqrt{xy} = (19 + 5 - 10)\sqrt{xy} = 14\sqrt{xy}$

7. $4\sqrt{2} - 5\sqrt{50} - 3\sqrt{98}$

$= 4\sqrt{2} - 5\sqrt{25}\sqrt{2} - 3\sqrt{49}\sqrt{2}$

$= 4\sqrt{2} - 5(5)\sqrt{2} - 3(7)\sqrt{2}$

$= 4\sqrt{2} - 25\sqrt{2} - 21\sqrt{2}$

$= -42\sqrt{2}$

8. $4\sqrt{2x} + \sqrt{18x} - 2\sqrt{125x} - 6\sqrt{20x}$

$= 4\sqrt{2x} + \sqrt{9}\sqrt{2x} - 2\sqrt{25}\sqrt{5x} - 6\sqrt{4}\sqrt{5x}$

$= 4\sqrt{2x} + 3\sqrt{2x} - 10\sqrt{5x} - 12\sqrt{5x}$

$= 7\sqrt{2x} - 22\sqrt{5x}$

9. $3x\sqrt[3]{54x^4} - 3\sqrt[3]{16x^7}$

$= 3x\sqrt[3]{27x^3}\sqrt[3]{2x} - 3\sqrt[3]{8x^6}\sqrt[3]{2x}$

$= 3x(3x)\sqrt[3]{2x} - 3(2x^2)\sqrt[3]{2x}$

$= 9x^2\sqrt[3]{2x} - 6x^2\sqrt[3]{2x}$

$= 3x^2\sqrt[3]{2x}$

7.4 Student Practice

1. $(-4\sqrt{2})(-3\sqrt{13x}) = (-4)(-3)\sqrt{2 \cdot 13x} = 12\sqrt{26x}$

2. $\sqrt{2x}(\sqrt{5} + 2\sqrt{3x} + \sqrt{8})$

$= (\sqrt{2x})(\sqrt{5}) + (\sqrt{2x})(2\sqrt{3x}) + (\sqrt{2x})(\sqrt{8})$

$= \sqrt{10x} + 2\sqrt{6x^2} + \sqrt{16x}$

$= \sqrt{10x} + 2\sqrt{x^2}\sqrt{6} + \sqrt{16}\sqrt{x}$

$= \sqrt{10x} + 2x\sqrt{6} + 4\sqrt{x}$

3. By FOIL: $(\sqrt{7} + 4\sqrt{2})(2\sqrt{7} - 3\sqrt{2})$

$= 2\sqrt{49} - 3\sqrt{14} + 8\sqrt{14} - 12\sqrt{4}$

$= 14 + 5\sqrt{14} - 24$

$= -10 + 5\sqrt{14}$

4. $(2 - 5\sqrt{5})(3 - 2\sqrt{2}) = 6 - 4\sqrt{2} - 15\sqrt{5} + 10\sqrt{10}$

5. By FOIL: $(\sqrt{5x} + \sqrt{10})^2 = (\sqrt{5x} + \sqrt{10})(\sqrt{5x} + \sqrt{10})$

$= \sqrt{25x^2} + \sqrt{50x} + \sqrt{50x} + \sqrt{100}$

$= 5x + 2\sqrt{25}\sqrt{2x} + 10$

$= 5x + 2(5)\sqrt{2x} + 10$

$= 5x + 10\sqrt{2x} + 10$

6. (a) $\sqrt[3]{2x}\left(\sqrt[3]{4x^2} + 3\sqrt[3]{y}\right)$

$= \left(\sqrt[3]{2x}\right)\left(\sqrt[3]{4x^2}\right) + 3\left(\sqrt[3]{2x}\right)\left(\sqrt[3]{y}\right)$

$= \sqrt[3]{8x^3} + 3\sqrt[3]{2xy}$

$= 2x + 3\sqrt[3]{2xy}$

(b) $\left(\sqrt[3]{7} + \sqrt[3]{x^2}\right)\left(2\sqrt[3]{49} - \sqrt[3]{x}\right)$

$= 2\sqrt[3]{343} - \sqrt[3]{7x} + 2\sqrt[3]{49x^2} - \sqrt[3]{x^3}$

$= 2(7) - \sqrt[3]{7x} + 2\sqrt[3]{49x^2} - x$

$= 14 - \sqrt[3]{7x} + 2\sqrt[3]{49x^2} - x$

7. (a) $\dfrac{\sqrt{75}}{\sqrt{3}} = \sqrt{\dfrac{75}{3}} = \sqrt{25} = 5$

(b) $\sqrt[3]{\dfrac{27}{64}} = \dfrac{\sqrt[3]{27}}{\sqrt[3]{64}} = \dfrac{3}{4}$

(c) $\dfrac{\sqrt{54a^3b^7}}{\sqrt{6b^5}} = \sqrt{\dfrac{54a^3b^7}{6b^5}} = \sqrt{9a^3b^2} = 3ab\sqrt{a}$

8. $\dfrac{7}{\sqrt{3}} = \dfrac{7}{\sqrt{3}} \cdot \dfrac{\sqrt{3}}{\sqrt{3}} = \dfrac{7\sqrt{3}}{\sqrt{9}} = \dfrac{7\sqrt{3}}{3}$

9. Method 1: $\dfrac{8}{\sqrt{20x}} = \dfrac{8}{\sqrt{4}\sqrt{5x}} = \dfrac{8}{2\sqrt{5x}} \cdot \dfrac{\sqrt{5x}}{\sqrt{5x}} = \dfrac{8\sqrt{5x}}{10x} = \dfrac{4\sqrt{5x}}{5x}$

Method 2: $\dfrac{8}{\sqrt{20x}} = \dfrac{8}{\sqrt{20x}} \cdot \dfrac{\sqrt{5x}}{\sqrt{5x}} = \dfrac{8\sqrt{5x}}{\sqrt{100x^2}} = \dfrac{8\sqrt{5x}}{10x} = \dfrac{4\sqrt{5x}}{5x}$

10. Method 1: $\sqrt[3]{\dfrac{6}{5x}} = \dfrac{\sqrt[3]{6}}{\sqrt[3]{5x}} = \dfrac{\sqrt[3]{6}}{\sqrt[3]{5x}} \cdot \dfrac{\sqrt[3]{25x^2}}{\sqrt[3]{25x^2}} = \dfrac{\sqrt[3]{150x^2}}{\sqrt[3]{125x^3}} = \dfrac{\sqrt[3]{150x^2}}{5x}$

Method 2: $\sqrt[3]{\dfrac{6}{5x}} = \sqrt[3]{\dfrac{6}{5x} \cdot \dfrac{25x^2}{25x^2}} = \sqrt[3]{\dfrac{150x^2}{125x^3}} = \dfrac{\sqrt[3]{150x^2}}{\sqrt[3]{125x^3}} = \dfrac{\sqrt[3]{150x^2}}{5x}$

11. $\dfrac{4}{2 + \sqrt{5}} = \dfrac{4}{2 + \sqrt{5}} \cdot \dfrac{2 - \sqrt{5}}{2 - \sqrt{5}}$

$= \dfrac{8 - 4\sqrt{5}}{2^2 - (\sqrt{5})^2}$

$= \dfrac{8 - 4\sqrt{5}}{4 - 5}$

$= \dfrac{8 - 4\sqrt{5}}{-1}$

$= -\left(8 - 4\sqrt{5}\right)$

$= -8 + 4\sqrt{5}$

12. $\dfrac{\sqrt{11} + \sqrt{5}}{\sqrt{11} - \sqrt{5}} = \dfrac{\sqrt{11} + \sqrt{5}}{\sqrt{11} - \sqrt{5}} \cdot \dfrac{\sqrt{11} + \sqrt{5}}{\sqrt{11} + \sqrt{5}}$

$= \dfrac{\sqrt{121} + 2\sqrt{55} + \sqrt{25}}{(\sqrt{11})^2 - (\sqrt{5})^2}$

$= \dfrac{11 + 2\sqrt{55} + 5}{11 - 5}$

$= \dfrac{16 + 2\sqrt{55}}{6}$

$= \dfrac{2\left(8 + \sqrt{55}\right)}{2 \cdot 3} = \dfrac{8 + \sqrt{55}}{3}$

7.5 Student Practice

1. $\sqrt{3x - 8} = x - 2$

$\left(\sqrt{3x - 8}\right)^2 = (x - 2)^2$

$3x - 8 = x^2 - 4x + 4$

$0 = x^2 - 7x + 12$

$0 = (x - 3)(x - 4)$

$x - 3 = 0 \quad \text{or} \quad x - 4 = 0$

$x = 3 \qquad\qquad x = 4$

Check.

For $x = 3$: $\sqrt{3(3) - 8} \stackrel{?}{=} 3 - 2$

$\sqrt{1} \stackrel{?}{=} 1$

$1 = 1$ ✓

For $x = 4$: $\sqrt{3(4) - 8} \stackrel{?}{=} 4 - 2$

$\sqrt{4} \stackrel{?}{=} 2$

$2 = 2$ ✓

The solutions are 3 and 4.

2. $-4 + \sqrt{x + 4} = x$

$\sqrt{x + 4} = x + 4$

$\left(\sqrt{x + 4}\right)^2 = (x + 4)^2$

$x + 4 = x^2 + 8x + 16$

$0 = x^2 + 7x + 12$

$0 = (x + 3)(x + 4)$

$x + 3 = 0 \quad \text{or} \quad x + 4 = 0$

$x = -3 \qquad\qquad x = -4$

Check.

For $x = -3$: $-4 + \sqrt{-3 + 4} \stackrel{?}{=} -3$

$-4 + \sqrt{1} \stackrel{?}{=} -3$

$-3 = -3$ ✓

For $x = -4$: $-4 + \sqrt{-4 + 4} \stackrel{?}{=} -4$

$-4 + \sqrt{0} \stackrel{?}{=} -4$

$-4 = -4$ ✓

The solutions are -3 and -4.

3. $\sqrt{2x + 5} - 2\sqrt{2x} = 1$

$\sqrt{2x + 5} = 2\sqrt{2x} + 1$

$\left(\sqrt{2x + 5}\right)^2 = \left(2\sqrt{2x} + 1\right)^2$

$2x + 5 = \left(2\sqrt{2x} + 1\right)\left(2\sqrt{2x} + 1\right)$

$2x + 5 = 8x + 4\sqrt{2x} + 1$

$-6x + 4 = 4\sqrt{2x}$

$-3x + 2 = 2\sqrt{2x}$

$(-3x + 2)^2 = \left(2\sqrt{2x}\right)^2$

$9x^2 - 12x + 4 = 8x$

$9x^2 - 20x + 4 = 0$

$(9x - 2)(x - 2) = 0$

$9x - 2 = 0 \quad \text{or} \quad x - 2 = 0$

$9x = 2$

$x = \dfrac{2}{9} \qquad\qquad x = 2$

Check. For $x = \dfrac{2}{9}$:

$\sqrt{2\left(\dfrac{2}{9}\right) + 5} - 2\sqrt{2\left(\dfrac{2}{9}\right)} \stackrel{?}{=} 1$

$\sqrt{\dfrac{4}{9} + 5} - 2\sqrt{\dfrac{4}{9}} \stackrel{?}{=} 1$

$\sqrt{\dfrac{49}{9}} - 2\left(\dfrac{2}{3}\right) \stackrel{?}{=} 1$

$\dfrac{7}{3} - \dfrac{4}{3} \stackrel{?}{=} 1$

$\dfrac{3}{3} \stackrel{?}{=} 1$

$1 = 1$ ✓

For $x = 2$:

$\sqrt{2(2) + 5} - 2\sqrt{2(2)} \stackrel{?}{=} 1$

$\sqrt{9} - 2\sqrt{4} \stackrel{?}{=} 1$

$3 - 4 \stackrel{?}{=} 1$

$-1 \neq 1$

The only solution is $\dfrac{2}{9}$.

4. $\sqrt{y - 1} + \sqrt{y - 4} = \sqrt{4y - 11}$

$\left(\sqrt{y - 1} + \sqrt{y - 4}\right)^2 = \left(\sqrt{4y - 11}\right)^2$

$\left(\sqrt{y - 1} + \sqrt{y - 4}\right)\left(\sqrt{y - 1} + \sqrt{y - 4}\right) = 4y - 11$

$y - 1 + 2\sqrt{(y - 1)(y - 4)} + y - 4 = 4y - 11$

$2\sqrt{(y - 1)(y - 4)} = 2y - 6$

$\sqrt{(y - 1)(y - 4)} = y - 3$

$\left(\sqrt{y^2 - 5y + 4}\right)^2 = (y - 3)^2$

$y^2 - 5y + 4 = y^2 - 6y + 9$

$y = 5$

Check. $\sqrt{5-1} + \sqrt{5-4} \stackrel{?}{=} \sqrt{4(5)-11}$
$\sqrt{4} + \sqrt{1} \stackrel{?}{=} \sqrt{9}$
$2 + 1 \stackrel{?}{=} 3$
$3 = 3$ ✓

The solution is 5.

7.6 Student Practice

1. (a) $\sqrt{-49} = \sqrt{-1}\sqrt{49} = (i)(7) = 7i$

(b) $\sqrt{-31} = \sqrt{-1}\sqrt{31} = i\sqrt{31}$

2. $\sqrt{-98} = \sqrt{-1}\sqrt{98} = i\sqrt{98} = i\sqrt{49}\sqrt{2} = 7i\sqrt{2}$

3. $\sqrt{-8}\cdot\sqrt{-2} = \sqrt{-1}\sqrt{8}\cdot\sqrt{-1}\sqrt{2}$
$= i\sqrt{8}\cdot i\sqrt{2}$
$= i^2\sqrt{16}$
$= -1(4) = -4$

4. $-7 + 2yi\sqrt{3} = x + 6i\sqrt{3}$
$x = -7, \quad 2y\sqrt{3} = 6\sqrt{3}$
$y = 3$

5. $(3 - 4i) - (-2 - 18i)$
$= [3 - (-2)] + [-4 - (-18)]i$
$= (3 + 2) + (-4 + 18)i$
$= 5 + 14i$

6. $(4 - 2i)(3 - 7i)$
$= (4)(3) + (4)(-7i) + (-2i)(3) + (-2i)(-7i)$
$= 12 - 28i - 6i + 14i^2$
$= 12 - 28i - 6i + 14(-1)$
$= 12 - 28i - 6i - 14 = -2 - 34i$

7. $-2i(5 + 6i)$
$= (-2)(5)i + (-2)(6)i^2$
$= -10i - 12i^2$
$= -10i - 12(-1) = 12 - 10i$

8. (a) $i^{42} = (i^{40+2}) = (i^{40})(i^2) = (i^4)^{10}(i^2) = (1)^{10}(-1) = -1$

(b) $i^{53} = (i^{52+1}) = (i^{52})(i) = (i^4)^{13}(i) = (1)^{13}(i) = i$

9. $\dfrac{4+2i}{3+4i} = \dfrac{4+2i}{3+4i}\cdot\dfrac{3-4i}{3-4i}$
$= \dfrac{12 - 16i + 6i - 8i^2}{9 - 16i^2}$
$= \dfrac{12 - 10i - 8(-1)}{9 - 16(-1)}$
$= \dfrac{12 - 10i + 8}{9 + 16}$
$= \dfrac{20 - 10i}{25}$
$= \dfrac{5(4 - 2i)}{5\cdot 5}$
$= \dfrac{4 - 2i}{5}$ or $\dfrac{4}{5} - \dfrac{2}{5}i$

10. $\dfrac{5-6i}{-2i} = \dfrac{5-6i}{-2i}\cdot\dfrac{2i}{2i}$
$= \dfrac{10i - 12i^2}{-4i^2}$
$= \dfrac{10i - 12(-1)}{-4(-1)}$
$= \dfrac{10i + 12}{4}$
$= \dfrac{2(5i + 6)}{2\cdot 2}$
$= \dfrac{6 + 5i}{2}$ or $3 + \dfrac{5}{2}i$

7.7 Student Practice

1. Let s = maximum speed,
h = horsepower.
$s = k\sqrt{h}$
Substitute $s = 128$ and $h = 256$.
$128 = k\sqrt{256}$
$128 = 16k$
$8 = k$
Now we know the value of k so $s = 8\sqrt{h}$.
Let $h = 225$.
$s = 8\left(\sqrt{225}\right)$
$s = 8(15) = 120$
The maximum speed is 120 miles per hour.

2. $y = \dfrac{k}{x}$
Substitute $y = 45$ and $x = 16$.
$45 = \dfrac{k}{16}$
$720 = k$
We now write the equation $y = \dfrac{720}{x}$.
Find the value of y when $x = 36$.
$y = \dfrac{720}{36}$
$y = 20$

3. Let w = weight,
h = height of column.
$w = \dfrac{k}{h^2}$
Substitute $w = 2$ and $h = 7.5$.
$2 = \dfrac{k}{(7.5)^2}$
$112.5 = k$
We now write the equation $w = \dfrac{112.5}{h^2}$.
Now substitute $h = 3$ and solve for w.
$w = \dfrac{112.5}{(3)^2}$
$w = \dfrac{112.5}{9} = 12.5$
The column can support 12.5 tons.

4. $y = \dfrac{kzw^2}{x}$
To find the value of k, substitute $y = 20$, $z = 3$, $w = 5$, and $x = 4$.
$20 = \dfrac{k(3)(5)^2}{4}$
$20 = \dfrac{75k}{4}$
$\dfrac{80}{75} = k$
$\dfrac{16}{15} = k$
We now substitute $\dfrac{16}{15}$ for k.
$y = \dfrac{16zw^2}{15x}$
We use this equation to find y when $z = 4$, $w = 6$, and $x = 2$.
$y = \dfrac{16(4)(6)^2}{15(2)} = \dfrac{2304}{30}$
$y = \dfrac{384}{5}$

Chapter 8

8.1 Student Practice

1. $x^2 - 121 = 0$
$x^2 = 121$
$x = \pm\sqrt{121}$
$x = \pm 11$

The two roots are 11 and −11.

Check:

$(11)^2 - 121 \stackrel{?}{=} 0$ $\quad$ $(-11)^2 - 121 \stackrel{?}{=} 0$
$121 - 121 \stackrel{?}{=} 0$ $\quad$ $121 - 121 \stackrel{?}{=} 0$
$0 = 0$ ✓ $\quad$ $0 = 0$ ✓

2. $x^2 = 18$
$x = \pm\sqrt{18} = \pm\sqrt{9 \cdot 2}$
$x = \pm 3\sqrt{2}$

The roots are $3\sqrt{2}$ and $-3\sqrt{2}$.

3. $5x^2 + 1 = 46$
$5x^2 = 45$
$x^2 = 9$
$x = \pm\sqrt{9}$
$x = \pm 3$

The roots are 3 and −3.

4. $3x^2 = -27$
$x^2 = -9$
$x = \pm\sqrt{-9}$
$x = \pm 3i$

The roots are $3i$ and $-3i$.

Check:

$3(3i)^2 \stackrel{?}{=} -27$ $\quad$ $3(-3i)^2 \stackrel{?}{=} -27$
$3(9i^2) \stackrel{?}{=} -27$ $\quad$ $3(9i^2) \stackrel{?}{=} -27$
$3(-9) \stackrel{?}{=} -27$ $\quad$ $3(-9) \stackrel{?}{=} -27$
$-27 = -27$ ✓ $\quad$ $-27 = -27$ ✓

5. $(2x + 3)^2 = 7$
$2x + 3 = \pm\sqrt{7}$
$2x = -3 \pm \sqrt{7}$
$x = \dfrac{-3 \pm \sqrt{7}}{2}$

The roots are $\dfrac{-3 + \sqrt{7}}{2}$ and $\dfrac{-3 - \sqrt{7}}{2}$.

6. $x^2 + 8x + 3 = 0$
$x^2 + 8x = -3$
$x^2 + 8x + (4)^2 = -3 + (4)^2$
$x^2 + 8x + 16 = -3 + 16$
$(x + 4)^2 = 13$
$x + 4 = \pm\sqrt{13}$
$x = -4 \pm \sqrt{13}$

The roots are $-4 + \sqrt{13}$ and $-4 - \sqrt{13}$.

7. $2x^2 + 4x + 1 = 0$
$2x^2 + 4x = -1$
$x^2 + 2x = -\dfrac{1}{2}$
$x^2 + 2x + (1)^2 = -\dfrac{1}{2} + 1$
$(x + 1)^2 = \dfrac{1}{2}$
$x + 1 = \pm\sqrt{\dfrac{1}{2}}$
$x + 1 = \pm\dfrac{1}{\sqrt{2}}$
$x = -1 \pm \dfrac{\sqrt{2}}{2}$ or $\dfrac{-2 \pm \sqrt{2}}{2}$

8.2 Student Practice

1. $x^2 + 5x = -1 + 2x$
$x^2 + 3x + 1 = 0$
$a = 1, b = 3, c = 1$
$x = \dfrac{-3 \pm \sqrt{3^2 - 4(1)(1)}}{2(1)}$
$x = \dfrac{-3 \pm \sqrt{5}}{2}$

2. $2x^2 + 7x + 6 = 0$
$a = 2, b = 7, c = 6$
$x = \dfrac{-7 \pm \sqrt{7^2 - 4(2)(6)}}{2(2)}$
$x = \dfrac{-7 \pm \sqrt{49 - 48}}{4} = \dfrac{-7 \pm \sqrt{1}}{4}$
$x = \dfrac{-7 + 1}{4} = -\dfrac{6}{4}$ or $x = \dfrac{-7 - 1}{4} = -\dfrac{8}{4}$
$x = -\dfrac{3}{2}$ $\qquad$ $x = -2$

3. $2x^2 - 26 = 0$
$a = 2, b = 0, c = -26$
$x = \dfrac{-0 \pm \sqrt{0^2 - 4(2)(-26)}}{2(2)}$
$x = \dfrac{\pm\sqrt{208}}{4} = \dfrac{\pm\sqrt{16}\sqrt{13}}{4} = \dfrac{\pm 4\sqrt{13}}{4} = \pm\sqrt{13}$

4. $0 = -100x^2 + 4800x - 52{,}559$
$a = -100, b = 4800, c = -52{,}559$
$x = \dfrac{-4800 \pm \sqrt{(4800)^2 - 4(-100)(-52{,}559)}}{2(-100)}$
$x = \dfrac{-4800 \pm \sqrt{23{,}040{,}000 - 21{,}023{,}600}}{-200}$
$x = \dfrac{-4800 \pm \sqrt{2{,}016{,}400}}{-200}$
$x = \dfrac{-4800 \pm 1420}{-200}$
$x = \dfrac{-4800 + 1420}{-200} = \dfrac{-3380}{-200} = 16.9 \approx 17$

or

$x = \dfrac{-4800 - 1420}{-200} = \dfrac{-6220}{-200} = 31.1 \approx 31$

For zero profit, they should manufacture 17 or 31 modems.

5. $\dfrac{1}{x} + \dfrac{1}{x - 1} = \dfrac{5}{6}$ $\quad$ The LCD is $6x(x - 1)$.

$6x(x - 1)\left(\dfrac{1}{x}\right) + 6x(x - 1)\left(\dfrac{1}{x - 1}\right) = 6x(x - 1)\left(\dfrac{5}{6}\right)$
$6(x - 1) + 6x = 5x(x - 1)$
$6x - 6 + 6x = 5x^2 - 5x$
$0 = 5x^2 - 17x + 6$

$a = 5, b = -17, c = 6$
$x = \dfrac{-(-17) \pm \sqrt{(-17)^2 - 4(5)(6)}}{2(5)}$
$x = \dfrac{17 \pm \sqrt{289 - 120}}{10} = \dfrac{17 \pm \sqrt{169}}{10} = \dfrac{17 \pm 13}{10}$
$x = \dfrac{17 + 13}{10} = \dfrac{30}{10} = 3,\quad x = \dfrac{17 - 13}{10} = \dfrac{4}{10} = \dfrac{2}{5}$

6. $2x^2 - 4x + 5 = 0$

$a = 2, b = -4, c = 5$

$$x = \frac{-(-4) \pm \sqrt{(-4)^2 - 4(2)(5)}}{2(2)}$$

$$x = \frac{4 \pm \sqrt{16 - 40}}{4} = \frac{4 \pm \sqrt{-24}}{4}$$

$$x = \frac{4 \pm 2i\sqrt{6}}{4} = \frac{2(2 + i\sqrt{6})}{4} = \frac{2 \pm i\sqrt{6}}{2}$$

7. $9x^2 + 12x + 4 = 0$

$a = 9, b = 12, c = 4$

$b^2 - 4ac = 12^2 - 4(9)(4) = 144 - 144 = 0$

Since the discriminant is 0, there is one rational solution.

8. (a) $x^2 - 4x + 13 = 0$

$a = 1, b = -4, c = 13$

$b^2 - 4ac = (-4)^2 - 4(1)(13) = 16 - 52 = -36$

Since the discriminant is negative, there are two complex solutions containing i.

(b) $9x^2 + 6x - 7 = 0$

$a = 9, b = 6, c = -7$

$b^2 - 4ac = 6^2 - 4(9)(-7) = 36 + 252 = 288$

Since the discriminant is positive, there are two different irrational solutions.

9. $x = -10 \qquad x = -6$

$x + 10 = 0 \qquad x + 6 = 0$

$(x + 10)(x + 6) = 0$

$x^2 + 16x + 60 = 0$

10. $x - 2i\sqrt{3} = 0 \qquad x + 2i\sqrt{3} = 0$

$$(x - 2i\sqrt{3})(x + 2i\sqrt{3}) = 0$$

$$x^2 + 2ix\sqrt{3} - 2ix\sqrt{3} - (2i\sqrt{3})^2 = 0$$

$$x^2 - 4i^2(\sqrt{9}) = 0$$

$$x^2 - 4(-1)(3) = 0$$

$$x^2 + 12 = 0$$

8.3 Student Practice

1. $x^4 - 5x^2 - 36 = 0$

Let $y = x^2$. Then $y^2 = x^4$.

Thus, our new equation is

$y^2 - 5y - 36 = 0.$

$(y - 9)(y + 4) = 0$

$y - 9 = 0 \quad$ or $\quad y + 4 = 0$

$y = 9 \qquad y = -4$

$x^2 = 9 \qquad x^2 = -4$

$x = \pm\sqrt{9} \qquad x = \pm\sqrt{-4}$

$x = \pm 3 \qquad x = \pm 2i$

2. $x^6 - 5x^3 + 4 = 0$

Let $y = x^3$. Then $y^2 = x^6$.

$y^2 - 5y + 4 = 0$

$(y - 1)(y - 4) = 0$

$y - 1 = 0 \quad$ or $\quad y - 4 = 0$

$y = 1 \qquad y = 4$

$x^3 = 1 \qquad x^3 = 4$

$x = \sqrt[3]{1} = 1 \qquad x = \sqrt[3]{4}$

3. $3x^{4/3} - 5x^{2/3} + 2 = 0$

Let $y = x^{2/3}$. Then $y^2 = x^{4/3}$.

$3y^2 - 5y + 2 = 0$

$(3y - 2)(y - 1) = 0$

$3y - 2 = 0 \quad$ or $\quad y - 1 = 0$

$y = \frac{2}{3} \qquad y = 1$

$x^{2/3} = \frac{2}{3} \qquad x^{2/3} = 1$

$(x^{2/3})^3 = \left(\frac{2}{3}\right)^3 \qquad (x^{2/3})^3 = 1^3$

$x^2 = \frac{8}{27} \qquad x^2 = 1$

$x = \pm\sqrt{\frac{8}{27}} \qquad x = \pm\sqrt{1}$

$x = \pm\frac{2\sqrt{2}}{3\sqrt{3}} \qquad x = \pm 1$

$x = \pm\frac{2\sqrt{2}}{3\sqrt{3}} \cdot \frac{\sqrt{3}}{\sqrt{3}}$

$x = \pm\frac{2\sqrt{6}}{9}$

Check: $x = \frac{2\sqrt{6}}{9}$:

$$3\left(\frac{2\sqrt{6}}{9}\right)^{4/3} - 5\left(\frac{2\sqrt{6}}{9}\right)^{2/3} + 2 \stackrel{?}{=} 0$$

$$3\left(\frac{4}{9}\right) - 5\left(\frac{2}{3}\right) + 2 \stackrel{?}{=} 0$$

$$\frac{4}{3} - \frac{10}{3} + 2 \stackrel{?}{=} 0$$

$$0 = 0 \ ✓$$

$x = -\frac{2\sqrt{6}}{9}$:

$$3\left(-\frac{2\sqrt{6}}{9}\right)^{4/3} - 5\left(-\frac{2\sqrt{6}}{9}\right)^{2/3} + 2 \stackrel{?}{=} 0$$

$$3\left(\frac{4}{9}\right) - 5\left(\frac{2}{3}\right) + 2 \stackrel{?}{=} 0$$

$$\frac{4}{3} - \frac{10}{3} + 2 \stackrel{?}{=} 0$$

$$0 = 0 \ ✓$$

$x = 1$:

$3(1)^{4/3} - 5(1)^{2/3} + 2 \stackrel{?}{=} 0$

$3(1) - 5(1) + 2 \stackrel{?}{=} 0$

$0 = 0 \ ✓$

$x = -1$:

$3(-1)^{4/3} - 5(-1)^{2/3} + 2 \stackrel{?}{=} 0$

$3(1) - 5(1) + 2 \stackrel{?}{=} 0$

$0 = 0 \ ✓$

4. $3x^{1/2} = 8x^{1/4} - 4$

$3x^{1/2} - 8x^{1/4} + 4 = 0$

Let $y = x^{1/4}$. Then $y^2 = x^{1/2}$.

$3y^2 - 8y + 4 = 0$

$(3y - 2)(y - 2) = 0$

$3y - 2 = 0 \quad$ or $\quad y - 2 = 0$

$y = \frac{2}{3} \qquad y = 2$

$x^{1/4} = \frac{2}{3} \qquad x^{1/4} = 2$

$(x^{1/4})^4 = \left(\frac{2}{3}\right)^4 \qquad (x^{1/4})^4 = (2)^4$

$x = \frac{16}{81} \qquad x = 16$

Check: $x = \frac{16}{81}$:

$$3\left(\frac{16}{81}\right)^{1/2} \stackrel{?}{=} 8\left(\frac{16}{81}\right)^{1/4} - 4$$

$$3\left(\frac{4}{9}\right) \stackrel{?}{=} 8\left(\frac{2}{3}\right) - 4$$

$$\frac{4}{3} = \frac{4}{3} \ ✓$$

$x = 16$:

$$3(16)^{1/2} \stackrel{?}{=} 8(16)^{1/4} - 4$$

$$3(4) \stackrel{?}{=} 8(2) - 4$$

$$12 = 12 \ ✓$$

5. $2x^{-2} + x^{-1} = 6$

$2x^{-2} + x^{-1} - 6 = 0$

Let $y = x^{-1}$. Then $y^2 = x^{-2}$.

$2y^2 + y - 6 = 0$

$(2y - 3)(y + 2) = 0$

$2y - 3 = 0$ or $y + 2 = 0$

$y = \frac{3}{2}$ $\quad y = -2$

$x^{-1} = \frac{3}{2}$ $\quad x^{-1} = -2$

$(x^{-1})^{-1} = \left(\frac{3}{2}\right)^{-1}$ $\quad (x^{-1})^{-1} = (-2)^{-1}$

$x = \frac{2}{3}$ $\quad x = -\frac{1}{2}$

Check: $x = \frac{2}{3}$: $\quad x = -\frac{1}{2}$:

$2\left(\frac{2}{3}\right)^{-2} + \left(\frac{2}{3}\right)^{-1} \stackrel{?}{=} 6$ $\quad 2\left(-\frac{1}{2}\right)^{-2} + \left(-\frac{1}{2}\right)^{-1} \stackrel{?}{=} 6$

$2\left(\frac{9}{4}\right) + \frac{3}{2} \stackrel{?}{=} 6$ $\quad 2(4) + (-2) \stackrel{?}{=} 6$

$6 = 6$ ✓ $\quad 6 = 6$ ✓

8.4 Student Practice

1. $V = \frac{1}{3}\pi r^2 h$

$\frac{3V}{\pi h} = r^2$

$\pm\sqrt{\frac{3V}{\pi h}} = r$ so $r = \sqrt{\frac{3V}{\pi h}}$

2. $2y^2 + 9wy + 7w^2 = 0$

$(2y + 7w)(y + w) = 0$

$2y + 7w = 0$ or $y + w = 0$

$2y = -7w$ $\quad y = -w$

$y = -\frac{7}{2}w$

3. $3y^2 + 2fy - 7g = 0$

Use the quadratic formula.

$a = 3, b = 2f, c = -7g$

$y = \frac{-2f \pm \sqrt{(2f)^2 - 4(3)(-7g)}}{2(3)}$

$y = \frac{-2f \pm \sqrt{4f^2 + 84g}}{6}$

$y = \frac{-2f \pm \sqrt{4(f^2 + 21g)}}{6}$

$y = \frac{-2f \pm 2\sqrt{f^2 + 21g}}{6}$

$y = \frac{-f \pm \sqrt{f^2 + 21g}}{3}$

4. $d = \frac{n^2 - 3n}{2}$

Multiply each term by 2.

$2d = n^2 - 3n$

$0 = n^2 - 3n - 2d$

Use the quadratic formula.

$a = 1, b = -3, c = -2d$

$n = \frac{-(-3) \pm \sqrt{(-3)^2 - 4(1)(-2d)}}{2(1)}$

$n = \frac{3 \pm \sqrt{9 + 8d}}{2}$

5. (a) $a^2 + b^2 = c^2$

$b^2 = c^2 - a^2$

$b = \sqrt{c^2 - a^2}$

(Only the positive root is used.)

(b) $b = \sqrt{c^2 - a^2}$

$b = \sqrt{(26)^2 - (24)^2} = \sqrt{676 - 576} = \sqrt{100}$

$b = 10$

6. $x + (x - 7) + c = 30$

$c = -2x + 37$

$a = x, b = x - 7, c = -2x + 37$

By the Pythagorean Theorem,

$x^2 + (x - 7)^2 = (-2x + 37)^2$

$x^2 + x^2 - 14x + 49 = 4x^2 - 148x + 1369$

$-2x^2 + 134x - 1320 = 0$

$x^2 - 67x + 660 = 0$

Use the quadratic formula.

$a = 1, b = -67, c = 660$

$x = \frac{-(-67) \pm \sqrt{(-67)^2 - 4(1)(660)}}{2}$

$x = \frac{67 \pm \sqrt{1849}}{2} = \frac{67 \pm 43}{2}$

$x = \frac{67 + 43}{2} = 55$ or $x = \frac{67 - 43}{2} = 12$

The only answer that makes sense is $x = 12$; therefore,

$x = 12$

$x - 7 = 5$

$-2x + 37 = 13$

The longest boundary of the piece of land is 13 miles. The other two boundaries are 5 miles and 12 miles.

7. $A = \pi r^2$

$A = \pi(6)^2 = 36\pi$

The area of the cross section of the old pipe is 36π square feet.

Let $x =$ the radius of the new pipe.

(area of new pipe) − (area of old pipe) = 45π

$\pi x^2 - 36\pi = 45\pi$

$\pi x^2 = 81\pi$

$x^2 = 81$

$x = \pm 9$

Since the radius must be positive, we select $x = 9$. The radius of the new pipe is 9 feet. The radius of the new pipe has been increased by 3 feet.

8. Let $x =$ width. Then $2x - 3 =$ the length.

$x(2x - 3) = 54$

$2x^2 - 3x = 54$

$2x^2 - 3x - 54 = 0$

$(2x + 9)(x - 6) = 0$

$x = -\frac{9}{2}$ or $x = 6$

We do not use the negative value.

Thus, width = 6 feet

length = $2x - 3 = 2(6) - 3 = 9$ feet

9. Let $x =$ his speed on the secondary road. Then $x + 10 =$ his speed on the better road.

	Distance	Rate	Time
Secondary Road	150	x	$\frac{150}{x}$
Better Road	240	$x + 10$	$\frac{240}{x + 10}$
Total	390	(not used)	7

$$\frac{150}{x} + \frac{240}{x+10} = 7$$

The LCD of this equation is $x(x + 10)$. Multiply each term by the LCD.

$$x(x+10)\left(\frac{150}{x}\right) + x(x+10)\left(\frac{240}{x+10}\right) = x(x+10)(7)$$

$$150(x + 10) + 240x = 7x(x + 10)$$
$$150x + 1500 + 240x = 7x^2 + 70x$$
$$0 = 7x^2 - 320x - 1500$$
$$0 = (x - 50)(7x + 30)$$
$$x - 50 = 0 \quad \text{or} \quad 7x + 30 = 0$$
$$x = 50 \qquad x = -\frac{30}{7}$$

We disregard the negative answer. Thus, $x = 50$ mph, so Carlos drove 50 mph on the secondary road and 60 mph on the better road.

8.5 Student Practice

1. $f(x) = x^2 - 6x + 5$
$a = 1, b = -6, c = 5$

Step 1 The vertex occurs at $x = \frac{-b}{2a}$. Thus, $x = \frac{-(-6)}{2(1)} = \frac{6}{2} = 3$

The vertex has an x-coordinate of 3.
To find the y-coordinate, we evaluate $f(3)$.

$$f(3) = 3^2 - 6(3) + 5$$
$$= 9 - 18 + 5$$
$$= -4$$

Thus, the vertex is $(3, -4)$.

Step 2 The y-intercept is at $f(0)$.

$$f(0) = 0^2 - 6(0) + 5$$
$$= 5$$

The y-intercept is $(0, 5)$.

Step 3 The x-intercepts occur when $f(x) = 0$.

$$x^2 - 6x + 5 = 0$$
$$(x - 5)(x - 1) = 0$$
$$x - 5 = 0 \quad x - 1 = 0$$
$$x = 5 \quad x = 1$$

Thus, the x-intercepts are $(5, 0)$ and $(1, 0)$.

2. $g(x) = x^2 - 2x - 2$
$a = 1, b = -2, c = -2$

Step 1 The vertex occurs at

$$x = \frac{-b}{2a}$$
$$x = \frac{-(-2)}{2(1)} = \frac{2}{2} = 1$$

The vertex has an x-coordinate of 1. To find the y-coordinate, we evaluate $g(1)$.

$$g(1) = 1^2 - 2(1) - 2$$
$$= 1 - 2 - 2$$
$$= -3$$

Thus, the vertex is $(1, -3)$.

Step 2 The y-intercept is at $g(0)$.

$$g(0) = 0^2 - 2(0) - 2$$
$$= -2$$

The y-intercept is $(0, -2)$.

Step 3 The x-intercepts occur when $g(x) = 0$. We set $x^2 - 2x - 2 = 0$ and solve for x. The equation does not factor, so we use the quadratic formula.

$$x = \frac{-b \pm \sqrt{b^2 - 4ac}}{2a} = \frac{-(-2) \pm \sqrt{(-2)^2 - 4(1)(-2)}}{2(1)}$$
$$= \frac{2 \pm \sqrt{12}}{2} = 1 \pm \sqrt{3}$$

The x-intercepts are approximately $(2.7, 0)$ and $(-0.7, 0)$.

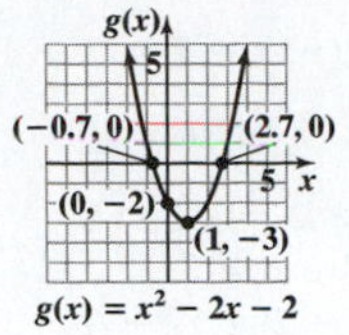

3. $g(x) = -2x^2 - 8x - 6$
$a = -2, b = -8, c = -6$
Since $a < 0$, the parabola opens downward.
The vertex occurs at

$$x = \frac{-b}{2a} = \frac{-(-8)}{2(-2)} = \frac{8}{-4} = -2.$$

To find the y-coordinate, evaluate $g(-2)$.

$$g(-2) = -2(-2)^2 - 8(-2) - 6$$
$$= -8 + 16 - 6$$
$$= 2$$

Thus, the vertex is $(-2, 2)$.
The y-intercept is at $g(0)$.

$$g(0) = -2(0)^2 - 8(0) - 6$$
$$= -6$$

The y-intercept is $(0, -6)$.
The x-intercepts occur when $g(x) = 0$.
Use the quadratic formula.

$$x = \frac{-(-8) \pm \sqrt{(-8)^2 - 4(-2)(-6)}}{2(-2)}$$
$$= \frac{8 \pm \sqrt{16}}{-4} = -2 \pm -1$$
$$x = -3, x = -1$$

The x-intercepts are $(-3, 0)$ and $(-1, 0)$.

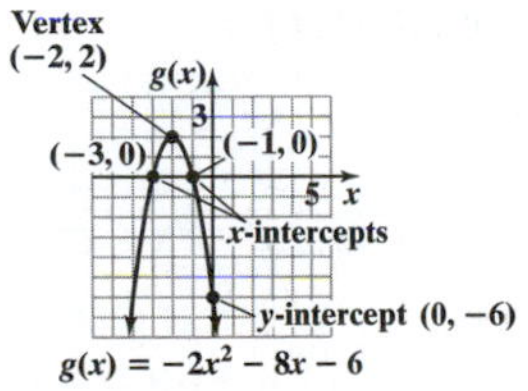

8.6 Student Practice

1. $x^2 - 2x - 8 < 0$
Replace the inequality symbol by an equals sign and solve.

$$x^2 - 2x - 8 = 0$$
$$(x + 2)(x - 4) = 0$$
$$x + 2 = 0 \quad \text{or} \quad x - 4 = 0$$
$$x = -2 \qquad x = 4$$

Region I $\quad x < -2 \quad$ Use $x = -3$
$(-3)^2 - 2(-3) - 8 = 9 + 6 - 8 = 7 > 0$

Region II $\quad -2 < x < 4 \quad$ Use $x = 0$
$0^2 - 2(0) - 8 = -8 < 0$

Region III $\quad x > 4 \quad$ Use $x = 5$
$(5)^2 - 2(5) - 8 = 25 - 10 - 8 = 7 > 0$

Thus, $x^2 - 2x - 8 < 0$ when $-2 < x < 4$.

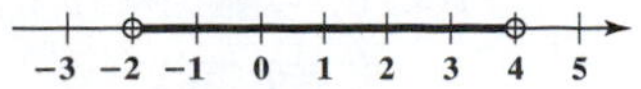

2. $3x^2 - x - 2 \geq 0$

$$3x^2 - x - 2 = 0$$
$$(3x + 2)(x - 1) = 0$$
$$3x + 2 = 0 \quad \text{or} \quad x - 1 = 0$$
$$x = -\frac{2}{3} \qquad x = 1$$

Region I $x < -\frac{2}{3}$ Use $x = -1$

$3(-1)^2 - (-1) - 2 = 3 + 1 - 2 = 2 > 0$

Region II $-\frac{2}{3} < x < 1$ Use $x = 0$

$3(0) - 0 - 2 = -2 < 0$

Region III $x > 1$ Use $x = 2$

$3(2)^2 - 2 - 2 = 12 - 2 - 2 = 8 > 0$

Thus, $3x^2 - x - 2 \geq 0$ when $x \leq -\frac{2}{3}$ or when $x \geq 1$.

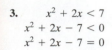

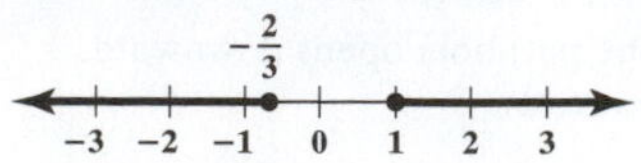

3. $x^2 + 2x < 7$

$x^2 + 2x - 7 < 0$

$x^2 + 2x - 7 = 0$

$$x = \frac{-2 \pm \sqrt{2^2 - 4(1)(-7)}}{2(1)} = \frac{-2 \pm \sqrt{4 + 28}}{2}$$

$$= \frac{-2 \pm \sqrt{32}}{2} = \frac{-2 \pm 4\sqrt{2}}{2} = -1 \pm 2\sqrt{2}$$

$-1 + 2\sqrt{2} \approx -1 + 2.828 \approx 1.828 \approx 1.8$

$-1 - 2\sqrt{2} \approx -1 - 2.828 \approx -3.828 \approx -3.8$

Region I Use $x = -5$

$(-5)^2 + 2(-5) - 7 = 25 - 10 - 7 = 8 > 0$

Region II Use $x = 0$

$(0)^2 + 2(0) - 7 = -7 < 0$

Region III Use $x = 3$

$(3)^2 + 2(3) - 7 = 9 + 6 - 7 = 8 > 0$

Thus, $x^2 + 2x < 7$ when $-1 - 2\sqrt{2} < x < -1 + 2\sqrt{2}$.

Approximately $-3.8 < x < 1.8$

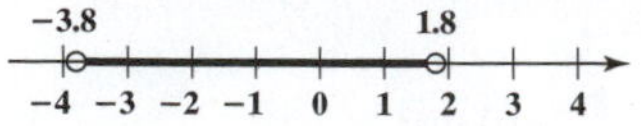

Chapter 9

9.1 Student Practice

1. Let $(x_1, y_1) = (-6, -2)$ and $(x_2, y_2) = (3, 1)$

$$d = \sqrt{(x_2 - x_1)^2 + (y_2 - y_1)^2}$$
$$= \sqrt{[3 - (-6)]^2 + [1 - (-2)]^2}$$
$$= \sqrt{(3 + 6)^2 + (1 + 2)^2}$$
$$= \sqrt{(9)^2 + (3)^2}$$
$$= \sqrt{81 + 9} = \sqrt{90} = 3\sqrt{10}$$

2. $(x + 1)^2 + (y + 2)^2 = 9$

To compare this to $(x - h)^2 + (y - k)^2 = r^2$, we write it in the form

$$[x - (-1)]^2 + [y - (-2)]^2 = 3^2.$$

Thus, we see the center is $(h, k) = (-1, -2)$ and the radius is $r = 3$.

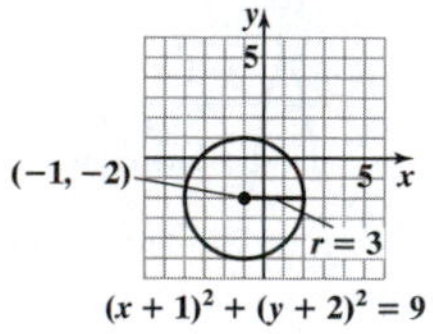

3. We are given that $(h, k) = (-5, 0)$ and $r = \sqrt{3}$. Thus, $(x - h)^2 + (y - k)^2 = r^2$ becomes

$$[x - (-5)]^2 + (y - 0)^2 = \left(\sqrt{3}\right)^2$$
$$(x + 5)^2 + y^2 = 3$$

4. To write $x^2 + 4x + y^2 + 2y - 20 = 0$ in standard form, we complete the squares.

$$x^2 + 4x + \_\_\_\_ + y^2 + 2y + \_\_\_\_ = 20$$
$$x^2 + 4x + 4 + y^2 + 2y + 1 = 20 + 4 + 1$$
$$x^2 + 4x + 4 + y^2 + 2y + 1 = 25$$
$$(x + 2)^2 + (y + 1)^2 = 25$$

The circle has its center at $(-2, -1)$ and the radius is 5.

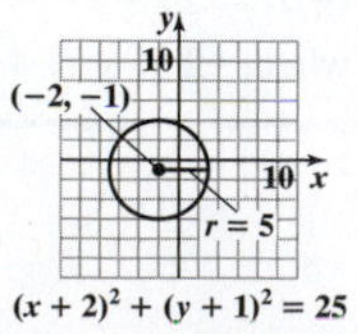

9.2 Student Practice

1. Make a table of values. Begin with $x = -3$ in the middle of the table because $(-3 + 3)^2 = 0$. Plot the points and draw the graph.

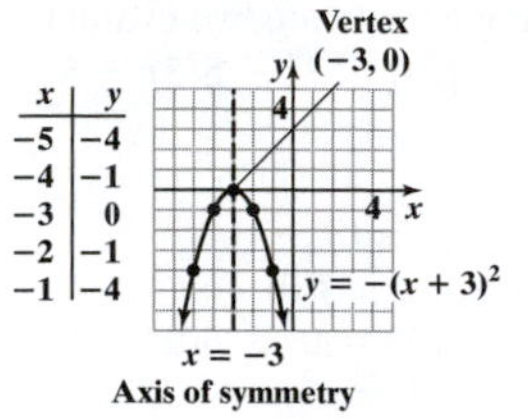

The vertex is $(-3, 0)$, and the axis of symmetry is the line $x = -3$.

2. This graph looks like the graph of $y = x^2$, except that it is shifted 6 units to the right and 4 units up.

The vertex is $(6, 4)$. The axis of symmetry is $x = 6$.

If $x = 0$, $y = (0 - 6)^2 + 4 = 36 + 4 = 40$, so the y-intercept is $(0, 40)$.

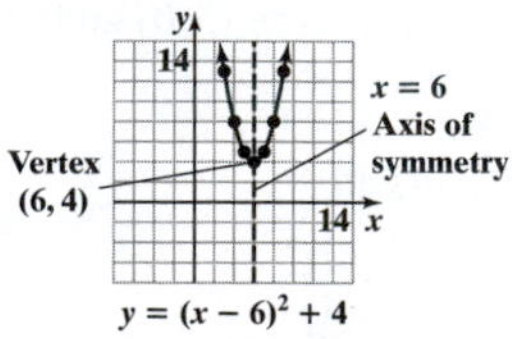

3. Step 1 The equation has the form $y = a(x - h)^2 + k$, where $a = \frac{1}{4}$, $h = 2$, and $k = 3$, so it is a vertical parabola.

Step 2 $a > 0$; so the parabola opens upward.

Step 3 We have $h = 2$ and $k = 3$. Therefore, the vertex is $(2, 3)$.

Step 4 The axis of symmetry is the line $x = 2$. Plot a few points on either side of the axis of symmetry. At $x = 4$, $y = 4$. Thus, the point is $(4, 4)$. The image from symmetry is $(0, 4)$. At $x = 6$, $y = 7$. Thus the point is $(6, 7)$. The image from symmetry is $(-2, 7)$.

Step 5 At $x = 0$,

$$y = \frac{1}{4}(0 - 2)^2 + 3$$
$$= \frac{1}{4}(4) + 3$$
$$= 1 + 3 = 4.$$

Thus, the y-intercept is $(0, 4)$.

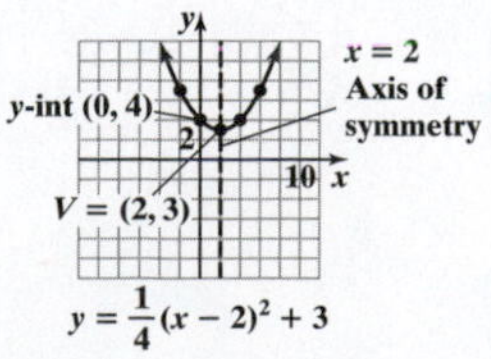

4. Make a table, plot points, and draw the graph. Choose values of y and find x. Begin with $y = 0$.

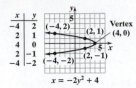

x	y
−4	2
2	1
4	0
2	−1
−4	−2

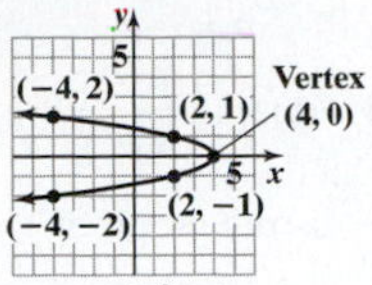

$x = -2y^2 + 4$

The vertex is at (4, 0). The axis of symmetry is the x-axis.

5. Step 1 The equation has the form $x = a(y - k)^2 + h$, where $a = -1, k = -1$, and $h = -3$, so it is a horizontal parabola.

Step 2 $a < 0$; so the parabola opens to the left.

Step 3 We have $k = -1$ and $h = -3$. Therefore, the vertex is $(-3, -1)$.

Step 4 The line $y = -1$ is the axis of symmetry. At $y = 0, x = -4$. Thus, we have the point $(-4, 0)$ and $(-4, -2)$ from symmetry. At $y = 1, x = -7$. Thus, we have the point $(-7, 1)$ and $(-7, -3)$ from symmetry.

Step 5 At $y = 0$,

$$\begin{aligned} x &= -(0 + 1)^2 - 3 \\ &= -(1) - 3 \\ &= -1 - 3 = -4 \end{aligned}$$

Thus, the x-intercept is $(-4, 0)$.

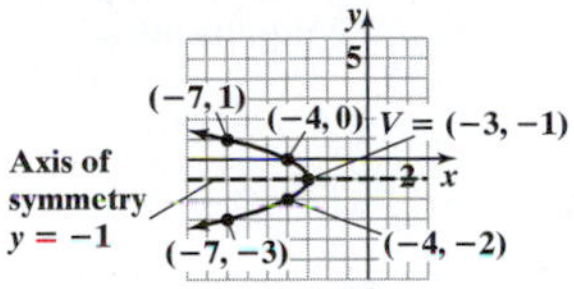

$x = -(y + 1)^2 - 3$

6. Since the y-term is squared, we have a horizontal parabola. The standard form is $x = a(y - k)^2 + h$.

$$\begin{aligned} x &= y^2 - 6y + 13 \\ &= y^2 - 6y + \left(\frac{6}{2}\right)^2 - \left(\frac{6}{2}\right)^2 + 13 \\ &= (y^2 - 6y + 9) + 4 \qquad \text{Complete the square.} \\ &= (y - 3)^2 + 4 \end{aligned}$$

Therefore, we know that $a = 1, k = 3$, and $h = 4$. The vertex is at (4, 3). The axis of symmetry is $y = 3$. If $y = 0$, $x = 13$. So the x-intercept is (13, 0).

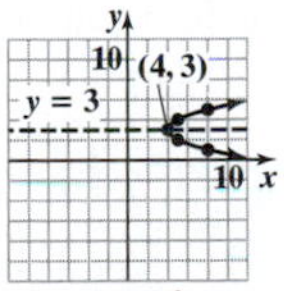

$x = (y - 3)^2 + 4$

7. $y = 2x^2 + 8x + 9$

Since the x-term is squared, we have a vertical parabola.

The standard form is

$$\begin{aligned} y &= a(x - h)^2 + k \\ y &= 2x^2 + 8x + 9 \\ &= 2(x^2 + 4x + \_\_\_\_) - \_\_\_\_ + 9 \\ &= 2(x^2 + 4x + 4) - 2(4) + 9 \qquad \text{Complete the square.} \\ &= 2(x + 2)^2 + 1 \end{aligned}$$

The parabola opens upward. The vertex is $(-2, 1)$, and the y-intercept is (0, 9). The axis of symmetry is $x = -2$.

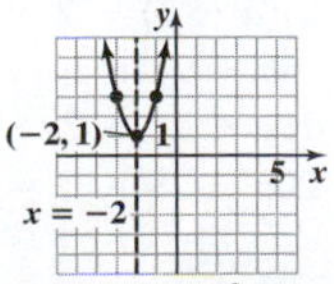

$y = 2(x + 2)^2 + 1$

9.3 Student Practice

1. Write the equation in standard form.

$$\begin{aligned} 4x^2 + y^2 &= 16 \\ \frac{4x^2}{16} + \frac{y^2}{16} &= \frac{16}{16} \\ \frac{x^2}{4} + \frac{y^2}{16} &= 1 \end{aligned}$$

Thus, we have:

$$a^2 = 4 \quad \text{so} \quad a = 2$$
$$b^2 = 16 \quad \text{so} \quad b = 4$$

The x-intercepts are (2, 0) and $(-2, 0)$ and the y-intercepts are (0, 4) and $(0, -4)$.

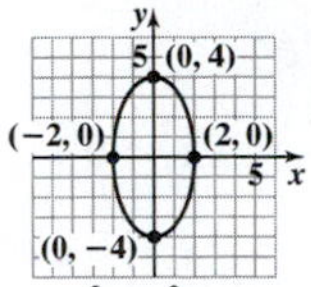

$\frac{x^2}{4} + \frac{y^2}{16} = 1$

2. $\frac{(x - 2)^2}{16} + \frac{(y + 3)^2}{9} = 1$

The center is $(h, k) = (2, -3), a = 4$, and $b = 3$. We start at $(2, -3)$ and measure to the right and to the left 4 units, and up and down 3 units.

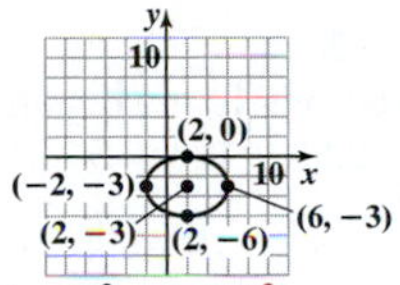

$\frac{(x - 2)^2}{16} + \frac{(y + 3)^2}{9} = 1$

9.4 Student Practice

1. $\frac{x^2}{16} - \frac{y^2}{25} = 1$

This is the equation of a horizontal hyperbola with center (0, 0), where $a = 4$ and $b = 5$. The vertices are $(-4, 0)$ and $(4, 0)$. Construct a fundamental rectangle with corners at (4, 5), $(4, -5)$, $(-4, 5)$, and $(-4, -5)$. Draw extended diagonals as the asymptotes.

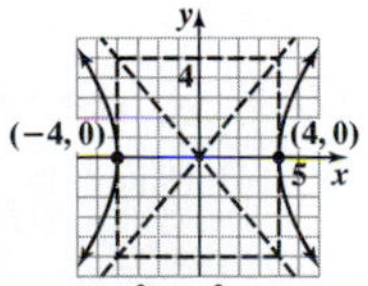

$\frac{x^2}{16} - \frac{y^2}{25} = 1$

2. Write the equation in standard form.

$$\begin{aligned} y^2 - 4x^2 &= 4 \\ \frac{y^2}{4} - \frac{4x^2}{4} &= \frac{4}{4} \\ \frac{y^2}{4} - \frac{x^2}{1} &= 1 \end{aligned}$$

This is the equation of a vertical hyperbola with center (0, 0), where $a = 1$ and $b = 2$. The vertices are (0, 2) and $(0, -2)$. The fundamental rectangle has corners at (1, 2), $(1, -2)$, $(-1, 2)$, and $(-1, -2)$.

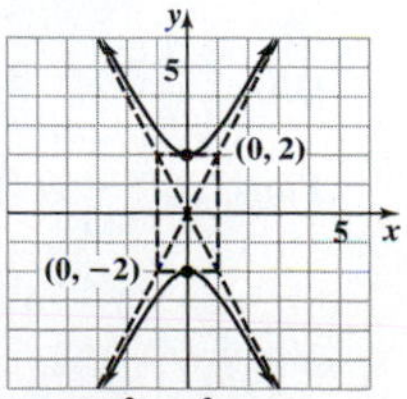

$\frac{y^2}{4} - \frac{x^2}{1} = 1$

3. $\frac{(y+2)^2}{9} - \frac{(x-3)^2}{16} = 1$

This is a vertical hyperbola with center at (3, −2), where $a = 4$ and $b = 3$. The vertices are (3, 1) and (3, −5). The fundamental rectangle has corners at (7, 1), (7, −5), (−1, 1), and (−1, −5).

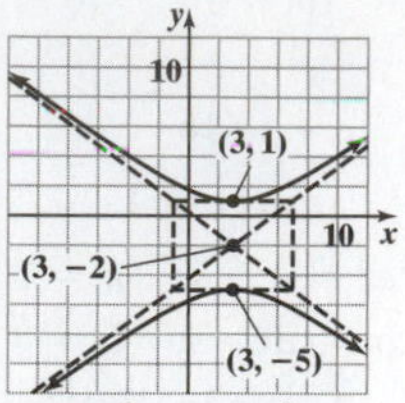

$\frac{(y+2)^2}{9} - \frac{(x-3)^2}{16} = 1$

9.5 Student Practice

1. $\frac{x^2}{4} - \frac{y^2}{4} = 1 \qquad x + y + 1 = 0$

$x^2 - y^2 = 4 \quad (1) \qquad y = -x - 1 \quad (2)$

Substitute (2) into (1).

$$x^2 - (-x-1)^2 = 4$$
$$x^2 - (x^2 + 2x + 1) = 4$$
$$x^2 - x^2 - 2x - 1 = 4$$
$$-2x - 1 = 4$$
$$-2x = 5$$
$$x = \frac{5}{-2} = -2.5$$

Now substitute the value for x in the equation $y = -x - 1$.

For $x = -2.5$: $y = -(-2.5) - 1 = 2.5 - 1 = 1.5$.

The solution is (−2.5, 1.5).

2. $2x - 9 = y \quad (1)$

$xy = -4 \quad (2)$

Equation (1) is solved for y.

Substitute (1) into equation (2).

$$x(2x - 9) = -4$$
$$2x^2 - 9x = -4$$
$$2x^2 - 9x + 4 = 0$$
$$(2x - 1)(x - 4) = 0$$
$$2x - 1 = 0 \quad \text{or} \quad x - 4 = 0$$
$$x = \frac{1}{2} \qquad x = 4$$

For $x = \frac{1}{2}$: $y = 2\left(\frac{1}{2}\right) - 9 = -8$

For $x = 4$: $y = 2(4) - 9 = -1$

The solutions are (4, −1) and $\left(\frac{1}{2}, -8\right)$.

The graph of $y = \frac{-4}{x}$ is the graph of $y = \frac{1}{x}$ reflected across the x-axis and stretched by a factor of 4. The graph of $y = 2x - 9$ is a line with slope 2 passing through the point (0, −9). The sketch shows that the points (4, −1) and $\left(\frac{1}{2}, -8\right)$ seem reasonable.

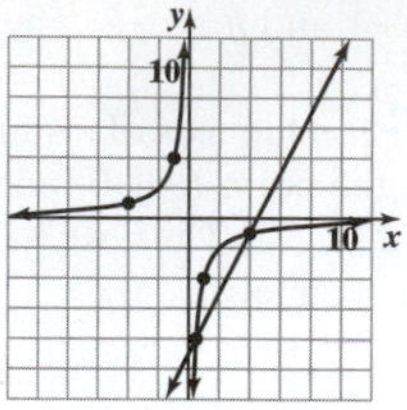

3. $x^2 + y^2 = 12 \quad (1)$

$3x^2 - 4y^2 = 8 \quad (2)$

$4x^2 + 4y^2 = 48$ Multiply (1) by 4.

$3x^2 - 4y^2 = 8$ Add the equations.

$$7x^2 = 56$$
$$x^2 = 8$$
$$x = \pm\sqrt{8}$$
$$x = \pm 2\sqrt{2}$$

If $x = 2\sqrt{2}$, then $x^2 = 8$. Substituting this value into equation (1) gives

$$8 + y^2 = 12$$
$$y^2 = 4$$
$$y = \pm\sqrt{4}$$
$$y = \pm 2$$

Similarly, if $x = -2\sqrt{2}$, then $y = \pm 2$.

Thus, the four solutions are $(2\sqrt{2}, 2)$, $(2\sqrt{2}, -2)$, $(-2\sqrt{2}, 2)$, $(-2\sqrt{2}, -2)$.

Chapter 10

10.1 Student Practice

1. (a) $g(a) = \frac{1}{2}a - 3$

(b) $g(a + 4) = \frac{1}{2}(a + 4) - 3$

$$= \frac{1}{2}a + 2 - 3$$
$$= \frac{1}{2}a - 1$$

(c) $g(a) = \frac{1}{2}a - 3$

$g(4) = \frac{1}{2}(4) - 3 = 2 - 3 = -1$

Thus, $g(a) + g(4) = \left(\frac{1}{2}a - 3\right) + (-1)$

$$= \frac{1}{2}a - 3 - 1$$
$$= \frac{1}{2}a - 4$$

2. (a) $p(-3) = -3(-3)^2 + 2(-3) + 4$

$$= -3(9) + 2(-3) + 4$$
$$= -27 - 6 + 4$$
$$= -29$$

(b) $p(a) = -3(a)^2 + 2(a) + 4$

$$= -3a^2 + 2a + 4$$

(c) $p(2a) = -3(2a)^2 + 2(2a) + 4$

$$= -3(4a^2) + 2(2a) + 4$$
$$= -12a^2 + 4a + 4$$

(d) $p(a - 3) = -3(a - 3)^2 + 2(a - 3) + 4$

$$= -3(a - 3)(a - 3) + 2(a - 3) + 4$$
$$= -3(a^2 - 6a + 9) + 2(a - 3) + 4$$
$$= -3a^2 + 18a - 27 + 2a - 6 + 4$$
$$= -3a^2 + 20a - 29$$

3. (a) $r(a + 2) = \frac{-3}{(a + 2) + 1}$

$$= \frac{-3}{a + 3}$$

(b) $r(a) = \dfrac{-3}{a+1}$

(c) $r(a+2) - r(a) = \dfrac{-3}{a+3} - \left(\dfrac{-3}{a+1}\right)$

$= \dfrac{-3}{a+3} + \dfrac{3}{a+1}$

$= \dfrac{(-3)(a+1)}{(a+3)(a+1)} + \dfrac{3(a+3)}{(a+1)(a+3)}$

$= \dfrac{-3a-3}{(a+1)(a+3)} + \dfrac{3a+9}{(a+1)(a+3)}$

$= \dfrac{-3a-3+3a+9}{(a+1)(a+3)}$

$= \dfrac{6}{(a+1)(a+3)}$

4. $g(x+h) = 2 - 5(x+h) = 2 - 5x - 5h$

$g(x) = 2 - 5x$

$g(x+h) - g(x) = (2 - 5x - 5h) - (2 - 5x)$

$= 2 - 5x - 5h - 2 + 5x$

$= -5h$

Therefore, $\dfrac{g(x+h) - g(x)}{h} = \dfrac{-5h}{h} = -5.$

5. (a) $S(r) = 16(3.14)r + 2(3.14)r^2 = 50.24r + 6.28r^2$

(b) $S(2) = 50.24(2) + 6.28(2)^2$

$= 50.24(2) + 6.28(4)$

$= 125.6$ square meters

(c) $S(e) = 50.24(2+e) + 6.28(2+e)^2$

$= 50.24(2+e) + 6.28(2+e)(2+e)$

$= 50.24(2+e) + 6.28(4 + 4e + e^2)$

$= 100.48 + 50.24e + 25.12 + 25.12e + 6.28e^2$

$= 125.6 + 75.36e + 6.28e^2$

(d) $S(e) = 125.6 + 75.36e + 6.28e^2$

$S(0.3) = 125.6 + 75.36(0.3) + 6.28(0.3)^2$

$= 125.6 + 22.608 + 0.5652$

≈ 148.77 square meters

Thus, if the radius of 2 meters was incorrectly measured as 2.3 meters, the surface area would be approximately $148.77 - 125.6 = 23.17$ square meters too large.

10.2 Student Practice

1. By the vertical line test, this relation is not a function.

2. $f(x) = x^2$

x	f(x)
−2	4
−1	1
0	0
1	1
2	4

$h(x) = x^2 - 5$

x	h(x)
−2	−1
−1	−4
0	−5
1	−4
2	−1

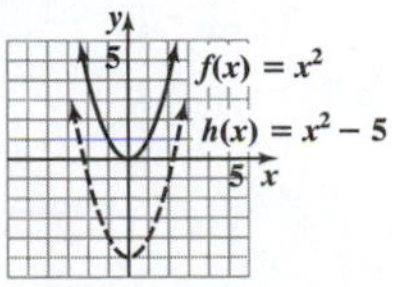

3. $f(x) = |x|$

x	f(x)
−4	4
−3	3
−2	2
−1	1
0	0
1	1
2	2

$p(x) = |x+2|$

x	p(x)
−4	2
−3	1
−2	0
−1	1
0	2
1	3
2	4

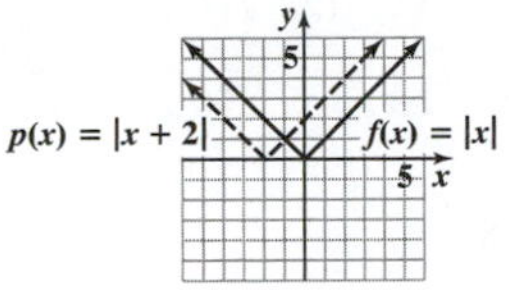

4. $f(x) = x^3$

x	f(x)
−2	−8
−1	−1
0	0
1	1
2	8

We recognize that $h(x)$ will have a similar shape, but the curve will be shifted 4 units to the left and 3 units upward.

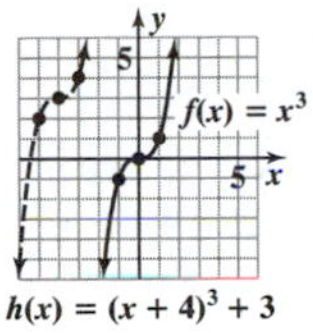

5. $f(x) = \dfrac{2}{x}$

x	f(x)
−4	$-\frac{1}{2}$
−2	−1
−1	−2
$-\frac{1}{2}$	−4
0	undefined
$\frac{1}{2}$	4
1	2
2	1
4	$\frac{1}{2}$

The graph of $g(x)$ is 1 unit to the left and 2 units below $f(x)$. We use each point on $f(x)$ to guide us in graphing $g(x)$.
Note: $g(x)$ is not defined for $x = -1$.

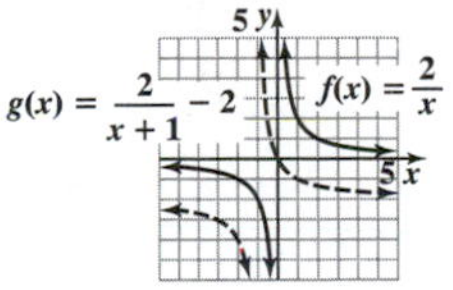

10.3 Student Practice

1. (a) $(f+g)(x) = f(x) + g(x)$

$= (4x + 5) + (2x^2 + 7x - 8)$

$= 4x + 5 + 2x^2 + 7x - 8$

$= 2x^2 + 11x - 3$

(b) Use the formula obtained in **(a)**.

$(f+g)(x) = 2x^2 + 11x - 3$

$(f+g)(4) = 2(4)^2 + 11(4) - 3$

$= 2(16) + 11(4) - 3$

$= 32 + 44 - 3$

$= 73$

2. (a) $(fg)(x) = f(x) \cdot g(x)$
$= (3x + 2)(x^2 - 3x - 4)$
$= 3x^3 - 9x^2 - 12x + 2x^2 - 6x - 8$
$= 3x^3 - 7x^2 - 18x - 8$

(b) Use the formula obtained in **(a)**.
$(fg)(x) = 3x^3 - 7x^2 - 18x - 8$
$(fg)(2) = 3(2)^3 - 7(2)^2 - 18(2) - 8$
$= 3(8) - 7(4) - 18(2) - 8$
$= 24 - 28 - 36 - 8$
$= -48$

3. (a) $\left(\frac{g}{h}\right)(x) = \frac{5x+1}{3x-2}$, where $x \neq \frac{2}{3}$

(b) $\left(\frac{g}{p}\right)(x) = \frac{5x+1}{5x^2+6x+1} = \frac{5x+1}{(5x+1)(x+1)} = \frac{1}{x+1}$,
where $x \neq -\frac{1}{5}, x \neq -1$

(c) $\left(\frac{g}{h}\right)(x) = \frac{5x+1}{3x-2}$
$\left(\frac{g}{h}\right)(3) = \frac{5(3)+1}{3(3)-2} = \frac{15+1}{9-2} = \frac{16}{7}$

4. $f[g(x)] = f(3x - 4)$
$= 2(3x - 4) - 1$
$= 6x - 8 - 1$
$= 6x - 9$

5. (a) $f[g(x)] = f(x + 2)$
$= 2(x + 2)^2 - 3(x + 2) + 1$
$= 2(x^2 + 4x + 4) - 3(x + 2) + 1$
$= 2x^2 + 8x + 8 - 3x - 6 + 1$
$= 2x^2 + 5x + 3$

(b) $g[f(x)] = g(2x^2 - 3x + 1)$
$= (2x^2 - 3x + 1) + 2$
$= 2x^2 - 3x + 1 + 2$
$= 2x^2 - 3x + 3$

6. (a) $(g \circ f)(x) = g[f(x)]$
$= g(3x + 1)$
$= \frac{2}{(3x+1) - 3}$
$= \frac{2}{3x + 1 - 3}$
$= \frac{2}{3x - 2}, x \neq \frac{2}{3}$

(b) $(g \circ f)(-3) = \frac{2}{3(-3) - 2} = \frac{2}{-9-2} = \frac{2}{-11} = -\frac{2}{11}$

10.4 Student Practice

1. (a) A is a function. No two pairs have the same second coordinate. Thus, A is a one-to-one function.

(b) B is a function. The pair $(1, 1)$ and $(-1, 1)$ share a common second coordinate. Therefore, B is not a one-to-one function.

2. The graphs of **(a)** and **(b)** do not represent one-to-one functions. Horizontal lines exist that intersect each graph more than once.

3. The inverse of B is obtained by interchanging x- and y-values of each ordered pair.
$$B^{-1} = \{(2, 1), (8, 7), (7, 8), (12, 10)\}$$

4. $g(x) = 4 - 6x$
$y = 4 - 6x$
$x = 4 - 6y$
$x - 4 = -6y$
$-x + 4 = 6y$
$\frac{-x+4}{6} = y$
$g^{-1}(x) = \frac{-x+4}{6}$

5. $f(x) = 0.75 + 0.55(x - 1)$
$y = 0.75 + 0.55(x - 1)$
$x = 0.75 + 0.55(y - 1)$
$x = 0.75 + 0.55y - 0.55$
$x = 0.55y + 0.2$
$x - 0.2 = 0.55y$
$\frac{x - 0.2}{0.55} = y$
$f^{-1}(x) = \frac{x - 0.2}{0.55}$

6. $f(x) = -\frac{1}{4}x + 1$
$y = -\frac{1}{4}x + 1$
$x = -\frac{1}{4}y + 1$
$x - 1 = -\frac{1}{4}y$
$-4x + 4 = y$
$f^{-1}(x) = -4x + 4$
Now graph each line.

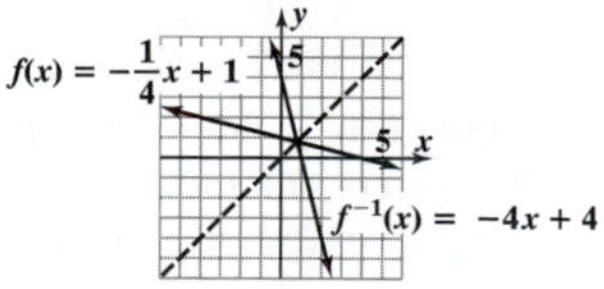

Chapter 11

11.1 Student Practice

1. $f(x) = 3^x$
$f(-2) = 3^{-2} = \left(\frac{1}{3}\right)^2 = \frac{1}{9}$
$f(-1) = 3^{-1} = \frac{1}{3}$
$f(0) = 3^0 = 1$
$f(1) = 3^1 = 3$
$f(2) = 3^2 = 9$

x	f(x)
−2	$\frac{1}{9}$
−1	$\frac{1}{3}$
0	1
1	3
2	9

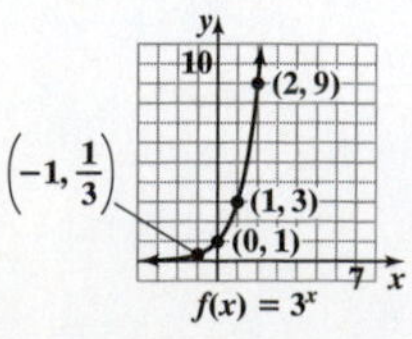

2. $f(x) = \left(\frac{1}{3}\right)^x$
$f(-2) = \left(\frac{1}{3}\right)^{-2} = 3^2 = 9$
$f(-1) = \left(\frac{1}{3}\right)^{-1} = 3^1 = 3$
$f(0) = \left(\frac{1}{3}\right)^0 = 1$
$f(1) = \left(\frac{1}{3}\right)^1 = \frac{1}{3}$
$f(2) = \left(\frac{1}{3}\right)^2 = \frac{1}{9}$

x	f(x)
−2	9
−1	3
0	1
1	$\frac{1}{3}$
2	$\frac{1}{9}$

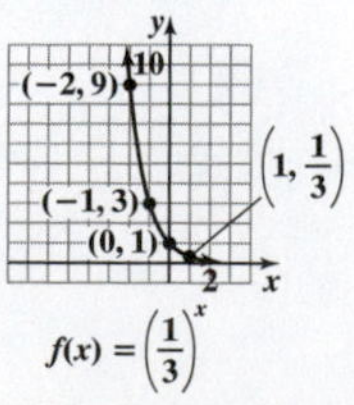

3. $f(x) = 3^{x+2}$

$f(-4) = 3^{-4+2} = 3^{-2} = \frac{1}{3^2} = \frac{1}{9}$

$f(-3) = 3^{-3+2} = 3^{-1} = \frac{1}{3}$

$f(-2) = 3^{-2+2} = 3^0 = 1$

$f(-1) = 3^{-1+2} = 3^1 = 3$

$f(0) = 3^{0+2} = 3^2 = 9$

x	f(x)
−4	$\frac{1}{9}$
−3	$\frac{1}{3}$
−2	1
−1	3
0	9

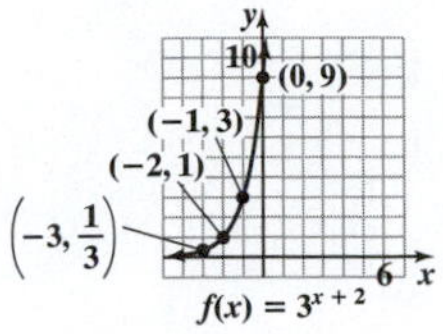

4. $f(x) = e^{x-2}$ (values rounded to nearest hundredth)

$f(4) = e^{4-2} = e^2 \approx 7.39$

$f(3) = e^{3-2} = e^1 \approx 2.72$

$f(2) = e^{2-2} = e^0 = 1$

$f(1) = e^{1-2} = e^{-1} \approx 0.37$

$f(0) = e^{0-2} = e^{-2} \approx 0.14$

x	f(x)
4	7.39
3	2.72
2	1
1	0.37
0	0.14

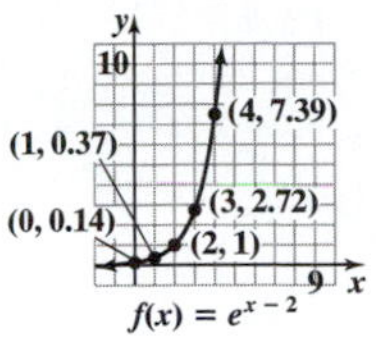

5. $2^x = \frac{1}{32}$

$2^x = \frac{1}{2^5}$ Because $2^5 = 32$.

$2^x = 2^{-5}$ Because $\frac{1}{2^5} = 2^{-5}$.

$x = -5$ Property of exponential equations.

6. Here $P = 4000$, $r = 0.11$, and $t = 2$.

$A = P(1 + r)^t$

$= 4000(1 + 0.11)^2$

$= 4000(1.11)^2$

$= 4000(1.2321)$

$= 4928.4$

Uncle Jose will have \$4928.40.

7. Here $P = 1500$, $r = 8\% = 0.08$, $t = 8$, and $n = 4$.

$A = P\left(1 + \frac{r}{n}\right)^{nt}$

$= 1500\left(1 + \frac{0.08}{4}\right)^{(4)(8)}$

$= 1500(1 + 0.02)^{32}$

$= 1500(1.02)^{32}$

$\approx 1500(1.884540592)$

≈ 2826.81089

Collette will have \$2826.81.

8. Here $C = 20$ and $t = 5000$.

$A = 20e^{-0.0016008(5000)}$

$A = 20e^{-8.004}$

$A \approx 20(0.0003341) = 0.006682 \approx 0.007$

Thus, 0.007 milligram of americium 241 would be present in 5000 years.

11.2 Student Practice

1. Use the fact that $x = b^y$ is equivalent to $\log_b x = y$.

(a) Here $x = 49$, $b = 7$, and $y = 2$. So $2 = \log_7 49$.

(b) Here $x = \frac{1}{64}$, $b = 4$, and $y = -3$. So $-3 = \log_4\left(\frac{1}{64}\right)$.

2. (a) Here $y = 3$, $b = 5$, and $x = 125$. Thus, since $x = b^y$, $125 = 5^3$.

(b) Here $y = -2$, $b = 6$, and $x = \frac{1}{36}$. So $\frac{1}{36} = 6^{-2}$.

3. (a) $\log_b 125 = 3$; then $125 = b^3$

$5^3 = b^3$

$b = 5$

(b) $\log_{1/2} 32 = x$; then $32 = \left(\frac{1}{2}\right)^x$

$2^5 = \left(\frac{1}{2}\right)^x$

$\frac{1}{2^{-5}} = \left(\frac{1}{2}\right)^x$

$\left(\frac{1}{2}\right)^{-5} = \left(\frac{1}{2}\right)^x$

$x = -5$

4. $\log_{10} 0.1 = x$

$0.1 = 10^x$

$10^{-1} = 10^x$

$-1 = x$

Thus, $\log_{10} 0.1 = -1$.

5. To graph $y = \log_{1/2} x$, we first write $x = \left(\frac{1}{2}\right)^y$. We make a table of values.

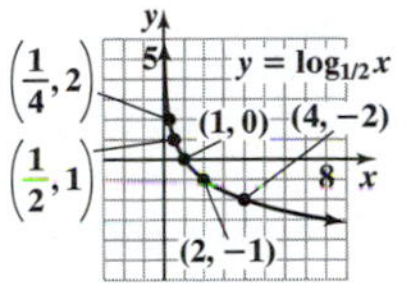

x	y
$\frac{1}{4}$	2
$\frac{1}{2}$	1
1	0
2	−1
4	−2

6. Make a table of values for each equation.

$y = \log_6 x$

x	y
$\frac{1}{6}$	−1
1	0
6	1
36	2

$y = 6^x$

x	y
−1	$\frac{1}{6}$
0	1
1	6
2	36

Ordered pairs reversed

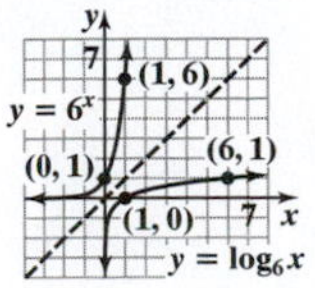

11.3 Student Practice

1. Property 1 can be extended to three logarithms.

$$\log_b MNP = \log_b M + \log_b N + \log_b P$$

Thus,

$$\log_4 WXY = \log_4 W + \log_4 X + \log_4 Y.$$

2. $\log_b M + \log_b N + \log_b P = \log_b MNP$
Therefore,
$$\log_7 w + \log_7 8 + \log_7 x = \log_7(w \cdot 8 \cdot x) = \log_7 8wx.$$

3. $\log_3\left(\frac{17}{5}\right) = \log_3 17 - \log_3 5$

4. $\log_b 132 - \log_b 4 = \log_b\left(\frac{132}{4}\right) = \log_b 33$

5. $\frac{1}{3}\log_7 x - 5\log_7 y = \log_7 x^{1/3} - \log_7 y^5$ By property 3.

$= \log_7\left(\frac{x^{1/3}}{y^5}\right)$ By property 2.

6. $\log_3\left(\frac{x^4y^5}{z}\right) = \log_3 x^4y^5 - \log_3 z$ By property 2.

$= \log_3 x^4 + \log_3 y^5 - \log_3 z$ By property 1.

$= 4\log_3 x + 5\log_3 y - \log_3 z$ By property 3.

7. $\log_4 x + \log_4 5 = 2$

$\log_4 5x = 2$ By property 1.

Converting to exponential form, we have

$4^2 = 5x$

$16 = 5x$

$\frac{16}{5} = x$

$x = \frac{16}{5}$

8. $\log_{10} x - \log_{10}(x + 3) = -1$

$\log_{10}\left(\frac{x}{x+3}\right) = -1$ By property 2.

$\frac{x}{x+3} = 10^{-1}$ Convert to exponential form.

$x = \frac{1}{10}(x + 3)$ Multiply each side by $(x + 3)$.

$x = \frac{1}{10}x + \frac{3}{10}$ Simplify.

$\frac{9}{10}x = \frac{3}{10}$

$x = \frac{1}{3}$

9. **(a)** $\log_7 1 = 0$ By property 5.

(b) $\log_8 8 = 1$ By property 4.

(c) $\log_{12} 13 = \log_{12}(y + 2)$

$13 = y + 2$ By property 6.

$y = 11$

10. $\log_3 2 - \log_3 5 = \log_3 6 + \log_3 x$

$\log_3\left(\frac{2}{5}\right) = \log_3 6x$ By property 1 and property 2.

$\frac{2}{5} = 6x$ By property 6.

$x = \frac{1}{15}$

11.4 Student Practice

1. **(a)** $4.36\,\boxed{\log} \approx 0.639486489$

(b) $436\,\boxed{\log} \approx 2.639486489$

(c) $0.2418\,\boxed{\log} \approx -0.616543703$

2. We know that $\log x = 2.913$ is equivalent to $10^{2.913} = x$.
Using a calculator we have
$2.913\,\boxed{10^x} \approx 818.46479$.
Thus, $x \approx 818.46479$.

3. To evaluate antilog(-3.0705) using a scientific calculator, we have
$3.0705\,\boxed{+/-}\,\boxed{10^x} \approx 8.5015869 \times 10^{-4}$.
Thus, antilog$(-3.0705) \approx 0.00085015869$.

4. **(a)** $\log x = 0.06134$ is equivalent to $10^{0.06134} = x$.
$0.06134\,\boxed{10^x} \approx 1.1517017$
Thus, $x \approx 1.1517017$.

(b) $\log x = -4.6218$ is equivalent to $10^{-4.6218} = x$.
$4.6218\,\boxed{+/-}\,\boxed{10^x} \approx 2.3889112 \times 10^{-5}$
Thus, $x \approx 0.000023889112$.

5. **(a)** $4.82\,\boxed{\ln} \approx 1.572773928$

(b) $48.2\,\boxed{\ln} \approx 3.875359021$

(c) $0.0793\,\boxed{\ln} \approx -2.53451715$

6. **(a)** If $\ln x = 3.1628$, then $e^{3.1628} = x$.
$3.1628\,\boxed{e^x} \approx 23.636686$
Thus, $x \approx 23.636686$.

(b) If $\ln x = -2.0573$, then $e^{-2.0573} = x$.
$2.0573\,\boxed{+/-}\,\boxed{e^x} \approx 0.1277986$
Thus, $x \approx 0.1277986$.

7. To evaluate $\log_9 3.76$, we use the change of base formula.
$$\log_9 3.76 = \frac{\log 3.76}{\log 9}$$
On a calculator, find the following.
$3.76\,\boxed{\log}\,\boxed{\div}\,\boxed{9}\,\boxed{\log} = 0.602769044$.

8. By the change of base formula,
$$\log_8 0.009312 = \frac{\log_e 0.009312}{\log_e 8} = \frac{\ln 0.009312}{\ln 8}$$
On a calculator, find the following.
$0.009312\,\boxed{\ln}\,\boxed{\div}\,8\,\boxed{\ln}\,\boxed{=}\ -2.24889774$
Thus, $\log_8 0.009312 \approx -2.24889774$.

9. Make a table of values.

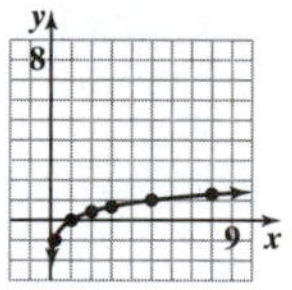

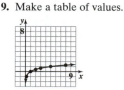

x	$y = \log_5 x$
0.2	−1
1	0
2	0.4
3	0.7
5	1
8	1.3

11.5 Student Practice

1. $\log(x + 5) = 2 - \log 5$

$\log(x + 5) + \log 5 = 2$

$\log[5(x + 5)] = 2$

$\log(5x + 25) = 2$

$5x + 25 = 10^2$

$5x + 25 = 100$

$5x = 75$

$x = 15$

Check. $\log(15 + 5) \stackrel{?}{=} 2 - \log 5$

$\log 20 \stackrel{?}{=} 2 - \log 5$

$1.301029996 \stackrel{?}{=} 2 - 0.698970004$

$1.301029996 = 1.301029996$ ✓

2. $\log(x + 3) - \log x = 1$

$\log\left(\frac{x+3}{x}\right) = 1$

$\frac{x+3}{x} = 10^1$

$\frac{x+3}{x} = 10$

$x + 3 = 10x$

$3 = 9x$

$\frac{1}{3} = x$

Check. $\log\left(\frac{1}{3}+3\right)-\log\frac{1}{3}\stackrel{?}{=}1$

$$\log\frac{10}{3}-\log\frac{1}{3}\stackrel{?}{=}1$$

$$\log\left(\frac{\frac{10}{3}}{\frac{1}{3}}\right)\stackrel{?}{=}1$$

$$\log 10\stackrel{?}{=}1$$

$$1=1\ \checkmark$$

3. $\log 5-\log x=\log(6x-7)$

$$\log\left(\frac{5}{x}\right)=\log(6x-7)$$

$$\frac{5}{x}=6x-7$$

$$5=6x^2-7x$$

$$0=6x^2-7x-5$$

$$0=(3x-5)(2x+1)$$

$3x-5=0$ or $2x+1=0$

$3x=5$ $\quad 2x=-1$

$x=\frac{5}{3}$ $\quad x=-\frac{1}{2}$

Check. $\log 5-\log x=\log(6x-7)$

$x=\frac{5}{3}$: $\log 5-\log\left(\frac{5}{3}\right)\stackrel{?}{=}\log\left[6\left(\frac{5}{3}\right)-7\right]$

$$\log 5-\log\frac{5}{3}\stackrel{?}{=}\log(10-7)$$

$$\log 5-\log\frac{5}{3}\stackrel{?}{=}\log 3$$

$$\log\left[\frac{5}{\frac{5}{3}}\right]\stackrel{?}{=}\log 3$$

$$\log 3=\log 3\ \checkmark$$

$x=-\frac{1}{2}$: $\log 5-\log\left(-\frac{1}{2}\right)\stackrel{?}{=}\log\left[6\left(-\frac{1}{2}\right)-7\right]$

Since logarithms of negative numbers do not exist, $x=-\frac{1}{2}$ is not a valid solution

The only solution is $\frac{5}{3}$.

4.

$$3^x=5$$

$$\log 3^x=\log 5$$

$$x\log 3=\log 5$$

$$x=\frac{\log 5}{\log 3}$$

5. Take the logarithm of each side.

$$2^{3x+1}=9^{x+1}$$

$$\log 2^{3x+1}=\log 9^{x+1}$$

$$(3x+1)\log 2=(x+1)\log 9$$

$$3x\log 2+\log 2=x\log 9+\log 9$$

$$3x\log 2-x\log 9=\log 9-\log 2$$

$$x(3\log 2-\log 9)=\log 9-\log 2$$

$$x=\frac{\log 9-\log 2}{3\log 2-\log 9}$$

Use the following keystrokes.

[(] 9 [log] [−] 2 [log] [)] [÷] [(] 3 [×] 2 [log] [−] 9 [log] [)] [=] −12.76989838

Rounding to the nearest thousandth, we have $x\approx -12.770$.

6. Take the natural logarithm of each side.

$$20.98=e^{3.6x}$$

$$\ln 20.98=\ln e^{3.6x}$$

$$\ln 20.98=(3.6x)(\ln e)$$

$$\ln 20.98=3.6x$$

$$\frac{\ln 20.98}{3.6}=x$$

On a scientific calculator, find the following.

20.98 [ln] [÷] 3.6 [=] 0.845436001

Rounding to the nearest ten-thousandth, we have $x\approx 0.8454$.

7. We use the formula $A=P(1+r)^t$, where $A=10{,}000$, $P=4000$, and $r=0.08$.

$$10{,}000=4000(1+0.08)^t$$

$$10{,}000=4000(1.08)^t$$

$$\frac{10{,}000}{4000}=(1.08)^t$$

$$2.5=(1.08)^t$$

$$\log 2.5=\log(1.08)^t$$

$$\log 2.5=t(\log 1.08)$$

$$\frac{\log 2.5}{\log 1.08}=t$$

On a scientific calculator,

2.5 [log] [÷] 1.08 [log] [=] 11.905904.

Thus, it would take approximately 12 years.

8.

$$A=A_0e^{rt}$$

$5000=1300e^{0.043t}$ Substitute known values.

$\frac{5000}{1300}=e^{0.043t}$ Divide each side by 1300.

$\ln\left(\frac{5000}{1300}\right)=\ln e^{0.043t}$ Take the natural logarithm of each side.

$$\ln\left(\frac{50}{13}\right)=(0.043t)\ln e$$

$$\ln 50-\ln 13=0.043t$$

$$\frac{\ln 50-\ln 13}{0.043}=t$$

Using a scientific calculator,

[(] 50 [ln] [−] 13 [ln] [)] [÷] 0.043 [=] 31.32729414

Rounding to the nearest whole year, the food shortage will develop in about 31 years.

9. Let $I_J=$ intensity of the Japan earthquake.

Let $I_S=$ intensity of the San Francisco earthquake.

$$8.9=\log\left(\frac{I_J}{I_0}\right)=\log I_J-\log I_0$$

Solving for $\log I_0$ gives $\log I_0=\log I_J-8.9$

$$7.1=\log\left(\frac{I_S}{I_0}\right)=\log I_S-\log I_0$$

Solving for $\log I_0$ gives $\log I_0=\log I_S-7.1$

Therefore, $\log I_J-8.9=\log I_S-7.1$

$\log I_J-\log I_S=8.9-7.1$

$$\log\left(\frac{I_J}{I_S}\right)=1.8$$

$$\frac{I_J}{I_S}=10^{1.8}$$

$$\frac{I_J}{I_S}\approx 63.09573445$$

$$\frac{I_J}{I_S}\approx 63$$

$$I_J\approx 63I_S$$

The earthquake in Japan was about sixty-three times more intense than the San Francisco earthquake.

Appendix B Student Practice

1. (a) $\begin{vmatrix} -7 & 3 \\ -4 & -2 \end{vmatrix} = (-7)(-2) - (-4)(3) = 14 - (-12) = 14 + 12 = 26$

(b) $\begin{vmatrix} 5 & 6 \\ 0 & -5 \end{vmatrix} = (5)(-5) - (0)(6) = -25 - 0 = -25$

2. (a) Since 3 appears in the first column and third row, we delete them.

$$\begin{vmatrix} 1 & 2 & 7 \\ -4 & -5 & -6 \\ 3 & 4 & -9 \end{vmatrix}$$

The minor of 3 is $\begin{vmatrix} 2 & 7 \\ -5 & -6 \end{vmatrix}$.

(b) Since -6 appears in second row and third column, we delete them.

$$\begin{vmatrix} 1 & 2 & 7 \\ -4 & -5 & -6 \\ 3 & 4 & -9 \end{vmatrix}$$

The minor of -6 is $\begin{vmatrix} 1 & 2 \\ 3 & 4 \end{vmatrix}$.

3. We can find the determinant by expanding it by minors of elements in the first column.

$$\begin{vmatrix} 1 & 2 & -3 \\ 2 & -1 & 2 \\ 3 & 1 & 4 \end{vmatrix} = 1\begin{vmatrix} -1 & 2 \\ 1 & 4 \end{vmatrix} - 2\begin{vmatrix} 2 & -3 \\ 1 & 4 \end{vmatrix} + 3\begin{vmatrix} 2 & -3 \\ -1 & 2 \end{vmatrix}$$

$$= 1[(-1)(4) - (1)(2)] - 2[(2)(4) - (1)(-3)] + 3[(2)(2) - (-1)(-3)]$$

$$= 1[-4 - 2] - 2[8 - (-3)] + 3[4 - 3]$$
$$= 1(-6) - 2(11) + 3(1)$$
$$= -6 - 22 + 3$$
$$= -25$$

4. $D = \begin{vmatrix} 5 & 3 \\ 2 & -5 \end{vmatrix}$
$= (5)(-5) - (2)(3)$
$= -25 - 6$
$= -31$

$D_x = \begin{vmatrix} 17 & 3 \\ 13 & -5 \end{vmatrix}$
$= (17)(-5) - (13)(3)$
$= -85 - 39$
$= -124$

$D_y = \begin{vmatrix} 5 & 17 \\ 2 & 13 \end{vmatrix}$
$= (5)(13) - (2)(17)$
$= 65 - 34$
$= 31$

$$x = \frac{D_x}{D} = \frac{-124}{-31} = 4, \qquad y = \frac{D_y}{D} = \frac{31}{-31} = -1$$

5. We will expand each determinant by the first column.

$$D = \begin{vmatrix} 2 & 3 & -1 \\ 3 & 5 & -2 \\ 1 & 2 & 3 \end{vmatrix} = 2\begin{vmatrix} 5 & -2 \\ 2 & 3 \end{vmatrix} - 3\begin{vmatrix} 3 & -1 \\ 2 & 3 \end{vmatrix} + 1\begin{vmatrix} 3 & -1 \\ 5 & -2 \end{vmatrix}$$
$$= 2[15 - (-4)] - 3[9 - (-2)] + 1[-6 - (-5)]$$
$$= 2(19) - 3(11) + 1(-1)$$
$$= 4$$

$$D_x = \begin{vmatrix} -1 & 3 & -1 \\ -3 & 5 & -2 \\ 2 & 2 & 3 \end{vmatrix} = -1\begin{vmatrix} 5 & -2 \\ 2 & 3 \end{vmatrix} - (-3)\begin{vmatrix} 3 & -1 \\ 2 & 3 \end{vmatrix} + 2\begin{vmatrix} 3 & -1 \\ 5 & -2 \end{vmatrix}$$
$$= -1[15 - (-4)] + 3[9 - (-2)] + 2[-6 - (-5)]$$
$$= -1(19) + 3(11) + 2(-1)$$
$$= 12$$

$$D_y = \begin{vmatrix} 2 & -1 & -1 \\ 3 & -3 & -2 \\ 1 & 2 & 3 \end{vmatrix} = 2\begin{vmatrix} -3 & -2 \\ 2 & 3 \end{vmatrix} - 3\begin{vmatrix} -1 & -1 \\ 2 & 3 \end{vmatrix} + 1\begin{vmatrix} -1 & -1 \\ -3 & -2 \end{vmatrix}$$
$$= 2[-9 - (-4)] - 3[-3 - (-2)] + 1[2 - 3]$$
$$= 2(-5) - 3(-1) + 1(-1)$$
$$= -8$$

$$D_z = \begin{vmatrix} 2 & 3 & -1 \\ 3 & 5 & -3 \\ 1 & 2 & 2 \end{vmatrix} = 2\begin{vmatrix} 5 & -3 \\ 2 & 2 \end{vmatrix} - 3\begin{vmatrix} 3 & -1 \\ 2 & 2 \end{vmatrix} + 1\begin{vmatrix} 3 & -1 \\ 5 & -3 \end{vmatrix}$$
$$= 2[10 - (-6)] - 3[6 - (-2)] + 1[-9 - (-5)]$$
$$= 2(16) - 3(8) + 1(-4)$$
$$= 4$$

$$x = \frac{D_x}{D} = \frac{12}{4} = 3;\ y = \frac{D_y}{D} = \frac{-8}{4} = -2;\ z = \frac{D_z}{D} = \frac{4}{4} = 1$$

Appendix C Student Practice

1. The matrix represents the equations $x + 9y = 33$ and $0x + y = 3$. Since we know that $y = 3$, we can find x by substitution.

$$x + 9y = 33$$
$$x + 9(3) = 33$$
$$x + 27 = 33$$
$$x = 6$$

The solution to the system is $(6, 3)$.

2. The augmented matrix for this system of equations is

$$\left[\begin{array}{cc|c} 3 & -2 & -6 \\ 1 & -3 & 5 \end{array}\right].$$

Interchange rows one and two so there is a 1 as the first element in the first row.

$$\left[\begin{array}{cc|c} 1 & -3 & 5 \\ 3 & -2 & -6 \end{array}\right] \quad R_1 \leftrightarrow R_2$$

We want a 0 as the first element in the second row. Multiply row one by -3 and add this to row two.

$$\left[\begin{array}{cc|c} 1 & -3 & 5 \\ 0 & 7 & -21 \end{array}\right] \quad -3R_1 + R_2$$

Now, to obtain a 1 as the second element of row two, multiply each element of row two by $\frac{1}{7}$.

$$\left[\begin{array}{cc|c} 1 & -3 & 5 \\ 0 & 1 & -3 \end{array}\right]. \quad \frac{1}{7}R_2$$

This represents the linear system $x - 3y = 5$, $y = -3$.

We know $y = -3$. Substitute this value into the first equation.

$$x - 3(-3) = 5$$
$$x + 9 = 5$$
$$x = -4$$

The solution to the system is $(-4, -3)$.

3. The augmented matrix is

$$\left[\begin{array}{ccc|c} 2 & 1 & -2 & -15 \\ 4 & -2 & 1 & 15 \\ 1 & 3 & 2 & -5 \end{array}\right]$$

$$\left[\begin{array}{ccc|c} 1 & 3 & 2 & -5 \\ 4 & -2 & 1 & 15 \\ 2 & 1 & -2 & -15 \end{array}\right] \quad R_1 \leftrightarrow R_3$$

$$\left[\begin{array}{ccc|c} 1 & 3 & 2 & -5 \\ 0 & -14 & -7 & 35 \\ 0 & -5 & -6 & -5 \end{array}\right] \quad \begin{array}{l} -4R_1 + R_2 \\ -2R_1 + R_3 \end{array}$$

$$\left[\begin{array}{ccc|c} 1 & 3 & 2 & -5 \\ 0 & 1 & \frac{1}{2} & -\frac{5}{2} \\ 0 & -5 & -6 & -5 \end{array}\right] \quad -\frac{1}{14}R_2$$

$$\left[\begin{array}{ccc|c} 1 & 3 & 2 & -5 \\ 0 & 1 & \frac{1}{2} & -\frac{5}{2} \\ 0 & 0 & -\frac{7}{2} & -\frac{35}{2} \end{array}\right] \quad 5R_2 + R_3$$

$$\left[\begin{array}{ccc|c} 1 & 3 & 2 & -5 \\ 0 & 1 & \frac{1}{2} & -\frac{5}{2} \\ 0 & 0 & 1 & 5 \end{array}\right] \quad -\frac{2}{7}R_3$$

We now know that $z = 5$. Substitute this value into the second equation to find y.

$$y + \tfrac{1}{2}(5) = -\tfrac{5}{2}$$
$$y + \tfrac{5}{2} = -\tfrac{5}{2}$$
$$y = -5$$

Now substitute $y = -5$ and $z = 5$ into the first equation.

$$x + 3y + 2z = -5$$
$$x + 3(-5) + 2(5) = -5$$
$$x - 15 + 10 = -5$$
$$x - 5 = -5$$
$$x = 0$$

The solution to this system is $(0, -5, 5)$.

Answers to Selected Exercises

Chapter 1

To Think About, page 4 1, since $a \cdot 1 = a$; sample answer: $5 \cdot \frac{1}{5} = 1$; $\frac{5}{2}$

1.1 Exercises **1.** Integers include negative numbers. Whole numbers are nonnegative integers. **3.** A terminating decimal is a decimal that comes to an end. **5.** natural, whole, integer, rational, real **7.** integer, rational, real **9.** rational, real **11.** rational, real **13.** irrational, real **15.** rational, real **17.** irrational, real **19.** $-25, -\frac{28}{7}$ **21.** $-25, -\frac{28}{7}, -\frac{18}{5}, -\pi, -0.763, -0.333\ldots$ **23.** $\frac{1}{10}, \frac{2}{7}, 9, 52.8, \frac{283}{5}$ **25.** $-\frac{18}{5}, -0.763, -0.333\ldots$ **27.** $1, 3, 5, 7, \ldots$ **29.** Commutative property of addition **31.** Inverse property of multiplication **33.** Identity property of addition **35.** Commutative property of multiplication **37.** Identity property of multiplication **39.** Distributive property of multiplication over addition **41.** Associative property of addition **43.** Commutative property of addition **45.** Associative property of multiplication **47.** 12 **49.** $-\frac{3}{5}$ **51. (a)** 37.5% **(b)** 62.5% **(c)** None **53.** There are several possible answers. The second and third numbers increase by \$100. So the \$500 does not seem part of the pattern. The fourth through the eleventh numbers fit a doubling pattern, so the \$125,000 does not seem part of the pattern. **55.** No, when the winnings reach down to \$15,625, the next smallest winning amount would be one half of that amount. This is \$7812.50, which is not a whole number. **57.** About 62 people

Quick Quiz 1.1 *See Examples noted with Ex.* **1.** $-2\pi, \sqrt{7}$ (Ex. 1) **2.** $-2\pi, -0.5333\ldots, -\frac{3}{11}$ (Ex. 1) **3.** Distributive property of multiplication over addition (Ex. 3) **4.** See Student Solutions Manual

To Think About, page 11 Because you get 0, the additive identity, when you add a number and its opposite.

1.2 Exercises **1.** To add two real numbers with the same sign, add their absolute values. The sum takes the common sign. To add two real numbers with different signs, find the difference of their absolute values. The answer takes the sign of the number with the larger absolute value. **3.** $\frac{2}{3}$ **5.** 8.3 **7.** 5 **9.** b **11.** -18 **13.** 2 **15.** $-\frac{5}{3}$ or $-1\frac{2}{3}$ **17.** 9 **19.** -0.03 **21.** 5.6 **23.** $-\frac{1}{36}$ **25.** -0.4 **27.** $\frac{7}{6}$ or $1\frac{1}{6}$ **29.** 19 **31.** 0 **33.** Undefined **35.** 0 **37.** 0 **39.** 1 **41.** $\frac{1}{9}$ **43.** $-\frac{10}{9}$ or $-1\frac{1}{9}$ **45.** 7 **47.** 5 **49.** 2 **51.** 13 **53.** 8 **55.** 1 **57.** 4 **59.** Approximately -0.678197169 **61.** One or three of the quantities a, b, c must be negative. In other words, when multiplying an odd number of negative numbers, the answer will be negative. **63.** Commutative property of addition **64.** Associative property of multiplication **65.** $-\frac{1}{2}\pi, \sqrt{3}$ **66.** $-16, 0, 9.36, \frac{19}{2}, 10.\overline{5}$

Quick Quiz 1.2 *See Examples noted with Ex.* **1.** -20 (Ex. 7) **2.** 9.4 (Ex. 4) **3.** 3 (Ex. 8) **4.** See Student Solutions Manual

To Think About, page 18 A negative number raised to an even power is a positive number. A negative number raised to an odd power is a negative number.

1.3 Exercises **1.** The base is a. The exponent is 3. **3.** Positive **5.** $+11$ and -11; $(+11)(+11) = 121$ and $(-11)(-11) = 121$ **7.** 9^4 **9.** $(-6)^5$ **11.** $x^6 \cdot y^3$ **13.** 32 **15.** 25 **17.** -36 **19.** -1 **21.** $\frac{4}{9}$ **23.** $\frac{1}{256}$ **25.** 0.49 **27.** 0.000064 **29.** 9 **31.** -4 **33.** $\frac{2}{3}$ **35.** 0.3 **37.** 4 **39.** 60 **41.** 1 **43.** Not a real number **45.** Not a real number **47.** -23 **49.** -33 **51.** 36 **53.** 6 **55.** 5 **57.** -46 **59.** 0 **61.** -28 **63.** 2 **65.** 3 **67.** 4 **69.** $\frac{2}{3}$ **71.** 1 **73.** Approximately 7685.702373 **75.** 192 **77.** Inverse property of multiplication **78.** Inverse property of addition **79.** 700% **80.** Three times greater **81.** 14% **82.** 15,852 field goals

Quick Quiz 1.3 *See Examples noted with Ex.* **1.** $\frac{32}{243}$ (Ex. 2) **2.** -2 (Ex. 11) **3.** 7 (Ex. 10) **4.** See Student Solutions Manual

How Am I Doing? Sections 1.1–1.3 **1.** $\pi, \sqrt{7}$ (obj. 1.1.1) **2.** $\sqrt{9}, -5, 3, \frac{6}{2}, 0$ (obj. 1.1.1) **3.** The irrational number set and the real number set (obj. 1.1.1) **4.** Associative property of addition (obj. 1.1.2) **5.** Inverse property of multiplication (obj. 1.1.2) **6.** 8 (obj. 1.2.3) **7.** 6 (obj. 1.2.3) **8.** $-\frac{5}{2}$ (obj. 1.2.3) **9.** $\frac{2}{3}$ (obj. 1.2.3) **10.** -18 (obj. 1.2.3) **11.** $\frac{4}{7}$ (obj. 1.3.2) **12.** 0.9 (obj. 1.3.2) **13.** 256 (obj. 1.3.1) **14.** 9 (obj. 1.3.3) **15.** -54 (obj. 1.3.3) **16.** $-\frac{7}{3}$ (obj. 1.3.3) **17.** 7 (obj. 1.3.3) **18.** 1 (obj. 1.3.3)

1.4 Exercises **1.** $\frac{1}{9}$ **3.** $\frac{1}{x^5}$ **5.** $\frac{1}{49}$ **7.** -9 **9.** x^{12} **11.** 17^5 **13.** $-6x^6$ **15.** $11x^6y^9$ **17.** $4y$ **19.** $7xy$ **21.** $24x^2y^3z$ **23.** $-\frac{3}{n}$ **25.** x^{11} **27.** $\frac{1}{a^5}$ **29.** 8 **31.** $\frac{2}{x^5}$ **33.** $-5x^3z$ **35.** $-\frac{10}{7}a^2b^4$ **37.** x^{16} **39.** $81a^{20}b^4$ **41.** $\frac{x^{12}y^{18}}{z^6}$ **43.** $\frac{9a^2}{16b^{12}}$

45. $\frac{1}{8x^{12}y^{18}}$ **47.** $\frac{4x^4}{y^6}$ **49.** $-\frac{27m^{13}}{n^5}$ **51.** $2a^4$ **53.** $\frac{27}{y^3}$ **55.** x^2y^6 **57.** $\frac{a^3}{b^5}$ **59.** $\frac{1}{4x^2}$ **61.** $\frac{7}{5}$ **63.** $-\frac{6x}{y}$
65. Approximately $-9460.906704x^{21}y^{28}$ **67.** 3.8×10^1 **69.** 1.73×10^6 **71.** 8.3×10^{-1} **73.** 8.125×10^{-4} **75.** 713,000
77. 0.307 **79.** 0.000000901 **81.** 4.65×10^{-6} **83.** 3×10^1 **85.** 204,282.9 m³/sec **87.** 7.65×10^6 **89.** 1.06×10^{-18} g
91. 5742 times greater **92.** 2×10^{10} calories **93.** 9 **94.** −28

Quick Quiz 1.4

See Examples noted with Ex. **1.** $\frac{4x^6}{5y^3}$ (Ex. 8) **2.** $\frac{9b^4}{a^{14}}$ (Ex. 11) **3.** 5.78×10^{-4} (Ex. 14) **4.** See Student Solutions Manual

1.5 Exercises

1. It is $-5x^2$, since y is multiplied by $-5x^2$. **3.** $5x^3; -6x^2; 4x; 8$ **5.** 1, −3, −8 **7.** 5, −3, 1 **9.** 6.5, −0.02, 3.05
11. $11ab$ **13.** $-5x - 2y$ **15.** $-3x^2 - 4$ **17.** $-ab$ **19.** $-0.4x^2 + 3x$ **21.** $\frac{1}{3}m + \frac{7}{6}n$ **23.** $a^2 - 6b$ **25.** $-7.7x^2 - 3.6x$
27. $18x^2 + 6xy$ **29.** $-y^3 + 3y^2 - 5y$ **31.** $-2a^3 + 6a^4 - 4a^3b$ **33.** $2x^3y - 6x^2y^2 + 8xy^3$ **35.** $6x^2 - 3x + \frac{3}{2}$ **37.** $x^3 - \frac{2}{5}x^2 + \frac{x}{5}$
39. $3a^5b^2 - 9a^3b + 3a^2b - 3ab^2$ **41.** $x^3 - 2x^2y + 1.5x^2y^2$ **43.** $2x^2 + x + 4$ **45.** $18x + 16$ **47.** $-4a^2 - 9ab$ **49.** $-3x^3 - 14x^2$
51. $20x - 88$ **53.** $3y^2 - 3xy + 15y$ **55.** 6 **56.** 9 **57.** −1.8 **58.** −249 **59.** Approximately 7.4547×10^{10} organisms
60. Approximately 27.3%

Quick Quiz 1.5

See Examples noted with Ex. **1.** $-6x^3y + 8x^2y^2 - 10x^2y$ (Ex. 6) **2.** $-7x^3 + 16x^2y$ (Ex. 8) **3.** $33x - 33$ (Ex. 9)
4. See Student Solutions Manual

To Think About, page 43

In part **(a)**, $(-2x)^2$ indicates that the product $-2x$ is being squared, while in part **(b)**, $-2x^2$ indicates that only x is being squared.

1.6 Exercises

1. 2 **3.** −1 **5.** 9 **7.** −55 **9.** 81 **11.** −3 **13.** 17 **15.** 9 **17.** 48.41439 **19.** −76°F **21.** 50°C
23. 6.28 sec **25.** \$5664 **27.** \$2552 **29.** 144 ft **31.** 784 ft **33.** $\frac{18}{5}$ or $3\frac{3}{5}$ **35.** 160 mg **37.** 0.785 in.² **39.** 56 cm² **41.** $\frac{7}{2}$ yd²
43. 24.705 cm² **45.** 36 cm² **47.** **(a)** 552.64 cm³ **(b)** 376.8 cm² **49.** $A = 200.96$ cm²; $C = 50.24$ cm **51.** $\frac{36y^6}{x^8}$ **52.** $\frac{8x^9}{27y^3}$
53. $20x + 22$ **54.** −4 **55.** 6,300,000 people **56.** 231 people were at the reunion

Quick Quiz 1.6

See Examples noted with Ex. **1.** 40 (Ex. 1) **2.** −56 (Ex. 2) **3.** 78.5 cm² (Ex. 8) **4.** See Student Solutions Manual

Use Math to Save Money

1. \$13,710 − \$12,360 = \$1350 **2.** \$1350/\$12,360 ≈ 0.11 = 11% **3.** \$12,360 × 0.05 = \$618
4. \$12,360 + \$618 = \$12,978 **5.** \$12,978 − \$500 = \$12,478

You Try It

1. (a) Identity property of addition **(b)** Associative property of addition **(c)** Identity property of multiplication **(d)** Commutative property of addition **(e)** Commutative property of multiplication **(f)** Associative property of multiplication **(g)** Inverse property of multiplication **(h)** Distributive property of multiplication over addition **(i)** Inverse property of addition **2. (a)** 11 **(b)** −14 **(c)** $-\frac{1}{2}$ **(d)** 7 **3. (a)** 25 **(b)** −10.5 **4. (a)** 48 **(b)** 4 **(c)** −18 **(d)** −8 **5. (a)** 7 **(b)** 10 **(c)** $\frac{3}{5}$ **(d)** 1.25 **6.** −10 **7. (a)** $\frac{1}{a^5}$ **(b)** $\frac{1}{27}$ **(c)** $\frac{b^4}{a^2}$ **8. (a)** $12x^7$ **(b)** $3x^5$ **9. (a)** 1 **(b)** 1 **(c)** 1 **10. (a)** 3^8 **(b)** $\frac{8}{a^{15}}$ **(c)** $\frac{64y^9}{x^6}$ **11. (a)** 3.124×10^3 **(b)** 1.825×10^7 **(c)** 2.78×10^{10} **(d)** 3.9×10^{-2} **(e)** 2.1×10^{-4} **(f)** 7×10^{-7} **12.** $-6a^2 - 7a + 15b$ **13.** $15x^2 - 5x + 25$ **14.** $-18x + 48$
15. 51 **16.** 40 ft²

Chapter 1 Review Problems

1. integer, rational, real **2.** rational, real **3.** natural, whole, integer, rational, real **4.** rational, real
5. irrational, real **6.** Commutative property of addition **7.** Associative property of multiplication **8.** Yes **9.** 5 **10.** −23.5 **11.** 48
12. $\frac{1}{3}$ **13.** $-\frac{7}{12}$ **14.** 4 **15.** 5.4 **16.** 0 **17.** Undefined **18.** 0 **19.** −1 **20.** 1 **21.** 10 **22.** 15 **23.** 1 **24.** $\frac{2}{3}$ **25.** 8
26. −0.064 **27.** $-24x^4y^6$ **28.** $-30a^5b^3c^2$ **29.** $\frac{1}{3b^3c^2}$ **30.** $\frac{81y^4}{16x^4z^8}$ **31.** $-8x^3y^{18}$ **32.** $-\frac{10x}{y^3}$ **33.** $\frac{x^7}{4y^7}$ **34.** $\frac{5x^2y^3}{64}$ **35.** $\frac{1}{125x^9y^{12}}$
36. 7.21×10^{-3} **37.** 1.06×10^{16} **38.** $6x^2 + 6x + 4$ **39.** $-5a^4b^2 - 10a^3b^3 + 15ab^3 + 20ab^2$ **40.** $2x^2 - 21x - 1$ **41.** $4x^2 - 5x - 2$
42. 28 **43.** 226.08 in.³ **44.** 2288 yd² **45.** 420 in.² **46.** −87 **47.** $14a^2b^4c^2$ **48.** $\frac{27y^{10}}{16x^7}$ **49.** $\frac{9}{4a^6b^6}$ **50.** 5.8×10^{-5} **51.** 89,500,000
52. $5x^3 - x^2 + 5x$ **53.** $10a^4b - 2a^4b^2 - 6a^3b$ **54.** $28x + 24y$ **55.** 19 **56.** 50.24 m² **57.** 25.12 sec **58.** $3x - 4y$ **59.** \$3776

How Am I Doing? Chapter 1 Test

1. $\pi, 2\sqrt{5}$ (obj. 1.1.1) **2.** $-2, 12, \frac{9}{3}, \frac{25}{25}, 0, \sqrt{4}$ (obj. 1.1.1) **3.** Commutative property of multiplication (obj. 1.1.2) **4.** 18 (obj. 1.3.3) **5.** 3 (obj. 1.3.3) **6.** $\frac{4x^4}{5y^4}$ (obj. 1.4.3) **7.** $-\frac{10}{y^5}$ (obj. 1.4.2) **8.** $\frac{25b^2}{a^6}$ (obj. 1.4.4)
9. $-17x^2$ (obj. 1.5.1) **10.** $-8a^2 - 2a + 5b$ (obj. 1.5.1) **11.** $12x^2y^2 - 9xy^3 + 6x^3y^2$ (obj. 1.5.2) **12.** 2.186×10^{-6} (obj. 1.4.5)
13. 2,158,000,000 (obj. 1.4.5) **14.** 1.52×10^{-6} (obj. 1.4.5) **15.** $10x^3 - 6x^2y - 4x$ (obj. 1.5.3) **16.** $36x - 56$ (obj. 1.5.3) **17.** 38 (obj. 1.6.1)
18. 9 (obj. 1.6.1) **19.** 78 m² (obj. 1.6.2) **20.** 113.04 m² (obj. 1.6.2) **21.** \$9200 (obj. 1.6.2)

Chapter 2

2.1 Exercises **1.** No; when you replace x by -20 in the equation, you do not get a true statement. **3.** Yes; when you replace x by $\frac{2}{7}$ in the equation, you get a true statement. **5.** Multiply each term of the equation by the LCD, 12, to clear the fractions. **7.** No; it would be easier to subtract 3.6 from both sides of the equation since the coefficient of x is 1. **9.** $x = 8$ **11.** $x = -7$ **13.** $x = 5$ **15.** $x = \frac{3}{2}$ or $1\frac{1}{2}$ or 1.5 **17.** $x = -3$ **19.** $x = 3$ **21.** $a = -2$ **23.** $y = -\frac{1}{3}$ **25.** $y = 1$ **27.** $x = 12$ **29.** $y = -\frac{23}{3}$ or $-7\frac{2}{3}$ **31.** $x = -2$ **33.** $x = 3$ **35.** $x = 0$ **37.** $x = 6$ **39.** $x = \frac{8}{5}$ or $1\frac{3}{5}$ or 1.6 **41.** $x = 19$ **43.** $x = -2$ **45.** $x = 8$ **47.** $x = \frac{3}{2}$ or $1\frac{1}{2}$ or 1.5 **49.** $x = \frac{1}{4}$ or 0.25 **51.** $x = 0$ **53.** $x = 2$ **55.** No solution **57.** Any real number is a solution. **59.** No solution **61.** Any real number is a solution. **63.** -5 **64.** 16 **65.** $\frac{27y^3}{8x^3}$ **66.** $\frac{1}{4x^6y^2}$

Quick Quiz 2.1 *See Examples noted with Ex.* **1.** $x = 5$ (Ex. 5) **2.** $x = \frac{11}{4}$ or $2\frac{3}{4}$ or 2.75 (Ex. 6) **3.** $x = \frac{4}{5}$ or 0.8 (Ex. 7) **4.** See Student Solutions Manual

2.2 Exercises **1.** $x = \frac{3 - 5y}{6}$ **3.** $x = \frac{18 - y}{7}$ **5.** $x = \frac{3y + 12}{2}$ **7.** $y = \frac{-12x + 8}{9}$ **9.** $\frac{A}{w} = l$ **11.** $B = \frac{2A - hb}{h}$ **13.** $r = \frac{A}{2\pi h}$ **15.** $b = \frac{3H - 2a}{4}$ **17.** $x = \frac{-6y}{a}$ **19. (a)** $b = \frac{2A}{a}$ **(b)** $b = 24$ **21. (a)** $n = \frac{A - a + d}{d}$ **(b)** $n = 2\frac{2}{3}$ or $\frac{8}{3}$ **23.** $x = \frac{y - 1.35}{0.076}$ or $x = \frac{1000y - 1350}{76}$; 2015 **25. (a)** $m = 1.15k$ **(b)** 33.35 mi/hr **27. (a)** $D = \frac{C - 5.8263}{0.6547}$ **(b)** \$5.7 billion **29.** $\frac{x^6}{4y^2}$ **30.** $\frac{y^{15}}{125x^{18}}$ **31.** -4 **32.** $7a + 4b - 6$ **33.** \$9610 **34.** 19.8 mi/gal

Quick Quiz 2.2 *See Examples noted with Ex.* **1.** $b = \frac{A - d - 3g}{g}$ (Ex. 3) **2.** $x = \frac{3V}{y}$ (Ex. 2) **3.** $w = \frac{4G - 16x - 3}{2}$ (Ex. 2) **4.** See Student Solutions Manual

To Think About, page 76 Solve $ax + b = 0$ because the only number with absolute value 0 is 0. No; the absolute value of a number is never negative.

2.3 Exercises **1.** It will always have 2 solutions. One solution is when $x = b$ and one when $x = -b$. Since $b > 0$ the values of b and $-b$ are always different numbers. **3.** You must first isolate the absolute value expression. To do this you add 2 to each side of the equation. The result will be $|x + 7| = 10$. Then you solve the two equations $x + 7 = 10$ and $x + 7 = -10$. The final answer is $x = -17$ and $x = 3$. **5.** $x = 30, -30$ **7.** $x = 6, -14$ **9.** $x = 9, -4$ **11.** $x = -\frac{3}{2}, 4$ **13.** $x = 10, 2$ **15.** $x = 2, 7$ **17.** $x = 6, -10$ **19.** $x = -2, \frac{10}{3}$ **21.** $x = -9, 15$ **23.** $x = -\frac{1}{5}, \frac{13}{15}$ **25.** $x = 5, -1$ **27.** $x = -\frac{7}{3}, -1$ **29.** $x = 3, 1$ **31.** $x = 0, 12$ **33.** $x \approx -0.59, -3.29$ **35.** $x = 0, -8$ **37.** $x = \frac{3}{5}$ **39.** No solution **41.** $x = \frac{11}{15}, -\frac{1}{15}$ **43.** $5xy^3$ **44.** $\frac{3}{5}$

Quick Quiz 2.3 *See Examples noted with Ex.* **1.** $x = \frac{53}{3}, -15$ (Ex. 1) **2.** $x = \frac{21}{2}, -\frac{27}{2}$ (Ex. 3) **3.** $x = -9, -\frac{1}{3}$ (Ex. 4) **4.** See Student Solutions Manual

2.4 Exercises **1.** -90 **3.** \$48 **5.** 18 lb **7.** 72 weeks **9.** 3 weeks **11.** The width is 44 yards. The length is 126 yards. **13.** Longest side = 23 ft; shortest side = 15 ft; third side = 21 ft **15.** Identity property of addition **16.** Associative property of multiplication **17.** 3 **18.** -94

Quick Quiz 2.4 *See Examples noted with Ex.* **1.** -92 **2.** First side = 46 yards; second side = 23 yards; third side = 31 yards (Ex. 3) **3.** 18 hr (Ex. 1) **4.** See Student Solutions Manual

How Am I Doing? Sections 2.1–2.4 **1.** $x = -\frac{37}{10}$ or $-3\frac{7}{10}$ or -3.7 (obj. 2.1.1) **2.** $x = -18$ (obj. 2.1.1) **3.** $x = -\frac{10}{7}$ or $-1\frac{3}{7}$ (obj. 2.1.1) **4.** $x = -100$ (obj. 2.1.1) **5.** $y = \frac{5x + 18}{9}$ or $\frac{5}{9}x + 2$ (obj. 2.2.1) **6.** $a = \frac{b + 24}{11b}$ (obj. 2.2.1) **7.** $r = \frac{A - P}{Pt}$ (obj. 2.2.1) **8.** $r = \frac{3}{50}$ or 0.06 (obj. 2.2.1) **9.** $x = -1, -\frac{11}{5}$ (obj. 2.3.1) **10.** $x = 6, 12$ (obj. 2.3.2) **11.** $x = 2.5, -5.5$ (obj. 2.3.1) **12.** $x = 5, 0.75$ (obj. 2.3.3) **13.** 60 in. × 80 in. (obj. 2.4.1) **14.** 26 (obj. 2.4.1) **15.** Cindi picked up 250 lb; Alan picked up 205 lb (obj. 2.4.1) **16.** Shortest side = 14.5 ft; longest side = 24 ft; third side = 23.5 ft (obj. 2.4.1)

2.5 Exercises **1.** Approximately 250.6 million people **3.** \$425 **5.** 1140 residents **7.** 230 hemlocks, 460 spruces, 710 balsams **9.** Walker earned \$260/week; Angela earned \$240/week **11.** Packet A = 5 mg; packet B = 3 mg **13.** \$270 **15.** \$232.50 **17.** She invested \$3900 at 5% and \$2500 at 8%. **19.** He invested \$12,000 in the CD and \$6000 in the fixed interest account. **21.** 12 g of cheese that contains 45% fat and 18 g of cheese that contains 20% fat **23.** 75 lb of the hamburger with 30% fat and 25 lb of hamburger with 10% fat **25.** 45 gal of 25% fertilizer and 105 gal of 15% fertilizer **27.** Her speed on secondary roads was 35 mph. **29.** They each used the treadmill for $\frac{3}{4}$ hour. **31.** 7 **32.** 15 **33.** 1 **34.** $\frac{1}{4}$

Quick Quiz 2.5 *See Examples noted with Ex.* **1.** \$2400 (Ex. 1) **2.** 150 gal of 50% fertilizer and 50 gal of 30% fertilizer (Ex. 4) **3.** \$1500 at 5% and \$2500 at 7% (Ex. 3) **4.** See Student Solutions Manual

2.6 Exercises **1.** True **3.** True **5.** False **7.** $>$ **9.** $<$ **11.** $>$ **13.** $<$ **15.** $<$ **17.** $<$

19. 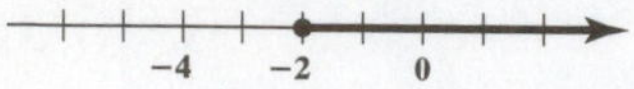

21. 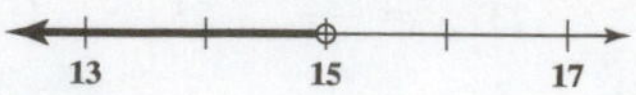

23. $x \leq 1$ 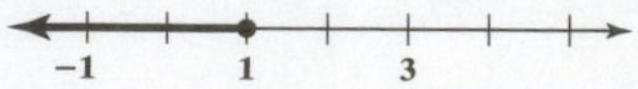

25. $x < -2$ 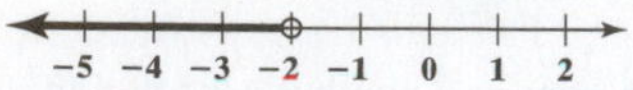

27. $x > -1$ 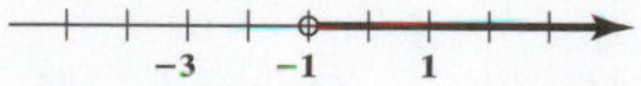

29. $x > 4$ **31.** $x \leq -4$ **33.** $x > -\frac{5}{3}$

35. $x < -3$ **37.** $x \leq 0$ **39.** $x \leq 4$ **41.** $x \leq 0$ **43.** $x \leq 1$ **45.** $x \leq -26$ **47.** More than 13 tables **49.** A maximum of 18 minutes **51.** A maximum of 13 computers **53.** $6xy - x^2y + 4x^2$ **54.** $4a^2b - \frac{4}{3}ab^2 + 6ab$ **55.** $\frac{64x^6w^3}{27y^3}$ **56.** $\frac{b^6}{9c^{10}}$

Quick Quiz 2.6 *See Examples noted with Ex.* **1.** $x > -8$ (Ex. 5) **2.** $x > -\frac{4}{9}$ (Ex. 6) **3.** $x \leq -7$ (Ex. 8) **4.** See Student Solutions Manual

2.7 Exercises **1.** 

3. 

5. 

7. 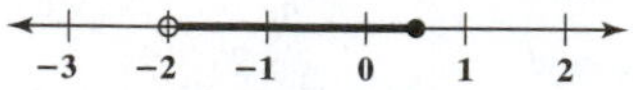

9. 

11. 

13. 

15. 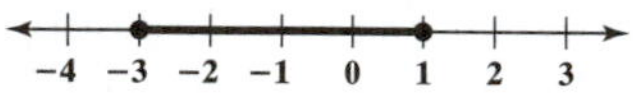

17. 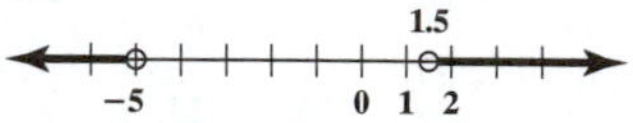

19. No solution **21.** $t < 10.9$ *or* $t > 11.2$ **23.** $5000 \leq c \leq 12{,}000$ **25.** $-4° \leq F \leq 51.8°$ **27.** $\$498.83 \leq d \leq \$844.51$ **29.** $-2 < x < 2$ **31.** $-3 \leq x \leq 1$ **33.** $x < -2$ *or* $x \geq 3$ **35.** $x \leq 2$ **37.** $x \geq 2$ **39.** No solution **41.** $x = 5$ **43.** $\frac{2}{3} < x \leq \frac{7}{3}$ **45.** $x - 17$ **46.** 28.26 in.^{2} **47.** $x = \frac{3y - 8}{5}$ **48.** $y = \frac{-7x - 12}{6}$

Quick Quiz 2.7 *See Examples noted with Ex.* **1.** $-\frac{16}{3} < x < 2$ (Ex. 9) **2.** $5 < x < 12$ (Ex. 9) **3.** $x \leq -8$ *or* $x \geq 1$ (Ex. 8) **4.** See Student Solutions Manual

2.8 Exercises **1.** $-8 \leq x \leq 8$ 

3. $-9.5 < x < 0.5$ 

5. $-2 \leq x \leq 8$ **7.** $-2 \leq x \leq 3$ **9.** $-\frac{2}{5} \leq x \leq 1\frac{1}{5}$ **11.** $-5 < x < 15$ **13.** $-32 < x < 16$ **15.** $-4 < x < 8$ **17.** $-3\frac{1}{3} < x < 4\frac{2}{3}$ **19.** $x < -5$ *or* $x > 5$ **21.** $x < -7$ *or* $x > 3$ **23.** $x \leq -1$ *or* $x \geq 3$ **25.** $x \leq -\frac{1}{2}$ *or* $x \geq 4$ **27.** $x < 10$ *or* $x > 110$ **29.** $x < -9\frac{1}{2}$ *or* $x > 10\frac{1}{2}$ **31.** $-13 < x < 17$ **33.** $-7\frac{1}{3} < x < 4$ **35.** $x < -2$ *or* $x > 2\frac{3}{4}$ **37.** $18.53 \leq m \leq 18.77$ **39.** $9.63 \leq n \leq 9.73$ **41.** 4.5×10^{-5} **42.** $x = -\frac{7}{2}$ or $x = \frac{9}{2}$ **43.** 29.83 m **44.** 62.8 ft

Quick Quiz 2.8 *See Examples noted with Ex.* **1.** $-12\frac{1}{2} < x < 11\frac{1}{2}$ (Ex. 3) **2.** $-2 \leq x \leq 3$ (Ex. 2) **3.** $x < -1\frac{4}{5}$ *or* $x > 1$ (Ex. 7) **4.** See Student Solutions Manual

You Try It **1.** $x = 7$ **2.** $b = \frac{2A - hB}{h}$ **3.** $x = 2$ or $x = -\frac{16}{3}$ **4.** \$4000 was invested at 6% and \$8000 was invested at 9%

5. (a) $x \geq -2$ 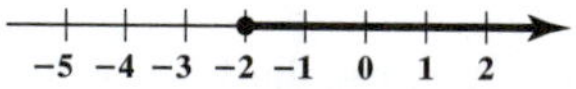

(b) $x < 22$ 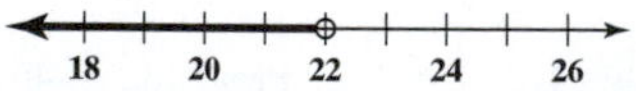

6. $x > -8$ and $x < 2$ 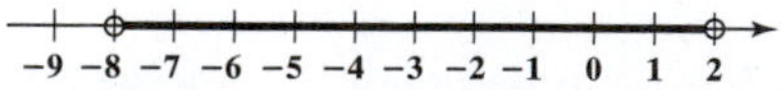

7. $x \leq -2$ *or* $x \geq 3$ 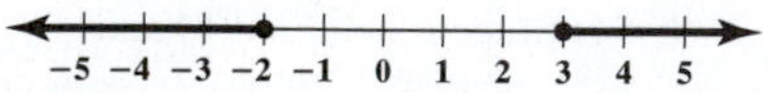

8. $-12 < x < 5$ 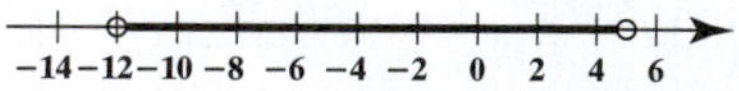

9. $x < -12$ *or* $x > -4$ 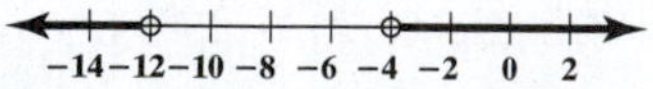

Use Math to Save Money **1.** Apt. 1: \$800 + \$110 + \$90 + \$90 + \$25 = \$1115; Apt. 2: \$850 + \$90 + \$90 + \$25 = \$1055; Apt. 3: \$900 + \$110 + \$25 = \$1035 **2.** ((\$1115 × 12) − \$800)/12 = \$1048.33 **3.** They should rent Apt. 3 since it has the lowest monthly expenses. **4.** \$1035 / 2 = \$517.50

Chapter 2 Review Problems **1.** $x = -\frac{5}{4}$ or $-1\frac{1}{4}$ or -1.25 **2.** $x = -28$ **3.** $x = \frac{5}{2}$ or $2\frac{1}{2}$ or 2.5 **4.** $x = 9$ **5.** $x = -3$ **6.** $x = -63$ **7.** $a = \frac{2P}{b}$ **8.** $a = -\frac{2y}{3x}$ **9. (a)** $F = \frac{9C + 160}{5}$ **(b)** $F = 50°$ **10. (a)** $W = \frac{P - 2L}{2}$ **(b)** 29.5 m **11.** $x = 8, -1$ **12.** $x = 1, -\frac{9}{5}$ **13.** $x = 2, \frac{8}{3}$ **14.** $x = 12, -\frac{4}{3}$ **15.** $x = 44, -20$ **16.** $x = \frac{13}{2}, \frac{3}{2}$ **17.** Length = 15 ft, width = 6 ft **18.** 120 men; 160 women **19.** 240 mi **20.** Retirement = \$10; state tax = \$23; federal tax = \$69 **21.** Nicholas sold 35 tickets; Emma sold 65 tickets; Jackson sold 80 tickets **22.** 2100 students **23.** \$5500 at 11%; \$3500 at 6% **24.** 8 liters of 2%; 16 liters of 5% **25.** 12 lb of \$4.25; 18 lb of \$4.50 **26.** 500 full-time students; 390 part-time students **27.** $x < -4$ **28.** $x > 1$ **29.** $x < -2$ **30.** $x \leq 1$ **31.** $x < 3$ **32.** $x > 4$

33. 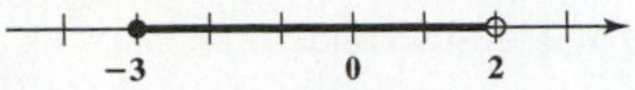 **34.**  **35.** 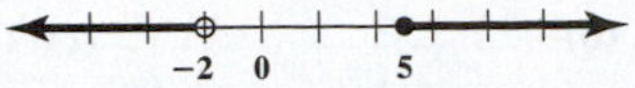

36.  **37.**  **38.** 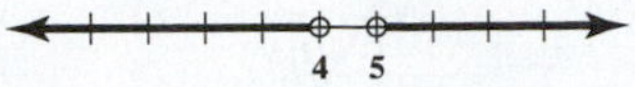

39. $x > 9$ *or* $x < -1$ **40.** No solution **41.** $-6 < x < 3$ **42.** $-4 \le x \le 1\frac{2}{3}$ **43.** $1\frac{4}{5} \le x < 5$ **44.** All real numbers

45. $-22 < x < 8$ **46.** $-27 < x < 9$ **47.** $-7\frac{1}{2} < x < -\frac{1}{2}$ **48.** $x \le -4$ *or* $x \ge 5$ **49.** $x \le -\frac{1}{3}$ *or* $x \ge 1$ **50.** $x \le 4$ *or* $x \ge 6$

51. 15 min **52.** 18 packages **53.** 7 yd^3 **54.** $2{,}404{,}480 \le x \le 3{,}025{,}240$ **55.** $x = -\frac{1}{2}$ or -0.5 **56.** $B = \frac{4H + 64}{3}$ *or* $\frac{4}{3}(H + 16)$

57. 80 g of 77% pure copper, 20 g of 92% pure copper **58.** $x > 6$ 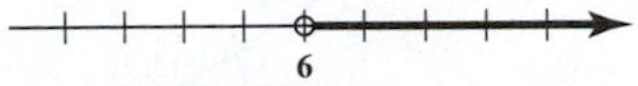

59. $x \ge 3$ 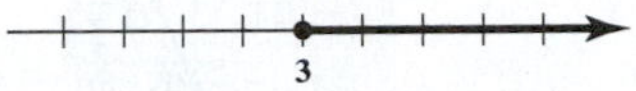 **60.** $-3 \le x \le 3$ 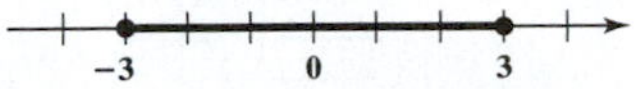

61. $x < -4$ *or* $x > 3$ 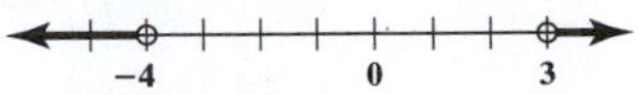 **62.** $x = 4, 3$ **63.** $-\frac{15}{4} \le x \le \frac{21}{4}$ **64.** $x < -3$ *or* $x > \frac{11}{5}$

How Am I Doing? Chapter 2 Test

1. $x = -\frac{2}{11}$ (obj. 2.1.1) **2.** $x = \frac{1}{2}$ or 0.5 (obj. 2.1.1) **3.** $x = 1$ (obj. 2.1.1) **4.** $x = \frac{17}{14}$ or $1\frac{3}{14}$ (obj. 2.1.1) **5.** $n = \frac{L - a + d}{d}$ (obj. 2.2.1) **6.** $b = \frac{2A}{h}$ (obj. 2.2.1) **7.** $b = 3$ cm (obj. 2.2.1) **8.** $r = \frac{4H - 12b + 1}{2}$ (obj. 2.2.1) **9.** $x = -7, \frac{39}{5}$ (obj. 2.3.1) **10.** $x = 6, -18$ (obj. 2.3.2) **11.** 1st side = 16 m; 2nd side = 32 m; 3rd side = 21 m (obj. 2.4.1) **12.** \$2620 (obj. 2.5.1) **13.** 2.5 gal of 90%, 7.5 gal of 50% (obj. 2.5.1) **14.** \$1800 at 6%, \$3200 at 10% (obj. 2.5.1)

15. $x > -2$ 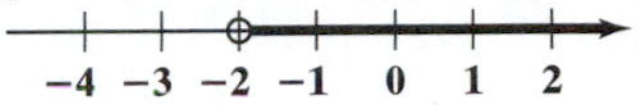 (obj. 2.6.3) **16.** $x \le -1$  (obj. 2.6.3)

17. $-5 < x \le -1$ (obj. 2.7.1) **18.** $x \le -2$ *or* $x \ge 1$ (obj. 2.7.2) **19.** $-\frac{15}{7} \le x \le 3$ (obj. 2.8.1) **20.** $x < -\frac{8}{3}$ *or* $x > 2$ (obj. 2.8.2)

Chapter 3

3.1 Exercises

1. variables **3.** To locate the point (a, b), assuming that $a, b > 0$, we move a units to the right and b units up. To locate the point (b, a), we move b units to the right and a units up. If $a \ne b$, the graphs of the points will be different. Thus, the order of the numbers in (a, b) matters. (1, 3) is not the same as (3, 1). **5.** -13 **7.** -4

The coordinates that are used to plot the points to graph each equation in the selected answers to exercises 11–35 are suggested coordinates. You may choose other replacements for x and y and hence generate a different set of points. The graph of the equation, however, will be the same.

9.

x	y
0	−3
2	1
4	5

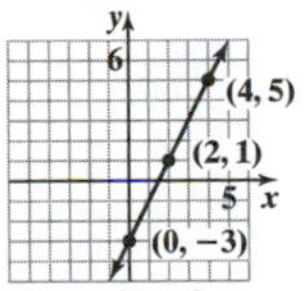

$y = 2x - 3$

11.

x	y
−1	6
0	4
2	0

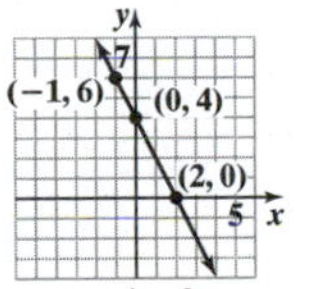

$y = 4 - 2x$

13.

x	y
−3	−6
0	−4
3	−2

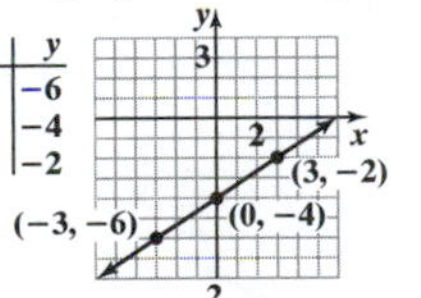

$y = \frac{2}{3}x - 4$

15.

x	y
−2	0
0	3
2	6

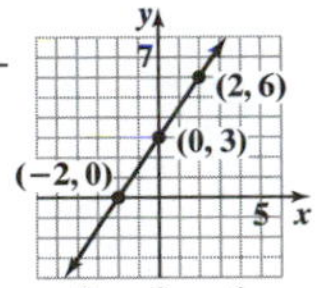

$2y - 3x = 6$

17.

x	y
0	−6
3	0
2	−2

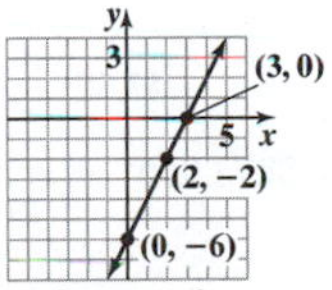

$2x - y = 6$

19.

x	y
$-1\frac{1}{2}$	0
0	−2
−3	2

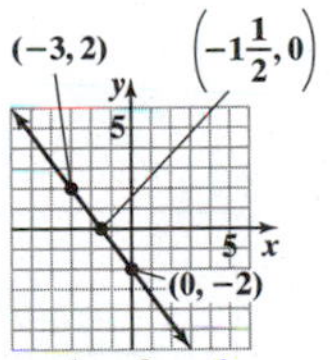

$-4x - 3y = 6$

21. $y = \frac{3}{5}x$

x	y
−5	−3
0	0
5	3

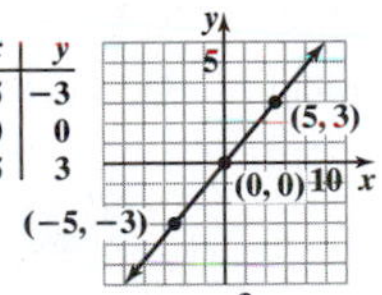

$y = \frac{3}{5}x$

23. $x = -5$; vertical line

x	y
−5	0
−5	5
−5	−2

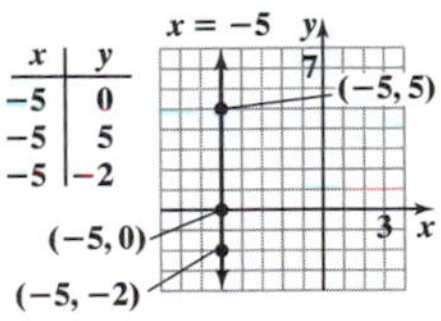

25. $x = 4$; vertical line

x	y
4	0
4	1
4	2

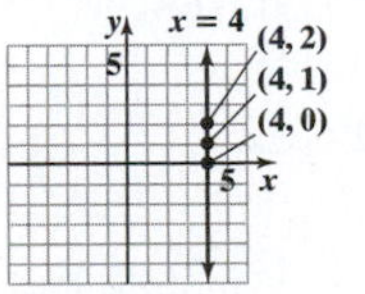

27. $y = -4$; horizontal line

x	y
0	−4
1	−4
2	−4

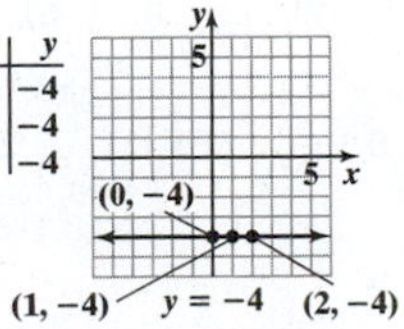

29.

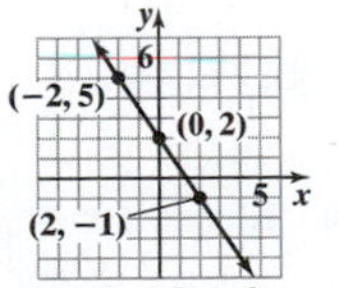

$y = -1.5x + 2$

31.

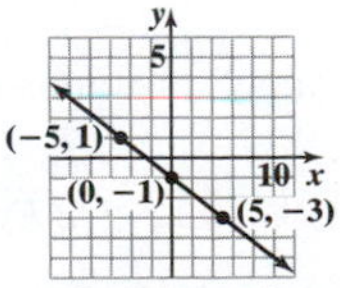

$2x + 5y = -5$

33. $3x - y = 4$

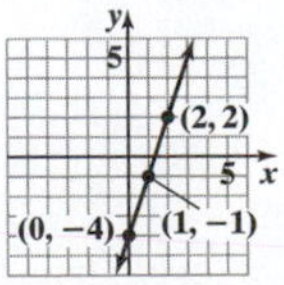

$3x - y = 4$

35. Vertical scale: 1 square = 50 units

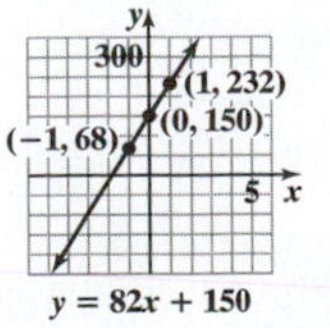

$y = 82x + 150$

37. Between 2007 and 2009 **39.** In 2009 **41.** 38.9%

43. **(a)** 120; 88; 56; 24; −8 **(b)**

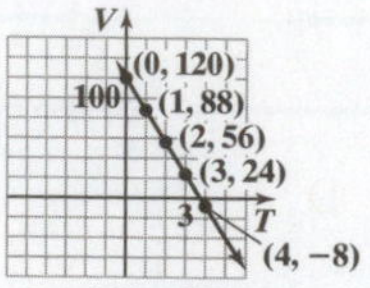

$V = 120 - 32T$

(c) The baseball is moving downward instead of upward at $T = 4$ seconds.

45. −3 **46.** $x \geq -3$ **47.** 520 red, 260 green, 1560 blue, 260 yellow, 130 white **48.** $170,000

Quick Quiz 3.1 *See Examples noted with Ex.*

1. (Ex. 1)

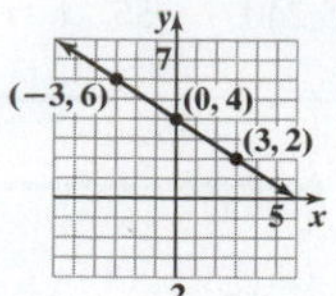

$y = -\frac{2}{3}x + 4$

2. (Ex. 3)

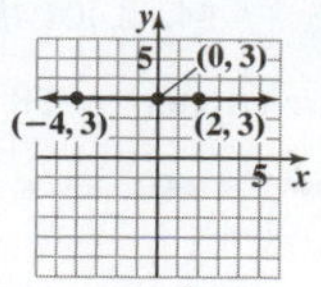

$y = 3$

3. (Ex. 2)

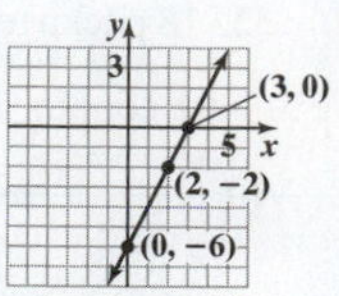

$-4x + 2y = -12$

4. See Student Solutions Manual

3.2 Exercises

1. vertical; horizontal **3.** 0 **5.** No, division by zero is undefined. **7.** $m = -2$ **9.** $m = \frac{8}{7}$ **11.** $m = \frac{2}{3}$ **13.** Undefined slope **15.** $m = 0$ **17.** $m = -\frac{3}{2}$ **19.** $\frac{3}{5}$ or 0.6 **21.** $\frac{1}{4}$ or 0.25 **23.** 42 ft **25.** $m_{\|} = -\frac{2}{9}$ **27.** $m_{\|} = -4$ **29.** $m_{\|} = \frac{3}{2}$ **31.** $m_{\perp} = -\frac{5}{3}$ **33.** $m_{\perp} = -\frac{1}{2}$ **35.** $m_{\perp} = -2$ **37.** Yes. Each line has the same slope. **39.** $m_{AD} = m_{BC} = -\frac{1}{6}; m_{AB} = m_{CD} = 1$ **41.** **(a)** $\frac{1}{12}$ **(b)** 2 ft **(c)** 20.4 ft **43.** $\frac{5x^{10}}{y^3}$ **44.** $8x^2 - 10x - 2y$ **45.** Length = 10 ft; width = 8 ft **46.** $x > 3$ *or* $x < -2$

Quick Quiz 3.2 *See Examples noted with Ex.*

1. $m = \frac{1}{4}$ (Ex. 2) **2.** $m = 0$ (Ex. 1) **3.** $m_{\perp} = -2$ (Ex. 4) **4.** See Student Solutions Manual

3.3 Exercises

1. First determine the slope from the coordinates of the points. Then substitute the slope and the coordinates of one point into the point-slope form of the equation of a line. You may rewrite the equation in standard form or in slope-intercept form. **3.** $y = \frac{3}{4}x - 9$ **5.** $3x - 4y = -2$ **7.** $m = \frac{1}{2}; (0, 3); y = \frac{1}{2}x + 3$ **9.** $m = -1; (0, 3); y = -x + 3$ **11.** $m = 3; (0, 0); y = 3x$ **13.** $m = -\frac{3}{2}; (0, 0); y = -\frac{3}{2}x$ **15.** $m = \frac{3}{4}; (0, -4); y = \frac{3}{4}x - 4$ **17.** $y = x - 5; m = 1; (0, -5)$ **19.** $y = \frac{5}{4}x + 5; m = \frac{5}{4}; (0, 5)$ **21.** $y = -\frac{8}{3}x - 4; m = -\frac{8}{3}; (0, -4)$

23.

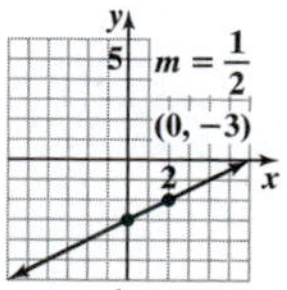

$y = \frac{1}{2}x - 3$

25.

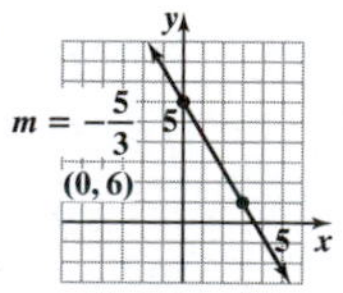

$5x + 3y = 18$

27. $y = -\frac{1}{2}x + 6$ **29.** $y = 5x + 33$ **31.** $y = -\frac{1}{5}x + \frac{6}{5}$ **33.** $y = \frac{5}{7}x + \frac{13}{7}$ **35.** $y = -\frac{2}{3}x - \frac{8}{3}$ **37.** $y = -3$ **39.** $5x - y = -10$ **41.** $x - 3y = 8$ **43.** $2x - 3y = 15$ **45.** $7x - y = -27$ **47.** Neither **49.** Parallel **51.** Perpendicular **53.**

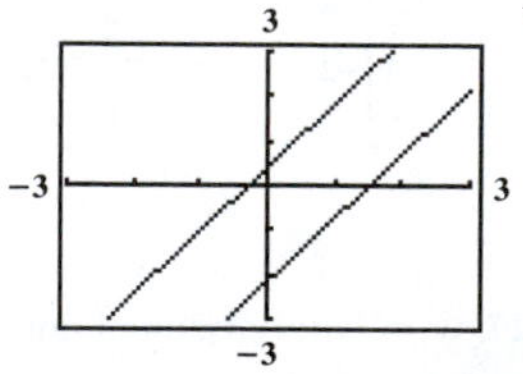

Yes **55.** $272,900 **57.** Answers may vary; approximately $250,000 **58.** 78.5 mm^2 **59.** $x = -4$ **60.** $x = 1$ **61.** $3000 at 4%; $6000 at 6%

Quick Quiz 3.3 *See Examples noted with Ex.*

1. $y = \frac{3}{7}x - \frac{9}{7};\ m = \frac{3}{7};\ \left(0, -\frac{9}{7}\right)$ (Ex. 3) **2.** $9x - y = -29$ (Ex. 5) **3.** $y = -\frac{3}{4}x - 5$ (Ex. 1) **4.** See Student Solutions Manual

How Am I Doing? Sections 3.1–3.3

1. $a = -\frac{24}{5}$ or −4.8 (obj. 3.1.1)

2. (obj. 3.1.1)

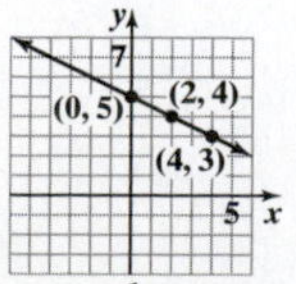

$y = -\frac{1}{2}x + 5$

3. (obj. 3.1.2)

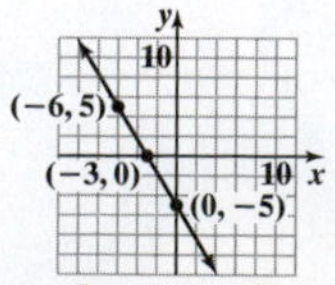

$5x + 3y = -15$

4. (obj. 3.1.1)

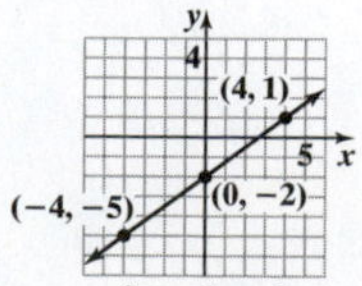

$3x - 4y = 8$

5. $m = -9$ (obj. 3.2.1) **6.** $m_{\|} = -36$ (obj. 3.2.2) **7.** $m_{\perp} = -\frac{9}{2}$ (obj. 3.2.2) **8.** 65 ft (obj. 3.2.1) **9.** $m = \frac{1}{2}$; y-intercept $= \left(0, -\frac{5}{2}\right)$ (obj. 3.3.1) **10.** $y = -2x + 11$ (obj. 3.3.2) **11.** $y = -\frac{5}{3}x - \frac{11}{3}$ (obj. 3.3.3) **12.** $y = 2x + 7$ (obj. 3.3.2)

3.4 Exercises

1. Use a dashed line when graphing a linear inequality that contains the $>$ symbol or the $<$ symbol. **3.** To graph the region $x > 5$, shade the region to the right of the line $x = 5$. **5.** The point (0, 0) is on the line; false.

7.
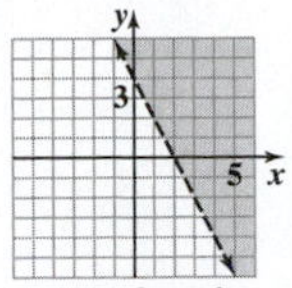

$y > -2x + 4$

9.
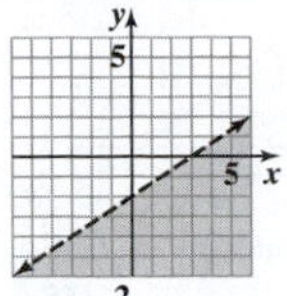

$y < \frac{2}{3}x - 2$

11.
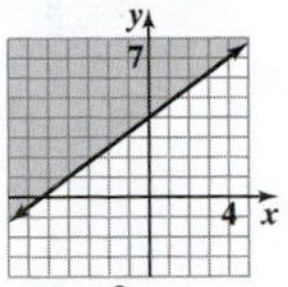

$y \geq \frac{3}{4}x + 4$

13.
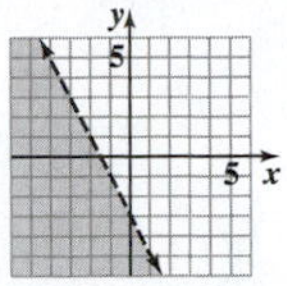

$-2x - y > 3$

15.
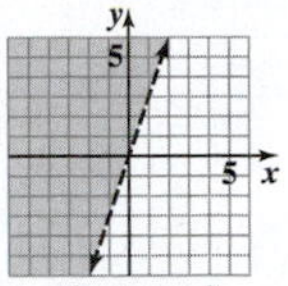

$-3x + y > 0$

17.
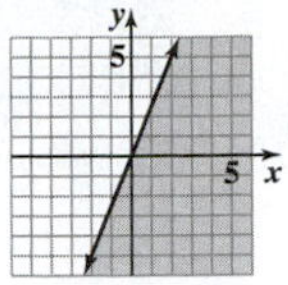

$5x - 2y \geq 0$

19.
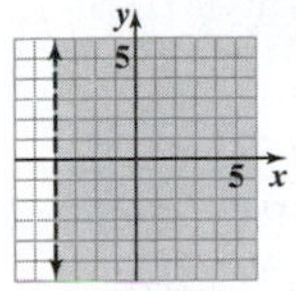

$x > -4$

21.
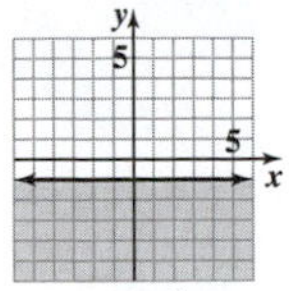

$y \leq -1$

23.
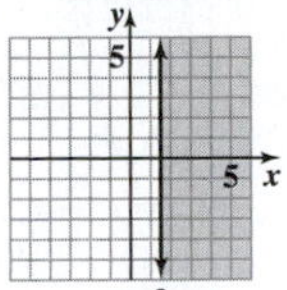

$x \geq \frac{3}{2}$

25.
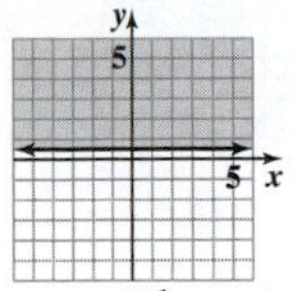

$y \geq \frac{1}{2}$

27. Length is 14 ft; width is 3 ft
28. $-1 < x < 7$
29. $x \leq 0$ *or* $x \geq 2$

Quick Quiz 3.4

See Examples noted with Ex. **1.** Above the line (Ex. 2) **2.**

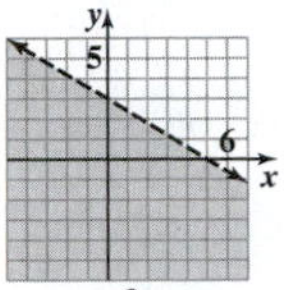

$y < -\frac{3}{5}x + 3$

(Ex. 3) **3.**

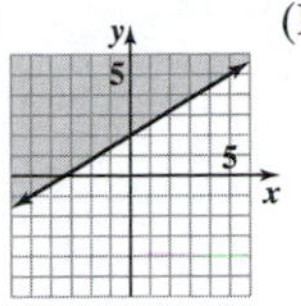

$3x - 5y \leq -10$

(Ex. 2)

4. See Student Solutions Manual

3.5 Exercises

1. A relation is any set of ordered pairs. A function is a set of ordered pairs in which no two different ordered pairs have the same first coordinate. **3.** As a set of ordered pairs; as an equation; as a graph **5.** Domain $= \{0, 5, 7\}$; Range $= \{0, 13, 11\}$; *not* a function **7.** Domain $= \{85, 16, -102, 62\}$; Range $= \{-12, 4, 48\}$; function **9.** Domain $= \{6, 8, 10, 12, 14\}$; Range $= \{38, 40, 42, 44, 46\}$; function **11.** Domain = {Jan., Feb., Mar., Apr., May, June, July, Aug., Sept., Oct., Nov., Dec.}; Range $= \{81, 80, 79\}$; function **13.** Domain = {Chicago, New York}; Range $= \{1454, 1350, 1250, 1136, 1127, 1046\}$; *not* a function **15.** Domain $= \{10, 20, 30, 40, 50\}$; Range $= \{11.51, 23.02, 34.53, 46.04, 57.55\}$; function **17.** Function **19.** Not a function **21.** Function **23.** Not a function **25.** Function **27.** -9.8 **29.** -4 **31.** -5 **33.** $\frac{13}{4}$ **35.** 7 **37.** 1.42 **39.** 57 **41.** -27 **43.** 2 **45.** 2 **47.** Range $= \{3, 4, 7\}$ **49.** Domain $= \{12, 8, 0, -3\}$ **50.** $x = 1, \frac{1}{3}$ **51.** $2 \leq x \leq 8$ **52.**

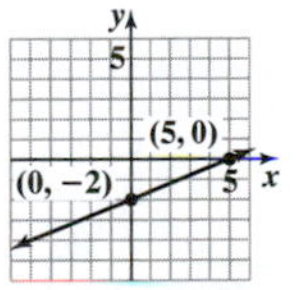

$2x - 5y = 10$

Quick Quiz 3.5

See Examples noted with Ex. **1.** Domain $= \{9, -2, 4, -4\}$; Range $= \{3, 5\}$ (Ex. 2) **2.** 8 (Ex. 4) **3.** 38 (Ex. 4)
4. See Student Solutions Manual

To Think About, page 172

The graphs of $f(x) = |x|$ and $h(x) = |x + 2|$ have the exact same shape, but different positions on the coordinate axes. To obtain the graph of $h(x) = |x + 2|$, take the graph of $f(x) = |x|$ and shift it 2 units to the left.

To Think About, page 173

1. The graphs of $r(x) = (x - 2)^2$ and $s(x) = x^2 + 2$ have the exact same shape, but different positions on the coordinate axes. The graph of $r(x) = (x - 2)^2$ has its vertex on the x-axis and the graph of $s(x) = x^2 + 2$ has its vertex on the y-axis. To obtain the graph of $q(x) = (x + 5)^2$, take the graph of $p(x) = x^2$ and shift it 5 units to the left. **2.** To obtain the graph of $s(x) = x^2 + 5$, take the graph of $p(x) = x^2$ and shift it 5 units up. To obtain the graph of $f(x) = x^3 - 2$, the graph of $g(x) = x^3$ is shifted 2 units down.

To Think About, page 174

To obtain the graph of $q(x) = -\frac{4}{x}$, the graph of $p(x) = \frac{4}{x}$ is reflected across the x-axis.

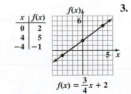

3.6 Exercises

1.

x	$f(x)$
0	2
4	5
−4	−1

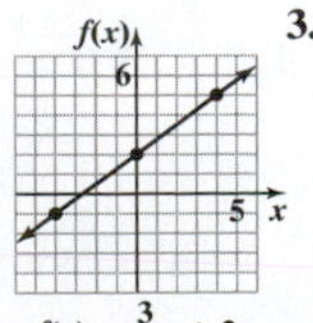

$f(x) = \frac{3}{4}x + 2$

3.

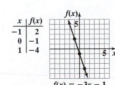

x	$f(x)$
−1	2
0	−1
1	−4

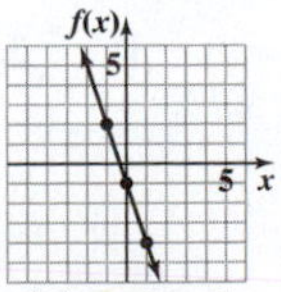

$f(x) = -3x - 1$

5. $C(x) = 25 + 0.15x$; $C(0) = 25$; $C(100) = 40$; $C(200) = 55$; $C(300) = 70$

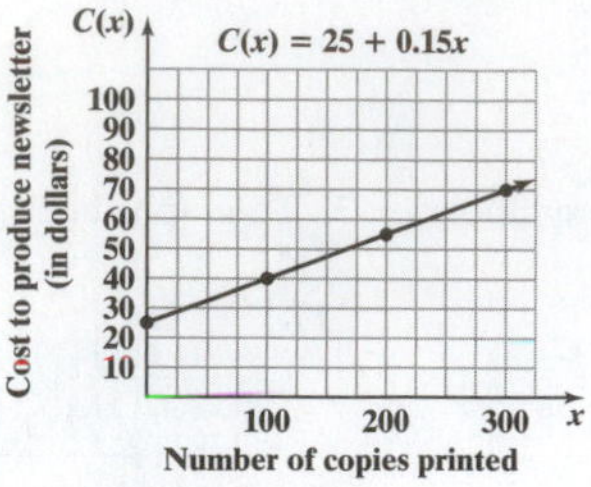

7. $P(x) = 45{,}000 - 1500x$; $P(0) = 45{,}000$; $P(10) = 30{,}000$; $P(30) = 0$
With 30 tons of pollutants, there will be no fish.

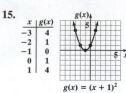

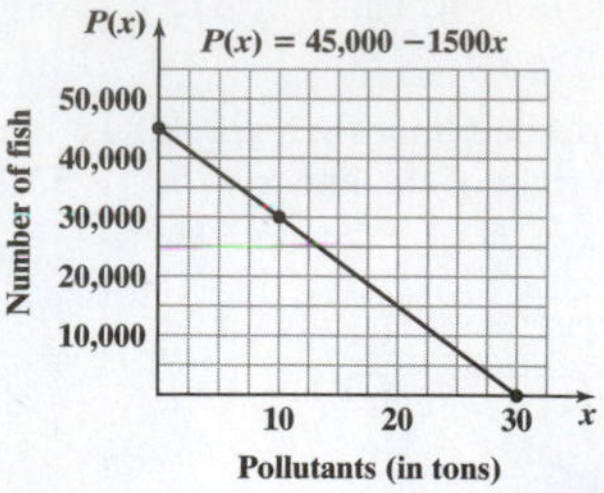

9.

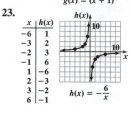

x	$f(x)$
−1	2
0	1
1	0
2	1
3	2

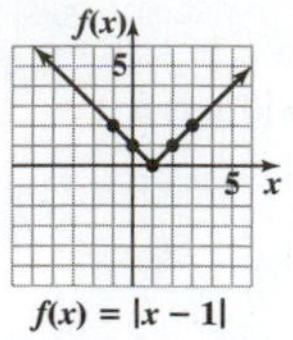

$f(x) = |x - 1|$

11.

x	$g(x)$
−2	−3
−1	−4
0	−5
1	−4
2	−3

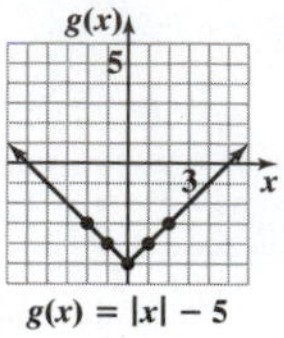

$g(x) = |x| - 5$

13.

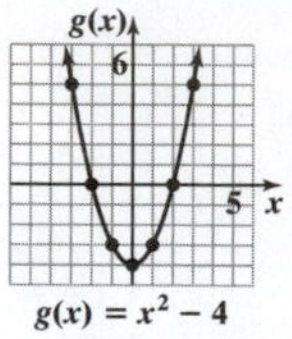

$g(x) = x^2 - 4$

15.

x	$g(x)$
−3	4
−2	1
−1	0
0	1
1	4

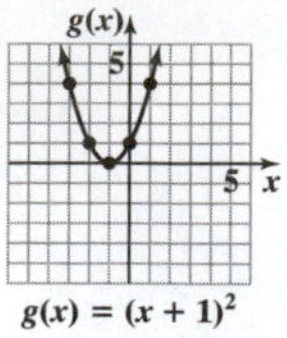

$g(x) = (x + 1)^2$

17.

x	$g(x)$
−1	−4
0	−3
1	−2
2	5

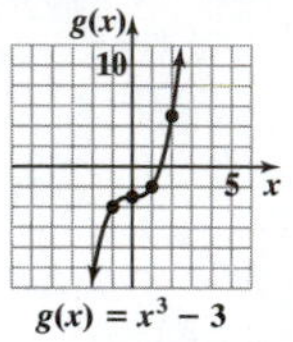

$g(x) = x^3 - 3$

19.

x	$p(x)$
−2	8
−1	1
0	0
1	−1
2	−8

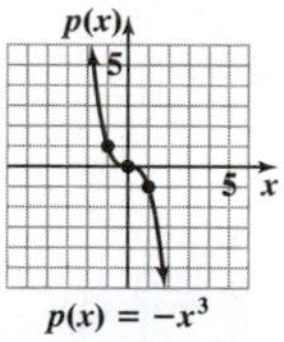

$p(x) = -x^3$

21.

x	$f(x)$
1	2
2	1
4	$\frac{1}{2}$
$\frac{1}{2}$	4
−1	−2
−2	−1
−4	$-\frac{1}{2}$
$-\frac{1}{2}$	−4

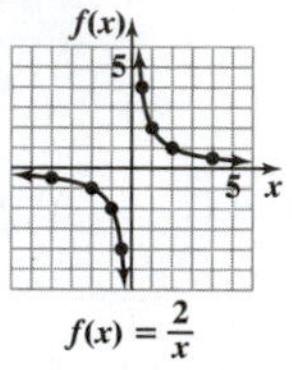

$f(x) = \frac{2}{x}$

23.

x	$h(x)$
−6	1
−3	2
−2	3
−1	6
1	−6
2	−3
3	−2
6	−1

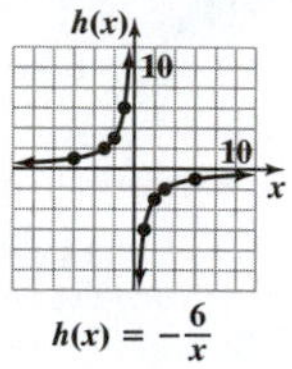

$h(x) = -\frac{6}{x}$

25. When the coefficient of x^2 is greater than 1, the curve is closer to the y-axis. When the coefficient is less than 1, the curve is farther away from the y-axis.

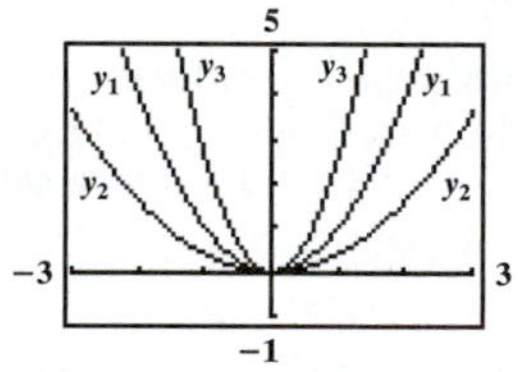

27.

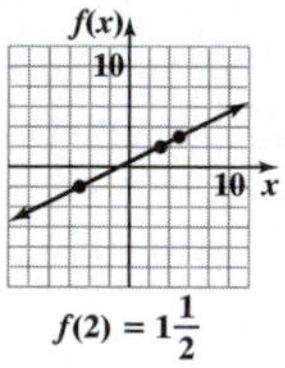

$f(2) = 1\frac{1}{2}$

29.

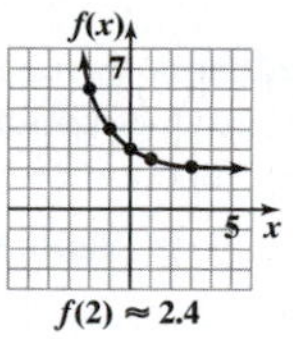

$f(2) \approx 2.4$

31.

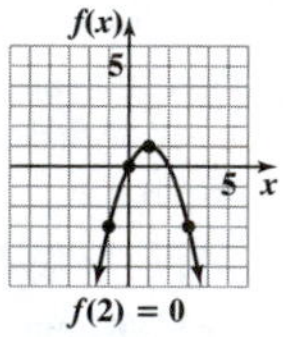

$f(2) = 0$

33. (a)

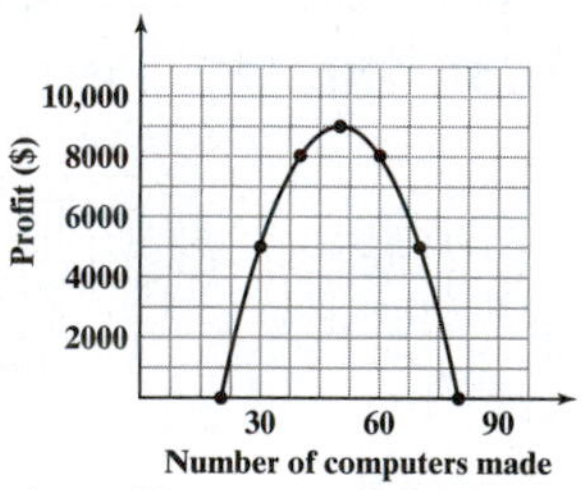

(b) 50 **(c)** Between 40 and 60 **(d)** They will operate at a loss. **(e)** \$8700

35. $x = \dfrac{8y + 15}{a}$ **36.** $x = 7.25$ **37.** $m = -2$ **38.** $x + 3y = -10$

Quick Quiz 3.6 *See Examples noted with Ex.*

1. (Ex. 2)

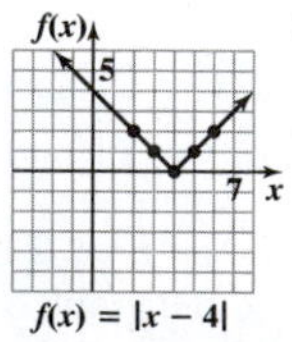

$f(x) = |x - 4|$

2. (Ex. 3)

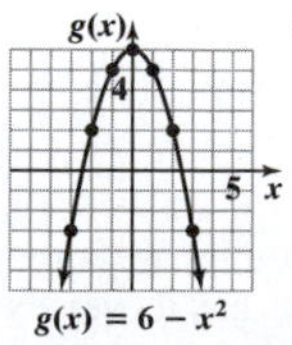

$g(x) = 6 - x^2$

3. (Ex. 4)

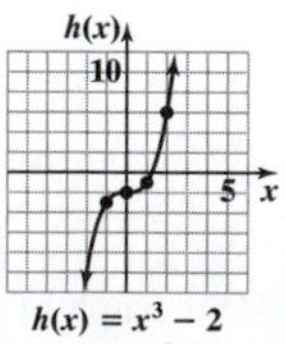

$h(x) = x^3 - 2$

4. See Student Solutions Manual

Use Math to Save Money

1. \$14,970.16 **2.** \$25,094.08 **3.** \$25,970.16 **4.** \$876.08 **5.** He will save \$281.22 each month in car payments. **6.** To get the best overall price, Louvy should buy the car because he will save \$876.08 on the total price. To get a lower monthly payment, Louvy should lease the car because he will save \$281.22 each month in car payments. **7.** Answers will vary. **8.** Answers will vary. **9.** Answers will vary.

You Try It

1.

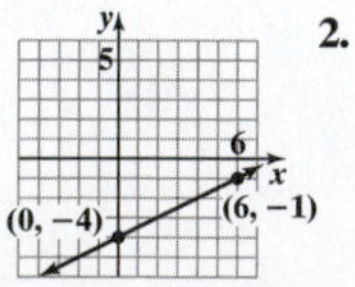

2. x-intercept = (5, 0); y-intercept = (0, −3) **3.** $m = -\frac{1}{2}$ **4.** 0 **5.** The slope is undefined. **6. (a)** $m = \frac{1}{2}$ **(b)** $m = -\frac{1}{2}$ **7.** $2x - y = 8$ **8. (a)** $m = \frac{1}{4}$; y-intercept $= \left(0, -\frac{3}{4}\right)$ **(b)** $m = 2$; y-intercept = (0, 5) **9.** $y = -\frac{2}{5}x + \frac{16}{5}$

10.

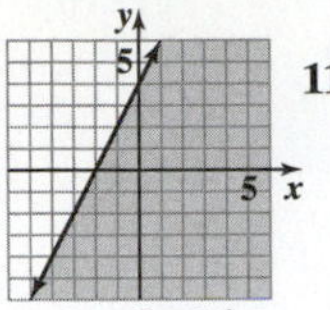

$y - 2x \le 4$

11. Domain $= \{0, -1, 4\}$; Range $= \{2, 3, 1, -1\}$ **12. (a)** Not a function **(b)** Function

13. (a) Not a function **(b)** Function **14.** 8 **15.**

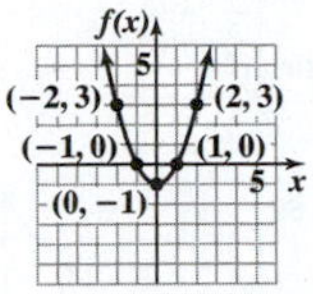

Chapter 3 Review Problems

1.

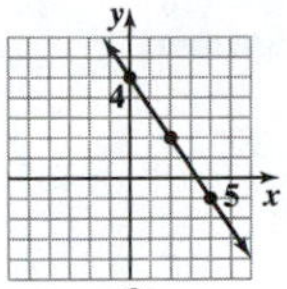

$y = -\frac{3}{2}x + 5$

2.

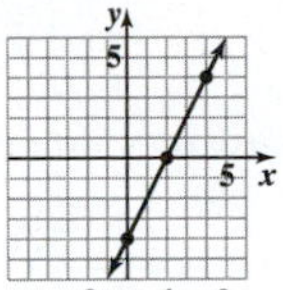

$y - 2x + 4 = 0$

3.

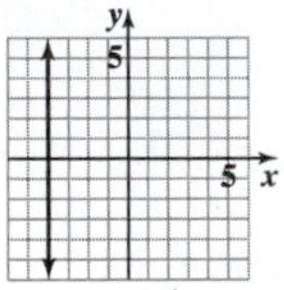

$x = -4$

4.

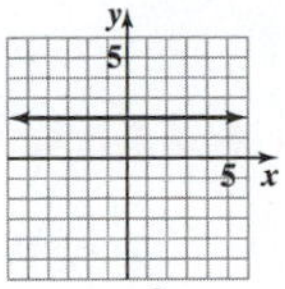

$y = 2$

5. $m = \frac{1}{10}$ **6.** $m = 0$ **7.** $m = \frac{1}{3}$; y-intercept $= (0, 4)$ **8.** $m = -2$ **9.** $x - 3y = -15$ **10.** $4x + y = 0$ **11.** $y = 1$

12. (a) The slope is 140. **(b)** At least 15 microcomputers **13.** $13x - 12y = -7$ **14.** $x = -6$ **15.** $8x - 7y = -51$ **16.** $3x - 2y = 13$

17. $m = -2$
y-intercept $= (0, 6)$
$y = -2x + 6$

18. $m = 0$
y-intercept $= (0, 2)$
$y = 2$

19.

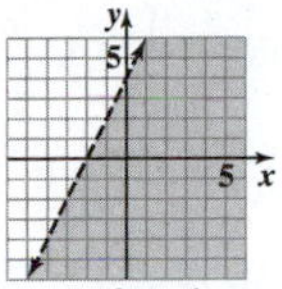

$y < 2x + 4$

20.

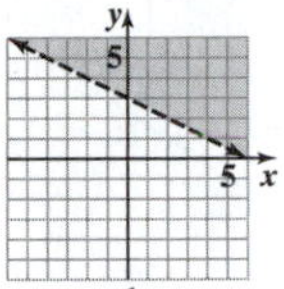

$y > -\frac{1}{2}x + 3$

21.

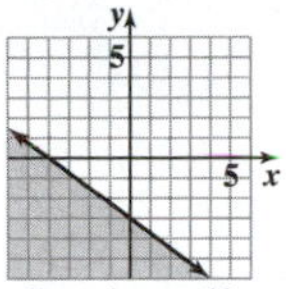

$3x + 4y \le -12$

22.

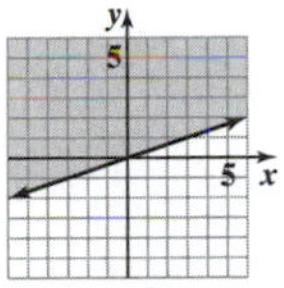

$x \le 3y$

23.

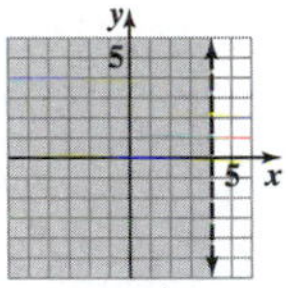

$x < 4$

24.

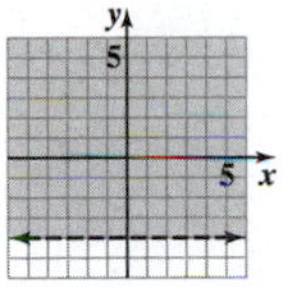

$y > -4$

25. Domain $= \{-20, -18, -16, -12\}$; Range $= \{18, 16, 14\}$; function
26. Domain $= \{0, 1, 2, 3\}$; Range $= \{0, 1, 4, 9, 16\}$; not a function **27.** Function **28.** Not a function
29. $f(-1) = 12$; $f(-5) = 20$ **30.** $h(-1) = 14$

31.

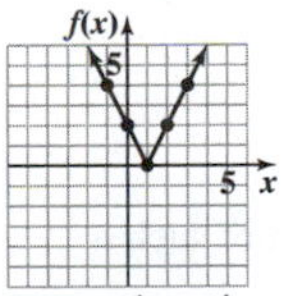

$f(x) = 2|x - 1|$

32.

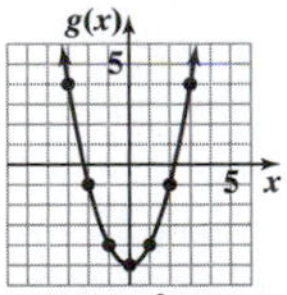

$g(x) = x^2 - 5$

33.

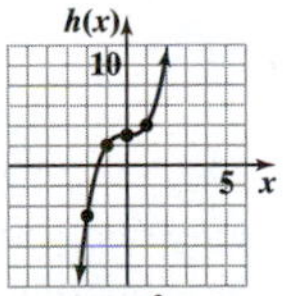

$h(x) = x^3 + 3$

34.

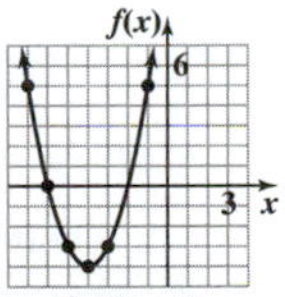

$f(-2) = 0$

35.

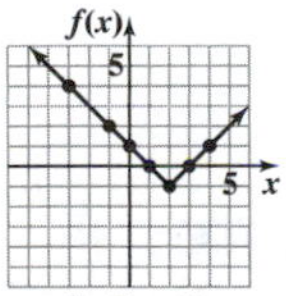

$f(-2) = 3$

36.

x	$f(x)$
−5	7
0	3
10	−5

37.

x	$f(x)$
−3	31
0	4
4	24

38. $m = \frac{3}{2}$; y-intercept $= \left(0, -\frac{9}{2}\right)$ **39.** $m = -\frac{4}{3}$ **40.** $m = 7$

41. $5x - 6y = 30$ **42.** $y = 5x - 10$ **43.** $y = \frac{1}{4}x + \frac{19}{4}$

44. $2x + y = -5$ **45.** $x = 5$ **46.** $f(x) = 40 + 0.20x$

47. $f(x) = 24{,}000 + 200x$ **48.** $f(x) = 18{,}000 - 65x$

How Am I Doing? Chapter 3 Test

1. (obj. 3.1.1)

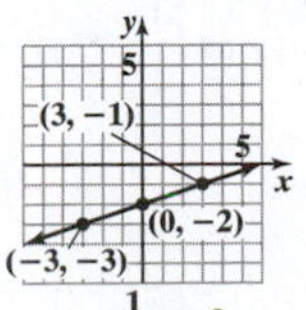

$y = \frac{1}{3}x - 2$

2. (obj. 3.1.3)

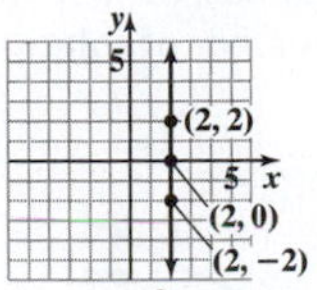

$x = 2$

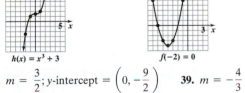

3. (obj. 3.1.1)

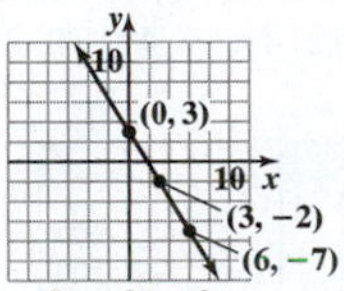

$5x + 3y = 9$

4. (obj. 3.1.1)

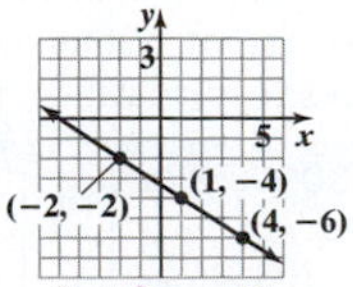

$2x + 3y = -10$

5. $m = 2$ (obj. 3.2.1) **6.** $m = 0$ (obj. 3.2.1) **7.** $m = \frac{2}{5}$; y-intercept $= (0, 3)$ (obj. 3.3.1) **8.** $7x + 6y = -12$ (obj. 3.3.3)

9. $x + 8y = -11$ (obj. 3.3.2) **10.** $y = 2$ (obj. 3.3.2) **11.** $y = -5x - 8$ (obj. 3.3.1)

12. (obj. 3.4.1)

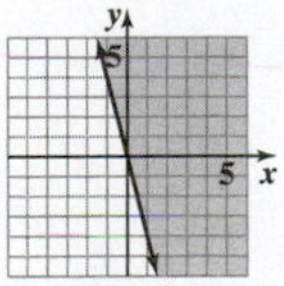

$y \ge -4x$

13. (obj. 3.4.1)

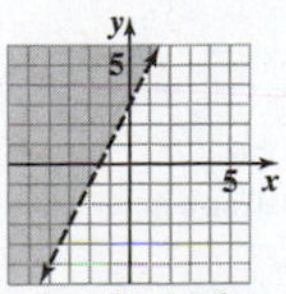

$4x - 2y < -6$

14. Domain = {0, 1, 2}; Range = {0, 1, −1, 4, −4} (obj. 3.5.1) **15.** $f\left(\frac{3}{4}\right) = -\frac{3}{2}$ (obj. 3.5.3) **16.** $g(-4) = 11$ (obj. 3.5.3)
17. $h(-9) = 10$ (obj. 3.5.3) **18.** $p(-2) = 22$ (obj. 3.5.3)

19. (obj. 3.6.1)

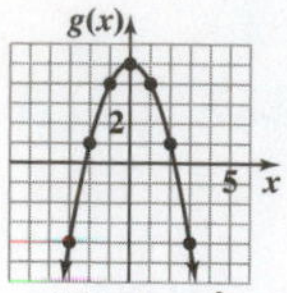

$g(x) = 5 - x^2$

20. (obj. 3.6.1)

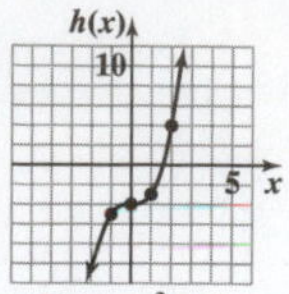

$h(x) = x^3 - 4$

21. 10 mi (obj. 3.6.2)

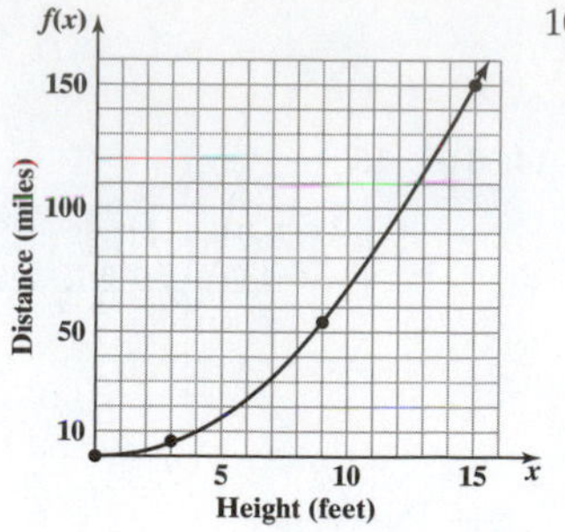

Cumulative Test for Chapters 1–3

1. Inverse property of addition **2.** 12 **3.** $\frac{a^6}{4b^8}$ **4.** $\frac{4a^3}{5b^2}$ **5.** $2x^2 + 10xy - 4$
6. −2 **7.** 4.37×10^{-4} **8.** 14.13 in.^{2} **9.** $x = 2$ **10.** $x = 1$ or $x = -\frac{5}{2}$ **11.** $-3 \le x \le 7$
12. $x < -2$ *or* $x > 4$

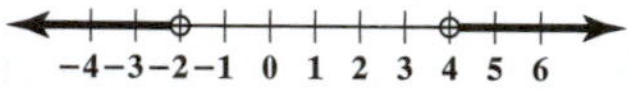

13. $x = \frac{6a + y}{9}$ **14.** Width = 15 cm; Length = 31 cm
15. \$3300 at 4%; \$3700 at 7% **16.**

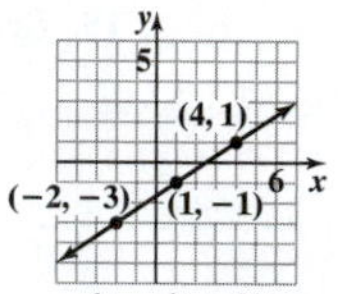

$4x - 6y = 10$

17. $m = \frac{1}{2}$ **18.** $2x + y = 11$ **19.** $3x - y = -7$
20. Domain = $\left\{3, 5, \frac{1}{2}, 2\right\}$; Range = {7, 8, −1, 2}; function

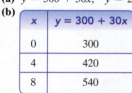

21.

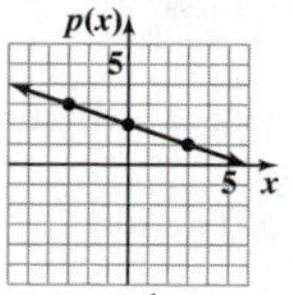

$p(x) = -\frac{1}{3}x + 2$

22.

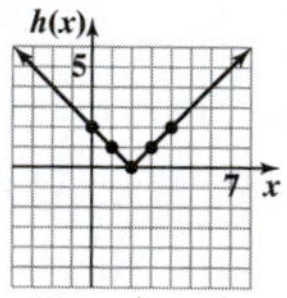

$h(x) = |x - 2|$

23.

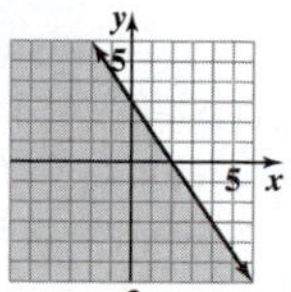

$y \le -\frac{3}{2}x + 3$

24.

x	f(x)
−2	20
0	4
3	−50

25. (a) $f(x) = 1.5x + 31.1$ **(b)** 49.1 million people

Chapter 4

4.1 Exercises

1. There is no solution. There is no point (*x*, *y*) that satisfies both equations. The graph of such a system yields two parallel lines. **3.** It may have one solution, it may have no solution, or it may have an infinite number of solutions.
5. $\left(\frac{3}{2}, -1\right)$ is a solution to the system. **7.**

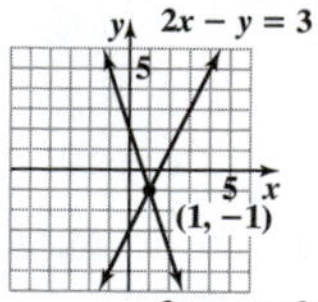

9.

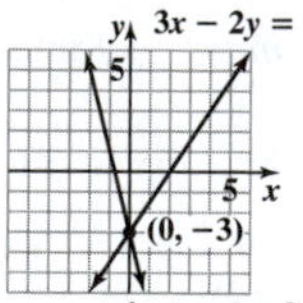

11.

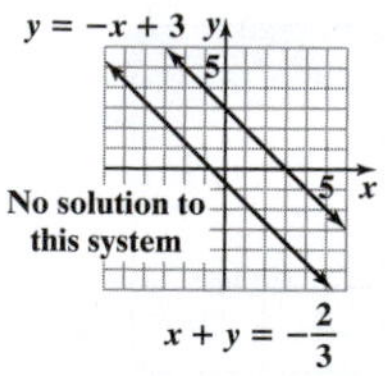

13.

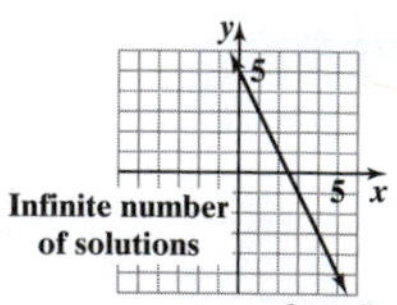

15. (18, −10) **17.** (2, −2) **19.** (−1, 2) **21.** (6, 0) **23.** (0, 1) **25.** $\left(2, \frac{4}{3}\right)$ **27.** (1, −3) **29.** (3, −2) **31.** (2, 8) **33.** (−4, 5)
35. (6, −8) **37.** No solution; inconsistent system of equations **39.** Infinite number of solutions; dependent equations **41.** (5, −3)
43. No solution; inconsistent system of equations **45.** (12, 4) **47.** (0, 2) **49.** Infinite number of solutions; dependent equations
51. (a) $y = 300 + 30x$, $y = 200 + 50x$
(b)

x	y = 300 + 30x
0	300
4	420
8	540

x	y = 200 + 50x
0	200
4	400
8	600

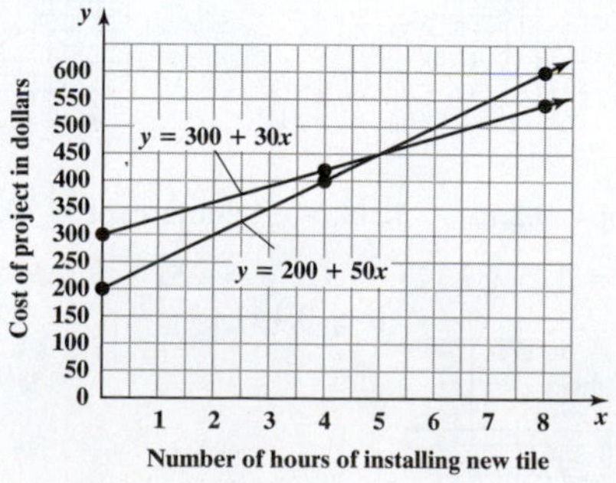

(c) The cost will be the same for 5 hours of installing new tile.
(d) The cost will be less for Modern Bathroom Headquarters.

53. (2.46, −0.38)

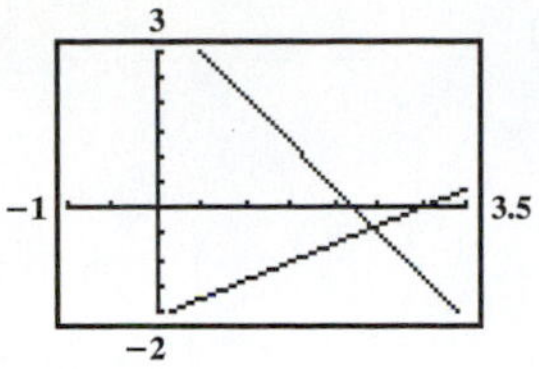

55. (−2.45, 6.11)

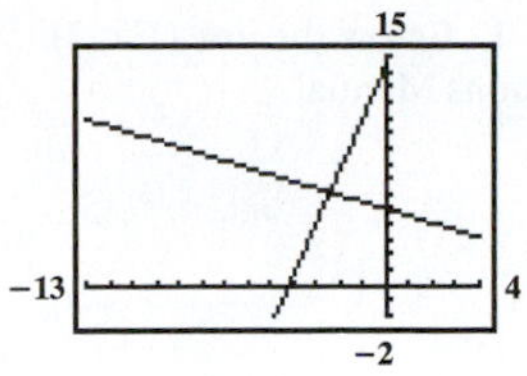

56. Approximately \$0.01/lb **57.** 341,889 cars

Quick Quiz 4.1 *See Examples noted with Ex.* **1.** (4, 2) (Ex. 3) **2.** (2, 2) (Ex. 6) **3.** (−12, −15) (Ex. 4) **4.** See Student Solutions Manual

4.2 Exercises **1.** Yes, it is. **3.** No, it is not. **5.** (1, 3, −2) **7.** (−2, 1, 5) **9.** (0, −2, 5) **11.** (1, −1, 2) **13.** (2, 1, −4) **15.** (4, 0, 2) **17.** (3, −1, −2) **19.** $x = 1.10551, y = 2.93991, z = 1.73307$ **21.** (4, −1, −3) **23.** $\left(\frac{1}{2}, \frac{2}{3}, \frac{5}{6}\right)$ **25.** (1, 3, 5) **27.** (4, −2, −3) **29.** Infinite number of solutions; dependent equations **31.** No solution; inconsistent system of equations **33.** $x = -3, x = 4$ **34.** 7.63×10^7 **35.** $m = \frac{4-3}{1+2} = \frac{1}{3}, y - 4 = \frac{1}{3}(x - 1), x - 3y = -11$ **36.** $m(\perp \text{ line}) = \frac{3}{2}, y - 2 = \frac{3}{2}(x + 4), 3x - 2y = -16$

Quick Quiz 4.2 *See Examples noted with Ex.* **1.** (−1, 2, 3) (Ex. 2) **2.** (1, 0, −1) (Ex. 2) **3.** (2, 1, −1) (Ex. 3) **4.** See Student Solutions Manual

How Am I Doing? Sections 4.1–4.2 **1.** (1, 5) (obj. 4.1.3) **2.** (2, −5) (obj. 4.1.4) **3.** (9, 9) (obj. 4.1.6) **4.** (3, −2) (obj. 4.1.6) **5.** Infinite number of solutions; dependent equations (obj. 4.1.5) **6.** (2, −2) (obj. 4.1.6) **7.** No solution; inconsistent system of equations (obj. 4.1.5) **8.** No, it is not. (obj. 4.2.1) **9.** (1, 3, 0) (obj. 4.2.2) **10.** $\left(-1, 1, \frac{2}{3}\right)$ (obj. 4.2.2) **11.** (−2, 3, 4) (obj. 4.2.3) **12.** (−1, 3, 2) (obj. 4.2.3)

4.3 Exercises **1.** 62 is the larger number; 25 is the smaller number **3.** 16 heavy equipment operators; 19 general laborers **5.** 51 tickets for regular coach seats; 47 tickets for sleeper car seats **7.** 25 experienced employees; 10 new employees **9.** 30 packages of old fertilizer; 25 packages of new fertilizer **11.** One scone costs \$1.89; one large coffee costs \$1.49 **13.** Speed of plane in still air is 216 mph; speed of wind is 36 mph **15.** Speed of boat is 14 mph; speed of current is 2 mph **17.** 28 free throws; 36 regular (2-point) shots **19.** 220 weekend minutes; 405 weekday minutes **21.** \$10,258 for a car; \$17,300 for a truck **23.** 7 pens, 5 notebooks, and 3 highlighters **25.** 80 adults, 100 high school students, and 120 children **27.** 800 senior citizens, 10,000 adults, and 1200 children under 12 **29.** 5 small pizzas, 8 medium pizzas, and 7 large pizzas **31.** 3 of Box *A*, 4 of Box *B*, and 3 of Box *C* **32.** $x = \frac{26}{7}$ or $3\frac{5}{7}$ **33.** $x = \frac{7}{18}$ **34.** $y = \frac{5}{3}$ or $1\frac{2}{3}$ **35.** $x = \frac{18ay - 3}{5a}$

Quick Quiz 4.3 *See Examples noted with Ex.* **1.** Speed of wind = 40 mph; speed of plane = 440 mph (Ex. 3) **2.** \$35/day; \$0.25/mi (Ex. 1) **3.** Drawings were \$5, carved elephants were \$15, and a set of drums was \$10. (Ex. 4) **4.** See Student Solutions Manual.

4.4 Exercises **1.** They would be dashed. The boundary lines are not included in the solution of a system of inequalities whenever the system contains only < or > symbols. **3.** She could substitute (3, −4) into each inequality. Since (3, −4) does not satisfy the second inequality, we know that the solution does not contain that ordered pair. Therefore, that point would not lie in the shaded region.

5.

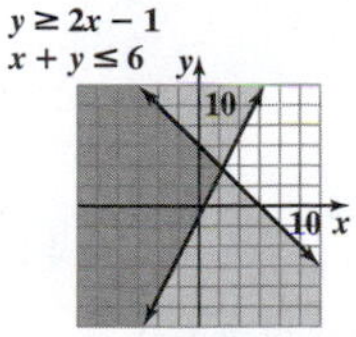

7.

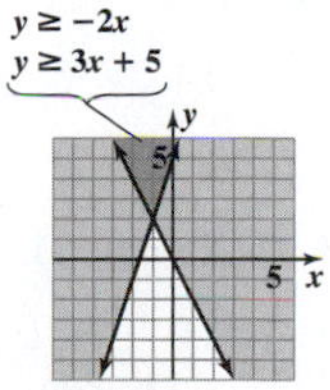

9.

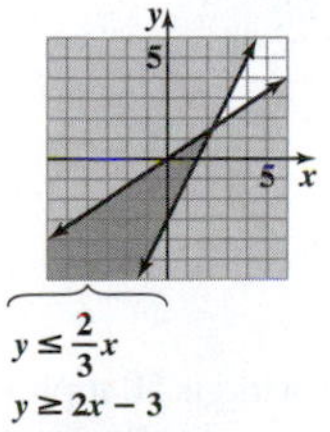

11.

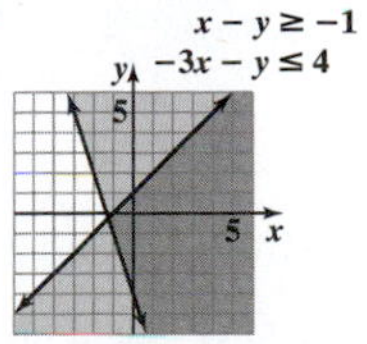

13.

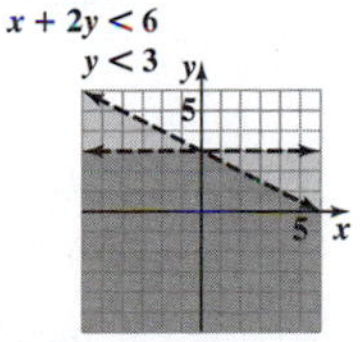

15.

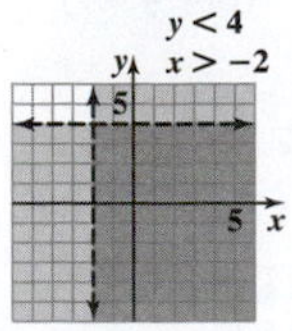

17.

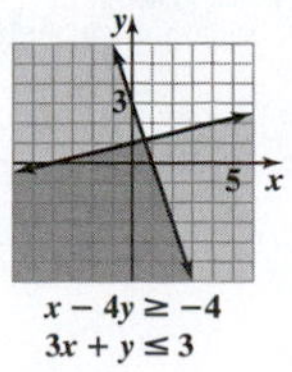

19.

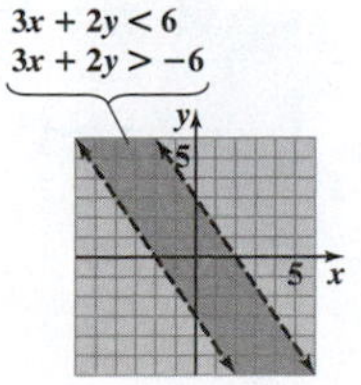

21.

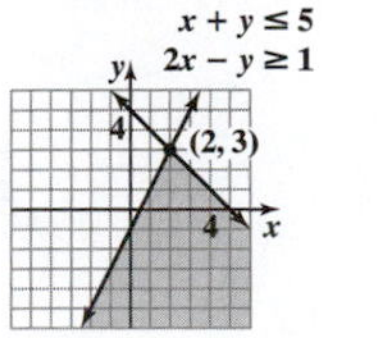

23.

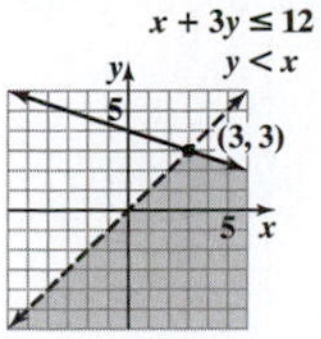

25.

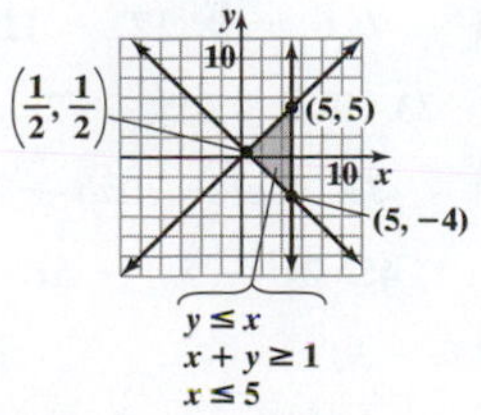

27.

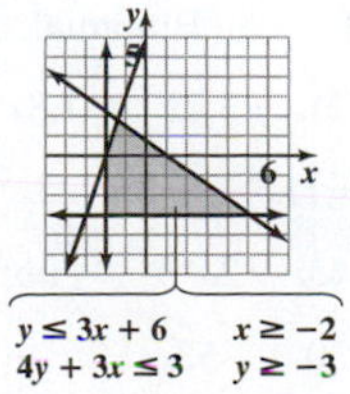

29. (a)

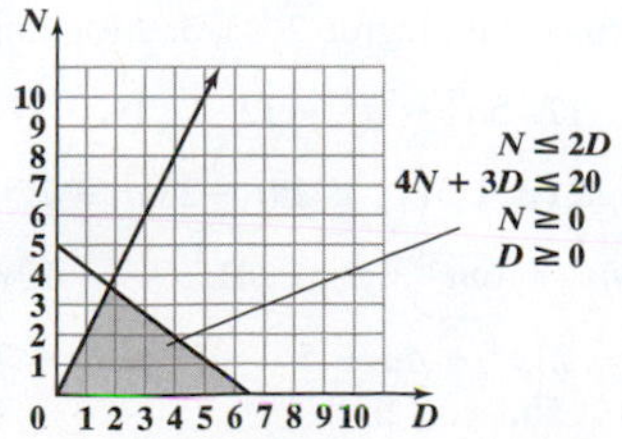

(b) Yes **(c)** No **31.** 13 **32.** $-4x + 4y$ **33.** The driving range takes in \$1250 on a rainy day and \$800 on a sunny day. **34.** One chicken sandwich costs \$3.25. One side salad costs \$2.50. One soda costs \$0.85.

Quick Quiz 4.4

See Examples noted with Ex. **1.** Below the line (Ex. 2) **2.** Solid lines (Ex. 1)
3. (Ex. 2)

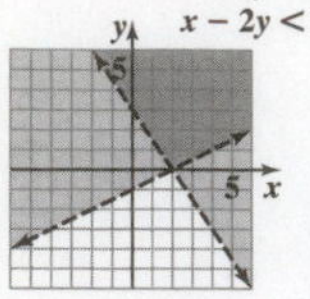

4. See Student Solutions Manual.

Use Math to Save Money

1. Job 1: \$50,310/52 = \$967.50; Job 2: \$20.50 × 30 = \$615 **2.** Job 1: \$967.50 × 0.15 ≈ \$145.13; Job 2: \$615 × 0.12 = \$73.80 **3.** Job 1: \$175 + \$75 + \$25 + \$50 + (\$0.35 × 45 × 5) = \$403.75; Job 2: \$150 + \$50 + \$25 + \$25 + (\$0.35 × 5 × 5) = \$258.75 **4.** Job 1: \$967.50 − (\$145.13 + \$403.75) = \$418.62; Job 2: \$615 − (\$73.80 + \$258.75) = \$282.45 **5.** Job 1: \$418.62/50 ≈ \$8.37; Job 2: \$282.45/31 ≈ \$9.11 **6.** She should take Job 2. **7.** She should take Job 1.

You Try It

1.

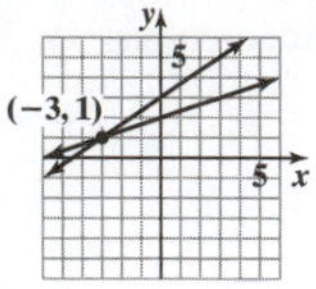

The solution is (−3, 1). **2.** (2, −1) **3.** (10, 7) **4.** No solution **5.** Infinite number of solutions **6.** (−4, −2, 5) **7.**

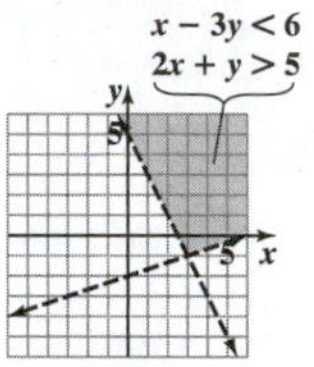

Chapter 4 Review Problems

1.

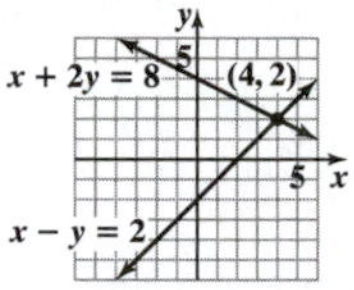

2.

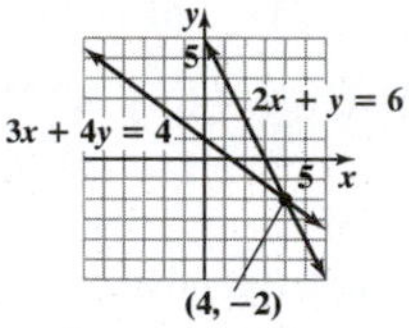

3. (−1, 3) **4.** (3, −2) **5.** (−4, 0) **6.** (1, −2) **7.** (2, 3) **8.** (5, −11) **9.** No solution; inconsistent system of equations **10.** Infinite number of solutions; dependent equations **11.** (1, −5) **12.** (0, 3) **13.** $\left(\frac{4}{3}, -\frac{1}{2}\right)$ **14.** $\left(0, \frac{2}{3}\right)$ **15.** No solution; inconsistent system of equations **16.** (5, 2) **17.** (1, 1, −2) **18.** (1, −2, 3) **19.** (5, −3, 8) **20.** $\left(7, \frac{1}{2}, -3\right)$ **21.** (4, −2, 0) **22.** (1, 2, −4) **23.** Speed of plane in still air = 264 mph; speed of wind = 24 mph **24.** 9 touchdowns; 4 field goals **25.** Laborers = 15; mechanics = 10 **26.** Children's tickets = 215; adult tickets = 115 **27.** Hats = \$3; shirts = \$15; pants = \$12 **28.** Jess scored 92; Nick scored 85; Chris scored 72 **29.** One jar of jelly = \$1.50; one jar of peanut butter = \$2.80; one jar of honey = \$1.90 **30.** Buses = 2; station wagons = 4; sedans = 3 **31.** $\left(\frac{4}{3}, \frac{1}{3}\right)$ **32.** (−3, 2) **33.** (0, 2) **34.** (0, 12) **35.** (2, 5) **36.** (11, 6) **37.** (3, 2) **38.** (−3, −2, 2) **39.** (3, −1, −2) **40.** (5, −5, −20)

41.

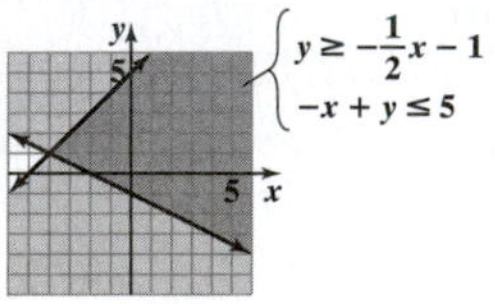

42.

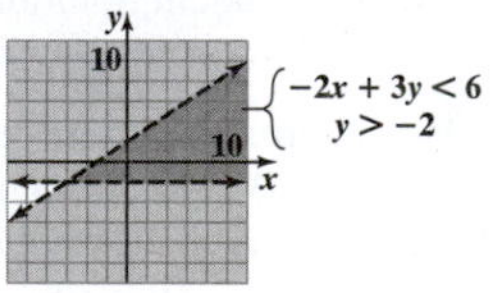

43.

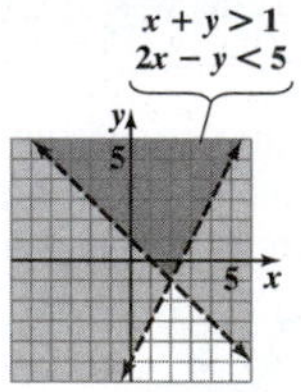

44.

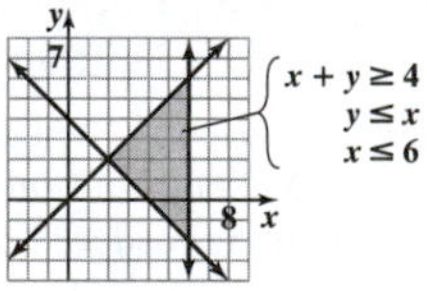

How Am I Doing? Chapter 4 Test

1. (10, 7) (obj. 4.1.3) **2.** (3, −4) (obj. 4.1.4) **3.** (3, 4) (obj. 4.1.3) **4.** $\left(\frac{1}{2}, \frac{3}{2}\right)$ (obj. 4.1.4) **5.** (1, 2) (obj. 4.1.4) **6.** No solution; inconsistent system of equations (obj. 4.1.5) **7.** (2, −1, 3) (obj. 4.2.2) **8.** (−2, 3, 5) (obj. 4.2.3) **9.** (−4, 1, −1) (obj. 4.2.2) **10.** Speed of plane in still air is 450 mph; speed of wind is 50 mph (obj. 4.3.1) **11.** Pen = \$1.50, mug = \$4.00, T-shirt = \$10.00 (obj. 4.3.2) **12.** \$30/day; \$0.20/mi (obj. 4.3.1) **13.** (obj. 4.4.1)

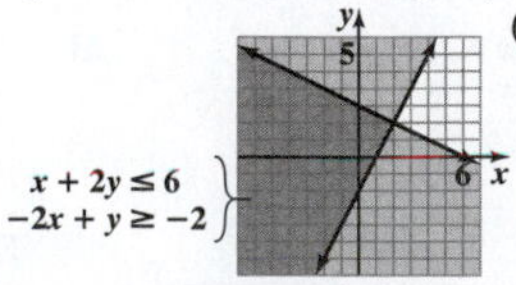

14. (obj. 4.4.1)

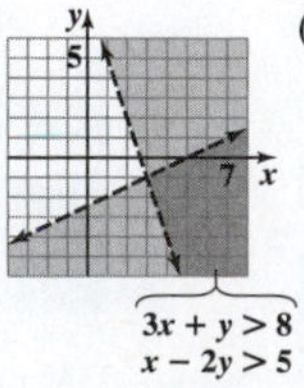

Chapter 5

5.1 Exercises

1. Trinomial, degree 2 **3.** Monomial, degree 8 **5.** Binomial, degree 4 **7.** 6 **9.** 17 **11.** 6 **13.** $-3x - 6$ **15.** $10m^3 + 4m^2 - 6m - 1.3$ **17.** $5a^3 - a^2 + 12$ **19.** $\frac{7}{6}x^2 + \frac{14}{3}x$ **21.** $-3.2x^3 + 1.8x^2 - 4$ **23.** $10x^2 + 61x + 72$ **25.** $15aw - 20bw + 6ad - 8bd$ **27.** $-12x^2 + 32xy - 5y^2$ **29.** $-28ar - 77rs^2 + 4as^2 + 11s^4$ **31.** $25x^2 - 64y^2$ **33.** $25a^2 - 20ab + 4b^2$ **35.** $49m^2 - 14m + 1$ **37.** $36 - 25x^4$ **39.** $9m^6 + 6m^3 + 1$ **41.** $6x^3 - 10x^2 + 2x$ **43.** $-2a^2b + \frac{5}{2}ab^2 + 5ab$ **45.** $2x^3 - 5x^2 + 5x - 3$ **47.** $6x^3 - 7x^2y - 10xy^2 + 6y^3$ **49.** $3a^4 - 11a^3 - 21a^2 + 4a + 5$ **51.** $2x^3 - 7x^2 - 7x + 30$ **53.** $-2a^3 - 3a^2 + 29a - 30$ **55.** $(4x^3 + 12x^2 + 13x + 20)$ ft^3 **57.** $x \geq 30$ **58.** $x = 0$

Quick Quiz 5.1 *See Examples noted with Ex.* **1.** $-3x^2 + 21x - 10$ (Ex. 4) **2.** $25x^2 - 20xy^2 + 4y^4$ (Ex. 8) **3.** $8x^3 - 8x^2 - 8x + 3$ (Ex. 9) **4.** See Student Solutions Manual

5.2 Exercises **1.** $6x^2 - 2x - 11$ **3.** $-3x^3 + x^2 - 7x$ **5.** $4b^2 - 3b - \frac{1}{2}$ **7.** $3ab^2 - 5ab + 1$ **9.** $5x - 2$ **11.** $3x + 4$ **13.** $5x - 3$ **15.** $x^2 - 2x + 13 - \frac{14}{x+1}$ **17.** $2x^2 + 3x + 6 + \frac{5}{x-2}$ **19.** $2x^2 - 4x + 2 - \frac{5}{2x+1}$ **21.** $x^3 + 4x + 6$ **23.** $2t^2 - 3t + 2$ **25.** $(3x - 1)$ m **27.** $m = -\frac{6}{5}$ **28.** $m = -\frac{3}{2}$ **29.** $x = \frac{11 - 4y}{2}$ **30.** $x = \frac{9y + 2}{2}$

Quick Quiz 5.2 *See Examples noted with Ex.* **1.** $-4x^2 + 5x + 2$ (Ex. 1) **2.** $x^2 - 5x + 2$ (Ex. 2) **3.** $x^3 - 2x^2 - 3x - 4$ (Ex. 4) **4.** See Student Solutions Manual

5.3 Exercises **1.** $2x + 1 + \frac{-2}{x-6}$ **3.** $3x^2 - 2x + 1 + \frac{3}{x+1}$ **5.** $x^2 + 4x + 5$ **7.** $4x^2 + 5x + \frac{5}{x-4}$ **9.** $x^2 - 4x + 8 + \frac{-8}{x+2}$ **11.** $6x^3 - 5x^2 + 15x - 10 + \frac{6}{x+3}$ **13.** $2x^3 + x^2 + 2 + \frac{3}{x+1}$ **15.** $2x^3 + 2x^2 + 2x - 3 + \frac{-6}{x-1}$ **17.** $2x^4 - 6x^3 + 5x^2 - 5x + 15 + \frac{-39}{x+3}$ **19.** $x^5 - x^4 + x^3 - 6x^2 + 7x - 7 + \frac{19}{x+1}$ **21.** $2x^2 - 6x + 9 + \frac{-21}{2x+3}$ **23.** 308,200 ft^3 **24.** 35 mi/hr **25.** 116

Quick Quiz 5.3 *See Examples noted with Ex.* **1.** $x^2 - 3x - 2$ (Ex. 1) **2.** $x^2 - 6x - 12$ (Ex. 2) **3.** $x^3 - 3x^2 + 2x - 4$ (Ex. 3) **4.** See Student Solutions Manual

5.4 Exercises **1.** $10(8 - y)$ **3.** $5a(a - 5)$ **5.** $4a(ab^3 - 2b + 8)$ **7.** $6y^2(5y^2 + 4y + 3)$ **9.** $5ab(3b + 1 - 2a^2)$ **11.** $10a^2b^2(b - 3ab + a - 4a^2)$ **13.** $(x + y)(3x - 2)$ **15.** $(a - 3b)(5b + 8)$ **17.** $(a + 5b)(3x + 1)$ **19.** $(3x - y)(2a^2 - 5b^3)$ **21.** $(5x + y)(3x - 8y - 1)$ **23.** $(a - 6b)(2a - 3b - 2)$ **25.** $(x + 5)(x^2 + 3)$ **27.** $(x + 3)(2 - 3a)$ **29.** $(b - 4)(a - 3)$ **31.** $(x - 4)(5 + 3y)$ **33.** $(3y - 2)(3 - x)$ **35.** $(ab + 1)(b^2 + c)$ **37.** $x\left(\frac{1}{3}x^2 + \frac{1}{2}x + \frac{1}{6}\right)$ or $\frac{1}{6}x(2x^2 + 3x + 1)$

39.

x	y
-2	0
-1	3
0	6

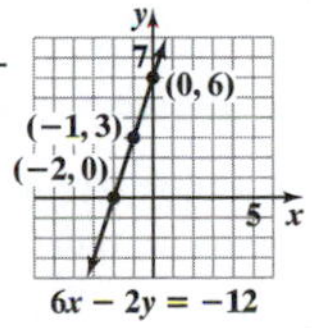

40.

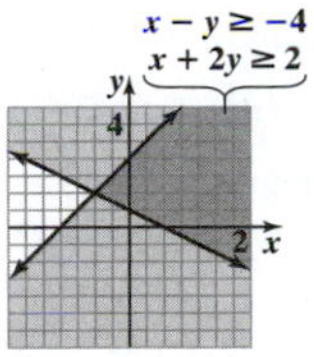

41. $m = -3$ **42.** $m = -3, \left(0, -\frac{3}{2}\right)$

Quick Quiz 5.4 *See Examples noted with Ex.* **1.** $5a^2b^3(3 - 2a - 5b)$ (Ex. 3) **2.** $(5x - 3y)(2x - 5)$ (Ex. 4) **3.** $(3c - 4b)(a - 5b)$ (Ex. 7) **4.** See Student Solutions Manual

5.5 Exercises **1.** $(x + 7)(x + 1)$ **3.** $(x - 5)(x - 3)$ **5.** $(x - 6)(x - 4)$ **7.** $(a + 9)(a - 5)$ **9.** $(x + 7y)(x + 6y)$ **11.** $(x - 14y)(x - y)$ **13.** $(x^2 - 8)(x^2 + 5)$ **15.** $(x^2 + 7y^2)(x^2 + 9y^2)$ **17.** $2(x + 11)(x + 2)$ **19.** $x(x + 5)(x - 4)$ **21.** $(2x + 1)(x - 1)$ **23.** $(3x - 5)(2x + 1)$ **25.** $(3a - 5)(a - 1)$ **27.** $(4a - 7)(a + 2)$ **29.** $(2x + 3)(x + 5)$ **31.** $(3x^2 + 1)(x^2 - 3)$ **33.** $(3x + y)(2x + 11y)$ **35.** $(7x - 3y)(x + 2y)$ **37.** $x(4x + 1)(2x - 1)$ **39.** $5x^2(2x + 1)(x + 1)$ **41.** $(x - 9)(x + 7)$ **43.** $(3x + 2)(2x - 1)$ **45.** $(x - 17)(x - 3)$ **47.** $(5x + 2)(3x - 1)$ **49.** $2(x + 8)(x - 6)$ **51.** $3(3x + 2)(2x + 1)$ **53.** $8a(5x - 1)(x + 2)$ **55.** $2x(3x - 2)(x + 5)$ **57.** $(7x^2 - 1)(x^2 + 2)$ **59.** $(3a + b)(3a - 7b)$ **61.** $(x^3 - 13)(x^3 + 3)$ **63.** $5xy(2x + 1)(x - 2)$ **65.** 28.26 in.^2 **66.** $\frac{3A - 4a}{3} = b$ **67.** 24 first-class seats and 160 coach seats

68.

x	y
-2	0
0	-3
2	-6

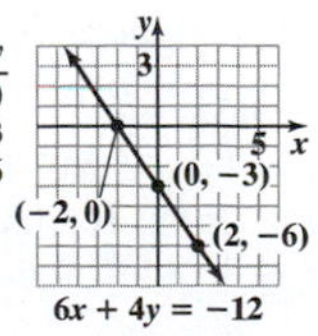

Quick Quiz 5.5 *See Examples noted with Ex.* **1.** $(x - 7)(x + 12)$ (Ex. 2) **2.** $(3x - 4)(2x - 3)$ (Ex. 8) **3.** $2x(x + 4)(2x - 5)$ (Ex. 9) **4.** See Student Solutions Manual

How Am I Doing? Sections 5.1–5.5 **1.** $x^2 - 11x + 4$ (obj. 5.1.3) **2.** $2x^3 - 7x^2 - 7x + 30$ (obj. 5.1.7) **3.** $5a^2 - 43a + 56$ (obj. 5.1.4) **4.** $9y^2 - 25$ (obj. 5.1.5) **5.** $9x^4 + 24x^2 + 16$ (obj. 5.1.6) **6.** $p(-3) = -80$ (obj. 5.1.2) **7.** $5x - 6y - 10$ (obj. 5.2.1) **8.** $3y^2 + y + 4 + \frac{7}{y-2}$ (obj. 5.2.2) **9.** $2x^3 - x^2 + 4x - 2$ (obj. 5.2.2) **10.** $2x^3 + 4x^2 - x - 3$ (obj. 5.3.1) **11.** $12a^3b^2(2 + 3a - 5b)$ (obj. 5.4.1) **12.** $(4x - 3y)(3x - 2)$ (obj. 5.4.1) **13.** $(5w + 3z)(2x - 5y)$ (obj. 5.4.2) **14.** $(3a + b)(6a - 5)$ (obj. 5.4.2) **15.** $(x - 5)(x - 2)$ (obj. 5.5.1) **16.** $(2y - 5)(2y + 3)$ (obj. 5.5.2) **17.** $(7x - 3y)(4x - y)$ (obj. 5.5.2) **18.** $(2x + 5)(x + 8)$ (obj. 5.5.2) **19.** $3(x + 4)(x - 6)$ (obj. 5.5.2) **20.** $4x(2x + 1)(3x + 4)$ (obj. 5.5.2)

5.6 Exercises **1.** The problem will have two terms. It will be in the form $a^2 - b^2$. One term is positive and one term is negative. The coefficients and variables for the first and second terms are both perfect squares. So each one will be of the form 1, 4, 9, 16, 25, 36, and/or x^2, x^4, x^6, etc. **3.** There will be two terms added together. It will be of the form $a^3 + b^3$. Each term will contain a number or variable cubed or both. They will be of the form 1, 8, 27, 64, 125, and/or x^3, x^6, x^9, etc. **5.** $(a + 8)(a - 8)$ **7.** $(4x + 9)(4x - 9)$ **9.** $(8x + 1)(8x - 1)$ **11.** $(7m + 3n)(7m - 3n)$ **13.** $(10y + 9)(10y - 9)$ **15.** $(1 + 6xy)(1 - 6xy)$ **17.** $3(3x + 5)(3x - 5)$ **19.** $3x(1 + 3x)(1 - 3x)$ **21.** $(3x - 1)^2$ **23.** $(7x - 1)^2$ **25.** $(9w + 2t)^2$ **27.** $(6x + 5y)^2$ **29.** $2(2x + 5)^2$ **31.** $3x(x - 4)^2$ **33.** $(x - 3)(x^2 + 3x + 9)$ **35.** $(x + 5)(x^2 - 5x + 25)$ **37.** $(4x - 1)(16x^2 + 4x + 1)$ **39.** $(2x - 5)(4x^2 + 10x + 25)$ **41.** $(1 - 3x)(1 + 3x + 9x^2)$ **43.** $(4x + 5)(16x^2 - 20x + 25)$ **45.** $(4s^2 + t^2)(16s^4 - 4s^2t^2 + t^4)$ **47.** $5(y - 2)(y^2 + 2y + 4)$ **49.** $2(5x + 1)(25x^2 - 5x + 1)$ **51.** $x^2(x - 2y)(x^2 + 2xy + 4y^2)$ **53.** $(5w^2 + 1)(5w^2 - 1)$ **55.** $(b^2 + 3)^2$ **57.** $(3m^3 + 8)(3m^3 - 8)$ **59.** $(6y^3 - 5)^2$ **61.** $5(3z^4 + 1)(3z^4 - 1)$ **63.** $(5m + 2n)(25m^2 - 10mn + 4n^2)$ **65.** $3(2a - b)(4a^2 + 2ab + b^2)$ **67.** $(2w - 5z)^2$ **69.** $9(2a + 3b)(2a - 3b)$ **71.** $(4x^2 + 9y^2)(2x + 3y)(2x - 3y)$ **73.** $(5m^2 + 2)(25m^4 - 10m^2 + 4)$ **75.** $(5x + 4)(5x + 1)$ **77.** $(7x - 4)(7x - 1)$ **79.** $A = (4x + y)(4x - y)$ ft^2 **81.** During the year 2020 **82.** Ryan paid \$45; Megan paid \$145; Grant paid \$175

Quick Quiz 5.6 *See Examples noted with Ex.* **1.** $(5x + 8y)(5x - 8y)$ (Ex. 2) **2.** $(7x - 4y)^2$ (Ex. 8) **3.** $(4x - 3y)(16x^2 + 12xy + 9y^2)$ (Ex. 11) **4.** See Student Solutions Manual

5.7 Exercises **1.** To factor out a common factor if possible **3.** You cannot factor polynomials of the form $a^2 + b^2$. All such polynomials that are the sum of two squares are prime. You can factor polynomials of the form $a^2 - b^2$, but you do NOT have that form in this problem. **5.** $3y(x - 2z)$ **7.** $(y - 2)(y + 9)$ **9.** $(3x - 5)(x - 1)$ **11.** $(2x + 5)^2$ **13.** $(2x - 5y)(4x^2 + 10xy + 25y^2)$ **15.** $a(a^2 - 3b - c)$ **17.** Prime **19.** $(8y + 5z)(8y - 5z)$ **21.** $(6x + 1)(x - 4)$ **23.** Prime **25.** $x(x + 7)(x + 2)$ **27.** $(5x - 4)^2$ **29.** $6(a - 3)(a + 2)$ **31.** $(x - 1)(3x - y)$ **33.** $(9a^2 + 1)(3a + 1)(3a - 1)$ **35.** $2x(x + 3)(x - 3)(x^2 + 1)$ **37.** $2ab(2a + 5b)(2a - 5b)$ **39.** $(a - 4y)(2x + w)$ **41.** $(x - 10)(5x + 8y)$ ft^2 $= (5x^2 - 50x + 8xy - 80y)$ ft^2 **43.** $x \le -9$ **44.** $-\frac{1}{5} < x < \frac{3}{5}$ **45.** $x < -\frac{7}{4}$ *or* $x > \frac{17}{4}$ **46.** $x \ge 11$ *or* $x \le 4$

Quick Quiz 5.7 *See Examples noted with Ex.* **1.** $x(7x + 6y)^2$ (Ex. 1) **2.** $(2x^2 + 3)(3x^2 - 4)$ (Ex. 1) **3.** $3(x + 2)(3x - 2)$ (Ex. 1) **4.** See Student Solutions Manual

5.8 Exercises **1.** $x = 3, x = -2$ **3.** $x = 0, x = \frac{6}{5}$ **5.** $x = -\frac{4}{3}, x = \frac{4}{3}$ **7.** $x = -\frac{4}{3}, x = 2$ **9.** $x = \frac{3}{4}, x = -\frac{1}{2}$ **11.** $x = \frac{3}{8}, x = 1$ **13.** $x = 0, x = -1$ **15.** $x = -\frac{1}{5}$, double root **17.** $x = 0, x = -4, x = -3$ **19.** $x = 0, x = 8, x = -6$ **21.** $x = 0, x = -3, x = 3$ **23.** $x = 0, x = -5, x = 3$ **25.** $x = -\frac{3}{7}, x = 1$ **27.** $x = -\frac{3}{2}, x = 4$ **29.** $x = 0, x = \frac{8}{5}$ **31.** $c = -2$, the other solution is $x = 2$ **33.** Altitude is 18 in.; base is 20 in. **35. (a)** Altitude is 8 ft; base is 26 ft **(b)** Altitude is $2\frac{2}{3}$ yd; base is $8\frac{2}{3}$ yd **37. (a)** Width is 28 cm; length is 32 cm **(b)** Width is 280 mm; length is 320 mm **39.** 12 feet **41.** Length is 9 in.; height is 11 in. **43.** Width is 6 mi; length is 9 mi **45.** \$311,000,000,000 **47.** 2004 **49.** $200x^{11}y^{10}$ **50.** $\frac{a^4}{2b^2}$ **51.** $(1, 2, -2)$ **52.** $y = -x + 7$

Quick Quiz 5.8 *See Examples noted with Ex.* **1.** $x = -8, x = 4$ (Ex. 1) **2.** $x = \frac{2}{5}, x = \frac{1}{3}$ (Ex. 2) **3.** Altitude = 16 yd; base = 20 yd (Ex. 6) **4.** See Student Solutions Manual

Use Math to Save Money **1.** $(4 \times \$8) + \$13 + (5 \times \$10) = \$95$ **2.** $10 \times \$3 = \$30$ **3.** $\$95 - \$30 = \$65$ **4.** $\$2275/\$65 = 35$; after 35 weeks.

You Try It **1. (a)** $5a^2 - 10a + 12$ **(b)** $-a^2 + 2ab - 12b^2$ **2. (a)** $15x^4 + 3x^3 - 6x^2 - 12x$ **(b)** $9x^2 - 21x + 6$ **(c)** $2x^3 + 7x^2 + 5x - 2$ **3.** $-6a^3 + a^2 + 3a - 3$ **4.** $4x^2 - x + 3 + \frac{5}{2x - 1}$ **5.** $2x^3 + 3x^2 + x + 3 + \frac{8}{x - 3}$ **6. (a)** $6a(2a^3 - 3a - 4)$ **(b)** $7xy^3(2xy + 3x^2y^2 + 4)$ **7.** $(4a - 3)(5b - c)$ **8. (a)** $(x + 6)(x + 3)$ **(b)** $3(x - 3)(x - 4)$ **(c)** $(x - 8)(x + 10)$ **(d)** $2(x - 5)(x + 4)$ **9. (a)** $(3x + 2)(x + 1)$ **(b)** $3(2a - 1)(a - 5)$ **(c)** $(6x - 5)(x + 1)$ **(d)** $y(4y + 1)(y - 4)$ **10. (a)** $(5 + 2a)(5 - 2a)$ **(b)** $3(4c + 1)(4c - 1)$ **11. (a)** $(2x + 3)^2$ **(b)** $3(5a + 2b)^2$ **(c)** $(3a - 8)^2$ **(d)** $x(6x - y)^2$ **12. (a)** $(3x + 1)(9x^2 - 3x + 1)$ **(b)** $2(y + 4x)(y^2 - 4xy + 16x^2)$ **(c)** $(a - 5)(a^2 + 5a + 25)$ **(d)** $3(2a - 3b)(4a^2 + 6ab + 9b^2)$ **13.** $x = -2, x = 5$

Chapter 5 Review Problems **1.** $-x^2 - 10x + 13$ **2.** $x^2y - 5xy - 8y$ **3.** $-11x^2 + 10xy + 6y^2$ **4.** $-11x^2 + 15x - 15$ **5.** $8x - 1$ **6.** $-2x^2 + 3x - 7$ **7.** -199 **8.** 2 **9.** 46 **10.** 3 **11.** -25 **12.** 5 **13.** $3x^3y - 3x^2y^2 + 3xy^3$ **14.** $6x^3 - 3x^2 + 2x - 1$ **15.** $25x^4 + 30x^2 + 9$ **16.** $2x^3 - 7x^2 - 7x + 30$ **17.** $-2x^4 + 7x^3 - 7x^2 + 7x - 2$ **18.** $9x^3 - 9x^2 - 22x + 20$ **19.** $25a^2b^2 - 4$ **20.** $18a^2 - 21ab^2 + 5b^4$ **21.** $-5x^2 + 3x + 20$ **22.** $4a^3 - 7a^2 - \frac{1}{2}a$ **23.** $4x + 1$ **24.** $x^2 - x + 1 - \frac{2}{2x + 3}$ **25.** $2x^3 + 2x^2 + x + 7 + \frac{10}{x - 1}$ **26.** $2x^3 - 3x^2 + x - 4$ **27.** $3x^3 - x^2 + x - 1$ **28.** $3y^2 + 9y + 25 + \frac{80}{y - 3}$ **29.** $5ab(3a + b - 2)$ **30.** $x^2(x^3 - 3x^2 + 2)$ **31.** $4m(3n - 2)$ **32.** $(x - 6)(y + 3)$ **33.** $(8y + b)(x^2 + 1)$ **34.** $(b - 5)(3a - 2)$ **35.** $(x - 11)(x + 2)$ **36.** $(4x + 3)(x - 2)$ **37.** $(3x + 7)(2x - 3)$ **38.** $(10x + 7)(10x - 7)$ **39.** $(2x - 7)^2$ **40.** $(2a - 3)(4a^2 + 6a + 9)$ **41.** $(3x + 11)(3x - 11)$ **42.** $(5x - 1)(x - 2)$ **43.** $x(x + 6)(x + 2)$ **44.** $(x + 4w)(x - 2y)$ **45.** Prime **46.** Prime

47. $x(3x - 1)(9x^2 + 3x + 1)$ **48.** $-a^2b^3(3a - 2b + 1)$ or $a^2b^3(-3a + 2b - 1)$ **49.** $(3x^2 + 1)(x^2 - 2)$ **50.** $b(3a + 7)(3a - 2)$ **51.** Prime **52.** $y^2(4y - 9)(y - 1)$ **53.** $5x^2(2y^2 - 4y + 1)$ **54.** $(a + b^3)(a + 4b^3)$ **55.** $(2x + 3)(x + 2)(x - 2)$ **56.** $2(x + 3)(x - 3)(x^2 + 3)$ **57.** $4(a + b)(2 - x)$ **58.** $2x(2x - 1)(x + 3)$ **59.** $(3x^2y - 5)^2$ **60.** $3xy(3x + 1)(3x - 1)$ **61.** $(5x + 4y)(b - 7)$ **62.** $x = -\frac{1}{5}, x = 2$ **63.** $x = \frac{3}{2}, x = 4$ **64.** $x = -\frac{4}{3}, x = -\frac{3}{2}$ **65.** $x = 0, x = 4$ **66.** $x = 0, x = -\frac{10}{3}$ **67.** $x = 0, x = -3, x = -4$ **68.** Base is 10 m; altitude is 15 m **69.** Width is 5 ft; length is 9 ft **70.** 5 calculators **71.** Old side is 1 yd; new side is 5 yd

How Am I Doing? Chapter 5 Test

1. $-4x^2y - 1$ (obj. 5.1.3) **2.** $5a^2 - 9a - 2$ (obj. 5.1.3) **3.** $-2x^2 - 6xy + 8x$ (obj. 5.1.7) **4.** $4x^2 - 12xy^2 + 9y^4$ (obj. 5.1.6) **5.** $2x^3 - 3x^2 - 3x + 2$ (obj. 5.1.7) **6.** $5x^2 + 4x - 7$ (obj. 5.2.1) **7.** $x^3 - x^2 + x - 2$ (obj. 5.2.2) **8.** $x^2 - 3x + 1$ (obj. 5.2.2) **9.** $x^3 - 1 - \frac{2}{x + 1}$ (obj. 5.3.1) **10.** $(11x + 5y)(11x - 5y)$ (obj. 5.6.1) **11.** $(3x + 5y)^2$ (obj. 5.6.2) **12.** $x(x - 2)(x - 24)$ (obj. 5.7.1) **13.** $4x^2y(x + 2y + 1)$ (obj. 5.4.1) **14.** $(x + 3y)(x - 2w)$ (obj. 5.4.2) **15.** Prime (obj. 5.7.2) **16.** $3(6x - 5)(x + 1)$ (obj. 5.5.2) **17.** $2a(3a - 2)(9a^2 + 6a + 4)$ (obj. 5.7.1) **18.** $x(3x^2 - y)^2$ (obj. 5.7.1) **19.** $(3x^2 + 2)(x^2 + 5)$ (obj. 5.7.1) **20.** $(x + 2y)(3 - 5a)$ (obj. 5.7.1) **21.** -18 (obj. 5.1.2) **22.** 17 (obj. 5.1.2) **23.** $x = -2, x = 7$ (obj. 5.8.1) **24.** $x = -\frac{1}{3}, x = 4$ (obj. 5.8.1) **25.** $x = 0, x = \frac{2}{7}$ (obj. 5.8.1) **26.** Base is 14 in.; altitude is 10 in. (obj. 5.8.2)

Chapter 6

6.1 Exercises

1. All real numbers except 3 **3.** All real numbers except -8 and 3 **5.** $-\frac{3x^2}{2y^5}$ **7.** $\frac{x^2}{2}$ **9.** $\frac{3x}{4x - 5}$ **11.** $\frac{y(x - 3)}{2x^2(1 - 2y)}$ **13.** $\frac{1}{x - 5}$ **15.** $y - 3$ **17.** $-\frac{x + 2}{x}$ **19.** $-\frac{2y + 5}{2 + y}$ **21.** $-\frac{4n^4}{m}$ **23.** $\frac{a(a - 2)}{a + 2}$ **25.** $\frac{3}{x - 3}$ **27.** $(x - 5y)(x + 6y)$ **29.** $\frac{y + 3}{2(y + 1)}$ **31.** $\frac{y(x - 5)}{x^2}$ **33.** $\frac{1}{5}$ **35.** $\frac{b - 3}{2b - 1}$ **37.** $\frac{x - 3y}{x(x^2 + 2)}$ **39.** $\frac{1}{3x^2y^2}$ **41.** $\frac{x}{3}$ **43.** $\frac{3(x + 7)}{xy}$ **45.** $\frac{x - 2}{xy^2(x + 1)}$ **47.** Cannot be simplified **49.** All real numbers except $x \approx -1.4$ and $x \approx 0.9$ **51.** 150 fish **53.** 225 fish

55.

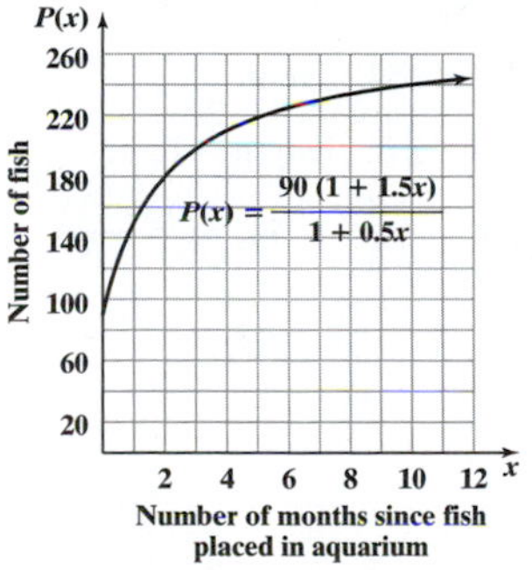

57. $x > -3$ **58.** $16x^2 - 8x + 1$ **59.** About 145%; 3121 people

Quick Quiz 6.1

See Examples noted with Ex. **1.** $\frac{x + 5}{3x - 1}$ (Ex. 4) **2.** 6 (Ex. 7) **3.** $-\frac{3(x - 2)}{x + 5}$ (Ex. 9) **4.** See Student Solutions Manual

6.2 Exercises

1. The factors are 5, x, and y. The factor y is repeated in one fraction three times. Since the highest power of y is 3, the LCD is $5xy^3$. **3.** $(x - 1)^2$ **5.** $2m^3n^2$ **7.** $(x + 1)(2x + 5)^3$ **9.** $x(3x + 2)(6x - 1)$ **11.** $\frac{3x - 10}{(x + 4)(x - 4)}$ **13.** $\frac{3(3y + x)}{4xy^2}$ **15.** $\frac{8x - 15}{x(x - 4)(x - 3)}$ **17.** $\frac{14x - 20}{2x - 5}$ **19.** $\frac{-5y^2 + 11y + 6}{(y + 1)(y - 1)^2}$ **21.** $\frac{a^2 + 9a + 5}{3(a + 2)(a - 2)}$ **23.** $\frac{2x + 17}{(x - 4)(x + 1)}$ **25.** $\frac{-2x + 7}{(x + 1)^2(x - 2)}$ **27.** $\frac{-y^2 + 6y + 3}{(y + 1)(y + 2)}$ **29.** $\frac{2a^2 - a - 7}{2a - 5}$ **31.** $P(x) = \frac{12x^2 - 68x + 120}{x^2 - 5x + 6}$ **33.** \$11,429 **35.** $-\frac{2}{7}$ **36.** 3.51×10^{-4} **37.** Amanda's car cost \$5500; Denise's car cost \$7500; Jamie's car cost \$11,000 **38.** 40 L of 15% and 20 L of 30%

Quick Quiz 6.2

See Examples noted with Ex. **1.** $\frac{x + 15}{x(x + 3)}$ (Ex. 5) **2.** $\frac{7x + 19}{(x + 1)(x + 2)(x + 4)}$ (Ex. 7) **3.** $\frac{7(3x - 5)}{3x - 7}$ (Ex. 4) **4.** See Student Solutions Manual

6.3 Exercises

1. $\frac{7y}{3}$ **3.** $\frac{2(x - 1)}{x(x + 5)}$ **5.** $\frac{3y - 4}{3(y + 2)}$ **7.** $\frac{y^2 - 3}{3}$ **9.** $\frac{2}{(y - 3)(y + 6)}$ **11.** $\frac{(x + 5)(x - 2)}{2(x^2 - 2x + 2)}$ **13.** $-\frac{4(x - 1)}{x + 2}$ **15.** $\frac{x(2x + 5)}{6(2x + 1)}$ **17.** $-\frac{12 + x + y}{3y}$ **19.** $\frac{1}{x(x - a)}$ **21.** $x = -\frac{2}{3}$ or $x = 2$ **22.** $-\frac{3}{5} < x < \frac{9}{5}$ **23.** \$12,000 at 9%; \$8000 at 5% **24.** 7

Quick Quiz 6.3

See Examples noted with Ex. **1.** $\frac{9y}{14x}$ (Ex. 1) **2.** $\frac{-2x-5}{4x+13}$ (Ex. 3) **3.** $\frac{2x-3}{5}$ (Ex. 4) **4.** See Student Solutions Manual

How Am I Doing? Sections 6.1–6.3

1. $\frac{7x+3}{x+1}$ (obj. 6.1.1) **2.** $\frac{x+6}{x-2}$ (obj. 6.1.1) **3.** $\frac{2x+1}{x-5}$ (obj. 6.1.1) **4.** $\frac{2(9a-8)}{2a-5}$ (obj. 6.1.2) **5.** $\frac{15(x-5y)}{2y^4(x+5y)}$ (obj. 6.1.3) **6.** 3 (obj. 6.1.2) **7.** $\frac{x^2-5x+10}{2x(x-2)}$ (obj. 6.2.2) **8.** $\frac{12x+5}{(x+5)(x-5)}$ (obj. 6.2.2) **9.** $\frac{5y+4}{(y+2)(y+4)(y-4)}$ (obj. 6.2.2) **10.** $\frac{-3x}{(x+4)(x-4)}$ (obj. 6.2.2) **11.** $\frac{3x}{2}$ (obj. 6.3.1) **12.** $\frac{x}{(2x-1)(6x+1)}$ (obj. 6.3.1) **13.** $\frac{5+3x}{6-2x}$ (obj. 6.3.1) **14.** $\frac{x^2+x+3}{x^2+5x+12}$ (obj. 6.3.1)

6.4 Exercises

1. $3 = x$ **3.** $x = \frac{3}{5}$ or 0.6 **5.** $x = 6$ **7.** $y = -2$ **9.** No solution **11.** $x = 2$ **13.** $x = 3$ **15.** $x = -1$ **17.** $x = -3$ **19.** $y = 1$ **21.** $z = \frac{34}{3}$ or $11\frac{1}{3}$ **23.** $x = 0$ **25.** No solution **27.** $x = -5$ **29.** When the solved value of the variable causes the denominator of any fraction to equal 0, or when the variable drops out and leaves a false statement. **31.** $7(x+3)(x-3)$ **32.** $2(x+5)^2$ **33.** $(4x-3y)(16x^2+12xy+9y^2)$ **34.** $(3x-7)(x-2)$ **35.** $x^2+6xy+9y^2$ **36.** $2x^2+x+5+\frac{6}{x-2}$

Quick Quiz 6.4

See Examples noted with Ex. **1.** $x = -\frac{3}{2}$ or $-1\frac{1}{2}$ or -1.5 (Ex. 3) **2.** $x = 3$ (Ex. 2) **3.** $x = 1$ (Ex. 3) **4.** See Student Solutions Manual

6.5 Exercises

1. $m = \frac{y-b}{x}$ **3.** $b = \frac{af}{a-f}$ **5.** $h = \frac{V}{lw}$ **7.** $\frac{3G}{b+c} = a$ **9.** $V = \frac{4\pi r^3}{3}$ **11.** $\frac{Er}{R+r} = e$ **13.** $S_1 = \frac{R_1S_2}{R_2}$ **15.** $w = \frac{S-2lh}{2h+2l}$ **17.** $T_1 = \frac{ET_2}{T_2-1}$ **19.** $x_1 = \frac{mx_2-y_2+y_1}{m}$ **21.** $D = \frac{2Vt+at^2}{2}$ **23.** $b = \frac{2A-hB}{h}$ **25.** $T_2 = \frac{-qT_1}{W-q}$ **27.** $v_0 = \frac{s-s_0-gt^2}{t}$ **29.** $T \approx T_0 + 0.1702\left(\frac{V}{V_0}\right) - 0.1656$ **31.** Approximately 119.2 km **33.** Approximately 54.55 mph **35.** 80 grizzly bears **37.** 26 officers, 91 seamen **39.** 44 people in marketing; 143 people in sales **41.** Width is 12 in.; length is 15 in. **43.** 60 people prefer the new software. **45.** 6.4 ft high **47.** 3.6 hr **49.** 6 hr **51.** 64.5 ft **53.** $m = \frac{7}{3}; b = -\frac{8}{3}$ **54.** $m = -\frac{1}{5}$ **55.** $(-4, 1)$ **56.** $(2,1,-4)$

Quick Quiz 6.5

See Examples noted with Ex. **1.** $H = \frac{4AW-3W}{10A}$ (Ex. 1) **2.** 288 students (Ex. 3) **3.** Width = 66 in.; length = 105 in. (Ex. 4) **4.** See Student Solutions Manual

Use Math to Save Money

1. 750, 468.75, and 250 gallons respectively **2.** \$2700.00, \$1687.50, and \$900.00 respectively **3.** \$1012.50 **4.** About six and a half years **5.** \$5062.50, \$10,125.00 **6.** \$1800 **7.** About one year and three months **8.** \$9000, \$18,000

You Try It

1. $\frac{x+4}{2(x-3)}$ **2.** $\frac{x}{2}$ **3.** $\frac{5x+1}{3x+1}$ **4.** $\frac{5}{5x-4}$ **5.** $\frac{3x^2+9x+4}{(x-4)(2x+3)}$ **6.** $\frac{x}{9x^2+12}$ **7.** $\frac{x}{9x^2+12}$ **8.** No solution **9.** $d = \frac{a^2-aw}{r}$ **10.** 96 faculty; 132 staff

Chapter 6 Review Problems

1. $\frac{x}{2}$ **2.** $\frac{2x^2}{3y^2}$ **3.** $\frac{2x-3}{3x+5}$ **4.** $\frac{a-b}{3(x-2)}$ **5.** $\frac{x(2x-1)}{x-2}$ **6.** $\frac{1}{8xy}$ **7.** $\frac{y-5}{y-4}$ **8.** $\frac{1}{5}$ **9.** $\frac{a^2(3x-1)}{2b^2(2x+1)}$ **10.** $\frac{2a+5}{2a(2a^2-7a-13)}$ **11.** $-\frac{x^2+10x-9}{(2x+1)(x-2)}$ **12.** $\frac{-7x+20}{4x(x+4)}$ **13.** $\frac{-6y-11}{36y}$ or $-\frac{6y+11}{36y}$ **14.** $\frac{y^2+4}{(y+2)^2(y-2)}$ **15.** $-\frac{y-3}{y+2}$ **16.** $\frac{a-2}{a+3}$ **17.** $\frac{-x^2+2}{(x+4)^2(2x+1)}$ **18.** $\frac{4x^2+6x+5}{2x}$ **19.** $\frac{x}{x-5}$ **20.** $\frac{4(x-2)}{2x+5}$ **21.** 4 **22.** $\frac{a^2+a+1}{a^2-a-1}$ **23.** $-\frac{1}{x+y}$ **24.** $-\frac{(x+3)(x+4)}{x(x-4)}$ **25.** $x = 6$ **26.** $x = 1$ **27.** $x = 1$ **28.** $a = -6$ **29.** $y = \frac{3}{4}$ or 0.75 **30.** $a = -\frac{5}{2}$ or $-2\frac{1}{2}$ or -2.5 **31.** No solution **32.** $x = 0$ **33.** $M = \frac{mV}{N} - N$ or $\frac{mV-N^2}{N}$ **34.** $x = \frac{y-y_0+mx_0}{m}$ **35.** $a = \frac{bf}{b-f}$ **36.** $t = \frac{2S}{V_1+V_2}$ **37.** $R_2 = \frac{dR_1}{L-d}$ **38.** $r = \frac{S-P}{Pt}$ **39.** $\frac{x+6}{x+5}$ **40.** $\frac{2x-1}{x+2}$ **41.** $\frac{4x^2+17x+11}{x+4}$ **42.** $\frac{1}{x+1}$ **43.** $x = \frac{5}{7}$ **44.** 224 graphing calculators; 96 scientific calculators **45.** 35 in. wide; 49 in. long **46.** 5.6 hr **47.** 500 rabbits **48.** 28 officers **49.** 182 ft tall **50.** 6 hr **51.** 6 min **52.** \$199 billion **53.** \$202 billion **54.** \$184 billion **55.** \$235 billion

How Am I Doing? Chapter 6 Test **1.** $\frac{x+2}{x-3}$ (obj. 6.1.1) **2.** $-\frac{5p^3}{9r^3}$ (obj. 6.1.1) **3.** $\frac{2(2y-1)}{3y+5}$ (obj. 6.1.2) **4.** $-\frac{6}{(2x-1)(x+1)}$ (obj. 6.1.3) **5.** $\frac{x+3}{x(x+1)}$ (obj. 6.2.2) **6.** $\frac{3x^2+8x+6}{(x+3)^2(x+2)}$ (obj. 6.2.2) **7.** $-\frac{2}{5}$ (obj. 6.3.1) **8.** $-x+1$ (obj. 6.3.1) **9.** $x = \frac{16}{3}$ or $5\frac{1}{3}$ (obj. 6.4.1) **10.** $x = -2$ (obj. 6.4.1) **11.** $y = 5$ (obj. 6.4.1) **12.** $x = 3$ (obj. 6.4.1) **13.** $W = \frac{S-2Lh}{2h+2L}$ (obj. 6.5.1) **14.** $h = \frac{3V}{\pi r^2}$ (obj. 6.5.1) **15.** 39 employees got the bonus; 247 did not. (obj. 6.5.2) **16.** 1500 ft wide; 2550 ft long (obj. 6.5.2)

Cumulative Test for Chapters 1–6 **1.** $\frac{9}{y}$ **2.** -2 **3.** 1.42×10^7 **4.** -22 **5.** $-x+15y$ **6.** $x = -\frac{42}{5}$ or $-8\frac{2}{5}$ or -8.4 **7.** $x = -6$ **8.** \$700 at 5% interest; \$6300 at 8% interest **9.** $x < -1$

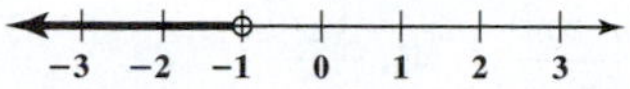

10. $-\frac{9}{2} \le x \le \frac{7}{2}$

11.

x	y
0	−6
1	−3
2	0

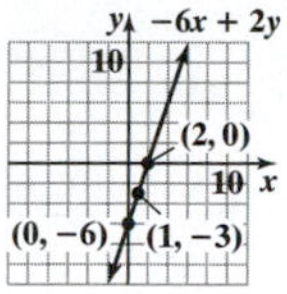

12. $5x - 6y = 13$ **13.** $y = -x + 3$ **14.** -5 **15.** 31 **16.** (5, 3) **17.** (4, −3,1) **18.** T-shirts = \$8: sweatshirts = \$12 **19.**

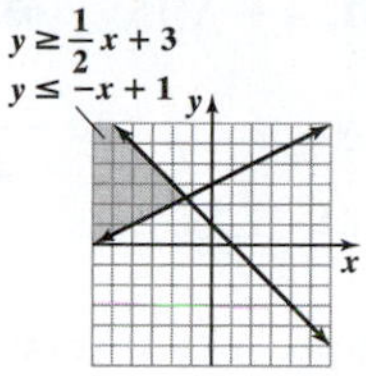

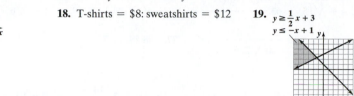

20. $4x^2 + 20x + 25$ **21.** $3x^2 + x + 4$ **22.** $(2x-5y)(4x^2+10xy+25y^2)$ **23.** $x(9x-5y)^2$ **24.** $x = -18, x = -2$ **25.** $\frac{x(x+1)}{2x+5}$ **26.** $\frac{2(x+1)(2x-1)}{16x^2-7}$ **27.** $\frac{7x-12}{(x-6)(x+4)}$ **28.** $x = -4$ **29.** $b = \frac{5H-2x}{3+4H}$ **30.** 693 on foot; 2541 in squad cars

Chapter 7

7.1 Exercises **1.** $\frac{81x^4}{y^4z^8}$ **3.** $\frac{16a^2}{9b^6}$ **5.** $\frac{y^3}{8x^6}$ **7.** $\frac{y^{10}}{9x^2}$ **9.** $x^{3/2}$ **11.** y^8 **13.** $x^{2/5}$ **15.** $x^{2/3}$ **17.** $a^{7/4}$ **19.** $x^{4/7}$ **21.** $a^{7/8}$ **23.** $y^{1/2}$ **25.** $\frac{1}{x^{3/4}}$ **27.** $\frac{b^{1/3}}{a^{5/6}}$ **29.** $\frac{1}{6^{1/2}}$ **31.** $\frac{3}{a^{1/3}}$ **33.** 243 **35.** 8 **37.** -32 **39.** 9 **41.** $xy^{1/6}$ **43.** $-14x^{7/12}y^{1/12}$ **45.** $6^{4/3}$ **47.** $2x^{7/10}$ **49.** $-\frac{4x^{5/2}}{y^{6/5}}$ **51.** $2ab$ **53.** $16x^{1/2}y^5z$ **55.** $x^2 - x^{13/15}$ **57.** $m^{3/8} + 2m^{15/8}$ **59.** $\frac{1}{2}$ **61.** $\frac{1}{125}$ **63.** 32 **65.** $\frac{3y+1}{y^{1/2}}$ **67.** $\frac{1+6^{4/3}x^{1/3}}{x^{1/3}}$ **69.** $2a(5a^{1/4} - 2a^{3/5})$ **71.** $3x(4x^{1/3} - x^{3/2})$ **73.** $a = -\frac{3}{8}$ **75.** Radius = 1.86 m **77.** Radius = 5 ft **79.** $x = -\frac{3}{2}$ **80.** $b = \frac{2A - ah}{h}$ or $b = \frac{2A}{h} - a$

Quick Quiz 7.1 *See Examples noted with Ex.* **1.** $-12x^{5/6}y^{3/4}$ (Ex. 3) **2.** $2x^{10/3}$ (Ex. 3) **3.** $125x^{3/8}$ (Ex. 1 and 5) **4.** See Student Solutions Manual

7.2 Exercises **1.** A square root of a number is a value that when multiplied by itself is equal to the original number. **3.** $\sqrt[3]{-8} = -2$ because $(-2)(-2)(-2) = -8$ **5.** 10 **7.** 13 **9.** $-\frac{1}{3}$ **11.** Not a real number **13.** 0.2 **15.** 4.6, 4.9, 6, 3; all real numbers x where $x \ge -7$ **17.** 0, 1, 2, 2.2; all real numbers x where $x \ge 10$ **19.**

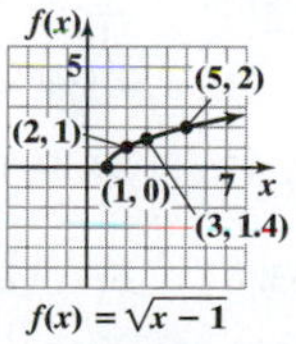

21.

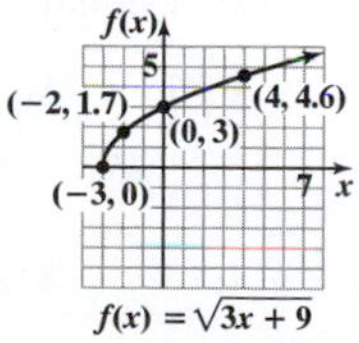

23. 4 **25.** -10 **27.** 2 **29.** 3 **31.** 8 **33.** 5 **35.** $-\frac{1}{2}$ **37.** $y^{1/3}$ **39.** $m^{3/5}$ **41.** $(2a)^{1/4}$ **43.** $(a+b)^{3/7}$ **45.** $x^{1/6}$ **47.** $(3x)^{5/6}$ **49.** 12 **51.** x^4y **53.** $6x^4y^2$ **55.** $2a^2b$ **57.** $\sqrt[7]{y^4}$ or $(\sqrt[7]{y})^4$ **59.** $\frac{1}{\sqrt[3]{49}}$ **61.** $\sqrt[4]{(a+5b)^3}$ or $(\sqrt[4]{(a+5b)})^3$ **63.** $\sqrt[5]{(-x)^3}$ or $(\sqrt[5]{(-x)})^3$ **65.** $\sqrt[5]{8x^3y^3}$ **67.** 27 **69.** $\frac{2}{5}$ **71.** 2 **73.** $\frac{1}{8x^2}$ **75.** $11x^2$ **77.** $12a^3b^{12}$ **79.** $5|x|$ **81.** $-2x^2$ **83.** x^2y^4 **85.** $|a^3b|$ **87.** $5x^6y^2$ **89.** $x = \frac{2y-6}{5}$ **90.** $y = \frac{3x-12}{2}$

Quick Quiz 7.2 *See Examples noted with Ex.* **1.** $\frac{8}{125}$ (Ex. 9) **2.** -4 (Ex. 1) **3.** $11x^5y^6$ (Ex. 7) **4.** See Student Solutions Manual

7.3 Exercises **1.** $2\sqrt{2}$ **3.** $3\sqrt{2}$ **5.** $2\sqrt{7}$ **7.** $5\sqrt{2}$ **9.** $3x\sqrt{x}$ **11.** $2a^3b^3\sqrt{10b}$ **13.** $3xz^2\sqrt{10xy}$ **15.** 2 **17.** $2\sqrt[3]{5}$ **19.** $3\sqrt[3]{2a^2}$ **21.** $3ab^3\sqrt[3]{a^2}$ **23.** $2x^4y^4\sqrt[3]{5y}$ **25.** $3p^5\sqrt[4]{kp^3}$ **27.** $-2xy\sqrt[5]{y}$ **29.** $a = 4$ **31.** $12\sqrt{5}$ **33.** $4\sqrt{3} - 4\sqrt{7}$

35. $11\sqrt{2}$ **37.** $11\sqrt{3}$ **39.** $-5\sqrt{2}$ **41.** 0 **43.** 0 **45.** $8\sqrt{3x}$ **47.** $19\sqrt{2x}$ **49.** $2\sqrt{11} - \sqrt{7x}$ **51.** $6x\sqrt{2x}$ **53.** $11\sqrt[3]{2}$ **55.** $4xy\sqrt[3]{x} - 3y\sqrt[3]{xy^2}$ **57.** $20.7846097 = 20.7846097$ **59.** 7.146 amps **61.** 3.14 sec **63.** $x(4x - 7y)^2$ **64.** $y(9x + 5)(9x - 5)$ **65.** 2.5 servings of scallops and 2 servings of skim milk **66.** 5 servings of scallops; 4 servings of skim milk

Quick Quiz 7.3

See Examples noted with Ex. **1.** $2x^3y^4\sqrt{30x}$ (Ex. 5) **2.** $2x^5y^3\sqrt[3]{2y}$ (Ex. 5) **3.** $10\sqrt{3}$ (Ex. 7) **4.** See Student Solutions Manual

7.4 Exercises

1. $\sqrt{35}$ **3.** $-30\sqrt{10}$ **5.** $-24\sqrt{5}$ **7.** $-3\sqrt{5xy}$ **9.** $-12x^2\sqrt{5y}$ **11.** $15\sqrt{ab} - 25\sqrt{a}$ **13.** $-3\sqrt{2ab} - 6\sqrt{5a}$ **15.** $-a + 2\sqrt{ab}$ **17.** $14\sqrt{3x} - 35x$ **19.** $22 - 5\sqrt{2}$ **21.** $4 - 6\sqrt{6}$ **23.** $14 + 11\sqrt{35x} + 60x$ **25.** $\sqrt{15} + 3 + 2\sqrt{10} + 2\sqrt{6}$ **27.** $29 - 4\sqrt{30}$ **29.** $81 - 36\sqrt{b} + 4b$ **31.** $3x + 13 + 6\sqrt{3x + 4}$ **33.** $3x\sqrt[3]{4} - 4x^2\sqrt[3]{x}$ **35.** $1 - \sqrt[3]{12} + \sqrt[3]{18}$ **37.** $\frac{7}{5}$ **39.** $\frac{2\sqrt{3x}}{7y^3}$ **41.** $\frac{2xy^2\sqrt[3]{x^2}}{3}$ **43.** $\frac{y^2\sqrt[3]{5y^2}}{3x}$ **45.** $\frac{3\sqrt{2}}{2}$ **47.** $\frac{2\sqrt{3}}{3}$ **49.** $\frac{\sqrt{5y}}{5y}$ **51.** $\frac{\sqrt{7ay}}{y}$ **53.** $\frac{\sqrt{3x}}{3x}$ **55.** $\frac{x(\sqrt{5} + \sqrt{2})}{3}$ **57.** $2y(\sqrt{6} - \sqrt{5})$ **59.** $\frac{\sqrt{6y} - y\sqrt{2}}{6 - 2y}$ **61.** $4 + \sqrt{15}$ **63.** $\frac{3x - 3\sqrt{3xy} + 2y}{3x - y}$ **65.** $11\sqrt{2}$ **67.** $-24 + \sqrt{6}$ **69.** $\frac{9\sqrt{2x}}{4x}$ **71.** $3 - \sqrt{5}$ **73.** 1.194938299; 1.194938299; yes; yes **75.** $-\frac{43}{4(\sqrt{2} - 3\sqrt{5})}$ **77.** \$78.58 **79.** $(x + 8\sqrt{x} + 15)$ mm^2 **81.** $(-3, 5)$ **82.** $(2, 0, -4)$ **83.** 72% **84.** 72,250 rings

Quick Quiz 7.4

See Examples noted with Ex. **1.** $8 + \sqrt{15}$ (Ex. 3) **2.** $\frac{3\sqrt{3x}}{x}$ (Ex. 9) **3.** $\frac{14 + 9\sqrt{5}}{11}$ (Ex. 12) **4.** See Student Solutions Manual

How Am I Doing? Sections 7.1–7.4

1. $\frac{6y^{5/6}}{x^{1/4}}$ (obj. 7.1.1) **2.** $\frac{b^3}{a^4}$ (obj. 7.1.1) **3.** $-8x^3y^{1/3}$(obj. 7.1.1) **4.** $\frac{9x^4}{y^6}$ (obj. 7.1.1) **5.** $\frac{1}{81}$ (obj. 7.2.3) **6.** -3 (obj. 7.2.1) **7.** 9 (obj. 7.2.1) **8.** $8a^4y^8$ (obj. 7.2.3) **9.** $3a^4b^2c^5$ (obj. 7.2.4) **10.** $(4x)^{5/6}$ (obj. 7.2.2) **11.** $2x^5y^7$ (obj. 7.3.1) **12.** $2x^2y^5\sqrt[3]{4x^2}$ (obj. 7.3.1) **13.** $3\sqrt{11}$ (obj. 7.3.2) **14.** $2y\sqrt{3y} + 5\sqrt[3]{2}$ (obj. 7.3.2) **15.** $-25 + 9\sqrt{10}$ (obj. 7.4.1) **16.** $\frac{5\sqrt{2x}}{6x}$ (obj. 7.4.3) **17.** $-5 - 2\sqrt{6}$ (obj. 7.4.3)

7.5 Exercises

1. Isolate one of the radicals on one side of the equation. **3.** $x = 3$ **5.** $x = 1$ **7.** $y = 2; y = 1$ **9.** $x = 3$ **11.** No solution **13.** $y = 7$ **15.** $y = -2, y = -3$ **17.** $x = 3, x = 7$ **19.** $x = 0, x = \frac{1}{2}$ **21.** $x = \frac{5}{2}$ **23.** $x = 7$ **25.** $x = 12$ **27.** $x = 0, x = 5$ **29.** $x = \frac{1}{4}$ **31.** $x = -3, x = -4$ **33.** $x = 0, x = 8$ **35.** No solution **37.** $x = 9$ **39.** $x \approx 3.3357, x \approx 0.9443$ **41. (a)** $S = \frac{V^2}{12}$ **(b)** 75 ft **43.** $x = 0.055y^2 + 1.25y - 10$ **45.** $c = 1$ **47.** $16x^4$ **48.** $\frac{1}{2x^2}$ **49.** $-6x^2y^3$ **50.** $-2x^3y$ **51.** Approximately 1.33 mph **52.** 16 dogs; 12 cats

Quick Quiz 7.5

See Examples noted with Ex. **1.** $x = 1, x = 4$ (Ex. 1) **2.** $x = 2$ (Ex. 2) **3.** $x = 5$ (Ex. 3) **4.** See Student Solutions Manual

7.6 Exercises

1. No. There is no real number that, when squared, will equal -9. **3.** No. To be equal, the real parts must be equal, and the imaginary parts must be equal. $2 \neq 3$ and $3i \neq 2i$ **5.** $5i$ **7.** $5i\sqrt{2}$ **9.** $\frac{5}{2}i$ **11.** $-9i$ **13.** $2 + i\sqrt{3}$ **15.** $-2.8 + 4i$ **17.** $-3 + 2i\sqrt{6}$ **19.** $-\sqrt{10}$ **21.** -12 **23.** $x = 5; y = -3$ **25.** $x = 1.3; y = 2$ **27.** $x = -6, y = 3$ **29.** $-5 + 11i$ **31.** $1 - i$ **33.** $1.2 + 2.1i$ **35.** -14 **37.** 42 **39.** $7 + 4i$ **41.** $6 + 6i$ **43.** $-10 - 12i$ **45.** $-\frac{3}{4} + i$ **47.** $-\sqrt{21}$ **49.** $12 - \sqrt{10} + 3i\sqrt{5} + 4i\sqrt{2}$ **51.** i **53.** 1 **55.** -1 **57.** $-i$ **59.** 0 **61.** $1 + i$ **63.** $\frac{1 + i}{2}$ or $\frac{1}{2} + \frac{i}{2}$ **65.** $\frac{1 + i}{3}$ or $\frac{1}{3} + \frac{i}{3}$ **67.** $\frac{-2 - 5i}{6}$ or $-\frac{2}{6} - \frac{5i}{6}$ **69.** $-2i$ **71.** $\frac{35 + 42i}{61}$ or $\frac{35}{61} + \frac{42i}{61}$ **73.** $\frac{11 - 16i}{13}$ or $\frac{11}{13} - \frac{16i}{13}$ **75.** $7i\sqrt{2}$ **77.** $9 - 8i$ **79.** $-12 - 14i$ **81.** $\frac{1 - 8i}{5}$ or $\frac{1}{5} - \frac{8i}{5}$ **83.** $Z = \frac{2 - 3i}{3}$ or $\frac{2}{3} - \frac{3i}{3}$ **85.** 18 hr producing juice in glass bottles; 25 hr producing juice in cans, 62 hr producing juice in plastic bottles **86.** \$96,030

Quick Quiz 7.6

See Examples noted with Ex. **1.** $32 - 9i$ (Ex. 6) **2.** $\frac{-2 + 11i}{5}$ or $-\frac{2}{5} + \frac{11i}{5}$ (Ex. 9) **3.** i (Ex. 8) **4.** See Student Solutions Manual

7.7 Exercises

1. Answers may vary. A person's weekly paycheck varies as the number of hours worked. $y = kx$, y is the weekly salary, k is the hourly salary, x is the number of hours. **3.** $y = \frac{k}{x}$ **5.** $y = 24$ **7.** 71.4 lb/in.^{2} **9.** 160 ft **11.** $y = 160$ **13.** 2993 gal **15.** 3.9 hr **17.** 400 lb **19.** Approximately 3.1 min **21.** $x = \frac{2}{3}, x = 2$ **22.** $x = 1, x = -8$ **23.** \$460 **24.** 55 gal

Quick Quiz 7.7 *See Examples noted with Ex.* **1.** $y = 4.5$ (Ex. 2) **2.** $y = 3$ (Ex. 4) **3.** 135.2 ft (Ex. 1) **4.** See Student Solutions Manual

Use Math to Save Money **1.** \$545.75 **2.** \$578.06 **3.** He did not deposit enough money to cover the checks he wrote for May. But the \$300.50 he already had in the bank will help to cover his expenses for May. **4.** \$268.19 **5.** Eventually Terry will be in debt. **6.** Answers will vary. **7.** Answers will vary.

You Try It **1. (a)** $x^{1/5}$ **(b)** $2^{1/3}a^{-1}b^{-1/12}$ **(c)** $\frac{5^{1/4}x^{-3/4}}{4^{-1/2}y^{1/2}}$ **2.** $-4a^{5/6}$ **3.** $-6x^{1/4}$ **4. (a)** $\frac{5}{a^3}$ **(b)** $\frac{a^4}{2}$ **(c)** $\frac{1}{32}$ **5.** $x^{7/6} - x$ **6.** 1 **7. (a)** 2 **(b)** -1 **(c)** Not a real number **8. (a)** $\sqrt[5]{x^4}$ or $\sqrt[5]{x}^{\,4}$ **(b)** $v^{9/4}$ **(c)** 81 **9. (a)** $|x|$ **(b)** y **10.** $-4m^6$ **11. (a)** $2x^2\sqrt{6x}$ **(b)** $3rs^3\sqrt[3]{2r}$ **12.** $30\sqrt{2}$ **13. (a)** $3\sqrt{30}$ **(b)** $18\sqrt{2} - 9\sqrt{5}$ **(c)** $1 - \sqrt{15}$ **14.** $\frac{\sqrt[4]{3}}{2}$ **15. (a)** $\frac{\sqrt{6}}{2}$ **(b)** $-4\sqrt{2} - 4\sqrt{3}$ **16.** $x = 9$ **17. (a)** $10i$ **(b)** $2i\sqrt{6}$ **18. (a)** $11 - 6i$ **(b)** $-1 + 2i$ **19.** $9 - 7i$ **20.** 1 **21.** $\frac{3 - 14i}{5}$ or $\frac{3}{5} - \frac{14}{5}i$ **22.** $y = 36$ **23.** $w = 0.5$

Chapter 7 Review Problems **1.** $\frac{15x^3}{y^{5/2}}$ **2.** $4a^3b^{5/2}$ **3.** $3^{2/3}$ **4.** $\frac{x^{1/2}y^{3/10}}{2}$ **5.** $\frac{x}{32y^{1/2}z^4}$ **6.** $7a^5b$ **7.** $2a^2$ **8.** 16 **9.** $\frac{2x + 1}{x^{2/3}}$ **10.** $3x(2x^{1/2} - 3x^{-1/2})$ **11.** -4 **12.** -2 **13.** $-\frac{1}{5}$ **14.** 0.2 **15.** Not a real number **16.** $-\frac{1}{2}$ **17.** 16 **18.** 625 **19.** $7x^2y^5z$ **20.** $4a^4b^{10}$ **21.** $-2a^4b^5c^7$ **22.** $7x^{11}y$ **23.** $a^{2/5}$ **24.** $(2b)^{1/2}$ **25.** $(5a)^{1/3}$ **26.** $(xy)^{7/5}$ **27.** $\sqrt{m}$ **28.** $\sqrt[5]{y^3}$ **29.** $\sqrt[3]{9z^2}$ or $\sqrt[3]{9z}^{\,2}$ **30.** $\sqrt[7]{8x^3}$ **31.** 8 **32.** 9 **33.** $\frac{1}{3}$ **34.** 0.7 **35.** 6 **36.** $125a^3b^6$ **37.** $11\sqrt{2}$ **38.** $13\sqrt{7}$ **39.** $25\sqrt{3}$ **40.** $8x\sqrt{5x}$ **41.** $11\sqrt{2x} - 5x\sqrt{2}$ **42.** $-6\sqrt[3]{2}$ **43.** $90\sqrt{2}$ **44.** $-24x\sqrt{5}$ **45.** **46.** $4\sqrt{3a} - 3a\sqrt{7}$ **47.** $2b\sqrt{7a} - 2b^2\sqrt{21c}$ **48.** $4 - 9\sqrt{6}$ **49.** $74 - 12\sqrt{30}$ **50.** $2x - \sqrt[3]{2xy} + 2\sqrt[3]{3x^2} - \sqrt[3]{6y}$ **51. (a)** $f(12) = 8$ **(b)** All real numbers x where $x \geq -4$ **52. (a)** $f(9) = 3$ **(b)** All real numbers x where $x \leq 12$ **53.** $\frac{y\sqrt{6x}}{x}$ **54.** $\frac{3\sqrt{5y}}{5y}$ **55.** $\sqrt{3}$ **56.** $2\sqrt{6} + 2\sqrt{5}$ **57.** $\frac{3x - \sqrt{xy}}{9x - y}$ **58.** $-\frac{\sqrt{35} + 3\sqrt{5}}{2}$ **59.** $\frac{2 + 3\sqrt{2}}{7}$ **60.** $\frac{\sqrt[3]{4x^2y}}{2y}$ **61.** $4i + 3i\sqrt{5}$ **62.** $x = \frac{-7 + \sqrt{6}}{2}$; $y = -3$ **63.** $-9 - 11i$ **64.** $-10 + 2i$ **65.** $21 + 9i$ **66.** $32 - 24i$ **67.** $-8 + 6i$ **68.** $-5 - 4i$ **69.** -1 **70.** i **71.** $\frac{13 - 34i}{25}$ **72.** $\frac{11 + 13i}{10}$ **73.** $\frac{-3 - 4i}{5}$ **74.** $\frac{18 + 30i}{17}$ **75.** $x = 9$ **76.** $x = 3$ **77.** $x = 4$ **78.** $x = 5$ **79.** $x = 5, x = 1$ **80.** $x = 1, x = \frac{3}{2}$ **81.** $y = 16.5$ **82.** 22.5 g **83.** 3.5 sec **84.** $y = 0.5$ **85.** $y = 100$ **86.** 432 cm^3

How Am I Doing? Chapter 7 Test **1.** $-6x^{5/6}y^{1/2}$ (obj. 7.1.1) **2.** $\frac{7x^{9/4}}{4}$ (obj. 7.1.1) **3.** $8^{3/2}x^{1/2}$ or $16(2x)^{1/2}$ (obj. 7.1.1) **4.** $\frac{8}{27}$ (obj. 7.1.1) **5.** -2 (obj. 7.2.1) **6.** $\frac{1}{4}$ (obj. 7.2.1) **7.** 32 (obj. 7.2.1) **8.** $5a^2b^4\sqrt{3b}$ (obj. 7.3.1) **9.** $7a^2b^5$ (obj. 7.3.1) **10.** $3mn\sqrt[3]{2n^2}$ (obj. 7.3.1) **11.** $6\sqrt{6} + 2\sqrt{2}$ (obj. 7.3.2) **12.** $2\sqrt{10x} + \sqrt{3x}$ (obj. 7.3.2) **13.** $-30y\sqrt{5x}$ (obj. 7.4.1) **14.** $18\sqrt{2} - 10\sqrt{6}$ (obj. 7.4.1) **15.** $12 + 39\sqrt{2}$ (obj. 7.4.1) **16.** $\frac{6\sqrt{5x}}{x}$ (obj. 7.4.3) **17.** $\frac{\sqrt{3xy}}{3}$ (obj. 7.4.3) **18.** $\frac{9 + 7\sqrt{3}}{6}$ (obj. 7.4.3) **19.** $x = 2, x = 1$ (obj. 7.5.1) **20.** $x = 10$ (obj. 7.5.1) **21.** $x = 6$ (obj. 7.5.2) **22.** $2 + 14i$ (obj. 7.6.2) **23.** $-1 + 4i$ (obj. 7.6.2) **24.** $18 + i$ (obj. 7.6.3) **25.** $\frac{-13 + 11i}{10}$ or $-\frac{13}{10} + \frac{11i}{10}$ (obj. 7.6.5) **26.** $27 + 36i$ (obj. 7.6.3) **27.** $-i$ (obj. 7.6.4) **28.** $y = 3$ (obj. 7.7.2) **29.** $y = \frac{5}{6}$ (obj. 7.7.3) **30.** About 83.3 ft (obj. 7.7.1)

Chapter 8

8.1 Exercises **1.** $x = \pm 10$ **3.** $x = \pm\sqrt{15}$ **5.** $x = \pm 2\sqrt{10}$ **7.** $x = \pm 9i$ **9.** $x = \pm 4i$ **11.** $x = 3 \pm 2\sqrt{3}$ **13.** $x = -9 \pm \sqrt{21}$ **15.** $x = \frac{-1 \pm \sqrt{7}}{2}$ **17.** $x = \frac{9}{4}, x = -\frac{3}{4}$ **19.** $x = 1, x = -6$ **21.** $x = \pm\frac{3\sqrt{2}}{2}$ **23.** $x = -5 \pm 2\sqrt{5}$ **25.** $x = 4 \pm \sqrt{33}$ **27.** $x = 6, x = 8$ **29.** $x = \frac{-3 \pm \sqrt{41}}{2}$ **31.** $y = \frac{-5 \pm \sqrt{3}}{2}$ **33.** $x = \frac{-5 \pm \sqrt{31}}{3}$ **35.** $x = -2 \pm \sqrt{10}$ **37.** $x = 4, x = -2$ **39.** $x = \frac{1 \pm i\sqrt{11}}{6}$ **41.** $x = \frac{1 \pm i\sqrt{7}}{2}$ **43.** $x = \frac{3 \pm i\sqrt{7}}{4}$ **45.** $(-1 + \sqrt{6})^2 + 2(-1 + \sqrt{6}) - 5 \stackrel{?}{=} 0$; $1 - 2\sqrt{6} + 6 - 2 + 2\sqrt{6} - 5 \stackrel{?}{=} 0$; $0 = 0$ ✓ **47.** $x = 16$ **49.** Approximately 1.12 sec **51.** 8 **52.** 7 **53.** 40 **54.** 4

Quick Quiz 8.1 *See Examples noted with Ex.* **1.** $x = \frac{3 \pm 2\sqrt{3}}{4}$ (Ex. 5) **2.** $x = 4 \pm 2\sqrt{11}$ (Ex. 6) **3.** $x = \frac{-5 \pm \sqrt{3}}{2}$ (Ex. 7) **4.** See Student Solutions Manual

8.2 Exercises

1. Place the quadratic equation in standard form. Find a, b, and c. Substitute these values into the quadratic formula. **3.** One real **5.** $x = \frac{-1 \pm \sqrt{21}}{2}$ **7.** $x = \frac{1 \pm \sqrt{13}}{6}$ **9.** $x = 0, x = \frac{2}{3}$ **11.** $x = 1, x = -\frac{2}{3}$ **13.** $x = \frac{-3 \pm \sqrt{41}}{8}$ **15.** $x = \pm\frac{\sqrt{6}}{2}$ **17.** $x = \frac{-1 \pm \sqrt{3}}{2}$ **19.** $x = \pm 3$ **21.** $x = \frac{2 \pm 3\sqrt{2}}{2}$ **23.** $x = 2 \pm \sqrt{10}$ **25.** $y = 6, y = 4$ **27.** $x = -2 \pm 2i\sqrt{2}$ **29.** $x = \pm\frac{i\sqrt{22}}{2}$ **31.** $x = \frac{4 \pm i\sqrt{5}}{3}$ **33.** Two irrational roots **35.** Two rational roots **37.** One rational root **39.** Two nonreal complex roots **41.** $x^2 - 15x + 26 = 0$ **43.** $x^2 + 12x + 27 = 0$ **45.** $x^2 + 16 = 0$ **47.** $2x^2 - x - 15 = 0$ **49.** $x \approx 1.4643, x \approx -2.0445$ **51.** 18 mountain bikes or 30 mountain bikes per day **53.** $-3x^2 - 10x + 11$ **54.** $-y^2 + 3y$

Quick Quiz 8.2

See Examples noted with Ex. **1.** $x = \frac{9 \pm 5\sqrt{5}}{22}$ (Ex. 1) **2.** $x = -\frac{8}{3}, x = 1$ (Ex. 5) **3.** $x = \frac{5 \pm i\sqrt{47}}{4}$ (Ex. 6) **4.** See Student Solutions Manual

8.3 Exercises

1. $x = \pm\sqrt{5}, x = \pm 2$ **3.** $x = \pm\sqrt{3}, x = \pm 2i$ **5.** $x = \pm\frac{i\sqrt{3}}{2}, x = \pm 1$ **7.** $x = 2, x = -1$ **9.** $x = 0, x = \sqrt[3]{3}$ **11.** $x = \pm 2, x = \pm 1$ **13.** $x = \pm\frac{\sqrt[4]{54}}{3}$; these are the only real roots. **15.** $x = -64, x = 27$ **17.** $x = \frac{1}{64}, x = -\frac{8}{27}$ **19.** $x = 81$ **21.** $\frac{1}{256}$ **23.** $x = 32, x = -1$ **25.** $x = \sqrt[3]{7}, x = \sqrt[3]{-2}$ **27.** $x = 3, x = -2, x = \frac{1 \pm \sqrt{17}}{2}$ **29.** $x = 9, x = 4$ **31.** $x = -\frac{1}{3}$ **33.** $x = \frac{1}{5}, x = -\frac{1}{2}$ **35.** $x = \frac{5}{6}, x = \frac{3}{2}$ **37.** $x = -2, y = 3$ **38.** $\frac{15x + 6}{7 - 3x}$ **39.** $3\sqrt{10} - 12\sqrt{3}$ **40.** $6 - 2\sqrt{10} + 6\sqrt{3} - 2\sqrt{30}$ **41.** High school graduate 37.3%; Associate's degree 36.4%; Bachelor's degree 40.0%; Master's degree 40.9% **42.** An additional 10.9 yr

Quick Quiz 8.3

See Examples noted with Ex. **1.** $x = \pm 4, x = \pm\sqrt{2}$ (Ex. 1) **2.** $x = \frac{1}{4}, x = -\frac{2}{5}$ (Ex. 5) **3.** $x = 125, x = -1$ (Ex. 3) **4.** See Student Solutions Manual

How Am I Doing? Sections 8.1–8.3

1. $x = \pm 3\sqrt{2}$ (obj. 8.1.1) **2.** $x = \frac{1 \pm 2\sqrt{7}}{3}$ (obj. 8.1.1) **3.** $x = 6, x = 2$ (obj. 8.1.2) **4.** $x = \frac{2 \pm \sqrt{10}}{2}$ (obj. 8.1.2) **5.** $x = \frac{4 \pm \sqrt{10}}{2}$ (obj. 8.2.1) **6.** $x = -2 \pm \sqrt{11}$ (obj. 8.2.1) **7.** $x = \frac{-3 \pm 2i\sqrt{2}}{2}$ (obj. 8.2.1) **8.** $x = -2, x = \frac{6}{5}$ (obj. 8.2.1) **9.** $x = 0, x = \frac{1}{5}$ (obj. 8.2.1) **10.** $x = \frac{3 \pm i\sqrt{87}}{8}$ (obj. 8.2.1) **11.** $x = -\frac{2}{3}, x = 3$ (obj. 8.2.1) **12.** $x = 2, x = -1$ (obj. 8.3.1) **13.** $w = \pm 8, w = \pm 2\sqrt{2}$ (obj. 8.3.2) **14.** $x = \pm\sqrt[4]{2}, x = \pm\sqrt[4]{3}$ (obj. 8.3.1) **15.** $x = -3, x = \frac{1}{4}$ (obj. 8.3.2)

8.4 Exercises

1. $t = \pm\frac{\sqrt{S}}{4}$ **3.** $r = \pm\frac{1}{3}\sqrt{\frac{S}{\pi}}$ **5.** $y = \pm\sqrt{\frac{10M}{b}}$ **7.** $y = \pm\frac{\sqrt{7R - 4w + 5}}{2}$ **9.** $M = \pm\sqrt{\frac{2cQ}{3mw}}$ **11.** $r = \pm\sqrt{\frac{V - \pi R^2 h}{\pi h}}$ **13.** $x = 0, x = \frac{3a}{7b}$ **15.** $I = \frac{E \pm \sqrt{E^2 - 4RP}}{2R}$ **17.** $w = \frac{-5t \pm \sqrt{25t^2 + 72}}{18}$ **19.** $r = \frac{-\pi h \pm \sqrt{\pi^2 h^2 + S\pi}}{\pi}$ **21.** $x = \frac{-5 \pm \sqrt{25 - 8aw - 8w}}{2a + 2}$ **23.** $b = 2\sqrt{5}$ **25.** $a = \sqrt{15}$ **27.** $a = \frac{12\sqrt{5}}{5}, b = \frac{24\sqrt{5}}{5}$ **29.** $5\sqrt{2}$ in. **31.** Width is 0.13 mi; length is 0.2 mi **33.** Width is 7 ft; length is 18 ft **35.** Base is 8 cm; altitude is 18 cm **37.** Speed in rain was 45 mph; speed without rain was 50 mph **39.** 30 mi **41.** 2,121,700 incarcerated adults **43.** In the year 2004 **45.** $\frac{4\sqrt{3x}}{3x}$ **46.** $\frac{\sqrt{30}}{2}$ **47.** $\frac{3(\sqrt{x} - \sqrt{y})}{x - y}$ **48.** $-2 - 2\sqrt{2}$

Quick Quiz 8.4

See Examples noted with Ex. **1.** $y = \pm\sqrt{\frac{5ab}{H}}$ (Ex. 1) **2.** $z = \frac{-7y \pm \sqrt{49y^2 + 120w}}{12}$ (Ex. 3) **3.** Width = 7 yd; length = 25 yd (Ex. 8) **4.** See Student Solutions Manual

8.5 Exercises

1. $V(1, -9)$; $(0, -8)$; $(4, 0)$, $(-2, 0)$ **3.** $V(-4, 25)$; $(0, 9)$; $(-9, 0)$, $(1, 0)$ **5.** $V(-2, -9)$; $(0, 3)$; $(-0.3, 0)$, $(-3.7, 0)$ **7.** $V\left(-\frac{1}{3}, -\frac{17}{3}\right)$; no x-intercepts; y-intercept $(0, -6)$ **9.** $V\left(-\frac{1}{2}, -\frac{9}{2}\right)$; $(0, -4)$; $(1, 0)$, $(-2, 0)$ **11.**

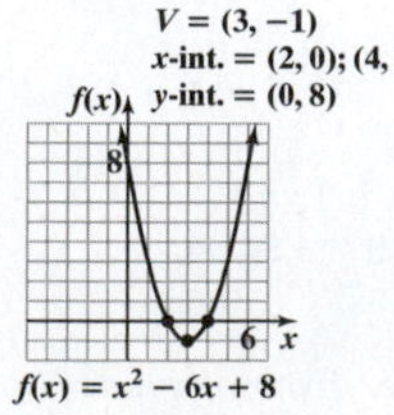

13. 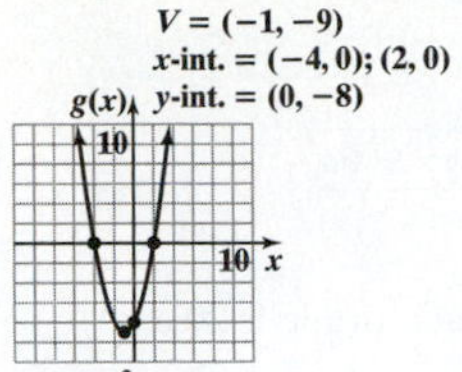

$g(x) = x^2 + 2x - 8$

15. 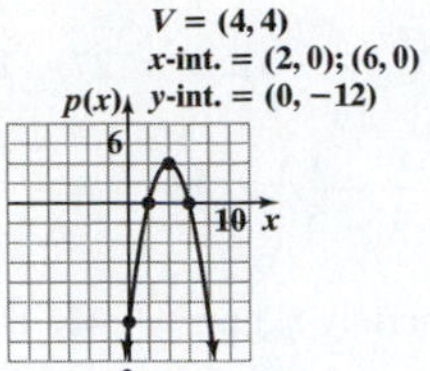

$p(x) = -x^2 + 8x - 12$

17. 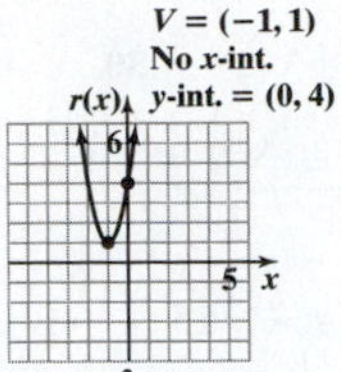

$r(x) = 3x^2 + 6x + 4$

19. 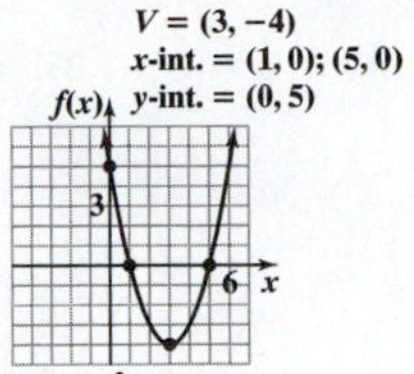

$f(x) = x^2 - 6x + 5$

21. 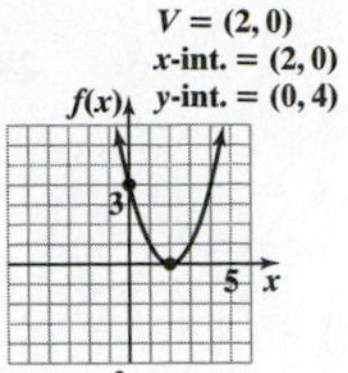

$f(x) = x^2 - 4x + 4$

23. 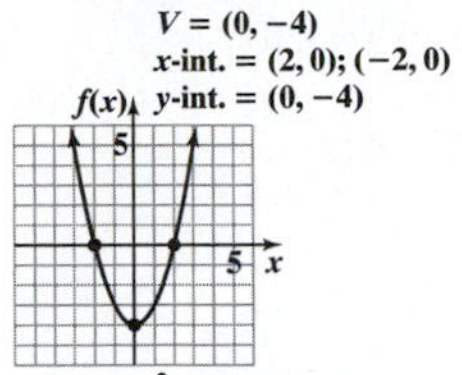

$f(x) = x^2 - 4$

25. $N(10) = 1229.8$; $N(30) = 2686.6$; $N(50) = 5735.4$; $N(70) = 10{,}376.2$; $N(90) = 16{,}609$ **27.** $N(60) \approx 8000$ Approximately 8,000,000 people participate in boating and have a mean income of \$60,000. **29.** Approximately 68. This means that 10,000,000 people who participate in boating have a mean income of \$68,000. **31.** $P(16) = -216$; $P(20) = 168$; $P(24) = 360$; $P(30) = 288$; $P(35) = -102$ **33.** 26 tables per day; \$384 per day **35.** 18 tables per day or 34 tables per day **37.** 56 ft; about 2.9 sec **39.** x-intercepts $(-0.3, 0)$ and $(2.6, 0)$

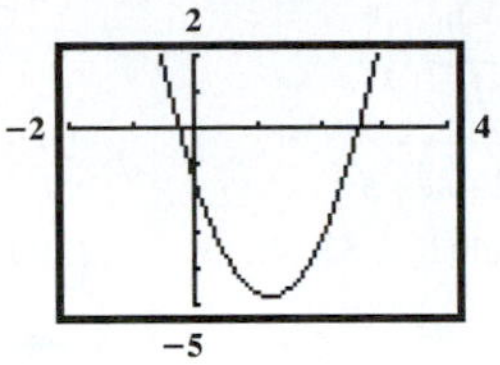

41. $a = 3, b = 8, c = -10$ **43.** $(3, 1, 2)$ **44.** $(3, -5, 8)$

Quick Quiz 8.5

See Examples noted with Ex. **1.** $(-1, 8)$ (Ex. 1) **2.** $(0, 6)$; $(-3, 0)$, $(1, 0)$ (Ex. 1) **3.** (Ex. 3)

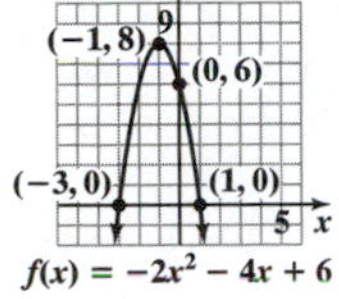

$f(x) = -2x^2 - 4x + 6$

4. See Student Solutions Manual

8.6 Exercises

1. The boundary points divide the number line into regions. All values of x in a given region produce results that are greater than zero, or else all the values of x in a given region produce results that are less than zero. **3.** $-4 < x < 3$

5. $x \le -2$ or $x \ge 2$

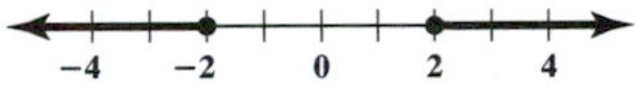

7. $-\frac{3}{2} < x < 1$

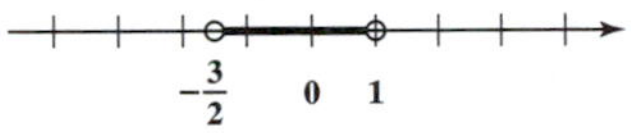

9. $x < -5$ or $x > 4$ **11.** $-\frac{1}{4} \le x \le 3$ **13.** $x < -\frac{2}{3}$ or $x > \frac{3}{2}$ **15.** $-10 \le x \le 3$ **17.** All real numbers **19.** $x = 2$ **21.** Approximately $x < -1.2$ or $x > 3.2$ **23.** Approximately $-0.8 < x < 1.8$ **25.** All real numbers **27.** No real number **29.** Greater than 15 sec but less than 25 sec **31. (a)** Approximately $9.4 < x < 190.6$ **(b)** \$57,000 **(c)** \$66,000 **33.** Any two test scores that total 167 will be sufficient to participate in synchronized swimming. **34.** 4.4 oz chocolate banana bites, 8.8 oz Dutch coffee, 2.8 oz Belgian chocolate **35.** \$400 for the 2-hour trip, \$483 for the 3-hour trip **36.** 9 adults, 12 children, 2 seniors

Quick Quiz 8.6

See Examples noted with Ex. **1.** $x < 1$ or $x > 6$ (Ex. 1) **2.** $-\frac{1}{2} < x < \frac{2}{3}$ (Ex. 2) **3.** $x \le -5.5$ or $x \ge 1.5$ (Ex. 3) **4.** See Student Solutions Manual

Use Math to Save Money

1. The Gold plan **2.** The Silver plan **3.** The Silver plan **4.** The Bronze plan **5.** \$30 **6.** \$60 **7.** \$90 **8.** The Silver plan **9.** \$2319

You Try It

1. $x = \pm 2\sqrt{5}$ **2.** $x = \dfrac{-3 \pm \sqrt{15}}{2}$ **3.** $2x^2 - 6x - 11 = 0$ **4.** $x = 4 \pm \sqrt{21}$ **5.** $x = 1$ **6. (a)** $x = \frac{-y}{2}, x = 3y$ **(b)** $x = \pm 2\sqrt{a - y^2}$ **(c)** $x = \dfrac{w \pm \sqrt{w^2 + 36w}}{4}$ **7.** $b = 5\sqrt{3}$ **8.**

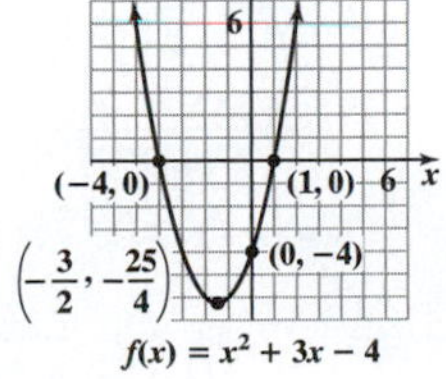

$f(x) = x^2 + 3x - 4$

9. $-\frac{3}{2} < x < 3$

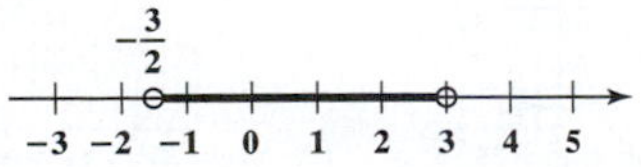

Chapter 8 Review Problems

1. $x = \pm 2$ **2.** $x = 1, -17$ **3.** $x = -4 \pm \sqrt{3}$ **4.** $x = \dfrac{2 \pm \sqrt{3}}{2}$ or $1 \pm \dfrac{\sqrt{3}}{2}$ **5.** $x = 2 \pm \sqrt{6}$ **6.** $x = \frac{2}{3}, 2$ **7.** $x = \frac{3}{2}$ **8.** $x = 7, -2$ **9.** $x = 0, \frac{9}{2}$ **10.** $x = \dfrac{5 \pm \sqrt{17}}{4}$ **11.** $x = \pm\sqrt{2}$ **12.** $x = \frac{2}{3}, 5$ **13.** $x = -1 \pm \sqrt{5}$ **14.** $x = \pm 2i\sqrt{3}$ **15.** $x = \dfrac{-5 \pm \sqrt{13}}{6}$ **16.** $x = -\frac{1}{2}, 4$ **17.** $x = \dfrac{-1 \pm i}{3}$ **18.** $x = -\frac{1}{4}, -1$ **19.** $y = -\frac{5}{6}, -2$ **20.** $y = -5, 3$ **21.** $y = 0, -\frac{5}{2}$ **22.** $x = -2, 3$ **23.** Two irrational solutions **24.** Two rational solutions **25.** One rational solution **26.** $x^2 - 25 = 0$

27. $x^2 + 9 = 0$ **28.** $8x^2 + 14x + 3 = 0$ **29.** $x = \pm 2, \pm\sqrt{2}$ **30.** $x = -\frac{\sqrt[3]{4}}{2}, \sqrt[3]{3}$ **31.** $x = 27, -1$ **32.** $x = \pm 1, \pm\sqrt{2}$ **33.** $A = \pm\sqrt{\frac{3MN}{2}}$ **34.** $t = \pm\sqrt{3ay - 2b}$ **35.** $x = \frac{3 \pm \sqrt{9 + 28y}}{2y}$ **36.** $d = \frac{x}{4}, -\frac{x}{5}$ **37.** $y = \frac{-2a \pm \sqrt{4a^2 + 6a}}{2}$ **38.** $x = \frac{2 \pm \sqrt{4 - 6y^2 + 3AB}}{3}$ **39.** $c = \sqrt{22}$ **40.** $a = 4\sqrt{15}$ **41.** Approximately 3.3 mi **42.** Base is 7 cm; altitude is 20 cm **43.** Width is 7 m; length is 29 m **44.** 20 mph for 80 mi, 10 mph for 10 mi **45.** 50 mph during first part, 45 mph during rain **46.** Approximately 2.4 ft wide **47.** 0.5 m wide **48.** Vertex $(3, -2)$; y-intercept $(0, -11)$; no x-intercepts **49.** Vertex $(-5, 0)$; x-intercept $(-5, 0)$; y-intercept $(0, 25)$ **50.**

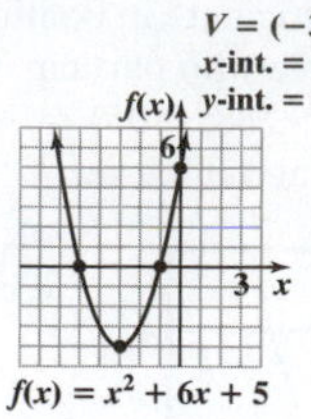

51.

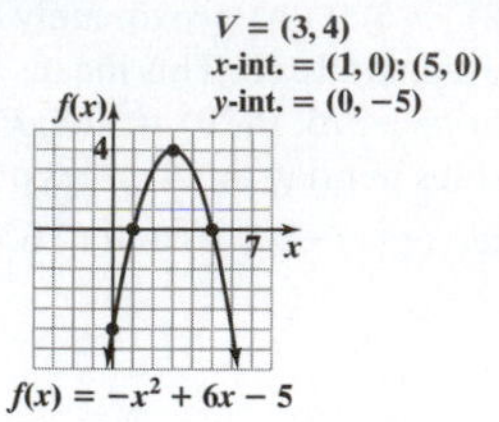

52. Maximum height is 2540 ft; 25.1 sec for complete flight **53.** $R(x) = x(1200 - x)$; the maximum revenue will occur if the price is \$600 for each unit. **54.** $-9 < x < 2$

55. $x < 4$ or $x > 5$

56. $-\frac{3}{2} \le x \le 2$ **57.** $\frac{4}{3} \le x \le 3$ **58.** $x < -\frac{2}{3}$ or $x > \frac{2}{3}$ **59.** $x \le -6$ or $x \ge -2$ **60.** $x < -8$ or $x > 2$ **61.** $x < 1.4$ or $x > 2.6$ **62.** No real solution **63.** $x < -4$ or $2 < x < 3$ **64.** $-4 < x < -1$ or $x > 2$

How Am I Doing? Chapter 8 Test

1. $x = 0, -\frac{9}{8}$ (obj. 8.1.2) **2.** $x = \frac{3 \pm \sqrt{33}}{12}$ (obj. 8.1.2) **3.** $x = 2, -\frac{2}{9}$ (obj. 8.2.1) **4.** $x = -2, 10$ (obj. 8.2.1) **5.** $x = \pm 2\sqrt{2}$ (obj. 8.1.1) **6.** $x = \frac{7}{2}, -1$ (obj. 8.2.1) **7.** $x = \frac{3 \pm i}{2}$ (obj. 8.2.1) **8.** $x = \frac{3 \pm \sqrt{3}}{2}$ (obj. 8.2.1) **9.** $x = \pm 3, \pm\sqrt{2}$ (obj. 8.3.1) **10.** $x = \frac{1}{5}, -\frac{3}{4}$ (obj. 8.3.2) **11.** $x = 64, -1$ (obj. 8.3.2) **12.** $z = \pm\sqrt{\frac{xyw}{B}}$ (obj. 8.4.1) **13.** $y = \frac{-b \pm \sqrt{b^2 - 30w}}{5}$ (obj. 8.4.1) **14.** Width is 5 mi; length is 16 mi (obj. 8.4.3) **15.** $c = 4\sqrt{3}$ (obj. 8.4.2) **16.** 2 mph during first part; 3 mph after lunch (obj. 8.4.3) **17.** $V = (-3, 4)$; y-int. $(0, -5)$; x-int. $(-5, 0), (-1, 0)$ (obj. 8.5.2)

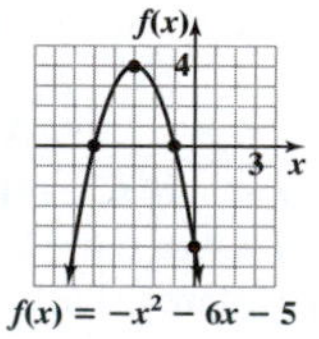

18. $x \le -\frac{9}{2}$ or $x \ge 3$ (obj. 8.6.1) **19.** $-2 < x < 7$ (obj. 8.6.1) **20.** $x < -4.5$ or $x > 1.5$ (obj. 8.6.2)

Chapter 9

9.1 Exercises

1. Subtract the values of the points and use the absolute value: $|-2 - 4| = 6$ **3.** Since the equation is given in standard form, we determine the values of h, k, and r to find the center and radius: $h = 1$, $k = -2$, and $r = 3$. Thus, the center is $(1, -2)$ and the radius is 3. **5.** $\sqrt{5}$ **7.** $3\sqrt{5}$ **9.** 10 **11.** $\frac{\sqrt{10}}{3}$ **13.** $\frac{2\sqrt{26}}{5}$ **15.** $5\sqrt{2}$ **17.** $y = 10, y = -6$ **19.** $y = 0, y = 4$ **21.** $x = 3, x = 5$ **23.** 4.4 mi **25.** $(x + 3)^2 + (y - 7)^2 = 36$ **27.** $(x + 2.4)^2 + y^2 = \frac{9}{16}$ **29.** $x^2 + \left(y - \frac{3}{8}\right)^2 = 3$

31.

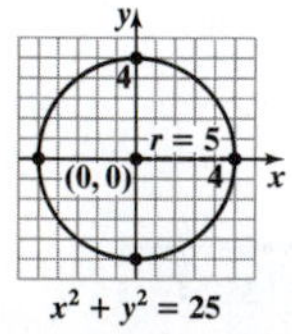

33.

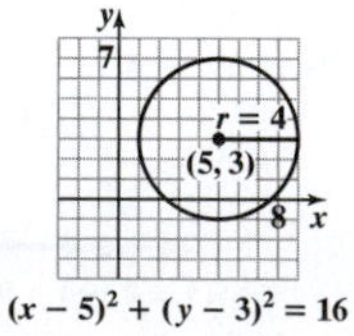

35.

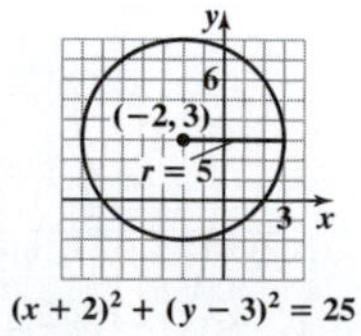

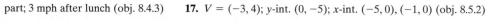

37. $(x + 4)^2 + (y - 3)^2 = 49$; center $(-4, 3)$, $r = 7$ **39.** $(x - 5)^2 + (y + 3)^2 = 36$; center $(5, -3)$, $r = 6$ **41.** $\left(x + \frac{3}{2}\right)^2 + y^2 = \frac{17}{4}$; center $\left(-\frac{3}{2}, 0\right)$, $r = \frac{\sqrt{17}}{2}$ **43.** $(x - 61.5)^2 + (y - 55.8)^2 = 1953.64$

45.

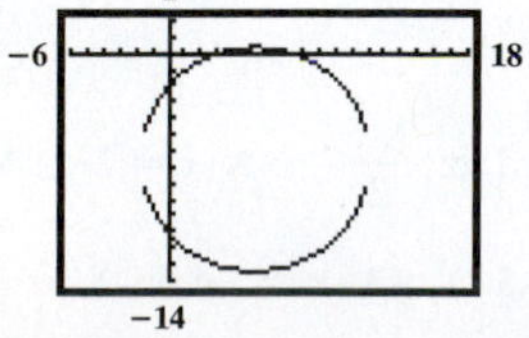

47. $x = \frac{-1 \pm \sqrt{5}}{4}$ **48.** $x = \frac{3 \pm 2\sqrt{11}}{5}$ **49.** Approximately 8.364×10^{10} ft^3 **50.** Approximately 81 sec

Quick Quiz 9.1

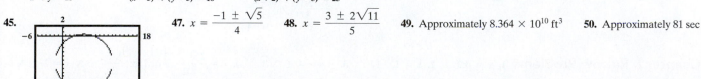

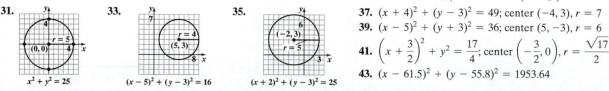

See Examples noted with Ex. **1.** $\sqrt{29}$ (Ex. 1) **2.** $(x - 5)^2 + (y + 6)^2 = 49$ (Ex. 3) **3.** $(x + 2)^2 + (y - 3)^2 = 9$; center $(-2, 3)$; $r = 3$ (Ex. 4) **4.** See Student Solutions Manual

9.2 Exercises

1. y-axis; x-axis **3.** If it is in the standard form $y = a(x - h)^2 + k$, the vertex is (h, k). So in this case the vertex is $(3, 4)$.

5.
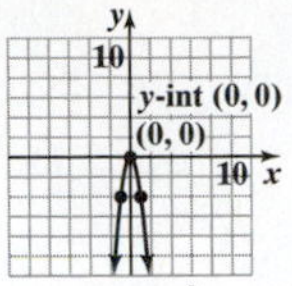

$y = -4x^2$

7.
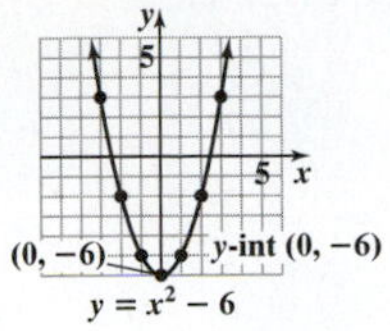

$y = x^2 - 6$

9.
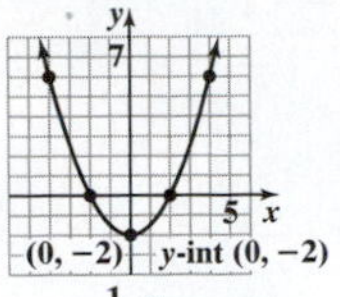

$y = \frac{1}{2}x^2 - 2$

11.
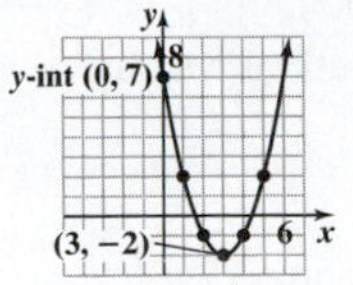

$y = (x - 3)^2 - 2$

13.
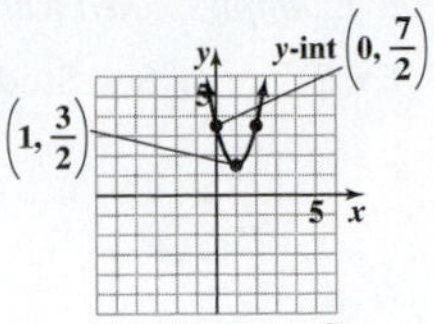

$y = 2(x - 1)^2 + \frac{3}{2}$

15.
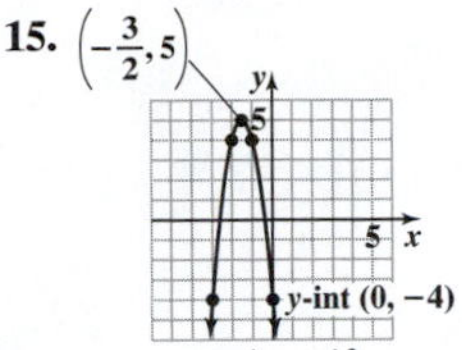

$y = -4\left(x + \frac{3}{2}\right)^2 + 5$

17.
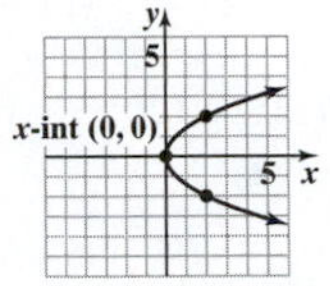

$x = \frac{1}{2}y^2$

19.
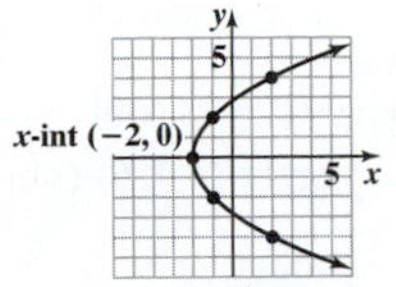

$x = \frac{1}{4}y^2 - 2$

21.
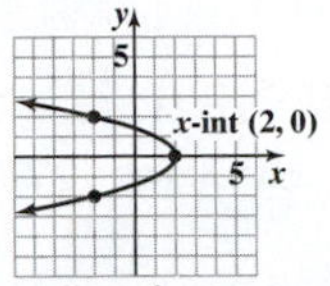

$x = -y^2 + 2$

23.
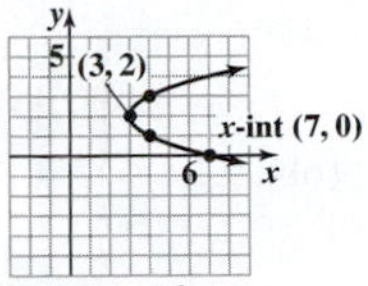

$x = (y - 2)^2 + 3$

25.
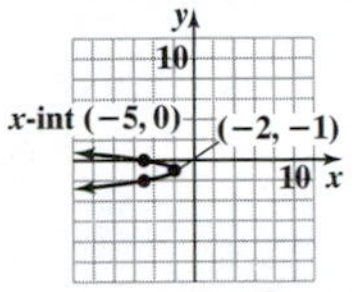

$x = -3(y + 1)^2 - 2$

27. $y = (x - 2)^2 - 5$ **(a)** Vertical **(b)** Opens upward **(c)** $(2, -5)$ **29.** $y = -2(x - 2)^2 + 7$ **(a)** Vertical **(b)** Opens downward **(c)** $(2, 7)$ **31.** $x = (y + 4)^2 - 7$ **(a)** Horizontal **(b)** Opens right **(c)** $(-7, -4)$ **33.** $y = \frac{1}{20}x^2$ **35.** 5 in. **37.** Vertex $(-1.62, -5.38)$; y-intercept $(0, -0.1312)$; x-intercepts are approximately $(0.02012195, 0)$ and $(-3.260122, 0)$ **39.** Maximum profit = \$55,000; number of watches produced = 100 **41.** Maximum yield = 202,500; number of trees planted per acre = 450 **43.** 28 mph **44.** $27\frac{1}{3}$ mi **45.** 7392 blooms **46.** Approximately 1098 buds

Quick Quiz 9.2

See Examples noted with Ex. **1.** Vertex $(-2, 5)$; y-intercept $(0, -7)$ (Ex. 2) **2.** $x = (y - 2)^2 + 3$ (Ex. 5)

3.
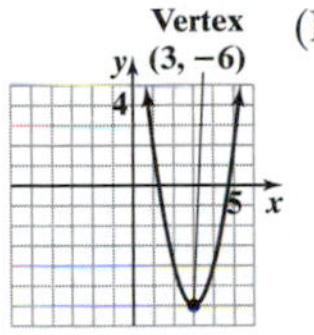

$y = 2(x - 3)^2 - 6$

(Ex. 7) **4.** See Student Solutions Manual

9.3 Exercises

1. In the ellipse $\frac{(x - h)^2}{a^2} + \frac{(y - k)^2}{b^2} = 1$ the center is at (h, k). In this case, the center of the ellipse is $(-2, 3)$.

3.
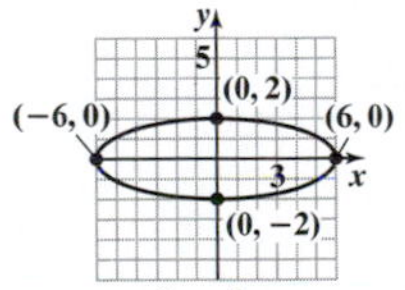

$\frac{x^2}{36} + \frac{y^2}{4} = 1$

5.
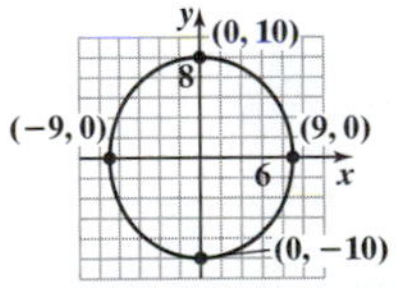

$\frac{x^2}{81} + \frac{y^2}{100} = 1$

7.
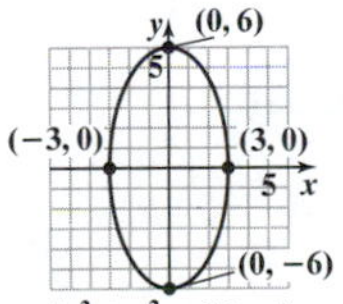

$4x^2 + y^2 - 36 = 0$

9.
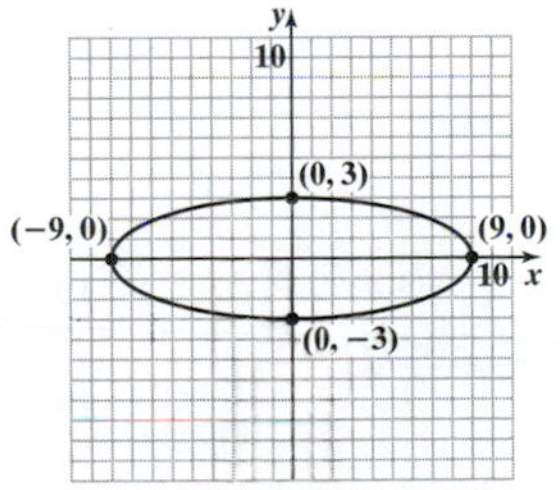

$x^2 + 9y^2 = 81$

11.
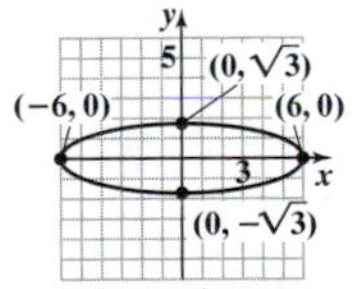

$x^2 + 12y^2 = 36$

13.
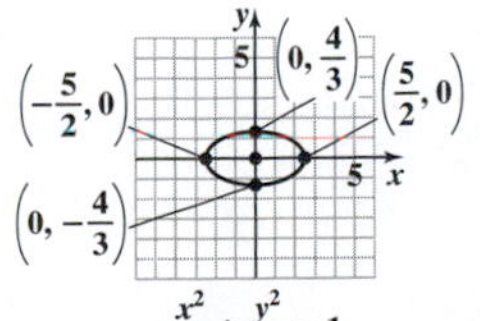

$\frac{x^2}{\frac{25}{4}} + \frac{y^2}{\frac{16}{9}} = 1$

15.
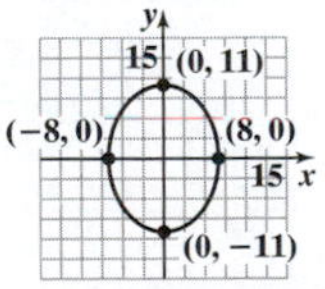

$121x^2 + 64y^2 = 7744$

17. $\frac{x^2}{169} + \frac{y^2}{144} = 1$ **19.** $\frac{x^2}{36} + \frac{y^2}{48} = 1$ **21.** 142 million mi

23.
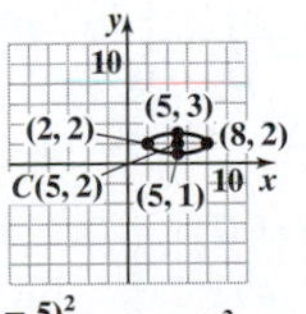

$\frac{(x - 5)^2}{9} + (y - 2)^2 = 1$

25.
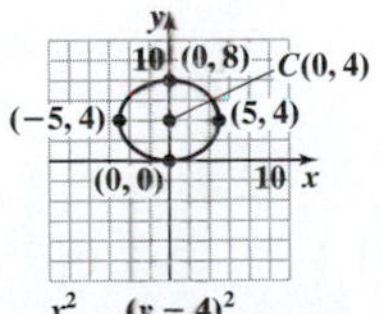

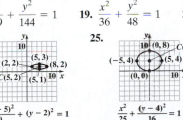

$\frac{x^2}{25} + \frac{(y - 4)^2}{16} = 1$

27.
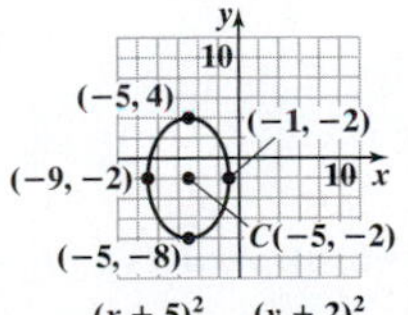

$\frac{(x + 5)^2}{16} + \frac{(y + 2)^2}{36} = 1$

29. $\frac{(x - 4)^2}{4} + \frac{(y - 3)^2}{16} = 1$ **31.** $\frac{(x - 30)^2}{900} + \frac{(y - 20)^2}{400} = 1$ **33.** $(0, 7.2768)$, $(0, 3.3232)$, $(4.2783, 0)$, $(2.9217, 0)$ **35.** $\frac{5(\sqrt{2x} + \sqrt{y})}{2x - y}$ **36.** $30\sqrt{2} - 2\sqrt{6} + 40\sqrt{3} - 8$ **37.** $22\frac{2}{3}$ weeks

Quick Quiz 9.3 *See Examples noted with Ex.* **1.** $\frac{x^2}{25} + \frac{y^2}{36} = 1$ (Ex. 1) **2.** $\frac{(x+3)^2}{16} + \frac{(y+2)^2}{9} = 1$ (Ex. 1)

3.

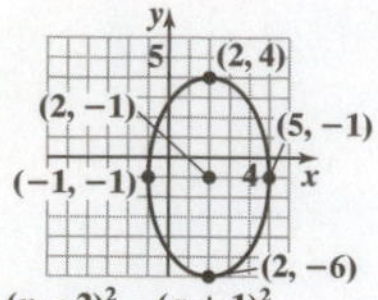

$\frac{(x-2)^2}{9} + \frac{(y+1)^2}{25} = 1$

(Ex. 2) **4.** See Student Solutions Manual

How Am I Doing? Sections 9.1–9.3 **1.** $(x+3)^2 + (y-7)^2 = 2$ (obj. 9.1.3) **2.** $3\sqrt{5}$ (obj. 9.1.1)

3.

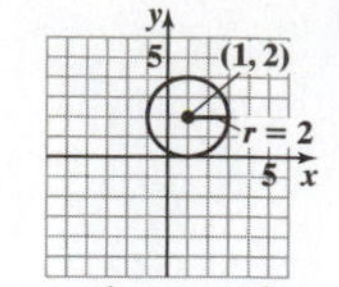

$(x-1)^2 + (y-2)^2 = 4$

(obj. 9.1.4) **4.** The line $x = 3$. (obj. 9.2.1) **5.** Vertex $(-4, 6)$ (obj. 9.2.1) **6.**

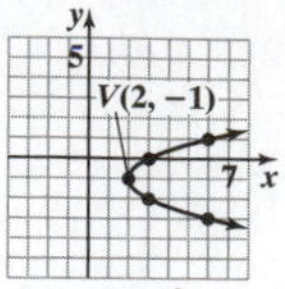

$x = (y+1)^2 + 2$

(obj. 9.2.2)

7.

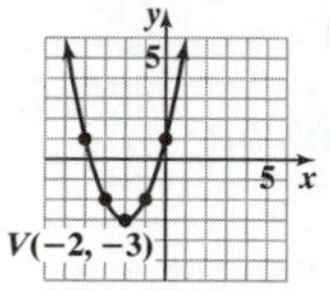

$y = (x+2)^2 - 3$

(obj. 9.2.3) **8.** $\frac{x^2}{100} + \frac{y^2}{49} = 1$ (obj. 9.3.1) **9.**

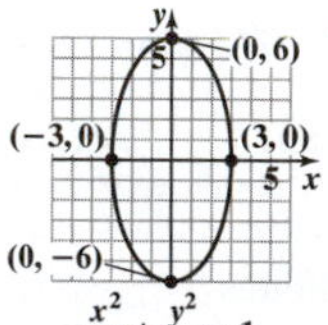

$\frac{x^2}{9} + \frac{y^2}{36} = 1$

(obj. 9.3.1) **10.**

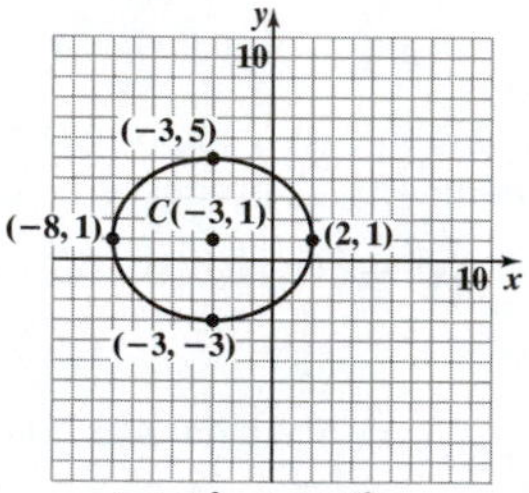

$\frac{(x+3)^2}{25} + \frac{(y-1)^2}{16} = 1$

(obj. 9.3.2)

9.4 Exercises **1.** The standard form of a horizontal hyperbola centered at the origin is $\frac{x^2}{a^2} - \frac{y^2}{b^2} = 1$ with a and b being positive real numbers.
3. This is a horizontal hyperbola, centered at the origin, with vertices at (4, 0) and (−4, 0). Draw a fundamental rectangle with corners at (4, 2), (4, −2), (−4, 2), and (−4, −2). Extend the diagonals of the rectangle as asymptotes of the hyperbola. Construct each branch of the hyperbola passing through a vertex and approaching the asymptotes. **5.**

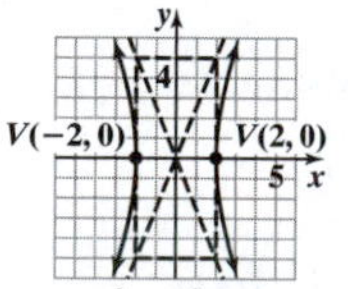

$\frac{x^2}{4} - \frac{y^2}{25} = 1$

7.

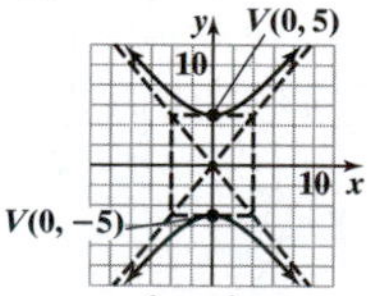

$\frac{y^2}{25} - \frac{x^2}{16} = 1$

9.

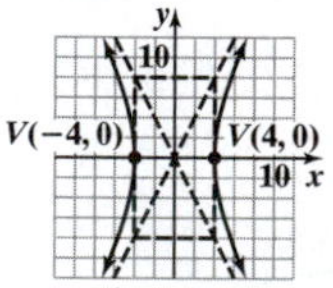

$\frac{x^2}{16} - \frac{y^2}{64} = 1$

11.

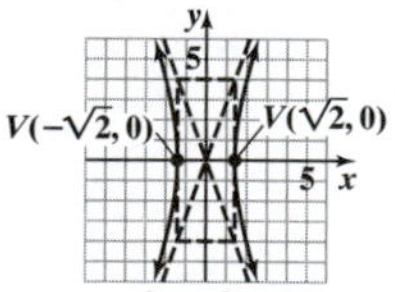

$\frac{x^2}{2} - \frac{y^2}{16} = 1$

13.

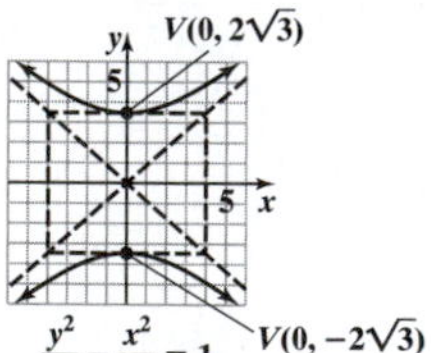

$\frac{y^2}{12} - \frac{x^2}{16} = 1$

15. $\frac{x^2}{9} - \frac{y^2}{16} = 1$ **17.** $\frac{y^2}{121} - \frac{x^2}{169} = 1$

19. $\frac{x^2}{14{,}400} - \frac{y^2}{129{,}600} = 1$, where x and y are measured in millions of miles

21.

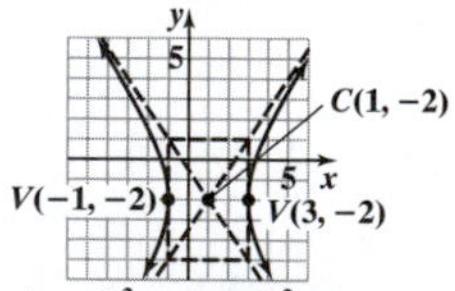

$\frac{(x-1)^2}{4} - \frac{(y+2)^2}{9} = 1$

23.

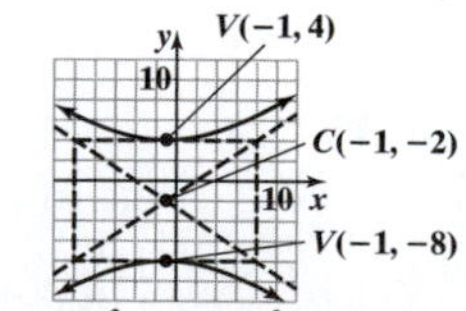

$\frac{(y+2)^2}{36} - \frac{(x+1)^2}{81} = 1$

25. Center $(-6, 0)$; vertices $(-6 + \sqrt{7}, 0), (-6 - \sqrt{7}, 0)$ **27.** $\frac{(y+7)^2}{49} - \frac{(x-4)^2}{16} = 1$ **29.** $y = \pm 9.055385138$

31. $\frac{5x}{(x-3)(x-2)(x+2)}$ **32.** $\frac{-x-6}{(5x-1)(x+2)}$ or $-\frac{x+6}{(5x-1)(x+2)}$ **33.** \$90.8 million **34.** 86.3 million barrels

Quick Quiz 9.4 *See Examples noted with Ex.* **1.** (0, 1), (0, −1) (Ex. 2) **2.** $\frac{x^2}{16} - \frac{y^2}{25} = 1$ (Ex. 1) **3.**

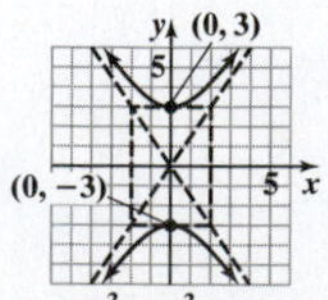

$\frac{y^2}{9} - \frac{x^2}{4} = 1$

(Ex. 1)

4. See Student Solutions Manual

9.5 Exercises

1. $\left(\frac{1}{2}, 1\right), (2, -2)$

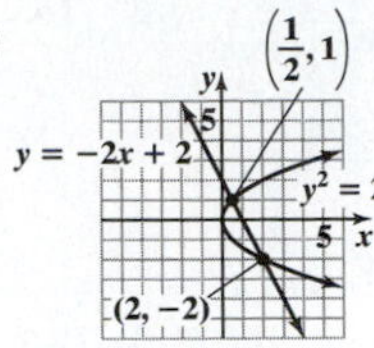

3. $(-4, 2), (4, -2)$

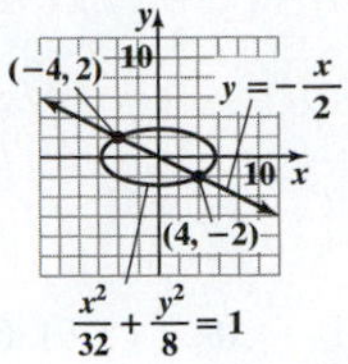

5. $(-2, 3), (1, 0)$ **7.** $(-5, 0), (4, 3)$ **9.** $\left(\frac{2}{3}, \frac{4}{3}\right), (2, 0)$ **11.** $\left(\frac{5}{2}, \frac{3}{2}\right)$

13. $(3, 2), (-3, 2), (3, -2), (-3, -2)$

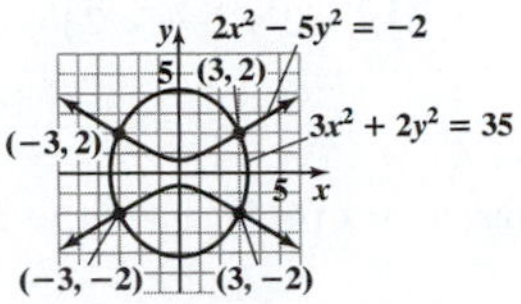

15. $(2, \sqrt{5}), (2, -\sqrt{5}), (-2, \sqrt{5}), (-2, -\sqrt{5})$

17. $\left(\frac{\sqrt{30}}{3}, \frac{\sqrt{21}}{3}\right), \left(-\frac{\sqrt{30}}{3}, \frac{\sqrt{21}}{3}\right), \left(\frac{\sqrt{30}}{3}, -\frac{\sqrt{21}}{3}\right), \left(-\frac{\sqrt{30}}{3}, -\frac{\sqrt{21}}{3}\right)$

19. $(2, \sqrt{3}), (-2, \sqrt{3}), (2, -\sqrt{3}), (-2, -\sqrt{3})$ **21.** $\left(5, \frac{1}{2}\right), \left(-2, -\frac{5}{4}\right)$

23. $(-3, 2), (1, -6)$ **25.** No real solution **27.** Yes, the hyperbola intersects the circle; (3290, 2280) **29.** $x^2 - 3x - 10$ **30.** $\frac{2x(x + 1)}{x - 2}$

Quick Quiz 9.5

See Examples noted with Ex. **1.** $(4, 4), (1, -2)$ (Ex. 1) **2.** $(0, -4), (\sqrt{7}, 3), (-\sqrt{7}, 3)$ (Ex. 3) **3.** $(-2, 4), (1, 1)$ (Ex. 2)
4. See Student Solutions Manual

Use Math to Save Money

1. Private loan: $I = (10{,}000)(0.0465)(20) = \$9300$. Total Amount $= 10{,}000 + 9300 = \$19{,}300$. Subsidized loan: $I = (10{,}000)(0.06)(20 - 4.5) = \$9300$. Total Amount $= 10{,}000 + 9300 = \$19{,}300$. **2.** The total amount that Alicia must pay back is the same for each loan. **3.** Total Amount ÷ Number of Payments = Monthly Payment. Private loan: 19,300 ÷ (12 months × 20 years) = 19,300 ÷ 240 = \$80.42. Subsidized loan: 19,300 ÷ (12 months × 15.5 years) = 19,300 ÷ 186 = \$103.76.
4. Private student loan **5.**

Type of Loan	Total Amount	Number of Monthly Payments	Monthly Payment Amount	When Payments Begin
Private loan	\$19,300	240	\$80.42	Immediately
Subsidized loan	\$19,300	186	\$103.76	6 months after graduation

6. Answers will vary.

You Try It

1. 5 **2.**

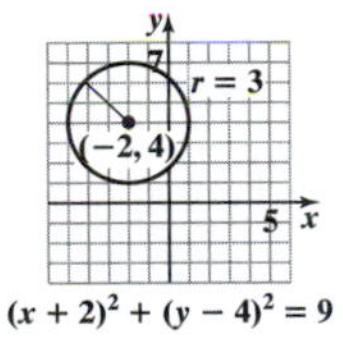

3.

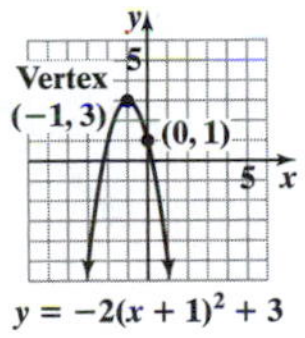

4.

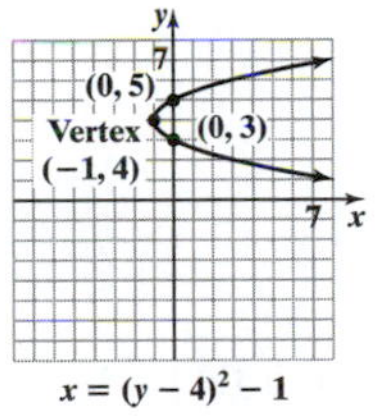

5.

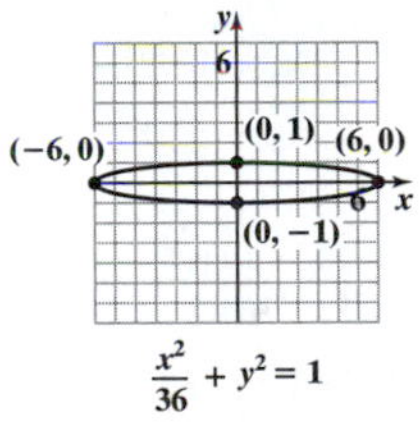

6.

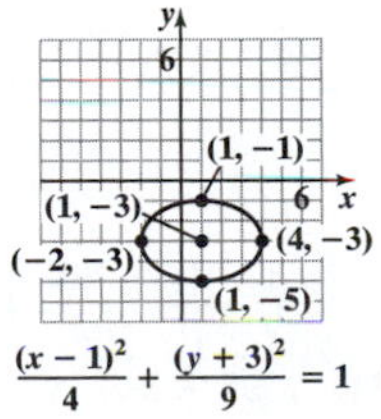

7.

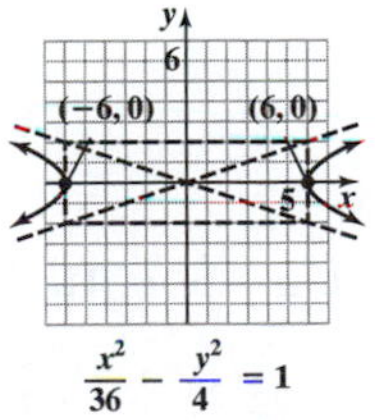

8.

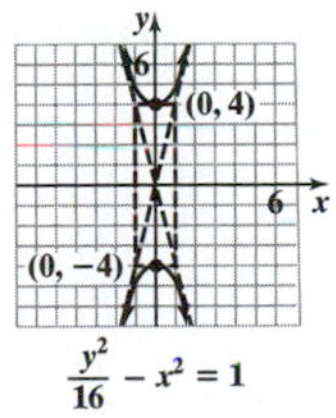

9.

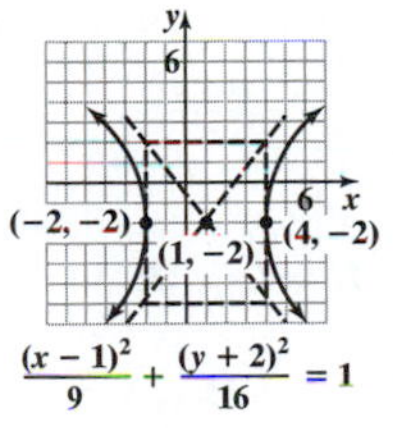

10.

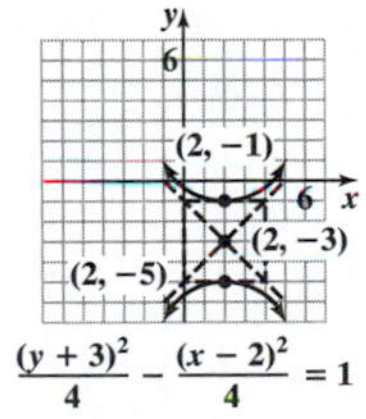

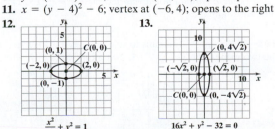

11. $(0, -2), \left(\frac{9}{7}, \frac{13}{7}\right)$

Chapter 9 Review Problems

1. $\sqrt{73}$ **2.** $\sqrt{41}$ **3.** $(x + 6)^2 + (y - 3)^2 = 15$ **4.** $x^2 + (y + 7)^2 = 25$
5. $(x + 1)^2 + (y - 3)^2 = 5$; center $(-1, 3)$, $r = \sqrt{5}$ **6.** $(x - 5)^2 + (y + 6)^2 = 9$; center $(5, -6)$, $r = 3$

7.

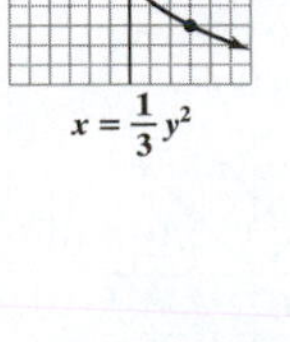

8.

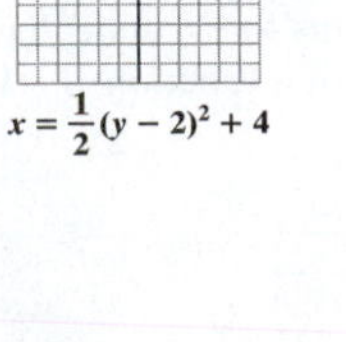

9.

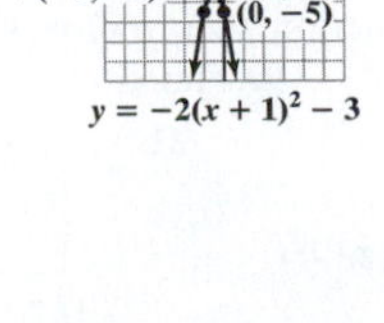

10. $y = (x + 3)^2 - 5$; vertex at $(-3, -5)$; opens upward
11. $x = (y - 4)^2 - 6$; vertex at $(-6, 4)$; opens to the right
12.

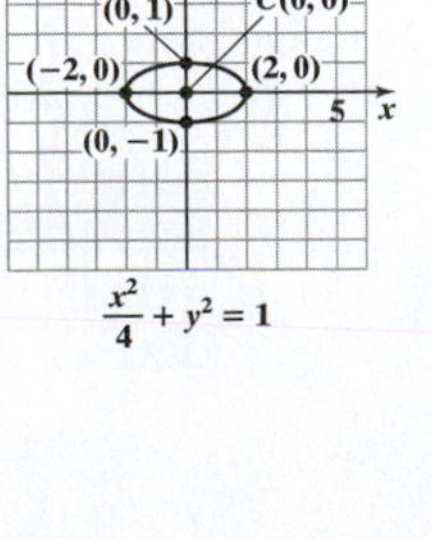

13.

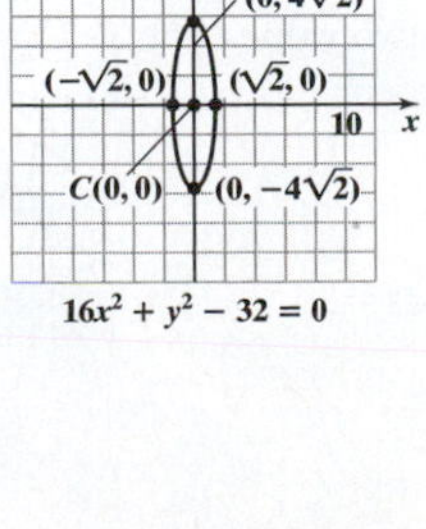

14. Center $(-5,-3)$; vertices are $(-3,-3)$, $(-7,-3)$, $(-5,2)$, $(-5,-8)$ **15.**

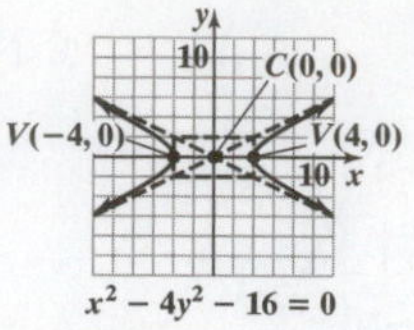

$x^2 - 4y^2 - 16 = 0$

16.

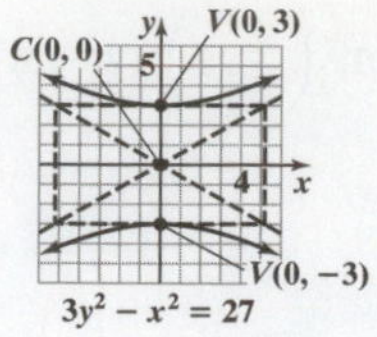

$3y^2 - x^2 = 27$

17. Vertices are $(0,-3)$, $(4,-3)$; center $(2,-3)$ **18.** $(-3,0)$, $(2,5)$ **19.** $(0,2)$, $(2,0)$ **20.** $(2,3)$, $(-2,3)$, $(2,-3)$, $(-2,-3)$
21. No real solution **22.** $(0,1)$, $(\sqrt{5},6)$, $(-\sqrt{5},6)$ **23.** $(1,4)$, $(-1,-4)$, $(2\sqrt{2},\sqrt{2})$, $(-2\sqrt{2},-\sqrt{2})$
24. $(1,2)$, $(-1,2)$, $(1,-2)$, $(-1,-2)$ **25.** $(2,2)$ **26.** 0.78 ft **27.** 1.56 ft

How Am I Doing? Chapter 9 Test

1. $\sqrt{185}$ (obj. 9.1.1) **2.** Parabola; vertex $(4,3)$; $x = (y-3)^2 + 4$ (obj. 9.2.3)

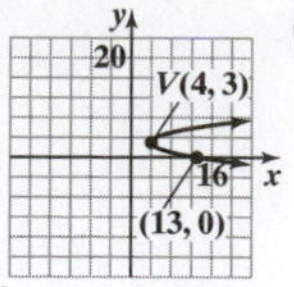

$y^2 - 6y - x + 13 = 0$

3. Circle; center $(-3,2)$; $(x+3)^2 + (y-2)^2 = 4$ (obj. 9.1.4)

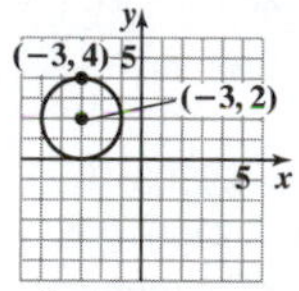

$x^2 + y^2 + 6x - 4y + 9 = 0$

4. Ellipse; center $(0,0)$ (obj. 9.3.1)

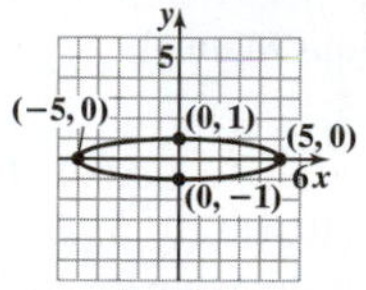

$\frac{x^2}{25} + \frac{y^2}{1} = 1$

5. Hyperbola; center $(0,0)$ (obj. 9.4.1)

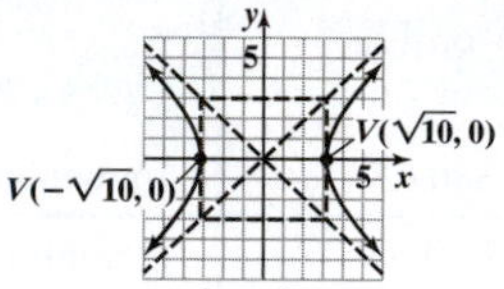

$\frac{x^2}{10} - \frac{y^2}{9} = 1$

6. Parabola; vertex $(-3,4)$ (obj. 9.2.1)

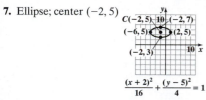

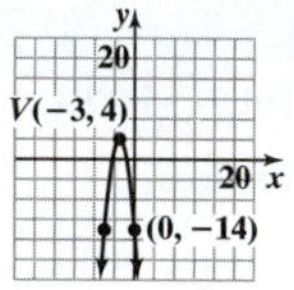

$y = -2(x+3)^2 + 4$

7. Ellipse; center $(-2,5)$ (obj. 9.3.2)

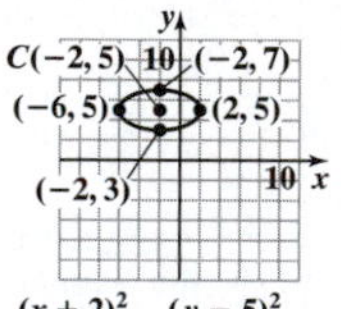

$\frac{(x+2)^2}{16} + \frac{(y-5)^2}{4} = 1$

8. Hyperbola; center $(0,0)$ (obj. 9.4.1)

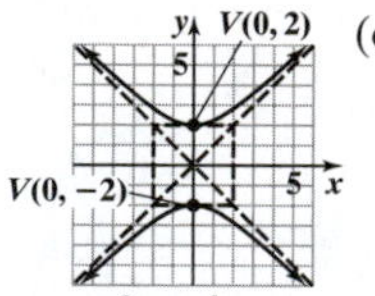

$7y^2 - 7x^2 = 28$

9. $(x-3)^2 + (y+5)^2 = 8$ (obj. 9.1.3) **10.** $\frac{x^2}{9} + \frac{y^2}{25} = 1$ (obj. 9.3.1) **11.** $x = (y-3)^2 - 7$ (obj. 9.2.3) **12.** $\frac{x^2}{9} - \frac{y^2}{25} = 1$ (obj. 9.4.1)
13. $(0,5)$, $(-4,-3)$ (obj. 9.5.1) **14.** $(0,-3)$, $(3,0)$ (obj. 9.5.1) **15.** $(1,0)$, $(-1,0)$ (obj. 9.5.2)
16. $(3,-\sqrt{3})$, $(3,\sqrt{3})$, $(-3,\sqrt{3})$, $(-3,-\sqrt{3})$ (obj. 9.5.2)

Chapter 10

10.1 Exercises

1. -7 **3.** $3a - 17$ **5.** $\frac{1}{2}a - 4$ **7.** $\frac{3}{2}a$ **9.** $a - 5$ **11.** $\frac{1}{2}a^2 - \frac{1}{5}$ **13.** 2 **15.** $\frac{3}{4}$
17. $3a^2 + 10a + 5$ **19.** $\frac{4a^2}{3} - \frac{8a}{3} - 2$ **21.** 3 **23.** $2\sqrt{3}$ **25.** $\sqrt{a^2+4}$ **27.** $\sqrt{-2b+5}$ **29.** $2\sqrt{a+1}$ **31.** $\sqrt{b^2+b+5}$
33. $\frac{7}{4}$ **35.** 14 **37.** $\frac{7}{a^2-3}$ **39.** $\frac{7}{a-1}$ **41.** $-\frac{7}{5}$ **43.** 2 **45.** $2x + h - 1$ **47. (a)** $P(w) = 2.5w^2$ **(b)** 1000 kilowatts
(c) $P(e) = 2.5e^2 + 100e + 1000$ **(d)** 1210 kilowatts **49.** The percent would decrease by 13; 26 **51.** $0.002a^2 - 0.120a + 1.23$ **53.** $x = -3$
54. $x = 5$ **55.** Approximately 6.7 times greater **56.** Approximately 14.4 times greater

Quick Quiz 10.1

See Examples noted with Ex. **1.** $\frac{3}{5}a + \frac{9}{5}$ (Ex. 1) **2.** $\frac{8}{9}a^2 - 2a + 4$(Ex. 2) **3.** $\frac{3}{a-2}$ (Ex. 3)
4. See Student Solutions Manual

10.2 Exercises

1. No, $f(x+2)$ means to substitute $x+2$ for x in the function $f(x)$. $f(x) + f(2)$ means to evaluate $f(x)$ and to evaluate $f(2)$ and then to add the two values. **3.** Up **5.** Not a function **7.** Function **9.** Function **11.** Not a function **13.** Function

15.

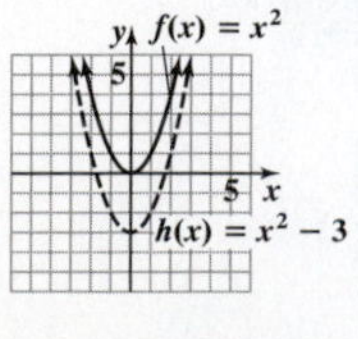

17.

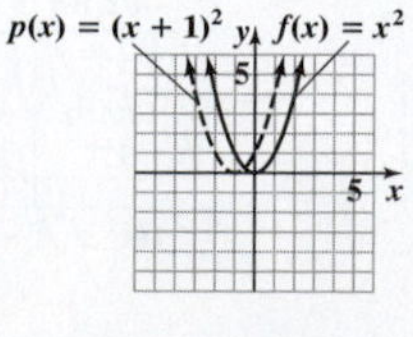

19.

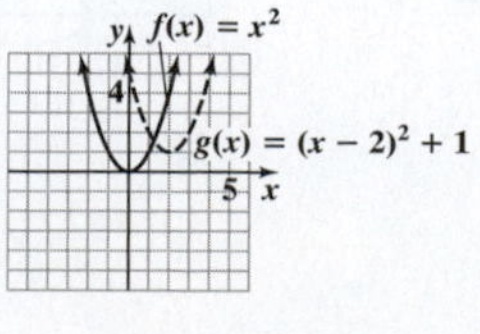

21.

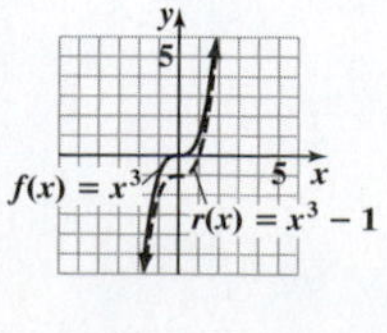

23.

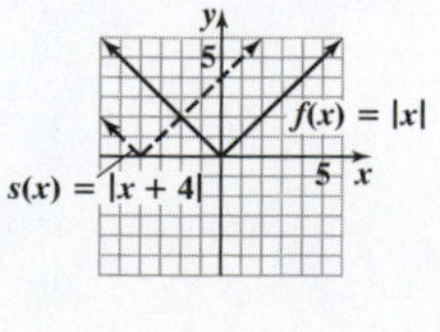

25. 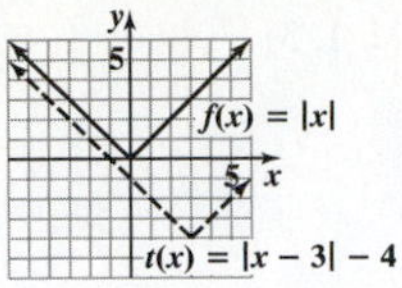

27. 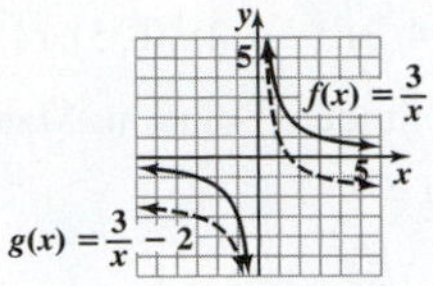

29. $15\sqrt{2} - 10\sqrt{3}$ **30.** $3x - 2\sqrt{3x} + 1$ **31.** $\frac{7 - 3\sqrt{5}}{4}$

Quick Quiz 10.2

See Examples noted with Ex. **1.** The graph of $h(x)$ is 3 units to the right of the graph of $f(x)$. (Ex. 3) **2.** The graph of $g(x)$ is 4 units below the graph of $f(x)$. (Ex. 2) **3.** 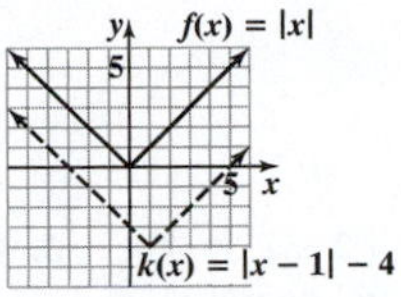

(Ex. 4) **4.** See Student Solutions Manual

How Am I Doing? Sections 10.1–10.2

1. -12 (obj. 10.1.1) **2.** $2a - 6$ (obj. 10.1.1) **3.** $4a - 6$ (obj. 10.1.1) **4.** $2a - 2$ (obj. 10.1.1) **5.** 13 (obj. 10.1.1) **6.** $5a^2 + 2a - 3$ (obj. 10.1.1) **7.** $5a^2 - 8a$ (obj. 10.1.1) **8.** $20a^2 - 4a - 3$ (obj. 10.1.1) **9.** $\frac{6(a^2 - 2)}{a(a + 2)}$ (obj. 10.1.1) **10.** $\frac{18(a - 1)}{5(3a + 2)}$ (obj. 10.1.1) **11.** Function (obj. 10.2.1) **12.** Not a function (obj. 10.2.1) **13.** 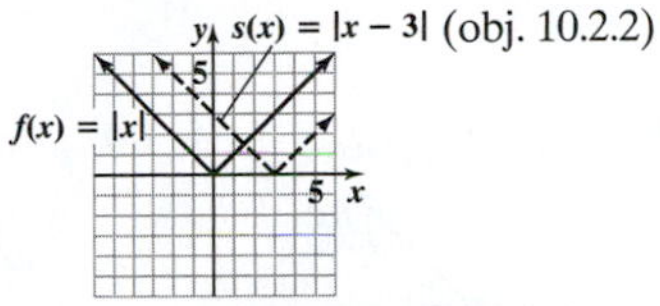

(obj. 10.2.2)

14. 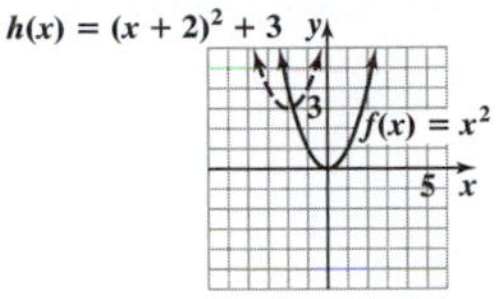

(obj. 10.2.2) **15.** 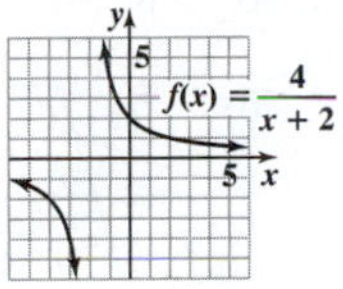

(obj. 10.2.2) **16.** The curve would be the same shape shifted 2 units down. (obj. 10.2.2)

10.3 Exercises

1. (a) $2x + 5$ **(b)** $-6x + 1$ **(c)** 9 **(d)** 7 **3. (a)** $3x^2 + 4x + 2$ **(b)** $3x^2 - 6x - 2$ **(c)** 22 **(d)** 7 **5. (a)** $x^3 + \frac{1}{2}x^2 + \frac{3}{4}x - 5$ **(b)** $x^3 - \frac{3}{2}x^2 + \frac{5}{4}x + 5$ **(c)** $\frac{13}{2}$ **(d)** $\frac{5}{4}$ **7. (a)** $3\sqrt{x + 6}$ **(b)** $-13\sqrt{x + 6}$ **(c)** $6\sqrt{2}$ **(d)** $-13\sqrt{5}$ **9. (a)** $-4x^3 + 11x - 3$ **(b)** 72 **11. (a)** $\frac{2(x - 1)}{x}, x \neq 0$ **(b)** $\frac{8}{3}$ **13. (a)** $-3x\sqrt{-2x + 1}$ **(b)** $9\sqrt{7}$ **15. (a)** $\frac{x - 6}{3x}, x \neq 0$ **(b)** $-\frac{2}{3}$ **17. (a)** $x + 1, x \neq 1$ **(b)** 3 **19. (a)** $x + 5, x \neq -5$ **(b)** 7 **21. (a)** $\frac{1}{x + 2}, x \neq -2, x \neq \frac{1}{4}$ **(b)** $\frac{1}{4}$ **23.** $-x^2 + 5x + 2$ **25.** $3x, x \neq 2$ **27.** -3 **29.** -3 **31.** $-6x - 13$ **33.** $2x^2 - 4x + 7$ **35.** $-5x^2 - 7$ **37.** $\frac{7}{2x + 1}, x \neq -\frac{1}{2}$ **39.** $|2x + 2|$ **41.** $9x^2 + 30x + 27$ **43.** $3x^2 + 11$ **45.** 11 **47.** $\sqrt{x^2 + 1}$ **49.** $\frac{3\sqrt{2}}{2} + 5$ **51.** $\sqrt{26}$ **53.** $K[C(F)] = \frac{5F + 2297}{9}$ **55.** $a[r(t)] = 28.26t^2$; 11,304 ft^2 **56.** $(6x - 1)^2$ **57.** $(5x^2 + 1)(5x^2 - 1)$ **58.** $(x + 3)(x - 3)(x + 1)(x - 1)$ **59.** $(3x - 1)(x - 2)$

Quick Quiz 10.3

See Examples noted with Ex. **1.** $5x^2 - 9x - 6$ (Ex. 1) **2.** $\frac{x^2}{4} - 2x + 1$ (Ex. 5) **3.** $\frac{1}{5}$ (Ex. 3) **4.** See Student Solutions Manual

10.4 Exercises

1. have the same second coordinate **3.** $y = x$ **5.** Yes, it passes the vertical line test. No, it does not pass the horizontal line test. **7.** Not one-to-one **9.** One-to-one **11.** One-to-one **13.** One-to-one **15.** Not one-to-one **17.** Not one-to-one **19.** $J^{-1} = \{(2, 8), (1, 1), (0, 0), (-2, -8)\}$ 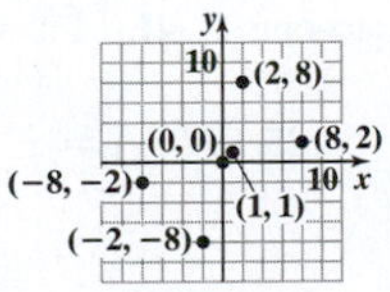

21. $f^{-1}(x) = \frac{x + 5}{4}$ **23.** $f^{-1}(x) = \sqrt[3]{x - 7}$ **25.** $f^{-1}(x) = -\frac{4}{x}$ **27.** $f^{-1}(x) = \frac{4}{x} + 5$ or $\frac{4 + 5x}{x}$ **29.** $f[f^{-1}(x)] = x$

31. $g^{-1}(x) = \frac{x - 5}{2}$ 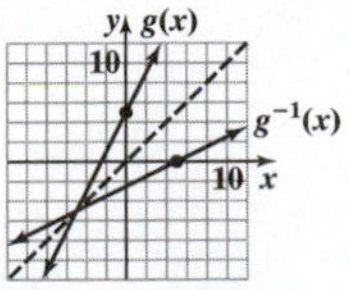

33. $h^{-1}(x) = 2x + 4$ 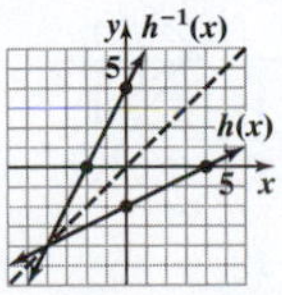

35. $r^{-1}(x) = -\frac{x + 1}{3}$ 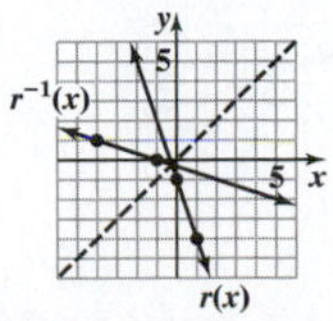

37. No, $f(x)$ is not one-to-one. **39.** $f[f^{-1}(x)] = 2\left[\frac{1}{2}x - \frac{3}{4}\right] + \frac{3}{2} = x$; $f^{-1}[f(x)] = \frac{1}{2}\left[2x + \frac{3}{2}\right] - \frac{3}{4} = x$ **41.** $x = 4$ **42.** $x = -64, x = -27$ **43.** 1,153,333 people **44.** 34%

Quick Quiz 10.4

See Examples noted with Ex. **1. (a)** Yes **(b)** Yes **(c)** $A^{-1} = \{(-4, 3), (-6, 2), (6, 5), (4, -3)\}$ (Ex. 1, 3)

2. $f^{-1}(x) = \dfrac{-x + 5}{2}$ (Ex. 6)

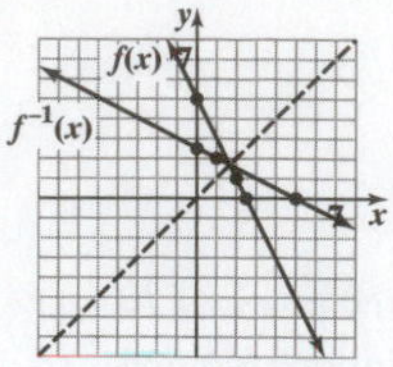

$f(x) = 5 - 2x$
$f^{-1}(x) = \dfrac{-x + 5}{2}$

3. $f^{-1}(x) = \sqrt[3]{2 - x}$ (Ex. 4) **4.** See Student Solutions Manual

Use Math to Save Money

1. December \$295.00, November \$355.00 **2.** \$1775.00 **3.** 4° lower = 8% savings, \$1775.00 × 0.08 = \$142 **4.** \$355 = 20% of \$1775; 20% = 10 degrees lower, 72 degrees − 10 degrees = 62 degrees **5.** Answers will vary. **6.** Answers will vary.

You Try It

1. (a) -5 **(b)** $-a^2 + a - 5$ **(c)** $-a^2 + 3a - 7$ **(d)** $-16a^2 + 4a - 5$ **2.** Yes **3. (a)**

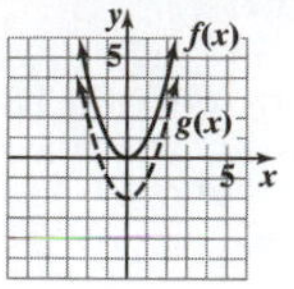

(b)

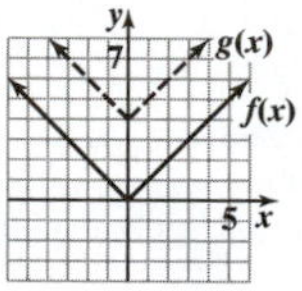

4. (a)

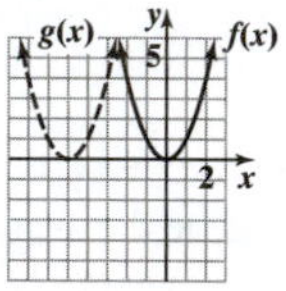

(b)

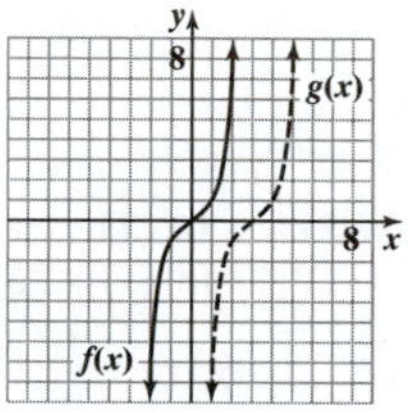

5. (a) $x^2 + 4x - 6$ **(b)** $x^2 + 2x + 6$ **(c)** $x^3 - 3x^2 - 18x$ **(d)** $\dfrac{x^2 + 3x}{x - 6}, x \neq 6$ **6. (a)** $-x^2 - 1$ **(b)** $x^2 - 4x + 7$ **7.** No **8.** No **9.** $\{(-2, 3),(4, 0),(1, 5)\}$ **10.** $f^{-1}(x) = 2x + 10$ **11.**

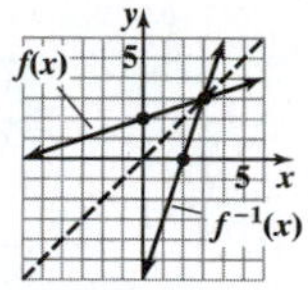

Review Problems

1. $\dfrac{1}{2}a + \dfrac{5}{2}$ **2.** $-\dfrac{1}{2}$ **3.** $\dfrac{1}{2}b^2 + \dfrac{3}{2}$ **4.** -28 **5.** $-8a^2 + 6a - 16$ **6.** $-2a^2 - 5a - 3$ **7.** 1 **8.** $\left|\dfrac{1}{2}a - 1\right|$ **9.** $|4a^2 - 6a - 1|$ **10.** $\dfrac{5}{3}$ **11.** $\dfrac{6a - 15}{2a - 1}$ **12.** $\dfrac{30a + 36}{7a + 28}$ **13.** 7 **14.** 6 **15.** $4x + 2h - 5$ **16. (a)** Not a function **(b)** Not one-to-one **17. (a)** Function **(b)** Not one-to-one **18. (a)** Not a function **(b)** Not one-to-one **19. (a)** Function **(b)** One-to-one

20.

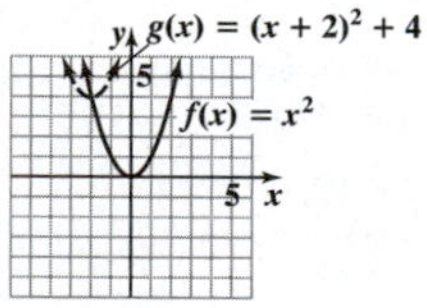

21.

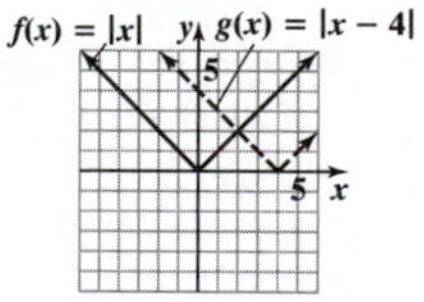

22.

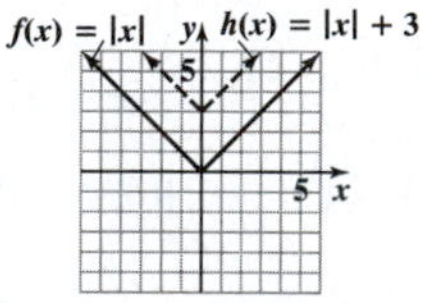

23.

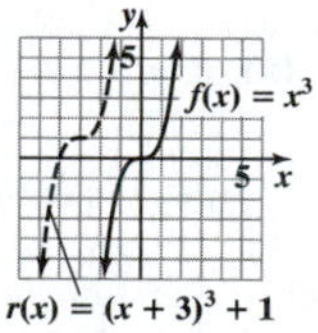

24.

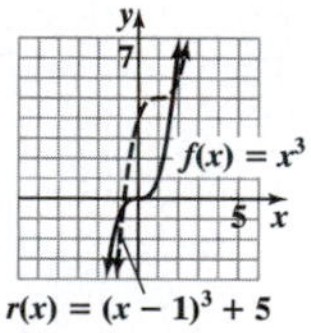

25.

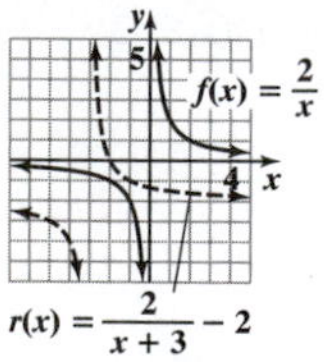

26. $\dfrac{5}{2}x + 2$ **27.** $2x^2 - 6x - 1$ **28.** -5 **29.** $\dfrac{6x + 10}{x}, x \neq 0$ **30.** $\dfrac{2x - 8}{x^2 + x}, x \neq 0, x \neq -1, x \neq 4$ **31.** -6 **32.** $18x^2 + 51x + 39$ **33.** $\sqrt{2x^2 - 3x + 2}$ **34.** 2 **35.** $f[g(x)] = \dfrac{6}{x} + 5 = \dfrac{6 + 5x}{x}$; $g[f(x)] = \dfrac{2}{3x + 5}$; $f[g(x)] \neq g[f(x)]$ **36.** $f^{-1}[f(x)] = x$ **37. (a)** Domain $= \{0, 3, 7\}$ **(b)** Range $= \{-8, 3, 7, 8\}$ **(c)** Not a function **(d)** Not one-to-one **38. (a)** Domain $= \{100, 200, 300, 400\}$ **(b)** Range $= \{10, 20, 30\}$ **(c)** Function **(d)** Not one-to-one **39. (a)** Domain $= \left\{-\dfrac{1}{3}, \dfrac{1}{4}, \dfrac{1}{2}, 4\right\}$ **(b)** Range $= \left\{-3, \dfrac{1}{4}, 2, 4\right\}$ **(c)** Function **(d)** One-to-one

40. (a) Domain $= \{0, 1, 2, 3\}$ **(b)** Range $= \{-3, 1, 7\}$ **(c)** Function **(d)** Not one-to-one **41.** $A^{-1} = \left\{\left(\dfrac{1}{3}, 3\right), \left(-\dfrac{1}{2}, -2\right), \left(-\dfrac{1}{4}, -4\right), \left(\dfrac{1}{5}, 5\right)\right\}$

42. $f^{-1}(x) = -\dfrac{4}{3}x + \dfrac{8}{3}$ **43.** $g^{-1}(x) = -\dfrac{1}{4}x - 2$ **44.** $h^{-1}(x) = \dfrac{6}{x} - 5$ or $\dfrac{6 - 5x}{x}$ **45.** $p^{-1}(x) = x^3 - 1$ **46.** $r^{-1}(x) = \sqrt[3]{x - 2}$

47. $f^{-1}(x) = -3x - 2$

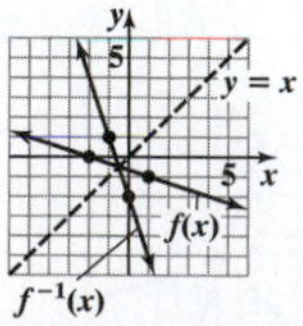

48. $f^{-1}(x) = -\dfrac{4}{3}x + \dfrac{4}{3}$

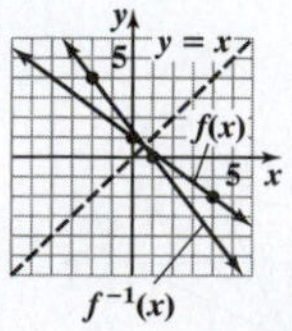

How Am I Doing? Chapter 10 Test

1. -8 (obj. 10.1.1) **2.** $\dfrac{9}{4}a - 2$ (obj. 10.1.1) **3.** $\dfrac{3}{4}a - \dfrac{3}{2}$ (obj. 10.1.1) **4.** 124 (obj. 10.1.1) **5.** $3a^2 + 4a + 5$ (obj. 10.1.1) **6.** $3a^2 - 2a + 9$ (obj. 10.1.1) **7.** $12a^2 + 4a + 2$ (obj. 10.1.1) **8. (a)** Function (obj. 10.2.1) **(b)** Not one-to-one (obj. 10.4.1) **9. (a)** Function (obj. 10.2.1) **(b)** One-to-one (obj. 10.4.1)

10. 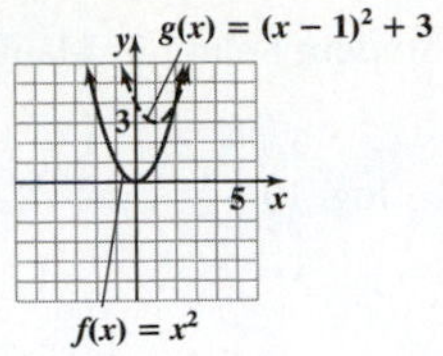

(obj. 10.2.2) **11.** 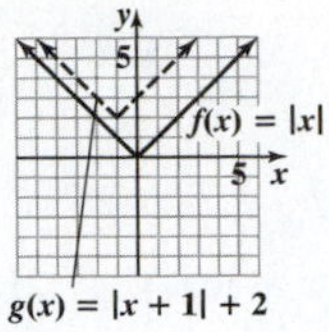

(obj. 10.2.2) **12. (a)** $x^2 + 4x + 1$ (obj. 10.3.1) **(b)** $5x^2 - 6x - 13$ (obj. 10.3.1)
(c) 19 (obj. 10.3.1) **13. (a)** $\frac{6x - 3}{x}, x \neq 0$ (obj. 10.3.1) **(b)** $\frac{3}{2x^2 - x}, x \neq 0, x \neq \frac{1}{2}$ (obj. 10.3.1) **(c)** $\frac{3}{2x - 1}, x \neq \frac{1}{2}$ (obj. 10.3.2)
14. (a) $2x - \frac{1}{2}$ (obj. 10.3.2) **(b)** $2x - 7$ (obj. 10.3.2) **(c)** 0 (obj. 10.3.2) **15. (a)** One-to-one (obj. 10.4.1)
(b) $B^{-1} = \{(8, 1), (1, 8), (10, 9), (9, -10)\}$ (obj. 10.4.2) **16. (a)** Not one-to-one (obj. 10.4.1) **(b)** Inverse cannot be found (obj. 10.4.2)
17. $f^{-1}(x) = \frac{x^3 + 1}{2}$ (obj. 10.4.2) **18.** $f^{-1}(x) = -\frac{1}{3}x + \frac{2}{3}$ 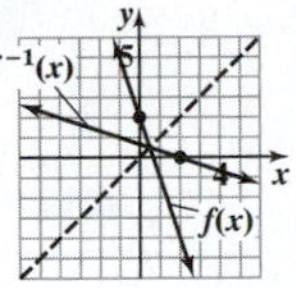

(obj. 10.4.3) **19.** $f^{-1}[f(x)] = x$ (obj. 10.3.2) **20.** -8 (obj. 10.1.1)

Chapter 11

To Think About, page 585 The graphs of $f(x) = 2^x$ and $f(x) = 2^{-x}$ are reflections of each other across the y-axis.

To Think About, page 586 **1.** To graph $f(x) = 3^{x+2}$, take the graph of $f(x) = 3^x$ and shift it two units to the left. **2.** The graph of $f(x) = e^x$ gets shifted two units to the right to obtain the graph of $f(x) = e^{x-2}$.

11.1 Exercises **1.** $f(x) = b^x$, where $b > 0, b \neq 1$, and x is a real number. **3.** 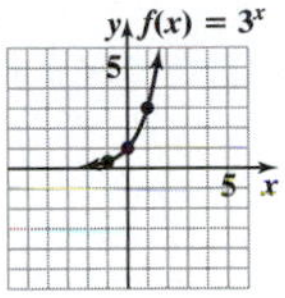

5. 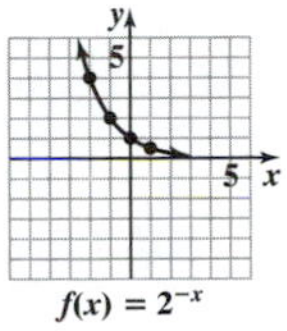

7. 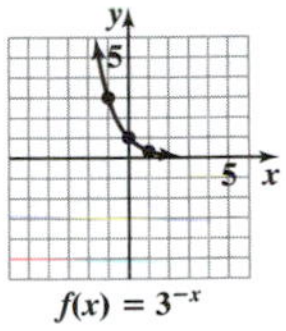

9. 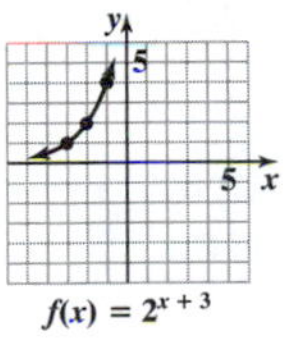

11. 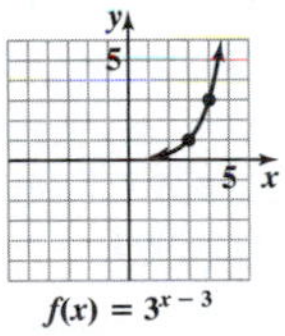

13. 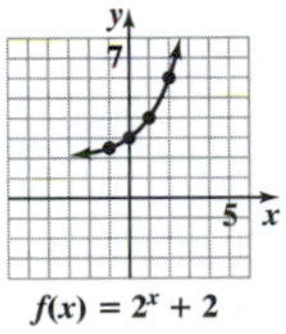

15. 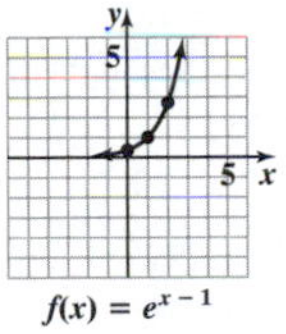

17. 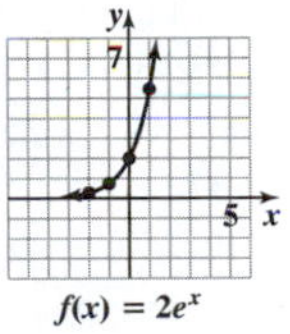

19. 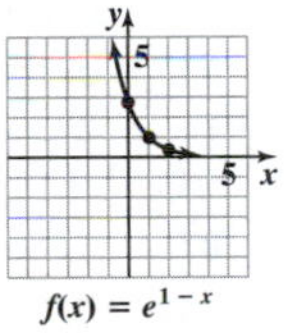

21. $x = 2$ **23.** $x = 0$ **25.** $x = -1$ **27.** $x = 4$ **29.** $x = 0$ **31.** $x = 2$ **33.** $x = 4$ **35.** $x = 2$ **37.** $x = 1$ **39.** \$2402.31 **41.** \$3632.24; \$3634.08 **43.** 32,000; 2,048,000 **45.** 37%; yes **47.** 3.91 mg **49.** 1.80 lb/in.^{2} **51.** \$1,191,000; about \$2,118,680; about 78% **53.** 1955 **55.** 10.1 billion people; about 8 billion people **57.** $x = -6$ **58.** $x = -2$

Quick Quiz 11.1 *See Examples noted with Ex.* **1.** 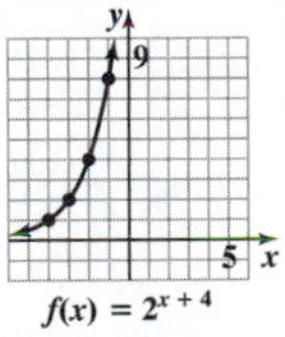

(Ex. 3) **2.** \$6975.33 (Ex. 7) **3.** $x = 2$ (Ex. 5) **4.** See Student Solutions Manual

11.2 Exercises **1.** exponent **3.** $x > 0$ **5.** $\log_7 49 = 2$ **7.** $\log_2 16 = 4$ **9.** $\log_{10}(0.001) = -3$ **11.** $\log_2\left(\frac{1}{32}\right) = -5$
13. $\log_e y = 5$ **15.** $3^2 = 9$ **17.** $17^0 = 1$ **19.** $16^{1/2} = 4$ **21.** $10^{-2} = 0.01$ **23.** $3^{-4} = \frac{1}{81}$ **25.** $e^{-3/2} = x$ **27.** $x = 16$
29. $x = \frac{1}{1000}$ **31.** $y = 3$ **33.** $y = -2$ **35.** $a = 11$ **37.** $a = 10$ **39.** $w = \frac{1}{2}$ **41.** $w = -1$ **43.** $w = 1$ **45.** $w = 9$
47. -3 **49.** 7 **51.** 0 **53.** $\frac{1}{2}$ **55.** 0 **57.** 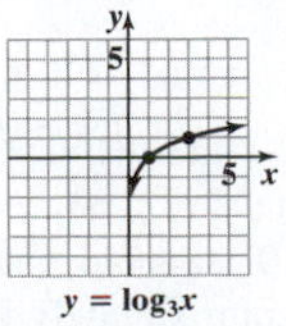

59. 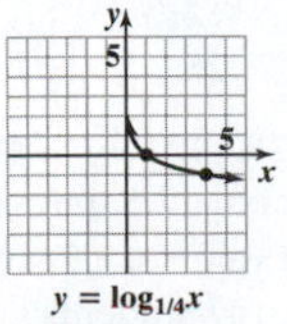

61. 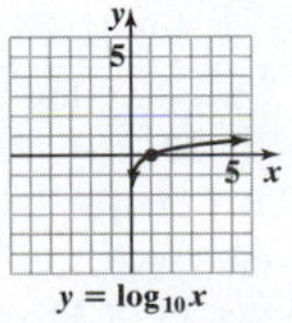

63. 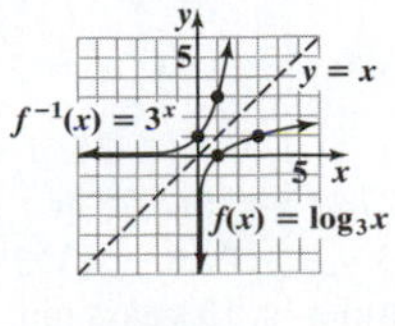

65. 10 **67.** $10^{-3.5}$ **69.** 2.957 **71.** 11,200 sets **73.** \$10,000,000 **75.** $y = \frac{3}{2}x + 7$ **76.** $m = -\frac{1}{5}$ **77. (a)** 36,000 cells **(b)** 36,864,000 cells **78. (a)** \$5353.27 **(b)** \$24,424.03

Quick Quiz 11.2 *See Examples noted with Ex.* **1.** 4 (Ex. 4) **2.** $x = 64$ (Ex. 3) **3.** $w = \frac{1}{3}$ (Ex. 3) **4.** See Student Solutions Manual

11.3 Exercises **1.** $\log_3 A + \log_3 B$ **3.** $\log_5 7 + \log_5 11$ **5.** $\log_b 9 + \log_b f$ **7.** $\log_9 2 - \log_9 7$ **9.** $\log_b H - \log_b 10$ **11.** $\log_a E - \log_a F$ **13.** $7 \log_8 a$ **15.** $-2 \log_b A$ **17.** $\frac{1}{2}\log_5 w$ **19.** $2 \log_8 x + \log_8 y$ **21.** $\log_{11} 6 + \log_{11} M - \log_{11} N$ **23.** $\log_2 5 + \log_2 x + 4 \log_2 y - \frac{1}{2}\log_2 z$ **25.** $\frac{4}{3}\log_a x - \frac{1}{3}\log_a y$ **27.** $\log_4 39y$ **29.** $\log_3\left(\frac{x^5}{7}\right)$ **31.** $\log_b\left(\frac{49y^3}{\sqrt{z}}\right)$ **33.** 1 **35.** 1 **37.** 0 **39.** 3 **41.** $x = 7$ **43.** $x = 11$ **45.** $x = 0$ **47.** $x = 1$ **49.** $x = 4$ **51.** $x = 21$ **53.** $x = 2$ **55.** $x = 5e$ **57.** $x = 3$ **59.** 6 **61.** Approximately 62.8 m^3 **62.** Approximately 50.2 m^2 **63.** $(3, -2)$ **64.** $(-1, -2, 3)$ **65.** \$7301; 17.22% **66.** \$7422; 16.74%

Quick Quiz 11.3 *See Examples noted with Ex.* **1.** $\frac{1}{3}\log_5 x - 4 \log_5 y$ (Ex. 6) **2.** $\log_6\left(\frac{x^3y}{5}\right)$ (Ex. 5) **3.** $x = 625$ (Ex. 10) **4.** See Student Solutions Manual

How Am I Doing? Sections 11.1–11.3 **1.**

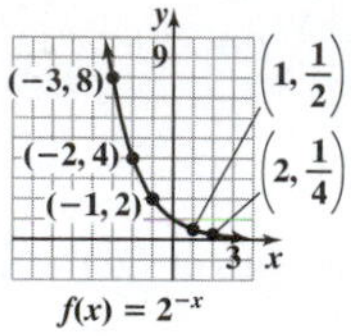

(obj. 11.1.1) **2.** $x = 2$ (obj. 11.1.2) **3.** $x = -5$ (obj. 11.1.2) **4.** $x = -\frac{1}{3}$ (obj. 11.1.2) **5.** \$15,735.19 (obj. 11.1.3) **6.** $\log_7\left(\frac{1}{49}\right) = -2$ (obj. 11.2.1) **7.** $10^{-3} = 0.001$ (obj. 11.2.2) **8.** $x = 125$ (obj. 11.2.3) **9.** $x = \frac{1}{7}$ (obj. 11.2.3) **10.** 4 (obj. 11.2.3) **11.** $2 \log_5 x + 5 \log_5 y - 3 \log_5 z$ (obj. 11.3.3) **12.** $\log_4\left(\frac{\sqrt{x}}{w^3}\right)$ (obj. 11.3.3) **13.** $x = \frac{81}{2}$ (obj. 11.3.4) **14.** $x = 8$ (obj. 11.3.4) **15.** $x = 0$ (obj. 11.3.4) **16.** $x = \frac{9}{2}$ (obj. 11.3.4) **17.** $x = \frac{4}{3}$ (obj. 11.3.4) **18.** $x = \frac{e}{3}$ (obj. 11.3.4)

11.4 Exercises **1.** Error. You cannot take the logarithm of a negative number. **3.** 1.089905111 **5.** 1.408239965 **7.** 0.903089987 **9.** 5.096910013 **11.** −1.910094889 **13.** $x \approx 103.752842$ **15.** $x = 0.01$ **17.** $x \approx 8519.223264$ **19.** $x \approx 2{,}939{,}679.609$ **21.** $x \approx 0.000408037$ **23.** $x \approx 0.027089438$ **25.** 41,831,168.87 **27.** 0.082679911 **29.** 1.726331664 **31.** 0.425267735 **33.** 11.82041016 **35.** −5.15162299 **37.** $x \approx 2.585709659$ **39.** $x \approx 11.02317638$ **41.** $x \approx 0.951229425$ **43.** $x \approx 0.067205513$ **45.** 472.576671 **47.** 0.1188610637 **49.** 2.020006063 **51.** 1.02507318 **53.** −1.151699337 **55.** 0.91759992 **57.** −1.846414 **59.** 1.996254706 **61.** 7.3375877 **63.** 0.1535084 **65.** $x \approx 3.6593167 \times 10^8$ **67.** $x \approx 3.3149796$ **69.**

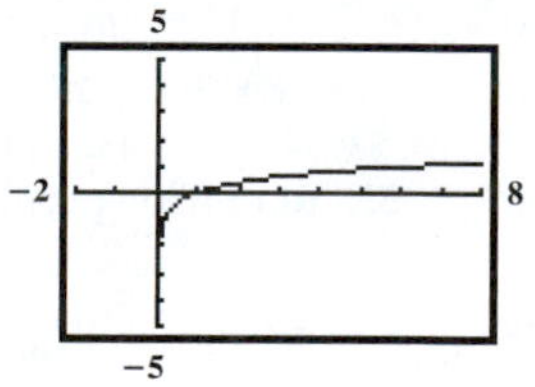

71.

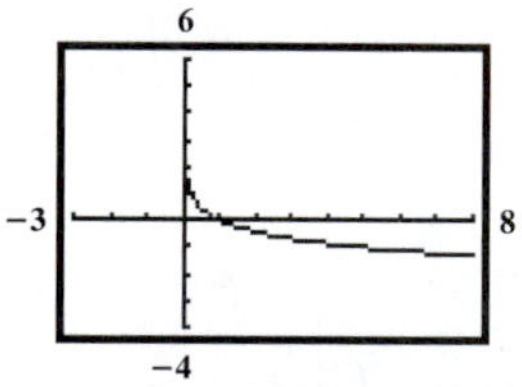

73. 35.27; 36.40; 3.2% **75.** $R \approx 4.75$ **77.** The shock wave was about 5,011,872 times greater than the smallest detectable shock wave. **79.** $x = \frac{11 \pm \sqrt{181}}{6}$ **80.** $y = \frac{-2 \pm \sqrt{10}}{2}$ **81.** 6 mi; 12 mi **82.** 21 mi; 24 mi

Quick Quiz 11.4 *See Examples noted with Ex.* **1.** 0.9713 (Ex. 1) **2.** $x \approx 1.68$ (Ex. 4) **3.** 1.3119 (Ex. 7) **4.** See Student Solutions Manual

11.5 Exercises **1.** $x = 20$ **3.** $x = 6$ **5.** $x = 4$ **7.** $x = 2$ **9.** $x = \frac{3}{2}$ **11.** $x = 2$ **13.** $x = 3$ **15.** $x = 5$ **17.** $x = \frac{5}{3}$ **19.** $x = 4$ **21.** $x = 5$ **23.** $x = \frac{\log 12 - 3 \log 7}{\log 7}$ **25.** $x = \frac{\log 17 - 4 \log 2}{3 \log 2}$ **27.** $x \approx 1.582$ **29.** $x \approx 6.213$ **31.** $x \approx 5.332$ **33.** $x \approx 1.739$ **35.** $t \approx 16$ yr **37.** $t \approx 19$ yr **39.** 4.5% **41.** 40 yr **43.** 31 yr **45.** 17,876 biomedical engineers **47.** 2015 **49.** 27 yr **51.** 55 hr **53.** 42,941 people **55.** About 12.6 times more intense **57.** About 31.6 times more intense **59.** $3\sqrt{2} - \sqrt{6} + 4\sqrt{3} - 4$ **60.** $7xy\sqrt{2x}$ **61.** 9 years old, 2 students; 10 years old, 1 student; 11 years old, 23 students; 12 years old, 39 students; 13 years old, 61 students; 14 or 15 years old, 147 students **62.** Approximately \$551,285,282 per mi

Quick Quiz 11.5 *See Examples noted with Ex.* **1.** $x = 4$ (Ex. 1) **2.** $x = 6$ (Ex. 2 and 3) **3.** $x \approx -0.774$ (Ex. 5) **4.** See Student Solutions Manual

Use Math to Save Money **1.** 8/3 or 2 2/3 cups **2.** 56 oz **3.** 3.5 lb **4.** \$9.31 **5.** \$6.23 **6.** Lucy saves approximately 67%. **7.** \$162 × 52 = \$8424; \$8424 × 67% = \$5644.08 **8.** \$1664

You Try It **1.**

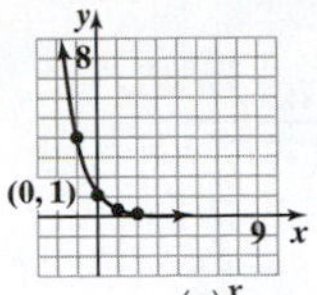

$f(x) = \left(\frac{1}{4}\right)^x$

2. $x = -4$ **3. (a)** $4^{0.5x} = 9$ **(b)** $\log_5 28 = x$ **(c)** $x = -3$ **4. (a)** $\log_2 a + \frac{1}{2}\log_2 b - 2\log_2 c$ **(b)** $\log_5\left(\frac{a^3b^{2/3}}{c^2}\right)$ **(c)** -2 **5. (a)** Approximately 1.2174839 **(b)** Approximately 3.4873751 **6. (a)** $x \approx 118.8502227$ **(b)** $x \approx 4.6089497$ **7.** Approximately 0.4406427 **8.** $x = 3$ **9.** $x \approx 0.1826583$

Chapter 11 Review Problems **1.**

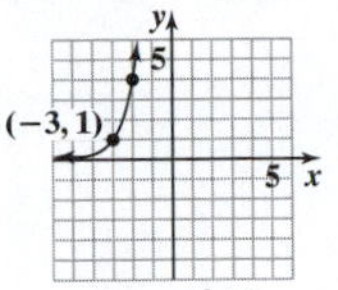

$f(x) = 4^{3+x}$

2.

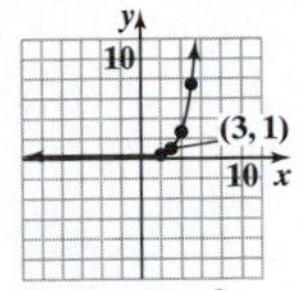

$f(x) = e^{x-3}$

3. $x = 1$ **4.** $x = -2$ **5.** $10^{-2} = 0.01$ **6.** $\log_4 8 = \frac{3}{2}$ **7.** $w = 2$ **8.** $x = 1$ **9.** $w = \frac{1}{7}$ **10.** $w = 4$ **11.** $w = 0.1$ or $\frac{1}{10}$ **12.** $x = 3$ **13.** $x = 6$ **14.** $x = -2$

15.

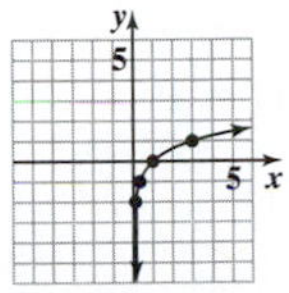

$y = \log_3 x$

16. $\log_2 5 + \log_2 x - \frac{1}{2}\log_2 w$ **17.** $3\log_2 x + \frac{1}{2}\log_2 y$ **18.** $\log_3\left(\frac{x\sqrt{w}}{2}\right)$ **19.** $\log_8\left(\frac{w^4}{\sqrt[3]{z}}\right)$ **20.** 6 **21.** 1.376576957 **22.** −1.087777943 **23.** 1.366091654 **24.** 6.688354714 **25.** $n \approx 13.69935122$ **26.** $n \approx 5.473947392$ **27.** 0.49685671 **28.** 3.084962501 **29.** $x = 25$ **30.** $x = 25$ **31.** $x = 25$ **32.** $x = 12$ **33.** $t = \frac{3}{4}$ **34.** $t = -\frac{1}{8}$ **35.** $x = \frac{\log 14}{\log 3}$ **36.** $x = \frac{\log 130 - 3\log 5}{\log 5}$ **37.** $x = \frac{\ln 100 + 1}{2}$ **38.** $x \approx -1.4748$ **39.** $x \approx 2.3319$ **40.** $x \approx 101.3482$ **41.** 9 yr **42.** \$6312.38 **43.** 41 yr **44.** 9 yr **45. (a)** Approximately 282 lb **(b)** 7.77 lb/in.^{3} **46.** 50.1 times more intense

How Am I Doing? Chapter 11 Test **1.**

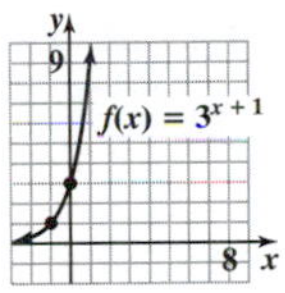

(obj. 11.1.1) **2.**

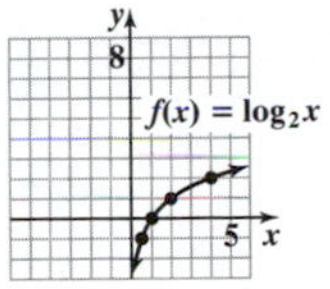

(obj. 11.2.4) **3.** $x = 0$ (obj. 11.1.2) **4.** $w = 5$ (obj. 11.2.2) **5.** $x = \frac{1}{64}$ (obj. 11.2.3) **6.** $\log_7\left(\frac{x^2y}{4}\right)$ (obj. 11.3.3) **7.** 1.7901 (obj. 11.4.3) **8.** 1.3729 (obj. 11.4.1) **9.** 0.4391 (obj. 11.4.5) **10.** $x \approx 5350.569382$ (obj. 11.4.2) **11.** $x \approx 1.150273799$ (obj. 11.4.4) **12.** $x = \frac{3}{7}$ (obj. 11.5.1) **13.** $x = \frac{16}{3}$ (obj. 11.5.1) **14.** $x = \frac{3 + \ln 57}{5}$ (obj. 11.5.2) **15.** $x \approx -1.4132$ (obj. 11.5.2) **16.** \$2938.66 (obj. 11.5.3) **17.** 14 yr (obj. 11.5.3)

Practice Final Examination **1.** 4 **2.** 3.625×10^7 **3.** $2a + 7b + 5ab + 3a^2$ **4.** $-9x - 15y$ **5.** $F = -31$ **6.** $y = -30$ **7.** $b = \frac{2A - ac}{a}$ **8.** $x = 3, x = 9$ **9.** $x < 1$ **10.** Length = 520 m; width = 360 m **11.** \$2600 at 12%; \$1400 at 14% **12.** $x \le -9$ *or* $x \ge -2$ **13.** $-\frac{5}{2} < x < \frac{15}{2}$ **14.** x-intercept $(-2, 0)$; y-intercept $(0, 7)$

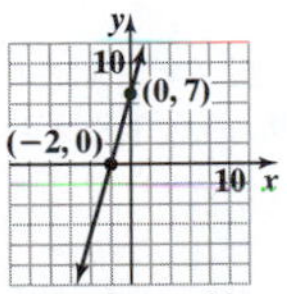

$7x - 2y = -14$

15.

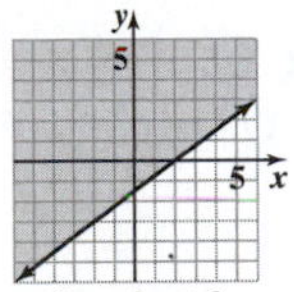

$3x - 4y \le 6$

16. $m = \frac{8}{3}$ **17.** $3x + 2y = 5$ **18.** $f(3) = 12$ **19.** $f(-2) = 17$ **20.**

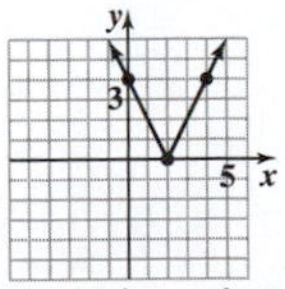

$f(x) = |2x - 4|$

21. (5, 5) **22.** (6, 4) **23.** (1, 4, −2) **24.** No solution **25.** 11 adults' tickets, 4 children's tickets **26.**

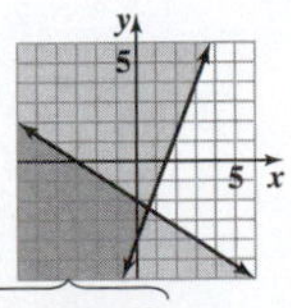

$3y \ge 8x - 12$
$2x + 3y \le -6$

27. $6x^3 - 16x^2 + 17x - 6$ **28.** $5x^2 - x + 2$ **29.** $(2x - 3)(4x^2 + 6x + 9)$ **30.** $(x + 2)(x + 2)(x - 2)$ **31.** $x(2x - 1)(x + 8)$ **32.** $x = -6, x = -9$ **33.** $\frac{x(3x - 1)}{x - 3}$ **34.** $\frac{x + 5}{x}$ **35.** $\frac{3x^2 + 6x - 2}{(x + 5)(x + 2)}$

36. $\dfrac{3-2x}{5x+2}$ **37.** $x = -1$ **38.** 64 **39.** $2a^2b^3\sqrt{11bc}$ **40.** $18\sqrt{2}$ **41.** $\dfrac{5(\sqrt{7}+2)}{3}$ **42.** $8i$ **43.** $x = -3$ **44.** $y = 33.75$

45. $x = \dfrac{1 \pm \sqrt{21}}{10}$ **46.** $x = 0, x = -\dfrac{3}{5}$ **47.** $x = 8, x = -343$ **48.** $x \le -\dfrac{1}{3}$ or $x \ge 4$ **49.**

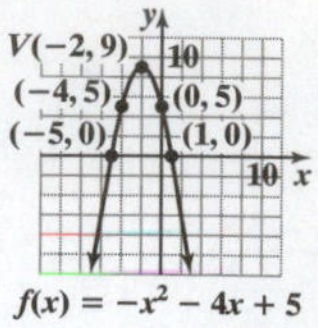

50. Width = 4 cm; length = 13 cm **51.** $(x+3)^2 + (y-2)^2 = 4$; center $(-3, 2)$; radius = 2 **52.** Ellipse

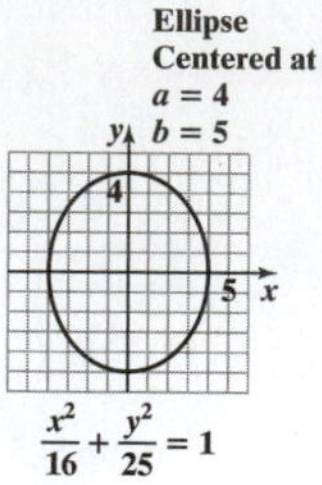

53. Hyperbola

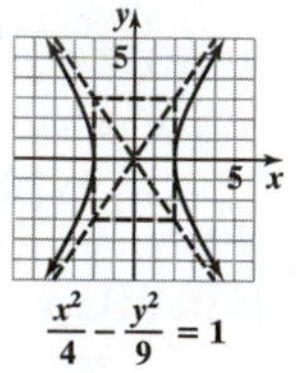

54. Parabola opening right

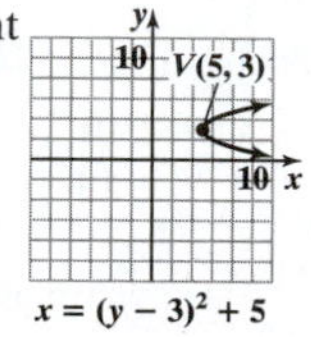

55. $(0, -4), (\sqrt{7}, 3), (-\sqrt{7}, 3)$ **56. (a)** $f(-1) = 10$

(b) $f(a) = 3a^2 - 2a + 5$ **(c)** $f(a+2) = 3a^2 + 10a + 13$ **57.** $f[g(x)] = 80x^2 + 80x + 17$ **58.** $f^{-1}(x) = 2x + 14$ **59.** No

60.

x	0	1	−2	4	−1
y	2	1	8	$\frac{1}{8}$	4

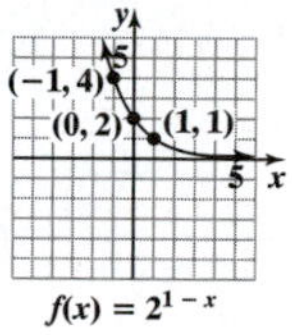

61. $x = 0$ **62.** $x = 21$ **63.** $y = -2$ **64.** $x = 8$

Appendix B

Exercises **1.** −7 **3.** 15 **5.** 2 **7.** 47 **9.** 18 **11.** 0 **13.** 0 **15.** −0.6 **17.** $-7a - 4b$

19. $\dfrac{11}{84}$ **21.** $\begin{vmatrix} 6 & 10 \\ -5 & 9 \end{vmatrix}$ **23.** $\begin{vmatrix} 3 & -4 \\ 1 & -5 \end{vmatrix}$ **25.** −7 **27.** −26 **29.** 11 **31.** −27 **33.** −8 **35.** 0 **37.** −3.179 **39.** 18,553

41. $x = 2; y = 3$ **43.** $x = -2; y = 5$ **45.** $x = 10; y = 2$ **47.** $x = 4; y = -2$ **49.** $x \approx 1.5795; y \approx -0.0902$

51. $x = 1; y = 1; z = 1$ **53.** $x = -\dfrac{1}{2}; y = \dfrac{1}{2}; z = 2$ **55.** $x = 4; y = -2; z = 1$ **57.** $x \approx -0.219; y \approx 1.893; z \approx -3.768$

59. $w \approx -3.105; x \approx 4.402; y \approx 15.909; z \approx 6.981$

Appendix C

Exercises **1.** (4, −1) **3.** (3, −9) **5.** (0, 3) **7.** (2, 2) **9.** (1.2, 3.7) **11.** (3, −1, 4) **13.** (1, −1, 3)
15. (0, −2, 5) **17.** (0.5, −1, 5) **19.** (3.6, 1.8, 2.4) **21.** (4.2, −3.6, 8.8, 5.4)

Intermediate Algebra Glossary

Absolute value inequalities (2.8) Inequalities that contain at least one absolute value expression.

Absolute value of a number (1.2) The distance between the number and 0 on the number line. The absolute value of a number x is written as $|x|$. The absolute value can be determined by

$$|x| = \begin{cases} x, & \text{if } x \geq 0 \\ -x, & \text{if } x < 0 \end{cases}$$

Algebraic expression (1.5) A collection of numerical values, variables, and operation symbols. $\sqrt{5xy}$ and $3x - 6y + 3yz$ are algebraic expressions.

Algebraic fraction (6.1) An expression of the form $\frac{P}{Q}$, where P and Q are polynomials and Q is not zero. Algebraic fractions are also called *rational expressions*. For example, $\frac{x + 3}{x - 4}$ and $\frac{5x^2 + 1}{6x^3 - 5x}$ are algebraic fractions.

Approximate value (1.3) A value that is not exact. The approximate value of $\sqrt{3}$, correct to the nearest tenth, is 1.7. The symbol $\approx$ is used to indicate "is approximately equal to." We write $\sqrt{3} \approx 1.7$.

Associative property of addition (1.1) For all real numbers a, b, and c: $a + (b + c) = (a + b) + c$.

Associative property of multiplication (1.1) For all real numbers a, b, and c: $a(bc) = (ab)c$.

Asymptote (9.4) A line that a curve continues to approach but never actually touches. Often an asymptote is a helpful reference in making a sketch of a curve, such as a hyperbola.

Augmented matrix (Appendix C) A matrix derived from a linear system of equations. It consists of the coefficients of each variable in a linear system and the constants. The augmented matrix of the system $\begin{aligned} -3x + 5y &= -22 \\ 2x - y &= 10 \end{aligned}$ is the matrix $\left[\begin{array}{cc|c} -3 & 5 & -22 \\ 2 & -1 & 10 \end{array}\right]$. Each row of the augmented matrix represents an equation of the system.

Axis of symmetry of a parabola (9.2) A line passing through the focus and the vertex of a parabola, about which the two sides of the parabola are symmetric. See the sketch.

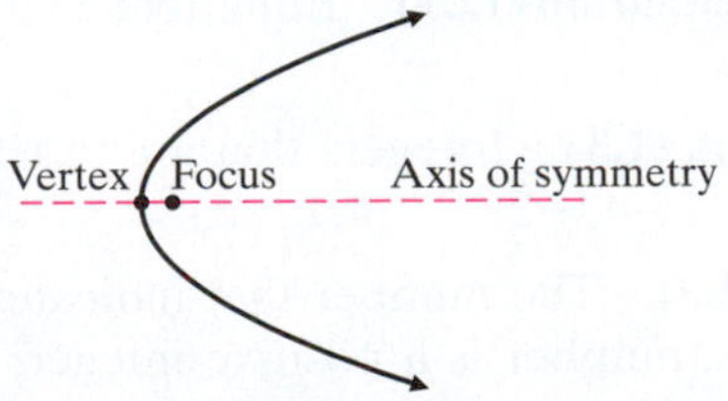

Base (1.3) The number or variable that is raised to a power. In the expression 2^3, the number 2 is the base.

Base of an exponential function (11.1) The number b in the function $f(x) = b^x$.

Binomial (5.1) A polynomial of two terms. For example, $z^2 - 9$ is a binomial.

Boundary points of a quadratic inequality (8.6) In a quadratic inequality of the form $ax^2 + bx + c > 0$ or $ax^2 + bx + c < 0$, those points where $ax^2 + bx + c = 0$.

Cartesian coordinate system (3.1) Another name for the rectangular coordinate system named after its inventor, René Descartes.

Circle (1.6) A geometric figure that consists of a collection of points that are of equal distance from a fixed point called the *center*.

Circumference of a circle (1.6) The distance around a circle. The circumference of a circle is given by the formulas $C = \pi d$ and $C = 2\pi r$, where d is the diameter of the circle and r is the radius of the circle.

Closure property of addition (1.1) For all real numbers a and b: the sum $a + b$ is a real number.

Closure property of multiplication (1.1) For all real numbers a and b: the product ab is a real number.

Coefficient (1.5) Any factor or group of factors in a term. In the term $8xy$, the coefficient of xy is 8. However, the coefficient of x is $8y$. In the term $abcd$, the coefficient of $abcd$ is 1.

Combine like terms (1.5) The process of adding and subtracting like terms. If we combine like terms in the expression $5x - 8y - 7x - 12y$, we obtain $-2x - 20y$.

Combined variation (7.7) Variation that depends on two on more variables. An example would be when y varies directly with x and z and inversely with d^2, written $y = \frac{kxz}{d^2}$, where k is the constant of variation.

Common denominator (1.2) The same number or polynomial in the denominator of two fractions. The fractions $\frac{4}{13}$ and $\frac{7}{13}$ have a common denominator of 13.

Common logarithm (11.4) The common logarithm of a number x is given by $\log x = \log_{10} x$ for all $x > 0$. A common logarithm is a logarithm using base 10.

Commutative property of addition (1.1) For all real numbers a and b: $a + b = b + a$.

Commutative property of multiplication (1.1) For all real numbers a and b: $ab = ba$.

Complex fraction (also called a Complex rational expression) (6.3) A fraction made up of polynomials or numerical values in which at least one fraction appears in the numerator or denominator or both. Examples of complex fractions are

$$\frac{\frac{1}{3} + \frac{1}{5}}{\frac{2}{7}} \quad \text{and} \quad \frac{\frac{1}{x} + 3}{2 + \frac{5}{x}}.$$

Complex number (7.6) A number that can be written in the form $a + bi$, where a and b are real numbers and $i = \sqrt{-1}$.

Compound inequalities (2.7) Two inequality statements connected together by the word *and* or by the word *or*.

Conjugate of a binomial with radicals (7.4) The expressions $a\sqrt{x} + b\sqrt{y}$ and $a\sqrt{x} - b\sqrt{y}$. The conjugate of $2\sqrt{3} + 5\sqrt{2}$ is $2\sqrt{3} - 5\sqrt{2}$. The conjugate of $4 - \sqrt{x}$ is $4 + \sqrt{x}$.

Conjugate of a complex number (7.6) The expressions $a + bi$ and $a - bi$. The conjugate of $5 + 2i$ is $5 - 2i$. The conjugate of $7 - 3i$ is $7 + 3i$.

Coordinates of a point (3.1) An ordered pair of numbers (x, y) that specifies the location of a point on a rectangular coordinate system.

Counting numbers (1.1) The counting numbers are the natural numbers. They are the numbers in the infinite set

$$\{1, 2, 3, 4, 5, 6, 7, \ldots\}.$$

The degree of a polynomial (5.1) The degree of the highest-degree term in the polynomial. The polynomial $5x^3 + 4x^2 - 3x + 12$ is of degree 3.

The degree of a term (5.1) The sum of the exponents of the term's variables. The term $5x^2y^2$ is of degree 4.

Denominator (6.1) The bottom expression in a fraction. The denominator of $\frac{5}{11}$ is 11. The denominator of $\frac{x - 7}{x + 8}$ is $x + 8$.

Descending order for a polynomial (5.1) A polynomial is written in descending order if the term of the highest degree is first, the term of the next-to-highest degree is second, and so on, with each succeeding term of less degree. The polynomial $5y^4 - 3y^3 + 7y^2 + 8y - 12$ is in descending order.

Determinant (Appendix B) A square array of numbers written between vertical lines. For example $\begin{vmatrix} 1 & 5 \\ 2 & 4 \end{vmatrix}$ is a 2×2 determinant. It is also called a *second-order determinant.* $\begin{vmatrix} 1 & 7 & 8 \\ 2 & -5 & -1 \\ -3 & 6 & 9 \end{vmatrix}$ is a 3×3 determinant. It is also called a *third-order determinant.*

Different signs (1.2) When one number is positive and one number is negative, the two numbers are said to have different signs. The numbers 5 and -9 have different signs.

Direct variation (7.7) When a variable y varies directly with x, written $y = kx$, where k represents some real number that will stay the same over a range of x-values. This value k is called the *constant of variation.*

Discriminant of a quadratic equation (8.2) In the equation $ax^2 + bx + c = 0$, where $a \neq 0$, the expression $b^2 - 4ac$. It can be used to determine the nature of the roots of the quadratic equation. If the discriminant is *positive,* there are two rational or irrational roots. The two roots will be rational only if the discriminant is a perfect square. If the discriminant is *zero,* there is only one rational root. If the discriminant is *negative,* there are two complex roots.

Distance between two points (9.1) The distance between point (x_1, y_1) and point (x_2, y_2) is given by the formula $d = \sqrt{(x_2 - x_1)^2 + (y_2 - y_1)^2}$.

Distributive property of multiplication over addition (1.1) For any real numbers a, b, and c: $a(b + c) = ab + ac$.

Dividend (5.2) The expression that is being divided by another. In $12 \div 4 = 3$, the dividend is 12. In $x - 5\overline{)5x^2 + 10x - 3}$, the dividend is $5x^2 + 10x - 3$.

Divisor (5.2) The expression that is divided into another. In $12 \div 4 = 3$, the divisor is 4. In $x + 3\overline{)2x^2 - 5x - 14}$, the divisor is $x + 3$.

Domain of a relation or a function (3.5) When the ordered pairs of a relation or a function are listed, all the different first items of each pair.

***e* (11.1)** An irrational number that can be approximated by the value 2.7183.

Elements (1.1) The objects that are in a set.

Ellipse (9.3) The set of points in a plane such that for each point in the set, the sum of its distances to two fixed points is constant. Each of the fixed points is called a *focus.* Each of the following graphs is an ellipse.

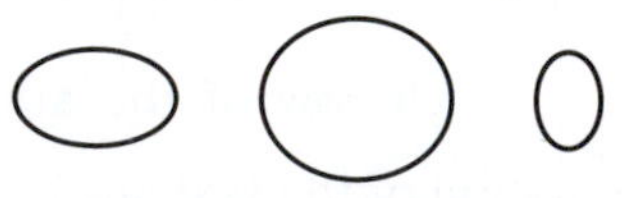

Equation (2.1) A mathematical statement that two quantities are equal.

Equilateral hyperbola (9.4) A hyperbola for which $a = b$ in the equation of the hyperbola.

Equivalent equations (2.1) Equations that have the same solution(s).

Even integers (1.3) Integers that are exactly divisible by 2, such as $\ldots, -4, -2, 0, 2, 4, 6, \ldots$.

Exponent (1.3) The number that indicates the power of a base. If the number is a positive integer, it tells us how many factors of the base occur. In the expression 2^3, the exponent is 3. The number 3 tells us that there are 3 factors, each of which is 2 since $2^3 = 2 \cdot 2 \cdot 2$. If an exponent is negative, use the property that $x^{-n} = \frac{1}{x^n}$. If an exponent is zero, use the property that $x^0 = 1$, where $x \neq 0$.

Exponential function (11.1) $f(x) = b^x$, where $b > 0$, $b \neq 1$, and x is any real number.

Expression (1.3) Any combination of mathematical operation symbols with numbers or variables or both. Examples of mathematical expressions are $2x + 3y - 6z$ and $\sqrt{7xyz}$.

Extraneous solution to an equation (6.4) A correctly obtained potential solution to an equation that when substituted back into the original equation does not yield a true statement. For example, $x = 2$ is an extraneous solution to the equation

$$\frac{x}{x-2} - 4 = \frac{2}{x-2}$$

An extraneous solution is also called an *extraneous root.*

Factor (1.5 and 5.4) Each of the two or more numbers, variables, or algebraic expressions that is multiplied. In the expression $5st$, the factors are 5, s, and t. In the expression $(x - 6)(x + 2)$, the factors are $(x - 6)$ and $(x + 2)$.

First-degree equation (2.1) A mathematical equation such as $2x - 8 = 4y + 9$ or $7x = 21$ in which each variable has an exponent of 1. It is also called a *linear equation.*

First-degree equation in one unknown (2.1) An equation such as $x = 5 - 3x$ or $12x - 3(x + 5) = 22$ in which only one kind of variable appears and that variable has an exponent of 1. It is also called a *linear equation in one variable.*

Focus point of a parabola (9.2) The focus point of a parabola has many properties. For example, the focus point of a parabolic mirror is the point to which all incoming light rays that are parallel to the axis of symmetry will collect. A parabola is a set of points that is the same distance from a fixed line called the *directrix* and a fixed point. This fixed point is the focus.

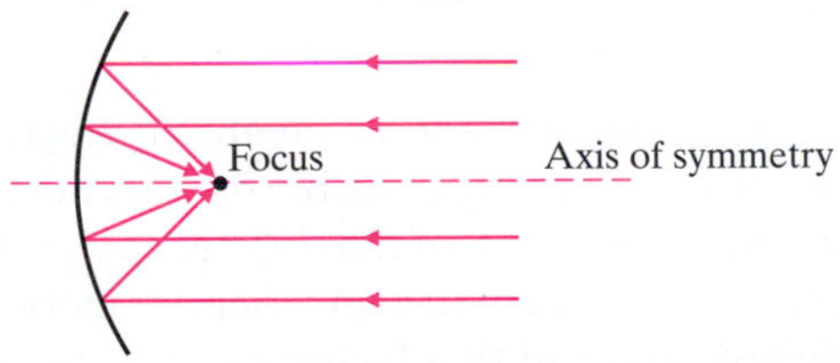

Formula (1.6) A rule for finding the value of a variable when the values of other variables in the expression are known. For example, the formula for finding the Fahrenheit temperature when the Celsius temperature is known is $F = 1.8C + 32$.

Fractional equation (6.4) An equation that contains a rational expression. Examples of fractional equations are

$$\frac{x}{3} + \frac{x}{4} = 7 \quad \text{and} \quad \frac{2}{3x-3} + \frac{1}{x-1} = \frac{-5}{12}.$$

Function (3.5) A relation in which no different ordered pairs have the same first coordinate.

Graph of a function (3.5, 10.2) A graph in which a vertical line will never cross in more than one place. The following sketches represent the graphs of functions.

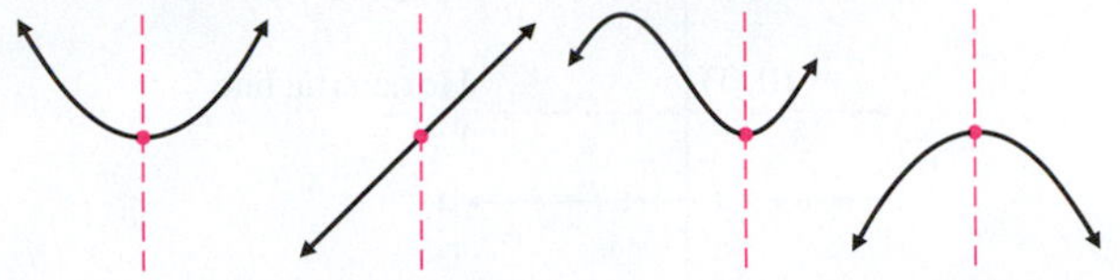

Graph of a linear inequality in two variables (3.4) A shaded region in two-dimensional space. It may or may not include the boundary line. If the line is included, the sketch shows a solid line. If it is not included, the sketch shows a dashed line. Two sketches follow.

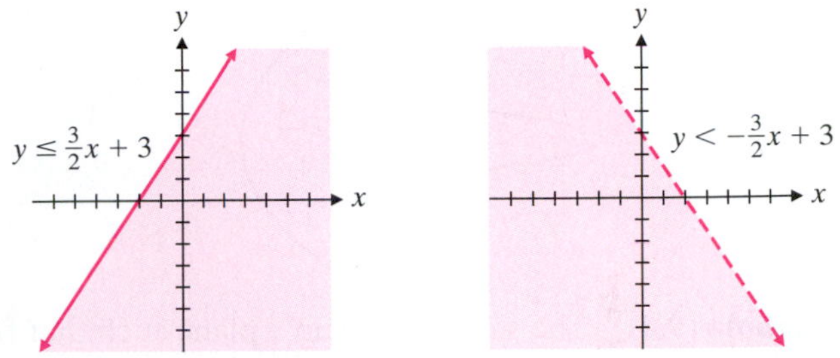

Graph of a one-to-one function (10.4) A graph of a function with the additional property that a horizontal line will never cross the graph in more than one place. The following sketches represent the graphs of one-to-one functions.

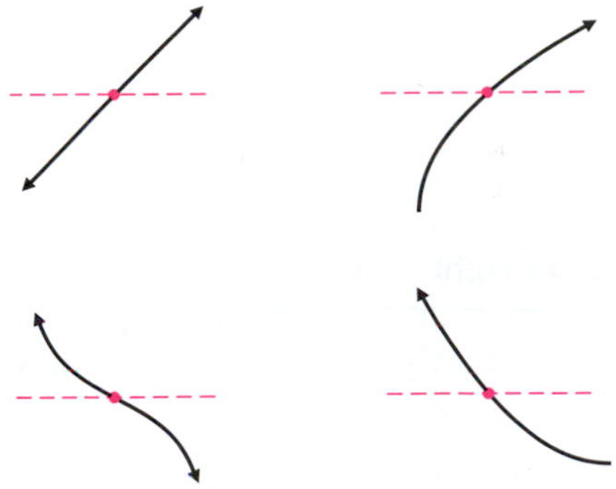

Greater than or equal to symbol (2.6) The $\geq$ symbol.

Greater than symbol (1.2, 2.6) The $>$ symbol. $5 > 3$ is read, "5 is greater than 3."

Greatest common factor of a polynomial (5.4) A common factor of each term of the polynomial that has the largest possible numerical coefficient and the largest possible exponent for each variable. For example, the greatest common factor of $50x^4y^5 - 25x^3y^4 + 75x^5y^6$ is $25x^3y^4$.

Higher-order equations (8.3) Equations of degree 3 or higher. Examples of higher-order equations are $x^4 - 29x^2 + 100 = 0$ and $x^3 + 3x^2 - 4x - 12 = 0$.

Higher-order roots (7.2) Cube roots, fourth roots, and all other roots with an index greater than 2.

Horizontal line (3.1) A straight line that is parallel to the x-axis. A horizontal line has a slope of zero. The equation of any horizontal line can be written in the form $y = b$, where b is a constant. A sketch is shown.

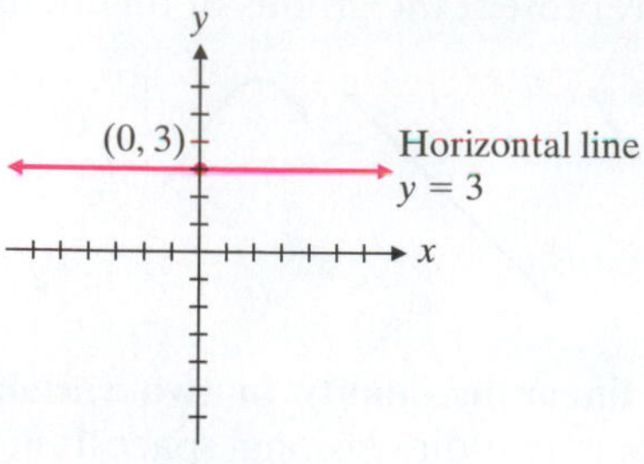

Horizontal parabolas (9.2) Parabolas that open to the right or to the left. The following graphs represent horizontal parabolas.

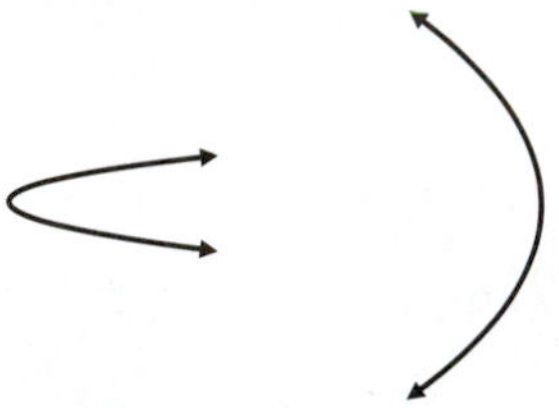

Hyperbola (9.4) The set of points in a plane such that for each point in the set, the absolute value of the difference of its distances to two fixed points is constant. Each of these fixed points is called a *focus*. The following sketches represent graphs of hyperbolas.

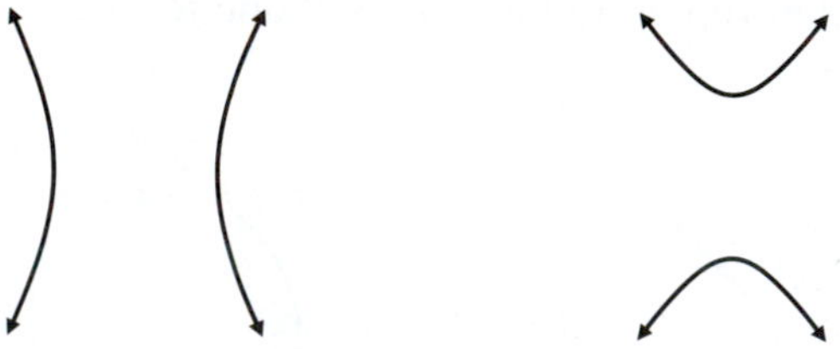

Hypotenuse of a right triangle (8.4) The side opposite the right angle in any right triangle. The hypotenuse is always the longest side of a right triangle. In the following sketch the hypotenuse is side c.

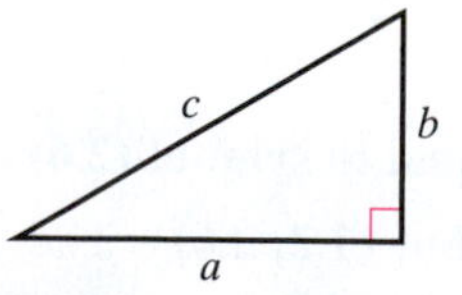

Identity property for addition (1.1) For any real number a: $a + 0 = a = 0 + a$.

Identity property for multiplication (1.1) For any real number a: $a(1) = a = 1(a)$.

Imaginary number (7.6) i, defined as $i = \sqrt{-1}$ and $i^2 = -1$.

Inconsistent system of equations (4.1) A system of equations for which no solution is possible.

Index of a radical (7.2) Indicates what type of a root is being taken. The index of a cube root is 3. In $\sqrt[3]{x}$, the 3 is the index of the radical. In $\sqrt[4]{y}$, the index is 4. The index of a square root is 2, but the index is not written in the square root symbol, as shown: $\sqrt{x}$.

Inequality (2.6) A mathematical statement expressing an order relationship. The following are inequalities:

$$x < 3, \quad x \geq 4.5, \quad 2x + 3 \leq 5x - 7,$$
$$x + 2y < 8, \quad 2x^2 - 3x > 0$$

Infinite set (1.1) A set that has no end to the number of elements that are within it. An infinite set is often indicated by placing an ellipsis (. . .) after listing some of the elements of the set.

Integers (1.1) The numbers in the infinite set

$$\{\ldots, -3, -2, -1, 0, 1, 2, 3, \ldots\}.$$

Interest (2.5) The charge made for borrowing money or the income received from investing money. Simple interest is calculated by the formula $I = prt$, where p is the principal that is borrowed, r is the rate of interest, and t is the amount of time the money is borrowed.

Inverse function of a one-to-one function (10.4) That function obtained by interchanging the first and second coordinates in each ordered pair of the function.

Inverse property of addition (1.1) For any real number a: $a + (-a) = 0 = (-a) + a$.

Inverse property of multiplication (1.1) For any real number $a \neq 0$: $a\left(\frac{1}{a}\right) = 1 = \left(\frac{1}{a}\right)a$.

Inverse variation (7.7) When a variable y varies inversely with x, written $y = \frac{k}{x}$, where k is the constant of variation.

Irrational numbers (1.1) Numbers whose decimal forms are nonterminating and nonrepeating. The numbers π, e, $\sqrt{2}$, and 1.56832574 . . . are irrational numbers.

Joint variation (7.7) Variation that depends on two or more variables. An example would be when a variable y varies jointly with x and z, written $y = kxz$, where k is the constant of variation.

Least common denominator of algebraic fractions (6.2) A polynomial that is exactly divisible by each denominator. The LCD is the product of all the *different prime factors*. If a factor occurs more than once in a denominator, we must use the highest power of that factor.

For example, the LCD of

$$\frac{5}{2(x + 2)(x - 3)^2} \quad \text{and} \quad \frac{3}{(x - 3)^4}$$

is $2(x + 2)(x - 3)^4$. The LCD of

$$\frac{5}{(x + 2)(x - 3)} \quad \text{and} \quad \frac{7}{(x - 3)(x + 4)}$$

is $(x + 2)(x - 3)(x + 4)$.

Least common denominator of numerical fractions (2.1) The smallest whole number that is exactly divisible by all the denominators of a group of fractions. The least common denominator (LCD) of $\frac{1}{7}$, $\frac{9}{21}$, and $\frac{3}{14}$ is 42. The number 42 is the smallest number that can be exactly divided by 7, 21, and 14. The least common denominator is sometimes called the *lowest common denominator*.

Leg of a right triangle (8.4) One of the two shorter sides of a right triangle. In the following sketch, sides a and b are the legs of the right triangle.

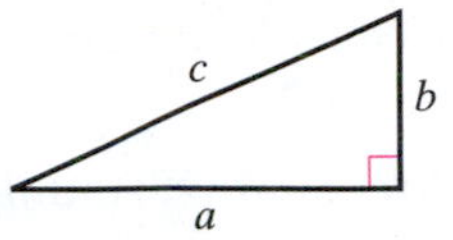

Less than or equal to symbol (2.6) The $\leq$ symbol.

Less than symbol (1.2) The $<$ symbol. $2 < 8$ is read, "2 is less than 8."

Like terms (1.5) Terms that have identical variables and identical exponents. In the mathematical expression $5x - 8syz + 7x + 15syz$, the terms $5x$ and $7x$ are like terms, and the terms $-8syz$ and $15syz$ are like terms.

Linear equation (2.1) A mathematical equation such as $3x + 7 = 5x - 2$ or $5x + 7y = 9$, in which each variable has an exponent of 1.

Linear inequality (2.6) An inequality statement in which each variable has an exponent of 1 and no variables are in the denominator. Some examples of linear inequalities are

$$2x + 3 > 5x - 6, \qquad y < 2x + 1, \quad \text{and} \quad x < 8.$$

Literal equation (2.2) An equation that has other variables in it besides the variable for which we wish to solve. $I = prt$, $7x + 3y - 6z = 12$, and $P = 2w + 2l$ are examples of literal equations.

Logarithm (11.2) For a positive number x, the power to which the base b must be raised to produce x. That is, $y = \log_b x$ is the same as $x = b^y$, where $b > 0$ and $b \neq 1$. A logarithm is an exponent.

Logarithmic equation (11.2) An equation that contains at least one logarithm.

Magnitude of an earthquake (11.5) The magnitude of an earthquake is measured by the formula $M = \log\left(\frac{I}{I_0}\right)$, where I is the intensity of the earthquake and I_0 is the minimum measurable intensity.

Matrix (Appendix C) A rectangular array of numbers arranged in rows and columns. We use the symbol [ ] to indicate a matrix. The matrix $\begin{bmatrix} 3 & 4 & 5 \\ 6 & 7 & 8 \end{bmatrix}$ has two rows and three columns and is called a 2×3 *matrix*.

Minor of an element of a third-order determinant (Appendix B) The second-order determinant that remains after we delete the row and column in which the element appears. The minor of the element 6 in the determinant $\begin{vmatrix} 1 & 2 & 3 \\ 7 & 6 & 8 \\ -3 & 5 & 9 \end{vmatrix}$ is the second-order determinant $\begin{vmatrix} 1 & 3 \\ -3 & 9 \end{vmatrix}$.

Monomial (5.1) A polynomial of one term. For example, $3a$ is a monomial.

Natural logarithm (11.4) For a number x, $\ln x = \log_e x$ for all $x > 0$. A natural logarithm is a logarithm using base e.

Negative integers (1.1) The numbers in the infinite set

$$\{-1, -2, -3, -4, -5, -6, -7, \ldots\}.$$

Nonlinear system of equations (9.5) A system of equations in which at least one equation is not a linear equation.

Nonzero (1.4) A nonzero value is a value other than zero. If we say that the variable x is nonzero, we mean that x cannot have the value of zero.

Numerator (6.1) The top expression in a fraction. The numerator of $\frac{3}{19}$ is 3. The numerator of $\frac{x + 5}{x^2 + 25}$ is $x + 5$.

Numerical coefficient (1.5) The numerical value multiplied by the variables in a term. The numerical coefficient of $-8xyw$ is -8. The numerical coefficient of abc is 1.

Odd integer (1.3) Integers that are not exactly divisible by 2, such as $\ldots, -3, -1, 1, 3, 5, 7, \ldots$.

One-to-one function (10.4) A function in which no two different ordered pairs have the same second coordinate.

Opposite of a number (1.2) That number with the same absolute value but a different sign. The opposite of -7 is 7. The opposite of 13 is -13.

Ordered pair (3.1) A pair of numbers represented in a specified order. An ordered pair is used to identify the location of a point. Every point on a rectangular coordinate system can be represented by an ordered pair (x, y).

Origin (3.1) The point determined by the intersection of the x-axis and the y-axis. It has the coordinates $(0, 0)$.

Parabola (9.2) The set of points that is the same distance from some fixed line (called the *directrix*) and some fixed point (called the *focus*) that is not on the line. The graph of any equation of the form $y = ax^2 + bx + c$ or $x = ay^2 + by + c$, where a, b, and c are real numbers and $a \neq 0$, is a parabola. Some examples of the graphs of parabolas are shown.

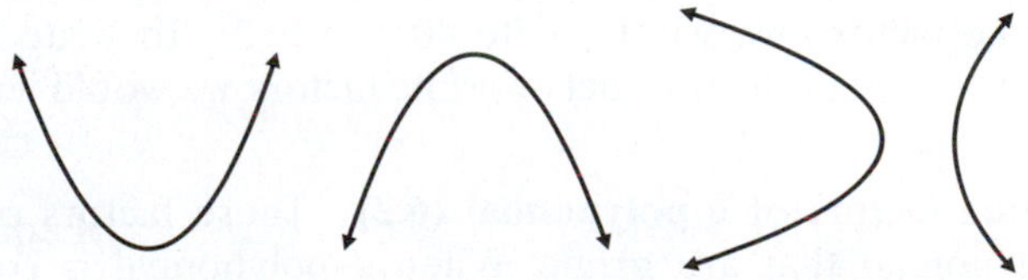

Parallel lines (3.2) Two straight lines that never intersect. Parallel lines have the same slope.

Parallelogram (1.6) A four-sided geometric figure with opposite sides parallel. The opposite sides of a parallelogram are equal.

Percent (1.6) Hundredths or "per one hundred"; indicated by the % symbol. Thirty-seven hundredths means thirty-seven percent: $\frac{37}{100} = 37\%$.

Perfect square (1.3) If x is an integer and a is a positive real number such that $a = x^2$, then x is a square root of a and a is a perfect square. Some numbers that are perfect squares are 1, 4, 9, 16, 25, 36, 49, 64, 81, and 100.

Perfect square trinomials (5.6) Trinomials of the form $a^2 + 2ab + b^2$ or $a^2 - 2ab + b^2$.

Perpendicular lines (3.2) Two straight lines that meet at a 90-degree angle. If two nonvertical lines have slopes m_1 and m_2, and m_1 and $m_2 \neq 0$, then the lines are perpendicular if and only if $m_1 = -\dfrac{1}{m_2}$.

pH of a solution (11.2) Defined by the equation $\text{pH} = -\log_{10}(\text{H}^+)$, where H^+ is the concentration of the hydrogen ion in the solution. The solution is an acid when the pH is less than 7 and a base when the pH is greater than 7.

Pi (1.6) An irrational number, denoted by the symbol π, that is approximately equal to 3.141592654. In most cases, 3.14 can be used as a sufficiently accurate approximation for π.

Point–slope form of the equation of a straight line (3.3) For a straight line passing through the point (x_1, y_1) and having slope m, $y - y_1 = m(x - x_1)$.

Polynomials (1.5 and 5.1) Variable expressions that contain terms with nonnegative integer exponents. A polynomial must contain no division by a variable. Some examples of polynomials are $5y^2 - 8y + 3$, $-12xy$, $12a - 14b$, and $7x$.

Positive integers (1.1) The numbers in the infinite set $\{1, 2, 3, 4, 5, 6, 7, \ldots\}$. The positive integers are the natural numbers.

Power (1.3) When a number is raised to a power, the number's exponent is that power. Thus, two to the third power means 2^3. The power is the exponent, which is 3. In the expression x^5, we say, "x is raised to the fifth power."

Prime factors of a number (6.2) Those factors of a number that are prime. To write the number 40 as a product of prime factors, we would write $40 = 5 \times 2^3$. To write the number 462 as the product of prime factors, we would write $462 = 2 \times 3 \times 7 \times 11$.

Prime factors of a polynomial (6.2) Those factors of a polynomial that are prime. When a polynomial is completely factored, it is written as a product of prime factors. Thus, the prime factors of $x^4 - 81$ are written as $x^4 - 81 = (x^2 + 9)(x - 3)(x + 3)$.

Prime number (6.2) A positive integer that is greater than 1 and has no factors other than 1 and itself. The first ten prime numbers are 2, 3, 5, 7, 11, 13, 17, 19, 23, and 29.

Prime polynomial (5.7) A polynomial that cannot be factored. Examples of prime polynomials are $2x^2 + 100x - 19$, $25x^2 + 9$, and $x^2 - 3x + 5$.

Principal (1.6) In monetary exercises, the original amount of money invested or borrowed.

Principal square root (1.3 and 7.2) The positive square root of a number. The symbol indicating the principal square root is $\sqrt{}$. Thus, $\sqrt{4}$ means to find the principal square root of 4, which is 2.

Proportion (6.5) An equation stating that two ratios are equal. For example, $\dfrac{a}{b} = \dfrac{c}{d}$ is a proportion.

Pythagorean Theorem (8.4) In any right triangle, if c is the length of the hypotenuse and a and b are the lengths of the two legs, then $c^2 = a^2 + b^2$.

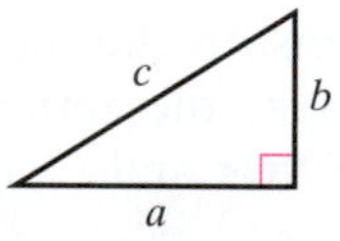

Quadrants (3.1) The four regions into which the x-axis and the y-axis divide the rectangular coordinate system.

Quadratic equation in standard form (5.8, 8.1) An equation of the form $ax^2 + bx + c = 0$, where a, b, and c are real numbers and $a \neq 0$. A quadratic equation is classified as a second-degree equation.

Quadratic formula (8.2) If $ax^2 + bx + c = 0$ and $a \neq 0$, then the roots to the equation are found by the formula

$$x = \frac{-b \pm \sqrt{b^2 - 4ac}}{2a}.$$

Quadratic inequalities (9.5) An inequality written in the form $ax^2 + bx + c > 0$, where $a \neq 0$ and a, b, and c are real numbers. The $>$ symbol may be replaced by a $<$, $\geq$, or $\leq$ symbol.

Quotient (1.4) The result of dividing one number or expression by another. In the equation $12 \div 4 = 3$, the quotient is 3.

Radical equation (7.5) An equation that contains one or more radicals. The following are examples of radical equations.

$$\sqrt{9x - 20} = x \quad \text{and} \quad 4 = \sqrt{x - 3} + \sqrt{x + 5}$$

Radical sign (1.3, 7.2) The symbol $\sqrt{}$, which is used to indicate the root of a number.

Radicand (1.3, 7.2) The expression beneath the radical sign. The radicand of $\sqrt{7x}$ is $7x$.

Radius of a circle (1.6) The distance from any point on the circle to the center of the circle.

Range of a relation or a function (3.5) When the ordered pairs of a relation or a function are listed, all of the different second items of each pair.

Ratio (6.5) The ratio of two values is the first value divided by the second. The ratio of a to b, where $b \neq 0$, is written as $\frac{a}{b}$, a/b, $a \div b$, or $a : b$.

Rational equation (6.4) An equation that has at least one variable in a denominator. Examples of rational equations are

$$\frac{x+6}{3x} = \frac{x+8}{5} \quad \text{and} \quad \frac{x+3}{x} - \frac{x+4}{x+5} = \frac{15}{x^2+5x}.$$

Rational exponents (7.2) When an exponent is a rational number, this is equivalent to a radical expression in the following way: $x^{m/n} = (\sqrt[n]{x})^m = \sqrt[n]{x^m}$. Thus, $x^{3/7} = (\sqrt[7]{x})^3 = \sqrt[7]{x^3}$.

Rational expressions (6.1) A fraction of the form $\frac{P}{Q}$, where P and Q are polynomials and Q is not zero. Rational expressions are also called *algebraic fractions*. For example, $\frac{7}{x-8}$ and $\frac{3x-5}{2x^2+1}$ are rational expressions.

Rational numbers (1.1) An infinite set of numbers containing all the integers and all exact quotients of two integers where the denominator is not zero. In set notation, the rational numbers are the set of numbers $\left\{\frac{a}{b} \mid a \text{ and } b \text{ are integers but } b \neq 0\right\}$.

Rationalizing the denominator (7.4) The process of transforming a fraction that contains one or more radicals in the denominator to an equivalent fraction that does not contain any radicals in the denominator. When we rationalize the denominator of $\frac{5}{\sqrt{3}}$, we obtain $\frac{5\sqrt{3}}{3}$. When we rationalize the denominator of $\frac{-2}{\sqrt{11}-\sqrt{7}}$, we obtain $-\frac{\sqrt{11}+\sqrt{7}}{2}$.

Rationalizing the numerator (7.4) The process of transforming a fraction that contains one or more radicals in the numerator to an equivalent fraction that does not contain any radicals in the numerator. When we rationalize the numerator of $\frac{\sqrt{5}}{x}$, we obtain $\frac{5}{x\sqrt{5}}$.

Real number line (1.2) A number line on which all the real numbers are placed. Positive numbers lie to the right of 0 on the number line, and negative numbers lie to the left.

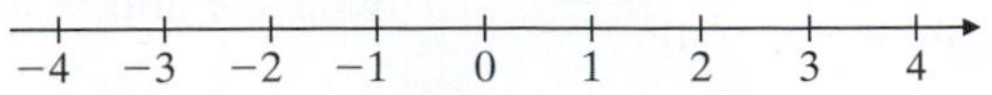

Real Number Line

Real numbers (1.1) The set of numbers containing the rational and irrational numbers.

Reciprocal (1.2) The reciprocal of a number is 1 divided by that number. Therefore, the reciprocal of 12 is $\frac{1}{12}$. The reciprocal of $\frac{3}{4}$ is $\frac{4}{3}$. The reciprocal of $-\frac{5}{8}$ is $-\frac{8}{5}$.

Rectangle (1.6) A four-sided figure with opposite sides parallel and all interior angles measuring 90 degrees. The opposite sides of a rectangle are equal.

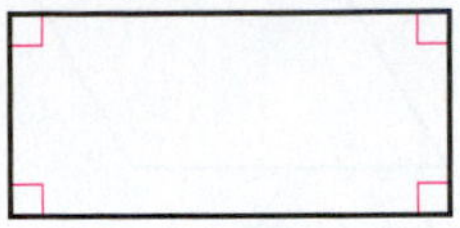

Rectangular solid (1.6) A three-dimensional object in which each side is a rectangle. A rectangular solid has the shape of a box.

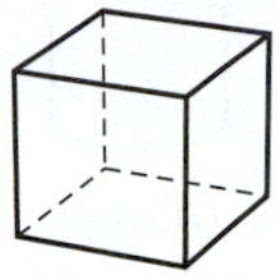

Reduced row echelon form (Appendix C) In the reduced row echelon form of an augmented matrix, all the numbers to the left of the vertical line are 1s along the diagonal from the top left to the bottom right. If there are elements below or above the 1s, these elements are 0s. Two examples of matrices in reduced row echelon form are

$$\left[\begin{array}{cc|c} 1 & 0 & 3 \\ 0 & 1 & 4 \end{array}\right] \quad \text{and} \quad \left[\begin{array}{ccc|c} 1 & 0 & 0 & 5 \\ 0 & 1 & 0 & 6 \\ 0 & 0 & 1 & 7 \end{array}\right].$$

Reducing a fraction (6.1) Using the basic rule of fractions to simplify a fraction. The basic rule of fractions is: For any polynomials a, b, and c, where $b \neq 0$ and $c \neq 0$, $\frac{ac}{bc} = \frac{a}{b}$. Reducing the fraction $\frac{x^2-16}{2x+8}$, we have $\frac{(x+4)(x-4)}{2(x+4)} = \frac{x-4}{2}$.

Relation (3.5) Any set of ordered pairs.

Remainder (5.2) The amount left after the final subtraction when working out a division problem. In the problem

$$\begin{array}{r} 2x - 3 \\ x - 2\overline{)2x^2 - 7x + 9} \\ \underline{2x^2 - 4x} \\ -3x + 9 \\ \underline{-3x + 6} \\ 3 \leftarrow \text{the remainder is 3.} \end{array}$$

Repeating decimal (1.1) A number that in decimal form has one or more digits that continue to repeat. The numbers 0.33333 . . . and 0.128128128 . . . are repeating decimals.

Reversing an inequality (2.6) When multiplying or dividing both sides of an inequality by a negative number, the greater than symbol changes to a less than symbol or the less than symbol changes to a greater than symbol. For example, to solve $-3x < 9$, we divide each side by -3. $\frac{-3x}{-3} > \frac{9}{-3}$, so $x > -3$. The $<$ symbol was reversed to the $>$ symbol.

Rhombus (1.6) A parallelogram with four equal sides and no right angle.

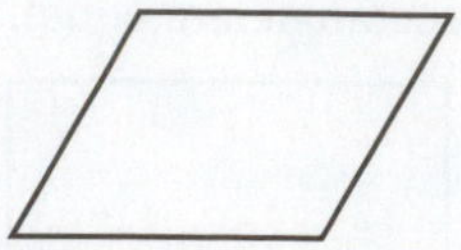

Right circular cylinder (1.6) A three-dimensional object shaped like a tin can.

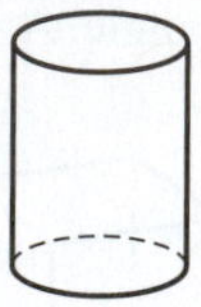

Right triangle (8.4) A triangle that contains one right angle (an angle that measures exactly 90 degrees). It is indicated by a small square at the corner of the angle.

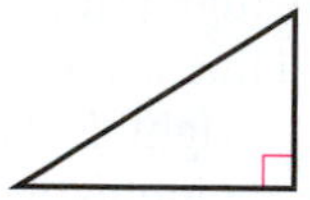

Root of an equation (2.1) A number that when substituted into a given equation yields a true mathematical statement. The root of an equation is also called the *solution of an equation*.

Scientific notation (1.4) A positive number written in the form $a \times 10^n$, where $1 \le a < 10$ and n is an integer.

Set (1.1) A collection of objects.

Signed numbers (1.2) Numbers that are either positive, negative, or zero. Positive signed numbers such as 5, 9, or 124 are usually written without a plus sign. Negative signed numbers such as -5, -3.3, or -178 are always written with a minus sign.

Similar radicals (7.3) Two radicals that are simplified and have the same radicand and the same index. $2\sqrt[3]{7xy^2}$ and $-5\sqrt[3]{7xy^2}$ are similar radicals. Usually similar radicals are referred to as *like radicals*.

Similar triangles (6.5) Two triangles whose corresponding sides are proportional. For example, the following two triangles are similar.

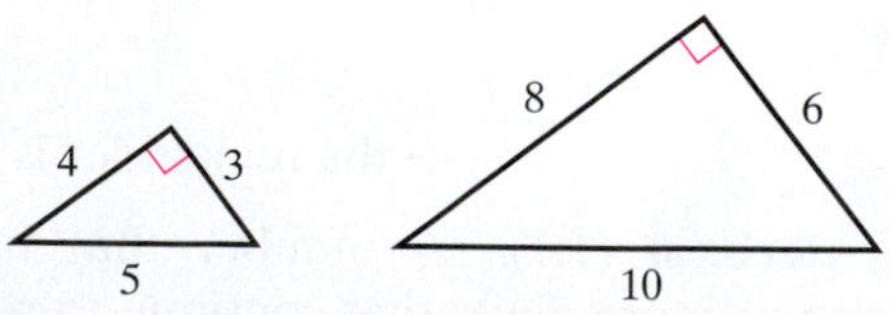

Simplifying imaginary numbers (7.6) Using the property that states for all positive real numbers a, $\sqrt{-a} = \sqrt{-1}\,\sqrt{a} = i\sqrt{a}$. Thus, simplifying $\sqrt{-7}$, we have $\sqrt{-7} = \sqrt{-1}\,\sqrt{7} = i\sqrt{7}$.

Simplifying a radical (7.3) To simplify a radical when the root cannot be found exactly, we use the product rule for radicals, $\sqrt[n]{ab} = \sqrt[n]{a}\sqrt[n]{b}$ for $a \ge 0$ and $b \ge 0$. To simplify $\sqrt{20}$, we have $= \sqrt{4}\sqrt{5} = 2\sqrt{5}$. To simplify $\sqrt[3]{16x^4}$, we have $= \sqrt[3]{8x^3}\,\sqrt[3]{2x} = 2x\sqrt[3]{2x}$.

Slope–intercept form of the equation of a straight line (3.3) $y = mx + b$, where m is the slope and $(0, b)$ is the y-intercept.

Slope of a straight line (3.2) A straight line that passes through the points (x_1, y_1) and (x_2, y_2) has

$$\text{slope} = m = \frac{y_2 - y_1}{x_2 - x_1}, \qquad \text{where } x_1 \neq x_2.$$

Solution of an equation (2.1) A number that when substituted into the equation yields a true mathematical statement. The solution of an equation is also called the *root of an equation*.

Sphere (1.6) A perfectly round three-dimensional object shaped like a ball.

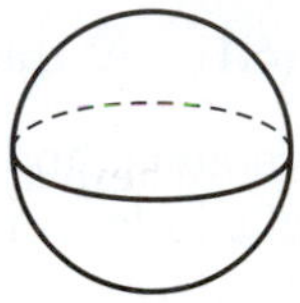

Square root (1.3 and 7.2) If x is a real number and a is a positive real number such that $a = x^2$, then x is a square root of a. One square root of 16 is 4 since $4^2 = 16$. Another square root of 16 is -4 since $(-4)^2 = 16$.

Standard form of the equation of a circle (9.1) For a circle with center at (h, k) and a radius of r,

$$(x - h)^2 + (y - k)^2 = r^2.$$

Standard form of the equation of an ellipse (9.3) For an ellipse with center at the origin,

$$\frac{x^2}{a^2} + \frac{y^2}{b^2} = 1, \quad \text{where } a \text{ and } b > 0.$$

This ellipse has intercepts at $(a, 0)$, $(-a, 0)$, $(0, b)$, and $(0, -b)$.

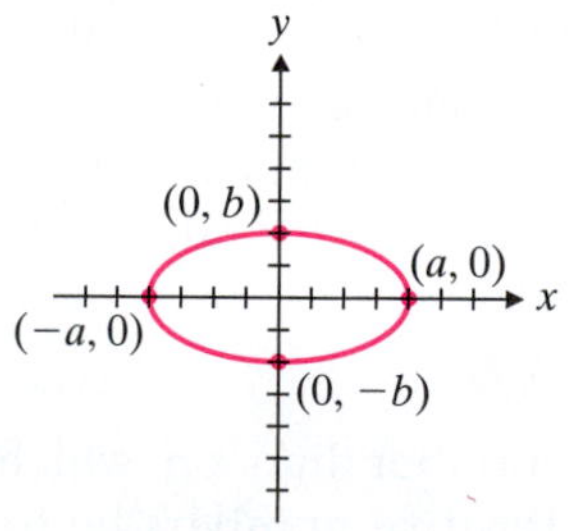

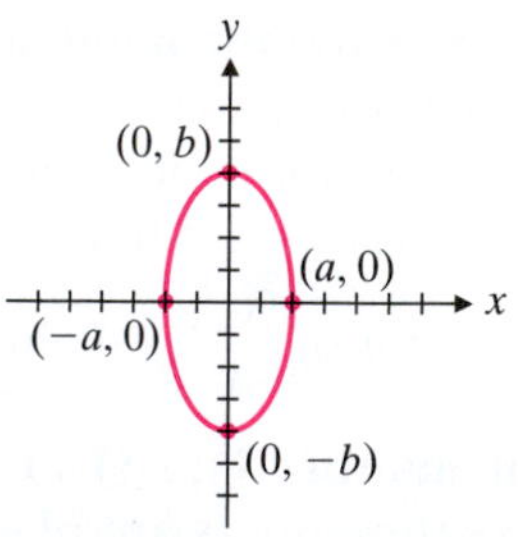

For an ellipse with center at (h, k),

$$\frac{(x - h)^2}{a^2} + \frac{(y - k)^2}{b^2} = 1, \quad \text{where } a \text{ and } b > 0.$$

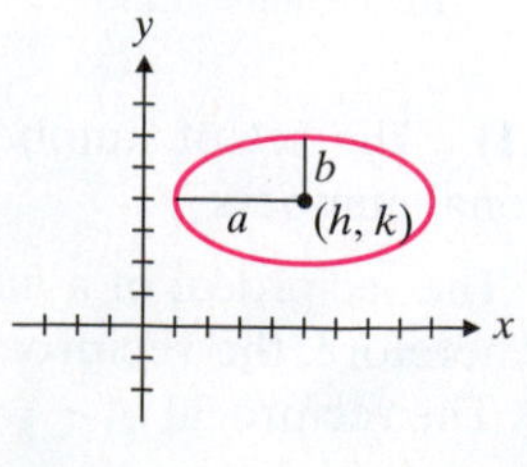

Standard form of the equation of a hyperbola with center at the origin (9.4) For a horizontal hyperbola with center at the origin,

$$\frac{x^2}{a^2} - \frac{y^2}{b^2} = 1, \quad \text{where } a \text{ and } b > 0.$$

The vertices are at $(-a, 0)$ and $(a, 0)$.

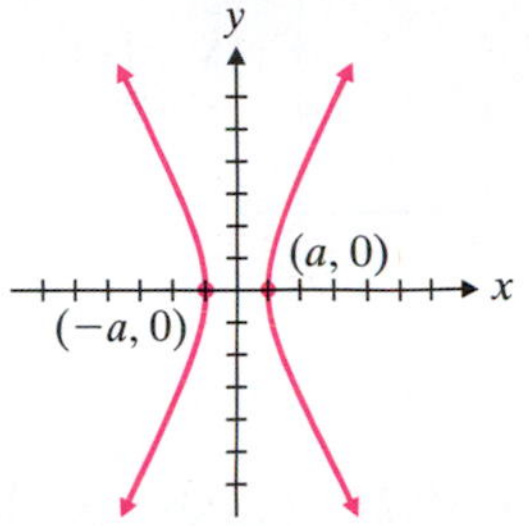

For a vertical hyperbola with center at the origin,

$$\frac{y^2}{b^2} - \frac{x^2}{a^2} = 1, \quad \text{where } a \text{ and } b > 0.$$

The vertices are at $(0, b)$ and $(0, -b)$.

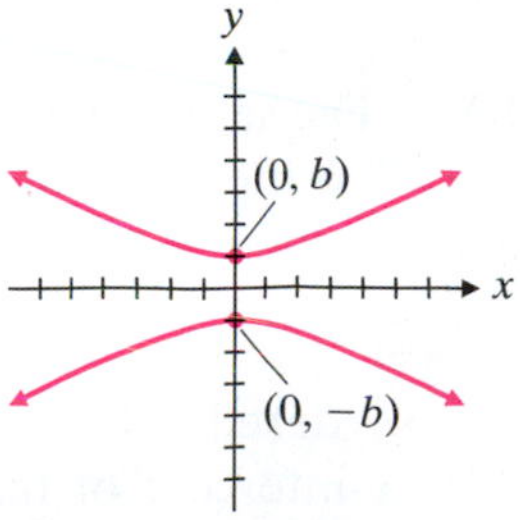

Standard form of the equation of a hyperbola with center at point (h, k) (9.4) For a horizontal hyperbola with center at (h, k),

$$\frac{(x-h)^2}{a^2} - \frac{(y-k)^2}{b^2} = 1, \quad \text{where } a \text{ and } b > 0.$$

The vertices are $(h - a, k)$ and $(h + a, k)$.

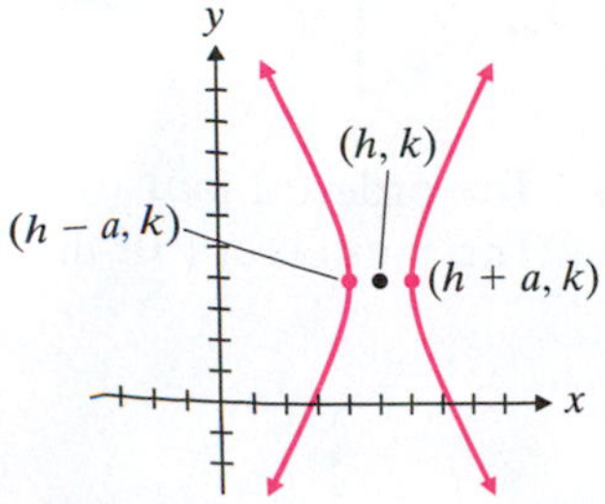

For a vertical hyperbola with center at (h, k),

$$\frac{(y-k)^2}{b^2} - \frac{(x-h)^2}{a^2} = 1, \quad \text{where } a \text{ and } b > 0.$$

The vertices are at $(h, k + b)$ and $(h, k - b)$.

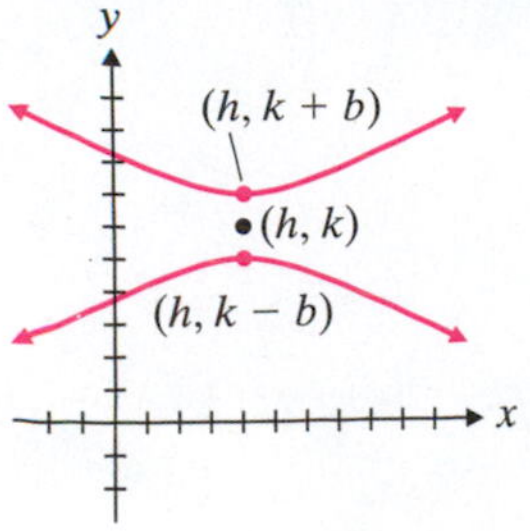

Standard form of the equation of a parabola (9.2) For a vertical parabola with vertex at (h, k),

$$y = a(x - h)^2 + k, \quad \text{where } a \neq 0.$$

For a horizontal parabola with vertex at (h, k),

$$x = a(y - k)^2 + h, \quad \text{where } a \neq 0.$$

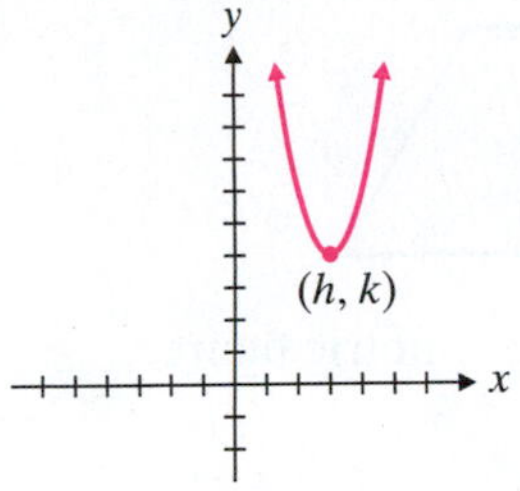

Vertical Parabola

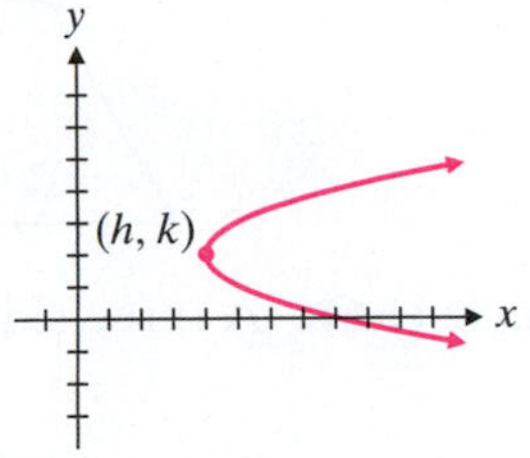

Horizontal Parabola

Standard form of the equation of a straight line (3.3) $Ax + By = C$, where A, B, and C are real numbers.

Standard form of a quadratic equation (5.8) $ax^2 + bx + c = 0$, where a, b, and c are real numbers and $a \neq 0$. A quadratic equation is classified as a second-degree equation.

Subset (1.1) A set whose elements are members of another set. For example, the whole numbers are a subset of the integers.

System of dependent equations (4.1) A system of n linear equations in n variables in which some equations are dependent. It does not have a unique solution but an infinite number of solutions.

System of equations (4.1) A set of two or more equations that must be considered together. The solution is the value for each variable of the system that satisfies each equation.

$$x + 3y = -7$$
$$4x + 3y = -1$$

is a system of two equations in two unknowns. The solution is $(2, -3)$, or the values $x = 2$, $y = -3$.

System of inequalities (4.4) Two or more inequalities in two variables that are considered at one time. The solution is the region that satisfies every inequality at one time. An example of a system of inequalities is

$$y > 2x + 1$$
$$y < \frac{1}{2}x + 2.$$

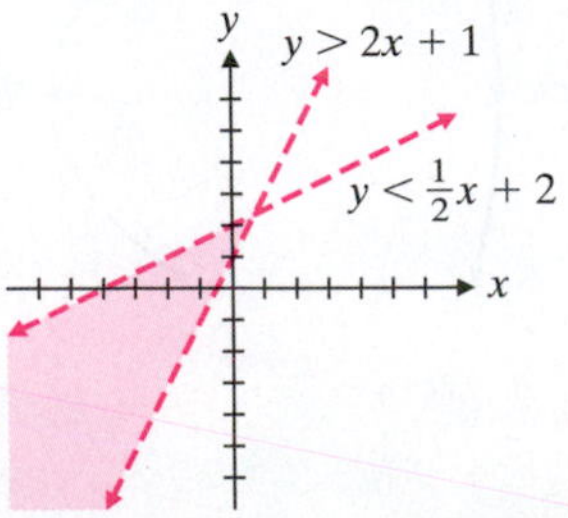

Term (1.5) A real number, a variable, or a product or quotient of numbers and variables. The expression $5xyz$ is one term. The expression $7x + 5y + 6z$ has three terms.

Terminating decimal (1.1) A number in decimal form such as 0.18 or 0.3462, where the number of nonzero digits is finite.

Trapezoid (1.6) A four-sided geometric figure with two parallel sides. The parallel sides are called the *bases of the trapezoid.*

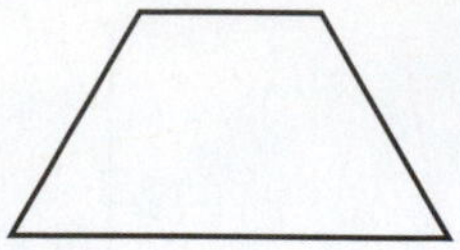

Triangle (1.6) A three-sided geometric figure.

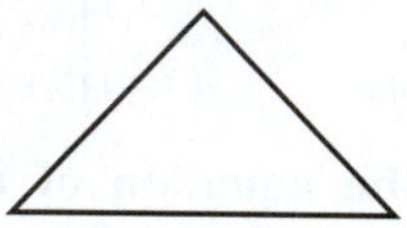

Trinomial (5.1) A polynomial of three terms. For example, $2x^2 + 3x - 4$ is a trinomial.

Unknown (2.1) A variable or constant whose value is not known.

Value of a second-order determinant (Appendix B) For a second-order determinant $\begin{vmatrix} a & b \\ c & d \end{vmatrix}$, $ad - cb$.

Value of a third-order determinant (Appendix B) For a third-order determinant $\begin{vmatrix} a_1 & b_1 & c_1 \\ a_2 & b_2 & c_2 \\ a_3 & b_3 & c_3 \end{vmatrix}$,

$$a_1b_2c_3 + b_1c_2a_3 + c_1a_2b_3 - a_3b_2c_1 - b_3c_2a_1 - c_3a_2b_1.$$

Variable (1.1) A letter used to represent a number.

Vertex of a parabola (9.2) In a vertical parabola, the lowest point on a parabola opening upward or the highest point on a parabola opening downward.

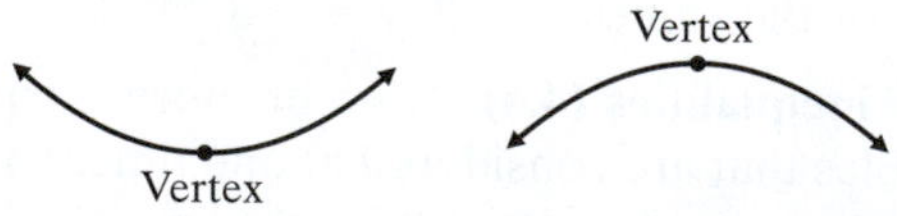

In a horizontal parabola, the leftmost point on a parabola opening to the right or the rightmost point on a parabola opening to the left.

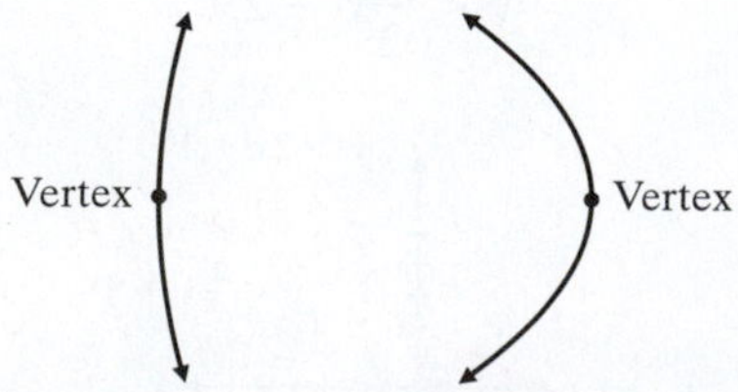

Vertical line (3.1) A straight line that is parallel to the y-axis. The slope of a vertical line is undefined. Therefore, a vertical line has no slope. The equation of a vertical line can be written in the form $x = a$, where a is a constant. A sketch of a vertical line is shown.

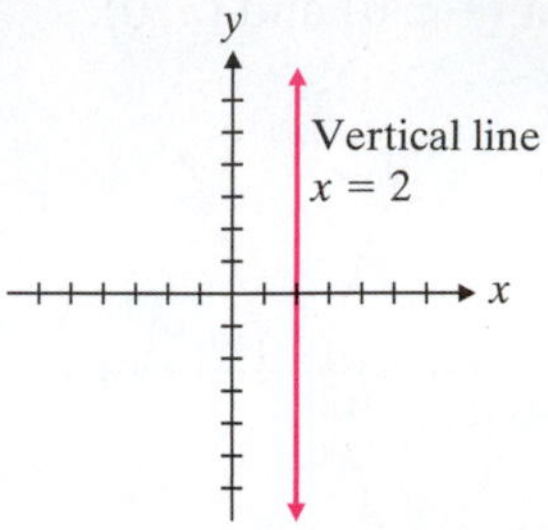

Vertical parabolas (9.2) Parabolas that open upward or downward. The following graphs represent vertical parabolas.

Whole numbers (1.1) The set of numbers containing the natural numbers as well as the number 0. The whole numbers can be written as the infinite set

$$\{0, 1, 2, 3, 4, 5, 6, 7, \ldots\}.$$

***x*-intercept (3.1)** The ordered pair $(a, 0)$ in the line that crosses the x-axis. The x-intercept of the following line is $(5, 0)$.

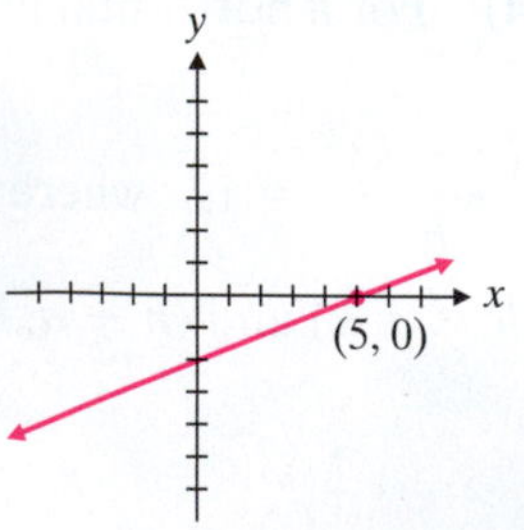

***y*-intercept (3.1)** The ordered pair $(0, b)$ in the line that crosses the y-axis. The y-intercept of the following line is $(0, 4)$.

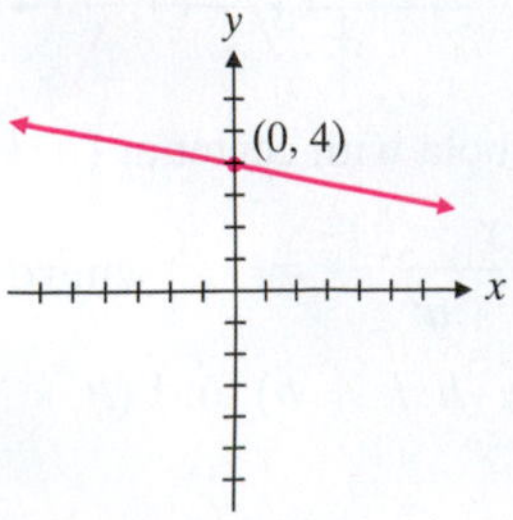

Subject Index

Photo Credits

CHAPTER 1 **CO** Tibor Bognar/AGE Fotostock **p. 7** Oleg Kozlov/Shutterstock **p. 36** Jacques Jangoux/Photo Researchers Inc. **p. 51** Barry Austin Photography/Thinkstock **p. 57** Courtesy of authors

CHAPTER 2 **CO** John Pitcher/iStockphoto **p. 72** Reglen Paassen/Shutterstock **p. 82** Imac/Alamy **p. 83** World History Archive/Alamy **p. 87** Forestphotoart/Dreamstime **p. 118** David Sacks/Thinkstock **p. 125** Courtesy of authors

CHAPTER 3 **CO** Eric Michaud/iStockphoto **p. 185** Eugene Choi/iStockphoto **p. 194** Courtesy of authors

CHAPTER 4 **CO** Laszio Podor/Alamy **p. 223** Belinda Images/SuperStock **p. 240** Joé/Fotolia **p. 247** Courtesy of authors

CHAPTER 5 **CO** Dobermaraner/Shutterstock **p. 261** Johnson Space Center/NASA **p. 293** Kirill Mikhirev/Shutterstock **p. 299** Exactostock/SuperStock **p. 307** Courtesy of authors

CHAPTER 6 **CO** Douglas Healey/AP Images **p. 317** Matt Jones/Shutterstock **p. 345** Getty Images/Thinkstock **p. 352** Courtesy of authors

CHAPTER 7 **CO** Kirill Putchenko/iStockphoto **p. 404** MikLav/Shutterstock **p. 409** Yegor Korzh/Shutterstock **p. 410** Creatas/Thinkstock **p. 419** Courtesy of authors

CHAPTER 8 **CO** Ronscall1/Dreamstime **p. 450** C. Daveney/Shutterstock **p. 452** Natalia Bratslavsky/Shutterstock **p. 473** webphotographeer/iStockphoto **p. 483** Courtesy of authors

CHAPTER 9 **CO** Michael Boyny/Alamy **p. 512** Songquan Deng/Shutterstock **p. 528** Ryan McVay/Thinkstock **p. 537** Courtesy of authors

CHAPTER 10 **CO** CB2/ZOB/Newscom **p. 558** Mikeledray/Shutterstock **p. 563** Strathroy/iStockphoto **p. 572** Tetra Images/Superstock **p. 581** Courtesy of authors

CHAPTER 11 **CO** Pavel Cheiko/Shutterstock **p. 587** Ford Smith/Corbis **p. 621** Robert Paul Van Beets/Shutterstock **p. 627** Andy Ward/Photolibrary New York **p. 633** Courtesy of authors

PROPERTIES OF THE REAL NUMBERS

If a, b, and c are real numbers:

Closure Properties	$a + b$ is a real number. ab is a real number.
Commutative Properties	$a + b = b + a$ $ab = ba$
Associative Properties	$a + (b + c) = (a + b) + c$ $a(bc) = (ab)c$
Identity Properties	$a + 0 = 0 + a = a$ $a \cdot 1 = 1 \cdot a = a$
Inverse Properties	$a + (-a) = -a + a = 0$ $a\left(\frac{1}{a}\right) = \left(\frac{1}{a}\right)a = 1$ (where $a \neq 0$)
Distributive Property	$a(b + c) = ab + ac$

QUADRATIC FORMULA

If $ax^2 + bx + c = 0$, where $a \neq 0$,

$$x = \frac{-b \pm \sqrt{b^2 - 4ac}}{2a}.$$

PYTHAGOREAN THEOREM

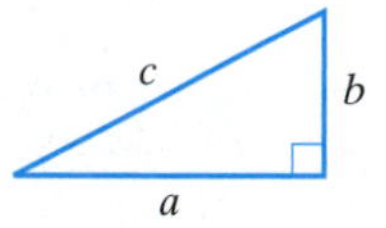

In a right triangle if c is the length of the hypotenuse and a and b are the lengths of the legs, then $c^2 = a^2 + b^2$.

PROPERTIES OF INEQUALITIES

If $a < b$, then $a + c < b + c$ and $a - c < b - c$.

If $a < b$ and c is a **positive** real number,

then $ac < bc$ and $\frac{a}{c} < \frac{b}{c}$.

If $a < b$ and c is a **negative** real number,

then $ac > bc$ and $\frac{a}{c} > \frac{b}{c}$.

If $|x| < a$, then $-a < x < a$.

If $|x| > a$, then $x > a$ or $x < -a$.

ABSOLUTE VALUE

$$|x| = \begin{cases} x & \text{if } x \geq 0 \\ -x & \text{if } x < 0 \end{cases}$$

EQUATIONS OF STRAIGHT LINES

Standard Form $Ax + By = C$

A, B, and C are real numbers.

Slope–Intercept Form $y = mx + b$

$m =$ slope $\quad (0, b) = y$-intercept

Point–Slope Form $y - y_1 = m(x - x_1)$

(x_1, y_1) is a point on the line, and $m =$ slope

PROPERTIES ABOUT POINTS AND STRAIGHT LINES

The **distance between two points** (x_1, y_1) and (x_2, y_2), d, is $\sqrt{(x_2 - x_1)^2 + (y_2 - y_1)^2}$.

The **slope of a line**, m, passing through (x_1, y_1) and (x_2, y_2) is $\frac{y_2 - y_1}{x_2 - x_1}$, where $x_2 \neq x_1$.

Parallel lines have the same slope: $m_1 = m_2$

Nonvertical **perpendicular lines** have slopes whose product is -1. This may be written $m_1 = -\frac{1}{m_2}$.

Horizontal lines have a slope of 0.

Vertical lines have no slope. (The slope is not defined for a vertical line.)

FACTORING AND MULTIPLYING FORMULAS

$a^2 - 2ab + b^2 = (a - b)^2$	**Perfect Square Trinomial**
$a^2 + 2ab + b^2 = (a + b)^2$	**Perfect Square Trinomial**
$a^2 - b^2 = (a + b)(a - b)$	**Difference of Two Squares**
$a^2 + b^2$ cannot be factored.	
$a^3 - b^3 = (a - b)(a^2 + ab + b^2)$	**Difference of Two Cubes**
$a^3 + b^3 = (a + b)(a^2 - ab + b^2)$	**Sum of Two Cubes**